AF876443

# GENESIS - IN THE BEGINNING

# Cellular Origin, Life in Extreme Habitats and Astrobiology

Volume 22

*Series Editor:*

Joseph Seckbach
*The Hebrew University of Jerusalem, Israel*

For further volumes:
http://www.springer.com/series/5775

# Genesis - In The Beginning

## Precursors of Life, Chemical Models and Early Biological Evolution

*Edited by*

Joseph Seckbach

*The Hebrew University of Jerusalem, Israel*

Springer

*Editor*
Joseph Seckbach
Hebrew University of Jerusalem
P.O. Box 1132, Mevo Hadas 20
90435 Efrat
Israel

ISSN 1566-0400
ISBN 978-94-007-2940-7 ISBN 978-94-007-2941-4 (eBook)
DOI 10.1007/978-94-007-2941-4
Springer Dordrecht Heidelberg New York London

Library of Congress Control Number: 2012933105



Printed on acid-free paper

Springer is part of Springer Science+Business Media (www.springer.com)

## TABLE OF CONTENTS



### PART 1:
## CHEMICAL EVOLUTION



*We thank our colleague Professor Julian Chela-Flores for reorganizing the Table of Contents

## PART 2: THE ROLE OF PHYSICS AND GEOPHYSICS IN THE ORIGIN AND EVOLUTION OF LIFE



## PART 3: THE ROLE OF SYMMETRY, INFORMATION, AND ORDER IN THE ORIGIN OF LIFE

## PART 4:
## PANSPERMIA



## PART 5:
## THE FIRST STEPS OF CELLULAR EVOLUTION



## PART 6:
## THE DISTRIBUTION OF LIFE IN THE SOLAR SYSTEM AND ELSEWHERE IN THE COSMOS

## PART 7: HISTORY AND FRONTIERS OF ASTROBIOLOGY AND THE HUMANITIES

## PATHWAYS TO "GENESIS – IN THE BEGINNING: PRECURSORS OF LIFE, CHEMICAL MODELS, AND EARLY BIOLOGICAL EVOLUTION"

This book is volume 22 of the *Cellular Origin, Life in Extreme Habitats and Astrobiology* ["*COLE*"] series, which covers several perspectives of the Origins: Genesis, Evolution and Diversity of Life. There are several approaches to the *origins of life* and genesis on our planet Earth and possibly elsewhere in the universe, but the final answer is still mysterious and under investigation. Definite answers are being sought for issues such as where or when the beginning of life took place? What were the initial building blocks of life? How did the biochemical polymers and macromolecules enter into a membranous system for the first cells? Then another mystery, how did the prokaryotic microorganisms evolve to higher forms?

Several ideas have been proposed for the initial prebiotic steps on the road of life. Among these scenarios are chemical-biological generation within a primordial soup, clay minerals as substrate for prebiotic polymerization of nucleic acids and their encapsulation in a membranous compartment to yield first cellular life [Gallori and Branciamore – in this volume], homochirality, or panspermia (a notion which claims that seeds of life have been transformed from outer space to Earth), and then there is the divine creation of living beings [see: www.worldscibooks.com/lifesci/6998.html]. The literature contains arguments about where the first steps of life took place, was it in thermophilic or rather in cryophilic environments ("warm ponds" or at the bottom of the ocean in volcanic hydrothermovents versus icy, chilled areas) or elsewhere in our solar system or galaxy. Other alternatives are that the creation of life took place in alkaline or acidic solution. Part of these above proposals are covered in articles in this volume, which also presents the history of life origins.

Eukaryogenesis, i.e., the transfer from the prokaryotic level (e.g., bacteria and/or archaebacteria) into nucleated eukaryotes has been explained in recent decades by endosymbiosis, and it is also dealt with in this volume. Similarly, the issue of the possibility of life having been imported from space (panspermia) was raised again in the literature (see Hoover, 2011, an article which made quite a stir). This hypothesis proposes the "import" of spores or whole microbes from outer space while their further evolution took place on Earth. We do not ask "how and when" these extraterrestrial life forms originated before they supposedly arrived on Earth. The proponents of this theory claim that such seeds of life originated from planets, such as meteorites, satellites, and comets from Mars; they supposedly fell and germinated on Earth [and perhaps also elsewhere], see McKay et al. (1996), McNichol and Gordon (2012), Gordon and McNichol (2012), Wickramasinghe (2012) (the latter three in this volume).

Evidence from the fossil record is also indicative of the early period of the life on earth (some of these arguments are found in the COLE series: volume 18 [Stromatolites, 2011] and volume 12 [*From Fossils to Astrobiology,* 2008]). Some dormant microorganisms have been revived in the laboratory after 40 million years, such as the bacteria from insects' abdomen embedded in amber or 250-million-year-old halophilic bacteria in salt mines.

The extremophiles [or polyextremophiles – volume 26 of COLE, 2012] also represent the initial prokaryotic and eukaryotic organisms that have been able to tolerate several harsh environments, such as anaerobic conditions (ANOXIA, see COLE volume 21, 2011), extreme ranges of temperature, extreme levels of pH solutions, pressure, UV radiation, or high salt concentration (see COLE volume 11, *Algae and Cyanobacteria in Extreme Environments*, 2007). Recently, it has been published that bacteria grow using toxic arsenic instead of phosphorus. These extremophiles could serve as analogues or models for extraterrestrial life on planets or satellites.

Some chapters in this volume deal with prebiotic conditions while others with aspects of astrobiology (present conditions and possible habitability of other planets and satellites). This volume is a continuation and complement to the *Origins: Genesis, Evolution and Diversity of Life* [COLE volume 6, 2004]. This current volume contains 50 chapters in eight sections, contributed from a total of 75 authors, who represent 21 countries, namely, Australia, Austria, Brazil, Canada, Cuba, France, Germany, Hungary, Israel, Italy, Japan, Mexico, Netherlands, Poland, Portugal, Romania, Russia, Slovenia, Spain, UK, and the USA. All chapters have been peer reviewed.

The editor thanks all the contributors and the many peer reviewers and external referees who assisted in creating *Genesis – In The Beginning*. Special acknowledgment is due to *Professor Richard Gordon* who invested a great deal of effort to bring this volume to its present status. Last but not least, we are grateful to the team from Springer (*Melanie van Overbeek* and *Maryse Walsh*) who work in full harmony with the editor in chief in all COLE books, as well as to *Mr. P. Saravanan* the typesetter who treated the text in this volume so diligently.

## References

Hoover RB (2011) Fossils of cyanobacteria in CI1 carbonaceous meteorites: implications to life on Comets, Europa, and Enceladus. J Cosmol 13:03–06

McKay DS et al (1996) Search for past life on mars: possible relic biogenic activity in Martian meteorite ALH84001. Science 273(5277):924–930

**Joseph Seckbach**
P.O. Box 1132 Efrat, 90435 Israel

Biodata of **Joseph Seckbach**, editor of this volume and the author (with Julian Chela-Flores) of the chapter "***Astrobiology: From Extremophiles to the Solar System.***" and the summary

**Professor Joseph Seckbach** is the founder and chief editor of book series ***Cellular Origins, Life in Extreme Habitats and Astrobiology*** **("COLE")**. See www.springer.com/sereis/5775. He is the author of several chapters in this series. Dr. Seckbach earned his Ph.D. from the University of Chicago, Chicago, IL (1965), and spent his postdoctoral years in the Division of Biology at Caltech (Pasadena, CA). Then he headed at the University of California at Los Angeles (UCLA) a team for searching for extraterrestrial life. He has been appointed to the faculty of the Hebrew University (Jerusalem, Israel) and spent his sabbatical periods in UCLA, Harvard University, and served at Louisiana State University (LSU, Baton Rouge, LA) as the first Chair for the Louisiana Sea Grant. Dr. Seckbach obtained two DAAD fellowships (German fellowships for exchange academicians) in Tübingen and in Ludwig Maximilians University in Munich.

Among his publications are books, scientific articles, and edited and translated popular books. Dr. Seckbach is the coauthor (with R. Ikan) of the *Chemistry Lexicon* (1991, 1999) and a coeditor of *Proceeding of Endocytobiology VII Conference* (Freiburg, Germany, 1998) and the *Proceedings of Algae and Extreme Environments meeting* (Trebon, Czech Republic, 2000). His edited volume (with Richard Gordon) entitled *Divine Action and Natural Selection: Science, Faith, And Evolution* has been published by World Scientific Publishing Company (2008). His recent interest is in the field of enigmatic microorganisms, life in extreme environments, and Astrobiology.

E-mail: **joseph.seckbach@mail.huji.ac.il**

## FOREWORD

Why is the "origin of life" question so difficult? The "origins question" has, of course, been recognized for the longest time. Nearly every culture has origins stories. Until lately, no one seems to have objected much to these. Indeed, large parts of humankind still today find cultural stories about origins reasonable. And nearly everyone, if asked to respond honestly, would admit that an answer to the "origin of life" question would most likely have no constructive impact on their daily lives.

As reviewed elsewhere in this volume, "origin of life" questions became challenging only after the Enlightenment, especially as it was influenced by the Age of Discovery. Leaving home is a powerful way to drive paradigm change. Discovery of new biology and new geology in new worlds certainly did this in post-Enlightenment Europe.

The Age of Discovery forced natural historians to organize the new biology, much of which proved to be incompatible with creation stories. For example, while the collection of animals known to the Hebrew fathers could conceivably fit into Noah's Ark (whose dimensions were specified in the Bible), the collection discovered in Africa, new Asia, Australia, and the Americas clearly could not.

Further, the "science" that emerged following the Enlightenment proved in the nineteenth century to be able to answer other classic questions. What is the nature of matter? What are the stars? What is electricity? What is light? The rationalism, spirit of inquiry, and reductionist philosophy that the Enlightenment made possible allowed all of these natural phenomena to be understood.

This "understanding" was clearly not just self-deception. The models and theories of nineteenth-century sciences proved to be technologically empowering. The empowerment was robust, extending into the twentieth century as classical sciences were developed and improved. For example, in the world of chemistry, we can today do most anything we want, and if we cannot, we can say *why* it cannot be done.

The success of science in so many areas has driven many to demand a "scientific" answer to the "origin of life" question. Here arose the first source of problems that make the "origins" problem so difficult: We are not so certain how to recognize a "scientific" solution to *any* problem, outside of an expedient that asks the science to show itself through technological empowerment.

The success of science made this fact slow to be realized. The positivist school in Vienna, populated mostly by physicists, proposed a century ago that scientific statements could be generated by formula, a logical structure that connected universal terms in mathematical relations to observation and experimental confirmation. When this model for "the scientific method" proved problematic, Popper set aside "proof" but rather proposed that a scientific statement was different from a non-scientific statement because it could be disproven.

Even though the Popperian view is still widely accepted by practicing scientists, it is well known to have problems. Quine and Duhem pointed out that disproof of single proposition was essentially impossible. An experimental result that might, at first glance, disprove a hypothesis might in fact simply contradict one of innumerable auxiliary hypotheses that are involved in any real measurement. Perhaps the instrument was not working, explaining why results expected under the hypothesis were not observed. Perhaps the reagents were impure.

In fact, scientific solutions are accepted in a community not because they are proven or because they have been tested and have not been disproven. Rather, solutions in science are accepted when *burdens of proof* are met (Galison, 1987; Benner, 2009). These are standards accepted by a community of scientists and transmitted through the culture through the training of scientists. When those standards are met, the community moves on.

This provides a second reason why the "origin of life" question is so difficult. The question is an archetype of a multidisciplinary puzzle. The puzzle draws from different fields of science as input to a solution. These include biology and chemistry, of course, but also physics, planetary science, and astronomy. And these different fields have very different standards of proof.

For example, the isolation of a natural product from a living system creates a "structure elucidation puzzle" in chemistry. This standard puzzle asks: "What is the structure of this natural product?" The community expects (and will accept) an answer in the form of a physical model for the product's constituent molecules, scaled in dimension by approximately eight orders of magnitude. That model may be rendered in three dimensions, using labeled balls to represent individual atoms and using sticks connecting those balls to represent bonds that hold those atoms together. Paper and computer renderings are also acceptable.

The culture of chemistry also understands what the solution to a standard puzzle need *not* include. For example, although the concept of a "bond" is deeply rooted within quantum mechanics, chemists need not go further than a ball-and-stick model for their proposed solution to the structure elucidation puzzle to be accepted.

If accepted by modest (and often flawed) peer review, the rendered physical model for a molecule can be published. It then enters the lexicon of other models for other natural product structures. These form a network interconnected by a logic that makes each structure a confirmation of all of the others. This collection also provides deep confidence that molecules, atoms, and bonds actually exist, even though no one has ever directly seen them.

"Atoms," "bonds," and "molecules" are terms in the *metalanguage* of chemistry. In addition to conveying community-acceptable answers to standard puzzles in chemistry, its vocabulary also limits the extent to which parts of the physical world can be "reduced" by the chemist. Our model for the structure of the natural product is reductionist in the sense that the product, a visible substance (a powder, for example) with a color and a melting point is "reduced" to a more fundamental model using terms that can be applied to other molecules (and, under chemical theory, all molecules). However, this metalanguage provides a platform beyond which reduction need not continue and, because of its vocabulary, cannot continue.

The metalanguage also defines how the solution might be challenged. Challenges to the solution might come when others examine the same powder and perhaps observe behaviors that are not consistent with the proposed arrangement of atoms and bonds. They do not come by questioning the lack of a quantum mechanical description of the bonds.

Likewise, standard chemistry holds many "mechanism" puzzles, which challenge chemists to say how a starting molecule is transformed into a product molecule. A mechanism puzzle begins, of course, only after the structures of the educt and product are "solved." Given these, the community accepts as a solution to a mechanism puzzle that comprises a graphical representation of the educt, showing where bonding electrons are located in its structure and then how these electrons move from their positions in the educt to their positions in the product. This movement is conveyed using "curved arrows" drawn on the representation of the educt that start with pairs of electrons in the starting molecule and end in the representation at positions where those electrons *will be* in the product.

Solutions to a mechanism problem might also propose the existence of one or more metastable states, molecules that are formed and decompose rapidly on the way from educt to product, molecules that are not isolated. Challenges to a mechanistic proposal might come as chemists attempt (but fail) to trap and observe those intermediates. The proposal may be challenged by observing unexpected behavior of the reaction as conditions are perturbed. But again, the language used to express the solution to the mechanistic problem provides a platform below which no chemist must go to call his/her mechanistic proposal acceptable. If such challenges fail, the mechanism will be accepted, adding to the network of mechanisms that collectively guide chemists to anticipate how molecules will (or will not) react under any set of specified conditions.

Natural historians have their own metalanguage. A planetary scientist will speak of strata from limestone and sandstone, drawing inferences about relative dates without bothering to reduce their components to the underlying metalanguage of chemistry. However, a proposed solution to a problem in planetary science need not be reduced to chemistry for it to be acceptable to the community. However, chemists need not find those solutions acceptable.

Biologists, at least those interested in organisms, have their own metalanguage. Organismic and evolutionary biologists speak using terms like "species" and "population." These terms have, of course, underlying chemical reality. However, as before, organismic and evolutionary biologists need not reference the underlying chemistry to have their proposed solutions to standard puzzles in the field be accepted by their communities. However, chemists need not find those solutions acceptable.

If we accept a common definition-theory of life, is a self-sustaining chemical system capable of Darwinian evolution, an answer using the metalanguage of chemistry might be expected for the question: How *might* life originate? Reforming the question given that definition-theory, the question would require as starting materials a collection of chemicals that did *not* have access to Darwinian evolution and generate from these a collection of chemicals that did.

Thus, a community of chemists would almost certainly accept an answer in the form of a set of curved arrow mechanisms showing the steps by which simple organic molecules that plausibly existed on early Earth might be transformed into those capable of supporting Darwinian evolution under conditions known to have existed on early Earth. That answer would be viewed as stronger if the chemists then went into a laboratory, did experiments that mixed those molecules under those conditions, and observed spontaneous emergence of Darwinian chemical systems. To chemists, the demonstration of the "possible" might be contradicted by denying that the requisite starting materials existed on early Earth or by denying the historical existence of the conditions necessary for the proposed transformation. It might be contradicted by an alternative model with equal support.

If such experiments were reported and widely reproduced, they need not satisfy the natural historians; however, who would point out that the question remained: *Did* life emerge on Earth in this way? Here, however, none of the tools routinely used by natural historians appear to be available that are used elsewhere to adjudicate historical hypotheses. There are no fossils from near the time of the origin of life. There are no strata from near the time of the origin of life. Indeed, the time for the origin of life is poorly constrained. As some argue in this volume, even the *place* where terran life originated is open to dispute.

This means that among the various communities that might contribute to this problem, no agreement exists on the semantic content of the question. No agreement exists on standards of proof. No agreement exists on acceptable forms of the answer. And for natural historians, the types of evidence that might satisfy their standards are missing.

Let us attempt to resolve these problems by ignoring them, accepting (perhaps arbitrarily) the chemist's question (How *might* life have originated?), its approach to answer the question, and the standards of proof that chemists accept. Are we done?

Unfortunately, no. Another problem arises because very few chemists will actually attempt to meet the community burden of proof, because very few chemists constructively believe that *any* combination of reagents and conditions will convert a set of molecules not supporting Darwinian evolution into a set that can.

Why is this? A reading of the literature identifies several reasons. First, while it has become widely accepted within the chemical community that the distinction between inanimate and animate matter lies in the access of the second (but not the first) to Darwinian processes, it has been less clear to the chemical community what chemical structures confer that access. Many decades of efforts have gone into research attempting to create molecular systems that are *unlike* DNA but nevertheless are capable of self-replicating. A few successes are in hand. However, to date, all of these have failed to also support replicability *with mutation* where the mutations are themselves *also* replicable. The replicability of mutations is an essential for Darwinian evolution.

This issue has been mitigated, at least in part, by efforts in "synthetic biology." Here, the focus has been on understanding better why nucleic acids (DNA and RNA) *are* able to support Darwinian evolution through imperfect replications where the imperfections are themselves replicable. Efforts to create alternative nucleic acid structures that support replication, mutation, and the replication of mutant forms have identified features of biopolymers that appear to be essential to genetics as we see it in the contemporary terran biosphere. In particular, a biopolymer with a repeating backbone charge (a *polyelectrolyte*) may be necessary for such processes in water (Benner, 2009).

Certainly, DNA and RNA are polyelectrolytes. Further, by inferences drawn from natural products structure analysis, we can safely infer that both DNA and RNA were present in the last common ancestor of all life on Earth. It also appears that the further we go backward in time, the greater the role that RNA plays in biocatalysis. Indeed, the paleogenetics experiment that led us to conclude that ancient bacteria lived at 65°C involved proteins that were intimately associated with RNA catalysts (Benner, 2009).

So far, so good. This line of reason has encouraged many from communities in biology to accept that the first form of life on Earth, the first chemical system capable of Darwinian evolution, was based on RNA that arose spontaneously from a prebiotic soup. Accordingly, communities accepting this reasoning have asked chemists to find curved arrow mechanisms that might create RNA from simple prebiotic molecules, together with environmental conditions where those mechanisms might operate.

Many in the chemical community have attempted to do just that. Without being exhaustive, chemists have now found conditions where five-carbon carbohydrates, nucleobases, nucleobase-ribose combinations, ribonucleosides attached to phosphorus, and even long oligomeric RNA can be produced from educts that are one or several steps earlier in the progression of a conjectural prebiotic metabolism.

So is the problem now "solved"? Here we face yet another problem. Even as separate solutions have been obtained that create ribose, adenine, and other components of RNA from precursors, none do so in a continuous pot. At the same time, these experiments also provide products that are "tars." And this leads to a general paradox: Organic material *not* having access to Darwinian evolution, given energy, *generally* becomes tar, not life.

You can yourself create another example of this. Go to your kitchen and pull off the shelf a sample of organic matter. It makes no difference which; a can of clam chowder or Cheerios® both will serve. Add some water (or not, your choice) and give that sample some energy by putting it into your oven and turning the temperature to 220°C. Then, return to reading this book until the stench from your kitchen is overwhelming, the smoke alarm turns on, or the fire company arrives. Then go see whether that complex chemical mixture ended up looking more or less like life.

Everything in the experience of any organic chemist suggests that your experiment will generate something that looks less like life than what you started with. That experience suggests that the mixture will have "evolved" to look like asphalt. The outcome, by physical appearance, will not depend much on what organic material you pulled from the pantry. Things will look pretty much the same if you started with chowder or Cheerios®.

You might try this in other ways. You might exclude oxygen (you probably should if you want to make the experimental environment more like early Earth). You might add methane (as on Titan). You might mix different organics. You might use energy from the household current instead of microwaves. But the outcome will be analogous. Without much energy, nothing happens (at least on a human time scale). With much energy, you get tar.

Analogous transformations happen on Earth as a planet, of course. Organic materials from previously living systems are often heated without oxygen. The products are petroleum, tar sands, and coal. Chemists can infer something about the organic material that fed the process if they examine the tar closely enough, just as chemists could infer whether the tar created in your kitchen began with chowder or Cheerios®. But the more energy put into the system, the less the outcome resembles life.

This commonplace observation explains why chemists complain when biologists suggest that chemistry is "pregnant with life," or mathematicians, physicists, and lawyers assert that life emerges "sensibly" by cooking complex mixtures of organic species. Maybe it can, and maybe it did, but to believe so would require us to disregard thousands of experiments of the type that you just did in the kitchen.

Now, let us consider another commonplace observation. Go to the pet store. Buy two guinea pigs, one male and one female. This pair exemplifies a chemical system that is capable of (and historically has had access to) Darwinian evolution.

Now, put the two pigs in a cage and give them some Cheerios® (we recommend Cheerios® over chowder, but that is a preference specific to guinea pigs) and a few other essentials (like water). Settle back to read this book. Do this for 40 days. The odds are that after this time, when you return to the cage, you will discover that the Cheerios® were converted not into asphalt but into more guinea pigs. Babies that display all of the attributes expected of the living state.

Our definition-theory suggests that the only relevant difference between the two systems is that the second has access to Darwinian evolution while the first has lost that access. But independent of that theory, the pair of observations suggests a paradox surrounding origins. If chemical systems spontaneously and intrinsically make tar when provided energy, how can they ever spontaneously generate any Darwinian system? RNA or otherwise?

These two commonplace observations (tar versus baby guinea pigs) seem to create a bona fide crisis in science, as discussed by Thomas Kuhn (1962).

The paradox is even recognized in the popular press. Consider this exchange recorded in The DNA Files (2001), a science education broadcast that pitted journalist John Hockenberry against a know-it-all computer named Mnemosyne. The exchange went like this:

> Mnemosyne: Well, chemists think that if you could recreate the conditions of the Earth about four and a half billion years ago, you'd see life happen spontaneously. You'd just see DNA just pop out of the mix. Hockenberry: Can they do that in a laboratory?
>
> Mnemosyne: Well, no. Actually, they've tried, but so far they can't seem to pull it off. In fact chemists have a little joke about that, you know: they say that life is impossible. Experience shows that it can't happen. That we're just imagining it. Ha ha ha ha....
>
> Hockenberry: Right. Those chemists. But even if these issues are resolved sufficiently to allow progress, we will encounter a different problem, one that might be called a "problem of aggregation." Let us assume an experiment that makes adenine from hydrogen cyanide, an experiment that makes ribose from formaldehyde, an experiment that makes adenosine from ribose and adenine, and experiment that makes adenosine monophosphate from adenosine and inorganic phosphate, and an experiment that makes oligomeric RNA polyadenylic acid from adenosine monophosphate. Does one have, then, a pathway from hydrogen cyanide and formaldehyde to RNA?

Certainly, if all of these reactions occurred in the same flask at the same time, permitting continuous synthesis of oligomeric RNA from hydrogen cyanide and formaldehyde, the community is likely to think that the burden of proof has shifted strongly to those who doubt this process as a solution to the origins of life.

Unfortunately, this is not the case. The conditions that make adenine from hydrogen cyanide are quite different from the conditions that make ribose from formaldehyde. Further, components from the reaction that makes adenine interfere with the reaction that makes ribose.

We can certainly conceive of the first reaction having occurred in one spot or at one time under favorable conditions, the second having occurred in a different spot or perhaps at a different time under favorable conditions, and the products having survived long enough to allow their remixing under a third set of conditions to make a ribonucleoside.

But many in the community of chemistry are not satisfied by this "relay" synthesis as the solution to the problem. This dissatisfaction is not mitigated by having reliable structures for each and every one of the molecules postulated or by a good curved arrow mechanism for every step.

Rather, this dissatisfaction arises from a view of probabilities, a view that has been eloquently expressed recently by Robert Shapiro. Each set of conditions, Shapiro argued, is contingent and not highly likely under most models for the early Earth. Therefore, the likelihood of having all sets of conditions is the product of several low probabilities. Still less likely is the probability that the various conditions would be met in the correct sequence in time, with sufficient rapidity so that sufficient amounts of product survived from the previous step to be available for each subsequent step. In the laboratory of the chemist, each of these conditions are engineered and their time sequence managed by a living chemist. But, axiomatically, no such chemist was available on prebiotic Earth to orchestrate the conditions and timing. Accordingly, Shapiro likened a relay synthesis to a golfer "who, having played a golf ball through an 18-hole course, then assumed that the ball could also play itself around the course in his absence."

This pain is felt especially by chemists. Indeed, chemists, observing the propensity of carbohydrates to form tar, wrote in 1995 that "stability considerations *preclude* the use of ribose and other sugars as prebiotic reagents .... It follows that ribose and other sugars were not components of the first genetic material...." No ribose, the "R" in RNA, means no RNA.

In this regard, the different standards of proofs found in communities of physicists and mathematicians are remarkable. For example, the physicist Freeman Dyson (1985) wrote that "Orgel *demonstrated* that nucleotide monomers will ... polymerize to form RNA if they are given an RNA template to copy without any polymerase enzyme. Orgel found that zinc ions in the solution are a good catalyst for the RNA synthesis." This, he felt, was evidence that the formation of life is easy.

Dyson went on: "Eigen *demonstrated* that a solution of nucleotide monomers will, under suitable conditions, give rise to a nucleic acid polymer molecule which replicates and mutates and competes with progeny for survival."

Of course, Orgel demonstrated only that a *few* nucleotide monomers will polymerize to give a *short* RNA product. And that RNA product spontaneously falls apart. Eigen's demonstration was done in the presence of a protein molecule that was itself a product of 3 billion years of terran biological evolution. But to a physicist, those details are minor. Life is easy.

Because of the strength of this paradox, the contrast between the behavior of organic material without and with access to Darwinian evolution, it may not be timely to try to create RNA in a flash. The tar-versus-pigs paradox may be so strong that neither funding agencies nor coworkers have a belief sufficiently constructive to enthusiastically fund or do experiments to answer those questions at this time. Enthusiasm and funding are near requirements for success in science.

Rather, a more timely question may be: How can the propensity of organic molecules to form tar be naturally constrained? The strategic plan is simple. Should an example of a natural constraint be provided, the constructive beliefs of the community might change. This would be a Kuhnian "paradigm shift." One example of a natural constraint on organic tar formation might create a constructive belief in individuals in a community that others exist. This might generate funding to find more examples of natural constraints on tar formation. This, in turn, might eventually uncover natural ways to assemble RNA itself from the complexity of organic molecules available on early Earth.

Alternatively, we might look at the problem in the opposite direction. What is the probability that an RNA molecule having useful properties can be found in a collection of RNA molecules with random sequence? Here, RNA molecules can be prepared using the full armamentarium of molecular biology; no need for spontaneous assembly. What might emerge from this is a better understanding of the enormity of the challenge that must be met for the "RNA first" hypothesis for the origin of life to prevail.

With these thoughts in mind, I invite you into this book. Keep alert for evidence as to what standards individual authors accept as burdens to prove or disprove hypotheses. See how well they match the standards of other fields discussed in other chapters. Evaluate their proposals in light of the paradoxes raised by natural chemistry. And above all, have fun. It is a key ingredient of enthusiasm, which is key to progress.

## References

Benner SA (2009) Life, the universe and the scientific method. FfAME Press, Gainesville
Dyson F (1985) Origins of life. Cambridge University Press, Cambridge
Galison PL (1987) How experiments end. University of Chicago Press, Chicago
Kuhn TR (1962) Structure of scientific revolutions. University of Chicago Press, Chicago

**Steven A. Benner**
Foundation for Applied Molecular Evolution
The Westheimer Institute for Science and Technology
P.O. Box 13174 Gainesville, FL 32604 USA

Biodata of **Steven A. Benner**, author of the "***Foreword***"

**Steven A. Benner** is a distinguished fellow at the Foundation for Applied Molecular Evolution and The Westheimer Institute for Science and Technology in Gainesville, Florida. He received his Ph.D. in Chemistry at Harvard University in the laboratories of Robert Woodward and Frank Westheimer. He was then on the faculty of Harvard University, the Swiss Federal Institute of Technology in Zurich, and the University of Florida, before establishing the Foundation and the Institute a decade ago. His research combines two broad traditions in science, the first from natural history and the second from the physical sciences. Toward this goal, his group works in fields such as organic chemistry, biophysics, molecular evolution, bioinformatics, geobiology, and planetary science. He contributed to the founding of several new fields, including synthetic biology, paleogenetics, and computational bioinformatics. His synthetic genetic systems improve the medical care of some 400,000 patients each year suffering from infectious diseases. He also cochaired with John Baross the National Research Committee's 2007 panel on the "Limits to Organic Life in the Solar System" and has contributed to our understanding how mineral species and organic molecules might have interacted to originate life.

E-mail: **sbenner@ffame.org**

# INTRODUCTION

The authors of individual chapters in this book describe a variety of experimental and theoretical approaches to the fundamental question of life's origin. Therefore, it seems appropriate in an introduction to try to define what we mean by "life." Surprisingly, there is no definition that is generally accepted by biologists. Even the simplest microorganisms are extraordinarily complex, and dictionary-style definitions of a sentence or two do not seem to encompass such complexity. However, in the next few years, it is likely that someone will claim to have fabricated artificial life in the laboratory, so they will need to match their claim to a satisfactory definition.

Because life is such a complex phenomenon, perhaps the best we can do is to state a minimal set of properties that, taken together, exclude anything that is not alive. Here is one such list that defines prokaryotes, the simplest form of life today that are descendents of the earliest life:

- Life is a system of polymers synthesized by chemical reactions (metabolism) that take place in membrane-bounded compartments called cells.
- The primary polymers of life today are nucleic acids and proteins. Nucleic acids have a unique ability to store and transmit genetic information. Proteins called enzymes have a unique ability to act as catalysts that increase the rates of metabolic reactions.
- Cells use nutrients and energy available in their environment to produce activated monomers and link the monomers into polymers. The primary chemical linkages are ester and peptide bonds, produced by condensation reactions in which the equivalent of a water molecule is removed to form the bond. Polymer synthesis is the fundamental process leading to growth of a living system.
- Life is characterized by a cyclic feedback system in which information in genetic polymers is used to direct the synthesis of catalytic polymers, which in turn catalyze the synthesis and replication of genetic polymers.
- During growth, the cyclic system of polymers reproduces itself, and the cellular compartment divides.
- Reproduction is not perfect, so that variations arise, resulting in differences between cells in a population. Because different cells have varying capacities to grow and survive in a given environment, individual cells undergo selection according to their ability to compete for nutrients and energy. As a result, populations of cells have the capacity for evolution.

This list would be obvious to biochemists and molecular biologists because it emerged from their research over the past 50 years. Each item in the list has been demonstrated experimentally in the laboratory, and the new tools of synthetic biology suggest that it may be possible to assemble systems of molecules that integrate all of these properties. If so, we will have succeeded in fabricating artificial living cells.

With this working description, we can ask how such a complex system of molecules could have emerged spontaneously on the surface of a sterile planet like the early Earth and perhaps Mars. Chemists perceive the origin of life as a chemical process, and it is indisputable that when the first microscopic organisms began to grow and reproduce on the early Earth, chemical reactions associated with growth, metabolism, and replication were central to the process. But how did the chemistry begin? It seems likely that life emerged from the interaction of physical and chemical processes that produce molecular systems of polymers capable of catalysis, replication, and evolution. If it becomes possible to fabricate such artificial systems, the next step will be to relate them to the origin of life by testing whether they could arise and survive in simulated early Earth conditions.

In the past, simulated prebiotic conditions have focused on chemical reactions such as the Miller-Urey experiment in which a gas mixture was exposed to an electrical discharge. Another approach has been to test whether a primitive version of metabolism could emerge from reactants adsorbed on the surfaces of minerals. However, I suspect that chemical reactions in solution or on surfaces are insufficient to account for the origin of life and that we need to incorporate other processes that were likely to have been present in the early Earth environment. For instance, geothermal and hydrothermal conditions are potential sources of chemical and thermal activation energy that can drive organic polymerization reactions. If membranous compartments are added to simulations, molecules in solution can be encapsulated and organized into interacting systems. Simulations incorporating cycles of temperature and hydration have the capacity to concentrate reactants and pump reactive molecules toward ever increasing complexity. However, for this to occur, we need to have a better understanding of the kinetic traps that can maintain polymers away from thermodynamic equilibrium.

In a sufficiently complex laboratory simulation, and presumably in the prebiotic environment, the result of this combination of conditions is vast numbers of microscopic compartments. Because every compartment will be different from all the rest in terms of the mix of potentially catalytic polymers they contain, each would be equivalent to a microscopic experiment in a primitive version of combinatorial chemistry. Most will be inert, but a few may have a specific, yet-to-be-discovered combination of interacting polymers that can catalyze their own growth and reproduction using energy and nutrients available in the environment.

Of course, not all would agree with this strategy for experimental approaches to the origin of life problem. It runs counter to reductionist methods which tacitly assume that we will be able to deduce how life began if we know enough about the specific chemical reactions that are the foundation of life today. However, reductionism by definition limits the potential for interacting systems to emerge. The experimental conditions of a typical test tube simulation may be insufficiently complex, while the origin of life clearly occurred under conditions that were sufficiently complex.

In my view, future progress in understanding the origin of life will involve simulations of prebiotic conditions that take this complexity into account. Some of the chapters in this book focus on specific aspects of the chemical and physical processes that would be involved in life's origins, while others incorporate increments in complexity such as mineral surfaces and compartments. In a sense, each chapter represents a piece of a puzzle, and readers who take the time to read all the chapters may see unexpected patterns emerge that will give clues to solving the puzzle.

**David W. Deamer**
Department of Biomolecular Engineering
and Department of Chemistry and Biochemistry
at the University of California,
Santa Cruz, CA. USA

Biodata of **David W. Deamer**, author of the "***Introduction***"

**David W. Deamer** is a research professor in the Department of Biomolecular Engineering and Department of Chemistry and Biochemistry at the University of California, Santa Cruz. His undergraduate B.Sc. degree was in Chemistry at Duke University, Durham, NC (1961), and his Ph.D. in Physiological Chemistry from the Ohio State University School of Medicine (1965). Following postdoctoral research at UC Berkeley, he joined the faculty at UC Davis in 1967. In 1994, he moved his laboratory to UC Santa Cruz.

Prof. Deamer's research interest concerns the process by which cellular life arose on the Earth nearly 4 billion years ago. This involves studies of meteorites that contain organic carbon compounds and self-assembly of complex lipid-protein structures that exhibit some of the properties of life. A second research area concerns DNA transport through nanoscopic pores in membranes. This work focuses on developing an instrument that can analyze nucleic acids as individual molecules. His recent books are:

Deamer, D.W. and Szostak, J. eds. 2010. *Origins of Life.* Cold spring Harbor Press

Deamer, D.W. *First Life.* University of California Press, June 2011

E-mail: **deamer@soe.ucsc.edu**

## ORIGINS OF LIFE – KEEPING OPTIONS OPEN

The origin of life is an immense subject and could be considered one of the most important single questions in science. It is also clear that the variety of possible paths which one might follow in approaching the subject matter is highly diverse, to say the least. This book reflects that breadth as well as the eclectic nature of the field. While a great deal has been achieved in the nearly 60 years since the publication of the first description of what has become universally known as the "Miller Experiment" (Miller, 1953), there remain many areas which have only been touched upon. One of these is the question of how any order could have emerged from the embarrassment of riches created in such a synthetic process (Schwartz, 2007). This question also creates a conundrum for astrobiology. What we cannot understand in scenarios for the early Earth is certainly no easier with respect to the emergence of life elsewhere. In searching for answers to the problem, we could do worse than to maintain open minds about the kinds of planetary conditions which might be considered as possible sources of complex chemistry, as well as potential abodes of life. Let me immediately say that I am not referring to "exotic biochemistries" but to the historical roots of the field of prebiotic synthesis.

The first published attempt to simulate prebiotic synthesis by Garrison et al. (1951) was a step which actually extended a long tradition of approaches toward an understanding of the origins of photosynthesis (Chittenden and Schwartz, 1981). It has been relegated to the background as a result of the much more productive reducing model simulation which Miller introduced. In a period when a strongly reducing atmosphere was *the* standard model, Hartman, in a perspicacious paper (1975), speculated on the advantages of non-reducing conditions in which mineral surfaces catalyzed a slow conversion of $CO_2$ to ever more complex products and, ultimately, to entire systems of metabolism. Of interest is the fact that the citric acid cycle was central to Hartman's prebiotic scenario, as it is to certain hypotheses today (Morowitz et al., 2000; Smith and Morowitz, 2004; Orgel, 2000, 2008). Although $CO_2$ was originally thought to be an unproductive source of carbon for prebiotic synthesis, this view has been revised and reactions in $CO_2/N_2$ can be surprisingly effective in producing amino acids (Cleaves et al., 2008). In addition, increased interest in models of carbon fixation (utilizing CO or $CO_2$) catalyzed by transition metals and mineral surfaces also seems to place Hartman's model in a new light (Cody et al., 2004). On the other hand, recent work on the synthetic activity of submarine hot springs has begun to suggest that the reducing level of Earth's primitive atmosphere may not have been the critical factor permitting organic synthesis. The process of serpentinization, which is a source of reducing potential, is attracting increasing interest (McCollom and

Seewald, 2006). It seems that our options keep changing as advances are made in this very uncertain field of early Earth geochemistry. Such a process should not be surprising, considering the general history of progress in science.

We have seen periods when reducing atmospheres were the fashion, we have seen periods when they went out of fashion, we have seen periods when they came back in fashion, and – certainly when we consider the significance of exogenic sources of organic compounds – we are apparently in a period when they no longer matter. What really matters is the determination to seek experimental answers to difficult questions involving chemical evolution in many kinds of environments. The search for other examples of life provides not only a stimulus for research but an opportunity to reexamine our models as well.

## References


Chittenden GJF, Schwartz AW (1981) Prebiotic photosynthetic reactions. BioSystems 14:75–32

Cleaves HJ, Chalmers JH, Lazcano A, Miller SL, Bada JL (2008) A reassessment of prebiotic organic synthesis in neutral planetary atmospheres. Orig Life Evol Biosph 38:105–115

Cody GD, Boctor NZ, Brandes JA, Filley TR, Hazen RM, Yoder Jr HS (2004) Assaying the catalytic potential of transition metal sulfides for abiotic carbon fixation. Geochim Cosmochim Acta 68:2185–2196

Garrison WM, Morrison DC, Hamilton JG, Benson AA, Calvin M (1951) Reduction of carbon dioxide in aqueous solutions by ionizing radiation. Science 114:416–418

Hartman H (1975) Speculations on the origin and evolution of metabolism. J Mol Evol 4:359–370

McCollom TM, Seewald JS (2006) Carbon isotope composition of organic compounds produced by abiotic synthesis under hydrothermal conditions. Earth Planet Sci Lett 243:74–84

Miller SL (1953) A production of amino acids under possible primitive earth conditions. Science 117:528–529

Morowitz HJ, Kostelnik JD, Yang J, Cody GD (2000) The origin of intermediary metabolism. Proc Natl Acad Sci 97:7704–7708

Orgel LE (2000) Self-organizing biochemical cycles. Proc Natl Acad Sci 97:12503–12507

Orgel LE (2008) The implausibility of metabolic cycles on the prebiotic earth. PloS Biology 6:e18

Schwartz AW (2007) Intractable mixtures and the origin of life. Chem Biodivers 4:656–664

Smith E, Morowitz HJ (2004) Universality in intermediary metabolism. Proc Natl Acad Sci 101:13168–13172



**Alan W. Schwartz**

Evolutionary Biology, RUN, Lindenheuvel 12, 1217 JX Hilversum, The Netherlands

Biodata of **Alan W. Schwartz** author of the "***Introduction to Origins: Genesis, Evolution and Diversity of Life (2nd Edition)***."

**Professor Alan Schwartz** is a member of the Faculty of Science at Radboud University Nijmegen, The Netherlands. He obtained his Ph.D. in Biochemistry at the Florida State University, Tallahassee, Florida, in 1965. Dr. Schwartz spent his postdoctoral (1965–1967) as a research appointee, Biomedical Research Group, in Los Alamos Scientific Laboratory, New Mexico. During 1967–1968, he was a national research council resident research associate at the NASA Ames Research Center, Moffett Field, California. Among his visiting positions were Chemical Evolution Laboratory, The Salk Institute for Biological Studies, La Jolla, California, USA, and University of Tsukuba, Tsukuba, Japan. Dr. Schwartz has been the managing editor of *BioSystems* (1972–1999) and the editor in chief of *Origins of Life and Evolution of Biospheres* (2000–present). He was awarded the H.C. Urey Medal, International Society for the Study of the Origin of Life [ISSOL] (1999), and serves in various advisory functions for ESA, NASA, the European Science Organization, and the Dutch Space Organization.

E-mail: **alan@sci.ru.ni**

## LIST OF AUTHORS AND THEIR ADDRESSES

**ABEL, DAVID L.**
DEPARTMENT OF PROTOBIOCYBERNETICS/PROTOBIOSEMIOTICS, THE GENE EMERGENCE PROJECT, THE ORIGIN-OF-LIFE SCIENCE FOUNDATION, INC. GREENBELT, MD, USA

**AGMON, ILANA**
SCHULICH FACULTY OF CHEMISTRY, INSTITUTE FOR ADVANCED STUDIES IN THEORETICAL CHEMISTRY, TECHNION – ISRAEL INSTITUTE OF TECHNOLOGY, HAIFA, ISRAEL
FRITZ HABER RESEARCH CENTER FOR MOLECULAR DYNAMICS, HEBREW UNIVERSITY, JERUSALEM, ISRAEL

**SCHWARTZ, ALAN W.**
EVOLUTIONARY BIOLOGY, RADBOUD UNIVERSITY NIJMEGEN, HILVERSUM, THE NETHERLANDS

**ARETXAGA-BURGOS, ROBERTO**
DEPARTMENT OF PHILOSOPHY, UNIVERSITY OF DEUSTO, BILBAO, SPAIN

**ARIGA, KATSUHIKO**
WORLD PREMIER INTERNATIONAL (WPI) RESEARCH CENTER FOR MATERIALS NANOARCHITECTONICS (MANA), NATIONAL INSTITUTE FOR MATERIALS SCIENCE (NIMS), TSUKUBA, JAPAN
JAPAN SCIENCE AND TECHNOLOGY AGENCY, CREST, TSUKUBA, JAPAN

**BADA, JEFFREY L.**
SCRIPPS INSTITUTION OF OCEANOGRAPHY, UNIVERSITY OF CALIFORNIA, LA JOLLA SAN DIEGO, CA, USA

**BARBALET, TOM**
NOBLE APE AND BIOTA.ORG, LAS VEGAS, NV, USA

**BEJ, ASIM K.**
DEPARTMENT OF BIOLOGY, UNIVERSITY OF ALABAMA AT BIRMINGHAM, BIRMINGHAM, AL, USA

**BENNER, STEVEN A.**
FOUNDATION FOR APPLIES MOLECULAR EVOLUTION,
THE WESTHEIMER INSTITUTE FOR SCIENCE
AND TECHNOLOGY, GAINESVILLE, FL, USA

**BRANCIAMORE, SERGIO**
DEPARTMENT OF PHYSICS AND ASTRONOMY,
UNIVERSITY OF FLORENCE, FLORENCE, ITALY

**CÁRDENAS, ROLANDO**
DEPARTMENT OF PHYSICS, UNIVERSIDAD
CENTRAL DE LAS VILLAS, SANTA CLARA, CUBA

**CARRAPIÇO, FRANCISCO**
DEPARTMENTO DE BIOLOGIA VEGETAL, FACULDADE DE
CIÊNCIAS, CENTRO DE BIOLOGIA AMBIENTAL E CENTRO
DE FILOSOFIA DAS CIÊNCIAS, UNIVERSIDADE DE LISBOA,
LISBON, PORTUGAL

**CHELA-FLORES, JULIAN**
THE ABDUS SALAM ICTP, TRIESTE, ITALY
INSTITUTO DE ESTUDIOS AVANZADOS, IDEA,
CARACAS, REPUBLICA BOLIVARIANA DE VENEZUELA

**CHEN, IRENE A.**
FAS CENTER FOR SYSTEMS BIOLOGY,
HARVARD UNIVERSITY, CAMBRIDGE, MA, USA

**CHOPRA, ADITYA**
PLANETARY SCIENCE INSTITUTE, RESEARCH SCHOOL
OF ASTRONOMY AND ASTROPHYSICS AND THE RESEARCH
SCHOOL OF EARTH SCIENCES, AUSTRALIAN NATIONAL
UNIVERSITY, CANBERRA, AUSTRALIA

**CIMPOIASU, VILY M.**
DEPARTMENT OF BIOLOGY, UNIVERSITY OF CRAIOVA,
CRAIOVA, ROMANIA

**CLEAVES II, H. JAMES**
THE CARNEGIE INSTITUTION FOR SCIENCE,
WASHINGTON, DC, USA

**COCKELL, CHARLES S.**
CENTRE FOR EARTH, PLANETARY AND SPACE RESEARCH,
OPEN UNIVERSITY, MILTON KEYNES, UK

**DAMER, BRUCE**
DIGITAL SPACE CORPORATION, SANTA CRUZ, CA, USA

**DEAMER, DAVID W.**
DEPARTMENT OF CHEMISTRY AND BIOCHEMISTRY,
UNIVERSITY OF CALIFORNIA, SANTA CRUZ, CA, USA

**DEL BIANCO, CRISTINA**
CIBIO, UNIVERSITY OF TRENTO, MATTARELLO, ITALY

**DI MAURO, ERNESTO**
DIPARTIMENTO DI GENETICA E BIOLOGIA MOLECOLARE,
UNIVERSITÀ DI ROMA "SAPIENZA", ROME, ITALY

**DUNKER, A. KEITH**
DEPARTMENT OF BIOCHEMISTRY AND MOLECULAR
BIOLOGY, CENTER FOR COMPUTATIONAL BIOLOGY
AND BIOINFORMATICS, INDIANA UNIVERSITY
SCHOOL OF MEDICINE, INDIANAPOLIS, IN, USA

**FERNANDO, CHRISANTHA**
DEPARTMENT OF INFORMATICS,
UNIVERSITY OF SUSSEX, FALMER, BRIGHTON, UK

**FITZ, DANIEL**
DIVISION OF THEORETICAL CHEMISTRY, INSTITUTE
OF GENERAL, INORGANIC AND THEORETICAL CHEMISTRY,
UNIVERSITY OF INNSBRUCK, INNSBRUCK, AUSTRIA

**FRIEDEN, B. ROY**
COLLEGE OF OPTICAL SCIENCES,
UNIVERSITY OF ARIZONA, TUCSON, AZ, USA

**GALLORI, ENZO**
DEPARTMENT OF PHYSICS AND ASTRONOMY,
UNIVERSITY OF FLORENCE, FLORENCE, ITALY

**GATENBY, ROBERT A.**
MATHEMATICAL ONCOLOGY,
MOFFITT CANCER CENTER, TAMPA BAY, FL, USA

**GORDON, RICHARD**
UNIVERSITY OF MANITOBA, WINNIPEG, MN, CANADA

**GRAHAM, JOHN**
SAN DIEGO STATE UNIVERSITY VISUALIZATION CENTER, CA, USA

**GRIFFITH, ROBERT W.**
DEPARTMENT OF BIOLOGY, UNIVERSITY OF MASSACHUSETTS DARTMOUTH, NORTH DARTMOUTH, MA, USA

**HARRIS, KELLEY**
UC BERKELEY MATHEMATICS DEPARTMENT, BERKELEY, CA, USA

**HENGEVELD, ROB**
VRIJE UNIVERSITEIT, RANDWIJK, AMSTERDAM, THE NETHERLANDS

**HILL, JONATHAN P.**
WORLD PREMIER INTERNATIONAL (WPI) RESEARCH CENTER FOR MATERIALS NANOARCHITECTONICS (MANA), NATIONAL INSTITUTE FOR MATERIALS SCIENCE (NIMS), TSUKUBA, JAPAN
JAPAN SCIENCE AND TECHNOLOGY AGENCY, CREST, TSUKUBA, JAPAN

**HORVATH, J. E.**
DEPARTMENT OF ASTRONOMY, IAG-UNIVERSIDADE DE SÃO PAULO, SÃO PAULO, BRAZIL

**ICHIHASHI, NORIKAZU**
ERATO PROJECT, JAPAN SCIENCE AND TECHNOLOGY AGENCY, SUITA, OSAKA, JAPAN

**IKEHARA, KENJI**
NARASAHO COLLEGE, NARA, JAPAN
INTERNATIONAL INSTITUTE FOR ADVANCED STUDIES, KYOTO, JAPAN
THE OPEN UNIVERSITY OF JAPAN, NARA STUDY CENTER, KITA-UOYA-HIGASHI-MACHI, NARA, JAPAN
EMERITUS PROFESSOR OF NARA WOMEN'S UNIVERSITY, NARA, JAPAN

**JAKSCHITZ, THOMAS**
DIVISION OF THEORETICAL CHEMISTRY, INSTITUTE OF GENERAL, INORGANIC AND THEORETICAL CHEMISTRY, UNIVERSITY OF INNSBRUCK, INNSBRUCK, AUSTRIA

**KALTENEGGER, LISA**
ASTRONOMY DEPARTMENT, HARVARD UNIVERSITY,
CAMBRIDGE, MA, USA

**KAWAMURA, KUNIO**
DEPARTMENT OF APPLIED CHEMISTRY,
OSAKA PREFECTURE UNIVERSITY, SAKAI, OSAKA, JAPAN

**KITA, HIROSHI**
ERATO PROJECT, JAPAN SCIENCE
AND TECHNOLOGY AGENCY, SUITA, OSAKA, JAPAN

**KLUSSMANN, MARTIN**
MAX-PLANCK-INSTITUT FUER KOHLENFORSCHUNG,
MUELHEIM AN DER RUHR, GERMANY

**KOMPANICHENKO, VLADIMIR**
DEPARTMENT OF GENETICS AND EVOLUTION,
INSTITUTE FOR COMPLEX ANALYSIS, BIROBIDZHAN, RUSSIA

**LATHE, RICHARD**
PIETA RESEARCH, EDINBURGH, UK
INSTITUTE OF BIO-ORGANIC CHEMISTRY, PUSHCHINO,
MOSCOW REGION, RUSSIAN FEDERATION

**LENTE, GÁBOR**
DEPARTMENT OF INORGANIC AND ANALYTICAL CHEMISTRY,
UNIVERSITY OF DEDRECEN, DEBRECEN, HUNGARY

**LINEWEAVER, CHARLES H.**
PLANETARY SCIENCE INSTITUTE, RESEARCH SCHOOL
OF ASTRONOMY AND ASTROPHYSICS AND THE RESEARCH
SCHOOL OF EARTH SCIENCES, AUSTRALIAN NATIONAL
UNIVERSITY, CANBERRA, AUSTRALIA

**MANSY, SHEREF S.**
CIBIO, UNIVERSITY OF TRENTO, MATTARELLO, ITALY

**MARTIN, OSMEL**
DEPARTMENT OF PHYSICS, UNIVERSIDAD CENTRAL
DE LAS VILLAS, SANTA CLARA, CUBA

**MATSUURA, TOMOAKI**
ERATO PROJECT, JAPAN SCIENCE AND TECHNOLOGY AGENCY,
SUITA, OSAKA, JAPAN

**MCNICHOL, JESSE C.**
NATIONAL RESEARCH COUNCIL OF CANADA, INSTITUTE FOR MARINE BIOSCIENCES, HALIFAX, NOVA SCOTIA, CANADA

**MOSQUEIRA, F. G.**
DIRECCIÓN GENERAL DE DIVULGACIÓN DE LA CIENCIA, UNIVERSIDAD NACIONAL AUTÓNOMA DE MÉXICO, MÉXICO, D.F., MEXICO

**MULLER, ANTHONIE W. J.**
SWAMMERDAM INSTITUTE FOR LIFE SCIENCES, UNIVERSITY OF AMSTERDAM, AMSTERDAM, THE NETHERLANDS

**NEGRÓN-MENDOZA ALICIA**
INSTITUTO DE CIENCIAS NUCLEARES, UNIVERSIDAD NACIONAL AUTÓNOMA DE MÉXICO, MÉXICO, D.F., MEXICO

**NEWMAN, PETER**
DIGITALSPACE CORPORATION, SANTA CRUZ, CA, USA

**NORKUS, RYAN**
DIGITALSPACE CORPORATION, SANTA CRUZ, CA, USA

**PAWLIKOWSKI, MACIEJ**
LABORATORY OF BIOMINERALOGY, CATHEDRAL MINERALOGY, PETROGRAPHY AND GEOCHEMISTRY, FACULTY GEOLOGY, GEOPHYSICS AND PROTECTION OF ENVIRONMENTS, AGH UNIVERSITY OF SCIENCE AND TECHNOLOGY, CRACOW, POLAND

**PEÑATE, LIUBA**
DEPARTMENT OF BIOLOGY, UNIVERSIDAD CENTRAL DE LAS VILLAS, SANTA CLARA, CUBA

**PEREIRA, LUÍSA**
ESCOLA SECUNDÁRIA FERNANDO LOPES-GRAÇA, PAREDE, PORTUGAL

**POLANCO-GONZALEZ, CARLOS**
CENTRO DE INVESTIGACIONES QUÍMICAS, UNIVERSIDAD AUTÓNOMA DEL ESTADO DE MORELOS, MORELOS, MEXICO

**POPA, RADU**
DEPARTMENT OF BIOLOGY, PORTLAND STATE UNIVERSITY, PORTLAND, OR, USA

**RAMOS-BERNAL, SERGIO**
INSTITUTO DE CIENCIAS NUCLEARES, UNIVERSIDAD NACIONAL AUTÓNOMA DE MÉXICO, MÉXICO, D.F., MEXICO

**RAULIN, FRANÇOIS**
LISA-IPSL, CNRS/UPEC AND UNIV. PARIS DIDEROT, CRETEIL, FRANCE

**RAULIN-CERCEAU, FLORENCE**
CENTRE ALEXANDRE KOYRE (UMR 8560/CNRS-/EHESS/MNHN) MUSEUM NATIONAL D'HISTOIRE NATURELLE, PARIS, FRANCE

**RAVEN, JOHN A.**
DIVISION OF PLANT SCIENCES, UNIVERSITY OF DUNDEE AT TJHI, THE JAMES HUTTON INSTITUTE, INVERGOWRIE, DUNDEE, UK
DIVISION OF PLANT SCIENCES, UNIVERSITY OF DUNDEE AT SCRI, SCOTTISH CROP RESEARCH INSTITUTE, INVERGOWRIE, DUNDEE, UK

**RICHARDS, GARY J.**
WORLD PREMIER INTERNATIONAL (WPI) RESEARCH CENTER FOR MATERIALS NANOARCHITECTONICS (MANA), NATIONAL INSTITUTE FOR MATERIALS SCIENCE (NIMS), TSUKUBA, JAPAN

JAPAN SCIENCE AND TECHNOLOGY AGENCY, CREST, TSUKUBA, JAPAN

**RIMOLA, ALBERT**
DEPARTAMENT DE QUÍMICA, UNIVERSITAT AUTÒNOMA DE BARCELONA, BELLATERRA, SPAIN

**RODE, BERND MICHAEL**
DIVISION OF THEORETICAL CHEMISTRY, INSTITUTE OF GENERAL, INORGANIC AND THEORETICAL CHEMISTRY, UNIVERSITY OF INNSBRUCK, INNSBRUCK, AUSTRIA

**RODRIGUES, TELMA**
ESCOLA SECUNDÁRIA DR. JOSÉ AFONSO, SEIXAL, PORTUGAL

**SAPP, JAN**
DEPARTMENT OF BIOLOGY, YORK UNIVERSITY,
TORONTO, ON, CANADA

**SECKBACH, JOSEPH**
P.O. BOX 1132, MEVO HADAS 20, EFRAT 90435, ISRAEL

**SHELDON, ROBERT B.**
NATIONAL SPACE SCIENCE & TECHNOLOGY CENTER,
HUNTSVILLE, AL, USA

**SIMAKOV, MICHAEL**
GROUP OF EXOBIOLOGY, INSTITUTE OF CYTOLOGY,
RUSSIAN ACADEMY OF SCIENCES, ST. PETERSBURG, RUSSIA

**SODUPE, MARIONA**
DEPARTAMENT DE QUÍMICA, UNIVERSITAT
AUTÒNOMA DE BARCELONA, BELLATERRA, SPAIN

**STRICK, JAMES E.**
DEPARTMENT OF EARTH AND ENVIRONMENT,
FRANKLIN AND MARSHALL COLLEGE, LANCASTER, PA, USA

**SUMMERS, DAVID P.**
CARL SAGAN CENTER, SETI INSTITUTE, C/O NASA
AMES RESEARCH CENTER, MOFFETT FIELD, CA, USA

**SUNAMI, TAKESHI**
ERATO PROJECT, JAPAN SCIENCE AND TECHNOLOGY AGENCY,
SUITA, OSAKA, JAPAN

**SUZUKI, HIROAKI**
ERATO PROJECT, JAPAN SCIENCE AND TECHNOLOGY AGENCY,
SUITA, OSAKA, JAPAN
DEPARTMENT OF BIOINFORMATIC ENGINEERING,
GRADUATE SCHOOL OF INFORMATION SCIENCE AND
TECHNOLOGY, OSAKA UNIVERSITY, SUITA, OSAKA, JAPAN

**SVETINA, SAŠA**
INSTITUTE OF BIOPHYSICS, FACULTY OF MEDICINE,
UNIVERSITY OF LJUBLJANA, LJUBLJANA, SLOVENIA
JOŽEF STEFAN INSTITUTE, LJUBLJANA, SLOVENIA

**TREVORS, JACK T.**
SCHOOL OF ENVIRONMENTAL SCIENCES, LABORATORY OF MICROBIOLOGY, UNIVERSITY OF GUELPH, GUELPH, ON, CANADA

**TRIFONOV, EDWARD N.**
GENOME DIVERSITY CENTER, INSTITUTE OF EVOLUTION, UNIVERSITY OF HAIFA, MOUNT CARMEL, HAIFA, ISRAEL
DEPARTMENT OF FUNCTIONAL GENOMICS AND PROTEOMICS, INSTITUTE OF EXPERIMENTAL BIOLOGY, FACULTY OF SCIENCE, MASARYK UNIVERSITY, BRNO, CZECH REPUBLIC

**UGLIENGO, PIERO**
DIPARTIMENTO CHIMICA IFM, UNIVERSITÀ DI TORINO & NIS CENTRE OF EXCELLENCE, TORINO, ITALY

**VAN ELSAS, J. D.**
LINNEAUSBORGH, UNIVERSITY OF GRONINGEN, GRONINGEN, THE NETHERLANDS

**VASAS, VERA**
DEPARTAMENT DE GENÈTICA I DE MICROBIOLOGIA, GRUP DE BIOLOGIA EVOLUTIVA (GBE), UNIVERSITAT AUTÒNOMA DE BARCELONA, BELLATERRA (BARCELONA), SPAIN

**WACEY, DAVID**
CENTRE FOR MICROSCOPY, CHARACTERISATION AND ANALYSIS, THE UNIVERSITY OF WESTERN AUSTRALIA, CRAWLEY, WA, AUSTRALIA
SCHOOL OF EARTH AND ENVIRONMENT, THE UNIVERSITY OF WESTERN AUSTRALIA, CRAWLEY, WA, AUSTRALIA

**WICKRAMASINGHE, CHANDRA**
BUCKINGHAM CENTRE FOR ASTROBIOLOGY, THE UNIVERSITY OF BUCKINGHAM, BUCKINGHAM, UK

**YOMO, TETSUYA**
ERATO PROJECT, JAPAN SCIENCE AND TECHNOLOGY AGENCY, SUITA, OSAKA, JAPAN
DEPARTMENT OF BIOINFORMATIC ENGINEERING, GRADUATE SCHOOL OF INFORMATION SCIENCE AND TECHNOLOGY, OSAKA UNIVERSITY, SUITA, OSAKA, JAPAN
GRADUATE SCHOOL OF FRONTIER BIOSCIENCES, OSAKA UNIVERSITY, SUITA, OSAKA, JAPAN

# PART I: CHEMICAL EVOLUTION

Cleaves II
Bada
Fernando
Vasas
Gallori
Branciamore
Harris
Chen
Hengeveld
Ikehara
Kawamura

Martin
Cárdenas
Peñate
Horvath
Mosqueira
Negrón-Mendoza
Ramos-Bernal
Polanco-Gonzalez
Pawlikowski
Simakov
Summers

Biodata of **H. James Cleaves II** and **Jeffrey L. Bada**, authors of "***The Prebiotic Chemistry of Alternative Nucleic Acids***."

**Dr. H. James Cleaves II** is currently a visiting scientist at the Geophysical Laboratory of the Carnegie Institution of Washington. He received his Ph.D. in Chemistry and Biochemistry from UC San Diego in 2001. He was later a research fellow at the Scripps Institution of Oceanography, and a lecturer in the Department of Chemistry and Biochemistry at UCSD. Dr. Cleaves' research interests include prebiotic organic chemistry and the interactions of organic compounds with mineral surfaces.

E-mail: **hjcleaves@ciw.edu**

**Professor Jeffrey L. Bada** joined the faculty of the Scripps Institution of Oceanography in 1970, where he is now a Distinguished Research Professor of Marine Chemistry. He received his Ph.D. in Chemistry in 1968 from UC San Diego. His research concerns the biogeochemistry of amino acids, the sources and stability of organic compounds on the primitive Earth, the origin of homochirality in biochemistry, the search for evidence of prebiotic chemistry and primitive life on Mars and other solar system bodies, and cosmogeochemical applications of state-of-the-art analytical techniques.

E-mail: **jbada@ucsd.edu**

**H. James Cleaves II**

**Jeffrey L. Bada**



J. Seckbach (ed.), *Genesis - In The Beginning: Precursors of Life, Chemical Models and Early Biological Evolution*, Cellular Origin, Life in Extreme Habitats and Astrobiology 22, 3–33
DOI 10.1007/978-94-007-2941-4_1, © Springer Science+Business Media Dordrecht 2012

# THE PREBIOTIC CHEMISTRY OF ALTERNATIVE NUCLEIC ACIDS

**H. JAMES CLEAVES II[1] AND JEFFREY L. BADA[2]**
*[1]The Carnegie Institution for Science, 5251 Broad Branch Rd. NW, Washington, DC 20015, USA*
*[2]Scripps Institution of Oceanography, University of California at San Diego, La Jolla, CA 92093-0212, USA*

## 1. Introduction

There are two complementary, though at times conflicting, approaches for investigating the chemical origins of life. The first, the top-down or retrospective approach searches for common attributes of modern organisms that likely existed in the last common ancestor (LCA) of all life on Earth. By extrapolation, it is supposed that these were also necessary for the origin of life. There may be, however, a large temporal or organizational distance between the LCA and the origin of life. Intermediary stages in evolution of life since its origin may have left no trace of their existence in contemporary biochemistry. The bottom-up or "prospective" approach considers what appears to be possible via prebiotic chemistry, the compounds that may have been available on the primitive Earth and the geochemical conditions that may have prevailed at the time of the origin of life. Unfortunately, due to the dearth of geological evidence from this time period, these conditions remain highly uncertain. This approach is clouded by uncertainties in the sources of prebiotic compounds, which would have depended on the oxidation state of the primitive atmosphere and the flux and nature of extraterrestrial organic material delivered to the primitive Earth, local and global environmental conditions, and the time available for simple organic compounds to accumulate, polymerize, and attain some level of chemical complexity.

There are two main models for the "prospective" approach. The "gene-first" model postulates that life began with a self-replicating polymer of RNA or an RNA-like material (Woese, 1967; Crick, 1968; Orgel, 1968). The "metabolist" model proposes that life began with a self-propagating network of mutually catalytic reactions (Kaufman, 1986; de Duve, 1991), which were not necessarily structurally informational in the way that nucleic acids are. Metabolist hypotheses were first proposed by Oparin (1938) and later by a number of other researchers (Morowitz, 1992; Dyson, 1999). Both models regard life as being divisible into two processes, metabolism and replication, but differ in regard to the order of their appearance. In some sense, a self-propagating chemistry predating the advent of genetic system is still prebiotic chemistry. Thus, the origin of a genetic system and heritable information may define the origin of life. The idea of autocatalytic nongenetic chemistry has some merit. However, if a set of reactions which self-propagate have several potential end points, some of which are non-genetic chemical systems and some of which are genetic chemical systems, only the ones which culminate in genetic molecules lead to the origin of life.

It seems logical that replication must have predated metabolism. The term metabolism is at times bandied about loosely, and as mentioned above, protometabolist concepts could be interpreted as forms of prebiotic chemistry. Strictly speaking, the only example of metabolism known is that of the reactions occurring inside living cells, which are directed by a complex host of genetically encoded enzymes. It seems unlikely that metabolic processes would be capable of continuous evolution without a genetic means of propagating their chemical composition (Orgel, 2000, 2008).

Although Hermann Muller in the early part of the twentieth century had posited the origin of life as beginning with the origin of a mutable genetic molecule (Lazcano, 2000), the structure and function of nucleic acids was not understood at that time. The elucidation of the structure of DNA in 1953 immediately suggested a physical basis for the role of nucleic acids in heredity (Watson and Crick, 1953). The impact of this discovery on the experimental study of the origins of life was not, however, quickly realized, with perhaps the first deliberate demonstration of a prebiotic synthesis of a nucleic acid component being Oró's adenine synthesis in 1960 (Oró and Kimball, 1961). The elucidation of the flow of biological information from DNA to RNA to protein led several scientists to suggest that RNA could have functioned as both information storage and catalyst in a primitive or perhaps original organism (Woese, 1967; Crick, 1968; Orgel, 1968), the so-called RNA World (Gesteland et al., 1999). This set off a spate of studies to understand how RNA could have arisen on the primitive Earth.

The general conclusion reached from these studies has been that while there are indeed possible abiotic routes to the synthesis of RNA, the low yields of the precursors such as purines, pyrimidines, and sugars, the specialized conditions required for the stepwise assembly of monomers, and the instability of RNA render the prebiotic origin of RNA "a near miracle" (Joyce and Orgel, 1993).

Cairns-Smith suggested that replicating clays may have been the first replicators and may have served as templates for RNA synthesis, though it seems he was also aware that logically another organic replicator may have preceded RNA (Cairns-Smith, 1977; Joyce and Orgel, 1993). Indeed, the idea of a genetic takeover is inherent in the notion of an RNA World: if life ever passed through an RNA World stage, every living thing on Earth has made the transition from using RNA to DNA as a long-term informational storage molecule. A seminal paper by Joyce et al. (1987) suggested a number of possible RNA precursor molecules. This idea was preceded by many years of biomedical research spawned by the elucidation of the mechanisms of RNA and DNA biosynthesis, which suggested that nucleic acid analogues could be useful treatments for various virally and bacterially caused diseases, due to differences in the ways host and pathogen enzymes process these molecules (Périgaud et al., 1992).

After several decades of research into the prebiotic synthesis of DNA and RNA, it is only relatively recently that the diversity of nucleic acid analogues developed for gene and antisense therapy has begun to have an impact on origins of life research. However, most of the structural alterations of nucleic acids of possible relevance to the origin of life remain uninvestigated, though this may prove to be one of the most *experimentally tractable* areas in prebiotic chemistry. We will attempt here to survey and analyze some of the literature on this problem to date, though any attempt to do so will naturally, and extremely regrettably, omit much relevant literature.

## 1.1. A SCHEME FOR GENETIC MOLECULE FORMATION

The overall scheme for the prebiotic synthesis of a genetic molecule hinges on the notion that organic molecules were available in the environment. A variety of sources of organic compounds are now considered plausible (Cleaves and Miller, 2007; Cleaves and Lazcano, 2009). Figure 1 broadly diagrams the steps likely involved in the prebiotic synthesis of a primordial genetic polymer.

Briefly, and importantly, it is likely that given a sufficiently reducing environment, a variety of energetic processes can induce carbon and nitrogen-containing gases to form reactive molecules such as HCHO, $NH_3$, and HCN. These can in turn further react, once delivered into an aqueous environment, to form more complex organic compounds such as amino acids, sugars, and heterocyclic compounds. Although the number of possible products grows exponentially as the number of atoms in a compound increases, prebiotic reactions are mechanistically constrained and tend to form a relatively small number of the possible low-molecular-weight organic compounds.

### *1.1.1. Searching Sequence Space*

Biological proteins are composed of linear polymers of 20 amino acid monomers which, based on the specific order of the monomers in the polymer, fold into discrete three-dimensional shapes with diverse and specialized catalytic activities.

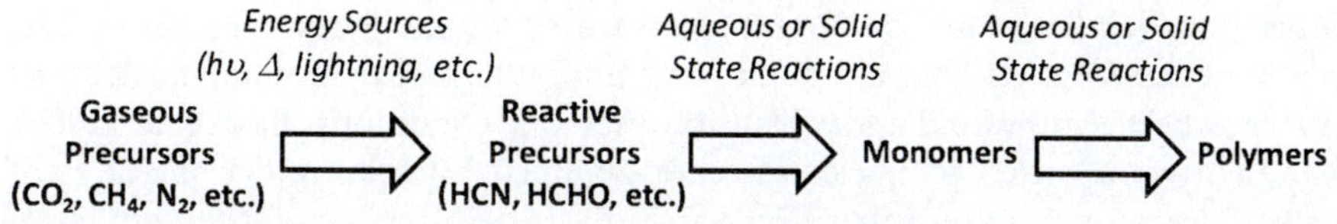


**Figure 1.** General scheme for the prebiotic origin of a genetic polymer.

This is also true of ribozymes evolved in the lab (Doherty and Doudna, 2000), and it seems reasonable that this would also be true of putative precursor genetic molecules. Given that the number of sequences grows as a function of $X^N$ where $N$ is the length of the polymer and $X$ is the number of monomer units, for even short polymers the number of possible sequences quickly becomes astronomical, though only a small fraction of these will likely have any given catalytic activity. It seems important then that the synthesis of prebiotic polymers, and their constituent monomers, must be fairly robust in order for them to collectively sample the vast potential of possible sequence space.

## 1.2. THE RNA WORLD

The idea that life began with a self-replicating RNA molecule has been bolstered by demonstrations that RNA can act as a catalyst (Cech et al., 1981; Guerrier-Takada et al., 1983; Cech, 1987), as well as "molecular fossil" evidence based on coenzyme structure and the role of RNA in translation and metabolic pathway organization (Brachet, 1959; Rich, 1962; Haldane, 1965; Woese, 1967; Crick, 1968; Orgel, 1968; White III, 1982; Gilbert, 1986). This has led to speculation that the "chicken-or-egg" problem of which came first, nucleic acids or proteins, has been solved (Gilbert, 1986). There are, however, numerous problems with the prebiotic synthesis of RNA (Shapiro, 1988; Schwartz, 1995).

For example, while the polymerization of formaldehyde produces ribose, it requires high pH and possibly unreasonably high concentrations of formaldehyde, yields a complex product mixture, and the synthetic conditions also catalyze the destruction of the products (Reid and Orgel, 1967). However, some progress has been made in the selective synthesis of ribose (Pitsch et al., 1995; Krishnamurthy et al., 1999; Ricardo et al., 2004; Lambert et al., 2010). Sugars are also notoriously unstable; for example, at pH 7 and 25°C, the half-life for ribose decomposition is estimated at 300 days (Larralde et al., 1995). Most other sugars and sugar phosphates have half-lives within an order of magnitude of this, thus sugars may be too unstable to have accumulated to significant concentrations in prebiotic environments. These stability estimates do not take into account the competing degradation reactions caused by congener amines, such as ammonia and amino acids, which would likely result in Maillard-type chemistry. Nucleic acid bases are also of limited, although much greater, stability under geochemical conditions (Levy and Miller, 1998).

Thus far, the demonstrated prebiotic syntheses of purine nucleosides are rather poor (Fuller et al., 1972) and much worse for the pyrimidine nucleosides (Sanchez and Orgel, 1970), although some nonbiological pyrimidine nucleosides have now been synthesized under plausibly prebiotic conditions (Bean et al., 2006). Although good yields of nucleotides are obtained by heating dry mixtures of nucleosides, ammonium oxalate (Schwartz et al., 1975) or urea (Osterberg et al., 1973), and naturally occurring phosphate minerals, the availability of activated phosphates for prebiotic phosphorylation reactions, however, has been questioned based on the insolubility and scarcity of most phosphate minerals and the lability of phosphate esters (Keefe and Miller, 1995). More recently, an ingenious synthesis of the biological pyrimidine nucleotides has been demonstrated directly from small molecule precursors (Sutherland and Weaver, 1994; Powner et al., 2009), and it remains possible that some as yet unexplored mechanism will easily explain the prebiotic synthesis of RNA. Such a synthesis would greatly simplify the origin of life. Nevertheless, the accumulated evidence suggests this is a difficult problem.

Enantiomeric cross-inhibition of nucleotide stereoisomers in template polymerizations of activated RNA monomers is another hurdle for prebiotic RNA synthesis (Joyce, 1987). It is worth noting that proposed RNA precursors which are achiral would not suffer from this problem (Schwartz and Orgel, 1985; Joyce et al., 1987), but this would postpone the problem of the introduction of chirality into biological systems until later, perhaps once catalysts capable of discriminating between enantiomers had evolved.

Overall, it appears that the problems with the prebiotic synthesis of RNA cannot be overestimated (Cairns-Smith, 1977; Shapiro, 1988), and thus the presence of RNA in biological systems is the result of a protracted period of abiological and biological evolution. According to this view, RNA is a specialized solution to a complex set of interlocking biochemical problems: for example, the fidelity of base-pairing, the ability of the monomers to be synthesized by available catalysts and from available metabolites, the ability of large nucleic acid molecules to be folded to fit into cells, and the kinetic and thermodynamic stability of the monomers and polymers, among others. What in this view then makes RNA unique was its ability to solve multiple functional problems as they arose evolutionarily. These could be problems which could have been solved by other polymers but were not due to the contingencies of stochastic evolutionary processes. Alternatively, RNA may represent the best solution to these multiple selection pressures.

The necessary length for a nucleic acid strand to be able to serve as a catalyst is estimated as ~40–80 nucleotides (Joyce and Orgel, 1993), while the error threshold for a prebiotic polymer without error-correction mechanisms is estimated to be ~80 bases (Eigen and Schuster, 1976; Biebricher and Eigen, 2005). The crux of the problem is then the synthesis of an oligomer of some 40 monomer units from plausibly prebiotic starting materials. This is not an experimentally well-constrained size range. It is possible that shorter polymers of more highly functionalized nucleic acid analogues could be highly functional catalysts.

## 1.3. ALTERNATIVES TO RNA

As noted above, the difficulty of prebiotic RNA synthesis suggests that RNA may not have been the first informational polymer (Benner et al., 1989), and RNA may have been preceded by a simpler molecule (Cairns-Smith, 1977; Joyce et al., 1987; Schwartz and Orgel, 1985; Schwartz, 1989). This idea is supported by studies conducted with peptide nucleic acid (PNA), which has been shown to be able to template the polymerization RNA (Böhler et al., 1995), and glycol nucleic acid (GNA), which has also been shown to base-pair with RNA (Zhang et al., 2005), among many other analogue structures.

The prebiotic plausibility of a primordial genetic polymer depends on several considerations, such as its ease and robustness of synthesis, stability, and functional adaptability. The relative plausibility of two or more alternative nucleic acid structures should be decided based on how these parameters compare for each molecule (Fig. 2 and Table 1).

Some of the most obvious points of comparison are shown in Table 1. This set of criteria is by no means exhaustive.

For example, first, the conditions of synthesis must be geochemically plausible, that is to say that there must be an easily demonstrable and environmentally plausible pathway for the synthesis of the monomers from small reactive intermediates such as HCN and HCHO, among others. Although they have not been investigated as extensively as RNA with regard to higher order chemistry such as evolvable and selectable catalytic activity, peptide nucleic acid (PNA) (Egholm et al., 1992) monomers may be more plausible from the standpoint of prebiotic chemistry (Nelson et al., 2000).

Unfortunately, as mentioned above, as the conditions on the primitive Earth remain controversial, what constitutes a geochemically plausible synthesis is debatable. One possible touchstone for prebiotic chemistry comes from carbonaceous chondrites which contain a host of organic compounds formed in the primitive solar system (Cronin et al., 1995). When compounds generated in prebiotic simulations are also found in these natural samples, their prebiotic plausibility is reinforced, although it is reasonable to suspect there were environments on the early Earth which provided conditions unlike those found in carbonaceous chondrites. Organic molecules within a compound class tend to decrease in abundance with increasing carbon number in the Murchison meteorite and in prebiotic simulations (Wolman et al., 1972; Cronin et al., 1995) (Fig. 3). Thus, smaller

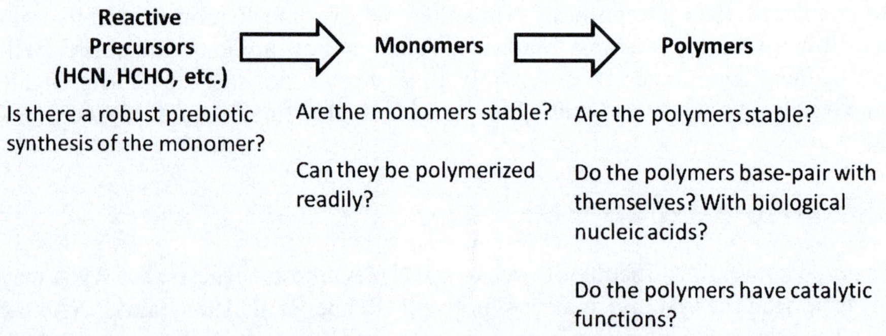


**Figure 2.** General scheme for evaluating the plausibility of an alternate primordial nucleic acid.

**Table 1.** A comparison of several genetic polymers with respect to some prebiotic selection criteria.

| Molecule | Prebiotic monomer synthesis | Monomer stability | Prebiotic polymer synthesis | Polymer stability | Functionality |
|---|---|---|---|---|---|
| *DNA* | Low for purines, unknown but likely poor for pyrimidines | Purines unstable to acid | Possible through use of activating reagents | Higher than RNA under neutral conditions | Base pairs with RNA and DNA, DNAzymes are known, appear to be poorer catalysts than RNA |
| *RNA* | Low for purines, can be poor for pyrimidines, with some recent exceptions | Nucleosides fairly stable but prone to dephosphorylation | Possible through use of activating reagents, also from 2′,3′-cyclic phosphates | Internal hydrolysis rapid through 2′-OH attack on backbone | Base pairs with RNA and DNA, many ribozymes known |
| *Aeg-PNA* | One-pot syntheses have been demonstrated for both purine and pyrimidine monomers | More stable than nucleosides | Activated dimers have been polymerized on templates | Prone to N-terminal-catalyzed strand scission and rearrangement unless capped | Base pairs with self, RNA, and DNA, catalytic potential unknown |
| *GNA* | Unknown | More stable than nucleosides? | Unknown | Higher than RNA | Unknown |
| *TNA* | Similar to RNA monomers | Similar to nucleosides? | Unknown, similar to RNA? | Higher than RNA | Self-base-pairing and base-pairing with DNA and RNA, catalytic potential unknown |
| *Alanyl-PNA* | Synthesis poor | Similar to nucleosides, rapidly decompose via $\beta$-elimination | Unknown, similar to amino acids? | Unknown, may be similar to a peptide, which is higher than RNA, but $\beta$-elimination problematic | Do not base-pair with DNA or RNA, evidence for self-pairing, catalytic potential unknown |
| *α−PNA* | One-pot syntheses have been demonstrated | Similar to nucleosides | Possible through use of activating reagents | Unknown, may be similar to a peptide, which is higher than RNA | Do not base-pair with DNA or RNA, evidence for self-pairing, catalytic potential unknown |
| *Others?* | ? | ? | ? | ? | ? |

*DNA* deoxyribonucleic acid, *RNA* ribonucleic acid, *aeg-PNA* aminoethylglycine-based PNA, *GNA* glycol nucleic acid, *TNA* threose nucleic acid, *alanyl-PNA* PNA based on α-alanine nucleobase analogues, *α-PNA* PNA based on homoserine nucleobase analogues.

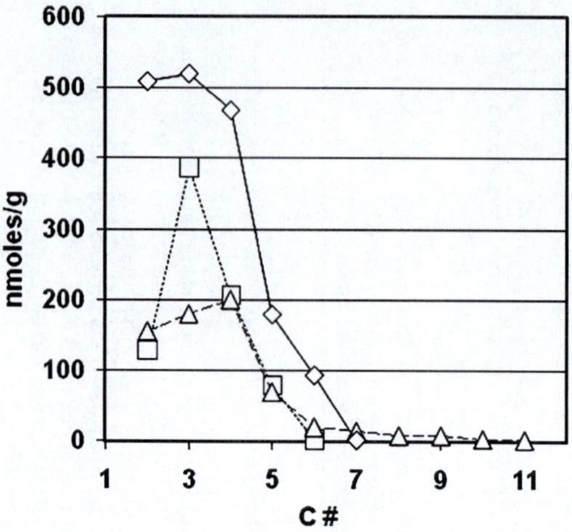


**Figure 3.** Abundance of compounds as a function of the number of carbon atoms in the Murchison meteorite. ◊: amino acids, □: hydroxy acids, □: straight-chain carboxylic acids (See Cronin et al. 1995).

compounds with fewer carbon atoms are likely to be more abundant, reflecting their synthesis from small precursors. For example, none of the nitrogenous bases found in RNA contain a continuous carbon chain longer than 3, although ribose contains a 5 carbon chain.

Second, the monomers must have some degree of stability in order for them to have accumulated to a degree to allow them to be oligomerized. Most organic compounds are more stable near neutral pH and at lower temperatures, which possibly says more about what types of environments are favorable than what types of molecules are (Cleaves and Chalmers, 2004). Nucleosides and nucleotides are by no means the most stable compounds in biochemistry, but they are stable enough in the context of their use in the cell. We cannot say how stable is stable enough, but we can compare analogues to each other, and stability might confer some initial advantage.

Third, the monomers would need some mechanism of being oligomerized. All biological nucleic acids are biosynthesized from activated precursors, nucleoside triphosphates, which overcomes the thermodynamic instability of the polymer. Various prebiotic nucleotide-activating agents have been proposed over the years, and many of these could also likely be used for other types of monomer. Alternatively, a reversible polymerization mechanism requiring no activating agent might have been used by the first genetic molecules.

In order for monomers to be polymerized on a template, reactions need to take place at temperatures low enough to allow for noncovalent interactions that stabilize self-recognition but high enough to overcome thermodynamic barriers to polymerization. Polymerization processes, such as template polymerizations

(Orgel, 1987; Kanavarioti et al., 2001; Trinks et al., 2005), may be more facile at low temperature, which enables Watson-Crick (WC) base-pairing and $\pi$–$\pi$ stacking interactions.

Fourth, the stability of the polymer itself could be an important consideration, although again it is impossible to say how stable is stable enough. We note, however, that DNA is considerably more stable than RNA under physiological conditions, perhaps the precursor polymer was less stable than RNA. Again, most polymers are likely more stable at lower temperatures (Cleaves and Chalmers, 2004).

Lastly, if life began with a one-polymer system, and that system was responsible for carrying out catalytic and informational roles, these functions would be selection criteria for the primordial genetic molecule. As shown in Table 1, many of the nonnatural nucleic acid analogues investigated to date have been shown to base-pair with biological nucleic acids, and many of them base-pair well with themselves as well, which is a more urgent property for a primordial molecule. It is a genuine, though understandable, tragedy for origins of life research that despite the intense interest in antisense analogue research, very little effort has been devoted to investigating the WC self-complementarity of nucleoside analogue polymers. Many RNA- and DNA-binding molecules have been discarded as drug targets without investigating whether they might be good self-binders or potentially good self-replicators. Far less is known about the ability of these molecules to serve as evolvable catalysts, which has been widely studied for RNA largely because the laboratory tools are available to conduct such studies. The difficulties of conducting such studies with alternate polymers will likely persist for some time.

It would seem essential that a prebiotic genetic polymer be able to base-pair with biological nucleic acids, as this would facilitate the transition to modern biochemistry. However, early genetic polymers may have passed their information to one or more intermediary molecules (Fig. 4). Logically, only the *penultimate* genetic polymer would need to be able to interact with RNA or DNA; no "molecular fossil" (White III, 1976) or palimpsest (Benner et al., 1989) evidence of the original or intermediary polymers might then remain in contemporary biochemistry.

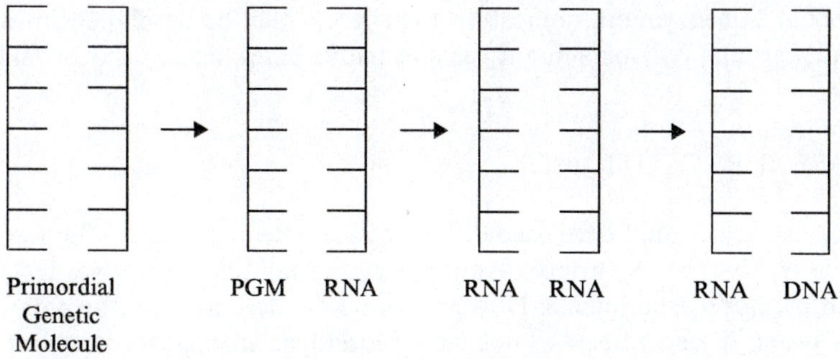


**Figure 4.** Possible transitions from earlier genetic molecules. *PGM* primordial genetic material.

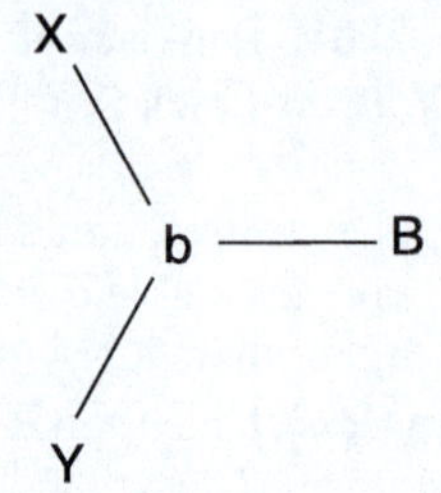


**Figure 5.** The minimal structural requirements for a nucleic acid monomer.

All that may have been required of the "original" genetic material is that it was able to interact with its own type, as well as putative intermediary molecules.

## 2. To Build a Primordial Nucleic Acid

In order to form a linear polymer with the ability to interact with a complementary strand, a monomer with three functionalities is required (Fig. 5): two reactive end groups (*X* and *Y*) which can be linked together, and an interacting motif, *B*, all linked to a central backbone scaffold, *b*.

Explanations have been offered as to why phosphates (Westheimer, 1987) (e.g., ionizability, resistance to leakage across a cell membrane, strand repulsion) and 3′,5′-linkages (Usher and McHale, 1976; Breslow and Shepard, 1996) (e.g., flexibility, stability) are used in biological nucleic acids and why monomers may need to be structured so as to avoid cyclization (Hill et al., 1988). These speculations may give clues to the structural requirements of prebiotic alternatives. For example, Benner and colleagues have argued that a nucleic acid must have the property of COSMIC-LOPER (Capable of Searching Mutation-space Independent of Concern over Loss of Properties Essential for Replication) (Benner et al., 1999), and for this, the polymer must have a repeating charge on the backbone.

Freed from the constraints of prebiotic plausibility, molecules that are more amenable to nonenzymatic replication than RNA may be developed in the lab. Primordial genetic polymers might use alternative bases, backbones, or linkers.

### 2.1. BASE SUBSTITUTIONS

Early nucleic acids could have used different bases (Benner et al., 1999; Levy and Miller, 1999) (Fig. 6). A variety of purine and pyrimidine derivatives have been found in prebiotic experiments. However, some of these may not be compatible with inclusion in a polymer, as some would lose their aromaticity when attached to a backbone, as in the case of 2,4-diaminopyrimidine attached via the $N^1$ nitrogen

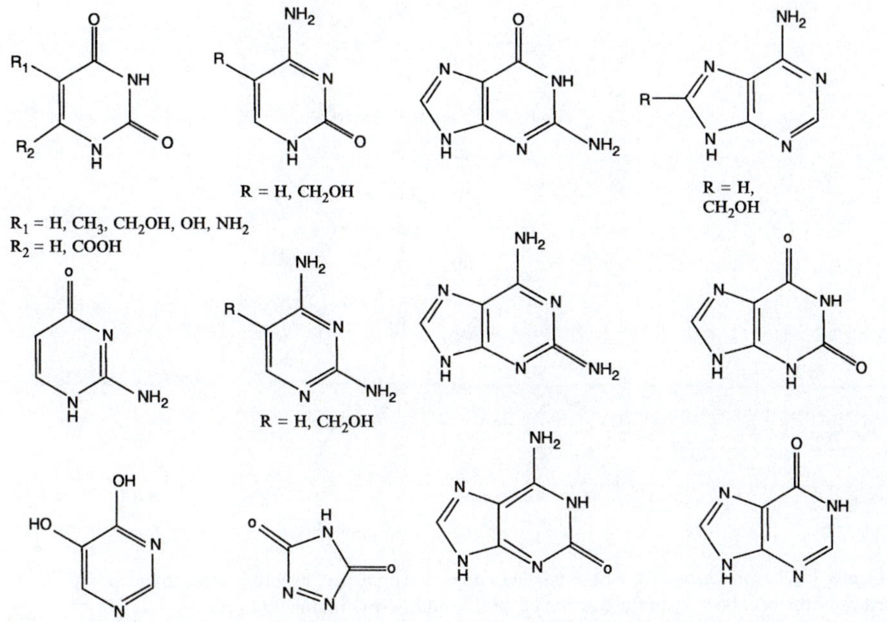


**Figure 6.** Alternative bases that may have been involved in the original informational polymer.

atom. Eschenmoser and coworkers have also shown that some of these are rather poor at base-pairing (Mittapalli et al., 2007a, b).

Considerable research has been conducted in this area (Kool, 2002; Benner, 2004). Mac Dónaill (2002) has addressed base-pairing from a theoretical perspective. Earlier structures could have included alternative pyrimidines (Ferris et al., 1978) or purines (Grzeskowiak et al., 1984) or other heterocycles that might have been more easily synthesized, more stable, or more prone to engage in self-organizational chemistry. Bases capable of forming more than three hydrogen bonds may have been used, although the advantage of these would depend on their availability as well as the interactions these might have with the backbone and the environmental conditions such as ionic strength, pH, and temperature, all of which influence the strength and fidelity of WC base-pairing. For example, the GC base pair, with three hydrogen bonds, is considerably stronger than the AU or AT base pairs, with two hydrogen bonds each (Saenger, 1984). Little attention has been given to the prebiotic synthesis of alternative bases, for what may be obvious reasons, with a few exceptions (Schwartz and Bakker, 1989).

Nucleic acid polymers based on different points of attachment rather than the N$^9$ position of the purines or the N$^1$ position of the pyrimidines have been proposed. Wächtershäuser (1988) suggested an all-purine nucleic acid linked at the N$^3$ position, although Orgel and coworkers (Hill et al., 1991) found this to work poorly in template-directed polymerizations. Schwartz proposed a polymer based

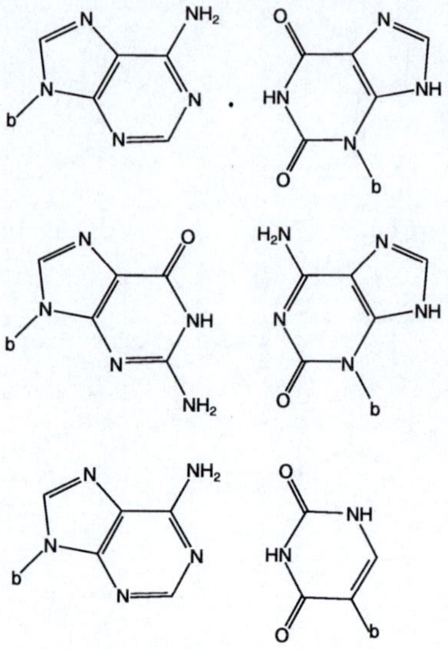


**Figure 7.** Watson-Crick (WC) base-pairing could be supported by other attachment points of the bases to the backbone than the normal $N^9$-purine and $N^1$-pyrimidine linkages.

on 8-hydroxymethyladenine and 5-hydroxymethyluracil, both of which have robust prebiotic syntheses (Schwartz, 1989; Robertson and Miller, 1995). Although these were proposed as part of a glycerolphosphate backbone system, they could be incorporated into other backbones. It should be noted that the transition from an all-purine to a mixed purine-pyrimidine base-pairing scheme may offer some difficulties with maintaining a constant interstrand distance (see below).

Many base substitutions in nucleic acids do not disrupt base-pairing and can be polymerized by enzymes (Lutz et al., 1998), and alternative attachment points for some bases, such as pseudouridine, still allow for (WC) base-pairing (Fig. 7).

The constant interstrand distance in DNA is maintained by pairing pyrimidines with purines (Fig. 8). This may have been an important early selection criterion and may have limited the degree of "play" available for the testing of alternative base-pairing molecules. A primordial polymer may have needed some form of simple size complementarity in the recognition element of the monomer unit.

Self-recognition is the most fundamental attribute of a self-replicating molecule (Rebek, 1991; Park et al., 1992; Bag and von Kiedrowski, 1996) since it is the first step in self-reproduction. Information transfer in biological nucleic acids depends upon the sequence of WC base pairs, mediated by the bases' ability to form hydrogen bonds. Nitrogen heterocycles seem uniquely fit for use in informational molecules. The bases' stacking interactions are greater than those of their isostructural aromatic hydrocarbons (Saenger, 1984). The nucleic acid bases' water solubility does not seem to be important: while indene and benzene are less soluble than adenine and uracil, respectively, 1,3-dihydroxybenzene is much more

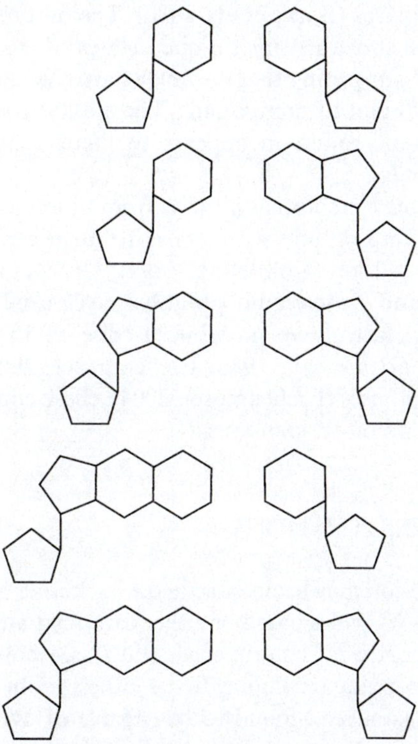

**Figure 8.** Various complementary ring systems could maintain constant interstrand distance. The outermost pentagonal shape represents a backbone moiety; the inner polygons represent various ring systems which allow for constant interstrand distance.

soluble than uracil. Guanine is extremely insoluble, due to the formation of hydrogen-bonded aggregates. In any event, even extremely insoluble molecules can be solubilized by attachment to backbone linker residues. For example, guanosine and 5′-GMP are much more soluble than guanine.

Although only weak evidence can be found for hydrogen bonding of the biological nucleic acid heterocycles in aqueous solution (Voet and Rich, 1970), stacking interactions are observable, and the two forces are quite strong once the monomers are joined covalently in a polynucleotide (Saenger, 1984). Many aromatic compounds stack well in a nucleic acid context, often better than the nucleobases themselves (Kool et al., 2000). However, these would not be able to hydrogen bond, and it is not obvious how they would react prebiotically to attach to a linker molecule.

Many recent experiments have demonstrated the remarkable, and perhaps unique, photostability of the biological nucleobases (Nir et al., 2000; Sobolewski and Domcke, 2010), which suggests that photochemistry may have been one of the earliest selection pressures for nucleic acid composition, and indeed experiments which include a photochemical selection pressure appear to preferentially yield biological nucleobases (Barks et al., 2010). The naturally occurring nucleosides have some of the shortest triplet excited states of many potentially useful nucleic acid analogues, suggesting that they may have been among the few to have survived from an initial glut of compounds. The ability to dissipate energy from absorbed electromagnetic radiation appears to increase in nucleosides, nucleotides, and their polymers.

Urazole has a robust prebiotic synthesis from biuret and hydrazine and has the advantage of reacting rapidly with ribose to form a pyrimidine nucleoside analogue in aqueous solution (Kolb et al., 1994). Urazole is also transparent at wavelengths where *U* and *T* are readily photolyzed (Cockell and Airo, 2002).

C-nucleosides are attractive possibilities due to their potentially facile reactions with ribose (Hill et al., 1988; Eschenmoser, 2004); however, despite suggested prebiotic syntheses (Eschenmoser, 2004), these compounds have not yet been demonstrated as prebiotic compounds.

## 2.2. BACKBONE SUBSTITUTIONS

A great deal of substitution may be possible in the backbone. Several molecules that base-pair according to WC rules based on other carbohydrate structures have been prepared (Eschenmoser, 1997; Hendrix et al., 1997). Glycol nucleic acids (GNA) based on a glycerolphosphate backbone, first synthesized in the early 1970s, have recently been reinvestigated and found to be capable of WC self-pairing (Zhang et al., 2005). Polymers based on ganciclovir *bis*phosphates also appear to be capable of WC base-pairing and template-directed synthesis (Schwartz and Orgel, 1985).

A six-atom repeat, as occurs in natural nucleic acids, PNA, GNA, and TNA, no longer appears necessary (Weller et al., 1991; Eschenmoser, 2004): longer and shorter repeats compatible with hybridization with natural nucleic acids have now been reported (Diederichsen and Schmitt, 1998; Brückner et al., 2003).

### *2.2.1. Alternative Sugar Backbones*

A number of alternative sugar-phosphate backbone nucleic acids have been investigated (Pitsch et al., 1993; Schoning et al., 2000). The prebiotic synthesis of these analogues may be as difficult as that of the natural nucleosides themselves, due to the complexity of the carbohydrate motif and the difficulty of attaching the bases. It is likely that any sugar phosphate–based chemistry will suffer from the same drawbacks as RNA (Shapiro, 1988). It is possible that subtle attributes of sugars, such as the interactions of *cis*- vs. *trans*-diol motifs, may have selected some sugars based on mineral interactions (Ricardo et al., 2004; Bielski and Tencer, 2007; Lambert et al., 2010).

### *2.2.2. Acyclic Nucleotides*

Polyvinyl nucleic acid analogues were among the first alternative nucleic acids investigated (Pitha and Pitha, 1970; Pitha et al., 1970). Heterocyclic bases react robustly and regiospecifically with acrolein (Nelsestuen, 1980; Cleaves, 2002).

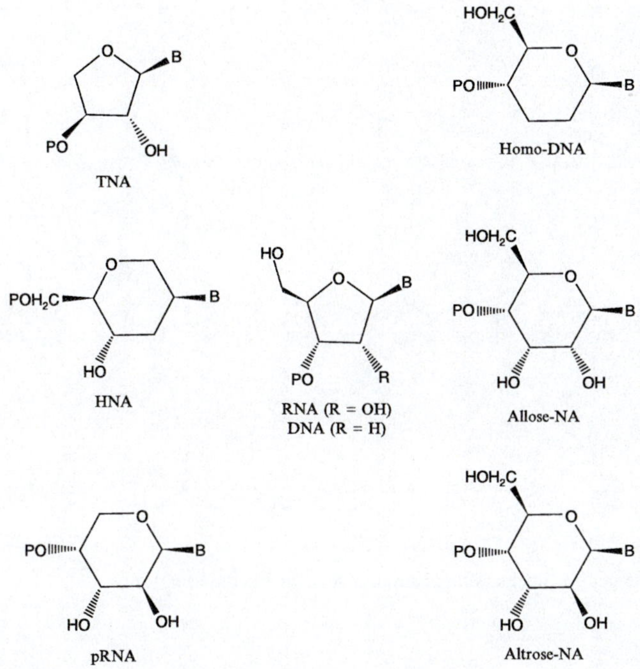


**Figure 9.** Some alternative sugar-based nucleic acid analogues which have been investigated.

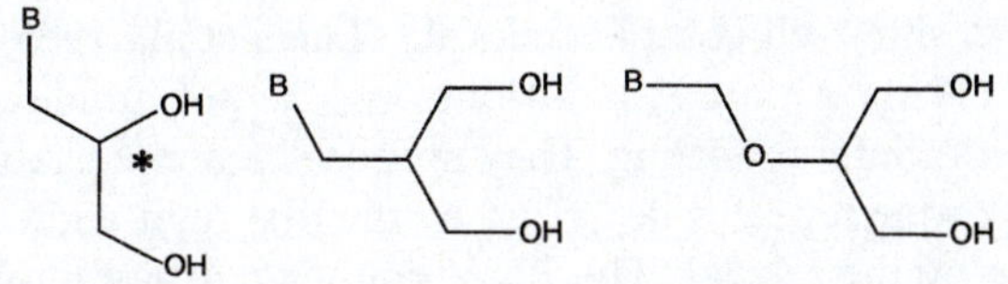


**Figure 10.** Some acyclic nucleotide analogues that have been investigated with regard to prebiotic chemistry (see also Schneider and Benner, 1990). The asterisk denotes a chiral center.

It has been suggested that this chemistry might selectively produce nucleoside analogues (Tohidi and Orgel, 1989). Acrolein is robustly synthesized from the reaction of HCHO with acetaldehyde (Cleaves, 2003), as well as from UV irradiation of HCHO as suggested by the high yields of pentaerythritol produced (Schwartz and de Graaf, 1993) (Figs. 9 and 10).

### *2.2.3. Peptide Nucleic Acids*

Homochiral polymers of the $\alpha$-amino acid-thymidine analogues of the bases (Fig. 11, top left) failed to show any complementary binding with polyadenosine (Buttrey et al., 1975). Alternating copolymers of homochiral forms of these and glycine neither base-paired with each other nor DNA (Lohse et al., 1996). However,

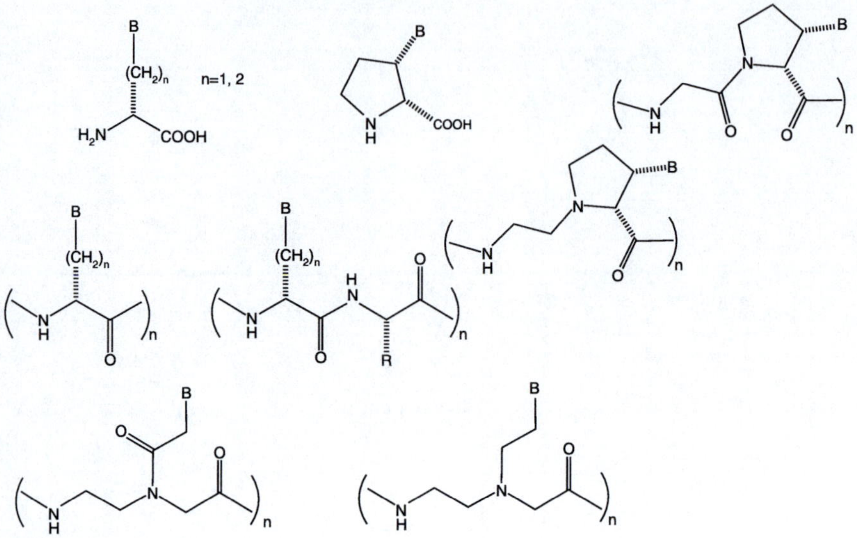


**Figure 11.** Some peptide nucleic acid analogues described in the literature.

polymers with alternating chirality monomers do in fact base-pair with each other (Diederichsen, 1996), again underscoring the subtlety of nucleoside analogue chemistry. Analogues with one more $CH_2$ unit in the linker moiety, termed $\alpha$–PNA, appear to show self-complementarity (Lenzi et al., 1995).

Weller et al. (1991) explored the various possible polyamide and polypeptide nucleic acids via computer modeling. They proposed a nomenclature in which the length of the backbone repeat is described by the first digit and the length of the linker to the base by the second. The PNA synthesized by Egholm et al. (1992) would be a 6(2) polymer, as would the natural nucleic acids. Weller et al. (1991) also considered the possible intramolecular chemistry that these analogues might be susceptible to, and suggested that the bases need to be kept at least 3 carbons away from the carboxyl groups of the backbone to avoid elimination or racemization reactions (Fig. 12).

We have observed precisely this behavior from these monomers in our laboratory (unpublished results). Weller et al. also suggested that analogues with the bases attached directly to the backbone were unlikely to form low-energy conformers due to steric interactions. Four-atom repeats were unable to be elongated into helical structures compatible with A or B forms of DNA. Dipeptide polymers (P6(1)) were found to be more conformationally restricted than polyamide polymers (A6(1)) by a factor of ~35. These authors and others[114] concluded that polypeptides are the ideal nucleic acid analogues because of the conformational restriction afforded by amide linkages.

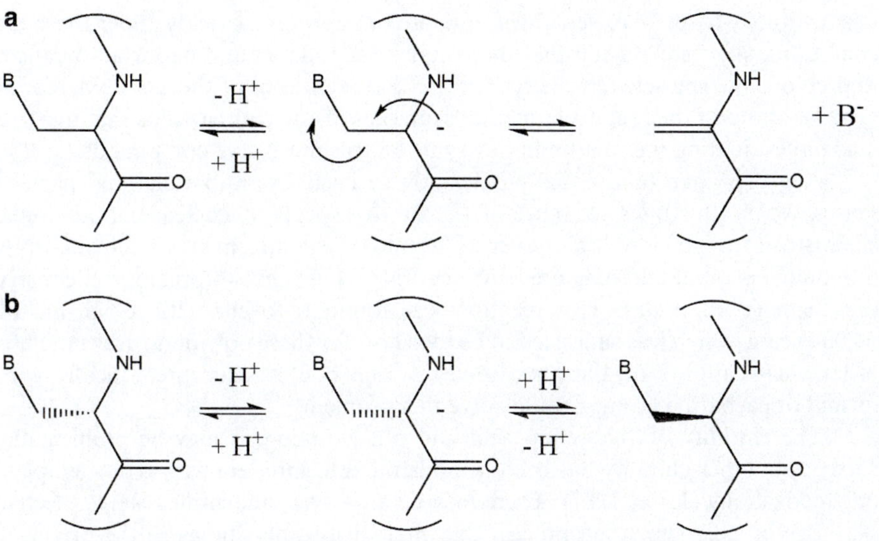


**Figure 12.** Base catalyzed elimination (**a**) and racemization (**b**) of monomers from $\alpha$-amino acid PNA monomers.

Interestingly, the PNA described by Egholm et al. (1992) provides an exception to the COSMIC-LOPER idea (Benner et al., 1999); these molecules are contrary to popular belief, are often surprisingly soluble, and are extremely good base pairers.

A large variety of alternative nucleic acids have now been investigated experimentally, with varying abilities to bind RNA/DNA. These have been reviewed elsewhere (Sanghvi and Cook, 1994). Structures based on a prochiral glycerol-formaldehyde backbone have been proposed (van Vliet et al., 1994). PNAs built from $\alpha$, $\gamma$ diaminobutyric acid or ornithine based on carbonic or glycolic acid linkages have also been proposed (Nielsen, 1993).

It has been suggested that peptide nucleic acids based on Nielsen and coworkers' 2-aminoethylglycine backbone (aeg-PNA) are capable of carrying out the functions that would be required of an early genetic material (Miller, 1997). Aeg-PNA can serve as a template for the polymerization of activated RNA monomers (Koppitz et al., 1998). The prebiotic synthesis of aeg-PNA has been demonstrated (Nelson et al., 2000) (see Fig. 15 below), but other structures may be more plausible and perhaps may have other properties more conducive to sequence space searching for catalytic structures.

Due to the surprising properties of aeg-PNA, a small but significant effort has been expended into the prebiotic synthesis of PNA (Nelson et al., 2000). The prebiotic synthesis of the components of aeg-PNA may offer insight into the prebiotic synthesis of alternative polymers. The N-acetic acid substituted- and unsubstituted-bases were generated in approximately equal yields under the same conditions, suggesting again that there may be significant and important nuances to prebiotic nucleic acid chemistry. The prebiotic synthesis of the aeg-PNA monomers was among the first experimental examples of the comparisons that must be made in evaluating the plausibility of synthesis of alternative compounds.

Aeg-PNA can bind equally efficiently to itself in antiparallel and parallel senses, which illustrates yet more of the subtle aspects of nucleic acid analogue chemistry to be explored. However, N-terminal rearrangement of the aeg-PNA monomers is rapid and may preclude aeg-PNA's relevance to prebiotic chemistry (Eriksson et al., 1998). This reaction was found to occur with a half-life of ~1,000 years, which was suggested to be too short for these polymers to accumulate. N-terminal capping of these molecules, which could occur prebiotically with formyl or carbamyl groups, could solve this problem.

The chirality of the amino acid and other analogues may be problematic. Mixed monomer chirality has been demonstrated to interfere with RNA template polymerizations (Joyce, 1987). There may be an as-yet-unidentified stereoselective synthetic or polymerization process that may give some clue as to the origin of chirality in biological systems, as there does now appear to be some real enantiomeric bias in the amino acids found in carbonaceous chondrites (Pizzarello, 2006).

## 2.3. LINKER SUBSTITUTIONS

A variety of linker substitutions are possible, including polyamides, polyesters (Harada and Orgel, 1990), polyanhydrides, polythioamides, sulfides (Huang et al., 1991; Wu and Orgel, 1991), sulfates (Huang et al., 1991), borates, silicates, thioethers (Uhlmann and Peyman, 1990), phosphoramidates, polyethers, polyurethanes, polycarbonates, sulfonates (Huang et al., 1991), carbamates (Stirchak et al., 1987), or phosphonates (de Graaf et al., 1998) (Fig. 13). Although sulfonates and phosphonates are extremely abundant in the Murchison meteorite (Botta and Bada, 2002), except for certain amide replacements, these modifications appear to generally destabilize helix formation (De Mesmaeker et al., 1994). Thus, compounds such as arsenates (Wolfe-Simon et al., 2009) and sulfonates (Huang et al., 1991) may pose some structural problems. Although there are many possible backbones, few are probably prebiotically accessible and many can probably be eliminated via stability arguments, although the inventory of prebiotically available small organic molecules is incompletely surveyed (Miller, 1992).

Oligonucleotides may need to be charged to ensure strand repulsion and solubility (Harada and Orgel, 1990; Benner and Hutter, 2002). Polyanhydrides are generally unstable in water, and polyphosphonates are unstable to oxidation and hydrolysis (Peyser and Ferris, 2001), although polyphosphates could be produced from the oxidation of polyphosphonates or phosphides present in meteorites (Pasek, 2008). Another interesting possibility is that phosphate could be replaced by glyoxylic acid (Bean et al., 2006), which would create an acetal linkage which would still be charged.

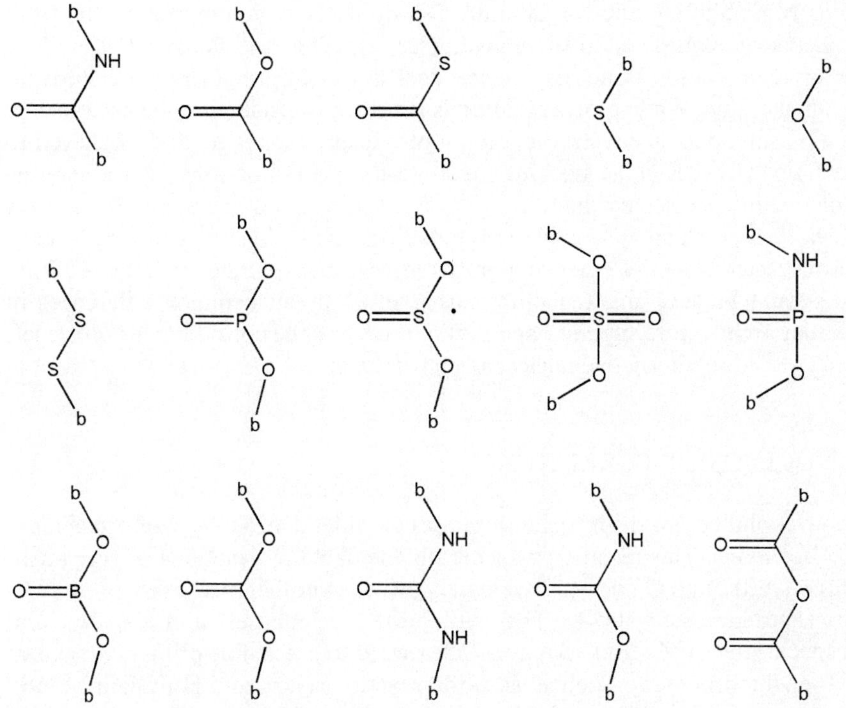


**Figure 13.** Some possible alternative linkages for primordial nucleic acids.

## 2.4. RECOGNITION SURFACES

Interstrand interactions in biological nucleic acids are mediated by hydrogen bonding; however, it is conceivable that another type of interaction could have carried out this function in earlier polymers. Duplex formation between ionic species such as positively and negatively charged amino acid side chains would likely be much weaker and less precise than that formed between complementary multiply hydrogen-bonded aromatic molecules. Hydrogen bonds may be better suited for molecular recognition than van der Waals forces due to their better-defined directionality (Whitesides et al., 1991). The relative entropic cost of self-assembly is much greater than the enthalpic contribution and can be as high as +6.2 kcal $mol^{-1}$ for the $-T\Delta S$ term. Molecular self-recognition is favored by rigidity in the interacting species as long as this does not interfere with the interacting recognition surfaces, and by the larger size of the interacting molecular surfaces.

The forces which contribute to nucleic acid duplex formation have been the subject of considerable study. The contributions of hydrogen bonding and $\pi$ stacking are estimated to be roughly equal in the formation of a helix between two nucleic acid strands (Kool et al., 2000), and the contributions due to cations such as $Mg^{2+}$ binding and the organization of water structure around the helix have also been considered in some detail (see, e.g., Hud and Feigon, 1997). Since there are many rotable bonds in a nucleic acid strand, duplex formation is entropically unfavorable. While protein folding is generally considered to be entropically driven by solvophobic effects, the folding of nucleic acids is generally believed to be enthalpically driven, as stacking interactions and H-bonding outcompete the loss of freedom caused by the restriction of backbone conformation. PNA offers an advantage in stabilizing stacking interactions as the backbone amide linkage has a restricted range of rotation compared with phosphodiester linkages. That there should be such marked differences caused by such minor differences in molecular architecture suggests, again, that there may be considerable nuance left to be explored in alternative nucleic acid structures.

## 2.5. NUCLEOSIDE FORMATION

The nucleophilicity of the nitrogenous bases provides a mechanism of attachment to the backbone. This reactivity is generally mediated by reactions of ring nitrogen atoms, although C-nucleosidation may provide another mechanism of attachment (Eschenmoser, 2004). The biological nucleobases add rapidly and regiospecifically to Michael acceptors (Cleaves, 2002). Adenine primarily reacts at the $N^9$ position in many Michael addition reactions (Lira and Huffmann, 1966). Guanine reacts at both $N^7$ and $N^9$ positions, although the reactions of guanine have not been studied as exhaustively, partly due to the insolubility of guanine. The $N^1$ adducts of the pyrimidines are often formed preferentially over the $N^3$ adducts; however, alkylation of pyrimidines often results in mixtures of $N^1$, $N^3$ and $N^1$, $N^3$-substituted products (Crippa et al., 1993). Cytosine and uracil also react appreciably at the $C^5$ position with electrophiles such as HCHO (Robertson and Miller, 1995). The reactivity of various heterocycles with the likely prebiotic compound acrolein (Cleaves, 2003; Hollis et al., 2004) has been studied in some detail. Such adducts could provide a method of synthesis of several monomeric compounds (Fig. 14).

Alternatively, the base/backbone linkage could be constructed during the synthesis of the monomer, as has been demonstrated with prebiotic pyrimidine nucleosides (Sanchez and Orgel, 1970; Osterberg et al., 1973; Schwartz et al., 1975; Sutherland and Weaver, 1994; Bean et al., 2006; Powner et al., 2009) and PNA monomers (Nelson et al., 2000) (Fig. 15).

## 2.6. THE CONSTRAINTS OF PREBIOTIC CHEMISTRY

Despite the diversity of possible analogue structures, there is likely a limited set of alternative nucleic acids that may be obtained prebiotically. Few monomer combinations that allow for appropriately spaced PNAs may be stable enough for

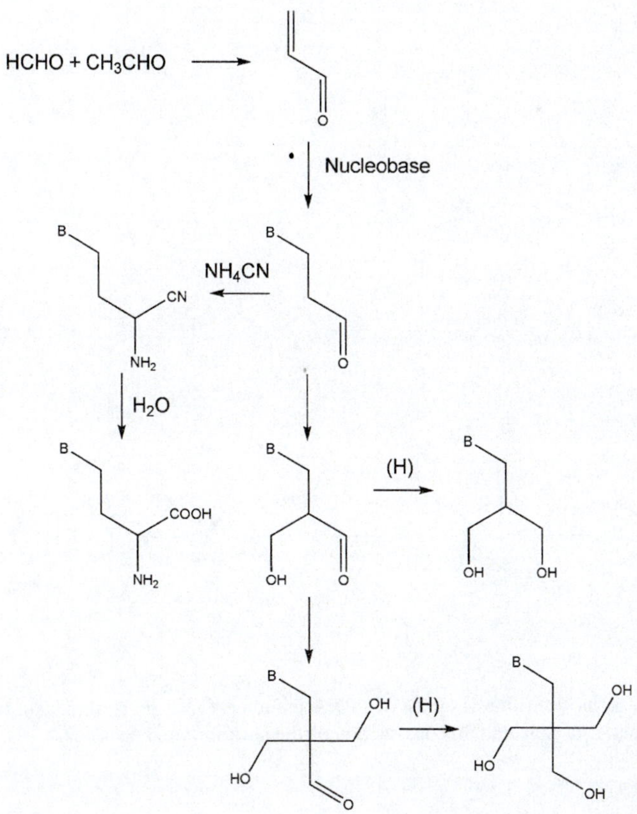


**Figure 14.** Possible routes to prebiotic monomers from base-acrolein adducts.

polymerization, and fewer still likely allow for the higher order supramolecular chemistry which might allow for replication and Darwinian selection.

Aqueous HCN spontaneously oligomerizes to produce a variety of bioorganic compounds (Ferris et al., 1978; Schwartz et al., 1984; Levy et al., 1999). The hydrolysis conditions of HCN polymers have a large impact on the products released (Schwartz et al., 1984; Miyakawa et al., 2002a; Borquez et al., 2005). Since it would be unlikely for a prebiotic process to allow for the synthesis of HCN but not HCHO, the products of mixed HCN/HCHO polymers are of considerable interest. These seemingly mundane reactions may harbor significant chemical surprises. For example, in the presence of HCHO, cyanide polymerizations produce more 8-hydroxymethyladenine than adenine (Schwartz and Bakker, 1989). The robust reaction of 5-hydroxymethylpyrimidines with amino acids has been demonstrated (Robertson and Miller, 1995).

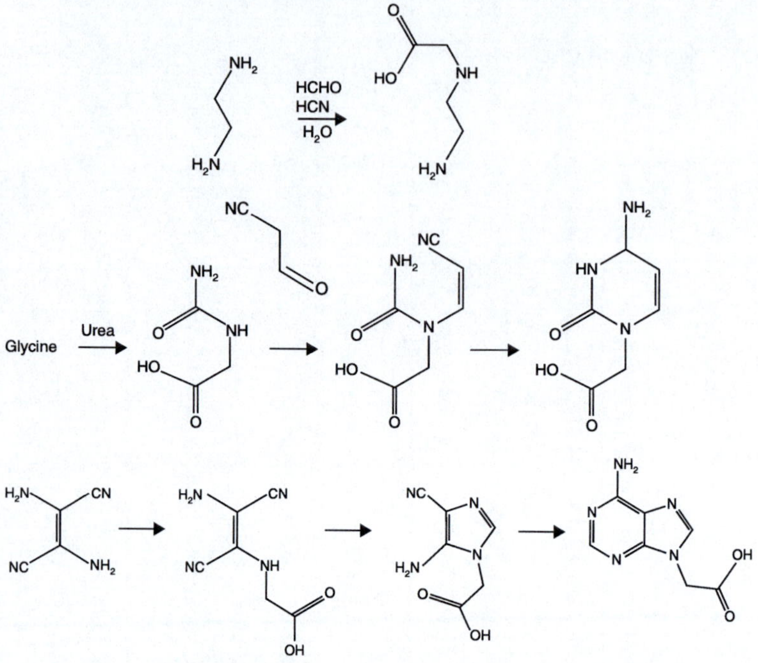


**Figure 15.** The prebiotic synthesis of the aeg-PNA monomers (Nelson et al., 2000), an example of monomer synthesis via concomitant construction of the backbone and nucleobase.

If the primordial genetic molecule included the principle of purine-pyrimidine base-pairing (see Wächtershäuser (1988) and Zubay (1996) for alternative views), it would be simpler if purines and pyrimidines were synthesized together. The oligomerization of aqueous HCN produces the biological purines (Schwartz et al., 1984; Levy et al., 1999), uracil (Voet and Schwartz, 1982), orotic acid (Ferris et al., 1978), and nonbiological pyrimidines such as 5-hydroxyuracil and 4,5-dihydroxypyrimidine (Ferris et al., 1978), which might be compatible with base-paring. What other small molecules are produced in these reactions? How might their structures inform our investigations of the origin of life? These are tractable questions worthy of future consideration. Bredereck et al. (1959) and Yamada and Okamoto (1972) and more recently Saladino and coworkers (2001) have also shown that a complex set of heterocycles is generated from heating neat formamide at high temperatures. Whether such high concentrations of formamide could ever have accumulated in one place is an open question (Miyakawa et al., 2002b).

The alanyl-PNA and α-PNA structures were recently investigated in our laboratory. These were selected based on their appearing to have plausible prebiotic syntheses and their predicted stability to polymerization (Orgel, 1998).

In alanyl-PNA, the nucleobases are substituted on the β-carbon of alanine (Lohse et al., 1996) (Fig. 11, top left, $n=1$); in α-PNA (Fig. 11, top left, $n=2$), the nucleobases are attached to the γ-carbon of a homoserine residue (Howarth et al., 1996). Alanyl-PNA monomers could potentially be synthesized prebiotically from dehydroalanine via Michael addition (as has been shown for tryptophan (Friedmann et al., 1971)). We were unable to demonstrate this, although we did find that mixtures of HCHO and HCN produced significant amounts of dehydroalanine hydantoin, and that this was a more stable molecule than serine hydantoin under a variety of conditions.

It has been suggested that α-PNA monomers could be derived prebiotically from acrolein, a nucleobase, and ammonium cyanide (Cheikh and Orgel, 1990) (Fig. 14). We were able to confirm this synthesis.

It was found that the alanyl-substituted bases were extremely prone to β-elimination and that the dipeptide homoserine analogues (glycyl-homoseryl analogue or alanyl-homoseryl analogue) did not polymerize readily via plausible prebiotic mechanisms (unpublished results). In addition, it seems that the necessary chirality of the polymers or strict necessity of nucleobase-analogue/amino acid alternation in the polymer would render them problematic prebiotic candidate molecules.

## 3. Future Directions

If a genetic polymer were required for the origin of life and if RNA is, as it appears to be, the product of extended biological evolution, then a search for plausible prebiotic genetic molecules may be rewarding. This could offer an almost limitless area for investigation, but numerous criteria likely limit the necessary searchable structure space considerably. We suggest that the most promising avenues for future research in this field are: investigating the polymerizability and stability of various monomers (lactones, lactams, etc.), with a concerted effort to constrain the kinetic and thermodynamic parameters, which might make some monomers' survival and polymerization more feasible, and the photochemical behavior of prebiotic organic mixtures, which might severely reduce the inventory of monomers available.

## 4. Conclusions

The remarkable diversity of nucleic acid structures synthesized in the laboratory underscores the fact that RNA and DNA are not unique in their ability to store genetic information. The difficulty of the prebiotic synthesis of RNA and the enormous unsurveyed structure space of prebiotic chemistry suggest it is at least possible that another type of molecule stored genetic information before RNA and was possibly important for the origin of life.

While the plethora of structures is daunting, certain design principles are likely to have been important in prebiotic nucleic acid construction. Prebiotic synthesis is likely to impose severe restrictions on which structures are possible, and the ability to base-pair likely imposes yet more restrictions. The investigation of the various possibilities with respect to these constraints is likely to be a rewarding avenue for future research.

## 5. References

Bag BG, von Kiedrowski G (1996) Templates, autocatalysis and molecular replication. Pure Appl Chem 68:2145–2152

Barks HL, Buckley R, Grieves GA, Di Mauro E, Hud NV, Orlando TM (2010) Guanine, adenine, and hypoxanthine production in UV-irradiated formamide solutions: relaxation of the requirements for prebiotic purine nucleobase formation. Chembiochem 11:1240–1243

Bean HD, Anet FA, Gould IR, Hud NV (2006) Glyoxylate as a backbone linkage for a prebiotic ancestor of RNA. Orig Life Evol Biosph 36:39–63

Benner SA (2004) Understanding nucleic acids using synthetic chemistry. Acc Chem Res 37:784–797

Benner SA, Hutter D (2002) Phosphates, DNA, and the search for non-terrean life: a second generation model for genetic molecules. Bioorg Chem 30:62–80

Benner SA, Ellington AD, Tauer A (1989) Modern metabolism as a palimpsest of the RNA world. Proc Natl Acad Sci U S A 86:7054–7058

Benner SA, Burgstaller P, Battersby TR, Jurczyk S (1999) Did the RNA world exploit an expanded genetic alphabet? In: Gesteland RF, Cech T, Atkins JF (eds) The RNA world, 2nd edn. Cold Spring Harbor Laboratory Press, Cold Spring Harbor

Biebricher CK, Eigen M (2005) The error threshold. Virus Res 107:117–127

Bielski R, Tencer M (2007) A possible path to the RNA world: enantioselective and diastereoselective purification of ribose. Orig Life Evol Biosph 37:167–175

Böhler C, Nielsen PE, Orgel LE (1995) Template switching between PNA and RNA oligonucleotides. Nature 376:578–581

Borquez E, Cleaves HJ, Lazcano A, Miller SL (2005) An investigation of prebiotic purine synthesis from the hydrolysis of HCN polymers. Orig Life Evol Biosph 35:79–90

Botta O, Bada JL (2002) Extraterrestrial organic compounds in meteorites. Surv Geophysics 23:411–467

Brachet J (1959) Les acides nucléiques et l'origène des protéines. In: The origin of life on Earth (Academy of Sciences of the U.S.S.R: Oparin AI, Pasynskii AG, Braunshtein AE, Pavlovskaya TE (eds), English-French-German edition, Clark F, Synge RLM (eds)). MacMillan, New York, pp 361–367

Bredereck H, Gompper R, Schuh HGV, Theilig G (1959) Neuere Methoden der präparativen organischen Chemie II. 16. Synthesen mit Säure-amiden, insbesondere mit Formamid. Angew Chem 71:753

Breslow R, Shepard TL (1996) Why natural DNA is based on 2′-deoxyribose, with 3′,5′- phosphodiester links. Pure Appl Chem 68:2037–2041

Brückner AM, Garcia M, Marsh A, Gellman SH, Diederichsen U (2003) Synthesis of novel nucleo-β-amino acids and nucleobase-functionalized β-peptides. Eur J Org Chem 18:3555–3561

Buttrey JD, Jones AS, Walker RT (1975) Synthetic analogues of polynucleotides. XIII. The resolution of DL-β-(thymin-1-yl) alanine and polymerization of the DL-β-(thymin-1-yl) alanines. Tetrahedron 31:73–75

Cairns-Smith A (1977) Takeover mechanisms and early biochemical evolution. Biosystems 9:105–109

Cech TR (1987) The chemistry of self-splicing RNA and RNA enzymes. Science 236:1532–1539

Cech TR, Zaug AJ, Grabowski PJ (1981) In vitro splicing of the ribosomal RNA precursor of tetrahymena: involvement of a guanosine nucleotide in the excision of the intervening sequence. Cell 27:487–496

Cheikh AB, Orgel LE (1990) Polymerization of amino acids containing nucleotide bases. J Mol Evol 30:315–321
Cleaves HJ (2002) The reactions of nitrogen heterocycles with acrolein: scope and prebiotic significance. Astrobiology 2:403–415
Cleaves HJ (2003) The prebiotic synthesis of acrolein. Monatsh Chem 134:585–593
Cleaves HJ, Chalmers JH (2004) Extremophiles may be irrelevant to the origin of life. Astrobiology 4:1–9
Cleaves HJ, Lazcano A (2009) Origin of biomolecules. In: Zaikowski L, Friedrich JM (eds) Chemical Evolution II: From origins of life to modern society, American Chemical Society Symposium Series. Oxford University Press, New York
Cleaves HJ, Miller SL (2007) Organic chemistry on the primitive Earth and beyond. In: Rigoutsos I, Stephanopoulos G (eds) Systems biology: Volume I: Genomics. Oxford University Press, New York
Cockell CS, Airo A (2002) On the plausibility of a UV transparent biochemistry. Orig Life Evol Biosph 32:255–274
Crick FHC (1968) The origin of the genetic code. J Mol Biol 38:367–379
Crippa S, Di Gennaro P, Lucini R, Orlandi M, Rindone B (1993) Characterization of adducts of nucleic acid bases and acrylic monomers. Gazz Chim Ital 123:197–203
Cronin JR, Cooper GW, Pizzarello S (1995) Characteristics and formation of amino acids and hydroxy acids of the Murchison meteorite. Adv Space Res 15:91–97
de Duve C (1991) Blueprint for a cell: the nature and origin of life. N. Patterson, Burlington
De Graaf RM, Visscher J, Schwartz AW (1998) Prebiotic chemistry of phosphonic acids: products derived from phosphonoacetaldehyde in the presence of formaldehyde. Orig Life Evol Biosph 28:271–282
De Mesmaeker A, Waldner A, Fritsch V, Lebreton J, Wolf RM (1994) Synthetic modifications of antisense oligonucleotides: novel backbone replacements with improved properties. Bull Soc Chim Belg 103:705–717
Diederichsen U (1996) Pairing properties of alanyl peptide nucleic acids containing an amino acid backbone with alternating configuration. Angew Chem Int Ed 35:445–448
Diederichsen U, Schmitt HW (1998) β-homoalanyl PNAs: synthesis and indication of higher ordered structures. Angew Chem Int Ed 37:302–305
Doherty EA, Doudna JA (2000) Ribozyme structures and mechanisms. Annu Rev Biochem 69:597–615
Dyson F (1999) Origins of life (Revised edition). Cambridge University Press, Cambridge/New York
Egholm M, Buchardt O, Nielsen PE, Berg RH (1992) Peptide nucleic acids (PNA) oligonucleotide analogues with an achiral peptide backbone. J Am Chem Soc 114:1895–1897
Eigen M, Schuster P (1976) The hypercycle: a principle of self-organization. Springer, Berlin
Eriksson M, Christensen L, Schmidt J, Haaima G, Orgel L, Nielsen PE (1998) Sequence dependent N-terminal rearrangement and degradation of peptide nucleic acid (PNA) in aqueous solution. New J Chem 22:1055–1059
Eschenmoser A (1997) Towards a chemical etiology of nucleic acid structure. Orig Life Evol Biosph 27:535–553
Eschenmoser A (2004) The TNA-family of nucleic acid systems: properties and prospects. Orig Life Evol Biosph 34:277–306
Ferris JP, Joshi PC, Edelson EH, Lawless JG (1978) HCN: a plausible source of purines, pyrimidines, and amino acids on the primitive Earth. J Mol Evol 11:293–311
Friedmann N, Haverland WJ, Miller SL (1971) Prebiotic synthesis of the aromatic and other protein amino acids. In: Buvet R, Ponnamperuma C (eds) Chemical evolution and the origin of life. North-Holland Publishing Company, Amsterdam/New York
Fuller WD, Sanchez RA, Orgel LE (1972) Solid-state synthesis of purine nucleosides. J Mol Evol 1:249–257
Gesteland R, Cech T, Atkins J (1999) The RNA world, 2nd edn, Cold Spring Harbor Monograph Series 37. Cold Spring Harbor Laboratory Press, Cold Spring Harbor

Gilbert W (1986) The RNA world. Nature 319:618
Grzeskowiak K, Webb TR, Orgel LE (1984) Template-directed synthesis with 2-aminoadenosine. J Mol Evol 21:81–83
Guerrier-Takada C, Gardiner K, Marsh T, Pace N, Altman S (1983) The RNA moiety of ribonuclease P is the catalytic subunit of the enzyme. Cell 35:849–857
Haldane JBS (1965) Data needed for a blueprint of the first organism. In: Fox SW (ed) The origin of prebiological systems and of their molecular matrices. Academic, New York, pp 11–18
Harada K, Orgel LE (1990) Template-directed oligomerization of 5′-deoxy-5′-nucleosideacetic acid derivatives. Orig Life Evol Biosph 20:151–160
Hendrix C, Rosemeyer H, Verheggen I, Seela F, Van Aerschot A, Herdewijn P (1997) 1′,5′-anhydro-hexitol oligonucleotides: synthesis, base pairing and recognition by regular oligodeoxyribonucleotides and oligoribonucleotides. Chem A Eur J 3:110–120
Hill AR Jr, Nord LD, Orgel LE, Robins RK (1988) Cyclization of nucleotide analogues as an obstacle to polymerization. J Mol Evol 28:170–171
Hill AR Jr, Kumar S, Patil VD, Leonard NJ, Orgel LE (1991) Which 3-ribofuranosyl-substituted purine 5′-phosphates undergo template-directed oligomerization? J Mol Evol 32:447–453
Hollis JM, Jewell PR, Lovas FJ, Remijan A, Møllendal H (2004) Green Bank Telescope detection of new interstellar aldehydes: propenal and propanal. Astrophys J 610:L21–L24
Howarth NM, Topham CM, Searcey M, Wakelin LPG (1996) Molecular simulation and synthesis of peptide nucleic acid analogues (α-PNA) of DNA. Irish J Med Sci 165(Supp 3):23
Huang Z, Schneider KC, Benner SA (1991) Building blocks for oligonucleotide analogues with dimethylene sulfide, sulfoxide, and sulfone groups replacing phosphodiester linkages. J Org Chem 56:3869–3882
Hud NV, Feigon J (1997) Localization of divalent metal ions in the minor groove of DNA A-tracts. J Am Chem Soc 119:5756–5757
Joyce GF (1987) Non-enzymatic template-directed synthesis of informational macromolecules. Cold Spring Harb Symp Quant Biol 52:41–51
Joyce GF, Orgel LE (1993) Prospects for understanding the origin of the RNA world. In: Gesteland RF, Atkins JF (eds) The RNA world. Cold Spring Harbor Laboratory Press, Cold Spring Harbor
Joyce G, Schwartz A, Miller S, Orgel L (1987) The case for an ancestral genetic system involving simple analogs of the nucleotides. Proc Natl Acad Sci U S A 84:4398–4402
Kanavarioti A, Monnard PA, Deamer DW (2001) Eutectic phases in ice facilitate nonenzymatic nucleic acid synthesis. Astrobiology 1:271–281
Kaufman SA (1986) Autocatalytic sets of proteins. J Theor Biol 119:1–24
Keefe AD, Miller SL (1995) Are polyphosphates or phosphate esters prebiotic reagents? J Mol Evol 41:693–702
Kolb VM, Dworkin JP, Miller SL (1994) Alternative bases in the RNA world: the prebiotic synthesis of urazole and its ribosides. J Mol Evol 38:549–557
Kool ET (2002) Replacing the nucleobases in DNA with designer molecules. Acc Chem Res 35:936–943
Kool ET, Morales JC, Guckian KM (2000) Mimicking the structure and function of DNA: insights into DNA stability and replication. Angew Chem Int Ed 39:990–1009
Koppitz M, Nielsen PE, Orgel LE (1998) Formation of oligonucleotide-PNA-chimeras by template-directed ligation. J Am Chem Soc 120:4563–4569
Krishnamurthy R, Arrhenius G, Eschenmoser A (1999) Formation of glycolaldehyde phosphate from glycolaldehyde in aqueous solution. Orig Life Evol Biosph 29:333–354
Lambert JB, Gurusamy-Thangavelu SA, Ma K (2010) The silicate-mediated formose reaction: bottom-up synthesis of sugar silicates. Science 327:984–986
Larralde R, Robertson MP, Miller SL (1995) Rates of decomposition of ribose and other sugars: implications for chemical evolution. Proc Natl Acad Sci U S A 92:8158–8160
Lazcano A (2000) Origins of life: history of ideas. In: Margulis L, Matthews C, Haselton A (eds) Environmental evolution: effects of the origin and evolution of life on planet Earth. The MIT Press, Cambridge, MA

Lenzi A, Reginato G, Taddei M, Trifilieff E (1995) Solid phase synthesis of a selfcomplementary (antiparallel) chiral peptidic nucleic acid strand. Tetahedron Lett 36:1717–1718

Levy M, Miller SL (1998) The stability of the RNA bases: implications for the origin of life. Proc Natl Acad Sci U S A 95:7933–7938

Levy M, Miller SL (1999) The prebiotic synthesis of modified purines and their potential role in the RNA world. J Mol Evol 48:631–637

Levy M, Miller SL, Oró J (1999) Production of guanine from $NH_4CN$ polymerizations. J Mol Evol 49:165–168

Lira EP, Huffmann CW (1966) Some Michael-type reactions with adenine. J Org Chem 31:2188–2191

Lohse P, Oberhauser B, Oberhauser-Hofbauer B, Baschang G, Eschenmoser A (1996) Chemie von α-aminonitrilen. XVII. Oligo(nukleodipeptamidinium)-salze. Croat Chem Acta 69:535–562

Lutz MJ, Horlacher J, Benner SA (1998) Recognition of a non-standard base pair by thermostable DNA polymerases. Bioorg Med Chem Lett 8:1149–1152

Mac Donaill D (2002) A parity code interpretation of nucleotide alphabet composition. Chem Commun 2002:2062–2063

Miller SL (1992) The prebiotic synthesis of organic compounds as a step toward the origin of life. In: Schopf JW (ed) Major events in the history of life. Jones and Bartlett Publishers, Boston, pp 1–28

Miller SL (1997) Peptide nucleic acids and prebiotic chemistry. Nat Struct Biol 4:167–169

Mittapalli G, Reddy K, Xiong H, Munoz O, Han B, De Riccardis F, Krishnamurthy R, Eschenmoser A (2007a) Mapping the landscape of potentially primordial Informational oligomers: oligodipeptides and oligodipeptoids tagged with triazines as recognition elements. Angew Chem Int Ed 46:2470–2477

Mittapalli GK, Osornio YM, Guerrero MA, Reddy KR, Krishnamurthy R, Eschenmoser A (2007b) Mapping the landscape of potentially primordial informational oligomers: oligodipeptides tagged with 2,4-disubstituted 5-aminopyrimidines as recognition elements. Angew Chem Int Ed Engl 46:2478–2484

Miyakawa S, Cleaves HJ, Miller SL (2002a) The cold origin of life: B. Implications based on pyrimidines and purines produced from frozen ammonium cyanide solutions. Orig Life Evol Biosph 32:209–218

Miyakawa S, Cleaves HJ, Miller SL (2002b) The cold origin of life: A. Implications based on the hydrolytic stabilities of hydrogen cyanide and formamide. Orig Life Evol Biosph 32:195–208

Morowitz HJ (1992) Beginnings of cellular life: metabolism recapitulates biogenesis. Yale University Press, New Haven

Nelsestuen GL (1980) Origin of life: consideration of alternatives to proteins and nucleic acids. J Mol Evol 15:59–72

Nelson K, Levy M, Miller S (2000) Peptide nucleic acids rather than RNA may have been the first genetic molecule. Proc Natl Acad Sci U S A 97:3868–3871

Nielsen PE (1993) Peptide nucleic acid (PNA): a model structure for the primordial genetic material? Orig Life Evol Biosph 23:323–327

Nir E, Kleinermanns K, de Vries MS (2000) Pairing of isolated nucleic-acid bases in the absence of the DNA backbone. Nature 408:949–950

Oparin AI (1938) The origin of life (translated by S Margulis). Macmillan, New York

Orgel LE (1968) Evolution of the genetic apparatus. J Mol Biol 38:381–393

Orgel LE (1987) The origin of self-replicating molecules. In: Eugene Yates F (ed) Self-organizing systems: the emergence of order. Plenum Press, New York

Orgel LE (1998) Polymerization on the rocks: theoretical introduction. Orig Life Evol Biosph 28:227–234

Orgel LE (2000) Self-organizing biochemical cycles. Proc Natl Acad Sci U S A 97:12503–12507

Orgel LE (2008) The implausibility of metabolic cycles on the prebiotic Earth. PLoS Biol 6:e18

Oró J, Kimball AP (1961) Synthesis of purines under primitive Earth conditions I. Adenine from hydrogen cyanide. Arch Biochem Biophys 92:221

Osterberg R, Orgel LE, Lohrmann R (1973) Urea-catalyzed phosphorylation reactions. J Mol Evol 2:231–234

Park TK, Feng Q, Rebek J Jr (1992) Synthetic replicators and extrabiotic chemistry. J Am Chem Soc 114:4529–4532

Pasek MA (2008) Rethinking early Earth phosphorus geochemistry. Proc Natl Acad Sci U S A 105:853–858

Périgaud C, Gosselin G, Imbach JL (1992) Nucleoside analogues as chemotherapeutic agents: a review. Nucleosides Nucleotides Nucleic Acids 11:903–945

Peyser JR, Ferris JP (2001) The rates of hydrolysis of thymidyl-3′,5′-thymidine-H-phosphonate: the possible role of nucleic acids linked by diesters of phosphorous acid in the origins of life. Orig Life Evol Biosph 31:363–380

Pitha J, Pitha PM (1970) Preparation and properties of poly-9-vinyladenine. Biopolymers 9:965–977

Pitha J, Pitha PM, Ts'o POP (1970) Poly (1-vinyluracil): the preparation and interaction with adenosine derivatives. Biochem Biophys Acta 204:39–48

Pitsch S, Wendeborn S, Juan B, Eschenmoser A (1993) Why pentose and not hexose-nucleic acids? Part VII. Pyranosyl-RNA ("p-RNA"). Helv Chim Acta 76:2161–2183

Pitsch S, Eschenmoser A, Gedulin B, Hui S, Arrhenius G (1995) Mineral induced formation of sugar phosphates. Orig Life Evol Biosph 25:297–334

Pizzarello S (2006) The chemistry of life's origin: a carbonaceous meteorite perspective. Acc Chem Res 39:231–237

Powner MW, Gerland B, Sutherland JD (2009) Synthesis of activated pyrimidine ribonucleotides in prebiotically plausible conditions. Nature 459:239–242

Rebek J Jr (1991) Molecular recognition and the development of self-replicating systems. Experientia 47:1096–1104

Reid C, Orgel LE (1967) Synthesis of sugars in potentially prebiotic conditions. Nature 216:455–456

Ricardo A, Carrigan MA, Olcott AN, Benner SA (2004) Borate minerals stabilize ribose. Science 303:196

Rich A (1962) On the problems of evolution and biochemical information transfer. In: Kasha M, Pullman B (eds) Horizons in biochemistry. Academic Press, New York, pp 103–126

Robertson MP, Miller SL (1995) Prebiotic synthesis of 5-substituted uracils: a bridge between the RNA world and the DNA-protein world. Science 268:702–705

Saenger W (1984) Principles of nucleic acid structure. Springer, New York

Saladino R, Crestini C, Costanzo G, Negri R, Di Mauro E (2001) A possible prebiotic synthesis of purine, adenine, cytosine, and 4(3H)-pyrimidinone from formamide: implications for the origin of life. Bioorg Med Chem 9:1249–1253

Sanchez RA, Orgel LE (1970) Studies in prebiotic synthesis V. Synthesis and photoanomerization of pyrimidine nucleosides. J Mol Biol 47:531–543

Sanghvi YS, Cook PD (1994) Carbohydrate modifications in antisense research, ACS Symposium series 580. American Chemical Society, Washington, DC

Schneider KC, Benner SA (1990) Oligonucleotides containing flexible nucleoside analogues. J Am Chem Soc 112:453–455

Schoning K, Scholz P, Guntha S, Wu X, Krishnamurthy R, Eschenmoser A (2000) Chemical etiology of nucleic acid structure: the α-threofuranosyl-(3′-2′) oligonucleotide system. Science 290:1347–1351

Schwartz AW (1989) Nucleic acid analogues and the origins of replication. Adv Space Res 9:77–81

Schwartz AW (1995) The RNA world and its origins. Planet Space Sci 43:161–165

Schwartz AW, Bakker CG (1989) Was adenine the first purine? Science 245:1102–1104

Schwartz A, de Graaf R (1993) The prebiotic synthesis of carbohydrates: a reassessment. J Mol Evol 36:101–106

Schwartz AW, Orgel LE (1985) Template-directed synthesis of novel, nucleic acid-like structures. Science 228:585–587

Schwartz AW, van der Veen M, Bisseling T, Chittenden GJF (1975) Prebiotic nucleotide synthesis – demonstration of a geologically plausible pathway. Orig Life Evol Biosph 6:163–168

Schwartz AW, Voet AB, Van der Veen M (1984) Recent progress in the prebiotic chemistry of HCN. Orig Life Evol Biosph 14:91–98

Shapiro R (1988) Prebiotic ribose synthesis: a critical analysis. Orig Life Evol Biosph 18:71–85
Sobolewski AL, Domcke W (2010) Molecular mechanisms of the photostability of life. Phys Chem Chem Phys 12:4897–4898
Stirchak EP, Summerton JE, Weller DD (1987) Uncharged stereoregular nucleic acid analogues. I. Synthesis of a cytosine-containing oligomer with carbamate internucleoside linkages. J Org Chem 52:4202–4206
Sutherland JD, Weaver GW (1994) Studies on a potentially prebiotic synthesis of RNA. Tetrahedron Lett 35:9105–9108
Tohidi M, Orgel L (1989) Some acyclic analogues of nucleotides and their template-directed reactions. J Mol Evol 28:367–373
Trinks H, Schröder W, Biebricher C (2005) Ice and the origin of life. Orig Life Evol Biosph 35:429–445
Uhlmann U, Peyman A (1990) Antisense oligonucleotides: a new therapeutic principle. Chem Rev 90:543–579
Usher DA, McHale AH (1976) Hydrolytic stability of helical RNA: a selective advantage for the natural 3′,5′-bond. Proc Natl Acad Sci U S A 73:1149–1153
Van Vliet MJ, Visscher J, Scwartz AW (1994) An achiral (oligo) nucleotide analog. J Mol Evol 38:438–442
Voet D, Rich A (1970) The crystal structures of purines, pyrimidines and their intermolecular complexes. Prog Nucleic Acid Res Mol Biol 10:183–265
Voet A, Schwartz A (1982) Uracil synthesis via hydrogen cyanide oligomerization. Orig Life 12:45–49
Wächtershäuser G (1988) An all-purine precursor of nucleic acids. Proc Natl Acad Sci U S A 85:1134–1135
Watson JD, Crick FHC (1953) Molecular structure of nucleic acids: a structure for deoxyribose nucleic acid. Nature 171:737–738
Weller DD, Daly DT, Olson WK, Summerton JE (1991) Molecular modeling of acyclic polyamide oligonucleotide analogues. J Org Chem 56:6000–6006
Westheimer FH (1987) Why nature chose phosphates. Science 235:1173–1178
White HB III (1976) Coenzymes as fossils of an earlier metabolic state. J Mol Evol 7:101–104
White HB III (1982) Evolution of coenzymes and the origin of pyridine nucleotides. In: Everse J, Anderson B, You K-S (eds) The pyridine nucleotide coenzymes. Academic, New York, pp 1–17
Whitesides GM, Mathias JP, Seto CT (1991) Molecular self-assembly and nanochemistry: a chemical strategy for the synthesis of nanostructures. Science 254:1312–1319
Woese CR (1967) The genetic code: the molecular basis for gene expression. Harper and Row, New York, p 186
Wolfe-Simon F, Davies PCW, Anbar AD (2009) Did nature also choose arsenic? Int J Astrobiol 8:69–74
Wolman Y, Haverland W, Miller S (1972) Nonprotein amino acids from spark discharges and their comparison with the Murchison meteorite amino acids. Proc Natl Acad Sci U S A 69:809–811
Wu T, Orgel LE (1991) Disulfide-linked oligonucleotide phosphorothioates: novel analogues of nucleic acids. J Mol Evol 32:274–277
Yamada H, Okamoto T (1972) Polycyclic N-hetero compounds XIV. Reactions of methylpyridines with formamide. Chem Pharm Bull 20:623
Zhang L, Peritz A, Meggers E (2005) A simple glycol nucleic acid. J Am Chem Soc 127:4174–4175
Zubay G (1996) Arguments in favor of an all-purine RNA first. Chemtracts: Biochem Mol Biol 6:251–260

Biodata of **Chrisantha Fernando** and **Vera Vasas**, authors of "***Cooptive Evolution of Prebiotic Chemical Networks.***"

**Dr. Chrisantha Fernando** is currently a postdoc at the Center for Computational Neuroscience and Robotics at Sussex University, England. Originally qualifying and practicing in medicine at Wadham College, Oxford, he obtained his Ph.D. from the Sussex University and completed a Marie Curie Fellowship at the Institute for Advanced Study in Budapest with Professor Eros Szathmary.

E-mail: **ctf20@sussex.ac.uk**

**Ms. Vera Vasas** is currently a Ph.D. student at the Autonomous University of Barcelona, Spain. She is currently working with Prof. Eros Szathmary and Prof. Mauro Santos on models of early evolution.

E-mail: **vvasas@yahoo.com**

**Chrisantha Fernando**

**Vera Vasas**



J. Seckbach (ed.), *Genesis - In The Beginning: Precursors of Life, Chemical Models and Early Biological Evolution*, Cellular Origin, Life in Extreme Habitats and Astrobiology 22, 35–53
DOI 10.1007/978-94-007-2941-4_2, 

# COOPTIVE EVOLUTION OF PREBIOTIC CHEMICAL NETWORKS

**CHRISANTHA FERNANDO[1] AND VERA VASAS[2]**
*[1]Department of Informatics, University of Sussex, Falmer, Brighton BN1 9RH, UK*
*[2]Departament de Genètica i de Microbiologia, Grup de Biologia Evolutiva (GBE), Universitat Autònoma de Barcelona, 08193 Bellaterra (Barcelona), Spain*

## 1. Introduction

Natural selection can act on units that are capable of multiplication, variation, and heredity, given these units have differential fitness and can be stably selected (Maynard Smith, 1986). The fundamental problem faced by any such unit of evolution is how information can be stored and reliably transmitted between generations. One solution that is generally agreed to have once been found by evolution is the RNA world. Here, RNA strands capable of template replication also possessed catalytic activity (Gilbert, 1986) and were most likely enclosed in compartments (Fernando et al., 2005).

However, the route to the RNA world is not known. RNA replication is template-based, see Fig. 1, meaning that the sequence of molecules on the parent strand is replicated (with mutation) to the child strand by a topographic mapping between the parent and child strand. The number of possible sequences that can be copied for a string of length $L$ is $n^L$ where $n$ is the number of nucleotide types and $L$ is the length of the string. Even for $n=2$ and $L=100$, there are $10^{30}$ possible unique sequences. As this is more than could be realized in any realistically sized system such as the universe, heredity is effectively unlimited (Szathmary, 2000). In addition, template replication has an important feature that adds power to natural selection, although is not a necessary feature for a unit of evolution as defined above: selection can, in principle, move from any sequence to any other sequence by a process of small mutations that produce *correlated variation* between parent and offspring if the fitness landscape allows it. Hogeweg has referred to information carried by template replicators as storage-based because all possible sequences are equally stable and transmissible, using the same copying mechanism (Hogeweg, 1998). While unlimited heredity and correlated variation seems necessary to explain the wondrous diversity of life, there are several problems with the spontaneous emergence of RNA templates or other molecules capable of template replication in the first place.

This chapter considers the mechanisms that could have led to the origin of nucleotides or, alternatively, of other monomers that made template-based

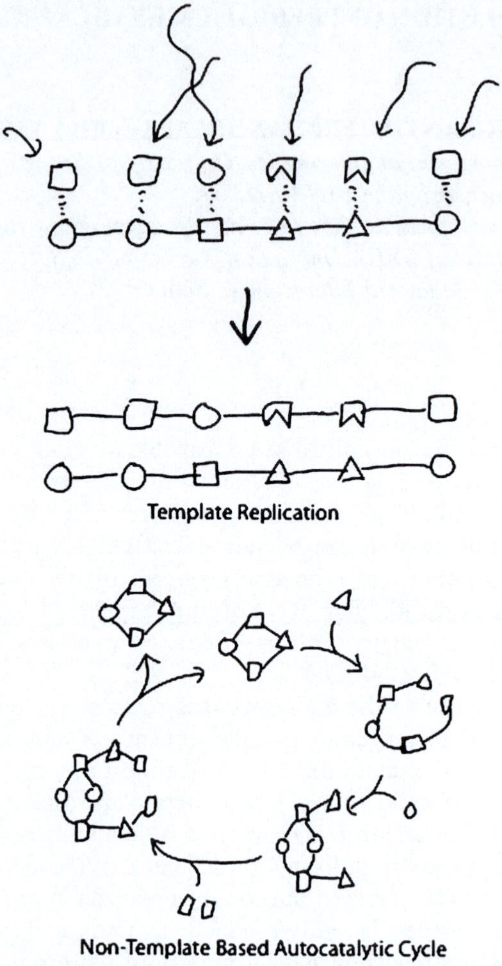


**Figure 1.** *Top*: *Template replication.* The parent strand contains a sequence of molecules (circle, circle, square, etc.) and is connected by strong bonds. Complementary monomers attach specifically to their counterpart by weak bonds and due to their proximity are connected together into a new strand by strong bonds. The power of template replication arises from the fact that variants can also be equally well replicated, for example, swapping the first circle with a triangle would not prevent the capacity for this novel sequence to replicate. *Bottom: Nontemplate replication.* In nontemplate-based autocatalysis, a copy is made, but variants are generally not capable of replication because a sequence-specific set of chemical reactions are involved in the autocatalytic cycle. For a more detailed discussion of the problem of the origin of nonenzymatic template replication, see Fernando et al., 2007.

replication possible. All such monomers are likely to be complex molecules that require an ordered set of chemical reactions for their synthesis. How could such a set of reactions have arisen prior to template-encoded catalytic enzymes thought to be necessary for maintaining a nucleotide synthesizing metabolism?

The difficulty of evolving functional nucleotide analogues can be compared to the difficulty of evolving feathers for flight. It is thought that feathers were originally evolved to solve the problem of keeping warm, and later exapted for flight. It is possible that the original function of nucleotides was not template replication, but something else.

Several solutions have been proposed, for example, a spontaneous generation type theory (chemical determinism) imagines a multitude of chemical environments on earth (or elsewhere), for example, black smokers (Wächtershäuser, 1990) or white nonsmokers (Martin and Russell, 2007) or a primordial soup as in Miller's famous experiment (Miller, 1953), or CO- and methane-rich atmospheres (Zahnle et al., 2011). It is perhaps wishful thinking, however, to suppose that some of these environments may well have been suitable for the spontaneous production of activated nucleotides in abundance, powered by some chemical or electromagnetic energy source. The extreme implausibility that an environment is perfectly suited to the production of activated nucleotides requires one to ask whether any other processes could have made the appearance of nucleotides more likely (Lifson, 1997; Shapiro, 2006; Wächtershäuser, 1990). We envision a scenario where chemical networks, enclosed in microscopic compartments, were capable of a limited type of evolution and gradually became more complex. Such networks might have been able to discover nucleotides and, eventually, template-encoded information transmission. Note that the primary chemical network that preceded nucleotides and the extant chemical networks of catalyzed metabolism today may differ greatly. This is discussed further elsewhere (Fernando et al., 2005).

## 2. Self-Organization Without Natural Selection

An influential set of theories that attempts to explain how pretemplate systems could have reached the levels of order necessary to explain the origin of nucleotides stems from Dyson's models (Dyson, 1986) and involves self-organization (in the absence of natural selection) (Kauffman, 1986; Kauffman, 1993; Fontana and Buss, 1994). These theories claimed that a process of random growth of a chemical network in a flow reactor would inevitably lead to increasingly complex chemical organizations capable of reproduction due inherently to scaling laws. They suggest that as the number of species grows, eventually, autocatalytic cycles arise that are capable of generating themselves from the food set of molecules. Stuart Kauffman proposed that a reactor supplied with a food set of short random peptides, where each peptide had some low probability $P$ of catalyzing ligation or cleavage reactions of other peptides, could result in the growth of a peptide catalytic network in which self-reproducing autocatalytic sets could form (Kauffman, 1993). Autocatalytic sets were defined by Kauffman as a set of species in which the production of each species in the set is catalyzed by at least one other species in the set. Note that the model is general enough to be applied with caution to ribozyme networks as well as other networks of molecules. It was demonstrated in a mathematical model that a supracritical reaction network that grows with an accelerating speed arose above a certain catalytic probability threshold (Kauffman, 1986). However, simulations that were undertaken failed to test whether these reaction networks were indeed truly sets of collectively autocatalytic molecules (Farmer et al., 1986).

Two main objections were made of Kauffman's model. Shinor Lifson pointed out that Kauffman's model assumed an unrealistically high probability (of one) that a peptide could act as a catalyst. This means that the probability that any randomly generated peptide would catalyze at least one reaction inevitably reaches one as the number of possible reactions increases (Lifson, 1997). However, Lifson's criticism was too strong in assuming that a catalyst is expected to catalyze only one out of the infinite possible reactions. Defining the probability that a catalyst catalyzes a given reaction as one divided by the number of existing reactions, he implicitly claimed that these probabilities are not independent – but why would considering a bigger reaction space make catalysis less likely? Also, it was proposed that if catalytic poisoning was included in Kauffman's model, autocatalytic sets would not have formed spontaneously due to the "paradox of specificity" (Szathmary, 2000): a high number of molecular types are required for the spontaneous emerge of autocatalytic sets, but the harmful effect of side reactions calls for a small system size. We have demonstrated in a reimplementation of the original simulations of Kauffman's theory (Farmer et al., 1986) that neither of these objections qualitatively prevents the formation of growing catalytic networks, see Fig. 2.

Large catalytic networks do form under various values of catalytic probability, even with strong poisoning of catalysis ($K = 0.01$). However, the further claims of Kauffman for the power of self-organization independent of natural selection to allow accumulation of adaptations in reproducing autocatalytic sets have been abandoned (Kauffman, 1993). Indeed, an early simulation showed that Kauffman's system was only capable of Spenserian and not true Darwinian evolution (Bagley et al., 1991; Bagley and Farmer, 1991), i.e., it was not capable of accumulating adaptations.

## 3. Compartmentalized Chemical Networks Undergoing Natural and Artificial Selection

An alternative theory to Kauffman's that has combined self-organization with natural selection prior to template replicators is Doron Lancet's GARD (Graded Autocatalytic Replication Domain) model. Lancet proposes that lipid-like molecules capable of self-assembly could produce vesicles that grow spontaneously and, due to environmental fluctuations, split (Segrè et al., 1998). Such vesicles possess the trinity needed to qualify as units of evolution: multiplication, variation, and heredity (Maynard Smith, 1986). The dynamical behavior of the system is governed by the matrix of catalytic interactions. Each element of the matrix describes the rate at which molecular species "A" enhances the inclusion of species "B" into the growing assembly. These rate enhancement factors are drawn randomly

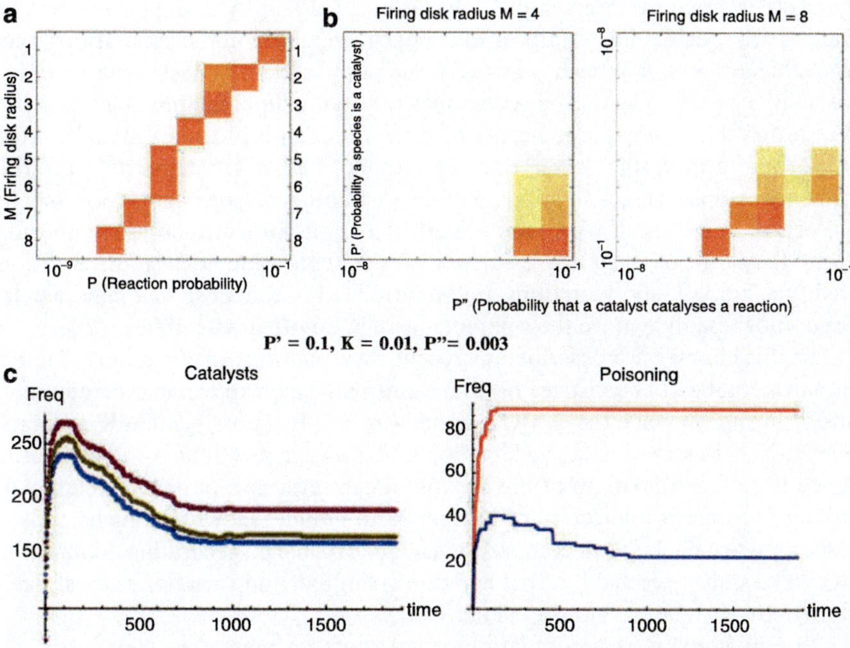


**Figure 2.** (**a**) There is a critical threshold (shown as *squares*) on the right of which reaction networks grow supracritically and on the *left* of which networks are subcritical. The *darkness* of a *square* reflects the proportion of 100 runs in which the network exceeded one of the following conditions: >20,000,000 reactions or >100,000 species. (Note that in any finite system, the reaction network cannot be explored infinitely due to mass constraints.) The critical threshold depends on the firing disc (maximum length of species which are to be provided as food) and the probability that a species catalyzes a specific reaction. (**b**) Instead of one $P$ value, two probabilities are defined, $P'$ = probability that a species can be catalytic. $P''$ = per reaction probability that this catalyst catalyzes a reaction. When $P'$ decreases, $P''$ must be considerably higher for reaction networks to keep growing. Nevertheless, there is a threshold above which catalytic networks grow supracritically. (**c**) Inhibition does not prevent formation of large catalytic reaction networks. For values of $P'$ and $P''$ that do produce catalytic network growth, strong noncompetitive inhibition is introduced by choosing with probability $K = 0.01$ that a species removes another species from the reactor completely if at least one molecule of the inhibiting species exists. *Left*: The number of catalysts present above the threshold concentration in the food set (*purple*, *top*) and nonfood set (*blue*, *bottom*). *Yellow* shows the number of all nonfood set molecules. *Right*: *Red* (*top*) = number of species that are inhibited (removed). *Blue* (*bottom*) = number of active inhibitors.

from a lognormal distribution and are always positive, meaning that all species catalyze the inclusion of themselves and all other species' (note that in Kauffman's model, catalysis was rare and specific). Variation occurs due to stochastic growth and splitting of the assembly at a certain size, thus giving the potential for evolution of "composomes" – compositional genomes in which the composition of the assembly is the heritable trait.

Recently, it was shown that the GARD model cannot sustain natural selection, i.e., the capacity for a particular assembly to be stably selected if it is of higher fitness than other assemblies (Vasas et al., 2010). Although it is true that vesicles with specific lipid compositions appear regularly in the simulations, they too frequently turn into each other, and the population of vesicles can be better characterized as a collection of assemblies randomly flipping among a few stable states rather than competing lineages of different compositions. Even if the fittest composome grows 10% faster than all others, it is still not possible for this composome type to reach fixation and persist. In other words, there is no capacity to select for one of multiple coexisting attractors, because attractors are unstable due to the large off-diagonal elements of the replication matrix. In addition, heredity is limited, and variations are uncorrelated in the sense that child assemblies do not resemble more their parents as they do other assemblies.

In this chapter, we describe our recent investigations that examine whether compartmentalized chemistries of the Kauffman type are capable of acting as units of evolution while the GARD model was not. It is possible that Kauffman's generative chemistry of catalyzed ligation and cleavage reactions is still of considerable interest for the origin of life because it can serve as a basis for pretemplate heredity. Our recent replication of Farmer et al.'s model shows that such catalytic networks, when enclosed in compartments, are capable of generating adaptations that can be stably selected but that heredity is limited, and variation is most likely only weakly correlated (Vasas et al., 2012).

In our simulations, a population of ten "compartments" each enclosing the generative chemistry was allowed to grow for a fixed period of time. These compartments model microscopic prebiotic vesicles. Fitness of a compartment was defined as the total mass of nonfood species present in a compartment at the end of the growth phase, just prior to division. Each compartment was modeled exactly as in Farmer et al.'s original paper, i.e., as a flow reactor in which food is input and materials leave at a fixed rate. The evolutionary algorithm used was generational fitness proportionate selection, i.e., compartments were chosen to populate the next generation in proportion to their fitness by using roulette wheel selection. If a compartment was chosen, it produced an offspring by taking a hypergeometric distribution of its molecular contents, thus modeling compartment splitting. Note that due to the fitness proportionate algorithm, a single compartment was, in theory, capable of producing more than two offspring. A prerequisite for accumulating adaptations is that it is possible to stably select for a particular attractor in a network that possesses *coexisting multiple attractors.* However, we found that original Kauffman-type networks of catalyzed cleavage and ligation reactions have only one attractor, meaning that the system is typically not capable of hereditary variation. Networks with inhibition, while able to spontaneously transition between attractors due to growth and division noise, (a) did not transition between correlated attractors, (b) typically transitioned periodically, or chaotically, and (c) attractors typically could not be stably selected, see Fig. 3. However, a different kind of evolution, which we can call evolution by chemical avalanches, was possible when variation was produced by allowing spontaneous novel species to arise from the "shadow" (uncatalyzed

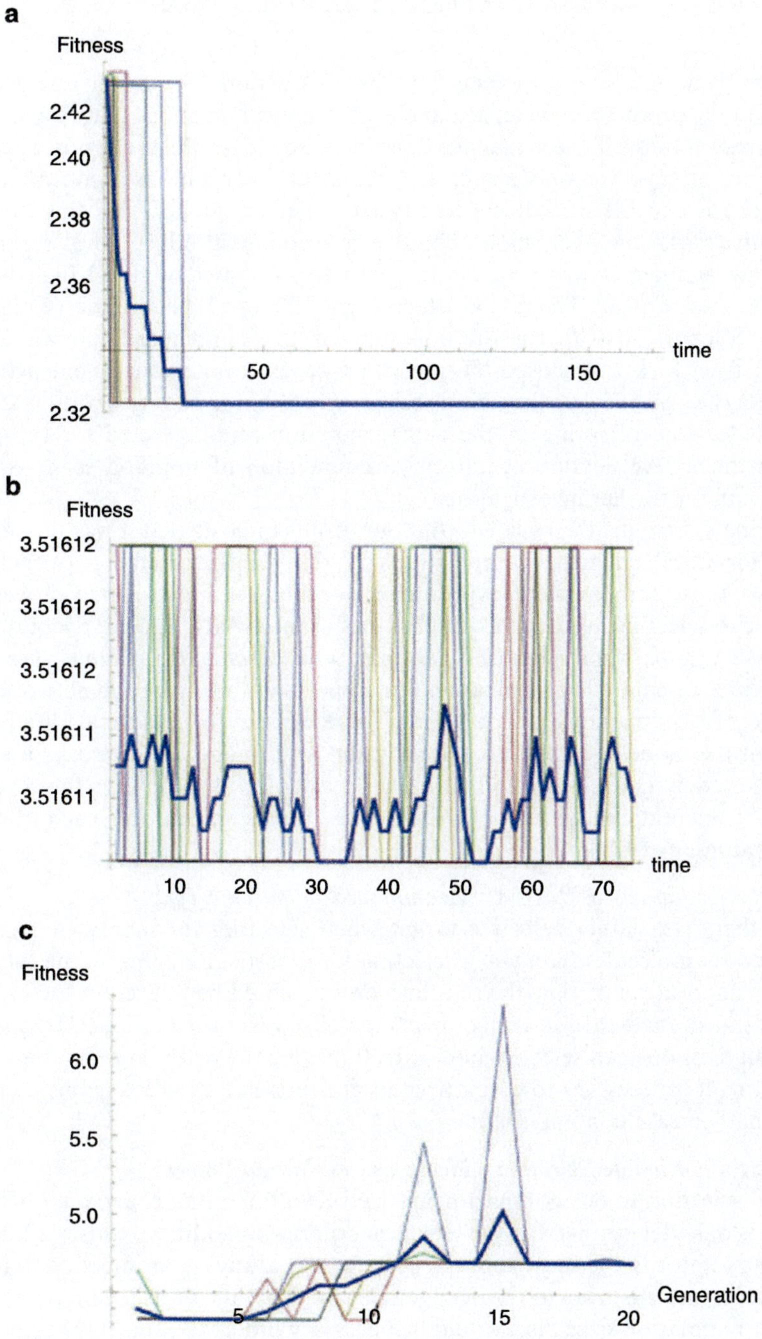


**Figure 3.** The fitness changes in the population resulting from artificial selection for more rapidly growing nonfood set mass. In the original Kauffman-type network, multiple attractors were not observed and selection was not possible. With inhibition included in the model, multiple stable states become possible; still, the population typically settled down into one equilibrium (**a**) or fluctuated stochastically or periodically between attractors (**b**). In both cases, selection had little effect. However, when spontaneous rare reactions that produce novel species were included in the model, the fitness graph showed the fixation of an adaptation (**c**). Each chemical attractor has a unique fitness because the fitness of a compartment was defined as the total mass of nonfood species present in a compartment at the end of the growth phase, just prior to division, and this was used to implement artificial selection.

reactions among existing molecules). The production of a novel rare species results in the production of an avalanche of chemical reactions that may or may not increase fitness. If the avalanche does increase fitness, then selection attempts to stabilize it; however, only a subset of chemical organizations produced in this avalanche is capable of stable heredity (see later section). Using this kind of generative variation when spontaneous rare reactions produce novel species in low copy number, it was possible to select for a subset of novel high fitness attractors, see Fig. 3c. The rate at which these "shadow" species emerge must be low in comparison with the division time of the compartment in which the chemical network is enclosed. However, as we used not natural, but artificial selection (i.e., we ensured that the fittest compartment in each generation always contributed one offspring to the next generation), and selected for the gross compartment level feature of rate of accumulation of nonfood set mass, the results call for further investigations.

Firstly, we must then ask why did our results indicate that it was possible to select for novel chemical compositions in the Kauffman-type networks, for example, those compartments with higher nonfood set mass, whereas it was not possible to select for particular composomes in the GARD model? Secondly, we must ask why *only some reaction networks were capable of sustaining heredity*. What kinds of catalytic organization that appears in these reaction networks are capable of accumulating adaptations? There appear to be several differences between the model we used for selecting for Kauffman-type networks and the models used by Lancet's group that make it easier for us to observe the accumulation of adaptations. Later work should test the significance of each of these assumptions.

*Selection for Increased Mass vs. Selection for Composome Type*

First, there is a larger selective target when selecting for increased mass of nonfood set molecules than when selecting for a particular composome type. To test the importance of this, the selection criteria could be relaxed in the GARD model and strengthened in the compartmentalized Kauffman model. In biological evolution, one can select in both directions: for the better and for the worse. Therefore, it is necessary to check whether it is possible to select against adaptations that increase nonfood set mass.

*Fitness Proportionate Selection with Elitism vs. Moran Process*

We call selection in our compartmentalized Kauffman model *artificial* because fitness is only determined by our selection criteria. In addition, it uses a kind of elitism operator in the sense that the best agent is always guaranteed to contribute at least one offspring to the next generation. This allows us to simulate much smaller population sizes than would be necessary otherwise. It is physically realizable but would not occur naturally. Natural selection could be used in the compartmentalized Kauffman model instead, and artificial selection could be introduced into the GARD model and the effects observed.

*Uniform Catalytic Activity Distribution vs. Lognormal Distribution*
In Kauffman's model, a catalyst either does or does not catalyze a particular reaction. If it does catalyze a reaction, then the catalytic velocity is chosen from a uniform distribution between 10 and 1,000. In the GARD model, all species catalyze the incorporation of all other species, but the catalytic velocities are drawn from a lognormal distribution. As the lognormal distribution of rate enhancement parameters in the GARD results in a few strong and many weak interactions, an analogy can be drawn if we interpret "not catalyzing" in Kauffman's model as rounding down weak interactions. The lognormal distribution of catalytic velocities in the GARD model can be replaced with a catalytic probability *P* that a lipid catalyzes another lipid, and catalytic velocities are taken from the same distribution chosen by Farmer et al. It has been shown (Eors Szathmary, personal communication) that introducing such variation into the GARD model does permit evolvability. The GARD model is based on a more realistic distribution of kinetics than Farmer (and our) model, and this certainly makes evolvability harder in their system. To allow Kauffman's model to play on a more similar playing field as the GARD model, one could set $P=1$ and draw catalytic velocities from the same lognormal distribution used in GARD.[1] It is not known whether this would prevent evolvability in our simulations. Another important modification to be made to the Kauffman model is to replace the random catalytic chemistry with one based on pattern matching, i.e., catalytic effects are determined by overlapping sequences. This would permit a more realistic distribution of catalytic effects.

*Ligation and Cleavage Chemistry vs. Incorporation Only Chemistry*
In GARD, there is only one level of catalyzed reactions, i.e., incorporation of lipids into the composome is catalyzed by other lipids in the composome, whereas in Kauffman's model, the catalytic network can be much more complex, involving many topologies of ligation and cleavage reactions. Similar reactions have already been incorporated into the GARD model (but have not been analyzed here) (Shenhav et al. 2005), and the generative potential of the compartmentalized Kauffman model could be restricted in further experiments.

*Rare Novel Reactions vs. No Rare Novel Reactions*
There were two sources of variation in our implementation of growing compartmentalized Kauffman's networks. The first is rare spontaneously occurring reactions among existing molecular species that produce novel species, and the second is noise at division. The GARD model possesses the later source of variation, as well as stochastic growth of the composome; however, it does not permit spontaneous novel reactions as may occur, for example, by the occasional addition of novel lipids with random catalytic potential, a feature that could be added to the

[1] It will be interesting in the future to consider the addition of inhibition into the GARD model, where a lipid can also prevent the incorporation of other lipids into the composome.

GARD model. Not permitting rare novel reactions in the Kauffman model removes any potential for continuing generation of adaptations and allows only exploration of multiple attractors in a fixed reaction network.

There are some serious technical difficulties in conducting some of these experiments. For example, there are severe computational constraints in running the compartmentalized Kauffman model with a sufficiently large chemical network with a large population size. This was the main motivation for incorporating artificial selection with an elitism-type operator. In the next section, we turn to the second question of what kind of catalytic organization can sustain heredity and present a classification of catalytic networks that makes sense of the often-confusing terminology used in papers on chemical evolution. We define a fundamental unit called the *viable strongly connected core* (VC). We propose this as the fundamental unit of chemical evolution.

## 4. Autocatalytic Sets and Viable Strongly Connected Cores

The broadest class of network is just a *reversible reaction network*. It consists of reactions where any number of reactants is converted to any number of products and vice versa. A more specific class of networks is a *catalyzed reversible reaction network*. Here, each reaction in the network must be catalyzed by another species in the network. Next, we consider specific types of such catalyzed networks and add to the system the notion of a *food set* (F) of species that are provided as input. Figure 4 shows our classification of chemical reaction networks of relevance for pretemplate evolution.

As Lifson has pointed out so clearly (Lifson, 1997), molecular replication by autocatalysis must have preceded biological reproduction, and as so, it was essential for the emergence of life. The uniqueness of autocatalysis relies in exponential, instead of linear, growth. For a single catalyst molecule with the velocity of one catalyzed reaction per microsecond, it would take more time than the estimated age of the universe to produce a mole of products – in contrast, for an autocatalyst with the same catalytic properties, this task would require 79 microseconds (Lifson, 1997). In addition to the simple case of elementary autocatalysis where a molecule catalyzes its own production, other setups of reflexively autocatalytic molecules are also possible, for example, A catalyzes B and vice versa. In the catalytic graph where nodes depict molecules and edges are catalytic actions, such *autocatalytic loops* occur as cycles.

Kauffman's original paper describes *autocatalytic sets* in relation to the food set. He states that they must be "catalytically closed," i.e., it is a set of molecules in which "every member of the autocatalytic set has at least one of the possible last steps in its formation catalyzed by some member of the set, and that connected sequences of catalyzed reactions lead from the maintained 'food set' to all members of the autocatalytic set" (Kauffman, 1986). This is more formally defined by Hordijk and Steel (2004) who state that reflexive autocatalytic (RA)

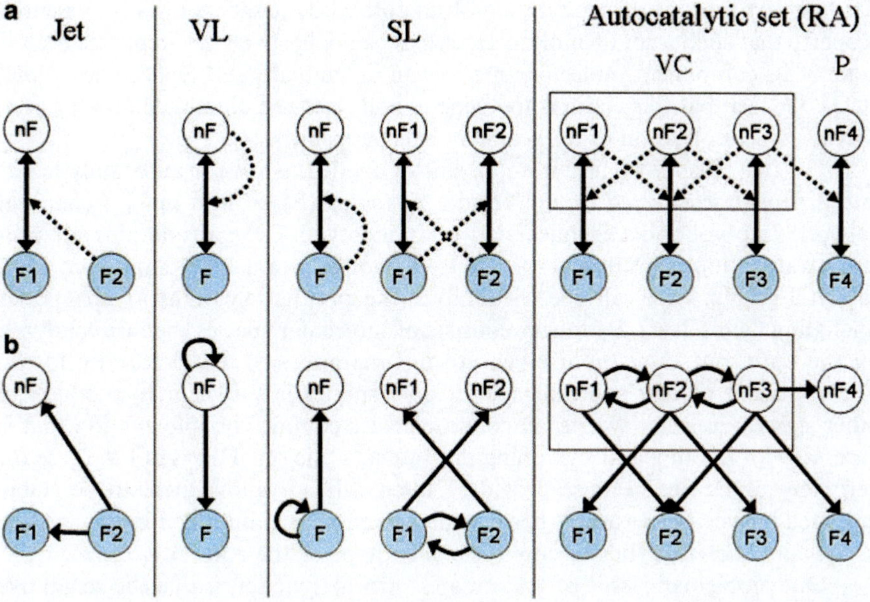


**Figure 4.** Our classification of reaction networks of relevance for heredity in chemical networks. Blue nodes indicate members of the food set (*F*); nonfood species (*nF*) are white. **(a)** Reaction networks, where *solid arrows* mark reactions, and *dashed arrows* catalysis. **(b)** Catalytic graph showing which species catalyzes which. Autocatalytic loops appear here as cycles of any size. A *jet* does not contain autocatalytic molecules, and its catalytic graph is acyclic. Viable loops (*VL*) are the necessary units needed for exponential growth of autocatalysts, in contrast to suicidal loops (*SL*) that require reactants that are only produced by the autocatalytic loop itself. *Autocatalytic sets* (*RA*s) contain a core of autocatalytic molecules (either viable or suicidal; here, a viable core – *VC* – made up by two VLs is shown) and a periphery (*P*) that is catalyzed by the core. As all core species catalyze the production of all others, any one molecule of core species is sufficient for igniting and sustaining the autocatalytic set. We propose that viable cores are the key to heritable adaptations in pretemplate chemical networks.

and food-generated (F-generated) *RAF sets* are as follows: given a network of catalyzed chemical reactions, a (sub)set *R* of such reactions is called:

1. Reflexively autocatalytic (RA) if every reaction in *R* is catalyzed by at least one molecule involved in any of the reactions in *R*
2. F-generated (F) if every reactant in *R* can be constructed from a small "food set" *F* by successive applications of reactions from *R*
3. Reflexively autocatalytic and F-generated (RAF) if both RA and F

Autocatalytic cycles, hypercycles, and (collectively) autocatalytic sets are all instances of RA sets. Note that Kauffman's definition and the above definition exclude autocatalytic sets that could not be constructed from the food set. Other work has defined autocatalytic sets more abstractly in terms of the catalytic graph: "an autocatalytic set is a graph, each of whose nodes has at least one incoming link from a node belonging to the same graph" (Jain and Krishna, 1998).

Krishna says that Kauffman's notion of an autocatalytic set captures the essential property that such a set of molecular species can collectively self-replicate even if none of its component molecular species can individually self-replicate (Krishna, 2003). We feel that this claim is too general and does not illuminate how, or even *whether*, true replication of autocatalytic sets happens.

For this reason, we find it important to divide the RA (or autocatalytic set) into a *strongly connected core* (C) and a *periphery* (P), or in a sense a chemical network genotype and a chemical network phenotype. The core contains adjacent autocatalytic loops and corresponds to a strongly connected component in a catalytic graph. Here, all species catalyze the production of all other species (including themselves). Periphery consists of molecular species that are catalyzed by the core, but they themselves are not members of autocatalytic loops. The provision of any one molecule of core species is sufficient to produce all other species; in other words, all core molecules contain the information that is necessary for igniting and sustaining the autocatalytic set. This is not the case for periphery molecules because providing them will not allow them to be stably inherited between generations because they cannot be maintained in the absence of the core, therefore they resemble the phenotype of the core.

One problematic issue of Kauffman's autocatalytic sets is that the generative chemistry is also capable of producing a growing network of nonfood set molecules in the absence of any autocatalytic sets whatsoever, because the food set can be catalytic, and thus can produce a *jet*. This is simply an acyclic directed catalytic graph of reactions that arises when the number of molecular species is small compared to the size of the food set. As the network grows, catalytic closure is reached.

However, if we take a closer look on the organization of a Kauffman-type network, we see that the vast majority of autocatalytic loops are not capable of exponential growth, the feature that made autocatalyis important in the first place. There is no pathway from the food set to the immediate reactants that are needed for sustaining autocatalysis; instead, the reactants are produced by the loop species from the food set. So, the loop catalyzes reactions that destroy itself, effectively it is a *suicidal loop* (SL). Eors Szathmary has pointed out (personal communication) the known existence of self-digesting autocatalytic loops in metabolism, for example, the Horecker cycle (Kresge et al., 2005). Indeed, only a small minority of autocatalytic species form *viable loops* (VL) and use valid reactants in the autocatalytic reactions. The GARD model, on the contrary, contains viable loops by definition because the only reaction structure considered is the transition between food monomers and nonfood – composome member – molecules catalyzed exclusively by the latter.

The critical units in this system are the viable loops that make up a *viable core* (VC). Also, we note that viable cores can be related to each other in various ways. For example, independent cores can exist in the same network, while some cores may be in the periphery of others, resulting in upstream and downstream cores. It is possible to reignite a downstream core from an upstream core, but not vice versa. Our notion of a viable core corresponds to the notion of an *autocatalytic seed* (Kun et al., 2008). Kun et al. showed that such cores actually exist in extant metabolism in the form of ATP/ADP and other small molecule cores that are necessary for the ignition of the extant metabolic network (in addition to encoded molecules). Thus, it seems that heredity based on viable cores still persists in life today! This concludes our classification.

Thus, we see that the notion of an autocatalytic set is a very broad one, encompassing many different kinds of chemical organization consisting of various types of autocatalytic loops, some of which may, and some of which may not, be capable of sustaining hereditary variation. It is important to note at this point that as there is only one platonic chemical network that encompasses all possible reactions, variation can only arise if the system has alternative stable attractors. This is not the case (in the absence of inhibition) in the original work of Farmer et al. (Farmer et al., 1986) because they examined exclusively food-generated autocatalytic sets – in other words, chemical networks that always appear when food species are provided. On the contrary, we find that heritable adaptations can appear when novel species that are produced by spontaneous rare reactions join the network as seeds of *nonfood-generated viable cores* of various sizes. They are, by definition, downstream of previously existing cores, as all previously existing molecules are members of the autocatalytic set. We propose that such *nonfood-generated viable cores constitute the source of heritable variation in pretemplate chemical networks and that adaptations appear as avalanches of novel molecular species.*

## 5. Cooptive Evolution in Catalytic Networks

Figure 3 showed that artificial selection of an adaptation is possible in compartmentalized Kauffman-type networks capable of rare spontaneous novel reactions that generate reaction avalanches.

We call this cooptive evolution because what happens here differs from Darwinian evolution in not requiring variation to be highly correlated; instead, it works by coopting a weak or uncorrelated novel extension to the network. This implies *limited heredity* and *weakly correlated variation* between parent and offspring compartments. What would this cooptive evolution by chemical avalanches look like if translated to template replication? It would be as if templates did not change by mutation but instead grew by randomly adding nucleotides to their ends. If the resulting strand had higher fitness it would be accepted, and if it had lower fitness it would be rejected. Add to this the caveat that only a small subset of these extending sequences could be stably transmitted between generations in the face of serial dilution due to division, and we have a "template" version of cooptive evolution. A classification of the various kinds of natural selection is shown in Fig. 5, based on correlated vs. uncorrelated variation, one

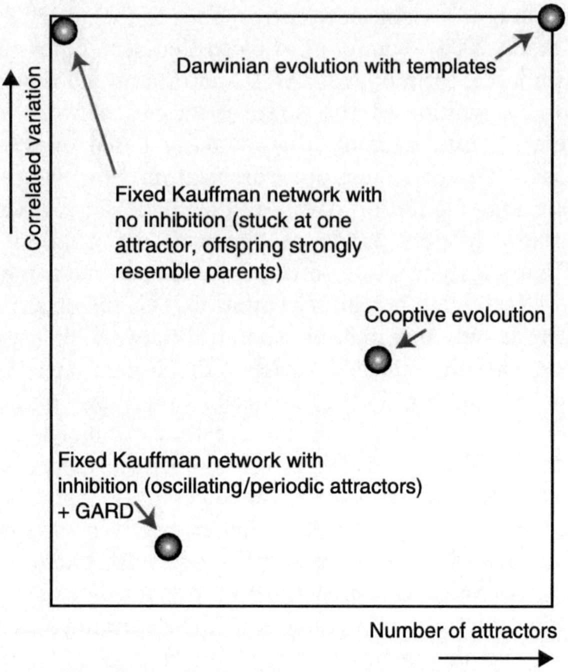


**Figure 5.** A classification of various kinds of natural selection. Standard Darwinian evolution with template replication has unlimited heredity, variants can be stably selected, and variation is correlated (offspring resemble their parents more than they do other parents). In fixed compartmentalized Kauffman-type networks, there may rarely be few coexisting alternative attractors, but these are not selectable because each attractor is *too stable* (i.e., there is no heritable variation between parents and offspring that selection can act on) in the face of noise during compartment division. When inhibition is introduced, there are more attractors, and there is some capacity to transition between them; however, selectability is still low because transitions are chaotic or periodic. The same is the case for GARD. However, cooptive evolution (by chemical avalanches) which involves expansion into the adjacent possible of reaction space does have the potential to produce several stable novel attractors (due to nonfood-generated viable cores); however, there is limited correlated variation, i.e., offspring may differ greatly from their parents if the avalanche size is large.

attractor vs. unlimited attractors, and unselectable attractors (periodic/chaotic) vs. selectable attractors.

In the absence of novel viable cores, it may be possible that another method based on compositional inheritance exists in which selection can stably maintain one of multiple stable concentration vectors (coexisting multiple attractors); however, our preliminary investigations revealed only very rare examples of this (probably due to removal of a food species in the concentration randomization procedure used to search for multiple attractors). In the GARD model, although transitions between multiple coexisting attractors were possible, stable selection was not possible due to the tendency for spontaneous flipping between composomes. However, this chapter has concentrated on the importantly different process of avalanche-based evolution. This appears to be a much more effective method of generating variation in chemical evolution than compositional hereditary variation.

Still, there are several limitations of cooptive evolution by chemical avalanches. Proper adjustment of the relative timescales of protocell division and chemical avalanches is essential (Fernando and Rowe, 2007). If there are too many avalanches per cell division, then it is not possible to weed out harmful avalanches from beneficial ones. Secondly, there must be sufficient variety of spontaneous novel reactions in the shadow of the existing set of species for novel species to arise in low copy number. Thirdly, because the benefit conferred to the cell by a species is not correlated with its autocatalytic rate or other factors relating to the probability of its persistence, it is possible that a selfish autocatalyst is capable of displacing another autocatalyst that benefited the cell more. Cell level selection will then have to act to deselect cheaters. However, it is not clear, as in the stochastic corrector model, how much information (i.e., how many distinct sequences with different replication rates or how many autocatalysts with different autocatalytic growth rates) could be sustained within the same protocell without competitive exclusion (Fontanari et al., 2006). The problem is more severe in chemical network evolution where there is no capacity in a random chemistry to fine-tune the concentration vector of autocatalysts to maximally benefit the cell, as variation is by macromutation and not micromutation. Fourthly, while template replication is "computational," i.e., involves the exploitation of a universal function that translates genotype to phenotype and therefore all sequences are equally heritable, avalanche-based evolution requires a novel autocatalytic mechanism for each new viable loop to ensure transmission across generations. While this can potentially give incremental adaptation, it is at a cost of an ever-increasing complexity of the molecular network. To solve this problem, it is necessary to understand the origin of micromutation from macromutation, i.e., the origin of highly correlated variation. This requires a more sophisticated model of chemical structure than merely a random chemistry to understand how autocatalysts could have been selected that were capable of varying smoothly in their replication rates and catalytic potentials.

## 6. Conclusions

Recent experiments by Ben Davis' group at Oxford University have managed to enclose the Formose autocatalytic cycle in compartments (Gardner et al., 2009). This is a fundamental component in Ganti's chemoton, a highly persuasive and coherent theory for the organization of the minimal unit of life (Gánti, 2003). Although they have not quite yet managed to obtain cell division with enclosed Formose cycles, they have been able to select for particular compositions. This is possibly an example in which seeding of a substrate/product or related molecule skews the output of the system; however, further analysis of the Formose networks would be needed to confirm whether those systems depended on viable cores for heredity as we have predicted they do here. Once cell division is obtained in these amazing experiments, we predict based on these simulations of Kauffman-type networks, of the GARD model, and on our previous simulations (Fernando and Rowe, 2008) that it will be possible to select for chemical organizations, i.e., viable loops (VLs) and viable strongly connected cores (VCs), thus for the first time demonstrating the possibility of pretemplate heredity in a chemical system. We suggest that to avoid having to use very large population sizes, artificial selection will have to be used in the laboratory to efficiently select for variants. Also, we think that techniques will have to be used to reduce the rate of avalanche production and avalanche size in relation to the protocell division period. Finally, we note that our proposal here of the mechanism of cooptive evolution is a mechanism which could also be responsible for evolution by symbiosis (Watson, 2006) and evolution of ecosystems (Williams and Lenton, 2007).

## 7. Acknowledgements

Funding was provided by the E-FLUX FET OPEN ICT FP7 grant. We thank Mauro Santos, Stuart Kauffman, and Eors Szathmary for invaluable discussions and guidance in this project.

## 8. References

Bagley RJ, Farmer JD (1991) Spontaneous emergence of a metabolism. In: Langton CG, Taylor C, Farmer JD, Rasmussen S (eds) Artificial life II: Santa Fe Institute studies in the sciences of complexity. Addison-Wesley, Redwood City

Bagley RJ, Farmer JD et al (1991) Evolution of a metabolism. In: Langton CG, Taylor C, Farmer JD, Rasmussen S (eds) Artificial life II, Studies in the sciences of complexity. Addison-Wesley, Redwood City

Dyson F (1986) Origins of life. Cambridge University Press, Cambridge

Farmer JD, Kauffman SA et al (1986) Autocatalytic replication of polymers. Phys D 22:50–67

Fernando C, Rowe J (2007) Natural selection in chemical evolution. J Theor Biol 247:152–167

Fernando C, Rowe J (2008) The origin of autonomous agents by natural selection. Biosystems 91:355–373

Fernando C, Santos M et al (2005) Evolutionary potential and requirements for minimal protocells. Top Curr Chem 259:167–211

Fernando C, Von Kiedrowski G, Szathmáry E (2007) A stochastic model of nonenzymatic nucleic acid replication: "elongators" sequester replicators. J Mol Evol 64:572–585

Fontana W, Buss LW (1994) What would be conserved if 'the tape were played twice'? Proc Natl Acad Sci U S A 91:757–761

Fontanari JF, Santos M et al (2006) Coexistence and error propagation in pre-biotic vesicle models: a group selection approach. J Theor Biol 239(2):247–256

Gánti T (2003) The principles of life. Oxford University Press, Oxford

Gardner PM, Winzer K et al (2009) Sugar synthesis in a protocellular model leads to a cell signalling response in bacteria. Nat Chem 1:377–383

Gilbert W (1986) Origin of life: the RNA world. Nature 319:618

Hogeweg P (1998) On searching generic properties of non generic phenomena: an approach to bioinformatic theory formation. In: Adami C, Belew RK, Kitano H, Taylor CE (eds) Artificial life VI. MIT Press, Cambridge/London, pp 285–294

Hordijk W, Steel M (2004) Detecting autocatalytic, self-sustaining sets in chemical reaction systems. J Theor Biol 227:451–461

Jain S, Krishna S (1998) Autocatalytic sets and the growth of complexity in an evolutionary model. Phys Rev Lett 81:5684–5687

Kauffman SA (1986) Autocatalytic sets of proteins. J Theor Biol 119:1–24

Kauffman SA (1993) The origins of order. Oxford University Press, New York

Kresge N, Simoni RD et al (2005) Bernard L. Horecker's contributions to elucidating the pentose phosphate pathway. J Biol Chem 280:e26

Krishna, S (2003) Formation and destruction of autocatalytic sets in an evolving network model. PhD thesis, Center for Theoretical Studies, Indian Institute of Science, Bangalore

Kun A, Papp B et al (2008) Computational identification of obligatorily autocatalytic replicators embedded in metabolic networks. Genome Biol 9:R51

Lifson S (1997) On the crucial stages in the origin of animate matter. J Mol Evol 44:1–8

Martin W, Russel M (2007) On the origin of biochemistry at an alkaline hydrothermal vent. Philos Trans R Soc Lond B Biol Sci 362:1887–1926

Maynard Smith J (1986) The problems of biology. Oxford University Press, Oxford

Miller SL (1953) Production of amino acids under possible primitive Earth conditions. Science 117:3046

Segrè D, Lancet D et al (1998) Graded Autocatalysis Replication Domain (GARD): kinetic analysis of self-replication in mutually catalytic sets. Orig Life Evol Biosph 28:501–514

Shapiro R (2006) Small molecule interactions were central to the origin of life. Q Rev Biol 81(2):105–125

Shenhav B, Bar-Even A, Kafri R, Lancet D (2005) Polymer GARD: computer simulation of covalent bond formation in reproducing molecular assemblies. Orig Life Evol Biosph 35(2):111–133

Szathmary E (2000) The evolution of replicators. Philos Trans R Soc Lond B 355:1669–1676

Vasas V, Szathmáry E et al (2010) Lack of evolvability in self-sustaining autocatalytic networks constraints metabolism-first scenarios for the origin of life. Proc Natl Acad Sci U S A 107(4):1470–1475

Vasas V, Fernando C, Santos M, Kauffman S , Szathmary E (2012) Evolution before genes. Biol Direct 7:1

Wächtershäuser G (1990) Evolution of the first metabolic cycles. Proc Natl Acad Sci U S A 87:200–204

Watson RA (2006) Compositional evolution: the impact of sex, symbiosis, and modularity on the gradualist framework of evolution. MIT Press, Cambridge

Williams H, Lenton T (2007) Artificial ecosystem selection for evolutionary optimisation. In: Almeida e Costa F et al (eds) Advances in artificial life: proceedings of the 9th European conference on artificial life. Springer, Berlin/Heidelberg, pp 93–102

Zahnle K, Schaefer L, Fegley B (2011) Earth's earliest atmospheres. Cold Spring Harb Perspect Biol 2010 2:a004895

Biodata of **Enzo Gallori** and **Sergio Branciamore**, authors of "***Origin and Evolution of Self-Replicating Polymers on Mineral Habitats.***"

**Enzo Gallori**, born on 1949, graduate *cum laude* in Chemistry in 1974 at the University of Florence, Italy, is a professor of Astrobiology and Genetics at the Department of Physics and Astronomy of the University of Florence. Member of several scientific associations, he has done researches at the Max Planck Institut für Molekulare Genetik, Berlin; the Department of Biology, University of Essex, UK; the Microbiology Institute of the Academy of Sciences, Prague; the Department of Biology, New York University; and the Institute of Geochemistry of the Russian Academy of Sciences, Moscow. His scientific activity, documented by around 200 publications as well as some books of genetics, is currently devoted to the study of the origin of life on Earth and in the universe. For these researches, Professor Gallori has been invited to lecture at the most important international symposia in this field and has been charged with organizing the XII ISSOL Meeting – XV International Conference on the Origin of Life – that has been held in Florence in 2008.

E-mail: **enzo.gallori@unifi.it**

J. Seckbach (ed.), *Genesis - In The Beginning: Precursors of Life, Chemical Models and Early Biological Evolution*, Cellular Origin, Life in Extreme Habitats and Astrobiology 22, 55–66
DOI 10.1007/978-94-007-2941-4_3, 

**Sergio Branciamore**, born on 1976, graduate *cum laude* in Biological Sciences in 1999, is a postdoc at the Department of Physics and Astronomy of the University of Florence. He earned his Ph.D. in Chemistry at the University of Florence in 2003, and has done researches at the University of Chicago, the Collegium Budapest, Institute for Advanced Study, Budapest, and he was Susumu Ohno Distinguished Fellow in the Department of Theoretical Biology at Beckman Research Institute, Duarte, California. He is a member of the International Society for Studies on Origins of Life (ISSOL), the Società Italiana di Astrobiologia (SIA), and the International Society for Artificial Life (ISAL). He is currently devoted to the study of the origin of life and the evolutionary dynamics of replicators in simulated systems.

E-mail: **sergio.branciamore@unifi.it**

# ORIGIN AND EVOLUTION OF SELF-REPLICATING POLYMERS ON MINERAL HABITATS

**ENZO GALLORI AND SERGIO BRANCIAMORE**
*Department of Physics and Astronomy, University of Florence, Largo E. Fermi 2, 50125 Florence, Italy*

## 1. Introduction

All life forms on Earth today, and all life for which there is fossil evidence, are made of cells and share the same biological program based on nucleic acids (DNA/RNA) and proteins (Ganti, 2003). The genetic information, recorded in the sequences of nucleotides (the structural unit of nucleic acids) along the DNA/RNA chains, supplies the language of life and, by the steps of transcription and translation, leads to the production of the amino acid sequences of a polypeptide, as asserted by the *Central Dogma of Biology* (Lewin, 2007). Moreover, the specific sequence of nucleotides in DNA is handed down from one generation to the next, by DNA replication process, thus ensuring the perpetuation of the genetic information on Earth.

These are the fundamental aspects of living organisms, and although other features are necessary to life, the presence of genetic material is absolutely essential.

The appearance in an ancestral era of our planet of a self-replicating polymer, able to undergo evolution through processes of replication, mutation, and natural selection, marks the turning point between the prebiotic chemistry and the beginning of life (Joyce, 2002).

The formation of a primordial genetic molecule, whatever it was, would have required specific conditions, as the synthesis and availability of precursors (i.e., nucleotides), the polymerization of these precursors to obtain larger molecules (polynucleotides), and their persistence in critical conditions, like those present in primeval terrestrial habitats, characterized by the presence of harsh UV and cosmic radiations, so ensuring their ability to evolve in a changing environment (Nisbet and Sleep, 2001).

The occurrence of the steps indicated above implies, necessarily, the presence of a credible, protected environment where the genetic molecule could originate and express its biological potential (Pace, 1991).

## 2. The Mineral Environment

Classical research in the field of chemistry of the origin of life has focused mainly on reactions in aqueous conditions (Delaye and Lazcano, 2005) in the belief that building blocks of the biopolymers could be readily obtained starting from simple organic molecules present on the surface of our planet. However, the formation of complex molecules, as nucleic acids, required the joining of precursors once synthesized. As biological polymers are generally formed by dehydration, it is difficult to conceive that complex macromolecules could have originated by random collisions in the presence of a high concentration of water, like that of a primordial ocean. In these conditions, hydrolysis is favored, not polymerization (Pace, 1991).

This problem is exemplified by the properties of RNA. It is currently believed that in an era indicated as the “RNA World,” the RNA molecule could have functioned both as genetic material and as an enzyme (“ribozyme”) (Joyce, 2002; Orgel, 2004). The presence of the 2′-OH group in ribose, which renders RNA catalytic, also makes the molecule particularly susceptible to hydrolysis (Doudna and Cech, 2002). Therefore, the very complex structure of RNA and its intrinsic instability make it very difficult to imagine the origin of a hypothetical RNA World in free solution.

In the last years, several alternative environments have been proposed to overcome the difficulties above mentioned (Bada and Lazcano, 2002). It has been suggested by various authors that mineral environments, as the surfaces of iron-sulfur minerals (pyrite, $FeS_2$) (Wachtershauser, 1988), the warm alkaline springs of ocean floor (Martin and Russell, 2003; Martin et al., 2008), or the sediments of clay minerals (Ferris and Hagan, 1986), were the sites of the prebiotic reactions that led to the first biomolecules. In particular, the last one, originally proposed by Bernal in the middle of last century (Bernal, 1951), suggested that clay minerals, for their chemical nature and crystal structure, could have bound organic molecules from the surrounding water habitat, concentrating them and protecting the molecules against destruction by high temperatures and strong radiations, so providing an ad hoc environment for the assembly of larger molecules.

In recent years, numerous observations have reinforced the hypothesis of a mineral-surface-mediated origin of life. These observations, at first, included the synthesis of the constituents of nucleotides, the pentose sugars (Pitsch et al., 1995) and nitrogen bases (Saladino et al., 2004), in the presence of clay minerals montmorillonite and kaolinite, phyllosilicates produced by the erosion of volcanic ash, and the possible role played by boron-bearing minerals in stabilizing sugar ribose in ancestral environments (Ricardo et al., 2004). On the other hand, results obtained in the field of astrobiology identify in extraterrestrial mineral bodies (i.e., meteorites, micrometeorites, comets, and cosmic dust particles) the principal agents in the enrichment of early Earth of the organic matter necessary for the synthesis of life molecules (Chyba and Sagan, 1992). Chemical analysis of cosmic objects fallen on our planet, as Murchison meteorite, shows the presence of large quantities of organic compounds, including carbohydrates, amino acids, and bases of nucleic acids (Martins et al., 2008; Pizzarello et al., 2008). In particular, it must be outlined the role played by the so-called cosmic dust in the dissemination of organic compounds throughout the space. The IDPs (interstellar dust particles), small grains of silicates and other materials, and aggregates of such grains permeate in fact the entire universe and, according to recent views, are considered the principal agents in the formation of the complex molecules mentioned earlier, thanks to their peculiar surface-catalytic activities (Monnard, 2005). The crucial role of mineral surfaces is also evidenced by their ability to promote the polymerization of nucleotides. The fact that right components were present in primeval habitats, in fact, is not sufficient by itself to explain the appearance of complex biomolecules. The formation of these macromolecules required the polymerization of single components. Studies performed by Ferris and coworkers (Ertem and Ferris, 1996; Ferris, 2002, 2006; Huang and Ferris, 2003) demonstrated the polymerization of oligonucleotides up to the length of a small ribozyme on montmorillonite clay, especially when the nucleotides are activated, with or without the need of a primer. More recently, the nonenzymatic synthesis of RNA-like polymers from nonactivated mononucleotides in lipid environment has been demonstrated (Rajamani et al., 2008). The theoretical aspects of the assembly and evolution of organic molecules on the surface of silica-rich minerals have been investigated at the end of last century (Parsons et al., 1998; Smith, 1998; Smith et al., 1999). More recently, the molecular aspects of the interaction between nucleotides and oligonucleotides and clay surface have been approached by using molecular dynamic techniques and computer simulations. Mathew and Luthey-Schulten provided information on the reaction of polymerization of nucleotides on clay mineral montmorillonite (Mathew and Luthey-Schulten, 2010), and Swadling and coworkers suggested a mechanism for the regiospecificity and elongation of RNA oligomers on clay surfaces (Swadling et al., 2010).

In addition to this, laboratory studies on the fate of genetic information in different habitats have demonstrated the role of clay minerals in environmental protection of genetic molecules (DNA, RNA) against biotic (i.e., nucleases) and abiotic (UV, X-ray radiation) degrading agents (Ciaravella et al., 2004; Gallori et al., 1994; Scappini et al., 2004). Indeed, nucleic acid molecules adsorbed on clays ("clay-nucleic acid complexes") can persist for a long time in hostile environments while still maintaining their biological activities, such as the ability to transform competent bacterial cells (Gallori et al., 1994; Stotzky et al., 1996), to transmit the genetic information contained in their sequences, and to interact with molecules present in the environment (Franchi and Gallori, 2005; Vettori et al., 1996).

All together, these observations suggest that mineral surfaces could have played a central role in the formation and evolution of ancestral genetic material by the establishing of an intimate association with it.

## 3. Clay-Nucleic Acid Complexes

Studies carried out in our laboratory have evidenced that clay-nucleic acid complexes possess peculiar physical-chemical characteristics that could account, at least partially, for the resistance to degradation in the natural habitats (Franchi et al., 1999).

Clay-nucleic acid complexes prepared by reacting nucleic acids of different chemical compositions and molecular structure (DNA, RNA), with two clay minerals, montmorillonite (M) and kaolinite (K), were investigated by different analytical techniques, including electron microscopy (TEM, SEM), X-ray diffractometry (X-RD), and Fourier transform infrared (FT-IR) spectroscopy. TEM and SEM observations (Fig. 1) indicated that DNA strands were associated with several points of the edges of clay particles, and the molecules appear to lie on the planar surface of mineral particles (Franchi et al., 1999). The lack of penetration of nucleic acid molecules into M and K was also suggested by X-RD. Diffractograms of the complexes showed that none of the nucleic acids significantly intercalated the Al-Si layers of the two clays, and the adsorption occurred primarily on the external surface of the clay mineral (Franchi et al., 1999). The adsorption/binding process of nucleic acid molecules on clay surfaces is

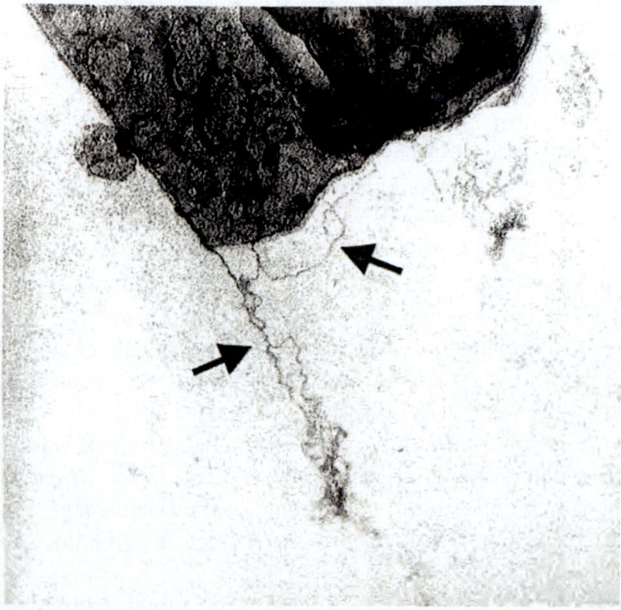

**Figure 1.** TEM picture of genomic DNA from *B. subtilis* on clay mineral montmorillonite. *Arrows* indicate the fibers of DNA.

promoted by cations. Interestingly, the presence of divalent cations, mainly $Mg^{2+}$ and $Ca^{2+}$, highly increases the adsorption of nucleic acids if compared to monovalent cations such as $Na^{+}$ (Franchi et al., 2003). Recent computer simulation studies on the interaction between RNA molecules and montmorillonite (Swadling et al., 2010) supported our previous observations, confirming the different roles of monovalent and divalent cations in the adsorption of nucleic acid molecule on clay. Moreover, simulations provided a new insight in the mechanism of adsorption of nucleic acid polymer, suggesting that divalent cations screen more efficiently the negative charges of phosphate groups of RNA, as well as the negative charges of sites associated with aluminum ions on clay surface.

According to the study, clay-RNA adsorption is mainly mediated by the interaction of planar purine nucleobases with clay surface through Lennard-Jones forces (Swadling et al., 2010). However, experimental observations (Franchi et al., 1999, 2003) suggested two different types of interactions between nucleic acid molecules and clay. During adsorption experiments, it is possible to recognize that a fraction of RNA molecules weakly interact with mineral surface, the adsorption/bound is reversible, and RNA molecules are easily removed by washing. On the contrary, another fraction of nucleic acids is tightly bound on montmorillonite and cannot be removed by simple washing. Cations are needed for the interaction between nucleic acid molecules with clay, and once the complexes are formed, the nucleic acid cannot be easily removed. This observation suggested that the interactions between nucleic acids and minerals are, in this case, higher than expected by simple physioadsorption, as described in simulations studies. The interaction of RNA tightly bound on clay implies higher adsorption energy and could involve electrostatic interactions such as assumed by the "cation bridge" model (Fig. 2), where divalent cations take part directly in the formation of nucleic acid-clay complexes (Franchi et al., 1999, 2003).

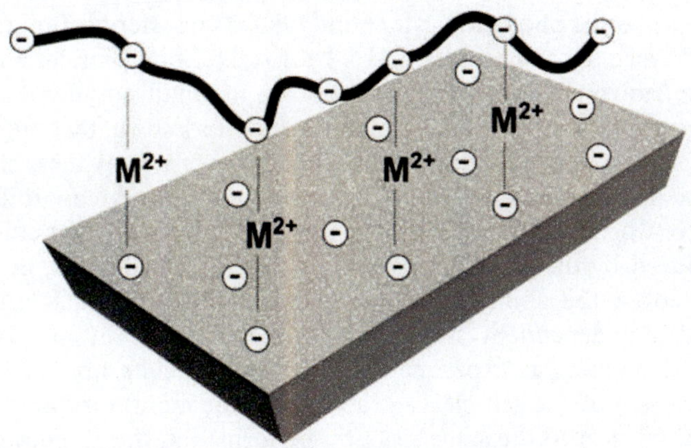


**Figure 2.** "Cation bridge" model (see Franchi et al., 2003, for details).

FT-IR spectra of clay complexes significantly differ from those of free molecules; this can indicate that the DNA and RNA conformations changed when the molecules were adsorbed on the clay. In particular, DNA showed a transition from the B and A form (Franchi et al., 1999), which could account for the increased resistance of clay complexes to degradation in the environment.

With the aim to answer the question if nucleic acid-like molecules (primitive genetic polymers) adsorbed on clay were in the rights conditions to express their biological "potentiality," we investigated the ability of clay-bound DNA/RNA molecules to perform specific reactions in a mineral environment. DNA and 16S RNA adsorbed on clay were still able to act as template for molecular replication by polymerase chain reaction (PCR) (Vettori et al., 1996) and reverse transcription and amplification (RT-PCR) (Franchi and Gallori, 2005). This indicates the possibility of transferring sequence information through different generations of molecules, corroborating the hypothesis that clay-bound complexes could have acted as a "storage" of genetic information on early Earth and, also, the possibility of the passage from a single-strand molecule to a double-strand one in a prebiotic context, supporting the hypothesis of the "RNA World" (Joyce, 2002). Moreover, clay-adsorbed poly(A) was able to interact with complementary polymers, poly(U), present in the surrounding medium and to establish a specific association with them, probably by the annealing of complementary helices (Franchi and Gallori, 2005).

These findings indicated that the adsorption/binding of nucleic acid molecules on clay does not hinder their ability to interact with other molecules present in the environment, allowing the occurrence of further more organized steps. The possibility that RNA molecules with catalytic activity (ribozymes) could perform their activities when complexed with clay minerals was assessed by studying the behavior of the self-cleavage reaction (Biondi et al., 2007b) of a so-called hammerhead ribozyme in the presence of montmorillonite. The hammerhead ribozymes are small RNA motifs which, in the presence of a divalent metal ion, self-cleave at a specific phosphodiester bond (via a transesterification reaction) to produce 2′,3′ cyclic phosphodiester and 5′ hydroxyl termini (Forster and Symons, 1987). These ribozymes are present in nature in small circular naked RNA molecules, named viroids, the smallest plant pathogens known, that are considered possible relics of the RNA world (Diener, 2001). The results of these studies indicated that the hammerhead ribozyme of the avocado sun blotch viroid (ASBVd) was still active when adsorbed on the clay surface, even though its efficiency was partially reduced by the adsorption/binding on montmorillonite (Biondi et al., 2007a). Moreover, the ability of clay to protect the catalytic RNA molecule ADHRI (adenine-dependent hairpin ribozyme I) (Meli et al., 2003) from UV-induced damages was experimentally checked. Results obtained showed a threefold retention of the self-cleavage activity of the montmorillonite-protected molecule with respect to the same reaction performed in the absence of the clay (Biondi et al., 2007a).

## 4. Evolution of Replicators on "Mineral Surfaces"

Altogether, the above observations suggest that mineral surfaces could have favored the first steps in chemical evolution promoting concentration/synthesis/protection of early genetic material. However, prebiotic synthesis of ancestral genetic molecules is a condition necessary but not sufficient by itself to trigger the molecular evolution that led to the first living cells. A living system must be able to replicate and undergo Darwinian evolution.

Determining how these steps occurred is one of the most difficult problems that prebiotic evolution has to face, in particular, with regard to the formation and preservation of a replicating system in the absence of efficient enzymatic systems of repair. In fact, as underlined by Eigen (1971), if the copy fidelity of a replicating system is not sufficiently high, it falls into the "error catastrophe," consequently losing its capability to transfer the information to the following generations and to evolve. To escape from "error catastrophe" at least two different scenarios have been proposed in the context of the origin of life: the "hypercycle" model (Fig. 3a) (Eigen and Schuster, 1977) and the "metabolic" model (Fig. 3b) (Czaran and

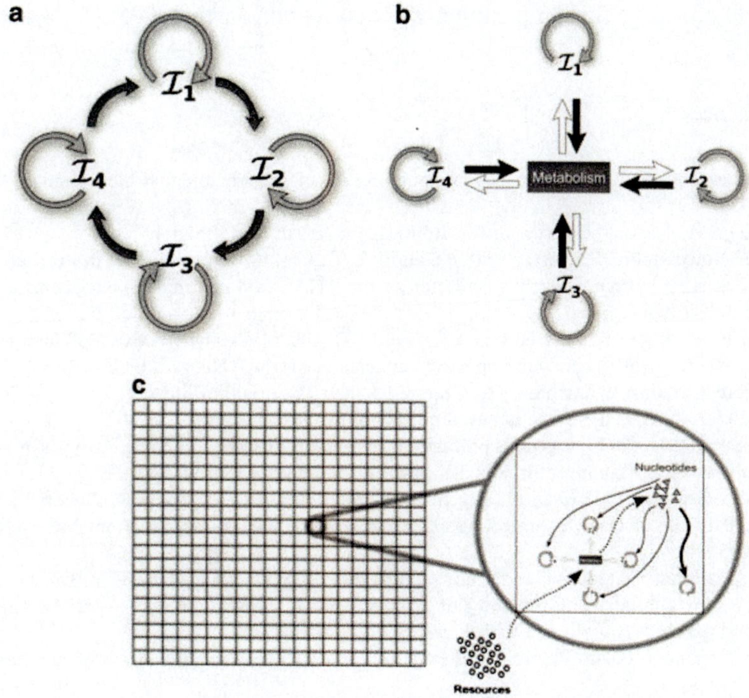


**Figure 3.** Schematic representations of **(a)** "hypercycle" model, **(b)** "metabolic" model, and **(c)** "mineral honeycomb" model (see Branciamore et al., 2009, for details).

Szathmary, 2000; Scheuring et al., 2003; Szathmary and Demeter, 1987). Both models try to overcome the basic ecological problem of keeping the assembly together without the fastest replicating member outcompete all the rest. The hypercycle and the metabolic replicator models postulate different mechanisms for that purpose. The hypercycle model assumes a cycle of direct heterocatalytic help among autocatalytic replicators, whereas the metabolic model relies on the indirect mutualistic interactions of autocatalytic replicators through their heterocatalytic contributions to a common metabolism that produces monomers for their own replication. Simulations of evolutionary dynamics (Branciamore et al., 2009; Czaran and Szathmary, 2000; Konnyu et al., 2008; Takeuchi and Hogeweg, 2008) have showed that "mineral environment" could promote the cooperation and evolution through temporarily sustaining local neighborhood structures in replicator assemblages. In particular, it was found that mineral surfaces are able to preserve the genetic diversity and the phenotypic functionality of the metabolic replicator system even in the presence of parasitic sequences (Branciamore et al., 2009). Moreover, it has been evidenced how the complicated network of interconnected pores and cavities ("mineral honeycomb model," Fig. 3c), inside the mineral, could further favored coexistence and evolution of replicating molecules (Branciamore et al., 2009), supporting the importance in prebiotic evolution of inorganic compartments (Koonin and Martin, 2005).

## 5. References


Bada JL, Lazcano A (2002) Origin of life. Some like it hot, but not the first biomolecules. Science 296:1982–1983

Bernal JD (1951) The physical basis of life. Routledge & Kegan Paul, London

Biondi E, Branciamore S, Maurel MC, Gallori E (2007a) Montmorillonite protection of an UV-irradiated hairpin ribozyme: evolution of the RNA world in a mineral environment. BMC Evol Biol 7(Suppl 2):S2

Biondi E, Branciamore S, Fusi L, Gago S, Gallori E (2007b) Catalytic activity of hammerhead ribozymes in a clay mineral environment: implications for the RNA world. Gene 389:10–18

Branciamore S, Gallori E, Szathmary E, Czaran T (2009) The origin of life: chemical evolution of a metabolic system in a mineral honeycomb? J Mol Evol 69:458–469

Chyba C, Sagan C (1992) Endogenous production, exogenous delivery and impact-shock synthesis of organic molecules: an inventory for the origins of life. Nature 355:125–132

Ciaravella A, Scappini F, Franchi M, Cecchi-Pestellini C, Barbera M, Candia R, Gallori E, Micela G (2004) Role of clays in protecting adsorbed DNA against X-ray radiation. Int J Astrobiol 3:31–35

Czaran T, Szathmary E (2000) Coexistence of replicators in prebiotic evolution. In: Dieckmann U, Law R, Metz J (eds) The geometry of ecological interactions: simplifying spatial complexity. Cambridge University Press, Cambridge, pp 116–134

Delaye L, Lazcano A (2005) Prebiological evolution and the physics of the origin of life. Phys Life Rev 2:47–64

Diener TO (2001) The viroid: biological oddity or evolutionary fossil? Adv Virus Res 57:137–184

Doudna JA, Cech TR (2002) The chemical repertoire of natural ribozymes. Nature 418:222–228

Eigen M (1971) Self-organization of matter and the evolution of biological macromolecules. Naturwissenschaften 58:465–523

Eigen M, Schuster P (1977) The hypercycle. A principle of natural self-organization. Part A: Emergence of the hypercycle. Naturwissenschaften 64:541–565

Ertem G, Ferris JP (1996) Synthesis of RNA oligomers on heterogeneous templates. Nature 379:238–240

Ferris JP (2002) Montmorillonite catalysis of 30–50 mer oligonucleotides: laboratory demonstration of potential steps in the origin of the RNA world. Orig Life Evol Biosph 32:311–332

Ferris JP (2006) Montmorillonite-catalysed formation of RNA oligomers: the possible role of catalysis in the origin of life. Philos Trans R Soc B361:1777–1786

Ferris JP, Hagan WJ Jr (1986) The adsorption and reaction of adenine nucleotides on montmorillonite. Orig Life Evol Biosph 17:69–84

Forster AC, Symons RH (1987) Self-cleavage of plus and minus RNAs of a virusoid and a structural model for the active sites. Cell 49:211–220

Franchi M, Gallori E (2005) A surface-mediated origin of the RNA world: biogenic activities of clay-adsorbed RNA molecules. Gene 346:205–214

Franchi M, Bramanti E, Bonzi LM, Orioli PL, Vettori C, Gallori E (1999) Clay-nucleic acid complexes: characteristics and implications for the preservation of genetic material in primeval habitats. Orig Life Evol Biosph 29:297–315

Franchi M, Ferris JP, Gallori E (2003) Cations as mediators of the adsorption of nucleic acids on clay surfaces in prebiotic environments. Orig Life Evol Biosph 33:1–16

Gallori E, Bazzicalupo M, Dal Canto L, Fani R, Nannipieri P, Vettori C, Stotzky G (1994) Transformation of *Bacillus subtilis* by DNA-bound on clay in nonsterile soil. FEMS Microbiol Ecol 15:119–126

Ganti T (2003) Chemoton theory: theory of living systems. Oxford University Press, Oxford

Huang WH, Ferris JP (2003) Synthesis of 35–40 mers of RNA oligomers from unblocked monomers. A simple approach to the RNA world. Chem Commun 12:1458–1459

Joyce GF (2002) The antiquity of RNA-based evolution. Nature 418:214–221

Konnyu B, Czaran T, Szathmary E (2008) Prebiotic replicase evolution in a surface-bound metabolic system: parasites as a source of adaptive evolution. BMC Evol Biol 8:267

Koonin EV, Martin W (2005) On the origin of genomes and cells within inorganic compartments. Trends Genet 21:647–654

Lewin B (2007) Genes IX. Jones & Barlett Learning, Sudbury

Martin W, Russell MJ (2003) On the origins of cells: a hypothesis for the evolutionary transitions from abiotic geochemistry to chemoautotrophic prokaryotes, and from prokaryotes to nucleated cells. Philos Trans R Soc Lond B Biol Sci 358:59–83; discussion 83–55

Martin W, Baross J, Kelley D, Russell MJ (2008) Hydrothermal vents and the origin of life. Nat Rev Microbiol 6:805–814

Martins Z, Botta O, Fogel ML, Sephton MA, Glavin DP, Watson JS, Dworkin JP, Schwartz AW, Ehrenfreund P (2008) Extraterrestrial nucleobases in the Murchison meteorite. Earth Planet Sci Lett 270:130–136

Mathew DC, Luthey-Schulten Z (2010) Influence of montmorillonite on nucleotide oligomerization reactions: a molecular dynamics study. Orig Life Evol Biosph 40:303–317

Meli M, Vergne J, Maurel MC (2003) In vitro selection of adenine-dependent hairpin ribozymes. J Biol Chem 278:9835–9842

Monnard PA (2005) Catalysis in abiotic structured media: an approach to selective synthesis of biopolymers. Cell Mol Life Sci 62:520–534

Nisbet EG, Sleep NH (2001) The habitat and nature of early life. Nature 409:1083–1091

Orgel LE (2004) Prebiotic chemistry and the origin of the RNA world. Crit Rev Biochem Mol Biol 39:99–123

Pace NR (1991) Origin of life-facing up to the physical setting. Cell 65:531–533

Parsons I, Lee MR, Smith JV (1998) Biochemical evolution II: origin of life in tubular microstructures on weathered feldspar surfaces. Proc Natl Acad Sci U S A 95:15173–15176

Pitsch S, Eschenmoser A, Gedulin B, Hui S, Arrhenius G (1995) Mineral induced formation of sugar phosphates. Orig Life Evol Biosph 25:297–334

Pizzarello S, Huang Y, Alexandre MR (2008) Molecular asymmetry in extraterrestrial chemistry: insights from a pristine meteorite. Proc Natl Acad Sci U S A 105:3700–3704

Rajamani S, Vlassov A, Benner S, Coombs A, Olasagasti F, Deamer D (2008) Lipid-assisted synthesis of RNA-like polymers from mononucleotides. Orig Life Evol Biosph 38:57–74

Ricardo A, Carrigan MA, Olcott AN, Benner SA (2004) Borate minerals stabilize ribose. Science 303:196

Saladino R, Crestini C, Ciambecchini U, Ciciriello F, Costanzo G, Di Mauro E (2004) Synthesis and degradation of nucleobases and nucleic acids by formamide in the presence of montmorillonites. Chembiochem 5:1558–1566

Scappini F, Casadei F, Zamboni R, Franchi M, Gallori E, Monti S (2004) Protective effect of clay minerals on adsorbed nucleic acid against UV radiation: possible role in the origin of life. Int J Astrobiol 3:17–19

Scheuring I, Czaran T, Szabo P, Karolyi G, Toroczkai Z (2003) Spatial models of prebiotic evolution: soup before pizza? Orig Life Evol Biosph 33:319–355

Smith JV (1998) Biochemical evolution. I. Polymerization on internal, organophilic silica surfaces of dealuminated zeolites and feldspars. Proc Natl Acad Sci U S A 95:3370–3375

Smith JV, Arnold FP Jr, Parsons I, Lee MR (1999) Biochemical evolution III: polymerization on organophilic silica-rich surfaces, crystal-chemical modeling, formation of first cells, and geological clues. Proc Natl Acad Sci U S A 96:3479–3485

Stotzky G, Gallori E, Khanna M (1996) Transformation in soil. In: Akkermans ADL, van Elsas JD, de Brujin FJ (eds) Molecular microbial ecology manual. Kluwer, Dordrecht, pp 1–28

Swadling JB, Coveney PV, Greenwell HC (2010) Clay minerals mediate folding and regioselective interactions of RNA: a large-scale atomistic simulation study. J Am Chem Soc 132:13750–13764

Szathmary E, Demeter L (1987) Group selection of early replicators and the origin of life. J Theor Biol 128:463–486

Takeuchi N, Hogeweg P (2008) Evolution of complexity in RNA-like replicator systems. Biol Direct 3:11

Vettori C, Paffetti D, Pietramellara G, Stotzky G, Gallori E (1996) Amplification of bacterial DNA bound on clay minerals by the random amplified polymorphic DNA (RAPD) technique. FEMS Microbiol Ecol 20:251–260

Wachtershauser G (1988) Before enzymes and templates: theory of surface metabolism. Microbiol Rev 52:452–484

Biodata of **Kelley Harris** and **Irene A. Chen**, authors of "***Mathematical Models of Prebiotic Replication of Informational Molecules.***"

**Kelley Harris** is a Ph.D. student at the University of California, Berkeley. She received an M.Phil in Biological Science in 2011 from Cambridge University, working at the Wellcome Trust Sanger Genome Campus, and earned an A.B. in mathematics from Harvard in 2009. Her current interests include population genetics, identity by descent in genome data, and the inference of human demographic history.

E-mail: **harris.kelley@gmail.com**

**Irene A. Chen** is currently a Bauer Fellow at the FAS Center for Systems Biology at Harvard University. She received an M.D. and Ph.D. in Biophysics from Harvard in 2007. Her scientific interests include experimental and theoretical studies on the origins of life, the structure of chemical activity landscapes, and the use of bacteriophages to control the dynamics of bacterial populations.

E-mail: **ichen@post.harvard.edu**

**Kelley Harris**

**Irene A. Chen**



J. Seckbach (ed.), *Genesis - In The Beginning: Precursors of Life, Chemical Models and Early Biological Evolution*, Cellular Origin, Life in Extreme Habitats and Astrobiology 22, 67–88
DOI 10.1007/978-94-007-2941-4_4, 

# MATHEMATICAL MODELS OF PREBIOTIC REPLICATION OF INFORMATIONAL MOLECULES

**KELLEY HARRIS[1] AND IRENE A. CHEN[2]**
[1]*UC Berkeley Mathematics Department, Berkeley, CA 94720, USA*
[2]*FAS Center for Systems Biology, Harvard University, 52 Oxford Street, Cambridge, MA 02138, USA*

## 1. Introduction

Anyone who studies the origin of life must contend with huge gaps in the available information. The earliest stages of life have left very few traces that survived the challenges of time. It is possible for natural history to degenerate into speculation when physical and chemical fossils cannot be found, but we can still make good use of universal consequences of physics and mathematics, which have not changed over time. Surprisingly, concrete claims about early life can be derived using chemical kinetics and information theory, in particular, and this chapter surveys major claims in this area. This chapter is not meant to review the entire literature of prebiotic models but to give a few examples of major frameworks as an introduction to interdisciplinary scientists.

Attempts to mimic prebiotic reactions, from Stanley Miller's classic experiments to recent advances in RNA synthesis, have done much to explain how life's building blocks could have appeared on earth (Miller and Urey, 1959; Ricardo et al., 2004; Powner et al., 2009; Orgel, 2004). These experiments may eventually be enough to explain the terrestrial presence of chemicals like amino acids and nucleic acid monomers. However, novelties of organization like metabolism and replication remain largely unexplained.

Few biomolecules are functional outside layers of native organization. To describe a protein means to consider structure at the levels of both sequence and fold. For example, precise folding is the difference between raw and denatured egg white, and once heating destroys the information contained in the fold, the native protein cannot be recovered. In the seventeenth century, this denaturation was attributed to the expulsion of an unknown "vital force," an understandable, if inappropriate, intuition given the biological importance of organization (Weisstein, 2007). Many proteins cannot assume their bioactive, water-soluble native fold without help from cellular chaperones. However, proteins with certain energy landscapes are capable of folding spontaneously (Jackson, 1998). This ability to self-organize would have been an essential feature of the earliest living systems.

In some remarkable cases, it is possible to engineer a peptide sequence that both folds spontaneously and has a rudimentary ability to self-replicate, catalyzing the ligation of shorter peptides to produce the same sequence (Lee et al., 1996). However, in contrast to nucleic acids, such peptides appear to be special cases. Because of the generalized templating ability of nucleic acids, any sequence can, in principle, readily catalyze the formation of a complementary sequence through template-directed ligation (James and Ellington, 1997; Rohatgi et al., 1996a, b). Among nucleic acids, RNA is of particular interest to the origins of life because of its ability to serve as genetic material and to catalyze a wide variety of chemical reactions (Guerrier-Takada et al., 1983; Kruger et al., 1982). Although an RNA polymerase ribozyme capable of true self-replication has not yet been discovered in the laboratory, work in this direction has demonstrated that ribozymes are capable of both oligonucleotide ligation and processive mononucleotide polymerization chemistry (Doudna and Szostak, 1989; Green and Szostak, 1992; Bartel and Szostak, 1993; Ekland and Bartel, 1995; Hager and Szostak, 1997; Jaeger et al., 1999; Johnston et al., 2001; Wochner et al., 2011).

While synthetic biologists make progress in building specific self-replicating systems, a complementary approach to understanding the origins of life is to look for mathematical constraints that must be satisfied by any replicating system. Several such constraints arise from the quasispecies model, presented in Sect. 2 and 3 (Eigen, 1971, 1977; Eigen and Schuster, 1978a, b; Eigen et al., 1988). These models were developed to capture the dynamics of self-replicating molecules and have been shown to describe viral evolution quite accurately (Eigen, 1993; Quer et al., 1996).

Section 4 will review progress in addressing the chicken-and-egg paradox of the error threshold, which suggests that early genomes might not have been capable of encoding enough information to encode replication machinery. This section includes a discussion of hypercycles, which Eigen and Schuster developed in the context of protein translation in a world of RNA quasispecies (Eigen, 1977; Eigen and Schuster, 1978a, b). Section 5 reviews recent efforts to understand how chemical kinetics become replicator dynamics in a solution of polymers that form more or less randomly (Nowak and Ohtsuki, 2008). This model is somewhat abstract, but it could describe nontemplated nucleic acid polymerization (e.g., catalyzed by charged clay surfaces, assisted by heating and drying cycles in lipids, or carried out in the eutectic phase (Ertem and Ferris, 1996; Ferris and Ertem, 1992, 1993; Ferris et al., 1996; Rajamani et al., 2008; Monnard and Deamer, 2001, 2002)) and templated polymerization (Sawai and Orgel, 1975; Orgel, 1992; Manapat et al., 2009; Ohtsuki and Nowak, 2009).

Experimental verification of theories related to origins of life is particularly important because of the plethora of interesting, plausible, but essentially untested ideas. Quasispecies theory has proven itself to be useful in understanding real systems and particularly where viruses are concerned. One approach to combating viral infection is to design drugs that increase the viral mutation rate beyond the error threshold (Eigen, 1993, 2002; Quer et al., 1996; Domingo and Holland, 1997;

Grande-Perez et al., 2002; Crotty et al., 2004; Vignuzzi et al., 2006). It has also been shown that artificially low mutation rates actually decrease virulence by reducing the rate at which a virus can adapt and evolve (Vignuzzi et al., 2006). Recently, more attention has focused on using experimental systems to test theoretical predictions, such as a study of the error threshold in nonenzymatic nucleic acid replication (Rajamani et al., 2010). Several experimental systems are under development, including an elementary hypercycle based on fragmentation of a ligase ribozyme (Hayden et al., 2005), a system of cooperating and competing ligase ribozymes (Kim and Joyce, 2004; Lincoln and Joyce, 2009) and model protocells encapsulating replicating nucleic acids (Mansy and Szostak, 2009). Although this chapter does not survey the entire literature on interesting simulations and models of prebiotic replication, our hope is that it will facilitate the dialogue between experimentalists and theorists by reviewing some major theories regarding information in early genomes whose predictions would benefit immediately from further experimental study.

## 2. Introducing Kinetics of Self-Replication

Let us imagine $n$ polymers $x_1,\ldots,x_n$ that coexist in a pool containing a limited supply of activated subunits (e.g., nucleotides or oligomers). For simplicity, we will refer to the subunits as monomers. Each polymer can use the monomers to reproduce, and we wish to predict the outcome of their struggle for existence. Enzymes and other catalysts might be present in solution, giving the polymers the ability to be copied, or each $x_i$ might incorporate one nucleotide at a time, begetting new sequences by elongating itself. Every molecule that is formed will eventually degrade so that no species can persist without regular replenishment.

The mechanism of replenishment will be a reaction of the form

$$a_1x_1+\cdots+a_nx_n \leftrightharpoons b_1x_1+\cdots+b_nx_n, \tag{1}$$

describing the net process of reproduction and death (Eigen, 1971). Letting $\vec{k}$ and $\overleftarrow{k}$ denote the forward and reverse reaction rate constants, we recall that the system is at steady state when the concentrations $[x_1],\ldots,[x_n]$ satisfy

$$\vec{k}[x_1]^{a_1}\ldots[x_n]^{a_n} = \overleftarrow{k}[x_1]^{b_1}\ldots[x_n]^{b_n} \tag{2}$$

such that the forward and reverse reaction rates remain constant and equal. (Hereafter, we will omit concentration brackets to simplify notation.) Since every population moves toward a steady state over time, the dominant species will be the fittest species, the survivors of the evolutionary contest.

Equation (1) is not a true chemical equilibrium, as it omits the activated monomers that drive reproduction. True chemical equilibrium is incompatible with life, and some biologists see this incompatibility as more definitive of life than reproduction, postulating that metabolism, not replication, was the first lifelike property to arise (Anet, 2004; Pross, 2004; Wachtershauser, 1988). However, we will focus on replication-first scenarios in which the precursors to life were passive replicators exploiting a reserve of digestible energy, racing against the clock to evolve metabolism before an energy crisis pulled the plug on their existence.

## 3. Quasispecies Theory

In this section, we present the basic theory about a model population of self-replicating molecules $x_1,\ldots,x_n$ (Eigen, 1977; Eigen et al., 1988). We first outline $n$ differential equations that specify their replication dynamics. We assume a constant influx of activated monomers, balanced by a uniform outflux of polymers and their degradation products, so that a steady state is reached and the "fitness" of each $x_i$ does not vary with time. In addition, we assume that $x_i$ forms and degrades at a rate that depends linearly on $x_1,\ldots,x_n$, as is the case when one polynucleotide templates the synthesis of another.

### 3.1. THE REPLICATOR EQUATION

In a system close to chemical equilibrium, reactions tend to be reversible, such that the reaction $C \rightarrow A+B$ proceeds by reversing each step of $A+B \rightarrow C$. In contrast, systems far from equilibrium tend to be dominated by irreversible reactions, meaning that self-replicating molecules form and degrade by pathways with different rate dependencies. For example, $x_i$ templates polymer synthesis at a rate $A_i x_i$ that depends more strongly on catalytic activity than does the rate $D_i x_i$ at which $x_i$ is degraded.

Another source of $x_i$ is the imperfect replication of competitors like $x_j$; when $x_i$ and $x_j$ differ by a simple monomer insertion, deletion, or substitution, $x_j$ will catalyze $x_i$'s production at a nonzero rate $w_{ij}x_j$. Conversely, there is a nonzero probability $1-Q_i$ that $x_i$ will produce a mutant when it tries to copy itself. Finally, the uniformizing outflux will carry away $x_i$ at rate $\Phi_i(\mathrm{X}) = \Phi_i(x_1,\ldots,x_n)$ that depends linearly on each of the polymer concentrations. If we neglect all other factors that affect the concentration of $x_i$, we obtain the following system of $n$ differential equations:

$$\dot{x}_i = (A_i Q_i - D_i)x_i + \sum_{k \neq i} w_{ik} x_k - \Phi_i(\vec{x}). \qquad (3)$$

We repress the time dependence of $x_1,\ldots,x_n$ for simplicity, using $\dot{x}_i$ to denote a first-order time derivative. When fitness values $W_i = A_i Q_i - D_i$ and mutation probabilities $w_{ik}$ do not vary with time, this system of equations is solvable. An exact solution is reported in Thompson and McBride (1974), and good approximations can be derived using perturbation theory (Eigen, 1971).

Assuming that $x$ includes the full array of mutants that can result from $x_i$'s imperfect replication, it must hold that

$$\sum_{i=1}^{n} A_i(1-Q_i) = \sum_{j \neq i} w_{ji}. \tag{4}$$

Finally, we maintain constant reaction conditions by requiring that

$$\sum_{i=1}^{n} \Phi_i = \sum_{i=1}^{n} A_i x_i - \sum_{i=1}^{n} D_i x_i. \tag{5}$$

## 3.2. THE QUASISPECIES

No primitive autocatalyst could have entirely eliminated its competitors. Even when there is a "master sequence" $x_i$ whose fitness $W_i = A_i Q_i - D_i$ is much higher than all competing $W_j$'s, $x_i$ continuously populates a range of nearby mutants, except in the unrealistic boundary case $Q_i = 1$. Eigen proved algebraically that the mutant mixture will stabilize with time, converging to an eigenvector of the *mutation matrix* $\mathbf{W}$ whose diagonal entries are the $W_j$'s and whose off-diagonal entries are the $w_{ij}$'s. If we let $\mathbf{x}_i$ denote the $n$-entry vector whose $i$th entry is $x_i \delta_{ii'}$, then by construction, the replicator population will evolve toward the stationary state $\lim_{k \to \infty} \mathbf{W}^k \mathbf{x}_i$, which is an eigenvector of $\mathbf{W}$ by the theory of finite-state Markov chains.

It is accurate to say that natural selection acts on these eigenvectors rather than on individual molecules since a dominant molecule can only persist in the context of the mutant mixture it generates. For this reason, Eigen refers to the eigenvectors of $\mathbf{W}$ as *quasispecies*. One may verify, using the Perron-Frobenius theorem (Horn and Johnson, 1991), that $\mathbf{W}$ has one positive, real eigenvalue with larger absolute value than any of its other eigenvalues; the corresponding eigenvector is the fittest quasispecies. One may also use the Perron-Frobenius theorem to prove that this dominant eigenvector has nonnegative entries, thus specifying a physically meaningful sequence distribution (Harris, 2009).

## 3.3. THE ERROR THRESHOLD

Theory and experiment show that the "master sequence" rarely makes up more than a few percent of a replicator population (Eigen, 1977; Quer et al., 1996). Though each individual mutant is much less abundant than the master sequence, the set of possible mutants is diverse enough that they dominate the population numerically. But despite the relatively low abundance of master copies, the fitness of a quasispecies depends very strongly upon the fitness of the master. Eigen (1977) used perturbation theory to show that the fitness of the dominant quasispecies is approximately

$$W_m + \sum_{k \neq m} \frac{w_{km} w_{mk}}{W_m - W_k}; \tag{6}$$

it follows that a dominant quasispecies replicating with high fidelity in a population with large fitness differences should have a fitness very close to $W_m$.

Master quasispecies proliferate because $x_m$'s productivity exceeds the population average, meaning that

$$A_m - D_m > \frac{\sum_{k \neq m} (A_k - D_k) x_k}{\sum_{k \neq m} x_k}. \tag{7}$$

However, it is possible for the concentration of $x_m$ to decrease at the same time that its descendents proliferate. The number of possible mutants is great enough that back mutation is a negligible source of perfect master copies, meaning that the concentration of master sequences will only increase with time if $x_m$ makes perfect copies of itself at a rate that exceeds the average population productivity. This is contingent on the inequality

$$A_m Q_m - D_m > \frac{\sum_{k \neq m} (A_k - D_k) x_k}{\sum_{k \neq m} x_k} \tag{8}$$

and a population that satisfies (7) but not (8) will succumb to an *error catastrophe* where resources disperse evenly through sequence space. In this situation, dominance drifts randomly as every species dies out soon after it appears (Eigen et al., 1988).

Inequality (8) dictates a sharp *error threshold* that is usually represented as the minimum replication accuracy that is required to preserve a certain amount of information. It is standard to assume that replicators are assembled by adding monomers one by one to the end of a growing chain, letting $q$ be the probability of incorporating the correct monomer during any given assembly step. If $x_1,\ldots,x_n$ are polymers of length $v$, then $Q_1 = \cdots = Q_n = q^v$. Thus, we can let

$$\sigma_m = \frac{A_m}{D_m + \frac{\sum_{k \neq m} (A_k - D_k) x_k}{\sum_{k \neq m} x_k}} \tag{9}$$

denote a measure of $x_m$'s superiority and see that an error catastrophe happens when

$$v > \frac{\log \sigma_m}{1-q}. \tag{10}$$

The error threshold is a very testable result. Experimentalists have created error catastrophes by sabotaging error correction *in vitro*, and some useful antivirals have this effect (Anderson et al., 2004; Crotty et al., 2004; Eigen, 2002).

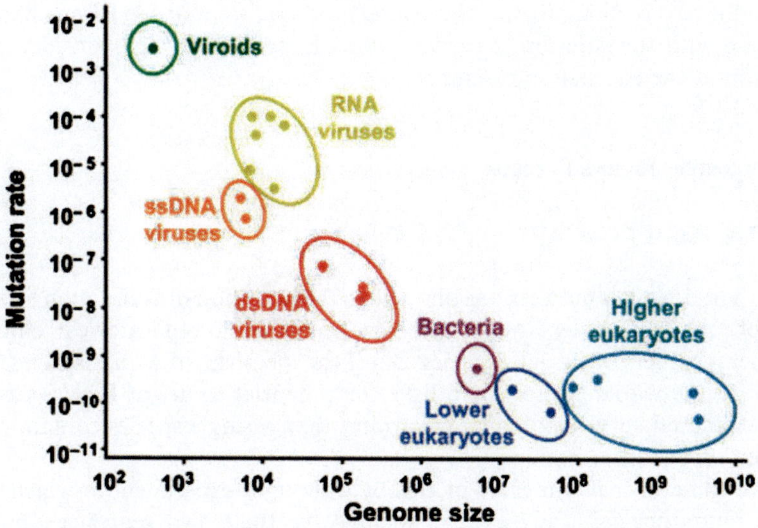


**Figure 1.** Inverse relationship between mutation rate and genome size, illustrating the extent to which the error threshold pressures large genomes to evolve high-fidelity replication mechanisms. (Reprinted with permission from a paper reporting that a 399-nucleotide viroid genome has the exceptionally high error rate of 1 mutation per 400 bases per generation (Gago et al., 2009).)

Researchers have also found a general inverse correlation between genome size and the effectiveness of error-correcting machinery (Gago et al., 2009; Fig. 1). Viral Qβ replicase, an RNA-dependent RNA polymerase, has a predicted error threshold around $10^4$ bases, which is the observed length of the virus's single-stranded RNA genome. Double-stranded genomes allow for specialized enzymatic error checks that reduce the mutation rate as much as 1,000-fold, and bacterial genomes can reach $10^7$ nucleotides (Eigen, 1977). Eukaryotes use error-correction mechanisms that are even more complex, some facilitated by diploidy, which permit even longer genomes.

Although the error threshold is a beautiful, nonobvious explanation for the inverse relationship between genome size and mutation rate, it is also true that error-correcting enzymes take up a lot of genome space and that the machinery for translating DNA into protein takes up even more space. A 10,000-base viral genome is too small to encode translation machinery (Eigen, 1977), yet it could not elongate without enzymatic stewardship. Even earlier in the emergence of life, ribozymes presumably propagated during the RNA world, but nonenzymatic template-directed polymerization appears to be too inaccurate to copy sequences of ribozyme length (error rate near 20%, while ribozymes are typically >30 bases) (Hagenbuch et al., 2005). Furthermore, a ribozyme that catalyzes template-directed RNA polymerization (Johnston et al., 2001) also appears to be too inaccurate to copy its own sequence (mutation rate of 3%, but length close to 200 bases). This apparent evolutionary barrier is known as *Eigen's paradox*.

In the next section, we describe several proposed solutions to Eigen's paradox. We begin with the solution that Eigen himself proposed when considering the evolution of the translation machinery.

## 4. Overcoming Eigen's Paradox

### 4.1. THE HYPERCYCLE

Eigen's paradox presents a serious obstacle to the invention of translation by a lone self-replicating molecule. However, we have seen that no replicator can eliminate the other members of its quasispecies. But if the members of a quasispecies could cooperate, the whole group would have the potential to defeat Eigen's paradox because it could survive at higher error rates than a single species containing the same amount of information.

We can convince ourselves of this by a thought experiment in which A, B, and C reproduce faithfully enough to preserve their own sequences, but not faithfully enough to replicate the concatenation ABC. If A, B, and C were concatenated by chance, perfect ABC master sequences would decline and disappear even as A, B, and C persisted as master subsequences. The expected frequency of perfect ABCs is proportional to the product of the frequencies of A, B, and C, and since A, B, and C are replicating so close to their own error thresholds, their abundances will be low enough that stochastic fluctuations would likely eliminate ABC entirely. But if A, B, and C are not concatenated and perfect copies have a mechanism for finding one another in solution, ABC cooperatives can persist at the concentration of their limiting component.

Eigen and Schuster used dynamical systems theory to model various catalytic replicator networks, varying the ways that participating species could catalyze one another's replication. They found that some networks dissolved as key species went extinct, while others stabilized and prospered (Eigen, 1977; Eigen and Schuster, 1978a, b). Specifically, they claim that the only stable self-replicating network is an unbranched circle they call a *hypercycle* (1978a), and they go on to describe a specific hypercycle that could account for the origin of translation (1978b).

The hypercycle is a network of $n$ species $E_1,\ldots,E_n$, where $E_i$ produces copies of $E_{i+1}$ and $E_n$ produces copies of $E_1$. DNA replication is a two-element hypercycle since each strand templates its complement rather than itself. In theory, if a hypercycle of ten RNA $10^4$-mers could replicate without error-correcting enzymes, this system could encode a translation system.

Since a dominant quasispecies drives competitors to extinction, the members of an early hypercycle would probably have to belong to the same quasispecies. One might object that ten RNA $10^4$-mers from the same quasispecies contain much less information than a typical RNA $10^5$-mer since they share many common residues. Whether ten related RNA sequences would be capable of performing different functions is an interesting experimental question, but one may argue that specialization of mutants from a quasispecies is plausible because it is analogous to gene duplication, which provides much raw material for modern evolution. Interestingly, examples of an RNA sequence that can perform two functions have been found (Schultes and Bartel, 2000; Vaidya and Lehman, 2009), supporting the idea that sequences from the same quasispecies could have different functions. However, such sequences may be relatively rare.

Eigen and Schuster proposed that early translation could proceed without ribosomes and synthetases by substituting nonspecific inorganic catalysts and modifying tRNA structure (1978b). They are left with an army of tRNAs that could, in principle, belong to one quasispecies. Phylogenetic analysis of modern tRNA sequences suggests that they could have diverged from a common ancestor within a quasispecies framework (Eigen and Winkler-Oswatitsch, 1981a, b). In addition, a computer simulation of RNA evolution that assigns a selective advantage to rows of stable internal base pairs has a very high probability of producing a cloverleaf structure that resembles tRNA (Eigen, 1971). These pieces of evidence add support to the idea that tRNA was the first part of the translation apparatus to evolve.

The scenario suggested by Eigen and Schuster is as follows. Once certain quasispecies members started to act as tRNAs, the RNA replicators $I_1,\ldots,I_n$ could template the formation of polypeptides $E_1,\ldots,E_n$. In turn, these polypeptides could catalyze nucleotide synthesis. If synthetases are not required and $E_i$ replicates $I_{i+1-\delta_{n1}n}$ more efficiently than it replicates any other $I_j$, we get a hypercyclic structure (Fig. 2). The relative strengths of catalysis are important; imbalances can cause a complex network to disintegrate to a simpler hypercycle and a parasitized network to go completely extinct (Fig. 3).

Implementing hypercyclic translation in the lab would involve overcoming significant challenges (e.g., lack of synthetases and ribosomes, existence of multifunctional tRNA-like molecules). However, these challenges must be weighed against the likelihood that branched networks are fated to collapse. If Eigen and Schuster's result is broadly applicable, then perhaps efforts to build a rudimentary translation apparatus should focus on constructing cyclic networks.

Some of the hypercycle's problems can be resolved by embedding it in a spatially organized ecology. Eigen and Schuster assume that the components of the hypercycle are well mixed with each other and with competing species, but hypercycles become more robust when new species take time to disperse from their places of origin. One problem with the classical hypercycle is its susceptibility to parasites that receive catalysis without offering any in return (Eigen and Schuster, 1978b), but when spatial organization forces parasites to attack the edges of a hypercycle instead of permeating it instantly, the hypercycle is capable of overcoming most challenges (Boerlijst and Hogeweg, 1991).

In addition, spatial organization makes hypercycles less likely to destroy one another. In contrast to replicator quasispecies, where low concentrations of less fit species persist indefinitely, dominant hypercycles drive all competing networks to extinction under conditions of instantaneous mixing. These unforgiving selection

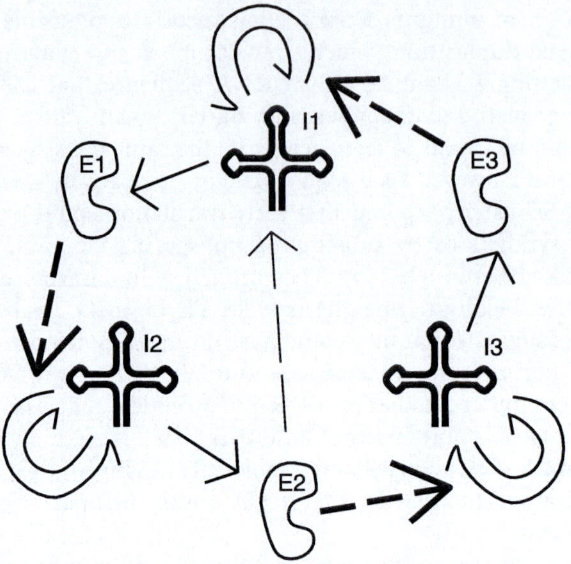


**Figure 2.** In this simple hypercyclic translation apparatus, $I_1,\ldots,I_3$ are tRNA-like molecules that double as templates for the peptide catalysts $E_1,\ldots,E_3$, which in turn facilitate production of the templates. A *dotted arrow* pointing to a *two-headed reaction arrow* denotes catalysis of a self-replication reaction, with *bold arrows* representing strong catalysis and the *thin arrow* representing weaker catalysis. *Solid arrows* denote the catalytic production of enzymes from monomers that are not shown. In theory, all molecular species survive indefinitely, stabilizing at finite equilibrium concentrations. If we complicate this hypercycle by introducing additional catalytic dependencies such as $E_2 \rightarrow I_1$ (*thin dotted arrow*), with $E_2$ catalyzing $I_1$'s self-replication, then as long as these dependencies have weaker rate constants than the dependencies $E_i \rightarrow I_{i+1}$, they will alter the position of equilibrium only slightly without jeopardizing the stable propagation of the hypercycle.

dynamics leave hypercycles vulnerable to parasites; the slightest fitness difference is enough to let a parasite extinguish an essential network link (see Fig. 3). They would also make it hard to escape a local fitness maximum, repressing the formation of modestly fit species that could evolve into dominant species over time.

Spatial organization mitigates the parasite threat by forcing competition to occur along a threshold in space. Such a threshold borders a parasite-free zone where the hypercycle reigns undisturbed. A similar threshold can separate regions that are dominated by two different coexisting hypercycles. Such partitioning could shield pockets of hypercycles that are less fit than their neighbors, and such hypercycles might evolve further to surpass the current fitness maximum.

Depending on the tendency of real catalytic networks to remain stuck in local fitness maxima, selection of a hypercycle precursor to translation could have been a "frozen accident" that persisted because of evolutionary inertia (Crick, 1963, 1968). The modern analogy is that any organism that attempts to alter its genetic code now would likely experience a severe fitness decline in the short run, due to the number of components that work together in the existing system.

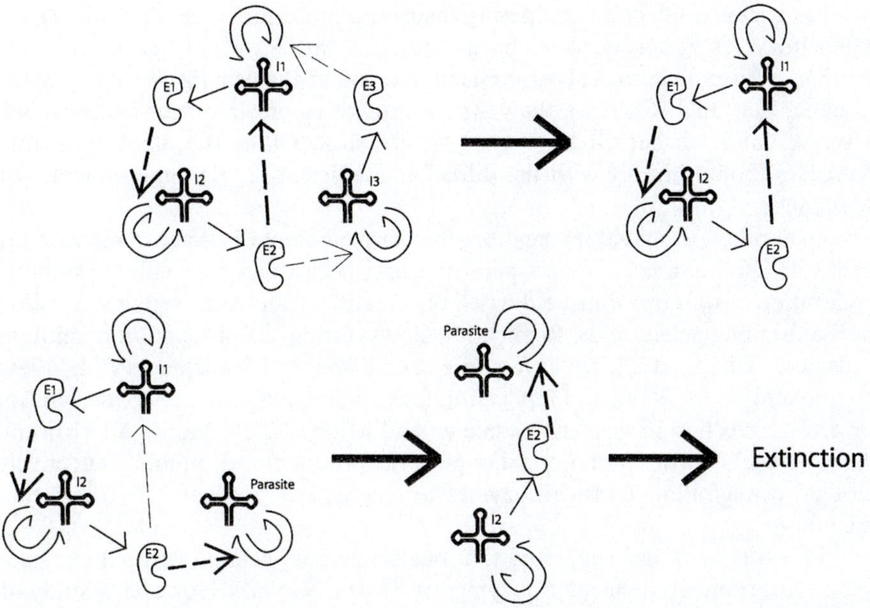


**Figure 3.** In Eigen and Schuster's treatment of well-mixed chemical systems (1978b), parasites invariably drive a network to extinction, while branched networks reduce to a single hypercyclic component.

However, some evidence suggests that the genetic code may actually be somewhat optimized. Plausibly, prebiotic amino acids are specified by GC-rich codons (e.g., glycine = GGC; alanine = GCC), which are the codons that would have functioned best without catalytic help (Eigen and Schuster, 1978b). Others have noted that the existing genetic code maximizes the likelihood that a point mutation will induce the substitution of an amino acid that is chemically similar to the wild-type residue (Freeland and Hurst, 1998; Freeland et al., 2000).

Historical questions such as these cannot always be satisfactorily resolved by examination of existing natural systems. Further insights may be gained through experimental efforts to implement a hypercycle. Lincoln and Joyce recently constructed a pair of cross-catalytic ribozymes, which form a kind of two-member hypercycle and can reproduce each other indefinitely in the absence of protein (2009). In addition, Hayden and Lehman were able to cleave the *Azoarcus* group I ribozyme into four parts that are inactive on their own but can find each other in solution to cooperatively cleave and recombine RNA (2006).

## 4.2. RELAXATION OF THE ERROR THRESHOLD

Researchers have come up with many scenarios that challenge Eigen's paradox without invoking translation or extreme hypercyclic cooperation. While spatial organization increases the robustness and believability of the hypercycle idea, it also has a similar effect on competing theories. For example, simulations suggest that ribozymes adhering to a charged mineral surface would encounter one another often enough to evolve significant reciprocal altruism (Szabo et al., 2002), an effect that might increase their fitness enough to prevent error catastrophes. Altmeyer and McCaskill have analytically shown that the error threshold decreases monotonically with the diffusion coefficient (2001), independently of hypercycles.

In addition to spatial segregation, there are other effects that can increase the error tolerance of simple, noncooperative quasispecies. One such effect is stalling, a relaxation of the error threshold based on a realistic feature of the polymerization mechanism of nucleic acids. Replication slows during the production of mutant sequences (Huang et al., 1992; Ichida et al., 2005a, b; Mendelman et al., 1990; Perrino and Loeb, 1989), and replication fork stalling is also a known mechanism for error correction in vivo (Krasilnikova and Mirkin, 2004; Mirkin and Mirkin, 2007). If stalling after mutations dampens the production of mutants effectively enough, it may allow useful ribozymes to emerge independently of translation machinery.

Previous work has suggested that nucleic acid replication without enzymes is too error prone to replicate ribozymes or deoxyribozymes. However, a study of nonenzymatic replication suggests that stalling could have significantly raised the prebiotic error threshold (Rajamani et al., 2010). Rajamani et al. determined rate constants for the incorporation of matched and mismatched nucleotides in a nonenzymatic polymerization reaction, finding that the mutation rate of a single nucleotide extension is 7.6%, which is too high to sustain a moderately fit ribozyme. However, a mismatch stalls further extension by more than two orders of magnitude, effectively curtailing the production of mutant sequences. Eigen's equations predict that a 7.6% mutation rate will cause an error catastrophe to destroy any replicator longer than 13 bases, but stalling pushes back this theoretical barrier to 39 bases, making their system accurate enough to propagate ribozyme-length sequences. This mechanism is essentially a way to propagate the very first ribozymes (e.g., able to cooperate with one another); such sequences could then evolve greater complexity through hypercyclic organization, cooperation in spatial structure, or other means.

Once stalling or similar effects allow ribozymes to arise through nonenzymatic polymerization, several mechanisms could lead to their diversification and fixation. Simulations involving "digital organisms" suggest that diverse quasispecies with flat fitness peaks tend to outcompete quasispecies with higher but narrower fitness peaks (Wilke et al., 2001). In an RNA world setting, such "survival of the flattest" might generate relatively diverse arrays of good catalysts that might later assemble themselves into hypercycles. For example, many nucleobases in a given ribozyme can be mutated without substantially altering its function (e.g., structural residues in a hairpin). Also, these sites contain essentially no information, so they would not "count" in the calculation of the error threshold. In other words, if 25% of sites are neutral (Kun et al., 2005), then the informative length of this ribozyme would be 75% of its physical length, allowing physically longer sequences to propagate.

We do not know how fit the earliest ribozymes could have been, and most scientists believe that early fitness was probably in the single digits (hence the approximation that $(\ln f)/\mu \approx 1/\mu$). However, invoking high fitness would increase the amount of information that could have predated the origin of translation. One study of mutagenesis data suggests that modern ribozymes have quite high chemical activity fitness, relative to nearby mutants (Kun et al., 2005).

The effects of stalling, spatial organization, and high fitness gradients take some pressure off of the hypercycle as the only way to circumvent Eigen's paradox. However, they could at most delay the need to develop higher-level organization. Hypercyclic structure is a kind of bridge between passive replicators and complex, metabolizing cells; Fontana et al. claim, based on λ-calculus models, that it is inevitable for replicators to eventually form hypercycles, particularly when a force begins to inhibit simple self-reproduction (1994).

In the final section of this chapter, we turn our attention to the origin of replication itself. Nowak and Ohtsuki have recently presented a model demonstrating how natural selection precedes replication (2008). In a sense, their work is a "prequel" to Eigen and Schuster's scenarios.

## 5. Prevolutionary Dynamics and the Origin of Replication

### 5.1. SELECTION BEFORE REPLICATION

Nowak and Ohtsuki's model of "prevolution" describes binary sequences of 0s and 1s that elongate by reacting with activated monomers. There is no black-box replication; for example, the species $x_{101}$ can be converted to the species $x_{1010}$ by addition of a single monomer but cannot make more copies of itself. Activated nucleotides can also be deactivated hydrolytically, producing the monomeric species $x_0$ and $x_1$. For constant population size, $x_i$ decays at the rate $Dx_i$. We let $x_{i'}$ denote the precursor of $x_i$ and let $a_i$ denote the rate at which $x_{i'}$ is transformed into $x_i$. Prevolution follows a linear set of growth equations:

$$\dot{x}_i = a_i x_{i'} - (D + a_{i0} + a_{i1}) x_i \tag{11}$$

Since the empty string is the immediate precursor to any 1-element string, it follows that $x_{0'} = x_{1'}$. Furthermore, since activated nucleotides become deactivated by reacting with superabundant water molecules, we assume that $x_{0'} = x_{1'} = 1$, i.e., that the empty string is more abundant than any other string.

Since (11) is linear with constant coefficients, prevolutionary dynamics are much simpler than the dynamics of quasispecies or catalytic networks. Letting $\mathbf{x} = (x_{0'} = x_{1'}, x_0, x_1, x_{00}, x_{01}, x_{10}, x_{11}, \ldots)$ be the infinite vector whose entries are the finite binary strings, we can write system (11) as $\dot{\mathbf{x}} = \mathbf{A}\mathbf{x}$ for a matrix $\mathbf{A}$ with constant entries. Its solutions are the vectors of the form $\mathbf{x} = \exp(\mathbf{A}t)\mathbf{x}_0$, where $\mathbf{x}_0$ is the constant vector of initial conditions. The endpoint of evolution is the solution of the equation system $\dot{\mathbf{x}} = \mathbf{0}$, whose vector entries satisfy

$$\frac{x_i}{x_{i'}} = \frac{a_i}{d + a_{i0} + a_{i1}} := b_i. \tag{12}$$

Given that $x_i$ begins with the nucleotide $\sigma \in \{0,1\}$, it follows that

$$x_i = \frac{x_i}{x_{i'}} \cdot \frac{x_{i'}}{x_{i''}} \cdots \frac{x_\sigma}{x_{\sigma'}} = b_i b_{i'} \ldots b_\sigma \tag{13}$$

at equilibrium. Thus, short sequences have exponentially greater abundance than long sequences. Such length distributions are typical of polymerization reactions (for a prebiotic example, see Lawless and Yuen, 1979).

Differences in reaction rate create further asymmetries that resemble the outcome of natural selection. The next step is to include a mechanism for self-replication.

## 5.2. SELECTION FOR REPLICATION

Transmission of genetic information could occur via an experimentally described process (Hill and Orgel, 1993; Mansy et al., 2008), where template-directed primer extension proceeds faster than the extension of an unpaired strand, even when no enzymes are involved. This rate asymmetry would select for master sequences that are particularly good at base pairing to form extendible primer-template complexes; the dynamics of this process should follow a scenario described by Nowak and Ohtsuki.

Working with the master sequence $0^n$ (assumed to be taken together with its complement), Nowak and Ohtsuki suppose that $0^k$ becomes $0^{k+1}$ at the rate $b$ whenever $k < n$, while all other extensions happen at the rate $a < b$. Fixing $a$ and letting $a_0 = a_1 = \alpha$, we can deduce from (12) that

$$x_m = \frac{\alpha}{a}\left(\frac{b}{a+b+d}\right)^{n-1}\left(\frac{a}{2a+d}\right), \tag{14}$$

such that the abundance of $x_m$ approaches $\alpha/(2a + d)$ as $b$ approaches infinity. In contrast, the abundance of $x_i$ approaches zero whenever $x_i$ is not a subsequence of $0^n$.

This computation indicates how a master sequence, once formed, could have avoided extinction and increased its abundance. However, it does not change the earlier result that sequence abundance decreases exponentially with length. Another mechanism is needed to shift the population toward long sequences, a prerequisite for ribozymes. Manapat et al. demonstrate how another chemical process, template-directed ligation, could bias prelife toward the production of long sequences (2010).

As an aside, papers about prelife avoid the 4-base complexity of the modern genetic code, but instead consider one or two bases and ignore complementarity (i.e., consider the entire duplex together). However, Sievers and von Kiedrowski report that cross-catalytic replication and self-catalytic replication should have similar efficiencies and obey similar dynamics (1994).

If mutations occur, certain assumptions make it possible to find the accompanying error threshold. Specifically, we assume that subsequences of $0^n$ will form at the rate $b(1-u)$, while mutants of $0^k 1$, with $0<k<n$, form at the rate $a+bu$. All other extensions proceed at the rate $a$. As before, the limiting master abundance is $\alpha/(2a+d)$, but $x_m$ cannot proliferate to $1/k$ times its maximum abundance unless

$$u<\frac{\log k}{n}-\frac{a+d}{b}. \tag{15}$$

Once true replicators start proliferating, they will obey the dynamics introduced in Sect. 3.1. Hybrid replicator equations describe the coexistence of "prevolution" with true evolution:

$$\dot{x}_i = a_i x_{i'} - (d + a_{i0} + a_{i1})x_i + r x_i (f_i - \phi). \tag{16}$$

The constant $r$ scales the relative speeds of template-directed replication and nontemplated sequence extension, while $f_i$ denotes the fitness of $x_i$. $\phi$ is the outflow that balances excess productivity.

Computer simulations show that the abundance of replicators shoots up immediately above the critical value of $r$ at which there are more copies of $x_i$ being created by replication than being converted to other sequences by random nucleotide addition (Nowak and Ohtsuki, 2008). This phase transition seems to denote the change from "prevolution" to evolution, when

$$-(d+a_{i0}+a_{i1})+r(f_i-\phi)>0, \tag{17}$$

such that

$$r>r_c:=\frac{d+a_{i0}+a_{i1}}{f_i-\phi}. \tag{18}$$

This threshold changes with the presence of replication errors, resulting in a different kind of error threshold.

Prelife as described by Manapat et al. (2009) has begun to converge with what experimentalists can make (Mansy et al., 2008): membrane-bound polynucleotides that base pair, dissociate, and extend, all without the help of protein catalysis. Lab results confirm that the theory is more than just a fantasy, while theory suggests that protocells in the lab could evolve in very interesting ways.

Every result in theoretical chemistry is conditional on simplifications, and these simplifications limit the degree to which anything can be proved to be "impossible." This is certainly the case with Eigen's paradox, which we have seen to admit a great many loopholes. However, it would have been much more difficult to find these loopholes without understanding Eigen's theory, looking at his assumptions and deliberately constructing situations where they break down.

Quasispecies theory and its offshoots predict that whole classes of experiments should fail, challenging lab scientists to look for success in less predictable setups. But just as hypercycle theory predicts that certain early translation mechanisms would have been inviable, Nowak and Ohtsuki's work predicts that certain chemical systems should be experimentally interesting. Similarly, experimentalists could contribute much to the direction of further theoretical treatments because realistic details that appear to be minor can have major consequences (e.g., the fitness landscape is an extremely important, but currently poorly understood, detail of replication models that is often lumped into the replicative rate "$r$").

## 6. Concluding Remarks

In this chapter, we have described a handful of major frameworks for modeling prebiotic replication of informational molecules mathematically, with the intention of connecting experimentalists to this fruitful area. However, the related literature is much larger than the focus of this chapter. Outside the domain of purely mathematical models, much recent effort has been devoted to computer simulation of evolving digital organisms, such as the Avida platform (Ofria and Wilke, 2004), which has also yielded interesting insights and predictions (e.g., about complexity (Adami et al., 2000)). In addition, the consideration of cellular compartmentalization, such as by lipid membranes, leads to interesting simulations of higher organization (e.g., see Hutton, 2007, and Sole et al., 2009). We have not reviewed this extensive and rich literature.

We might never know exactly how our earliest history unfolded, but we can hope to learn the features of that history that were necessary, not contingent Fontana and Buss (1994). Even scholars of time periods that did leave fossils behind work to distinguish contingency from necessity, a goal that was popularized in Stephen Jay Gould's *Wonderful Life* (1989); after piecing together some of the history of the animals preserved in the Burgess Shale, Gould wonders how much of that history would stay the same if "the tape were played twice." The theories we have described are essentially attempts to delineate necessary constraints on early life, with the hope that whatever we learn would hold during any playback of Gould's "tape." Determining whether this is actually the case will require many joint adventures in experiment and theory.

## 7. References


Adami C, Ofria C, Collier TC (2000) Evolution of biological complexity. Proc Natl Acad Sci 97(9):4463–4468

Altmeyer S, McCaskill JS (2001) Error threshold for spatially resolved evolution in the quasispecies model. Phys Rev Lett 86:5819–5822

Anderson JP, Daifuku R, Loeb LA (2004) Viral error catastrophe by mutagenic nucleosides. Annu Rev Microbiol 58:183–205

Anet FAL (2004) The place of metabolism in the origin of life. Curr Op Chem Biol 8(6):654–659

Bartel DP, Szostak JW (1993) Isolation of a new ribozymes from a large pool of random sequences. Science 261:1411–1418

Boerlijst MC, Hogeweg P (1991) Spiral wave structure in pre-biotic evolution: hypercycles stable against parasites. Phys D: Nonlinear Phenomena 48(1):17–28

Crick FHC (1963) On the genetic code: deductions about the general nature of the code are drawn from results of biochemical experimentation. Science 8:461–464

Crick FHC (1968) The origin of the genetic code. J Mol Biol 38(3):367–379

Crotty S, Cameron CE, Andino R (2004) RNA virus error catastrophe: direct molecular test by using ribavirin. Proc Natl Acad Sci U S A 98(12):6895–6900

Domingo E, Holland JJ (1997) RNA virus mutations and fitness for survival. Annu Rev Microbiol 51:151–178

Doudna JD, Szostak JW (1989) RNA-catalyzed synthesis of complementary-strand RNA. Nature 339:519–522

Eigen M (1971) Selforganization of matter and the evolution of biological macromolecules. Naturwissenschaften 33a:465–523

Eigen M (1977) The hypercycle: a principle of natural self-organization; Part A: emergence of the hypercycle. Naturwissenschaften 64:541–565

Eigen M (1993) The origin of genetic information: viruses as models. Gene 135:37–47

Eigen M (2002) Error catastrophe and antiviral strategy. Proc Natl Acad Sci U S A 99(21):13374–13376

Eigen M, Schuster P (1978a) The hypercycle: a principle of natural self-organization; Part B: The abstract hypercycle. Naturwissenschaften 65:7–41

Eigen M, Schuster P (1978b) The hypercycle: a principle of natural self-organization; Part C: The realistic hypercycle. Naturwissenschaften 65:341–369

Eigen M, Winkler-Oswatitsch R (1981a) Transfer RNA: the early adaptor. Naturwissenschaften 68:217–228

Eigen M, Winkler-Oswatitsch R (1981b) Transfer RNA, an early gene? Naturwissenschaften 68:282–292

Eigen M, McCaskill J, Schuster P (1988) Molecular quasi-species. J Phys Chem 92:6881–6891

Ekland EH, Bartel DP (1995) The secondary structure and sequence optimization of an RNA ligase ribozyme. Nucleic Acids Res 23(16):3231–3238

Ertem G, Ferris JP (1996) Synthesis of RNA oligomers on heterogeneous templates. Nature 379:238–240

Ferris JP, Ertem G (1992) Oligomerization of ribonucleotides on montmorillonite: reaction of the 5′-phosphorimidazolide of adenosine. Science 257(5075):1387–1389

Ferris JP, Ertem G (1993) Montmorillonite catalysis of RNA oligomer formation in aqueous solution. A model for the prebiotic formation of RNA. J Am Chem Soc 115(26):12270–12275

Ferris JP, Hill AR, Liu R, Orgel LE (1996) Synthesis of long prebiotic oligomers on mineral surfaces. Nature 381:59–61

Fontana W, Buss LW (1994) What would be conserved if "the tape were played twice"? Proc Natl Acad Sci U S A 91:757–761

Fontana W, Wagner GP, Buss LW (1994) Beyond digital naturalism. Artif Life 1:211–227

Freeland SJ, Hurst LD (1998) The genetic code is one in a million. J Mol Evol 47:238–248

Freeland SJ, Knight RD, Landweber LF, Hurst LD (2000) Early fixation of an optimal genetic code. Mol Biol Evol 17:511–518

Gago S, Elena SF, Flores R, Sanjuan R (2009) Extremely high mutation rate of a hammerhead viroid. Science 323:1308

Gould SJ (1989) Wonderful life. Norton, New York

Grande-Perez A, Sierra S, Castro MG, Domingo E, Lowenstein PR (2002) Molecular indetermination in the transition to error catastrophe: systematic elimination of lymphocytic choriomeningitis virus through mutagenesis does not correlate linearly with large increases in mutant spectrum complexity. Proc Natl Acad Sci U S A 99(20):12938–12943

Green R, Szostak JW (1992) Selection of a ribozyme that functions as a superior template in a self-copying reaction. Science 258(5090):1910–1915

Guerrier-Takada C, Gardiner K, Marsh T, Pace N, Altman S (1983) The RNA moiety of ribonuclease P is the catalytic subunit of the enzyme. Cell 35:849–857

Hagenbuch P, Kervio E, Hochgesand A, Plutowski U, Richert C (2005) Chemical primer extension: efficiently determining single nucleotides in DNA. Angew Chem Int Ed 44:6588–6592

Hager AJ, Szostak JW (1997) Isolation of novel ribozymes that ligate AMP-activated RNA substrates. Chem Biol 4(8):607–617

Harris K (2009) On the dynamics of prebiotic evolution. A.B. honors thesis, Harvard University, http://brauer.harvard.edu/theses/senior/harris/harris.pdf

Hayden EJ, Lehman N (2006) Self-assembly of a group I intron from inactive oligonucleotide fragments. Chem Biol 13(8):909–918

Hayden EJ, Riley CA, Burton AS, Lehman N (2005) RNA-directed construction of structurally complex and active ligase ribozymes through recombination. RNA 11:1678–1687

Hill AR Jr, Orgel LE (1993) The limits of template-directed synthesis with nucleoside-5′-phosophoro (2-methyl)imidazolides. Orig Life Evol Biosph 23:285–290

Horn RA, Johnson CR (1991) Matrix analysis. Cambridge University Press, New York

Huang M-M, Arnheim N, Goodman MF (1992) Extension of base mispairs by *Taq* DNA polymerase: implications for single nucleotide discrimination in PCR. Nucleic Acids Res 20(17):4567–4573

Hutton TJ (2007) Evolvable self-reproducing cells in a two-dimensional artificial chemistry. Artif Life 13(1):11–30

Ichida JK, Horhota A, Zou K, McLaughlin LW, Szostak JW (2005a) High fidelity TNA synthesis by Therminator polymerase. Nucleic Acids Res 33(6):5219–5225

Ichida JK, Zou K, Horhota A, Yu B, McLaughlin LW, Szostak JW (2005b) An *in vitro* selection system for TNA. J Am Chem Soc 127(9):2802–2803

Jackson SE (1998) How do small single-domain proteins fold? Fold Des 3:R81–R91

Jaeger L, Wright MC, Joyce GF (1999) A complex ligase ribozyme evolved *in vitro* from a group I ribozyme domain. Proc Natl Acad Sci U S A 96(26):14712–14717

James KD, Ellington AD (1997) Surprising fidelity of template-directed chemical ligation of oligonucleotides. Chem Biol 4(8):595–605

Johnston WK, Unrau PJ, Lawrence MS, Glasner ME, Bartel DP (2001) RNA-catalyzed RNA polymerization: accurate and general RNA-templated primer extension. Science 292:1319–1325

Kim D-E, Joyce GF (2004) Cross-catalytic replication of an RNA ligase ribozyme. Chem Biol 11(11):1505–1512

Krasilnikova MM, Mirkin SM (2004) Replication stalling at Friedrich's ataxia $(GAA)_n$ repeats *in vivo*. Mol Cell Biol 24(6):2286–2295

Kruger K, Grabowski PJ, Zaug AJ, Sands J, Gottschling DE, Cech TR (1982) Self-splicing RNA: autoexcision and autocyclization of the ribosomal RNA intervening sequence of Tetrahymena. Cell 31:147–157

Kun A, Santos M, Szathmary E (2005) Real ribozymes suggest a relaxed error threshold. Nat Genet 37(9):1008–1011

Lawless JG, Yuen GU (1979) Quantification of monocarboxylic acids in the Murchison carbonaceous meteorite. Nature 282:396–398

Lee DH, Granja JR, Martinez JA, Severin K, Ghadiri MR (1996) A self-replicating peptide. Nature 382:525–528

Lincoln TA, Joyce GF (2009) Self-sustained replication of an RNA enzyme. Science 323(5918): 1229–1232

Manapat ML, Ohtsuki H, Burger R, Nowak MA (2009) Originator dynamics. J Theor Biol 256: 586–595

Manapat ML, Chen IA, Nowak MA (2010) The basic reproductive ratio of life. J Theor Biol 263(3):317–327

Mansy SS, Szostak JW (2009) Reconstructing the emergence of cellular life through the synthesis of model protocells. Cold Spring Harb Symp Quant Biol 74:47–54

Mansy SS, Schrum JP, Krishnamurthy M, Tobe S, Treco DA, Szostak JW (2008) Template-directed synthesis of a genetic polymer in a model protocell. Nature 454:122–125

Mendelman LV, Petruska J, Goodman MC (1990) Base mispair extension kinetics: comparison of DNA polymerase alpha and reverse transcriptase. J Biol Chem 265:2338–2346

Miller SL, Urey HC (1959) Organic compound synthesis on the primitive Earth: several questions about the origin of life have been answered, but much remains to be studied. Science 130(3370): 245–251

Mirkin EV, Mirkin AM (2007) Replication fork stalling at natural impediments. Microbiol Mol Biol Rev 71(1):13–35

Monnard PA, Deamer DW (2001) Nutrient uptake by protocells: a liposome model system. Orig Life Evol Biosph 31(1):147–155

Monnard PA, Deamer DW (2002) Membrane self-assembly processes: steps toward the first cellular life. Anat Rec 268:196–207

Nowak MA, Ohtsuki H (2008) Prevolutionary dynamics and the origin of evolution. Proc Natl Acad Sci U S A 105(39):14924–14927

Ofria C, Wilke CO (2004) Avida: a software platform for research in computational evolutionary biology. Artif Life 10(2):191–229

Ohtsuki H, Nowak MA (2009) Prelife catalysts and replicators. Proc R Soc B 279:3783–3790

Orgel L (1992) Molecular replication. Nature 358:203–209

Orgel L (2004) Prebiotic chemistry and the origin of the RNA world. Crit Rev Biochem Mol Biol 39:99–123

Perrino FW, Loeb LA (1989) Differential extension of 3′ mispairs is a major contribution to the high fidelity of calf thymus DNA polymerase-alpha. J Biol Chem 264:2898–2905

Powner MW, Gerland B, Sutherland JD (2009) Synthesis of activated pyrimidine ribonucleotides in prebiotically plausible conditions. Nature 459:239–242

Pross A (2004) Causation and the origin of life: metabolism or replication first? Orig Life Evol Biosph 34:3307–3321

Quer J, Huerta R, Novella IS, Tsimring L, Domingo E, Holland JJ (1996) Reproducible nonlinear population dynamics and critical points during replicative competitions of RNA virus quasispecies. J Mol Biol 264:465–471

Rajamani S, Vlassov A, Benner S, Coombs A, Olasagasti F, Deamer D (2008) Lipid-assisted synthesis of RNA-like polymers from mononucleotides. Orig Life Evol Biosph 38(1):57–74

Rajamani S, Ichida JK, Antal T, Treco DA, Leu K, Nowak MA, Szostak JW, Chen IA (2010) Effect of stalling after mismatches on the error catastrophe in non-enzymatic nucleic acid replication. J Am Chem Soc 132(16):5880–5885

Ricardo A, Carrigan MA, Olcott AN, Benner SA (2004) Borate minerals stabilize ribose. Science 303(5655):196

Rohatgi R, Bartel DP, Szostak JW (1996a) Kinetic and mechanistic analysis of nonenzymatic, template-directed oligoribonucleotide ligation. J Am Chem Soc 118(14):3332–3339

Rohatgi R, Bartel DP, Szostak JW (1996b) Nonenzymatic, template-directed ligation of oligoribonucleotides is highly regioselective for the formation of 3′-5′ phosphodiester bonds. J Am Chem Soc 118(14):3340–3344

Sawai H, Orgel LE (1975) Oligonucleotide synthesis catalyzed by the $Zn^{2+}$ ion. J Am Chem Soc 97:3532–3533

Schultes EA, Bartel DP (2000) One sequence, two ribozymes: implications for the emergence of new ribozyme folds. Science 289(5478):448–452

Sievers D, von Kiedrowski G (1994) Self-replication of complementary nucleotide-based oligomers. Nature 369:221–224

Sole RV, Macia J, Fellermann H, Munteanu A, Sardanyes J, Valverde S (2009) Models of protocell replication. In: Rasmussen S, Bedau MA, Chen L, Deamer D, Krakauer DC, Packard NH, Stadler PF (eds) Protocells: bridging nonliving and living matter. MIT Press, Cambridge

Szabo P, Scheuring I, Czaran T, Szathmary E (2002) *In silico* simulations reveal that replicators with limited dispersal evolve towards higher efficiency and fidelity. Nature 420:340–343

Thompson CJ, McBride JL (1974) On Eigen's theory of the self-organization of matter and the evolution of biological macromolecules. Math Biosci 21:127

Vaidya N, Lehman N (2009) One RNA plays three roles to provide catalytic activity to a group I intron lacking an endogenous internal guide sequence. Nucleic Acids Res 37(12):3981–3989

Vignuzzi M, Stone JK, Arnold JJ, Cameron CE, Andino R (2006) Quasispecies diversity determines pathogenesis through cooperative interactions within a viral population. Nature 439(7074): 344–348

Wachtershauser G (1988) Before enzymes and templates: theory of surface metabolism. Microbiol Rev 52(4):452–484

Weisstein EW (2007) Vitalism theory—from Eric Weisstein's world of chemistry. Accessed at http://scienceworld.wolfram.com/chemistry/VitalismTheory.html on 14 Jan 2009

Wilke CO, Wang JL, Ofria C, Lenski RE, Adami C (2001) Evolution of digital organisms at high mutation rates leads to survival of the flattest. Nature 412:331–333

Wochner A, Attwater J, Coulson A, Holliger P (2011) Ribozyme-catalyzed transcription of an active ribozyme. Science 332(6026):209–212

Biodata of **Rob Hengeveld**, author of "***Life Without Carbon***."

I have been working as a theoretical biologist all my scientific life. Using real-life data, my interest shifted from ecology to biogeography and invasion biology and more recently, using an ecological approach, to biogenesis. The last 4 years were spent working on *Why We are Wasting our World* (in press) (University of Chicago Press), an analysis of the deteriorating conditions for life on Earth. In 1982, I received my Ph.D. from the University of Leiden. Developing the subject of the thesis led to *The Dynamics of Biological Invasions* (1989) (Chapman and Hall) and *Dynamic Biogeography* (1990) (Cambridge University Press). Still in preparation, *Autecology*, written in collaboration with Dr. G. H. Walter, Brisbane, will be published by Cambridge University Press. In 1994, I received the Distinguished Statistical Ecologist Award from the International Association for Ecology.

E-mail: **rhengeveld@wish.net**

J. Seckbach (ed.), *Genesis - In The Beginning: Precursors of Life, Chemical Models and Early Biological Evolution*, Cellular Origin, Life in Extreme Habitats and Astrobiology 22, 89–106
DOI 10.1007/978-94-007-2941-4_5, 

# LIFE WITHOUT CARBON

**ROB HENGEVELD**
*Vrije Universiteit, Knoppersweg 11, 6668 AT Randwijk, Amsterdam, The Netherlands*

## 1. Introduction

Applying a deductive approach has constraints to the way life on Earth and elsewhere in the universe could have originated and evolved, and the conditions under which that took place.

A systematic investigation into biogenesis started with Miller's (1953) experiments although earlier attempts have been made. This half century of intense work has not yet led to a homogeneous body of theory; what remains is a multitude of scattered experiments based on individual conceptions and definitions of core processes of life (e.g., Fry, 2000; Lahav, 1999). This is surprising as the inductive methodology can only generate hypotheses. The inductive approach is certainly needed for exploring a new research field, but the deductive approach should be used for testing the hypotheses generated and for formulating theory. The explorative, hypothesis generating phase should be kept as short as possible. At present, by not appreciating the function of the inductive approach, time is lost to define life in a limitless number of a priori criteria working hypotheses (e.g., Popa, 2004). This leaves the door open for the formulation of an equally limitless number of biogenetic scenarios, so misguiding the direction and depth of biogenetic research (Hengeveld, 2010b).

Experimental testing of these inductive scenarios is only possible using a deductive methodology. For example, the deductive approach is seen in experiments involving the precipitation of crusts around trickles of alkaline water into acid seawater that envelop a cavity as the most primitive protocell (Russell and Hall, 1997). Another example concerns an investigation into how a succession of elements of increasing electronegativity took part in the early chemistries on Earth (Hengeveld and Fedonkin, 2007). Thus, the crusts allow for the generation of energy, and the succession of elements in the cavity inside the crust allows the energy to be processed. These processes are essential for the upstart of a chemical system that brings about an increase in the energy dissipation rate (Prigogine and Stengers, 1984). In this way, favorable tests put constraints on the number of biogenetic scenarios left, both for the origin of life here on Earth and elsewhere in the universe.

Systems thinking, in turn, puts further constraints on the number of possible scenarios as the processes concerned are assumed to operate in concert.

Only certain kinds of processes can occur immediately after crust formation, and later in evolution, others may either be unnecessary, impossible, or essential components for their functioning or their formation may still be missing. Furthermore, systems are collections of components interacting with each other such that it is the interactions that determine the dynamic of the system more than does the nature of those components itself. This applies especially when the interactions evolve into strict interdependencies, whereby the components remain interchangeable. This also holds for a biochemical system. For example, iron in a heme group can be substituted by copper without significantly changing the heme group and its functioning; it improves this functioning under altered environmental conditions. However, a set of tuned interactions must remain as it is because any change in even a single interaction would disrupt all others, often with dire effects. As soon as biochemical systems increase in complexity, the number of changes therein or alternative systems becomes severely restricted.

The tuning of interactive processes between various system components depends particularly on physical factors, such as temperature. When the temperature changes, the rates of various processes usually change differently, which in turn disrupts or destroys their tuning, which can be disastrous for sustaining the system. Conversely, when we find apparently ancient systems still operating today, we can justifiably assume that temperature has remained the same during the intervening time. This also constrains our assumptions concerning the conditions under which upstarting systems occurred. Also, temperatures on other planets should be similar to those on Earth for biological systems to originate and to continue for long periods of evolutionary time as well.

We now turn to a more detailed discussion of each of these four constraining consequences of a deductive approach to biogenesis. Thus, this chapter extends the idea of habitable planets, which restricts the possibility of life occurring to a narrow belt around the Sun, and this to a planet with certain chemical and physical properties only (Broecker, 1987; Kasting, 2010).

## 2. Crust Formation

Russell and Hall (1997) were the first to recognize the need for a mechanism for generating energy at the very beginning of biogenesis. Energy is essential in any kind of organization, static or dynamic, nonbiological or biological. In order to keep working, dynamic organizations need a continuous flow of energy, similar to the hydrological cycle on Earth which is kept running by solar energy. At present, solar energy also keeps all life on Earth going, which raises the question whether this has always been the case: how and when did this continuous flow start up, and in what form? Obviously, at some point, a thermodynamic disequilibrium must have originated, which has continued ever since, although it may have evolved into different forms. For this permanent disequilibrium to exist on a microscale, two media were separated from each other by a crust, a film, or a membrane, as in a battery. For a crust to form, a precipitate could have been formed around a trickle of alkaline fluid flowing from the basalt ocean floor into acid seawater, just as happens in chemical gardens. This precipitate then separates two kinds of water with a different electric charge, which then generates the thermodynamic disequilibrium required: just as in a battery, energy flows from one compartment, the seawater, into the other, the tiny cavity within the precipitate. The difference between this system and a real battery is that the precipitate separating the two compartments is permeable to material carrying the energy from the seawater to within the cavity, whereas it is impermeable in a battery. Russell and Hall (1997) suggested that the assumed precipitate would have consisted of FeS, an ancient molecule which in the form of ferredoxins would still be found in the cellular membrane. By connecting the crust with these compounds that are still central to the energetics of the cell, they had their deduced mechanism for the generation of the essential energy.

The step they took differs from those made in most current biogenetic models in that it deduces a physical mechanism for generating the energy needed for chemical reactions to take place within the cavity. It thus shifts the attention away from the formation of specific chemical compounds to the thermodynamics of a system (Hengeveld, 2007). By accounting for a continuous energy flow from a charge disequilibrium, it also avoids the difficulty of reaching an untimely chemical equilibrium which is unavoidable in the organic soup scenarios.

## 3. Trends in the Periodic Table of Elements and in Biological Evolution

In their groundbreaking article, Russell and Hall (1997) stayed with the generation of energy through the prebiotic initiation of a charge disequilibrium. Of course, a permanent energy flow thus generated assumed the existence of a permanent energy sink within the tiny cavity. And it also meant that the crust prevented an exchange of a significant amount of material from occurring between the cavity and the outside world. Therefore, right from the beginning, some cyclic process(es) had to develop within the cavity in order that the enclosed material did not need to be renewed continually. The flow of energy through the system, though, remained linear. Since enzymes were not available at that time, the question is which chemical compounds can form and break down again easily in a chemical cycle without needing complex enzymes. Carbon compounds, mentioned here because they are usually considered essential to any form of life on Earth, would have been unsuitable because of their great stability; their formation and breakdown require enzymes. This holds for nearly all six elements at present essential for the operation of living systems (CHONSP), namely, nitrogen and oxygen, and to a lesser extent, sulfur and phosphorus. Those elements, all nonmetals, typically occur in the top right-hand corner of the Periodic Table of Elements (PTE); they are all elements with a great electron pulling power, that is, with a strong electronegativity. Their strong electronegativity allows for their stability, which explains why particularly carbohydrates, and to a lesser extent also proteins and phosphates, serve as energy-storing molecules. Similarly, the stable water molecule can only be split into hydrogen and oxygen by using a large amount of solar energy which is released as chemical energy by the subsequent reconstitution of water during respiration. Yet, the heme group of chlorophyll, the molecule that receives the solar radiation energy and transforms it into the chemical energy needed for the splitting of water, is extremely complex. This use of solar energy is therefore the result of a long evolutionary development. Biogenetic scenarios starting with the formation of carbon, nitrogen, and oxygen compounds, therefore, cannot work, and probably neither those based on sulfur and phosphorus. Using amino acids rained down from outer space or formed by strong electric discharges is no solution, because we have to explain the mechanism of their biochemical formation, not their chance capture from the environment. Apart from this, they are the building blocks of intricately folded protein chains, parts of complex biochemical systems, too complex for life to start with.

The electronegativity is strongest among the nonmetals of the top row and especially the halogens to the right of them, whereas it gets weaker toward the metals, which occur in the rows underneath. Selenium, occurring immediately below sulfur, but in the same period, has a weak electronegativity, almost as weak as that of hydrogen, whereas these two elements can both accept electrons as well as donate them. Therefore, Hengeveld and Fedonkin (2007) proposed that they might have formed the first cycle, in which $H_2Se$ was formed and broken down again, forming chains over evolutionary time. The hydrogen could have been supplied by a crust of FeNi or by the still existing ancient hydrogenase FeNiSe (Cammack et al., 2001). From there and under the gradually more oxidizing conditions of the environment, chemical evolution could have resulted in the substitution of Se, first by sulfur to form $H_2S$, which was eventually substituted by oxygen forming $H_2O$. In this process of elemental substitution, the biochemical structures and processes only needed relatively minor adjustments as all three elements occur in the same period of the PTE having similar chemical properties. However, at some point, oxygen was included, and it may be that its late date was a consequence of its strong electronegativity: $H_2O$ requires much energy to be split, which can only be done by Photosystem II, a slightly altered version of Photosystem I needed for splitting $H_2S$ for the same reason. Later, it gives much of this energy back during respiration. Still, oxygen was largely kept out of the more recent and the evolutionarily more developed biochemistries, although hydrogen remains central in them. This exclusion possibly happens because of its strong electronegativity and because of its more recent addition to the biochemistry as a consequence of this strong electronegativity.

According to this scheme, carbon, nitrogen, and oxygen were among the later elements to be inserted into the developing biochemistry of the living systems rather than involved in their evolutionary upstart. The RNA world in which phosphorus played an important part would have been an intermediate evolutionary stage before one dominated by carbohydrates, proteins, and DNA.

Similar trends were deduced to have followed by other nonmetals, as well as by metals as electron-supplying catalysts. The final phase could have been reached by the multicellular stages of evolution at the end of the Precambrian when animals, for example, developed a nervous system operating with elements that donate electrons easily, sodium and potassium, and a hormonal system operating with strong electron acceptors in this case halogens at the opposite end of the Table.

Operating with strong electron donors and acceptors puts great demands both on the energy supply as on the enzymes and reaction mechanisms. These enzymes and reaction mechanisms not only have to be specific, elaborate, and strong, but they also have to be formed, whereas the mechanisms forming them have to be formed themselves as well. And this, in turn, puts demands on the stability of the structure as a whole, which requires a stable regulating mechanism in the form of DNA, which should, on top of all this, also be able to transfer the structure from one generation to the next without too many mistakes being made. This genetic transfer mechanism itself is highly intricate and complex. This enormous increase in complexity, continuous organization, and maintenance costs energy and materials. It, therefore, seems unlikely that the complex suite of mechanisms, each of which already highly complex, coevolved. Rather, each mechanism evolved independently, after which they were put together during a succession of endosymbiotic events. Kooijman and Hengeveld (2005) suggested such a succession series for the evolution of metabolism. Martin and Muller (1998) proposed that at a later stage of evolution, the eukaryotic cell resulting from a similar process occurring that ensured a regular supply of hydrogen to the host cell. And recently, Lane and Martin (2010) suggested that eukaryotic endosymbiosis brought in the energy needed for the increased genome complexity underlying the enormous complexity of both the metabolism and the genetic mechanism. All this suggests that the shift to the present organic chemistry of the cell has a very long history, consisting of many steps. It would have been impossible for living systems to have originated straightaway from elements with strong electronegativities.

As seen from the perspective given in this chapter, the window of life's origin is rather small; much depends on the very specific qualitative and quantitative composition of the chemical environment allowing the formation of an energy-generating crust with feasible chemical cycles. Chemical soup models are less defined in this respect. Furthermore, the initial, reducing environmental conditions gradually changed to oxidizing ones which necessitated a further trend in the chemical conditions during early and later evolutionary stages (e.g., Anbar and Knoll, 2002). Yet, the exact adaptations made also depended on the rapidity of the environmental changes, on spatial variation in those conditions, on the properties of the developing systems available, as well as on a large stochastic component, to name the most significant determinants. The chemical conditions put down by thermodynamics restrict the possibilities of life originating, both here on Earth and elsewhere in the universe. Starting inductively using an almost free choice of criteria for defining life, such as the possession of macromolecules, of DNA, of amino acids, of carbohydrates, and so on, does not put any restrictions on the initial conditions under which life might have originated. The only condition generally thought to restrict the occurrence of life is the presence of water. Therefore, the possibility of life continuing under those changing conditions and the direction evolution has taken over almost 4 billion years, partly as a consequence of these changes, make the end result unique; life as we know it cannot be found anywhere else in the universe.

## 4. Building an Integrated System

Starting from, say, amino acids or from adenine formed in some way elsewhere in the universe and raining down from outer space onto the surface of the Earth concentrates on individual molecules rather than on their role within an evolutionarily developing chemical system. Also, amino acids or adenine cannot have begun operating straightaway after their inception without an already elaborate system being present within which to operate and function. Moreover, students choose such molecules for their biogenetic research with regard to the significance they have in the present biochemical system. In fact, however, the system itself has had to develop a mechanism for making such compounds rather than depend on an external source for their supply in order to operate. Furthermore, such a mechanism would make them only when needed not to run short relative to other processes, or not to waste energy and material, each mechanism needing in turn to evolve. This supply and fine tuning contrast with chemical soup models that implicitly assume that at one moment, all chemical cell components of present relevance are scrambled together and put within a membrane. A well-tuned system would then be formed with all components in place, accounting for the generation and processing of chemicals and energy. But those components and mechanisms in the original unbounded organic soup cannot have coevolved, operated, and recycled in isolation without a regular and organized supply of energy. Instead, it is their role within the biochemical system that defines their presence, their shape, their interactive functioning, and their further evolution within a changing environment and a system that is becoming more and more complex. If we are to understand systems as we find them today, we need to understand how they have evolved as integrated, dynamic systems. They have to be built up from scratch, initially without enzymes, although certain metals could have catalyzed some reactions from early stages onward as electron donors. We have to reconstruct those systems from scratch, bit by bit, component by component, mechanism by mechanism.

Because of the limited, or initially even absent interchange with the environment, the earliest systems had to form chemical reaction cycles comprising a succession of well-adjusted components and running with well-tuned reaction rates. If those rates had not been tuned, delays would have resulted whereby either shortages would soon have killed the system or the system would have drowned in its own products. At a later evolutionary stage, when chemical interchange with the environment became feasible, mechanisms of the uptake and rejection of chemicals had to evolve. Such mechanisms respond to the relative concentrations of the chemical elements or compounds within the cell, which, according to some preset standards, are either be too low or too high. Those concentrations therefore had to be measured and interpreted, for which a homeostasis mechanism with reaction norms had to develop, allowing mechanisms to be switched on or off for making or breaking down particular molecules.

Such integrated, homeostatic systems based on recycling with a minimum of waste, which, moreover, maintain a permanent thermodynamic disequilibrium for their continuous energy supply, can only develop within an envelope that has separated them from their surroundings from scratch. The system components and their interactions together form a structure of dissipating energy and are, as such, shaped by the flow of energy. As this energy is gradually degrading, new energy has to be injected continually at certain points into the biochemical pathways and cycles to keep them running. Ultimately, it is this flow of energy through the system that increased in speed and volume relative to other flows in the environment; it is this flow and the resulting structured dynamic system that has to be explained and understood. In fact, it is the origin and maintenance of this energy flow that makes living systems; the materials are the carriers of the energy circulating in those systems. This makes it impossible to start research in biogenesis by producing individual molecules or other properties abiotically without a biologically functional context, certainly when they define present-day life, such as highly specialized macromolecules like DNA (cf. Popa (2004) for a list of almost 100 definitions). Concentrating research on molecules and mechanisms of later significance hampers our understanding of the very early stages of life's beginning on Earth (Hengeveld, 2010b).

As the environment changed, living systems had to follow suit and not run short of chemicals or energy. However, as systems, they had to change as little as possible, which means that certain chemical elements that were system components were replaced by other ones with similar chemical properties. Molecules or molecular mechanisms altered, but only gradually and slightly, adding to variations on the initial theme that kept the dynamic going. This meant that elements of the same period within the PTE were often selected, although of increasing electronegativity.

Thus, the initial selenium may have been replaced by sulfur, and sulfur was eventually replaced by oxygen. The last replacement may have happened when, for example, Photosystem I evolved into Photosystem II splitting $H_2S$ and $H_2O$, respectively, or some thioester-RNA world, evolved into an organic one (de Duve, 1991). FeS kernels and metals and the nucleotides may thus have obtained their protein chains transforming them into ferredoxins (Eck and Dayhoff, 1966) and other metalloproteins, whereas coenzymes and the nucleotides may have evolved into a large variety of RNAs, these in turn evolving into ribosomes (Chech, 2000), spliceosomes (Reanny, 1979), and DNA. The nucleotides could themselves have started as phosphate strings (Wald, 1962; Westheimer, 1977; Baltscheffsky and Baltscheffsky, 1994), or before that, in principle, even from arsenates because of the relatively weak bonds of arsenic (Wolfe-Simon et al., 2009), a weakness which later became a drawback. This need for increasing stability was also why yet later DNA replaced RNA, so that it is not to be expected that arsenic can still have occurred in DNA as Wolfe-Simon et al. (2010) suggested or even in the earlier forms of RNA. Step by step, the complexity of the cell, and later that of multicellular systems, was built up. For energy storage and mechanical reasons, this required the selection of the strongly electronegative elements of the top row to the right of the PTE, such as carbon. Similarly, nitrogen could have been selected for proteins in order to enhance and to specify the catalytic functions of metals and coenzymes by orienting and polarizing the reactants (e.g., Pullman, 1972), and water as the ultimate hydrogen donor. Apart from that, the increasing versatility of bonding was enhanced in the same direction by the double and multiple bonds made by these elements (Pullman, 1972). Halogens in the top rows of the PTE were later attributed with mechanical functions and with chemical communication (Hengeveld and Fedonkin, 2007). And so on.

In spite of some changes, the operation and biochemical functioning of many mechanisms would have remained the same since entirely new mechanisms would have disrupted the interactive system and the tuning of its defining interactions. This implies that the direction evolution takes is limited from the start at the beginning by local conditions and by chance. Local conditions and chance continue to determine the direction of evolutionary progress. There is therefore no possibility of turning back and trying again. This process is typical for all developing systems, also including systems in computer science, where it is known as backward compatibility. According to this principle, a newly evolving system should be compatible with the previous one from which it originated. Therefore, general rules for system development and behavior determine the margin of variation of evolutionary development. However, what happens within that margin remains indeterminate.

The restrictions imposed by systems thinking regarding life as a system mean that the probability of occurring and developing elsewhere in the universe decreases accordingly (cf. also Wald, 1962). Reconstructing how life began and applying this model to extraterrestrial conditions are now even more restricted than for chemical and thermodynamic reasons. Developing and maintaining any organized system require a continuous input of energy which has to be channeled the easiest way. If we regard life as a special form of organization, we need to focus on this flow of energy, how it has shaped the organization and its components, and how the energy flow has selected these components during its upstart and subsequent evolutionary development (Wald, 1962; Williams and Fraústo da Silva, 1996, 2006). Moreover, our task is primarily to understand life here on Earth rather than on other planets (e.g., Hengeveld, 2010a).

## 5. Constancy of Environmental Conditions

Those working within the field of biogenesis not only discuss the compounds and processes that might have accounted for the upstart of life but also the conditions under which this may have happened. For example, could it have happened at the same pressure and temperature, under which life mostly operates at present, or at higher pressures or temperatures? What were the chemical conditions of the environment in which life originated and evolved? Did they remain the same? If not, how did the systems respond to the changes? Can those conditions be deduced from the general approach we have taken so far? How much freedom do we have to assume that higher pressures and temperatures were dominant at the time?

Concerning pressure, we need to explain life as we know it, however ancient or recent it is. Sooner or later, a system developed that operated at a pressure similar to the one at present, when it was not dependent on a high pressure. Systems that might once have originated from strong electric discharges (Miller, 1953), under conditions of shock (Hazen, 2005), or have operated under much higher pressures – whatever their cause – must have died out when these conditions ceased to exist. The fact that amino acids can be formed as a result of electric discharges does not help our research because such discharges do not occur in any of the existing biochemical systems known. In contrast to chemical conditions which have changed over geological time, and to which systems had and were able to adapt, we have to explain the present systems by the physical conditions we are experiencing today. The same reasoning applies to the problem of whether or not the temperature was the same at life's origin as it is now.

Many biochemical processes involve the transfer of hydrogen, a process which often requires the use of enzymes to overcome the energy barrier. There is a possibility that the enzymes lower the energy threshold, so that less energy is needed for the transfer. An alternative is to "tunnel through" the barrier which would not cost any energy at all; this is called quantum tunneling (cf. Dutton et al., 2006), a quantum mechanical process. According to quantum mechanics, hydrogen can be conceived both as a particle as well as a wave of a particular length. This wave length is defined by the size of the "particle"; the larger the "particle," the shorter the length of this wave. This wave can also be conceived in statistical terms as a probability density distribution of the presence of the particle, which has the highest values in the central parts of the wave and which tapers from this center in all directions. Therefore, when two molecules, one containing a hydrogen atom and the other lacking one, come close to each other during vibrations, at certain probabilities, the hydrogen can be shared at a certain probability. When the molecules separate again, the hydrogen staying in either the one or the other molecule, it may have been transferred.

This process is only significant for the transfer of hydrogen: all other elements are too large for this tunneling process to be able to happen. For example, deuterium, an isotope of hydrogen, and only twice as large since its nucleus contains a proton and a neutron instead of just a proton, has a wave length that is 100 times shorter. This implies that life would not be possible in a deuterium world (e.g., Greener, 2005). Obviously, however, the vibrations and intermolecular distances are affected by both temperature and pressure. Increases in temperature decrease the flexibility of proteins, and increases in pressure decrease the distance. For the same reason, the pH must have remained constant over time. The reaction rates of hydrogen with other chemical elements would therefore be affected, which cannot be simply turned back when those physical conditions change; they have to be redefined such that a new, yet different set of interactions – and thereby a different system – results: the present, Earthly system. And this is the only kind of system that we know.

Because of the significance of quantum tunneling, we can exclude the possibility of life having originated under a temperature much higher than those at present: on average, the distances to be covered by hydrogen may have been too long and too short-lived for quantum tunneling to have obtained its significance in presently known organisms. Recent geological information seems to confirm this (e.g., Hren et al., 2009; Kasting, 2010). Also, hyperthermophilic bacteria adapted to high temperatures appear to have built in special structures that give their protein molecules thermal stability (Jaenicke, 1996; Rice et al., 1996). Being special structures, they may have evolved later.

Therefore, concerning the possibility of life originating on our planet under high temperatures and pressures, the probability is low that systems use hydrogen in their basic biochemical processes in the same way as present living systems do. Should hydrogen be used as a central biochemical element, processes will be much slower and more complex, making one wonder if life could be feasible under such conditions anyway. This also applies to other planets within or outside our solar system.

Contrary to the physical conditions which probably have remained constant during the time that life has been on Earth, chemical conditions are known to have changed considerably over geological time (e.g., Anbar and Knoll, 2002). When the Earth was formed, the environment may have been reducing, gradually becoming oxidizing (e.g., Kasting, 2010). This means that, at first, protons and electrons would have been relatively easily available but that their availability decreased. This implies that as time went on, the nonmetals with a stronger electronegativity at the upper right-hand corner of the PTE had to replace the ones used in earlier life in the lower rows of the Table that have a weak electronegativity. It implies too that ever more robust biochemical mechanisms had to evolve. Apart from these adaptations, there must also have been mechanisms keeping the system running as it had always done before, a matter of backward compatibility. These highly complex adaptations can indeed be found both in the membrane and in the cytosol: both maintain a low redox potential, possibly at the level it had during the upstart of life.

According to Russell and Hall (1997), the membrane separates the reducing outside world from the oxidizing world inside the membrane. This difference accounted for the charge disequilibrium on which the inward-bound energy flow depended. However, as the environment shifted from reducing to oxidizing, the disequilibrium decreased. At some point, it could have flattened out altogether. The energy flow would have stopped with the total extinction of life as a result. However, it may be that the diminution of the energy flow prompted the development of the proton-motive force. Protons are first pumped from the inside to the outside of the membrane, after which they reenter the cell, thus maintaining the charge disequilibrium. The energy this process needs is generated in the same membranes by the conversion of radiant solar energy into chemical energy.

A similar response to changing redox conditions outside the cell may have happened in the cytosol. Here, we find the Z-scheme, whereby electrons generated in the membrane at relatively high, positive redox values – water holding on to its electrons at +820 mV – enter the biochemical system in the cytosol in two steps at high negative values – NADPH holding on to its electrons loosely at −320 mV – as they initially may have been. Interestingly, the first step is taken by splitting $H_2O$ by Photosystem II which evolved last, that is, under relatively oxidizing environmental conditions. The second step is taken by the older Photosystem I that splits $H_2S$. During this process, protons are pumped across the membrane, thereby generating the charge or electrochemical gradient just mentioned. This gradient drives the synthesis of ATP which, when hydrolyzed, releases its bond energy for further biochemical reactions downstream in the cytosol.

Both counter-intuitive processes can be deduced directly from the need of systems to remain the same under known changing conditions. However, because of the large stochastic component involved, the exact form the adaptations will take cannot be deduced. In that sense, the direction the evolution of life takes – or its probability of extinction – remains unpredictable. To take an extreme, we cannot predict with certainty where, when, and how a meteorite of a certain size will strike the Earth, what geological and meteorological consequences the impact will have, what part of life will be affected, and what direction evolution will take after the impact and after the extinction wave caused. As all these processes vary stochastically, we can only calculate the chance occurrence of their simplest components, such as chance impacts dependent on the number of meteorites, or whether the impact will be somewhere in the ocean or on land depending on the relative surface areas of water and land. Similarly, we can expect evolutionary trends in living systems to follow the structure of the PTE, but it is senseless to try and reconstruct the exact evolution of, say, adenine, cytosine, DNA, or the genetic information system. And it is equally senseless to search for similar life forms on other planets (e.g., Mayr, 1985).

## 6. Discussion

The inductive way followed so far by biogenetic research began by defining life in terms of its present-day chemical properties, that is, in terms of the origination of certain carbohydrates, of simple proteins, or of DNA. Popa (2004) enlisted no less than 98 such criteria given in the literature for defining life, many of which have served as starting points for biogenetic research. His list, though, is certainly not exhaustive; among many other criteria, the presence of DNA or of other macromolecules, for example, are still missing. This huge number by itself already indicates that this is not an objective way forward; anything can be taken to define life, and anything can be taken as a basis for research for further sharpening the hypotheses generated. Moreover, many of the criteria suggested are vague and nonoperational, and all of them are circular. "Life is alive if it can die" (Lauterbur, 2002) is one of the 98, one which would be pretty useless in biogenetic research. Other texts concentrate on chemical reaction mechanisms, such as the one by Zubay (2000) who put the existence of covalent bonds central, as formed by the six lightest elements most dominant in present-day organisms, the so-called CHONSP elements: carbon, hydrogen, oxygen, nitrogen, sulfur, and phosphorus. Among these elements are those forming the strongest bonds with which, according to deductive reasoning, it would have been impossible to have started the cycles first needed for life. The metabolic energetics is phrased in terms of energy transfer by ATP and other nucleotides, which, however, would have evolved well after the first beginnings of an increased flow of energy.

However, instead of carbon having been at the cradle of life, as an element forming very strong covalent bonds, hydrogen may have taken the central place from early on to the present time. This means that we need to reorient our approach to metabolism, from the generally accepted carbon-based one to a metabolism based on hydrogen as the pivotal element transferring energy through the system. Hydrogen also has physical properties such as the possibility of quantum tunneling that enhance reaction rates without costing energy. Other elements have substituted each other during evolution, but hydrogen has not lost its initial importance. Furthermore, in that way, the mechanism of system maintenance becomes the focus of attention: the energy flow becomes essential in the origin and maintenance of biological systems within and over the generations. Then, there is nothing weird in a system made up of different elements, as it is the system that maintains as it goes through various stages of evolutionary complexification, remaining the same; we only need to look at it differently, that is, from a basically thermodynamic viewpoint rather than from an organic-chemical one.

One of the great advantages of deductive reasoning over inductive reasoning is that it excludes several alternative explanations from further consideration. However, deductive reasoning can be applied to relatively simple systems only because of the very rapid increase of the number of interactions that are possible between an increasing number of components. The form interactions take depends largely on chance. Worse, the number of interactions – the number of permutations between the system's components – depends on the number of components, and this increases factorially. This factorial increase is expressed as n!, which means the number of components n times this number minus 1, times $n-2$, and so on. For 3 components, therefore, we get $3 \times 2 \times 1 = 6$, but for 8 components, the number of permutations already amounts to 40,320. Given the vastly larger number of atoms within even a very confined space of a single bacterial cell separated from its environment, the number of permutations readily becomes enormous, incalculable in fact (Elsasser, 1987). But for their survival, bacterial cells are dependent on each other and on their immediate chemical and physical environment. Moreover, this large number of permutations only concerns the number of molecules that may be formed by the atoms of the various elements available, but these molecules repeat the same permutation process on a higher level; they form a reaction pathway or cycle. And these in turn repeat the same story once again on a third level of interaction and so on. Apart from these processes happening on those different levels, the levels themselves are also integrated, implying that they can interact as well. Of course, not all permutations are equally viable, and we also have to account for backward compatibility as well as for repair mechanisms, etc., all of which reduce the final number. But it still remains incredibly high.

On the other hand, we also need to account for the repeatability of the historical process of life. Evolving in a changing environment means that the chance reduces incredibly fast. To obtain the chance of some process happening twice in two successive steps, namely, we cannot add the chances of a particular result of each other steps together, but we have to multiply them. For example, to throw, say, 4 with a dice gives a chance of 1 in 6; to throw it twice consecutively and independently, the chances add together, which gives 1 in 3. Over time, they converge to 1; it is impossible never to throw 4 in a great number of games. But to throw a 4 twice in succession, the chance becomes 1 in 36. These chances rapidly get smaller. Such multiplications of in themselves incalculably small chances happen on all temporal and spatial scales, from the atomic and molecular scales to the geological and astronomical ones of some large meteorite impacting the Earth, and this over almost 4 billion years. It is this chance we are talking about when it concerns the evolution of other beings of similar intelligence wanting to communicate with us from elsewhere in the universe.

The number of permutations itself, independently of the small chance of the repeatability of a historical process, is unimaginably large, larger in principle than the number of experiments that can be done for testing a hypothesis following a deductive reconstruction of an initial biochemical system. This is already the case after even the first steps having been made during an assumed origination of life (Elsasser, 1987). Therefore, for any further progress along these lines, one has to compartmentalize the process of biochemical evolutionary change into simple systems. On the other hand, without a deductive way of system reconstruction, that is, without restrictions imposed by this kind of reasoning on the number of possible permutations, this number is obviously even larger. From this perspective, inductive reasoning without a deductive framework leads nowhere. Moreover, as the hypotheses it generates cannot be tested within this way of reasoning, its results remain inconclusive.

In connection with biogenesis, it is imperative to start with system thinking: it is the biochemical system that has to be reconstructed from scratch. This means that the energetics of the system is understood with regard to the formation and maintenance of its interactions. It is the energy flow which is getting structured in ever more complex forms. This energy flow puts demands and restrictions on the nature of the elements and the intensity of their selection, as well as on those of the selection of molecules and molecular mechanisms as the material basis of the system. It also puts demands and restrictions on the stability of the physical conditions of the environment. In turn, these environmental conditions and the variation and changes therein decide the viability of the system, not only its internal functioning, but also whether it can originate and whether it can follow the external changes in living conditions adequately to keep going.

Obviously, understanding the origin of life on Earth in these terms will take time, but it is the shortest, quickest, and most certain way possible. Within this perspective, it seems premature to begin thinking about the possibility of similar processes happening elsewhere, particularly in such detail.

## 7. Conclusions

We cannot expect the earliest biological system to be based on carbon, nor can we expect it to occur under present-day oxidizing conditions. It might exist in reducing environments as rich in phosphorus, sulfur, and metals as at the time of biogenesis. In a reducing environment, simple systems may be able to occur instead of systems with highly developed, energy-hungry biochemical and genetic mechanisms that contain strong electronegative elements; such elements could have been redundant or disrupting under reducing conditions. However, it is doubtful whether primitive life forms could have persisted under such reducing conditions for several billions of years.

## 8. Acknowledgments

Hereby, I thank my wife Claire Hengeveld-Nicholls for correcting the English and the logic of this chapter.

## 9. References

Anbar AD, Knoll AH (2002) Proterozoic ocean chemistry and evolution: a bioinorganic bridge? Science 297:1137–1142

Baltscheffsky H, Baltscheffsky M (1994) Molecular origin and evolution of early biological energy conversion. In: Bengtson S (ed) Early life on Earth. Columbia University Press, New York, pp 81–90

Broecker WS (1987) How to build a habitable planet. Eldigio Press, Palisades

Cammack R, Frey M, Robinson R (eds) (2001) Hydrogen as a fuel. Taylor & Francis, New York

Chech TR (2000) The ribosome is a ribozyme. Science 289:878–879

de Duve C (1991) Blueprint for a cell: the nature and origin of life. Patterson, Burlington

Dutton L, Scrutton N, Sutcliffe M, Munrow A (eds) (2006) Quantum catalysis in enzymes – beyond the transition state theory. Philos Trans R Soc B 361:1291–1455
Eck R, Dayhoff MO (1966) Evolution of the structure of ferredoxin based on living relics of primitive amino acid sequences. Science 152:363–366
Elsasser WM (1987) Reflections on a theory of organisms: holism in biology. Johns Hopkins University Press, Baltimore
Fry I (2000) The emergence of life on Earth: a historical and scientific overview. Free Association Books, London
Greener M (2005) Did enzymes evolve to capitalize on quantum tunneling? Scientist 19:17–29
Hazen RM (2005) Genesis: the scientific quest for life's origin. Joseph Henry Press, Washington, DC
Hengeveld R (2007) Two approaches to the study of the origin of life. Acta Biotheor 55:97–131
Hengeveld R (2010a) Weird life in a shadowy environment. J Cosmol 13:3627–3630
Hengeveld R (2010b) Definitions of life are not only unnecessary, but they can do harm to understanding. Found Sci. doi:10.1007/s10699-010-9208-5
Hengeveld R, Fedonkin MA (2007) Bootstrapping the energy flow in the beginning of life. Acta Biotheor 55:181–226
Hren MT, Tice MM, Chamberlain CP (2009) Oxygen and hydrogen isotope evidence for a temperate climate 3.42 billion years ago. Nature 416:205–208
Jaenicke R (1996) Glyceraldehyde-3-phosphate dehydrogenase from *Thermatoga maritima*: strategies of protein stabilization. FEMS Microbiol Rev 18:215–224
Kasting J (2010) How to find a habitable planet. Princeton University Press, Princeton
Kooijman SALM, Hengeveld R (2005) The symbiotic nature of metabolic evolution. In: Reydon TAC, Hemerik L (eds) Current themes in theoretical biology: a Dutch perspective. Springer, Dordrecht, pp 159–202
Lahav N (1999) Biogenesis: theories of life's origin. Oxford University Press, New York
Lane N, Martin W (2010) The energetics of genome complexity. Nature 467:929–934
Lauterbur PC (2002) Short definitions of life. In: Palyi G, Zucchi C, Caglioti L (eds) Fundamentals of life. Elsevier, New York, p 41
Martin W, Muller M (1998) The hydrogen hypothesis for the first eukaryote. Nature 392:37–41
Mayr E (1985) The probability of extraterrestrial life. In: Regis E (ed) Extraterrestrials: science and alien intelligence. Cambridge University Press, Cambridge, pp 23–30
Miller SL (1953) A production of amino acids under possible primitive earth conditions. Science 117:528–529
Popa R (2004) Between necessity and probability: searching for the definition and origin of life. Springer, Berlin
Prigogine I, Stengers I (1984) Order out of chaos: man's new dialogue with nature. Heinemann, London
Pullman B (1972) Electronic factors in biochemical evolution. In: Ponnamperuma C (ed) Exobiology. North-Holland Publishing Company, Amsterdam, pp 136–169
Reanney D (1979) RNA splicing and polynucleotide evolution. Nature 277:598–600
Rice DW et al (1996) Insights into the molecular basis of thermal stability from the structure determination of *Pyrococcus furiosis* glutamate dehydrogenase. FEMS Microbiol Rev 18:105–117
Russell MJ, Hall AJ (1997) The emergence of life from iron monosulphide bubbles at a submarine hydrothermal redox and pH front. J Geol Soc Lond 154:377–402
Wald G (1962) Life in the second and third periods; or why phosphorus and sulfur for high-energy bonds? In: Kasha M, Pullman B (eds) Horizons in biochemistry. Academic, New York, pp 127–142
Westheimer FH (1977) Why nature chose phosphate. Science 235:1173–1178
Williams RJB, Fraústo da Silva JJR (1996) The natural selection of the chemical elements: the environment and life's chemistry. Oxford University Press, Oxford
Williams RJP, Fraústo da Silva JJR (2006) The chemistry of evolution: the development of our ecosystem. Elsevier, Amsterdam
Wolfe-Simon F, Davies PCW, Anbar AD (2009) Did nature also choose arsenic? Int J Astrobiol 8:69–74

Wolfe-Simon F, Switzer Blum J, Kulp TR, Gordon GW, Hoeft SE, Pett-Ridge J, Stolz J, Webb SM, Weber PK, Davies PCW, Anbar AD, Oremland RS (2010) A bacterium that can grow by using arsenic instead of phosphorus. Science 324:36–44
Zubay G (2000) Origins of life on the Earth and in the Cosmos. Academic, San Diego

Biodata of **Kenji Ikehara**, author of "***[GADV]-Protein World Hypothesis on the Origin of Life.***"

**Professor Kenji Ikehara** is currently Director of Nara Study Center of the Open University of Japan. He obtained his Ph.D. from Kyoto University in 1976 and continued his studies and research at the University of Tokyo and Nara Women's University. Professor Kenji Ikehara's scientific interests are in the area of origins of genes, the genetic code, proteins, and life.

E-mail: **ikehara@cc.nara-wu.ac.jp**

J. Seckbach (ed.), *Genesis - In The Beginning: Precursors of Life, Chemical Models and Early Biological Evolution*, Cellular Origin, Life in Extreme Habitats and Astrobiology 22, 107–121
DOI 10.1007/978-94-007-2941-4_6, 

# [GADV]-PROTEIN WORLD HYPOTHESIS ON THE ORIGIN OF LIFE

**KENJI IKEHARA**[1,2,3,4]
[1]*Narasaho College, Nara 630-8566, Japan*
[2]*International Institute for Advanced Studies, Kyoto 619-0225, Japan*
[3]*The Open University of Japan, Nara Study Center, Kita-uoya-higashi-machi, Nara 630-8589, Japan*
[4]*Emeritus Professor of Nara Women's University, Nara, Japan*

## 1. Introduction

### 1.1. BACKGROUND OF THE PRESENT STUDIES ON THE ORIGIN OF LIFE

At present, many researchers studying on the origin of life believe that life originated from RNA world, which was formed by RNA self-replication, since Gilbert proposed the hypothesis about 25 years ago (Gilbert, 1986). The main reasons why RNA world hypothesis is widely accepted are that it has been generally considered that acquisition of genetic information or genes must precede the formation of proteins with catalytic functions, because proteins composed of 20 kinds of amino acids are too complex to be produced without the support of genetic information, and that genes must be firstly selected from a pool of RNA accumulated through RNA self-replication. It is described in recent ScienceDaily Feb. 24, 2010, that Blumenthal said, "it now appears that the first catalytic macromolecules could have been RNA molecules, since they are somewhat simpler and RNA molecules were likely to exist early in the formation of the first life forms, and are capable of catalyzing chemical reactions without proteins being present."

But the RNA world hypothesis has many weak points, as follows (Ikehara, 2005, 2009): (1) It is very difficult to prebiotically synthesize nucleotides, building blocks of RNA, since nucleotides are rather complex organic compounds composed of sugar, nucleobase, and phosphate (Fig. 1). In addition, nucleotides are unstable under the conditions on the primitive earth. In fact, nucleotides have not been detected in meteorites. (2) RNA could not be self-replicated due to the self-discrepancy between RNA catalysts with stable tertiary structures and RNA templates without structures. (3) Even if RNA were self-replicated, it is also quite difficult for the self-replicated RNA to carry genetic information because genes would never be formed by random polymerization of nucleotides. It is not statistically expected for RNA randomly polymerized in the RNA world to encode a functional water-soluble globular protein since genes are organized by a linear arrangement of triplet base sequences or codons encoding the corresponding amino acids. (4) It is also quite difficult to explain evolutionary processes of the

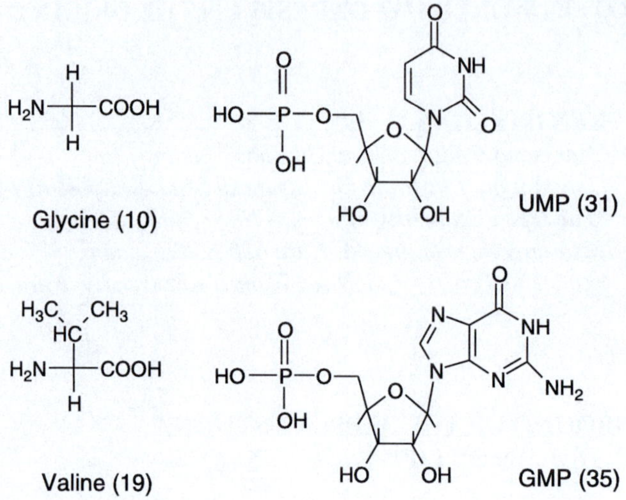


**Figure 1.** Chemical structures of amino acids glycine ([G]) and valine ([V]) and of nucleotides UMP and GMP which are the simplest and the most complex organic compounds among four [GADV]-amino acids and four ribonucleotides. *Numbers in parentheses* indicate those of atoms in the amino acids and nucleotides.

fundamental life system, composed of genes, genetic code, and proteins according to the RNA world hypothesis because formation of the first genetic code also could not be explained based on the RNA world hypothesis, since the capability of RNA for self-replication is not relevant to the genetic code or triplet codon sequences for protein synthesis (Ikehara, 2009).

## 1.2. PROPERTIES OF [GADV]-AMINO ACIDS

It is well known that amino acids, especially as glycine [G], alanine [A], aspartic acid [D], and valine [V], which are encoded by GNC codons, are simple organic compounds having one hydrogen atom, positive amino and negative carboxylic groups, and a side chain at α-carbon atom (α-amino acids) (Fig. 1), and all of them are easily synthesized in Miller's discharge experiments (Higgs, 2009; van der Gulik, 2009). Here square brackets are used to discriminate one-letter symbols of amino acids from nucleobases. In addition, enough amounts of [GADV]-amino acids could be accumulated on the primitive earth, since the amino acids are enough stable for a long time under prebiotic heat conditions (Vallentyne, 1964). Moreover, peptide bond, which is formed between amino group of one amino acid and carboxyl group of the other amino acid, has a planar character, which is favorable for folding polypeptide chain into regular secondary structures and successively into tertiary structures, prerequisitely required to form catalytic center on surface of the tertiary structure.

It must be required to elucidate the origin of the genetic code and to introduce one or two new concepts in order to solve the difficult problem on the origin of life from a new point of view. In addition to the our original GNC-SNS primitive genetic code hypothesis (Ikehara et al., 2002), where N and S means either of four nucleobases (G, C, A, and U (T)) and guanine (G) or cytosine (C), respectively, we introduced two new concepts, protein 0th-order structure composed of roughly equal amounts of [GADV]-amino acids and pseudo-replication of [GADV]-proteins. As a result, we have proposed GADV hypothesis on the origin of life (Ikehara, 2005, 2009). In this chapter, I will describe and discuss the [GADV]-protein world hypothesis or GADV hypothesis.

## 2. The Origin of the Genetic Code

### 2.1. GENETIC CODE IN THE FUNDAMENTAL LIFE SYSTEM

Genetic code occupies a core position relating genetic function to catalytic function in the fundamental life system (Fig. 2a). Therefore, the genetic code is not only a simple representation of triplet base sequence with an amino acid, but also it should be a key point when one understands the formation processes of the life system, which is composed of genetic function, genetic code, and catalytic function, or the origin of life (Fig. 2b).

It is considered that the establishment of the most primitive genetic code triggered to produce the "chicken–egg relationship" connecting between genes and proteins. However, it seems to me that previous researches on the origin of life have ignored the important role or the origin of the genetic code.

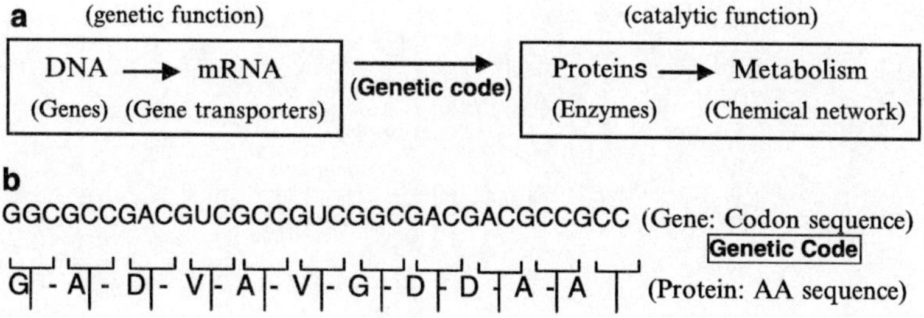


**Figure 2.** The role of genetic code in the fundamental life system. (**a**) Genetic code connects genetic function with catalytic function. (**b**) Genetic code correlates triplet codons in genes to amino acids (AA) in proteins.

## 2.2. GC-NSF(A) HYPOTHESIS ON FORMATION OF ENTIRELY NEW GENES UNDER THE PRESENT LIFE SYSTEM

We accidentally started from a study on formation of entirely new genes, that is, the first ancestor genes in gene families consisting of homologous genes, independent of the origins of the genetic code and life (Fig. 3). From analyses of microbial genes and proteins obtained from the GenomeNet database, we found that the entirely new genes could be produced from nonstop frames on antisense strands of not AT-rich but GC-rich microbial genes (GC-NSF(a)) (Ikehara et al., 1996) (Fig. 4).

This conclusion was obtained mainly based on the facts that hypothetical proteins encoded by GC-NSF(a)s well satisfy six conditions (hydropathy, formabilities of $\alpha$-helix, $\beta$-sheet, and turn/coil structures, acidic amino acid, and basic amino acid compositions) for folding of polypeptide chains into water-soluble globular structures (Ikehara et al., 1996).

The six conditions were obtained as each of the six average values plus/minus the respective standard deviations, which were obtained by calculation using amino acid structural indices described in Stryer's textbook (Berg et al., 2002), and the corresponding amino acid compositions of extant proteins encoded by seven microbial genomes with different GC contents (Table 1) (Ikehara et al., 2002).

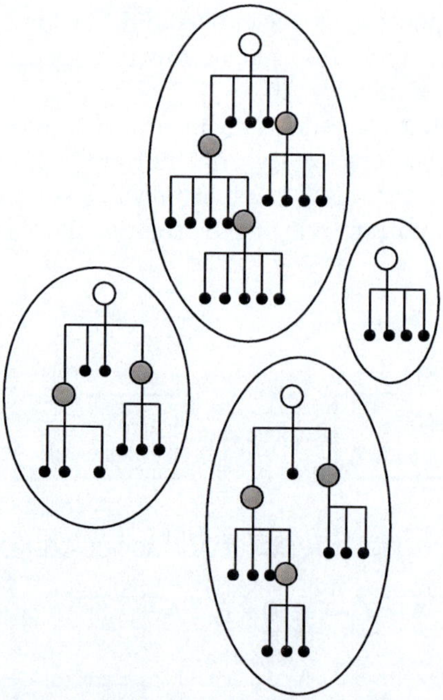

**Figure 3.** Large ellipsoids indicate gene/protein families. *Open* and *gray circles* and *black dots* represent the first ancestor genes/proteins of the families, ancestral genes/proteins, and progeny genes/proteins originated from the first ancestor genes/proteins, respectively.

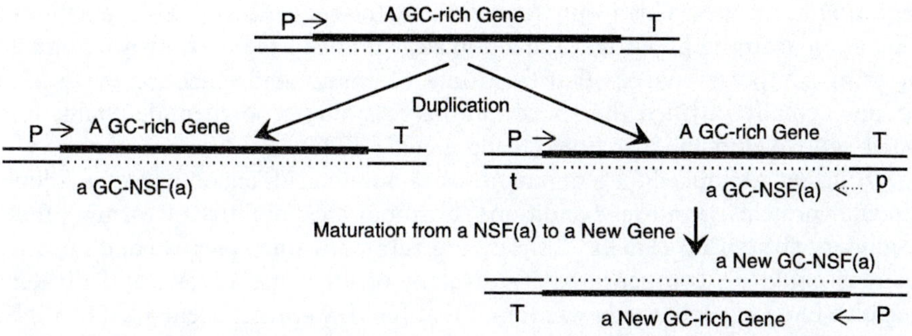


**Figure 4.** GC-NSF(a) hypothesis on the formation of entirely new genes, suggesting that entirely new genes were created from base sequences on GC-rich nonstop frames on the antisense strands (GC-NSF(a)). *Thick black lines* and *dotted lines* show sense sequences encoding functional proteins and their antisense sequences, respectively.

**Table 1.** Mean values and their standard deviations (SD) of hydropathy, secondary structure formabilities (α-helix, β-sheet, and turn/coil formabilities), and acidic and basic amino acid contents, which were obtained by analyses of extant proteins from seven microorganisms.

| Condition | Mean | SD | Top 10 | Bottom 10 |
|---|---|---|---|---|
| Hydropathy | −1.282 | 0.749 | 2.33 | −5.07 |
| α-Helix | 1.025 | 0.036 | 1.244 | 0.817 |
| β-Sheet | 1.011 | 0.032 | 1.192 | 0.795 |
| Turn/coil | 0.956 | 0.050 | 1.258 | 0.644 |
| Acidic A.C. | 0.120 | 0.032 | – | – |
| Basic A.C. | 0.141 | 0.031 | – | – |

Hydrophobicity and secondary structure indices of amino acids were obtained from Stryer's textbook of "Biochemistry" (Berg et al., 2002). Top 10 and Bottom 10 indicate the respective average values of 10 amino acids from the top and from the bottom, respectively. Those mean values plus/minus SD values of the six structural indices of proteins were used as the six conditions for judging whether a hypothetical polypeptide chain is folded into water-soluble globular structure.

The reason why we convinced that the six conditions can be used for judgment of folding ability of a polypeptide chain is that those values of most proteins held nearly constant levels, regardless of variable GC contents, which causes wide variation of half number of natural 20 amino acids. Sufficiently small probability of stop codon appearance is also favorable to produce nonstop frames on the GC-NSF(a)s at a high probability (Ikehara et al., 2002).

## 2.3. GNC-SNS PRIMITIVE GENETIC CODE HYPOTHESIS

Why can proteins encoded by GC-NSF(a)s, not AT-NSF(a)s, well satisfy the six conditions? One reason is that base sequences on both strands on GC-rich genes (around 60–70%) are rather symmetrical and are quite similar to SNS repeating sequences. This result suggests that the sequences of SNS repetitions might hold a strong potential to function as genes. We have previously confirmed by using computer analysis that even only 10 amino acids encoded by 16 SNS codons actually satisfied the six conditions for folding polypeptide chains into water-soluble globular structures (Ikehara et al., 2002).

Further, we looked for a minimum set of amino acids encoding water-soluble globular proteins with four conditions (hydropacy, capabilities of forming three secondary structures; $\alpha$-helix; $\beta$-sheet; and turn/coil) since positive and negative charges could be compensated by divalent positive metal ions and divalent negative ions, respectively. It was found that [GADV]-proteins encoded by GNC code satisfied the four conditions when about equal amounts of [GADV]-amino acids were contained in the proteins (Fig. 5a) (Ikehara et al., 2002), but all four amino acids encoded by other four codons in rows and columns in the universal genetic code table did not satisfy at least one of the four conditions, except for the GNG code, a slightly modified form of the GNC code (Ikehara et al., 2002). The results of this search indicate that random synthesis in a pool of four [GADV]-amino acids could produce water-soluble globular proteins that are basically comparable in their potential to contemporary proteins as abilities for forming secondary and tertiary structures.

Table 2 indicates that [GADV]-amino acids have adequately favorable characteristics necessary to fold polypeptide chains composed of them into water-soluble globular proteins. That is, the four properties are allotted for four [GADV]-amino acids, as Gly, Ala, and Val are turn/coil-, $\alpha$-helix-, and $\beta$-sheet-forming amino acids, respectively, and Asp and Val are hydrophilic and

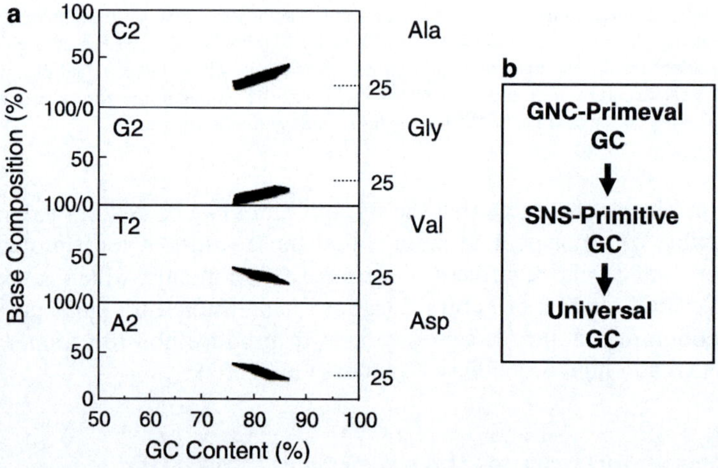


**Figure 5.** (**a**) Amino acid compositions of hypothetical [GADV]-proteins satisfying the four conditions for folding of a polypeptide chain into water-soluble globular structure. (**b**) The origin and evolutionary process of the genetic code (**GC**).

**Table 2.** Structural indices (hydropathy and *α*-helix, *β*-sheet, and turn/coil propensities) of [GADV]-amino acids used to calculate the formability of water-soluble globular proteins were obtained from Stryer's textbook of "Biochemistry" (Berg et al., 2002).

| | Gly [G] | Ala [A] | Asp [D] | Val [V] |
|---|---|---|---|---|
| Hydropathy | 1.0 | 1.6 | −9.2 | **2.6** |
| *α*-Helix | 0.56 | **1.29** | 1.04 | 0.91 |
| *β*-Sheet | 0.92 | 0.90 | 0.72 | **1.49** |
| Turn/coil | **1.64** | 0.78 | **1.41** | 0.47 |
| Acidic A.A. | | | (−) | |

Acidic amino acid is shown with minus (−) symbol. Bold numbers and underlined numbers indicate that the indices are in 5 from the top and from the bottom, respectively.

hydrophobic amino acids, respectively (Table 2). Therefore, it is considered that the amino acids were selected for primitive protein synthesis, based on unique characters of the amino acids from a pool of amino acids easily accumulated on the primitive earth. Here, we concluded that the genetic code originated from GNC code and evolved to the universal genetic code through SNS code (Fig. 5b).

2-Aminobutylic acid (2-ABA) having ethyl group on the side chain could be accumulated on the primitive earth similarly as [GADV]-amino acids, but it was not used as a natural amino acid since 2-ABA without a branched chain or bulky group as benzene ring on *β*-carbon atom is an α-helix-forming amino acid similarly as Ala, Leu, Met, and so on (Berg et al., 2002). Therefore, Ala, not 2-ABA, was used for primitive protein synthesis since Ala with methyl group has simpler structure than 2-ABA.

## 2.4. RECENT STUDIES ON THE ORIGIN OF THE GENETIC CODE

Recently, van der Gulik et al. (2009) have reported that the early functional peptides could be short (3–8 amino acids long) and were made of [GADV]-amino acids because they were abundantly produced in many prebiotic synthesis experiments and observed in meteorites, and the neutralization of Asp's negative charge is achieved by metal ions. Furthermore, they conjectured that the abundance of prebiotic [GADV]-oligopeptides is tightly connected with the appearance of the genetic code.

In parallel, Higgs and Pudritz (2009) have also obtained a similar conclusion, suggesting that the genetic code originated from [GADV]-amino acids based on the order of amino acid amounts synthesized by Miller's atmospheric discharge experiments and in realistic simulations of hydrothermal vents and those detected in meteorites.

Di Giulio (2008) also has reported a similar conclusion, suggesting that the first amino acids to evolve along the biosynthetic pathways are predominantly [GADV]-amino acids and Glu codified by codons of the type GNN, and this observation is found to be statistically significant, and therefore the close biosynthetic relationships between the sibling amino acids Ala-Ser, Ser-Gly, Asp-Glu, and Ala-Val are not random in the genetic code table, based on an extension of the coevolution theory that attributes a crucial role to the first amino acids.

Biro (2009) also have obtained a similar conclusion on the origin of the genetic code, based on his Proteomic Code theory, which determines how individual amino acids interact with each other during folding and in specific protein–protein interactions. He described in his book that the GNC-SNS primitive genetic code hypothesis, which has been proposed by us, is similar to the Proteomic Code and that both concepts agree with each other regarding.

As described above, van der Gulik, Higgs, di Giulio, and Biro have obtained a conclusion similar to the GNC-SNS primitive genetic code hypothesis, suggesting that the first proteins must be composed of only four kinds of [GADV]-amino acids, although they did not discuss on the origin of life. The facts that the same conclusion was obtained by several independent investigations indicate that the genetic code probably originated from the first genetic code, GNC, encoding four [GADV]-amino acids. The conclusion supports GADV hypothesis on the origin of life.

Ser is easily synthesized as similarly as Asp and Val in [GADV]-amino acids in Miller-type discharge experiments using primitive atmosphere, as shown in Table 1 of Higgs's paper (Higgs and Pudritz, 2009). But it is supposed that the amino acid was not used from the beginning of the emergence of life for natural protein synthesis through the genetic code, but the use of Ser in protein synthesis was delayed because Ser is much more unstable than Asp and Val under heat conditions on the primitive earth due to the existence of hydroxyl group on the side chain. Therefore, enough amount of the amino acid required for the primitive protein synthesis could not accumulate on the primitive earth (Vallentyne, 1964).

Similarly, nucleotides having many hydroxyl groups on ribose ring would not be accumulated on the primitive earth because of the instability under the conditions of primitive earth. From this reason too, RNA world must not be realized on the primitive earth since enough amounts of RNA, which is necessary for formation of the RNA world, could not be prebiotically synthesized. We therefore conclude that it is impossible to explain the emergence of life according to the RNA world hypothesis.

Some amino acids used for protein synthesis in extant organisms have been actually detected in meteorites. The kind and the ratio of amino acid detected in meteorites are rather similar to those of observed by Miller's atmospheric discharge experiments and in realistic simulations of hydrothermal vents (Higgs and Pudritz, 2009). This indicates that life was not brought directly into the earth by meteorites and/or comets, and that the amino acids, which led to the emergence of life, had been produced in exterior space as well as on the primitive earth.

## 3. [GADV]-Protein World Hypothesis

### 3.1. NEW CONCEPTS LEADING TO THE GADV HYPOTHESIS

It is generally considered that amino acid sequence, which is a primary structure of a protein encoded by one-dimensional genetic information on DNA or RNA, is always the most important when folding of a polypeptide chain into three-dimensional structure is discussed (Berg et al., 2002). Therefore, it has not been recognized by many researchers that amino acid composition is important for folding of a polypeptide chain. It is recognized that the amino acid composition is obtained simply after hydrolysis of polypeptide chain or protein. But unique amino acid composition containing about equal amounts of [GADV]-amino acids, which satisfy the four conditions (hydropathy, $\alpha$-helix, $\beta$-sheet, and turn formabilities) obtained from calculated values with amino acid compositions of extant proteins and structure indices of the corresponding amino acids, is important for effective production of functional proteins through random processes before formation of the first genetic information. The reasons are as follows. Since structure formability of one protein is the same as others randomly assembled in the same amino acid composition, every polypeptide chain randomly synthesized among amino acids in the amino acid composition could be folded into water-soluble globular structure similar to presently existing proteins. But the polypeptide chains should be folded into different, not the same, structures because the proteins have the same amino acid composition but different amino acid sequences from each other. We have named such a specific amino acid composition favorable for formation of water-soluble globular protein structure as protein 0th-order structure (Ikehara, 2009).

Amino acid composition of roughly equal amounts of [GADV]-amino acids is one of the protein 0th-order structure. This indicates that various kinds of water-soluble globular [GADV]-proteins could be created by random polymerization of [GADV]-amino acids even in the absence of any genetic function or before creation of the first gene because individual [GADV]-amino acids are functional units for protein structure formation, and the amino acid composition composed of roughly equal amounts of [GADV]-amino acids satisfies the four conditions for formation of water-soluble globular proteins (Ikehara et al., 2002). The new second concept which we call as "pseudo-replication of [GADV]-proteins" came from the consideration on the first concept of the protein 0th-order structure (Ikehara, 2009).

We reached to the GADV hypothesis on the origin of life based on the above two new concepts, "[GADV]-protein 0th-order structure" and "pseudo-replication of [GADV]-proteins." The GADV hypothesis suggests that life originated from [GADV]-protein world, which was formed by pseudo-replication of [GADV]-proteins in a pool composed of roughly equal amounts of [GADV]-amino acids (Ikehara, 2009). Another important point for solving the riddle on the origin of life was that even four kinds of amino acids were enough for formation of water-soluble globular proteins, which is necessary as a premise to produce functional enzymes. This could produce weak but significantly high catalytic activities on surfaces of [GADV]-proteins, when the activities were compared with the level without any proteinous enzyme on the primitive earth. The catalytic functions of the proteins would open a new gate for production of functional proteins in [GADV]-protein world before appearance of genetic information. Oligonucleotides and RNA synthesis carried out in the [GADV]-protein world led the formation of next RNA-[GADV]-protein world, in which the first life was created on the primitive earth.

## 3.2. POSSIBLE STEPS TO THE EMERGENCE OF LIFE

Possible evolutionary processes to the emergence of life are described below (Ikehara, 2002, 2005; Oba et al., 2005). Enough amounts of [GADV]-amino acids were synthesized and accumulated on the primitive earth. Successively, [GADV]-proteins were produced, for example, by repeated heat-drying processes of [GADV]-amino acids in tide pools on the primitive earth (Fig. 6a). [GADV]-proteins were further accumulated by pseudo-replication to form [GADV]-protein world (Figs. 6b and 7a). Nucleotides and oligonucleotides as well as [GADV]-amino acids could be synthesized through enough catalytic activities of [GADV]-proteins in the protein world (Fig. 6b). In these cases too, places such as tide pools, where chemical compounds could be concentrated by heat-drying process, were necessary for the syntheses. In addition, divalent metal cations would be used to compensate negative charges on [GADV]-proteins and to accelerate association of [GADV]-protein with nucleotide for the oligonucleotide and RNA syntheses. Judging from

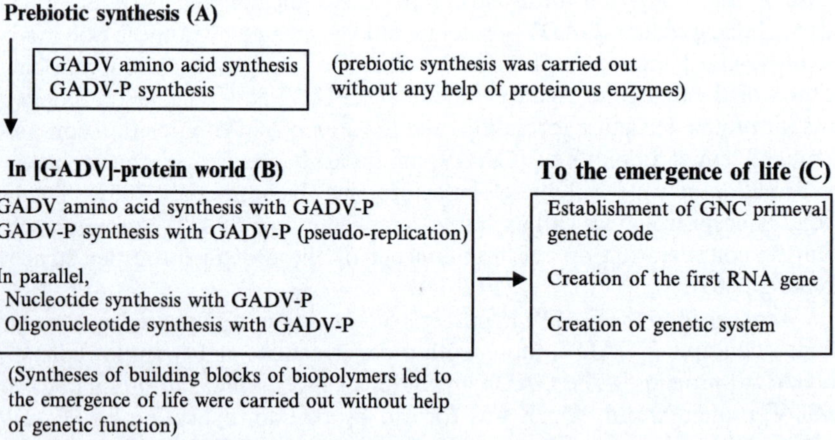


**Figure 6.** Possible steps from prebiotic synthesis (**a**) to the emergence of life (**c**). Roles of [GADV]-P ([GADV]-peptides and/or [GADV]-proteins) world (**b**) in the emergence of life.

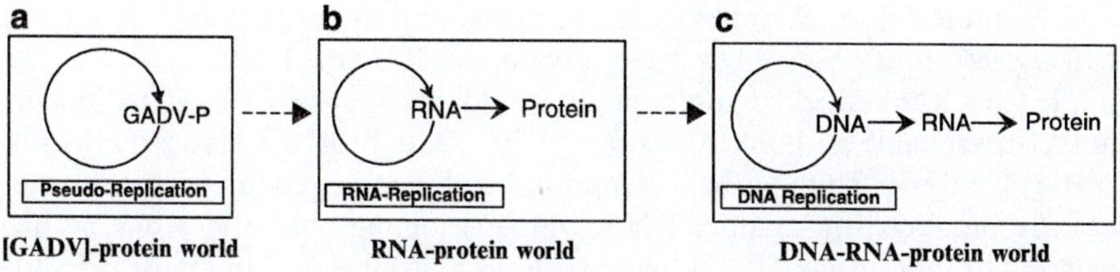


**Figure 7.** The modern fundamental life system (**c**), which was composed of DNA, RNA, and protein, evolved from [GADV]-protein world (**a**), which was formed by pseudo-replication of [GADV]-proteins, through RNA-protein world (**b**), which was established by RNA replication with proteinous enzymes.

the fact that [GADV]-amino acids are encoded by GNC primeval genetic code and that nucleotides are favorable for base pair formation and for transmittance of genetic information from parents to progenies, it is considered that the accumulation of GNC-containing oligonucleotides triggered establishment of the first genetic code or GNC primeval genetic code through stereospecific interaction among four [GADV]-amino acids and the corresponding GNC-containing oligonucleotides. The GNC-containing oligonucleotides or proto-tRNAs might be charged with the corresponding [GADV]-amino acids, and the amino acids were delivered to the peptide bond formation with a [GADV]-protein or a proto-rRNA ribozyme. The proto-tRNA could play roles of both extant tRNA and mRNA by binding with each other through base pair formation between two complementary proto-tRNAs. More efficient [GADV]-protein synthesis with the proto-tRNA-[GADV]-amino acid complexes than direct [GADV]-protein synthesis among individual [GADV]-amino acids could assist to establish the GNC primeval genetic code (Fig. 6c). In parallel, translation system could be evolved through formation of the peptide bonds with complexes of proto-rRNA molecules and [GADV]-proteins to form primitive ribosomal small and large subunits. Next, GNC-repeating sequences were produced by random phosphodiester bond formation among GNC codons or anticodons in the complexes of GNC-containing oligonucleotides and [GADV]-amino acids. Thus, it is assumed that the first single-stranded $(GNC)_n$ gene was created, when one $(GNC)_n$ sequence encoding a [GADV]-protein with a required function was selected out from a pool of versatile $(GNC)_n$ polynucleotides, leading to the emergence of the first life. At this point, [GADV]-protein world ceased a role and transformed to RNA-protein world, which was formed by RNA replication with [GADV]-protein enzymes (Fig. 7b).

On the way of the evolutionary process of the life system, genetic functions on RNA strands were transferred to base sequences on DNA strands, which were produced by DNA replication (Fig. 7c). Even a reasonable process, how the "chicken and egg relationship" between genes and proteins was formed on the primitive earth, can be explained from a standpoint of the GADV hypothesis. That is, it could be established as going up from the lower (pseudo-replication of

[GADV]-proteins) to the upper stream (creation of double-stranded DNA genes) of the genetic flow observed in extant organisms (Ikehara, 2009).

In the RNA world hypothesis, it would be impossible even to imagine another reasonable and concrete strategy for creation of the first gene because genetic function with unique base compositions at the three codon positions would never be produced by random RNA synthesis joining nucleotides one by one. Contrary to that, in the GADV hypothesis, establishment of the most primitive code or GNC primeval genetic code gave a clue for creation of the first genetic function and to proposition of GADV hypothesis on the origin of life (Fig. 6).

## 4. Summary and Conclusions

RNA world hypothesis was proposed only for solving the "chicken–egg dilemma" between genes and proteins without consideration of formation processes of the fundamental life system, especially of the origin of the genetic code, which should be the most important point when the origin of life is considered. Therefore, in the RNA world hypothesis, it would be impossible even to imagine another reasonable and concrete strategy for creation of the first gene, genetic function with unique base compositions at three codon positions. Genetic sequences would be never produced by RNA synthesis joining nucleotides one by one because the genetic sequences are not simple nucleotide sequences but are codon sequences. Of course, I believe that the formation processes of "chicken–egg dilemma" between genes and proteins have been explained by GADV hypothesis more rationally than the RNA world hypothesis, as described above. What I would like to emphasize here is that the RNA world, which was formed by RNA self-replication, never existed on the evolutionary way from era of the chemical evolution to the emergence of life.

## 5. References

Berg JM, Tymoczko JL, Stryer L (2002) Biochemistry, 5th edn. W. H. Freeman and Company, New York

Biro JC (2009) The Proteomic Code. Homulus Foundation, Los Angeles, Chap. 2

Di Giulio M (2008) An extension of the coevolution theory of the origin of the genetic code. Biol Direct 3:37–57

Gilbert W (1986) The RNA world. Nature 319:618

Higgs PG, Pudritz RE (2009) A thermodynamic basis for prebiotic amino acid synthesis and the nature of the first genetic code. Astrobiology 5:483–490

Ikehara K (2002) Origins of gene, genetic code, protein and life: comprehensive view of life system from a GNC-SNS primitive genetic code hypothesis. J Biosci 27:165–186

Ikehara K (2005) Possible steps to the emergence of life: the [GADV]-protein world hypothesis. Chem Rec 5:107–118

Ikehara K (2009) Pseudo-replication of [GADV]-proteins and origin of life. Int J Mol Sci 10:1527–1537

Ikehara K, Amada F, Yoshida S, Mikata Y, Tanaka A (1996) A possible origin of newly-born bacterial genes: significance of GC-rich nonstop frame on antisense strand. Nucleic Acids Res 24:4249–4255

Ikehara K, Omori Y, Arai R, Hirose A (2002) A novel theory on the origin of the genetic code. A GNC-SNS hypothesis. J Mol Evol 54:530–538

Oba T, Fukushima J, Maruyama M, Iwamoto R, Ikehara K (2005) Catalytic activities of [GADV]-peptides. Orig Life Evol Biosph 35:447–460

Vallentyne JR (1964) Biogeochemistry of organic matter II. Thermal reaction kinetics and transformation products of amino compounds. Geochim Cosmochim Acta 28:157–188

van der Gulik P, Massar S, Gilis D, Buhrman H, Rooman M (2009) The first peptides: the evolutionary transition between prebiotic amino acids and early proteins. J Theor Biol 261:531–539

Biodata of **Kunio Kawamura**, author of "***Reality of the Emergence of Life-like Systems from Simple Prebiotic Polymers on Primitive Earth.***"

**Professor Kunio Kawamura** is currently the assistant professor of Department of Applied Chemistry in Osaka Prefecture University, Japan. He obtained his Ph.D. from Tohoku University in 1988 on the kinetic investigations of water-soluble porphyrins and the applications for trace metal analyses. He continued his studies at Tohoku University for inorganic synthesis and coal oxidation under hydrothermal conditions. His interests moved to the origin of life problem from applied chemistry and joined with Professor James P. Ferris at Rensselaer Polytechnic Institute in 1992–1995 as postdoctoral fellowship. He moved to Osaka Prefecture University in 1995. Professor Kawamura's scientific interests are in the areas of analytical tools for hydrothermal reactions, hydrothermal reactions of peptides, chemical evolution of RNA, primitive evolutionary molecular system, industrial applications of his hydrothermal systems, green analytical techniques, and unified theory for chemical evolution–biological evolution–human society.

E-mail: **kawamura@shudo-u.ac.jp**

J. Seckbach (ed.), *Genesis - In The Beginning: Precursors of Life, Chemical Models and Early Biological Evolution*, Cellular Origin, Life in Extreme Habitats and Astrobiology 22, 123–144
DOI 10.1007/978-94-007-2941-4_7, 

# REALITY OF THE EMERGENCE OF LIFE-LIKE SYSTEMS FROM SIMPLE PREBIOTIC POLYMERS ON PRIMITIVE EARTH

**KUNIO KAWAMURA**
*Department of Human Environmental Studies, Hiroshima Shudo University, 1-1-1, Ozuka-higashi, Asaminami-ku, Hiroshima 731-3195, Japan*

## 1. Introduction

Life would have emerged between 4.6 and 3.8 billion years ago (Mojzsis et al., 1996, 2001). The environments during the emergence of life would be quite different from the present Earth, where strong cosmic rays, ultraviolet light, electric discharges, meteorite impacts, and high-$CO_2$ atmosphere were present (Maher and Stevenson, 1988; Sleep et al., 1989; Bada et al., 1994). Water is considered to have been essential for the emergence of life, and the primitive ocean was considered to be present before 4 billion years ago (Frey, 1977). However, the temperature of ocean would be much higher than the present (Walker, 1985; Karhu and Epstein, 1986; Kasting and Ackerman, 1986). Nevertheless, the extreme environments would have been suitable for chemical evolution, where strong energy sources have resulted in chemical evolutions. On the other hand, the progress of molecular biology led to the RNA world hypothesis (Gilbert, 1986). In addition, the discovery of hyperthermophilic organisms led to the hydrothermal origin-of-life hypothesis (Pace, 1991) although this hypothesis has been widely disputed (Forterre, 1996). In a typical scenario of the origin of life, self-replication system has formed from the mixture of prebiotic molecules (prebiotic soup) and finally resulted a life-like system. However, how did a prebiotic soup including self-replicating materials, such as RNA molecules, evolve to a real life? Although we have gained knowledge about the functions of prebiotic polymers, the process from the biopolymers possessing self-replication and enzyme-like activities to a real life is still unclear. If we try to approach to this problem, it would be necessary to evaluate the origin-of-life hypotheses from a variety of viewpoints. In particular, the evaluation of candidate molecular systems would be important from the viewpoint "what are the essential characteristics of life" (Fig. 1).

In this chapter, prior to the evaluation of life-like systems from the viewpoint of the essential characteristics of life, RNA-based and protein-based life-like systems will be evaluated from the viewpoint of hydrothermal origin-of-life hypothesis (Kawamura, 2004). The evaluation suggests the importance of the stability of systems including the stability of biopolymers. In addition, the essential characteristics of life have been studied to clarify the unified theory of chemical evolution – biological evolution – civilization, the relationship between individual organisms

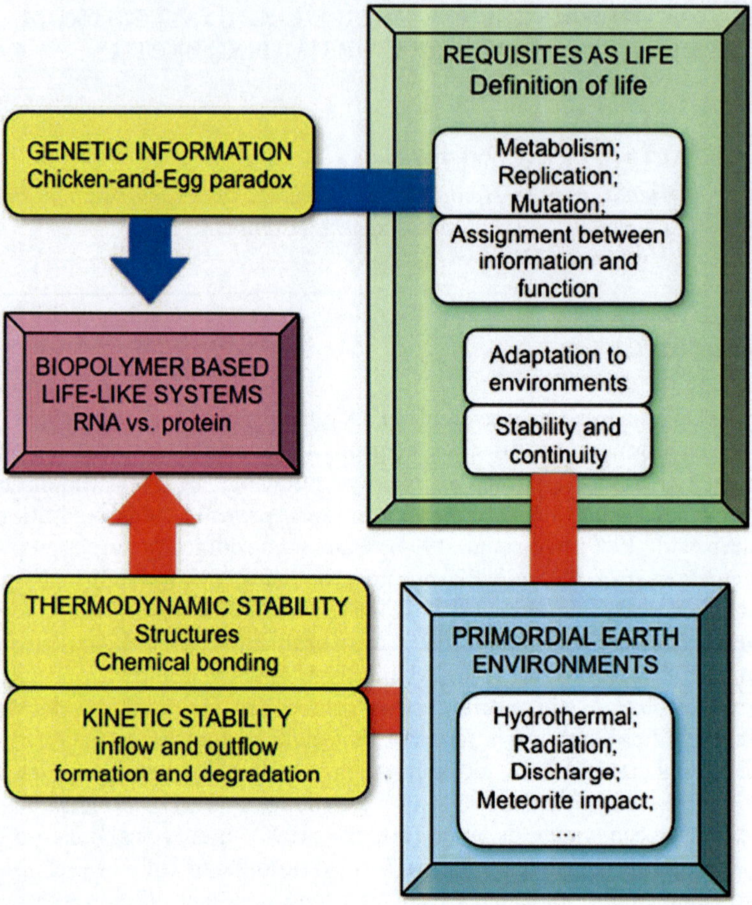


**Figure 1.** Evaluation strategies for the primitive life-like systems.

and their environments, and the analogies among life-like systems at different hierarchical levels (Kawamura, 2002a, 2003a, 2007). These studies led to a proposal on the realistic life-like systems from the mixture of biopolymers.

## 2. Chemical Evolution Models from Small Molecules to Biopolymers

### 2.1. PREBIOTIC LIFE-LIKE SYSTEM CONSISTING OF RNA MOLECULES

The RNA world hypothesis was first proposed in 1986 (Gilbert, 1986), while pioneer works had appeared many years ago (Eigen, 1971; Orgel and Crick, 1993). The RNA world hypothesis is now considered as the one most important

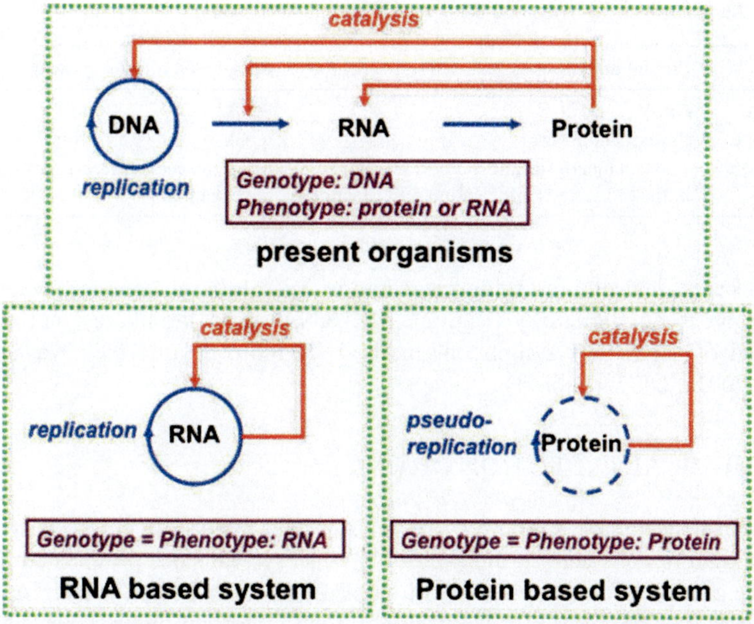


**Figure 2.** The relationships between genotype and phenotype molecules in different life-like systems. *Top*: present organisms, *bottom left*: possible RNA-based life-like system, *bottom right*: possible protein-based life-like system. *Dashed circle* indicates the pseudoreplication by GADV proteins.

concerning the origin-of-life hypotheses (Cech et al., 2005). This is mainly due to the potential of this hypothesis to solve the "egg and chicken problem" regarding the relationship between genetic information and protein functions. The mechanism of the information flow from DNA to proteins is very complicated (Fig. 2, top). Thus, the emergence of the information flow in primitive life-like systems has been the most important issue on the origin-of-life problem.

The discovery of ribozyme (Kruger et al., 1982) suggests that the RNA or RNA-like molecules play important roles for the emergence of life on primitive Earth, where RNA molecules had preserved both information and enzymatic activities instead of DNA and proteins. Thus, the information flow in the RNA-based life-like system can be illustrated (Fig. 2, bottom left). According to my previous work, the relationship between the genetic information and protein functions (genotype and phenotype) was generalized as the relationship between information and function of life-like systems (Kawamura, 2002a, 2003a, 2007) so that the RNA-based life-like system can be considered as a system, of which the RNA molecules preserve both information and function.

Here, the importance of the principle regarding the assignment of information and function of life-like systems should be pointed out as well as other requisite for life, that is, metabolism, replication, and mutation (Kuppers, 1985). For the cell-type system, a whole set of molecular machinery for the assignment method is preserved in cells. In a possible RNA-based life-like system, RNA molecules act

**Table 1.** Assignment of genotype and phenotype in the life-like systems.

| | **Present organisms** | **RNA-based life-like systems** |
|---|---|---|
| Genotype | DNA | RNA |
| Phenotype | Protein and RNA | RNA |
| Assignment | The machinery for information transfer in the cell assigns information and function | RNA molecules preserve information and function |

as molecular devices preserving information by the base sequence, where the translation is not necessary (Fig. 2). The RNA-based life-like system can be regarded of a simplest assignment method (Nemoto et al., 1997; Kawamura, 2002a, 2003a, 2007) (Table 1).

## 2.2. CHEMICAL EVOLUTION OF RNA

Several evidences have supported the RNA world hypothesis. First, it is much simpler than to consider a protein-based life-like system since proteins could not preserve genetic information in the modern organisms. Second, the RNA world hypothesis has been supported by continuous investigations of the chemical evolution of RNA and *in vitro* selection studies of functional RNA molecules. The plausible pathways are shown in Fig. 3. These investigations support that these molecules could have evolved under primitive Earth conditions. Besides, the origin of chirality remains unclear.

Phosphorimidazolide of 5′-nucleoside, which could be formed under primitive conditions (Lohrmann and Orgel, 1973), has been used as prebiotic-activated nucleotide monomers. There have been successful studies of the formation of oligonucleotides in the presence of template polynucleotides without enzymes (Inoue and Orgel, 1982, 1983). Spontaneous formation of oligonucleotides occurs in the presence of metal ions (ME reaction) (Sawai, 1976; Sawai et al., 1981) or clay minerals (CL reaction) (Ferris and Ertem, 1992; Kawamura and Ferris, 1994). By these reactions, 10-mer and higher oligonucleotides form by one pot reactions. Oligonucleotide with 50 nucleotide units forms in the presence of clay mineral under the continuous supply of the activated nucleotide (Ferris et al., 1996). Besides, the formation of oligonucleotides with 40 nucleotide units in the presence of a polynucleotide template has been studied as a primitive RNA polymerase model (Inoue and Orgel, 1982, 1983; Joyce et al., 1987), which could compose a replication system of RNA (template-directed formation of RNA (TD reaction)). It has been shown that the mixtures of activated nucleotide monomers indeed form oligonucleotides involving different bases, which support that a partial replication occurs without enzymes. Furthermore, TD reaction of oligoguanylate proceeds in the presence of oligocytidylate template, which is formed by the spontaneous formation in the presence of clay mineral (Ertem and Ferris, 1996). Thus, the TD reaction is considered as a candidate replication system.

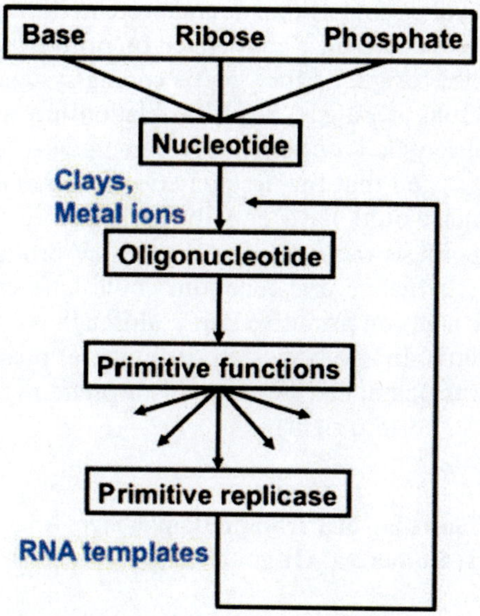


**Figure 3.** Pathways for the formation of a replication system of RNA.

## 2.3. *IN VITRO* SELECTION OF FUNCTIONAL RNA

The principle of *in vitro* selection techniques for RNA has been developed by two independent groups (Tuerk and Gold, 1990; Ellington and Szostak, 1990). These techniques consist of the formation of a random pool of RNA, amplification of RNA, and selection of RNA. Thus, it is analogical to the evolution of organisms as the natural selection, replication, and mutation of organism are involved. By these methods, several kinds of functional RNA molecules have been created (Wilson and Szostak, 1999; Gold et al., 1999; Szostak et al., 2001). If similar prebiotic mechanisms have existed on primitive Earth, Darwinian evolution of RNA molecules could have proceeded. Thus, the presence of functional RNA and this speculation strongly support the RNA world hypothesis, although the primitive machinery of replication and mutation should have been essential.

## 2.4. PROTEIN-BASED LIFE-LIKE SYSTEM

It seems to be difficult that the protein-based life-like system could have preserved genetic information while proteins would be essential for metabolism and enzymatic activities. Naturally, it would not be necessary to assume that nucleotides or proteins should have evolved independently since nucleic acids and proteins cooperatively work in modern organisms. A necessary process for the origin-of-life study is to identify the roles of nucleic acids and proteins during chemical evolution and how these molecules could have cooperated on primitive Earth.

Recently, a possible origin of the genetic coding system has been proposed on the basis of continuous investigations of the relationship between the structures of proteins and the nucleotide-base contents within modern genes (Ikehara, 2005). This analysis has suggested that the first genetic coding system generated from a simple system which could have coded only glycine, alanine, valine, and aspartic acid; the hypothesis was named as the GADV protein world hypothesis. The hypothesis suggests that several functions could have emerged from merely G, A, D, and V amino acids on primitive Earth, although is not yet experimentally verified (Oba et al., 2005). In addition, the possibility for preserving genetic information by proteins was postulated by assuming a pseudoreplication mechanism (Ikehara, 2009) (Fig. 2, bottom right).

## 3. Life-like System Consisting of Prebiotic Biopolymers from the Viewpoint of Hydrothermal Origin-of-Life Hypothesis

### 3.1. HYDROTHERMAL ORIGIN-OF-LIFE HYPOTHESIS

Estimation of the temperatures under the primitive Earth environments would be important to evaluate the emergence of a life-like system. The meteorite impacts could have eventually risen up the Earth's temperature (Maher and Stevenson, 1988; Sleep et al., 1989; Bada et al., 1994) around 3,800 million years ago. On the contrary, some evidences suggest that the primitive ocean was frozen since the solar luminosity at that time was relatively less than the present luminosity (Sagan and Mullen, 1972; Newman and Rood, 1977; Gough, 1981). Thus, the temperature of the primitive ocean in which life originated cannot be readily justified (Walker, 1985; Karhu and Epstein, 1986; Kasting and Ackerman, 1986).

Despite that, the hydrothermal origin-of-life hypothesis was proposed based on continuous studies on thermophilic organisms (Corliss et al., 1981; Baross and Hoffman, 1985; Nisbet, 1986) and the phylogenetic analysis of the present generation of organisms. The last common ancestor (LCA) is considered as a thermophilic organism (Pace, 1991), while its nature is still disputed (Miller and Bada, 1988; Forterre, 1996; Galtier et al., 1999; Bada and Lazcano, 2002). Thus, the formations of prebiotic monomers and polymers have been verified under the simulated hydrothermal vent conditions in our group.

### 3.2. STABILITY OF BIOPOLYMERS

Few systematic studies regarding the chemical evolution of biopolymers regarding the prebiotic formation, the stabilities, prebiotic functions, and three-dimensional structures were conducted on the chemical evolution of nucleic acids under hydrothermal conditions (White, 1984; Larralde et al., 1995). This was due to the lack of practical experimental techniques to investigate hydrothermal reactions. Our group has been studying new methods for monitoring hydrothermal reactions (Kawamura, 1998, 2000, 2002b). The system consists of a high-pressure pump, a sample injector, a hydrothermal reactor, a quenching unit, a back-pressure regulator, and a sampling port. The hydrothermal heater involves a narrow capillary of different materials, such as stainless-steel tubing and fused-silica capillary with 0.015–0.25 mm inner diameter and 5–200 cm length. This method enables monitoring hydrothermal reactions within 2 ms–200 s at temperatures up to 400°C at 30 MPa. The samples injected through the sample injector can be withdrawn at the sampling port and then analyzed by conventional analytical techniques, such as HPLC and GC. In addition, this system enables in situ monitoring the hydrothermal reactions within millisecond to second time scale at temperatures up to 400°C using an optical device, which is attached on the window of the fused-silica capillary tubing. By using our techniques, the investigations have been carried out regarding the stability of monomers and polymers, the formation of biopolymers, and the three-dimensional structures of biopolymers in detail under the hydrothermal environments. While the fastest process for the degradation of amino acids is racemization (Kawamura and Yukioka, 2001) and that of nucleic acids is cleavage of phosphodiester bond, the cleavage of phosphodiester bonding is approximately 10,000 times faster than that of racemization of amino acids. Moreover, the cleavage of phosphodiester bond of RNA is approximately 100 times weaker than that of peptide bonding (Kawamura, 2003b, c; Kawamura et al., 2005). Furthermore, the ribonuclease loses the catalytic activity within 30 s at 275°C (Kawamura et al., 2009). Although proteins and amino acids are much stable as compared to RNA and nucleotide monomers, it should be noted that these reactions proceed within much shorter time scale than the geological time scale. In the modern organisms, DNA is used as a universal informational device due to the greater stability of DNA than RNA. However, the half-life of DNA molecules is also much shorter than the geological time scale. According to these discussions, the time scales of degradation of RNA, DNA, and proteins are much shorter than the geological time scale. Such direct comparison of the degradation rate of these molecules with the geological time scale would deduce the implication that biologically important molecules would not possess sufficient stabilities to make life. This paradox can be solved if a kinetic model is considered where the formation and degradation (or inflow and outflow) of biomolecules are balanced as discussed in a later section.

## 3.3. FORMATION OF BIOPOLYMERS

The prebiotic formation of oligonucleotides could have occurred from activated monomers, which are considered on primitive Earth (Lohrmann and Orgel, 1973; Lohrmann, 1977). The activated nucleotide monomers have been used for the formation of oligonucleotides with and without a polynucleotide template in the presence of metal ions and clay catalysts. From the viewpoint of hydrothermal origin-of-life hypothesis, we have carried out kinetic analyses of prebiotic formation models of RNA using the activated nucleotide monomers or water-soluble carbodiimide as a condensation reagent at temperatures up to 100°C. The following models have been successfully analyzed regarding (1) the template-directed formation of oligoguanylate on a polycytidylate template (TD reaction) (Kawamura and Umehara, 2001), (2) the cyclization of oligonucleotides (CY reaction) (Kawamura et al., 2003), (3) the oligocytidylate formation in the presence of $Pb^{2+}$ (ME reaction) (Kawamura and Maeda, 2007), and (4) the oligocytidylate formation in the presence of montmorillonite clay (CL reaction) (Kawamura and Maeda, 2008).

Here, it is noted that the accumulation of oligonucleotides is determined by the relative magnitude of the formation and degradation processes. The kinetic analyses of the four types of RNA formation models suggested that the low efficiency of oligonucleotide formation at high temperatures is mainly due to the weak association between an activated nucleotide monomer and an elongating oligonucleotide at high temperatures since hydrogen bonding and hydrophobic interaction decrease with increasing temperature. This trend was observed for all the four types of prebiotic reactions. For the cases of TD, ME, and CL reactions, it is generally found that the association between an activated monomer and a monomer (or another activated monomer) for the formation of 2-mer becomes weak and the relative rate of the formation of 2-mer decreases notably as compared to 3-mer and 4-mer formations. According to these data, it was conclusively implied that the oligonucleotides could have formed at high temperatures if the association between the activated nucleotide monomer and the elongation oligonucleotide is facilitated by additives, such as protein-like molecules, mineral surfaces, and metal ions.

On the other hand, the formation of protein-like molecules is also possible under the hydrothermal conditions in the absence of condensation reagent (Imai et al., 1999), while the efficiency is lower than that of the dry model. Actually, the formation of proteins from amino acids under hydrothermal conditions is not so easy, where the yield of oligopeptides formation is typically 0.1–1%. One reason is that the dehydration of amino acids is generally difficult in aqueous solution. By using the hydrothermal flow reactor, we have discovered the oligoalanine formation within 10–30 s at 250–330°C using 4-mer oligoalanine and longer (Kawamura et al., 2005) with 10% yield and the one-step formation of oligopeptides including 20 amino acid units from Asp and Glu within 3 min at 275°C (Kawamura and Shimahashi, 2008). However, the oligopeptides could not have survived for a long time under hydrothermal conditions because they are thermodynamically unstable. Thus, it has been frequently assumed that oligopeptides could have accumulated in the surrounding cool ocean once the peptides are evacuated from the hydrothermal vent (Imai et al., 1999).

## 3.4. INTERACTIONS OF BIOPOLYMERS UNDER HYDROTHERMAL CONDITIONS

According to our recent investigations regarding the interactions of biopolymers under hydrothermal conditions, double-stranded DNA becomes single-stranded DNA at temperatures below 100°C and single-stranded DNA becomes insoluble in aqueous medium at temperatures higher than 100°C. At above 200°C, DNA is decomposed to short oligonucleotides or monomeric nucleotides (Kawamura and Nagayoshi, 2007). On the other hand, the stability of proteins and the interaction between proteins with chromogenic reagents have been investigated. It was concluded that (1) the solubility of proteins likely decreases with increasing temperature and (2) the association of a protein with chromogenic reagent becomes weak mainly due to the conformational change of the protein (Kawamura et al., 2010).

## 3.5. POSSIBILITY OF LIFE-LIKE SYSTEMS UNDER HYDROTHERMAL CONDITION

Conclusively, prebiotic biopolymers could have kinetically accumulated under hydrothermal conditions, as shown in Fig. 4. (Kawamura, 2004, 2009). In addition, although biological interactions become weak with increasing temperature, it seems not effective as compared to the increase of degradation rates of biomolecules. Furthermore, the solubility of biomolecules should be taken into account while this has not been focused for the evaluation of the hydrothermal origin-of-life hypothesis.

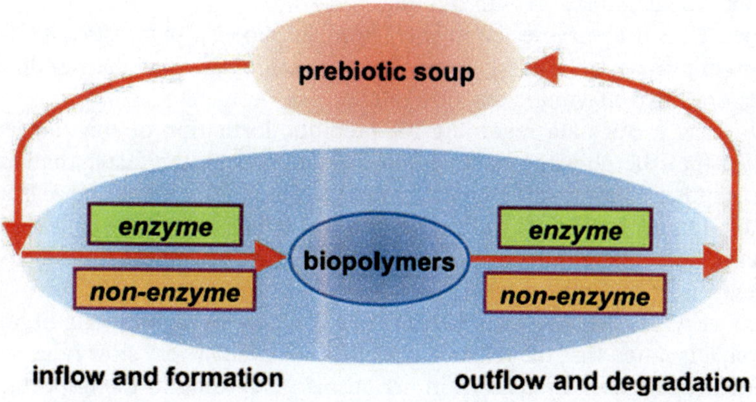


**Figure 4.** Accumulation of biopolymers determined by formation + inflow and degradation + outflow on the primitive Earth.

The RNA world hypothesis and the GADV protein world hypothesis have been independently constructed from different evidences. From the viewpoint of the assignment between information and function, RNA-based life-like system would possess an advantage since RNA preserves both informational and catalytic functions, while proteins are not capable to preserve information in modern organisms. However, if these hypotheses were evaluated from the viewpoint of hydrothermal conditions, protein-based life-like system would be suitable rather than the RNA-based system because of the stability of proteins. However, it is hardly determined which of nucleotides or proteins are more advantageous under such extreme conditions since (1) the stabilities of these biopolymers are much shorter than the geological time scale and (2) the temperature effect for weak interactions would act for both RNA and proteins. Thus, it is reasonable to assume such corporative cochemical evolution on primitive Earth. However, there are few investigations on the cooperative chemical evolution of RNA and proteins. Thus, we have investigated the TD reaction of RNA and the stability of RNA in the presence of protein-like molecules. It was found that the influence of protein-like molecules is very limited, and we found merely weak activities of protein-like molecules for the formation and degradation of RNA (Kawamura et al., 2004). On contrary, the formation of oligonucleotides under hydrothermal conditions is not yet studied since the moieties of nucleotides are less stable than amino acids, peptides, and proteins.

The viewpoints to determine whether biopolymers are stable or not have been discussed in the previous investigations (Kawamura, 2004). The following two views were applied to evaluate the possibility of the accumulation of biopolymers.

View I: The accumulation of prebiotic biopolymers should be evaluated from the viewpoint of kinetics of the accumulation of prebiotic polymers (Fig. 4). The accumulation of biopolymers in a cell is determined by the formation with inflow and the degradation with outflow.

View II: Since enzymes control reactions in modern organisms, the rate of reactions in primitive life-like systems should be evaluated from the standpoint of possible primitive enzymatic reaction rates.

On view I, our data regarding the prebiotic formation of oligonucleotides suggested that the phosphodiester bond formation could be faster than that of the decomposition even at high temperatures if potential prebiotic catalysts, such as protein-like molecules, clay minerals, and metal ions, could have facilitated the association of the monomer and the elongating oligomers for RNA and peptides.

On view II, enzymes control biological reactions in modern organisms. However, it is noted that the reactions can proceed even at very slow rates without enzymes as background reactions in organisms. According to the importance of this principle (Radzicka and Wolfenden, 1995; Kawamura, 2004), we have deduced that this principle would provide a temperature limit for the primitive

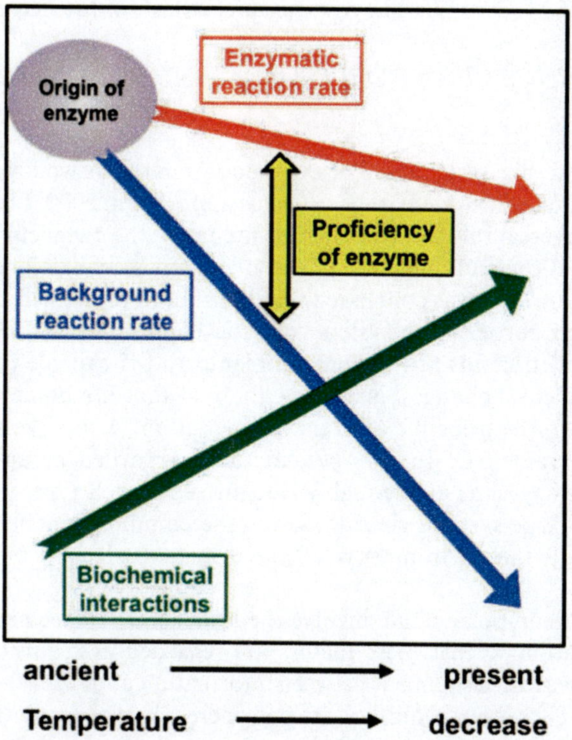


**Figure 5.** The limit temperatures for the primitive enzymes deduced from the comparison between formation and degradation rates.

life-like system, where primitive enzymes could have facilitated the target reactions with faster rates than the background reactions (Fig. 5). Even a small difference between the rates with enzyme-like molecule and background rates could be considered as candidate primitive enzyme activity. The large difference between the enzymatic rates and the background rates even at very high temperatures might reflect that the primitive enzymatic activities would have emerged at high temperatures. Besides, it is well known that weak interactions for biomolecules, such as hydrogen bonding, become strong with decreasing of temperature. The fact that the strength of biologically important interactions becomes weak can be illustrated in Fig. 5. These investigations imply that the chemical evolution of primitive enzymes would have been synchronized with the decrease of temperature of primitive Earth. Furthermore, the evaluation from the viewpoint of hydrothermal origin-of-life hypothesis emphasizes the importance of stabilities of biopolymers. The importance of the stability should be applied to the evaluation of a life-like system to determine if the system could have survived under primitive Earth environments.

## 4. Evaluation of Modern Life-like Systems on the Basis of the Requisites of Life

### 4.1. IMPORTANCE OF THE DEFINITION AND REQUISITES FOR LIFE

Now, modern life-like systems will be evaluated from the viewpoints of the essential characteristics of a system to be called as a life. From 2000, I have attempted to deduce a universal rule, which would be hidden in the chemical evolution, the biological evolution, and the development of human's civilizations (Kawamura, 2002a, 2003a, 2007). It was concluded that life should be defined on the basis of the relationship between the system and the environments. Life possesses a nature to adapt itself toward the environments and is capable to survive even under environmental changes. Naturally, the fact that life holds such nature is not in conflict to the principle of Darwin's evolution. A life-like system should satisfy this characteristic for the system to be regarded as alive. While the modern life-like systems are probably categorized as cell-type, virus-type, and possibly viroid-type systems since these use the common genetic materials, the activities toward the environments of these systems depend on the types of the systems.

Many of definitions of life involve the importance of metabolism, replication, and mutation so that these factors are regarded as requisites for life. As mentioned above, the assignment between information and function can be added as a requisite. These four requisites are considered as the elemental activities of life-like systems. On the other hand, if one focuses the functions and/or behavior of life-like system as a whole system, different requisites can be proposed to define life as shown in Table 2. Thus, further requisites have been discussed from this viewpoint. First, the relationship between the living system and environments is important. To adapt the life-like system to the environments, the stability of life would be essential. Second, present organisms possess the different hierarchical levels, such as individual organisms, species, and ecosystem. These are applied as requisites for the evaluation of life-like systems.

**Table 2.** Characteristics of present organisms and primitive life-like systems.

| Requisites | Present life-like systems | Primitive life-like systems |
|---|---|---|
| Boundary | Cell or virus | Compartment or particle |
| Metabolism | Mainly by proteins | Ribozymes of primitive enzymes |
| Replication and mutation | DNA-RNA-protein system | RNA replication or pseudoreplication by proteins |
| Environments | Adapt itself to the environments | Weak adaptation |
| Species | All of organisms have species | Replicated copy, pseudospecies |
| Ecosystem | Multiple hierarchical levels | Limited hierarchical levels |

## 4.2. THE STABILITY OF LIFE-LIKE SYSTEMS

For a living system, the stability and continuity of the system is essential. Different types of life-like systems would have different stabilities. First, the cell-type systems keep the molecules inside of a micrometer-sized bag, and this method would enhance the stability of the molecules. The stability would involve both static and kinetic stabilities. Static stability of a system is caused from the inherent stability of the biomolecules, of which some of molecules would possess longer lifetime than the lifetime of cell. The kinetic stability indicates that the population of molecule is dynamically controlled though the formation and decomposition and the inflow and outflow.

Besides, viruses are less stable as compared to the cell-type organisms. Most of viruses possess RNA (or DNA) molecules covered by proteins, and viroids are circular RNA molecules, where the proteins could play a role in preventing degradation of virus gene. In addition, for the present viruses and viroids, the machinery for amplification of virus or viroids is stored in host cells. Thus, viruses can keep the stability and continuity with the action of host cells. This reflects the difference of the stabilities among these life-like systems. The rapid evolution of viruses as compared to cell-type organisms might reflect a trend that viruses are less stable than cell-type systems, where the rapid life cycle is necessary on the basis of the principle regarding the formation and decomposition mentioned above. It would be generally true that the compartment or integration of biopolymers is important for the life-like systems from the viewpoint of the stability and continuity of the system.

## 4.3. EVALUATION OF POSSIBLE PRIMITIVE LIFE-LIKE SYSTEMS ON THE BASIS OF THE STABILITY

The RNA world hypothesis has given us a strong impression that the origin of life can be solved by this hypothesis. However, the details of the first living system based on RNA molecules are not sufficiently understood whether the first system consisted of cell, what kinds of functions of RNA were evolved, and how the first system evolved to the last common ancestor. Naturally, the same viewpoint is applied for the protein-based life-like system so that the essential characteristics or requisites are needed to apply to these life-like systems.

A primitive life-like system would have hardly adapted to small environmental changes. However, if the postulation that RNA- or protein-based life-like systems were the first living systems is correct, such system should have possessed the characteristic to adapt itself toward the environments. Thus, an analogical compartment and/or integration of molecules to the present life-like systems would have been necessary to keep the continuity and stability for a primitive system. As described above, the environments for the primitive life-like system would have been extreme conditions so that the influence from environments should have

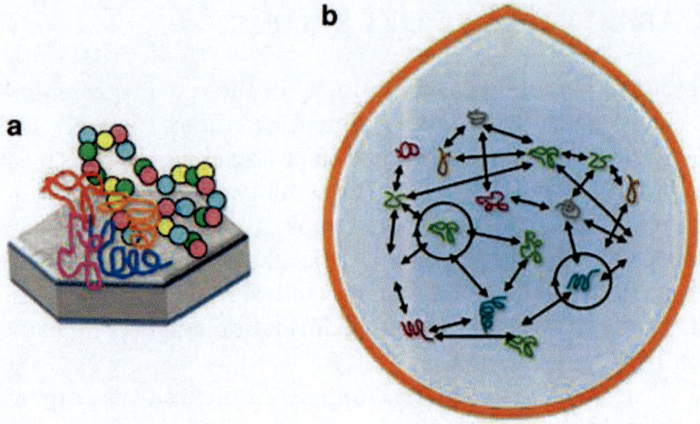


**Figure 6.** Imaginable particle-type (**a**) and cell-type (**b**) life-like systems.

more seriously damaged the primitive life-like systems. Here, primitive life-like systems are going to evaluate on the basis of analogies of cell-type and virus-type systems, as shown in Fig. 6. Here, these are named as cell- and particle-type life-like systems.

## 4.4. PARTICLE-TYPE LIFE-LIKE SYSTEM

Particle types of life-like systems are assumed as possible strategy to keep the stability and continuity as analogy of present viruses (and viroids) (Fig. 6a). The figure indicates a possible combination of RNA and protein-like molecules adsorbed on a mineral particle, where mineral particle might be important to protect the biomolecules. Actually, minerals serve as catalysts for the formation of RNA oligomers (Ferris and Ertem, 1992; Kawamura and Ferris, 1994; Ferris et al., 1996). Viruses and viroids require host cells so that these ancient particle types of life-like system would have required some corresponding host. This would be very important since it is reasonable that such small particle could not possess complicated functions because of the limitation of size. Actually, modern viruses and viroids could possess limited functions, and this is probably due to primarily the limitation of size. The metabolism for the particle-type life-like system should have proceeded in the ancient host material. Host material would not be necessary as cell-type life-like system; prebiotic soup would be also a candidate host (Fig. 7). The RNA-based life-like systems would have possessed similar structures to those of viruses or viroids. Actually, viroids are circular RNA molecules which could not amplify by themselves. Primitive protein-like molecules and minerals would be the candidate materials, which protect or hold the particle-type life-like system.

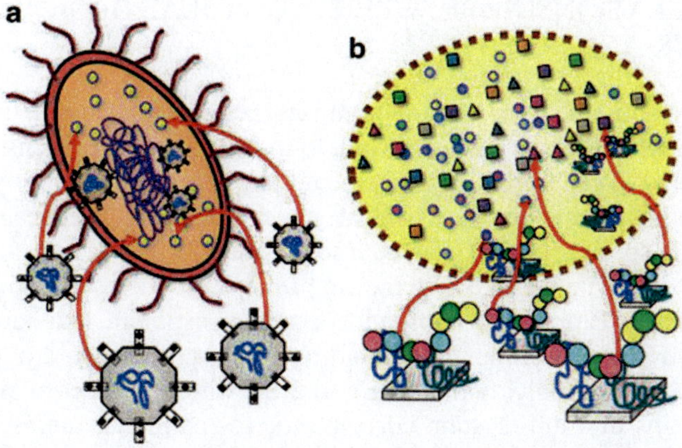


**Figure 7.** Analogies between viruses and a cell (**a**) vs. primitive particle-type life-like systems and a prebiotic soup (**b**).

## 4.5. CELL TYPE OF LIFE-LIKE SYSTEM

There is no extensive discussion concerning the RNA-based life-like system could have possessed cell-type compartment. The system involved RNA molecules in a cell-type compartment can be assumed (Fig. 6b), where low-efficiency functional molecules were packed in the cell. Naturally, the system would have possessed low-efficiency metabolic systems. In addition, the primitive method for the cell division should be taken into account. On the other hand, by cell-type compartment, the reactions in the system would be accelerated since molecules can be concentrated and the diffusion of molecules readily proceeds. On contrary, if life-like systems do not possess boundary with the surrounding ocean, a primitive enzyme has hardly met its target substrates.

By the way, the main materials for boundary of organisms are phospholipids and proteins. There is a trend that the amount of membrane proteins is high for primitive organisms. It is imaginable that by increasing the size of inner sphere of the cell-type system, it becomes similar to the particle-type life-like system. In other words, a cell-type system of which the inner volume is zero can be equated to a particle-type system. This implication suggests that the particle-type system and the cell-type system are not readily classified with decreasing of size. Although a medium system between viruses and prokaryote has not yet been observed in the real biota, such system would connect between the simplicity of viruses and complexity of prokaryote.

The verification of the primitive life-like systems suggests the importance of compartment and integration so that the experimental evaluations of the postulation should be taken into account of RNA-based or protein-based life-like systems.

### 4.6. THE RELATIONSHIP BETWEEN INDIVIDUAL ORGANISM, SPECIES, AND ECOSYSTEM

There is no exception that the present organisms belong to its species. Even, viruses and viroids possess the one corresponding to the species of organisms. This is due to the capability of replication of the present life-like systems. Similarly, the origin of this relationship and the existence of hierarchical levels between individual organisms and species should be taken into account.

For particle-type life-like system, the principle of the pseudospecies (Eigen et al., 1988) can be applied for a pseudospecies consisting of individual particle-type life-like system. Each particle including RNA, proteins, and minerals, etc., would correspond to individual element, and the whole member of particle-type life-like systems in a unit system corresponds to the higher hierarchical level. The presence of different types of RNA molecules, such as tRNA, rRNA, mRNA, ribozymes, and micro-RNA, would reflect the fossil quasispecies of RNA members. Similarly, for the case of GADV protein-based life-like systems, each protein corresponds to the individuals and the member of proteins corresponds to the species. The pseudospecies as a higher hierarchical level system can be automatically formed since the replication primarily occurs in these systems. Naturally, the pseudospecies could not be identified as a living unit without compartment or integration since such group would be readily destroyed in the presence of certain influence from the outside.

On the contrary, if RNA or protein-based life-like systems consisted of a cell, the cell corresponds to individual and the whole members of cells correspond to the species. Similarly, the relationship between individual members and the ecosystem should be taken into account for the verification of primitive life-like system.

Furthermore, the relationship between individual life-like systems and the surrounding ecosystems is also applied to estimate the emergence of life. Now, the estimation of such relationship is limited as very speculative. Thus, in this chapter, this viewpoint will not be discussed.

## 5. Conclusions

This chapter summarizes the possibility of RNA and/or protein-based primitive life-like systems on the basis of chemical evolution of these biopolymers, the evaluation of these systems under hydrothermal conditions, and the restriction conditions what system can be called as life. These considerations imply that the first primitive system should have possessed some mechanisms and/or materials to stabilize itself, as found in modern life-like systems, that is, cells and viruses. Fortunately, the implication shown in this paper would be experimentally verified in future by simulation experiments under the primitive Earth conditions.

The origin-of-life hypotheses regarding RNA and protein molecules, that is, the RNA world hypothesis and GADV protein hypothesis have problems from the viewpoint of hydrothermal origin-of-life hypothesis. In other words, these hypotheses have not provided practical models, which could have survived under the extreme conditions on primitive Earth. Even if life could have originated under much milder conditions, the question how such system initially consisting of simple biopolymers could have evolved to a primitive system to regard as life has not been extensively focused, where the definition or essential characteristics for life should be clarified to investigate such system. The stability and continuity of life-like system is essential to possess the nature that living system is capable to adapt itself to and survive against the change of environments. Thus, in this chapter, the stability and continuity of life-like system was mainly focused to draw the nature and structure of the first life-like systems, which can be regarded as life. The analysis showed possible implication of such pathways and the scenario from molecules to a system possessing the essential characteristics for life.

## 6. References


Bada JL, Lazcano A (2002) Origin of life – some like it hot, but not the first biomolecules. Science 296:1982–1983

Bada JL, Bigham C, Miller SL (1994) Impact melting of frozen oceans on the early Earth – implications for the origin of life. Proc Natl Acad Sci U S A 91:1248–1250

Baross JA, Hoffman SE (1985) Submarine hydrothermal vents and associated gradient environments as sites for the origin and evolution of life. Orig Life Evol Biosph 15:327–345

Cech TR, Atkins JF, Gesteland RF (eds) (2005) The RNA world, 3rd edn, Cold Spring Harbor monograph series. Cold Spring Harbor Laboratory Press, New York

Corliss JB, Baross JA, Hoffman SE (1981) An hypothesis concerning the relationship between submarine hot springs and the origin of life on Earth. Oceanol Acta 4(Suppl):59–69

Eigen M (1971) Selforganization of matter and the evolution of biological macromolecules. Naturwissenschaften 58:465–523

Eigen M, McCaskill J, Schuster P (1988) Molecular quasi-species. J Phys Chem 92:6881–6891

Ellington AD, Szostak JW (1990) *In vitro* selection of RNA molecules that bind specific ligands. Nature 346:818–822

Ertem G, Ferris JP (1996) Synthesis of RNA oligomers on heterogeneous templates. Nature 379:238–240

Ferris JP, Ertem G (1992) Oligomerization of ribonucleotides on montmorillonite: reaction of the 5′-phosphorimidazolide of adenosine. Science 257:1387–1389

Ferris JP, Hill AR, Liu JR, Orgel LE (1996) Synthesis of long prebiotic oligomers on mineral surfaces. Nature 381:59–61

Forterre P (1996) A hot topic: the origin of hyperthermopiles. Cell 85:789–792

Frey H (1977) Origin of the Earth's ocean basins. Icarus 32:235–250

Galtier N, Tourasse N, Gouy M (1999) A nonhyperthermophilic common ancestor to extant life forms. Science 283:220–221

Gilbert W (1986) The RNA world. Nature 319:618–618

Gold L, Polisky B, Uhlenbeck O, Yarus M (1999) Diversify of oligonucleotide functions. Annu Rev Biochem 68:763–797

Gough DO (1981) Solar interior structure and luminosity variations. Sol Phys 74:21–34

Ikehara K (2005) Possible steps to emergence of life: the [GADV]- protein world hypothesis. Chem Rec 5:107–118
Ikehara K (2009) Pseudo-replication of [GADV]-proteins and origin of life. Int J Mol Sci 10:1525–1537
Imai E, Honda H, Hatori K, Brack A, Matsuno K (1999) Elongation of oligopeptides in a simulated submarine hydrothermal system. Science 283:831–833
Inoue T, Orgel LE (1982) Oligomerization of (guanosine 5′-phosphor)-2-methylimidazolide on poly(C). An RNA polymerase model. J Mol Biol 162:201–217
Inoue T, Orgel LE (1983) A nonenzymatic RNA polymerase model. Science 219:859–862
Joyce GF, Shwartz AW, Miller SL, Orgel LE (1987) The case for an ancestral genetic system involving simple analogues of the nucleotides. Proc Natl Acad Sci U S A 84:4398–4402
Karhu J, Epstein S (1986) The implication of the oxygen isotope records in coexisting cherts and phosphates. Geochim Cosmochim Acta 50:1745–1756
Kasting JF, Ackerman TP (1986) Climatic consequences of very high-carbon dioxide levels in the Earth's early atmosphere. Science 234:1383–1385
Kawamura K (1998) Kinetic analysis of hydrothermal reactions by flow tube reactor -hydrolysis of adenosine 5′-triphosphate at 398–573 K. Nippon Kagaku Kaishi 255–262
Kawamura K (2000) Monitoring hydrothermal reactions on the millisecond time scale using a micro-tube flow reactor and kinetics of ATP hydrolysis for the RNA world hypothesis. Bull Chem Soc Jpn 73:1805–1811
Kawamura K (2002a) The origin of life from the life of subjectivity. In: Palyi G, Zucchi C, Caglioti L (eds) Fundamentals of life. Elsevier, Paris, pp 563–574
Kawamura K (2002b) In situ UV–VIS detection of hydrothermal reactions using fused-silica capillary tubing within 0.08–3.2 s at high temperatures. Anal Sci 18:715–716
Kawamura K (2003a) The relative importance of genes, subjectivity, and self-organization for the origin and evolution of life. In: Levit GS, Popov IY, Hossfeld U, Breidbach O (eds) In the shadow of Darwinism: alternative evolutionary theories in the 20th century. Fineday-Press, St-Petersburg, pp 218–239
Kawamura K (2003b) Kinetics and activation parameter analyses of hydrolysis and interconversion of 2′,5′- and 3′,5′-linked dinucleoside monophosphate at extremely high temperatures. Biochim Biophys Acta 1620:199–210
Kawamura K (2003c) Kinetic analysis of cleavage of ribose phosphodiester bond within guanine and cytosine rich oligonucleotides and dinucleotides at 65–200°C and its implications on the chemical evolution of RNA. Bull Chem Soc Jpn 76:153–162
Kawamura K (2004) Behavior of RNA under hydrothermal conditions and the origins of life. Int J Astrobiol 3:301–309
Kawamura K (2007) Civilization as a biosystem examined by the comparative analysis of biosystems. Biosystems 90:139–150
Kawamura K (2009) Realistic life-like systems consisting of RNA under the hydrothermal environments on primitive Earth. Viva Origino 37:31–42
Kawamura K, Ferris JP (1994) Kinetics and mechanistic analysis of dinucleotide and oligonucleotide formation from the 5′-phosphorimidazolide of adenosine on $Na^+$-montmorillonite. J Am Chem Soc 116:7564–7572
Kawamura K, Maeda J (2007) Kinetic analysis of oligo(C) formation from the 5′-monophosphorimidazolide of cytidine with Pb(II) ion catalyst at 10–75°C. Orig Life Evol Biosph 37:153–165
Kawamura K, Maeda J (2008) Kinetics and activation parameter analysis for the prebiotic oligocytidylate formation on $Na^+$-montmorillonite at 0–100°C. J Phys Chem A 112:8015–8023
Kawamura K, Nagayoshi H (2007) Behavior of DNA under hydrothermal conditions with $MgCl_2$ additive using an in situ UV-visible spectrophotometer. Thermochim Acta 466:63–68
Kawamura K, Shimahashi M (2008) One-step formation of oligopeptide-like molecules from Glu and Asp in hydrothermal environments. Naturwissenschaften 95:449–454
Kawamura K, Umehara M (2001) Kinetic analysis of the temperature dependence of the template-directed formation of oligoguanylate from the 5′-phosphorimidazolide of guanosine on a poly(C) template with $Zn^{2+}$. Bull Chem Soc Jpn 74:927–935

Kawamura K, Yukioka M (2001) Kinetics of the racemization of amino acids at 225–275°C using a real-time monitoring method of hydrothermal reactions. Thermochim Acta 375:9–16

Kawamura K, Nakahara N, Okamoto F, Okuda N (2003) Temperature dependence of the cyclization of guanine and cytosine mix hexanucleotides with water-soluble carbodiimide at 0–75°C. Viva Origino 31:221–232

Kawamura K, Kuranoue K, Nagahama M (2004) Prebiotic inhibitory activity of protein-like molecules to the template-directed formation of oligoguanylate from guanosine 5′-monophosphate 2-methylimidazolide on a polycytidylic acid template. Bull Chem Soc Jpn 77:1367–1375

Kawamura K, Nishi T, Sakiyama T (2005) Consecutive elongation of alanine oligopeptides at the second time range under hydrothermal condition using a micro flow reactor system. J Am Chem Soc 127:522–523

Kawamura K, Nagayoshi H, Yao T (2009) Stability of ribonuclease A under hydrothermal conditions in relation to the origin-of-life hypothesis: verification with the hydrothermal micro-flow reactor system. Res Chem Intermed 35:879–891

Kawamura K, Nagayoshi H, Yao T (2010) In situ analysis of proteins at high temperatures mediated by capillary-flow hydrothermal UV–vis spectrophotometer with a water-soluble chromogenic reagent. Anal Chim Acta 667:88–95

Kruger K, Grabowski PJ, Zaug AJ, Sands J, Gottschling DE, Cech TR (1982) Self-splicing RNA: autoexcision and autocylization of the ribosomal RNA intervening sequence of *Tetrahymena*. Cell 31:146–157

Kuppers BO (1985) Molecular theory of evolution. Outline of a physico-chemical theory of the origin of life. Springer, Berlin

Larralde R, Robertson MP, Miller SL (1995) Rates of decomposition of ribose and other sugars: implications for chemical evolution. Proc Natl Acad Sci U S A 92:8158–8160

Lohrmann R (1977) Formation of nucleoside 5′-phosphorimidates under potentially prebiological conditions. J Mol Evol 10:137–154

Lohrmann R, Orgel LE (1973) Prebiotic activation processes. Nature 244:418–420

Maher KA, Stevenson DJ (1988) Impact frustration of the origin of life. Nature 331:612–614

Miller SL, Bada JL (1988) Submarine hot springs and the origin of life. Nature 334:609–611

Mojzsis SJ, Arrhenius G, McKeegan KD, Harrison TM, Nutman AP, Friend CRL (1996) Evidence for life on Earth before 3,800 million years ago. Nature 384:55–59

Mojzsis SJ, Harrison TM, Pidgeon RT (2001) Oxygen-isotope evidence from ancient zircons for liquid water at the Earth's surface 4,300 Myr ago. Nature 409:178–181

Nemoto N, Miyamoto-Sato E, Husimi Y, Yanagawa H (1997) In vitro virus: bonding of mRNA bearing puromycin at the 3′-terminal end to the C-terminal end of its encoded protein on the ribosome in vitro. FEBS Lett 414:405–408

Newman MJ, Rood RT (1977) Implications of solar evolution for the Earth's early atmosphere. Science 198:1035–1037

Nisbet EG (1986) RNA and hot-water springs. Nature 322:206–206

Oba T, Fukushima J, Maruyama M, Iwamoto R, Ikehara K (2005) Catalytic activities of [GADV]-peptides: formation and establishment of [GADV]-protein world for the emergence of life. Orig Life Evol Biosph 35:447–460

Orgel LE, Crick FHC (1993) Anticipating an RNA world some past speculations on the origin of life: where are they today? FASEB J 7:238–239

Pace NR (1991) Origin of life – facing up to the physical setting. Cell 65:531–533

Radzicka A, Wolfenden R (1995) A proficient enzyme. Science 267:90–93

Sagan C, Mullen G (1972) Earth and Mars: evolution of atmospheres and surface temperatures. Science 177:52–56

Sawai H (1976) Catalysis of internucleotide bond formation by divalent metal ions. J Am Chem Soc 98:7037–7039

Sawai H, Shibata T, Ohno M (1981) Preparation of oligoadenylates with 2′-5′ linkage using $Pb^{2+}$ ion catalyst. Tetrahedron 37:481–485

Sleep NH, Zahnle KJ, Kasting JF, Morowitz HJ (1989) Annihilation of ecosystems by large asteroid impacts on the early Earth. Nature 342:139–142

Szostak JW, Bartel DP, Luisi PL (2001) Synthesizing life. Nature 409:387–390

Tuerk C, Gold L (1990) Systematic evolution of ligands by exponential enrichment: RNA ligands to bacteriophage T4 DNA polymerase. Science 249:505–510

Walker JCG (1985) Carbon-dioxide on the early Earth. Orig Life Evol Biosph 16:117–127

White RH (1984) Hydrolytic stability of biomolecules at high temperatures and its implication for life at 250°C. Nature 310:430–432

Wilson DS, Szostak JW (1999) *In vitro* selection of functional nucleic acids. Annu Rev Biochem 68:611–647

Biodata of **Osmel Martín, Liuba Peñate, Rolando Cárdenas,** and **J.E. Horvath,** authors of "***The Photobiological Regime in the Very Early Earth and the Emergence of Life.***"

**Osmel Martín, M.Sc.**, is currently assistant of the Department of Physics of the Faculty of Mathematics, Physics, and Computing Science in the Central University of Las Villas (CULV), Santa Clara, Cuba. He is in the final stage of his Ph.D., which he develops under the advice of Professors Rolando Cárdenas (CULV) and Jorge E. Horvath at the University of São Paulo, Brazil. His scientific interests are in the areas of: influence of cosmic and solar radiations on planetary atmospheres and hydrospheres and their influence on biosphere, and mathematical modeling of complex biological systems.

E-mail: **osmel@uclv.edu.cu**

**Liuba Peñate, M.Sc.,** is currently assistant of the Biology Department in the Central University of Las Villas (CULV), Santa Clara, Cuba. In 2006, she obtained her M.Sc. from the Laboratory for Plant Cell and Tissue Culture at the Bioplant Centre of the University of Ciego de Avila, Cuba, and is working for her Ph.D. at the Interdisciplinary Environmental Modeling group mainly hosted in the Physics Department of CULV. Her scientific interests are in the areas of effects of cosmic and solar radiation on oceanic photosynthesis, emerging and evolution of the early life in Earth and exoplanets, and cell and molecular biology.

E-mail: **liubapa@uclv.edu.cu**

**Osmel Martín**

**Liuba Peñate**



J. Seckbach (ed.), *Genesis - In The Beginning: Precursors of Life, Chemical Models and Early Biological Evolution*, Cellular Origin, Life in Extreme Habitats and Astrobiology 22, 145–155
DOI 10.1007/978-94-007-2941-4_8, 

**Dr. Rolando Cárdenas** is currently associate professor of the Department of Physics of the Faculty of Mathematics, Physics, and Computing Science in the Central University of Las Villas (CULV), Santa Clara, Cuba. He obtained his Ph.D. in 2003 as a cross appointment of the Department of Physics of CULV and the Nuclear Physics Department at the Higher Institute for Nuclear Science and Technology at Havana, Cuba. His main scientific interests are in the areas of influence of cosmic and solar radiations on planetary atmospheres and hydrospheres and their influence on biosphere, and mathematical modeling of complex biological systems.

E-mail: **rcardenas@uclv.edu.cu**

**Professor J.E. Horvath** is appointed at the IAG-USP, Brazil, since 1992. He obtained his Ph.D. from the Universidad Nacional de La Plata, Argentina, in 1989 and visited the U of A, USA, and other research institutions to collaborate with several colleagues. Professor Horvath scientific interests are in the areas of relativistic and high-energy astrophysics and many mathematical and physical aspects of astrobiology.

E-mail: **foton@astro.iag.usp.br**

**Rolando Cárdenas**

**J.E. Horvath**

# THE PHOTOBIOLOGICAL REGIME IN THE VERY EARLY EARTH AND THE EMERGENCE OF LIFE

**OSMEL MARTÍN[1], LIUBA PEÑATE[2], ROLANDO CÁRDENAS[1], AND J.E. HORVATH[3]**
*[1]Department of Physics, Universidad Central de Las Villas, Santa Clara, Cuba*
*[2]Department of Biology, Universidad Central de Las Villas, Santa Clara, Cuba*
*[3]Department of Astronomy, IAG-Universidade de São Paulo, São Paulo, Brazil*

## 1. A General Model for the Emergence of Life

Life as we know it is an extremely complex phenomenon. However, one might wonder whether it is possible to define general environmental conditions for the emergence of life elsewhere in the universe. Such an endeavor seems to be within the reach of astrobiologists. It is recognized that at least three conditions are to be fulfilled:

- The presence of biogenic chemical elements. Example: C, H, O, N in Earth
- Solvent where biogenic chemical elements can readily mix and react to form complex biomolecules. Example: water in Earth
- A source of energy guiding abovementioned biochemistry. Example: solar radiation and the process of photosynthesis in primary producers on Earth

Some alternatives to above example conditions on Earth have been proposed, such as life based on silicon instead of carbon, with hydrocarbons (methane, ethane) as solvent. However, alternative biochemistries still have to be very carefully worked out to avoid important shortcomings. Along the same lines, ecosystems living around hydrothermal vents in marine depths on Earth do not need solar radiation, but they represent a very small portion of terrestrial biosphere. Therefore, in this work, we will assume that carbon-based life can arise on any planetary surface where the following conditions are accomplished:

- Biogenic chemical elements are available.
- There is liquid water at surface.
- Light from a parent star can be used for the process of photosynthesis.

Thus, we investigate the constraints on life imposed by the photobiological regime at the planetary surface, and in that part of the ocean where light is enough for the key process of photosynthesis to be done. That part of the ocean is called the *photic zone* and is roughly defined as the zone extending from the surface down to the depth where light intensity or irradiance (in $W \cdot m^{-2}$) is 1% of the intensity of incident light at surface.

## 2. Origin of Photosynthetic Life on Earth

The timing for the origin of life of Earth is an open issue. Many authors locate it namely at the beginning of the Archean, around 3,800 millions of years (Ma) ago, presumably the age of the first stromatolites or colonies of cyanobacteria. Others prefer a less controversial dating of 3,500 Ma, well inside the Early Archean. But, could life be older? Could it have a Hadean origin? Answering the latter question is more or less equivalent to answering the following ones: When was water available on Earth? When were the oceans formed?

Classically, the Hadean Earth has been considered an inhospitable world with magma oceans and heavy asteroid bombardment, high surface temperatures not allowing surface water oceans to exist. However, recent study of minerals called zircons has provided a hypothesis for a Cool Early Earth during what we call "Mid-Hadean," roughly between 4.4 and 4.0 Ga ago (Valley et al., 2002). Several kinds of zircons need liquid water to be formed. This scenario would allow liquid oceans at surface, possibly amenable for life. However, the Hadean eon has left very little fingerprints as to infer characteristics of a potential ancient life at those times.

If the Cool Early Earth scenario existed, perhaps the photobiological regime during Mid-Hadean was not very different from the one existing in Early Archean: the Sun would had already passed the T-Tauri phase (thus, its ultraviolet and visible emissions would not differ much in both eras), while the exact composition of those atmospheres is a subject under debate, but basically levels of $N_2$ similar to current ones seem reasonable and also important quantities of greenhouse gases $CO_2$ and $CH_4$. In both eras, the faster rotation of the planet would result in strong surface winds of more than 150 km/h, causing very active vertical circular currents in the ocean extending down to tens of meters (Langmuir circulation). As we will see in the next section, this last fact would have a key role in the efficiency of an extremely important biological process of simple unicellular organisms: photosynthesis.

## 3. Photosynthetic Rates of Early Archean Life

The range of wavelengths useful for photosynthesis is roughly the same that we can see with our eyes, from 400 to 700 nm. This band of light is usually called photosynthetic active radiation (PAR). In this work, we will assume that all this light favors photosynthesis, although in some (not frequent) cases, high PAR intensities can be rather inhibitory. On the other hand, the ultraviolet radiation (UVR) causes inhibition of photosynthesis due to the following main reasons:

- It causes damages in photosystem II and in the photosynthetic enzyme Rubisco (Cockell, 2000; Hader and Worrest, 1991; Neale et al., 1993; Cullen and Neale, 1994; Vincent and Roy, 1993).
- It damages nucleic acids, making the cell to spend energy to repair them. This indirectly affects the efficiency of the photosynthetic process.

Cells have the possibility of repairing part of the damages done by ultraviolet radiation. Concerning this skill, several models have been developed to account for real biological damage of UVR; they are classified according to repair capabilities: intermediate, good, and no repair at all (Fritz et al., 2008).

Without an atmosphere with UVR screeners such as ozone, and allowing for the usual land-habitat arguments, it is likely that all Early Archean life was aquatic. This relies on the water column attenuation of radiations, although some life could colonize the rather small area of the planet covered by land, provided the potential protection of some wet minerals (Cockell and Raven, 2007). Also, during Early Archean, maybe there was chemosynthetic life at hydrothermal vents in oceanic depths, well protected from all kinds of surface radiations and asteroid impacts. Indeed, such scenario nowadays looks attractive as the potential first site where life appeared on Earth. As the aim of our work is the emergence and early evolution of photosynthetic life, we will focus on oceanic life living in the photic zone, where there is enough light for the photosynthetic process.

The upper part of the photic zone of the ocean is usually influenced by Langmuir circulation, provided winds with speed greater than 3 m/s blew, a condition expected to happen practically all the time during Early Archean. Langmuir circular vertical currents mix the waters of the upper part of the ocean, thus making its physical properties, such as temperature and density, constant. That is why this layer is called the mixed layer of the ocean. Below it, circulation is stopped by density and temperature gradients (pycnoclines and thermoclines) and the so-called stratification of the water settles: negligible circulation.

Considering the crude photobiological regime during Early Archean Earth (Cockell, 2000), it is reasonable to assume that biota living in the mixed layer of the ocean would be radiation resistant, especially considering Langmuir circulation. These circular currents would expose periodically any organism to the surface, where it would receive high doses of ultraviolet irradiation. Thus, we would expect in the mixed layer of the Early Archean ocean a biota perhaps numerically abundant, but with low diversity due to the ultraviolet constraint. In addition, these organisms should have had good repair capabilities, so we employ the so-called $E$ model appearing in Fritz et al. (2008) to account for rates of photosynthesis in the mixed layer of the Archean ocean, and by extension in the entire photic zone.

The photosynthetic rate $P$ in this model can be calculated by

$$P = P_{pot}\left(\frac{1}{1+E^{*}_{inh}}\right) \quad (1)$$

where $E^{*}_{inh}$ is the nondimensional inhibitory irradiance, given by the ultraviolet radiation, while $P_{pot}$ is the photosynthetic rate in the absence of photoinhibition, given by

$$P_{pot} = P_s\left(1-e^{-E_{PAR}/E_S}\right). \quad (2)$$

In the above expression, $P_s$ is the light saturated rate of photosynthesis in the absence of inhibition, $E_{PAR}$ (in $W\cdot m^{-2}$) is the irradiance of visible light, while $E_s$ (in $W\cdot m^{-2}$) is the characteristic irradiance for light saturation. The inhibitory nondimensional irradiance of ultraviolet radiation is given by

$$E^{*}_{inh} = \sum_{\lambda=200nm}^{400nm} E(\lambda)\,\varepsilon_E(\lambda)\,\Delta\lambda \quad (3)$$

where $\varepsilon_E(\lambda)$ (in $W^{-1}\cdot m^{2}$) are the biological weightings quantifying effectiveness of spectral exposure $E(\lambda)$ (in $W\cdot m^{-2}\cdot nm^{-1}$), that is, $\varepsilon_E(\lambda)$ represent the inhibition of photosynthesis caused by ultraviolet radiation of wavelength $\lambda$.

Substituting Eq. 2 in Eq. 1 and normalizing to $P_s$, we obtain

$$\frac{P}{P_s} = \frac{1-e^{-E_{PAR}/E_S}}{1+E^{*}_{inh}}. \quad (4)$$

It is clear that photosynthesis rate is a combination of two factors: the numerator in Eq. 4 favors photosynthesis (as the irradiance of photosynthetic active radiation (PAR) is there), while the denominator inhibits it because of the presence of the ultraviolet inhibitory irradiance factor. The question of when oxygenic photosynthesis evolved is still unanswered, but it seems reasonable to assume the same photosynthetic routes and approximately similar rates during all life history. This allows us to estimate the influence of the photobiological regime on photosynthesis, that is, the influence of different solar irradiances and atmospheric transmittances.

We split the photic zone in two layers to estimate photosynthetic rates: the mixed layer and the rest of the photic zone beneath it. The depth of the mixed layer, where Langmuir circulation exists, depends on the speed of the wind blowing at surface. For the Early Archean, it is reasonable to assume depths equal or greater than 30 m (Cockell, 2000). In this work, as an example for our calculations, we assume a mixed layer of 40 m depth, largely because in the above reference, plenty of data are available up to this depth. Indeed, to calculate the photosynthetic rates in the mixed layer, in Eq. 4, we use $E_{PAR}$ and $E^*_{inh}$ as reported in Cockell (2000), and average $E_s$ as given in Fritz et al. (2008). On the other hand, to estimate photoinhibition $E_{PAR}$ and $E^*_{inh}$ beneath the mixed layer, we followed two approaches:

(a) We calculated $E_{PAR}$ and $E^*_{inh}$ beneath the mixed layer, as these are not previously reported in literature. To do that, we use the irradiances at 40 m depth as given in Cockell (2000) and rewrite Eq. 3 as

$$E^*_{inh}(z) = \sum_{\lambda=200nm}^{400nm} E(\lambda, z)\, \varepsilon_E(\lambda)\Delta\lambda \tag{5}$$

$$E^*_{PAR}(z) = \sum_{\lambda=400nm}^{700nm} E(\lambda, z)\Delta\lambda \tag{6}$$

where

$$E(\lambda, z) = E(\lambda, 40)e^{-K_d(z-40)}. \tag{7}$$

(b) We extrapolated the results in Cockell (2000) until depths of 200 m.

The differences between the two approaches above are within 1%, thus giving a consistency check to our calculations. In the rest of this chapter, we then report results obtained with the (b) approach.

When the irradiance of photosynthetic active radiation $E_{PAR}$ takes the value of the parameter $E_s$, then the numerator of the right hand side of Eq. 4 becomes 0.63, that is, the parameter $E_s$ represents the irradiance of photosynthetic active radiation $E_{PAR}$ which ensures 63% of maximum photosynthetic rate ($P/P_s = 63\%$). Thus, the smaller the value of $E_s$, the more efficient is the organism performing the photosynthesis, as it reaches 63% of maximum rate with smaller PAR irradiance.

We might also wonder whether appreciable differences in the photosynthetic potential would be possible between the Early Archean and today's unicellular photosynthetic organisms. To account for that, we considered not only the value $E_s \approx 20 W.m^{-2}$ obtained by averaging the 16 values reported in Fritz et al. (2008), but we also used $E_s = 15 W.m^{-2}$ and $E_s = 25 W.m^{-2}$. The value of $E_s$ for plenty of current photosynthetic organisms, including multicellular ones, falls in this range (from 15 to 25 $W \cdot m^{-2}$). An example of that can be inferred from the upper panel of Fig. 2 of Biber et al. (2003), where photosynthesis irradiance curves are shown for three groups of macroalgae (bear in mind that the irradiance units in that paper are not $W \cdot m^{-2}$).

As we said above, during the Early Archean, a complete turn of the Earth was completed in around 15 h. Such a fast rotation would have implied strong winds, and, consequently, Langmuir circular currents should have developed down to tens of meters. These currents would sink and upwell unicellular organisms captured in the mixed layer so that the average irradiances received by a cell during a complete sinking-upwelling cycle would be

$$\overline{E^*_{inh}} = \frac{\sum_{z=0^-}^{z_L} \sum_{\lambda=200nm}^{400nm} E(\lambda,z)\, \varepsilon_E(\lambda)\, \Delta\lambda\, \Delta z}{z_L} \tag{8}$$

$$\overline{E_{PAR}} = \frac{\sum_{z=0^-}^{z_L} \sum_{\lambda=400nm}^{700nm} E(\lambda,z)\Delta\lambda\, \Delta z}{z_L} \tag{9}$$

where $z_L$ is the maximum depth at which Langmuir circulation extends. As we mentioned above, beneath this mixed layer, water stratification appears as a result of temperature and or density gradients (thermoclines and pycnoclines), which do not allow water circulation and mixing.

Below we present plots of percent relative photosynthetic rate P/Ps vs. depth z in the Archean ocean for three values of the parameter $E_s$. As in Cockell (2000), we consider solar zenith angles of 0° and 60°. The former provides maximum irradiation with the Sun directly overhead, while the latter gives irradiances similar to current ones at noon in high latitudes.

As expected, in both cases, the lower the value of $E_s$, the greater the maximum rate of photosynthesis. As can be inferred from Eq. 4, this parameter is the value of PAR irradiance for which 63% of maximum photosynthetic rate is achieved if there was not inhibition, hence a measure of the effectiveness of the idealized photosynthetic process.

We also see that the conditions for maximum photosynthetic rate are given at depths of around 75 m for solar zenithal angle (sza) of 0°, and at 50–60 m for sza of 60°, as the smaller ultraviolet pressure allows going closer to sea surface to gather more photosynthetic active radiation (PAR).

However, we should stress that the above plots can be misleading if the role of Langmuir circulation in the ocean is neglected. The plots give photosynthetic rates as if the unicellular planktonic organisms of early life would be at fixed depth all the time. However, those cells living in the mixed layer could be easily captured in Langmuir circular currents, thus suffering cyclic sinking and upwelling.

As an example, we consider such a circulation with maximum depth $z_L$ of 40 m and calculate the average irradiances received by an organism trapped in it using Eqs. 8 and 9. Then we substitute in Eq. 4 to obtain average photosynthetic rate during a full cycle. Results are in Table 1.

**Table 1.** Average values, in a complete cycle, of photosynthetic rate of a cell trapped by a Langmuir circular current.

| $E_s$; W·m$^{-2}$ | 15 | 20 | 25 |
|---|---|---|---|
| $P/P_s$; % ($sza=0°$) | 7.12 | 7.10 | 7.07 |
| $P/P_s$; % ($sza=60°$) | 15.8 | 15.2 | 14.5 |

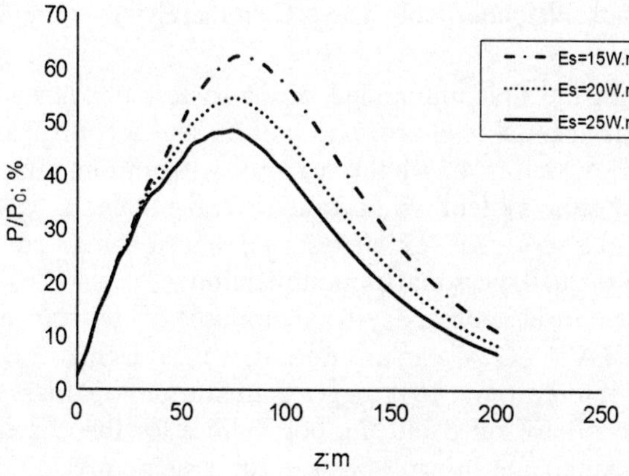


**Figure 1.** Relative photosynthetic rate vs. depth in the Archean ocean, for a solar zenith angle of 0°.

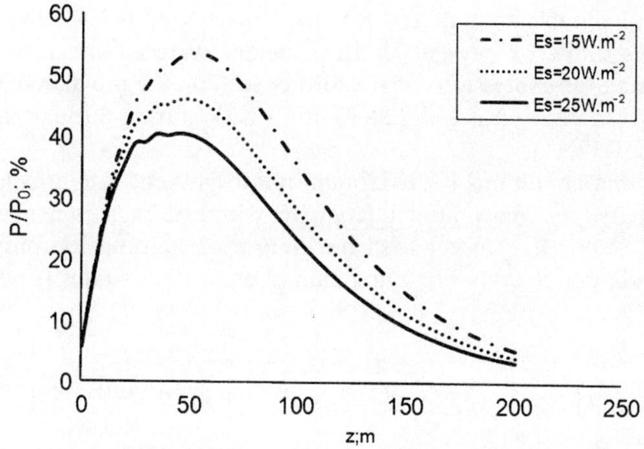


**Figure 2.** Relative photosynthetic rate vs. depth in an Archean ocean, for a solar zenith angle of 60°.

We obtain small average photosynthetic rates of around 7% for all the three values of $E_s$ used in this work when the Sun is at zenith and values of 15% for a solar zenith angle of 60°. As we can see, comparing with Figs. 1 and 2, to be captured in the Langmuir circulation considerably reduces the photosynthetic possibilities of organisms living in the mixed layer, especially when there is very intense irradiance. Of course, this is due to the crude exposition to ultraviolet radiation while the organism is circulating in the upper part of the ocean. Beneath the mixed layer, photosynthesis would be reasonably good up to 150 m and would extend even deeper than 200 m.

## 4. Life in Our Cosmic Neighborhood: The $\alpha$ Centauri System as an Example

It is very likely that the first unmanned probe to test the existence of photosynthetic life in our cosmic neighborhood will be sent to our nearest planetary system: $\alpha$ Centauri, which is 4.37 light years away from our Sun. With present technology, reaching the system would take several millennia, but forthcoming new technologies like space sail or nuclear pulse fusion may cut this down to decades (Hearnshaw, 2010, personal communication).

Actually, $\alpha$ Centauri is a binary system composed of two stars quite similar to our Sun. $\alpha$ Centauri A is a G2V star, as our Sun, while $\alpha$ Centauri B is a K1V star. The mass of the A star is around 10% bigger than Sun's, and B mass is around 10% smaller. Observations have ruled out the possibilities for the existence of Jupiter mass planets (gas giants) and brown dwarfs in this system, making more likely the existence of the smaller Earth-like ones. These are more difficult to detect, though. Computer simulations for planet formation around $\alpha$ Centauri B predict one to fourth Earth mass planets in stable orbits Guedes et al. (2008), one or two inside the stellar habitable zone (i.e., where liquid water can exist at planet's surface).

A rough calculation of photosynthetic rate for a hypothetical Earth-mass planet in the habitable zone of $\alpha$ Centauri B (located at 0.7 AU from this star, which ensures similar to present Earth planetary surface temperatures) can be attempted using the expressions of the former section. We use data of ultraviolet irradiances given in Segura et al. (2003), for a K2V star, as a rough model for $\alpha$ Centauri B, a K1V star.

We see that conditions for maximum photosynthetic rate are closer to the surface compared to Archean Earth (compare with Fig. 2). This can be explained because $\alpha$ Centauri B is cooler than our Sun, thus emitting less proportion of ultraviolet radiation. Comparing maximum photosynthetic rates (Figs. 2 and 3),

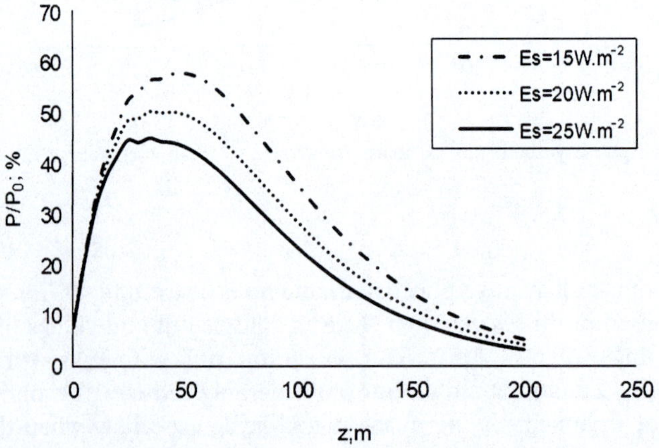


**Figure 3.** Relative photosynthetic rate vs. depth in the ocean, for a solar zenith angle of 60°, of an Earth mass planet at 0.7 AU of $\alpha$ Centauri B.

we could infer that there are better conditions for the emergence of photosynthetic life in an Earth mass planet located at 0.7 AU from $\alpha$ Centauri B than in Archean Earth, other environmental conditions being similar.

## 5. References

Biber PD, Harwell MA, Cropper WP (2003) Modeling the dynamics of three functional groups of macroalgae in tropical seagrass habitats. Ecol Model 175:25–54

Cockell CS (2000) Ultraviolet radiation and the photobiology of Earth's early oceans. Orig Life Evol Biosph 30:467–499

Cockell CS, Raven JA (2007) Ozone and life on the Archean Earth. Philos Trans R Soc A 365:1889–1901

Cullen JJ, Neale PJ (1994) Ultraviolet radiation, ozone depletion, and marine photosynthesis. Photosynth Res 39:303–320

Fritz JJ, Neale PJ, Davis RF, Pelloquin JA (2008) Response of Antarctic phytoplankton to solar UVR exposure: inhibition and recovery of photosynthesis in coastal and pelagic assemblages. Mar Ecol Prog Ser 365:1–16

Guedes J, Rivera EJ, Davis E, Laughlin G, Quintana EV, Fischer DA (2008) Formation and detectability of terrestrial planets around α Centauri B. Astrophys J 679:1582–1587

Hader DP, Worrest RC (1991) Effects of enhanced solar ultraviolet radiation on aquatic ecosystems. Photochem Photobiol 53:717–725

Neale PJ, Cullen JJ, Lesser MP, Melis A (1993) Physiological bases for detecting and predicting photoinhibition of aquatic photosynthesis by PAR and UV radiation. In: Yamamoto HY, Smith CM (eds) Photosynthetic responses to the environment. American Society of Plant Physiologists, Rockville, pp 61–77

Segura A, Krelove K, Kasting JF, Sommerlatt D, Meadows V, Crisp D, Cohen M, Mlawer E (2003) Ozone concentrations and ultraviolet fluxes on Earth-like planets around other stars. Astrobiology 3:689–708

Valley JW, Peck WH, King EM, Wilde SA (2002) A cool early Earth. Geology 4:351–354

Vincent WF, Roy S (1993) Solar ultraviolet-B radiation and aquatic primary production: damage, protection and recovery. Environ Rev 1:1–12

Biodata of **F.G. Mosqueira**, **Alicia Negrón-Mendoza**, **Sergio Ramos-Bernal,** and **Carlos Polanco-Gonzalez**, authors of "***Prebiotic Formation of Peptides: A Markov Chain Approach.***"

**Dr. F.G. Mosqueira** is Biochemistry engineer from Instituto Politecnico Nacional. He was at Faculté des Sciences, Université Libre de Bruxelles. He obtained his Ph.D. in 1981; his studies were in the group of Prof. Ilya Prigogine (a Nobel laurate in 1977). Currently, Dr. Mosqueira is working at the Science Museum Universum of the Universidad Nacional Autonoma de Mexico (UNAM). His research area is connected with the origin of life.

E-mail: **gmosque@universum.unam.mx**

**Dr. Alicia Negrón-Mendoza** was born in Mexico City. She studied at Universidad Nacional Autonoma de Mexico (UNAM). She got her Ph.D. in 1980 under the direction of Prof. Ponnamperuma, at the University of Maryland, USA. Currently, she works in radiation chemistry applied to problems in chemical evolution at Instituto de Ciencias Nucleares, UNAM, in Mexico City. Her research is related to the role of minerals and ionizing radiation in chemical evolution studies. She is pioneering in experimental studies related to chemical evolution at UNAM, Mexico.

E-mail: **negron@nucleares.unam.mx**

**F.G. Mosqueira**

**Alicia Negrón-Mendoza**



J. Seckbach (ed.), *Genesis - In The Beginning: Precursors of Life, Chemical Models and Early Biological Evolution*, Cellular Origin, Life in Extreme Habitats and Astrobiology 22, 157–169
DOI 10.1007/978-94-007-2941-4_9, 

**Dr. Sergio Ramos-Bernal** is researcher at the Instituto de Ciencias Nucleares, UNAM. He studied physics at UNAM, and he obtained his Ph.D. from the University of Manchester, England, in 1973. His research interest is solid state physics and problems related to chemical evolution and the origin of life. He has many publications in refereed journals.

E-mail: **ramos|@nucleares.unam.mx**

**Dr. Carlos Polanco-Gonzalez** was born in Mexico City. He studied mathematics at the Universidad Nacional Autonoma de Mexico (UNAM) and computer science at Universidad Autonoma Metropolitana in Mexico City. Later, in 2009, he obtained his Ph.D. in Sciences at Facultad de Ciencias, UNAM. His research area is bioinformatics and the origin of life. He has a postdoctoral position at Universidad Autonoma del Estado de Morelos, Mexico.

E-mail: **polanco@unam.mx**

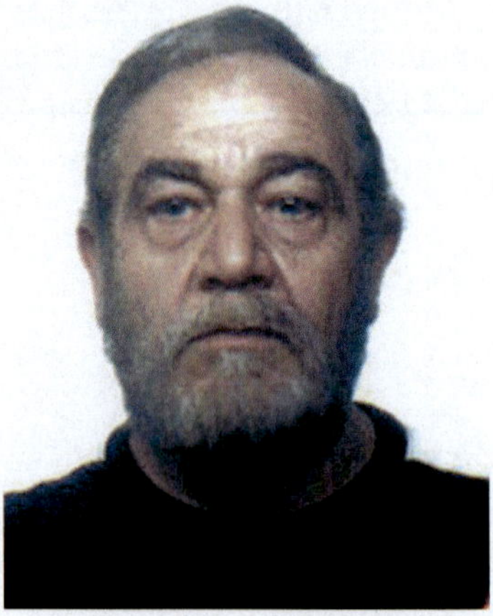

**Sergio Ramos-Bernal**

**Carlos Polanco-Gonzalez**

# PREBIOTIC FORMATION OF PEPTIDES: A MARKOV CHAIN APPROACH

**F.G. MOSQUEIRA[1], ALICIA NEGRÓN-MENDOZA[2], SERGIO RAMOS-BERNAL[2], AND CARLOS POLANCO-GONZALEZ[3]**
*[1]Dirección General de Divulgación de la Ciencia, Universidad Nacional Autónoma de México, Cd. Universitaria, AP. 70-487, 04510 México, D.F., Mexico*
*[2]Instituto de Ciencias Nucleares, Universidad Nacional Autónoma de México, Cd. Universitaria, A.P. 70-543, 04510 México, D.F., Mexico*
*[3]Centro de Investigaciones Químicas, Universidad Autónoma del Estado de Morelos, Av. Universidad 1001 A.P., 62209 Morelos, Mexico*

## 1. Introduction

We have been studying the synthesis of short sequences of amino acids as produced in simulated prebiotic conditions (anhydrous poly-condensation of amino acids). To that end, we have proposed a simple probabilistic model to more thoroughly explore oligomerization (Mosqueira et al., 2000, 2002, 2008). A main premise of such studies is that the reactivity among monomers varies. With such a view, we expect to see a bias in the sequences of the oligomers produced. A biased mechanism would result in a much *smaller* set of sequences than the set of sequences produced, assuming an *equal* probability of reaction among amino acids. That condition may favour the persistence of a given set of oligopeptides that might carry out complex functions relevant to the origin of life (Mosqueira, 1988).

We adhere to the interpretation of Sidney W. Fox and collaborators in respect to the anhydrous poly-condensation of amino acids. They found that the reactivity between different amino acids is uneven (for an overview, see Fox and Dose, 1977). Fox called this experimental finding "the principle of self-ordering of amino acids."

In our model, we adopted the Dickerson and Geis (1969) classification of amino acids, i.e. polar positive ($p^+$), polar negative ($p^-$), neutral (n), and non-polar (np), which is an electromagnetic classification. Due to our focus on possible chemical reactions among amino acids, such an electromagnetic classification is important. In chemical kinetics, we must consider the electromagnetic nature of the reacting species. For example, we may have a reaction between an ion and a molecule but quite a different situation between two non-polar molecules. With these principles in mind, we adhered to this classification of amino acids.

## 2. The Model

Let us define a finite Markov chain (Moran, 1984; Érdi and Janos, 1989; Bremaud, 1999). First, we must consider events that can occur at successive discrete stages and denote them by a variable, $k$, which can take the values 0, 1, …, $n$ …. At each stage, a finite number of events $E_1, E_2, \ldots, E_n \ldots$ can occur. These events represent the possible states of the system.

At each stage $k+1$, we suppose that the events $E_1, \ldots, E_n$ occur with certain probabilities, which depend only on the events that occurred at stage $k$ and not on anything that had happened previously. We express $p_{ij}$ for the probability of $E_j$ to occur at stage $k+1$, conditional on $E_i$ having occurred at stage $k$.

The set of quantities, $p_{ij}$, $i=1, \ldots, n$, $j=1, \ldots, n$, known as the transition probabilities, are non-negative and satisfy the following conditions:

$$\sum_j p_{ij} = 1, i = 1, \ldots, n. \tag{1}$$

Additionally, $\mathbf{P} = (p_{ij})$ is an $n \times n$ matrix, known as the transition probability (or reactivity, or stochastic) matrix of the system.

If the probabilities of the events $E_1, \ldots, E_n$ at any stage $k$ are denoted by $p_1(k), \ldots, p_n(k)$, for this matrix state after $k$ stages, we arrive at the following:

$$p_j(k+1) = \sum_i p_i(k) p_{ij}, \tag{2}$$

These equations can be written in the matrix form:

$$\mathbf{p}(k+1) = \mathbf{p}(k)\mathbf{P} \tag{3}$$

where $\mathbf{p}(k)$ is a row vector (or $1 \times n$ matrix) whose elements are $p_1(k), \ldots, p_n(k)$. Let us define a $1 \times n$ initial state matrix (or an initial state row vector) $\mathbf{p}(0)$. By applying (3) repeatedly, we see the following:

$$\mathbf{p}(k) = \mathbf{p}(0)\mathbf{P}^k \tag{4}$$

where $k$ is an integer.

We assume different electromagnetic interactions among the reacting monomers (amino acids). To that end, in accordance to Dickerson and Geis (1969), we classify amino acids into four groups: polar positive ($p^+$), polar negative ($p^-$), neutral (n), and non-polar (np). Therefore, we arrange the four possible electromagnetic interactions between amino acids into a $4 \times 4$ matrix as follows:

$$\begin{pmatrix} p^+p^+ & p^+p^- & p^+n & p^+np \\ p^-p^+ & p^-p^- & p^-n & p^-np \\ np^+ & np^- & nn & nnp \\ npp^+ & npp^- & npn & npnp \end{pmatrix} = \mathbf{P} = (p_{ij}) \tag{5}$$

Thus, the element $p_{13}$ is equal to $p^{+}n$, for example. The state of the system also is represented at any stage $k$ by a matrix of the state of the system that is a row matrix with four elements:

$$\left(p^{+}p^{-}n\,np\right) \tag{6}$$

Finally, we should make a succinct comment on the interpretation that we give to ($p_{ij}$) in equation (5), which differs slightly from an orthodox interpretation of a Markov chain transition matrix. In a Markov chain, a matrix element $p_{ij}$ signifies the probability that an entity $i$ becomes an entity $j$. In our approach, we interpret it as the probability of a chemical reaction between entities $i$ and $j$. Up to this point is the summary of the model.

## 3. Relevant Chemical Aspects in the Oligomerization of Amino Acids Under Prebiotic Conditions

Before going further, there are several chemical aspects to revisit. This process should be undertaken as our model is applied at every stage of the reaction mechanism. In general, the electromagnetic character of the reacting species changes as the chemical reaction proceeds.

### 3.1. OLIGOMERIZATION UNDER DRY AND MODERATE THERMAL CONDITIONS, ASSISTED BY THE ESSENTIAL PARTICIPATION OF TRI-FUNCTIONAL AMINO ACIDS

A set of $\alpha$-amino acids is thermally oligomerized under anhydrous conditions to simulate certain prebiotic conditions (e.g. the vicinity of volcanic areas). Though other conditions have been used, the most common temperatures used in such experiments are 160–200°C for approximately 9–12 h. It has been established that tri-functional amino acids (such as glutamic acid, aspartic acid or lysine) must be present in order to oligomerize ordinary amino acids with two functional groups (bi-functional amino acids); otherwise, the heating of purely bi-functional amino acids is recognized as a destructive treatment in which no oligomerization occurs (for a review, see Fox and Dose, 1977).

The chemical transformation of tri-functional and bi-functional amino acids using thermal energy is quite different. Hartmann et al. (1981) have studied its mechanism of reaction using glutamic acid as the tri-functional amino acid and tyrosine and glycine as the bi-functional amino acids. We will comment on each stage of the reaction mechanism in the following sections.

#### *3.1.1. Formation of the Initiator*

In a first rapid stage, pyroglutamic acid **1** is formed from glutamic acid (see Fig. 1). In terms of our model, we interpret it as follows: A glutamic acid molecule has

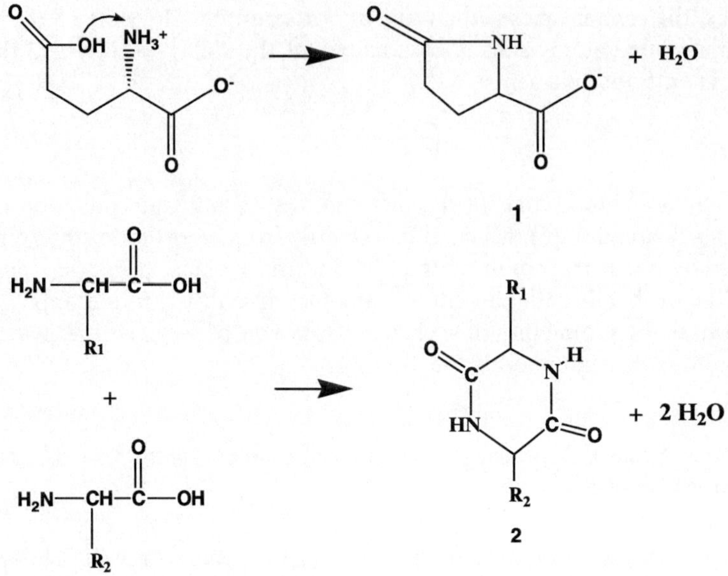


**Figure 1.** Formation of pyroglutamic acid (1) and a diketopiperazine (2).

three centres of charge (two negatives and one positive), neither having predominance. The first stage, the formation of pyroglutamic acid (pyrGlu), is an *internal cyclization* reaction that proceeds readily, because in the same molecule we have $p^+$ and $p^-$ nearby. Through internal rotation of the molecule, they react rapidly to get pyrGlu (see Fig. 1). The product of this intra-reaction has a concentrated negative charge on it, giving rise to a powerful initiator for the polymerization reaction. This characteristic previously has been remarked upon based on chemical analysis (Fox et al., 1977).

### *3.1.2. Formation of the Diketopiperazine*

In a second stage, slower than the first, diketopiperazine **2** is formed from the cyclization of two amino acids, such as tyrosine and glycine (see Fig. 1). We call this stage the *external cyclization* reaction. We may classify them as np for glycine and n for tyrosine (Dickerson and Geis, 1969). In the same experiment, another pair of amino acids may cyclize to form different diketopiperazine molecules, such as trace amounts of Gly-Gly and Tyr-Tyr.

A special question arises here because a diketopiperazine molecule contains traits of the two molecules from which it is formed. Sometimes it is easy to resolve the electromagnetic nature of the produced diketopiperazine, including when both residues of the amino acids ($R_1$ and $R_2$) have the same electromagnetic character. If $R_1$ and $R_2$ are both neutral (n), then of course the diketopiperazine that is formed will be neutral as well. If both residues are $p^+$, then the diketopiperazine that is formed will be $p^+$. When residues are from different groups, then one must choose the most influential residue to assign the character to the diketopiperazine. We have analyzed all possible combinations of residue groups and found the emergence of a new electromagnetic species: a dipole (*d*), constituted by the cyclization of a $p^+$ residue and a $p^-$ residue (Mosqueira et al., 2008).

### *3.1.3. Oligomerization*

Finally, in a third stage, pyroglutamic acid reacts with the diketopiperazine to form (in this example) two tripeptides: pyroGlu-Tyr-Gly and pyroGlu-Gly-Tyr (or in general, pyroGlu-$R_1$-$R_2$ and pyroGlu-$R_2$-$R_1$). We emphasize that both trimers now have a negative charge at one end of the oligomer. This charge arises because pyroGlu breaks the cyclic diketopiperazine molecule and the resulting trimers end with their acid group, bearing a negative charge ($p^-$). Then, the negative charge of the initiator pyroGlu ($p^-$) is reconstituted in both trimers (see Fig. 2). The reconstitution of $p^-$ continues indefinitely because the oligomer always contains a terminal amino acid with its exposed acidic group (see Fig. 3).

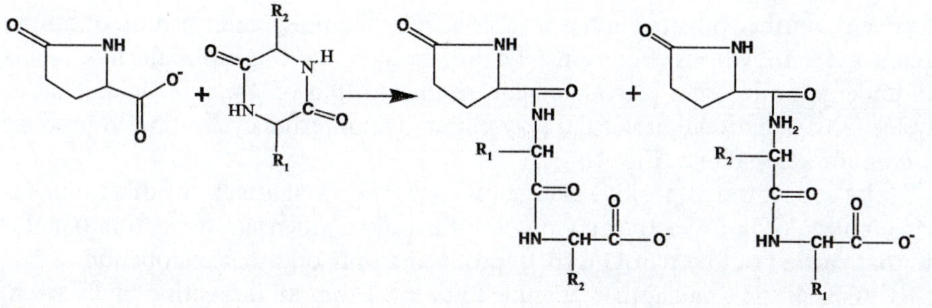

**Figure 2.** Reaction of a diketopiperazine with pyroglutamic acid to synthesize two tripeptides with different sequence.

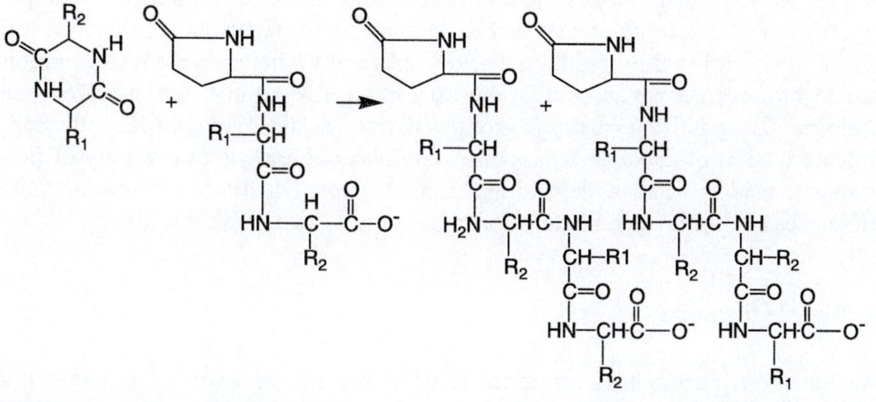

**Figure 3.** Synthesis of two pentapeptides with different sequence from one possible tripeptide and a given diketopiperazine molecule.

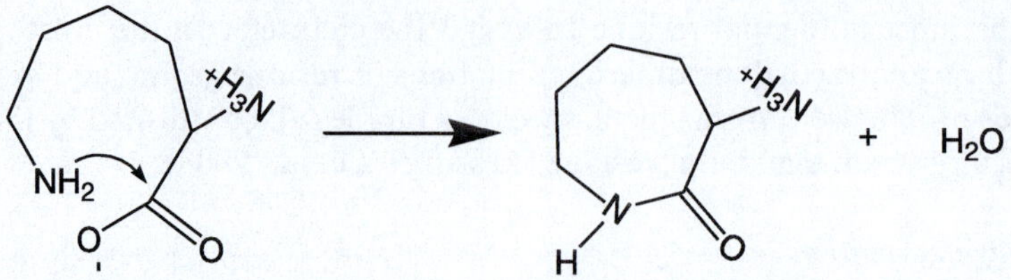


**Figure 4.** Cyclization of lysine.

Another feature of this reaction is that odd-oligomers always are produced (only 3, 5, 7, 9, ... monomers in the oligomers are formed). We call the above mechanism as the *diketopiperazine reaction mechanism* (Mosqueira et al., 2008).

### *3.1.4. Other Initiators*

Lysine is another possible initiator molecule for the poly-condensation of amino acids under anhydrous conditions. We may assume that this molecule, too, is able to undergo an internal rearrangement, similar to that of glutamic acid. That is, lysine (a tri-functional molecule) may undergo an internal cyclization to produce a monofunctional one (Fig. 4).

Its concentrated positive charge ($p^+$) may very well attack the diketopiperazine molecules in order to carry on with the poly-condensation reaction (similar to that carried out by pyroGlu ($p^-$)), producing only odd-mer oligopeptides.

Aspartic acid is another possible initiator. From an inspection of its structural formula, it is clear that it has one methyl group less than glutamic acid. It is tempting to suppose that aspartic acid may also undergo a molecular rearrangement to allow internal cyclization similar to that of glutamic acid (using its amino and carboxyl groups). Such an assumption would lead to a speculated cyclic species of four atoms, which seems to be too unstable to participate in the chemical reaction. (Recall that pyroglutamic acid is a five-atom heterocyclic nitrogen molecule.) However, aspartic acid is able to form a five-atom heterocyclic oxygen molecule using its two carboxyl groups (Kovacs et al., 1961). In fact, the self-polymerization of aspartic acid is more complicated and yields a variety of polyaspartic acid molecules, depending on its degree of hydration. Some of these molecules exhibit nitrogen heterocyclic monomeric units with five atoms.

## 4. Biased Oligomers

We had been particularly interested in studying the generation of biased oligopeptides. Such conditions may favour the persistence of a given set of oligopeptides that may carry out complex functions relevant to the origin of life (Mosqueira, 1988).

Fox and Dose (1977) and collaborators have extensively investigated and interpreted experimental findings. These studies indicate a relatively reduced number of thermal proteins as compared to the much larger number of such polymers that would be expected, assuming an even probability of reaction between amino acids. Fox named this result *"the principle of self-ordering of amino acids."* We have adopted and reinforced this view by proposing a probabilistic model—by means of Markov chains—to delve into the study and consequences of assuming *unequal* probability of reaction among amino acids.

In the next sections, we briefly review experimental results showing this aspect. Then, we present some predictions on the nature of such biased oligopeptides derived from our simple probabilistic model.

## 4.1. BIASED TRIPEPTIDES OBTAINED BY EXPERIMENT

A highly biased composition of thermal tyrosine-containing tripeptides has been reported (Nakashima et al., 1977). Due to experimental facilities, researchers only sought trimers containing tyrosine. The researchers expected to find 36 trimers in their experiment based on bias-free randomness. However, they found only two tyrosine-containing tripeptides out of 36 expected. Hartman et al. (1981) further studied the reaction mechanism of this reaction. Therefore, our simple probabilistic model was able to account for the experimental finding of only two trimers versus 36 possible trimers (Mosqueira et al., 2000).

The experimental results are limited to these simple oligopeptides. To our knowledge, there are no reports for the sequencing of larger oligopeptides produced by chance under prebiotic conditions.

## 4.2. PREDICTIONS OF THE MODEL: BEHAVIOUR OF PARTICULAR STEADY STATES

We have applied the reactivity matrix (5) to particular reacting systems in which only two different species participate. This situation is reflected in matrix (5), which allows only two rows with elements different from zero. However, the other two rows are all elements equal to zero (Mosqueira et al., 2002).

Three different situations may arise depending on the specificity of the reactants. The following sections detail each scenario.

### *4.2.1. Diagonal Interactions of the Reactivity Matrix Are Neglected*

This case illustrates the situation in which both reactants interact with high specificity. Neither is able to self-react. The appearance of the reactivity matrix (5) is a row with one element equal to one, and the other elements in the same row forcefully are equal to zero. It also has another row with the same appearance. The other two rows have zeros everywhere. Such systems present a sustained oscillatory steady state.

#### 4.2.2. *Equal Symmetrical Interactions with Diagonal Interactions Non-null and Small*

This case illustrates a situation in which both reactants interact with less specificity. Both reactants are able to self-react smoothly. The appearance of the reactivity matrix (5) contains one row with an element equal to $a_{ij}$ and another equal to $1 - a_{ij}$ (we recall that $i, j = 1, 2, 3, 4$). The other two elements in the same row are equal to zero. Another row has an element equal to $a_{ji}$ and another equal to $1 - a_{ji}$. The other two rows have zeros everywhere.

Under such conditions, we find a series of transient state matrices characterized by a damped oscillatory behaviour that approaches the steady-state matrix (0 0.5 0.5 0).

#### 4.2.3. *Symmetrical and Diagonal Interactions Equal to 0.5*

The reaction matrix (5) in this case has two elements equal to 0.5 in one row and another row with the same structure. All other elements are equal to zero. If we start with an initial matrix, for example (0 $x$ $y$ 0), where $x$ and $y$ are non-negative and satisfy the condition $x+y=1$, we reach the steady-state matrix (0 0.5 0.5 0) in a single stage (i.e. in $k=1$).

In summary, with the exception of the case in Sect. 4.3.1, all steady states arrive to the steady-state matrix with two matrix elements equal to 0.5. This peculiar situation arises from the symmetrical form of **P**. To assign $a_{ij} = a_{ji}$ means that we make no distinction in the interaction of object $i$ to object $j$, in comparison to the interaction of object $j$ to object $i$. For this reason, either after a short transient (i.e. $k=1$) or a longer one ($k \approx 12$), we arrive to a steady-state matrix with two elements equal to 0.5. This result corroborates the "*initial concentration independence*" mentioned in Sect. 4.2.1.

These examples represent a sample from the set including all pairwise electromagnetic interactions among four classes of amino acids. The assignment of more specific numerical values to the transition probabilities and electromagnetic kinds is open to future work.

### 4.3. PREDICTIONS OF THE MODEL: LIMITED DIVERSITY

There are several factors that contribute to reducing the variety of oligopeptides in the sequence space. The most important factors are as follows: (1) the unequal probability of reaction among amino acids, (2) the existence of a Markov chain steady state, (3) an observed independence of the initial conditions of the system, and (4) production of only odd-mer peptides. In other words, all even-mer peptides are excluded from the resulting products. These factors will be reviewed in the following sections.

#### 4.3.1. *Unequal Probability of Reaction Among Amino Acids*

This condition appears to be self-evident. The probability cannot be the same to cause a reaction among pair $p^+$ with $p^-$ than another pair of species like neutral (n) and non-polar (np). The consequences of considering *equal* probabilities of reaction among amino acids have been studied in detail elsewhere (Mosqueira, 1988).

*4.3.2. Convergence to the Steady State and Independence of Initial Conditions*

The state of the system—in concentrated variables—is represented at any stage $k$ by a row matrix with four elements represented in Eq. 6. The reactivity matrix (5) entails the unequal probability of reaction among amino acids. On the other hand, Eq. 3 is the description of the oligomerization reaction; it describes how the state of the system changes from the state $k$ to the state $k+1$.

Similarly, as a differential equation attains its steady state under appropriate conditions, a Markov chain also attains a steady state. To calculate such a steady state of the system, we set (Moran, 1984):

$$\mathbf{p}(k)=\mathbf{p}(k)\mathbf{P} \tag{7}$$

In words, this equation states that the state of the system changes from state $k$ to state $k$. The row matrix ceases to change when we apply the reactivity matrix **P**. This is, by definition, the steady state of the system. Solving Eqs. 7 and 1 allows one to find the row matrix representing the concentration of the steady state. We have seen from our experience that once $k$ has attained a few stages of transitions (i.e. around $5<k<12$), it has attained the steady state of the system. Furthermore, this condition persists at all subsequent stages, as long as the process is sustained, i.e. in our case, as long as the chemical process of polymerisation proceeds.

The attainment of a steady state is already an important mechanism that limits variability in polymer sequencing. The state matrix is fixed in its steady state (see Eq. 7). This condition prevents a state matrix from roaming about over a huge sequence space that has been shown to exist that would introduce a greater variability in oligomer sequencing (Mosqueira, 1988). However, a quantitative evaluation of the extent of variability limitation has not been done so far.

Let us look at the variables on which the steady state depends. From Eqs. 4 and 7, it is obvious that the steady state depends on two factors: the initial state matrix **p**(0) and the transition matrix **P**. However, from our analysis, it appears that steady-state dependence is mostly on **P**. This dependence exists because we have verified the independence of the initial conditions on the steady state from our previous works. Regardless of the initial conditions (concentrations of amino acids), we arrive at the same steady state. In fact, this situation also occurs in differential equations.

Then, the independence of initial conditions—in conjunction with the attainment of the steady state—contributes to achieving biased oligopeptides.

*4.3.3. Production of Only Odd-Mer Peptides*

In Sect. 3.1.3, we outlined the reaction mechanism via the diketopiperazine molecule, which is a cyclic dehydrate condensation of two bi-functional amino acids. This mechanism allows only the production of odd-mer oligopeptides, with the exclusion of all even-mer oligopeptides. This fact surely is another aspect that reinforces the production of biased oligopeptides.

#### 4.3.4. *In the Prebiotic World, Heterooligopeptides Were More Common than Homooligopeptides*

It is instructive to give a *qualitative* figure for the interactions between different pairs of amino acids. Possible combinations are as follows:

$$\begin{pmatrix} p^+p^+ & p^+p^- & p^+n & p^+np \\ p^-p^+ & p^-p^- & p^-n & p^-np \\ np^+ & np^- & nn & nnp \\ npp^+ & npp^- & npn & npnp \end{pmatrix} = \begin{pmatrix} 0.01 & 0.79 & 0.15 & 0.05 \\ 0.79 & 0.01 & 0.15 & 0.05 \\ 0.4 & 0.4 & 0.15 & 0.05 \\ 0.4 & 0.4 & 0.15 & 0.05 \end{pmatrix} \quad (8)$$

Notice that the sum of values of the elements in each row is unity, according to Eq. 1. Besides, the higher values of matrix elements correspond to interactions that we know to be more intense from physical chemistry, including $p^+p^-$ or $p^-p^+$. On the other hand, lower values are given for interactions that are known to be much weaker, including nnp and $p^+p^+$.

From this perspective, it can be envisaged that contiguous alike charges or monomers will not be favoured in a polymerization process. On the contrary, it would be easier to unite contiguous charges of different polarity. With this background, we predict that for oligopeptides produced by purely prebiotic means, the heterooligopeptides would be more prevalent than the homooligopeptides. Such conditions would be welcome in the prebiotic environment because heterooligopeptides likely would have more pre-catalytic activities than homooligopeptides.

## 5. Conclusions

In this work, we have proposed a simple probabilistic model to describe the oligomerization of amino acids under prebiotic conditions. We have emphasized the role of unequal probabilities of reaction among amino acids to produce biased oligopeptides within the frame of the reaction mechanism "*diketopiperazine reaction.*" From this perspective, our probabilistic model has been able to justify some experimental results in respect to the synthesis of particular tripeptides. In addition, several predictions derived from our model have been put forward and are waiting to be experimentally verified.

Of particular relevance to us is one of the predictions related to the nature of primordial oligopeptides. In the prebiotic world, in anhydrous environments with a steady source of heat, it would be more likely to have heterooligopeptides than homooligopeptides. This idea may be unexpected as it might be thought that primitive oligopeptides were highly monotonous, with monomers being repeated throughout the sequence with little variation. This model instead suggests a primitive world with not so much monotonous sequences of oligopeptides, and with an implied catalytic potential.

## 6. Acknowledgements

This work was supported by PAPIT-UNAM grant IN104109-3 and CONACyT grant 82937.

## 7. References

Bremaud P (1999) Markov chains: Gibbs fields, Monte Carlo simulation, and queues, vol 31, Series texts in applied mathematics. Springer, Heidelberg, 444p

Dickerson RE, Geis I (1969) The structure and action of proteins. Harper and Row Publishers, New York

Érdi P, Tóth J (1989) Mathematical models of chemical reactions. Theory and applications of deterministic and stochastic models. Manchester University Press, Manchester, pp 91–161

Fox SW, Dose K (1977) Molecular evolution and the origin of life. Marcel Dekker, Inc., New York, 370p

Fox SW, Melius P, Nakashima T (1977) N-Terminal pyroglutamyl residues in proteins and thermal peptides. In: Matsubara H, Yamanaka T (eds) Proceedings of the symposium on evolution of proteins molecules. Scientific Society Press, Tokyo

Hartmann J, Brand MC, Dose K (1981) Formation of specific amino acid sequences during thermal polymerization of amino acids. Biosystems 13:141–147

Kovacs J, Kovacs HN, Könyves I, Császár J, Vajda T, Mix H (1961) Chemical studies of polyaspartic acids. J Org Chem 26:1084–1091

Moran PAP (1984) An introduction to probability theory. Clarendon, Oxford, 542p

Mosqueira FG (1988) On the origin of life event. Orig Life Evol Biosph 18:143–156

Mosqueira FG, Ramos-Bernal S, Negrón-Mendoza A (2000) A simple model of the thermal prebiotic oligomerization of amino acids. Biosystems 57:67–73

Mosqueira FG, Ramos-Bernal S, Negrón-Mendoza A (2002) Biased polymers in the origin of life. Biosystems 65:99–103

Mosqueira FG, Ramos-Bernal S, Negrón-Mendoza A (2008) Prebiotic thermal polymerization of crystals of amino acids via the diketopiperazine reaction. Biosystems 91:195–200

Nakashima T, Jungck JR, Fox SW, Lederer E, Das BC (1977) A test in randomness in peptides isolated from a thermal polyamino acid. Int J Quant Chem v Quant Biol Symp 4:65–72

Biodata of **Maciej Pawlikowski**, author of "***Atomic Structural Templates of the Earliest Life on Earth: Vibration and Lightning Experiments with Quartz and Amino Acids.***"

**Professor Maciej Pawlikowski** is currently a research scientist of the Faculty of Geology, Geophysics and Protection of Environment in AGH—University of Science and Technology, Cracow, Poland. He obtained his Ph.D. from AGH—University of Science and Technology (in 1975) and continued his research at the same university. Professor Pawlikowski's scientific interests are in the areas of life creation by minerals as well as in biomineralogy, i.e. minerals present in human tissues. Part of his work is devoted for investigation of archaeological objects and artefacts. He teaches mineralogy, biomineralogy and petroarchaeology.

E-mail: **mpawlik@agh.edu.pl**

J. Seckbach (ed.), *Genesis - In The Beginning: Precursors of Life, Chemical Models and Early Biological Evolution*, Cellular Origin, Life in Extreme Habitats and Astrobiology 22, 171–177
DOI 10.1007/978-94-007-2941-4_10, 

# ATOMIC STRUCTURAL TEMPLATES OF THE EARLIEST LIFE ON EARTH: VIBRATION AND LIGHTNING EXPERIMENTS WITH QUARTZ AND AMINO ACIDS

MACIEJ PAWLIKOWSKI
*Laboratory of Biomineralogy, Cathedral Mineralogy, Petrography and Geochemistry, Faculty Geology, Geophysics and Protection of Environments, AGH University of Science and Technology, Cracow, Poland*

## 1. Introduction

The atomic structure of some organic molecules and mineral chains is remarkably similar (Fig. 1) (Pawlikowski, 1995; Pawlikowski and Palka–Zielińska, 2000). This has led to recent investigations into how the earliest organic life formed on Earth. Most of them are devoted to organic compounds (Miller, 1953; Oparin, 1953; Cairns-Smith, 1982; Orgel, 1994; Muller, 1995; Robertson and Miller, 1995).

## 2. Methodology

Experiments were performed to assess the alternation of amino acid structures by vibrating piezoelectric quartz. The experiments consisted of two phases.

### 2.1. PHASE I

Plates of piezoelectric quartz of various thickness (therefore different vibration frequencies) were excited using a specially prepared generator. The quartz plates were coated with gold, and a small window on the surface of the plates was coated with various amino acids.

The plates were recorded on an infrared spectrophotometer under the following parameters: Bio-Rad and Win-IR. Two IR analyses of amino acids were obtained, the first analysis when the quartz plate was still and the second analysis when the plate was vibrated.

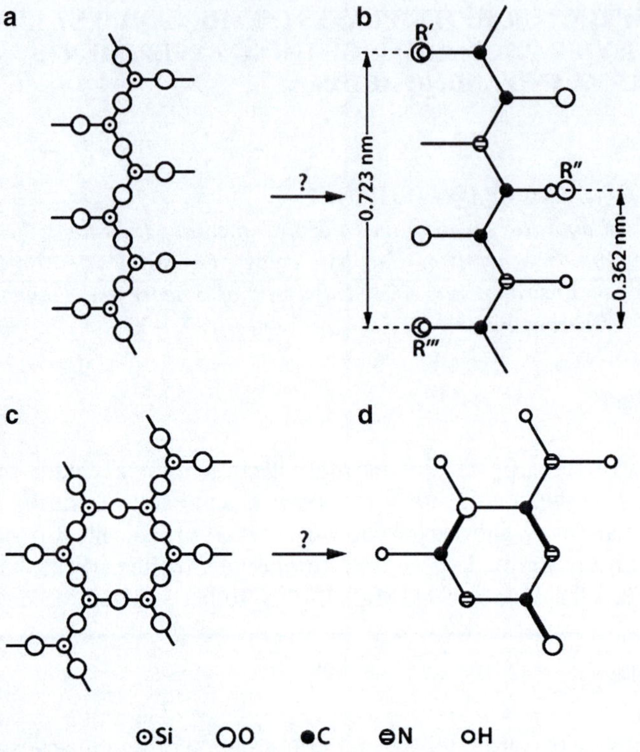


**Figure 1.** Similarity of selected mineral and organic structures. (**a**) The structure of chain silicate (Bolewski, 1965), (**b**) structure of peptide chain (Lasek, 1978), (**c**) structure of layered silicate (Bolewski, 1965), and (**d**) structure of cytozine, one of most important DNA component (Jerzmanowski, 1994).

## 2.2. PHASE II

This experiment was performed with quartz, artificially created lightning and amino acids dissolved in $H_2O$ (4% liquids). The lightning generator was specially constructed for these experiments (Fig. 2). It houses a rotating container within which was placed a mixture of crushed pure quartz and a liquid solution of amino acids (uracil, glycine, glutamic acid, alanine, L-alanine, L-tryptophan, DL-arginine). The surface of the quartz and amino acid mixture was struck by artificial lightning (at about 50 kV). Two IR curves of amino acids were obtained, the first with no lightning and the second after the mixture was activated by artificial lightning.

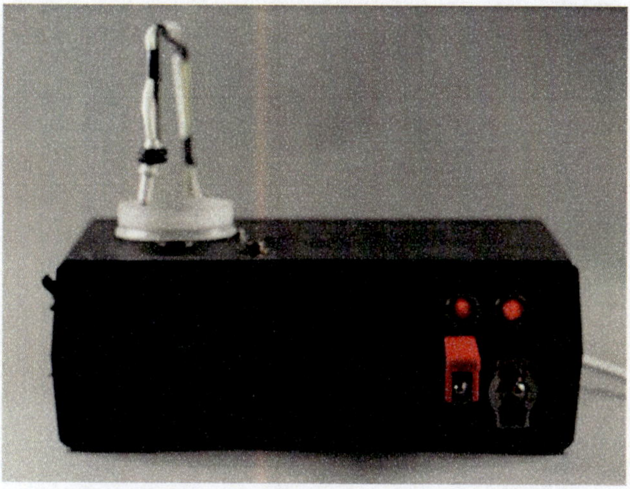

**Figure 2.** Apparatus used for phase II experiments.

## 3. Results

### 3.1. PHASE I

On the IR spectral analysis, the positions of the amino acid absorption peaks change during vibration.

### 3.2. PHASE II

The analyses of both IR curves suggest that the position of selected IR absorption peaks was changed by a reaction with artificial lightning. The spectral data are also presented in Table 1.

## 4. Interpretation

### 4.1. PHASE I

The change in IR absorption frequencies is interpreted to be a direct result of changes in the amino acid structure during vibration. It is likely that the primary structures of the amino acid are modified.

**Table 1.** Data from phase II experiments. The IR absorption peak data was obtained after 24 h of interaction between quartz, artificial lightning, and glycine. The position of the peaks is measured in $cm^{-1}$.

| Natural glycine | Glycine after experiment | |
|---|---|---|
| **Primary position** | **Secondary position** | **Various bands of absorption** |
| 752.935 | – | × |
| 891.328 | 891.328 | |
| 997.985 | 997.985 | |
| 1034.09 | 1034.09 | |
| 1132.49 | 1132.49 | |
| 1331.09 | 1330.85 | × |
| 1412.94 | 1412.56 | × |
| 1513.43 | 1507.41 | × |
| 1602.00 | 1597.00 | × |

### 4.2. PHASE II

Table 1 shows that the amino acid was slightly transformed during activation by lightning.

Though experiments performed at phases I and II are various because at phase I we have selected frequency while at phase II quartz grains produce a spectrum of frequency, both methods showed possibility of modification of amino acids structure. Unfortunately, what was not observed was the vibration of amino acids after stopping the quartz vibration, but this phenomena may be active because of the very short time that is impossible to observe using the methods described in this chapter.

## 5. Discussion

In relation to the hypothesis concerning the creation of life on Earth, the effect of lightning and vibration of quartz is to alter the amino acid structure from stable to vibrating. Later, it is possible that vibrating molecules of amino acids bonded one with another and modified, leading to more and more complicated biological forms.

Unfortunately, the evidence for new and primitive forms of amino acid compounds in the rock record has not been preserved, due to the age of these molecules and the effect of geological processes.

In relation to presented hypothesis of early life on Earth, it is understandable that there is no life (as we know it) on the nearest astral bodies (planets, moons, etc.), as the surface geological layers do not contain quartz.

## 6. Conclusion

It was observed that vibrating quartz may alter the internal structure of selected amino acids. This phenomena is observed as a change in IR peak position.

## 7. References


Bolewski A (1965) Mineralogia szczegółowa (Detailed mineralogy). Wyd. Geol, Warszawa

Cairns-Smith AG (1982) Genetic takeover and the mineral origins of life. Cambridge University Press, Cambridge

Jerzmanowski A (1994) Geny i ludzie (Gens and Man). Wyd. Szkolne i Pedagogiczne, Warszawa

Lasek W (1978) Kolagen – chemia i wykorzystanie (Collage – chemistry and use). Wyd. Nauk.-Techn., Warszawa

Miller SL (1953) A production of amino acids under possible primitive Earth conditions. Science 117(3046):528–529

Muller AWJ (1995) Were the first organisms heat engines? A new model for biogenesis and the early evolution of biological energy conversion. Prog Biophys Mol Biol 63(2):193–231

Oparin AI (1953) Origin of life. Dover Publications, New York

Orgel L (1994) The origin of life on Earth. Sci Am 271(4):81

Pawlikowski M (1995) Sekrety mineralizacji tkanek (Secrets of tissues mineralization). Wyd. Centrum PPGSMiE PAN, Kraków

Pawlikowski M, Palka–Zielińska E (2000) Mineralogical research of quartz – organic compounds interaction: possible factor in life creation phenomena. In: Materiały Konferencji pt.: Mineralogy and life. Syktyvkar, Republic of Komi, Russia, June 5–8, pp 142–144

Robertson MP, Miller SL (1995) An efficient prebiotic synthesis of cytosine and uracil. Nature 375(6534):772–774

Biodata of **Michael Simakov**, author of "***The Chemical Evolution in Open Space: A Link to the Origin of Life***."

**Dr. Michael Simakov** is a scientific researcher in the Group of Exobiology, Institute of Cytology, Russian Academy of Sciences, St. Petersburg, Russia. He obtained his Ph.D. (1991) in organic chemistry at St. Petersburg State University. His current research interests include (1) the origins of life, chemical evolution, abiogenic synthesis of biologically important compounds; (2) the role of energy sources of open space (ultraviolet radiation of different wavelengths, high-charged particles) in abiogenic synthesis of nucleotides and oligopeptides; (3) exobiology—possible sites of exobiological activity on Titan. He designed and conducted the exobiological scientific experiments on board of space satellites "Bion" and Russian module of International Space Station. Dr. Simakov is a member of Scientific Council of Russian Astrobiology Center as well as a member of other societies such as ISSOL and IAU Commission 51-Bioastronomy.

E-mail: **exobio@mail.cytspb.rssi.ru**

J. Seckbach (ed.), *Genesis - In The Beginning: Precursors of Life, Chemical Models and Early Biological Evolution*, Cellular Origin, Life in Extreme Habitats and Astrobiology 22, 179–200
DOI 10.1007/978-94-007-2941-4_11, 

# THE CHEMICAL EVOLUTION IN OPEN SPACE: A LINK TO THE ORIGIN OF LIFE

MICHAEL SIMAKOV
*Group of Exobiology, Institute of Cytology, Russian Academy of Sciences, Tikhoretsky Av., 4, St. Petersburg 194064, Russia*

## 1. Introduction

It is now well known that carbonaceous chondrites contain large quantities of prebiotic molecules, including amino acids, carbohydrates, and heterocyclic bases of nucleic acids. It has become evident that the chemical evolution proceeds under conditions of open space on the small bodies of the Solar system, which surface serves as a huge chemical reactor. The extraterrestrial origin of these prebiotic molecules has been established on the basis of diversity, molecular structure, and isotopic composition. The molecular composition of a meteorite's organics resembles abiotic processes governed by physicochemical processes that occurred on the parent bodies, including asteroids. The synthesis of soluble compounds in carbonaceous chondrites consists of a multistep process in which organic precursors were formed in the interstellar cloud, incorporated with many volatiles into icy planetesimals, and upon aqueous processing produced the variety of compounds which can form very complex organics in such harsh environments. We show experimentally that the solid mixtures of biologically significant compounds (nucleosides and amino acids) produce more complex compounds when they are exposed to either vacuum UV photons or ionizing radiation.

## 2. The Places of Chemical Evolution in Space

The standard scenario of planetary formation of the terrestrial planets and the cores of giant planets are formed through accretion of many small bodies called planetesimals (Kokubo and Ida, 2000). The lifetime of protoplanetary disks span the range of $10^6$–$10^7$ years. And at this stage, the complex chemical processes could take place on the surface of the planetesimals. A typical photon flux for diffuse interstellar environments is estimated to be $8 \times 10^7$ photons $cm^{-2}s^{-1}$ for photons with energies higher than 6 eV (Mathis et al., 1983). In more opaque dense interstellar clouds, fluxes are expected to be 3 (Shen et al., 2004) to 5 (Prasad and Tarafdar, 1983) orders of magnitude lower. Prior to the onset of star formation, dust grains effectively shield molecules from interstellar UV photons; however, cosmic rays can penetrate throughout and drive a rich ion–molecule chemistry, supplemented by radical processes, in which many complex organic species may be produced. Interstellar cloud chemistry, including radical and ion chemistry, leads for organic compounds synthesized at low density and low temperature to abundances different from those that would result from the solar nebula chemistry (neutral chemistry at warmer temperature and higher density). Star formation has a profound effect on the chemistry of the surrounding medium. During and after accretion, the processed materials are exposed to different energetic processes (thermal, action of the cosmic rays, UV with different wavelengths, X-ray fluxes) all of which can act to alter the original starting organic material. The young stars emit to $10^4$ times more UV than the present Sun, and of course, the inner part of planetary nebula was sterilized by such light. But models of protoplanetary disks suggest a multilayer structure, and the chemical evolution could take place in midplane regions where UV radiation is not such harsh. As dust grains with ice mantles are transported to warm, dense, and active protostellar regions, the processes such as ultraviolet radiation and cosmic ray bombardment may become important and alter the grain composition. Laboratory experiments show that irradiation processes of ice analogs lead to the formation of radicals, complex molecules, and complex organic refractory material (Cottin et al., 1999). As we can see, the main attention in experimental works devotes now to energetic chemistry in "simulated ice conditions," i.e., interstellar grains and cometary ices. However, the part of grain could lose their highly volatile mantles, but have less volatile species yet. The reaction of complex organic compounds can take place on the solid surface of such grains. When dust–gas mixtures of the solar nebula segregate into large-sized objects, the solar wind and flares can reach the planetary region and may have a significant role in synthesis of organic compound. So, all young solar system objects are subjected to energetic processing by photons and ions. As a result, the chemical and physical properties of the materials composing these objects will change over time significantly. Energetic processing of organic compounds into more complex species can be driven by significantly enhanced UV field in star-forming regions, high-energy particle bombardment and UV radiation from the T Tauri phase in stellar birth, and UV radiation of different wavelengths, protons of the Solar wind, and flares at early stage of evolution and at present days. The surviving remnants of the population of planetesimals which filled the inner early solar system during the late stages of collapsing solar nebula had a tremendous total surface, and the reactions of complex organic compounds could take place on the solid surface of those bodies. It was important to test experimentally how far the process of chemical evolution could take place on the surface of space bodies under action of energy sources available at that period. It is now well known that carbonaceous chondrites contain large quantities of prebiotic molecules (Botta and Bada, 2002), including amino acids (Martins et al., 2007), carbohydrates (Cooper et al., 2001), and heterocyclic bases of nucleic acids (Martins et al., 2008). Evidently, the chemical evolution proceeds under conditions of the open space, which serves as a huge chemical reactor. The extraterrestrial origin of the meteoritic organics is established on the basis of diversity, molecular structure, and isotopic composition (Pizzarello et al., 2004). Primitive chondritic meteorites contain up to 2 wt.% organic, and interplanetary dust particles (IDPs) can contain even more. This organic is generally divided into soluble (SOM) and insoluble (IOM) organic matter fractions, the IOM being the dominant fraction (75–99%). Both fractions can have large D and $^{15}N$ enrichments. In the SOM, there are considerable isotopic variations among individual compounds within and between meteorites (Pizzarello et al., 2008). It is almost universally accepted that the large D and $^{15}N$ enrichments in the organics are the products of low-temperature chemistry in the interstellar medium (ISM) or in the early outer Solar system. The amino acids share a deuterium enrichment found in other meteoritic organic extracts, such as kerogen, strongly hinting at synthesis by processes operating in the interstellar medium or the Solar nebula and subsequently modification in the meteorites' asteroidal parent bodies (Herd et al., 2011). The chondrites contain a much wider variety of amino acids as normally found in terrestrial organisms, as never used in terrestrial biochemistry.

The comets compose the second class of space bodies interesting from astrobiological point of view. Comets are often considered as the last witness of the formation of our solar system (Cottin et al., 1999). There are roughly 80 species firmly identified in comets, but this is almost certainly a very incomplete list (Despois and Cottin, 2005). Ethylene glycol, a simple sugar-like compound, is the largest molecule discovered in comets (Crovisier et al., 2004). Almost all of them are also seen in the interstellar medium (Charnley and Rodgers, 2008). And it is now quite obvious that comets have brought to the Earth a certain amount of volatile compounds such as water and other inorganic and organic compounds (Owen, 2008).

The NASA Stardust mission successfully collected material from Comet 81P/Wild 2, including authentic cometary grains. Analysis of these samples indicates the presence of oxygen-rich and nitrogen-rich organic materials, which contain a broad variety of functional groups (carbonyls, C C bonds, aliphatic chains, amines, amides, etc.). Stardust samples were also shown to contain glycine, the smallest biological amino acid, and alanine (Sandford et al., 2006). Organic residues produced from the UV irradiation of astrophysical ice analogs are already known to contain a large suite of organic molecules including amino acids, amphiphilic compounds (fatty acids), and other complex species (Cottin and Despois, 2009).

The "simulated space ice conditions" experiments have shown the synthesis of simple biochemical compounds in the form of amino acid's precursors (Bernstein et al., 2002; Munoz Caro et al., 2002; Hudson et al., 2008; Nuevo et al., 2008) and pyrimidine bases (uracil, cytosine, and thymine) of the nucleic acids (Kobayashi et al., 2004; Nuevo et al., 2009). UV radiation probably causes significant photochemistry within the mixed molecular ices found in space. This process has been simulated in the laboratory using high-tech gadgetry. By simulating the photochemistry of interstellar and cometary ices in the laboratory, we were able to produce a host of organic compounds (i.e., compounds composed primarily of carbon, the kinds of molecules from which we and all living things are made). We suggest that this process may be responsible for the richness of the organics seen in the diffuse and dense interstellar medium, comets, and meteorites.

Simple organic compounds were formed originally as products of low-temperature reactions that occurred in the interstellar molecular clouds, under radiation processing of astrophysical ices, and by gas-phase reactions in the primordial solar nebula (Fisher–Tropsch-type reactions), and they entered the parent bodies at the time of their accretion.

Such reactions may have been involved in the production of interstellar molecules, organic compounds in meteorites, and prebiotic organic matter on planets. The production of organic compounds on asteroids as parent bodies of the meteorites may have been catalyzed by the minerals found on the surface of these bodies. There are many environments in space in which it is possible to find solid bodies subjected to UV irradiation and particle bombardment, such as interplanetary dust particles, asteroids, comets, meteorites, and planetary satellites.

McDonald et al. (1996) reported the production of a large diversity of organic compounds by experimental simulation. Alkanes, ketones, esters, and other complex molecules have been detected by gas chromatography/mass spectrometry following the irradiation of the mixture $H_2O$:$CH_3OH$:$CO_2$:$C_6H_6$ (80:16:3.2:0.8) with helium plasma. A wide variety of the initial amino acids were also found following the irradiation of an ice mixture of propane (or methane, or carbon monoxide), ammonia, and water with 3 MeV protons. Dr. Kobayashi with coworkers identified such compounds as glycine (the most abundant), alanine, and α–amino–*n*–butyric acid (Takahashi et al., 1999). Amino acids were formed from carbon monoxide, nitrogen, and water, which are possible constituents of extraterrestrial atmospheres, following irradiation with high-energy particles (protons, helium nuclei, and electrons, all components of cosmic rays). Glycine; aspartic acid; serine; alanine; threonine; glutamic acid; α–, β–, and γ–aminobutyric acids; α– and β–aminoisobutyric acids; valine; norvaline; sarcosine; and proline were detected (Kobayashi et al., 1998). These compounds were identified after irradiation of different gas mixtures ($CH_4$–$N_2$–$H_2O$; CO–$NH_3$–$H_2O$; CO–$N_2$–$H_2O$) by X-ray (Takahashi et al., 1999) and vacuum ultraviolet (VUV) irradiation (Takahashi et al., 2005). Kobayashi et al. (1995) showed that some amino acids, such as glycine and alanine, were detectable in hydrolysates when a mixture of simple inorganic compounds (CO, $NH_3$, and $H_2O$) was irradiated in a cryostat at 10 K (simulated cometary ices) with high-energy protons. The synthesis of simple oligopeptides (Gly–Trp and Gly–Tyr) from a solid mixture of single amino acids under the action of vacuum UV radiation ($\lambda < 200$ nm) was reported by Simakov et al. (1996).

Among several energy sources available for abiogenic synthesis of biomolecules in open space, UV light with different wavelengths and cosmic rays are two of the most abundant. All extraterrestrial material exhibits strong evidence that irradiation by H and He ions is an important process on solids throughout the interstellar and interplanetary medium (Bradley, 1994) and includes a rich variety of materials with different histories. The chemical anomalies accumulate in grains exposed to ionizing radiation.

Organics synthesized by the processing of interstellar materials could have been incorporated into cometary ice and then delivered to the early Earth. The chemical evolution of the simple molecules would continue on the surface of these bodies under the action of the energy sources available in open space. Simple oligopeptides can be formed on solid state by proton radiation, vacuum UV light, heat, and $\gamma$–radiation (Simakov et al., 1996). Thus, it can be assumed that the chemical evolution would have taken place during the early stage of the Solar system origin and reached the stage of polymerization before the end of planet accretion. Such complex biomolecules as oligopeptides would have been brought onto the primordial Earth with comets, carbonaceous chondrites, and interplanetary dust particles (Chyba and Sagan, 1992). The next stage of evolution on surface of our planet would be the origin of the first protoliving organisms.

Our investigation dealt with further reaction of nucleosides to nucleotides—main components of RNA and DNA—and single amino acids to oligopeptides. And we investigated two types of reactions: (1) abiogenic synthesis of nucleotides from mixtures of nucleoside + inorganic phosphate and (2) abiogenic synthesis of dipeptides from mixtures of simple amino acids. In order to simulate the process of chemical evolution that takes place in the interstellar and interplanetary medium, some experimental mixtures have been placed into the space environment onboard of the Russian spacecrafts (Kuzicheva and Gontareva, 2003). The full range of the solar light spectrum and the primary components of cosmic rays are available in space at high density. In contrast, the flux and energy range of light radiation are limited in laboratory exposure experiments on the ground. UV light and cosmic rays can work together on organic molecules in space, but it is difficult to apply these two energy sources together in laboratory.

Hydrated minerals are found within a chondrite matrix of meteorites together with significant amounts of organic matter. Clays are likely to have formed when water was present on parent meteoritic bodies (Brearly, 2006). In order to verify the existence of a relationship between abiogenic synthesis of nucleotides and inorganic components of the meteorites, we have investigated possible abiogenic reactions associated with different clay (montmorillonite, kaolinite) and a basaltic one (Tyatya's volcanic ash) under action of open space energy sources as a model of different exobiological environments on the surface of small space bodies. The abiogenic synthesis of natural adenine nucleotides from a mixture of adenosine and inorganic phosphate has been observed following irradiation with VUV light in the presence of different mineral samples. The yields of the products (5′AMP, 2′AMP, 3′AMP, 2′3′cAMP, and 3′5′cAMP) depended on irradiation time and kinds of minerals used. The discovery that meteoritic organic compounds may be trapped and protected within a clay mineral matrix has implications for our understanding of prebiotic molecular evolution in the early Solar system.

Clay minerals may also have concentrated organic compounds thereby promoting polymerization reactions. An adsorption/binding of nucleic acids components by clay crystals could change the electron distribution and/or the conformation of the molecules. The remnant water molecules in the clay sheets also could influence the course of the reaction. Clay immobilization of phosphate could play an important role in this reaction. Chondritic material could have been a common component of the inner Solar system shortly after its formation, and the biologically useful products of clay mineral–organic matter interactions could have also widespread and delivered to planetary surfaces through the accretion of carbonaceous asteroids.

## 3. Abiogenic Synthesis

### 3.1. NUCLEOTIDES

Nucleotides are the building blocks of RNA (ribonucleic acid) and DNA (deoxyribonucleic acid), the genetic material used by all living organisms on Earth. They consist of three different components: nucleobases, ribose (in case of RNA) and deoxyribose (in case of DNA), and phosphate.

The nucleobases molecular structures are based on two N-heterocyclic aromatic compounds, namely, pyrimidine ($C_4H_4N_2$) and purine ($C_5H_4N_4$). Among the biological nucleobases, three are pyrimidine-based: uracil, thymine, and cytosine; and two are purine-based: adenine and guanine. The molecular structures of pyrimidine, purine, and their derived nucleobases are given in Fig. 1.

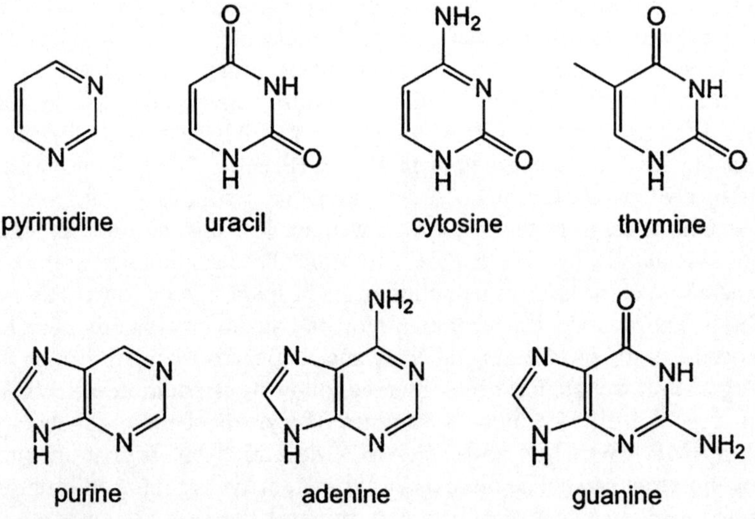


**Figure 1.** Molecular structure of nucleobases.

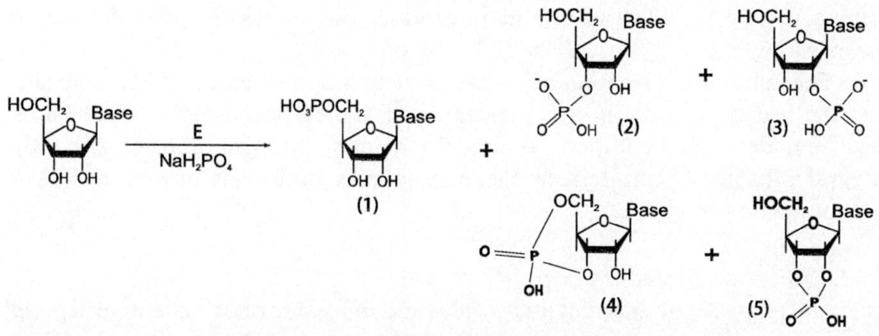


**Figure 2.** Abiogenic synthesis of nucleotides from mixture of nucleoside and phosphate.

The scientific aim of our experiment was to expose simulated organic mixtures in solid state to interstellar dust and meteoritic (asteroid) surfaces. To this end, (1) the efficiency of abiotic synthesis of complex organic compounds, such as nucleotides and dipeptides, from nucleosides and simple amino acids deposited on grains was evaluated in a quantitative manner; (2) the role of each energy source in this synthesis was assessed; and (3) the role of mineral surface on abiogenic synthesis was evaluated.

The investigated compounds were irradiated as solid mixtures prepared by air-drying an aqueous solution (300 mL) of an equimolar mixture of nucleoside with inorganic phosphate ($NaH_2PO_4$). The area of the film was ~1 $cm^2$. The films contained 1.3 mM nucleoside and 1.3 mM $NaH_2PO_4$. The $H_2PO_4^-$ ion was chosen as a source of phosphate based on the results of Kuzicheva et al. (1993) that showed that the ability to phosphorylate decreases in the order $H_2PO_4^- > HPO_4^{2-} > PO_4^{3-}$ but the process takes place in all cases. When the effect of a substrate is being studied, to aqueous solution of the reagents was added 3 mg of mineral component, and the suspension was dried as above. The following minerals were used as a substrate: kaolinite $(Al_2)(Si_2)O_5(OH)_4$, montmorillonite $(M_x(Al_2{-}xMg_x)(Si_4)O_{10}(OH)_2{\times}nH_2O)$, and ash from the Tyatya volcano. Homoinic $Na^+$ −montmorillonite has been prepared by titration (Banin et al., 1985). The exchange capacity (CEC) of the cations $Na^+$ − montmorillonite used in this study was 92 meq/100 g clay.

VUV irradiation of the mixture of nucleoside and inorganic phosphate resulted in the production of the natural monophosphates of the corresponding nucleosides. The main products were nucleoside–5′–monophosphates (5′–NMP) (**1**) and some small amount of by-products such as 3′–monophosphate (**2**), 2′–monophosphate (**3**), 2′,3′–cyclomonophosphate (**5**), and 3′,5′–cyclomonophosphate (**4**) (Fig. 2). The yields of products were small, within a few percentages; however, the effectiveness of the abiogenic synthesis with VUV irradiation was higher than that with UV radiation ($\lambda = 254$ nm) and heat. When the investigated films were irradiated with protons, the full mixture of nucleotides was also found. The maximal summary yield was 9.43%. The main product was 5′–AMP under the driving action of VUV.

The dominant formation of 5′–adenosinemonophosphate (5′AMP) indicates that the hydroxyl group in the 5′–position of the carbohydrate residue is most reactive under these conditions. A similar regularity was also observed under the integral influence of extraterrestrial energy sources on dry mixtures of nucleoside and inorganic phosphate.

### *3.1.1. Abiogenic Synthesis in Open Space*

The solid mixtures of different nucleosides and inorganic phosphate were exposed to space conditions during the space flight experiment onboard the BION–11 satellite. The abiogenic synthesis of the full set of the natural nucleotides was observed, with yields of up to 6%. After returning back to Earth, the experimental films were dissolved in water and analyzed by HPLC. The abiogenic synthesis of nucleotides was observed. Such nucleotides as 5′–, 3′–, and 2′–monophosphates of adenosine, deoxyadenosine, cytidine, and thymidine were detected together with cyclomonophosphates of these nucleosides. In the case of purines and pyrimidines, the main products of the reactions were 5′–monophosphates of the corresponding nucleosides. The preference to 5′–monophosphate formation is indicative of the more advantageous spatial position of the 5′–hydroxyl group in the carbohydrate residue in comparison with the 2′– and 3′–hydroxyl groups. The summary nucleotide yields were for adenosine, 5.8%; for cytidine, 4.7%; for deoxyadenosine, 2.3%; and for thymidine, 2.0%. The yield of products was calculated in relation to the initial amount of nucleoside. The yield of phosphorylated nucleoside is dependent on the kind of sugar moiety. Compounds which have ribose in its structure form 5′–nucleotides twofold more frequently than those with deoxyribose. For example, the yield of 5′AMP was 3.23%, while for that 5′dAMP was 1.87%. This result is related to the higher reaction capability of the ribose in the ribonucleosides. Purine nucleosides react slightly more easily than the pyrimidine ones. Thus, the yield of 3′AMP was 0.71% while that of 3′CMP—0.55%. In general, however, it can be concluded that the formation of nucleotides in the space environment does not depend significantly on the nature of the nucleic acid base.

The 2′,3′–cyclomonophosphates were formed on the solid mixtures under open space energy sources as well (if it was possible based on the structure of nucleosides). It is interesting to note that the yield of 2′,3′–cyclomonophosphates is more than those of 2′– and 3′–monophosphates. There have been previous reports of an interaction between all three nucleotides under VUV radiation (Kuzicheva et al., 1996; Simakov et al., 1997a) and γ–radiation (Kuzicheva et al., 1993). This possibility can be extended to the open space condition as well. It is possible that the temperature variations (which were about 130°C) on the satellite surface during its rotation also contribute noticeably to abiogenic synthesis. Despite the fact that UV radiation is the most abundant energy source in space, the heat used in the ground experiments was the most soft and plausible factor for the abiogenic synthesis of biochemical compounds. The radiation level on the surface of the spacecraft was as much as 6 Gy (600 rad) per day and up to 80 Gy (8,000 rad) per 15-day mission (depending on solar activity), and solar electromagnetic radiation was on the order of 40 solar constant hours per mission. The total dose of solar light (including UV) radiation during the pace mission was $1.83 \times 10^9$ J/m$^2$ at a UV photon flow of $1.03 \times 10^{21}$ photons/cm$^2$.

Both the destruction of the initial compounds as well as the synthesis of natural products were observed. The ribose residue was more resistant to the action of destructive influences than the deoxyribose one. This was observed for both the purine as well as the pyrimidine nucleotides. This is confirmed by earlier received data for adenosine and deoxyadenosine (Kuzicheva et al., 1996). On the other hand, the double ring of purine nucleotides is more stable than the pyrimidine ring. Pyrimidine is more stable against UV photodestruction also when mixed within $H_2O$ ice than when it is isolated (Peeters et al., 2005), and its photoproducts are stable enough to survive at room temperature. During the flight time, 66% of the cytidine was destroyed, which is the maximal value of decomposition of all nucleosides. Adenosine was the most stable nucleotide in the open space environment, with 54% of the initial nucleoside surviving the flight time. Reactions involving the decomposition of the heterocyclic rings, cleavage of the N–glycoside bond, and destruction of the sugar moiety were observed in the investigated mixtures. About 20% of each nucleoside was decomposed with a cleavage of the glycoside bond and the elimination of a free nucleic acid base. In such reactions, VUV radiation at wavelengths below 200 nm has the greatest effect (Kuzicheva et al., 1996). UV light at wavelengths of about 254 nm could be observed to be a factor in the destruction of the heterocyclic rings. Broad spectrum UV light from the Sun is the principal source of energy for the destruction of organic molecules. In the ground experiments, quanta yields of nucleosides as a result of photochemical destruction have been found. For cytidine, these are $4 \times 10^{-3}$ for VUV radiation and $4 \times 10^{-5}$ for UV radiation (254 nm). The 5′–nucleotides are the dominant products of the nucleoside reaction with inorganic phosphate with all sources of energy, both in the lab (Simakov et al., 2002) and in the space flight (Kuzicheva and Simakov, 1999). In addition to nucleotide formation, the destruction of the initial adenosine, such as depurination (base release) and decomposition of the base moiety, also may take place.

The results of our laboratory experiments and experiments in open space show that many kinds of energy were effective in the chemical evolution on the surface of asteroids.

### *3.1.2. Influence of the Mineral on the Abiogenic Synthesis of Nucleotides*

Hydrated minerals are found within a chondrite matrix of meteorites together with significant amounts of organic matter (Pearson et al., 2002). Clays are likely to have formed when water was presented on the parent meteoritic bodies (Brearly, 2006). In order to verify the existence of a relationship between the abiogenic synthesis of nucleotides and the inorganic components of the meteorites, possible abiogenic reactions in the presence of different clay minerals (montmorillonite, kaolinite) and the basaltic sample (Tyatya's volcanic ash) were investigated under action of the open space energy sources for modeling of different exobiological environments on the surface of small space bodies.

Clay minerals may have concentrated organic compounds, thereby also promoting polymerization reactions. The main works on polymerization on clays were done on water solution (Mathew and Luthey-Schulten, 2010; Swadling et al., 2010) instead of our experiments in solid state. An adsorption/binding of nucleic acid components by clay crystals could change the electron distribution and/or the conformation of the molecules. The irradiation of dry adenosine and inorganic phosphate films with VUV in the presence of mineral substrates (kaolinite, montmorillonite, volcanic ash) produced 5′–, 2′–, and 3′–adenosine phosphates along with cyclic products, such as 2′,3′– and 3′,5′–cycloadenosinemonophosphates. Under the maximal used irradiation dose ($4 \times 10^5$ J/m$^2$), the total yields of nucleotides were 5.96%, 5.55%, and 3.88% in the presence of montmorillonite, Tyatya's ash, and kaolinite, respectively. In the absence of the mineral substrate, the yield was 2.12%.

The results clearly demonstrate that, depending on the mineral substrate used, increases in the yield of nucleotides can vary by 1.8–2.8 times. Kaolinite and montmorillonite are the clay minerals, whereas the Tyatya's ash is compositionally very similar to terrestrial and extraterrestrial basalts, e.g., Martian soil (ten Kate et al., 2006). The whole spectrum of natural nucleotides is present in the reaction products. The maximal yield of the main product, 5′AMP, was 3.48% after irradiation in the presence of montmorillonite. The yield increased with increasing irradiation dose in the presence or absence of the mineral substrate. The second major reaction product was 2′,3′–cycloadenosine-monophosphate (2′,3′cAMP). The maximal yields were 0.33% without the mineral substrate and 0.89% with montmorillonite; i.e., the mineral component increases the nucleotide yield as in the case of 5′AMP synthesis. It is interesting to note the yield of 2′,3′cAMP is greater than those of the 2′– and 3′–monophosphates, respectively. The yields from nucleoside decay are 11% (Tyatya), 13% (kaolinite), 15% (without the mineral), and 18% (montmorillonite). The volcanic ash has more shielding effect along with others minerals whereas montmorillonite has some catalytic action on nucleoside decay.

Here we can see a shielding effect of the volcanic ash and catalytic effect of the montmorillonite. The clay facilitates the destruction of the initial adenosine. In the case of using the mineral substrate, a maximal influence on the reaction yield is typical of the montmorillonite. This clay gave the highest yields, both in terms of nucleotide formation and adenosine destruction. The mineral is composed of units made up of two silica tetrahedral sheets and an alumina octahedral sheet (Fig. 3). There are isomorphic substitutions in both tetrahedral and octahedral sheets, which result in charge deficiency. This charge deficiency is counterbalanced by the metal ions, called interlayer cations or exchangeable metal ions, held between the layers. In contemporary clays, these are mainly $Na^+$ and $Ca^{2+}$.

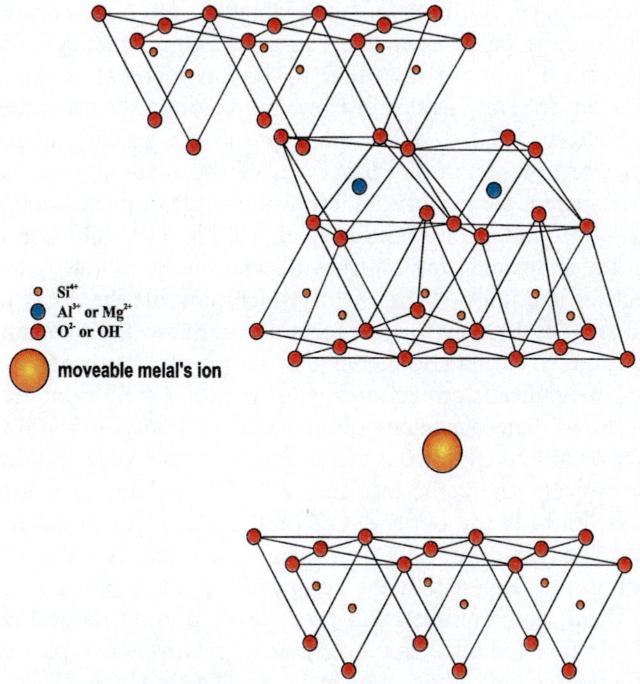


**Figure 3.** The structure of montmorillonite.

The hydration or dissolving of these metal ions result in the expansion of the layers (increase of the interlayer distance), thereby permitting the entrance of organic molecules into the interlayer. The isomorphic substitutions are much less in kaolinite than in montmorillonite, resulting in fewer numbers of interlayer cations and, therefore, less expansion and a shorter interlayer distance. This structure can absorb not only inorganic ions but also organic material. For example, 1 g of this mineral can absorb up to $5 \times 10^{-10}$ moles of adenine (El Amri et al., 2002).

It is precisely this property that makes montmorillonite different from kaolinite, another clay mineral used in our experiment. Kaolinite possesses layers consisting of a 1:1 pair of octahedral and tetrahedral sheets. The shorter distance between the layers prevents any accumulation of organic molecules into the kaolinite. Consequently, nucleoside molecules can be adsorbed only by its surface. The specific surface area of both clays is also different: montmorillonite has 78 $m^2/g$ and kaolinite, 8 $m^2/g$. The higher catalytic activity of montmorillonite, relative to kaolinite, has also been established by several other researchers studying the synthesis of abiogenic biologically significant compounds (BSCs) such as amino acids (Yushkin, 1999). Ash from the Tyatya volcano represents an erupted silicate product with an extremely porous texture and various oxides of different elements. The abiogenic synthesis of many organic compounds occurs precisely at the surface of ash particles in the ash-gas volcanic channels (Basiuk and Navarro-Gonzalez, 1996). On the other hand, this material could serve as an analog of Martian soil, and ash from Hawaiian volcanoes is often used for such purpose (ten Kate et al., 2006). When adenosine associates with the mineral substrate, its behavior depends on the properties of this material. Ash particles are almost nontransparent to VUV radiation, whereas clay minerals are sufficiently permeable to VUV radiation (Boillot et al., 2002). The influence of mineral substrates on the abiogenic synthesis may depend on the following mechanisms: (1) the protective function of the mineral, which prevents the decay of both the initial components and newly formed reaction products; the screening effect of clays has been studied earlier (Bonner et al., 1984), and it has been found that the clays provided a significant protection against photolysis; (2) catalytic properties of the mineral in heterogeneous photochemical phosphorylation reactions (Otroshchenko et al., 2002); (3) the presence of residual water molecules, which can generate radicals under the influence of VUV radiation, in the interlayer space of clays; and (4) the dependence of the reactivity of both adenosine and inorganic phosphate molecules on their absorption by mineral substrates. The results obtained suggest that the action of the VUV radiation on films of BSCs makes them more complex and produces important natural compounds. One can assume that the abiogenic synthesis of relatively complex compounds can proceed under conditions of open space. The discovery that meteoritic organic compounds may be trapped and protected within a clay mineral matrix has implications for our understanding of prebiotic molecular evolution in the early Solar system (Pearson et al., 2002). Clay minerals may have concentrated organic compounds, thereby also promoting polymerization reactions. Organic molecules can then enter the interlayer region and interact with the negatively charged basal surfaces of montmorillonite (Ferris, 2005). The interaction between the adsorbed molecules and the clay surface may be via simple electrostatic attraction, van der Waals forces, the formation of hydrogen bonding, or a combination of any two or three of the above. Adsorption of the compounds in the interlayer brings the molecules into close proximity of each other. Furthermore, it may orient them into a favorable position, thereby enabling intermolecular interactions. An adsorption/binding of nucleic acids components by clay crystals could change the electron distribution and/or the conformation of the molecules. In addition, an immobilization of phosphate on clay also could play an important role in our reaction (Goldberg and Sposito, 1985). Chondritic material would have been a common component of the inner Solar system shortly after its formation, and the biologically useful products of clay mineral–organic matter interactions would have also been widespread and ultimately delivered to planetary surfaces through the accretion of carbonaceous asteroids. The mineral components selected for substrates can serve as a model for the surface of minor bodies in the Solar system.

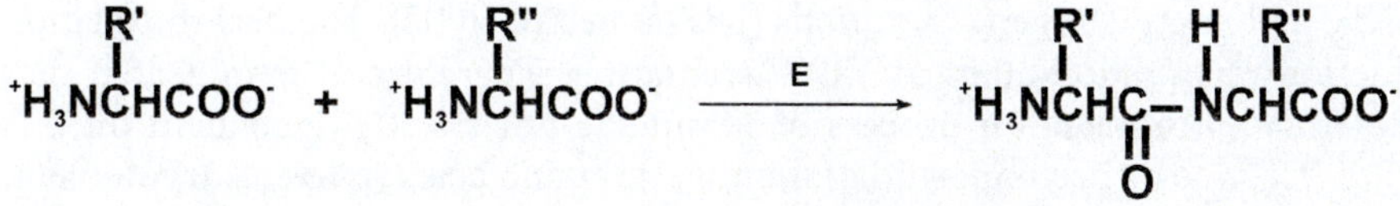


**Figure 4.** Abiogenic synthesis of dipeptides from mixture of amino acids.

## 3.2. OLIGOPEPTIDES

The reactions of the amino acids in solid mixtures were the primary objective of our investigation—primarily, the abiogenic synthesis of dipeptides from mixtures of simple amino acids (Fig. 4). Four mixtures of aromatic (tyrosine or tryptophan) and aliphatic (glycine or alanine) amino acids were investigated.

These mixtures were prepared by air-drying aqueous solutions and have been irradiated with different sources of energy: (1) VUV light of 145 nm, (2) UV light of 254 nm, and (3) high-energy protons (2–6 MeV). Films containing a mixture of amino acids have been shown to yield various oligopeptides after being exposed to energetic action (Simakov et al., 1996, 1997b). Such dipeptides as Trp–Trp, Trp–Gly, Gly–Trp, Ala–Trp, Trp–Ala, Tyr–Tyr, Tyr–Gly, and Gly–Tyr were detected as products of irradiation with summary yields of ~2% in the case of VUV irradiation and ~2.5% for proton irradiation. When solid films of single amino acids—tyrosine or tryptophan—were irradiated with protons, dipeptides, such as tryptophyl–tryptophan and tyrosyl–tyrosine, were found also. The yields of dipeptides were 2.43% for Trp–Trp and 2.38% for Tyr–Tyr under a maximal dose of irradiation ($5 \times 10^{11}$ protons/cm$^2$). In the case of a mixture consisting of aromatic and aliphatic amino acids (tyrosine + glycine), the yields of synthetic reactions did not essentially differ from those for single amino acid. Dipeptides, Gly–Tyr and Tyr–Gly, were formed with the equal probability, together with Tyr–Tyr and Gly–Gly.

The control experiments without irradiation did not yield any trace of oligopeptides. These values suggest that dipeptide formation does not depend on the type of side-chain residue. Di–, tri–, and tetraglycine were formed under VUV irradiation of solid glycine (Nakagawa et al., 2009). The super-excited state that is produced during the interaction of high-energy particles with amino acids may play a major role in these processes. Free radicals are formed due to the collision of protons with molecules, and these free radicals could react with each other.

The initial amino acids are able to undergo processes of destruction as well as the process of abiogenic synthesis of oligopeptides. Of the amino acids tested, glycine is the most resistant to decomposition by UV photons. For alanine in the solid phase irradiated with VUV, both the decarboxylation and deamination reactions were major processes, and the methyl group hardly dissociated. This result was reasonably consistent with that at the gas phase (Jochims et al., 2004) and acidic and neutral aqueous solutions (Nishino et al., 2002). The decarboxylation reaction was predominant at VUV irradiation. Therefore, it is indicated that the VUV photoreaction process of alanine is not strictly dependent on this state. The rate of decomposition for aromatic amino acids (tyrosine, tryptophan, and phenylalanine) is significantly accelerated under these conditions, mostly due to the strong UV absorption of the aromatic group (Orzechowska et al., 2007).

#### *3.2.1. Abiogenic Synthesis in Open Space*

Flight experiments were performed in terms of French–Russian MIR mission “PERSEUS 1999” on a special exposure hardware placed onboard of the orbital station. Flight duration was 113 days—from April 16, 1999 to August 7, 1999. Experimental hardware was designed by Didier Chaput (CNES, Toulouse) and manufactured by Steel Industries (Toulouse, France).

Dry films were loaded in the cavity face of the $MgF_2$ windows 1.1-mm thick. This kind of glass is transmissive to UV radiation. Each strip has two cavity levels: the top cavities (9 mm in diameter) are exposed to the radiation, whereas the bottom ones serve as in-flight controls. The temperature was recorded inside the cavities by two thermal sensors fixed in one exposed top cavity and in the corresponding shielded bottom cavity. Lunar soil was delivered to the Earth by automatic station “Luna-16” (probe 1640). Amino acids were impregnated inside the mineral bed in order to investigate its catalytic and protective properties against the destructive action of the solar radiation. Lunar soil samples consist of dark-gray regolith particles 0.02 mm in diameter, electronegative, and hydrophobic. The amount of each amino acid exposed corresponds to 750 μg for Gly and 408 μg for Trp (amino acid mixture) and 816 μg tryptophan for single amino acid. Mixtures were loaded by dropping 300 μl of stock sample solution on the cavity face of the $MgF_2$ windows and dried under $CaCl_2$ to afford a thin film. When the amino acids were loaded with regolith, 150 μl of stock sample solution loaded preliminarily on the cavity face was covered by 3 mg of powder and later on, the rest of amino acid solutions (150 μl) were added.

After the delivery, exposed dry films were analyzed using high-performance liquid chromatography (HPLC) method. Identification of the reaction products was made against standard solution curves with accuracy not less than 98%. Yield of the polypeptides was calculated relative to Trp.

After the exposure of Gly+Trp and Trp dry mixtures during flight experiments, some polypeptides have been found. For the Gly+Trp mixtures, data reveal the following polypeptides synthesized, ranging from Gly–Gly (0.97%) to Trp–Trp–Trp (0.004%). Although the amounts of di- and tripeptides are rather small, they are significant, since this reaction take place in such harsh environments.

After the exposure of tryptophan films during space flight, only the formation of dipeptide with a small yield of 0.004% was revealed. This low amount of dipeptide as well as the absence of tripeptide could be related to the high destruction rate of the basic tryptophan (85%). The presence of mineral bed during film exposure in space conditions leads to the increasing of the yield. Total yield of the products analyzed at the exposure of Gly+Trp dry films had valued up to 1.31% in absence and 2.06% in presence of Lunar soil, i.e., increased in 1.5 times. Similar films were irradiated by vacuum ultraviolet (145 nm) and ultraviolet (245 nm) light in the parallel laboratory experiments. Following peptides were identified as the reaction products: Gly–Gly, Trp–Gly, Gly–Trp, Trp–Trp (in case of Gly+Trp mixture), and Trp–Trp, Trp–Trp–Trp (for tryptophan only). The protective properties of lunar soil could be clearly detected also after the orbital exposure of tryptophan films.

Presence of regolith increases Trp–Trp yield from 0.004% up to 0.007%. Such small amounts of products synthesized could be described by the destructive action of radiation on the newly synthesized peptides. Apart from activating the synthesis processes, UV and VUV radiation could activate the decay of polypeptides as well.

This matter could be illustrated by our previous investigations on tryptophan and its dipeptide destruction after proton treatment (6.6 MeV). The results obtained revealed that energy flux of $4 \times 10^9$ protons/cm$^2$ causes 28% tryptophan decay and 31% dipeptide decay (Simakov et al., 1997b). VUV radiation is characterized by selectivity of peptide bond absorption leading to its destruction. This kind of selectivity is absent with γ–radiation—the excitation of electron states occurs due to the hits of molecules with high-energy ionization electrons, appearing after the absorption of γ–quanta.

Breakage of peptide bond and the formation of di- and tripeptides at VUV radiation of tyrosine in solid state were discussed by Khoroschilova et al. (1991). The results presented in the paper showed that the destruction of Tyr–Tyr peptide in solid state comes along with peptide bond breakage as a result of a single-quanta process of UV absorption.

Apart from the VUV- and proton-induced synthesis processes, the possibility of peptide synthesis in solid state activated by γ–radiation was demonstrated earlier. In their paper (Kuzicheva et al., 1993), the possibility of Gly–Trp dipeptide formation was discussed after γ–irradiation ($10^6$ J/m$^2$) of tyrosine, tryptophan, and glycine solid films with maximal yield of 0.4%. Further increasing of the radiation dose led to the degradation of synthesized product, for the processes of synthesis and decay occurred in a competitive condition.

Comparing results on VUV and γ–radiation, we could come to the conclusion that Gly–Trp peptide yields of 0.04% could be obtained at different values of the energy absorbed. These values are equal to $10^6$ J/m$^2$ and $10^2$ J/m$^2$ for VUV and γ–radiation, respectively. Comparatively, higher contents of Gly–Gly and Trp–Gly dipeptides after exposure in flight and laboratory data could be described by higher rate of their formation and lower rate of the reverse destruction.

Polymerization is an essential step in prebiological evolution, and the results reported here show that this process probably could take place even at an early stage of the Solar system formation, before planet accretion, on the surface of small bodies.

## 4. Exobiological Implications

The problem of the origins of life is a very intriguing question of contemporary science. Geological data indicate that complex photosynthetic organisms possibly related to cyanobacteria existed 3.5 billion years (Gyr) ago (Schopf, 1993). There is also evidence of biologically mediated carbon isotope fractionation in the 3.85-Gyr-old Isua metasediments, which indicate that the photoautotrophy biochemical processes must have existed at that time (Mojzsis et al., 1996). So, we have to consider an additional source of complex organic compounds, for chemical evolution took place on our planet.

All Solar system objects, such as planets, satellites, rings, comets, asteroids, meteorites, and interplanetary dust particles (IDPs), are subjected to energetic processing by different kinds of open space energy. The study of the role of charged particles, UV radiation of different wavelengths, and other energy sources in the abiogenic synthesis of BSCs is closely related to exobiological investigations in the Earth's orbit related to chemical evolution and the issue of the origin of life on Earth and throughout the universe. According to recent views, the majority of biologically significant compounds may have been introduced to the primeval Earth by micrometeorites, meteorites, and comets (Chyba and Sagan, 1992). These cosmic objects contain large quantities of complex organic compounds, including carbohydrates, amino acids, and heterocyclic bases of nucleic acids. Evidently, the chemical evolution proceeds under conditions of open space, which serves as a huge chemical reactor. "Simulated space ice conditions" experiments have shown the synthesis of simple BSC in the form of the precursors of amino acids (Bernstein et al., 2002; Munoz Caro et al., 2002) and bases of the nucleic acids (Kobayashi et al., 2004). The present investigation dealt with the subsequent reaction of the nucleic acid components to nucleotides, which are the main components of RNA and DNA. Abiogenic synthesis of nucleotides has been reported to take place under conditions of space flight (Kuzicheva and Simakov, 1999) in the presence of different types of energy: proton beams (Simakov et al., 2002), VUV (Kuzicheva et al., 1996; Simakov et al., 1997a,b), UV, and $\gamma$–radiation (Kuzicheva et al., 1993). In contrast, just before that period, the surface of Earth may have been melted by high intensive comet and meteoritic bombardment, and any structures related to life may have been killed off by such catastrophic events (Sleep et al., 1989). Life would therefore arise rapidly on Earth once conditions became clement, and this process could have been very rapid, almost instantaneous on the geological time scale. Only 300 million years (from the end of the late heavy bombardment, 4–3.8 billion years ago, up to the first well-evolved living organisms at 3.5 billion years ago) appear to be left for terrestrial chemical evolution to lead to primitive yet fully developed organisms. There is a very short time interval for producing the biochemical compounds in the atmosphere and on the surface of planet, but this type of matter evolution had to occur before the biological one.

It can be suggested that the chemical evolution which had proceeded at the early stages of the Solar system origin would lead to the appearance of very complex organic compounds. The compounds synthesized under these conditions could have survived inside comets and meteorites, the last witnesses of the formation of our Solar system, and have been delivered to early Earth at the stage of the "late heavy bombardment" and later. The comets, carbonaceous chondrites, and other primitive bodies of the Solar system, including interplanetary and interstellar dust particles, are responsible for essentially all carbon compounds and water currently present in our planet. Therefore, these highly complex organic molecules have been proposed as the starting material for early biological activity on Earth and on other planets in the universe (Oro et al., 1992; Flinn, 1996). Anders (1989) estimated the flux of intact organic matter reaching the contemporary Earth in IDPs. Our planet is currently accreting ~$3.2 \times 10^5$ kg/year of intact organics. An additional source of exogenous organics on early Earth would have been the terrestrial accretion of dust as the Solar system passed through interstellar clouds. Greenberg (2002) estimated that during its first $7 \times 10^8$ years, Earth should have passed through approximately four to five such clouds, thereby accreting organic molecules during each approximately $6 \times 10^5$ year passage at a rate $10^6$–$10^7$ kg/year.

The mechanism proposed by Chyba and Sagan (1992) for the transport of biologically significant compounds synthesized in space seems sufficiently plausible. Phosphorylated and nonphosphorylated nucleosides transported by micron-sized interplanetary dust particles could resist temperatures up to 500°C generated during atmospheric entry (Marcano et al., 2004). So, the dust and micrometeorites could have contributed significantly to the availability of nucleotides in the early Earth and thereby could have allowed the arising of an early biological activity.

## 5. References

Anders E (1989) Prebiotic organic matter from comets and asteroids. Nature 342:255–256

Banin A, Lawless JG, Mazzurco J, Church FM, Margulies L, Orenberg JB (1985) pH profile of the adsorption of nucleotides onto montmorillonite. Orig Life Evol Biosph 15:89–101

Basiuk V, Navarro-Gonzalez R (1996) Possible role of volcanic ash-gas clouds in the Earth's prebiotic chemistry. Orig Life Evol Biosph 26:173–194

Bernstein MP, Dworkin JP, Sandford SA, Cooper GW, Allamandola LJ (2002) Racemic amino acids from the ultraviolet photolysis of interstellar ice analogues. Nature 416:401–403

Boillot F, Chabin A, Bure C, Venet M, Belsky F, Bertrand-Urbaniak M, Delmas F, Brack A, Barbier B (2002) The perseus exobiology mission on Mir: behaviour of amino acids and peptides in Earth orbit. Orig Life Evol Biosph 32:359–385

Bonner WA, Hall H, Liang Y (1984) Racemization of amino acids on clays. Orig Life Evol Biosph 15:103–114

Botta O, Bada JL (2002) Extraterrestrial organic compounds in meteorites. Surv Geophys 23:411–467

Bradley JP (1994) Chemically anomalous, preaccretionally irradiated grains in interplanetary dust from comets. Science 265:925–929

Brearly AJ (2006) The action of water. In: Lauretta DS, McSween HY (eds) Meteorites and the early solar system II. University of Arizona Press, Tucson, pp 587–624

Charnley SB, Rodgers SD (2008) Interstellar reservoirs of cometary matter. Space Sci Rev 138:59–73

Chyba CF, Sagan C (1992) Endogenous production, exogenous delivery, and impact-shock synthesis of organic molecules: An inventory for the origins of life. Nature 355:125–132

Cooper GW, Kimmich N, Belisle W, Sarinana J, Brabham K, Garrel L (2001) Carbonaceous meteorites as a source of sugar-related organic compounds for the early Earth. Nature 414:879–883

Cottin H, Despois D (2009) Comets and astrobiology. In: Tze-Fei Wong J, Lazcano A (eds) Prebiotic evolution and astrobiology. Landes Bioscience, Austin, pp 38–45

Cottin H, Gazeau MC, Raulin F (1999) Cometary organic chemistry: a review from observation, numerical and experimental simulation. Planet Space Sci 47:1141–1162

Crovisier J, Bockelee-Morvan D, Biver N, Colom P, Despois D, Lis DC (2004) Ethylene glycol in comet C/1995 O1 (Hale-Bopp). Astron Astrophys 418:L35–L38

Delsemme A (2006) The origin of the atmosphere and of the oceans. In: Thomas PJ et al (eds) Comets and the origin and evolution of life. Springer, Berlin, pp 23–68

Despois D, Cottin H (2005) Comets: potential sources of prebiotic molecules for the early Earth. In: Gargaud M, Barbier B, Martin H, Reisse J (eds) Lectures in astrobiology I. Springer, Berlin

El Amri AG, Baron M-H, Maurel MC (2002) Adenine in mineral samples: development of a methodology based on surface enhanced Raman and Raman microscopy for picomole detection. In: Lacoste H (ed) Proceedings of the 2nd European workshop on exo/astrobiology. ESA Publication Division, Noordwijk, pp 449–450

Ferris JP (2005) Mineral catalysis and prebiotic synthesis: montmorillonite-catalyzed formation of RNA. Elements 1:145–149

Flinn GJ (1996) The delivery of organic matter from asteroids and comets to the early surface of Mars. Earth Moon Planet 72:469

Goldberg S, Sposito G (1985) On the mechanism of specific phosphate adsorption by hydroxylated mineral surfaces: a review. Commun Soil Sci Plant Anal 16:801–821

Greenberg JM (2002) Cosmic dust and our origins. Surf Sci 500:793–822

Herd CDK et al (2011) Origin and evolution of prebiotic organic matter as inferred from the Tagish Lake meteorite. Science 332:1304–1307

Hudson RL, Moore MH, Dworkin JP, Martin MP, Pozun ZD (2008) Amino acids from ion-irradiated nitrile-containing ices. Astrobiology 8:1–9

Jochims H-W, Schwell M, Chotin J-L, Clemino M, Dulieu F, Baumgartel H, Leach S (2004) Photoion mass spectrometry of five amino acids in the 6–22 eV photon energy range. Chem Phys 298:279–297

Joshi P, Aldersley M, Ferris J (2011) Homochiral selectivity in RNA synthesis: Montmorillonite-catalyzed quaternary reactions of D, L–purine with D, L–pyrimidine nucleotides. Orig Life Evol Biosph 41:213–236

Khoroschilova EV, Tsyganenko NM, Petrov MYu, Dodonova NYa (1991) Photosynthesis of peptides at tyrosine exposure by vacuum ultraviolet radiation. Doklady Akad Nauk SSSR 319:1244–1247

Kobayashi K, Kaneko T, Saito T, Oshima T (1998) Amino acid formation in gas mixtures by high energy particle irradiation. Orig Life Evol Biosph 28:155–165

Kobayashi K, Kasamatsu T, Kaneko T, Koike J, Oshima T, Saito T, Yamamoto T, Yanagawa H (1995) Formation of amino acid precursors in cometary ice environments by cosmic radiation. Adv Space Res 16:21–26

Kobayashi K, Takano Y, Masuda H, Tonishi H, Kaneko T, Hashimoto H, Saito T (2004) Possible cometary organic compounds as sources of planetary biospheres. Adv Space Res 33:1277–1281

Kokubo E, Ida S (2000) Formation of protoplanets from planetesimals in the Solar nebula. Icarus 143:15–27

Kuzicheva EA, Gontareva NB (2003) Exobiological investigations on Russian spacecrafts. Astrobiology 3:253–261

Kuzicheva EA, Simakov MB (1999) Abiogenic synthesis of nucleotides in conditions of space flight of the biosputnik "BION-11". Adv Space Res 23:387–391

Kuzicheva EA, Zhukova VF, Simakov MB (1993) Role of UV and γ-radiation in the abiogenic synthesis of nucleotides in solid state. Zhurn Evol Biokhimii i Fisiologii 29:339–343 (in Russian)
Kuzicheva EA, Simakov MB, Malko IL, Dodonova NYa, Gontareva NB (1996) Role of vacuum ultraviolet (VUV) radiation in abiogenic synthesis of adenine nucleotides. Adv Space Res 18:65–68
Marcano V, Benitez P, Campins J, Matheus P, Cedeno C, Falcon N, Palacios-Pru E (2004) Pyrolisis of phosphorylated molecules and survivability limits during the atmospheric passage in earthlike planets. Planet Space Sci 52:613–621
Martins Z, Alexander CMO'D, Orzecowska GE, Fogel ML, Ehrenfreund P (2007) Indigenous amino acids in primitive CR meteorites. Meteorit Planet Sci 42:2125–2136
Martins Z, Botta O, Fogel ML, Sephton MA, Glavin DP, Watson JS, Dworkin JP, Schwartz AW, Ehrenfreund P (2008) Extraterrestrial nucleobases in the Murchison meteorite. Earth Planet Sci Lett 270:130–136
Mathew DC, Luthey-Schulten Z (2010) Influence of montmorillonite on nucleotide oligomerization reactions: a molecular dynamics study. Orig Life Evol Biosph 40:303–317
Mathis JS, Mezger PG, Panagia N (1983) Interstellar radiation field and dust temperatures in the diffuse interstellar matter and in giant molecular clouds. Astron Astrophys 128:212–229
McDonald GD, Whited LJ, Deruiter C, Khare BN, Patnaik A, Sagan C (1996) Production and chemical analysis of cometary ice tholins. Icarus 122:107–117
Mojzsis SJ, Arrhenius G, McKeegan KD, Harrison TM, Nutman AP, Friend CRL (1996) Evidence for life on Earth by 3800 million years ago. Nature 384:55–59
Munoz Caro GM et al (2002) Amino acids from ultraviolet irradiation of interstellar ice analogues. Nature 416:403–406
Nakagawa K, Matsui T, Izumi Y, Agui A, Tanaka M, Muro T (2009) Radiation-induced chemical evolution of biomolecules. Radiat Phys Chem 78:1198–1201
Nishino H, Kosaka A, Hembury GA, Aoki F, Miyauchi K, Shitomi H, Onuki H, Inoue Y (2002) Absolute asymmetric photoreactions of aliphatic amino acids by circularly polarized synchrotron radiation: critically pH-dependent photobehavior. J Am Chem Soc 124:11618–11627
Nuevo M, Auger G, Blanot D, d'Hendecourt L (2008) A detailed study of the amino acids produced from the vacuum UV irradiation of interstellar ice analogs. Orig Life Evol Biosph 38:37–56
Nuevo M, Milam SN, Sandford SA, Elsila JE, Dworkin JP (2009) Formation of uracil from the ultraviolet photo-irradiation of pyrimidine in pure $H_2O$ ices. Astrobiology 9:683–695
Oro J, Mills T, Lazcano A (1992) Comets and the formation of biochemical compounds on the primitive Earth—a review. Orig Life Evol Biosph 21:267
Orzechowska GE, Goguen JD, Johnson PV, Tsapin A, Kanik I (2007) Ultraviolet photolysis of amino acids in a 100 K water ice matrix: application to the outer solar system bodies. Icarus 187:584–591
Otroshchenko VA, Alekseev VA, Ryabchuk VK (2002) Nonequilibrium processes of organic substances synthesis in interstellar gas dust clouds. In: Simakov MB, Pavlov AK (eds) Astrobiology in Russia, Proceedings of International Workshop, St. Petersburg, Russia, 25–28 March 2002, pp 63–92
Owen T (2008) The contributions of comets to planets, atmospheres, and life: insights from Cassini-Huygens, Galileo, Giotto, and inner planet missions. Space Sci Rev 138:301–316
Pearson VK, Sephton MA, Kearsley AT, Bland PA, Franchi IA, Gilmor I (2002) Clay mineral—organic matter relationships in the early Solar system. Meteor Planet Sci 37:1829–1833
Peeters Z, Botta O, Charnley SB, Kisiel Z, Kuan Y-J, Ehrenfreund P (2005) Formation and photostability of N-heterocycles in space. I. The effect of nitrogen on the photostability of small aromatic molecules. Astron Astrophys 433:583–590
Pizzarello S, Huang Y, Fuller M (2004) The carbon isotopic distribution of Murchison amino acids. Geochim Cosmochim Acta 68:4963–4969
Pizzarello S, Huang Y, Alexandre MR (2008) Molecular asymmetry in extraterrestrial chemistry: insights from a pristine meteorite. Proc Natl Acad Sci USA 105:3700–3704

Prasad SS, Tarafdar SP (1983) UV radiation field inside dense clouds—its possible existence and chemical implications. Astrophys J 267:603–609

Sandford SA et al (2006) Organics captured from comet 81P/Wild 2 by the Stardust spacecraft. Science 314:1720–1724

Schopf JW (1993) Microfossils of the early Archean Apex chert: new evidence of the antiquity of life. Science 260:640–644

Shen CJ, Greenberg JM, Schutte WA, van Dishoeck EF (2004) Cosmic ray induced explosive chemical desorption in dense clouds. Astron Astrophys 415:203–215

Simakov MB, Kuzicheva EA, Mal'ko IL, Dodonova NYa (1996) Abiogenic synthesis of oligopeptides in solid state under action of vacuum ultraviolet light (100–200 nm). Adv Space Res 18:61–64

Simakov MB, Kuzicheva EA, Malko IL (1997a) Abiogenic synthesis of pyrimidine nucleotides in solid state by vacuum ultraviolet radiation. Adv Space Res 19:1059–1062

Simakov MB, Kuzicheva EA, Dodonova NY, Antropov AE (1997b) Formation of oligopeptides on the surface of small bodies in Solar system by cosmic radiation. Adv Space Res 19:1063–1066

Simakov MB, Kuzicheva EA, Antropov AE, Dodonova NYa (2002) Abiogenic synthesis of nucleotides on the surface of small space bodies with high energy particles. Adv Space Res 30:1489–1494

Sleep NH, Jahnle KJ, Kasting JF, Morowitz HJ (1989) Annihilation of ecosystems by large asteroids impacts on the early Earth. Nature 342:139–142

Swadling JB, Coveney PV, Greenwell HC (2010) Clay minerals mediate folding and regioselective interactions of RNA: a large-scale atomistic simulation study. J Am Chem Soc 132:13750–13764

Takahashi J, Hosokawa T, Masuda H, Kaneko T, Kobayashi K, Saito T, Utsumi U (1999) Abiotic synthesis of amino acids by X-ray irradiation of simple inorganic gases. Appl Phys Lett 74:877–879

Takahashi J, Masuda H, Kaneko T, Kobayashi K, Saito T, Hosokawa T (2005) Photochemical abiotic synthesis of amino-acid precursors from simulated planetary atmospheres by vacuum ultraviolet light. J Appl Phys 98:024907

ten Kate IL, Garry JRC, Peeters Z, Foing B, Ehrenfreund P (2006) The effects of Martian near surface conditions on the photochemistry of amino acids. Planet Space Sci 54:296–302

Yushkin NP (1999) Hydrocarbon crystallization of life. Earth Sci Frontiers 6:71–78

Biodata of **David P. Summers**, author of "***The Prebiotic Chemistry of Nitrogen and the Origin of Life***."

**Dr. David P. Summers** is currently a principal investigator at the Carl Sagan Center of the SETI Institute. Under cooperative agreement with NASA, he has labs and office at the Ames Research Center in Mountain View, CA, USA. He obtained his Ph.D. from the Massachusetts Institute of Technology in 1984. He then worked at SRI International studying fuel cells and electrochemical fuel production. He was a Senior Research Fellow under the NRC at NASA Ames before taking his current position. Dr. Summers' scientific interests are in the areas of prebiotic chemistry and the origin of life, the search for life, planetary habitability, and mission instrumentation.

E-mail: **David.P.Summers@nasa.gov**

J. Seckbach (ed.), *Genesis - In The Beginning: Precursors of Life, Chemical Models and Early Biological Evolution*, Cellular Origin, Life in Extreme Habitats and Astrobiology 22, 201–216
DOI 10.1007/978-94-007-2941-4_12, 

# THE PREBIOTIC CHEMISTRY OF NITROGEN AND THE ORIGIN OF LIFE

**DAVID P. SUMMERS**
*Carl Sagan Center, SETI Institute, c/o NASA Ames Research Center, Mail Stop 239-4, Bldg 239, Room 365, Moffett Field, CA 94035-1000, USA*

## 1. Nitrogen and the Origin of Life

When considering how the origin of life may have occurred, attention tends to focus first on the chemistry of carbon. While there are certainly good reasons for this, it should be remembered that nitrogen is also essential to life as we know it. Nitrogen, and its chemistry, is essential for the peptide linkages that are used to build proteins up from their component amino acids. Without nitrogen, protein chemistry as we know it would be impossible. Similarly, the base-base interactions in both RNA and DNA depend on nitrogen functional groups, and genetics, as we know them, would not be possible without nitrogen. It is simply impossible to construct anything that resembles life as we know it without nitrogen. Here we seek to provide a very brief review of the major known sources of nitrogen for the origin of life.

What are the likely sources of prebiotic nitrogen and how could they contribute the origin of life? As with carbon, these sources can be broken down into three general categories:

- Reactions in the atmosphere and in/on condensed phases at the surface, leading to fixed/reduced nitrogen (which we will call "atmospheric fixation").
- The reduced nitrogen present in meteorites as they are delivered to the Earth (which is often called "exogenous delivery").
- The formation of reduced nitrogen in reactions as seawater is swept through geological hydrothermal systems (often called "hydrothermal formation").

This review will, as one would expect, focus primarily on the Earth as the only known example of where prebiotic chemistry led to life. However, all of these processes will generally apply to a wide range of terrestrial planets.

## 2. Atmospheric Fixation

Perhaps the best studied source of prebiotic nitrogen is the formation of nitrogen compounds by terrestrial processes that start from elemental nitrogen, or "dinitrogen" ($N_2$). The earliest work on prebiotic synthesis assumed atmospheres that already contained reduced nitrogen (in the form of ammonia, $NH_3$). That meant that one only needed to understand the subsequent formation of the organonitrogen compounds directly necessary for the formation of life (Miller and Urey, 1959). However, the current view is that nitrogen was mostly likely released into the atmosphere as dinitrogen. According to current hypotheses, terrestrial atmospheres are formed from the release of gases originally trapped in the solid interior of the planets during their final stages of accretion. The composition of these gases depends on the redox state of the mantle/crust they are released from. Geochemical and geological evidence suggest that iron migrated to the core very early in Earth's history and volatiles outgassed from a relatively oxidized mantle, much like that of the modern Earth (Delano, 2001; Hunten, 1993; Mattioli and Wood, 1986; Walker, 1985; Wood and Vigo, 1989). Based on these models, $CO_2$ and $N_2$ were the major atmospheric constituents outgassed.

Dinitrogen is one of the least reactive compounds found in nature. This is because of the very high strength of its triple, N≡N, bond. This gives it a very high dissociation energy (at 3,000°C, at standard pressure, there is no significant dissociation) and makes chemical interactions with its π system (electrons involved in the triple bond) very weak (Cotton and Wilkinson, 1980). This allows it to remain as an atmospheric component after other species have been removed. For example, $N_2$ is not removed from the atmosphere by weathering, like $CO_2$. For this reason, the first step in atmospheric chemistry to fix nitrogen involves processes such as lightning and meteors. These produce very high temperatures, >10,000 K (Boyd, 2000; Jebens et al., 1992; Jenniskens and Stenbaek-Nielsen, 2004; Sobral et al., 2000) which can dissociate dinitrogen and allow it to react.

The products of such heating depend on the composition of the atmosphere being shocked. This has been modeled both theoretically and experimentally (Borucki and Chameides, 1984; Chameides and Walker, 1981; Fegley et al. 1986; Kasting, 1990; Miller and Schlesinger, 1984; Miller and Urey, 1959; Navarro–González et al., 2001b; Nna Mvondo et al., 2001; Prinn and Fegley 1987; Scattergood et al., 1989; Zahnle, 1990) (see Fig. 1). In neutral atmospheres (i.e., atmospheres with only a few percent hydrogen or CO or less, Miller and Schlesinger, 1983; Miyakawa et al., 2002a), processes such as lightning, meteors, and/or coronal discharges cause the formation of NO and CO (Borucki and Chameides, 1984; Chameides, 1979; Chameides and Walker, 1981; Fegley et al. 1986; Kasting, 1990; Navarro–González et al., 1998, b; Nna Mvondo et al., 2001; Prinn and Fegley 1987; Yung and McElroy, 1979; Zahnle, 1990). (Here we define a "neutral" atmosphere as all compositions that produce primarily NO and CO from high temperature processing.) The NO/CO formation in such atmospheric compositions has been well studied experimentally (Navarro–González et al., 1998, 2001b; Nna Mvondo et al., 2001, 2005). Cleaves et al. (2008) did find hydrogen cyanide (HCN), ammonia, and amino acid production in small quantities by spark discharge in a neutral atmosphere. The contrast with other studies (Navarro–González et al., 2001b; Nna Mvondo et al., 2001) may be a result of a closed system, allowing compounds to be shock-processed multiple times, and the

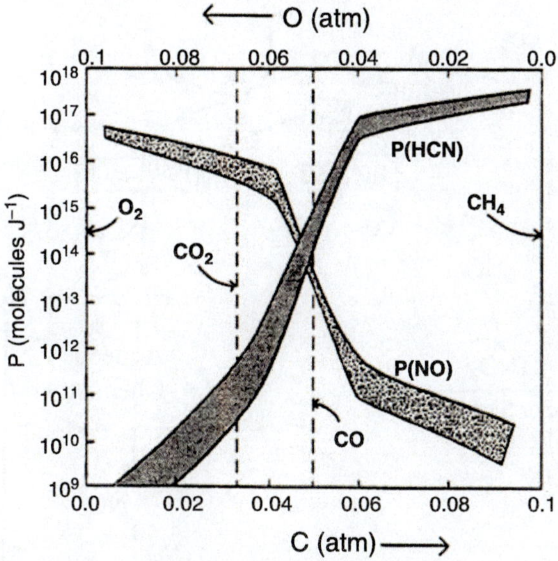


**Figure 1.** Yields of HCN and NO for lightning-produced shocks as theoretically predicted by a "freeze-out" model in the original paper by Chameides and Walker (1981).

presence of a water phase selectively segregating polar species. Plankensteiner et al. (2004) found amino acids under similar conditions, though in this case with added copper ions.

In a mildly reducing atmosphere (Hydrogen > ~10% and/or CO > ~5%), one observes the production of significant amounts of HCN (Chameides and Walker, 1981; Miller and Schlesinger, 1983; Navarro–González et al., 2001a). HCN is a form of reduced nitrogen that can enter directly into prebiotic chemistry, and we will say little more about it until we discuss the formation of amino acids and purines/pyrimidines. A reducing atmosphere (large amounts of $CH_4$, $NH_3$, and $H_2$ or CO as major constituents) produces complex organic species that can directly lead to such compounds as amino acids though, as mentioned above, recent results make such a composition unlikely. It has been proposed that the early terrestrial planets were able to evolve significant compositions of $H_2$ and CO, though questions remain and amounts are uncertain (Catling, 2006; Kharecha et al., 2005; Kulikov et al., 2007; Rosing et al., 2010; Tian et al., 2005). There has been only limited study of what hydrogen sinks might be present in the crust and oceans, which would affect the concentrations (Rosing et al., 2010). Enough hydrogen could make the atmosphere mildly reducing. It is worth noting that the atmosphere of Titan is also mildly reducing (here we consider it a "mildly reducing," rather than a Miller-Urey type "strongly reducing," atmosphere since it produces large amounts of HCN rather than amino acids, Navarro–González et al., 2001a). Contemporary Venus and Mars contain little (<700 ppm) CO and almost no hydrogen.

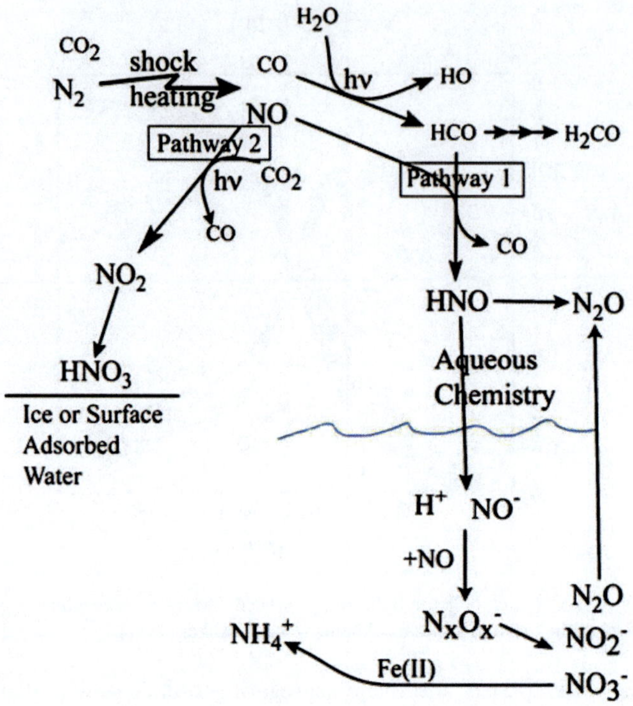


**Figure 2.** Summary of chemistry leading form NO generation by shock heating to formation of fixed and/or reduced nitrogen.

In a neutral atmosphere, as we have seen, shock heating produces NO. While this is a more oxidized form of nitrogen, it is also a "step forward" toward the formation of reduced nitrogen, since breaking the triple bond of $N_2$ is more important than the oxidation state. It was first proposed that the formation of NO would lead to nitrogen fixation by Mancinelli and McKay (1988). They suggested that, after the photochemical formation of HNO and its dissolution in aqueous phases, a series of reactions would lead to the formation of species such as $NO^-$, $N_2O_2^-$, and $N_3O_3^-$ which would then decompose to form $NO_2^-$, $NO_3^-$, and $N_2O$ (see Fig. 2).

Subsequent experimental work showed that irradiation of a $CO_2/N_2$ gas mixture containing CO and NO, by 185 nm light in the presence of liquid water, can cause a reaction forming $NO_2^-$, $NO_3^-$, and $N_2O$ (Summers and Khare, 2007). This work also showed that the reaction of NO was connected to the photochemistry of CO (which is formed along with NO during shock heating). CO is consumed photochemically in the presence of water. This reaction, which proceeds by the formation of HCO, leads to formaldehyde and other organics (Bar–Nun and Chang, 1983; Bar–Nun and Hartman, 1978; Hubbard et al., 1971; Mancinelli and McKay, 1988; Pinto et al., 1980; Wen et al., 1989). What was observed by Summers and Khare was that CO will not react in the presence of NO. When the mixture is irradiated, consumption of the CO is not seen while the NO is still present. This is consistent with the theoretical understanding of such chemistry. It is thought that HNO is not formed directly from NO but instead is formed by a hydrogen transfer from HCO (1, 2, 3). This means that the presence of CO is catalytic for the formation of HNO (and hence for the formation of nitrite and nitrate), and that NO chemistry is parasitic on the formation of formaldehyde through HCO.

$$H_2O + \text{light} \rightarrow H + OH \tag{1}$$

$$H + CO + M \rightarrow HCO + M \tag{2}$$

$$NO + HCO \rightarrow CO + HNO \tag{3}$$

Another NO reaction pathway was also observed (Summers and Khare, 2007). This involves the photochemical transfer of an oxygen atom from a $CO_2$ molecule to a NO molecule, forming $NO_2$ and CO (4). This chemistry is slower than the pathway

$$NO + CO_2 + \text{light} \rightarrow NO_2 + CO \tag{4}$$

through HNO and only becomes important when liquid water is not present. This reaction produces, rather than consumes, CO. $NO_2$ is itself quite reactive with water and will react with liquid water, or absorbed or frozen water, to form nitric acid. In the complete absence of water, it is not possible to form HNO, and reactions through $NO_2$ are the only ones observed. In the presence of water vapor, but no liquid water, the formation of both $NO_2$ and HNO occurs. The HNO cannot be removed from the gas phase by dissolution in water, and, instead, a slower dimerization to $N_2O$ is observed.

This chemistry makes species such as nitrate and nitrite available for prebiotic chemistry. But how does reduced nitrogen get formed? It is likely that the early oceans contained significant amount of ferrous iron (Beukes and Klein, 1992; Braterman and Cairns-Smith, 1987; Braterman et al., 1983; Cloud, 1973; Derry and Jacobsen, 1990; Drever, 1974; Holland, 1973, 1984; Towe, 1983; Veizer, 1978; Walker and Brimblecombe, 1985). Dissolved ferrous ions can reduce nitrite and nitrate to ammonia (Summers and Chang, 1993). Since then, work has also shown that suspended ferrous-containing minerals such as green rust (which is also involved in reduction by ferrous ions), FeO, $FeCO_3$, and FeS will all reduce nitrate and nitrite to ammonia (Hansen and Koch, 1996; Hansen et al., 1996; Ottley et al., 1997; Rakshit et al., 2008, 2005; Summers, 2005; Summers and Chang, 1993). The presence of other cations has an important effect on these reactions (Hansen and Koch, 1996; Ottley et al., 1997; Summers and Chang,

1993). The reduction of nitrite has been observed with sugars, with and without Fe(III) (and presumably also Fe(II) which would be formed by sugar reduction) as a catalyst (Weber, 2010). Similarly, nitrite and nitrate can be reduced in hydrothermal systems (see below). A study of the kinetics of the reduction of nitrite indicates that these species will be reduced in the upper ocean as quickly as they are formed (Summers, 1999). It was estimated that atmospheric fixation could produce ~$5 \times 10^{10}$ mol reduced nitrogen per year or ~$7 \times 10^{11}$ g reduced N year$^{-1}$ (Summers, 1999; Summers and Chang, 1993). This estimate, however, did not include any contribution by NO produced by the lightning generated by volcanoes (Navarro–González et al., 1998).

It should also be noted that ammonia could also potentially be formed by the photochemical action of titanium dioxide containing desert sands (Schrauzer et al., 1983). However, the importance of this source is unclear, since the ammonia formed would be subject to photolysis and reaction with OH radicals.

## 3. Exogenous Delivery

The Earth is being continuously showered by meteorites and interplanetary dust particles (IDPs). Many of these objects, particularly the carbonaceous chondrites, contain significant amounts of organic material and their delivery to the Earth has been postulated as an important source of organic compounds for the origin of life (Anders, 1989; Chyba and Sagan, 1992, 1997; Delsemme, 1992; Jenniskens et al., 2000). The presence of reduced nitrogen in these objects means that they are also a source of prebiotic nitrogen. The nitrogen in carbonaceous chondrites (the most organic rich of the chondrites, a common class of meteorites) is present primarily as organic nitrogen. In C1 and C2 chondrites, typical chondrites, nitrogen makes up 2% of all the soluble organic matter and 300–450 ppm of the meteorite overall (Dodd, 1981; Kung and Clayton, 1978; Pizzarello, 2004). Nitrogen comprises 3% of the insoluble organic matter, 500–700 ppm. The insoluble organic matter is present in the form of large macromolecular species (Pizzarello et al., 2006). Much of the nitrogen in that phase is probably present in heteroaromatic rings.

It has been estimated that the flux of organics, mostly of chondritic composition, was (from IDPs and micrometeorites) about $4 \times 10^{11}$ g of carbon year$^{-1}$ (Chyba and Sagan, 1997; Maurette et al., 2000). This same flux would have delivered about $1 \times 10^{10}$ g of reduced N year$^{-1}$. This represents material that is not ablated by heating in the atmosphere before it reaches the surface. If organics (and reduced nitrogen) can survive ablation (Jenniskens, 2001; Jenniskens et al., 2004), then that could increase these numbers. It has also been proposed that nitrogen present in the iron meteorites (presumably as nitrides) could have been delivered to the Earth and released as ammonium upon weathering (Smirnov et al., 2008). This could have delivered a flux as high as ~$1 \times 10^{6}$ g of nitrogen year$^{-1}$.

Perhaps most important nitrogen compounds in the soluble fraction, in both terms of abundance (60 ppm in Murchison) and in terms prebiotic importance,

**Table 1.** Nitrogen-containing soluble organics from Murchison meteorite (From Pizzarello et al., 2006, and references therein).

| Class of compound | Concentration (ppm) | # Compounds identified |
|---|---|---|
| Amino acids | 60 | 74[a] |
| Dicarboximides | >50 | 2 |
| Pyridine carboxylic acids | >7 | 7 |
| N-heterocycles | 7 | 31 |
| Amines | 13 | 20 |
| Amides | ND[b] | 27[a] |
| Imino acids | ND[b] | 10 |

[a]This group contained significantly more compounds than were unequivocally identified.
[b]*ND* not determined.

are the amino acids (Pizzarello et al., 2006). They are present in an enormous array of different compounds. For example, at least 74 different amino acids are present in the Murchison meteorite. This set includes many common biological amino acids and includes what are postulated to have been the five most common amino acids in the last universal ancestor (Brooks et al., 2002).

Also of interest are reports of the presence of nucleobases in carbonaceous chondrites. While these species are present in much smaller quantities than the amino acids (<1 ppm), they are one of the fundamental building blocks of RNA and DNA. One pyrimidine compound (uracil) and four purines (adenine, guanine, hypoxanthine, and xanthine) have been found (Hayatsu et al., 1975; Martins et al., 2008; Stoks and Schwartz, 1981; van der Velden and Schwartz, 1977). This source is especially relevant to an early atmosphere that was not reducing enough to produce hydrogen cyanide by shock heating. Since HCN polymerization is the best route that has been discovered for the terrestrial synthesis of these species (see below), exogenous delivery would provide a source of those compounds in the absence of routes for terrestrial synthesis. Exogenous delivery of pyrimidines could also obviate the need for a prebiotic source of cyanoacetylene (see below).

There are a number of other species that are less directly of prebiotic importance but which could provide a feedstock for organic nitrogen compounds to support prebiotic chemistry (Pizzarello et al., 2006). These are, in decreasing abundance, dicarboxyimides, amine, pyridine carboxylic acids, N-heterocyclic compounds, amides, and the imino acids (Table 1).

## 4. Hydrothermal Formation

In hydrothermal systems, water is heated (often by close proximity to magma) while in the presence of mineral catalysts. Significant effort is being directed at determining if synthesis in hydrothermal systems could have been an important source of organics (e.g., see Holm and Andersson, 2005; Simoneit, 2004). However, work on the formation of reduced nitrogen has been more limited, perhaps in part because of the "difficulties in the realistic simulation of submarine hydrothermal environments in the laboratory considering their diversity and the chemically and physically complex nature of natural hydrothermal systems" (Holm and Andersson, 2005). The formation of amino acids from aldehydes and ammonia and/or HCN as a starting material has been reported (which will be covered in the next section).

Smirnov et al. (2008) found that dinitrogen can be reduced in hydrothermal systems with metallic iron, metallic nickel, or alloys of iron and nickel. They postulated that these metals would have been formed in highly reducing environments produced by serpentinization processes similar to those occurring today. Passivation is an issue with the reactions at ferrous metal surfaces. Oxidation of the iron produces iron oxides which can block further reaction. Depending on what assumptions one considers appropriate, production rates were estimated to be as low as $10^9$ g reduced nitrogen year$^{-1}$ or as high as $10^{13}$ g year$^{-1}$. It was also found that this system will reduce nitrate and nitrite to ammonium (Smirnov et al., 2008). Such a process would be important if amounts of Fe(II) in the ocean were too low to reduce all the nitrite/nitrate before it reaches hydrothermal systems.

Schoonen and Xu (2001) looked at reduction of dinitrogen in an $H_2S$/FeS system (hydrogen sulfide makes the system more reducing than a system with FeS alone). They found that reduction of $N_2$ to ammonia occurs but concluded that it could not have produced enough ammonia to sustain the formation of prebiotic organonitrogen compounds.

## 5. Formation of Amino Acids and Nucleobases

We have looked at a number of processes that can produce, or deliver, "reduced nitrogen." In some cases, these provide compounds directly relevant to the origin of life, and in other cases they provide precursors. It is worth looking at how we get to some of the key organonitrogen compounds thought to be needed for the origin of life; the amino acids that are the building blocks of proteins and the nucleobases that, once attached to sugars and phosphate linkages, are strung together to make up RNA/DNA and the genetic code.

### 5.1. AMINO ACIDS

One prebiotic source that directly supplies amino acids is the delivery of organonitrogen compounds by meteorites. One of the reasons that such delivery has been considered such an important potential source of prebiotic compounds, and has received such study, has been the large number of amino acids that are found in meteorites.

The other sources described generally produce ammonia or, if the atmosphere was reducing enough, hydrogen cyanide. As mentioned above, the atmosphere is not considered to have been reducing enough for shock heating to produce amino acids as easily or abundantly as with Miller-Urey-type experiments (Miller and Urey, 1959). So how do we get to amino acids from ammonia and/or hydrogen cyanide? The most accepted route is the Strecker synthesis (5) where ammonia, hydrogen cyanide, and an aldehyde or ketone react to form amino acids (Chang et al., 1983; Enders and Shilvock, 2000; Miller, 1957; Miller and Trump, 1981; Summers and Lerner, 1996). This reaction is a likely candidate for how the amino acids in the meteorites were formed in the first place (Lerner et al., 1993).

It has also been shown that hydrothermal systems will cause the formation of amino acids from formaldehyde (where R=H in Eq. 5), ammonia, and cyanide. Unlike the conventional Strecker synthesis, where the amino acid formed depends on the starting aldehyde (different aldehydes would be formed from condensation of formaldehyde), this system produces a range of amino acids from just formaldehyde. Aubrey et al. (2009) found that amino acids could be formed from just ammonia and formaldehyde (but not when nitrate or nitrite is used in place of the ammonia). Again, a number of different amino acids were formed. This would obviate the need to have a source of hydrogen cyanide for the formation of amino acids.

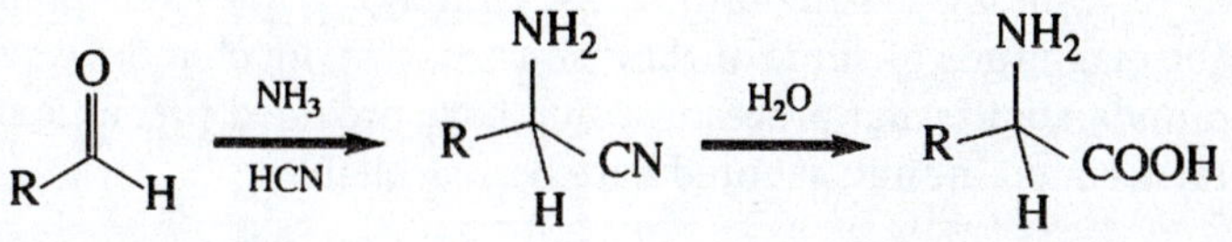


(5)

## 5.2. NUCLEOBASES

As mentioned above, it is known that nucleobases could have been delivered by meteorites. The concentrations are, however, low. The best known route for terrestrial synthesis of purines involves reactions of hydrogen cyanide with ammonia or with photochemical activation (Borquez et al., 2005; Hill and Orgel, 2002; Levy et al., 1999; Miyakawa et al., 2002b; Orgel, 2004; Zubay and Mui, 2001; and references therein). The route to pyrimidines involves reactions with cyanoacetylene (Nelson et al., 2001; Orgel, 2004; Powner et al., 2009). It has been suggested that activated pyrimidine ribonucleotides could be formed directly, bypassing the formation of the nucleobase, in a process that uses cyanamide and cyanoacetylene as the nitrogen-containing starting materials (Powner et al., 2009).

The use of cyanoacetylene as a starting material is problematic. The proposed route for its formation requires spark discharges in a reducing atmosphere containing methane (Nelson et al., 2001; Orgel, 2002). Cyanoacetaldehyde, the other alternative, is itself formed from cyanoacetylene by hydrolysis. The steps in the synthesis of purines and pyrimidines are complex and not completely understood. For a good summary, see the review by Orgel (2004).

Miyakawa et al. (2002b) found that eutectic concentration of ammonium cyanide in water ice can yield both purines and pyrimidines upon hydrolysis. While this was presented as a solution to the problem of low HCN concentrations that may have existed in the prebiotic environment, it would also eliminate the need to have either cyanoacetylene or cyanoacetaldehyde. (This reaction does not produce cytosine but does produce orotic acid.) It has also been proposed that purines and pyrimidines could have been formed from pure formamide and inorganic catalysts (Saladino et al., 2001). Formamide could have been formed from HCN hydrolysis, though it then also has to be separated to form a neat organic solvent.

## 6. Summary

Nitrogen, like carbon, is essential for the formation of life. However, in many ways the sources of prebiotic nitrogen are more of an open question than sources of prebiotic carbon. The difficulty in abiotically fixing nitrogen is one reason why the removal of nitrogen from the Earth's atmosphere is currently so heavily dominated by biology (and by a surprisingly few types of microorganisms). All of the proposed routes for the formation of prebiotic nitrogen compounds find some way of overcoming the chemical inertness of $N_2$ (or, in the case of exogenous delivery, a way of avoiding passing through $N_2$ entirely). While some preliminary estimates of the magnitude of some of these sources exist, much more work needs to be done to understand what processes could have provided prebiotic nitrogen and in what forms and amounts it could have been available.

## 7. References

Anders E (1989) Pre-biotic organic matter from comets and asteroids. Nature 342:255–257

Aubrey AD, Cleaves HJ, Bada JL (2009) The role of submarine hydrothermal systems in the synthesis of amino acids. Orig Life Evol Biosph 39:91–108

Bar-Nun A, Chang S (1983) Photochemical reactions of water and carbon monoxide in Earth's primitive atmosphere. J Geophys Res 88:6662–6672

Bar-Nun A, Hartman H (1978) Synthesis of organic compounds from carbon monoxide and water by UV photolysis. Orig Life 9:93–101

Beukes NJ, Klein C (1992) 4.3 Models for iron-formation deposition. In: Schopf JW, Klein C (eds) The proterozoic biosphere: a multidisciplinary study. Cambridge University Press, Cambridge, pp 147–151

Borquez E, Cleaves HJ, Lazcano A, Miller SL (2005) An investigation of prebiotic purine synthesis from the hydrolysis of HCN polymers. Orig Life Evol Biosph 35:79–90

Borucki WJ, Chameides WL (1984) Lightning: estimates of the rates of energy dissipation and nitrogen fixation. Rev Geophys Space Phys 22:363–372

Boyd ID (2000) Computation of atmospheric entry flow about a Leonid meteoroid. Earth Moon Planet 82–83:93–108

Braterman PS, Cairns-Smith AG (1987) Iron photoprecipitation and the genesis of the banded iron-formations. In: Appel PWU, LaBerge GL (eds) Precambrian iron-formations. Theophrastus Publications, Athens, pp 215–245

Braterman PS, Cairns-Smith AG, Sloper RW (1983) Photooxidation of hydrated $Fe^{2+}$: the significance for banded iron formations. Nature 303:163–164
Brooks DJ, Fresco JR, Lesk AM, Singh M (2002) Evolution of amino acid frequencies in proteins over deep time: inferred order of introduction of amino acids into the genetic code. Mol Biol Evol 19:1645–1655
Catling DC (2006) Comment on "a hydrogen-rich early earth atmosphere". Science 311:38a
Chameides WL (1979) Effect of variable energy input on nitrogen fixation in instantaneous linear discharges. Nature 277:123–125
Chameides WL, Walker JCG (1981) Rates of fixation by lightning of carbon and nitrogen in possible primitive atmospheres. Orig Life 11:291–302
Chang S, DesMarais D, Mack R, Miller SL, Strathearn GE (1983) Prebiotic organic synthesis and the origin of life. In: Shopf JW (ed) Earth's earliest biosphere: its origin and evolution. Princeton University Press, Princeton, pp 53–92
Chyba C, Sagan C (1992) Endogenous production, exogenous delivery and impact-shock synthesis of organic molecules: an inventory for the origins of life. Nature 355:125–132
Chyba CF, Sagan C (1997) Comets as a source of prebiotic organic molecules for the early Earth. In: Thomas PJ, Chyba CF, McKay CP (eds) Comets and the origin and evolution of life. Springer, Berlin-Heidelberg, pp 147–173
Cleaves HJ, Chalmers JH, Lazcano A, Miller SL, Bada JL (2008) A reassessment of prebiotic organic synthesis in neutral planetary atmospheres. Orig Life Evol Biosph 38:105–115
Cloud P (1973) Paleoecological significance of the banded iron-formation. Econ Geol 68:1135–1143
Cotton FA, Wilkinson G (1980) Advanced inorganic chemistry: a comprehensive text. Wiley, New York, pp 412–413, Chapter 13–3
Delano JW (2001) Redox history of the Earth's interior since 3900 Ma: implications for prebiotic molecules. Orig Life Evol Biosph 31:311–341
Delsemme AH (1992) Cometary origin of carbon, nitrogen, and water on the Earth. Orig Life Evol Biosph 21:279–298
Derry LA, Jacobsen SB (1990) The chemical evolution of Precambrian seawater: evidence from REEs in banded iron formations. Geochem Cosmochem Acta 54:2965–2975
Dodd RT (1981) Chapter 2: The chondrites: chemistry and classification. In: Meteorites: a petrologic-chemical analysis. Cambridge University Press, Cambridge, pp 18–23
Drever JI (1974) Geochemical model of the origin of Precambrian banded iron formations. Geol Soc Am Bull 85:1099–1106
Enders D, Shilvock JP (2000) Some recent applications of α-amino nitrile chemistry. Chem Soc Rev 29:359–373
Fegley B Jr, Prinn RG, Hartman H, Watkins GH (1986) Chemical effects of large impacts on the Earth's primitive atmosphere. Nature 319:305–308
Hansen HCB, Koch CB (1996) Reduction of nitrate to ammonium by sulphate green rust: activation energy and reaction mechanism. Clay Miner 33:87–101
Hansen HCB, Koch CB, Nancke-Krogh H, Borggaard OK, Sorensen J (1996) Abiotic nitrate reduction to ammonium: key role of green rust. Environ Sci Technol 30:2053–2056
Hayatsu R, Studier MH, Moore LP, Anders E (1975) Purines and triazines in the Murchison meteorite. Geochim Cosmochim Acta 39:471–488
Hill A, Orgel LE (2002) Synthesis of adenine from HCN tetramer and ammonium formate. Orig Life Evol Biosph 32:99–102
Holland HD (1973) The oceans: a possible source of iron in iron-formations. Econ Geol 68:1169–1172
Holland HD (1984) The chemical evolution of the atmosphere and oceans. Princeton University Press, Princeton, pp 387–388, 396–397
Holm NG, Andersson E (2005) Hydrothermal simulation experiments as a tool for studies of the origin of life on Earth and other terrestrial planets: a review. Astrobiology 5:444–460
Hubbard JS, Hardy JP, Horowitz NH (1971) Photocatalytic productions of organic compounds from CO and $H_2O$ in a simulated Martian atmosphere. Proc Natl Acad Sci USA 68:574–578

Hunten DM (1993) Atmospheric evolution of the terrestrial planets. Science 259:915–920
Jebens DS, Lakkaraju HS, McKay CP, Borucki WJ (1992) Time resolved simulation of lighting by LIP. Geophys Res Lett 19:273–276
Jenniskens P (2001) Meteors as a delivery vehicle for organic matter to the early earth. In: Proceedings of the Meteoroids 2001 conference. European Space Agency, ESA SP-495. Swedish Insititute of Space Physics, Kiruna, Sweden, pp 247–254
Jenniskens P, Stenbaek-Nielsen HC (2004) Meteor wake in high frame-rate images—implications for the chemistry of ablated organic compounds. Astrobiology 4:95–108
Jenniskens P, Wilson MA, Packan D, Laux CO, Krüget CH, Boyd ID, Popova OP, Fonda M (2000) Meteors: a delivery mechanism of organic matter to the early earth. Earth Moon Planet 82–83:57–70
Jenniskens P, Schaller EL, Laux CO, Wilson MA, Schmidt G, Rairden RL (2004) Meteors do not break exogenous organic molecules into high yields of diatomics. Astrobiology 4:67–79
Kasting JF (1990) Bolide impacts and the oxidation state of carbon in the Earth's early atmosphere. Orig Life Evol Biosph 20:199–231
Kharecha P, Kasting J, Siefert J (2005) A coupled atmosphere–ecosystem model of the early Archean Earth. Geobiology 3:53–76
Kulikov YN, Lammer H, Lichtenegger HIM, Penz T, Breuer D, Spohn T, Lundin R, Biernat HK (2007) A comparative study of the influence of the active young Sun on the early atmospheres of Earth, Venus, and Mars. Space Sci Rev 129:207–243
Kung C-C, Clayton RN (1978) Nitrogen abundances and isotopic compositions in stony meteorites. Earth Planet Sci Lett 38:421–435
Lerner NR, Peterson E, Chang S (1993) The strecker synthesis as a source of amino acids in carbonaceous chondrites: deuterium retention during synthesis. Geochim Cosmochim Acta 57:4713–4723
Levy M, Miller SL, Oró J (1999) Production of guanine from $NH_4CN$ polymerizations. J Mol Evol 49:165–168
Mancinelli RL, McKay CP (1988) The evolution of nitrogen cycling. Orig Life Evol Biosph 18: 311–325
Martins Z, Botta O, Fogel ML, Sephton MA, Glavin DP, Watson JS, Dworkin JP, Schwartzg AW, Ehrenfreund P (2008) Extraterrestrial nucleobases in the Murchison meteorite. Earth Planet Sci Lett 270:130–136
Mattioli GS, Wood BJ (1986) Upper mantle oxygen fugacity recorded by spinel lherzolites. Nature 322:626–628
Maurette M, Duprat J, Engrand C, Gounelle M, Durat G, Matrajt G, Toppani A (2000) Accretion of neon, organics, $CO_2$, nitrogen, and water from large interplanetary dust particles on the early Earth. Planet Space Sci 48:1117–1137
Miller SL (1957) The mechanism of synthesis of amino acids by electric discharges. Biochim Biophys Acta 23:480–487
Miller SL, Schlesinger G (1983) The atmosphere of the primitive Earth and the prebiotic synthesis of organic compounds. Adv Space Res 3:47–53
Miller SL, Schlesinger G (1984) Carbon and energy yields in prebiotic syntheses using atmospheres containing $CH_4$, CO and $CO_2$. Orig Life Evol Biosph 14:83–90
Miller SL, Trump JEV (1981) The Strecker Synthesis in the primitive ocean. In: Wolmann A (ed) Origins of life. D. Reidel Publishing, Dordrecht, pp 135–141
Miller SL, Urey HC (1959) Organic compound synthesis on the primitive Earth. Science 130:245–251
Miyakawa S, Yamanashi H, Kobayashi K, Cleaves HJ, Miller SL (2002a) Prebiotic synthesis from CO atmospheres: implications for the origins of life. Proc Natl Acad Sci USA 99:14628–14631
Miyakawa S, Cleaves HJ, Miller SL (2002b) The cold origins of life: B. Implications based on pyrimidines and purines produced from frozen ammonium cyanide solutions. Orig Life Evol Biosph 32:209–218
Navarro-González R, Molina MJ, Molina LT (1998) Nitrogen fixation by volcanic lightning in the early Earth. Geophys Res Lett 25:3123–3126

Navarro–González R, Ramires SI, de la Rosa JG, Coll P, Raulin F (2001a) Production of hydrocarbons and nitriles by electrical processes in titan's atmosphere. Adv Space Res 27:271–282

Navarro–González R, Villagran-Muniz M, Sobral H, Molina LT, Molina MJ (2001b) The physical mechanism of nitric oxide formation in simulated lightning. Geophys Res Lett 28:3867–3870

Nelson KE, Robertson MP, Levy M, Miller SL (2001) Concentration by evaporation and the prebiotic synthesis of cytosine. Orig Life Evol Biosph 31:221–229

Nna Mvondo D, Navarro-Gonzgdez R, McKay CP, Coll P, Raulin F (2001) Production of nitrogen oxides by lightning and coronae discharges in simulated early Earth, Venus and Mars environments. Adv Space Res 27:217–223

Nna-Mvondo D, Navarro-González R, Raulin F, Coll P (2005) Nitrogen fixation by corona discharge on the early Precambrian Earth. Orig Life Evol Biosph 35:401–409

Orgel LE (2002) Is cyanoacetylene prebiotic? Orig Life Evol Biosph 32:279–281

Orgel L (2004) Prebiotic chemistry and the origin of the RNA world. Crit Rev Biochem Mol Biol 39:99–123

Ottley CJ, Davison W, Edmunds WM (1997) Chemical catalysis of nitrate reduction by iron (II). Geochim Cosmochim Acta 61:1819–1828

Pinto JP, Gladstone GR, Yung YL (1980) Photochemical production of formaldehyde in Earth's primitive atmosphere. Science 210:183–185

Pizzarello S (2004) Chemical evolution and meteorites: an update. Orig Life Evol Biosph 34:25–34

Pizzarello S, Cooper GW, Flynn GJ (2006) The nature and distribution of the organic material in carbonaceous chondrites and interplanetary dush particles. In: Lauretta DS, McSween HY Jr (eds) Meteorites and the Early Solar System II. The University of Arizona Press, Tuscson, pp 625–651

Plankensteiner K, Reiner H, Schranz B, Rode BM (2004) Prebiotic formation of amino acids in a neutral atmosphere by electric discharge. Angew Chem Int Ed 41:1886–1888

Powner MW, Gerland B, Sutherland JD (2009) Synthesis of activated pyrimidine ribonucleotides in prebiotically plausible conditions. Nature 459:239–242

Prinn RG, Jr. Fegley B (1987) Bolide impacts, acid rain, and biospheric traumas at the Cretaceous-Tertiary Boundary. Earth Planet Sci Lett 83:1–15

Rakshit S, Motocha CJ, Hazier GR (2005) Nitrate reduction in the presence of wusite. J Environ Qual 34:1286–1292

Rakshit S, Matocha CJ, Coyne MS (2008) Nitrite reduction by siderite. Soil Sci Am J 72:1070–1077

Rosing MT, Bird DK, Sleep NH, Bjerrum CJ (2010) No climate paradox under the feint young sun. Nature 464:744–747

Saladino R, Crestini C, Costanzo G, Negri R, Mauro ED (2001) A possible prebiotic synthesis of purine, adenine, cytosine, and 4(3H)-pyrimidinone from formamide: implications for the origin of life. Bioinorg Med Chem 9:1249–1253

Scattergood TW, McKay CP, Borucki WJ, Giver LP, Van Ghyseghem H, Parris JE, Miller SL (1989) Production of organic compounds in plasmas: a comparison among electric sparks, laser-induced plasmas, and UV light. Icarus 81:413–428

Schoonen MAA, Xu Y (2001) Nitrogen reduction under hydrothermal vent conditions: implications for the prebiotic synthesis of C-H-O-N compounds. Astrobiology 1:133–141

Schrauzer GN, Strampach N, Hui LN, Palmer MR, Salehi J (1983) Nitrogen photoreduction on desert sands under sterile conditions. Proc Natl Acad Sci 80:3873–3876

Simoneit BRT (2004) Prebiotic organic synthesis under hydrothermal conditions: an overview. Adv Space Res 33:88–94

Smirnov A, Hausner D, Laffers R, Strongin DR, Schoonen MA (2008) Abiotic ammonium formation in the presence of Ni-Fe metals and alloys and its implications for the Hadean nitrogen cycle. Geochem Trans 9:5–25

Sobral H, Villagràn-Muniz M, Navarro-Gonzàlez R, Raga AC (2000) Temporal evolution of the shock wave and hot core air in laser induced plasma. Appl Phys Lett 77:3158–3160

Stoks PG, Schwartz AW (1981) Nitrogen-heterocyclic compounds in meteorites: significance and mechanisms of formation. Geochim Cosmochim Acta 45:563–569

Summers DP (1999) Sources and sinks for ammonia and nitrite on the early Earth and the reaction of nitrite with ammonia. Orig Life Evol Biosph 29:33–46

Summers DP (2005) Ammonia formation by the reduction of nitrite/nitrate by FeS: ammonia formation under acidic conditions. Orig Life Evol Biosph 35:299–312

Summers DP, Chang S (1993) Prebiotic ammonia from reduction of nitrite by iron(II) on the early Earth. Nature 365:630–633

Summers DP, Khare B (2007) Nitrogen fixation on early Mars and other terrestrial planets. Experimental demonstration of abiotic fixation reactions to nitrite and nitrate. Astrobiology 7: 333–341

Summers DP, Lerner N (1996) Strecker synthesis using ammonia from iron(II) reduction of nitrite: do iron(II) and cyanide interfere with each other? Orig Life Evol Biosph 26:221–222

Tian F, Toon OB, Pavlov AA, Sterck HD (2005) A hydrogen-rich early Earth atmosphere. Science 308:1014–1017

Towe KM (1983) Precambrian atmospheric oxygen and banded iron formations: a delayed ocean model. Precambrain Res 20:161–170

van der Velden W, Schwartz AW (1977) Search for purines and pyrimidines in the Murchison meteorite. Geochim Cosmochim Acta 41:961–968

Veizer J (1978) Secular variations in the composition of sedimentary carbonate rocks, II. Fe, Mn, CA, Mg, Si and minor constituents. Precambrian Res 6:381–413

Walker JCG (1985) Carbon dioxide on the early Earth. Orig Life 16:117–127

Walker JCG, Brimblecombe P (1985) Iron and sulfur in the pre-biological ocean. Precambrian Res 28:205–222

Weber AL (2010) Sugar-driven prebiotic synthesis of ammonia from nitrite. Orig Life Evol Biosph 40:245–252

Wen J-S, Pinto JP, Yung YL (1989) Photochemistry of CO and $H_2O$: analysis of laboratory experiments and applications to the prebiotic Earth's atmosphere. J Geophys Res E Planet 94:14957–14970

Wood BJ, Vigo D (1989) Upper mantle oxidation state: ferric iron contents of lherzolite spinels by $^{57}$Fe Mossbauer spectroscopy and resultant oxygen fugacities. Geochim Cosmochim Acta 53:1277–1989

Yung YL, McElroy MB (1979) Fixation of nitrogen in the prebiotic atmosphere. Science 203:1002–1004

Zahnle KJ (1990) Atmospheric chemistry by large impacts. In: Sharpton VL, Wards PD (eds) Global catastrophes in Earth history: an interdisciplinary conference on impacts, volcanism and mass mortality. Geological Society of America, Boulder, pp 271–287

Zubay G, Mui T (2001) Prebiotic synthesis of nucleotides. Orig Life Evol Biosph 31:87–102

# PART II: THE ROLE OF PHYSICS AND GEOPHYSICS IN THE ORIGIN AND EVOLUTION OF LIFE


**Abel**
**Ariga**
**Richards**
**Hill**
**Barbalet**
**Damer**
**Newman**
**Norkus**
**Graham**
**Gordon**
**Ichihashi**
**Matsuura**
**Kita**
**Sunami**
**Suzuki**
**Yomo**
**Kompanichenko**
**Muller**
**Rimola**
**Sodupe**
**Ugliengo**
**Sheldon**
**Wacey**

Biodata of **David L. Abel**, author of "***Moving "Far from Equilibrium" in a Prebiotic Environment: The Role of Maxwell's Demon in Life Origin***."

**Dr. David L. Abel** majored in Zoology and studied at the University of Maryland (College Park), Virginia Tech (Blacksburg), and the University of Georgia (Athens) (Doctorate 1972). Dr. Abel has lectured at international scientific conferences in Europe and the United States. He has published numerous peer-reviewed science journal papers and life-origin book chapters. He is a theoretical biologist focusing on ProtoBioCybernetics and ProtoBioSemiotics. He is the Program Director of The Gene Emergence Project, an international consortium of scientists investigating the natural-process derivation of initial biocybernetic/biosemiotic programming and control in a primordial environment.

E-mail: **life@us.net**

J. Seckbach (ed.), *Genesis - In The Beginning: Precursors of Life, Chemical Models and Early Biological Evolution*, Cellular Origin, Life in Extreme Habitats and Astrobiology 22, 219–236
DOI 10.1007/978-94-007-2941-4_13, 

# MOVING "FAR FROM EQUILIBRIUM" IN A PREBIOTIC ENVIRONMENT: THE ROLE OF MAXWELL'S DEMON IN LIFE ORIGIN

**DAVID L. ABEL**
*Department of ProtoBioCybernetics/ProtoBioSemiotics, The Gene Emergence Project, The Origin-of-Life Science Foundation, Inc., 113-120 Hedgewood Dr., Greenbelt, MD 20770-1610, USA*

## 1. How Could Usable Energy Arise Out of Equilibrium?

Maxwell was among the first to ponder how usable energy could spontaneously arise out of equilibrium (Maxwell, 1871). Even the simplest heat engine requires a kinetic energy differential, or "potential," to accomplish "useful work." Maxwell was searching for a natural process that could produce not only this energy potential but a spontaneously forming nontrivial heat engine capable of harnessing that energy potential and transducing it into usable energy suitable for performing intuitive pragmatic "work."

The proper definition of "work," at least as it relates to abiogenesis, embodies formal utility ("useful work"), not just the mere thermodynamic transfer of energy from one body to another. The physics definition of "work" has its shortcomings too. If an object's kinetic energy changes as a result of a force acting upon it, we glibly say "work" is done. Physics defines work as a force causing an object to be displaced. Each unit of energy (joule) incorporates a newton of force moving an object 1 m. But 1 m in what direction for what purpose? Is there any meaning or function to this movement?

The above two questions are formal queries that cannot be answered by chance or physicodynamic factors alone. The only way the physics definition of work acquires the intuitive, everyday semantic meaning of "work" is that our supposedly naturalistic models smuggle in through the back door a lot of background information along with agent motives, values, and pursuits. $W = \Delta E = Fd$ and other "work"-related equations of physics by themselves tell us absolutely nothing about utility or how to achieve it. No such agent-valued and pursued factors existed in a prebiotic environment. Yet without achieving abundant practical functions in and between many cooperative systems, not even a protocell could have organized. Life is the most integrated, sophisticated interaction of functional processes known to science or engineering. Thus, the problem for life-origin specialists is daunting. Mere spontaneous self-replication is the simplest problem that needs solving.

Far more troublesome is explaining the source of prescriptive information (PI) (Abel, 2009a) needed to design, prescribe, and engineer a sustained heat engine. In the history of human observation (the cornerstone of science), creating sustained and harnessable "useful work" has always required formal engineering choices at true decisions nodes, logic gates, and configurable switch-settings. Goals such as algorithmic optimization must be pursued at each organizational step. Such pursuits are just not common to spontaneous physicodynamics in any amount of cosmic time. Not even evolution has any goals. Plausibility evaluations of life-origin models and hypotheses must take such factors into consideration (Abel, 2009e, 2010b).

Maxwell could not even find a natural mechanism capable of distinguishing, let alone separating, the hotter, faster-moving inert gas molecules from the cooler, slower-moving inert gas molecules. Once equalized on both sides of a compartmental partition, neither chance nor necessity could create a heat differential. The only model that seemed able to accomplish the task consisted of a thought experiment. A "finite being" had to purposefully choose when to open and close a trap door in the partition. Lord Kelvin later called this finite being, "Maxwell's Demon." When the Demon saw a hotter, faster-moving molecule approaching the trap door from the left compartment, he deliberately opened the trap door to let it pass into the right compartment. When he saw a colder molecule approaching the open trap door from the left compartment, he purposefully closed the trap door to keep that colder particle in the left compartment. His choices were reversed with inert gas molecules approaching the trap door from the right compartment. The end result was a kinetic energy potential waiting to be harnessed, transduced, stored, and later called up by sophisticated mechanisms and molecular machines capable of accomplishing useful work (Fig. 1).

## 2. Why Do the Gas Molecules Have to Be Ideal?

In order to avoid what is known as the "Gibbs Paradox," the "point particles" must be *indistinguishable* and immune to phase changes and chemical reactions. Ideal gas molecules are inert. They have no physicochemical preference for either compartment or for each other. Except for temperature and its kinetic relation to heat agitation, each particle's destiny is physicodynamically indeterminate (decoupled and incoherent from physical causation, physicodynamically inert) (Rocha, 2001; Rocha and Hordijk, 2005). The heat agitation of inert gas molecules is devoid of physicochemical affinities. As explained in Sect. 5 below, this feature is essential to the instantiation of nonphysical prescriptive information (PI) (Abel, 2009a) into physicality. Decision nodes, logic gates, and configurable switch-settings must be free from physicodynamic determinism in order to program formal function into physical media (Abel, 2008a, b, c, 2009b).

**Figure 1.** Maxwell's Demon has to purposefully choose when to open and close the trap door to accomplish his goal. Because the gas molecules are inert and ideal, no physicodynamic mechanism seems apparent to explain spontaneous self-ordering, let alone formal self-organization, to the utilitarian end of creating a nontrivial heat engine.

Suppose that these gas molecules were *not* ideal, not physicodynamically inert. If these molecules were charged particles, for example, they might have natural tendencies to dichotomize themselves from an equilibrium state (Dalenoort, 1989; Demetrius, 1984; Eigen, 1971a, b; Eigen and Schuster, 1977, 1982; Epstein and Eigen, 1979; Gánti, 1975, 1980; Goldbeter and Decroly, 1983; Kuhn, 1972, 1976; Küppers, 1979; Lifson, 1987; Morowitz, 1977, 1981; Morowitz et al., 1988; Nicolis and Prigogine, 1977; Rokhsar et al., 1986). They might spontaneously self-order due to charge attraction/repulsion. Or they might have differential tendencies to undergo phase changes or to chemically react. Such physicodynamic factors could spontaneously produce heterogeneity, moving them locally and temporarily away from equilibrium.

Could not physicodynamic interaction and self-ordering tendencies be a mechanism important to life origin? Many publications have promoted this perspective (Agrawal, 2002; Barab et al., 1999; Batten et al., 1991; Bedau, 2003; Comazine et al., 2001; Dalenoort, 1989; Deacon, 2006a, b; Demetrius, 1984; Eigen and Winkler, 1992; Feltz et al., 2006; Galimov, 2009; Gershenson, 2007; Harold, 2005; Ito and Gunji, 1994; Kauffman, 1993, 1995; Kurakin, 2005; Lehn, 2002a, b; Lozneanu and Sanduloviciu, 2009; Luisi, 2003; Moreno and Ruiz-Mirazo, 2009; Nicholis, 1989; Orgel, 1995; Ponnamperuma et al., 1995; Rokhsar et al., 1986; Ruiz-Mirazo et al., 1998, 2010; Schiffmann, 2007; Seeley, 2002; Stewart, 2003; Swenson, 1989; Takeuchi and Hogeweg, 2009a; Takeuchi and Hogeweg, 2009b; Umerez, 2001; Vesterby, 2008; Weber and Depew, 1996; Wills, 1993). Despite all of these publications, and many more, the answer to the question of whether physicodynamic interaction and self-ordering tendencies could generate abiogenesis is, surprisingly, "NO!" Mere heterogeneity and order do not even begin to satisfy the necessary and sufficient conditions for life. Self-ordering tendencies provide no mechanism for self-organization, let alone abiogenesis. All sorts of physical astronomical "clumping," weak-bonded molecular alignments, phase changes, and outright chemical reactions occur spontaneously in nature that have nothing to do with life. Life is organization-based, not order-based. As we shall see below in Sect. 6, order is poisonous to organization.

Stochastic ensembles of nucleotides and amino acids can polymerize naturalistically (with great difficulty). But *functional* sequencing of those monomers cannot be determined by any fixed physicodynamic law. It is well known that only one 150-mer polyamino acid string out of $10^{74}$ stochastic ensembles folds into a tertiary structure with any hint of protein function (Axe, 2004). This takes into full consideration the much publicized substitutability of amino acids without loss of function within a typical protein family membership. The odds are still only one functional protein out of $10^{74}$ stochastic ensembles. And 150 residues are of minimal length to qualify for protein status. Worse yet, spontaneously condensed levo-only peptides with peptide-only bonds between only biologically useful amino acids in a prebiotic environment would rarely exceed a dozen mers in length. Without polycodon prescription and sophisticated ribosome machinery, not even polypeptides form that would contribute much to "useful biological work."

## 3. The Trap Door Is Not Constrained; It Is *Demon-Controlled*

Since 1871, many attempts have been made to refine Maxwell's search. Quantum level models of heat engine generation have been suggested (Kieu, 2004; Quan et al., 2006, 2007; Scully, 2001). Muller and Schulze-Makuch wondered whether the first organisms were sorption heat engines (Muller, 1995). Szilard (1964) argued that Maxwell's Demon must be "informed." But to be informed requires semantic information. This raises the sticky problem of formal "knowledge of being" as opposed to objective physicodynamic being itself. Worse yet, it brings to the forefront the Demon's informed *choices*. He must not only know when to open or close the trap door, he must have the freedom to choose from among real options. He must have the power to control thermodynamic and physical destiny.

Semantic information has always represented a major problem for philosophic naturalism. Many have struggled with it, perhaps the most prominent being Bar-Hillel and Carnap (1953), Dretske (1981, 1995), Barwise and Perry (1983), Barwise and Etchemendy (1990), and Devlin (1991). Some have questioned the very existence of information, attempting to reduce it to mere thermodynamics and complex physicality (Barham, 1996; Boniolo, 2003; Godfrey-Smith, 2003; Griffiths, 2001; Kauffman, 1993; Sarkar, 1996, 2003; Stent, 1981; Stonier, 1996; Toussaint and Schneider, 1998). Deacon has attempted to provide a naturalistic basis for the emergence of teleology (Deacon and Sherman, 2008; Deacon, 2006a, b, c, 2010; Sherman and Deacon, 2007). Jacob (1974), Wolpert (2002), Stegmann (2005), Barbieri (2004), and Abel (2002, 2005) tend to view genetic and epigenetic information as real. Norbert Wiener argued that "Information is information, not matter or energy. Any materialism that fails to take account of this will not survive one day" (Wiener, 1961). Genetic cybernetics inspired Turing's (1936), von Neumann's (1950), and Wiener's (1948) development of computer science. Had it not been for their observation of an objective, heritable, linear digital genetic message and control system that predated human consciousness, computers might never have been invented.

## 4. "Far from Equilibrium" Is Not Unique to Life or the Key to Life

Chaos theory deals with many self-ordering phenomena that spontaneously move events far from equilibrium. But candle flames, vortices at bathtub drains, sand piles, and hurricanes have absolutely nothing to do with life. If anything, the "dissipative structures" of chaos theory (e.g., tornadoes) tend to destroy life and any other form of formal organization that they encounter.

The bottom line is that merely "moving far from equilibrium" is not the key to the life-origin problem as supposed. Most would agree that life is indeed "far from equilibrium." But so are hurricanes. It should be readily apparent from hurricane thermodynamics that merely "moving far from equilibrium" does not necessarily have anything to do with "life."

If there is any one feature of life that distinguishes it from nonlife, it is the exercise of control and regulation through time and extreme environmental insults. And this control is specifically mediated through a unique linear digital material symbol system found nowhere in the inanimate world. This linear digital symbol system communicates, genetically prescribes, and epigenetically regulates the most exquisite interactions and cooperation of pathways and cycles. The selection of each nucleotide "token" serves to write linguistic-like instructions. Each nucleotide polymerization corresponds to a quaternary configurable switch-setting (Abel, 2000, 2002, 2006, 2008a, b). Life is ultimately programmed by decision-node and logic-gate settings found in nucleic acid linear digital sequencing.

## 5. Order $\neq$ Organization

Order results from physicodynamic constraints working on otherwise random distributions (e.g., heat agitation) to reduce combinatorial possibilities. Order binds free molecules or individual entities into fewer conglomerates or categories. Structure and patterns arise that reduce disorder (increase order). The physical constraints that produce greater order simultaneously reduce combinatorial possibilities, thereby simultaneously reducing complexity. The higher the self-ordering tendency of physicodynamic interactions, the less complexity remains in hopes of explaining naturalistic abiogenesis.

It is ironic that many investigators simultaneously appeal to "complexity" AND to some yet-to-be-discovered "law" of self-organization. Complexity and law-like behavior are not only incompatible, they are in opposition (Abel and Trevors, 2005). The more law-like phenomena are, the less complex they are. Maximum complexity is random behavior, not highly ordered behavior that can be compressed down to a simple algorithmic formula or law. We cannot simultaneously appeal to complexity and law as explanations for life origin.

Organization, unlike law-like behavior, is highly complex and informational (Abel, 2008a, 2010a). Organization requires selection at bona fide decision nodes (Abel, 2008b, c). In addition, organization requires selection *for potential function*, not just selection of the best existing function (The GS Principle) (Abel, 2009a). Without prior organization, no sophisticated phenotypic function would exist for the environment to favor. Natural selection does not select at the programming level, and it does not select for isolated function (The GS Principle) (Abel, 2009c, d). Selection pressure preserves only the fittest already-programmed, already-living phenotypes (holistic organisms) (Gabora, 2006).

Organization is fundamentally formal, not physical (Abel, 2007, 2008b, 2009b, 2010a). Organization can be instantiated into physicality, but physicality itself is never the source of organization (Abel, 2008b, c). Physicality can self-order, but it cannot self-organize. Organization requires traversing The Cybernetic Cut (Abel, 2008b, c) across the one-way CS Bridge (configurable switch bridge) (Abel, 2008b, c). This unidirectional bridge provides the only passageway across a great ravine (The Cybernetic Cut) that runs through the only objective reality that science has ever observed. On the near side of the ravine are all those phenomena that can be adequately explained with nothing but physicodynamics. On the far side of this great ravine lie all those phenomena that can only be explained by formal choice contingency (e.g., purposeful choices at true decision nodes, circuit integration through the wise setting of logic gates, all forms of programming, computational success achieved through obeying formal rules rather than forced physical laws, artificial intelligence). Much of the confusion between order and organization was fostered by Eigen (1971a, b) and reinforced by Kuhn (1972), Ganti (1975), and Nicholis and Progogine (1977). Their papers and books purporting to explain "Self-Organization" in fact explained nothing more than self-ordering phenomena. Their confusion and conflation of organization with order is never more apparent than in the very title of Nicholis and Prigogine's famous book, *Self-Organization in Nonequilibrium Systems: From Dissipative Structures to Order Through Fluctuations* (Nicolis and Prigogine, 1977). Their title should have been simply: *Self-ordering into Nonequilibrium Dissipative Structures.* Neither bona fide "organization" nor algorithmic "systems" can be generated by the self-ordered dissipative structures of chaos theory. Prigogine's later title was far more accurate: *Order out of Chaos* (Prigogine and Stengers, 1984).

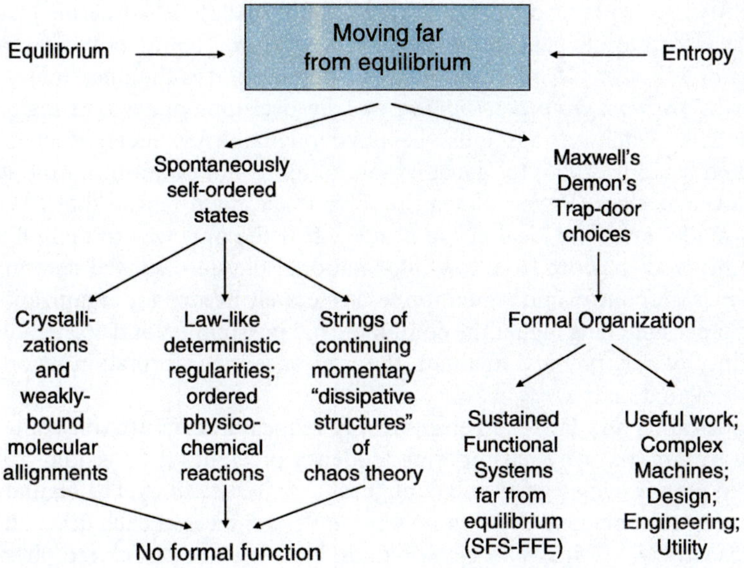


**Figure 2.** Diagram showing why "order" and merely moving "far from equilibrium" are not synonymous with organization, function, or life.

In addition to the null hypothesis found in this chapter's abstract, we also vigorously invite falsification of the following null hypothesis: *"Self-ordering physicodynamic phenomena alone cannot generate formal organization* (Abel, 2008b, 2009a, b, 2010a; Abel and Trevors, 2005, 2006b)." "Order's" capabilities are no match for those of "organization" (Abel, 2008a). Abiogenesis requires bona fide formal organization, not mere order (Abel and Trevors, 2006b) (Trevors and Abel, 2004). Crystallization produces abundant order, but crystals are not alive. While self-ordering phenomena might be able to locally and temporarily move events away from equilibrium (Galimov, 2009), self-ordering phenomena have never once been observed to algorithmically optimize, program, compute, or organize cooperative formal function. This is one reason Cairns-Smith's intriguing model of clay life fell by the wayside (Cairns-Smith, 1966, 1990) (Fig. 2).

## 6. Self-Ordering Precludes Organization

There are other reasons why merely "moving far from equilibrium" is *not* the key to life as seems so universally supposed. Disequilibrium stemming from mere physicodynamic constraints and self-ordering phenomena would actually be poisonous to life origin (Abel, 2009b). The price of such constrained and self-ordering tendencies in nature is the severe reduction of Shannon informational uncertainty in any physical medium (Abel, 2008b, 2010a). Self-ordering processes preclude information generation because they force conformity and reduce freedom of selection. If information needs anything, it is the uncertainty made possible by *freedom from determinism* at true decisions nodes and logic gates. Configurable switch-settings must be physicodynamically inert (Rocha, 2001; Rocha and Hordijk, 2005) for genetic programming and evolution of the symbol system to take place (Pattee, 1995a, b). This is the main reason that Maxwell's Demon model must use ideal gas molecules. It is the only way to maintain high uncertainty and freedom from low-informational physicochemical determinism. Only then is the control and regulation so desperately needed for organization and life origin possible. The higher the combinatorial possibilities and epistemological uncertainty of any physical medium, the greater is the information recordation potential of that matrix.

Constraints and law-like behavior only reduce uncertainty (bit content) of any physical matrix. Any self-ordering tendency precludes the freedom from law needed to program logic gates and configurable switch-settings. The regulation of life requires not only true decision nodes, but wise choices at each decision node. This is exactly what Maxwell's Demon does. No yet-to-be-discovered physicodynamic law will ever be able to replace the Demon's wise choices or explain the exquisite linear digital PI programming and organization of life (Abel, 2009a; Abel and Trevors, 2007). Organization requires choice contingency rather than chance contingency or law (Abel, 2008b, 2009b, 2010a). This conclusion comes via deductive logical necessity and clear-cut category differences, not just from best-thus-far empiricism or induction/abduction.

Programming lies in a formal category, the same as mathematics, language, logic theory, and cybernetics. Logic gates cannot be programmed by chance or physicodynamic necessity (Trevors and Abel, 2004). Even the simplest life form is the most highly controlled and regulated system known to science. It puts our finest hardware and operating systems to shame. No low-informational physical law is going to explain such abstract, conceptual, integrated, logic-gate selections at every nucleotide polymerization event (a quaternary [four-way] configurable switch-setting) (Abel, 2008b).

Appeals to complexity as an explanation for life origin are better founded, in some respects, than appeals to some yet-to-be-discovered physical law. But maximum complexity is randomness. Randomness has never generated a single complex machine or organized formal "system." Life's genetic and epigenetic programming decision nodes cannot be reduced to mere "bifurcation points." Biogenic controls require wise programming choices to be made at bona fide decision nodes, not just "coin flips" at bifurcation points. No nontrivial computational program has ever been written by a random number generator.

The nodes of a neural network cannot just be dumb "buttons" connected by "threads" or "strings." One node firing cannot create an all-or-none depolarization of all other nodes if one expects computational and functional success. Such a lack of selectivity at each node makes fine-tuned control and regulatory mechanisms impossible. A neural net must be organized and controlled to accomplish sophisticated formal tasks.

Neither long periods of time nor an imagined multiverse helps to circumvent the apparent need for deliberate purposeful selections by Maxwell's Demon at the trap door. It is the Demon's purposeful choices in deciding when to open and close the trap door, not chance, necessity, patterns, or order, that alone make organization and nontrivial energy utilization possible. An energy potential alone is not enough to explain life. Energy must be harnessed, transduced, stored, and later utilized in a highly coordinated and cooperative manor. Without such formal organization, metabolism cannot be integrated into the Aristotelian "final cause" of generating and maintaining life.

## 7. "Disorganization" Is a Better Description of Entropy than "Disorder"

The increase in entropy in accord with the second law primarily describes events and objects progressing toward disorganization, not necessarily toward disorder. To understand how and why this is, we need only return to Maxwell's diagram. Any attempt to deny the Demon his ability to purposefully choose when to open and close the trap door will only result in increased entropy and movement back toward equilibrium. The key to understanding entropy, equilibrium, and "moving far from equilibrium" centers around the Demon's *agency*, not mere order or disorder.

No Demon choice contingency is needed for inorganic crystallization or a hurricane to self-order. Not surprisingly, the overall long-term effects of crystallization and hurricanes do not generate an algorithmically or cybernetically organized "system" or *keep* it away from equilibrium for very long. Despite short-term self-ordering, both crystallization and hurricanes ultimately promote the loss of usable energy and only accelerates the progression toward eventual heat death. Sooner or later crystals will dissolve and release the remaining energy trapped temporarily in their bonds.

It is the Demon's ability to formally categorize and purposefully select for "*potential* usefulness" that makes both "organization" and a *sustained* "far from equilibrium" true "system" possible. "Weather systems" are not true systems. They are merely physicodynamic interfaces, fronts, and self-ordered phenomena with no pragmatic algorithmic component. Weather pseudosystems cannot compute, integrate, or program any formal function. They are purely physicodynamic.

*Potential* usefulness cannot be perceived or pursued by physicodynamics. Potential usefulness exists only as an abstract concept, desire, and goal. Programming choices lie in the formal realm, not in the physicodynamic realm.

*It is only the organizational version of moving far from equilibrium that has any relevance to life origin,* not the highly constrained and self-ordered redundancies of mere physicodynamics. Law-like self-ordering makes the instantiation of nontrivial formal PI into physicality impossible (Abel, 2008b). Neither chance nor necessity can program instructive linear digital DNA strings or the protein folding they prescribe. The organizational version of moving far from equilibrium necessarily involves cybernetics—bona fide control mechanisms and true regulation—not just formally indifferent, blind physicodynamic constraints (Abel, 2010a).

## 8. Sustained Functional Systems Require Organization

Maxwell's illustration and Demon are crucial to understanding how any system could have moved not only far from equilibrium but specifically toward a *sustained functional system (SFS)*. Inherent in any SFS is bona fide *organization*. The role of the Demon reduces the problem of both SFS formation, organization, and maintenance to its rock-bottom level. The Demon's *choice contingency* of when to open and close the trap door so as to accomplish the goal of a sustained energy potential goal represents the very first true decision-node instantiation into physicality. The Demon's first choice is the birth of engineering and the artificial intelligence movement. Deciding when to open and close the trap door is the very first logic gate—the very first configurable switch-setting. The Demon's voluntary (arbitrary) trap-door operation represents the birth of integrated circuits, computational cybernetics, and life's regulatory mechanisms.

Try to metaphysically exclude choice contingency from one's mental construct of reality, and that purely physicalistic worldview immediately becomes a cement body suit. Just too many repeatedly observable phenomena arise from undeniable choice contingency. The very practice of science itself is impossible apart from the reality of physicodynamically indeterminate choice at bona fide decision nodes. We cannot hypothesize, falsify, measure, sort data, dichotomize mere association from causation, evaluate results, or draw conclusions without choice contingency. Deny the Demon's choice of trap door control and all hope of organization and algorithmic optimization dies. When the physicalist seeks to eliminate choice contingency from physicalistic reality, he only shoots himself in the foot. He cannot even argue for his metaphysical belief system without employing the choice contingency he denies. Without choice contingency, both his materialistic worldview and naturalistic science are dead.

Organization is what makes sustained functional systems (SFS) far from equilibrium possible. Without SFSs, life is impossible. No finer example of sustained control, organization, function, and utility exists than life itself. We cannot begin to understand "systems biology" without first appreciating the reality of organization and its contribution to holistic metabolism. But life's organization depends wholly upon such factors as monomeric sequencing decisions at the primary structure level of biopolymer formation. "Messenger molecules" cannot be reduced to mere linguistic metaphor (Abel and Trevors, 2006a). No fundamental difference exists between the Demon's trap-door choices and those sequencing decisions that program future folding and binding function. They both are a form of undeniable cybernetic programming of *potential* (not-yet-existent) function.

Selection pressure cannot favor something that does not yet exist (The GS Principle) (2009c; 2009d).

Each purposeful opening and closing of the trap door by the Demon is a true cybernetic "operation." Each operation is a wise binary programming decision. The Demon's trap-door decisions to merely dichotomize hot from cold would not be sufficient to create a heat engine. Additional more anticipatory, sophisticated, and integrative decisions would be required to generate a heat engine to take advantage of the created energy potential.

"Operations" alone are what successfully compute and organize. Operations alone are what design and engineer machines. It is operations that perform "useful work." Useful work is what creates and keeps sustained functional systems (SFS) far from equilibrium—not only physical work but formal cybernetic work. Thus, we invite falsification of still another null hypothesis: "Without formal cybernetics, the 'work' addressed by thermodynamics and physics will not become organized into Aristotelian 'final' utility." Mere heat transfer from one particle to another does not measure up. A force moving a mass through a distance does not measure up either. Only the Demon's purposeful trap-door operations produce sustained formal function and utility in the form of SFSs.

## 9. Summary

1. Redundant, low-informational, pragmatically blind "Order" is not the same as imaginative, highly informational, pragmatic, choice-based "Organization." Physicodynamic self-ordering in nature precludes both organization and instantiation of prescriptive information (PI) into physical media such as RNA and DNA.
2. Constraints are not synonymous with controls. The two terms must not be used interchangeably. Constraints reduce options and impede potential controls.
3. Both self-ordering phenomena (such as the dissipative structures of chaos theory) and true organization can move events away from equilibrium. But only organization, not order, can steer, coordinate, and integrate sustained sustained functional systems (SFS) (e.g., biochemical pathways and cycles) into holistic metabolic schemes. Merely being far from equilibrium, therefore, is not a sufficient definition or cause of life.
4. No hint of formal function or pragmatism is inherent in any of the thermodynamic or physics equations relating to work
5. Life is not merely physicodynamically constrained. Life is formally controlled with expedient decision node, logic gate, and configurable switch-settings that cybernetically program computational success and that organize Aristotelian "final" utility (integrated holistic metabolism).
6. The birth of ProtoBioCybernetics was Maxwell's Demon's first choice of when to open or close his trap door for the purpose of moving away from equilibrium specifically toward organization and control, not necessarily order.

7. Prediction: Deny Maxwell's Demon his ability to purposefully select when to open and close the trap door, and no sustained functional system far from equilibrium (the key to life origin) will occur in nature.
8. All known life is cybernetic—controlled and regulated by formal computational logic-gate and configurable switch-settings. Chance and necessity cannot explain the arbitrary assignments of the genetic material symbol system, the code bijection (translation, correspondence of amino acid to codonic representational prescription), the noise-reducing Hamming block code described by the codon table, the error-correction mechanisms, the epigenetic regulation of already existing instructions, or the linear digital programming of prescriptive information (PI) that instructs and organizes life.
9. The best understanding of entropy is not "disorder" but "disorganization."

## 10. References


Abel DL (2000) Is life reducible to complexity? Paper presented at the Workshop on life: a satellite meeting before the Millennial World Meeting of University Professors, Modena, Italy

Abel DL (2002) Is life reducible to complexity? In: Palyi G, Zucchi C, Caglioti L (eds) Fundamentals of life. Elsevier, Paris, pp 57–72

Abel DL (2006) Life origin: the role of complexity at the edge of chaos. Paper presented at the Washington Science 2006, Headquarters of the National Science Foundation, Arlington, VA

Abel DL (2007) Complexity, self-organization, and emergence at the edge of chaos in life-origin models. J Wash Acad Sci 93(4):1–20

Abel DL (2008a) The capabilities of chaos and complexity. Paper presented at the Society for Chaos Theory: Society for Complexity in Psychology and the Life Sciences, International Conference at Virginia Commonwealth University, Richmond, VA

Abel DL (2008b) 'The Cybernetic Cut': progressing from description to prescription in systems theory. Open Cybern Syst J 2:234–244, Open access

Abel DL (2008c) The Cybernetic Cut (Scirus Topic Page). http://www.scitopics.com/The_Cybernetic_Cut.html. Last accessed Jan 2012

Abel DL (2009a) The biosemiosis of prescriptive information. Semiotica 2009(174):1–19

Abel DL (2009b) The capabilities of chaos and complexity. Int J Mol Sci 10(1 Special Issue on Life Origin):247–291, Open access at http://mdpi.com/1422–0067/1410/1421/1247

Abel DL (2009c) The GS (Genetic Selection) Principle. Front Biosci 14(January 1):2959–2969, Open access at http://www.bioscience.org/2009/v2914/af/3426/fulltext.htm

Abel DL (2009d) The GS (Genetic Selection) Principle (Scirus Topic Page). http://www.scitopics.com/The_Genetic_Selection_GS_Principle.html. Last accessed Jan 2012

Abel DL (2009e) The Universal Plausibility Metric (UPM) and Principle (UPP). Theor Biol Med Model 6(1):27

Abel DL (2010a) Constraints vs. controls. Open Cybern Syst J 4:14–27

Abel DL (2010b) The Universal Plausibility Metric (UPM) and Principle (UPP) [Scirus SciTopic Page], Jan 2012. http://scitopics.com/The_Universal_Plausbility_Metric_UPM_and_Principle_UPP.html

Abel DL, Trevors JT (2005) Three subsets of sequence complexity and their relevance to biopolymeric information. Theor Biol Med Model 2, Open access at http://www.tbiomed.com/content/2/1/29

Abel DL, Trevors JT (2006a) More than metaphor: genomes are objective sign systems. J BioSemiotics 1(2):253–267

Abel DL, Trevors JT (2006b) Self-organization vs. self-ordering events in life-origin models. Phys Life Rev 3:211–228

Abel DL, Trevors JT (2007) More than metaphor: genomes are objective sign systems. In: Barbieri M (ed) BioSemiotic research trends. Nova, New York, pp 1–15

Agrawal H (2002) Extreme self-organization in networks constructed from gene expression data. Phys Rev Lett 89(26):268702

Axe DD (2004) Estimating the prevalence of protein sequences adopting functional enzyme folds. J Mol Biol 341(5):1295–1315

Barab SA, Cherkes-Julkowski M, Swenson R, Garrett S, Shaw RE, Young M (1999) Principles of self-organization: learning as participation in autocatakinetic systems. J Learn Sci 8(3/4): 349–390

Barbieri M (2004) Biology with information and meaning. Hist Philos Life Sci 25(2 (June)):243–254

Barham J (1996) A dynamical model of the meaning of information. Biosystems 38(2–3):235–241

Bar-Hillel Y, Carnap R (1953) Semantic information. Br J Philos Sci 4:147–157

Barwise J, Etchemendy J (1990) Information, infons, and inference. In: Cooper R, Mukai K, Perry J (eds) Situation theory and its applications, vol 1. CSLI, Stanford, pp 33–78

Barwise J, Perry J (1983) Situations and attitudes. MIT Press, Cambridge

Batten M, Salthe S, Boschetti F (1991) Visions of evolution: self-organization proposes what natural selection disposes. Biol Theor 3(1):17–29

Bedau MA (2003) Artificial life: organization, adaptation and complexity from the bottom up. Trends Cogn Sci 7(11):505–512

Boniolo G (2003) Biology without information. Hist Philos Life Sci 25:255–273

Cairns-Smith AG (1966) The origin of life and the nature of the primitive gene. J Theor Biol 10(1): 53–88

Cairns-Smith AG (1990) Seven clues to the origin of life. Canto ed. Cambridge University Press, Cambridge

Comazine S, Deneubourg J-L, Franks N, Sneyd J, Theraulaz G, Bonabeau E (2001) Self-organization is biological systems. Princeton University Press, Princeton

Dalenoort GJ (1989) The paradigm of self-organization: current trends in self-organization. Gordon and Breach Science Publishers, New York

Deacon TW (2006a) Emergence: the hole at the wheel's hub. In: Clayton P, Davies PCW (eds) Re-emergence of emergence. Oxford University Press, Oxford

Deacon TW (2006b) Reciprocal linkage between self-organizing processes is sufficient for self-reproduction and evolvability. Biol Theor 1(2):136–149

Deacon TW (2006c) Reciprocal linkage between self-organizing processes is sufficient for self-reproduction and evolvability. Biol Theor 1(2):136–149

Deacon TW (2010) 8: What's missing from theories of information? In: Davies P, Gregersen NH (eds) Information and the nature of reality: from physics to metaphysics. Cambridge University Press, Cambridge

Deacon T, Sherman J (2008) The pattern which connects pleroma to creatura: the autocell bridge from physics to life. In: Hoffmeyer J (ed) A legacy for living systems: Gregory Bateson as precursor to biosemiotics. Springer, Dordrecht

Demetrius L (1984) Self-organization in macromolecular systems: the notion of adaptive value. PNAS 81(19):6068–6072. doi:10.1073/pnas.81.19.6068

Devlin K (1991) Logic and information. Cambridge University Press, New York

Dretske F (1981) Knowledge and the flow of information. MIT Press, Cambridge

Dretske F (1995) Naturalizing the mind. MIT Press, Cambridge

Eigen M (1971a) Molecular self-organization and the early stages of evolution. Experientia 27(11):149–212

Eigen M (1971b) Self-organization of matter and the evolution of biological macromolecules. Naturwissenschaften 58:465–523 (In German)

Eigen M, Schuster P (1977) The hypercycle. A principle of natural self-organization. Part A: emergence of the hypercycle. Naturwissenschaften 64(11):541–565

Eigen M, Schuster P (1982) Stages of emerging life – five principles of early organization. J Mol Evol 19(1):47–61

Eigen M, Winkler R (1992) Steps towards life: a perspective on evolution. Oxford University Press, Oxford/New York

Epstein IR, Eigen M (1979) Selection and self-organization of self-reproducing macromolecules under the constraint of constant flux. Biophys Chem 10(2):153–160

Feltz B, Crommelinck M, Goujon P (2006) Self-organization and emergence in life sciences. Springer, Dordrecht

Gabora L (2006) Self-other organization: why early life did not evolve through natural selection. J Theor Biol 241(3):443–450

Galimov EM (2009) Concept of sustained ordering and ATP-related mechanism of life's origin. Int J Mol Sci 10:2019–2030

Gánti T (1975) Organization of chemical reactions into dividing and metabolizing units: the chemotons. Biosystems 7(1):15–21

Gánti T (1980) On the organizational basis of the evolution. Acta Biol 31(4):449–459

Gershenson C (2007). Design and control of self-organizing systems. Ph.D., Vrije Universiteit Brussel, Brussels. Retrieved from http://complexity.vub.ac.be/~carlos/thesis.pdf

Godfrey-Smith P (2003) Genes do not encode information for phenotypic traits. In: Hitchcock C (ed) Contemporary debates in philosophy of science. Blackwell, London, pp 275–289

Goldbeter A, Decroly O (1983) Temporal self-organization in biochemical systems: periodic behavior vs. chaos. Am J Physiol 245(4):R478–483

Griffiths PE (2001) Genetic information: a metaphor in search of a theory. Philos Sci 68:394–412

Harold FM (2005) Molecules into cells: specifying spatial architecture. Microbiol Mol Biol Rev 69(4):544–564

Ito K, Gunji YP (1994) Self-organisation of living systems towards criticality at the edge of chaos. Biosystems 33(1):17–24

Jacob F (1974) The logic of living systems – a history of heredity. Allen Lane, London

Kauffman SA (1993) The origins of order: self-organization and selection in evolution. Oxford University Press, Oxford

Kauffman S (1995) At home in the universe: the search for the laws of self-organization and complexity. Oxford University Press, New York

Kieu TD (2004) The second law, Maxwell's demon, and work derivable from quantum heat engines. Phys Rev Lett 93(14):140403

Kuhn H (1972) Self-organization of molecular systems and evolution of the genetic apparatus. Angewandte Chemie Int Edn 11(9):798–820

Kuhn H (1976) Model consideration for the origin of life. Environmental structure as stimulus for the evolution of chemical systems. Naturwissenschaften 63(2):68–80

Küppers B (1979) Towards an experimental analysis of molecular self-organization and precellular Darwinian evolution. Naturwissenschaften 66(5):228–243

Kurakin A (2005) Self-organization vs Watchmaker: stochastic gene expression and cell differentiation. Dev Genes Evol 215(1):46–52

Lehn J-M (2002a) Toward complex matter: supramolecular chemistry and self-organization. PNAS, 072065599

Lehn J-M (2002b) Toward self-organization and complex matter. Science 295(5564):2400–2403

Lifson S (1987) Chemical selection, diversity, teleonomy and the second law of thermodynamics. Reflections on Eigen's theory of self-organization of matter. Biophys Chem 26(2–3):303–311

Lozneanu E, Sanduloviciu M (2009) Self-organization scenario grounded on new experimental results. Chaos, Solitons and Fractals 40(4):1845–1857

Luisi PL (2003) Autopoiesis: a review and a reappraisal. Naturwissenschaften 90(2):49–59

Maxwell JC (1871) Theory of heat. Dover, New York, reprinted (2001)

Moreno A, Ruiz-Mirazo K (2009) The problem of the emergence of functional diversity in prebiotic evolution. Biol Philos 24(5):585–605

Morowitz HJ (1977) Perspectives on thermodynamics and the origin of life. Adv Biol Med Phys 16:151–163

Morowitz HJ (1981) Phase separation, charge separation and biogenesis. Biosystems 14(1):41–47

Morowitz HJ, Heinz B, Deamer DW (1988) The chemical logic of a minimum protocell. Orig Life Evol Biosph 18(3):281–287
Muller AW (1995) Were the first organisms heat engines? A new model for biogenesis and the early evolution of biological energy conversion. Prog Biophys Mol Biol 63:193–231
Nicholis G (1989) Physics of far-from-equilibrium systems and self-organization. In: Davies P (ed) The new physics. Cambridge University Press, Cambridge, pp 316–347
Nicolis G, Prigogine I (1977) Self-organization in nonequilibrium systems: from dissipative structures to order through fluctuations. Wiley-Interscience, New York
Orgel LE (1995) Unnatural selection in chemical systems. Acc Chem Res 28(3):109–118
Pattee HH (1995a) Artificial life needs a real epistemology. In: Moran F (ed) Advances in artificial life. Springer, Berlin, pp 23–38
Pattee HH (1995b) Evolving self-reference: matter, symbols, and semantic closure. Commun Cognit-Artif Intel 12:9–28
Ponnamperuma C, Hobish MK, Wickramasinghe N (1995) The physicochemical origins of the genetic code. In: Chela-Flores J, Chadha M, Negron-Mendoza A, Oshima T (eds) Chemical evolution: self-organization of the macromolecules of life. Deepak Publishing, Hampton, pp 3–18
Prigogine I, Stengers I (1984) Order out of chaos. Heinemann, London, pp 285–287, 297–301
Quan HT, Wang YD, Liu YX, Sun CP, Nori F (2006) Maxwell's demon assisted thermodynamic cycle in superconducting quantum circuits. Phys Rev Lett 97(18):180402
Quan HT, Liu YX, Sun CP, Nori F (2007) Quantum thermodynamic cycles and quantum heat engines. Phys Rev E Stat Nonlin Soft Matter Phys 76(3 Pt 1):031105
Rocha LM (2001) Evolution with material symbol systems. Biosystems 60:95–121
Rocha LM, Hordijk W (2005) Material representations: from the genetic code to the evolution of cellular automata. Artif Life 11(1–2):189–214
Rokhsar DS, Anderson PW, Stein DL (1986) Self-organization in prebiological systems: simulations of a model for the origin of genetic information. J Mol Evol 23(2):119–126
Ruiz-Mirazo K, Moreno A, Moran F (1998) Merging the energetic and the relational-constructive logic of life. In: Adami C, Belew R, Kitano H, Taylor C (eds) Artificial life VI. MIT Bradford Books, Cambridge, pp 448–451
Ruiz-Mirazo K, Pereto J, Moreno A (2010) Defining life or bringing biology to life. Orig Life Evol Biosph 40(2):203–213
Sarkar S (1996) Biological information: a skeptical look at some central dogmas of molecular biology. In: Sarkar S (ed) The philosophy and history of molecular biology: new perspectives. Kluwer, Dordrecht, pp 187–231
Sarkar S (2003) Genes encode information for phenotypic traits. In: Hitchcock C (ed) Contemporary debates in philosophy of science. Blackwell, London, pp 259–274
Schiffmann Y (2007) Self-organization in and on biological spheres. Prog Biophys Mol Biol 95(1–3): 50–59
Scully MO (2001) Extracting work from a single thermal bath via quantum negentropy. Phys Rev Lett 87(22):220601
Seeley T (2002) When is self-organization used in biological systems? Biol Bull 202(3):314–318
Sherman J, Deacon T (2007) Teleology for the perplexed: how matter began to matter. Zygon 42: 873–901
Stegmann UE (2005) Genetic information as instructional content. Philos Sci 72:425–443
Stent GS (1981) Strength and weakness of the genetic approach to the development of the nervous system. Annu Rev Neurosci 4:163–194
Stewart I (2003) Self-organization in evolution: a mathematical perspective. Philos Trans R Soc Lond 361:1101–1123
Stonier T (1996) Information as a basic property of the universe. Biosystems 38(2–3):135–140
Swenson R (1989) Emergent attractors and the law of maximum entropy production. Syst Res 6:187–197
Szilard L (1964) On the decrease of entropy in a thermodynamic system by the intervention of intelligent beings. Behav Sci 9(4):301–310

Takeuchi N, Hogeweg P (2009a) Multilevel selection in models of prebiotic evolution II: a direct comparison of compartmentalization and spatial self-organization. PLoS Comput Biol 5(10):e1000542

Takeuchi N, Hogeweg P (2009b) Multilevel selection in models of prebiotic evolution II: a direct comparison of compartmentalization and spatial self-organization. PLoS Comput Biol 5(10):e1000542

Toussaint O, Schneider ED (1998) The thermodynamics and evolution of complexity in biological systems. Comp Biochem Physiol A Mol Integr Physiol 120(1):3–9

Trevors JT, Abel DL (2004) Chance and necessity do not explain the origin of life. Cell Biol Int 28:729–739

Turing AM (1936) On computable numbers, with an application to the entscheidungs problem. Proc R Soc Lond Math Soc 42(Ser 2):230–265, [correction in 243, 544–546]

Umerez J (2001) Howard Pattee's theoretical biology – a radical epistemological stance to approach life, evolution and complexity. Biosystems 60(1–3):159–177

Vesterby V (2008) Origins of self-organization, emergence and cause. ISCE Publishing, Goodyear

von Neumann J (1950) [Letter to physicist George Gamow (first scientist to elucidate *triplet* codons) on July 25, 1950. Cited by Heims SJ (1980) John von Neumann and Norbert Wiener: from mathematics to the technologies of life and death. MIT Press, Cambridge]

Weber BH, Depew DJ (1996) Natural selection and self-organization. Biol Philos 11(1):33–65

Wiener N (1948) Cybernetics. Wiley, New York

Wiener N (1961) Cybernetics, its control and communication in the animal and the machine, 2nd edn. MIT Press, Cambridge

Wills PR (1993) Self-organization of genetic coding. J Theor Biol 162(3):267–287

Wolpert L, Smith J, Jessell T, Lawrence P (2002) Principles of development. Oxford University Press, Oxford

Biodata of **Katsuhiko Ariga**, **Gary J. Richards**, and **Jonathan P. Hill**, authors of "***Life from Interface***."

**Dr. Katsuhiko Ariga** is currently principal investigator of World Premier International (WPI) Research Center for Materials Nanoarchitectonics (MANA), National Institute for Materials Science (NIMS). He also works as a director of the Supermolecules Group at NIMS and professor of Tokyo University of Science. He obtained his Ph.D. from Tokyo Institute of Technology in 1990. His major interests are the fabrication of novel nanostructures based on molecular recognition and self-assembly, including Langmuir–Blodgett films, layer-by-layer films, and mesoporous materials.

E-mail: **ARIGA.Katsuhiko@nims.go.jp**

**Dr. Gary J. Richards** is a postdoctoral researcher in the Supermolecules Group at the National Institute for Materials Science (NIMS). He obtained his Ph.D. from Hull University (UK) in 2001. He worked as a postdoctoral researcher at Oxford University (UK) before starting work at NIMS. His scientific interests are in the areas of liquid crystals and supramolecular chemistry together with their application to organic electronics, OLEDs, and OPVs.

E-mail: **RICHARDS.Gary@nims.go.jp**

**Katsuhiko Ariga**

**Gary J. Richards**



J. Seckbach (ed.), *Genesis - In The Beginning: Precursors of Life, Chemical Models and Early Biological Evolution*, Cellular Origin, Life in Extreme Habitats and Astrobiology 22, 237–252
DOI 10.1007/978-94-007-2941-4_14, 

**Dr. Jonathan P. Hill** is a senior researcher in the Supermolecules Group at the National Institute for Materials Science (NIMS). He obtained his Ph.D. from Brunel University (UK) in 1995. Prior to working in NIMS, he occupied postdoctoral positions in Tokyo, University of Karlsruhe, and Osaka. His research interests include supramolecular science, chemistry of the tetrapyrroles, molecular self-assembly in the bulk state and at interfaces, and the sensing, optical, and electronic properties of molecules.

E-mail: **Jonathan.HILL@nims.go.jp**

# LIFE FROM INTERFACE

**KATSUHIKO ARIGA[1,2], GARY J. RICHARDS[1,2], AND JONATHAN P. HILL[1,2]**
*[1] World Premier International (WPI) Research Center for Materials Nanoarchitectonics (MANA), National Institute for Materials Science (NIMS), 1-1 Namiki, Tsukuba 305-0044, Japan*
*[2] Japan Science and Technology Agency, CREST, 1-1 Namiki, Tsukuba 305-0044, Japan*

## 1. What Is Essential for Life?

When we consider "life" from a scientific point of view, the origins and evolution of its components are important research subjects. Moreover, the origin of functionalities of biomolecules exemplified by the operations of proteins and nucleic acids during biochemical processes has attracted the attention of many scientists. The discovery of *ribozymes*, which are ribonucleic acids (RNA) with enzyme-like catalytic functions, introduced a powerful RNA world hypothesis (Eddy, 2001), which suggests that RNA first evolved and fulfilled two of the major functions of life—genetic and enzymatic activities at the initial stages of evolution. Although these discussions are very attractive, consideration of individual molecular functions does not address all the fundamental problems of the origins of the complex biochemical reactions involved in living processes. Genetic processes, which transmit information from one generation to the next or render genetic information into biomolecules such as proteins, are crucial for life. Also, highly selective and efficient catalytic activities incorporated in enzyme functions are indispensable to life processes. However, if we consider their operation under ambient unconfined conditions, they can be simply regarded as complex combinations of molecular interactions. Operation within the confines of defined microscopic reactors, i.e., cells, has enabled their integration into harmonized and sophisticated biochemical systems. Therefore, the origin of cell membranes must be seriously addressed. It would appear to be significantly more important than the evolution of individual components including nucleic acids and proteins.

The basic construction of cell membranes is surprisingly simple although current cell membranes are formed with complicated combinations of several components. The basic structure of a cell membrane is a lipid bilayer. Lipid molecules copossess hydrophilic and hydrophobic portions in a single molecule. They assemble in such a way as to minimize energetically unfavorable contact between hydrophobic parts and any aqueous media resulting in the formation of a lipid bilayer membrane. Lipid bilayer membranes can be further organized to form hollow spherical microobjects, which are sometimes known as liposomes or vesicles. The latter structures can be regarded as models of cell membranes. Bangham demonstrated the formation of liposomes through a very simple process, where dispersion of phospholipid molecules spontaneously resulted in liposome formation (Bangham and Horne, 1964). Later, Kunitake and Okahata proved that similar hollow object formation can be achieved with nonbiological surfactants such as dialkylammonium salts (Kunitake and Okahata, in 1977). The latter implies that formation of cell-membrane-like structures is based on a simple physicochemical process in which assistance from specific biological components is not necessary.

Some hypotheses on the origin of life suggest that the contribution of cell membranes may come later than the evolution of fundamental biological components such as RNA. However, it would be quite difficult to imagine that structures formed through such simple processes cannot be evolved in the early stages of life. Even if the natural evolution of phospholipids with complicated structures is a difficult process, much simpler alternative lipid analogues could form cell membrane structures in the early stage of life. In this chapter, we discuss possible scenarios for the origin of cell membranes based on our collaboration with research teams at the Louis Pasteur University led by Ourisson and Nakatani. We assume that life comes from the interface.

## 2. Origin of Lipid Analogues

In order to establish hypotheses on the origin of the cell membrane, several plausible processes are available. One of the most important problems is to make reasonable speculation on plausible structures for lipid ancestors. Currently existing phospholipids, glycerol esters with nonbranched fatty acids and phosphate ester derivatives, have structures which are too complex. Such structures can only be formed as a result of enzyme reactions, and therefore would be created in life systems in the latter stages of life evolution. Consideration of more plausible structures based on organic chemistry is necessary to specify structures of the oldest lipids. Another important question to be answered concerns the mechanisms of assembly of the primitive lipids into cell-membrane-like structures. Amphiphilic molecules have an intrinsic capability to form cell-like structures if they are confined in a closed volume. Consideration of an appropriate environment to promote primitive lipid assembly is an essential process for the development of hypotheses on the origin of cell membrane.

Lipids of eukaryotes contain ester-linked fatty acid chains even at an evolutionarily early stage. On the other hand, those of prokaryotes are composed of chains of ether-linked isoprene units. Although chemical evolution of straight fatty acids through Fischer–Tropsch-like reactions has been proposed (Simoneit et al., 2007), prebiotic synthesis of isoprenoids could be synthetically easier. These facts indicate that ether-type lipids with branched chains with isoprene

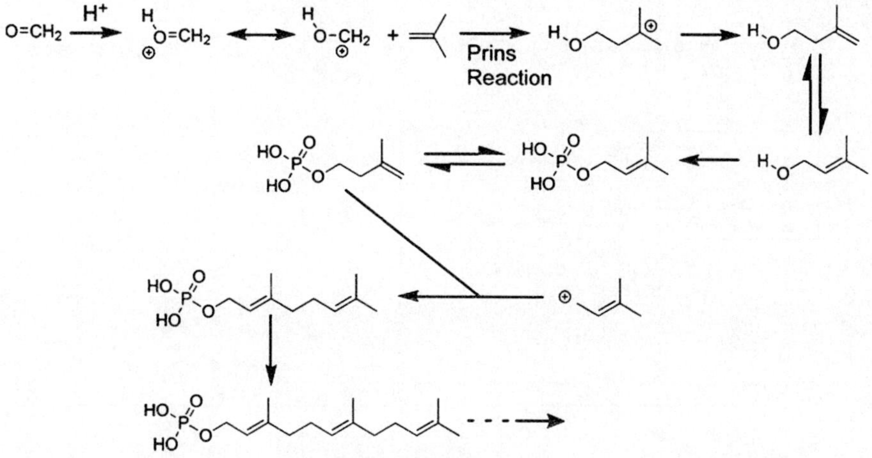


**Figure 1.** Chemical evolution of polyprenyl phosphate.

units are more probable structures of lipids at the early stages of life as compared with ester-type lipids with nonbranched straight chains. Ourisson and Nakatani proposed polyprenyl phosphates as possible lipid ancestors by showing a possible route for prebiotic chemical evolution of the ancient polyprenyl phosphates (Fig. 1) (Ourisson and Nakatani, 1999). According to their proposal, the $C_5$ isoprene unit (isopentenol) can be formed through acid-catalyzed Prins reaction of formaldehyde and isobutene. Further extension of chains into polyprenols can be made through acid-catalyzed condensation. The resulting polyprenyl alcohol can be easily converted to polyprenyl phosphates through phosphorylation. This scheme is reasonable because formaldehyde and phosphonic acid would be present even under prebiotic conditions. In addition, isobutene is also present in various media such as comets, volcanic solfataric gases, and fluidic inclusions in ancient geological formations. Formation of higher polyprenol ($C_{10}$ and $C_{15}$) from $C_5$ prenol units under prebiotic conditions was experimentally demonstrated by Désaubry et al. (2003). They reacted prenol (or dimethyl vinyl carbinol) and isopentenol (10 equiv.) in the presence of montmorillonite K10 at room temperature for 9 days, resulting in geraniol and isomers ($C_{10}$). Condensation of geraniol with isopentenol also provided farnesol and isomers ($C_{15}$). These demonstrations based on simple reactions suggest a high possibility for the chemical evolution of polyprenyl phosphates under prebiotic conditions.

It is well known that the cell membrane contains reinforcers to strengthen the lipid membrane structures. A typical example in currently existing biomembranes is cholesterol. Therefore, chemical evolutions of membrane reinforcers have to be included in a hypothesis reasonable for cell membrane evolution. Ourisson and Nakatani proposed transformation of several polyterpenoids as a possible origin of

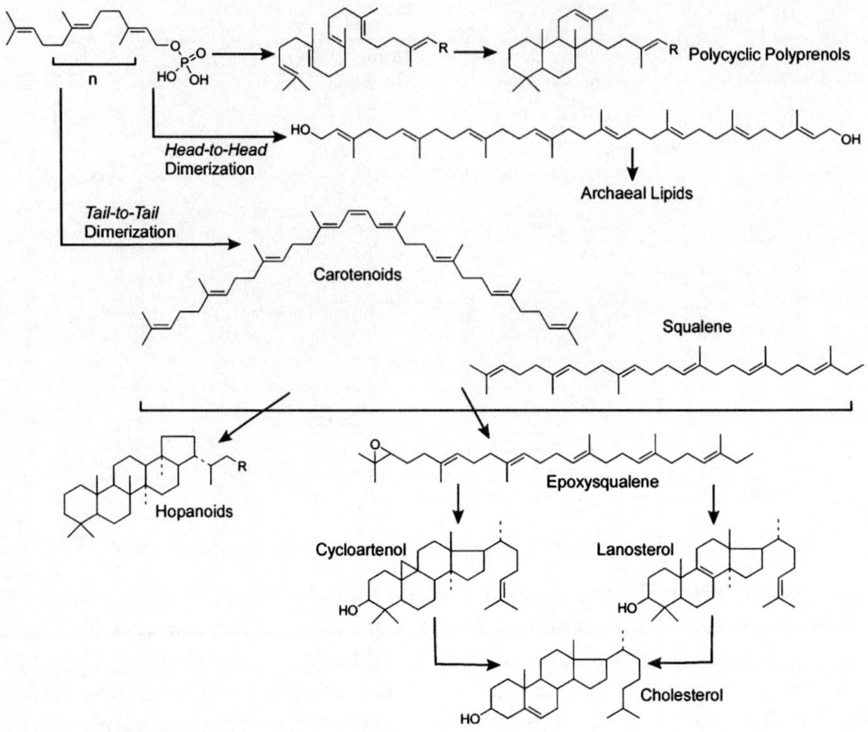


**Figure 2.** Further steps of chemical evolution.

the membrane reinforcers as well as transformation of the chain part of the lipids (Fig. 2). According to their hypothesis, several reactions which create useful membrane components are possible. Cyclization of polyprenyl phosphate leads to formation of polycyclopolyprenols. Head to head ω,ω-dimerization of the polyprenyl phosphate resulted in precursors of lipids, dipolar tetraether-type lipids. On the other hand, tail to tail α,α-dimerization results in formation of carotenoids and squalene, the latter of which can be cyclized into hopanoids. Addition of molecular oxygen to squalene also leads to formation of epoxysqualene. Further reactions produce cholesterol via cycloartenol and lanosterol. However, cholesterol is believed to have appeared much later than the hopanoids because the concentration of molecular oxygen in the atmosphere of the prebiotic Earth increased only after the appearance of advanced photosynthetic bacteria.

In order to verify the abovementioned hypothesis, Nakatani, Ourrison, and coworkers actually synthesized various polyprenyl phosphates, which were subjected to experiments of their aqueous dispersion for demonstration of formation of cell-like structures. Bisphosphonic esters of dimethylallyl alcohol, isopentenol, geraniol, farnesol, and geranylgeraniol were first investigated (Plobeck et al., 1992). Except the first two phosphates, sodium and calcium salts of the other phosphates easily form closed vesicles. Next, simpler compounds, single-chain phosphate and pyrophosphates of geraniol, farnesol, geranylgeraniol, phytol, and phytanol, were examined with respect to their capability of bilayer vesicle formation (Pozzi et al., 1996). Although phosphate and pyrophosphate with short geranyl chains are completely soluble in water at any pH and farnesyl phosphate gave only small vesicles, the other phosphates and pyrophosphates form giant vesicles under appropriate pH conditions. As mimics of lipids, phosphates with branched chains and cyclic chains were synthesized, and vesicle formation with these phosphates was also confirmed. Interestingly, much simpler compounds, polyprenyl alcohols, which are precursors of polyprenyl phosphates can form bilayer vesicle membranes upon mixing with polyprenyl phosphates (Streiff et al., 2007). The presence of the alcohols in the phosphate membranes reinforced the membrane structures and suppressed water permeability. Addition of bicycloprenols and tricycloprenols to the lipid bilayer of polyprenyl phosphate similarly induced reinforcing effects (Ribeiro et al., 2007). These species can be regarded as ancestors of current membrane reinforcers such as cholesterol.

## 3. Anomalies of Lipid Membranes at Interfaces

Prior to discussion of behaviors of polyprenyl phosphate derivatives, several specific properties of lipid membranes placed at interfaces will be considered. Kunitake and coworkers researched molecular recognition in biological systems. Molecular recognition based on many kinds of interactions including hydrogen bonding and electrostatic interaction is disadvantageous in aqueous media. These interactions are originated from an uneven distribution of electron density that is compensated and/or quenched through hydration. Surprisingly, most bioactive molecules are water soluble, and their recognition is still possible in the presence of water in biological systems. Kunitake pointed out that the key to solve this mystery is the interface. Most biological recognition occurs at certain kinds of interfacial environments such as the reaction pockets of enzymes, at the surface of biomembranes, and in the interior of the DNA double helix, and they hardly ever occur in a molecularly dispersed state. Against this background, Kunitake and coworkers investigated various molecular recognition systems at the air–water interface and related aqueous interfaces. They eventually found anomalies of aqueous interfaces (Ariga and Kunitake, 1998).

After pioneering work on the molecular recognition of aqueous nucleic acid base by Kitano and Ringsdorf (Kitano and Ringsdorf, 1985), direct evidence for hydrogen bond formation in a wide variety of systems including recognition of nucleotides, nucleic acid bases, and their analogues was obtained by Kunitake and coworkers using analytical techniques such as Fourier transform IR (FT-IR) and X-ray photoelectron spectroscopy (XPS). In order to get a better idea of molecular recognition events at aqueous interfaces, they compared binding efficiencies between guanidinium and phosphate in three different environments

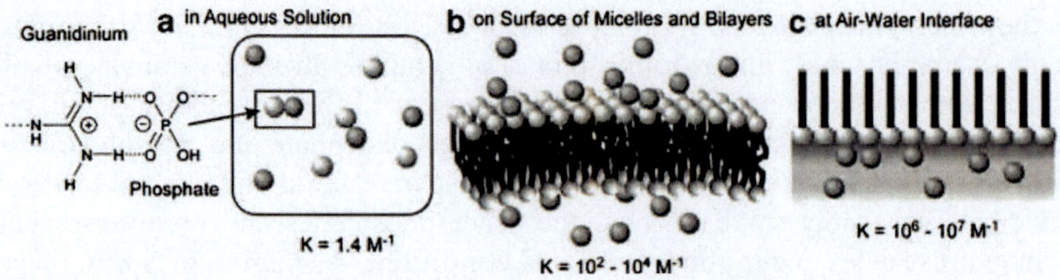


**Figure 3.** Binding constant between guanidinium and phosphate at different media.

(Fig. 3) (Onda et al., 1996). Binding constants of adenosine monophosphate (AMP) to the guanidinium functionality in aqueous aggregates such as micelle or bilayer vesicles were evaluated at $10^2$–$10^4$ $M^{-1}$ (11–25 kJ $mol^{-1}$). The values obtained are significantly larger than those between molecularly dispersed guanidinium and phosphate in water (1.4 $M^{-1}$, 0.9 kJ $mol^{-1}$). This result strikingly demonstrates enhancement of molecular recognition efficiencies at interfacial media. More surprisingly, much larger enhancement in binding constant was reported for binding of AMP and related molecules to guanidinium groups when embedded at a water surface. The values of binding constants become $10^6$–$10^7$ $M^{-1}$ (34–42 kJ $mol^{-1}$) (Sasaki et al., 1991). The latter facts indicate that molecular interactions are significantly influenced by types of interfaces.

For further understanding of these findings, Sakurai and coworkers investigated theoretical aspects of molecular recognition at the air–water interface using a quantum chemical approach including reaction field calculations combined with AM1 molecular orbital methods (Fig. 4) (Sakurai et al., 1997; Tamagawa et al., 1997). In their calculations, a guanidinium host and phosphate guest were placed in various positions at a model interface consisting of a lipid layer (low dielectric, $\varepsilon = 2$) and water phase (high dielectric, $\varepsilon = 80$). The calculated binding energies depended significantly on the position of the binding site relative to the two-phase boundary. When the binding site was located in the lipid phase, a large binding constant was calculated. In contrast, positioning the guanidinium group deep in the water phase resulted in a very small binding energy. Interesting insights were obtained in cases where interacting groups were located near interfacial regions. The binding sites are affected electronically by the low dielectric lipid layer even when the hydrogen bonding site was positioned in the water phase. In such situations, intermolecular hydrogen bonding and electrostatic interactions can be strengthened, resulting in significant increase of the binding constants at the border between the lipid and aqueous phases. These quantum chemical calculations suggested advantageous aspects of interfacial environments for efficient molecular recognition of substances in aqueous media. Interfaces of micelles and lipid bilayers are more dynamic than those given at the air–water interface. Water penetration and structural fluctuation at the interfaces may occur

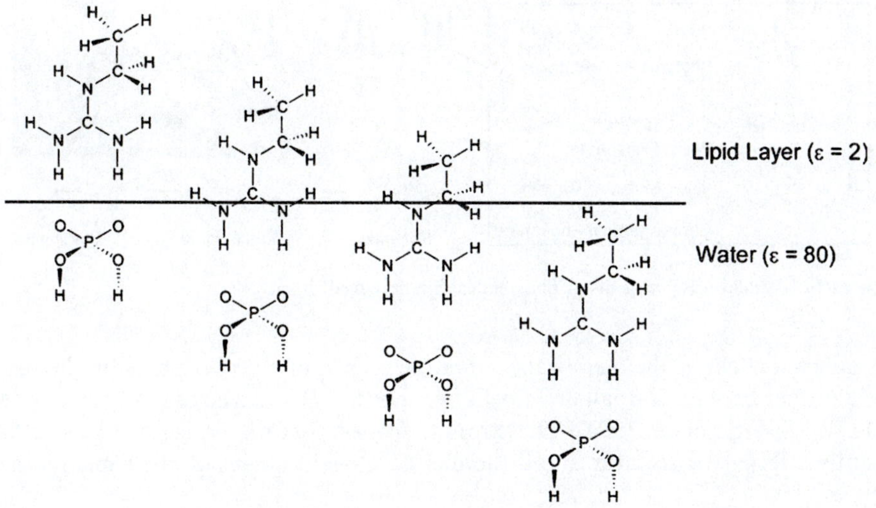


**Figure 4.** Model systems to estimate molecular recognition at the lipid–water interface.

in the former cases, probably leading to rather lower binding constants as compared with those observed at the air–water interface.

These anomalies observed at interfaces suggest that membrane properties are strongly dependent on interfacial structures. Therefore, consideration of individual molecules of lipid ancestors is rather meaningless. The origin of cell membranes needs consideration based on the interfacial environment.

## 4. Properties of Polyprenyl Phosphate Analogues at Interfaces

Properties of polyprenyl phosphates at interfaces may be different from those in bulk media. In particular, dissociation of the phosphate group and interaction between phosphate and external cations can govern stability of their membranes, which can be affected by the nature of the interfacial environment. According to calculations by Sakurai and coworkers, functional groups located in the aqueous phase near a hydrophobic lipid phase are strongly influenced by the low dielectric phase. Dissociation of neutral species to form charged ones would be suppressed in this situation, and this would work more effectively at a less disordered air–water interface as compared with the dynamic interface of an aqueous bilayer. We have actually demonstrated a significant difference in acid dissociation constants, $pK_a$, of amino acid residues between a lipid bilayer surface and Langmuir monolayer at the air–water interface (Ariga et al., 2005a). Smart and McCammon also calculated the $pK_a$ value of a long-chain alkylamine at the air–water interface, based on the nonlinear Poisson–Boltzmann equation, and expected significant shift of $pK_a$ value (Smart and McCammon, 1996). Swairjo et al. pointed out the

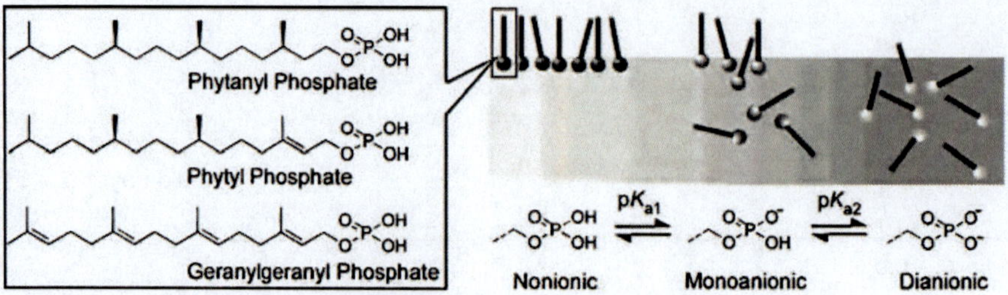


**Figure 5.** Monolayer experiments of phosphates with branched chain.

significant effect of the membrane curvature on p$K_a$ of the phosphate on the vesicle surface between a small unilamellar vesicle (SUV) and a large unilamellar one (LUV) (Swairjo et al., 1994). These results suggest that p$K_a$ values can be significantly affected by structures and the dielectric environment of their membrane states.

Therefore, we examined p$K_a$ values of phosphate groups of several polyprenyl phosphate analogues at the air–water interface and on aqueous vesicles (Fig. 5). The p$K_a$ values of single-chain polyprenyl phosphate were first examined at the air–water interface (Ariga et al., 2005b). Phytanyl phosphate and phytyl phosphate form stable monolayers under selected conditions although geranyl phosphate was water soluble regardless of pH. Therefore, the former two phosphates were subjected to further detailed analyses as shown below. In these cases, monolayers in their nonionic (R–$PO_4H_2$), monoanionic (R–$PO_4H^-$), and dianionic (R–$PO_4^{2-}$) forms are insoluble, partially soluble, and well soluble, respectively. The apparent molecular area of the monolayer decreases as phosphates dissociate. Relations of the observed molecular areas and two kinds of p$K_a$ values are summarized by the following equations:

$$A_{obsd} = (1 - x - y)A_{nonionic} + xA_{monoanionic} + yA_{dianionic} \tag{1}$$

$$A_{obsd} = \left(10^{-2pH} A_{nonionic} + 10^{-(pKa1 + pH)} A_{monoanionic} + 10^{-(pKa2 + pH)} A_{dianionic}\right) \Big/ \left(10^{-2pH} + 10^{-(pKa1 + pH)} + 10^{-(pKa2 + pH)}\right) \tag{2}$$

$$A_{obsd} = \left(10^{-2pH} A_{nonionic} + 10^{-(pKa1 + pH)} A_{monoanionic}\right) \Big/ \left(10^{-2pH} + 10^{-(pKa1+pH)} + 10^{-(pKa2+pH)}\right) \tag{3}$$

As described in Eq. 1, the observed molecular area ($A_{obsd}$) in the monolayer can be determined from the molar ratio of these three species as described in the equation using $A_{nonionic}$, $A_{monoanionic}$, and $A_{dianionic}$ for molecular areas of nonionic, monoanionic, and dianionic phosphate species at the air–water interface, respectively, upon assumption of negligible interactions between these species.

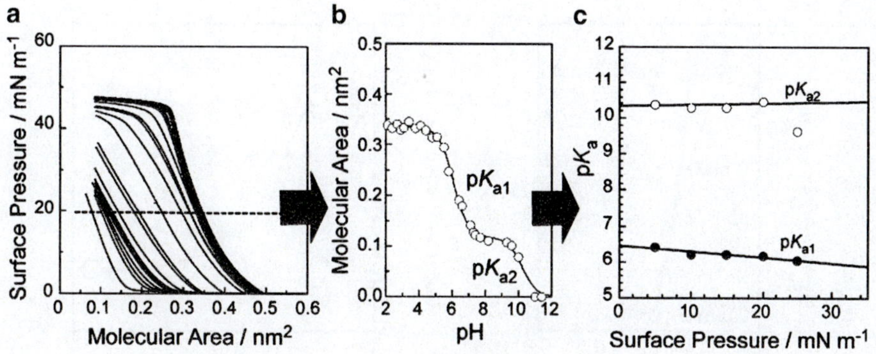


**Figure 6.** Analyses on phytyl phosphate monolayer. **(a)** π–A isotherms at various pH, **(b)** molecular areas at $20 mNm^{-1}$, and **(c)** p$K_a$ values at different surface pressures.

Molar fractions, $x$ (monoanionic form) and $y$ (dianionic form), can be determined by two acid dissociation constants (p$K_{a1}$ for nonionic/monoanionic and p$K_{a2}$ for monoanionic/dianionic) and subphase pH, resulting in conversion of Eq. 1 into Eq. 2. Because dianionic forms of phytanyl phosphate and phytyl phosphate are well soluble, $A_{dianionic}$ can be neglected. With this assumption, Eq. 2 can be simplified into Eq. 3. Curve fitting of the experimental data provides p$K_{a1}$ and p$K_{a2}$ values using this Eq. 3.

As shown in π–A isotherms of phytyl phosphate at pH values from 1.3 to 10 (Fig. 6a), the isotherms shifted toward lower molecular area as the pH value in the subphase decreases due to promoted dissolution of monolayer components into the water phase. The molecular area obtained at constant surface pressure was plotted as a function of pH value in the subphase, and curve fitting using the Eq. 3 provides values of p$K_{a1}$ and p$K_{a2}$ at this surface pressure (Fig. 6b). The obtained p$K_{a1}$ and p$K_{a2}$ values are then plotted against surface pressures, and their extrapolation to 0 mN $m^{-1}$ provides dissociation constants under strain-free condition (Fig. 6c). The p$K_{a1}$ and p$K_{a2}$ values of phytyl phosphate at the air–water interface were determined as 6.45 and 10.33, respectively. The p$K_{a1}$ and p$K_{a2}$ values of phytanyl phosphate were similarly obtained as 7.13 and 9.35, respectively, at the air–water interface. Separately, p$K_a$ values of the single-chain phosphates in their bilayer vesicle form in bulk water were determined by Ourisson, Nakatani, and coworkers. Phytanyl phosphate in a bilayer assembly has p$K_{a1}$ and p$K_{a2}$ values of 2.9 and 7.8, respectively, and p$K_{a1}$ and p$K_{a2}$ values of phytyl phosphate bilayer were determined as 2.8 and 6.6, respectively. The p$K_a$ values of aqueous bilayers of these phosphates are much lower than the corresponding values observed at the air–water interface. Such large differences in p$K_a$ values between Langmuir monolayer and aqueous vesicles could lead to the interesting phenomenon that stable forms of membranes at certain pH values depend significantly on the type of interfaces and assemblies. This point is discussed later in this chapter.

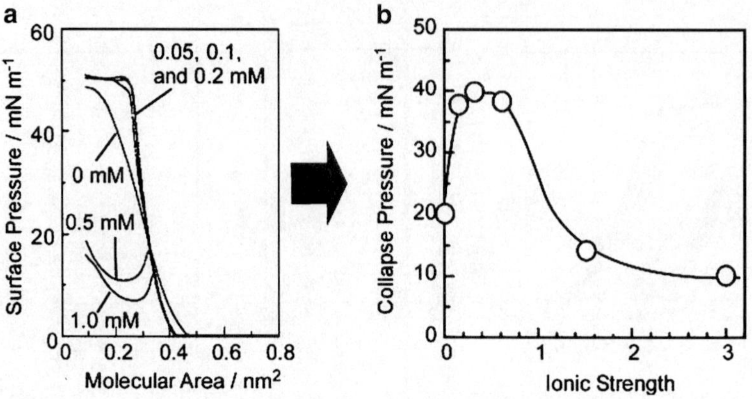


**Figure 7.** Analyses on interaction of $Ca^{2+}$ to phytanyl phosphate. (**a**) π–A isotherms at various $Ca^{2+}$ concentrations and (**b**) collapse pressure at various $Ca^{2+}$ concentrations.

Phosphate groups are known to interact with cations, and the properties of their salts such as solubility significantly depend on the cation species. Monolayer behaviors of phosphate compounds can be strongly influenced by the presence of cations in the subphase. Investigation of the Langmuir monolayers of polyprenyl phosphates in the presence of various cations can give important information on their behaviors. Therefore, effects of three typical cations, $Na^+$, $K^+$, $Mg^{2+}$, and $Ca^{2+}$ on monolayers of phytanyl phosphate were examined. Addition of large quantities of NaCl (ca. 60 or 150 mM) suppresses its dissolution although phytanyl phosphate is partially soluble in water at ambient pH. A phytanyl phosphate monolayer exhibited similar behavior when another monovalent cation $K^+$ was used. In the case of divalent $Mg^{2+}$, drastic effects on the solubility of the phytanyl phosphate monolayer were observed. The presence of a very small amount (0.05 mM) of $Mg^{2+}$ makes the phytanyl phosphate monolayer completely insoluble, although the insoluble nature of the phytanyl phosphate monolayer was not altered by further addition of $Mg^{2+}$ at neutral pH. Interestingly, another divalent cation $Ca^{2+}$ has a totally different effect on the stability of the phytanyl phosphate monolayer. π–A isotherms of phytanyl phosphate at various concentration of $CaCl_2$ in the subphase showed unusual effects on their collapse (Fig. 7a). Collapse pressure was estimated from surface pressure, giving maximum two-dimensional elasticity $[-d\pi/(dA/A)]$ values calculated from the π–A isotherms. The collapse pressure as a function of ionic strength indicates that the phytanyl phosphate monolayers are stabilized through addition of small amounts of $Ca^{2+}$ ($[CaCl_2]=0.05$–$0.2$ mM), but addition of further $Ca^{2+}$ makes the monolayer mechanically fragile as seen in reduced collapse pressures (Fig. 7b). It is generally known that $Mg^{2+}$ has greater hydration ability, attested to by the difference of solubility of their ($Mg^{2+}$ and $Ca^{2+}$) phosphate salts in water (Jallot, 2003). Less hydrated salts of $Ca^{2+}$ could cause easier collapse of the monolayer at the air–water interface.

## 5. Possible Scenario on Evolution of Cell Membranes

In this section, we will attempt to draw a possible scenario for the origin of cell-like membranes based on the research discussed above (Ariga et al., 2009). Even if chemically evolved lipid forerunners were homogeneously dispersed in the ocean, cell-like vesicle assemblies cannot be spontaneously formed due to their low concentrations. The chemically evolved lipid molecules should be somehow accumulated in a certain environment. For example, surfactants can accumulate as monolayers on stagnant water. Polyprenyl phosphates can be accumulated as insoluble monolayers at appropriate interfaces. The accumulated polyprenyl phosphates as monolayers require subsequent conversion into aqueous vesicles. This transformation requires the stable coexistence of the monolayer on the surface and aqueous vesicles under the same conditions. This bistable condition can be easily achieved judging by the difference between $pK_a$ values at the interfaces. Langmuir monolayers are generally stable when the nonanionic form of polyprenyl phosphate is the major component of membrane. On the other hand, the monoanionic form of the polyprenyl phosphates has an optimum amphiphilic balance to form bilayer vesicle structures in water. In fact, formation of vesicle structures was experimentally confirmed at pH regions where their monoanionic form is dominant. According to these findings, pH regions for membrane states are summarized in Fig. 8. A pH range over which both monolayer and vesicular structures are stable can be attained under slightly acidic conditions. This slightly acidic condition might have existed near sites of volcanic activity often associated with the emergence of life.

A trigger for the transformation from monolayer structure into aqueous vesicle is another important factor. Monolayer stability experiments demonstrated that the polyprenyl phosphates behave sensitively to certain cations such as $Ca^{2+}$. It is not difficult to imagine the presence of these ions in ancient oceans. These cations can stabilize or destabilize the monolayers of polyprenyl phosphates on water. Therefore, accumulated monolayers of the polyprenyl phosphates could be stabilized by cations and support lipid accumulation. However, the monolayer structures can also collapse into three-dimensional structures when concentration of the cations exceeds critical values. It was separately suggested that the monolayer at the air–water interface can be transformed into lipid bilayer vesicles. Therefore, the monolayer collapse triggered by appropriate cations can be induced. Some conditions promote both stabilization of lipid assembly and transformation into vesicles of polyprenyl phosphate in water, which can exist at slightly acidic conditions.

Finally, our proposed scenario for the origin of cell membranes is illustrated in Fig. 9. After ancient lipids (polyprenyl phosphates and their derivatives) had evolved chemically, they assembled as monolayers at a stagnation point. The subsequent transformation of the monolayer into vesicles under slightly acidic conditions can be triggered by ion binding and results in lipid bilayer vesicles as cell membrane precursors. We note that the entire process should not be in equilibrium.

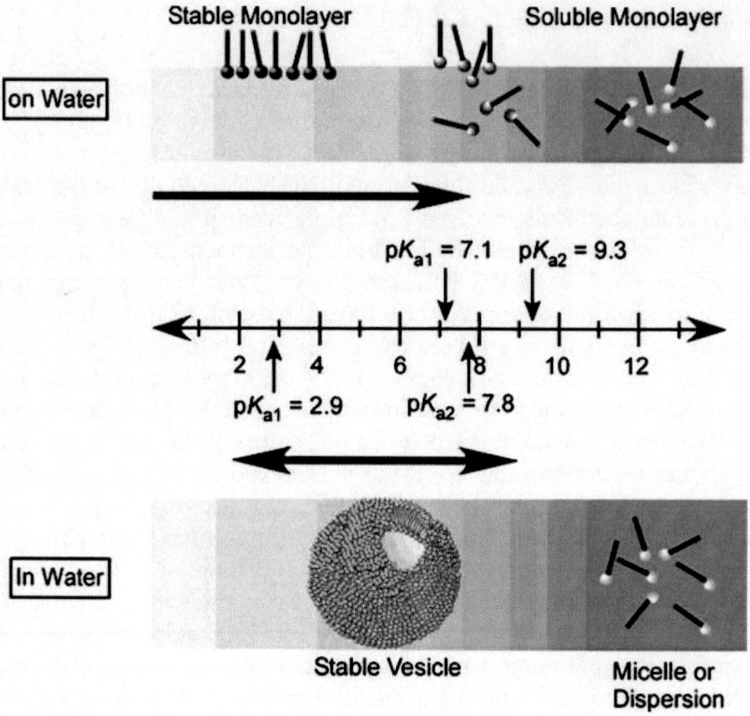


**Figure 8.** Summary of pH effect on polyprenyl phosphate (p$K_a$ values of phytanyl phosphates are used for explanation).

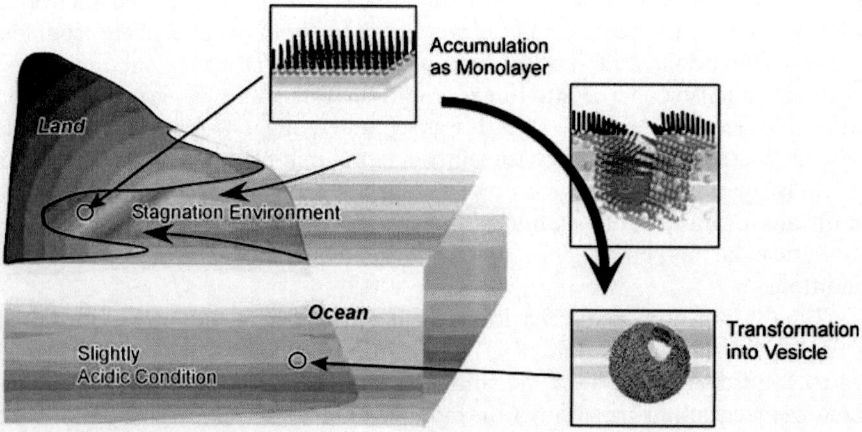


**Figure 9.** Plausible mechanism of origin of cell membrane at the air–ocean interface.

Otherwise, the vesicles formed may redisperse as monomeric phosphate into the bulk ocean. Therefore, a nonequilibrium process including poststabilization of the formed vesicles must be also considered.

As we mentioned earlier, formation of confined systems is more fundamental for the origin of life than the evolution of individual functional molecules. This process creates hierarchical assemblies of biological functions composed of several individual chemical reactions. Some theories for cell membrane evolution, such as the coacervate and marigranule hypotheses, have been developed. Although we do not disagree with them, the contribution of lipid components in the early stages of life evolution should be more seriously considered. It is much more likely that ancient lipid precursors consisted of simple polyprenyl phosphates rather than complex phospholipids since these compounds can be easily synthesized under natural conditions. It would be unwise to ignore them and their assemblies when considering the origins of life.

## 6. References


Ariga K, Kunitake T (1998) Molecular recognition at air-water and related interfaces: complementary hydrogen bonding and multisite interaction. Acc Chem Res 31:371–378

Ariga K, Nakanishi T, Hill JP, Shirai M, Okuno M, Abe T, Kikuchi J (2005a) Tunable pK of amino acid residues at the air-water interface gives an L-zyme (Langmuir enzyme). J Am Chem Soc 127:12074–12080

Ariga K, Yuki H, Kikuchi J, Dannemuller O, Albrecht-Gary AM, Nakatani Y, Ourisson G (2005b) Monolayer studies of single-chain polyprenyl phosphates. Langmuir 21:4578–4583

Ariga K, Lee MV, Labuta J, Okamoto K, Hill JP (2009) Studies on Langmuir monolayers of polyprenyl phosphates towards a possible scenario for origin of life. Colloid Surf B-Biointerface 74:426–435

Bangham AD, Horne RW (1964) Negative staining of phospholipids and their structural modification by surface active agents as observed in the electron microscope. J Mol Biol 8:660–668

Désaubry L, Nakatani Y, Ourisson G (2003) Toward higher polyprenols under ‘prebiotic’ conditions. Tetrahedron Lett 44:6959–6961

Eddy SR (2001) Non-coding RNA genes and the modern RNA world. Nat Rev Gen 2:919–929

Jallot E (2003) Role of magnesium during spontaneous formation of a calcium phosphate layer at the periphery of a bioactive glass coating doped with MgO. Appl Surf Sci 211:89–95

Kitano H, Ringsdorf H (1985) Surface behaviors of nucleic-acid base-containing lipids in monolayer and bilayer systems. Bull Chem Soc Jpn 58:2826–2828

Kunitake T, Okahata Y (1977) Totally synthetic bilayer membrane. J Am Chem Soc 99:3860–3861

Onda M, Yoshihara K, Koyano H, Ariga K, Kunitake T (1996) Molecular recognition of nucleotides by the guanidinium unit at the surface of aqueous micelles and bilayers. A comparison of microscopic and macroscopic interfaces. J Am Chem Soc 118:8524–8530

Ourisson G, Nakatani Y (1999) Origins of cellular life: molecular foundations and new approaches. Tetrahedron 55:3183–3190

Plobeck N, Eifler S, Brisson A, Nakatani Y, Ourisson G (1992) Sodium di-polyprenyl phosphates form primitive membranes. Tetrahedron Lett 33:5249–5252

Pozzi G, Birault V, Werner B, Dannenmuller O, Nakatani Y, Ourisson G, Terakawa S (1996) Single-chain polyprenyl phosphates form ‘primitive’ membranes. Angew Chem Int Ed Engl 35:177–180

Ribeiro N, Streiff S, Heissler D, Elhabiri M, Albrecht-Gary AM, Atsumi M, Gotoh M, Desaubry L, Nakatani Y, Ourisson G (2007) Reinforcing effect of bi- and tri-cyclopolyprenols on ‘primitive’ membranes made of polyprenyl phosphates. Tetrahedron 63:3395–3407

Sakurai M, Tamagawa H, Inoue Y, Ariga K, Kunitake T (1997) Theoretical study of intermolecular interaction at the lipid-water interface.1. Quantum chemical analysis using a reaction field theory. J Phys Chem B 101:4810–4816

Sasaki DY, Kurihara K, Kunitake T (1991) Specific, multiple-point binding of ATP and AMP to a guanidinium-functionalized monolayer. J Am Chem Soc 113:9685–9686

Simoneit BRT, Rushdi AI, Deamer DW (2007) Abiotic formation of acylglycerols under simulated hydrothermal conditions and self-assembly properties of such lipid products. Adv Space Res 40:1649–1656

Smart JL, McCammon JA (1996) Surface titration: a continuum electrostatics model. J Am Chem Soc 118:2283–2284

Streiff S, Ribeiro N, Wu Z, Gumienna-Kontecka E, Elhabiri M, Albrecht-Gary AM, Ourisson G, Nakatani Y (2007) "Primitive" membrane from polyprenyl phosphates and polyprenyl alcohols. Chem Biol 14:313–319

Swairjo MA, Seaton BA, Roberts MF (1994) Effect of vesicle composition and curvature on the dissociation of phosphatidic acid in small unilamellar vesicle. A $^{31}P$ NMR study. Biochim Biophys Acta 1191:354–361

Tamagawa H, Sakurai M, Inoue Y, Ariga K, Kunitake T (1997) Theoretical study of intermolecular interaction at the lipid-water interface.2. Analysis based on the Poisson-Boltzmann equation. J Phys Chem B 101:4817–4825

Biodata of **Tom Barbalet**, author of "***Breaking the Paradigm: The Pattern of Life***."

**Tom Barbalet** created the Noble Ape Simulation in 1996 and continues its development to this day. Noble Ape is used by Apple and INTEL as well as a number of universities to teach biodiversity, multimedia education, vector processing, real-time graphical interfaces, and a number of other technologies. He has been the editor of Biota.org (since 2005), a leading community resource for artificial life developers. He is the host of the Biota Live internet radio show where he discusses a variety of topics relating to artificial life, artificial intelligence, and simulation philosophy with a variety of guests. He is also the chair of the International Game Developers' Association's Intellectual Property Rights Special Interest Group.

E-mail: **tom@nobleape.com**

J. Seckbach (ed.), *Genesis - In The Beginning: Precursors of Life, Chemical Models and Early Biological Evolution*, Cellular Origin, Life in Extreme Habitats and Astrobiology 22, 253–266
DOI 10.1007/978-94-007-2941-4_15, 

# BREAKING THE PARADIGM: THE PATTERN OF LIFE

**TOM BARBALET**
*Noble Ape and Biota.org, Campbell, CA, USA*

How can there be a coherent understanding of the origin of life when life is so poorly understood? This chapter explores the perceptual boundaries of life with the result that the structure of living systems (outside the problems of the perception of life) provides an improved definition of what life is and how to find new forms of life. This is done with the implicit assumption that the origin of life existed outside the domain of human perception and thus needs to be determined through a new paradigm: the pattern of life.

## 1. The Impossible Zoo

Consider for a moment that you were invited to a zoo that contained animals you had never seen previously. These animals had only recently been discovered, perhaps on a tropical island, somewhere in some unreachably hilly jungle or living close to a hazardous volcano. As you pay your entry fee, you are warned that the things you will see in the zoo will be far from anything you have imagined previously.

What would be your considerations entering the zoo?

For a start, you would assume initially that you would be safe entering the zoo. You would be walking along delimited paths and the animals would be presented to you in a fashion where you could observe them and get a sense of what they might be like in their native environment. You would also assume that nothing you would see in the zoo would change your definitions associated with what animals were and possibly that you could roughly phylogenetically group what you saw. This phylogenetic grouping has after all been reinforced since your birth. You know where you stand in the phylogenetic order. It would be impossible to imagine this somehow being altered by walking through a zoo.

You entered the zoo with a number of tourists from all over the world: some from Japan, some from Germany, some from Swaziland, and some from Argentina. You believe this from their collections of clothing, accents, and languages spoken.

The first exhibit you come to appears to recreate a metropolitan apartment. It is a relatively small room with rugs, a couch, a bookshelf, and a television set. You notice a wide variety of cat toys and various climbing frames ideally suited for house cats.

All of a sudden you see two house cats with their backs towards you walk over to one of the climbing frames in the corner of the exhibit. For a moment you feel cheated. These are just regular house cats, not some fascinating creature of wonder.

One of the cats turn and you notice it has human eyes. It gestures to its friend and they both turn and stare at your group. These are in fact some kind of human-cat hybrid creatures. You are stunned. Due to the constant movement of people through the zoo, you eventually have to move on to the next exhibit.

The next exhibit appears to be a bull walrus sitting on a large rock, with a water mote between you and it. Again your emotions return to the sense that you have been cheated. This is just a walrus. But in the back of your mind you remember the last exhibit.

"What are you staring at?!" the walrus exclaims. The angry statement then appears to be repeated in Spanish, Japanese, and SiSwati (the language of Swaziland).

"Oh, you think I'm a parrot," the walrus continues. "Okay, you Swazis say something and I'll translate it to Japanese! C'mon." He then repeats this phrase again in Spanish, Japanese, and SiSwati to confirm to all the groups of tourists that he is capable of such a feat.

After about ten minutes being entertained by the walrus including a number of culturally idiomatic jokes, your time with him has ended and you have to again move on to the next exhibit.

This appears to be a cluster of computer mainframes interconnected with a series of screens showing what appears to be a line tracing screen saver. Now you definitely feel cheated. There is no way this is a new creature. The zoo has failed your expectations.

This thought experiment is designed to identify three implicit biases in our understanding of living systems (and intelligence). In order to convey to us a sense of life or intelligence there needs to be either appearance or action. Pure existence does not convey intelligence or life to us.

Interactive proof of intelligence or life is always a superior form of proof. While questions might have been left on the cats with human eyes, the talking walrus provided direct feedback of its competent intelligence and its ability not only to do something which is considered difficult by human standards – speaking multiple languages – but also offered characteristics of playfulness. Playfulness might be easily epistemologically devalued, but it offers an insight into the intelligence of the participant. Consider for example someone who has rote-learnt a language but whose command of the language appears to be stilted. This somewhat robotic grasp of a language devalues the appearance of the speaker's intelligence. But if the same speaker manipulates the language with humor, this indicates more of the speaker's intelligence.

When confronted by a series of interconnected machines, all the skills used to establish intelligence and life fall flat. We have not been trained with the skills needed to determine whether a complex human-created system is alive or intelligent. This is a multipart problem.

The purpose of this chapter is to propose an analytical method to solve this problem by offering a simple structural model that conveys either current life or the presence of prior life.

## 2. Disassembling Perception

As humans, we have very limited perceptual skills. We exist as captives to our senses, and our perception is reinforced both through our physical domain and our domain of understanding (Barbalet et al., 2008).

Let us analyze the claim: to determine if something is alive, what we lack in actual sense perception (both immediate and reinforced through years of existing as humans with all the social constraints and intellectual limitations) should be vastly improved by sensory-enhancing equipment (microscopes, satellites, scalpels, etc.) and academia. To paraphrase, while our blunt senses and social understanding may be lacking in establishing what is living and what is not living, we have tools and a knowledge base that allow us to understand living systems outside our limited senses.

Consider the elements individually.

Our senses are based on our size and our being. We are lumbering ape creatures with senses that are better designed for manual creative labor than thinking about or exploring the very small, the very large, or vast complexity. In fact, our relationship with the sublime is by definition unintelligible (Kant, 1764).

The author is sympathetic to the idea that senses do not and cannot exist independent of socially agreed-upon understanding. While it is possible to imagine a human completely isolated from all other humans, it is not clear if this creature would actually be a human in any meaningful sense. Our perceptions are not just a collection of images, smells, feelings, and sounds from the external world. Our perceptions are a tapestry of our part in a society and every element of our upbringing. The ability to speak a language and map our senses back onto a language indicates that the words we choose have been agreed upon by others (Wittgenstein, 1953). For this reason, we cannot consider our senses independent of our social understanding of what life is and what it means to be alive. In this regard, our senses are meaningless for identifying life outside what has already been socially agreed upon based on thousands of years of flawed pack ape cognition.

The author's view of the social perception of life is far more critical (Barbalet et al., 2010). Consider only human life for an instant. What is the value of human life? A positive analysis may say human life has meaningful value. There are murder laws and various other laws that stop humans from killing each other. This may be observed to endorse the value of human life. The laws governing corporate killing (death through corporate negligence) and societal killing (the death penalty) have a lower bar on mandating death than murder. Human-caused poverty – in particular starvation and limited access to food, shelter, nominally also education and health care – results in state-sanctioned death frequently

(Barbalet and Clune, 2010). While not wanting to get into the political issues associated with the solutions to these problems, the lack of criminal culpability to humans that cause these problems directly indicate a social acceptance that human life is in no way sacred.

This relates to a societal acceptance that human life has limited value for the state or state-comparable corporate interests. This is within groups of humans united by nation status. Even within nation states, there is segregation based on race, religion, social standing, and a variety of other factors that show humans within nations do not see the meaningful value of human life in a demonstrable fashion.

In even greater contrast to this is the practical nature of war in terms of the complete devaluing of nations of people where their lives are effectively meaningless. Here, certain humans may have the intellectual proclivity to isolate themselves from the collective reality of concepts like ongoing war, perhaps asserting that conflicts of nation states have very little to do with what appears down a microscope or in a computer simulation. The author would argue that rather than being a distant noise, the proponents of wars fund contemporary academia in such a symbiotic fashion that the epistemic result must be a source of concern for any analysis of the value of life and finding new life.

If the basis of understanding life has a cultural or social understanding of life that is fundamentally devalued, flawed, or perturbed by the examples provided, it will be difficult to establish new forms of life. Even established forms of life with the perception that is passed to us through our senses (or the description of these senses that has been socially agreed upon) have limited value for the understood, defined, and agreed-upon forms of life.

In terms of non-human life, humans have broadly no consideration of such things. The thousands of years of social evolution has given no value or benefit in understanding these kinds of life except for basic functional purposes usually relating to eradicating, utilizing, or consuming these forms of life. While these functional purposes have required specialized humans to learn a great deal about non-human life, this is fractured knowledge in no way shown over entire societies and has very minor to no feedback into social perception.

In presenting this analysis, the author has no analytical criticism of the tools currently used to determine life. The issue with the tools is not with the actual tools themselves, it is with their use and the understanding of the information they present. In this gang of four, the tools are the silent partner. The tools are there to allow interrogation or observation based on the flaws of the other three elements. It is fallacious to assume that the tools improve the ability to determine life as the definition comes from the sensorial, societal, and (in particular circumstances) the academically agreed-upon definition.

There is a vast dichotomy between the social perception of life and the academic perception of life. Rather than exploring value or pinnacle claims associated with the academic perception of life, the methodological problems associated with the academic perception are grounds enough for concern. Aside from the flaws already noted, the academic insights that have come outside these constraints have been limited and epistemologically flawed based on the need for a descriptive physics that is lacking a detailed metaphysics (the natural sciences) or a descriptive metaphysics that is lacking a detailed physics (the social sciences).

In terms of meaningful progress for finding new forms of life or a definition of life that can include vastly complex systems, it is like an explorer requiring a whole world map before he could begin on his journey. The divide between the traditional natural sciences and the social sciences is contrasted by the emerging simulation sciences (Barbalet, 2008), a field that is useful to both branches of traditional science precisely because the results are devoid of scale (topological) or disciplinary paradigmatic idioms (Barbalet and Guerin, 2009).

Rather than a means to understand living systems, all these elements – actual senses, social understanding of senses, equipment used to assist senses, and academia – inhibit rather than assist our understanding of living systems.

## 3. In Many Places All at Once

The problem with complex systems – particularly those created by humans – is they clearly exist and function. Our perception is limited. So far, this has been described in very general terms – the flaws in perception – but the crux of this is the inability to describe complex systems as whole environments linking the small interactions over the entire system: understanding both the physics and the metaphysics of the system. Fundamentally, this is a property of our size and our understanding of complexity. The tools aid in terms of the size issue. They enable us to see very small or very large; however, the underlying analysis remains flawed.

The understanding of life (and consequently intelligence) needs to be removed from perceived properties.

The author has previously proposed the use of survival as the primary metric of judging the intelligence of a complex system (Barbalet, 2008). This is a good overview of the scale problem, but it does not yield much structural insight. The survival metric holds two implicit assumptions: things that are larger tend through this metric to be more intelligent, and survival is a multilayered process. These two assumptions are positively correlated: the larger a vastly complex system is, the more likely there is to be redundancy in the system.

There are three structural concepts that constitute the basis for understanding life in vastly complex systems. They form a structural pattern that can be used to describe life at any level from single-celled organisms to the United Nations.

*Layers* describe effectively autonomous systems that are represented by underlying sub-systems. Each sub-system constitutes another layer, and it is assumed that there are multiple layers – tens of layers potentially – in any living system that is visible to the human eye. This could be mapped to very distinct biological properties, but for the purpose of this initial analysis, layers will be described without system and sub-system references. The term layer is used to add a visual description and the appearance of an almost armor-like quality when describing the remaining two concepts. It is also important to note that the concept of layers needs to be scalable from the atomic level to the size of large ecological systems or in the case of human-based or human-made systems, nation states (hence the lack of biological references at this stage).

*Redundancy* describes the components of a system that are interchangeable and reuseable. It does not necessarily mean the components are identical. Consider from a human perspective the number of vessels that could carry water. Among all the vessels, very few would even be physically similar, but they are fundamentally objects of redundancy for the purpose of carrying water.

*Criticality* is the antithesis of redundancy both in scope of use and in property. Unlike redundancy, criticality has the property of localized uniqueness and also critical importance to a particular system. If a point of criticality in the system itself fails, the system fails. In order to maintain survival, obviously these points of criticality are minimized. This is an interesting descriptor for living systems that will be explored in greater detail in this chapter.

It may seem strange to consider scale or the ambiguity of scale when constructing a definition of life for offering insight into the origins of life. The ideas discussed here are based on computational simulation and taking the ideas generated from computational simulation and mapping them back into scientific discourse (Barbalet and Guerin, 2009). This is loosely described as the burgeoning field of simulation science. Life within these systems knows no meaningful scale. The properties of scale in terms of atomic or human size are all arbitrary simulation constraints. This property within simulation science circles is noted frequently in terms of the insights linking the traditional natural sciences and the social sciences. It is worth noting this explicitly here.

It is important to note the independent analysis offered by Clune et al. (2010) which give the critical elements as regularity, modularity, and hierarchy. These appear orthogonal to layers, redundancy, and criticality (Barbalet and Clune, 2010). The expansion of this linked analysis will no doubt be the topic of future work.

## 4. In Search of a New Field

Aside from the limits to our perception, as far as being able to assess and understand life and intelligence, the combination of our intellectual development has been sorely lacking in this area too. While the three components (redundancy, criticality, and layers) have been described with particular reverence to the simulation sciences, the simulation sciences and their precursory contributors do not give a coherent starting point to explore the three components.

In terms of the academic origins of the structural means of describing life, the well-travelled path would begin with systems theory, specifically the sub-discipline living systems theory (Miller, 1978). As a descriptive and non-biologically-biased starting point, living systems theory offers a good basis. It is, however, not an analytical mechanism to understand living systems or add predictive analysis. As emphasized by the author and Guerin (2009), the numerical method of simulation science (also described as the new mathematics (Barbalet, 2008)) is not only incomplete, it does not exist in a coherent form.

Without a viable mathematics to offer an analytical basis, living systems theory is a descriptive analysis of many parts that cannot scale productively. In short, it is unnecessarily complicated. Through its complexity, it is beset by a problem given to most theories in contemporary discourse. Occam's razor applies strongly here: ideas however valid will be discarded if the means to obtain the ideas outweighs the benefits in gathering them. This is a truism that has relegated living systems theory to referential footnotes and continuation by very few acolytes of primary contact.

## 5. Human-Made, Not Made by Humans

Taking a step back from the broader analysis offered in this chapter, let us consider a network of computers. What do these computers have in terms of the properties of life so far discussed?

It may be difficult to understand how an obviously human-made system would have any properties of life. Each part in every layer has been explicitly designed by humans. Computer systems (by design) crash or fail if even a small piece of information is stored or added incorrectly to their environment. How then are ideas of redundancy and criticality used to describe a network of computers?

There are two important points to note – one explicit and one implicit.

Even though every component in a computer (at some point through its history of development) has been designed by a human with some emphasis on a priori, a posteriori, and luck, the elements are (by their very nature) mass-produced. These elements from the point of creation fit the criteria of redundancy perfectly. It is not that a human has labored to create each part uniquely. Every modern computer by necessity of its creation is a thing of mass production.

The implicit point to this is that modern computers – in particular the components associated with network communication – are highly fault tolerant. While traditional or observable redundancy is not part of that tolerance, it produces a property like redundancy. Computers have critically defined points that are mitigated by some elements of design but are primarily mitigated by creating an additional layer (or more) to maintain survival.

The network of computers does (by structure) have a form that would be considered to be alive by the structural analysis methods offered here.

## 6. The Structural Form of Life

Suppose you set up a series of digital cameras peering into a Petri dish awaiting the arrival of life. What would distinguish life from dust or noise from the digital cameras? What would distinguish something as being alive through this digital information?

This is a problem that seems to be more frequently posed within artificial life simulations (Barbalet et al., 2009) and the new emerging field of the computational origins of life (COoL): a linking of artificial life and artificial chemistry (Damer et al., 2010).

The most trivial analysis would say that there are a certain number of acceptable forms and patterns that match these forms with the output of the camera. In this way, life is defined as a fixed quantity that must be identical or near identical to other previously found forms. This represents the description of the traditional analytical biology paradigm contrasted here. In this method, the academy would need to agree that every new possible form was in fact alive before the camera could independently recognize life.

In contrast, the approach offered by the author is based on fractal dimensions (Hausdorff, 1919). Rather than the traditional use of fractal descriptions in nature (Mandelbrot, 1982), consider the structure of life as a particular ratio between criticality and redundancy. To define the concept of fractal dimensions, it is that between a point (zero dimensions), a line (one dimension), a plane (two dimensions), a solid (three dimensions), and the physical world (four dimensions), there are fractional dimensions that indicate the complexity of the entity between those basic geometric forms. For the purpose of this chapter, allow this definition to stand descriptively rather than mathematically.

Consider what is occurring with the cameras filming the Petri dish. In the artificial life simulation soup example, what is occurring in the simulated space that could be scanned for simulated life?

The structure of life is distinctly identified through a method similar to fractal dimensions. The limits imposed by perfect structural shapes in the life-searching fractal dimension algorithm represent the unattainable limits of structure that life will not reach. Put explicitly, the combination of redundancy and criticality means the fractal dimension analysis will always be less than pure redundancy defined by perfect structural shapes. Moreover, the result of imposing criticality and redundancy at multiple layers further diminishes the fractal dimension of life.

Rather than life existing in perfect fractals, structural analysis informs the life detection algorithm that when an environment can be interrogated for fractal dimensions, the positive detection of life is a definable ratio between noise and order. This ratio exists down to the lowest layer with the influence of criticality limiting the fractal dimensions even further. The benefit of the fractal dimension analysis method is that it works in two-, three-, and four-dimensional searches for life where the algorithm will return slightly less than the perfect structural shape solution locally to show instances of life.

The mathematical derivation of such a life-detecting algorithm is outside the scope of this chapter. It would probably require a derivation roughly three or more times the length of this chapter. However, with the field of artificial life (Barbalet et al., 2009) and the emerging field of the computational origins of life (COoL), such algorithms will without question be developed over the next decade for specific simulation uses (Damer et al., 2010).

## 7. Objections: Ubiquitous Structure and Negative Entropy

Two possible objections with the redundancy and criticality layered structural analysis of life offered relate to arbitrary structure and energy-use models. These objections: any complex structure would adhere to the properties of life described, and the negative entropy (Schrödinger, 1944) (or negative Gibbs free energy) methods of describing life (Avery, 2003) are superior to a structural analysis.

By offering a structural analysis method of describing life, there could be clear criticism of the relatively arbitrary and self-serving nature of establishing structural requirements that could be found in systems that clearly were not alive. Two possible counter examples are a crystal lattice structure and a dead cat.

The crystal lattice shows vast-ordered complexity and (in real-world terms) a variety of patterns that show redundancy and probably also occasional points that appear to be points of criticality. The crystal lattice is clearly not living. The flaw in the crystal lattice example is that the observation of both the criticality and redundancy are perceived rather than an actual property of the crystal lattice.

The lattice even with minor faults is too perfect to be alive. The relationship between redundancy and criticality, and the principle of layering means that living entities have properties – underlying and subtle ratios – that do not exist in crystal lattices. In short, flawed order (even when imperfect) cannot be living.

Consider the dead cat.

Here you have all the criteria for observation of a living system described in terms of redundancy and criticality. Assuming the cat expired recently, most of these could be observed through dissection analysis. This is an adequate criticism of structural analysis. At this point in the discussion, this example should be bookmarked to indicate that structural analysis provides a means of establishing whether something is alive or has ever been alive. In this regard, it could be used to identify fossils in rock as well as talking walruses and dead cats.

It is important to state explicitly that the structural definition of life and the description of a pattern for life are not arbitrary in their descriptive or analytical use. In fact, the ability to use this method to explicitly search for new and previously defined forms of life give it an applied utility outside any broader philosophical discourse.

To avoid any exact syntactical concerns, let us use the descriptive term "negative entropy" to describe the phenomenon where life observably appears to work against the second law of thermodynamics or more importantly that life fights entropy in a very measurable fashion. Whether the actual phenomenon relates to entropy or Gibbs free energy in the context of this broad discussion is unimportant. This idea is referred to here as negative entropy.

The negative entropy methods of establishing and testing for life cannot be faulted. They are useful methods of identifying biological and non-biological life in any observable system. The environment for finding the quantitative data required to establish life through negating these quantities makes it difficult to use in cases where energy tests are not possible.

Where possible, these negative entropy methods may be preferred for testing the presence of and identifying life; however, the structural analysis method allows basic observation to give a classification of life. As it does not require contact or any means of invasive testing, it provides a novel approach that is useful for methods like the observing camera on the Petri dish or scanning through a computational soup in the search for artificial life (Barbalet et al., 2009).

## 8. Origin of Structure

If life is described by a very specific structure, the origin of life becomes reframed in the origin of this form of structure. The ability to create this structure is self-perpetuating as this structure has a well-defined robustness through the use of both redundancy and layers to minimize the impact of criticality and allow for survival. Structural analysis does not give any temporality in terms of working backwards to find a point in time where these structures could have originated. Removed from the question of when life originated, the origination of life occurred at the building of the first layer.

As this analysis is removed from a discussion of specific chemical compounds, it may seem this has little to do with current discourse on the topic of the origin of life. This precursory glance is incorrect as it gives an attribution of life that is considerably further developed than the chemical origins. As noted, the structural analysis of life is also applicable to systems that are independent of the origins of biological life on Earth.

For the purpose of discussing the origins of the structure that describes life, let us consider two possible descriptions of the creation of this structure: perpetuation and design.

Perpetuation simply states that the criteria of redundancy mean that life as defined through structure will continue to multiply until it no longer can multiply. This expanse does not mean single entities continue to grow but that the property of redundancy expands layer by layer. In this regard, redundancy determines layers will be created, and these layers will in turn create more redundancy.

Design, here, does not mean some intelligent entity creates the structure of life but that the structures of life through their pattern of criticality, redundancy, and layers require some assembly instructions that are meta to their form.

The analysis associated with the description of the three components of life or the follow-up on meta-consequences of perpetuation and design do not answer any questions of why these things occur, and the concept of the origins of structure demands further analysis of these points. These questions are the nexus of multiple analytical methods in artificial life (Barbalet and Clune, 2010).

Consider these ideas as part of the premise of life as negative entropy (Avery, 2003) rather than in contrast with this premise. The structure as proposed offers reinforced and robust means for providing continued life. This method could be replaced eventually by a complete removal of criticality (in theory). The points of criticality tend also to be associated directly either with reproducing or maintaining redundancy. This is fundamentally the paradox of life that it is possible in the future that these points of criticality will be further minimized but it is inconceivable almost by definition that a living entity would not die from a critical failure.

The origin of life is the origin of the structural pattern in survival through redundancy, criticality, and layers to maximize the effects of redundancy and minimize the effects of criticality. Any other structural pattern does not represent a living system. It is clearly possible to create this pattern outside traditional biological systems. The understanding of these systems paradoxically is still in its infancy even as the development of these systems continues with the same vigor as the momentum of life.

## 9. References


Avery J (2003) Information theory and evolution. World Scientific, New Jersey

Barbalet TS (2008) Welcome to the simulation. In: Seckbach J, Gordon R (eds) Divine action and natural selection: science, faith and evolution. World Scientific, Singapore

Barbalet TS, Clune J (2010) Jeff Clune. Biota Podcast. Retrieved June 12, 2010, from http://www.archive.org/download/biota_podcasts/biota_061110.mp3

Barbalet TS, Guerin S (2009) Roots of simulation science. Biota Podcast. Retrieved May 8, 2010, from http://www.archive.org/download/biota_podcasts/biota_112109.mp3

Barbalet TS, Gordon R, de Jong G, Ventrella JJ (2008) The Cathedral and the Spider. Biota Podcast. Retrieved May 8, 2010, from http://www.archive.org/download/biota_podcasts/biota_103108.mp3

Barbalet TS, Damer B, Gordon R (2009) Summoning the EvoGrid. Biota Podcast. Retrieved May 8, 2010, from http://www.archive.org/download/biota_podcasts/biota_032909.mp3

Barbalet TS, Gordon R, Swan E, Buckley WR (2010) Do you like your turing strong or weak? Biota Podcast. Retrieved May 8, from http://www.archive.org/download/biota_podcasts/biota_010810.mp3

Clune J, Beckmann BE, McKinley PK, Ofria C (2010) Investigating whether HyperNEAT produces modular neural networks. Genetics and Evolutionary Computation Conference, GECCO 2010 Proceedings, Oregon

Damer B, Newman P, Gordon R, Barbalet TS, Deamer D, Norkus R (2010) The EvoGrid: a framework for distributed artificial chemistry cameo simulations supporting the computational origins of life endeavors. Proceedings of the Twelfth International Conference on the Synthesis and Simulation of Living Systems, Artificial Life XII, MIT Press, Massachusetts.

Hausdorff F (1919) Dimension und äußeres Maß. Mathematische Annalen 79(1–2):157–179

Kant I (1764). Observations on the feeling of the beautiful and the sublime (trans: Goldthwait JT). University of California Press, California

Mandelbrot B (1982) The fractal geometry of nature, Lecture notes in mathematics 1358. W. H. Freeman, New York

Miller JG (1978) Living systems. McGraw-Hill, New York

Schrödinger E (1944) What is life – a physical aspect of the living cell. Cambridge University Press, Cambridge

Wittgenstein L (1953) Philosophical investigations (trans: Anscombe GEM). Blackwell Publishing, Oxford

Biodata of **Bruce Damer**, **Peter Newman**, **Ryan Norkus**, **John Graham**, **Richard Gordon**, and **Tom Barbalet**, authors of "***Cyberbiogenesis and the EvoGrid: A Twenty-First Century Grand Challenge.***"

**Bruce Damer** (Ph.D., (2011) from University College, Dublin) is an independent designer, researcher, and historian. His design career ranges from the creation of some of the first graphical interfaces for personal computers in the 1980s to pioneering avatar multiuser virtual worlds in the 1990s and the use of 3D environments for mission design visualization for NASA in the 2000s. His research interest spans from optical computing, practical applications of 3D graphics, to his current focus on origin of life efforts supported by simulation environments such as the EvoGrid. He is also a historian of the origins of personal computing, having built a substantial collection of vintage computer hardware in his DigiBarn Computer Museum. Damer is the author of the seminal 1997 book *Avatars* as well as numerous publications in the field of virtual worlds, space exploration, and simulation in biology. Damer also founded Biota.org and initiated the Digital Biota conference series.

E-mail: **bdamer@digitalspace.com**

**Peter Newman** is the chief architect of the EvoGrid platform and has worked for several years with DigitalSpace, developing its real-time 3D platform used for mission visualization and design for NASA. Peter is a native Australian and currently resides in Melbourne. His other interests include massively multiplayer games and software development in digital materialization.

E-mail: **peter.newman@digitalspace.com**

**Bruce Damer**

**Peter Newman**



J. Seckbach (ed.), *Genesis - In The Beginning: Precursors of Life, Chemical Models and Early Biological Evolution*, Cellular Origin, Life in Extreme Habitats and Astrobiology 22, 267–288
DOI 10.1007/978-94-007-2941-4_16, © Springer Science+Business Media Dordrecht 2012

**Ryan Norkus** is the chief modeler and animator for DigitalSpace and has worked for almost a decade in support of its mission visualization and design projects for NASA and other customers. Ryan is a native Australian and currently resides in rural Victoria. Recently, Ryan has been responsible for rendering the EvoGrid and protocell concept movies which provided the public an understanding of the vision for Cyberbiogenesis. His other interests include 3D modeling for massively multiplayer games and avatar environment.

E-mail: **ryan.norkus@digitalspace.com**

**John Graham** Putting the Earth online is one of John Graham's most important ambitions as Chief Scientist of the San Diego State University's Visualization Center, a sister facility to the Scripps Institution of Oceanography Visualization Center at UCSD.

E-mail: **johng@telascience.org**

**Ryan Norkus**

**John Graham**

**Dr. Richard Gordon** is an itinerant theoretical biologist with B.Sc. (Math) from the University of Chicago and Ph.D. (Chemical Physics) from the University of Oregon. He is presently Theoretical Biologist at the Embryogenesis Center, Gulf Specimen Marine Laboratory (http://www.gulfspecimen.org/), recently retired from the University of Manitoba. He organized Books With Wings (http://bookswithwings.net) which sends books to universities in Afghanistan, and the Embryo Physics Course online in Second Life® (http://embryogenesisexplained.com/course). He writes a blog in Science 2.0 (http://www.science20.com/cosmic_embryo) and is Principal Scientific Advisor for the EvoGrid project (http://www.evogrid.org/), an attempt to simulate the origin of life. He has also worked in diatom nanotechnology, HIV/AIDS prevention, and algorithms for computed tomography.

E-mail: **DickGordonCan@gmail.com**

**Tom Barbalet** created the Noble Ape Simulation in 1996 and continues its development to this day. Noble Ape is used by Apple and INTEL as well as a number of universities to teach biodiversity, multimedia education, vector processing, real-time graphical interfaces, and a number of other technologies. He has been the editor of Biota.org (since 2005), a leading community resource for artificial life developers. He is the host of the Biota Live internet radio show where he discusses a variety of topics relating to artificial life, artificial intelligence, and simulation philosophy with a variety of guests. He is also the chair of the International Game Developers' Association's Intellectual Property Rights Special Interest Group.

E-mail: **tom@nobleape.com**

**Richard Gordon**

**Tom Barbalet**

# CYBERBIOGENESIS AND THE EVOGRID: A TWENTY-FIRST CENTURY GRAND CHALLENGE

**BRUCE DAMER[1], PETER NEWMAN[1], RYAN NORKUS[1], JOHN GRAHAM[2], RICHARD GORDON[3], AND TOM BARBALET[4]**
[1]*DigitalSpace Corporation, 343 Soquel Ave, Suite 70, Santa Cruz, CA 95062, USA*
[2]*San Diego State University Visualization Center, CA, USA*
[3]*Gulf Specimen Marine Laboratory, Panacea, FL, USA*
[4]*Noble Ape and Biota.org, Las Vegas, NV, USA*

## 1. Introduction

The modern quest for the understanding of possible mechanisms behind the origin of life, or in other words the *transformation of nonliving matter into living matter,* has been passed down to us from chemistry's precursors, Middle Ages alchemists (O'Connor, 1994). The mathematician Rene Descartes wrote in the seventeenth century of the then prevalent theory of spontaneous generation that "it is certainly not surprising that so many animals, worms, and insects form spontaneously before our eyes in all putrefying substances" (Margulis and Sagan, 2000, p. 64). Charles Darwin challenged the assertion of spontaneous generation in his seminal volume *On the Origin of Species* (Darwin, 1859), arguing that species evolved from previous generations through a process of natural selection. In a letter to botanist Joseph Hooker (1871), Darwin contemplated a chemical origin for life in "some warm little pond." The early twentieth-century work of Oparin (Oparin and Morgulis, 1938) and J.B.S. Haldane (1927) regarding the formation of cells and chemical conditions on the early Earth set the stage for the 1953 experiment by Miller and Urey (Miller, 1953) which synthesized amino acids within a laboratory model of the prebiotic environment.

In the same year of the Miller-Urey experiments, explorations into the origins and evolution of life entered the new medium of digital, electronic computation. In the spring of 1953, researcher Nils Aal Barricelli (1953) coded one of the first scientific computer programs onto punched cards and fed them into the promethean prototype of all electronic computers just put into operation at the Institute for Advanced Study in Princeton, New Jersey (Dyson, 1997). Calling it an *experiment in bionumeric evolution*, Barricelli was investigating the role of symbiosis in the origin of life and came to believe that while his five-kilobyte universe of *numerical symbioorganisms* exhibited criteria of a living, evolving system they would "never become anything more complex than plain numbers" (Barricelli, 1962, p. 73).

Decades later in the 1980s, John von Neumann's original design for the electronic computer at Princeton had come to dominate the computing world and began appearing on desktops as microcomputers. These tiny machines allowed intellectual successors to Barricelli such as Chris Langton to again work late into the night and code their own renditions of *life as it could be* and along the way creating a new field: *artificial life* (Alife) (Langton, 1986; Levy, 1993). Alife's close cousin, *artificial intelligence* (AI) had a parallel lineage from the 1950s but was aimed at representing conscious thought. Alife focused on a bottom-up approach, hoping to explore the dynamics of living systems through algorithmic techniques that generated emergent phenomena including the physics of motion (Sims, 1991) and the evolving of competing artificial genomes (Ray, 1991). Great promise was held that in the 1990s, increasing computing power would soon support teeming simulated ecosystems which biologists would come to recognize as true living systems. However, it was clear that by the turn of the past century, Alife development had stalled as the virtual worlds containing early examples of protobiota proved too simplistic for the phenomena of open-ended growth of complexity to be observed (Rasmussen et al., 2003a).

## 2. A New Synthesis: Cyberbiogenesis

In the early 2000s, interest was again growing in creating chemically based (*in vitro*) experiments in origins of life endeavors which would lead to the formation of so-called *protocells*, chemical structures exhibiting at least some properties of living systems (Rasmussen et al., 2008). In parallel, massively distributed computation and large-scale centralized grids and special-purpose hardware were hosting viable realistic simulations of very small volumes of interacting molecules over short but biologically significant timescales (Shaw and Dror, 2008). In 2011, a true synthesis of *in silico* simulation as a tool to design and predict the outcomes of *in vitro* experimentation seems to be beckoning to us from just over the horizon. This synthesis holds the promise of new tools for chemistry akin to computer-aided design (CAD) enjoyed by other fields such as product manufacturing and architecture. At a not-so-distant date in the future, biochemists should be able to simulate larger biomolecular structures such as proteins yielding a good measure of predictability of outcomes in the test tube (or the Petri dish). Such a synthesis also brings up a new and tantalizing possibility:

> Could we actually one day digitally simulate a complete step-by-step chemical scenario for an origin of life on Earth? And if we could carry out such a simulation while remaining faithful to the chemistry, could we then reproduce this particular pathway to life from nonlife on the chemistry workbench?

The computing part of this challenge was perhaps most definitively issued by Richard Gordon in the recently published book *Divine Action and Natural Selection: Questions of Science and Faith in Biological Evolution* (Seckbach and Gordon, 2008). In his chapter titled "Hoyle's Tornado Origin of Artificial Life: A Programming Challenge," Gordon challenges the Alife community to develop a computational environment to simulate an origin of artificial life from artificial nonlife (pp. 354–367):

> I would like to suggest that artificial life (Alife) enthusiasts take up Fred Hoyle's (Hoyle, 1984) challenge, that in a way they simulate a tornado going through a junkyard of parts, and come up with something we would all agree is alive, in the Alife sense, from components that are not alive in the Alife sense...

This author's response to Gordon's challenge was detailed in another chapter "The God Detector" in the same volume (pp. 66–85):

> What I am proposing is to engage all of the best programmers, artists and philosophers of our generation to create a gigantic network of software and computers, working to create a sort of "Evolution Grid" or "EvoGrid". This EvoGrid would start out as God the Mechanic (like Karl Sims' creatures) in which we build the simulation, set the initial conditions and then let the artificial ecosystem go from there.

The chemical fabrication part of this challenge is perhaps best represented by the field of synthetic biology. The recent successful *in vitro* substitution of a synthetically created genome into a living cell (Venter et al., 2001) seems to suggest that the fabrication of significant additional parts of living cells might also be possible.

The term *cyberbiogenesis* might suffice to capture the complete closing of this circuit from digital simulation to atomic realization. This term could be thought of as a kind of cousin to Mereschcowsky's word *symbiogenesis* (Mereschcowsky, 1909) which Margulis argues was a primary driver of evolution (Margulis, 1997). The scope of constructing an end-to-end cyberbiogenesis system would dwarf the recently completed Human Genome Project (Watson and Cook-Deegan, 1991) but is possibly realizable within this century. To compute an origin of life faithful enough to physical laws of chemistry to be reproducible, *in vitro* is perhaps one of the most audacious applications of technology in the history of our species.

The remainder of this chapter will explore cyberbiogenesis in a thought experiment, then detail one initial attempt at implementing an early computer prototype, and finally enumerate and illuminate some of the many scientific, technological, religious, and philosophical conundrums uncovered by such an effort. We hope that these words will lend some shape to cyberbiogenesis as a grand challenge for the coming century for those who might choose to take it up. With apologies to readers, we will leave it to others in this volume to provide a decent treatment of the major schools of thought regarding the origin of life.

## 3. Cyberbiogenesis: A Visually Rendered Thought Experiment

In mid-2008, the author (BD) engaged in a *Gedankenexperiment* (a thought experiment), drew storyboards (Fig. 1), and requested a collaborator to produce a short animated movie (Damer et al., 2008) designed to illustrate the concept of cyberbiogenesis.

Figures 2, 3, 4, 5, 6, 7, 8, 9, 10, 11 below depict and describe scenes from the movie which provides a visual cartoon of the thought experiment which imagines a completely realized cyberbiogenesis system.

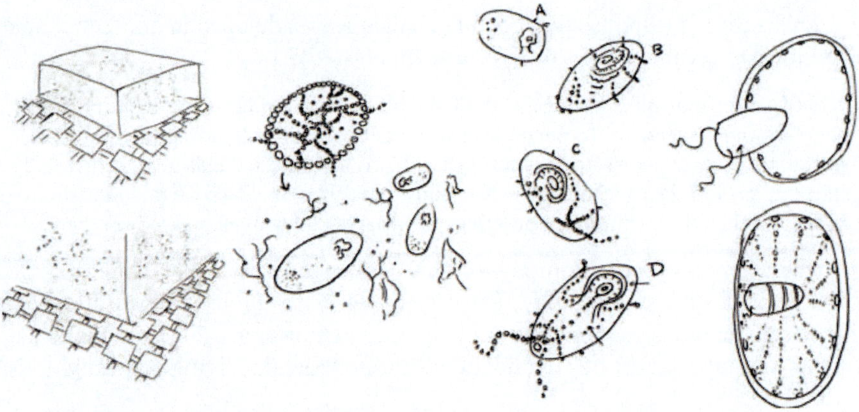


**Figure 1.** The author's original sketches of the thought experiment.

**Figure 2.** The conceptual cyberbiogenesis setup: on the right is the *in silico* molecular simulation space underlain and powered by numerous microprocessors; on the left is the molecular assembler and *in vitro* test beaker.

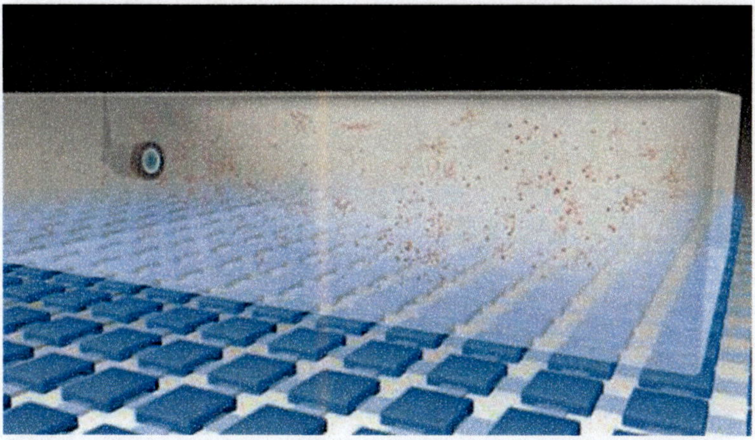

**Figure 3.** The simulation space renders the physics of an aqueous chemical environment.

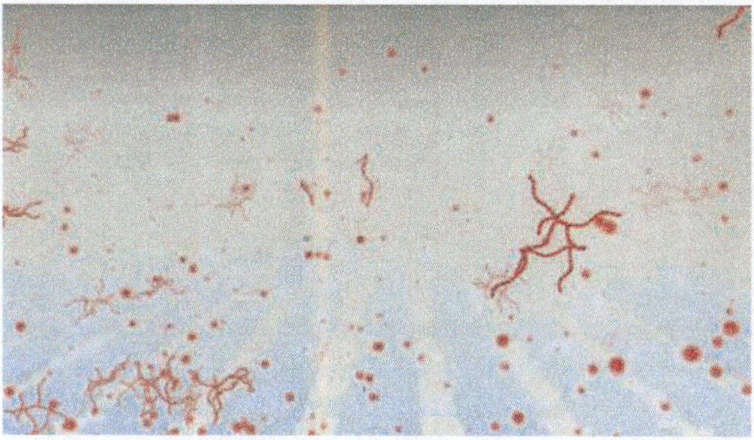

**Figure 4.** The formation of virtual molecules and self-organization occurs in the simulation space.

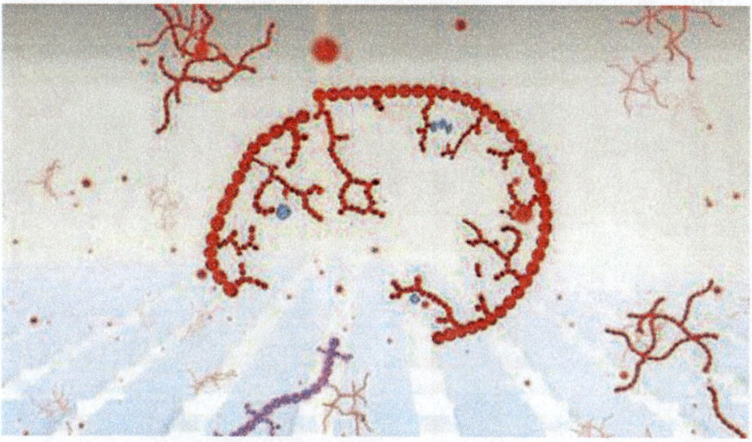

**Figure 5.** The formation of a vesicle is observed with the accidental capture of some other molecular machinery (on the *lower left center*).

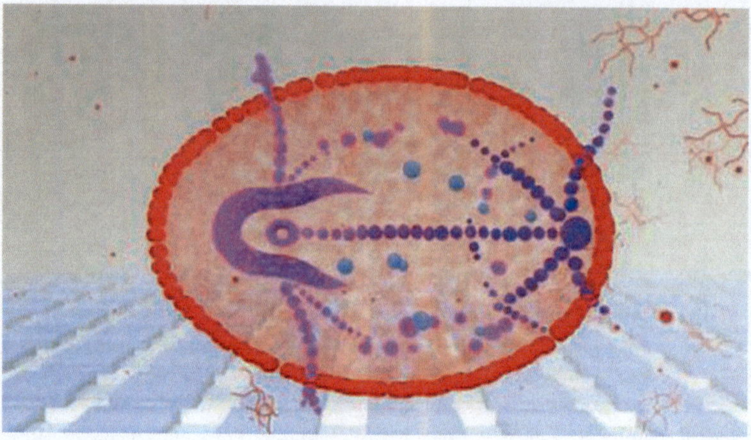

**Figure 6.** The virtual symbiotic entity is capable of a sufficient ensemble of lifelike behaviors including compartmentalization, metabolism, and replication with a mechanism for genetic heredity such that Darwinian natural selection has led to its growing sophistication.

**Figure 7.** A sufficiently evolved entity is selected for digital decomposition perhaps at its embryonic phase and transmitted from the *in silico* simulation to the molecular assembler.

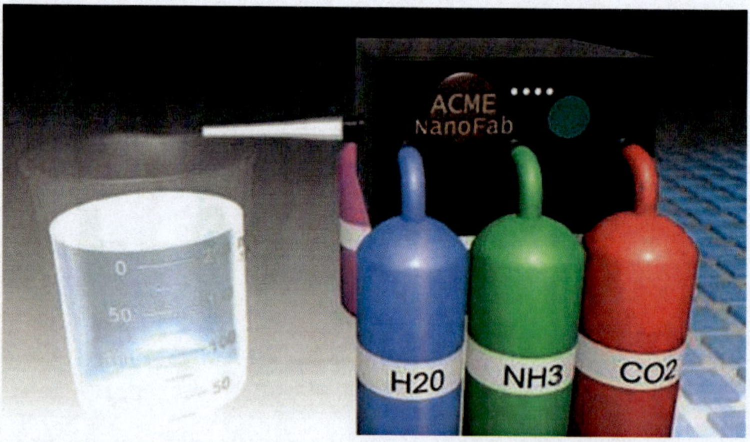


**Figure 8.** The hypothetical molecular assembler carries out a process akin to 3D printing and combines basic chemical elements to synthesize a molecular rendition of the virtual entity.

**Figure 9.** The fabricated entity emerges to drop into the beaker of formulated chemicals matching the environment in the original digital simulation.

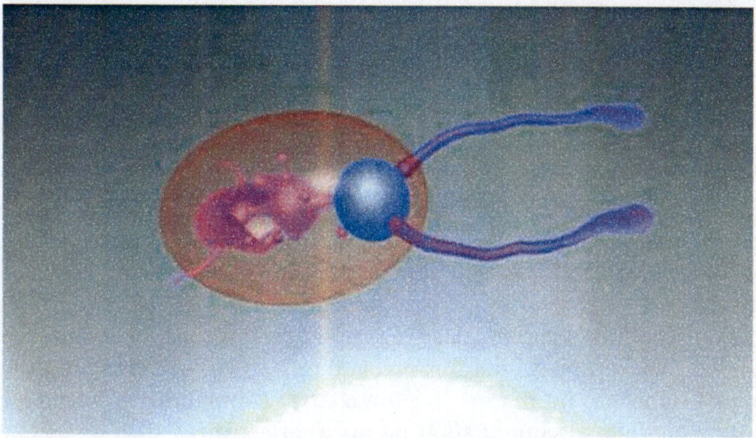

**Figure 10.** Within the *in vitro* environment, the molecular version of the entity starts to function as a new form of "living" entity.

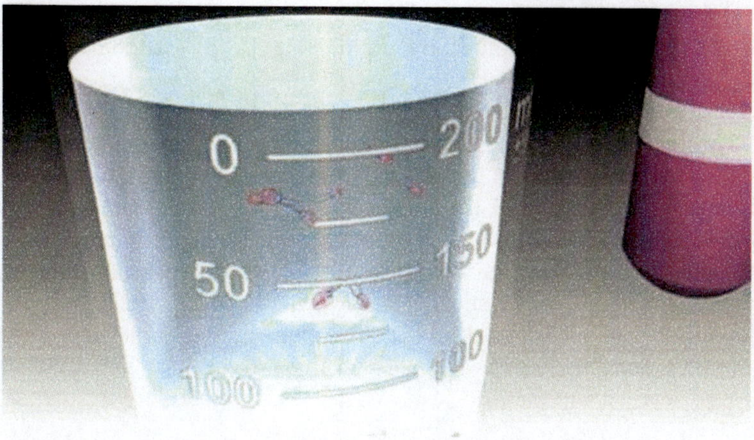


**Figure 11.** The entity advances further and begins to reproduce in the new environment, completing the cyberbiogenesis cycle.

To conclude this thought experiment, we might now ask why have the molecular assembler to fabricate the entity when instead we could simply recreate the same chemical environment as that from which the *in silico* entity emerged? It may be that the in vivo emergence of the parallel chemical entity may not so be easily achieved or may take substantially longer without the shortcuts provided to its computational *in silico* counterpart. We might also ask how realistic and realizable is this thought experiment? According to Nobel laureate Richard E. Smalley, a "black box" nanoscale molecular assembler is nowhere near to becoming a reality (Baum, 2003). However recent progress in functional representation of 3D virtual objects (Pasko et al., 2008) paired with digital materialization made possible by universal desktop fabrication (Vilbrandt et al., 2008) shows promise in this direction. In any case, near-term progress toward this goal would have to be made in the domain of computational simulation. Given the computing resources available in 2011, it was thought that a reasonable goal might be to produce a prototype that would reach the step depicted in Fig. 4 above: a few simple molecules forming from an atomistic soup.

## 4. One Naïve and Early Cyberbiogenesis Prototype: The EvoGrid

How would one begin to construct the computational part of a cyberbiogenesis system? The author's company, DigitalSpace, working with the guidance of an energetic international team of advisors, undertook to build a first prototype of the EvoGrid. The prototype began preliminary trial operations at the end of 2009 and has engaged in several longer experimental runs in 2010 and 2011 at the University of California at San Diego. Figure 12 depicts the architectural components of the EvoGrid in its first implementation. The EvoGrid is built upon an open source, published framework to allow a larger community to extend it and to permit fully distributed computation on donated networks such as the BOINC environment that supports the "@Home" experiments (Anderson, 2004). The salient innovation of the EvoGrid is the execution of a large number of small volumes of simulated molecular dynamics over short periods of time. Each of the small volumes is sampled and processed by a search function, which directs whether the simulation should become the starting "seed" for the next branch of simulations or abandoned.

It was hypothesized that the combination of search with small simulations would drive the overall system more rapidly toward the emergence of complex molecular structures within branched simulations. As shown at (B) in Fig. 13, the storage of the full tree state of the system would permit time-based analysis and backtracking at (C) to retry promising branches with the hopes to reaching a more complex end state at (D) than would be possible through linear simulation as depicted in (A). This is a stochastic hill-climbing method (E) used commonly in AI (Russell and Norvig, 2003, pp. 111–114) but applied here to molecular dynamics.

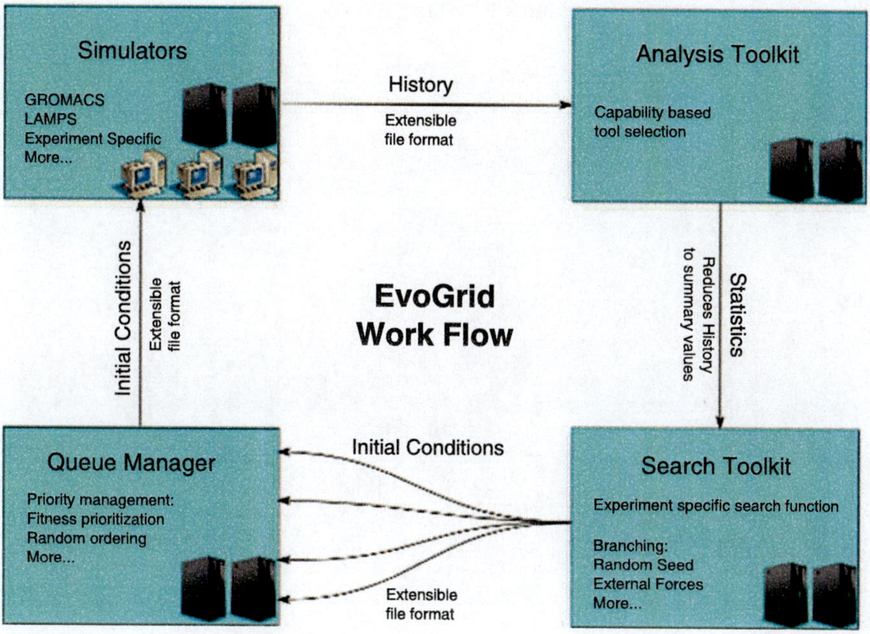


**Figure 12.** The architecture of the EvoGrid.

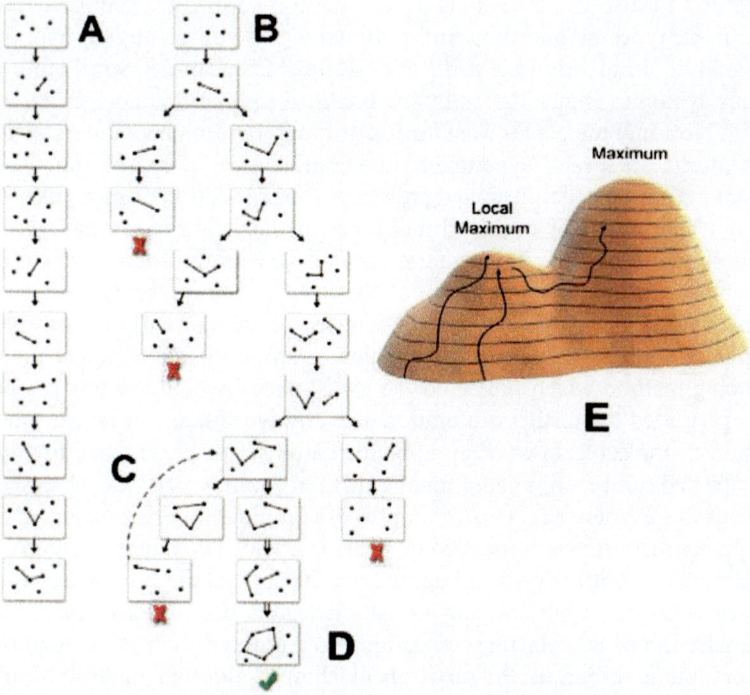


**Figure 13.** Illustration of the hill-climbing search tree method employed by the EvoGrid.

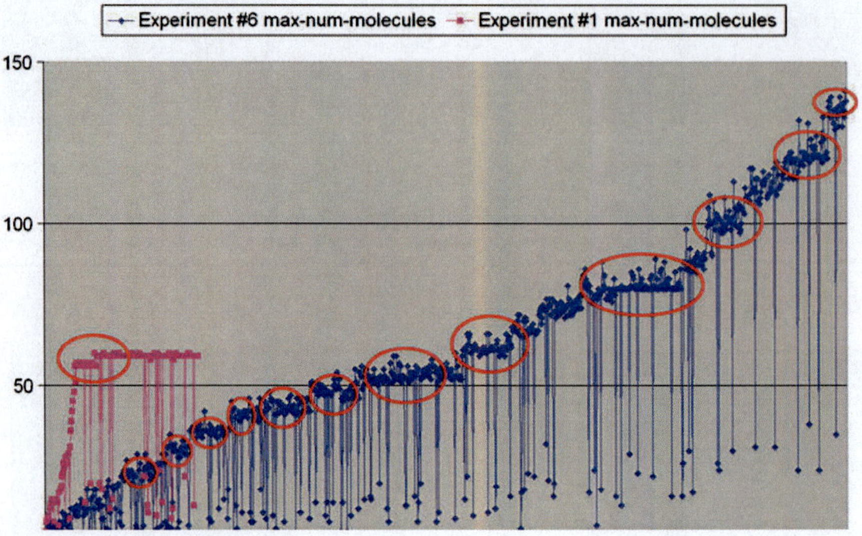


**Figure 14.** Plot of increasing molecular complexity.

Figure 14 illustrates the results of this method applied to an initially random volume of six types of one thousand notional atoms run through a hypothetical nanosecond of simulation time using the GROMACS (Van der Spoel et al., 2005) molecular dynamics engine driven by the EvoGrid simulation manager and search functions. Notional molecules were formed through random encounters within the soup of atoms. In several experiments the combination of search functions and back-tracking supported a sustained growth of the population of molecules, shown as a graph of number of observed molecules or yield (Y) versus the number of executed simulations (X). Momentary declines (the intermittent vertical drops) were overcome through abandonment of these less promising branches.

The results of two experiments are depicted in the figure. Both used the same starting conditions but Experiment #1 (lighter graph on left side) applied a simple hill-climbing method which preserved the gains made by high yielding simulations but only permitted the further exploration of pathways of equal or higher yield. The population of molecules (two or more bonded atoms) quickly climbed to a plateau of sixty (circled) and then no additional growth in "complexity" was observed. The experiment was terminated after two months of computation on a small grid of four servers. Experiment #6 was processed on a larger grid of two dozen cores running at 100% for a two month period ending in June, 2011. This time the search function implemented stochastic hill climbing permitting degradation off local optima and the random selection of new starting simulations from a wider pool. While requiring far more computing resources, this experiment climbed through multiple thresholds

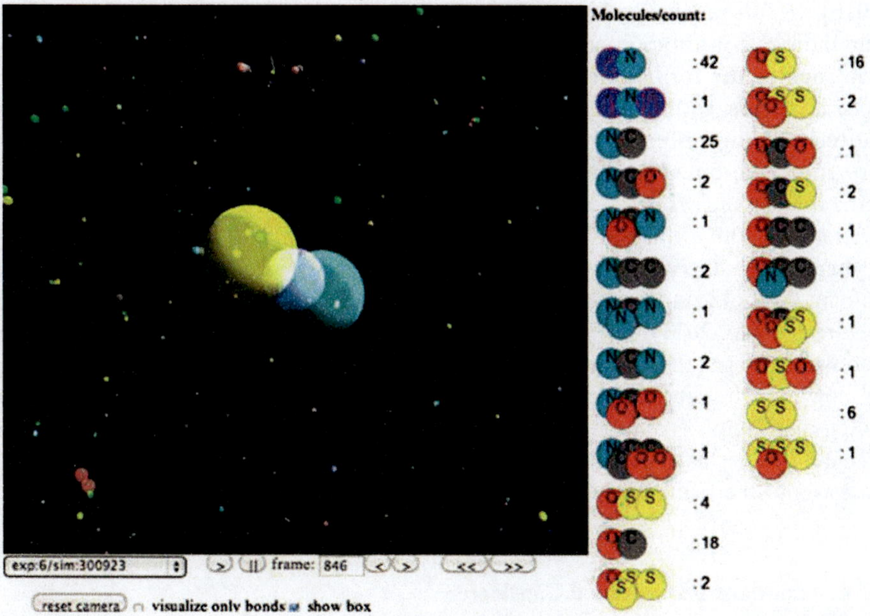


**Figure 15.** 3D view of a simulation volume showing molecular products.

(shown circled) arriving at what appeared to be a global maximum of 189 molecules (not shown).

At this point, the richest "soups" had sequestered over half of their total populations of 1,000 free atoms within molecular bonds. Thus the slight change in the search function permitted a "breakthrough" to ever higher complexity through the exploration of a highly rugged fitness landscape. The observed hill-climb in experiment #6 suggests that this landscape is not truly random but is in fact correlated (Kauffman, 1995). A 3D view of one higher yielding simulation is shown in Figure 15. The notional end products of the depicted simulation, which was terminated with over two thousand processed simulations and over two hundred thousand unprocessed simulations in abandoned branches, consisted of 23 kinds of molecules in a total population of 185 in the most populated volume. A control experiment (not shown) involving purely unsearched linear processing produced far fewer types of molecules (6) in fewer quantities (26) with no sustained trend.

The EvoGrid prototype showed that a grid of molecular dynamics simulations could be coupled with a carefully tuned search function such that the yields of experiments would grow rapidly to "interesting" maxima. These simulations were modeled very crudely on the diffuse atomistic regime of interstellar space, where organic and other bio-essential molecules are formed (Deamer and Damer,

2010). In future EvoGrid implementations it is conceivable that more ambitious simulations could be tuned to search their way biologically interesting phenomena such as the formation of vesicles, informational molecules, or autocatalytic sets. It is envisaged that this discovery system of small but rapidly searched simulations working in concert with larger centralized grid-based simulations could provide a combined computing environment rich enough to permit experiments surrounding the origin of life question (Damer et al., 2010). The EvoGrid is only a first step on a very long journey to a full system capable of undertaking Cyberbiogenesis. As more realistic chemical interactions are simulated they could inform a whole range of parallel "wet" laboratory experiments not conceivable before (Damer, 2011). The use of combinatorial chemistry techniques automating thousands of parallel experiments with computerized search and robotic reseeding of new experiments could yield a sort of chemical EvoGrid or, more poetically, a "Genesis Engine". With a commitment and investment like the Human Genome Project of the last Century, the Genesis Engines of the 21st Century may well carry us to the brink of a second abiogenesis on the Earth.

## 5. Conundrums Exposed and Considered

Given the likely distant prospects for any successful cyberbiogenesis effort, a valuable activity in the interim is to consider the societal impact of such a project on a range of human endeavors. Any enterprise that sets as its goal the emergence of an *artificial origin of life*, testable in chemistry and therefore ultimately realizable as a new form of chemical life is likely to draw controversy from many quarters. This controversy will in turn expose a number of conundrums that lie at the basis of science, technology, religion, ethics, and philosophy.

### 5.1. SCIENTIFIC CONUNDRUMS

The goals of cyberbiogenesis beg many basic questions in science including:

1. How might science define a living entity or indeed an entire living system? When asked to measure aliveness, would scientists simply "know it when they see it" when presented candidate lifelike systems?
2. In simulating steps to an origin of life, what is the experiment to be undertaken and at what point does it start? Do experiments begin with prebuilt components of some complexity but not considered to be living and proceed from there, as suggested by Gordon (2008)? Alternately, should simulation experiments be initiated much further down the ladder with simpler artificial precursor molecules or even farther down from basic atoms assembling precursor molecules within an *ab initio* primal soup?

3. How much influence is allowed or required to induce a sufficient measure of emergence? In other words, how much "intelligent design" is required in the setting up and operating of cyberbiogenesis experiments? What degree of ongoing human guidance should be permitted in both the virtual and chemical experiments which follow?
4. Would an entirely artificially evolved entity pose a current or future threat to any part of the natural environment in the Earth's biosphere or to technological or biological regimes within human civilization? How could such a threat be mitigated? If such a threat were plausible, would it be grounds for not pursuing this line of research?

## 5.2. TECHNOLOGICAL CONUNDRUMS

A decade ago, the artificial life community took stock of their field and proposed a set of "open problems in artificial life" (Bedau et al., 2000) which provide a clear look at the brickwork of the technological foundations of any serious cyberbiogenesis effort. The authors set a challenge in the second open problem to solve the challenge of abiogenesis in an artificial chemistry and identified that "[b]etter algorithms and understanding may well accelerate progress… [and] combinations of… simulations… would be more powerful than any single simulation approach" (pp. 367–368). The authors also pointed out that while the digital medium is very different from molecular biology, it "has considerable scope to vary the type of 'physics' underlying the evolutionary process" and that this would permit researchers to "unlock the full potential of evolution in digital media" (p. 369). Ten years later as projects such as the EvoGrid take aim an *in silico* abiogenesis, many technological conundrums have come to the fore including:

1. What level(s) do you simulate at and at what scale? Is molecular dynamics a sufficient level or are quantum dynamical effects required? Alternatively, is a more abstract artificial chemistry which can exhibit desired properties a better starting point than aiming at high fidelity to chemistry?
2. Nature operates in parallel at multiple scales with multiple physical properties emerging at these scales. Therefore, how can von Neumann computers (essentially serial processors) be adapted to meet this substantial computational challenge, or does this challenge belong to the domain of special-purpose hardware or an amalgam of digital and chemical computing?
3. What computational corners can be cut but still retain plausibility in nature and viability in experimental chemistry? Related to this is the claim by Abel (2009) that any computational simulation is formulaic, subject to predicative knowledge and not based on *physicodynamic* factors so may never be representative of solutions *in vitro*. In addressing this question, Gordon presents the following possibility for future EvoGrid implementations:

> Consider having the EvoGrid simulate a less plausible approximation to chemistry. Allow a more abstract chemistry to be tested which also might be subject to a proof

> by construction in mathematics. The components will be decent approximations of real chemistry. Allow yourself to introduce all the bias that you want but as long as the program constrains you to do things that are physically realistic then you might argue that you have something artificially alive. You just don't have the pathway to the end point but you know there is a way back. Decomposed parts could be markers on many paths to life. The important point is proving that one such path exists. (Gordon et al., 2010)

4. How much do the search functions and human-designed initial conditions and sought-after end points to experiments limit their ultimate *creativity*? This is the problem of systems employing a teleological approach: bias toward the sought-after goals limits the power of the system as an open-ended discovery mechanism. As suggested by Dawkins (1986) and others, evolution does not strive toward goals. Even though nature cannot be praised for the best "designed" solutions to problems, it also cannot be faulted for teleological bias. Gordon also makes the following points along this line:

> Examine the case of the EvoGrid where you act as the intelligent designer and use the tools to determine the minimal artificial organism. If you could then put one together then you could look for the properties and potential pathways to that minimal artificial organism. You could also consider an experiment where you start with bigger building blocks that are considered to be non alive and see if they assemble into something you would consider to be alive. (Gordon et al., 2010)

## 5.3. RELIGIOUS, ETHICAL, AND PHILOSOPHICAL CONUNDRUMS

Any seriously undertaken cyberbiogenesis endeavor will attract the following questions and controversy from the triumvirate magisteria of religion, ethics, and philosophy:

1. Does a successful cyberbiogenesis disprove the need for a supernatural creator as an agent in the origin of life and for the guiding of life's development?
2. What is the consequence for the world's religions of the creation of an artificially alive (in computer simulations) or a chemically alive entity?
3. Would an artificially sourced living entity be protected as an endangered species? Would only the chemical entity be protected or the virtual one as well?
4. Does the enterprise of cyberbiogenesis represent a willful achievement of human innovation or is it an inevitable expression of the entire biosphere and life itself, with humans as mere agents forwarding life into a new mechanism of evolution? Is this the means by which life is expanding itself out into other parts of our solar system or the universe? Are we willing or unwilling agents of this expansion?

In a discussion of ethical concerns in (Rasmussen et al., 2003b, p. 67), the authors echo some of the above points:

> Generating life de novo will create public reactions. The reactions will probably be along two lines: (i) Environmental concerns that the life-producing technology could "get out of control", and (ii) Religious and moral concerns, based on the beliefs that humankind must restrain from certain endeavors on grounds that they are fundamentally amoral.

Ethical questions arising around the possible creation of cells from existing biology or completely new molecular constructions have a storied history. Nontechnical press reaction from announcements in biotechnology and genomics such as the research on minimal cells (Fraser et al., 1995) and the announcement of the sequencing of the human genome (Venter et al., 2001) often turns to talk of "Frankencells." Concerns about more artificial nanostructures able to survive and reproduce in natural environments have also been discussed in the nanotechnology community (Merkle, 1992). It is clear that future work on the EvoGrid or any cyberbiogenesis-oriented system will have to address the above issues. Initially, the response to concerns might be to state that lifelike objects in a simulation are merely objects in a simulation and of no threat to people or the biosphere. The argument might be made that these objects might be a threat to computer networks, however, akin to computer viruses. However, virtual environments required to sustain an artificially alive system would be so complex and large that the system would effectively be an isolated island, similar to large multiplayer game systems. Once there is an active effort to reproduce these lifelike objects in physical chemistry, the alarm will be raised with many in the public and scientific community.

## 5.4. AN ORIGIN OF ARTIFICIAL LIFE TURING TEST

Related to concerns about Frankencells is the question: how does the observer know when something is lifelike enough in a computer simulation to declare it "alive"? In his 1950 paper *Computing Machinery and Intelligence* (Turing, 1950), Alan Turing wrote "I propose to consider the question 'Can machines think?'" (p. 433) and defined a test of machine intelligence, one variation of which consisted of a human judge conversing through a text interface with an unseen human and a similarly hidden computer (p. 442). If the human judge could not tell the difference between the human and computer conversant, then the machine would be deemed to have reached some sort of cognitive equivalence to the human participant.

At some day in the distant future, a test group of biologists, engineers, philosophers, and others may assemble in an online virtual space. In another access-controlled space, project specialists would be assembled observing a rich array of lifelike objects moving about in a virtual world running on a greatly advanced EvoGrid. The esteemed guests would then be asked to undertake a kind of *origin of artificial life Turing test*, wherein they may read descriptions of the objects and environments being observed but not revealed by the staff. They would also be exposed to descriptions of similar yet real, living entities and their terrestrial environments, such as a termite colony. Over many days or weeks, the test group would be able to ask the staff questions about both environments, perhaps even specifying experiments to be performed. If at the end of a long enough period a majority of the test group cannot consistently tell which environment is in fact the real, biological one and which is the one witnessed in the simulation then the simulated environment will have passed this new variant of the Turing test. Of course, there will likely be strong arguments for and against the "aliveness" of these virtual entities, especially if they exist solely in an abstract universe far from the norms of chemistry. If at some later date a chemistry-faithful entity is fabricated with molecules and survives to breed within a physical setting, then the concerns of the doubters may be quelled.

### 5.5. A VISIONARY VIEW: A LENS ON LIFE IN THE UNIVERSE

Let us now roll forward to some much farther future date when working cyberbiogenesis systems abound and when a further closure of the loop is in place whereby the observation of *in vitro* adaptations feeds directly back to changes in the *in silico* ecosystem. With this full closure, numerous variants on viable biologies could be generated, extending our ability to model origins of life in alien habitats and to cast light onto *life as it might be* in the universe. Such a system could be used to seed life forms in alien habitats such as the Martian ice cap or the surfaces of near-earth objects. Of course, there may well be a substantial range of viable artificial living systems for which there would exist *no physical medium in which they could be instantiated.* In this case, the only universe in which that these creatures could be chemically rendered out of the simulation into physical reality is a parallel one possessed of exotic physics. Taking this one step further, we might see that nearly infinitely endowed future cyberbiogenesis systems could serve as a lens into where in this universe or others life might arise and projecting how far it might evolve. Indeed, in the unlikely event that an intelligent higher form of life should arise within a cyberbiogenesis simulation, would we choose to instantiate it physically or seek out where its naturally evolved cousins might be resident? Presumably at that point the new form of life might have its own say in the matter.

## 6. Conclusions

Recent progress in synthetic biology, the creation of prototypical chemical protocells, and models of pathways to the origin of life together with increases in computing power and the ability to accurately model small volumes of molecules suggest that this century may see a solution to life's origins emerge *in silico* and be testable *in vitro*. A concerted effort or "grand challenge" to create a viable *cyberbiogenesis* system might therefore be well worth the effort. The construction of a shared, open, and extensible "primal soup Internet" may serve not only as an experimental environment for origins of life and complexity theory but also provide valuable CAD-type capabilities for biomedical research, ultimately simulating the structure and functions of an entire cell.

## 7. Acknowledgments

The lead author (BD) thanks his team at DigitalSpace and Elixir Technologies Corporation for funding support of this work and numerous additional contributors including Richard Gordon, Tom Barbalet, and David Deamer for their critical input to his recently completed Ph.D. thesis from which this chapter was drawn.

## 8. References

Abel DL (2009) The capabilities of chaos and complexity. Int J Mol Sci 10:247–291

Anderson DP (2004) Boinc: a system for public-resource computing and storage. In: Proceedings of the 5th IEEE/ACM international workshop on grid computing, international conference on grid computing. IEEE Computer Society, Washington, DC

Barricelli NA (1953) Experiments in bionumeric evolution executed by the electronic computer at Princeton, NJ. Archives of the Institute for Advanced Study, Princeton, NJ

Barricelli N (1962) Numerical testing of evolution theories: Part II. Acta Biotheoretica 16.

Baum R (2003) Nanotechnology drexler and smalley make the case for and against 'molecular assemblers' (online). Chemical & Engineering News. Available: http://pubs.acs.org/cen/coverstory/8148/8148counterpoint.html. Accessed 21 Mar 2011

Bedau MA, Mccaskill JS, Packard NH, Rasmussen S, Adami C, Green DG, Ikegami T, Kaneko K, Ray TS (2000) Open problems in artificial life. Artif Life 6:363–376

Damer B (2011) The evogrid: an approach to computational origins of life endeavours. PhD, University College Dublin

Damer B, Norkus R, Dhiren D (2008) Evogrid "the movies" and script and storyboards concept (online). Available at http://www.evogrid.org/evogrid-movie/index.html Accessed 16 June 2011

Damer B, Newman P, Gordon R, Barbalet T, Deamer D, Norkus R (2010) The evogrid: a framework for distributed artificial chemistry cameo simulations supporting computational origins of life endeavors. Proceedings of the 12th conference on artificial life. Published by MIT Press, Cambridge MA.

Darwin C (1859) On the origin of species by means of natural selection. J. Murray, London

Darwin C (1871) Letter to hooker. Darwin online library

Dawkins R (1986) The blind watchmaker: why the evidence of evolution reveals a universe without design. W.W. Norton & Company, New York

Dyson G (1997) Darwin among the machines: the evolution of global intelligence. Addison-Wesley Pub. Co, Reading

Fraser CM, Gocayne JD, White O, Adams MD, Clayton RA, Fleischmann RD, Bult CJ, Kerlavage AR, Sutton G, Kelley JM, Fritchman RD, Weidman JF, Small KV, Sandusky M, Fuhrmann J, Nguyen D, Utterback TR, Saudek DM, Phillips CA, Merrick JM, Tomb JF, Dougherty BA, Bott KF, Hu PC, Lucier TS, Peterson SN, Smith HO, Hutchison CA 3rd, Venter JC (1995) The minimal gene complement of mycoplasma genitalium. Science 270:397–403

Gordon R (2008) Hoyle's tornado origin of artificial life, a computer programming challenge. In: Gordon R, Seckbach J (eds) Divine action and natural selection: science, faith and evolution. World Scientific, Singapore

Gordon R, Damer B, Newman P (2010) Re: conversation between Richard Gordon, Bruce Damer and Peter Newman prior to the implementation of the full Evogrid design. Silver Bog, Second Life virtual world platform

Haldane JBS (1927) Possible worlds and other essays. Chatto & Windus, London

Hoyle F (1984) The intelligent universe. Holt Rinehart and Winston, New York

Langton CG (1986) Studying artificial life with cellular automata. Physica D 22:120–149

Levy S (1993) Artificial life: the quest for a new creation. Vintage Books, New York

Margulis I (1997) Beliefs and biology: theories of life and living. ISIS 88:522–523

Margulis L, Sagan D (2000) What is life? University of California Press, Berkeley

Mereschcowsky K (1909) The theory of two plasms as the basis of symbiogenesis, a new study or the origins of organisms. Biologisches Centralblatt XXX(8), 278–288

Merkle R (1992) The risks of nanotechnology. In: Lewis BCJ (ed) Nanotechnology -research and perspectives. MIT Press, Cambridge

Miller SL (1953) A production of amino acids under possible primitive earth conditions. Science 117:528–529

O'Connor K (1994) The alchemical creation of life (takwin) and other concepts of genesis in medieval Islam (PhD dissertation, UPenn 1994). PhD, University of Pennsylvania

Oparin AI, Morgulis S (1938) The origin of life. The Macmillan Company, New York

Pasko A, Adzhiev V, Comninos P (eds) (2008) Heterogeneous objects modelling and applications. Springer, Berlin/Heidelberg

Rasmussen S, Bedau MA, Raven M, Keating G (2003a) Collective intelligence of the artificial life community on its own successes, failures, and future. Artif Life 9:207–235

Rasmussen S, Chen L, Nilsson M, Abe S (2003b) Bridging nonliving and living matter. Artif Life 9:269–316

Rasmussen S, Bedau MA, Chen L, Deamer D, Krakauer DC, Packard NH, Stadler PF (eds) (2008) Protocells: bridging nonliving and living matter. MIT Press, Cambridge

Ray TS (1991) An approach to the synthesis of life. In: Langton C, Taylor C, Farmer JD, Rasmussen S (eds) Artificial life II, Santa Fe Institute Studies in the sciences of complexity. Addison-Wesley, Redwood City

Russell SJ, Norvig P (2003) Artificial intelligence: a modern approach. Prentice Hall/Pearson Education, Upper Saddle River

Shaw DE, Dror R (2008) Anton, a special-purpose machine for molecular dynamics simulation. Commun ACM 51:91–97

Sims K (1991) Artificial evolution for computer graphics. ACM Comput Graphics 25:319–328

Turing AM (1950) Computing machinery and intelligence. Mind 59:433–460

Van der Spoel D, Lindahl E, Hess B (2005) Gromacs: fast, flexible, and free. J Comput Chem 26:1701–1718

Venter JC, Adams MD, Myers EW, Li PW et al (2001) The sequence of the human genome. Science 291:1304–1351

Vilbrandt T, Malone E, Hod Lipson H, Pasko A (2008) Universal desktop fabrication. In: Pasko A, Adzhiev V, Comninos P (eds) Heterogeneous objects modelling and applications. Springer, Berlin/Heidelberg

Watson JD, Cook-Deegan RM (1991) Origins of the human genome project. FASEB J 5:8–11

Biodata of **Norikazu Ichihashi**, **Tomoaki Matsuura**, **Hiroshi Kita**, **Takeshi Sunami**, **Hiroaki Suzuki**, and **Tetsuya Yomo**, authors of "***Constructive Approaches for the Origin of Life.***"

**Dr. Norikazu Ichihashi** is a group leader of Exploratory Research for Advanced Technology project of Japan Science and Technology Agency. He obtained his Ph.D. from the University of Tokyo in the field of Pharmaceutical Science in 2007.

E-mail: **ichihashi@ist.osaka-u.ac.jp**

**Tomoaki Matsuura** is an associate professor at Osaka University and a group leader of Exploratory Research for Advanced Technology project of Japan Science and Technology Agency. He obtained his Ph.D. from Osaka University in the field of Engineering in 1999.

E-mail: **matsuura_tomoaki@bio.eng.osaka-u.ac.jp**

**Norikazu Ichihashi**

**Tomoaki Matsuura**



J. Seckbach (ed.), *Genesis - In The Beginning: Precursors of Life, Chemical Models and Early Biological Evolution*, Cellular Origin, Life in Extreme Habitats and Astrobiology 22, 289–303
DOI 10.1007/978-94-007-2941-4_17, © Springer Science+Business Media Dordrecht 2012

**Hiroshi Kita** is a researcher of Exploratory Research for Advanced Technology project of Japan Science and Technology Agency. He obtained his Ph.D. from Osaka University in the field of Engineering in 2008.

E-mail: **kita3@bio.eng.osaka-u.ac.jp**

**Takeshi Sunami** is a researcher of Exploratory Research for Advanced Technology project of Japan Science and Technology Agency. He obtained his Ph.D. from Osaka University in the field of Information Science in 2008.

E-mail: **sunami_takeshi@bio.eng.osaka-u.ac.jp**

**Hiroshi Kita**

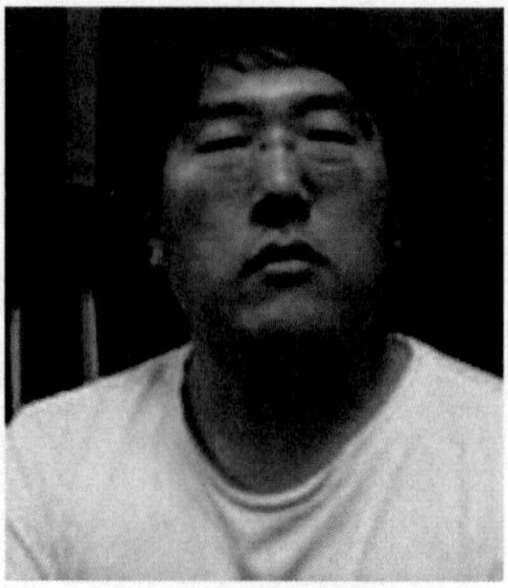

**Takeshi Sunami**

**Hiroaki Suzuki** is an associate professor at Osaka University and a group leader of Exploratory Research for Advanced Technology project of Japan Science and Technology Agency. He obtained his Ph.D. from the University of Tokyo in the field of Mechanical Engineering in 2003.

E-mail: **suzuki@ist.osaka-u.ac.jp**

**Tetsuya Yomo** is a professor at Osaka University and a research director of Exploratory Research for Advanced Technology project of Japan Science and Technology Agency. He obtained his Ph.D. from Osaka University in the field of Engineering in 1991.

E-mail: **yomo@ist.osaka-u.ac.jp**

**Hiroaki Suzuki**

**Tetsuya Yomo**

# CONSTRUCTIVE APPROACHES FOR THE ORIGIN OF LIFE

**NORIKAZU ICHIHASHI[1], TOMOAKI MATSUURA[1], HIROSHI KITA[1], TAKESHI SUNAMI[1], HIROAKI SUZUKI[1,2], AND TETSUYA YOMO[1,2,3]**

[1]*ERATO Project, Japan Science and Technology Agency, 1-5 Yamadaoka, Suita, Osaka 565-0871, Japan*

[2]*Department of Bioinformatic Engineering, Graduate School of Information Science and Technology, Osaka University, 1-5 Yamadaoka, Suita, Osaka 565-0871, Japan*

[3]*Graduate School of Frontier Biosciences, Osaka University, 1-5 Yamadaoka, Suita, Osaka 565-0871, Japan*

## 1. Introduction

Life is composed of a vast number of molecules, including polynucleotides, proteins, lipids, polysaccharides, and other low-molecular-weight compounds. These molecules are encapsulated within a microsized membrane structure termed "cell" and participate in a myriad of enzymatic reactions. Different sets of these reactions are part of functions essential for life, such as replication and translation of genetic information, metabolism, membrane growth, and division. These essential functions operate in a coordinated manner without inhibiting each other to sustain the cell in a "living state." This integrated cellular organization is surprising because the current technology cannot fabricate such a small multifunctional micromachinery. How could such a cell have arisen on earth? What conditions were required for this purpose? These questions remain to be addressed.

The route that led to the origin of the cell can be divided into the following three stages:

1. Origin of biological molecules
2. Origin of biological functions involving several of these biological molecules
3. Origin of a cell comprising several of these functions

The first stage was the nonbiological (chemical) origin of biological molecules, such as nucleotides, sugars, and amino acids. These molecules have been extensively reviewed in other chapters of this book series. However, the emergence of these biological molecules on earth did not guarantee the emergence of life. Several functions, such as replication of genetic information, transcription and translation of the information into functional molecules, metabolism, self-reproduction, and evolution, are required for nonliving molecules to be considered living. To answer the question regarding the origin of life, we need to understand the origins of these functions, which is the second stage in the origin of cell.

Here, we focus on the second stage. Starting from biological molecules, such as amino acids, nucleic acids, DNA, RNA, lipids, and proteins, we attempt to understand how these molecules constitute cellular functions. A straightforward means to clarify this point is to construct systems to perform cellular functions in vitro, and then examine what systems for such cellular functions we can and cannot construct. This type of approach to understand functions of interest by constructing artificial systems is called "constructive approach." In this chapter, we first define what we mean by constructive approach and then review studies that are regarded to have used constructive approaches to study several cellular functions. We will also identify their achievements and limitations.

## 2. Constructive Approach

In the constructive approach, biological molecules are combined based on existing knowledge to achieve a cellular function of interest. If the function could be performed in the constructed system, the properties of the molecules and the method(s) of combining them are concluded as conditions sufficient for performing the function in such a system (Kaneko, 2006). Such sufficient conditions can be an answer to the question of how biological molecules constitute a cellular function. However, if the function cannot be performed as expected or is inefficient compared to a native cellular function, which is often the case, it indicates the existence of missing conditions and also highlights the inadequacy of our knowledge. Consequently, a constructive approach tells us what is lacking in our understanding of the cell and provides hints to improve our knowledge. A similar approach is bottom-up approach or in vitro synthetic biology (Pohorille and Deamer, 2002; Luisi et al., 2006; Forster and Church, 2007; Liu and Fletcher, 2009).

The constructive approach has two requirements. The first requirement is to use only identified materials, which allows precise determinations of sufficient conditions. The use of a crude material whose composition is unknown hinders the determination of components in the material that are important for a function of interest. The second important requirement is to construct a simple system to perform the function. The simplest system allows the determination of the minimum sufficient condition.

This approach includes both in vitro reconstitutions of native systems operating within the cell and constructions of nonnative systems that do not exist in the cell to perform the functions of interest. DNA replication, for example, can be accomplished both by reconstituting the genomic DNA replication system of *Escherichia coli* (native system) and by performing polymerase chain reaction (PCR) (nonnative system). Reconstitution of a native system is a promising approach to accomplish a function although its complexity and the possible existence of unknown factors hinder its reconstitution. The construction of a nonnative system, in contrast, has no guarantee of accomplishing the function, but it can be used to pursue the simplest mechanism, which is an advantage for understanding the minimum sufficient conditions for the function. As shown in the next section, DNA replication is achieved in a much simpler manner by PCR than by the DNA replication system of *E. coli.*

A number of functions that constitute a cell have been partially performed in vitro, such as replication of genetic information, translation of proteins, and formation of membrane structures encapsulating enzymatic reactions. Further, we provide an overview of previous studies from the view point of the constructive approach.

## 3. Construction of Replication System of Genetic Information

Replication of genetic information is a central function of life. The first system to study replication of genetic information was constructed by Kornberg's group. They combined 12 enzymes and 1 polymerase holoenzyme (a complex of about 10 proteins) to reconstitute the genomic DNA replication system of *Escherichia coli*. This reaction system accomplished the in vitro replication of a plasmid harboring a chromosomal origin of DNA replication (Funnell et al., 1986). The functions of these enzymes were wide ranging and included DNA synthesis, primer RNA synthesis, primer degradation, DNA helix unwinding, helix recoiling, and synthesized DNA ligation. This reconstitution clearly demonstrated that this set of enzymes is sufficient for constructing the DNA replication system of *E. coli*.

In contrast, nonnative systems, which do not exist in the cell, have achieved DNA replication in a much simpler manner. An example of a nonnative system is PCR. To replicate a target DNA, PCR requires one polymerization enzyme, two oligonucleotides as primers, and an artificial thermal cycle (Saiki et al., 1985). In this series of reactions, artificial components or physicochemical manipulation substitute for the roles of enzymes required by the *E. coli* system. For example, heating during the thermal cycle and addition of chemically synthesized oligonucleotides substitute for DNA helix unwinding by helicase and primer synthesis by primase, respectively, in the *E. coli* system.

Another example of a nonnative system used for DNA replication is referred to as self-sustained sequence replication (3SR), which allows isothermal replication via RNA synthesis (Guatelli et al., 1990). 3SR involves three enzymes, namely RNA polymerase, RNA-dependent DNA polymerase (reverse transcriptase), and RNA degradation enzyme (RNaseH), along with two oligonucleotides used as primers. In this reaction, RNA is synthesized from a template DNA by RNA polymerase, after which complementary DNA is synthesized from the RNA by reverse transcriptase to form a DNA/RNA heteroduplex. The RNA strand in the duplex is degraded by RNaseH, and the residual single-stranded DNA is used as a template to produce double-stranded DNA by the DNA-dependent DNA polymerase activity of reverse transcriptase. This results in template DNA replication under isothermal conditions.

A much simpler means to replicate genetic information is to use RNA as a genetic molecule both in native and nonnative systems. In native RNA replication, the replication system of an RNA phage in which neither primers nor thermal cycling is required is constructed in vitro (Mills et al., 1967). An RNA-dependent RNA polymerase (RNA replicase, a complex of four proteins) originating from RNA phage Qβ synthesizes a complementary strand RNA from a template single-stranded RNA that harbors certain RNA sequences. The replicase initiates RNA polymerization without any primers, and the synthesized complementary RNA spontaneously dissociates from the template RNA to become a single strand. Thus, a template RNA autocatalytically replicates under isothermal conditions in the presence of the replicase.

An example of nonnative RNA replication is replication of a self-ligating ribozyme, which is the simplest form of RNA self-replication currently known (Lincoln and Joyce, 2009). This reaction requires four RNA fragments, one template ribozyme, and no proteins. Two of the RNA fragments form the original template ribozyme after ligation, and the other two form RNA that is partially complementary to the template after ligation. In this reaction, the ribozyme ligates the two RNA fragments that are partially complementary to itself. In addition, the ligated RNA also has ligation activity that can ligate the other two RNA fragments that are partially complementary to the initially ligated RNA, which results in copying the original ribozyme.

These examples demonstrated that the system for replication of genetic information could be constructed in a manner that is simpler than the DNA replication system of *E. coli* (Table 1). However, the simpler nonnative systems, such as PCR, 3SR, and RNA replication, are known to have a common problem; the appearance of undesired self-replicating molecules. For example, a nonspecific, smaller DNA fragment is predominantly amplified by the inappropriate annealing of primers or the formation of a primer dimer in PCR. In addition, a "selfish"

**Table 1.** Construction of replication system of genetic information.

| Names | Number of polymers |
|---|---|
| *Native system* | |
| Genomic DNA replication system of *Escherichia coli* (Funnell et al., 1986) | 22 Proteins |
| *Non-native systems* | |
| Polymerase chain reaction (Saiki et al., 1985) | 1 Protein<br>2 Oligonucleotides |
| Self-sustained sequence replication (Guatelli et al., 1990) | 3 Proteins<br>2 Oligonucleotides |
| RNA replication by Qβ replicase (Mills et al., 1967) | 4 Proteins |
| Replication of a self-ligating ribozyme (Lincoln and Joyce, 2009) | 4 Oligonucleotides |

RNA called RNA-Z, which shortcuts the process, is known to appear in 3SR (Breaker and Joyce, 1994). In RNA replication, self-replicating small RNAs called "6S" or "RQ" appear from several sources (Biebricher et al., 1986). These undesired self-replicators replicate more rapidly than the original template due to their small size and consume resources and dominate the population, and thus can be referred to as "parasites" in the replication system. Repressing such parasites is essential for the construction of a replication system that is as continuous as the native system in the cell.

How can a native system repress a parasitic replicator? We can find a hint for this question from the *E. coli* system in which the DNA replication system is encapsulated within a microscale compartment, the cell. In a bulk solution, such as in a test tube, even if a single parasite molecule appears, it replicates rapidly and prevails throughout the entire solution, finally inhibiting the entire target template replication. In contrast, if the reaction is encapsulated within a compartment of an appropriate size, a parasite that appears is confined to a minor fraction of the compartment, and thus, the other major fraction of the compartment is free of the parasite.

This type of compartmental strategy has been successfully applied to PCR (Ghadessy et al., 2001) and RNA replication (Urabe et al., 2010) in which efficient replications of long template DNA and RNA were achieved using water-in-oil emulsions as compartments. These successes of compartmentalization in vitro imply that the native compartmental structure, i.e., the cell membrane, may also contribute to the sustainability of internal replication of genetic information by reducing the possible appearance of parasites.

## 4. Construction of Translation System

Genetic information must be transcribed and translated into functional molecules such as proteins to play its roles. The system is one of the most complex reaction systems in a cell. In 2001, Ueda's group reported the construction of the translation system of *E. coli* using all identified materials (PURE system) (Shimizu et al., 2001), which consisted of ribosomes, NTPs, tRNAs, and 37 proteins, including T7 RNA polymerase, translational initiation proteins, elongation proteins, termination proteins, aminoacyl synthetase, and ATP recycling factors. Although this study clearly demonstrated that these molecules were sufficient to construct the translation system, the translation rate (160 μg/ml/h, corresponding to 1.8 aa/s/ribosome, assuming that all ribosomes were active) (Swartz, 2001) was much lower than that in the translation system in *E. coli* (17 aa/s/ribosome) (Young and Bremer, 1976). The lower translation rate implied the existence of components or conditions that were lacking in the in vitro system.

Our group recently attempted to identify the components that were lacking by introducing all gene products of *E. coli* into the translation system from which we identified several enhancing factors for the in vitro translation system

(Kazuta et al., 2008). In addition, we optimized the concentration of each component by taking into account nonlinear interactions of all components (Matsuura et al., 2009). These comprehensive optimizations have improved the translation rate to a certain extent, but the rate is still less than that of the translation system in *E. coli*. Further studies on other functional molecules and further optimizations will be required to construct a translation system as efficient as that in the cell.

## 5. Construction of Cell Membranes

Cell membranes are primarily composed of phospholipids, cholesterol, and membrane proteins. In the 1960s, phospholipids dispersed in water were reported to spontaneously form bilayer membranes, and these bilayers formed spherical structures called liposomes under certain conditions (Bangham and Horne, 1964). Using these liposomes as a model of the cell membrane, researchers have partially reconstructed several functions of cell membranes in vitro, including membrane transport, cell division, and membrane growth.

The transport of a small compound across the membrane was performed by introducing alpha-hemolysin, a channel protein that is permeable to molecules smaller than 10 kDa, into liposomes (Noireaux and Libchaber, 2004). This transport machinery enabled prolonged internal translation for more than 3 days. Liposome division was partially performed by arranging the *E. coli* FtsZ protein on the liposome to form a contractile circle, which shrank partially, after which liposome constriction was observed (Osawa et al., 2008). Constructing systems to perform cell membrane growth coupled with lipid synthesis is in progress. Enzymes that participate in lipid synthesis, i.e., two acyltransferases, were expressed within liposomes, and phospholipids were synthesized to a certain level. However, the amounts synthesized were insufficient to observe liposome growth (Kuruma et al., 2009).

Under appropriate pH conditions, fatty acids are also known to form bilayered structures and liposome-like vesicles. Luisi's group reported spontaneous vesicle growth by incorporating fatty acids that were dispersed outside the vesicles (Walde et al., 1994b). Szostak's group reported on the coupling of cycles of growth and division using these vesicles (Zhu and Szostak, 2009). Vesicle growth induced a pH gradient across the membrane, the same phenomenon as observed with cell membranes (Chen and Szostak, 2004). Sugawara's group constructed a self-reproducing vesicle using artificial lipids and performed the condensation reaction of lipids on the membrane (Takakura et al., 2003). These constructions demonstrated that a simpler membrane composed of fatty acids could achieve several functions of the cell membrane.

In summary, several systems have been constructed using liposomes and fatty acid vesicles. However, it will be necessary to construct systems for performing other important membrane functions, including selective or active transport and the growth and division of membranes coupled with enzymatic reactions. Native systems perform these functions via a number of membrane-integrated and membrane-associated proteins, such as transporter proteins that form channels in the membrane, and enzymes that catalyze lipid synthesis as described above. Therefore, a plausible method to construct such systems is to isolate these membrane proteins from the cell and to introduce them into liposomes in their functional forms. However, this is usually quite difficult because membrane proteins are insoluble and fragile, and therefore readily aggregate into inactive forms. A new method is needed to overcome these difficulties and to accelerate the construction of systems to perform membrane functions. One candidate approach is the construction of a translocation system involving specialized machinery in the cell to solubilize and insert newly synthesized membrane proteins into the membrane in functional form. The construction of such a system may allow the spontaneous insertion of translated membrane proteins into liposomes, and would therefore increase the availability of membrane proteins for construction of membrane functions (Shimizu et al., 2006).

## 6. Construction of Reactions Inside Liposome

Several systems have been constructed for performing enzymatic reactions, including RNA synthesis (Chakrabarti et al., 1994), PCR (Oberholzer et al., 1995a; Shohda and Sugawara, 2006), protein translation (Oberholzer et al., 1999; Yu et al., 2001; Nomura et al., 2003; Sunami et al., 2010), and a cascading reaction consisting of two successive gene expressions (Ishikawa et al., 2004), inside liposomes. In fatty acid vesicles, several nucleotides have been synthesized by RNA replication (Oberholzer et al., 1995b), template-independent RNA synthesis coupled with vesicle growth (Walde et al., 1994a), and nonenzymatic replication of RNA using activated NTP that entered the membrane from outside the vesicle (Mansy et al., 2008).

To date, several simple enzymatic reactions have been performed inside liposomes or fatty acid vesicles, indicating that conditions sufficient for these reactions are already known. The next step will be the development of reactions for more complex cellular functions by combinations with other functions. Next, we present an example of the recent progress in the constructions of systems for performing more complex cellular functions—replication of genetic information coupled with translation.

As an example of a system for complex cellular function, we constructed a replication system for replication of genetic information coupled with translation of a replication enzyme. Previous constructions of systems to perform replication reactions, such as DNA replication of *E. coli* or PCR, 3SR, and RNA replication, were all performed using extraneously supplied enzymes. The use of extraneously added replication enzymes differentiates these in vitro replication systems from the native replication systems in the cell, where a replication enzyme itself is translated from genetic information.

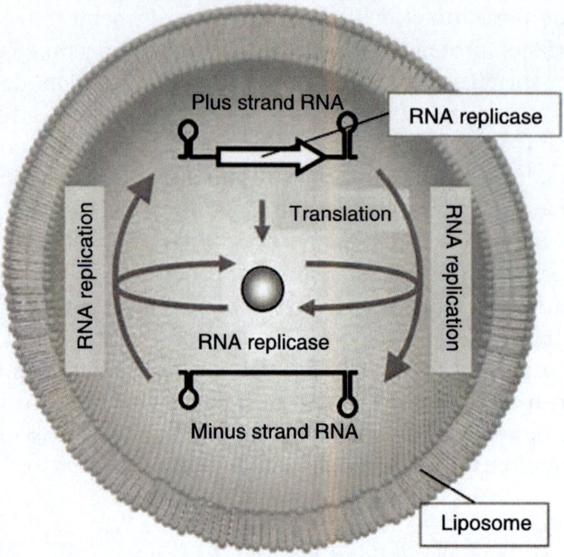


**Figure 1.** An RNA replication system coupled with translation of a replication enzyme in a liposome. The plus strand RNA was encapsulated into liposomes with the reconstituted translation system.

We attempted to combine an RNA replication system with a translation system to construct a more native, cell-like replication system for replication of genetic information. This system consisted of a reconstituted translation system, template RNA encoding RNA-dependent RNA polymerase (RNA replicase), and encapsulating liposomes (Kita et al., 2008). In this reaction system, RNA replicase was translated from the template RNA, which then bound to the recognition sequence on the original template RNA and synthesized the complementary strand RNA. The complementary strand RNA also became a template for the replicase, and the same RNA as the original template RNA was synthesized, resulting in template RNA replication (Fig. 1).

This replication system coupled with a translation system had difficulties in combining different systems. Before starting the construction, we simply thought that increased translation of replicase would result in increased replication and thus increased the ribosome concentration in the translation system. In contrast, the increased ribosome concentration inhibited RNA replication almost completely due to the competition between ribosomes and the replicase for template RNA. This competition was inevitable because both ribosomes and the replicase used the same template RNA and each polymerization reaction—RNA synthesis and translation—progressed in opposite directions. To minimize the negative effect of this competition, we characterized the behavior of the system using a mathematical model that took into account the competition effect, and thus, we determined the optimum balance between translation and replication (Ichihashi et al., 2008).

This result showed us that appropriate balancing of all reactions would be essential to construct a complex system comprised of several subsystems.

The next challenge is inclusion of evolvability into this coupled replication system. We introduced a β-galactosidase gene into the complementary strand RNA to detect RNA synthesis by determining the fluorescence produced by galactosidase activity. We can select a template RNA that has a higher ability to synthesize the complementary strand by collecting liposomes with strong fluorescence by FACS. Since mutations are introduced at the replication step, repeated cycles of replication and selection will let the template RNA evolve.

How can RNA evolve? There would be two evolutionary routes: evolution of activity of the replicase encoded by RNA and evolution of RNA template activity (the activity to be replicated by replicase). We know almost nothing about the evolution of such a multifunctional genetic molecule although natural genetic molecules (e.g., the genomic RNA of an RNA virus) evolve in the same manner. The construction of an evolvable replication system would contribute to broadening our knowledge regarding the evolutionary processes of extant natural systems.

## 7. Conclusion

The constructive approach has performed a number of cellular functions in vitro using identified materials. These approaches demonstrated that the components or their properties that were used to the construction were sufficient for performing the cellular functions of interest. However, functions performed in artificial systems have several shortcomings compared to those performed in the native system of the cell. For example, replication of genetic information is unsustainable due to parasitic replicators, translation rates of in vitro systems are less, growth and division of liposomes have not been accomplished, and undesired competition exists between replication and translation in the replication system coupled with a translation system. These problems would have been the same types of problems encountered by primordial life during the earliest times on earth. Therefore, finding solutions to these problems will contribute to our understanding of what conditions are required for the evolution of extant life forms.

Two possible strategies exist to solve these problems. The first is a rational design in which a researcher describes the behavior of a multifunctional system using a mathematical model to identify the problems to be solved, and then rationally designs a better system. The second is evolutionary engineering in which a researcher selects the best system from a large pool of systems with mutated genetic molecules. The repeated cycles of improvements by the use of these strategies and further comparisons corresponding native systems will provide knowledge regarding how native cells organize multiple systems into an integrated system. This knowledge will finally lead to the construction of an artificial cell or a minimal cell that will reveal the conditions necessary for the emergence of life (Szostak et al., 2001; Forster and Church, 2006; Luisi et al., 2006).

## 8. References


Bangham AD, Horne RW (1964) Negative staining of phospholipids and their structural modification by surface-active agents as observed in the electron microscope. J Mol Biol 8:660–668

Biebricher CK, Eigen M, Luce R (1986) Template-free RNA synthesis by Q beta replicase. Nature 321(6065):89–91

Breaker RR, Joyce GF (1994) Emergence of a replicating species from an *in vitro* RNA evolution reaction. Proc Natl Acad Sci USA 91(13):6093–6097

Chakrabarti AC, Breaker RR, Joyce GF, Deamer DW (1994) Production of RNA by a polymerase protein encapsulated within phospholipid vesicles. J Mol Evol 39(6):555–559

Chen IA, Szostak JW (2004) Membrane growth can generate a transmembrane pH gradient in fatty acid vesicles. Proc Natl Acad Sci USA 101(21):7965–7970

Forster AC, Church GM (2006) Towards synthesis of a minimal cell. Mol Syst Biol 2:45

Forster AC, Church GM (2007) Synthetic biology projects *in vitro*. Genome Res 17(1):1–6

Funnell BE, Baker TA, Kornberg A (1986) Complete enzymatic replication of plasmids containing the origin of the Escherichia coli chromosome. J Biol Chem 261(12):5616–5624

Ghadessy FJ, Ong JL, Holliger P (2001) Directed evolution of polymerase function by compartmentalized self-replication. Proc Natl Acad Sci USA 98(8):4552–4557

Guatelli JC, Whitfield KM, Kwoh DY, Barringer KJ, Richman DD, Gingeras TR (1990) Isothermal, *in vitro* amplification of nucleic acids by a multienzyme reaction modeled after retroviral replication. Proc Natl Acad Sci USA 87(5):1874–1878

Ichihashi N, Matsuura T, Kita H, Hosoda K, Sunami T, Tsukada K, Yomo T (2008) Importance of translation-replication balance for efficient replication by the self-encoded replicase. Chembiochem 9(18):3023–3028

Ishikawa K, Sato K, Shima Y, Urabe I, Yomo T (2004) Expression of a cascading genetic network within liposomes. FEBS Lett 576(3):387–390

Kaneko K (2006) Life: an introduction to complex systems biology. Springer, Berlin

Kazuta Y, Adachi J, Matsuura T, Ono N, Mori H, Yomo T (2008) Comprehensive analysis of the effects of Escherichia coli ORFs on protein translation reaction. Mol Cell Proteomics 7(8):1530–1540

Kita H, Matsuura T, Sunami T, Hosoda K, Ichihashi N, Tsukada K, Urabe I, Yomo T (2008) Replication of genetic information with self-encoded replicase in liposomes. Chembiochem 9(15):2403–2410

Kuruma Y, Stano P, Ueda T, Luisi PL (2009) A synthetic biology approach to the construction of membrane proteins in semi-synthetic minimal cells. Biochim Biophys Acta 1788(2):567–574

Lincoln TA, Joyce GF (2009) Self-sustained replication of an RNA enzyme. Science 323(5918): 1229–1232

Liu AP, Fletcher DA (2009) Biology under construction: *in vitro* reconstitution of cellular function. Nat Rev Mol Cell Biol 10(9):644–650

Luisi PL, Ferri F, Stano P (2006) Approaches to semi-synthetic minimal cells: a review. Naturwissenschaften 93(1):1–13

Mansy SS, Schrum JP, Krishnamurthy M, Tobe S, Treco DA, Szostak JW (2008) Template-directed synthesis of a genetic polymer in a model protocell. Nature 454(7200):122–125

Matsuura T, Kazuta Y, Aita T, Adachi J, Yomo T (2009) Quantifying epistatic interactions among the components constituting the protein translation system. Mol Syst Biol 5:297

Mills DR, Peterson RL, Spiegelman S (1967) An extracellular Darwinian experiment with a self-duplicating nucleic acid molecule. Proc Natl Acad Sci USA 58(1):217–224

Noireaux V, Libchaber A (2004) A vesicle bioreactor as a step toward an artificial cell assembly. Proc Natl Acad Sci USA 101(51):17669–17674

Nomura SM, Tsumoto K, Hamada T, Akiyoshi K, Nakatani Y, Yoshikawa K (2003) Gene expression within cell-sized lipid vesicles. Chembiochem 4(11):1172–1175

Oberholzer T, Albrizio M, Luisi PL (1995a) Polymerase chain reaction in liposomes. Chem Biol 2(10):677–682

Oberholzer T, Wick R, Luisi PL, Biebricher CK (1995b) Enzymatic RNA replication in self-reproducing vesicles: an approach to a minimal cell. Biochem Biophys Res Commun 207(1):250–257
Oberholzer T, Nierhaus KH, Luisi PL (1999) Protein expression in liposomes. Biochem Biophys Res Commun 261(2):238–241
Osawa M, Anderson DE, Erickson HP (2008) Reconstitution of contractile FtsZ rings in liposomes. Science 320(5877):792–794
Pohorille A, Deamer D (2002) Artificial cells: prospects for biotechnology. Trends Biotechnol 20(3):123–128
Saiki RK, Scharf S, Faloona F, Mullis KB, Horn GT, Erlich HA, Arnheim N (1985) Enzymatic amplification of beta-globin genomic sequences and restriction site analysis for diagnosis of sickle cell anemia. Science 230(4732):1350–1354
Shimizu Y, Inoue A, Tomari Y, Suzuki T, Yokogawa T, Nishikawa K, Ueda T (2001) Cell-free translation reconstituted with purified components. Nat Biotechnol 19(8):751–755
Shimizu Y, Kuruma Y, Ying BW, Umekage S, Ueda T (2006) Cell-free translation systems for protein engineering. FEBS J 273(18):4133–4140
Shohda K, Sugawara T (2006) DNA polymerization on the inner surface of a giant liposome for synthesizing an artificial cell model. Soft Matter 2(5):402–408
Sunami T, Hosoda K, Suzuki H, Matsuura T, Yomo T (2010) Cellular compartment model for exploring the effect of the lipidic membrane on the kinetics of encapsulated biochemical reactions. Langmuir (in press)
Swartz J (2001) A PURE approach to constructive biology. Nat Biotechnol 19(8):732–733
Szostak JW, Bartel DP, Luisi PL (2001) Synthesizing life. Nature 409(6818):387–390
Takakura K, Toyota T, Sugawara T (2003) A novel system of self-reproducing giant vesicles. J Am Chem Soc 125(27):8134–8140
Urabe H, Ichihashi N, Matsuura T, Hosoda K, Kazuta Y, Kita H, Yomo T (2010) Compartmentalization in a water-in-oil emulsion repressed the spontaneous amplification of RNA by Qβ replicase. Biochemistry 49(9):1809–1813
Walde P, Goto A, Monnard PA, Wessicken M, Luisi PL (1994a) Oparins reactions revisited – enzymatic-synthesis of poly(adenylic acid) in micelles and self-reproducing vesicles. J Am Chem Soc 116(17):7541–7547
Walde P, Wick R, Fresta M, Mangone A, Luisi PL (1994b) Autopoietic self-reproduction of fatty acid vesicles. J Am Chem Soc 116:11649–11654
Young R, Bremer H (1976) Polypeptide-chain-elongation rate in Escherichia coli B/r as a function of growth rate. Biochem J 160(2):185–194
Yu W, Sato K, Wakabayashi M, Nakaishi T, Ko-Mitamura EP, Shima Y, Urabe I, Yomo T (2001) Synthesis of functional protein in liposome. J Biosci Bioeng 92(6):590–593
Zhu TF, Szostak JW (2009) Coupled growth and division of model protocell membranes. J Am Chem Soc 131(15):5705–5713

Biodata of **Vladimir Kompanichenko**, author of "***Origin of Life by Thermodynamic Inversion: A Universal Process.***"

**Professor Vladimir Kompanichenko** is currently the Head of the Biogeochemical Group in the Institute for Complex Analysis of Regional Problems of the Russian Academy of Sciences in Birobidzhan, Russia. He obtained his Ph.D. from the Far Eastern Geological Institute in Vladivostok in 1983. Professor Kompanichenko's scientific interests are in the areas of the origin of life, astrobiology, and biogeochemical and geophysical exploration of hydrothermal systems.

E-mail: **kompanv@yandex.ru**

J. Seckbach (ed.), *Genesis - In The Beginning: Precursors of Life, Chemical Models and Early Biological Evolution*, Cellular Origin, Life in Extreme Habitats and Astrobiology 22, 305–320
DOI 10.1007/978-94-007-2941-4_18, 

# ORIGIN OF LIFE BY THERMODYNAMIC INVERSION: A UNIVERSAL PROCESS

**VLADIMIR KOMPANICHENKO**
*Department of Genetics and Evolution, Institute for Complex Analysis, 4 Sholom Aleyhem St., Birobidzhan 67016, Russia*

## 1. Introduction

It follows from the inversion approach described here that a necessary condition for the origin of life—fluctuations of thermodynamic and physicochemical parameters in the prebiotic environment—must be added to the consensus requirement of liquid water, availability of organic matter, and a source of energy. Such fluctuations are essential to drive the transition of prebiotic microsystems into the first living organisms.

## 2. Thermodynamic Background: The Inversion Approach to the Origin of Life

It is the consensus view that the general thermodynamic trend of biological evolution can be expressed in terms of entropy (*S*), free energy (*F*), and information (*I*). Free energy drives the course of evolution in the biosphere (f.i.: Vernadsky, 1980; de Duve, 2002). Entropy changes occur within the range of negative values because the entropy export from biological systems is more efficient than the internal entropy production, and a variety of negative entropy changes (negentropy) can be correlated with accumulating genetic information. Thus, Ashby (1959) and Brullien (1966) construe information as negentropy based on the similarities of "entropy" in thermodynamics and information theory. The thermodynamic trend corresponds with the ability of living systems to concentrate free energy and information, which is considered a fundamental biological property. (See Kompanichenko, 2008, for review.) This ability allows biological systems to accumulate an excess of free energy and information with respect to the environment as well as maintaining the corresponding positive F and I gradients.

The thermodynamic trend of biological evolution is unique. Only biological systems are able to extract free energy and information from the environment at the expense of its own activity. Some natural systems, for instance, a volcano or star, can accumulate free energy and keep an excess of free energy with respect to the surroundings. However, both of these natural systems enrich the surroundings with energy, while a biological system actively extracts energy (Kompanichenko, 2003). A living organism is able to be active within its environment because of the internal location of its entropy pump (Ebeling et al., 1990), whereas the entropy pump that sustains a geological system like a volcano or the entire Earth is external. That means that geological or cosmic systems do not possess internal (low entropy) structures, but instead, the export of entropy is possible due to the direct inflow of energy from a molten planetary interior, such as a volcanic system, or because a vast amount of energy is produced inside the system (a star).

Biological evolution proceeds in a thermodynamic direction that is opposite to evolution of nonbiological natural systems. This direction consists in concentration (accumulation) of free energy and information that is correlated with displacement of entropy in the area of negative values. Therefore, the start of biological evolution (through the transformation of nonbiological organic microsystems into initial living units) implies the *inversion* of the general trends of free energy change (from dissipation to continuous increase), information change (from spontaneous loss to continuing accumulation), and entropy change (from positive to negative values). This thermodynamic inversion can be mathematically expressed as follows:

$$(n-bs)\,F^{+} < F^{-} \rightarrow \left(F^{+} \approx F^{-},\ \text{inversion moment}\right) \rightarrow F^{+} > F^{-}\,(bs), \qquad (1)$$

where $F^{+}$ is the input of free energy in a system, both at the expense of internal production and extraction from the outside world; $F^{-}$ represents loss of free energy in a nonbiological or biological system, both at the expense of internal devalues and dissipation in the outside world; *n–bs* represents nonbiological systems, and *bs*, biological systems:

$$(n-bs)\,I^{+} < I^{-} \rightarrow \left(I^{+} \approx I^{-},\ \text{inversion moment}\right) \rightarrow I^{+} > I^{-}\,(bs), \qquad (2)$$

where $I^{+}$ is the input of information into a system, and $I^{-}$ is the loss of information from a system:

$$\begin{aligned} &(n-bs)\,S > 0 \rightarrow \left(S \approx 0,\ \text{inversion moment}\right) \rightarrow S < 0\,(bs),\\ &\text{or: } (n-bs)\,S^{+} \rightarrow \left(S^{+} \approx S^{-},\ \text{inversion moment}\right) \rightarrow S^{-}\,(bs), \end{aligned} \qquad (3)$$

where $S$ is entropy, $S^{+}$ is the input of (positive) entropy in a system, and $S^{-}$ is the input of negentropy in a system (both at the expense of internal production and extraction from the outside world).

The inversion can be approximately expressed through the balance "free energy contribution plus information contribution to entropy contribution" ($F_c + I_c/S_c$). The prevalence of free energy and information contributions over the entropy contribution allows the existence of a biological system. So, the negentropy trend of biological evolution is related with the positive balance $F_c + I_c/S_c$. Nevertheless, the balance $F_c + I_c/S_c$ is relative: the absolute equality in this ratio is unattainable because free energy, information, and entropy are measured in different units. It follows that a certain level of uncertainty exists close to the inversion point $F_c + I_c \approx S_c$.

There exist two opposite types of universal processes: *spontaneous*, or *basic*, and *nonspontaneous*, or *coupled*. In the long run, spontaneous processes lead to disappearance of various gradients (heat, substance, electric charge, etc.) in a system and its transition into the state that is characterized by maximum entropy and disorder. The counter nonspontaneous processes are responsible for transference of heat, substance, electric charge, etc. against the gradients. Contribution of (positive) entropy results in strengthening of spontaneous processes, while inputs of free energy and information support nonspontaneous ones. The positive balance $F_c + I_c/S_c$ means that efficiency of the nonspontaneous processes in a system exceeds efficiency of the spontaneous ones.

The negentropy trend of biological evolution correlates with four key biological properties. They are (1) the ability to concentrate free energy and information (by means of their active extraction from the environment); (2) the ability to exhibit an intensified counteraction to external influences; (3) expedient behavior; and (4) regular self-renovation on the various levels, including self-reproduction (Kompanichenko, 2008, 2009). The essence of the properties is that a viable living system concentrates free energy and information because it can react to external changes and extend its existence through self-renovation. In other words, a living organism is able to maintain positive F and I gradients with respect to the environment. By means of the positive free energy gradient, a viable living system transforms the environment more efficiently than the environment transforms it. Due to the positive information gradient, a viable biological system knows about the environment more than the environment "knows" about it. So, the first key property directly expresses sense of the negentropy trend, while the second and third properties indicate the mechanism of concentration of free energy and information.

The inversion of the thermodynamic trend during the transition from nonbiological to biological organization can be interpreted as follows. The inversion is the transformation of a system from a passive position in respect to the surroundings to the active one. At the point in time when prebiotic organic microsystems transform into initial biotic units, they arise as "centers of activity" in the medium. The primary living units and their communities begin to enhance the excess of free energy and information with respect to the surroundings by increasing the positive energetic and informational gradients. The inversion of the thermodynamic trend must be related to the formation of specific internal structures. To indicate such specific structures, E. Galimov (2006) used the term *low-entropy structures* that produce low-entropy products. These low-entropy structures were organized to maintain the existence of the entropy pump that supported the efficient entropy export. As cited above, entropy export must exceed a certain critical value in order to begin. Such overcritical impulse must be provided by the pumping of free energy (with or without information) into a system (Ebeling et al., 1990). Such an impulse was constructive in itself because it was directed against disorganization (entropy rise) and connected with nonspontaneous processes. In the framework of biology, the notion about radical constructive transformations of biological systems has been formulated by Baltscheffsky (1997). He called such constructive impulses *anastrophe* (opposite to *catastrophe*). It seems that the question concerning the inversion of the thermodynamic trend (i.e., inversion of the balance between basic and coupled processes) boils down to the nature of the first anastrophe that laid the foundation of life on Earth.

The field of uncertainty close to the relative balance $F_c + I_c \approx S_c$ leads to the existence of a heterogeneous system in which simultaneously some structures mainly produce entropy, and others, free energy and information (with the total negative balance $F_c + I_c < S_c$). This situation is characteristic of self-organizing systems that exist far from equilibrium. In accordance with the proposed concept, only far-from-equilibrium conditions might generate the powerful negentropy impulse that could overcome the entropy barrier and initiate the entropy pump in prebiotic microsystems. Behavior of a chemical system under conditions far from equilibrium radically differs from behavior under conditions near equilibrium. Nonequilibrium conditions are responsible for transformations that radically change a system's structure. The mechanism of radical transformations of natural systems under conditions far from equilibrium is well explored in the framework of the theory of dissipative structures and synergetics (Nicolis and Prigogine, 1977; Haken, 1978, 2003; Prigogine and Stengers, 1984; Ebeling et al., 1990). This mechanism is a bifurcation. A bifurcation occurs when a system cannot exist further in a given condition due to changes in the surroundings or inner mutations. A bifurcate transition is shown on Fig. 1.

The universal scheme of bifurcate transition of a system includes the following features: stable existence of a system → rise of instability through the powerful fluctuations → the highest point of instability (bifurcation, or critical point), radical change of the system's structure → choice of new way of development → the next period of its stable existence (Fig. 1a). At the bifurcation point, a system undergoes multiple accidental changes that may influence the direction of further development. That is why a system's development can bifurcate at the moment of its highest instability. Finally, the system has followed one of many permissible directions, which can be understood in terms of two principal trends: (1) to increasing complexity through self-organization (Trend A) and (2) to simplification and degradation (Trend B), up to full destruction (Trend B′). Reverse changes may initiate the back transition of the system in accordance with the trends C′ or C″ (Fig. 1b).

When approaching the bifurcation point, a system acquires some specific properties that occur only at the bifurcate transition: there appear heterogeneous structure, continuous macrofluctuations, cooperative events, and constant exchange of matter and energy with the surroundings. Such properties lie in the deep background of life. To keep these properties, a prebiotic microsystem must continuously maintain its own nonequilibrium state through oscillations around the bifurcation point (Kompanichenko, 2008, 2009). In this case, the system takes the intermediate position between the initial and potential advanced states (Fig. 1). Such oscillating type of a natural system was called *bistate* system (Kompanichenko, 2004).

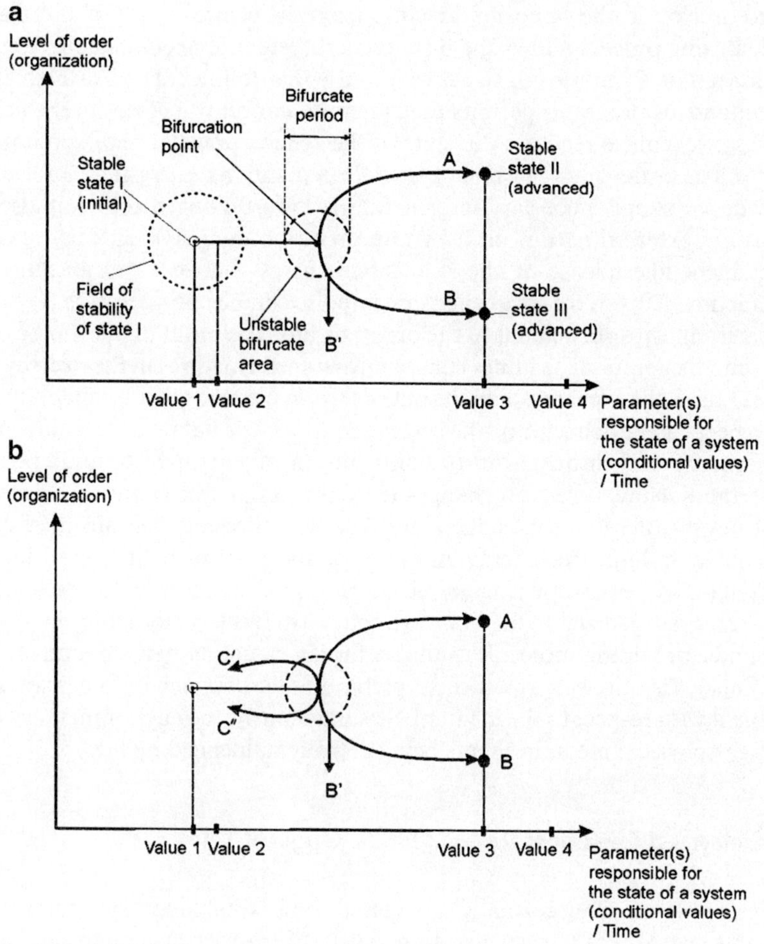


**Figure 1.** Principal scheme of bifurcate transition of a natural system under far-from-equilibrium conditions. (**a**) Direct transition from the initial stable state into one of permissible advanced stable states due to changed conditions in the outside world; (**b**) direct and reverse transitions (in case oscillating conditions in the outside world). *A*—trend to advanced higher-organized state; *B*—trend to advanced lower-organized state; *B'*—trend to complete destruction; and *C'* and *C''* —reverse trend to the approximately initial state.

A prebiotic bistate microsystem composed of diverse organic matter is a suitable candidate for the inversion process from several points of view. The microsystem is an open oscillating unit in which incessant reorganization of the molecules takes place. Its heterogeneous structure allows spontaneous processes to occur which decrease the energetic gradients and nonspontaneous ones that restore the gradients through self-organization. Also, the tendency to dichotomy appears in a prebiotic system having a bistate status: it simultaneously tends to develop in the opposite direction from the central point of bifurcation. The organizing process allows the bistate microsystem to accumulate free energy and information from the constructive impulse that follows. Thus, availability of nonspontaneous processes permits the system to concentrate free energy in local networks of chemical reactions, in spite of the general prevalence of spontaneous processes. The bistate prebiotic microsystems can pump free energy and information to provide the supply necessary for generating powerful negentropy impulses.

Strong external action on a bistate prebiotic microsystem may result in the avalanche-like release of the stored free energy that leads to an *intensified* counteraction. This overcritical energetic impulse can be considered as a basis for the greatest anastrophe related to the origin of life. The input of the stored information into the impulse facilitates constructive transformations in the microsystem. The input turns the exchange with the outside world into the most profitable way for initiating expedient behavior of the system, in which the balance of "contribution of free energy and information to contribution of entropy" become positive. The thermodynamic inversion changes the direction of free energy, information, and entropy transfer in the interchanging processes between the bistate microsystem and its surroundings: free energy and information start to be imported into the microsystem, while entropy is exported outside. Simultaneously, the exchange of substances is reorganized to facilitate extraction from the surrounding energy-rich compounds, providing molecules suitable for constructing new structures in the microsystem. In this way, the inverse prebiotic microsystem becomes an active constituent with respect to the medium. Simultaneously, the environment becomes part of the physical medium that is being actively influenced by life.

## 3. Hydrothermal Conditions Relevant for the Origin of Life

Two alternative media are usually considered to be suitable for the origin of life: ocean and hydrothermal systems. Both satisfy the consensus conditions for the origin of life: liquid water, availability of organic matter, and a source of energy. The fourth necessary condition follows the approach described here, that the medium must be undergoing fluctuations in order to drive increased complexity of organic structures in which prebiotic microsystems transition from the stable state into the unstable (bifurcate) state.

The scale of thermodynamic and physicochemical fluctuations depends on values of the corresponding gradients. The Earth's oceans (beyond the region of hydrothermal vents) are characterized by very low gradients. For instance, the minimal and maximal values of salinity are 30 and 38 g/l, and the pH ranges between 8.0 and 8.4. The low gradients provide only small amplitudes of fluctuations in the ocean. The scale of fluctuations in hydrothermal systems is diverse and ranges from very low to extraordinarily high, corresponding to extremes of pH (1–12), salinity (1–500 g/l), and temperature (30–500°C). To obtain information about such fluctuations in nature, we have mathematically processed changeability of pressure of hot water and water-steam mixture in Mutnovsky and Pauzhetsky hydrothermal fields in the Kamchatka peninsula. The period of pressure micro-oscillations in these fields ranges from 10 to 60 min. In Mura hydrothermal field, Slovenia, the microoscillations of thermodynamic and other physicochemical parameters of hot water are correlated with a period of about 70 min (Kralj and Kralj, 2000). Unlike the terrestrial hydrothermal systems, the changeability in submarine vents is usually minor but can increase dramatically due to nearby earthquakes or volcanic eruptions. In this connection, the author agrees with other investigators that hydrothermal conditions are the most probable medium for the origin of life on the early Earth (Corliss et al., 1981; Holm and Andersson, 2005; Russell et al., 2005; Washington, 2000). The preferable geological settings for the origin of life are defined as the following: hydrothermal systems themselves (beneath the Earth's surface), high-temperature discharges on oceanic floor, littoral discharges of hot water (where both the geothermal energy and sunlight are available), and hydrothermal discharges in groundwater aquifers within the continents.

## 4. Prebiotic Chemistry and Inversion Chemistry

In accordance with the inversion approach, it is reasonable to subdivide prebiotic chemical reactions into two categories:

(A) The prebiotic chemistry related to synthesis, accumulation, and degradation of organic matter in the context of the origin of life, including formation of various prebiotic microsystems

(B) The inversion chemistry connecting prebiotic chemistry and early biochemistry in the framework of which the chemical transformation (inversion) of prebiotic microsystems into the simplest living units are investigated

### 4.1. PREBIOTIC CHEMISTRY IN NATURAL AND MODEL HYDROTHERMAL ENVIRONMENTS

Hydrocarbons (including aromatics), lipid compounds, and simple amino acids under certain conditions can be synthesized and are relatively stable in high-temperature aqueous medium. They may exist in soluble and insoluble forms in a fluid with the temperature 200–300°C and below. The capacity of these insoluble organics to form three-dimensional microsystems is experimentally proven. Microdroplet emulsions (composed mainly of hydrocarbons) appear as a result of phase separation of oil from water in the process of temperature reduction of the fluid to about 200–300°C (Simoneit, 2003). The temperature range of 150–200°C is suitable for self-assembly of amino acids into proteinoid microspheres and lipids into liposomes if concentrations of these molecules reach a certain critical value (Fox and Dose, 1977; Deamer et al., 2002; Simoneit, 2004). Recent reviews of the research on prebiotic chemistry in natural and simulating hydrothermal environments have been published (Simoneit, 2004; Holm and Andersson, 2005).

Synthesis of ribose and other sugars, nucleotides, and nucleic acids in hot aqueous medium is much more difficult and demands very specific conditions. Ribose, an essential component of nucleotides, is unstable in hot aqueous solutions. Its half-life at pH 7 and 100°C is 73 min (Larralde et al., 1995). Furthermore, abiotic condensation reactions required to produce polymers in aqueous solutions face a very high thermodynamic barrier because hydrolysis is favored. The upper temperature limit for synthesis of nucleic acids is usually 50–60°C, in the rare cases up to 90°C. Thus, heating adenine and pyruvic aldehyde in concentrated aqueous solution to 60°C led to the complete disappearance of adenine in few hours (Vergne et al., 2000).

Laboratory experiments on prebiotic chemistry are essential for better understanding of origin-of-life processes. However, to make advancement in this field, we should take into consideration at least two distinctions between nature and laboratory simulations. First, the high-temperature limit for synthesis of nucleotide chains (50–60°C) concerns the processes in vitro, i.e., in a chemical system that does not possess even primary properties of the living state. The same synthesis in vivo, i.e., in a biochemical system in a living microorganism, sometimes proceeds at much higher temperatures. Thus, some species of Archaea can grow at 105–110°C and even higher (Stetter, 1995). Second, the researchers exploring the synthesis and autoreplication of RNA came to the conclusion that this cannot occur without the aid of an experimenter (Joyce et al., 1987; Horgan, 1991). However, according to the inversion approach described here, high-temperature synthesis of nucleotides could be possible if it is governed by the negentropy motive power that appeared in the microsystems just at the inversion moment. This nonspontaneous motive power might play the role of a "natural experimenter."

In summary, three-dimensional aggregates composed mainly of hydrocarbons, lipids, and simple amino acids (and/or their precursors) in various proportions and combinations should be considered as the most suitable initial prebiotic microsystems. It follows that such initial organic microsystems might assemble within the approximate range of temperatures near 100°C. The ability of organic substances to segregate from the aqueous phase under certain thermodynamic and physicochemical conditions is a characteristic feature of hydrothermal fluid.

## 4.2. INVERSION CHEMISTRY

In a living cell, a signal (energy impulse) intensifies cascades of biochemical reactions in which one product of a chemical reaction catalyzes synthesis of another product, which catalyzes synthesis of the next and so on (Fedonkin, 2008). According to the concept proposed here, the original cascade of biochemical reactions appeared in the course of avalanche-like release of the stored free energy in prebiotic bistate microsystems during their jump over the entropy barrier.

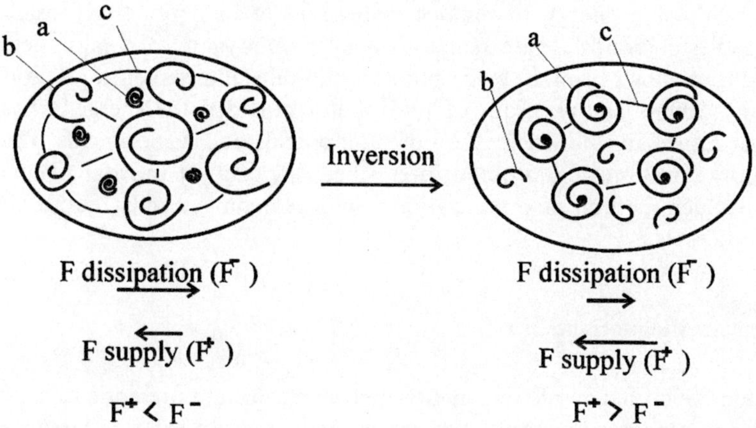


**Figure 2.** Inversion of the balance between the contribution of spontaneous processes and contribution of nonspontaneous processes in a chemical system. *Left* a system with prevalence of spontaneous processes, *right* a system with prevalence of nonspontaneous processes. *a*—areas with prevalence of nonspontaneous processes containing low-entropy structures that produce free energy and information, *b*—areas with prevalence of spontaneous processes containing high-entropy structures, *c*—associative links between high-entropy structures and low-entropy structures. *F*—free energy, $F^+$—input of free energy, and $F^-$—loss (dissipation) of free energy. .

The multistep release of free energy available in the network of prebiotic chemical reactions changed the ratio between spontaneous and nonspontaneous processes in favor of the latter. The additional inflow of energy tended to turn autocatalytic reactions into "ascending" cascades, which produced more and more free energy, and the universal effect of intensification that is characteristic of all living systems started from this moment. Under such transformation, the local low-entropy structures expanded and organized into the primary biological unit—probiont (Fig. 2). The surplus free energy began to circulate within the network of biochemical reactions and maintained the positive free energy gradient and information content of probiont with respect to the environment.

The extracted high-energy material was incorporated into the network of biochemical reactions and used to produce free energy that was then stored in the living system. The necessity to develop biochemical systems responsible for preservation and the release of surplus energy led to the expanded synthesis of carbohydrates, which are the most suitable molecules for this purpose. Simultaneously, the structure and functions of the lipid bilayers were reorganized to facilitate selective extraction of high-energy molecules, thus protecting the probionts from spontaneous loss of energy. Continuing transfer of protons gave rise to more efficient reactions of organic phosphates that were coupled to redox reactions, perhaps including pyrophosphate bonds of ATP. Hydrolysis of ATP proceeds with release of energy and uptake of water. In contrast, condensation reactions producing biopolymers (peptides, polysaccharides, polynucleotides) proceed by using energy to remove water and form peptide, glycoside, and phosphodiester bonds. These reactions are thermodynamically and chemically coupled that allows primary living units to minimize internal entropy producing (Galimov, 2006). The reactions of phosphates provided the energetic basis for oligomerization and polymerization of amino acids and nucleic acids. The integrative organization of a probiont prescribed any catalyst to catalyze a certain reaction. Such method of organization could be arranged only by bioinformational processes.

## 5. Origin of Bioinformation

The intensive exchange of information between a bistate prebiotic microsystem and its surroundings represents the highest sensitivity to external changes. Any change of the external conditions exerts a certain influence upon the microsystem, deforming its structure and the network of chemical reactions. The deformation may consist of modifying organic molecules, appearance of new pathways of chemical reactions, etc. In effect, this deformation reflects the change that occurred in the surroundings. In fact, the incessant external influences on an oscillating microsystem represent flow of information with the changing outside world. The influences are fixed in the heterogeneous structure of the bistate microsystem. The flow of structural changes initiated by external fluctuations travels and waves from the periphery to the center of the microsystem (Fig. 3a). Oligomeric and polymeric chains of amino acids possessing the highest lability seem the most appropriate molecules to maintain structural information.

When a probiont forms a prebiotic bistate microsystem during an inversion, it must accumulate bioinformation in order to develop further. The accumulation

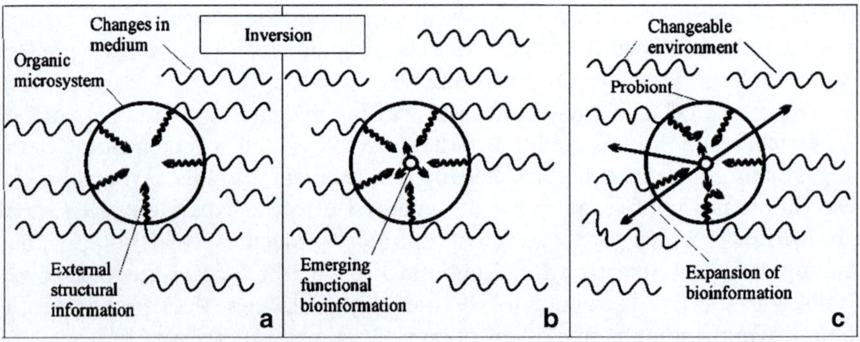


**Figure 3.** Emergence of functional bioinformation in a probiont. (**a**) Penetration of spontaneous flows of structural information in a prebiotic bistate microsystem from the changeable surroundings; (**b**) formation of the circulative informational core in a probiont and arise of the counter nonspontaneous bioinformational flows; and (**c**) expansion of bioinformation in the environment.

implies a succession of new (bio)information over the previous one. Informational capacity of oligomers or polymers of amino acids is low because of their lability, while chains of nucleic acids are more suitable for storage of bioinformation. Therefore, the existence and further evolution of a probiont depended on the availability of nucleic acids or their precursors. There were three possible sources of these molecules: they could be present in the composition of earlier prebiotic microsystems, extracted from the outside world through the lipid bilayer membrane (Deamer, 2004), or appeared as coproducts of (bio)chemical reactions (Fox and Dose, 1977).

High concentrations of (bio)information in a probiont could be achieved by means of algorithmic compression. Primary (bio)information contained in the initial sequences of amino acids were in some way transmitted into the initial sequences of nucleic acids (or their precursors) for storage. Step by step, the primary chains of amino acids catalyzed these molecules and transmitted the (bio) information in its own sequences into the nucleic acids. This (bio)information was conserved in the nucleotide sequences and then reflected back as the instructions for synthesis of the corresponding amino acid sequences. The rise of the bioinformation in a probiont led to accretion of the nucleoprotein interaction through the formation of increasingly complex intermediates. The dissymmetrical heterogeneous structure of a probiont facilitated formation of the stereo specificity—combination of L-amino acids and D-sugars that enhanced bioinformational capacity. In the case of spontaneous alternation of D- and L-sugars in the nucleic acids, the complementary pairs could not be formed. Due to the dissymmetrical macrostructure, the bioinformation concentrated in the core was divided by the plane of dissymmetry; in present cells, this dissymmetrical core has shaped the DNA helix. Growth of a probiont in size was associated with accumulation of high-energy compounds and bioinformation, then completed by its final division that began with the breakage of weak chemical bonds along the plane of dissymmetry.

From the thermodynamic point of view, bioinformation is characterized by the positive balance of "contribution of information/contribution of informational entropy" in a biological system. The surplus information provides bioinformation with a new quality in comparison to physicodynamical (or structural) information of nonliving systems. Compression of information is the algorithmic process. The reverse flow is reflected from the most compressed bioinformation located in the core of a probiont (Fig. 3b–c). It follows that the principal peculiarity of bioinformation consists in its *reflex,* or *inverse,* nature. The expanding circulation of bioinformation through its multifold reflection was related to the development of nucleoprotein interactions. The circulation proceeded on the background of the general bioinformational flow emerging from the most compressed block of bioinformation (in the direction "DNA → RNA → proteins" in modern cells).

The inverse deployment of the most algorithmically compressed bioinformation naturally proceeded under algorithmic control, through programming and prescriptive instructions. Functional processes arose from the algorithmic control and connected the main types of biopolymers—sequences of nucleotides and proteins. In this way, a functional sequence could be a gene or a protein. Unlike a polynucleotide or proteinoid chain artificially synthesized in vitro, a functional sequence is meaningful. Functional sequences produce function at their destination (binding site or ribosomal translation site). Function is optimized algorithmically. Algorithmic optimization (selection of each symbol in the sequence specifically for function) occurred in the course of internal circulation of bioinformation, through continuous recombination and selection. The relevant specificity of biological information has been considered by previous authors (Abel and Trevors, 2005; Durston et al., 2007).

## 6. Universal Aspects of the Inversion Approach

Most of the scientific knowledge in background of the proposed inversion approach to the origin of life applies not only to our planet but also to processes in habitable planets beyond the Earth. The key notions of "free energy" and "entropy" are universal and characterize the ability of a natural system to carry out work. For instance, eruptions of active volcanoes on Earth and on Jupiter's satellite Io can be approximately evaluated in terms of the free energy expenditure. The opposite spontaneous and nonspontaneous processes are universal as well. They can be used to describe the transition of a natural system to more or less probable corresponding state. Cooling of a lava flow on Earth and cooling of the entire Mars planet are similar phenomena related to the universal spontaneous process of heat radiation. All stars, planets, and other space bodies in the universe are composed of about one hundred elements. In this context, the behavior of chemical systems elsewhere in the universe should comply with fundamental laws of chemistry and physics, in particular with the Le Chatelier principle, and the theory of dissipative structures and synergetics, which are incorporated in the inversion approach. Organic microsystems are considered as prebiotic for the early Earth and are composed of diverse organic compounds that are also present in dense interstellar molecular clouds and in meteorites. Taking these considerations into account, the transition from organic microsystems to simplest living units in principle can be understood in the context of a general scenario of the origin of life in the universe.

Based on the universal aspects of the inversion approach, fluctuating hydrothermal conditions could serve as a potential cradle of life on any habitable planet. Volcanic activity and the associated hydrothermal processes occur on Earth and other planets in the Solar system. Thus, the diapirs in the ice crust of the Jupiter's satellite Europa represent themselves the intrusions of warm ice in the circumjacent cold ice (Rathbun and Squyres, 1998). Such diapirs may appear at the expense of intensive heat flow rising from the liquid ocean that is probably located beneath the ice cover. Another example is the discovered signs of relatively recent (10–100 Ma ago or less) volcanic and water flows in Elysium and

Amazonis Planitia regions of Mars, as reported by Sakimoto (2001). Some ancient thermal springs were found and explored in Vernal Crater, Arabia Terra on Mars (Allen and Oehler, 2008).

In conclusion, volcanic processes are universal and are accompanied by continuous thermodynamic and physicochemical fluctuations. According to the inversion approach, hydrothermal conditions associated with volcanism can provide a significant source of energy to drive chemical reactions that inevitably lead to the complex organic microsystems necessary for the origin of life.

## 7. References


Abel DL, Trevors JT (2005) Three subsets of sequence complexity and their relevance to biopolymeric information. Theor Biol Med Model 2:29. doi:10.1186/1742-4682-2-29

Allen CC, Oehler DZ (2008) A case for ancient springs in Arabia terra, Mars. Astrobiology 8(6):1093–1112

Ashby WR (1959) Introduction in cybernetics. Mir, Moscow

Baltscheffsky H (1997) Major “anastrophes” in the origin and early evolution of biological energy conversion. J Theor Biol 187:495–501

Brullien L (1966) Scientific uncertainty and information. Mir, Moscow

Corliss JB, Baross JA, Hoffman SE (1981) An hypothesis concerning the relationship between submarine hot springs and the origin of life on the Earth. Oceanol Acta SP 4:59–69

De Duve C (2002) Life is what is common to all living beings. In: Palyi G, Zucci C, Caglioti L (eds) Fundamentals of life. Elsevier, Paris, pp 26–27

Deamer DW (2004) Prebiotic amphiphilic compounds. In: Seckbach J (ed) Origins. Kluwer, Amsterdam, pp 75–89

Deamer D, Dworkin JP, Sandford SA, Bernstein MP, Allamandola LJ (2002) The first cell membranes. Astrobiology 2:371–382

Durston KK, Chiu DKJ, Abel DL, Trevors JT (2007) Measuring the functional sequence complexity of proteins. Theor Biol Med Model 4:47. doi:10.1186/1742-4682-4-47

Ebeling W, Engel A, Feistel R (1990) Physik der evolutionsprozesse. Akademie, Berlin (In German)

Fedonkin MA (2008) Ancient biosphere: the origin, trends and events. Russ J Earth Sci 10. doi:10.2205/2007ES000252

Fox S, Dose K (1977) Molecular evolution and the origin of life. Dekker, New York

Galimov EM (2006) Phenomenon of life: between equilibrium and nonlinearity: origin and principles of evolution. Editorial URSS, Moscow (In Russian)

Haken H (1978) Synergetics. Springer, Berlin/New York

Haken H (2003) Nonlinear Phenomena in Complex Systems. Special issue 5(4)

Holm N, Andersson E (2005) Hydrothermal simulation experiments as a tool for studies for the origin of life on Earth and other terrestrial planets: a review. Astrobiology 5(4):444–460

Horgan J (1991) Near cradle of life. Sci Am 4:68–79

Joyce GF, Schwartz AW, Miller SL, Orgel LE (1987) The case for an ancestral genetic system involving simple analogs of nucleotides. Proc Natl Acad Sci USA 84:4398–4402

Kompanichenko VN (2003) Distinctive properties of biological systems: the all-round comparison with other natural systems. Front Perspect 12(1):23–35

Kompanichenko VN (2004) Systemic approach to the origin of life. Front Perspect 13(1):22–40

Kompanichenko VN (2008) Three stages of the origin-of-life process: bifurcation, stabilization and inversion. Int J Astrobiol 7(1):27–46

Kompanichenko VN (2009) Changeable hydrothermal media as a potential cradle of life on a planet. Planet Space Sci 57:468–476

Kralj P, Kralj P (2000) Thermal and mineral waters in north-eastern Slovenia. Environ Geol 39(5):488–498

Larralde R, Robertson M, Miller S (1995) Rates of decomposition of ribose and other sugars: implications for chemical evolution. Proc Natl Acad Sci USA 92:8158–8160

Nicolis G, Prigogine I (1977) Self-organization in nonequilibrium systems. Wiley, New York

Prigogine I, Stengers I (1984) Order out of chaos. Bantam, New York

Rathbun JM Jr, Squyres S (1998) Ice Diapirs on Europa: implications for liquid water. Geophys Res Lett 25(22):4157–4160

Russell MJ, Hall AJ, Boyce AJ, Fallick AE (2005) On hydrothermal convection and the emergence of life. Econ Geol 100:419–438

Sakimoto S (2001) Volcanoes still active on Mars? New evidence for ongoing volcanism and water release. Abstracts of the Geological Society of America annual meeting, Boston

Simoneit BRT (2003) Petroleum generation, extraction and migration and abiogenic synthesis in hydrothermal systems. In: Ikan R (ed) Natural and laboratory simulated thermal geochemical rocesses. Kluwer, Amsterdam, pp 1–30

Simoneit BRT (2004) Prebiotic organic synthesis under hydrothermal conditions: an overview. Adv Space Res 33:88–94

Stetter KO (1995) Microbial life in hyperthermal environments. ASM News 61(6):328–340

Vergne J, Dumas L, Decout J-L, Maurel M-C (2000) Possible prebiotic catalysts formed from adenine and aldehyde. Planet Space Sci 48:1139–1142

Vernadsky VI (1980) Problems of biogeochemistry. Nauka, Moscow (In Russian)

Washington J (2000) The possible role of volcanic aquifers in prebiotic genesis of organic compounds and RNA. Orig Life Evol Biosph 30:53–79

Biodata of **Anthonie W.J. Muller**, author of "***Life Explained by Heat Engines.***"

**Dr. Anthonie W.J. Muller** is a biophysicist (postdoc) at the University of Amsterdam. For 30 years, he has been investigating the possibility of biological heat engines. He has a bachelor's degree in physical chemistry and a master's degree in experimental physics from the Free University of Amsterdam. His Ph.D. thesis (University of Amsterdam, 1990) was on the corrosion of dental NiCrMo alloys. Outside academia, he has worked as a software developer in the Netherlands and Scotland. He has investigated thermosynthesis—mainly theoretically—as independent researcher in the labs of Karel van Dam at the University of Amsterdam, Michael Russell at the University of Glasgow, David Apps at the University of Edinburgh, and Roel van Driel at the University of Amsterdam. From 2004 to 2006, he tried to obtain experimental evidence of thermosynthesis in the lab of Dirk Schulze-Makuch at Washington State University. Thermosynthesis has been applied in theoretical models for the origin of life and the emergence of the chemiosmotic machinery, photosynthesis, and the genetic system. Currently, he studies the possible relation between thermosynthesis and the emergence of the metazoans during the Snowball Earths/early Cambrian.

E-mail: **a.w.j.muller@uva.nl**

J. Seckbach (ed.), *Genesis - In The Beginning: Precursors of Life, Chemical Models and Early Biological Evolution*, Cellular Origin, Life in Extreme Habitats and Astrobiology 22, 321–344
DOI 10.1007/978-94-007-2941-4_19, 

# LIFE EXPLAINED BY HEAT ENGINES

**ANTHONIE W.J. MULLER**
*Swammerdam Institute for Life Sciences, University of Amsterdam, 1098XH Amsterdam, The Netherlands*

## 1. Introduction

It is generally agreed that life as it currently exists has emerged through the process of evolution. It is, however, also generally agreed that evolution is far from being completely understood. This incompleteness is illustrated by the absence of universally accepted models for the origin of life and for the emergence of the metazoans. Evolutionary considerations are moreover of little help in unraveling the complexity of today's life. Any search undertaken for extraterrestrial life will be all the more difficult if we do not know what early terrestrial life looked like. Clearly, our understanding of evolution could be improved.

According to the Thermosynthesis theory, early life was driven by temperature differences (Muller, 1985, 1995a, 2005, 2009). Similar ideas involving heat by classical authors such as Empedocles and Aristotle (Guthrie, 1957) have previously been discarded after it was found that today's life is powered by redox energy (Florkin, 1972):

> Modern biochemists have realized… that redox reactions are the ultimate source of all the energy of life, and that in all cells a large part of the oxidation-reduction energy is converted into phosphate-bond energy, the universal "go" of life, the remaining part being dispersed as waste in the form of heat. Heat became a waste product and lost the overvalorization which had originated in the mind of man by the tragic opposition of the cold of death and of the warmth of life.

Early life may, however, have differed greatly from the life we know today. Here, we show that a comprehensive scenario for the origin and evolution of life, including today's chemiosmotic machinery (Mitchell, 1979a), can be given based on heat engines. In particular, it is proposed that life's "First Protein" (FP) was (1) a progenitor of today's ATP Synthase and (2) a heat engine that powered the RNA World (Muller, 1995a, 2005). The bacterial photosynthetic membrane evolved from a biomembrane with thermosynthesis capabilities (Muller, 1995b).

Heat engines could have also played a role in the emergence of metazoans. Today, many invertebrates live in colonies (Beklemishev et al., 1969). The most complex colonies are found in the Siphonophora (Fig. 1) (Dunn, 2009), which include the Portuguese man-of-war that floats on the tropical seas. It is generally accepted that a colony can become a solitary new organism by "individuation," a process used to explain the emergence of the metazoans during the Cambrian

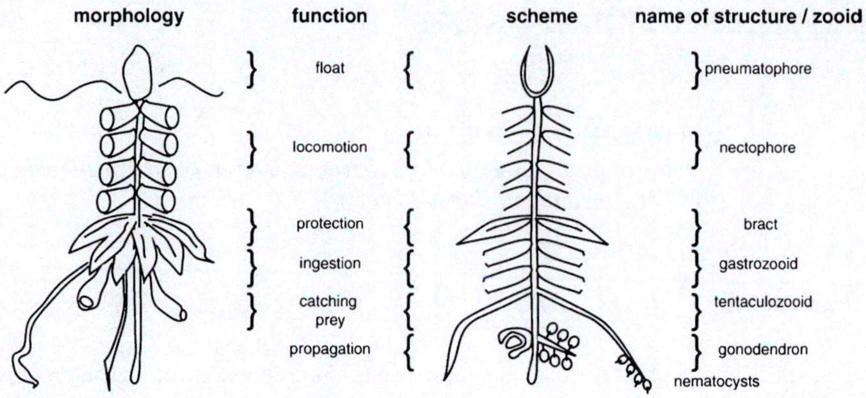


**Figure 1.** A siphonophore is a colony of invertebrates. The contributing organisms are called "zooids"; these strongly resemble other independently living invertebrates (Dunn, 2009). In colonies, the concept of individuality becomes diffuse. Organisms that contribute to the colony can merge to form a larger organism in a process called "individuation."

(Dewel, 2000; Dewel et al., 2001). According to the Thermosynthesis theory, the Cambrian zooids descended from organisms that acquired their large size while living by thermosynthesis above submarine hydrothermal vents during the Proterozoic Snowball Earths.

## 2. Thermodynamics

### 2.1. HEAT ENGINES

Man-made heat engines originated with the steam engine and evolved by the Otto and Diesel combustion engines to today's gas turbine engine (Reed, 2006). The high temperature is the result of combustion, the term for the process where an oxidant draws electrons from a fuel. The fuel cells that are coming into increasing use (Tollefson, 2010; Kunze and Stimming, 2009) circumvent this high temperature. Enhancing efficiency, the motor is directly driven by the current generated by electrons donated by fuel to and accepted by oxygen from electrodes.

Respiration in our body is based on fuel cells as well (Mitchell, 1979b). The mitochondria work on the organic reductants in the food that we eat and the oxygen in the air that we breathe.

In the past, it has been proposed that the muscle operates by a heat engine mechanism, but the efficiency of the muscle is too high. Therefore, the idea of the ancient Greeks that organisms live on heat was discarded, and it became accepted that the muscle works on nonequilibrated redox couples (Florkin, 1972). *The key notion of the Thermosynthesis theory is that although life may* not invoke *heat engines at present, life may have* evolved *from heat engines in the past*. Two significant unresolved evolutionary processes, the origin of life and the emergence of the metazoans, can be modeled in terms of heat engines.

## 2.2. ENTROPY AS GENERALIZED MACROSCOPIC FRICTION

Heat engines relate to thermodynamics, which in turn relates to entropy. Today, the concept of entropy is unfortunately often regarded as mysterious and difficult to understand (Lambert, 2002; Carson and Watson, 2002). This author finds it convenient to visualize entropy as a generalization of friction (Muller, 1998), which resembles the phenomenological point of view as formulated by Clausius that is still given in physical chemistry textbooks (Moore, 1962; Atkins, 1990).

We give some examples. A perpetual motion machine is impossible since every engine shows the irreversible decay of movement called friction when objects moving at different speed make contact. After Mayer's formulation in 1842 of the conservation of energy in the First Law of Thermodynamics, it became quickly accepted that during friction, kinetic energy is converted into heat.

Although friction cannot be avoided, its absence can be imagined: a *perfectly* elastic ball bounces back forever to the height from which it is dropped; a *perfectly* greased spinning wheel rotates forever. In practice, perpetual movement does not occur and kinetic energy is irreversibly converted into heat $Q$. It has been found advantageous to use the generated entropy, symbol $S$, as measure of friction:

$$\mathrm{d}S = Q / T \tag{1}$$

where $T$ is the temperature in degrees Kelvin. With both $Q$ and $T$ being well defined and measurable, the value of the increase in entropy $\mathrm{d}S$ is easily obtained.

Not all heat generation is associated with friction. Liquid evaporation and condensation under a piston in a cylinder occur reversibly: when the piston is slowly pushed outward and the head space increases, the evaporating liquid takes up heat at the environment. When the piston is slowly moved back again and the head space decreases, the vapor condenses and the heat previously entered is returned to the environment. Although there always must be some friction in the process, a *perfect* system of such reversible heat-work transfer can be imagined.

In a steam engine or, more generally, a heat engine, the heat $Q_h$ is taken from the high temperature $T_h$, and the heat $Q_l$ is donated to a reservoir (for instance, cooling water) at the low temperature $T_l$, with work $W$ being done. The First Law of Thermodynamics states that energy is conserved:

$$Q_h = W + Q_l \tag{2}$$

Just as in any engine, friction in a heat engine leads to entropy production. But just as one can imagine a *perfect machine* without friction, one can imagine a *perfect heat engine*, called the *Carnot engine*, which reversibly converts heat into work *without producing entropy*: hence, $Q_h/T_h = Q_l/T_l$. The relations

$$W = Q_h - Q_l = Q_h (T_h - T_l) / T_h = Q_l (T_h - T_l) / T_l \tag{3}$$

apply. *These relations are* fundamental, *and* define *in physics the temperature.*

Real heat engines are not perfect, and entropy is therein not only produced by friction but by heat conduction as well, as conducted heat is lost for conversion into work. During the conduction of heat $Q$ from a high temperature $T_h$ to a low temperature $T_l$, there is a partial entropy *decrease* of $-Q/T_h$ while heat is lost at $T_h$ and a partial entropy *increase* of $+Q/T_l$. The net change is an increase:

$$dS = -Q/T_h + Q/T_l = Q(1/T_l - 1/T_h) > 0 \tag{4}$$

Just as heat is wasted during heat conduction, the energy of pressurized gas is wasted when it leaks. The entropy then increases as well. Where an expanding gas, however, "perfectly" and reversibly drives a piston, no entropy is produced.

Hence, (1) friction, (2) heat conduction, and (3) the leakage of pressurized gas all increase the entropy. The Second Law of Thermodynamics generalizes these three processes: *in an isolated system the entropy only increases.*

## 2.3. ENTROPY AS GENERALIZED MICROSCOPIC DISORDER

As entropy is such an important tenet, some feel that it should not be fully described in engineering terms (Jammer, 1973). This author disagrees. Fundamental physical concepts such as time, distance, mass, inertia, force, work, energy, temperature, heat, and in particular friction, all play a role in daily life; the notion of an always increasing entropy involves a simple extension of these fundamental concepts—no further explanation is called for.

Sentiments similar to those espoused by Jammer may have motivated Boltzmann in the 1870s to search for a microscopic explanation of the entropy of a gas in terms of the statistical mechanics of molecules. His investigations resulted in a modified Second Law that states that a decrease in entropy is merely *very improbable* instead of *impossible*.

Applying statistical mechanics to systems different from a large set of point particles is difficult, however, as application requires a detailed molecular model (Einstein, 1910). The history of science shows that for real systems such models are either unknown, have evolved unpredictably, or require extensive use of quantum mechanics (Fast, 1968).

Entropy, on the molecular scale, is important for biochemistry, especially due to (1) the hydrophobic effect (Tanford, 1973), which stabilizes membranes and proteins, and (2) the high-energy phosphate bond (De Meis, 1989); both involve the interaction of biomolecules with water. A full understanding of these processes is elusive, as no adequate molecular model for water exists (Stanley et al., 2007).

In statistical mechanics, the entropy production during friction is often attributed to a dispersal of the energy from a macroscopic movement involving one degree of freedom to uncorrelated molecular movements involving many degrees of freedom. Some say that the increase of entropy is due to a natural tendency for order to decrease, but others object to this view (Lambert, 2002), order, and as a consequence, disorder, being difficult to define (Bohm, 2009).

Order and disorder are especially difficult to recognize in complex systems such as living organisms. Cardwell (1971) gives a convincing example of misuse:

> A distinguished modern writer has tried to explain the second law of thermodynamics to the lay reader by likening its implications to the empirically established principle that one cannot unscramble omelettes. Such an oversimplification must surely generate more confusion than edification.

It seems best to apply the concept of order only carefully, if at all (Lambert, 2002).

Nevertheless, in spite of these caveats, it seems reasonable to state, given the Second Law, that the emergence of a novel *order on a large scale* far from equilibrium requires a continuous input of energy. Such structures have been named "dissipative structures" (Nicolis and Prigogine, 1977). The local entropy decrease, an increase in order, is considered to be only possible by a compensating increase in entropy of the surroundings.

## 2.4. SELF-ORGANIZATION IN GEOLOGY

Convection occurs when a fluid is heated from below and cooled from the top. Thermal expansion decreases the fluid's density, causing the fluid to rise. After cooling, the fluid contracts, sinks, and returns to the heat source. The convecting fluid circulates as a whole. Entropy is produced as heat moves from high to low temperature; convection is an example of a "self-organizing dissipative structure."

Convection is ubiquitous (Muller, 2009). The convection cells in the Earth's mantle (Ogawa, 2008) sustain continental drift. In the thermal gradient between the Earth's interior and exterior, four dissipative structures are of interest:

1. *Convection in the Earth's mantle and in volcanic hot springs.* On the Earth's surface, volcanoes emerge both where convection currents in the mantle swell and where they descend. In volcanic areas, convection in volcanic hot springs—which occurs on a much smaller scale—causes thermal cycling of suspended organisms.
2. *Submarine hydrothermal vents.* Where the convection cells in the Earth's mantle swell, the Earth's surface is pulled apart, and new surface is created on the bottom of the ocean. Here, submarine hydrothermal vents are found (Baross and Hoffman, 1985), which combined with the cold ocean water create a large thermal gradient.
3. *The global carbon cycle and the Snowball Earths.* The sediments on the ocean floor contain the remains of dead organisms and precipitated carbonates. After subduction by convection currents descending in the mantle and being heated to high temperature, the organic remains and carbonates decompose to form $CO_2$. The $CO_2$ returns through the aforementioned volcanoes to the atmosphere, from where it is fixed again by photosynthesis.

   During the Neoproterozoic Snowball Earths (Kirschvink, 1992; Hoffman et al., 1998), the global carbon cycle was interrupted. A large withdrawal of the greenhouse gas $CO_2$ from the atmosphere triggered a global glaciation. The resulting ocean ice cover impeded photosynthesis, which is detected by the low $^{13}C$:$^{12}C$ ratio of precipitated carbonates (Kaufman et al., 1997). Reflection of sunlight stabilized glaciation but could not halt the accumulation of volcanic $CO_2$ in the atmosphere. Eventually, a global greenhouse put an end to the glaciation, and a new cycle could start.

The last Snowball Earth was the Gaskiers (~585–582 Ma (million years ago)) (Stern et al., 2006), which was followed by the Ediacaran period, during which the first metazoans appeared (Knoll and Carroll, 1999).

Similar global glaciations occurred much earlier as well. The Huronian glaciations (~2,300 Ma) (Papineau et al., 2007) are attributed to the oxidation of the greenhouse gas methane by biogenic oxygen (Kasting and Ono, 2006).

4. *Formation and breakup of the continents.* During subduction, some solids can remain behind near the Earth's surface, associate, and form continents or, sometimes, a single supercontinent. These supercontinents break up with a period of roughly 500–1,000 million years (My). In the absence of substantial heat transport, generally carried by fluid movement, such as occurs during convection, the temperature under the continent rises, which generates forces that eventually lead to its fragmentation. The breakup ~700 Ma of the Rodinia continent (Li et al., 2008) was followed by the formation of the Gondwana continent (Meert and Lieberman, 2008), right at the beginning of the Cambrian.

## 2.5. HEAT ENGINES APPLICABLE TO BIOLOGY

The convection cell is a heat engine. We propose life has been powered by four additional heat engines, which resemble heat engines known from engineering:

1. The *sorption heat engine* (*SHE*) (Fig. 2) (Muller and Schulze-Makuch, 2006; Aristov et al., 2008). The sorption heat engine has not yet found practical application as forward engine. The reverse engine is, however, well known and is widely applied for cooling purposes.
2. The *thermodielectric energy converter* (*TDEC*) (Fig. 3) (Sklar, 2005). A capacitor contains a dielectric with a temperature-dependent polarization, i.e., a pyroelectric. During thermal cycling, the voltage across the capacitor varies due to temperature-induced changes in polarization. The free energy gained in this process is applied to the load by timely switching (Muller, 1995b). A similar device that works on light–dark cycling and that uses the "photodielectric effect" has also been described (Glazebrook and Thomas, 1982; Muller, 1995b).
3. The *rubber heat engine* (*RHE*) (Fig. 4) (Wiegand and Snyder, 1934; Mullen et al., 1975). The rubber heat engine makes use of the temperature-induced expansion or contraction in a material such as metal, rubber, or protein. Heat engines based on shape memory metal (Wekjira, 2001) are variations of this type of engine.
4. The *thermogalvanic energy converter* (*TGEC*) (Fig. 5). Thermoelectricity is the generation of a voltage difference in a conductor by a thermal gradient

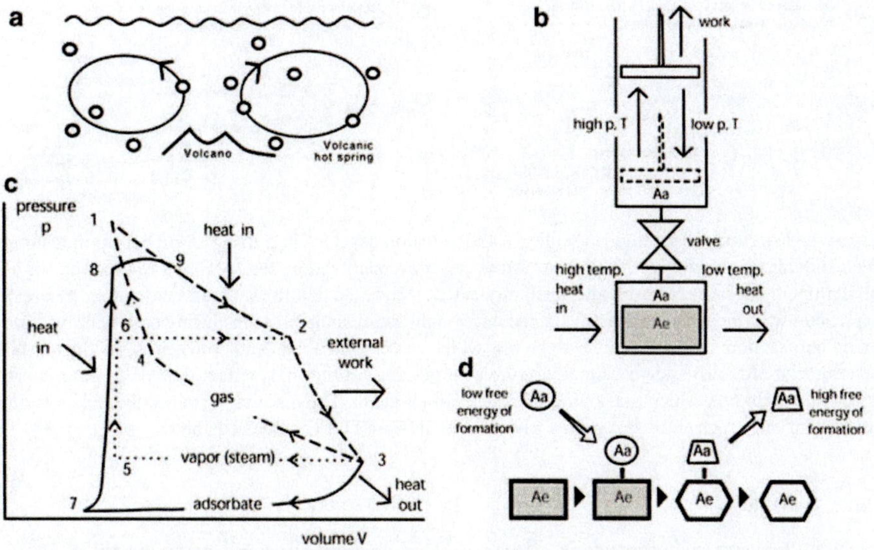


**Figure 2.** The sorption heat engine (Muller and Schulze-Makuch, 2006) makes use of thermal cycling such as in objects suspended in a convecting volcanic hot spring (**a**). In the working cycle (**b**), the pressure in the adsorbent (*Ae*) chamber strongly increases upon thermally stimulated desorption. After opening of the valve, the released adsorbate (*Aa*) pushes a piston, and work is done. The *p-V* diagrams of a gas engine (*1–2–3–4*), a vapor (steam) engine (*2–3–5–6*), and the sorption heat engine (*7–8–9–2–3*) are shown (**c**). A proposed sorption heat engine that does chemical work on the adsorbate is shown in (**d**); the mechanism could involve a thermal transition in the adsorbent.

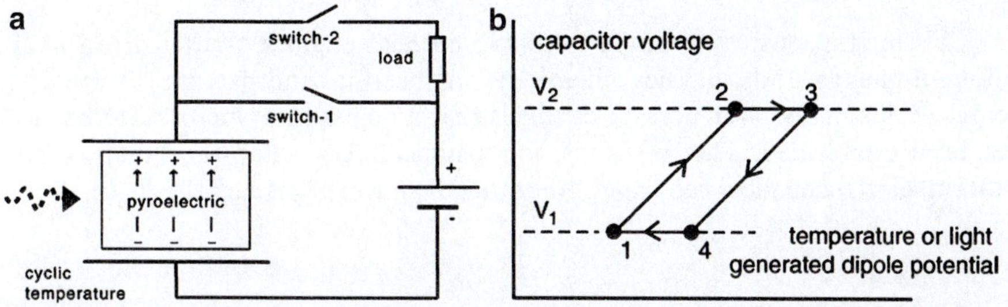


**Figure 3.** The thermodielectric energy converter (Muller, 1995b) makes use of thermal cycling, just as the sorption heat engine. The converter consists of a battery, a capacitor, two switches, and a load (**a**). By a variation in the polarization of the dielectric in the capacitor, thermal cycling causes a voltage variation (**b**). Timely opening and closing of switches drives a current through the load, on which work is done. Where light absorption creates metastable dipoles in the dielectric, an engine could also work on light–dark cycling. Combined light–dark and thermal cycling could also be made to work.

(Riffat and Ma, 2003). It is an example of nonequilibrium thermodynamics (de Groot and Mazur, 1962; Haase, 1963). Its application is well known in metals, with electrons acting as the charge carriers, but it occurs also in ionic solutions, with ions as charge carriers (Agar, 1963; Quickenden and Mua, 1995).

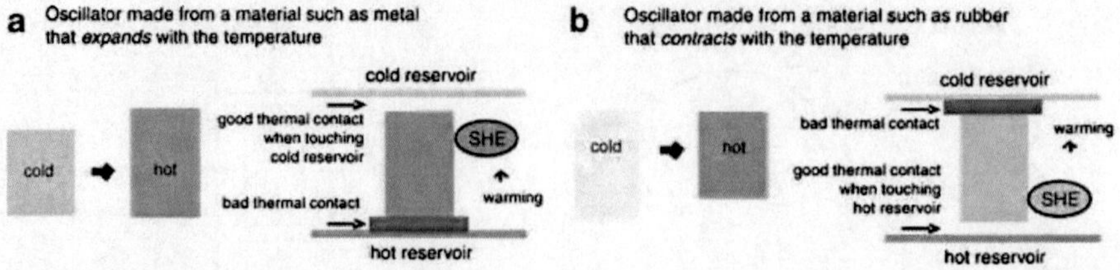


**Figure 4.** The rubber heat engine (Muller, 2009). Tension develops in a string of rubber upon heating. By a suitable combination with pulleys, a rotating movement can be set up. Here, we describe simple oscillators in which linear back and forth movement is induced in a thermal gradient, either by use of expansion with the temperature (**a**), as occurs in metals and some proteins during cold denaturation, or by contraction with the temperature (**b**), which occurs in rubber and some proteins during hot denaturation. In both systems, a temperature change causes a thermal contact that yields an opposite temperature change, which breaks the thermal contact again. The material is thermally cycled during the oscillation, which could be applied to drive the SHE or TDEC described above.

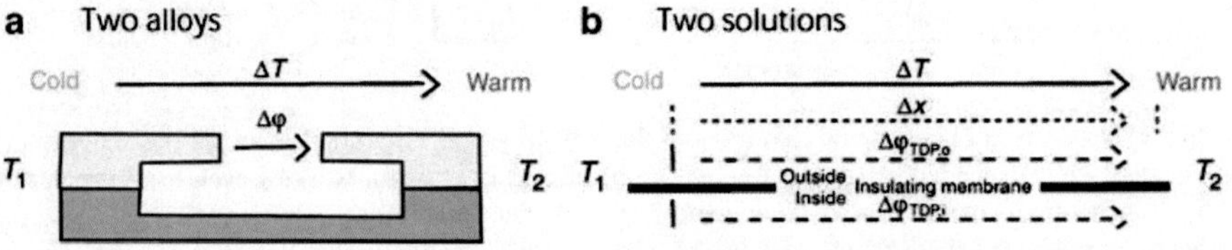


**Figure 5.** Thermogalvanic energy converter. In the well-known thermocouple (**a**), a voltage is generated when two different alloys make an electrical contact in two places at different temperatures. In the hardly known thermogalvanic energy converter (**b**), a voltage is generated in solution or in molten salt, with ions as charge carriers.

There is extensive literature available on heat engines, written from many different backgrounds: physics, chemistry, engineering, and patents. An Internet search yields many references. A caveat is that a common nomenclature has not yet been established. The TGEC is, for instance, also called the "thermocell," "thermoelectrochemical cell," and "thermogalvanic cell" (Hu et al., 2010).

## 3. Early Life

### 3.1. THE CHEMIOSMOTIC MACHINERY

Today's biological energy conversion comprises two basic types: ATP can be generated by (1) fermentation of simple small organic compounds by a set of enzymes and (2) chemiosmosis, which works on a voltage/proton gradient across a membrane; this gradient is generated during respiration and photosynthesis (Fig. 6), with ATP synthesis occurring within the ATP Synthase enzyme, which functions through the binding change mechanism (Fig. 7). Chemiosmosis, with its multiple components, is often seen as too complex to have been the first biological energy converter. The Thermosynthesis theory proposes that the chemiosmosis machinery descends from simple heat engines that powered the origin of life.

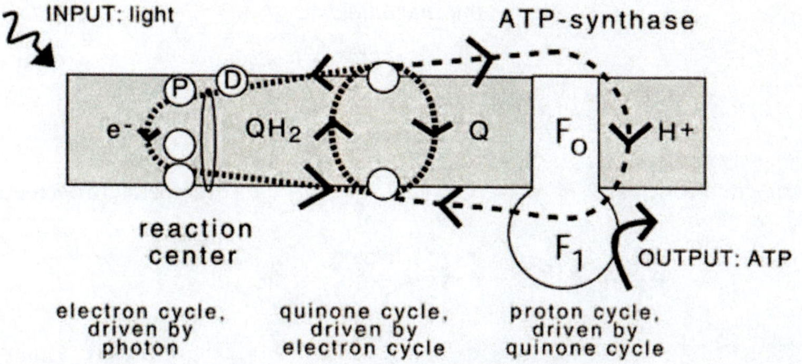


**Figure 6.** Chemiosmosis as it powers bacterial photophosphorylation today. Upon absorption of photons by a pigment *P*, excited electrons cross the membrane and combine with quinones *Q*, which also pick up protons from one side of the membrane. The resulting quinol $QH_2$ diffuses across the membrane and releases its electron and protons at the other side of the membrane. The energy of the protons is transduced into ATP while the translocated protons return through ATP Synthase.

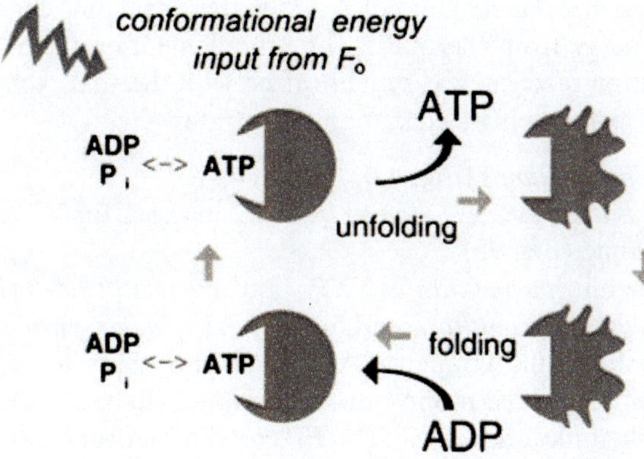


**Figure 7.** The binding change mechanism of ATP Synthase (Muller, 1995a). In the β subunit of the water-soluble $F_1$ moiety of the enzyme, ADP and phosphate are bound and react to form bound ATP, either because in the enzymatic cleft water is absent or because the water therein has a diminished activity (Boyer, 1993). The energy for the release of the bound ATP is obtained from the protons that cross the energized membrane through the $F_o$ part of the enzyme as depicted in Fig. 6.

## 3.2. THE FIRST PROTEIN AND THE FIRST CELL: THE LIPID WORLD

ATP Synthase performs a condensation reaction while immersed in water, with the condensed product having a higher free energy than the substrates. In the context of the origin of life, this is a significant and fundamental capability that is quite puzzling (Calvin, 1969). Moreover, it has been recognized that the hypothetical "first enzyme" probably had the capability to process many different substrates (Black, 1970).

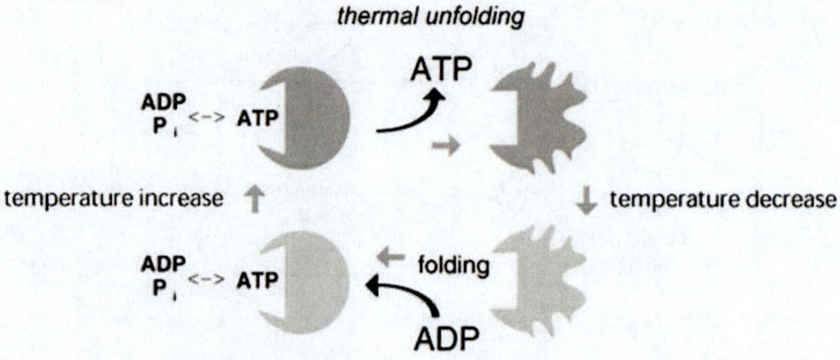


**Figure 8.** The proposed First Protein (*FP*) and its thermal binding change mechanism (Muller, 1995a). The enzyme functioned similar to today's mechanism of ATP Synthase (Fig. 7), the difference being that instead of using translocated high-energy protons, the FP released ATP upon a temperature change.

The Thermosynthesis theory postulates that life started with a First Protein (FP) (Fig. 8) that:

1. Functioned by a thermal variation of ATP Synthase's binding change mechanism
2. Gained free energy from thermal cycling, implementing the SHE of Fig. 2d. The condensation reaction was synchronized with thermal cycling
3. Could condense, following Black, many substrates:
   (a) *Lipids → phospholipids* (Fig. 9a)
   (b) *Amino acids* and *peptides → longer peptides*, in short, the FP could synthesize a peptide bond (Fig. 9b)
   (c) *Pyrimidine* and *purine bases, NXPs,* and *nucleic acids → ribonucleosides, phosphorylated nucleotides*, and *longer nucleic acids*: similar to ATP Synthase, the FP could condense ADP and phosphate to ATP. Analogous condensations yielded mono-, di-, and triphosphates of adenine, guanine, cytosine, thymine, and uracil. The FP could moreover create the phosphodiester bond of RNA (Fig. 9c).

Convection was the process that drove the thermal cycling of the FP and thus constituted the *inanimate self-organizing dissipative structure* that the origin of life requires.

During the origin of life, lipids spontaneously formed cell membranes. The stability of these membranes would have decreased at higher temperatures (Budin et al., 2009). Phospholipids, however, stabilize membranes. In early protocells, lipid phosphorylation by the FP would provide a selective advantage. These early cells in turn protected the FP from proteolysis by external proteins.

Over a long period of time, peptides synthesized by the FP would constitute a library of peptides/proteins (Fig. 9b). Their composition would be random as there was as yet no genetic apparatus to guide the synthesis of specific protein sequences. A very small fraction of this library might have the same capabilities

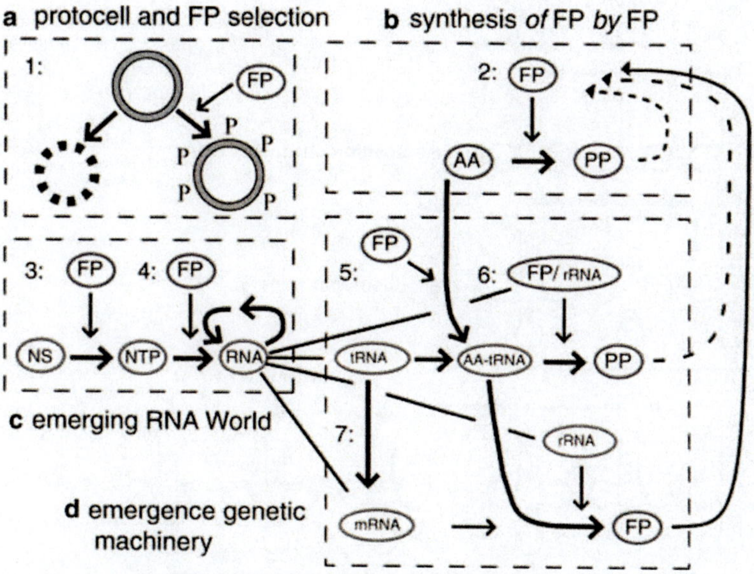


**Figure 9.** Model of the emergence of the genetic machinery (Muller, 2005). **(a)** *Protocell selection*: (*1*) the FP stabilized a protocell by membrane lipid phosphorylation. **(b)** *Synthesis* of *FP* by *FP*: (*2*) in the synthesized library of proteins, a small fraction functioned as FP: functional FP propagation. **(c)** *Emerging RNA World*: (*3*) nucleotide triphosphates (*NTPs*) were synthesized from nucleosides (*NS*) and (*4*) synthesis of RNA, including self-replicating RNA (Johnston et al., 2001). From NTPs, FPs synthesized RNA both without a template and by copying of RNA. RNA was selected that increased the synthesis of FPs. **(d)** *Emergence genetic machinery:* (*5*) tRNAs were charged with amino acids by ribozymes, or self-charged, (*6*) tRNAs assisted in FP synthesis by the FP or by rRNA, and (*7*) mRNA emerged from a tRNA (see Fig. 10); rRNA, mRNA, and tRNAs yielded the genetic machinery.

as an FP but would be "inexactly replicated." Similar ideas have previously been proposed by Dyson (1985) and Woese (1987).

## 3.3. RNA AND THE RNA WORLD: EMERGENCE OF THE GENETIC CODE

Figures 9 and 10 portray a hypothetical scenario for the emergence of the ribosome, the genetic code, and the genome in the "RNA World" that is proposed to have followed the "Lipid World." In this scenario, the FP synthesized the first RNA and its building blocks (Fig. 9c).

For details on this scenario, for the emergence of the genetic machinery and the genetic code, we refer the reader to a previous study (Muller, 2005). A single energy-generating FP containing a short conserved amino acid sequence that formed the FP's enzymatic cleft makes this model possible and plausible.

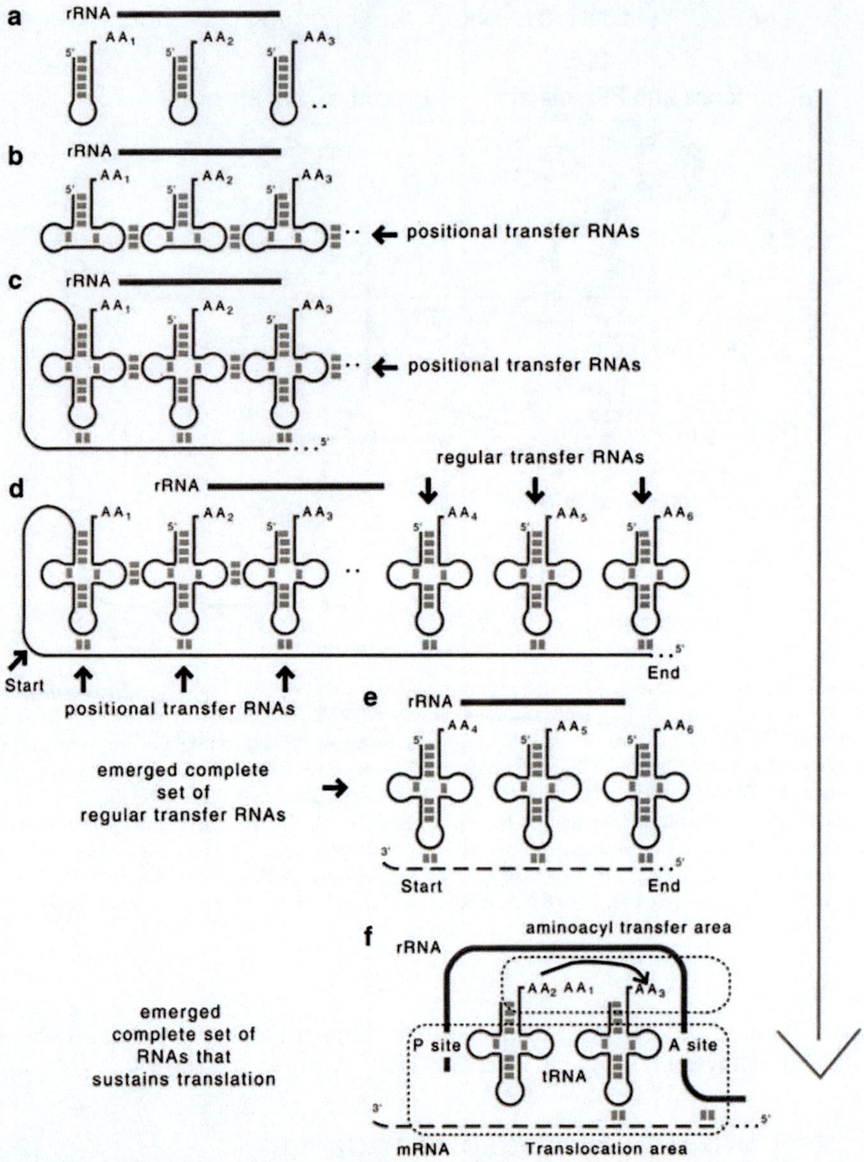


**Figure 10.** The proposed evolution of tRNA and emergence of mRNA (Muller, 2005). **(a)** The first tRNAs were minihelices that connected to amino acids and yielded charged tRNAs. The charged tRNAs formed proteins with random composition, a process catalyzed by an rRNA progenitor or possibly an FP. **(b)** Using acquired arms, positional transfer tRNAs recognized a predecessor and a follower, which resulted in a tRNA sequence that yielded a small protein sequence that was a key motif of the FP. **(c)** The first tRNA of the sequence extended to a template that bound all other motif tRNAs. The positional tRNAs acquired anticodons that bound to codons on the extension. **(d)** The extension was disconnected from the first tRNA (*arrow*) and yielded an mRNA (start at 3′ end). Mutual tRNA recognition disappeared but tRNA:mRNA recognition remained: *this transition from mutually recognizing positional tRNAs to a set of regular tRNAs that recognize the codons of mRNA constituted the emergence of the genetic code.* **(e)** Mutual tRNA recognition had disappeared. **(f)** A ribosome particle containing rRNA emerged that performed (1) aminoacyl transfer from tRNAs in the sequence given by mRNA to a growing peptide chain and (2) translocation along mRNA in the 3′ to 5′ direction.

After the emergence of system capable of a translating the first single mRNA coding for the FP, additional mRNAs would have quickly emerged by subsequent gene duplication and mutation, yielding specialized FPs adapted to specific substrates, allowing the original FP with the general condensing ability to vanish. These mRNAs would have contributed to the first set of genes, the first genome. In addition, this nascent genome would have contained the RNA that facilitated the whole process of translation: self-replicating RNA, a set of tRNAs, and rRNA.

## 4. Prokaryotes

In the model presented above, the first organisms lived on thermosynthesis and required thermal cycling. In order for "life as we know it" to arise, the development of DNA was necessary; this was synthesized by reverse transcriptase from RNA. In addition, life had to adapt to living at constant temperature. This involved two steps:

1. The acquisition of an isothermal energy source in the form of *bacterial photosynthesis*, which yields ATP (Fig. 11). The TDEC would have played an intermediate role in this development.
2. The use of ATP to mimic thermal cycling by cyclic phosphorylation/dephosphorylation of those enzymes that previously had a thermal cycling requirement for the release of their product by folding/unfolding (Urry, 1992).

After the acquisition of *bacterial photosynthesis*, diversification led to the two types of reaction centers seen in the green and the purple bacteria. Their merging yielded the cyanobacteria that function according to the Z-scheme (Duysens et al., 1961), in which NADPH is reduced after the absorption of two photons by electrons taken from water. This process powers *oxygenic photosynthesis*.

Since respiration and photosynthesis make use of similar enzymes, it is hard to decide on what came first. *Anaerobic respiration* was probably already present before *oxygenic photosynthesis* (Schäfer et al., 1996), but it was plausibly preceded by *bacterial photosynthesis* to which it donated and from which it accepted electrons. Figure 11 shows that *bacterial photosynthesis* may in turn descend from thermosynthesis. This suggests that thermosynthesis preceded respiration, i.e., *the biological heat engine preceded the biological fuel cell.*

## 5. Eukaryotes

Early *oxygenic photosynthesis* created oxidants and reductants that could remain separated for a very long time. Before predators or scavengers existed, the death of an organism resulted in reduced biomass that ended in ocean sediments, to be

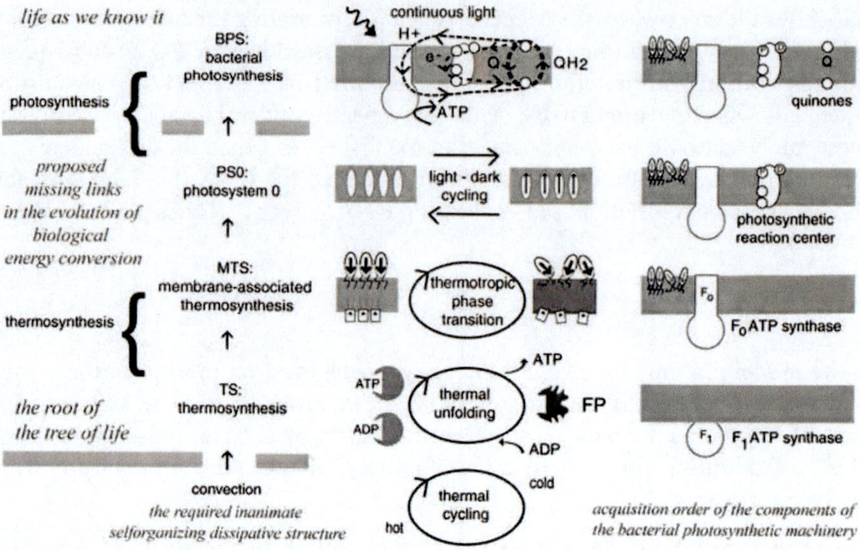


**Figure 11.** The thermosynthesis-based stepwise model for the emergence of bacterial photosynthesis (Fig. 6) (Muller, 2003). The components were acquired in the sequence: (1) the $F_1$ part of ATP Synthase which relates to the FP (Fig. 5). (2) The $F_o$ part of ATP Synthase and an asymmetric biomembrane were acquired during the emergence of MTS, which also worked on thermal cycling and implemented the TDEC (Fig. 3). Use was made of changes in the dipole potential of the membrane (Haydon and Hladky, 1972) during a thermotropic phase transition (Muller, 1993). Net ATP was gained by an ATP Synthase that charged and discharged the membrane with different $H^+$/ATP ratios (Muller, 1993). (3) The photosynthetic reaction center was acquired during the emergence of Photosystem 0; during light–dark cycling, the dipoles of metastable states, induced by light, enhanced the dipole potential of the membrane (Muller, 1995b). (4) The addition of membrane-diffusible quinols/quinones permitted bacterial photosynthesis that worked in continuous light.

transported down the lithosphere by mantle convection. Oxygen, in contrast, moved upward into the atmosphere. The Huronian global glaciations of the early Proterozoic between 2,450 and 2,320 Ma have been explained by a breakdown through emerged oxygen of the greenhouse gas methane, which had kept the Earth warm in spite of a much dimmer Sun (Bekker et al., 2004).

This model argues that when glaciation impeded photosynthesis, life fell back to relying on thermosynthesis. A "thermotether" that oscillated in the thermal gradient above hydrothermal vents, reminiscent of the RHE shown in Fig. 4a, pumped protons that were converted by ATP Synthase into ATP (Muller, 2009). When the proton pump at the end of the glaciation became redundant due to the return of photosynthesis, the thermotether developed into the prokaryotic flagellum, in which a proton gradient produced by respiration drives the flagellum.

This scenario is of interest as it presents a solution to the problem of the emergence of active biological movement. Emergence cannot start small (Muller, 2009), as the smaller the scale, the stronger the impediment of motion due to

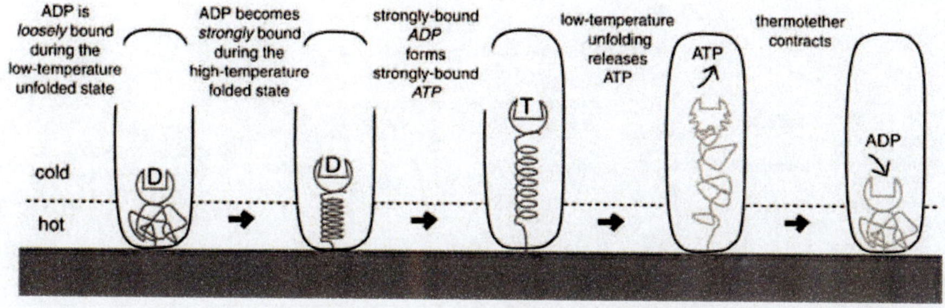


**Figure 12.** An ATP-producing progenitor of muscle that generated ATP using in a cell a combination of (1) an internal thermotether oscillating in the thermal gradient just as an RHE and (2) an FP.

viscosity (Purcell, 1977). We call this the "Purcell barrier" and introduce the term "Purcell barrier breaking mechanism" (PBBM) for the emergence of large moving organs driven by ATP from ATP generators driven by a thermal gradient.

The PBBM of an organ consists of the following steps: (1) it originates during a glaciation as a small organ that *generates* ATP by oscillating in the thermal gradient above a submarine hydrothermal vent; (2) its size increases, spanning—while remaining oscillating—a larger part of the thermal gradient; (3) during the long glaciation, the enhanced efficiency in the mechanisms causes approach of reversibility; and (4) at the end of the glaciation, the mechanism reverses, with ATP being *used* for oscillation. The ATP required for the last stage is generated by a respiring or photosynthesizing endosymbiont (Keeling, 2010).

Figure 12 gives an example of the ATP-generating stage of a PBBM during the emergence of muscle. A thermotether present inside a cell is connected to an FP that yields ATP, with the cell placed in thermal gradient. Reversal led to the emergence of actin and myosin, which evolved from the thermotether and the FP, respectively.

## 6. Metazoan Diploblasts of the Ediacaran

During the "boring billion," between 2,300 and 800 Ma, life saw little change. Four Snowball Earths occurred from 800 Ma until 580 Ma (Stern et al., 2006), events that may have been triggered by the breakup of Rodinia (Li et al., 2008).

The period between the last Snowball, the Gaskiers, and the Ediacaran witnessed the emergence of the first animals (Fig. 13) (Knoll and Carroll, 1999).

The PBBM that emerged during the Gaskiers-Ediacaran transition eventually yielded macroscopic organisms with oscillating fractal fans (fronds), large body cavities (sponges), and tentacles (anemones) (Narbonne, 2005). The collagen in the hanging tentacles of the ancestors of the anemones contracted at high temperature. The individuation of colonies composed of a large frond and small sponges would lead to the polyp.

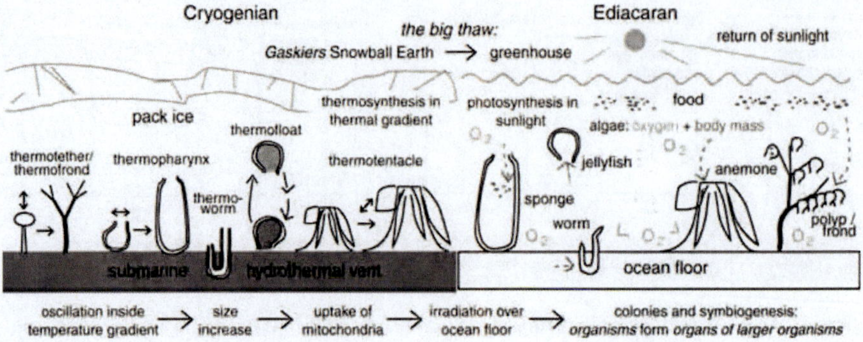


**Figure 13.** Emergence of the Ediacaran diploblast biota by the PBBM. During the Gaskiers, thermosynthesizers emerged that gained ATP from oscillators in the thermal gradient above a hydrothermal vent that either had the shape of a tether (thermotether) or of a hollow tube (thermopharynx). The thermoworm and the thermofloat are variations of these shapes, with the thermofloat using the shape for transport, just as a hot air balloon. In the thermopharynx, energy was gained by the RHE mechanism of Fig. 4a. With time, organism size increased, and a larger part of the thermal gradient became spanned. The thermopharynx acquired hanging thermotentacles that permitted it to gain free energy from the RHE mechanism of Fig. 4b. At the end of the glaciation, after acquisition of ATP-yielding endosymbionts, the first large animals emerged, which fed on algae and used oxygen for respiration.

## 7. Metazoan Triploblasts of the Cambrian

Fossils of animals from the Cambrian (542–505 Ma) are abundant, with most of today's animal phyla already being represented (Valentine, 2004). The decrease in the $^{13}C$:$^{12}C$ ratio of marine carbonate deposits from the beginning of this period indicates that global photosynthesis had decreased, just as during the Snowball Earths. Evidence for global glaciations is, however, absent. Instead, it appears that the ocean had turned anoxic (Kimura and Watanabe, 2001). The decrease in photosynthesis can be explained by intense global volcanism associated with the formation of the Gondwana supercontinent (Meert and Lieberman, 2008).

We propose a model for the emergence in the Cambrian of the triploblast metazoans based on colony individuation (Dewel, 2000; Dewel et al., 2001). We also postulate the origin of a novel zooid, which was not present in the Ediacaran period, the nerve, which could have descended from a thermosynthesizing "thermonerve" (Fig. 14).

This thermonerve would have consisted of a filamentous organism that spanned a major part of the thermal gradient above a submarine hydrothermal vent. Thermal diffusion potentials (TDPs) were generated inside and outside the organism. Although the TDP is small at room temperature, ~1 mV/C, above submarine hydrothermal vents, the attainment of temperature differences of ~100 C along the organisms would be feasible. This would have resulted in TDPs of ~100 mV. The different composition of the medium inside and outside the organism would have resulted in different TDPs values. Moreover, the TDP strongly increases as the critical temperature and pressure of the solvent is approached

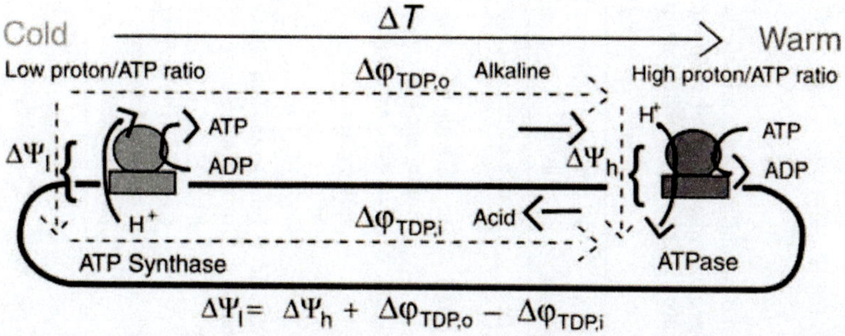


**Figure 14.** The thermonerve is the proposed progenitor of the nerve. The thermonerve gained energy from a biological implementation of the TGEC of Fig. 5. The difference in the internally and externally generated thermal diffusion potentials, $\Delta\varphi_{TDP,i}$ and $\Delta\varphi_{TDP,o}$, was sufficient to gain ATP from simultaneous activity of (1) an ATP Synthase that synthesized ATP while active at low $H^+$/ATP ratio in one part of the membrane and (2) an ATPase that pumped protons and therefore consumed ATP at high $H^+$/ATP ratio in another part of the membrane, at a different temperature.

(Haase, 1963), which is indeed the case for water near submarine hydrothermal vents. A large difference in TDP is therefore plausible, which could lead to the production of net ATP through the activities of an ATP Synthase and an ATPase. When conditions amenable to photosynthesis returned, the system would have adapted and reversed—just as during a PBBM—and would begin to function as the Na/K-ATPase that today sustains the inner composition of the nerve and enables data communication by the action potential (Barnett and Larkman, 2007).

In colonies, the nerve zooid would have integrated the activities of the other zooids through this communication, which made colonies with a larger size possible. Our collagen-containing skin may be a relic of a zooid containing collagen (related to the Ediacaran thermotentacle that contracted with the temperature (Fig. 13)) with a large free internal space that could contain several small zooids and eventually became the coelom of the triploblast (Clark, 1964; Strickberger, 2000). Other anemone-like and spongelike zooids that caught and digested food remained, however, in contact with the outside world. Upon individuation of the colony, the internal zooids turned into organs such as the heart, the gonads, and the kidney.

## 8. Discussion

Table 1 summarizes the proposed scenario that integrates geological phenomena, inanimate heat engines, and models for early life. As a result, Simon's (1971) idea on the role of machines in biology can be worked out in further detail:

> Biology's subject matter thus consists of a class of *naturally occurring machines*, machines that are assumed to be descended from *primitive protomachines*, the *original progenitor* of which was *self-assembled* [my italics].

**Table 1.** Overview of the proposed relations between geology, heat engine mechanisms, emerged biological systems, and organisms, according to the Thermosynthesis theory. Geology and biology are integrated: geology made biology possible, but biology also affected geology when removal by life of greenhouse gases such as methane and $CO_2$ resulted in global glaciations. These glaciations in turn had large consequences for biology, enabling major evolutionary steps forwards.

| Timing | Global geology | Derived small scale geology | Heat engine mechanism known from engineering | Emerged biosystem based on heat engine | Resulting emerged organisms |
|---|---|---|---|---|---|
| End of *Late Heavy Bombardment* 3,800 Ma – | Convection in Earth's mantle causing volcanism | Within volcanic hot springs, thermal cycling by convection | Sorption heat engine (SHE) | ATP Synthase from *First Protein* | Origin of life |
| | | | Thermodielectric energy converter (TDEC) | Energy-converting biomembrane | Prokaryotic *bacterial* and (at end) *oxygenic photosynthesizers* |
| Huronian glaciations 2,300 Ma | Removal of greenhouse gas methane by biogenic oxygen | Thermal gradient above submarine hydrothermal vent | Rubber heat engine (RHE) | Flagellum from thermotether | Eukaryotes |
| Late Proterozoic glaciations 770–582 Ma | Snowball Earths due to $CO_2$ oscillation in atmosphere | | | Pharynx and tentacle from thermopharynx thermotentacle | Ediacaran diploblasts: fronds, anemones, jellyfish |
| Start of Cambrian 542 Ma | Formation of Gondwana supercontinent | | Thermogalvanic energy converter (TGEC) | Nerve from thermonerve | Cambrian triploblasts: most of the animal phyla we know today |

His "naturally occurring machines" are identified as (1) the chemiosmotic machinery and (2) enzymes that work on ATP. The "primitive protomachines" are identified as the thermosynthesizers that (1) eventually yielded bacterial photosynthesis (Fig. 11) or (2) by PBBM yielded the large ATP-consuming systems conspicuous in macroscopic life. The "original progenitor" that "was self-assembled" is identified as an FP carried by the self-organizing convection cell.

Thermosynthesis may be simply testable, e.g., by the use of radioactive tracers or the use of $CO_2$ fixation (Muller, 1995a, 2001, 2003). Potential niches for thermosynthesizers are ubiquitous (Muller, 1985) and include many environments discussed in previous volumes of the COLE series. Many extraterrestrial environments also seem suitable: (1) bodies in outer space rotating in the sunlight such as meteorites, asteroids, and comets; (2) convecting water underneath surface ice on the Moon, Mars, and the satellites of the outer planets; and (3) convecting atmospheres of these outer planets and of Venus (Muller, 1985, 1996, 2001, 2003). The possibility of similar "Carnot creatures" in the convecting stellar atmospheres of brown dwarf stars has independently been proposed (Jones, 1997).

## 9. Conclusion

The presented theory on how life was powered during early evolution should be of great interest, as it solves long standing problems, is testable, and has the potential for wide application. Thermosynthesis is firmly based in physics, chemistry, geology, and engineering. In these disciplines, thermodynamics holds a central place; the theory puts thermodynamics in the center of biology as well.

## 10. Acknowledgments

Wolter Kaper, Steph Menken, and Roel van Driel are thanked for their comments on the manuscript; Kevin Crosby is thanked for extensive proofreading.

## 11. References

Agar JN (1963) Thermogalvanic cells. Adv Electrochem Electrochem Eng 3:31–121

Aristov YI, Vasiliev LL, Nakoryakov VE (2008) Chemical and sorption heat engines: state of the art and development prospects in the Russian Federation and Republic of Belarus. J Eng Phys Thermophys 81:17–47

Atkins PW (1990) Physical chemistry, 4th edn. Oxford University Press, Oxford

Barnett MW, Larkman PM (2007) The action potential. Pract Neurol 7:192–197

Baross JA, Hoffman SE (1985) Submarine hydrothermal vents and associated gradient environments as sites for the origin and evolution of life. Orig Life Evol Biosph 15:327–345

Bekker A, Holland HD, Wang P-L, Rumble D, Stein HJ, Hannah JL, Coerzee LL, Beukes NJ (2004) Dating the rise of atmospheric oxygen. Nature 427:117–120

Beklemishev WN, Maclennan JM, Kabata Z (1969) Principles of comparative anatomy of invertebrates I: promorphology, II: organology. Oliver and Boyd, Edinburgh

Black S (1970) Pre-cell evolution and the origin of enzymes. Nature 226:754–755

Bohm D (2009) Some remarks on the notion of order. In: Waddington CH (ed) Sketching theoretical biology: towards a theoretical biology, vol 2. Transaction Publishers, New Brunswick, pp 18–40

Boyer PD (1993) The binding change mechanism for ATP synthase—some probabilities and possibilities. Biochim Biophys Acta 1140:215–250

Budin I, Bruckner RJ, Szostak JW (2009) Formation of protocell-like vesicles in a thermal diffusion column. J Am Chem Soc 131:9628–9629

Calvin M (1969) Chemical evolution. Oxford University Press, Oxford

Cardwell DSL (1971) From Watt to Clausius. Iowa State University Press, Iowa City

Carson EM, Watson JR (2002) Undergraduate students' understanding of entropy and Gibbs free energy. Univ Chem Educ 6:4–12

Clark RB (1964) Dynamics in metazoan evolution. The origin of the Coelom and segments, Clarendon Press, Oxford

de Groot SR, Mazur P (1962) Non-equilibrium thermodynamics. NHPC, Amsterdam

De Meis L (1989) Role of water in the energy of hydrolysis of phosphate compounds—energy transduction in biological membranes. Biochim Biophys Acta 973:333–349

Dewel RA (2000) Colonial origin for Eumetazoa: major morphological transitions and the origin of bilaterian complexity. J Morphol 243:35–74

Dewel RA, Dewel WC, McKinney FK (2001) Diversification of the Metazoa: ediacarans, colonies, and the origin of eumetazoan complexity by nested modularity. Hist Biol 15:193–218

Dunn C (2009) Siphonophores. Curr Biol 19:R233–R234

Duysens LNM, Amesz J, Kamp BM (1961) Two photochemical systems in photosynthesis. Nature 190:510–511

Dyson F (1985) Origins of life. Cambridge University Press, Cambridge

Einstein A (1910) Theorie der Opaleszenz von homogenen Flüssigkeiten und Flüssigkeitsgemischen in der Nähe des kritischen Zustandes. Ann Phys 33:1275–1298

Fast JD (1968) Entropy, 2nd edn. Centen, Hilversum

Florkin M (1972) From forces-of-life to bioenergetics. Compre Biochem 30:215–249

Glazebrook RW, Thomas A (1982) Solar energy conversion via a photodielectric effect. J Chem Soc Faraday Trans II 78:2053–2065

Guthrie WKC (1957) In the beginning. Some Greek views on the origins of life and the early state of man. Methuen, London

Haase R (1963) Thermodynamik der irreversiblen Prozesse. Dietrich Steinkopff, Darmstadt

Haydon DA, Hladky SB (1972) Ion transport across thin lipid membranes: a critical discussion of mechanisms in selected systems. Q Rev Biophys 5:187–282

Hoffman PF, Kaufman AJ, Halverson GP, Schrag DP (1998) A Neoproterozoic Snowball Earth. Science 281:1342–1346

Hu R, Cola BA, Haram N, Barisci JN, Lee S, Stoughton S, Wallace G, Too C, Thomas M, Gestos A, DelaCruz ME, Ferraris JP, Zakhidov AA, Baughman RH (2010) Harvesting waste thermal energy using a carbon-nanotube-based thermo-electrochemical cell. Nano Lett 10:838–846

Jammer M (1973) Entropy. In: Wiener PP (ed) Dictionary of the history of ideas, vol 2. Scribner, New York, pp 112–120

Johnston WJ, Unrau PJ, Lawrence MS, Glasner ME, Bartel DP (2001) RNA-catalyzed RNA polymerization: accurate and general RNA-templated primer extension. Science 292:1319–1325

Jones D (1997) The dark is light enough. Nature 385:401

Kasting JF, Ono S (2006) Palaeoclimates: the first two billion years. Philos Trans R Soc Lond B Biol Sci 361:917–929

Kaufman AJ, Knoll AH, Narbonne GM (1997) Isotopes, ice ages, and terminal Proterozoic earth history. Proc Natl Acad Sci USA 94:6600–6605

Keeling PJ (2010) The endosymbiotic origin, diversification and fate of plastids. Philos Trans R Soc Lond B Biol Sci 365:729–748

Kimura H, Watanabe Y (2001) Oceanic anoxia at the Precambrian-Cambrian boundary. Geology 29:995–998

Kirschvink JL (1992) Late Proterozoic low-latitude global glaciation: the Snowball Earth. In: Schopf JW, Klein C (eds) The Proterozoic biosphere: a multidisciplinary study. Cambridge University Press, Cambridge, pp 51–52

Knoll AH, Carroll SB (1999) Early animal evolution: emerging views from comparative biology and geology. Science 284:2129–2137

Kunze J, Stimming U (2009) Electrochemical versus heat-engine energy technology: a tribute to Wilhelm Ostwalds's visionary statements. Angew Chem Int Ed 48:9230–9237

Lambert FL (2002) Disorder—a cracked crutch for supporting entropy discussions. J Chem Educ 79:187–192

Li ZX, Bogdanova SV, Collins AS, Davidson A, De Waele B, Ernst RE, Fitzsimons ICW, Fuck RA, Gladkochub DP, Jacobs J, Karlstrom KE, Lu S, Natapov LM, Pease V, Pisarevsky SA, Thrane K, Vernikovsky V (2008) Assembly, configuration, and break-up history of Rodinia: a synthesis. Precambrian Res 160:179–210

Meert JG, Lieberman BS (2008) The Neoproterozoic assembly of Gondwana and its relationship to the Ediacaran-Cambrian radiation. Gondwana Res 14:5–21

Mitchell P (1979a) Keilin's respiratory chain concept and its chemiosmotic consequences. Science 206:1148–1159

Mitchell P (1979b) Compartmentation and communication in living systems. Ligand conduction: a general catalytic principle in chemical, osmotic and chemiosmotic reaction systems. Eur J Biochem 95:1–20

Moore WJ (1962) Physical chemistry, 4th edn. Longmans, London

Mullen JG, Look GW, Konkel J (1975) Thermodynamics of a simple rubber-band heat engine. Am J Phys 43(4):349–353

Muller AWJ (1985) Thermosynthesis by biomembranes: energy gain from cyclic temperature changes. J Theor Biol 115:429–453

Muller AWJ (1993) A mechanism for thermosynthesis based on a thermotropic phase transition in an asymmetric biomembrane. Physiol Chem Phys Med NMR 25:95–111

Muller AWJ (1995a) Were the first organisms heat engines? A new model for biogenesis and the early evolution of biological energy conversion. Prog Biophys Mol Biol 63:193–231

Muller AWJ (1995b) Photosystem 0: a postulated primitive photosystem that generates ATP in fluctuating light. Available on the internet. http://dare.uva.nl/document/175059

Muller AWJ (1996) Life on Mars? Nature 380:100

Muller AWJ (1998) Thermosynthesis: where biology meets thermodynamics. In: Lugowski W, Matsuno K (eds) Uroboros, or biology between mythology and philosophy. Arboretum, Wroclaw, pp 139–167

Muller AWJ (2001) The thermosynthesis model for the origin of life: implications for Solar System exploration. Marsbugs 8(15):3–6

Muller AWJ (2003) Finding extraterrestrial organisms living on thermosynthesis. Astrobiology 3:555–564

Muller AWJ (2005) Thermosynthesis as energy source for the RNA world: a model for the bioenergetics of the origin of life. Biosystems 82:93–102

Muller AWJ (2009) Emergence of animals from heat engines—part 1. Before the Snowball Earths. Entropy 11:463–512

Muller AWJ, Schulze-Makuch D (2006) Sorption heat engines: simple inanimate negative entropy generators. Physica A 362:369–381

Narbonne GM (2005) The Ediacara Biota: Neoproterozoic origin of animals and their ecosystems. Annu Rev Earth Planet Sci 33:421–442

Nicolis G, Prigogine I (1977) Self-organization in non-equilibrium systems. Wiley, New York

Ogawa M (2008) Mantle convection: a review. Fluid Dyn Res 40:379–398

Papineau D, Mojzsis SJ, Schmitt AK (2007) Multiple sulfur isotopes from Paleoproterozoic Huronian interglacial sediments and the rise of atmospheric oxygen. Earth Planet Sci Lett 255:188–212

Purcell EM (1977) Life at low Reynolds number. Am J Phys 45:3–11

Quickenden TI, Mua Y (1995) A review of power-generation in aqueous thermogalvanic cells. J Electrochem Soc 142:3985–3994
Reed RC (2006) The superalloys. Fundamentals and applications. Cambridge University Press, Cambridge
Riffat SB, Ma X (2003) Thermoelectrics: a review of present and potential applications. Appl Therm Eng 23:913–935
Schäfer G, Purschke W, Schmidt CL (1996) On the origin of respiration: electron transport proteins from archaea to man. FEMS Microbiol Rev 18:173–188
Simon MA (1971) The matter of life. Philosophical problems of biology. Yale University Press, New Haven, p 167
Sklar AA (2005) A numerical investigation of a thermodielectric power generation system. Thesis, Georgia Institute of Technology, Atlanta, Georgia
Stanley HE, Kumar P, Xu L, Yan Z, Mazza MG, Buldyrev SV, Chen S-H, Mallamace F (2007) The puzzling unsolved mysteries of liquid water: some recent progress. Physica A 386:729–743
Stern RJ, Avigad D, Miller NR, Beyth M (2006) Evidence of the Snowball Earth hypothesis in the Arabian-Nubian shield and the East African Orogen. J Afr Earth Sci 44:1–20
Strickberger MW (2000) Evolution, 3rd edn. Jones and Bartlett, Sudbury, p 337
Tanford C (1973) The hydrophobic effect. Wiley, New York
Tollefson J (2010) Fuel of the future? Nature 464:1262–1264
Urry DW (1992) Free energy transduction in polypeptides and proteins based on inverse temperature transitions. Prog Biophys Mol Biol 57:23–57
Valentine JW (2004) On the origin of phyla. University of Chicago Press, Chicago
Wekjira JF (2001) The VT1 shape memory alloy heat engine design. Thesis, Virginia Polytechnic Institute, Blacksburg, VA
Wiegand WB, Snyder JW (1934) The rubber pendulum, the Joule effect, and the dynamic stress strain curve. Inst Rubber Ind Trans Proc 10:234–262
Woese CR (1987) Bacterial evolution. Microbiol Rev 51:221–271

Biodata of **Albert Rimola**, **Mariona Sodupe**, and **Piero Ugliengo**, authors of "***Computational Simulations of Prebiotic Processes.***"

**Dr. Albert Rimola** received his B.S. in Chemistry from the Autonomous University of Barcelona (UAB), where he also completed in 2007 his Ph.D. in Theoretical and Computational Chemistry under the supervision of Prof. Mariona Sodupe. Then he continued his studies as a postdoctoral researcher at the University of Torino for 2 years with Prof. Piero Ugliengo, focusing his research on modeling the interaction and reactivity of biomolecules with inorganic biomaterials for interests in nanobiotechnology and prebiotic chemistry. In 2010, he returned to UAB to work on ab initio modeling and reactivity of interstellar dust particles.

E-mail: **albert@klingon.uab.es**

**Prof. Mariona Sodupe** received her Ph.D. in Chemistry from the Autonomous University of Barcelona (UAB) in 1990. Then she continued her studies as postdoctoral researcher at NASA Ames Research Center (California) for 2 years with Prof. C.W. Bauschlicher. In 1992, she returned to UAB where she is now full professor at the Department of Chemistry. Research areas of interest are metal-ligand systems of biological interest, structure and reactivity of radical cations, and surface properties of silica-based materials. Recent projects focus on the role of mineral surfaces in prebiotic chemistry. She is coauthor of about 130 papers on peer-reviewed international journals.

E-mail: **mariona@klingon.uab.es**

**Albert Rimola**

**Mariona Sodupe**



J. Seckbach (ed.), *Genesis - In The Beginning: Precursors of Life, Chemical Models and Early Biological Evolution*, Cellular Origin, Life in Extreme Habitats and Astrobiology 22, 345–362
DOI 10.1007/978-94-007-2941-4_20, 

**Prof. Piero Ugliengo** received his Laurea degree (5 years course and an experimental work) in chemistry from the University of Torino where he is associate professor of physical chemistry working at the Department of Chemistry IFM. He collaborated with Daresbury Laboratory (UK), Humboldt University of Berlin (DE), Universitat Autonoma de Barcelona (ES), and UPMC Paris (FR). The main scientific interests are on ab initio modeling of hydrogen-bond interactions in solids and at their surfaces with focus on silica and zeolites. Recent projects are on modeling the role of minerals in prebiotic chemistry and the adsorption process at the surface of inorganic biomaterials. He is coauthor of about 160 papers on peer-reviewed international journals.

E-mail: **piero.ugliengo@unito.it**

# COMPUTATIONAL SIMULATIONS OF PREBIOTIC PROCESSES

**ALBERT RIMOLA[1], MARIONA SODUPE[1], AND PIERO UGLIENGO[2]**
*[1]Departament de Química, Universitat Autònoma de Barcelona, 08193 Bellaterra, Spain*
*[2]Dipartimento Chimica IFM, Università di Torino & NIS Centre of Excellence, Via P. Giuria 7, 10125 Torino, Italy*

## 1. Introduction

One of the most intriguing and fascinating topics in science is to figure out how the complexity of life arose. It is known that planet Earth is 4.5 billion years old with the first evidence of a cyanobacterium dating back to 3.5 billion years (Schopf, 1993). During the years, a lot of experiments have been carried out, and theories have been proposed in an attempt to clarify the elementary chemical steps that led to the appearance of life, which cover different stages according to the degree of complexity of the entities involved: (a) synthesis of basic building blocks in manners compatible with the hostile environment in the early Earth, (b) formation of biopolymers (nucleic acids, proteins), and (c) emergence of supramolecular architectures. Understanding these complex plots involve an interdisciplinary approach in which chemistry, geology, catalysis by minerals, and radiation effects play the key role.

Different scenarios have been proposed for the abiotic synthesis of relevant prebiotic molecules, from inert chemical compounds. Early Earth atmosphere (Miller, 1953,1955), the interstellar medium (Bernstein et al., 2002; Muñoz-Caro et al., 2002; Nuevo et al., 2008) or comets (Pizzarello, 2006), and deep-sea hydrothermal vents (Huber et al., 1998) have been proved to be powerful factories of important prebiotic molecules. Since the pioneering work of Stanley Miller and Harold Urey (Miller, 1953), the synthesis of most basic building blocks, and other organic molecules, has been demonstrated in different studies including experiments using different gas mixtures (Cleaves et al., 2008; Miller, 1955; Miyakawa et al., 2002; Oró, 1961), ultraviolet radiation in interstellar ice analogues (Bernstein et al., 2002; Muñoz-Caro et al., 2002), or powdered minerals (Cody et al., 2000; Huber and Wächtershäuser, 1998; Saladino et al., 2008).

More complex is the second stage by which the basic molecules of life are selected, concentrated, and assembled into key biological macromolecules, including proteins, carbohydrates, and DNA. British biophysicist J. D. Bernal suggested, as early as 1949, that mineral surfaces may have played a special role (Bernal, 1949) since they have the potential to select and promote polymerization by (a) providing adsorption sites which can immobilize, protect, and concentrate the monomer building blocks from the dilute soup and (b) by lowering the activation barrier of the condensation reaction due to the catalytic effect of the surface active sites. Since then, it has been recognized that rocks and minerals provide attractive alternative surfaces for molecular concentration and assembly, and subsequent experiments have supported Bernal's hypothesis (Ferris, 2006; Ferris et al., 1996).

Despite the experimental evidences on the formation of prebiotic molecules in icy particles and on the catalytic role of mineral surfaces, very little is known about the mechanistic steps occurring at the surface of these materials. In this context, quantum chemical studies can be very useful since they can provide information at a molecular level on the mechanism of these reactions. In this chapter, we will review the contributions of our groups on this topic. First, we will consider the formation of amino acid precursors in the interstellar and circumstellar medium, in which the catalytic role played by icy particles will be evidenced. Second, we will discuss the role of silica defective surface rings in the capture and exogenous delivery of molecular building blocks. Third, we will address the adsorption of amino acids on silica materials. Finally, we will discuss the role of active sites (Brønsted and Lewis) of a feldspar surface, one of the most abundant aluminosilicates on the Earth crust, on the polymerization of amino acids.

## 2. Computational Framework

As mentioned, a number of different scenarios have been put forward as conceivable "prebiotic factories" for building up complex biopolymers from very simple molecular building blocks (Hazen, 2005). This environmental variability envisages reactions occurring in gas-phase, in water excess, and at the surface of either minerals or icy particles, these latter acting as effective catalysts. Also, a large and diverse pool of reactive processes may occur, involving proton/electron transfer, nucleophylic/electrophylic attacks, and so on. Accurate modeling of these processes should, therefore, rely on quantum mechanical methods. Within this context, approaches based on the refinement of the wave function such as Møller-Plesset (MP2) and coupled cluster (CC) theory have become the methods of choice when high accuracy is needed, with the CCSD(T) method being capable to achieve chemical accuracy (Sherrill, 2010). Applicability of these methods is, however, hampered by the size of the system to model as, for instance, the simulation of the prebiotic synthesis of biopolymers, which involves rather large molecular building blocks (*i.e.*, amino acids, RNA/DNA bases). However, since the late 1990s, approaches based on electron density rather than on the wave function have become computationally cheaper alternatives to MP2 and CCSD(T), while achieving acceptable accuracy by means of well-designed electron density functionals (Sousa et al., 2007). Nevertheless, methods based on gradient-corrected density functional (*e.g.*, PBE, PW91, BLYP) as well as hybrid methods containing a fraction of exact Hartree-Fock exchange (*e.g.*, B3LYP), while being reasonable for chemical reactivity, are unable to cope with purely London forces, which are essential to properly describe attachment of molecular building blocks onto mineral and icy surfaces. A suitable solution has been recently proposed by Grimme (2006) and adopted by us in some of our simulations (Rimola et al., 2007, 2009a). When coupled to a given functional, *e.g.*, B3LYP, the Grimme's correction is refereed as B3LYP-D and leaves unaltered the original accuracy of the B3LYP functional for short-range interactions (bond breaking/making), while improving dramatically the long-range tail of the interaction (dominated by London forces).

As aforementioned, quantum mechanical simulations based on DFT may deal with reactions catalyzed by mineral surfaces (Bernal, 1949) or by the water at the surface of icy particles (Rimola et al., 2010a). In these cases, one has to find a way to simulate the catalytic effects exerted by the active sites present at a given surface. Basically, two methods are at hand: (a) the adoption of periodic boundary conditions to simulate an infinite 2D surface, whose unit cell contains the catalytic sites, and (b) the adoption of a subset of atoms cut out from model (a), which is large enough to contain the catalytic sites. Approach (a) appears the most natural one as, indeed, catalysis at the mineral surfaces envisages a crystalline surface exposing active sites. Computer codes apt to properly treat long-range electrostatic effects due to the infinite nature of a crystalline surface have been developed since long ago, either based on expanding the wave function in terms of plane waves (Hafner, 2008) or by means of localized Gaussian-type orbitals (Dovesi et al., 2005). The pros/cons of both approaches have been recently summarized by some of us (Tosoni et al., 2007). In general, the limitations behind approach (a) are due to the paucity of quantum mechanical methods that can be used to describe an infinite system. Indeed, all computer codes based on periodic boundary conditions are limited to DFT methods which, for the case of plane-waves basis set, rarely include the most accurate hybrid functionals, contrarily to Gaussian-type basis set calculations. Exceptions in which MP2 has been used to treat bulk or surface systems have recently appeared (Pisani et al., 2005) and can be considered the frontier of the simulation for crystalline materials. One further technical drawback of approach (a) is the far less-developed technology to search and characterize transition state structures, which are essential for the prediction of reaction energy barriers, a crucial point in the present context. Despite the recent progress in that direction (see Rimola et al., (2010b) and references therein), algorithms developed within the periodic boundary conditions paradigm are less powerful than those encoded in programs apt to treat molecular systems quantum mechanically and are normally used to study rather simple reactions at the surface of regular metals. Approach (b) does not suffer from the above drawbacks as the infinite nature of the system is reduced to a "finite cluster" that, when properly healed at its boundaries, becomes a molecule to be calculated with the variety of quantum molecular programs (Sauer et al., 1994). In this case, complex reactions can be studied by fully characterizing both minima and transition states.

Additionally, quantum level of theory is not limited to DFT so that, depending on the size of the cluster, even CCSD(T) can be used for high accuracy. This approach is plagued by the need of cutting covalent/ionic bonds present in the surface of the extended system with the need to "heal" the dangling bonds with proper terminal atoms. However, the size of the adopted cluster should be large enough to guarantee that the chemical features of the catalytic sites represented by the cluster are the same of those contained in the original infinite system. This implies that the size of the cluster may become prohibitively large, reducing the abovementioned advantages when adopting a molecular computer code. A possible way out is to adopt embedding techniques like the ONIOM method, proposed by Morokuma and coworkers (Morokuma, 2003). The trick is to treat the region close to the catalytic center by a highly accurate method (MP2, CCSD(T)), whereas the surrounding region is treated at a cheaper level (DFT, semiempirical or even molecular mechanics). This approach has been evolved to a point in which complex reactions of the kind of those relevant for the present topic can be studied within a controllable accuracy, and it is the method of choice adopted for many of the calculations reviewed here.

## 3. Results and Discussion

### 3.1. FORMATION OF AMINO ACID PRECURSORS IN THE DEEP SPACE

There is a wide variety of complex organic molecules in the clouds of gas and dust which pervade the galaxies. Universe, therefore, might have been an important source of molecules for the primordial Earth since some of the terrestrial prebiotic molecules could have firstly been synthesized in the interstellar and circumstellar medium (ICM) and then transported to the early Earth by comets, asteroids, and meteorites. In this line, careful analysis of carbonaceous meteorites and dust cometary grains reveals the presence of several amino acids (Elsila et al., 2009; Pizzarello, 2006), although they have not been directly detected in ICM yet (Jones et al., 2007; Snyder et al., 2005). Additionally, several experiments simulating ICM conditions in Earth laboratories resulted in the successful synthesis of amino acids,(Bernstein et al., 2002; Muñoz-Caro et al., 2002; Nuevo et al., 2008), in which the catalytic role played by icy dust grain particles is evidenced. In this sense, theoretical works focused on thermally activated astrochemical reactions show the strong catalytic effect of the ice water ($H_2O$-ice) due to the proton mobility at the ice surface itself (Koch et al., 2008), similarly to that observed for heterogeneous reactions occurring on stratospheric ices (Abbatt, 2003).

One of the possible routes to form amino acids is the Strecker synthesis, and it has been postulated to occur in ICM, since the raw molecules are available in relative abundance. Glycine (Gly) formation adopting Strecker-type reactions can be achieved following three steps: (a) condensation of $NH_3$ and $H_2C = O$ followed

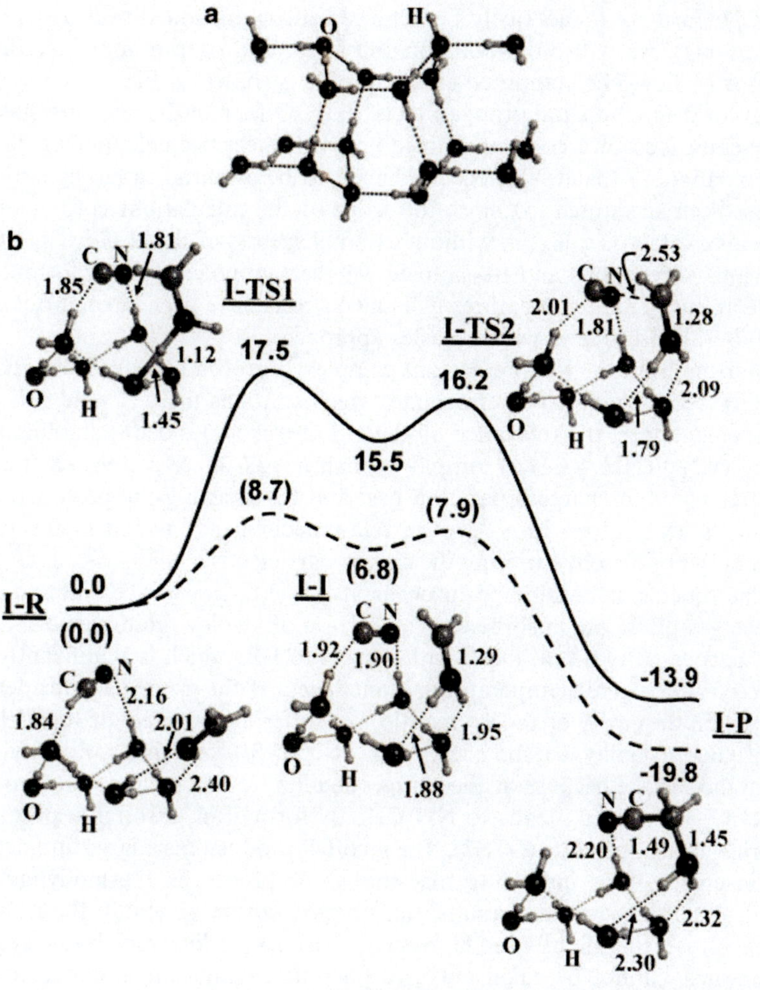


**Figure 1.** (**a**): cluster model of the $H_2O$-ice employed to simulate an icy grain particle. (**b**): ZPE-corrected B3LYP/6-31+G(d,p) potential energy profile for the reaction of HCN + NH = $CH_2$ → $NH_2CH_2CN$ in presence of $H_2O$-ice. *Dashed lines* refer to reactions occurring in a dielectric medium of $\varepsilon = 78$. Energies in kcal $mol^{-1}$, bond distances in Å.

by $H_2O$ elimination to form NH=$CH_2$; (b) C – C bonding between NH=$CH_2$ and HCN to give aminoacetonitrile, $NH_2CH_2CN$; and (c) hydrolysis of $NH_2CH_2CN$ to yield Gly. In this section, quantum mechanical results obtained for the second step, *i.e.*, formation of the aminoacetonitrile precursor of Gly, in the presence of a $H_2O$-ice cluster model as catalyst (Fig. 1a) are presented. It should be noted that the first and third step of the Strecker reaction were characterized to have too high energy barriers, even in the presence of the $H_2O$ ice model, to occur at cryogenic temperatures (see Rimola et al., 2010a for details). Thus, and assuming that

$NH=CH_2$ formation can easily be achieved through radical-radical reactions (Woon, 2002), we will only focus on the formation of the aminoacetonitrile precursor of Gly. The computed energy profile is shown in Fig. 1b. In order to account for the ice bulk long-range effects, the $H_2O$-ice *plus* the reactants has been further embedded in a continuum medium with dielectric constant $\varepsilon=78$. It is worth mentioning that UV effects, assumed to be essential in lab experiments, have not been accounted for, since the scope of the calculations is to determine the intrinsic catalytic role (*i.e.*, without external agents) of the $H_2O$-ice mantles in astrochemical reactions and to estimate whether aminoacetonitrile formation is feasible at cryogenic temperatures. All calculations have been carried out at the B3LYP/6-31+G(d,p) level with a cluster approach.

Starting from the **I-R** prereactant complex, a proton transfer from HCN to $NH=CH_2$ leads to the **I-I** intermediate, which contains the $CN^-$ and $NH_2CH_2^+$ ionic species. Here, the relevance of the $H_2O$-ice is twofold: (a) stabilizing the charged $(CN^-)/(NH_2=CH_2^+)$ ion pair, which in gas-phase cannot exist due to the extremely high charge separation cost and (b) assisting the proton transfer through its icy protons in a "proton relay mechanism," which facilitates the proton mobility thereby lowering the energy barrier.

The process, if considered to occur at the $H_2O$-ice surface, *i.e.*, molecules from the gas-phase are adsorbed at the surface of the icy grain particle, has an energy barrier of 17.5 kcal $mol^{-1}$ with respect to **I-R**, which is significantly high to proceed at cryogenic temperatures. Nonetheless, if the process is considered to occur within the cavity of the icy mantle, *i.e.*, suffering the effect of the dielectric response due to bulky ice, the energy barrier is of 8.7 kcal $mol^{-1}$, notably lower than on the surface because of the charges stabilization of the ion pair. From **I-I**, a direct C – C bonding leads to $NH_2CH_2CN$ formation, through a practically barrierless transition state (**I-TS2**). The final **I-P** product has a large and negative reaction energy, thus indicating that the whole process is thermodynamically favored. According to these results, the reaction occurring within the ice water, although presenting slightly too high energy barriers (~9 kcal $mol^{-1}$) for cryogenic temperatures, cannot be ruled out as a possible channel for aminoacetonitrile formation. This seems to be consistent with the detected aminoacetonitrile signature found in the ICM (Belloche et al., 2008). Accordingly, it is then conceivable that aminoacetonitrile could have been accumulated as a reservoir of Gly precursor in ICM, carried on cometary and interplanetary dust particles, and seeded the early Earth by microasteriodal bombardment, behaving as potential source for Gly formation.

## 3.2. EXOGENOUS DELIVERY OF MOLECULAR BUILDING BLOCKS

The main body of interplanetary dust particles (IDPs), besides organic fractions and ices, mainly contains inorganic silicates. Because of the several energy inputs in the deep space, one can find phyllosilicates and anhydrous silicates, both in

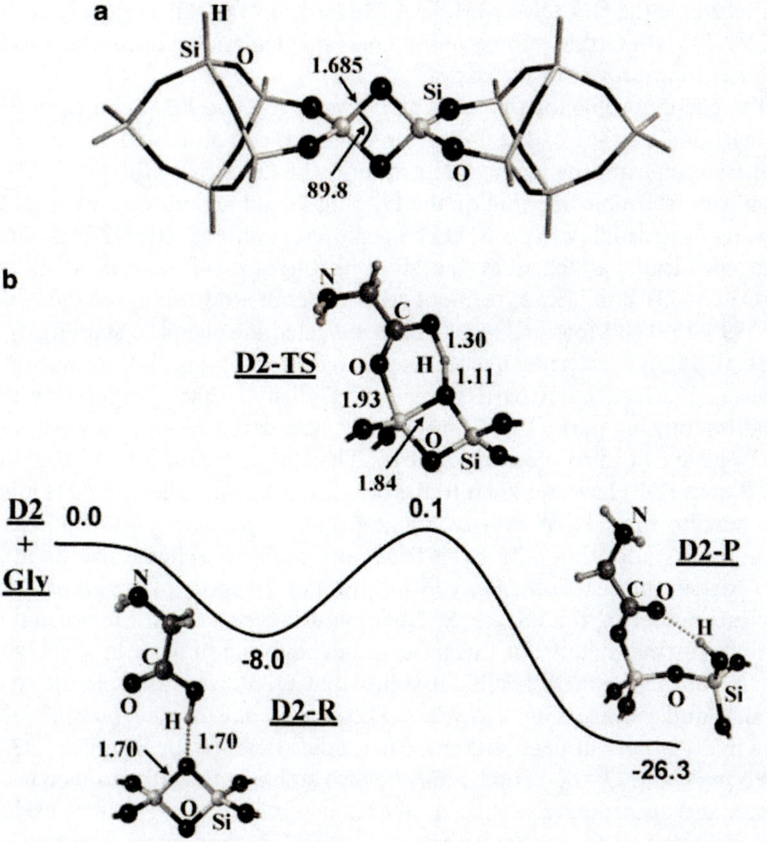


**Figure 2.** Section (**a**): cluster model employed for the D2 defect. Spheres/sticks treated at B3LYP/6-31+G(d,p)/MNDO levels with ONIOM2 approach. Section (**b**): ZPE-corrected B3LYP//ONIOM2 potential energy profile for the reaction of the D2 defect with Gly. Relative energies in kcal mol$^{-1}$; bond distances in Å; angles in degrees.

amorphous and crystalline phases, the latter one only possible by T annealing of the former one (*e.g.*, when orbiting close to stars, T > 1,000°C). Silica IDPs may thus undergo condensation processes such as Si–OH + Si–OH → Si–O–Si + $H_2O$, which may end up with some strained surface defects. Among them, the $(SiO)_2$ defective surface rings (so-called D2) are of particular interest because they are extremely reactive toward molecules (Bolis et al., 1991; Morrow et al., 1976) but seem to exhibit great stability in space due to the very low pressures and gas densities (Rimola and Ugliengo, 2008). The question to be addressed here is: might D2 defects have played some role in capturing and transporting molecular building blocks from deep space to our planet? To provide some answers, in this section, we present quantum chemical results for the reaction between Gly and a D2 cluster model (shown in Fig. 2a). Calculations have been carried out optimizing the structures at the ONIOM2[B3LYP/6-31+G(d,p):MNDO] level and performing full B3LYP/6-31+G(d,p) single-point energy calculations onto the ONIOM2 geometries to improve the accuracy.

The energy profile for the reaction between Gly and D2 is shown in Fig. 2b. The nucleophilic attack of the CO group toward the Si atom with a simultaneous proton transfer from the acidic OH group to the O atom of the Si–O–Si bond yields a very favorable opening of the D2 ring via a barrierless process (**D2-TS**), evidencing the feasibility to occur at cryogenic temperatures. The **D2-P** product has glycine chemically attached at the silica through a surface mixed anhydride $Si_{surf}$–O–C(=O) bond, in agreement with experimental measurements (Basiuk et al., 1990–1991).Theoretical results have indicated the chemical stability of **D2-P** against aggressive external agents (i.e., single molecules, UV radiation, and cosmic rays) present in ICM (Rimola and Ugliengo, 2009), suggesting that its population forming part of IDPs might have been large enough to seed the early Earth by means of micrometeoritic entries. The key point now is to address the fate of **D2-P** once IDPs have landed into the primordial ocean. Theoretical results (not shown here for the sake of brevity) suggest that the participation of at least four $H_2O$ molecules adopting a "proton transport catalyst" scheme results in a free energy barrier and reaction energy (T = 298 K) of 21 and −14 kcal $mol^{-1}$, respectively, thus rendering the release of Gly a plausible process in the normal ocean conditions (further details for this process can be found in Rimola and Ugliengo, 2009). According to these results, it seems that D2 defects are suitable to react, transport, and release Gly. However, D2 is just one of the possible defects among many others. Indeed, other surface silica defects, for instance, D3 rings, Si = O bonds, and $O=Si_3$ centers, may be also stable in the anhydrous conditions of space, and accordingly, a variety of organic molecules might have undergone similar reactions to those showed to occur with D2, thus, suggesting the significant role of silica surface defects in capturing and delivering prebiotic organic molecules onto the primitive Earth.

### 3.3. ADSORPTION OF MOLECULAR BUILDING BLOCKS ON OXIDE MINERALS

In a sequence of organizational events, the emergence of molecular building blocks (MBB) is followed by their polymerization. However, MBB must be previously concentrated to be in close proximity in order to facilitate condensation reactions. Bernal's hypothesis (Bernal, 1949) advocates the special role of oxide minerals in these processes, since they are ubiquitous materials capable to adsorb, capture, and concentrate MBB. Thus, studies oriented to the adsorption of MBB on oxide mineral surfaces are of relevance since they can provide quantitative data (*i.e.*, structural and energetic features) of the MBB in their previous stage to polymerization. On the other hand, defining a scale for the surface affinity of amino acids is a major issue since adsorption selectivity from amino acids mixtures

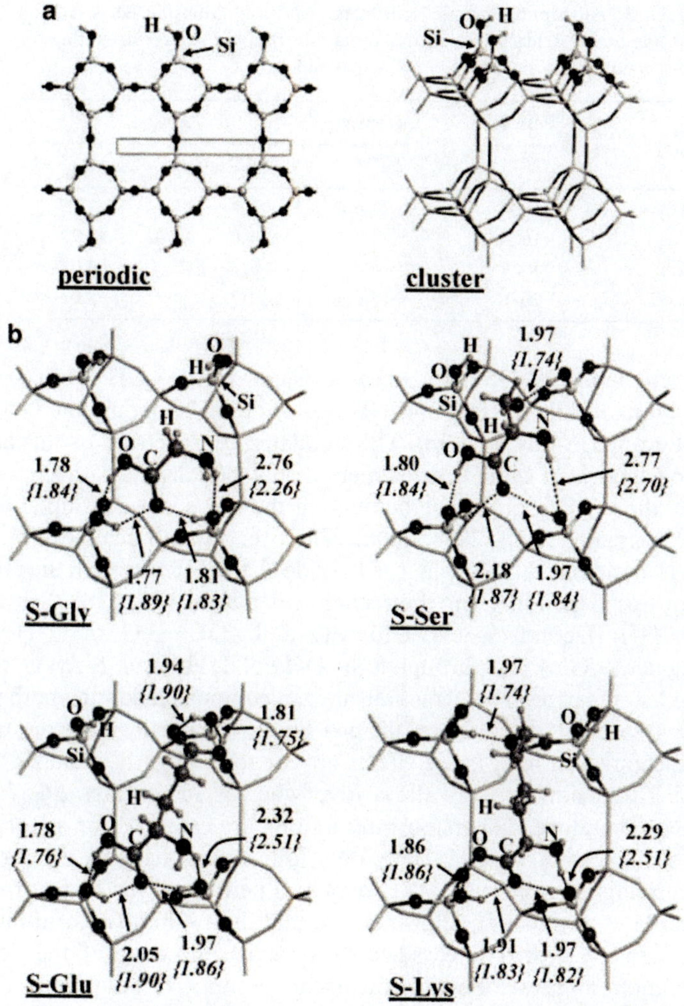


**Figure 3.** Section (**a**): periodic (showing unit cell) and cluster (atoms as balls/sticks represent high-/low-level zones in the ONIOM2 calculation) models for the moderately hydroxylated silica surface. Section (**b**): optimized geometries of amino acids interacting with the surface models. Bare values refer to cluster calculations, in *curl brackets* to periodic calculations. Bond distances in Å.

would be related with the selection of amino acids used in protein building. We have considered the adsorption of 15 amino acids with a moderately hydroxylated (*i.e.,* containing Si–OH silanol groups) silica surface by means of periodic (B3LYP/6-311G(d,p)) and cluster ONIOM2[B3LYP/6-311++G(d,p): MNDO] calculations (surface models shown in Fig. 3a). As an example, Fig. 3b shows the most stable adducts of some selected amino acids: glycine (Gly), serine (Ser), glutamic acid (Glu), and lysine (Lys).

**Table 1.** Adsorption energies (electronic $\Delta E^C$, and Gibbs at 298 K $\Delta G^C_{298}$) of the amino acids/silica systems. Final adsorption energies accounting for dispersion interactions $\Delta G_{298} + D$. In kcal mol$^{-1}$.

| | Periodic | Cluster | | |
|---|---|---|---|---|
| | $\Delta E^C$ | $\Delta E^C$ | $\Delta G_{298}{}^C$ | $\Delta G_{298}{}^C + D$ |
| Gly | −12.3 | −11.6 | 4.4 | −3.1 |
| Ser | −16.1 | −11.3 | 6.0 | −4.9 |
| Glu | −19.9 | −18.3 | 1.8 | −8.6 |
| Lys | −20.0 | −17.6 | 1.7 | −8.9 |

The interaction of Gly with silica envisages two Si–OH···O=C hydrogen bonds (H-bonds) while the OH and $NH_2$ groups are H-bonding the O atoms of the silanol groups (S-Gly, Fig. 3b). The simulated IR spectrum of this adduct is in very good agreement with the experimental one, specially the bathochromic shift suffered by the C=O group when contacting the OH surface group (Lomenech et al., 2005; Rimola et al., 2008, 2006). The other amino acids exhibit similar backbone H-bonding features, whereas the side chains interact with the remaining Si–OH groups depending on their chemical nature (Fig. 3b): Ser through a Si–OH···OH H-bond (S-Ser); Glu with a Si–OH···O C and OH···O(H)Si H-bonding net (S-Glu); Lys through a Si–OH···$NH_2$ H-bond (S-Lys).

Table 1 summarizes the interaction energies computed adopting both periodic and cluster treatments. Periodic results provide systematically more negative $\Delta E^C$ values, indicating that long-range effects appear to be slightly stabilizing. $\Delta G_{298}{}^C$ results (*i.e.*, dispersion free) are all positive, whereas, in contrast, $\Delta G_{298}{}^C + D$ are all negative. Therefore, dispersion interactions are essential for adsorption at the silica surface to take place. Here, only four amino acids have been shown. Notwithstanding, as mentioned, the study was performed for 15 natural amino acids (Rimola et al., 2009a), allowing to establish an amino acid affinity scale toward moderately hydroxylated silica surfaces as a function of the side chain chemical functionalities following the trend: nonpolar ~ polar/neutral < polar amidic ~ polar/acidic < polar/basic. Moreover, it has been found that the computed adsorption energies are reasonably well correlated with the hydropathy index, a number obtained experimentally reflecting the hydrophobic and hydrophilic character of natural amino acids (see Fig. 4).

## 3.4. POLYMERIZATION OF MBB ACTIVATED BY MINERAL SURFACES

Formation of biopolymers from the polymerization of their MBB envisages condensation reactions in which water molecules are released. The major concern of these processes is that they are disfavored in excess of water (Lambert, 2008), *i.e.*, hydrolysis is thermodynamically favored. For instance, it is estimated that two alanines condensate with a $\Delta G_{310}{}^0 \approx 4$ kcal mol$^{-1}$ in water solution at neutral pH

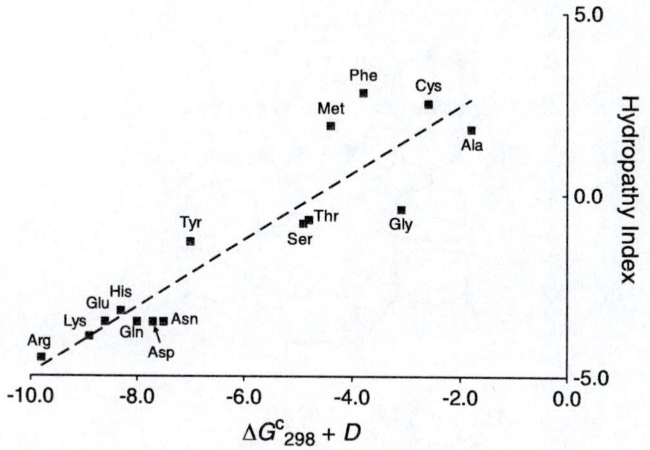


**Figure 4.** Plot of the final adsorption energies ($\Delta G_{298}{}^{C} + D$, in kcal $mol^{-1}$) vs hydropathy index of amino acids (see text for further details) obtained by Kyte and Doolittle (1982).

(Brack, 2007). In addition to that, theoretical calculations showed that the same reaction has a free energy barrier $\Delta G_{298}{}^{\neq} \approx 50$ kcal $mol^{-1}$, indicating that polymerization is also kinetically hindered. The Bernal's hypothesis (Bernal, 1949) invoked mineral surfaces to help in: (a) activating MBB through adsorption at surface active sites and lowering the energy barriers and (b) protecting the newly formed biopolymers from hydration and hydrolysis as they are firmly adsorbed at the surface. In this section, peptide bond formation processes between three glycine molecules have been studied quantum mechanically in the presence of a model of aluminosilicate sanidine, quite common at the Earth's crust, and suggested to be relevant as active mineral in the biopolymerization (Smith, 1998). This surface model contains both a Lewis (coordinatively unsaturated Al atom) site and an acidic Brønsted (Si–O(H)–Al) site in close proximity (see Fig. 5a), allowing for a synergic action (Rimola et al., 2007). It is worth mentioning that this model simulates just one of the many possible situations. Calculations have been carried out optimizing the structures at the ONIOM2[B3LYP/6-31+G(d,p):MNDO] level and performing full B3LYP/6-31+G(d,p) single-point energy calculations onto the ONIOM2 geometries to improve the accuracy. Dispersion forces have also been accounted for using the Grimme's correction.

The energy profile and the stationary points located for Gly polymerization are shown in Fig. 5b. The prereactant **F-G** complex involves one Gly chemically attached at the Lewis site through its N atom, whereas the Brønsted site H-bonds Gly through the COOH group. The condensation of the adsorbed Gly with a second one envisages three simultaneous steps: (a) a nucleophilic attack from the amino N atom of the incoming Gly toward the carboxyl C atom of the adsorbed one, (b) a proton transfer from the Brønsted site to the adsorbed Gly OH group, and (c) a proton transfer from the incoming amino group to a surface oxygen atom.

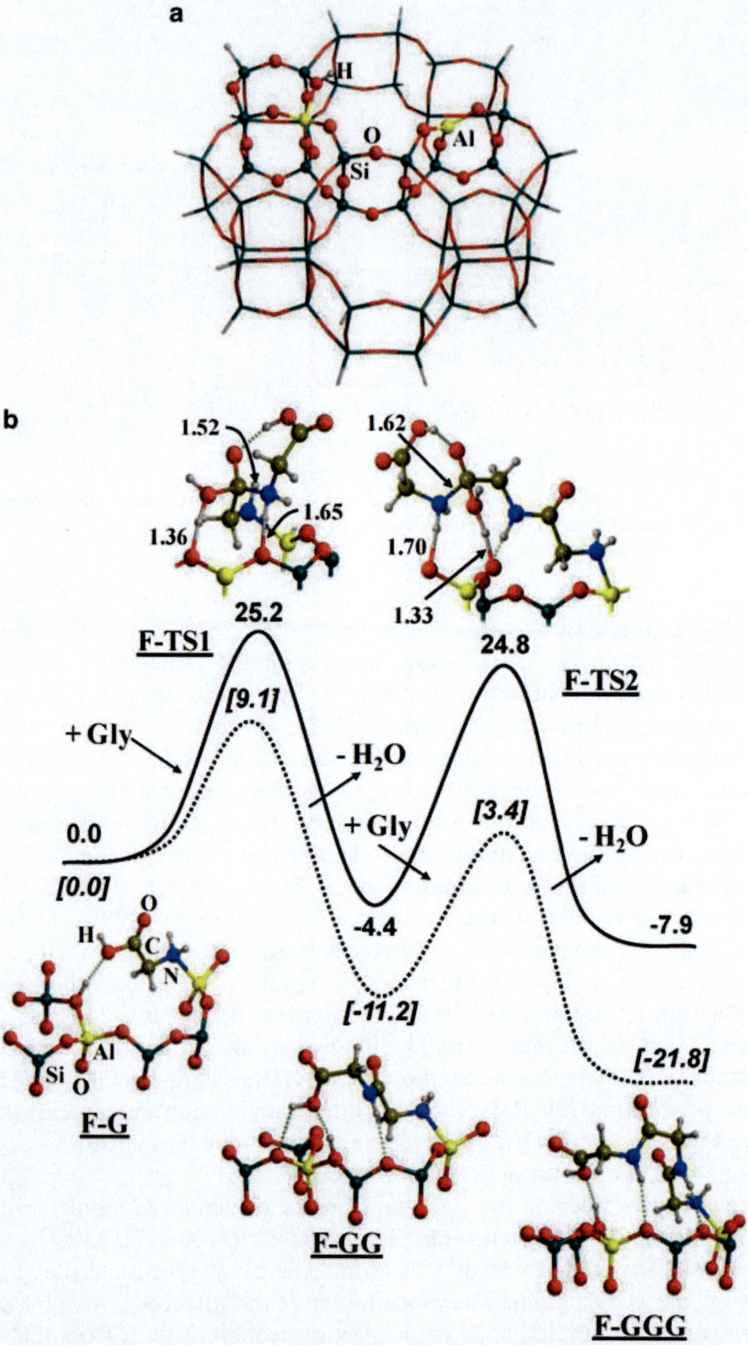


**Figure 5.** Section (**a**): cluster model of the aluminosilicate sanidine surface. Spheres/sticks as high-/low-level zones in the ONIOM2 calculations. Section (**b**): B3LYP//ONIOM2 free energy profiles for the condensation of three glycine molecules. *Solid line* refers to relative $\Delta G_{298}$, *dotted lines* to $\Delta G_{298} + D$. Relative energies in kcal mol$^{-1}$, bond distances in Å.

These processes form an eight-membered ring in the transition state **(F-TS1)** rendering a low-constrained structure that lowers the energy barrier to 25 kcal mol$^{-1}$ (half of the gas-phase uncatalyzed process). Furthermore, if dispersive forces are accounted for, the energy barrier decreases to 12 kcal mol$^{-1}$, resulting in an even more dramatic lowering. The resulting **F-GG** product remains attached to the surface giving reaction energies of −4 ($\Delta_r G_{298}$) and −11 ($\Delta_r G_{298} + D$) kcal mol$^{-1}$. Therefore, the mineral surface helps in stabilizing both the transition structure and the reaction product, which is further enhanced by the dispersion forces occurring between the surface and the biopolymer. In this particular mechanism, the Lewis site acts as a Gly concentrator whereas the Brønsted site activates the condensation via a double proton transfer (Rimola et al., 2007, 2009b).

The point to address now is: may **F-GG** undergo further condensation reactions with other incoming amino acids to be further elongated? The second part of the energy profile focuses on this aspect. The reaction of a third Gly molecule proceeds similarly to the first condensation, *i.e.*, formation of an eight-membered transition state ring due to the simultaneous proton transfers and the C-N formation **(F-TS2)**. The energies of this second process results in a $\Delta G_{298} + D = 14$ kcal mol$^{-1}$ and a $\Delta_r G_{298} + D = -10$ kcal mol$^{-1}$ with respect to **F-GG**, *i.e.*, the reaction is still favored by the presence of the feldspar surface. This fact suggests that the occurrence of amino acids polymerization on the surface may be an iterative process (namely, a successive elongation of the peptide) if the surface presents enough Brønsted acidic sites. Furthermore, since $\Delta_r G_{298} + D$ values are larger and more negative at each step of the polymerization, the newest peptides are expected to stay more and more tightly attached to the surface at each elongation step. These results are in agreement with the suggestion of L. E. Orgel, who stated that oligomers may be elongated indefinitely by repeated cycles at the mineral surfaces and that the affinity of a mineral surface for an oligomer increases with its length, rendering the adsorption essentially irreversible for sufficiently long oligomers (Orgel, 1998, 2004).

## 4. Conclusions

This chapter reviews our computational studies on the formation of amino acids precursors in icy particles, the adsorption of amino acids at a fully hydroxylated silica surface, and the role of feldspar active sites on the polymerization of amino acids. These studies have provided molecular insights on relevant prebiotic processes. Basically, they show that (a) synthesis of glycine precursor in icy particles is a plausible process at the interstellar medium conditions, mainly due to the catalytic role of ice, which lowers the energy barrier through a proton relay mechanism and stabilizes charged separated species; (b) defective surface rings in silica particles can play a major role in capturing glycine and further delivering it onto the primitive Earth; (c) amino acids adsorption on an hydroxylated surface is driven by hydrogen bond and dispersive interactions between the amino acid and the surface; and (d) Bronsted and Lewis sites in a feldspar surface play a synergic role in the catalysis of amino acid polymerization, i.e., the Lewis site strongly attaches glycine to the surface whereas the Bronsted site allows for an easy polymerization. Moreover, surface-biomolecule interactions render the condensation reaction energy to be more exoergonic, thus, favoring further polymerization. Overall, it is shown that, at present, quantum chemical calculations on well-designed models can provide a fundamental understanding on the mechanistic aspects involved in chemical evolution. It should be noted, however, that interpretation of theoretical results based on simplified models should be made with caution and that further improvements to describe these complex systems, including better surface models, role of the environment, and dynamical effects, are major goals to be achieved in the following years.

## 5. References


Abbatt JPD (2003) Interactions of atmospheric trace gases with ice surfaces: adsorption and reaction. Chem Rev 103:4783–4800

Basiuk VA, Gromovoy TY, Golovaty VG, Glukhoy AM (1990–1991) Mechanisms of amino acid polycondensation on silica and alumina surfaces. Orig Life Evol Biosph 20:483–498

Belloche A, Menten KM, Comito C, Müller HSP, Schilke P, Ott J, Thorwirth S, Hieret C (2008) Detection of amino acetonitrile in Sgr B2(N). Astron Astrophys 482:179–196

Bernal JD (1949) The physical basis of life. Proc Phys Soc B 62:597–618

Bernstein MP, Dworkin JP, Sandford SA, Cooper GW, Allamandola LJ (2002) Racemic amino acids from the ultraviolet photolysis of interstellar ice analogues. Nature 416:401–403

Bolis V, Fubini B, Marchese L, Martra G, Costa D (1991) Hydrophilic and hydrophobic sites on dehydrated crystalline and amorphous silicas. J Chem Soc Faraday Trans 87:497–505

Brack A (2007) From interstellar amino acids to prebiotic catalytic peptides: a review. Chem Biodivers 4:665–679

Cleaves HJ, Chalmers JH, Lazcano A, Miller SL, Bada JL (2008) A reassessment of prebiotic organic synthesis in neutral planetary atmospheres. Orig Life Evol Biosph 38:105–115

Cody GD, Boctor NZ, Filley TR, Hazen RM, Scott JH, Sharma A, Yoder HS Jr (2000) Primordial carbonylated iron-sulfur compounds and the synthesis of pyruvate. Science 289:1337–1340

Dovesi R, Orlando R, Civalleri B, Roetti C, Saunders VR, Zicovich-Wilson CM (2005) CRYSTAL: a computational tool for the ab initio study of the electronic properties of crystals. Z Kristallogr 220:571–573

Elsila JE, Glavin DP, Dworkin JP (2009) Cometary glycine detected in samples returned by Stardust. Meteorit Planet Sci 44:1323–1330

Ferris JP (2006) Montmorillonite-catalysed formation of RNA oligomers: the possible role of catalysis in the origins of life. Phil Trans R Soc B 361:1777–1786

Ferris JP, Hill-Jr AR, Liu R, Orgel LE (1996) Synthesis of long prebiotic oligomers on mineral surfaces. Nature 381:59–61

Grimme S (2006) Semiempirical GGA-type density functional constructed with a long-range dispersion correction. J Comput Chem 27:1787–1799

Hafner J (2008) Ab-initio simulations of materials using VASP: density-functional theory and beyond. J Comput Chem 29:2044–2078

Hazen R (2005) Genesis: the scientific quest for life's origins. Joseph Henry Press, Wahington, DC

Huber C, Wächtershäuser G (1998) Peptides by activation of amino acids with CO on (Ni, Fe)S surfaces: implications for the origin of life. Science 281:670–672

Jones PA, Cunningham MR, Godfrey PD, Cragg DM (2007) A search for biomolecules in Sagittarius B2 (LMH) with the Australia Telescope Compact Array. Mon Not R Astron Soc 372:579–589

Koch DM, Toubin C, Peslherbe GH, Hynes JT (2008) A theoretical study of the formation of the aminoacetonitrile precursor of glycine on icy grain mantles in the interstellar medium. J Phys Chem C 112:2972–2980

Kyte J, Doolittle RF (1982) A simple method for displaying the hydropathic character of a protein. J Mol Biol 157:105–132

Lambert J-F (2008) Adsorption and polymerization of amino acids on mineral surfaces: a review. Orig Life Evol Biosph 38:211–242

Lomenech C, Bery G, Costa D, Stievano L, Lambert JF (2005) Theoretical and experimental study of the adsorption of neutral glycine on silica from the gas phase. Chemphyschem 6:1061–1070

Miller SL (1953) A production of amino acids under possible primitive Earth conditions. Science 117:528–529

Miller SL (1955) Production of some organic compounds under possible primitive Earth conditions. J Am Chem Soc 77:2351–2666

Miyakawa S, Yamanashi H, Kobayashi K, Cleaves J, Miller SL (2002) Prebiotic synthesis from CO atmospheres: implications for the origins of life. Proc Natl Acad Sci USA 99:14628–14631

Morokuma K (2003) ONIOM and its applications to material chemistry and catalysis. Bull Korean Chem Soc 24:797–801

Morrow BA, Cody IA, Lee LSM (1976) Infrared studies of reactions on oxide surfaces. 7. Mechanism of the adsorption of water and ammonia on dehydroxylated silica. J Phys Chem 80:2761–2767

Muñoz-Caro GM, Meierhenrich UJ, Schutte WA, Barbier B, Segovia AA, Rosenbauer H, Thiemann WH-P, Brack A, Greenberg JM (2002) Amino acids from ultraviolet irradiation of interstellar ice analogues. Nature 416:403–406

Nuevo M, Auger G, Blanot D, d'Hendecourt L (2008) A detailed study of the amino acids produced from the vacuum UV irradiation of interstellar ice analogs. Orig Life Evol Biosph 38:37–56

Orgel LE (1998) Polymerization on the rocks: theoretical introduction. Orig Life Evol Biosph 28:227–234

Orgel LE (2004) Prebiotic chemistry and the origin of the RNA world. Crit Rev Biochem Mol Biol 39:99–123

Oró J (1961) Mechanism of synthesis of adenine from hydrogen cyanide under possible primitive Earth conditions. Nature 191:1193–1194

Pisani C, Busso M, Capecchi G, Casassa S, Dovesi R, Maschio L, Saunders VR, Zicovich-Wilson CM, Schütz M (2005) Local-MP2 electron correlation method for non conducting crystals. J Chem Phys 122:094113

Pizzarello S (2006) The chemistry of life's origin: a carbonaceous meteorite perspective. Acc Chem Res 39:231–237

Rimola A, Ugliengo P (2008) A quantum mechanical study of the reactivity of (SiO)2-defective silica surfaces. J Chem Phys 128:204702

Rimola A, Ugliengo P (2009) The role of defective silica surfaces in exogenous delivery of prebiotic compounds: clues from first principles calculations. Phys Chem Chem Phys 11:2497–2506

Rimola A, Sodupe M, Tosoni S, Civalleri B, Ugliengo P (2006) Interaction of glycine with isolated hydroxyl groups at the silica surface: first principles B3LYP periodic simulation. Langmuir 22:6593–6604

Rimola A, Sodupe M, Ugliengo P (2007) Aluminosilicate surfaces as promoters for peptide bond formation: an assessment of Bernal's hypothesis by ab initio methods. J Am Chem Soc 129:8333–8344

Rimola A, Civalleri B, Ugliengo P (2008) Neutral vs. zwitterionic glycine forms at the water/silica interface. Structure, energies and vibrational features from B3LYP periodic simulations. Langmuir 24:14027–14034

Rimola A, Sodupe M, Ugliengo P (2009a) Affinity scale for the interaction of amino acids with silica surfaces. J Phys Chem C 113:5741–5750

Rimola A, Ugliengo P, Sodupe M (2009b) Formation versus hydrolysis of the peptide bond from a quantum-mechanical viewpoint: the role of mineral surfaces and implications for the origin of life. Int J Mol Sci 10:746–760

Rimola A, Sodupe M, Ugliengo P (2010a) Deep-space glycine formation via Strecker-type reactions activated by ice water dust mantles. A computational approach. Phys Chem Chem Phys 12:5285–5294

Rimola A, Zicovich-Wilson CM, Dovesi R, Ugliengo P (2010b) Search and characterization of transition state structures in crystalline systems using valence coordinates. J Chem Theory Comput 6:1341–1350

Saladino R, Neri V, Crestini C, Costanzo G, Graciotti M, Mauro ED (2008) Synthesis and degradation of nucleic acid components by formamide and iron sulfur minerals. J Am Chem Soc 130:15512–15518

Sauer J, Ugliengo P, Garrone E, Saunders VR (1994) Theoretical-study of van der Waals complexes at surface sites in comparison with the experiment. Chem Rev 94:2095–2160

Schopf WJ (1993) Microfossils to the early Archean Apex Chert: new evidence of the antiquity of life. Science 260:640–646

Sherrill CD (2010) Frontier in electronic structure theory. J Chem Phys 132:110902

Smith JV (1998) Biochemical evolution. I. Polymerization on internal, organophilic silica surfaces of dealuminated zeolites and feldspars. Proc Natl Acad Sci 95:3370–3375

Snyder LE, Lovas FJ, Hollis JM, Friedel DN, Jewell PR, Remijan A, Ilyushin VV, Alekseev EA, Dyubko SF (2005) A rigorous attempt to verify interstellar glycine. Astrophys J 619:914–930

Sousa SF, Fernandes PA, Ramos MJ (2007) General performance of density functionals. J Phys Chem A 111:10439–10452

Tosoni S, Tuma C, Sauer J, Civalleri B, Ugliengo P (2007) A comparison between plane wave and Gaussian-type orbital basis sets for hydrogen bonded systems: formic acid as a test case. J Chem Phys 127:154102

Woon DE (2002) Pathways to glycine and other amino acids in ultraviolet-irradiated astrophysical ices determined via quantum chemical modeling. Astrophys J 571:L177–L180

Biodata of **Robert B. Sheldon**, author of "***Comets, Information, and the Origin of Life***."

**Robert B. Sheldon** received his Ph.D. from the University of Maryland, College Park in 1990, modeling the Earth's radiation belts with data from a time-of-flight mass spectrometer on the National Aeronautic and Space Administration (NASA) Active Magnetospheric Particle Tracer Experiment (AMPTE) mission. He invented an improved isochronous time-of-flight mass spectrometer that subsequently flew on three more satellite missions and made the first measurements of solar wind isotopes. After appointments at the University of Bern in Switzerland, Boston University, and the University of Alabama in Huntsville where he designed spacecraft instrumentation, he is currently consulting with NASA's Marshall Space Flight Center in the areas of space physics, dusty plasmas, particle instrumentation, astrobiology, nuclear propulsion, and the newly selected solar probe plus mission (http://rbsp.info/rbs/RbS/).

E-mail: **rbs@rbsp.info**

J. Seckbach (ed.), *Genesis - In The Beginning: Precursors of Life, Chemical Models and Early Biological Evolution*, Cellular Origin, Life in Extreme Habitats and Astrobiology 22, 363–389
DOI 10.1007/978-94-007-2941-4_21, 

# COMETS, INFORMATION, AND THE ORIGIN OF LIFE

**ROBERT B. SHELDON**
*National Space Science & Technology Center, 320 Sparkman Dr, Huntsville, AL 35805, USA*

## 1. Introduction

The origin-of-life (OOL) problem has been traditionally viewed as a informational barrier, whereas comets, when they have been considered at all, have been treated as passive, information-neutral carriers of life. In this chapter, we attempt to show how OOL and comets form a synergistic system, involving both information and transportation.

### 1.1. THE ORIGIN-OF-LIFE PROBLEM

After Darwin's success (Darwin, 1859) at reviving Lucretius' materialism (Lucretius Carus, 1921) with its rejection of teleological or vitalist explanations for evolution (Davies, 2000), there arose a paradox on the origin of that first life. On the one hand, Darwin rejected any inherent property of matter that made it alive; it had to be a naturalistic spontaneous generation from nonlife. But on the other hand, Pasteur demonstrated that life always came from life, that spontaneous generation did not easily occur (Pasteur, 1861; Farley, 1974). Darwin acknowledged the problem but merely expressed a belief that under the right conditions (a warm pond) and with sufficient time (eternity), spontaneous generation could still be likely (Darwin, 1887; Peret et al., 2009).

Subsequent discoveries have not been kind to Darwin's estimate. The age of the universe has shrunk from eternity to 13.7 Ga (Komatsu et al., 2011), and the complexity of the first living cell has grown astronomically from the "protoplasm" imagined by Darwin to the complexity of modern biochemistry (Meyer, 2009). Despite early evidence of the liquid water environment, a complete set of cellular nanomachines needed for life would require extensive assembly and dynamic initialization (Polanyi, 1968). To expect a proper assortment of pieces to randomly assemble is estimated in various places to have a probability of less than one in $10^{41,000}$ (Hoyle and Wickramasinghe, 1981a; Hoyle, 1999). These are not even astronomical, these are cosmological improbabilities, as illustrated by the following example.

Suppose that the J. Craig Venter Institute (www.jcvi.org) is successful in producing a stripped-down *Mycoplasma* with a mere 1,000 codons describing a minimally functional 1,000-amino-acid protein set (and substantially shorter than their synthetic *M. mycoides JCVI-syn1.0* genome with 991,920 codons, Gibson et al., 2010a, b; Wickramasinghe et al., 2003; Wickramasinghe, 2011). Further supposing one had a computer that generated random arrangements of 1,000 codons and then tested each for possible "life," how long would it likely take to find the right arrangement, where "likely" is a probability of one-half? Since there are 20 possible amino acids, a 1,000-long chain has $20^{1,000} = 10^{1,301}$ permutations. Supercomputers today are capable of "petaflops" or a million billion instructions per second (Oak Ridge National Laboratory, 2009). There are ~31 million seconds in a year, so if each instruction is a test of a random sequence, we have about $10^{22}$ evaluations in a year. At this rate, $10^{1,301}$ tests would take $10^{1,279}$ years, or much longer than the $10^{10}$ years that the universe has existed.

We know that computer chips are getting faster and smaller, so could such a computer be built in the future, even if it is impossible today? There are physical limits on speed and size, the most rigorous physics limit being the "Planck time" given by quantum mechanics for the shortest interval of time that has any meaning, or about $10^{-43}$ s. Then the maximum number of time intervals from the beginning of the universe is 13.7 Gyr * $3.1 \times 10^7$ s/year * $10^{43}$ intervals/s = $10^{61}$ intervals. We will further assume that at least one electron or elementary particle has to be involved in a calculation so the maximum number of computable bits must be no greater than the number of particles in the universe, or about $10^{80}$. Their product is $10^{141}$ maximum computer calculations, assuming the entire universe were an atomic computer (Dembski, 1998). More careful estimates applying the limitations of general relativity on the quantum physics give the computational capacity, or stochastic resources of the universe, to be about $10^{120}$ operations on $10^{120}$ bits (Lloyd, 2002).

This result sums up the current impasse in OOL research: randomly generating the proper arrangement of even a 1,000-peptide enzyme is outside the computational abilities of the universe, much less the 991,920 long minimal genome of *Mycoplasma*. So not only has it proved difficult to create life in the laboratory, or even find a mechanism to spontaneously generate it, but theoretically, it appears hugely improbable that a random search can ever find it.

One counterargument to these cosmological improbabilities is to argue that there are many more arrangements of the basic building blocks that are alive. That is, just because the minimal life form we chose has a specific arrangement of 1,000 amino acids representing a one in $10^{1,301}$ probability does not mean that there are no $10^{1,270}$ other arrangements of those 1,000 amino acids that are also "alive." Thus, the OOL computation need only find one of those other possible permutations, which would increase the odds and make spontaneous generation feasible.

The counter-counterargument is that the putative ubiquity of "living" permutations should cause spontaneous generation to be observed frequently, which it has not (Pasteur, 1861; Hoyle and Wickramasinghe, 1981a, Hoyle, 1999). Or it should leave behind a body of alternate forms for these basic proteins and gene coding, which it has not (Meyer, 2009). Furthermore, a laboratory that randomly permutes enzymes and genomes should frequently produce viable organisms, which also has not happened (Barrick et al., 2009). Rather, mapping out the viable organisms in "permutation space" reveals a tremendous desert of nonviable arrangements (Axe, 2004). Life appears to be highly specific, ordered, and particular, which puts severe limits on the ease of randomly making life.

An alternative counterargument suggests that we live in a "space" of many universes, each one birthed by random fluctuations of the vacuum. Since a fluctuation has a certain "Planck-length" volume, then at any given instant, the universe can be divided into cubes of size $10^{-35}$ m, or about $[10^{61}]^3$ cubes today. If we argue that a new universe can form in one Planck time, and each universe can spawn more universes, then we have an exponential series from our Big Bang onward (1 the first instant, $2^3$ the next, up to $[10^{61}]^3$), for a total less than $10^{243}$ "similar" universes (Guth, 2007; Susskind, 2007). String theory with its 11 dimensions achieves a larger number of about $[10^{61}]^{11}$, though Linde believes this number to be surpassed by entropy considerations of the cosmological constant, or about 10 to the power $10^{82}$ universes (Linde and Vanchurin, 2010). Of course, our universe might have begun unmeasured years after some "original" Big Bang, so in principle, the number of multiverses in the "landscape" may be infinite and time eternal.

There are numerous difficulties with the scenario sketched out above. For example, once infinite solutions are posited, it is difficult to find any single solution because the fastest growing solution becomes dominant, so the discussion changes from finding our universe to the difficulty of demonstrating why ours has the fastest growth rate in an infinite set. And since a directed search is faster than a random search, the fastest growing solution would also be the most teleological, which contradicts the materialist assumptions. For example, suppose in one of those infinite universes there comes into existence a being with the ability to communicate between multiverses giving it immense computational resources and allowing it to inject information into any particular universe much faster than random chance can produce it. By the usual definitions, this behavior would be supernatural, thus making the entire justification of the multiverse hypothesis—the naturalistic production of life—impossible. This argument can be made more rigorous using Gödel's method of enumerating the infinite worlds (Gödel, 1931), demonstrating any subset of "materialist multiverses" is still incomplete.

Ignoring this incoherence of infinities, there is significant doubt that universes can actually appear in the vacuum as hypothesized, since energy is not conserved in this model (Pitts, 2010). Furthermore, the minimal life recorded by Venter has combinatorial information at least $20^{991,920}(=10^{1,290,517})$ not including the dynamical information and the permutations of "fine-tuned" physical constants that add at least another factor of $10^{1,000}$. Nor is it clear that each multiverse would sample the solution space evenly, or even whether dynamically interacting systems can be constructed from noninteracting random steps. Thus, it would appear that the multiverse solution of reintroducing Democritus' infinities produces more problems than it solves.

The arguments and counterarguments do not agree on the density of viable arrangements in protein phase space; they do not agree on the minimum codon length needed for life, or even the nature of the universe. But this lack of agreement should not distract us from recognizing two common characteristics of the debate: first, the OOL problem involves astronomical probabilities in which incremental progress is measured in factors greater than ten billion ($10^{10}$), and second, a successful OOL theory hinges on ways to bring these astronomical probabilities down to Earth. In this chapter, we argue that comets can address both goals, though not without cost.

## 1.2. THE COMETARY HYDROSPHERE

Over the past century, it has become increasingly apparent that life does not reside solely on this planet Earth (e.g., Arrhenius, 1908) but probably exists throughout the Solar System wherever there is liquid water as on moons or planets (Levin and Straat, 1976; Coates et al., 2010; Strobel, 2010; NASA/Jet Propulsion Laboratory, 2010), but more importantly, on numerous, small icy bodies, called comets when they cross Mars' orbit, melt, and acquire visible tails (Hoyle and Wickramasinghe, 1980; 1981a; 2000). Recent discoveries of fossilized life on carbonaceous chondrite meteorites thought to be extinct comets also support comets as unique (Hoover et al., 2004; Hoover, 2008, 2011). Comets distinguish themselves in several ways from rocky bodies: they have short "summers" when they come near the Sun and melt followed by long "winters" far from the Sun when they refreeze; they explore a much larger volume of space in their orbit; they accrete material along their orbit; their "life span" is much shorter with a "death" involving disintegration into many smaller fragments; and they are frequently ejected from the Solar System gravitational well (Sheldon and Hoover, 2005, 2006, 2007). These properties of the cometary "life cycle" are so different from the gravitationally bound rocky bodies that we call this wide range of sizes, temperatures, and orbits "the cometary hydrosphere."

Supposing that a significant fraction of melted comets become infected with life, then the cometary hydrosphere is also a cometary biosphere that is able to survive, spread, and transport life across the galaxy, possibly from the moment when stars began to form 12 billion years ago (Sheldon and Hoover, 2008). This potential cometary biosphere can then interact with the OOL problem in several important ways. For example, since comets can transport life between rocky bodies, then if life is found on different planets, it is not necessary to hypothesize that life began independently twice, for it could have begun on either rocky body and spread to the other, or even begun on comets and spread to the rocky bodies. Therefore, transportation changes the OOL problem by permitting a much larger volume of space to be involved. But transportation does more than allow a greater volume of the universe to be involved, it also allows a greater timespan (distance divided by comet velocity) to be involved. Thus, comets integrate the entire volume of the galaxy into the OOL problem (with a smaller probability for multiple galaxies) over nearly the entire time since the Big Bang.

If we hypothesize that the OOL probability (OOLP) scales as the probability of a rare event times the number of locations and the amount of time, then this inclusion of a galaxy of locations existing over the entire time since the Big Bang increases OOLP by approximately $10^{24}$ compared to the probability of forming only on Earth. This number can be estimated very roughly by calculating the ratio of Earth OOLP to cometary OOLP, which means comparing the ratio of time intervals and volumes available for life to begin on Earth to that on comets.

Calculating the maximum amount of time available for Earth OOL gives a time interval between the molten-rock Hadean Age at the end of the planetary bombardment of 3.85 Gyr BP and the first appearance of biofractionated carbon at 3.65 Gyr BP (Mojzsis et al., 2003), or about 200 million years. A similar calculation for cometary OOLP starts from star formation some 12 Gyr BP to the same spot or about 8 billion years. Then the ratio of time intervals for cometary/Earth OOLP is about 40 times larger.

Likewise, an estimate for the volume of cometary water around the planet Earth since the Hadean is approximately equal to the volume of ocean water in the planet Earth today (Sheldon and Hoover, 2007). If each star system has a similar amount of cometary water, then we can multiply by the number of stars in the galaxy (and assuming no other rocky planets with oceans) which gives about 100 billion times more volume in galactic comets than on the Earth. If we further assume that other galaxies were accessible by comet (which is uncertain because the high velocities of intergalactic comets needed to cover the distance preclude capture into the gravitational well of a target solar system), we can increase this number by another 100 billion to account for the number of observable galaxies in the cosmos. Then the ratio of volume cometary water over Earth water increases the cometary OOLP by about $10^{22}$. Finally, combining the time and volume ratios gives a rough estimate of a $10^{24}$ increase in cometary over Earth OOLP.

We could increase this slightly by assuming a distribution of rocky bodies with liquid water oceans throughout the galaxy, but all these refinements hardly change this number by more than one order of magnitude, which when compared to estimates of Earth OOLP $< 10^{-1,301}$, provide insufficient progress in solving the OOLP puzzle. Note that these refinements are all "linear adjustments" to the OOLP calculation, scaling directly with volume and time interval. As we discuss later, it is enticing to consider whether the 70% "dark matter" of the universe is composed entirely of comets, in which case, we would have to increase our estimate of the cometary hydrosphere by another factor of about $10^7$, and yet would have made little progress on raising OOLP close to 1/2. That is, assuming the most radical changes to cosmology that incorporates every cubic centimeter of potential water into the OOLP calculation would hardly move the resulting probability.

These sorts of considerations suggest that the OOL problem will not be solved by tweaking linear factors. If time and space are only linearly correlated to OOLP, there will be no solution; however, there may yet be nonlinear corrections to OOLP made possible by the discovery of the cometary hydrosphere.

## 2. Origin-of-Life Probability and Linearity

One of the many difficulties in discussing the OOL problem is that we confuse the theory with the practical, or the immaterial with the material. The combinatorial problem of OOLP is theoretical because no chemical reaction, no cellular biochemistry proceeds in the manner described. For example, when Venter announced his synthetic bacteria (Gibson et al., 2010a, b) using nonorganic deoxyribonucleic acid (DNA) machine-manufactured from biologically derived reagents, the DNA fragments by themselves were useless. So they inserted these fragments into a living yeast cell so as to reassemble the 1,078 pieces, and injected that repaired DNA into a related bacterial species whose own DNA had been removed. One hurdle that took many months to solve was a result of a single missing codon. Nowhere in this experiment was there a theoretical problem similar to the combinatorial math of the OOLP problem, rather, all the biological protein machines were running and operational when the sleight of hand to change out the DNA occurred. Venter's success was not randomly finding a sequence but rather converting the immaterial logical sequences into living biological material.

The combinatorics assumes that there is a fixed target that we are to search for blindly, like a needle in a haystack, whereas the Venter problem was to swap one DNA for another in a living organism, reminiscent of electricians who rewire factories without turning off the power. The Venter approach began with a living chemical environment and tries to change it without killing it, whereas the combinatorial approach began with a dead chemical environment and hopes to enliven it without trying. The latter is an attempt to find life without a driver, whereas the former is an attempt to keep life going while switching drivers.

If we know the sequence we are after, then like the Venter Institute, we can produce that DNA after due attention to quality control. But if we do not know the sequence, it will take a very long time to find it. The OOLP problem has been stated as the difficulty of randomly finding the right sequence. Many computational biology approaches have been proposed as "smart" algorithms for finding the "living" sequence, but as Dembski argues, all these programs—*Weasel* (Dawkins, 1986), *Ev*, or *Avida*—smuggle in information that helps with the search (Dembski and Marks, 2010). In fact, the "No Free Lunch" theorem proves that without prior information, there is no "smart" algorithm that can outperform a random search, which is where we started our discussion (Dembski, 2002).

But perhaps the problem is assuming some sort of maximally random "warm pond" as the starting point and attempting life in one step. If an information-rich substrate, perhaps a clay, or a coacervate, permits the addition of information that leads to OOL, then the formation of life is much closer to Venter's problem, that of adding information without losing what is already there. That is, the "smuggling" of information, which is the bane of Dembski's algorithmic analysis, represents the pinnacle of Venter's experimental accomplishment.

For the OOLP calculation does not need to begin at zero and in one jump make it to *Mycoplasma*, rather, it may be possible to combine two information-rich subsets—say coacervates and ribonucleic acid ribozymes—to produce life. So both "smuggling" and "finding" contribute to OOLP. We are not saying that breaking down an improbable string into substrings changes the probability of forming the final string, only that smuggling or "adding up substrings" possesses probabilities as important to OOLP as finding the final string. The OOL problem is exacerbated, not reduced, by including the probability of experimentally adding information. For as the Venter Institute reported, it was quite difficult to add information, requiring real reagents manipulated in vitro with real organisms, rather than the manipulation of abstract symbols on a computer.

Since OOLP is proportional to the probability of finding the minimal sequence multiplied by the probability of the method producing that sequence, the information from experimental production of the proper sequence is just as important as the information in the sequence. Since there are a great many ways to make substrings and add them together, each with its own probability, OOLP must be the most probable method selected from all the possible paths to that destination. That is to say, while we cannot make the discovery of a long string of peptides more probable by breaking it into substrings, we can make the manufacture of that string of peptides more probable by breaking it into substrings. And nonlinear production mechanisms have the potential to be the most probable.

If it is comets that transport the reactants for OOL, then just as a dimerization reaction proceeds at a quadratic or non-linear power of the density of reactants, so also the density of comets functions as a non-linear factor both in the abiotic (purely chemical) OOL pathway as well as in the biotic or quasi-biotic evolution pathway. Debate over where to place the prebiotic versus biotic boundary is irrelevant, for whatever nonlinear mechanism we invoke should still generate sufficient probability to overcome the difficulties of addition. Comets fulfill this role nicely, providing the nonlinear delivery of reactants for an abiotic OOL synthesis or the nonlinear delivery of genes for biotic evolution. In both cases, it is a density-dependent nonlinear function that has the potential to approach an improbable OOLP.

## 3. Information Restatement of OOL

Despite the simplicity of modeling life on the molecular reactions needed to produce a living sequence, it would be inaccurate to quantify OOLP only by the density of reactants, which ignores the hidden effort of the biochemist, who, when abiotically synthesizing an important biochemical, carefully isolates the products from the reactants, performing several purifying steps for every synthesis step. Quantifying these actions of the chemist is analogous to a physicist's calculation of entropy. That is, purification produces no new products but does reduce the entropy of the products at hand. The inverse of entropy is information, so whenever making an OOLP calculation, it would seem convenient to keep track of the information content, whether added by purification or added by increasing codon length.

In fact, it would be mathematically advantageous to cast the entire OOLP problem as a problem of information, where life is assumed to be a highly informational state of matter. Then one could calculate OOLP quantitatively over the entire space from the low-information dilute chemicals to the high-information life, from the beginning of the time interval to the end.

This recasting of the OOL problem as a change in information content has several other advantages as well, making it independent of material details (viruses versus cyanobacteria) or temporal details (RNA-world versus metabolism-first). We merely set some informational threshold and argue that when the information in the system exceeds that threshold, we have OOL. Since life also concentrates that information into a small volume, we should restate the threshold as an information density spike achieved somewhere in our volume. This is still not quite right, because a dead bacterium may have the same information density as a live bacteria, yet be completely unable to propagate, and therefore not "alive." So we should further refine our threshold to include time or information density flow, where spatial derivatives are used to establish the density and temporal derivatives determine the flow.

If this information density is so very improbable, then the exact level of the threshold is unimportant because the gradients in space and time should be so very steep. If information is measured in "probability units," we could set it at $10^{150}$/cubic micron or at $10^{15,000}$/cubic micron with no real difference to the outcome. Likewise, the diffusive entropy flow should be enormous at sharp gradients, so the mere fact that a cell does not rapidly dissipate with time is a signature of strong informational flow. The entropic dissipation is a function of the strength of the gradient and the local temperature, so for the ease of computation, we normalize the information flow to the expected gradient-driven dissipation flow, with life demonstrating a flow of opposite sign to dissipation, and slightly greater than the expected dissipation flow. Note that for freeze-dried or lyophilized bacteria, the entropy flow is so very small that the countercurrent of living information flow may be virtually undetectable but nevertheless exist even in a state of suspended animation.

This discussion has been necessarily qualitative, but considering the many orders of magnitude involved in the OOLP calculation, we do not think we have oversimplified the problem yet. The OOL problem can then be restated as the appearance of very high informational density that also has an informational (negative entropic) flow slightly larger than the expected positive entropic decay rate. Calculating this quantity, then, will require a calculation of information density over all space and its time evolution (or temporal derivative). In the next section, we discuss this calculation mathematically.

### 3.1. SHANNON INFORMATION IN SPACETIME

In a series of groundbreaking papers, Shannon developed "information theory" from scratch, developing it to describe the carrying capacity of telephone cables and then applying it to the English language as a paradigm case (Shannon, 1948; Shannon and Weaver, 1949). In the ensuing development of the mathematical theory, there tend to be two simplifying directions of his research: calculating the information capacity (spatial derivative of the telephone cable) and calculating the informational flow (temporal derivative of the signal).

As an example to clarify the difference between capacity and flow, consider the coaxial cable used to bring cable television into a house, which has a higher capacity than copper twisted pair. While possessing less capacity, over time, the bit rate of twisted pair went from a 300-bits-per-second (bps) acoustic modem, to a 1,200-bps digital modem, to 9,600 bps "maximum" for vocal frequencies, to 56 kbps for digital compression. Each time a new modem arrived, we were told this was the theoretical maximum for twisted pair, yet today, we have twisted pair carrying DSL at 1–4 Mbps. This increase in bandwidth is not a function of time-independent geometry, as seen in the coax versus twisted-pair comparison, but a function of frequency and compression algorithms that are able to make each bit carry more information by relating it to the bits before and after it. Making a graphical analogy to water pipes, the coax is a wider pipe than twisted pair, whereas the improvement in twisted-pair modems is a faster flow or greater pressure. Therefore, information theory involves both a spatial and a temporal component, which are related by the speed of the information carrier: electricity for telephones, sound waves for liquids, chemical waves for biochemistry, and comets for astrophysics.

What exactly is this information which Shannon described? Shannon began by characterizing the noise on the telephone line as a binary bit stream. Noise comes from fluctuations, which may be described by their frequency dependence: a Gaussian distribution is thermal, or a flat distribution is "white," and a 1/frequency distribution is "brown." Whatever the noise distribution, the signal is what remains when the noise is subtracted out, which means that the signal is strongest on the "wings" of the Gaussian, where the more improbable the noise, the better the signal/noise ratio. As the number of particles gets larger, and a mole of molecules is already $10^{24}$particles, these Gaussians get extremely steep, making it much easier to manipulate the logarithm than the quantity itself. Boltzmann defined entropy, $S$, to be his constant, $k$, times the logarithm of the number of states in distribution, $\ln(\Omega)$ (and had the equation $S = k\ln(\Omega)$ engraved on his tombstone). Shannon's definition of information, $I$, is just the negative of Boltzmann's $S$, or what he called "negentropy."

Relating this negentropy to the distribution of chemicals in a warm pond, a smooth and dilute distribution is the most probable, and hence the "noisiest" or most entropic distribution, whereas a concentrated spot of chemicals is the least probable and therefore the higher negentropy information content. Using the appropriate scale length, we might say that the information in a particular dissolved chemical is its local concentration divided by the expected average concentration. But note that this is a time-independent measure, this is the "width" of the information pipe, not its "pressure."

To calculate the information "pressure" of this chemical concentration gradient, we have to compare it to the state immediately before and the state immediately after. If the states are describable by a simple law, say, the diffusive motion of a chemical gradient, then the entropy increases and the information decreases. If, however, there is no physical law that connects these states, or more precisely, the greater the deviation from the physical law of diffusion ($df/dt = D\, d^2f/dx^2$), the greater the information content in these adjacent states.

Shannon does this calculation in a 1951 paper on the information content of written English (Shannon, 1993). To calculate the spatial information of written English, the same statistics as cryptographers is employed, looking for the occurrence of specific letters, pairs of letters, triplets of letters, and so forth, which is a static analysis independent of global position, and the standard cryptographic technique for cracking a substitution cipher. But Shannon wanted to know how correlated are the letters for people who know the code. That is, a computer can tell us that the letter "q" is highly correlated with a following "u," but could it, say, determine that this rule is violated for Chinese names? A human could, so Shannon asked them to read texts that had letters removed in order to determine the information encoded in these longer range correlations. In our example, how many letters does it take to decide the word is likely a Chinese name instead of a Latin-root language? Although Shannon worked with written texts, these same rules apply to spoken texts, which make this experiment also a study of time-dependent information content.

Are the spatial and temporal ways of measuring information really independent? A recent paper on the undeciphered pictograms of the Picts demonstrates their independent character (Lee et al., 2010). The question posed by these carvings was whether they represented a picture, a hieroglyphic/pictogram script, or a syllabic language. Examples of each script type were collected, and the single-symbol frequency statistics (spatial) were plotted on one axis against the statistics of the following symbol (temporal) on the other axis. Each type of communication occupied a distinct cluster on the graph, and the Pictish symbols were adjacent to syllabic languages, suggesting a communication form midway between hieroglyphics and a syllabic language. The key point is that a circular cluster for each communication method, rather than a long ellipse or line, indicates that the two axes are relatively independent so that time and space correlations carry different information.

Applying this to our definition of life, we argue that both spatial and temporal correlations carry information. Not only does the cell exist as a distinct arrangement in space, but this arrangement persists in time, unlike random features seen in clouds or tea leaves. Now in order to avoid prejudicing one type of information over the other, we use Einstein four-vector notation to lump the temporal and spatial components together. Then our generalized Shannon information looks like: $S^\alpha = k \ln(\Omega^\alpha)$, $\alpha = (0,1,2,3)$, with Greek indices carrying the usual meaning of four-vector space-time where $\alpha = 0$ indicates the time and $\alpha = 1,2,3$ are $x,y,z$ spatial components. Note that the information is proportional to the deviation from the diffusion equation, for example, the magnitude of the negative diffusion coefficient needed to keep the structure from dissipating. Since Boltzmann calculated entropy in units of energy per Kelvin, the units on his constant $k_0$ in this four-vector notation includes the speed of light.

Since $S^\alpha$ is a function of the density of states, and density depends on whether the observer is moving with respect to the particles, we define a relativistic invariant for the information: $I = S^\alpha S_\alpha$ where $k \; k'/2$ to keep the normalization.

## 3.2. FOURIER SPACE INFORMATION

An analysis of the Shannon information above reveals that it is a local quantity. It depends upon sharp gradients in space and the maintenance of these gradients in time. But all these descriptions depend on nearest neighbors; they do not incorporate any global knowledge. By these criteria, a vat of beer yeast has no more information than the man who shovels it out. We need a global measure that indicates when diffuse information is correlated.

In Shannon's 1951 paper, he looked at long-range interactions in English words, how a letter two places removed from the missing letter influenced the prediction, or how a letter three places removed influenced it. In standard communication textbooks, these correlations are referred to as second order, third order, etc. (Cover and Thomas, 1991). We can generalize this long-range correlation as a Fourier transform, where second-order terms connect every other point, and third-order terms connect every third point, etc. It is not necessary to use sines and cosines as Fourier did, only that there be a transform with a basis set that covers all possible long-range correlations. Then just as nearest neighbors can have information in this density of states, so too can nonnearest neighbors have information in the transform of the density of states.

Note that the zeroth-order term in such a transform is just the same local term we described above. Thus, the information quantity we are interested in looks something like $I = S^\alpha S_\alpha + \Sigma_1^L F_i(S^\alpha S_\alpha) = \Sigma_0^L F_i(S^\alpha S_\alpha)$, where $F_i()$ is the transform at some spatial scale $i$ and $L$ is the limiting scale size.

Does the information in these various modes add, as we have assumed? From physics, we know that the entropy, $S$, is usually additive for volumes, so the information in different volumes is also additive. We have made a weak argument that information is additive for temporal constancy (a negative diffusion coefficient), which seems odd that something that does not change is increasing in information.

But the time dimension has different units, and what is important is that the information does not disappear at the diffusion speed. Like the Red Queen in Lewis Carroll's masterpiece, it takes energy to maintain homeostasis, and that energy expenditure (divided by temperature) is a negative entropy flow which is information. So the spatial and temporal entropies add.

Do the Fourier components of information add as well? Yes, it would seem natural that they are additive, though the units (inverse space and time) are not the same, nor the magnitudes equal. In Shannon's 1951 work, the information per added letter in an English sentence dropped from some 4.8 bits for the second letter to 1.2 bits for the tenth and following letters like an autocorrelation function, which Shannon estimated by subtracting the information in the $(n-1)$th letter from the $n$th letter, making the assumption that the information in the longer correlation lengths was additive.

So if the information in all these different modes is additive, and they all correspond to a logarithm of a density function, then we can create a density function for each mode and multiply these densities together. That is, if the transform of the logarithm is the same as the logarithm of the transform, then this sum can be replaced with a product, $I = k' \ln[\prod^{L}_{0} (\Omega^{\alpha}_{i}\, \Omega_{i\alpha})]$, where $i$ signifies the basis vectors of the transform space.

How much do these higher order terms add to the total information? Generalizing from Shannon's estimate, where the fifth-order term drops to one quarter of the zeroth-order term, we estimate that each decade of $L$ contains the same amount of information, giving a power-law dependence of information on scale size. Then starting at an atomic scale of $10^{-12}$ m = 1 fm, we would have about 38 decades up to the scale of the universe. This should probably be done for relativistic four volumes instead of lengths, so that the information in the Fourier components is about 152 times greater than that in the zeroth order. If we reexpressed the logarithm as a density of states, then we would say it is equivalent to a very high power of density, $I = k' \ln[\Omega^{\alpha}_{i}\, \Omega_{i\alpha}]^{152}$.

This rather heuristic approach can be physically motivated by considering a series of abiotic chemical steps that can hypothetically make life in a test tube. The reactions $k_1[a][b] \rightarrow [c]$ and $k_2\,[c][a] \rightarrow [d]$ taking place in a single flask could be written as $k_1 k_2\,[a]^2[b] \rightarrow [d]$ where the reaction rate or probability is nonlinear in [a]. Of course, some reactions may destroy the products much like atmospheric chemistry, so the expected output is found by solving a large matrix of coupled equations. This matrix is just a more accurate physical description of the independent and equal probabilities we had used in our earlier description of searching for arrangements that are "alive." The key difference is that now the different arrangements are not equally probable, nor are they independent. This means we have to abandon our linear approach of adding probabilities and consider the impact of nonlinear terms.

But if we lack the computational resources to find the correct sequence, how does it help if we add in the lack of experimental ability to even produce the correct reaction pathway? It helps because there are potentially both linear and nonlinear synthesis pathways, but if a nonlinear pathway exists, then under some set of conditions, it will dominate over the linear path. This means that two probabilities collapse down to the one which is more likely. It is a method of improving OOLP, by picking out pairs of probabilities and replacing them with a better single one.

Is it not even more speculative to talk of nonlinear synthesis without evidence? No, because we can examine the end product for examples of duplication, which would be the result of a nonlinear input. And duplicates can occur at any size scale, they can be "aa" or "abab" or "aaabbb" and so forth. Finding such items in a data set uses tools like autocorrelation functions which are calculated with Fourier transforms, or fractal analysis over "wavelet" basis vectors. The specific technique is not as important as the concept that information about duplicates and their compression of the linear probabilities can be found in "transform space." Just as structure can be found locally by taking local gradients, so too can duplicates be found globally by taking "Fourier" components, which appear in the calculation as nonlinear exponents on the densities.

## 4. OOL Detection

What does it mean that there is information in those Fourier components, how does nonlocal information contribute to micron-sized life? Consider "rogue" waves on the ocean. Most ocean waves are a meter or so high, but occasionally, with no warning, 10-, 20-, or even 30-m high waves can topple a ship. Oceanographers suggest that they form spontaneously as the reconstruction of many smaller waves that all arrive in phase. In the same way, each of these Fourier components of information can arrive "in phase" with other information, so as to add up to a greater sum.

Such an interpretation of Fourier components assumes that there can exist "an information wave" that propagates through space. This is precisely what comets represent in the universe, carrying water, carrying chemicals that have been processed by heat and liquid water, carrying genes that are being transported in bacteriophages and cyanobacteria, even carrying entire ecosystems identical to bacterial mats (Hoover, 2011). So the OOLP premise is that at some time, $t_0 = (t-1)$, no cubic micron in the universe had information content above the threshold, but a "collision" at time $t$ combined the information from two cubic microns to be above the OOL threshold.

Mathematically then, our OOL detector is a large calculation of the four-vector entropy density flow (where the time component is measuring antidiffusive flow), summed over all Fourier components of interest. When this registers a spike above the threshold, we have the OOL. Then OOLP is found by doing the same calculation summing over the entire universe in space and time for that first spike.

Now the reason for using four vectors becomes apparent. In order to calculate the best possible number for OOLP, we will need to include all the time between the first star formation (which could then melt comets) and the 3.65 Gyr isotopic identification of life on Earth. Using Einstein's block universe with time being just another dimension like the other three spatial dimensions, we can calculate OOLP as a four-vector information over the expanding universe for those 8 Gyr. As we described before, this additional volume and time barely changes the linear probabilities, adding a mere 24 zeroes to OOLP, but it does add information in the Fourier components, raised to the 152nd power. So a power-law in scale length provides a modest increase in the Shannon information exhibited, though not enough to change our OOLP calculation by very much. However, it does indicate a way in which comets can contribute nonlinear information to OOLP.

## 4.1. THE COMET ADVANTAGE

How does this "comet information wave" model differ from Darwin's warm pond? Darwin had all his chemicals in solution, a high-entropy and low-information situation, whereas comets keep all their chemicals locked in a deep freeze until the last moment (near perihelion), which is a low-entropy high-information system. Darwin added sunlight and heat to his pond to provide the energy for life, a high-entropy energy source, whereas comets provide inhomogeneous chemicals, often lingering at the melting point of ice, a low-entropy energy source. One way to characterize the suitability of energy sources for work is by calculating the Gibbs free energy or exergy, $G = H - TS$, where $H$ is the enthalpy and $T$ the temperature. Since $G$ is proportional to negative temperature, life prefers it cool, which is why trees evaporate 99% of the water they take in at their roots, a feat consuming 66% of sunlight energy, just to increase their exergy by cooling their leaves (Schneider and Sagan, 2005). Since comets linger near the melting point of water as they melt, they have the maximum exergy possible.

But most importantly, Darwin had no way for warm ponds to communicate. All the information had to be available locally, there was no method of communicating information, collecting information, or distributing it. Comets provide a mechanism for all these things, and in so doing, provide the network that permits Fourier space to influence real space, because Fourier space does more than communicate information, it also stores it. Another example demonstrates the importance of distributed information within a network.

Because the human brain has proportionately the same density of neurons as any other primate (Herculano-Houzel, 2009), it would seem that only brain size matters for intelligence. But most brain researchers argue that it is not size but the number of cross connects that make the human brain so versatile. The information lies not in the number of nodes (spatial complexity) but in the number of dendrites, the number of cross connects (Fourier complexity). The 30 billion cells of the human brain, with its 10,000 cross connects possesses about $10^{15}$ synapses. But many of those synapses form loops, which may be important to memory, so intelligence likely scales nonlinearly as well, to some power in synapse number.

This nonlinear behavior is a consequence not just of the connections between neurons but of order in which connections are made. This means we count permutations ($n!$) rather than combinations of synapses ($\sim n$), where the number of states or the information bits per synapse is an exponentially growing function of synapse number, just as in a maximally entangled coherent quantum state (Abrams and Lloyd, 1999). That is, instead of $10^{15}$ bits, we have perhaps 10 to the power $10^{15}$ bits per brain, all because of the coherency of the cross connects (Linde and Vanchurin, 2010).

Therefore, comets may provide a distributed but connected web of information flow in the solar system, in the galaxy, and possibly in the universe. This would permit the information content of the whole to be greater than the linear sum of the parts. In our terminology, this permits the Fourier space information to dominate over the local and linear information content. By analogy to the problem of the entire universe having only about $10^{120}$ computational bits, we have achieved much greater computational resources of the universe by replacing the serial computation of a (local) silicon chip with the parallel processing of a quantum (nonlocal) computer. It is precisely because a quantum computer incorporates entanglement between bits, the nonlocal and nonlinear correlations, that it outperforms local linear silicon-based computers (Shor, 1995).

But can even Fourier space provide enough information processing? Supposing the entire universe were a computer, with its 120 decades of information, we would need a nonlinearity of the 10th power to get it up to 1,200 decades for a moderately complex molecule and nonlinearities of 100th or 1,000th power to achieve a minimal life form. If simple loops produce a quadratic power, then how many "cross connects" are needed to get 10th power or 100th power? Isn't that asking a lot from comets?

It is—especially because the number of comets in the galaxy is not expected to be more than $10^{21}$, or in the universe, $10^{31}$. Collisions or cross connects between comets are not expected to be more than 10, so we are not really asking comets to provide the information storage and processing, only the distributed information network which connects information-rich regions. Once again, comets are macroscopic objects with some $10^{39}$ water molecules, so they stand, logarithmically, about halfway between atoms and the universe in scale size. Their purpose, then, is to provide the mechanism that connects information at the large scales of galaxies and stars with the information at the small scales of cells and organisms. Without them, the Fourier space of large scales would be devoid of information, or at best, there would be no information flow between the smallest and largest spatial scales. Comparing the lost connections, we have the Earth being a volume of about one part in $10^{59}$ of the universe, which logarithmically is roughly double the one part in $10^{32}$ for the ratio of a microbe to the volume of the Earth. Therefore, comets open up 200% more log-space volume for Fourier components.

That is, comets provide a mechanism to connect the universe of Darwin's warm ponds together, so as to provide a unified information system greater than the linear sum of the parts. Comets, in addition to their linear importance in adding to the total number of Darwinian ponds, also provide the nonlinear Fourier space information that connects information-rich regions together, both at larger and at smaller scales. Without this connection, the Fourier series would truncate early, unable to connect the information on one planet with another, much less the information from the whole galaxy.

### 4.2. COMETARY ABUNDANCES

The claim that comets connect large with small spatial scales should be elaborated, lest we fall back into the panspermia idea that comets merely transport microbes from one world to another, without providing an information source of their own (e.g., Arrhenius, 1908). We distinguish our model where comets are an integrating complex information system necessary for OOL from the linear panspermia model by calling ours panzooia, where the prefix "pan" refers to its nonlocality and the root "zooia" refers to all life (Sheldon and Hoover, 2007).

Astronomical measurements of the motion of the stars in the Andromeda galaxy reveal that they are orbiting the center, but with non-Keplerian speeds of the sort found for planetary orbits around the sun. Rather, the stars seem to orbit as a rigid body, as if they are embedded in an invisible sphere (Volders and van de Hulst, 1959; Rubin and Ford, 1970). The distribution of matter that permits such motion is proportional to distance from the center of the galaxy, such that the "funnel-shaped" gravitational potential of a stellar source of matter is broadened into a flattened well, usually attributed to "dark matter," or massive material that cannot be seen with astronomical telescopes. These stellar rotation curves do not require modifications to Newtonian gravity, or invocation of nonbaryonic matter (e.g., heavy neutrinos), they merely require a radially dependent star/mass ratio, where the galaxy becomes progressively more "dusty" with radius (Gallo and Feng, 2010). Since high stellar densities "heat" a cometary velocity through gravitational slingshots and jetting of gases on the comet, one expects this radial profile for cometary density in galaxies if the cometary kinetic and potential and starlight energy are "virialized" to the same $1/r^2$ dependence.

We refer to "dusty" as indicative of dark matter that has not yet been observed by telescope. If it were actual micron dust grains, we could observe them in the infrared frequency range. If it were neutral hydrogen, we could observe them in the radio, or if heated, in the UV range. If it were compact objects—black holes, neutron stars, brown dwarfs—we could observe their gravitational micro-lensing or their occultation of background stars. As it is, we only detect them from large-scale gravitational effects of changing the rotation curves of galaxies, or at the galaxy-wide level, lensing the background galaxies. Therefore, we are looking for dark matter that is neither too finely divided that it extinguishes light nor too highly clumped that it can be seen gravitationally; it has neither a large photon cross section nor a large gravitational cross section. This means it has to be larger than a sand grain, but smaller than a Jupiter. Comets fit that description.

The best support for large numbers of galactic comets comes from observations of the "bullet cluster" of colliding galaxies. The collision produced a distinct shock wave in the heated hydrogen gas clouds and a perceptible offset between the bright stellar center-of-mass and the gravitational lensing center-of-mass. With sophisticated modeling, the "dark matter" ratio of cross-sectional area to mass can be computed from these data (Randall et al., 2008). An upper limit puts the ratio at 0.7 $cm^2/g$. If we calculate this ratio for comets, and assuming a spherical comet of radius $r$, we have mass $m = 4/3\pi r^3 \rho$ and the cross-sectional area, $A = \pi r^2$, giving a ratio, $A/m = 3/(4r\rho)$. Plugging in a typical comet density of 0.5 $g/cm^3$, we get $r \sim 2$ cm or about 32 g. This is a bit small for solar system comets, which tend to have radii about 2,000 m, or $10^5$ larger than this. However, the bullet cluster merely sets an upper limit, and the smaller this ratio, the better it fits the comet model. It also illustrates very nicely the interpolation of cross sections between gas with radii $10^{-9}$ m and brown dwarfs with radii $\sim 10^8$ m.

There is one other objection to galactic comets fulfilling the role of "dark matter," and that is the assertion that 70% of the matter in the bullet cluster or in the universe is "dark" (Clowe et al., 2006; Angus et al., 2007). This would make comets and their associated carbon and oxygen more abundant than stars and their constituent hydrogen and helium, which would violate the 75:25:0.01 mass ratio of cosmological hydrogen:helium:metals production in the Big Bang nucleosynthesis (BBN) models. This is a serious problem for our galactic comet hypothesis, which can be resolved by either (a) following the current paradigm where 90% or more of dark matter is nonbaryonic with small admixtures of comets consistent with Solar System abundances; (b) positing some early stage of galactic formation that burns H and He to C and O, which later form comets (Gibson et al., 2010a); or (c) arguing that BBN models have not properly taken into account the "plasma" age of the universe, between nucleosynthesis and neutralization of atoms.

Our preference is (c), for if strong Big Bang magnetic fields exist, then magnetized plasma modes can provide degrees of freedom not available to the hot-gas models of BBN, prolonging the ~20-min era of giga-Kelvin temperatures and providing nonthermal channels for nucleosynthesis to continue. This may have changed the H:He:C:O ratios, whereupon later condensation into comets would have "hidden" the CO from spectroscopic discovery, since C and O are both "sticky" elements, likely to form interstellar solids that are not easily detected spectroscopically. Furthermore, their volatility in the protosolar nebula would have caused them to migrate antisunward during the accretion phase, so that they are underrepresented in stellar composition, and hence in spectroscopic observations of stars. In a now-discredited theory, Frank argues for the ubiquity of meter-sized comets in the solar system, making many of the same sort of "invisibility" arguments for "cometesimals" (Frank et al., 1986; Frank, 1990).

Nevertheless, we find this baryon-density argument to be a formidable objection, requiring a comprehensive plasma-BBN model to address this issue quantitatively.

## 4.3. COMET LOOPS

Making the assumption that comets are not just ubiquitous but numerous in the early universe, we can now see how they can connect the large and small scales of the universe. Since the atoms of C, O, or their simple hydrogenized forms—$CH_4$, $H_2O$, CO, and $CO_2$—are all easily condensed, they would nucleate much sooner than the giant hydrogen and helium clouds. As a consequence, gas clouds in the early universe would go unstable to accelerated gravitational collapse much sooner and provide the superstructure of galactic clusters and voids observed today, as is provided by "dark matter" in all the cosmological models.

Galaxy clusters, such as the Coma supercluster, must form before the galaxy begins stellar formation; they require a dark matter seed (Zwicky, 1937), which can be provided by comets since they are the first to condense out of the protogalactic nebula. The same is true at the smaller subgalactic scale of globular clusters, whose stars are generally much older than the galactic disk. Since globular clusters have higher average stellar velocities than galaxies and cluster galaxies higher than field galaxies, we would expect these clusters to evaporate comets with much higher relative velocities, enabling these high-speed comets to seed neighboring galactic nebulae. This self-seeding or catalytic character of comets is similar to diffusion limited growth and may account for some of the cosmic galactic structure such as "the great wall" which is presently attributed to unspecified "dark matter."

As stellar formation began in protogalaxies, the immediate heat flux would drive the comets away due to gas jetting on the surfaces of comets. Thus, comets have a built-in repulsion for stars, which we may be observing in the galactic rotation curves discussed earlier. The greater the repulsion, the more likely that comets will "evaporate" from galaxies and not contribute to star formation. Not coincidently, this "repulsive force" depends on the spectral reflectivity of comets, and the big surprise in the past 25 years was the discovery that "old" comets are blacker than carbon soot. This makes them maximally sensitive to thermal radiation and may be a consequence of cyanobacterial biofilms forming on the outside of the comet (Sheldon and Hoover, 2006, 2007). That is, the spectral characteristic of comets that makes them more efficient galactic messengers is itself a consequence of life.

Therefore, just as Gaia theory argues that Earth climate is stabilized by life, so it may be possible that galaxy formation was itself catalyzed by life, making this universe with all its anthropic contingency merely a consequence of biological homeostasis. Whether this corollary hypothesis bears up or not, we present it as an example of how the largest scales observed in cosmology can be connected to the smallest scales of biology through the mediation of comets.

We began this discussion on comets by describing them as the messengers of the universe, much like the neurons in the brain, connecting the spacetime pixels of static spatial information to produce dynamic information, populating the matrix of Fourier transform information. We ended by arguing that life could modulate galaxy formation in such a way as to make the universe hospitable for advanced life, a combination of strong anthropic principle and Gaia hypothesis (Carter, 1974; Lovelock and Margulis, 1974). But perhaps a better way to view this emphasis on Fourier space information is to recognize that the macrocosm mirrors the microcosm, that the universe bears more than passing resemblance to the cell, with comets providing an analog of the tubulin proteins that give shape and structure to the cell and are the highways for nondiffusive information transport. Thus, science may recover the medieval golden chain of being that connects the earth and the heavens (Lewis, 1964).

## 5. Discussion and Conclusion

The OOL problem arose in the nineteenth century when materialism replaced theism as the metaphysics of science. (Some might argue that science is defined by its materialist metaphysics, and therefore deny the existence of science until the Enlightenment, but this narrow definition does disservice to the contributions of Aristotle, Archimedes, and the countless "giants" on whose shoulders Newton stood.) In this more restricted scientific metaphysics, Aristotle's material causes trump his final causes, and life is to be described by "how" rather than "why." With the discovery by Pasteur that spontaneous generation is highly unlikely, and with the twentieth century advances in biochemistry that made spontaneous generation impossible in a finite universe (Meyer, 2009), the OOL problem crystallized all the metaphysical objections to materialism that had been raised by Aristotle and subsequent generations of philosophers.

Elsewhere in the twentieth-century science, materialism posed less of an impediment, and progress was made in information theory, astronomy, and cosmology that led to several important discoveries of conservation laws. The late 1800s saw the development of thermodynamics and its twin concepts of energy conservation and entropy growth, despite neither being a material property of matter envisioned by Democritus. Thermodynamics was brought back into the fold of materialism by Boltzmann, who gave it a particle (statistical mechanics) interpretation, and along the way defined entropy as a probabilistic ordering of these particles. A half-century later, Shannon laid the foundation of the computer revolution by demonstrating how the flip side of entropy is information and by showing how machines can process that information digitally.

The implications from Shannon's new field of information theory rippled outward into all of the sciences, especially physics. Quantum mechanics reported experiments whose outcome depended purely on information (e.g., Kim et al., 2000). Hawking began to consider the effects of entropy and energy on astrophysics, concluding that entropy (and therefore information) is conserved even as black holes devour the matter of materialism (Hawking, 2005; Susskind, 2008), so in some sense, the immaterial information is more permanent than the material matter. This progress toward immaterialism is no more evident than in the career of the late physicist John Wheeler, who described his life as composed of three phases: "Everything is Particles; Everything is Fields, and Everything is Information." His memorable aphorism to describe this final phase was "It from Bit" (Wheeler and Ford, 1999)—existence comes from information—which is the exact opposite of the materialist-inspired aphorism of the mid-twentieth century "existence precedes essence" (Sartre, 1943). For the first time since Aquinas, scientists are now seriously considering not just the inadequacy of materialism but the prior necessity of immaterialism.

This shift may explain the 2010 publication of a curious paper by Verlinde, in which he argues that conservation of energy and conservation of information, with some mathematical machinery of 4D spacetime, cannot only reproduce Einstein's general theory of relativity but also Newton's laws of motion (Verlinde, 2010). That is, the materialist assumptions of point particles traveling through the void which were so ably quantified by Newton's calculus have now been derived, not assumed, from conservation laws of energy and information. Two completely immaterial concepts have been combined so as to derive the material. Materialism is not the basis of science but a corollary of science.

This information, including that in the Fourier realm which we argue is necessary to explain the origin of life, is now thought by many to be a permanent feature of the universe, which from a physics standpoint, means a contingent feature of the Big Bang. The anthropic principle, which paled at the prospect of a finely tuned explosion to one part in $10^{60}$ such that one grain of sand more or less would have made the universe devoid of life, must now contend with contingent information of far greater magnitudes.

This conservation of information from the Big Bang is often misunderstood as "front loading," or as the British deists described it, as the winding up of a watch (Paley, 1809). This description is inadequate if time is treated as a separate dimension as in, for example, a wound-up watch with a spatial arrangement of springs and levers that deterministically evolve in time; a boundary condition in space that sets $x_0$ and $v_0$ and then follows $F=ma$ in time. Free choice seems to be missing from the equation, and likewise, entropy appears to be growing as the spring gets hot, destroying information. However, the watch in four-dimensional spacetime not only has a boundary condition in space, but a boundary condition in time. Thus, information is continually propagating to the watch from that temporal boundary condition as it unwinds, just as information is imparted to the watch from its spatial boundary condition as it unwinds. That information may include, for example, instructions to rewind the watch. These information flows in spacetime mean that the system is not "closed" to outside influence, which would then lead to entropic information loss, but capable of becoming more complex.

Applying this to our OOL problem, the appearance of life is not explained as an internal law of complexification (a.k.a. vitalism), or an internal production of information (a.k.a. materialist evolution), but as a consequence of external information flow (e.g., comets), bringing information from the 4D spacetime boundary condition that accompanied the Big Bang (Sheldon and Hoover, 2008).

What would this OOL scenario look like to an observer within the system? Making an analogy to Paley's watch, we can imagine labeling all the atoms of the watch and then running the movie back 1 year to see how that watch came to be. We would see tagged atoms of copper and tin and zinc coming from ores, being purified and concentrated, melted and mixed, and shaped and cut, and polished. Then from locations all over the Earth, these components would arrive and concentrate into subassemblies, which further transport would bring to the watch factory and suddenly they would all assemble and the watch would begin to function. Distributed information is displayed by the sequencing of these events, where at each moment of the movie, information is being added in the form of concentrating, shaping, and structuring, and where at no point in the movie would there be an entropic or information destroying event.

In the same way, the movie of OOL might show a highly diffuse and distributed information system that concentrated, altered, and structured the organic molecules. Not only would OOL involve more than a warm pond on the Earth, it would likely involve more than all the warm ponds in the Solar System and galaxy. As the universe expands and as the galaxies contract, the necessary information would likely concentrate, moving from the 4D boundary of the Big Bang toward the middle, toward a spacetime volume perhaps on a comet, where a living organism could then appear. How does this miracle occur? By adding slightly less information-rich systems together in improbable, but not wildly improbable steps, until the OOL threshold is crossed.

In conclusion, we have attempted to show that the OOL problem runs aground on the metaphysical shoals of materialism and its assumption of incoherence. Information theory provides a way forward, but must be expanded to include nonlocal or Fourier space information to accommodate the vast amounts of information encoded by life. This required capacity, when generalized to Einstein's spacetime, is claimed to be a conserved quantity of the universe that must also incorporate time, which makes information flow a necessary consequence of information capacity. When combined with the conservation laws, this information flow is from the 4D boundary conditions of the Big Bang inward, toward the volume that includes the Earth. Materially, we find that comets have all the properties to mediate this information flow, and the cometary hydrosphere can therefore be the physical realization of this mathematical necessity. We provide some weak justification for singling out comets for this monumental task, incidentally suggesting that they may also provide the solution to the missing "dark matter" problem.

When we examine the solution we have derived, we find that it has led several prominent physicists to propose the priority of information and the derivative nature of materialism. Thus, the OOL problem may be solved by turning materialism on its head. Rather than finding life to be a difficult accomplishment for a materialist, we find instead that materialism may be a trivial accomplishment for life.

## 6. References


Abrams DS, Lloyd S (1999) Quantum algorithm providing exponential speed increase for finding eigenvalues and eigenvectors. Phys Rev Lett 83:5162–5165. doi:10.1103/PhysRevLett.83.5162

Angus GW, Shan HY, Zhao HS, Famaey B (2007) On the proof of dark matter, the law of gravity, and the mass of neutrinos. Astrophys J Lett 654:L13–L16. doi:10.1086/510738

Arrhenius S (1908) Worlds in the making: the evolution of the universe. Harper & Row, New York

Axe DD (2004) Estimating the prevalence of protein sequences adopting functional enzyme folds. J Mol Biol 341:1295–1315

Barrick JE, Yu DS, Yoon SH, Jeong H, Oh TK, Schneider D, Lenski RE, Kim JF (2009) Genome evolution and adaptation in a long-term experiment with Escherichia coli. Nature 461:1243–1247

Carter B (1974) Large number coincidences and the anthropic principle in cosmology. In: IAU symposium 63: confrontation of cosmological theories with observational data, pp. 291–298, Reidel, Dordrecht

Clowe D, Brada M, Gonzalez AH, Markevitch M, Randall SW, Jones C, Zaritsky D (2006) A direct empirical proof of the existence of dark matter. Astrophys J Lett 648:L109–L113. doi:10.1086/508162

Coates AJ, Jones GH, Lewis GR, Wellbrock A, Young DT, Crary FJ, Johnson RE, Cassidy TA, Hill TW (2010) Negative ions in the Enceladus plume. Icarus 206:618–622. doi:10.1016/j.icarus.2009.07.013

Cover TM, Thomas JA (1991) Elements of information theory. Wiley, New York

Darwin C (1859) On the origin of species by means of natural selection, or the preservation of favoured races in the struggle for life. John Murray, London

Darwin F (1887) The life and letters of Charles Darwin, including an autobiographical chapter, vol 3. John Murray, London, 18 pp

Davies P (2000) The fifth miracle: the search for the origin and meaning of life. Simon & Schuster, New York

Dawkins R (1986) The blind watchmaker. Oxford University Press, Oxford

Dembski W (1998) The design inference: eliminating chance through small probabilities. Cambridge University Press, Cambridge

Dembski W (2002) No free lunch: why specified complexity cannot be purchased without intelligence. Rowman and Littlefield Pub. Inc., Lanham

Dembski W, Marks RJ II (2010) The search for a search: measuring the information cost of higher level search. J Adv Comput Intell Intell Inform 14(5):475–486

Farley J (1974) The spontaneous generation controversy from Descartes to Oparin. Johns Hopkins University Press, Baltimore

Frank LA (1990) The big splash: a scientific discovery that revolutionizes the way we view the origin of life, the water we drink, the death of the dinosaurs,…. Birch Lane Press, New York

Frank LA, Sigwarth JB, Craven JD (1986) On the influx of small comets into the earth's upper atmosphere. I—Observations. II— Interpretation. Geophys Res Lett 13:303–310. doi:10.1029/GL013i004p00303

Gallo CF, Feng JQ (2010) Galactic rotation described by a thin-disk gravitational model without dark matter. J Cosmol 6:1373–1380

Gibson CH, Wickramasinghe NC, Schild RE (2010a) First life in primordial-planet oceans: the biological big bang. ArXiv e-prints

Gibson DG, Glass JI, Lartigue C, Noskov VN, Chuang R-Y, Algire MA, Benders GA, Montague MG, Ma L, Moodie MM, Merryman C, Vashee S, Krishnakumar R, Assad-Garcia N, Andrews-Pfannkoch C, Denisova EA, Young L, Qi Z-Q, Segall-Shapiro TH, Calvey CH, Parmar PP, Hutchison CA III, Smith HO, Venter JC (2010b) Creation of a bacterial cell controlled by a chemically synthesized genome. Science 329(5987):52–56

Gödel K (1931) Über formal unentscheidbare Sätze der Principia Mathematica und verwandter Systeme, I. Monatshefte für Mathematik und Physik 38:173–198

Guth AH (2007) Eternal inflation and its implications. J Phys A Math Gen 40:6811–6826. doi:10.1088/1751-8113/40/25/S25

Hawking SW (2005) Information loss in black holes. PhysRevD 72(8):084, 013. doi: 10.1103/PhysRevD.72.084013

Herculano-Houzel S (2009) The human brain in numbers: a linearly scaled-up primate brain. Front Hum Neurosci 3:31

Hoover RB (2008) Comets, carbonaceous meteorites, and the origin of the biosphere. In: Dobretsov N, Kolchanov N, Rozanov A, Zavarzin G (eds) Biosphere origin and evolution. Springer, New York, pp 55–68

Hoover R (2011) Fossils of cyanobacteria in CI1 carbonaceous meteorites. J Cosmol 13:1+

Hoover RB, Pikuta EV, Wickramasinghe NC, Wallis MK, Sheldon RB (2004) Astrobiology of comets. In: Hoover RB, Levin GV, Rozanov AY (eds) Instruments, methods, and missions for astrobiology VII. Proceedings of SPIE, vol 5555. SPIE, Bellingham, pp 93–106

Hoyle F (1999) The mathematics of evolution. Acorn Enterprises, Memphis

Hoyle F, Wickramasinghe C (1980) In: Ponnamperuma C (ed) Comets and the origin of life. D. Reidel Publishing Co, Dordrecht, p 227

Hoyle F, Wickramasinghe C (1981a) Evolution from space. J.M. Dent, London

Hoyle F, Wickramasinghe C (1981b) Proofs that life is cosmic. Memoirs of the Institute of Fundamental Studies, Sri Lanka, No. 1. http://www.panspermia.org/proofslifeiscosmic.pdf

Hoyle F, Wickramasinghe C (2000) Astronomical origins of life. Kluwer Academic, Dordrecht

Kim Y-H, Yu R, Kulik SP, Shih Y, Scully MO (2000) Delayed 'choice' quantum eraser. Phys Rev Lett 84(1):1–5. doi:10.1103/PhysRevLett.84.1

Komatsu E et al (2011) Seven-year Wilkinson Microwave Anisotropy Probe (WMAP) observations: cosmological interpretation, Astrophys J Suppl 192:17. doi:10.1088/0067-0049/192/2/17

Lee R, Jonathan P, Ziman P (2010) Pictish symbols revealed as a written language through application of Shannon entropy. Proc R Soc A Math Phys Eng Sci 466(2121):2545–2560. doi:10.1098/rspa.2010.0041

Levin GV, Straat PA (1976) Viking labeled release biology experiment—interim results. Science 194:1322–1329

Lewis CS (1964) The discarded image: an introduction to medieval and renaissance literature. Cambridge University Press, Cambridge

Linde A, Vanchurin V (2010) How many universes are in the multiverse? Phys Rev D 81(8): 083,525–+. doi: 10.1103/PhysRevD.81.083525

Lloyd S (2002) Computational capacity of the universe. Phys Rev Lett 88(23): 237, 901. doi: 10.1103/PhysRevLett.88.237901

Lovelock JE, Margulis L (1974) Atmospheric homeostasis by and for the biosphere: the gaia hypothesis. Tellus 26:2

Lucretius Carus T (1921) Of the nature of things (De Rerum Natura). E. P. Dutton and Co. Inc., New York

Meyer SC (2009) Signature in the cell: DNA and the evidence for intelligent design. HarperCollins Pub, New York

Mojzsis SJ, Harrison TM, Manning CE (2003) The oldest known sediments on Earth: implications for exobiology. Geochim Cosmochim Acta Suppl 67:300

NASA/Jet Propulsion Laboratory (2010) Life on titan? New clues to what's consuming hydrogen, acetylene on Saturn's Moon. ScienceDaily. Retrieved January 5, 2012, from http://www.sciencedaily.com/releases/2010/06/100606103125.htm
Oak Ridge National Laboratory (2009) Oak ridge 'Jaguar' supercomputer is world's fastest. ScienceDaily. Retrieved January 5, 2012, from http://www.sciencedaily.com /releases/2009/11/091116204229.htm
Paley W (1809) Natural theology: or, evidences of the existence and attributes of the deity, 12th edn. J. Faulder, London
Pasteur L (1861) Mémoire sur les corpuscules organisés qui existent dans l'atmosphère. Examen de la doctrine des générations spontanées. Ann Sci Naturelles (partie zoologique) (Sér. 4) 16:5–98
Peret J, Bada J, Lazcano A (2009) Charles Darwin and the origin of life. Orig Life Evol Biosph 39:395–406. doi:10.1007/s11084-009-9172-7
Pitts JB (2010) Gauge-invariant localization of infinitely many gravitational energies from all possible auxiliary structures. General Relativity and Gravitation 42:601–610. doi:10.1007/s10714-009-0874-y
Polanyi M (1968) Life's irreducible structure. Science 160:1308–1312
Randall SW, Markevitch M, Clowe D, Gonzalez AH, Brada M (2008) Constraints on the self-interaction cross section of dark matter from numerical simulations of the merging galaxy cluster 1E 0657–56. Astrophys J 679:1173–1180. doi:10.1086/587859
Rubin VC, Ford WK Jr (1970) Rotation of the Andromeda Nebula from a spectroscopic survey of emission regions. Astrophys J 159:379. doi:10.1086/150317
Sartre J-P (1943) L'Être Et Le Neant. Gallimard, Paris
Schneider ED, Sagan D (2005) Into the cool: energy flow, thermodynamics, and life. University of Chicago Press, Chicago
Shannon CE (1948) A mathematical theory of communication. Bell Syst Tech J 27(379–423):623–656
Shannon CE (1993) Prediction and entropy in written English. In: Wyner AD, Sloane NJA (eds) Claude Shannon collected papers. IEEE Press, Piscataway, p 583
Shannon CE, Weaver W (1949) The mathematical theory of communication. University of Illinois Press, Urbana
Sheldon RB, Hoover RB (2005) Evidence for liquid water on comets. In: Hoover RB, Levin GV, Rozanov AY (eds) Instruments, methods, and missions for astrobiology VIII. Proceedings of SPIE, vol 5906A. SPIE, Bellingham, WA, pp 196–206
Sheldon RB, Hoover RB (2006) Implications of cometary water: deep impact, stardust and hayabusa. In: Hoover RB, Levin GV, Rozanov AY (eds) Instruments, methods, and missions for astrobiology IX. Proceedings of SPIE, vol 6309. SPIE, Bellingham, WA, pp 6309–0L
Sheldon RB, Hoover RB (2007) The cometary biosphere. In: Hoover RB, Levin GV, Rozanov AY (eds) Instruments, methods, and missions for astrobiology X. Proceedings of SPIE, vol 6694. SPIE, Bellingham, WA, pp 6694–0H
Sheldon RB, Hoover RB (2008) Cosmological evolution: spatial relativity and the speed of life. In: Hoover RB, Levin GV, Rozanov AY (eds) Instruments, methods, and missions for astrobiology XI. Proceedings of SPIE, vol 7097. SPIE, Bellingham, WA, pp 7097–41
Shor PW (1995) Quantum computers, factoring, and decoherence. Science 270:1633–1635
Strobel DF (2010) Molecular hydrogen in Titan's atmosphere: implications of the measured tropospheric and thermospheric mole fractions. Icarus 208:878–886. doi:10.1016/j.icarus.2010.03.003
Susskind L (2007) The anthropic landscape of string theory. Cambridge University Press, Cambridge, pp 247–+
Susskind L (2008) The black hole war: my battle with Stephen Hawking to make the world safe for quantum mechanics. Little, Brown, & Co, New York
Verlinde EP (2010) On the origin of gravity and the laws of Newton. High Energy Physics 4:29+
Volders L, van de Hulst HC (1959) Neutral hydrogen in extra-galactic systems. In: Bracewell RN (ed) URSI symposium 1: Paris symposium on radio astronomy. IAU symposium, vol. 9. Stanford University Press, Stanford, CA, pp 423–430
Wheeler JA, Ford KW (1999) Geons, black holes, and quantum foam: a life in physics. Norton, New York

Wickramasinghe C (2011) Viva Panspermia. The Observatory, 131:130–134

Wickramasinghe NC, Wainwright M, Narlikar JV, Rajaratnam P, Harris MJ, Lloyd D (2003) Progress towards the vindication of panspermia. Astrophys Space Sci 283:403–413. doi:10.1023/A:1021677122937

Zwicky F (1937) On the masses of nebulae and of clusters of nebulae. Astrophys J 86:217. doi:10.1086/143864

Biodata of **David Wacey**, author of "***In Situ High Spatial Resolution Techniques in the Search for the Origin of Life***."

**David Wacey** is a biogeochemist at the University of Western Australia. He obtained his D.Phil. from Oxford University studying the effects of sulfate-reducing bacteria on biomineralization. His current research focuses on morphological, chemical, and isotopic tracers of primitive Archean life, using cutting-edge techniques such as NanoSIMS and transmission electron microscopy. He is also the author of a new introductory textbook on Archean life entitled "Early Life on Earth: A Practical Guide," released in 2009.

E-mail: **David.Wacey@uwa.edu.au**

J. Seckbach (ed.), *Genesis - In The Beginning: Precursors of Life, Chemical Models and Early Biological Evolution*, Cellular Origin, Life in Extreme Habitats and Astrobiology 22, 391–411
DOI 10.1007/978-94-007-2941-4_22, 

# IN SITU HIGH SPATIAL RESOLUTION TECHNIQUES IN THE SEARCH FOR THE ORIGIN OF LIFE

**DAVID WACEY**[1,2]
[1]*Centre for Microscopy, Characterisation and Analysis, The University of Western Australia, Crawley, WA 6009, Australia*
[2]*School of Earth and Environment, The University of Western Australia, Crawley, WA 6009, Australia*

## 1. Introduction

Identifying the transition from a prebiotic world to one containing life is a complex task. The early Archean rock record is extremely fragmentary with few examples of sedimentary rocks older than 3,000 million years (Ma). Of these, even fewer have escaped the significant post-depositional deformation and metamorphism that would likely have destroyed any former signs of life. Add to this the fact that the earliest life on Earth would have been very small, morphologically simple and likely only subtly different from co-occurring nonbiological organic material, and the extent of the task becomes clear. Fortunately, modern technology gives researchers a fighting chance to distinguish biological signals from nonbiological artifacts in these ancient rocks. In this contribution, I will describe four types of instrumentation that have been introduced to the origins of life field in the past decade or so in an attempt to provide more robust in situ evidence for life in Earth's oldest rocks: (1) transmission electron microscopy (TEM) using focused ion beam (FIB)–milled ultra-thin sections; (2) secondary ion mass spectrometry (SIMS and NanoSIMS) incorporating elemental mapping and isotopic analysis; (3) laser-based techniques including laser Raman micro-spectroscopy and confocal laser scanning microscopy (CLSM); (4) synchrotron-based techniques including scanning transmission X-ray microscopy (STXM) and near-edge X-ray absorption fine structure (NEXAFS) spectroscopy. I will outline the types of high spatial resolution data that can now be obtained, and how this has furthered our understanding of Earth's earliest life. A prerequisite to the use of all such instrumentation is that one must have a firm understanding of the geological context (age, environmental setting, post-depositional history) of a given sample so that data obtained on the micrometer (μm) to nanometer (nm) scale may be interpreted correctly and integrated with the bigger picture.

## 2. Transmission Electron Microscopy

Transmission electron microscopy (TEM) is a broad term that encompasses numerous techniques that can be performed in a transmission electron microscope (TEM – note that the same acronym is used for both the instrument and the technique) and can be used to image and characterize the structure and chemistry of a sample down to the nanometer or even atomic scale. Details concerning the principles of TEM can be found in Williams and Carter (2009). Briefly, a beam of electrons is accelerated at high voltage (typically 200–300 KeV) through an ultra-thin (~100 nm) sample. As the electrons pass though the sample, they interact with the sample atoms and are scattered by varying amounts dependent upon how close an electron gets to an atomic nucleus and upon the charge associated with said nucleus. A standard TEM image results from variable electron scattering at different points within the sample. TEM image intensity can also be affected by diffraction contrast, whereby crystalline materials show different contrast depending upon their orientation to the electron beam. In some cases, this may add an extra layer of complexity to image interpretation. Nonetheless, in its simplest mode of operation, the TEM performs as a high-resolution imaging tool, able to resolve objects separated by distances as small as ~0.1 nm (i.e. a typical atomic spacing in a crystalline material). However, the real power of TEM comes with the ability to combine information from imaging with additional parameters including elemental composition, chemistry (bonding and oxidation states), crystal structure and crystal orientation.

The elemental composition and local chemistry of a sample can be analyzed using an electron spectrometer attached to the TEM. Electron energy-loss spectroscopy (EELS) uses the fact that the interaction of electrons with different elements causes them to lose different amounts of energy (Williams and Carter, 2009). Hence, a spectrum is produced showing the number of electrons that have passed through the sample as a function of the amount of energy they have lost, with specific core-loss peaks corresponding to specific elements. The area under a core-loss peak can be used to determine how much of a given element is present. The fine detail of a core-loss peak (the energy-loss near-edge structure or ELNES) can, in turn, be used to identify the bonding characteristics and/or oxidation state of a given element within a sample. For example, the transition of carbon from an unordered amorphous state to graphite during heating can be traced using ELNES (Daniels et al., 2007); similarly, $Fe^{3+}$ can be distinguished from $Fe^{2+}$ (e.g. Calvert et al., 2005).

ELNES has been used to analyze carbon within the ~3,465-Ma Apex chert and the ~3,430-Ma Strelley Pool Formation of Western Australia (De Gregorio and Sharp, 2006; Wacey et al., 2009), two rock units that have been at the forefront of the search for Earth's earliest life (Schopf, 1993; Brasier et al., 2002; Wacey et al., 2006; Wacey, 2010). The carbon ELNES spectra show that both the Apex and Strelley carbon have little order and lack the characteristic $\sigma^*_1$ peak of graphitization. The carbon spectra are also very similar to that measured from bona fide microfossils from the 1,900-Ma Gunflint chert (Moreau and Sharp, 2004). Unfortunately, ELNES cannot distinguish disordered biological carbon from disordered carbon produced from nonbiological reactions such as Fischer-Tropsch-type synthesis (Brasier et al., 2002) so it can only provide data consistent with biology, not unique to biology. Nevertheless, it is extremely useful when applied as part of a suite of complementary techniques to attempt to determine what constitutes Earth's earliest life (De Gregorio et al., 2009).

The EELS technique can be taken a step further to produce images of the elemental distribution within a sample by isolating (filtering) specific energy windows (energy-filtered TEM or EFTEM) from an EELS spectrum. This has been employed by Wacey et al. (2011) to investigate the distribution of potential biological elements and biominerals associated with pyrite in a sandstone from the aforementioned ~3,430-Ma Strelley Pool Formation. EFTEM was used to identify carbonaceous coatings on pyrite grains (Fig. 1a) that were intimately associated with microbial trace fossils on pyrite surfaces. EFTEM combined with X-ray microanalysis and selected area electron diffraction (see below) also identified nano-grains of iron oxide within the carbonaceous coatings and on pyrite surfaces that Wacey et al. (2011) interpreted as the metabolic products of iron-oxidizing microbes feeding on the pyrite. This work has provided the earliest evidence of iron and sulfur-oxidizing microbial metabolisms in the rock record.

An alternative method to determine the elemental composition of a sample in the TEM is to use energy-dispersive X-ray spectroscopy (EDS). As electrons pass through a sample, they can ionize atoms resulting in the emission of X-rays, with the energies of these X-rays characteristic of the elements from which they are emitted. Both qualitative and quantitative elemental analyses are possible using EDS. If elemental maps are required using X-rays, the TEM must be operated in scanning mode (STEM) so that X-rays can be collected at specific points as the beam scans across the sample. Elemental mapping using STEM is complementary to elemental mapping using EFTEM and may help to resolve problems caused by overlapping energy peaks encountered in the latter technique.

A true high-resolution image in the TEM (HRTEM or HREM) is a phase-contrast image with atomic-scale resolution, allowing the visualization of the arrangement of atoms within a sample. This provides information about the crystallinity of a sample, its lattice structure and any defects it may have. This has been particularly useful, in combination with EELS, in the analysis of carbon within ancient rocks. For example, graphite can be quickly visually distinguished from more poorly ordered carbon (Buseck and Bo-Jun, 1985; Buseck et al., 1988). HRTEM is not able to distinguish biological from nonbiological carbon, but it can provide vital information on whether the carbon is likely syngenetic with the host rock and whether its structure is at least consistent with a biological precursor (Fig. 1b; Wacey et al., 2009, 2011). Attempts have also been made to quantify HRTEM data from ancient rocks. Rouzaud et al. (2005) imaged the polyaromatic layers within carbon from materials ranging from meteorites to the ~3,465-Ma Apex chert, quantifying the length, interlayer spacings and stacking arrangement

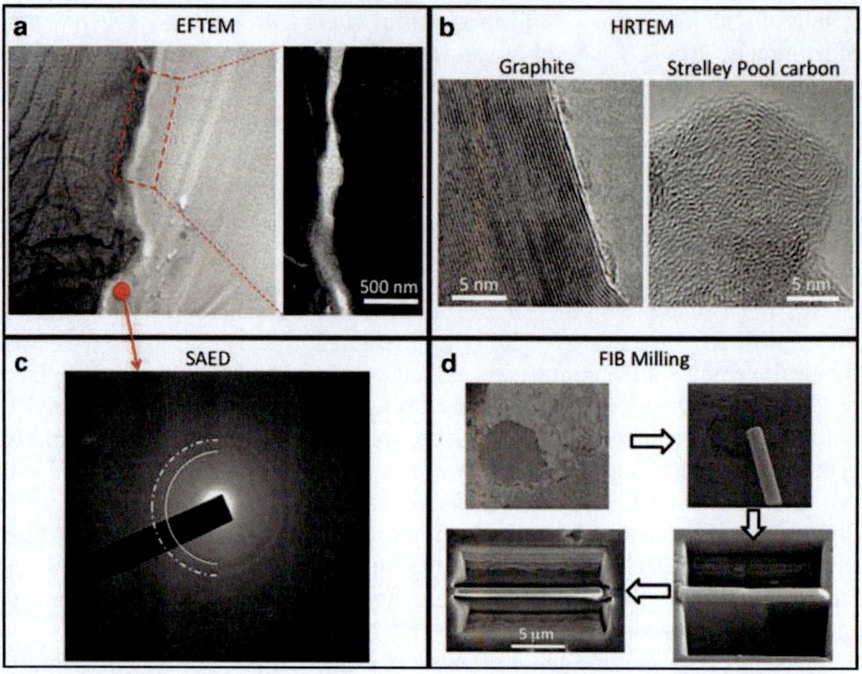


**Figure 1.** Examples of the application of TEM to investigating the origin and evolution of life. (**a**) Bright field TEM image showing pyrite (*dark grey*, *left*), silica-rich cement (*light grey*, *right*) and narrow band of low mass material (*white*, *center*) at the boundary of a pyrite grain from the ~3,430-Ma Strelley Pool Formation, Western Australia. Enlarged area is an energy-filtered TEM (EFTEM) elemental map of carbon distribution from the pyrite boundary. (**b**) High-resolution TEM (HRTEM) images demonstrating the degree of order within carbon. In the ~3,430-Ma Strelley Pool carbon (*right*), a mixture of short-range ordered and disordered carbons is present. Ordered regions exhibit the 0.34-nm interlayer spacing of graphite but do not approach the degree of order found in true graphite (*left*). (**c**) Selected area electron diffraction (SAED) in the TEM applied to the identification of mineral phases. This diffraction pattern comes from the area indicated in (**a**) and shows the presence of both fine-grained polycrystalline material (*sharp rings*) and amorphous material (diffuse background). The two semi-circles mark the two most intense polycrystalline rings at atomic planar spacings of 0.30 nm (± 0.01 nm) and 0.25 nm (± 0.01 nm), consistent with spacings present in the closely related iron oxides magnetite and maghemite. (**d**) Focused ion beam (*FIB*) sample preparation for the TEM. Features of interest are located in standard geological thin sections (*top left*, rounded cell-like feature). A protective platinum strip is deposited across the area of interest prior to ion beam milling (*top right*). Trenches are milled on either side of the area of interest, and the user can check that the feature continues below the sample surface (*bottom right*). Using reduced beam currents, the wafer is gradually thinned and cleaned to ~100 nm (*bottom left*). The wafer can then be removed from the sample for TEM analysis whilst retaining the context of the feature of interest (**a** and **c** modified from Wacey et al. (2011)).

of these layers on the nm scale (cf. Rouzaud and Clinard, 2002). Once a sufficient database of standard materials is built up, such data may be able to distinguish biological from nonbiological carbon.

Selected area electron diffraction (SAED) within the TEM (Williams and Carter, 2009) provides quantitative information on the distances between atomic planes in crystalline materials, permitting mineral identification. Wacey et al. (2011) used SAED to show that nano-crystals composed of iron and oxygen within the ~3,430-Ma Strelley Pool Formation (see above) were iron oxides belonging to the magnetite-maghemite group of minerals (Fig. 1c) and were therefore likely biologically mediated oxidation products of primary pyrite grains. SAED also allows the orientation of crystals to be studied. One potential application of this is to show the orientation of microquartz grains in the vicinity of putative carbonaceous microfossils, in order to determine whether the carbon is housed between grain boundaries, enclosed within a single grain or within fractures.

Correct sample preparation is vitally important for TEM work. Not only must the sample be ultra-thin (≤100 nm) in order to extract the maximum structural and chemical information, but the context of the structure to be analyzed must also be retained. Traditional TEM sample preparation involved grinding up a bulk sample and depositing it onto a copper grid; this was far from ideal for ancient samples where contamination is always a serious issue. Only relatively recently have researchers been able to prepare samples of uniform ~100-nm thickness directly from rock chips or thin sections using focused ion beam (FIB) milling. The FIB preparation technique uses a highly focused beam of heavy ions, typically gallium, to sputter ions from the sample. This essentially isolates a small portion of the sample at high spatial resolution and with great accuracy, leaving a rectangular wafer (typically about 15 μm × 10 μm × 0.1 μm) that can be lifted out of the rock chip or thin section and mounted on a TEM holder. Details of the FIB procedure are given in Wirth (2009) and references therein.

In the origins and evolution of life field, this preparation technique has been successfully utilized by Kempe et al. (2005) to investigate well-preserved 650-Ma cell walls, providing a valuable baseline for studies of older material. Wacey et al. (2011) also used FIB milling to prepare ~3,430-Ma Strelley Pool material for TEM analysis so that the relationships between pyrite, carbon, iron oxides and trace fossils (see above) could be visualized in their original context. An alternative use of FIB milling is to simply cut through specific features in ancient rock samples so that structures can be visualized in three dimensions. Westall et al. (2006) used FIB to mill through a small portion of a putative microbial mat from the ~3,400-Ma Josefsdal chert, South Africa. This allowed the internal structure of the mat to be imaged using scanning electron microscopy (SEM). This methodology can be extended to incorporate elemental analysis in the SEM and 3-D reconstructions of multiple slices of a sample (cf. Wirth, 2009). The major drawback to this technique is its destructive nature, making it unsuitable for samples with very rare microfossils; in these cases, X-ray tomography (see Sect. 5.1.1) may be preferable.

## 3. Secondary Ion Mass Spectrometry

Secondary ion mass spectrometry (SIMS) can be used to determine the elemental, isotopic or molecular composition of a sample surface in situ and at high spatial resolution. Details of the principles and variety of applications of SIMS are given elsewhere (Ireland, 1995; McKibben et al., 1998; Orphan and House, 2009). Briefly, the surface of a specimen is sputtered with a primary ion beam, typically cesium or oxygen, and ejected secondary ions are collected and analyzed using a mass spectrometer. SIMS has excellent sensitivity, being able to detect many elements present in the parts-per-billion (ppb) range.

### 3.1. LARGE RADIUS SIMS

In the origins and evolution of life field, traditional large radius SIMS instruments have been used mostly for isotopic work, both for dating rocks and more recently for investigating isotopic fractionations of key biological elements, especially carbon and sulfur (Mojzsis et al., 1996; House et al., 2000; Ueno et al., 2001; Philippot et al., 2007; Wacey et al., 2010a). The latest generation of large radius SIMS (CAMECA IMS 1280) are able to analyze objects as small as ~10–20 μm, with a precision generally better than 0.5 parts per thousand (‰). The measurement of isotopic fractionations, in situ, direct from putative microfossils, microbial mats or biominerals represents a huge step forward from bulk rock analyses because the possibility of contamination (especially when dealing with carbon) is greatly reduced.

A number of isotopic studies have made use of SIMS to increase our understanding of the earliest biosphere. House et al. (2000) analyzed individual bona fide microfossils from the Proterozoic Bitter Springs and Gunflint Formations. They found a range of light $\delta^{13}C$ values consistent not only with biological processing but with specific metabolic pathways (e.g. Calvin cycle or acetyl-CoA). Extending this idea to controversial putative microfossils, Ueno et al. (2001) measured the isotopic composition of individual carbonaceous filaments from the ~3,490-Ma Dresser Formation of Western Australia. A similar range of light $\delta^{13}C$ values ($-42$ to $-32‰_{PDB}$; where PDB is the international Pee Dee Belemnite reference standard) was interpreted as evidence of biological fractionation of carbon by autotrophic bacteria. Moving even further back into the rock record, Mojzsis et al. (1996) reported SIMS $\delta^{13}C$ values from carbonaceous inclusions in ~3,700-Ma (mean $-30‰_{PDB}$) and ~3,850-Ma (mean $-37‰_{PDB}$) rocks from Greenland, again interpreting such fractionations as evidence for biological reactions. More recent work on the fractionations of carbon isotopes associated with nonbiological reactions (see below) has, however, complicated the interpretation of these data.

SIMS has also been used to demonstrate the existence of a light and heterogenous $\delta^{13}C$ carbon reservoir ($-5$ to $-58‰_{PDB}$) prior to the oldest intact rock record preserved on Earth (Nemchin et al., 2008). These light $\delta^{13}C$ values were measured from tiny slivers of diamond and graphite within detrital ~4,250-Ma zircons ($ZrSiO_4$) from Jack Hills in Western Australia. Attributing these data to biological reactions is extremely controversial given that: (1) the host rock has not been preserved, so there is no way of knowing the true geological context of the putative biological signal; (2) a mechanism is required to transport "biological carbon", which presumably existed somewhere near the surface of the Earth, to the great depths and pressure for it to be transformed into diamond. This would imply that plate tectonics and subduction zones were already in existence at ~4,250 Ma; (3) Light carbon isotope values are by no means a uniquely biological signal. A "null hypothesis" that needs to be falsified for each potentially biological signal is the alternative nonbiological origin for the light carbon, for example, from Fischer-Tropsch-type reactions between CO and metals (Horita and Berndt, 1999; McCollom and Seewald, 2006) and/or from the metamorphic reduction of siderite (van Zuilen et al., 2003). These can both result in carbon isotope fractionations that lie within the traditional biological domain (i.e. lighter than $-15‰_{PDB}$). This latter point must also be considered when evaluating the biogenicity of $\delta^{13}C$ values reported from Greenland (Mojzsis et al., 1996), Australia (Ueno et al., 2001) and, in general, for Archean rocks.

Sulfur isotope fractionations on the early Earth have recently also been investigated using SIMS. With relevance to the origin and evolution of life, Philippot et al. (2007) measured the sulfur isotope fractionations of co-occurring pyrite and barite from the ~3,490-Ma Dresser Formation, Western Australia. They attributed the significant mass-dependent $\delta^{34}S$ fractionations (up to ~25‰) to microbial reactions, with a positive $\Delta^{33}S$ signature indicating that microbial elemental sulfur disproportionation was the most likely reaction pathway. This provides supporting evidence for the theory that sulfur-based metabolisms are some of the most deeply rooted on the tree of life (Shen and Buick, 2004 and references therein).

## 3.2. NANOSIMS

In the last 5 years, a new generation of dynamic SIMS has been introduced to the origin and evolution of life field. NanoSIMS enables isotopic measurements to be made from objects as small as ~1 μm and brings with it the additional capability of elemental mapping on the nm scale. For elemental mapping, the NanoSIMS is capable of sub-50-nm lateral resolution whilst imaging negatively charged secondary ions and ~150-nm lateral resolution for positive secondary ions. Up to six movable and one static detector record ion counts from the same sputtered sample volume at up to seven masses (elements or isotopes) simultaneously. This parallel measurement is extremely useful in the analysis of low concentration elements in organic materials that are rapidly sputtered away by the ion beam. For isotopic measurements, a precision of ~1‰ has been obtained from objects as small as

~5 μm in diameter (Fletcher et al., 2008) and ~3‰ from objects as small as ~1 μm in diameter (Wacey et al., 2010a). Whilst this is much poorer than the precision of conventional isotopic techniques, it is often sufficient to trace key biological fractionations in cases where spatial resolution is more critical than precision.

Significant advances have been made in the study of more modern microbial material using NanoSIMS that may have direct applications in the origins of life field. These include the detection of distinct cell wall and sheath structures enriched in carbon, nitrogen and sulfur in ~850-Ma microfossils (Oehler et al., 2006), the μm-scale mapping of changes in sulfide $\delta^{34}S$ with depth within modern microbial mats (Fike et al., 2008) and the μm-scale mapping of the relationship between individual microfossil components (e.g. sheaths, trichomes, extracellular substance), organic laminae and biomineralization in modern stromatolites (Fig. 2a; Wacey et al., 2010b).

In the origins and evolution of life field itself, Wacey (2010) extended previous NanoSIMS elemental mapping studies of modern microbialites (Kilburn and Wacey, 2011) back to some of the earliest known stromatolites from the ~3,430-Ma Strelley Pool Formation, interpreting the nm-scale co-occurrence of carbon, nitrogen and sulfur as evidence for a biological component to their formation (Fig. 2b). This reinforced previous biological interpretations of these stromatolites obtained from macro-scale studies (Hofmann et al., 1999; Allwood et al., 2006, 2009).

Wacey et al. (2008a, b) combined nm-scale mapping of carbon, nitrogen and biologically important trace elements (e.g. Co, Fe, Ni, Zn), with carbon isotope analysis to show that there was likely a biological component to the formation of ambient inclusion trails (AIT) in the ~3,430-Ma Strelley Pool Formation, and potentially other early Archean rocks. This provided the first geochemical data to support a long-standing biological hypothesis for AIT formation (Knoll and Barghoorn, 1974) and introduced AIT as a new potential biosignal in ancient rocks. Also from the Strelley Pool Formation, Wacey et al. (2010a) combined NanoSIMS $\delta^{34}S$ analysis of single pyrite grains with large radius SIMS $\delta^{34}S$ and $\Delta^{33}S$ to show that two coexisting sulfur-based metabolisms (microbial sulfate reduction and microbial sulfur disproportionation) were operational at ~3,430 Ma. This was able to resolve some of the controversy surrounding previous works that had provided conflicting data regarding the origin of these two metabolisms (Philippot et al., 2007; Ueno et al., 2008; Shen et al., 2009).

## 4. Laser-Based Techniques

### 4.1. LASER RAMAN MICRO-SPECTROSCOPY

Laser Raman micro-spectroscopy allows the collection of molecular and structural data, in situ, at high sensitivity and μm-scale spatial resolution. It is nonintrusive and nondestructive, and can be used on bulk samples or standard uncovered thin

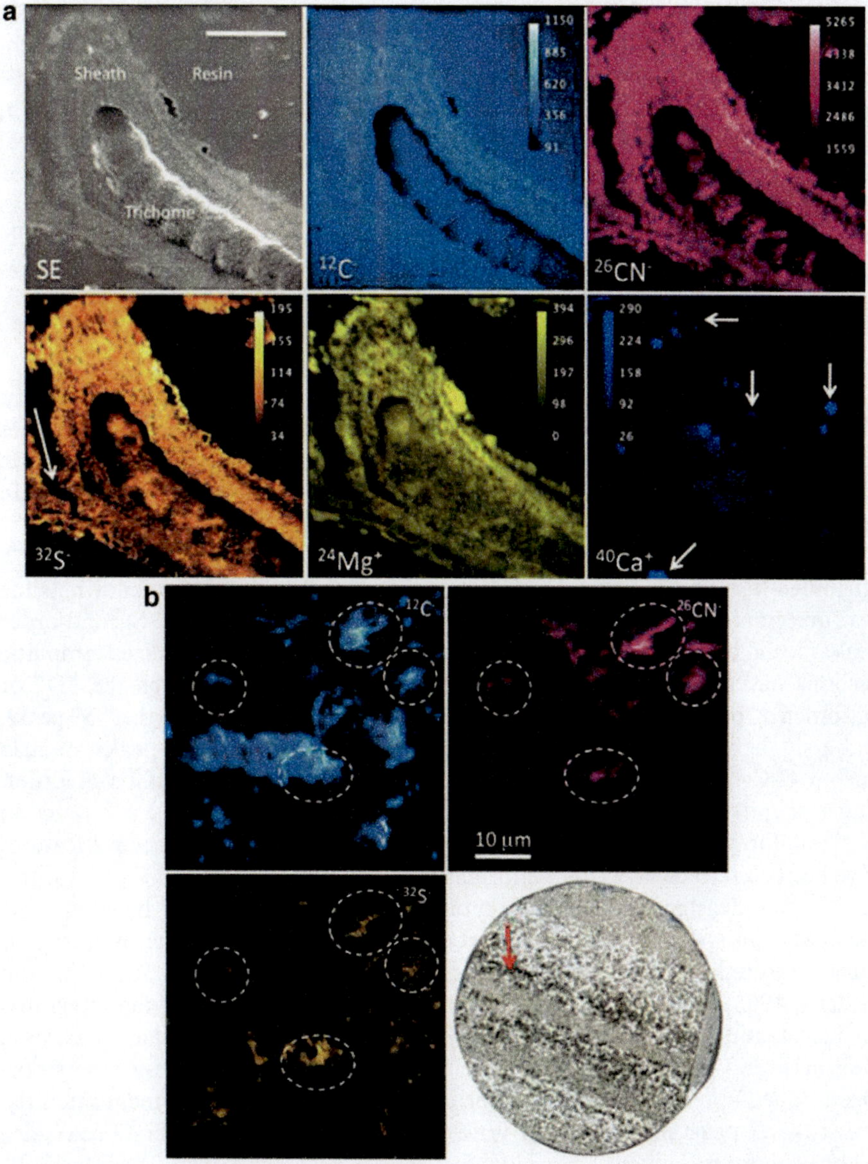


**Figure 2.** Ion images highlighting the information that can be obtained from modern and ancient stromatolites using NanoSIMS. **(a)** NanoSIMS ion images from a fossil cyanobacterium from a modern microbialite, Lake Clifton, Western Australia. Carbon, nitrogen, sulfur and magnesium co-occur in the cyanobacterial sheath and, to a lesser extent, within the trichome. Nano-grains of aragonite are seen to precipitate close to the sheath surface (*arrowed* in the $^{40}Ca^{+}$ image). A tear in the sheath is also visible (*arrowed* in the $^{32}S^{-}$ image) (Modified from Wacey et al., 2010b). Note that nitrogen does not form secondary ions so the $CN^{-}$ complex is used to map nitrogen distribution. **(b)** NanoSIMS ion images showing the co-occurrence of carbon, nitrogen and sulfur in organic material (*arrowed in thin section*) aligned along a stromatolite lamination from the ~3,430-Ma Strelley Pool Formation, Western Australia (Modified from Wacey (2010)).

sections, making it an ideal technique for the origin and evolution of life field where samples may be unique, and may have to be subjected to additional techniques subsequent to Raman. Analyses can be made from objects both at the surface and at depth (although the data quality deteriorates somewhat with depth) within a thin section, and Raman systems are often attached to a standard optical microscope so that imaging in transmitted and reflected light at a variety of magnifications can be performed in parallel to Raman data acquisition. Raman data are acquired by laser excitation of the samples using wavelengths typically ranging between the blue and infrared parts of the spectrum. Data can be presented as both point spectra and 2-D or 3-D images.

#### *4.1.1. Raman Point Spectra*

Raman point spectra give information on molecular bonding and crystallinity. One application of this is to quickly and simply identify different mineral phases within a sample. Of particular interest in the origin of life field is the identification and characterization of carbon-rich material associated with putative microfossils, trace fossils, laminae or biominerals.

In perfectly crystalline graphite, a single first order peak occurs at 1,582 $cm^{-1}$, attributed to stretching of the C-C bonds in basal graphite planes (known as the "G" or "graphite" peak) (Jehlicka et al., 2003; Schopf et al., 2005). Second-order peaks occur at ~2,695 $cm^{-1}$ and 2,735 $cm^{-1}$. Imperfectly crystallized graphitic carbons have additional peaks at ~1,355 $cm^{-1}$ (Fig. 3; known as the "D" or "disordered" peak) and ~1,620 $cm^{-1}$ ("D", occurring as a shoulder to the "G" peak), and a single broad second-order peak around 2,700 $cm^{-1}$. These peaks are also due to C-C bonds but arise because the lattice structure contains disorder. The exact position, height and width of these peaks can vary a little as a function of crystallite size and degree of ordering of the carbon. Raman spectral features have been used to try to extract information about both the biogenicity and antiquity of putative signs of early life. Antiquity has been investigated by comparing the change in spectral features of carbon to the change in thermal maturity of carbon through different metamorphic grades (Tice et al., 2004; Wopenka and Pasteris, 1993). In general, both the Raman D/G peak area (i.e. the integrated intensity) ratio and the D/G peak width (i.e. the full peak width at half maximum intensity) ratio of the carbonaceous material decrease with progressive metamorphism. The D peak becomes narrower, sharper and of more prominent intensity, whilst the G peak shifts to lower wave numbers. The second-order G peak also becomes more prominent.

Putative carbonaceous microfossils should exhibit the same Raman spectral features as other carbonaceous material in the same rock specimen because both should have undergone the same thermal maturation and geochemical alteration processes. Their spectra should also be consistent with known spectra from equivalent maximum metamorphic grade to the host rock unit and should contain little evidence of hydrated minerals that would not survive metamorphism. However, care must be taken when making comparisons with carbon housed in

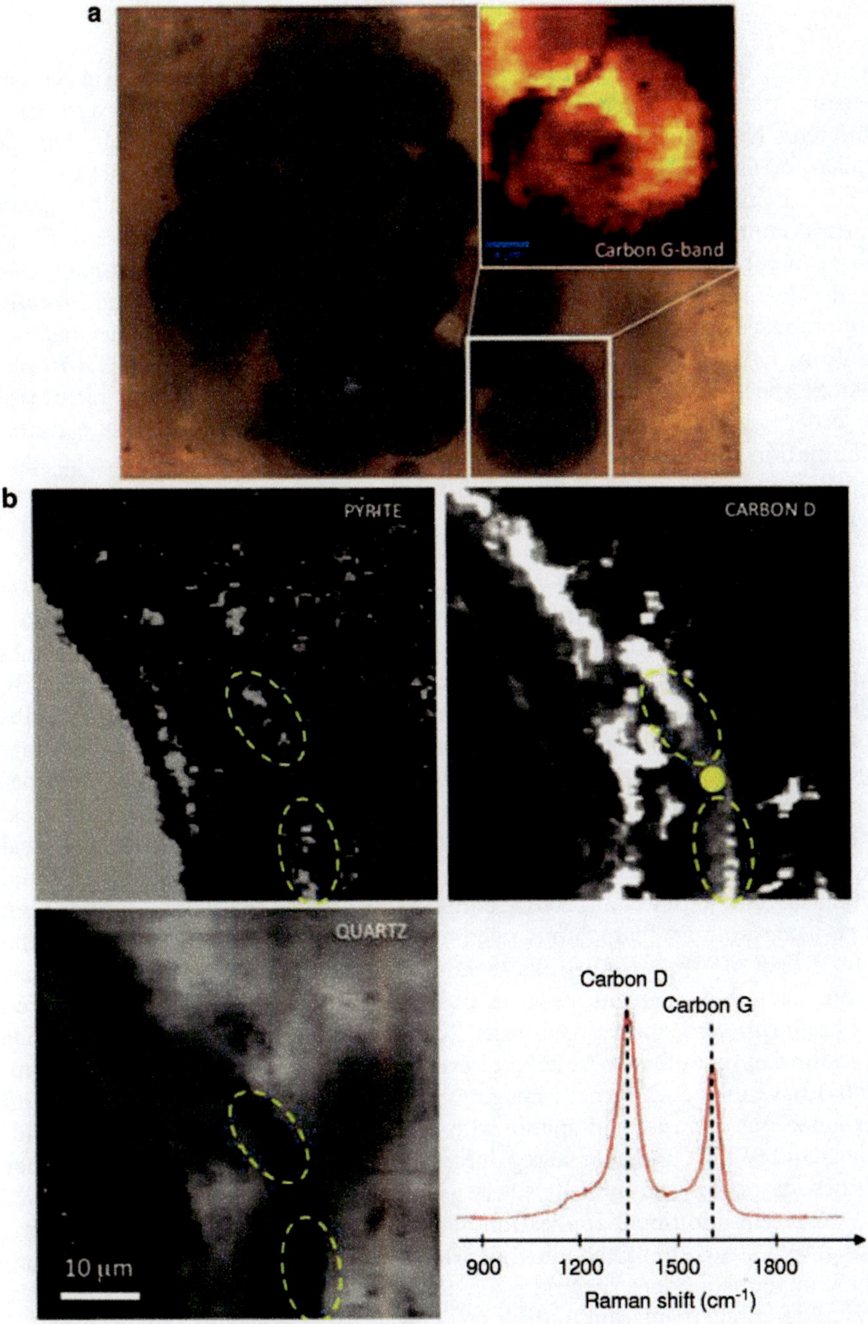


**Figure 3.** Examples of the application of laser Raman to the origin and evolution of life. (**a**) Cluster of well-preserved bona fide coccoid microfossils from the ~750-Ma Cailleach Head Formation of NW Scotland. The Raman image clearly indicates the carbonaceous composition of the cells (sample courtesy of L. Battison). (**b**) Raman images showing the association of pyrite, carbon and quartz within ~3,430-Ma Strelley Pool sandstone. *Dashed ovals* show equivalent areas in each image where pyrite and disordered carbon co-occur, and the *filled circle* indicates the area from which the Raman spectrum was obtained (Modified from Wacey et al. (2011)).

rocks of equivalent metamorphic grade but from different geographical regions because the starting composition and behavior of the carbon may have been different. Nevertheless, if serious inconsistencies exist, then it is likely that the "microfossils" are younger contaminants.

In complex biological carbonaceous compounds, additional spectral peaks corresponding to bonds such as C=O, C–H and C–N might be expected. These have not yet been observed in ancient samples using laser Raman probably due to the degree of thermal/geochemical alteration. Indeed, similar Raman spectra are produced from synthesized nonbiological disordered carbonaceous material as from *bona fide* biological kerogens (Pasteris and Wopenka, 2003). Hence, Raman spectra cannot be used to prove the biogenicity of a putative microfossil or other carbonaceous material. However, Raman can be used as a useful elimination test for biogenicity; if a carbonaceous microstructure does not give the Raman spectra of structurally disordered carbon, then it can generally be disregarded for further study.

### *4.1.2. Raman Imaging*

Raman imaging allows the micro-mapping of the distribution of certain spectral parameters within a sample. For example, how the ratio of D to G band varies over a given area, or which areas of a sample contain a certain mineral (Fig. 3). Three-dimensional maps are also obtainable by taking images at various depths within the thin section and then stacking them together using a 3-D software package (McKeegan et al., 2007). This allows a one-to-one correlation of carbonaceous molecular structural chemistry with optically recognizable morphological features. Most simply, this can be used to show the co-occurrence of biological morphology, such as putative cell walls, with disordered carbonaceous material (Fig. 3a; Kudryavtsev et al., 2001; Schopf and Kudryavtsev, 2005). This has been done using putative filamentous microfossils from a ~3,465-Ma chert from the Apex Basalt of Western Australia (Schopf et al., 2007; Schopf and Kudryavtsev, 2009). They show filamentous structures composed of hollow quartz-filled "cell cavities" with kerogen-like "cell walls", closely comparable to younger *bona fide* Precambrian microfossils. Moving to even more ancient rocks, McKeegan et al. (2007) have used 3-D Raman imaging to demonstrate the existence of ~5 μm carbonaceous inclusions in apatite within >3,800-Ma rocks from Akilia Island, Greenland. Whilst Raman imaging does not provide proof of biogenicity in either of these cases, it is undoubtedly a very useful technique.

A recent additional application of Raman imaging is to look at the distribution and structure of the minerals which host purported biological material. Many minerals produce spectral peaks that vary in intensity depending on their crystallographic orientation to the laser. This feature can be used, for example, to image the distribution of quartz crystallographic axes to see whether the biological material occurs between grain boundaries, is enclosed by entire grains or occurs in cracks (cf. SAED in the TEM).

### 4.2. CONFOCAL LASER SCANNING MICROSCOPY

Confocal laser scanning microscopy (CLSM) allows rapid and precise in situ characterization of 3-D morphology and taphonomic preservation. It is also nonintrusive and nondestructive, and its advantage over laser Raman for morphological investigations lies in the speed of acquisition and slightly higher spatial resolution (<100 nm is possible). However, it has currently only been applied to younger (micro)fossils freed from their rock matrix or to Proterozoic permineralized microfossils in chert (Schopf et al., 2006; Schopf and Kudryavtsev, 2009). It relies on the laser excitation of auto-fluorescent organic material, so it faces limitations when objects are either a long way below the surface of the thin section or have thick opaque walls. More importantly, it cannot be used on samples that are composed of highly geochemically altered carbon that does not fluoresce. This limits its use in the origin of life field since most samples have undergone significant geochemical alteration during metamorphism to greenschist facies and above.

## 5. Synchrotron-Based Techniques

Synchrotron radiation is a source of X-rays with high intensity that can be harnessed across a large energy range. Synchrotron-based techniques are, for the most part, nondestructive, require little sample preparation, have flexible spatial resolution (mm to ~50-nm scale) and are sensitive to at least the parts-per-million (ppm) level. Details of the techniques using synchrotron-generated X-rays and their applications can be found in Fenter et al. (2002), Hitchcock et al. (2008), Templeton and Knowles (2009) and references therein. Here, I will outline the two main techniques that have been used in the origins and evolution of life field in recent years, namely, scanning transmission X-ray microscopy (STXM) and near-edge X-ray absorption fine structure (NEXAFS) spectroscopy. These are essentially linked techniques that can be performed at the same time on the same areas of sample within a scanning transmission microscope attached to a synchrotron beamline and are often referred to together as X-ray spectromicroscopy.

### 5.1. X-RAY SPECTROMICROSCOPY

#### *5.1.1. Scanning Transmission X-ray Microscopy*

STXM uses soft X-rays (i.e. those with energies <4 KeV) to obtain both images and spectral data (NEXAFS spectra) at the nm scale. It uses the principle of absorption contrast around elemental absorption edges (e.g. C K-edge). Samples are rastered through the X-ray beam at stepwise-increasing incident X-ray energies, achieved by shortening the wavelength, to cover the absorption edges of the elements of interest (Templeton and Knowles, 2009).

With relevance to the origins and evolution of life field, Benzerara and colleagues have investigated microbe-mineral interactions using STXM. For example, they found concentrations of organic carbon in association with minute tubes in ~100-Ma volcanic glass (Benzerara et al., 2007) that have close morphological similarity to the earliest reported putatively biogenic microtubes in ~3,500-Ma volcanic glass from the Barberton region of South Africa (Furnes et al., 2004). Benzerara et al. (2006) also used STXM to investigate modern microbialites and showed that functional groups commonly associated with microbes could be mapped within mineralized samples and, hence, could be used as a proxy for fossilized microbes in ancient rocks. It is unknown, however, exactly which functional groups may survive in Archean rocks.

Other synchrotron-based imaging techniques are also emerging in this field. X-ray fluorescence (XRF) is another way of obtaining semi-quantitative element specific maps over flexible spatial scales (μm to mm) and can be particularly useful for an overview of the elemental composition of a given sample before focusing on a specific smaller area. Lemelle et al. (2008) used this approach to map and approximately quantify the amounts of sulfur within the cell walls of ~800-Ma coccoid microfossils, before using XANES (see below) at the sulfur K-edge to determine sulfur speciation. They showed that the sulfur was a reduced organic form most likely belonging to a thiophene-like compound. Thus, the combination of XRF and XANES can identify, map, quantify and distinguish inorganic from organic forms of sulfur and may be applicable to controversial fossils at the prebiotic-biotic boundary in the early Archean.

Synchrotron radiation X-ray tomography (SRXTM) is a nondestructive imaging technique that creates stacks of images at sub-μm resolution that can be combined into 3-D visualizations. SRXTM has recently been successfully applied to the study of fossil embryos preserved as calcium phosphate in the late Precambrian (Fig. 4a; Donoghue et al., 2006). Undoubtedly, this technique will soon be applied in the earlier rock record.

#### *5.1.2. Near-Edge X-ray Absorption Fine Structure Spectroscopy*

NEXAFS is a spectroscopic technique using soft X-rays whereby an element absorbs a photon at or above a characteristic energy, exciting a core electron to an unoccupied orbital, or completely out of the atom (Templeton and Knowles, 2009). The spectrum produced can be compared to standard spectra to provide information regarding the co-ordination chemistry and valence of the element of interest. It provides much of the same information as ELNES within the TEM but has a better energy resolution (~0.1 eV compared to ~0.5–1 eV), allowing very closely spaced peaks to be resolved. When hard X-rays (i.e. energies >4 KeV) are used, the technique is usually referred to as X-ray absorption near-edge structure (XANES).

In the origins and evolution of life field, NEXAFS is often used to investigate the chemical binding state of carbon, so spectra are recorded within a specific energy range corresponding to the carbon K-edge (1s-shell electrons).

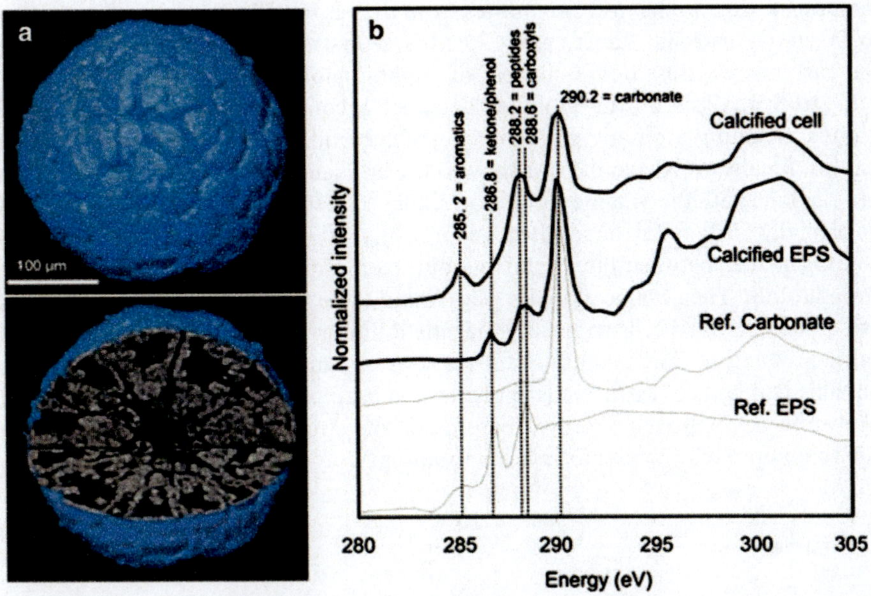


**Figure 4.** Examples of data obtained using synchrotron-based techniques. (**a**) X-ray tomographic 3-D reconstruction of a ~540-Ma embryo from Southern China enabling internal structure of the blastomeres to be visualized (Donoghue et al., 2006. Reproduced with permission from Nature Publishing). (**b**) NEXAFS data showing how the chemical binding state of carbon atoms can be investigated. Here, spectra from a modern calcified cell and calcified extracellular polymeric substances (EPS, *bold lines*) are compared to reference spectra from carbonate and EPS (*faint lines*). Carbon bound in aromatic, ketone, peptide, carboxyl and carbonate groups can be distinguished (Modified from Benzerara et al. (2006)).

Different absorption peaks can be seen for carbon bound, for example, in aromatic groups, aliphatic groups, ketones, peptides, carbonyls, carboxyls and carbonate (Fig. 4b). This type of spectrum may be used to distinguish between cellular and extracellular components and to strip out signals from carbonate minerals. These spectral data can also be combined with STXM images to build up maps of specific functional groups within potential microfossils. De Gregorio et al. (2009) used NEXAFS spectra to compare the carbon K-edge of kerogen from the ~3,465-Ma Apex chert with that from the fossiliferous 1,900-Ma Gunflint chert. NEXAFS spectra revealed peaks corresponding to carboxylic acid groups and possible phenol groups in addition to polyaromatic domains within the Apex kerogen; similar peaks were observed in the Gunflint kerogen. The complexity of the Apex NEXAFS spectrum and its similarity with the Gunflint spectrum was taken as supporting evidence for the biogenicity of the Apex kerogen (De Gregorio et al., 2009).

The power of combining NEXAFS/STXM with other high-resolution techniques such as FIB-TEM, Raman and CLSM was demonstrated by Lepot et al. (2008). These authors investigated the association of organic carbon and carbonate minerals within Archean stromatolites from the ~2,700-Ma Tumbiana Formation of Western Australia. Raman and CLSM were first used to characterise and map the carbon, revealing dense clusters of organic globules with the stromatolites. NEXAFS analysis of the globules detected a complex mixture of functional groups including aromatic, aliphatic, carboxyl, and possible ketone and phenol groups. Finally, STXM and TEM showed the intimate association of nano-crystals of aragonite with the organic globules, a similar pattern to that observed in modern microbially mediated stromatolites. Lepot et al. (2009) extended this study to show heterogeneities between the organic globules and organic matter elsewhere in the stromatolites. The organic globules were found to be sulfur-rich and interpreted as preserved microbial cells with early diagenetic sulfurization promoted by bacterial sulfate reduction. The remaining sulfur-poor organic material was interpreted as possible remnants of extracellular polymeric substances. These types of studies will likely now be applied to putative microfossils and stromatolites of earlier Archean age that approach the prebiotic-biotic boundary.

## 6. Summary

This chapter highlights some of the best options now available for researchers in the origins and evolution of life field to obtain high-resolution chemical and structural data, whilst at the same time, maintaining the original context of the structure of interest. The choice of instrumentation depends on a number of factors including ease of access, type of sample, required spatial resolution and sensitivity. A number of the techniques outlined above can provide similar information. For example, nm-scale elemental mapping can be achieved using NanoSIMS, EFTEM, STEM, laser Raman or STXM. However, if elemental mapping combined with isotopic data is required, NanoSIMS is the obvious choice. Elemental mapping combined with mineral identification at the <50-nm scale would favor the TEM. A truly unique sample where a one-to-one correlation of microfossil morphology and chemistry is required and would be best suited to the nondestructive nature of laser Raman, whereas complex organic material with numerous functional groups would be best analyzed by STXM and NEXAFS.

## 7. Acknowledgments

The author is supported by a University Postdoctoral Research Fellowship from The University of Western Australia. I acknowledge the facilities and the scientific and technical assistance of the Australian Microscopy and Microanalysis Research Facility at the Centre for Microscopy, Characterisation and Analysis, The University of Western Australia, a facility funded by the University, State and Commonwealth Governments. Special thanks go to Martin Saunders and Matt Kilburn for TEM and NanoSIMS data.

## 8. References

Allwood AC, Walter MR, Kamber BS, Marshall CP, Burch IW (2006) Stromatolite reef from the Early Archaean era of Australia. Nature 441:714–718

Allwood AC, Grotzinger JP, Knoll AH, Burch IW, Anderson MS, Coleman ML, Kanik I (2009) Controls on development and diversity of early Archean stromatolites. Proc Natl Acad Sci USA 106:9548–9555

Benzerara K, Menguy N, Lopez-Garcia P, Yoon T-H, Kazmierczak J, Tyliszczak T, Guyot F, Brown GED Jr (2006) Nanoscale detection of organic signatures in carbonate microbialites. Proc Natl Acad Sci USA 103:9440–9445

Benzerara K, Menguy N, Banerjee NR, Tyliszczak T, Guyot F, Brown GE Jr (2007) Alteration of submarine basaltic glass from the Ontong Java Plateau: a STXM and TEM study. Earth Planet Sci Lett 260:187–200

Brasier MD, Green OR, Jephcoat AP, Kleppe AK, Van Kranendonk MJ, Lindsay JF, Steele A, Grassineau NV (2002) Questioning the evidence for Earth's oldest fossils. Nature 416:76–81

Buseck PR, Bo-Jun H (1985) Conversion of carbonaceous material to graphite during metamorphism. Geochim Cosmochim Acta 49:2003–2016

Buseck PR, Bo-Jun H, Miner B (1988) Structural order and disorder in Precambrian kerogens. Org Geochem 12:221–234

Calvert CC, Brown A, Brydson R (2005) Determination of the local chemistry of iron in inorganic and organic materials. J Electron Spectrosc Relat Phenom 143:173–187

Daniels H, Brydson R, Rand B, Brown A (2007) Investigating carbonization and graphitization using electron energy loss spectroscopy (EELS) in the transmission electron microscope (TEM). Philos Mag 87:4073–4092

De Gregorio BT, Sharp TG (2006) The structure and distribution of carbon in 3.5 Ga Apex chert: implications for the biogenicity of Earth's oldest putative microfossils. Am Mineral 91:784–789

De Gregorio BT, Sharp TG, Flynn GJ, Wirick S, Hervig RL (2009) Biogenic origin for Earth's oldest putative microfossils. Geology 37:631–634

Donoghue PCJ, Bengtson S, Dong X, Gostling NJ, Huldtgren T, Cunningham JA, Yin C, Yue Z, Peng F, Stampanoni M (2006) Synchrotron X-ray tomographic microscopy of fossil embryos. Nature 442:680–683

Fenter P, Rivers M, Sturchio N, Sutton S (eds) (2002) Applications of synchrotron radiation in low-temperature geochemistry and environmental sciences. Rev Mineral Geochem 49

Fike DA, Gammon CL, Ziebis W, Orphan VJ (2008) Micron-scale mapping of sulfur cycling across the oxycline of a cyanobacterial mat: a paired NanoSIMS and CARD-FISH approach. ISME J 2:749–759

Fletcher IR, Kilburn MR, Rasmussen B (2008) NanoSIMS μm-scale in situ measurement of $^{13}C/^{12}C$ in early Precambrian organic matter, with permil precision. Int J Mass Spectrom 278:59–68

Furnes H, Banerjee NR, Muehlenbachs K, Staudigel H, de Wit M (2004) Early life recorded in Archean pillow lavas. Science 304:578–581

Hitchcock AP, Dynes JJ, Johansson G, Wang J, Botton G (2008) Comparison of NEXAFS microscopy and TEM-EELS for studies of soft matter. Micron 39:311–319

Hofmann HJ, Grey K, Hickman AH, Thorpe R (1999) Origin of 3.45 Ga coniform stromatolites in Warrawoona Group, Western Australia. Bull Geol Soc Am 111:1256–1262

Horita J, Berndt ME (1999) Abiogenic methane formation and isotopic fractionation under hydrothermal conditions. Science 285:1055–1057

House CH, Schopf JW, McKeegan KD, Coath CD, Harrison TM, Stetter KO (2000) Carbon isotopic composition of individual Precambrian microfossils. Geology 28:707–710

Ireland TR (1995) Ion microprobe mass spectrometry: techniques and applications in cosmochemistry, geochemistry, and geochronology. In: Hyman M, Rowe M (eds) Advances in analytical geochemistry. JAI Press, Greenwich, pp 1–118

Jehlicka J, Urban O, Pokorny J (2003) Raman spectroscopy of carbon and solid bitumens in sedimentary and metamorphic rocks. Spectrochim Acta A 59:2341–2352

Kempe A, Wirth R, Altermann W, Stark RW, Schopf JW, Heckl WM (2005) Focused ion beam preparation and in situ nanoscopic study of Precambrian acritarchs. Precambrian Res 140:36–54

Kilburn MR, Wacey D (2011) Elemental and isotopic analysis by NanoSIMS: insights for the study of stromatolites and early life on Earth. In: Tewari V, Seckbach J (eds) Stromatolites: interaction of microbes with sediments. Cellular origin, life in extreme habitats and astrobiology, vol 18. Springer, Dordrecht, 742p

Knoll AH, Barghoorn ES (1974) Ambient pyrite in Precambrian chert: new evidence and a theory. Proc Natl Acad Sci USA 71:2329–2331

Kudryavtsev AB, Schopf JW, Agresti DG, Wdowiak TJ (2001) In situ laser-Raman imagery of Precambrian microscopic fossils. Proc Natl Acad Sci USA 98:823–826

Lemelle L, Labrot P, Salome M, Simionovici A, Viso M, Westall F (2008) In situ imaging of organic sulfur in 700–800 My-old Neoproterozoic microfossils using X-ray spectromicroscopy at the S K-edge. Org Geochem 39:188–202

Lepot K, Benzerara K, Brown GE Jr, Philippot P (2008) Microbially influenced formation of 2,724-million-year-old stromatolites. Nat Geosci 1:118–121

Lepot K, Benzerara K, Rividi N, Cotte M, Brown GE Jr, Philippot P (2009) Organic matter heterogeneities in 2.72 Ga stromatolites: alteration versus preservation by sulfur incorporation. Geochim Cosmochim Acta 73:6579–6599

McCollom TM, Seewald JS (2006) Carbon isotope composition of organic compounds produced by abiotic synthesis under hydrothermal conditions. Earth Planet Sci Lett 243:64–84

McKeegan KD, Kudryavtsev AB, Schopf JW (2007) Raman and ion microscopic imagery of graphitic inclusions in apatite from older than 3830 Ma Akilia supracrustal rocks, west Greenland. Geology 35:591–594

McKibben MA, Shanks WC III, Ridley WI (1998) Applications of microanalytical techniques to understanding mineralizing processes. SEG Rev Econ Geol 7:263p

Mojzsis SJ, Arrenhius G, McKeegan KD, Harrison TM, Nutman AP, Friend CRL (1996) Evidence for life on Earth 3,800 million years ago. Nature 384:55–59

Moreau JW, Sharp TG (2004) A transmission electron microscopy study of silica and kerogen biosignatures in 1.9 Ga Gunflint microfossils. Astrobiology 4:196–210

Nemchin AA, Whitehouse MJ, Menneken M, Geisler T, Pidgeon RT, Wilde SA (2008) A light carbon reservoir recorded in zircon-hosted diamond from the Jack Hills. Nature 454:92–95

Oehler DZ, Robert F, Mostefaoui S, Meibom A, Selo M, McKay DS (2006) Chemical mapping of Proterozoic organic matter at submicron spatial resolution. Astrobiology 6:838–850

Orphan VJ, House CH (2009) Geobiological investigations using secondary ion mass spectrometry: microanalysis of extant and paleo-microbial processes. Geobiology 7:360–372

Pasteris JD, Wopenka B (2003) Necessary, but not sufficient: Raman identification of disordered carbon as a signature of ancient life. Astrobiology 3:727–738

Philippot P, van Zuilen M, Thomazo C, Farquhar J, Van Kranendonk MJ (2007) Early Archaean microorganisms preferred elemental sulfur, not sulfate. Science 317:1534–1537

Rouzaud J-N, Clinard C (2002) Quantitative high-resolution transmission electron microscopy: a promising tool for carbon materials characterization. Fuel Process Technol 77–78:229–235

Rouzaud J-N, Skrzypczak A, Bonal L, Derenne S, Quirico E, Robert F (2005) The high resolution transmission electron microscope: a powerful tool for studying the organization of terrestrial and extra-terrestrial carbons. LPSC XXXVI:A1322

Schopf JW (1993) Microfossils of the early Archean Apex chert: new evidence for the antiquity of life. Science 260:640–646

Schopf JW, Kudryavtsev AB (2005) Three-dimensional Raman imagery of Precambrian microscopic organisms. Geobiology 3:1–12

Schopf JW, Kudryavtsev AB (2009) Confocal laser scanning microscopy and Raman imaging of ancient microscopic fossils. Precambrian Res 173:39–49

Schopf JW, Kudryavtsev AB, Agresti DG, Czaja AD, Wdowiak TJ (2005) Raman imagery: a new approach to assess the geochemical maturity and biogenicity of permineralized Precambrian fossils. Astrobiology 5:333–371

Schopf JW, Tripathi AB, Kudryavtsev AB (2006) Three-dimensional confocal optical imagery of Precambrian microscopic organisms. Astrobiology 6:1–16

Schopf JW, Kudryavtsev AB, Czaja AD, Tripathi AB (2007) Evidence of Archean life: stromatolites and microfossils. Precambrain Res 158:141–155

Shen Y, Buick R (2004) The antiquity of microbial sulfate reduction. Earth Sci Rev 64:243–272

Shen Y, Farquhar J, Masterson A, Kaufman AJ, Buick R (2009) Evaluating the role of microbial sulfate reduction in the early Archean using quadruple isotope systematics. Earth Planet Sci Lett 279:383–391

Templeton A, Knowles E (2009) Microbial transformations of minerals and metals: recent advances in geomicrobiology derived from synchrotron-based x-ray spectroscopy and x-ray microscopy. Annu Rev Earth Planet Sci 37:367–391

Tice MM, Bostick BC, Lowe DR (2004) Thermal history of the 3.5–3.2 Ga Onverwacht and Fig Tree Groups, Barberton greenstone belt, South Africa, inferred by Raman microspectroscopy of carbonaceous material. Geology 32:37–40

Ueno Y, Isozaki Y, Yurimoto H, Maruyama S (2001) Carbon isotopic signatures of individual Archean microfossils(?) from Western Australia. Int Geol Rev 43:196–212

Ueno Y, Ono S, Rumble D, Maruyama S (2008) Quadruple sulfur isotope analysis of ca. 3.5 Ga Dresser Formation: New evidence for microbial sulfate reduction in the early Archean. Geochim Cosmochim Acta 72:5675–5691

van Zuilen MA, Lepland A, Teranes J, Finarelli J, Wahlen M, Arrhenius G (2003) Graphite and carbonates in the 3.8 Ga old Isua Supracrustal Belt, southern West Greenland. Precambrian Res 126:331–348

Wacey D (2010) Stromatolites in the 3,400 Ma Strelley Pool Formation, Western Australia: examining biogenicity from the macro- to the nano-scale. Astrobiology 10(4):381–395

Wacey D, McLoughlin N, Green OR, Parnell J, Stoakes CA, Brasier MD (2006) The 3.4 billion-year-old Strelley Pool sandstone: a new window into early life on Earth. Int J Astrobiol 5:333–342

Wacey D, Kilburn MR, McLoughlin N, Parnell J, Stoakes CA, Brasier MD (2008a) Use of NanoSIMS to investigate early life on Earth: ambient inclusion trails in a *c*. 3400 Ma sandstone. J Geol Soc Lond 165:43–53

Wacey D, Kilburn MR, Stoakes CA, Aggleton H, Brasier MD (2008b) Ambient inclusion trails: their recognition, age range and applicability to early life on earth. In: Dilek Y, Furnes H, Muehlenbachs K (eds) Links between geological processes, microbial activities and evolution of life. Springer, Dordrecht, pp 113–133

Wacey D, Kilburn MR, Saunders M (2009) A combined SIMS and TEM study of potential biosignals from Precambrian rocks. Geochim Cosmochim Acta 73:A1397

Wacey D, Gleeson D, Kilburn MR (2010a) Microbialite taphonomy and biogenicity: New insights from NanoSIMS. Geobiology 8(5):403–416

Wacey D, McLoughlin N, Whitehouse M, Kilburn MR (2010b) Two co-existing sulfur metabolisms in a 3,430 Ma sandstone. Geology 38(12):1115–1118

Wacey D, Saunders M, Brasier MD, Kilburn MR (2011) Earliest microbially mediated pyrite oxidation in 3.4 billion-year-old sediments. Earth Planet Sci Lett 301:393–402

Westall F, de Ronde CEJ, Southam G, Grassineau N, Colas M, Cockell C, Lammer H (2006) Implications of a 3.472–3.333 Gyr-old subaerial microbial mat from the Barberton greenstone belt, South Africa for the UV environmental conditions on the early Earth. Philos Trans R Soc Lond B Biol Sci 361:1857–1875

Williams DB, Carter CB (2009) Transmission electron microscopy: a textbook for materials science, 2nd edn. Springer, New York, 832p

Wirth R (2009) Focused Ion Beam (FIB) combined with SEM and TEM: Advanced analytical tools for studies of chemical composition, microstructure and crystal structure in geomaterials on a nanometre scale. Chem Geol 261:217–229

Wopenka B, Pasteris JD (1993) Structural characterisation of kerogens to granulite-facies graphite: applicability of Raman microprobe spectroscopy. Am Mineral 78:533–557

# PART III: THE ROLE OF SYMMETRY, INFORMATION AND ORDER IN THE ORIGIN OF LIFE

**Mauro**
**Dunker**
**Trifonov**
**Frieden**
**Gatenby**
**Jakschitz**
**Fitz**
**Rode**
**Klußmann**
**Lente**
**Popa**
**Cimpoiasu**

Biodata of **Ernesto Di Mauro**, **A. Keith Dunker**, and **Edward N. Trifonov**, authors of "***Disorder to Order, Nonlife to Life: In the Beginning There Was a Mistake***."

**Ernesto Di Mauro** is professor of Molecular Biology at the University "Sapienza" of Rome. His research activity has been consistently focused on the study of the structure and of the function of genetic materials, in both prokaryotes and eukaryotes.

Contributions were made in the fields of DNA topology and of topology-related functions, in the biophysics and biochemistry of gene expression. Special interest was devoted to the kinetic and topological aspects of gene regulation. Part of his research has dealt with the study of plausible prebiotic routes to the evolution of informational macromolecules. This latter approach has lead to the discovery of reactions affording the prebiotic synthesis of a complete set of nucleic acid precursors. The study of novel chemical mechanisms in an origin-of-life scenario is at present one of the main interests of his research group, recently leading to the discovery of the abiotic generation of long RNA molecules in water.

E-mail: **ernesto.dimauro@uniroma1.it**

J. Seckbach (ed.), *Genesis - In The Beginning: Precursors of Life, Chemical Models and Early Biological Evolution*, Cellular Origin, Life in Extreme Habitats and Astrobiology 22, 415–435
DOI 10.1007/978-94-007-2941-4_23, © Springer Science+Business Media Dordrecht 2012

**Prof. A. Keith Dunker** is the Director of the Center for Computational Biology and Bioinformatics at the Indiana University School of Medicine. His life-long interests have been in protein structure-function relationships using virus or phage structure, assembly and cell penetration as model systems for study. In the mid-1990s, from clues provided by mechanisms of host cell penetration by filamentous phage, Dr. Dunker and his collaborators began to use computational approaches to study proteins that utilize lack of structure rather than structure to carry out biological function. This work, which is continuing up to the present, suggests that the sequence-to-structure-to-function paradigm applies mainly to enzymes and membrane transport proteins while a new paradigm, sequence-to-unstructured, dynamic protein ensemble-to-function applies to proteins and regions of proteins that carry out control, regulation, and signaling.

E-mail: **kedunker@iupui.edu**

**Professor Edward N. Trifonov**, Head of Genome Diversity Center at the University of Haifa, Israel, and Professor at Masaryk University, Brno, Czech Republic. World known expert in Sequence Biology, the branch of Computational Biology devoted to deciphering biologically meaningful sequence codes (often called Bioinformatics). Edward and his collaborators were the first to detect major 10–11 base periodicity in DNA sequences (1981) – predicted attribute of curved DNA and of nucleosome positioning pattern. With many different sequence codes at hand, of which several were his own discoveries, Edward promoted the notion of the overlapping and interactions between the multiple codes (1989) – the unique property of genetic sequences.

Ed Trifonov is co-discoverer of modular, loop-fold structure of proteins (with I. N. Berezovsky and A. Grosberg, 2000). The closed-loop modules of standard size (25–30 amino-acid residues) make the basis of a proteomic code that allows to describe and compare proteins in the new language of the modules. This development got a new strong impetus with introduction of formatted protein sequence space (With Z. M. Frenkel, 2007).

The most recent passion of Ed Trifonov is reconstruction of the earliest events in the evolution of the triplet code, derivation of the sequence and structure of the very first small proteins.

M.Sc. and Ph.D., Moscow Physico-Technical Institute. Signal processing, biophysics, DNA. Assoc. Prof. and Professor, Weizmann Institute. Professor, University of Haifa. Professor, Masaryk University, Brno, Czech Republic.

E-mail: **trifonov@research.haifa.ac.il**

# DISORDER TO ORDER, NONLIFE TO LIFE: IN THE BEGINNING THERE WAS A MISTAKE

**ERNESTO DI MAURO[1], A. KEITH DUNKER[2], AND EDWARD N. TRIFONOV[3,4]**
[1]*Dipartimento di Genetica e Biologia Molecolare, Università di Roma "Sapienza", P.le Aldo Moro, 5, 00185 Rome, Italy*
[2]*Department of Biochemistry and Molecular Biology, Center for Computational Biology and Bioinformatics, Indiana University School of Medicine, Indianapolis, IN 46202, USA*
[3]*Genome Diversity Center, Institute of Evolution, University of Haifa, Mount Carmel, Haifa 31905, Israel*
[4]*Central European Institute of Technology and Faculty of Science, Masaryk University, Kamenice 5, Brno CZ-62500, Czech Republic*

## 1. Introduction

Adherence to some form of order is an all-natural trend in all human walks. First of all, this is the search for rules governing the world we live in. The observed reality with all its complexity is thereby reduced to the learned and canonized foci of the order-rules in the sea of the disordered rest. Since the discovery of the classical gas laws, we also know that the disorder itself may follow strict rules, thus becoming a legal component of the ordered world.

One may classify the order recognized in the living matter as structures and as the rules of interactions between them. The structures of DNA of viruses are obvious examples of the order, while the whole chart of biochemical pathways presents the order in the interactions. In such directional processes as individual life (ontogenesis) and evolution of species or of a whole biosphere, one recognizes consecutive emergence of key structures corresponding to different stages of a process. The structures can be considered, again, as foci of the order, and the evolutionary (developmental) journey from one key structure to the next ones in the process can be viewed as alternation of order and disorder, without a clear starting priority. It is this view of alternation of order/disorder in evolution that brought together the authors of this essay in their attempt to draw a desirably consistent picture of early molecular evolution. We will try to outline possible main stages/transitions from one key structure in the chemical/biomolecular evolution to the next one, all the way from clearly chemical world to clearly living matter, with full realization that this, probably, is a naive venture.

The authors also share the view (a hypothesis, as some would prefer to phrase it) that life has an abiotic origin so that the order–disorder alternation, in principle, could spread far back in the physical/chemical evolution stages of life's journey from none to present. For certainty, one can start with the order of the periodic table of elements. This beautiful key structure of chemical universe, with all its inherent valencies, is the basis of tremendous potential variety (disorder) of chemical substances, including, at our end, nucleic acids and proteins.

In the harsh conditions of early Earth (or of any other Earth-like planet), only rather simple substances could be formed via elementary chemical reactions, with the synthetic rates exceeding degradation rates of the substances. One can imagine gradual diversification of this primordial chemical arsenal, along with the moderation of conditions. That chemical variety (disorder) reigned until the first new key structures opened new chemical opportunities, leading to new classes of variety. We, thus, arrive at the most exciting stage, the origin of the structures leading to or already constituting what could be called life at its very origin.

The fuzzy definition of disorder and of its progression to order given above entails that disorder and order are two aspects of the same ensemble of properties. Passing from disorder to order is a continuum, possibly not separated by any hiatus. The problem can thus be defined as to in which frame(s) of reference these properties and their definitions apply, namely, do they apply to *life*? Assuming that the origin and the progression of order takes place from disorder in a local (uni)directional manner, and that some form of order is necessary for life to organize itself, it is worth considering that the logics space that separates *disorder* from *order* is the same as the one that separates various stages of *nonlife* and *life*, again crediting the uncertainty as to where exactly that magic *nonlife* to *life* spacer cuts the chain. The point thus is made that the rules and definitions which apply to the disorder to order category should also apply to the events on the way from nonlife to life.

## 2. About the Passage from Nonlife to Life

The development of modern thought on the two categories disorder–order and nonlife–life has historically followed different paths. Physicists and mathematicians on one side, biologists and physicians on the other, have made use of differing rational approaches.

At first sight, this lead to apparently different conclusions and to the difficulty in understanding the unitary nature of the problem. The pitfalls on the road to the unequivocal definitions of life are well known (Luisi, 1998; Damiano and Luisi, 2010). The current operational definition is: "*Life is a selfsustained chemical system capable of undergoing Darwinian evolution*" (Joyce, 1994). Rather than a definition, this is a description of the phenomenon, partially unsatisfactory because the definiendum is obtained based on its ability to change. Life is more a process than a system, the acceptable definition of a process being a natural phenomenon marked by gradual changes that lead toward a particular result. Darwinian evolution is one of the pillars of this definition of life and is a process not aiming to any specific goal. In this purport, the definition given above is closely related to Lewis Carroll's Red Queen paradox (Carroll, 1865) and suffers from the contradictions that characterized the Classic efforts to develop a criterion for classification, from Aristotle to Porfirius, all the way up to the contradictions that still plague Lynnean rules. The Red Queen was forced to run faster and faster in order to stay in the same place in a constantly accelerating world; similarly, the standard classification is growing heavier and heavier.

If one accepts the argument "system versus process" and the lack of semantic precision of the term "Darwinian evolution," the consequence is that no strict logical definition of life is possible, as well as rigorous classifications are out of reach. Does something that cannot be defined, actually exist as such? But do not we actually live and explore life itself even without having the definition? The vortex of the rigor is circumvented by necessary pragmatism. We do perceive a living entity as something that produces an entity very similar to itself, mostly with help of a sexual partner. Perhaps, the simplest and more general definition with the same Darwinian spirit would be "Life is self-reproduction with variations" (Trifonov, 2009). This is, of course, a *process*, accurate and inaccurate at the same time, and this magically balanced uncertainty, probably, reflects the very essence of life. We are to follow this process, watching its recognizable stages, irrespective of whether the satisfactory rigorous definition is at all ever to come.

Our parents derived from theirs, the process repeating itself all the way to the origins of the first organized self-replicating structures. The resolution of the *egg versus chicken* paradox is at the (always retreating from the explorer) start of these repetitions all the way down, close to the first atoms and to their interactive properties and whereabouts. And the thread could continue, back to the Big Bang and through it. Our point of view is that all these apparently different oppositions are resumed in the perpetual disorder-to-order passage. Somewhere in this chain, however, reside those disorder-to-order steps that are at the, imaginary so far, border beyond which we are already on life's territory.

Is it appropriate to transfer this reasoning on pregenetics into the disorder-to-order frame of reference? And does the disorder–order domain allow for discontinuities?

An immense complexity of observed life suggests to many that life could not even emerge to begin with. How possibly did the complex amount of chemical pathways that make us up come about? The solution is usually humbly formulated in Darwinian terms: once the appropriate precursors were synthesized and allowed to accumulate, the step-by-step integration of new reactions and their fitness-oriented controlled cooperation with the preexisting ones started. The very first integration was possibly that of the protometabolism leading to peptides with the protometabolism leading to oligo nucleic acids, hence, the possibility of proactive cooperation and booming evolution. This functional fitness-enhancing peptide-nucleic acids encounter, however, was necessarily preceded by the self-invention of the resolution of a more basic problem, that of the physical survival, of the integrity of the information-bearing molecules and macromolecules. The principle that hampers the accumulation of chemical information is defined by the standard-state Gibbs free energy (ΔG°') change for condensation reactions, essentially stating that the polymerization reactions that occur by release of water do not take place in water (which is intuitively understandable). It was noted by van Holde (1980) and Alberty (2006) that the only ways out from this paradox are provided by high concentration of the reactants or by their preactivation to higher energetic levels by other means. The synthesis of low-information molecules may well be conceived in solvents different than water, but their higher-order organization necessarily occurred in water.

Let us thus insert into this general frame the contemporary physical-chemical reality, as it is being shaped by recent findings.

## 3. The Generation of Nucleic Acid Precursors (and More) from the One-Carbon Compound Formamide

The production of all the nucleic bases in pure formamide (i.e., formamide-based polymerizations occurring in formamide) was described, as reviewed (Saladino et al., 2009; Ciciriello et al., 2009).

Formamide $NH_2COH$ is made of the four most abundant atoms in the universe (neglecting inert He) and is easily obtained from hydrogen cyanide HCN reacting with water $H_2O$. HCN is the most abundant three-atom organic compound in interstellar clouds; $H_2O$ is the most abundant three-atom inorganic compound. Formamide chemistry is simple and robust (Saladino et al., 2005a, b, c, 2007, 2009; Ciciriello et al., 2009). When heated between 80°C and 160°C in the presence of mineral catalysts, formamide yields a number of compounds encompassing all the nucleic bases present in extant nucleic acids, related compounds as imidazoles, and monocarboxilic acids. A list of the observed products of formamide-based syntheses encompasses purine, 2-aminopurine, adenine, hypoxanthine, N9-formylpurine, N9,N6-diformyl adenine, cytosine, isocytosine, hydroxypyrimidine, 2(1H)-and 4(3H)-pyrimidinone, uracil, thymine, and 5-hydroxymethyluracil. Compounds were also observed that are related to present-day nucleic acids biosynthetic or degradative pathways (as various imidazoles and parabanic acid) or that are known to react with amino acids and nucleotides, as carbodiimide. Guanine is missing in the list of compounds obtained at higher temperatures. Its instability in the conditions in which synthetic reactions are run, the presence of its more stable cognate compound hypoxanthine and of compounds related to its degradation, taken together, indicate that in heat, guanine is made and degraded. Guanine can be obtained from formamide by UV irradiation (Barks et al., 2010). The catalytic abilities and capacities of a large number of potentially prebiotic (i.e., of nonbiotic origin) minerals were analyzed, namely, silica, alumina, zeolites, $CaCO_3$, $TiO_2$, common clays, kaolin, montmorillonites, olivines, phosphate minerals [reviewed in (Saladino et al., 2005a, b, c, 2007)], sulfur and iron minerals (Saladino et al., 2008), and zirconium-based (Saladino et al., 2010) and boron-based minerals (Saladino et al., 2011). In the absence of catalysts, only purine is obtained, while yields of the other compounds largely vary according to the conditions (Saladino et al., 2005a, b, c, 2007, 2009; Ciciriello et al., 2009). More than the completeness of the list of obtained products, what is of possible prebiotic relevance is the fact that whatever catalyst is provided, a rich panel of nucleic acid chemistry–related compounds is obtained. Thus, the chemical key to the origin of nucleic precursors, not confined to any special mineral-related catalytic process, is formamide itself. This observation lends relevance to formamide chemistry in the origin of nucleic acids, as anyhow its chemical formula, resembling a fragment of a pyrimidine or of a purine, suggests at first glimpse. The complexity of the composition of the mixture of obtained products, the good yields, and, mostly, the nonfastidious requirements for efficient catalysis to occur all point to an intrinsic tendency of formamide to assist polymerization. What could this intrinsic tendency mean in terms of disorder-to-order and nonlife-to-life transitions? As stated in simple words by Stenger: *something came from nothing because it was more stable than nothing* (Stenger, 2006).

## 4. Polymerization

The polymerization (and the heteropolymerization) has been, definitely, a key structural stage in what we could imagine as the chain of events on the way to life. From a mess of the monomer products generated under the rule of the periodical table, including amino acids and nucleobases, the new order emerged – polymer chains. Life as we know it does not occur in formamide. Possibly, it did not even start in formamide. Organic solvents as formamide are likely to have been instrumental only for the very first reactions, those leading to the formation and accumulation of the basic precursors. However, given that we are made of (and live in) water, at some point, an evolution-friendly resolution of the $\Delta G°'$ paradox for polymerization must have been found. In line with Stenger's aphorism, some of us proposed (Saladino et al., 2005a, b, c, 2006a; Ciciriello et al., 2009) that the answer is to be found in the higher stability of polymers relative to monomers. This apparently paradoxical assertion relies on robust experimental ground (Saladino et al., 2005a, b, c, 2006a, b). The repeating unit in the ribooligomer chain is made of six covalent bonds, connecting as many atoms. The fragile site in this constituent unit of the polymer is the phosphoester bond at 3′ of the D-ribose moiety, easily undergoing hydrolysis and possible transphosphorylation. That is, the integrity of the polymers depends on the stability of this molecular Achilles' heel. The measurement of the stability of this bond in the monomer and in the polymer established that physical-chemical conditions exist in which the fragile site is more stable when it is part of the polymer (part of the bridge connecting two nucleotides) than when in the monomer, connecting the phosphate group to the 3′ carbon atom. These physical-chemical conditions define a thermodynamic niche in which that very chemical information survives longer if it had the chance to encounter similar molecules and with them form a polymer. The same occurs on DNA (Saladino et al., 2005a, b, c), in which a different bond (the 5′ phosphoester) is the key weak site. Also for DNA, the differential stability of the given bond in monomer versus polymer establishes that the transition from the disorder of a mix of nucleotides to the order of the polymer is an important Darwinian parameter for selection, on the possible route to higher complexity, which in this case corresponds to higher order. The evolution of RNA sequence complexity from homogeneous sequences as a favorable Darwinian character was also established (Ciciriello et al., 2008).

The higher stability of the key bonds (3′ phosphoester in RNA and 5′ phosphoester in DNA) in the polymers may provide longer half-life. However, this survival edge is in water between one and two orders of magnitude only (Ciciriello et al., 2009) and would not possibly explain the formation of polymers long enough to be endowed of (pre)genetic potential. Consequently, mechanisms must have come into play allowing facile polymerization and sequence expansion.

Mechanisms explaining the abiotic origin of nucleotides from nucleosides, and the accumulation of their 2′-3′ and 3′-5′ cyclic forms were discovered (Costanzo et al., 2007). Following this, the generation of long RNA chains in water was obtained (Costanzo et al., 2009), leading to the synthesis of polyG oligomers from 3′-5′ cyclic GMP and of polyA from 3′-5′ cyclic AMP. The nonenzymatic formation of polyA oligomers was also described, obtained by terminal ligation through canonical 3′-5′ phosphodiester bonds to yield correctly oriented multimers (Pino et al., 2008) up to >100mers. 3′-5′cyclic GMP showed to be particularly reactive, yielding longer than 30mers in a time frame in the range of minutes, in nondemanding pH and temperature conditions.

Together, these nonenzymatic sequence extension reactions account for the possibility of generation of RNA molecules long and complex enough to provide the (pre)genetic raw material for evolutionary exercises. The series of events potentially, and well within a plausible prebiotic frame, leading from H to the emergence of complexity are schematically indicated in Fig. 1.

Which were the first sequences? How can we reconstruct the steps that followed, leading to those functional vesicles or coacervates (Oparin, 1952) able to reproduce and transmit the molecular solutions that serendipitously turned out to be fertile? When did the Darwinian rules started to play in an otherwise futile exchange of phosphoester bonds?

## 5. Complementarity Enters the Scene

The abiotic polymerization has been, likely, biased toward heteropolymerization – formation of mixed RNA sequences. This would have been a stability-driven trend, as the mixed-sequence polymers are more resistant to spontaneous degradation

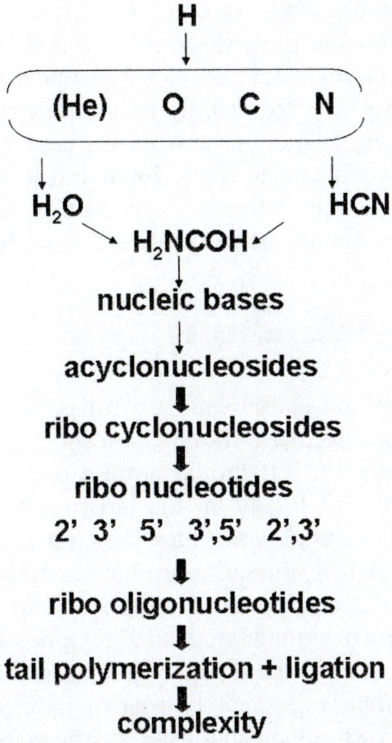


**Figure 1.** Every step, from hydrogen to complex RNA sequences, is abiotically possible, as detailed in the text. The *thicker arrows* highlight the passages (from acyclonucleosides to cyclonucleosides – Bean et al., 2007; Powner et al., 2009) not based on formamide chemistry.

(Ciciriello et al., 2008). This favorable development should have brought the events closer to the lifelike reality – a variety of sequences competing for monomers in the balance of mass law and upper hand of condensation versus degradation. The emerging sequence complexity, a disorder of mixed sequences, should have lead inevitably to the appearance of sequence pairs that matched each other partially or entirely by sterical fit stabilized by hydrogen bonds and van der Waals interactions, which is in original sense of molecular complementarity, as suggested first by Delbrück and Pauling in 1940 (as referred to in Rich, 1994). Nucleic acids, of course, have remarkable structural potential for inter- and intramolecular folding in various types of complementary structures (e.g., Bloomfield et al., 2000). If one of such structures would only have a catalytic activity, assisting further syntheses, one would be yet one step closer to the modern-day Watson-Crick complementary replication and transcription. One such possible natural design is folding of polyA on itself (Brahms, et al., 1966) that, apparently, is the main cause of virtually unlimited elongation of polyA in water (Pino et al., 2008), template copying of itself. Perhaps, a whole spectrum of similar replicating complementary structures may exist. Among them, undoubtedly, the Watson-Crick complementarity.

The molecular complementarity of RNA, thus, could well be the next-ordered key structure on the path we dare to follow. One would rather say "complementarities" in plural as the complexes complementary in the Delbrück-Pauling sense could have been formed, as well between abiotically synthesized peptides (e.g., Lahav et al., 1978) and between the peptides and RNA. Catalytic properties of peptides (Kelly and Roberts, 2006) and RNA (Doudna and Cech, 2002) only add to the opening of the vast spectrum of molecular competitions, making the emergence of complementary replication of RNA more real.

## 6. In the Beginning There Was… a Mistake

The above narrative from abiotic elementary syntheses to polymerization and to complementary replication went emotionless, except for dullness of skepticism: "well, it could go this way, too." It should not escape attention of a thoughtful reader, however, that we are already on the territory of what is called "RNA world." We even could say that this would have been purely RNA world, as proteins had not yet appeared as part of the self-reproduction cycles, except for, probably, passive sequence-nonspecific supportive presence. The undetected emotional component is hidden there behind the weathered textbook knowledge. Go back to the thrilling comment in the historical paper by Watson and Crick (1953): "It has not escaped our notice that the specific pairing we have postulated immediately suggests a possible copying mechanism for the genetic material." The magic thing happened in our order–disorder story. We did come to the self-reproduction, an undeniable major attribute of life. At the moment, however, it is only a part without the whole. Returning to Darwin, we need *variation*, to consider the process as natural selection. The variation in the replicating RNA would naturally come as the error of replication – mistaken incorporation of wrong base in place of the one prescribed by the complementarity. If such error that sneaked in the complementary strand were faithfully copied in the next round of replication, the new molecules will have the now propagating change, and if the change, following Darwin, turned out to be useful, "assuredly individuals thus characterized will have the best chance of being preserved in the struggle for life; and from the strong principle of inheritance, these will tend to produce offspring similarly characterized" (Darwin, 1859). We have just passed the border between nonlife and life, is it not ? It is now both self-reproduction and variation. What is next for this *simplissimus*? The natural selection of such simple organism is selection of the best replicator. Very much like in seminal experiments of Sol Spiegelman in which one such RNA replicator had been trained to the highest of its efficiency (Kacian et al., 1972). The mixed sequence of high complexity is not a good trait. Much better replicator would be a simple tandemly repeating sequence. The proof is in the highest polymorphism of tandem repeats in eukaryotic species. The change in the copy number of the repeats may even lead to pathology, of which triplet expansion diseases are the best examples. The expansion of $GCC_n$ and $GCU_n$

sequences is the most vicious (Trifonov and Bettecken, 1997). This would be the target for evolution of the *simplissimus*. Moreover, as speculated (Trifonov and Bettecken, 1997), the repeats $GCC_n$ and $GCU_n$ may represent the very first messenger RNAs, and the GCC (GCU) would be then the very first codons. This speculation, via succession of predictions and confirmations, developed in a whole new theory of early molecular evolution, with evolutionary chart of codons its culmination and a fundamental starting ground for further studies (Trifonov, 2000, 2004).

## 7. Evolutionary Chart of Codons: A Launch Pad for Reconstruction of the Earliest Events

The evolutionary chart has been preceded by derivation of consensus temporal order of appearance of amino acids in evolution (Trifonov, 2000, 2004). The order is a combined speculation based on now more than 100 different criteria (opinions), each suggesting its own version of the order, all speculative as well. The striking thing about the consensus order is that the codon-anticodon pairs which correspond to amino acids *today* display descending order of stability along the amino acid "chronology" (Trifonov, 2000, 2004). Such a fundamental correspondence could only be observed if (1) amino acid chronology consensus reflects an evolutionary reality, much beyond mere speculation, (2) the codon-anticodon stability has been crucial in the evolutionary history of the code, and (3) the codon assignments *today* are essentially the same as *in the early times* of origin and maturation of the triplet code, some 3–4 billion years back. After this fundamental observation, the reconstruction of the whole chart literally avalanched along a unique route revealing the evolutionary tree (chart) of an amazing order (Trifonov, 2000, 2004). The chart demonstrates the appearance of incoming codons in (complementary) pairs, GCC (alanine) and GGC (glycine) first, the most stable pair. The second step in the chart – appearance of GUC (valine) and GAC (aspartate), due to transition C to U or G to A in already engaged codons GCC and GGC. All subsequent steps, consecutively assigning next stablest codon pairs and, respectively, amino acids next in chronology, proceed via mutations in third positions of already assigned codons, and complementary copying the new codon, which results also in the change in the first position (e.g., GGG for glycine and CCC for proline). At all stages, all the way to the last, least stable codon pair (AAA and UUU), the middle purines of the incoming codons remain purines, and middle position pyrimidines remain pyrimidines. This suggests (predicts) that the 20 amino acids should split in two families – those encoded by mYn codons (Ala-family) and those encoded by mRn codons (Gly-family), which at least during formation of the codon table, and probably sometime later on, kept their identities in the events of amino acid substitutions. This prediction is confirmed since, indeed, the known $20 \times 20$ substitution matrices do split in two independent boxes (Trifonov, 2006; Gabdank et al., 2006), again, a fundamental fact not known before. This means that the modern sequences still carry a memory of their past that, perhaps, even spreads to the very beginning.

## 8. The Earliest Minigenes

The binary protein sequence organization allowed to evaluate the size of the very first encoded peptides consisting either from Ala-family residues only or Gly-family. The alternation of these two types of runs in a hidden form is still present in modern sequences so that the size of the first minigenes could be determined. They encoded peptides of seven residues (Trifonov et al., 2001). The very first peptides, thus, should have been $(Ala)_7$ and $(Gly)_7$ encoded in $(GCC)_7$ and $(GGC)_7$, two strands of thc earliest RNA duplex gene. This is well in line with Sydney Brenner, who hypothesized in 1988 that "the earliest proteins were small peptides of about ten amino acids, and specified by small primitive genes, probably made of RNA." The seven-residue units are visible also in the reconstruction of the first genes by derivation of binary form consensus (Trifonov, 2009) of omnipresent protein sequence motifs (Sobolevsky et al., 2007), likely descendants of LUCA (last universal common ancestor) proteins.

## 9. Mistakes Again?

The first round of variation, from the first replicating complementary sequences (no reconstruction succeeded so far) to their more efficient variants, according to the reasoning above has eventually lead to the simple repetitive sequences $(GCC)_n$ and $(GGC)_n$. From a disorder of presumably mixed replicating sequences, it has boiled down to the order of the unique, presumably best replicating sequence. No further changes would be justified after reaching that stage of the duplex perfectly replicating itself. The variation component of life is lost! To stay consistent with the logic of the chapter, we have to admit that the life (self-reproduction with variations) that just emerged with the complementarity came to the dead end. The evolutionary chart of the codons, however, suggests that life got another chance due to the occasional mutational changes of C to U and G to A in the duplex $(GCC)_n \bullet (GGC)_n$ with formation of GUC and GAC triplets. This change could stay only if respective loss of replication efficiency of the RNA was compensated by some new advantage. At this point, we only can speculate that the possible advantage would come from the protein component of the triplet-based composite replicon (Trifonov, 2009). The four-part replicon, $(GGC)_7$ , $(GCC)_7$, $ala_7$ , $gly_7$ – as speculated, has the ability to reproduce itself so that the peptides are somehow translated, while the replication of RNA is assisted by the peptides. These details may well become revealed by later reconstructions and experimentation.

## 10. The Evolution of Order from Disorder in Early Proteins

When one looks at a biological sequence, no matter how complex it is, even random-like, one realizes that the complexity is the result of massive evolutionary effort. Conservation and propagation of these sequences, even when we do not know what exactly they are good for, attests to their importance for continuation of life. The apparent disorder in the sequences is, as apparently, to the advantage for the species carriers of the sequences. Thus, there is nothing wrong in the sequence disorder, which is, actually, well-encoded order of a wealth of biomolecular events, interactions, and structures, indispensable for the advanced life we are part of.

As the proteins gained length and complexity, surely they gained structure and function in this RNA-dominated world. But what types of structures? And what types of functions? And what was the ordering of the events leading to this gain of structure and function? It is generally assumed that 3D structure is a prerequisite for protein function. However, whether or not prior structure is required depends on which functions are being discussed. For example, catalytic function, whether from RNA, protein, or an inorganic material, almost always requires prior formation of structure. Catalytic materials speed up their cognate reactions by using the binding energy for this purpose (Polanyi, 1921). But in particular, catalysis requires tighter binding to the transition state compared to the ground state (Pauling, 1946; Zalatan and Herschlag, 2009) which in turn leads to the need for prior formation of structure for the catalyst so that the chemical complementarity with the transition state is better than the complementarity with the ground state. This preferential binding requires fairly precise structure and leads to rather limited flexibility for the key catalytic moieties. On the other hand, mere association does not require the protein fold in advance. Functions that required only binding could have been carried out by early, even fairly short, unstructured polypeptides. Such unstructured peptides likely played key roles, even very early in the RNA world by, say, helping RNA molecules to fold correctly and then, once folded, by associating with and stabilizing the functional RNA molecule. Similar ideas that early polypeptides could have helped to fold and stabilize RNA molecules have been proposed by Poole et al. (1998), but without our emphasis that polypeptides containing just the early amino acids likely could not form structured proteins. Noller (2004) added the insight that small peptides, examples of which are known to cause substantial rearrangements of particular RNA molecules, could very substantially enlarge the structure-function space of RNA.

Let us consider RNA structure and folding a little more deeply. Given the multiplicity of ways of forming hydrogen bonds and the limited information content of RNA primary structure, RNA has a highly rugged energy landscape and therefore frequently becomes kinetically trapped in nonproductive folds (Treiber and Williamson, 1999, 2001; Solomatin et al., 2010). Some proteins, known as RNA chaperones, greatly speed up the correct folding of functional RNA molecules. An important feature of these RNA chaperone proteins is that they have large regions that lack stable 3D structure (Ivanyi-Nagy et al., 2005). That is, these proteins contain regions that exist as unstructured, dynamic intrinsically disordered ensembles. Thus, a very early protein function not requiring prior structure was very likely to assist the RNA molecules to fold correctly. In addition to chaperone function, a number of intrinsically disordered proteins have been shown to bind to catalytic RNAs and thereby help to stabilize the catalytic conformations of these molecules (Poole et al., 1998; Treiber and Williamson, 2001; Noller, 2004). To give a well-studied specific example, most of the ribosomal proteins contain idiosyncratically folded extensions that bind to several RNA domains and reach into the interior when in the ribosome (Nissen et al., 2000). Indeed, ribosome assembly likely involves cooperative cofolding of the RNA and protein components particularly involving these protein extensions (Klein et al., 2004), which are almost certainly intrinsically disordered in the absence of the RNA.

Besides being RNA chaperones and stabilizers, intrinsically disordered proteins and regions are heavily involved in signaling and regulation in modern cells. When two rigid molecules bind to each other, the free energy from the interactions of the contacting surfaces less the motional entropy equals the binding free energy. On the other hand, if binding involves a flexible molecule, much of the free energy from the interactions of the contacting surfaces is spent to rigidify the flexible molecule upon binding, and this rigidification free energy does not contribute to the affinity (Schulz, 1979; Dunker et al., 1998). The end result is typically weak, reversible binding coupled with high specificity, which is an ideal pair of properties for signaling molecules (Daughdrill et al., 2004). Thus, it is likely that intrinsically disordered proteins have been used for signaling and regulation from very early times, likely providing important capacities in the conversion from nonlife to life in the RNA world.

According to the evolutionary chart of codons, some amino acids occurred early and others came into being much later (Trifonov, 2009). The structural ramifications of the early versus late amino acids have not been considered previously. We developed a scale of the disorder propensity of the 20 amino acids (Campen et al., 2008). We noticed that the putatively early amino acids derived from the earlier codons generally promote lack of 3D structure or intrinsic disorder, while those associated with the later codons generally promote structure (Dunker et al., 2009). This rough correspondence of those amino acids that came later with those amino acids that most strongly promote structure is shown in Fig. 2.

As the amino acid chains increased in length, but with only the early amino acids, these chains could evolve to become very good at helping to fold and stabilize RNA structures and also very useful for signaling within the RNA world protocells, as discussed above, but they almost certainly could not form the highly stable enzymes observed in the modern world. Randomly generated amino acid sequences containing just three amino acids, namely, V, Q, and R, form fairly stable structure an amazingly high 5% of the time (Davidson and Sauer, 1994). However, such structures tend to be loose with poorly packed, mobile side chains and are more akin to molten globules, which were originally described as protein folding intermediates (Dolgikh et al., 1981, 1984), than to tightly structured enzymes. Indeed, even sequences that are carefully designed to fold into a particular 3D structure often fold correctly but yield stable molten globules that fail to progress into tightly packed proteins (see, e.g., Tanaka et al., 1994). In addition, a tightly structured enzyme was converted into a catalytically active molten globule

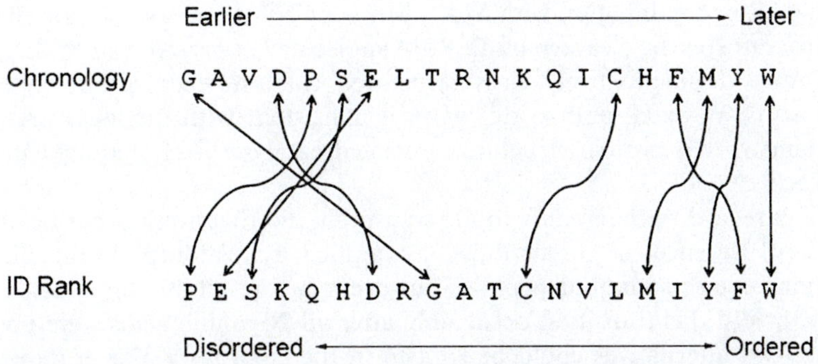


**Figure 2.** The correlation of 20 amino acids in various orders. In the *top line*, the 20 amino acids are aligned in the chronology of evolution from earlier to later times by building a consensus among the various theories of the evolution of the genetic code (Trifonov, 2009). In the *bottom line*, the 20 amino acids are ranked according to their increasing propensity to promote structure as estimated by comparing their grouped frequencies of occurrence in structured versus disordered regions of modern, well-characterized proteins (Campen et al., 2008).

by genetic engineering, and from this observation, it was proposed that modern enzymes might have evolved from molten globular precursors (Vamvaca et al., 2004). The important point here is the proteins made from just the earlier amino acids can likely form molten globules but not tightly packed structures.

The systematic conversion from molten globular enzymes into their modern counterparts, we believe, required the addition of the late amino acids, particularly the aromatics, phenylalanine, tyrosine, and tryptophan, as well as histidine, which is typically not counted as an aromatic. The conversion from molten globules into more tightly packed proteins would have been accompanied by very large increases in protein stability and substantial improvements in catalytic activity, both of which would lead to a significant selective advantage for those cells that were capable of incorporating these amino acids.

A major uncertainty regards when and the extent to which proteins appeared in the RNA world. For example, Waldrop (1992) and Cech (2009) present the idea that RNA alone could have carried out information storage, catalysis, and the various other necessities of life. In this view, matter could have passed the threshold into life before proteins were on the scene. Focusing on different chemistries, both Di Giulio (1997) and Carny and Gazit (2005) argue that the RNA world was really a ribonucleoprotein world from the start. Of course, the work of Poole et al. (1998) and Noller (2004) described above both argue that protein was crucial for the development of the RNA world even if not present at the very beginning. While we and these others have emphasized noncatalytic protein functions at early times, protein-based catalysis might have also developed quite early. For example, simple polypeptides formed of the amino acids K, I, and A, with a single H or W at the terminus, fold into four-helix bundles; furthermore, these KIA-H or KIA-W polypeptides and even short, unstructured peptides that are rich in K and H, in combination with $Mg^{++}$, $Mn^{++}$, or $Ni^{++}$ (but not $Cu^{++}$ or EDTA), can carry out specific cleavage of an RNA molecule (Lopez-Alonso et al., 2010). Notice that these putatively early structured catalysts require late-appearing amino acids, W or H, and so this work is consistent with the ideas presented here, namely, that protein structure-associated catalysis likely required the late amino acids.

With regard to the nonlife to life transition, did that event occur before the significant presence of polypeptides, as implied by Waldrop (1992), did this transition require some polypeptides but occur before all 20 amino acids were present, or did this transition occur only after all 20 amino acids were present? The last two alternatives could be studied in the laboratory. For example, one could determine whether functional ribosomes could be assembled using proteins in which the most modern amino acids were replaced by their more ancient counterparts. Secondly, one could determine whether catalytically competent enzymes would result if all of the modern amino acids were replaced by their ancient counterparts. Finally, the ultimate experiment would be to see whether a synthetic cell could survive if all of its codons for the modern amino acids were replaced with the codons for their more ancient counterparts.

The coexistence and transformations of the ordered and disordered proteins and parts thereof, probably, can be traced all the way back to the very first fusion products of the earliest minigenes. This is illustrated by the structures of the presumably earliest protein segments containing the omnipresent LUCA motifs (Sobolevsky et al., 2007). Among 20 most conserved elements (modules) harboring the omnipresent motifs, there is one (in RNA polymerases) that unlike other modules does not possess typical closed-loop structure ("molten globule"), being rather of extended shape, different in all crystal structures where it has been located.

The order and disorder march in evolution side-by-side, from one innovation to another, to the delight of those who chose to trace the elusive path from nonlife to life.

## 11. Acknowledgments

Ernesto Di Mauro was supported by "Istituto Pasteur – Fondazione Cenci Bolognetti." Thanks also to Silvia Lopizzo for helpful contributions. Keith Dunker was supported by: NIH 5R01GM071714-04 "Mining Structural Genomics Initiative for Disorder"; NIH 5R13LM006766-13 Pacific Symposium on Biocomputing; NIH 2R56LM007688-05A1 Bioinformatics – "Linkage of Protein Disorder and Function"; NSF 0849803 "DisProt Database: A Central Repository of Information on Intrinsically Disordered Proteins."

Thanks also to Maya Wagle for helpful contributions. Edward N. Trifonov is supported by Czech Ministry of Education (grant SM0021622415) and by SoMoPro (South Moravia Project).

## 12. References


Alberty RA (2006) Thermodynamic properties of enzyme-catalyzed reactions involving guanine, xanthine, and their nucleosides and nucleotides. Biophys Chem 121:157–162

Barks H, Buckley R, Grieves GA, Di Mauro E, Hud N, Orlando T (2010) Guanine, adenine, and hypoxanthine production in UV-irradiated formamide solutions: relaxation of the requirements for prebiotic purine nucleobase formation. Chembiochem 11:1240–1243

Bean HD, Sheng Y, Collins JP, Anet FA, Leszczynski J, Hud NV (2007) Formation of a beta-pyrimidine nucleoside by a free pyrimidine base and ribose in a plausible prebiotic reaction. J Am Chem Soc 129:9556–9557

Bloomfield VA, Crothers DM, Tinoco I (2000) Nucleic acids: structures, properties, and functions. University Science, Sausalito

Brahms J, Michelson AM, Van Holde KE (1966) Adenylate oligomers in single- and double-strand conformation. J Mol Biol 15:467–488

Brenner S (1988) The molecular evolution of genes and proteins: a tale of two serines. Nature 334:528–530

Campen A, Williams RM, Brown CJ, Meng J, Uversky VN, Dunker AK (2008) TOP-IDP-scale: a new amino acid scale measuring propensity for intrinsic disorder. Protein Pept Lett 15:956–963

Carny O, Gazit E (2005) A model for the role of short self-assembled peptides in the very early stages of the origin of life. FASEB J 19L:1051–1055

Carroll L (1865) Alice's adventures in Wonderland. Macmillan, London

Cech TR (2009) Crawling out of the RNA world. Cell 136:599–602

Ciciriello F, Costanzo G, Pino S, Crestini C, Saladino R, Di Mauro E (2008) Molecular complexity favors the evolution of ribopolymers. Biochemistry 47:2732–2742

Ciciriello F, Costanzo G, Pino S, Di Mauro E (2009) Spontaneous generation revisited at the molecular level. In: Pontarotti P (ed) Evolutionary biology: concept, modeling and application. Springer, Berlin/Heidelberg, pp 3–22

Costanzo G, Saladino R, Crestini C, Ciciriello F, Di Mauro E (2007) Nucleoside phosphorylation by phosphate minerals. J Biol Chem 282:16729–16735

Costanzo G, Pino S, Ciciriello F, Di Mauro E (2009) Generation of long RNA chains in water. J Biol Chem 284:33206–33216

Damiano L, Luisi PL (2010) Towards an autopoietic redefinition of life. Orig Life Evol Biosph 40:145–149

Darwin C (1859) Origin of species. John Murray, London, Chapter 4

Daughdrill GW, Pielak GJ, Uversky VN, Cortese MS, Dunker AK (2004) Natively disordered proteins. In: Buchner J, Kiefhaber T (eds) Protein folding handbook. Wiley-VCH, Weinheim, pp 271–353

Davidson AR, Sauer RT (1994) Folded proteins occur frequently in libraries of random amino acid sequences. Proc Natl Acad Sci USA 91:2146–2150

Di Giulio M (1997) On the RNA world: evidence in favor of an early ribonucleoprotein world. J Mol Evol 45:571–578

Dolgikh DA, Gilmanshin RI, Brazhnikov EV, Bychkova VE, Semisotnov GV, Venyaminov SYu, Ptitsyn OB (1981) Alpha-lactalbumin: compact state with fluctuating tertiary structure? FEBS Lett 136:311–315

Dolgikh DA, Kolomiets AP, Bolotina IA, Ptitsyn OB (1984) 'Molten-globule' state accumulates in carbonic anhydrase folding. FEBS Lett 165:88–92

Doudna JA, Cech T (2002) The chemical repertoire of natural ribozymes. Nature 418:222

Dunker AK, Garner E, Guilliot S, Romero P, Albrecht K, Hart J, Obradovic Z, Kissinger C, Villafranca JE (1998) Evolution of molecular recognition. Pac Symp Biocomput 3:473–484

Dunker AK, Chan WS, Karn SHB, Uversky VN, Brooks DJ, Oldfield CJ, White J, Perumal N, Romero P (2009) Protein folding as a transition step from ancient to modern life forms. Biophys J 96:318a

Gabdank I, Barash D, Trifonov EN (2006) Tracing ancient mRNA hairpins. J Biomol Struct Dyn 24:163–170

Ivanyi-Nagy R, Davidovic L, Khandjian EW, Darlix JL (2005) Disordered RNA chaperone proteins from functions to disease. Cell Mol Life Sci 62:1409–1417

Joyce J (1994) In: Deamer DW, Fleischaker GR (eds) The foreword of 'Origins of Life: The Central Concepts'. Jones and Bartlett, Boston. The original reference is a 1992 internal NASA document entitled 'Exobiology: Discipline Science Plan'.

Kacian DL, Spiegelman S, Mills DR, Kramer FR (1972) Replicating RNA molecule suitable for a detailed analysis of extracellular evolution and replication. Proc Natl Acad Sci USA 69:3038–3042

Kelly DR, Roberts SM (2006) Oligopeptides as catalysts for asymmetric epoxidation. Pept Sci 84:74–89

Klein DJ, Moore PB, Steitz TA (2004) The roles of ribosomal proteins in the structure assembly and evolution of the large ribosomal subunit. J Mol Biol 340:141–177

Lahav N, White D, Chang S (1978) Peptide formation in prebiotic era – thermal condensation of glycine in fluctuating clay environments. Science 201:67–69

Lopez-Alonso JP, Pardo-Cea MA, Gomez-Pinto I, Fernandez I, Chakrabarty A, Pedroso E, Gonzalez C, Laurents DV (2010) Putative one-pot prebiotic polypeptides with ribonuclease activity. Chem Eur J 16:5314–5323

Luisi PL (1998) About various definitions of life. Orig Life Evol Biosph 28:613–622

Nissen BN, Hansen J, Moore PB, Steitz TA (2000) The complete atomic structure of the large ribosomal subunit at 2.4 A resolution. Science 289:905–920

Noller HF (2004) The driving force for the evolution of translation. RNA 10:1833–1837

Oparin AI (1952) The origin of life. Dover, New York

Pauling L (1946) Molecular architecture and biological reactions. Chem Eng News 24:1375–1377

Pino S, Ciciriello F, Costanzo G, Di Mauro E (2008) Nonenzymatic RNA ligation in water. J Biol Chem 283:36494–36503

Polanyi M (1921) On adsorption catalysis. Z Eleckrochem 27:142–150

Poole AM, Jeffares DC, Penny D (1998) The path from the RNA world. J Mol Evol 46:1–17

Powner MW, Gerland B, Sutherland JD (2009) Synthesis of activated pyrimidine ribonucleotides in prebiotically plausible conditions. Nature 459:239–422

Rich A (1994) Pauling Linus (1901–1994) – Obituary. Nature 371:285

Saladino R, Crestini C, Busiello V, Ciciriello F, Costanzo G, Di Mauro E (2005a) Origin of informational polymers. Differential stability of 3′- and 5′-phosphoester bonds in deoxy monomers and oligomers. J Biol Chem 280:35658–35669

Saladino R, Crestini C, Neri V, Brucato JR, Colangeli L, Ciciriello F, Di Mauro E, Costanzo G (2005b) Synthesis and degradation of nucleic Acid components by formamide and cosmic dust analogues. Chembiochem 6:1368–1374

Saladino R, Crestini C, Costanzo G, Di Mauro E (2005c) On the prebiotic synthesis of nucleobases, nucleotides, oligonucleotides, pre-RNA and pre-DNA molecules. In: Walde P (ed) Topics in current chemistry, "Prebiotic chemistry", vol 259. Springer, Berlin, pp 29–68R

Saladino R, Crestini C, Ciciriello F, Di Mauro E, Costanzo G (2006a) Origin of informational polymers: differential stability of phosphoester bonds in ribomonomers and ribooligomers. J Biol Chem 281:5790–5796

Saladino R, Crestini C, Neri V, Ciciriello F, Costanzo G, Di Mauro E (2006b) Origin of informational polymers: the concurrent roles of formamide and phosphates. Chembiochem 711:1707–1714

Saladino R, Crestini C, Ciciriello F, Costanzo G, Di Mauro E (2007) Formamide chemistry and the origin of informational polymers chemistry & biodiversity. Helvetica Chimica Acta 4:694–720

Saladino R, Neri V, Crestini C, Costanzo G, Graciotti M, Di Mauro E (2008) Synthesis and degradation of nucleic acid components by formamide and iron sulfur minerals. J Am Chem Soc 130:15512–15518

Saladino R, Crestini C, Ciciriello F, Pino S, Costanzo G, Di Mauro E (2009) From formamide to RNA: the roles of formamide and water in the evolution of chemical information. Special issue on "the origin of life and microbiology". Res Microbiol 160:441–448

Saladino R, Neri V, Crestini C, Costanzo G, Graciotti M, Di Mauro E (2010) The role of the formamide/zirconia system in the synthesis of nucleobases and biogenic carboxylic acid derivatives. J Mol Evol 71:100–110

Saladino R, Barontini M, Cosetti C, Di Mauro E, Crestini C (2011) The effects of borate minerals on the synthesis of nucleic acid bases, amino acids and biogenic carboxylic acids from formamide. Orig Life Evol Biosph 41:317–330

Schulz GE (1979) Nucleic acid binding proteins. In: Balaban M (ed) Molecular mechanisms of biological recognition. Elsevier/North Holland Biomedical Press, Amsterdam, pp 79–94

Sobolevsky Y, Frenkel ZM, Trifonov EN (2007) Combinations of ancestral modules in proteins. J Mol Evol 65:640–650

Solomatin SV, Greenfeld M, Chu S, Herschlag D (2010) Multiple native states reveal persistent ruggedness of an RNA folding landscape. Nature 463:681–684

Stenger VJ (2006) The comprehensible cosmos: where do the laws of Physics come from? Prometheus, Amherst

Tanaka T, Hayashi M, Kimura H, Oobatake M, Nakamura H (1994) De novo design and creation of a stable artificial protein. Biophys Chem 50:47–61

Treiber DK, Williamson JR (1999) Exposing the kinetic traps in RNA folding. Curr Opin Struct Biol 9:339–345

Treiber DK, Williamson JR (2001) Beyond kinetic traps in RNA folding. Curr Opin Struct Biol 11:309–314

Trifonov EN (2000) Consensus temporal order of amino acids and evolution of the triplet code. Gene 261:139–151

Trifonov EN (2004) The triplet code from first principles. J Biomol Struct Dyn 22:1–11

Trifonov EN (2006) Theory of early molecular evolution: predictions and confirmations. In: Eisenhaber F (ed) Discovering biomolecular mechanisms with computational biology. Landes Bioscience, Georgetown, pp 107–116

Trifonov EN (2009) Origin of the genetic code and of the earliest oligopeptides. Res Microbiol 160:481–486

Trifonov EN, Bettecken T (1997) Sequence fossils, triplet expansion, and reconstruction of earliest codons. Gene 205:1–6

Trifonov EN, Kirzhner A, Kirzhner VM, Berezovsky IN (2001) Distinct stages of protein evolution as suggested by protein sequence analysis. J Mol Evol 53:394–401

Vamvaca K, Vogeli B, Kast P, Pervushin K, Hilvert D (2004) An enzymatic molten globule: efficient coupling of folding and catalysis. Proc Natl Acad Sci USA 101:12860–12864

Van Holde K (1980) In: Halvorson HO, van Holde KE (eds) The origins of life and evolution. Alan R. Liss, Inc., New York, p 31

Waldrop MM (1992) Finding RNA makes protein gives 'RNA world' a big boost. Science 256: 1396–1397

Watson JD, Crick FH (1953) Molecular structure of nucleic acids; a structure for deoxyribose nucleic acid. Nature 171:737–738

Zalatan JG, Herschlag D (2009) The far reaches of enzymology. Nat Chem Biol 5:516–520

Biodata of **B. Roy Frieden** and **Robert A. Gatenby**, authors of "***The Role of Information and Order in the Origin of Life.***"

**B. Roy Frieden** is a professor emeritus in the College of Optical Sciences of the University of Arizona, Tucson, Arizona. He obtained his Ph.D. from the University of Rochester in 1966 and has since been a professor at the University of Arizona. His principal research was in image processing and restoration until about 1990, when an investigation of noise fluctuations serendipitously uncovered the strong ability of Fisher information to subdue and control it. Further analysis showed the information to be a measure of order in all systems, regardless of their nature, that obey well-defined probability laws. All such laws display maximum Fisher information transfer from source to observer. This principle, called extreme physical information (EPI), has succeeded in rederiving many known laws of physics, biology, chemistry, etc. Given these verifications of the approach, EPI is now being used to derive growth and depletion laws in *unknown* areas of biology (ecological monitoring, cancer growth and its treatment), chemistry (many-body problems), economics (optimum investment programming), and social organization (optimizing their stability).

E-mail: **roy.frieden@optics.arizona.edu**

J. Seckbach (ed.), *Genesis - In The Beginning: Precursors of Life, Chemical Models and Early Biological Evolution*, Cellular Origin, Life in Extreme Habitats and Astrobiology 22, 437–468
DOI 10.1007/978-94-007-2941-4_24, © Springer Science+Business Media Dordrecht 2012

**Robert A. Gatenby, M.D.**, is the chairman of the Department of Radiology and codirector of Integrated Mathematical Oncology at H. Lee Moffitt Cancer Center. Following professorships at the University of Arizona in the departments of Radiology and Applied Mathematics, he joined Moffitt in 2008. He received a B.S.E. in Bioengineering and Mechanical Sciences from Princeton University and an M.D. from the University of Pennsylvania in 1977. He completed his residency in radiology at the University of Pennsylvania where he served as chief resident. Bob remains an active clinical radiologist specializing in body imaging. While working at the Fox Chase Cancer Center after residency, Bob perceived that cancer biology and oncology were awash in data but lacked coherent frameworks of understanding to organize this information and integrate new results. Since 1990, most of Bob's research has focused on exploring mathematical methods to generate theoretical models for cancer biology and oncology. His current modeling interests include (1) the tumor microenvironment and its role in tumor biology; (2) evolutionary dynamics in carcinogenesis, tumor progression, and therapy; and (3) information flow in living systems and its role in maintaining thermodynamic stability.

E-mail: **robert.gatenby@moffitt.org**

# THE ROLE OF INFORMATION AND ORDER IN THE ORIGIN OF LIFE

**B. ROY FRIEDEN[1] AND ROBERT A. GATENBY[2]**
*[1]College of Optical Sciences, University of Arizona,*
*8572 Tucson, AZ 85721, USA*
*[2]Mathematical Oncology, Moffitt Cancer Center,*
*FL 33612, Tampa Bay, FL 33612, USA*

## 1. Introduction

*Notes on terminology*: We use the terms "order," "structure," and "complexity" interchangeably. Unless otherwise qualified, the term "disorder" is used in the sense of a lack of the "order" measure $R$ defined here. We also distinguish between the terms "order" and "information." The latter is a property of data, the former a property of the system that forms the data. The precise relation between order and information is derived as Eq. (11c) in Frieden and Hawkins (2010b). This work is summarized and applied in Sects. 3.1–3.6. Meanwhile, we use the word "order" in its intuitive sense, freely quoting results from those sections.

The general question we address is, what are the fundamental physical characteristics of a living system? Carefully addressing these issues can only advance our understanding of how life originated. In a nutshell, our view is that its necessary characteristics are:

1. A boundary separating it from the environment
2. A stable, local entropy value using external energy
3. Capability of self replication

Addressing these issues in depth forms the major part of this work. We address them briefly here. To satisfy the 2nd law of thermodynamics, stable entropy in a living system over time can only be achieved by export of entropy from the system into the environment. This must be via a system that is robust and stable to a range of environmental perturbations. Such stability requires the system to be in a state of stable information. This emphasizes the concept of information as a concept that is central to cell stability and function. It also requires the system probability law $p(x)$, $x$ the position coordinate, to be stable to perturbations in the probabilities due to random intracell and extracell events.

The information about the process and system state is also to serve as a memory or template. This allows the system to continuously restore the primary state even as entropy is released due to normal cell functions. Finally, the system must be able to reproduce itself.

Regarding requirement 1 above, the living system must have distinct boundary conditions separating its entropy space from the environment. This is defined by an entropy gradient (Haynie, D.T.). Although the gradient may be steep, it is not a delta function. That is, there should be changes in the environment adjacent to a living system due to its export of entropy and changes in the living system adjacent to the border that reflect the environmental conditions. By requirement 2, the living system should be stable to these changes.

Thus, the living system should have locally ordered structure in a state of dynamic equilibrium. To satisfy the 2nd law of thermodynamics, this can only occur at the expense of decreased order elsewhere, i.e., outside the living system. However, living systems are not unique in accomplishing this trade-off in order. For example, a growing crystal is not alive and, yet, likewise increases its local order in this manner. It is interesting that these phenomena are structural in nature – both result from polymerization of long molecules. We know the niches that crystal growth occupies in nature, but what niche does growth of a living system fill? How does it use *its* local increase of order to promote life? Correctly answering this question can ultimately shed light on the principle question posed by this book, of how life originated.

The third requirement, of cell reproduction, will be seen to be a tactic enabling the living system to require smaller energy input demands than the same system in the absence of reproduction. A system containing many small living units, i.e., cells, is in a lower state of order than is a corresponding system containing but a few very large such units. Maintaining lower order requires less energy input than does maintaining high order.

Cell mitosis also avoids a problem of overly long carbon chains that are detrimental to cell function (Sect. 3.5).

## 1.1. WHY THE MEMBRANE?

In obeying the 2nd law of thermodynamics, the increased order within the living system is at the expense of a *compensating loss of order, i.e., increased disorder, elsewhere* – outside the living system. But a state of disorder tends to spatially spread, e.g., in the form of diffusing heat or some other form of waste energy. Thus, in order to maintain its locally ordered structure, the living system must form a barrier to the extra amount of disorder it shipped outside of it. This takes the form of *a membrane* – another ordered structure – that encases the system.

### *1.1.1. Role of Communication*

Aside from acting as a physical barrier to the loss of system order, the membrane also acts as a vital *communication link* between its inside and outside. Maintaining the system order requires that it be robust to a wide range of environmental perturbations. This requires information to serve as a *memory or template* of the *fundamental state* of the system (as, e.g., B cells preserve a memory of past pathogens so as to more quickly suppress them when they reappear) (Allman et al., 2004). As order is lost due to normal cell functions, the system must continuously act to restore it, i.e., restore the primary state. For the cell membrane to act quickly and efficiently for this purpose, it makes sense for the membrane to actually *contain* the needed memory of the fundamental state. An analogous example occurs in a neuron cell membrane that contains slowly inactivating calcium channels (Shen et al., 2000). Finally, the system must be able to reproduce itself, requiring further properties of the membrane.

### *1.1.2. Role of Carbon Chains*

A cell membrane is largely composed of long chains of carbon atoms (Lodish et al., 2004). About 30% of the membrane consists of three classes of amphipathic lipids. Of these, phospholipids are the most abundant, and in both these and glycolipids, the fatty chains usually contain an even number of carbon atoms, typically between 16 and 20. How could such long chains arise? Remarkably, they are long known to exist widely in interstellar space, even away from galactic centers (Winnewisser and Walmsley, 1979). Also, ordinary methane gas $CH_4$ when irradiated light with ultraviolet line forms such chains in the laboratory (Wu and Cheng, 2010). These, then, fill one specific need of cell creation.

## 1.2. INFORMATION AND ORDER NEEDS

A living system consists of mutually interacting components that send and receive information – Fisher information in particular, as we find below. The analogous crystal reacts to "messages" consisting of the concentrations of compounds around it, by growing or dissolving. These represent *changes in the level of order* of the crystal. Thus, these changes in order give rise to the messages and their information contents. In fact, we will find that, mathematically, the information arises out of a *change in the level of order* of the system, increasing linearly with that change. Thus, for a living system to continue living, it must continually change its level of order, e.g., grow. Conversely, a dead system is incapable of changing its order on demand.

In analogy with the crystal case, there should be changes in the environment just outside the membrane due to its export of disorder and analogous changes just within the membrane that reflect the environmental conditions. These changes are quantified as messages, in the form of molecular agents that are carried by protein ions from both directions.

## 1.3. VITAL ROLE OF FEEDBACK

The incoming messages, in particular, provide positive feedback since the messages are utilized to form changes (usually, increases) in the level of order of the membrane and the interior cell system. In this way, a living system exhibits many of the attributes of a communication channel with *positive* (reinforcing) feedback. Life is impossible without such feedback.

In the development to follow, we form the statistical principles needed to develop more specific properties of the channel. As one would hope, there are, in fact, direct, quantitative correspondences between how living and nonliving systems convey the Fisher information they need to exist. In particular, *both types of system evolve according to the same statistical principle of information transfer*. See Sects. 2.2–2.16. Indeed, many feedback activities arise as *evolutionary tactic* functions in the Lotka-Volterra equations (Sect. 2.10d).

## 2. What Was the First Life Form?

What was the first form of life that achieved the above requirements of local increase of order and mitosis? The first cell was formed out of fundamental processes that were either classical or quantum in nature, or somewhere in between. In fact, the processes that are most basic to a living cell occur on the molecular and, therefore, at least semiquantum level (Schroedinger, 1944; Davydov, 1982). However, any quantum system requires stringent conditions of coherence, and coherence is rapidly lost as outside or environmental inputs enter it.

Hence, more realistically, we have to seek an origin for the living cell that is *semiclassical,* i.e., lying somewhere between the quantum and classical limits. In fact, just such a system obeys the Hartree or "mean field" approximation, as discussed below. The Hartree equations for a certain medium will be seen to give rise to the known equations of biological population growth, those of Lotka and Volterra (Frieden and Gatenby, 2008). This is a new and rather startling connection between living and quantum processes.

### 2.1. GENERALIZED LOTKA-VOLTERRA EQUATIONS

The development of life follows the rules of natural selection first laid down by Darwin (1859). Natural selection leads inevitably to the growth and decay of competing cell types. How can this be quantified?

There are fundamental equations (Lotka, 1932; Volterra, 1926) defining this competition that are known to be correct to first order. These are the generalized (to $K$ species) Lotka-Volterra (L-V) equations:

$$\frac{dp_k}{dt} = p_k(t)G_k(t), \quad G_k(t) = \left[ g_k(t) + \sum_j g_{jk}(t)p_j(t) \right], \quad k = 1,\ldots,K, \tag{1}$$
$$p_j(t) = \int dr_j p_j(r,t).$$

These equations show that the rate of population change $dp_k / dt$ of the $k$th type of species in a system containing $K$-interacting species obeys a set of coupled, first-order, generally quadratic, differential equations. The nature of the growth is defined by an overall growth rate $G_k(t)$ for each species. These are defined by coefficients $g_k(t), g_{jk}(t)$ representing known *fitness coefficients* (Maynard-Smith, 1989). These $g$ coefficients are defined as the average number of new species members per generation that survive long enough to themselves produce a new generation. Note that the $g$ coefficients have no dependence upon spatial position, instead depending explicitly upon the time. There is also an implicit dependence upon all the particle (i.e., *species*) abundances, e.g., in a predator–prey system, where the fitness of the predator increases with the population of the prey species.

An example of growth obeying Eq. (1) is RNA replication, thought to characterize the growth of the earliest forms of life (Bernstein et al., 1983). At a general RNA replication number $k$, the $g$ coefficients are *constant* in time, obeying

$$g_k(t) = b_k \frac{A}{A_k + A} - d_k, \quad \text{with all } g_{jk}(t) = 0. \tag{2}$$

The latter means there are no interactions between different replications (the system lacks memory). Here, $A$ is a constant amount of resource available at all replications; $b_k$ is the maximum rate of replication of $k$, occurring when $A \gg A_k$; from the form of (2), $A_k$ is the resource level giving rise half the maximum rate $b_k / 2$ of replication; and $d_k$ is the attrition, or loss, rate.

## 2.2. PRINCIPLE OF EXTREME PHYSICAL INFORMATION (EPI)

It is long known, since Plato, Kant, Schopenhauer, and, more recently, Heisenberg, that any observation is generally imperfect. It is impossible to observe absolute truth. The engineer would say that random noise must always enter into the measurement. The physicist claims that the 2nd law, coarse graining, or quantum fluctuation, enters in. In fact, the philosopher Schopenhauer notably despaired of this fact. This lack of perfect observation is probably best explained by example.

### *2.2.1. Viewing a Microscope Slide*

Consider a "source," such as an illuminated microscope slide specimen. A viewer attempts to assess a property of the specimen, say its width. How well may he do this? At the microscope slide (the *source*), let there be a level $J$ of information about the unknown width. The specific *type* of information will be defined below. Since this is right at the source, before the information about the width flows (as photons) to the observer, there is not yet any deterioration in the information. Hence, $J$ defines the total available amount of information about the source. By comparison, once the photons reach the observer, the information about width deteriorates due to all possible interceding (or environmental) effects that affect the photon trajectories en route. This includes heat-induced turbulence effects, aberrations, diffraction, the imperfect human visual system, and others. Hence, the level $I$ that the observer receives is generally less than or equal to the full amount $J$. So far, Plato, Kant et al. would agree. However, the following would probably surprise them.

### 2.2.2. *Extreme Physical Information*

Nature decrees that, despite the imperfect nature of any observation, the received level $I$ of information about an observed parameter (here, the specimen width) is *maximally close* to the ideal value $J$. It obeys a principle

$$I - J = \text{extremum}, \quad \text{where } J \to I. \tag{3}$$

The arrow denotes the direction of the information flow, i.e., from source to observer. Fortunately for us, the extremum or extreme value is usually *a minimum*. Consequently, $I \to J$ in magnitude, so that $I$ = maximum of the available information. Also, informations $I$, $J$ are specifically *Fisher* informations (Fisher, 1922, 1959), (Frieden, 2001). These are as defined in (4a) and (4b) below. In summary, *the fundamental activity of observing, i.e., making a measurement of a parameter, excites a maximum transfer of Fisher information from any phenomenon (called a system) to the observer.*

Principle (3) is called that of extreme physical information (EPI). In a nutshell, the EPI principle states that despite the depredations of nature, it nevertheless cooperates in a *statistical sense* with the observer, supplying him/her with data containing the *maximum possible amount of information about* the parameter. Perhaps, Schopenhauer should not have despaired so.

Principle (3) is a statement that *we live in a universe that cooperates with the observer*. Other universes might not have this property. We can have confidence that this one does because the many past uses of principle (3) (as enumerated in references below) result in derivation of nearly all known laws of physics and many of biology and other sciences.

### 2.2.3. *Ubiquitous Nature of Information*

Why does EPI (3) have such a wide scope of application? Basically, this is because its working concept is "information," and:

*Every observation of a well-defined natural event involves a flow of information, from a source to an output/observer. In general, the "source" is the thing that is (indirectly, through data) observed, and the "observer" can be a person or a mechanism that naturally utilizes the physical output.*

Note that the concept of "information" is even more general than that of "energy" since an information value can be proportional to an energy, or, alternatively, to the reciprocal of a mean-squared fitness value, or a position value, or depending upon the application of (3). Hence, EPI applies to every statistically well-defined process in nature. It has tremendous scope.

Thus, aside from its past extensive use in deriving laws of physics, e.g., in Frieden and Soffer (1995, 2009, 2010), Frieden (1998, 2004), and Frieden et al. (1999), EPI has been used to predict effects nominally considered to be exterior to physics, e.g., population genetics (Frieden et al., 2001), in situ cancer growth (Gatenby and Frieden, 2002, 2004), cancer growth suppression (Gatenby and Frieden, 2008), the allometric laws of biology (Frieden and Gatenby, 2005, 2007), intracell protein dynamics (Gatenby and Frieden, 2010), and (even) optimum investment programs (Frieden and Hawkins, 2010a).

*2.2.4. Property of Similarity*

The ubiquitous nature of information, and principle (3), implies a property of *similarity* in use of EPI: Suppose some phenomenon (a) obeys an output law $p(x)$ of (3). Then *any other, apparently unrelated, phenomenon* (b) *will obey the same output law provided the source information* $J$ *for* (b) *has the same mathematical form* as did (a). This property of similarity is a powerful one. As an example, it will be used below to derive the Lotka-Volterra equations (1) of population growth out of a parallel EPI application to a *physical* phenomenon of particle scattering.

## 2.3. DEFINITION OF INFORMATION $I$

The working concept in principle (3) is the Fisher information:

$$I = \int_a^b dx \frac{p'^2(x)}{p(x)}, \quad p'(x) \equiv \frac{dp}{dx}, \tag{4a}$$

or

$$I = 4\int_a^b dx q'^2(x), \quad q'(x) \equiv \frac{dq}{dx}, \quad q(x) \equiv +\sqrt{p(x)}. \tag{4b}$$

The latter expresses $I$ in terms of the "probability amplitude function" $q(x)$ of the system. Useful cases of $I$ are next given.

Both forms (4a) and (4b) show that data whose probability density function (PDF) $p(x)$ contains steep slope (gradient) values tend to have high information content. Since the total area under $p(x)$ is fixed at value 1 by normalization, the existence of steep slope values means that the curve of $p(x)$ is squashed inward toward its peak value. That is, random values $x$ have significantly high probabilities $p$ only over a small range, and determinism is being approached. This effectively small range implies a small variance $\sigma^2$, which then implies a high level of information $I$. In fact, consider the simplest case of a Gaussian $p(x)$, i.e., Gaussian with variance $\sigma^2$,

$$p(x) = \frac{1}{\sqrt{2\pi}\sigma} e^{-x^2/2\sigma^2}, \quad a = -\infty, \quad b = +\infty. \tag{5a}$$

Its use in (4a) gives

$$I = \frac{1}{\sigma^2}. \tag{5b}$$

The received information is the reciprocal of the variance. The smaller the variance is, the higher is the information. Remarkably, an entire family of PDFs has the same information value (5b). This is the exponential family, which includes the law (5a), the simple exponential form $\exp(-kx)$, $k$ = constant, the Poisson law, Rayleigh law, and many others. However, *not all* PDFs belong to this family. Hence, (5b) *is not* the general answer for $I$. That is reserved for the full calculations (4a) or (4b).

## 2.4. COMPUTING $J$

EPI principle (3) cannot be used unless the source information level $J$ is also known. Conceptually, the two informations differ markedly: $I$ is a property (4a) and (4b) of the PDF, i.e., of the system. By hypothesis, the physics governing the source is not known to the observer, so that the acquired information level $I$ mirrors only this property, through the PDF $p(x)$, as mirrored in the data. On the other hand, as we saw, information $J$ is directly a property of the *source phenomenon*. As one might expect, different phenomena have different information levels $J$. Thus, the choice of input $J$ in principle (3) is crucial to getting the correct answer $p(x)$ for the phenomenon. For example, the user has to be confident that the chosen $J$ describes the required scenario and not some other.

The following section goes into details on finding $J$.

## 2.5. LEVELS OF PRIOR KNOWLEDGE

Since $J$ is a property of the source phenomenon, to construct $J$ requires a level of prior knowledge about that phenomenon. Thus, rules for constructing $J$ have been devised (Frieden and Gatenby, 2007) that correspond to different levels of the quality of such prior knowledge.

It was found by the nineteenth-century American philosopher-astronomer Charles S. Peirce (1903) that all prior knowledge scenarios fall into three general categories. These identify uses of principle (3) with descending orders of accuracy:

(a) *Abductive* – giving 100% accurate outputs $p(x)$, where the phenomena are in fact quantum in nature
(b) *Deductive* – giving high, but not complete, accuracy and for generally classical or macroscopic (e.g., biological) phenomena
(c) *Inductive* – giving a smooth or unbiased output $p(x)$ that usually well approximates the truth

Abductive outputs (a), having perfect accuracy, follow from use of a $J \equiv I$ numerically (but not functionally). Here, $J$ is computed in a data space that is unitary to the source space, so that the two are entirely equivalent in representing the phenomenon. An example is where EPI is used to derive the Schrodinger wave equation of quantum mechanics. Here, the two spaces are Fourier transforms of one another, and $J$ is $I$ expressed as the mean-squared momentum of the particle.

In general, regardless of phenomenon, source information $J$ is both a Fisher information and a measurable characteristic of the phenomenon, such as the mean-squared-momentum (i.e., mean energy) in the example preceding. This is one basic reason why Fisher information pervades natural phenomena in general. The other is that all well-defined phenomena yield data whose information level $I$ may always be computed by Eqs. (4a) or (4b) preceding.

Deductive outputs (b), having imperfect but generally high accuracy, follow as *a consequence* of the correct $p(x)$. An example of deductive prior knowledge is knowledge that $p(x)$ obeys continuity of flow. These applications give rise to classical (nonquantum) effects.

Inductive outputs (c), the least accurate, follow as a consequence of *knowledge of some data* from $p(x)$. An example is where biological populations are known at a few times.

## 2.6. APPLICATIONS OF EPI

### *2.6.1. Generic Nature*

As was mentioned, principle (3) is a generic one. It applies wherever a phenomenon defines a flow of any manifestation of information (whether it be of particles, photons, protein molecules, etc.). It is often used for finding the relative occurrence $p_k(x,t), k = 1,\ldots,N$, of $N$ kinds of "particle" in a given medium. The "particles" are either inanimate, e.g., electrons, or organic as in a living system such as an ecology. Hence, the term "particle" is used in this generic sense. The medium is defined by its sources and sinks of "resource." The particles are generally competing with one another for maximal use of the resource. The unknown effect $p_k(x,t), k = 1,\ldots,N$ is assumed to be activated or driven by inputs of *system resource*. The term "resource" is used in a generic sense (as was "particle" in the above).

In inanimate systems, the resources of the medium are the energy inputs such as charges exerting $1/r$-type forces on the particles. By comparison, in living systems, the resources are energy or nutrients supplied by all possible sources, such as the environmental inputs into the system of sunshine, heat, and any other "particles" (potential prey) of the system.

*2.6.2. Qualitative Difference Between Informations I and J*

In general, the source information $J$ is some function of the resource. Thus, there are two specific roles for $I$ and $J$ – the former is a *generic* quantity always computed as (4a) and (4b), and the latter is a resource effect that is *specific to* the particular scenario under observation. It is *numerically* equal to one of the forms (4a) and (4b) but has a functional form that depends upon the particular resource.

## 2.7. DETERMINING POPULATION GROWTH USING EPI

The Lotka-Volterra equations (1) were discussed as models for determining population growth effects in an ecology. These are next derived out of purely *information* flow considerations, via principle (3).

The EPI principle has been shown to derive the Schrodinger wave equation (SWE) (Frieden, 1998, 2004). This becomes quite complicated in the case of many interacting particles. However, there is an approximation called the "Hartree self-consistent fields" approach that allows good approximate solutions to these scenarios. Accordingly, our starting point is these equations.

We have shown (Frieden and Gatenby, 2008) that, in fact, under certain conditions, the SWE for the system of *interacting particles* become identical to the Lotka-Volterra equations for, instead, *interacting (or competing) species.* Since this has already been shown, we merely summarize here the assumptions and result.

We have previously distinguished between living and nonliving resources. In the Hartree derivation, the resources are nonliving and are energies. These take the form of kinetic energies and a total energy potential acting on a $k$ th *particle* that is the sum of its potentials – nonliving resources:

$$V_k(r_k,t), v_k(r_k,t) \text{ and } v_{jk}(r_{jk},t). \tag{6a}$$

Position $r_k$ is that of particle $k$, and $r_{jk}$ is its distance to the $j$ th particle. Potential $V_k$ is the mean potential felt by the particle due to some fixed source such as a charge; $v_k$ is the mean potential felt by the particle due to all other particles, and $v_{jk}$ is the interaction potential between the particle $k$ and some other particle $j$.

EPI states that Fisher information has the same generic definitions and properties regardless of system. Hence, we appropriately seek to describe a living system that has parallel properties (of resource, in particular) to the above nonliving system. Here, the concept of "fitness" replaces the above resource concept of "potential." The reasoning is as follows: The fitness of a species (particle) is defined as the average number of offspring it will produce in its lifetime that will survive to themselves reproduce. Now, these offspring are potential prey, i.e., potential resource for some other species.

Thus, for the corresponding living system, the resources are *fitness functions* – living resources:

$$G_k(r_k,t), g_k(r_k,t) \text{ and } g_{jk}(r_{jk},t). \tag{6b}$$

Positions $r_k, r_{jk}$ are as above; $G_k$ is the average fitness of particle $k$ due to resources aside from those of the other particles; $g_k$ is the average fitness of the particle due to the presence of all other particles; and $g_{jk}$ is the fitness of particle $k$ due specifically to the presence of the $j$ th particle.

## 2.8. L-V GROWTH FOLLOWS FROM HARTREE EQUATIONS

Let the species growth rates $G_k(t), g_k(t), g_{jk}(t)$ of Eq. (6b) relate linearly to corresponding interaction *potentials* $V_k(t), v_k(t), v_{jk}(t)$ of Eq. (6a), as

$$\begin{aligned} &V_k(t) \equiv i\frac{\hbar}{2}G_k(t) - \beta, \quad v_k(t) \equiv i\frac{\hbar}{2}g_k(t) - \beta, \quad \text{and} \\ &v_{jk} \equiv i\frac{\hbar}{2}g_{jk}(t), \quad k = 1,\ldots,K, \quad i = \sqrt{-1}. \end{aligned} \tag{7}$$

Here, $\hbar$ is Planck's constant divided by $2\pi$ and $\beta =$ a known constant. Note the requirement that all potentials $V_k, v_k$ and $v_{jk}$ be generally complex functions of the time.

Such complex potentials define a *scattering* system, as was assumed.

## 2.9. SIGNIFICANCE OF THE CORRESPONDENCE

The correspondence (7) between biological fitness coefficients $G, g$ and interactive particle potentials $V, v$ is an important statement. It shows a commonality in the growth characteristics for living and dead (inanimate) systems. In principle, *any* first-order population *growth scenario* obeys the same mathematics regardless of whether the growing populations are *biological* or *inanimate* in nature.

## 2.10. A PHYSICAL SCATTERING SYSTEM THAT MIMICS A BIOLOGICAL ONE

We next seek a Hartree solution to a particular scattering problem. This is one that mimics a corresponding biological system. Under what circumstances can this take place? As was shown (Frieden and Gatenby, 2008), these conditions are:

(a) *Well-mixed medium*, where

$$V_k(r,t) = V_k(t), \quad v_k(r,t) = v_k(t), \quad v_{jk}(r_{jk},t) = v_{jk}(t). \tag{8}$$

The indicated time dependences are arbitrary.

(b) *Populations $p_k(t)$ varying in time.*

(c) *Some finite level of coherence* in the system.

(d) *Allowance for evolutionary tactics*. The constant $\beta$ and new potential functions $G_k, g_k, g_{jk}$ in Eq. (7) (the $g$'s standing for "growth") *can even represent evolutionary tactics* in the corresponding biological system (Vincent, 2004). Since these potentials do not depend upon position, and assuming none of them are of magnetic origin, the particles are not acted upon by forces.

(e) *Open system requirement*. The two systems must further correspond in being open to the environment and, thus, not energy conserving.

(f) *State of low coherence.* The Hartree approach requires the system to be in a state of low coherence. Studies by Salle et al. (2001) verify these trends, showing that the Hartree approximation holds well over relatively large time scales.

Under these conditions, competing species with given *fitness interaction* functions $G_k(t)$, $g_k(t)$, $g_{jk}(t)$ will obey the same population growth laws as do interacting particles with *potentials* $V_k(t), v_k(t), v_{jk}(t)$ that equal these fitness interaction functions. Mayer et al. (2007) give examples of biological predator–prey systems, defined by given fitness interaction functions, whose growth laws may also be achieved by interacting particles with corresponding potential functions.

## 2.11. BIOLOGICAL AND PHYSICAL EFFECTS WITH PARALLEL STRUCTURES

Nature is remarkable in its symmetrical description of disparate phenomena. In fact, it is through such symmetry that we introduced the EPI principle. Here, it is, specifically, symmetry in the *mathematics.* The same mathematics describe many phenomena.

For example, the EPI-based derivation of the Schrodinger wave equation (SWE) and other wave equations of quantum theory (Frieden, 1998, 2004) follow simply from the existence of a space that is unitary, or *symmetric*, with that of measurement space. Recall that the latter was the ordinary position space of the observed particle, the former its momentum-energy space. This correspondence is also a statement that the two spaces are equivalent in describing the phenomenon. There is equivalence under the symmetry.

Therefore, we can expect *other phenomena as well* to be described by the same mathematical answer, the wave equation. A past example from EPI is the recent derivation of a well-known program of optimum investment programming over time in economics (Hawkins and Frieden, 2004). Another is the finding that *classica*l thermodynamics obeys the mathematical form of the SWE (Plastino and Plastino, 2004).

More pertinent to this exposition is the above derivation of the L-V equations of biological growth, where the "quantum" aspect of interactive potentials is replaced 1:1 by corresponding biological fitness functions. Are there further such correspondences?

### *2.11.1. Parallel Quantities of Physics and Biology*

Solutions to the quantum SWE, as well as the Heisenberg uncertainty principle, depend upon the existence of irreducible fixed parameters, either of energy or of "action." Thus, electromagnetic radiation of frequency $\nu$ carries energy of amount $h\nu$, where $h$ is Planck's constant of action. Then by information-based symmetry, biology should likewise include such a discrete parameter.

From the standpoint of EPI, this occurs when analyzing the dynamics of biological system whose resource, of whatever nature (energy, or fitness, or ...), is a function $V(x) = (1/2)kx^2, k = \text{constant}$, with $x$ the system coordinate. In physics, this "resource" represents the potential function of a simple harmonic oscillator. Use of this resource in the SWE gives, as is well known, quantized resource states $E$ of the system:

$$E = (n + 1/2)h\nu, \quad n = 0, 1, 2, \ldots \tag{9}$$

In physics, this represents quantized levels of energy, while in biology it represents discrete levels of resource.

## 2.12. ARE THERE BIOLOGICAL "QUANTA"?

Equation (9) describes total system resource $E$, including inputs and their consumption, as occurring only as discrete multiples ($n$+1/2) of an irreducible resource "chunk" $h\nu$. Since this development is on the basis of information, and therefore independent of particular application, we cannot expect the smallest energy chunk size $h\nu$ to equal the specifically quantum mechanical one, $h = 6.63 \times 10^{-34}$ J s, which is exceedingly small. Rather, since the biological level is generally macroscopic, these chunks ought to be much larger than the atomistic level of quantum mechanics.

### 2.12.1. *Do ATP Units Amount to Biological "Quanta" of Energy Resource?*

For example, in a eukaryotic cell, where ATP is used by intracellular proteins to at least partially fuel their motion, the irreducible resource chunk would be the amount of energy that an intracellular protein absorbs via ATP at each encounter with a mitochondrion.

### 2.12.2. *Dependence of "Quantum" Unit on System Size*

But also, cell size may be something to consider – bacteria are much smaller than eukaryotic cells, suggesting that the most basic life processes in bacteria occur only over very small distances. These processes might, then, likewise depend upon a correspondingly smaller biological quantum of energy.

Long ago, Walker (1983) postulated such finite quantities, in fact, calling them $E$ and $h$, apparently thinking along similar lines as ours. In a follow-up paper (Walker, 1988), he denotes them as $R_i, i = 1,2,\ldots$ These represent the mean resource amount needed for survival during a small time interval $\Delta t$. The estimated $R_i$ for observed phytoplankton are reported as lying somewhere on the range 0.19–0.24 (no units are cited).

## 2.13. POSSIBILITY OF LIFE'S ORIGIN AS AN INORGANIC POLYMER

There has been conjecture that the first living system was a polymer of some kind. As discussed in (Orgel, 2000), such conjecture is evidence-backed. The role of the polymer is thought to have been filled by TNA, which consists of threose nucleic acids. The TNA growth presumably acted as a template for the first RNA growth. The motivation for this template effect is two major effects:

1. Pairs of complementary TNA polymers are known to form stable helices of the Watson-Crick double-helix form.
2. Metal ions, such as those of magnesium or zinc, are capable of interaction.

These corresponding effects give rise to the following possibility for genesis (Frieden, 2004, p. 388): The metal ions in the growing TNA template interact with corresponding ions in the RNA, forcing upon it the same double-helix growth pattern. The interaction could, e.g., obey a shielded Coulomb law. In this way, life initiates as a synchronicity between corresponding physical and a biological growth processes.

Next, evolution takes over: As the first instance of natural selection, those RNA molecules that do not bond in this way with a complementary threose chain do not grow, allowing those that do bond to dominate over time. This is also, of course, basic natural selection as predicted by the L-V equations. It requires the metal ions to act as genetic traits, with their number varying with some degree of randomness from one "generation" to the next. Whether this requirement has been verified is unknown to the authors.

### *2.13.1. Compliance with Hartree Approximation*

Assume that the constituent ions of the RNA (a) move relatively slowly (a so-called breathing molecule effect), obeying what is called adiabatic motion; (b) move relatively independently, obeying the so-called "Born-Oppenheimer" approximation (Born and Oppenheimer, 1927); and (c) the level of temporal coherence is not high. Then, the motion of the particles is well described by the quasi-classical Hartree mean field approximation (Schiff, 1955). Here, each ion is, by property (b), separately affected by the mean motion of all the others.

Regarding the polymer possibility, assume that their molecules have the preceding properties. Then by Sects. 2.8–2.10, their growth characteristics obey the Hartree approximation. This is under conditions discussed in Glotzer and Paul (2002). Also, in applications to metallic nanostructures, the approximation was found to hold well over small times ~1 ns, and less well out to possible macroscopic times ~50 ms = 0.05 s (Burki, 1999).

Even the largest time scale mentioned in the preceding is small (0.05 s) This defines a brief event. Such an event must then have had a relatively high temporal concentration of energy so as to provide the energy required by the initial Hartree process. This describes an impulse-type event, e.g., due to *an explosion*.

## 2.14. POSSIBLE CELL FORMATION AS MICELLES

*Note*: A micelle is an aggregate of molecules dispersed in a liquid that lower its surface tension. A typical micelle in solution tends to aggregate molecules via hydrophilic and hydrophobic regions, the latter in the micelle center. Spontaneous ordering, as in crystal formation, can result from releases of energy and entropy. For example, "the ability of colloidal particles to self-organize suggests that colloidal particles could be used as precursors for advanced materials via the generation of complex microstructures" (Sturtevant and Corti, 2008). Of course, a cell is a complex microstructure. As discussed next, the micelle center defines an effective intracellular space.

Micelles will spontaneously form from carbon chains of sufficient length. These are ordered structures of lower energy. So, if we assume that the primitive oceans contained a high concentration of small carbon sources (2- or 3-carbon chains), they might under some conditions form longer carbon chains, with the latter catalyzing the addition of smaller carbons to themselves or others. This would result in micelle formation which, in turn, defines an *intracellular space*. If small carbons are used in that space to form carbon chains, then a gradient will be formed that promotes flow of the carbon precursors into the cell. But the carbon chains are too large to pass through the micelle, so that they accumulate. This would cause the cell to increase in size, were it not for the mechanism of Sects. 3.8 and 3.9, whereby many small cells are preferred on the basis of order. Hence a splitting into two cells becomes the favorable choice. Perhaps this represents the prototype living system.

## 2.15. QUORUM SENSING IN LIVING SYSTEMS OF CELLS

We have previously emphasized conditions giving rise to life within a single cell. Now, we consider a living *system* consisting of many cells. As a single cell required local ordering within it, a living system requires ordering to occur over many neighboring cells. As is discussed, this is permitted by a process called *quorum sensing*.

An example of environmental ordering is found in quorum sensing (Miller and Bassler, 2001), a widely used tactic of bacteria. In quorum sensing, each bacterium excretes a continuous quantity of some metabolite. Regarding the bacterium as the message source in a communication system, the message conveyed by the metabolite is simply "I am here." This message is received by other bacteria of the same species, which also excrete the metabolite. This initiates a kind of "information chain reaction," where each bacterium is both a receiver and a new source of information. This process also permits the bacteria to change their behavior in response to the information in cues that are placed in the environment by their neighbors.

### *2.15.1. Possible Application of EPI Principle*

This process describes an information channel and, hence, fits well into the EPI framework as previously described. With some work, it could well be quantitatively described by it.

When the concentration of the metabolite in one or more bacteria exceeds some threshold, they simultaneously change their behavior, resulting in dispersal or some other function that allows them to survive. Alternatively, when there are too many bacteria, they use up all of the substrate and die off. This demarks the limit of the living system. Notice that although this living system must, by the 2nd law of thermodynamics, increase the *total* entropy of its environment, it also sends out messages that partially offset this. These contribute increases in local order or, conversely, decreases in entropy.

### *2.15.2. An Extrapolation to General Cell Systems*

*We now propose that these properties qualitatively hold as well for any growing system of cells.* Thus, each such cell changes its behavior in response to cues from information its neighbors have placed in the environment. In turn, it relays the information to its neighbors, etc. This process allows us to further address the overall question of how living organisms initially grew (and grow).

A living cell is of finite size and, as it forms and grows, represents a local increase of order (decrease of entropy). As we saw at the beginning of Sect. 1, a local increase of order is possible because the cell still exports sufficient disorder to its surroundings to satisfy the 2nd law.

This trade-off permits life to exist, but does not describe how it grows. The process of quorum sensing attempts to accomplish this. Regarding a growing organism as an information relay system, it postulates that the growth of a multicellular organism is ultimately dictated by the type and amount of information that is present in a single cell.

Also, *each cell acts effectively as a two-edged sword: It exports order as well as disorder to its surroundings.*

That is, it ships order to outlying molecules such that these become ordered, morphing into new cells that become part of the growing organic system. Also, to satisfy the 2nd law, it likewise ships some disorder outward, since incorporation of the metabolite into the new cell cannot be thermodynamically perfect. Some energy goes into heat and entropy, which is wasted to the outside of the growing mass. Also, some cells die, further wasting entropy.

Of course, the process ends when there are so many cells present that they exhaust all available resources (the substrate provided by the environment). This demarks the temporal limiting state of the living system in two ways, i.e., as an equilibrium mass state and an equilibrium information state. By the latter, the organism, which has heretofore been increasing its local level of order or information content, now ceases such growth. Here, the total information received by each cell from all others and from the substrate that have enabled it to grow is, finally, balanced by the total level of disorder it receives from these.

In this way, *a multicellular organism that has stopped growing not only maintains a constant mass but also maintains a constant level of total information content.*

## 2.16. WERE THE FIRST CELLS FORMED DURING ACQUISITION OF OUR MOON?

The fossil record seems to say that life did not exist on Earth until about 3.8 billion years ago. But this is also about when the Earth purportedly collided with a planetoid orbiting somewhere between the Earth and Mars (Hartmann et al., 1986). This, then, could be the violent event required in Sect. 2.13 (see end). Hence, we propose that life could have originated at the explosive formation of the moon. Most of the dust resulting from the collision either fell back to Earth or, in time, accreted into the sphere that became our moon.

### *2.16.1. Possibility of Life on the Moon*

This suggests a substantial likelihood for life to also exist on the moon. Hence, the credibility of the moon-acquisition thesis would be furthered if such life were found there. If it does, this seems most likely to be in its craters, especially near the permanently shadowed poles, where abundant water remains; the latter, presumably from comets that struck the moon billions of years ago (O'Hanlon, 2010).

Also, it was recently discovered (preceding reference) that water uniformly comprises about 5% of the sand on the moon. Such water was presumably formed by reaction between incoming hydrogen from the sun and oxygen in the sand.

### *2.16.2. Extremophiles*

Although conditions for life on the moon would be harsh for the commonest forms of life on Earth, some existing forms of life are "extremophiles" (Cavicchioli and Thomas, 2000; Seckbach, 2006). These can survive in very warm, very cold, and/or very dry environments, even at ocean depths large enough to receive no sunlight. Their presence suggests that functioning life could exist on the moon. Or, in the absence of functioning life, the fossil remains of such life might still exist.

There is a possibility that life propagated as well to Mars or another relatively nearby planet. However, we regard these to be of much lower probability than life on the moon. This is principally because we now know, from a purposeful satellite crash into its north pole (NASA, 2009), that the moon has the necessary water.

### *2.16.3. Hartree Model Requirements*

We propose that the first life form resulted from either actual, physical Hartree scatter, as described in Sects. 2.9 and 2.10, or from its biological analog, as in Sects. 2.11 and 2.12. For either of these to be true, the requisite energy requirements must be verified, possibly from known data on releases of energy from asteroid-Earth collisions (Chapman and Morrison, 1994).

Other requirements of the Hartree analysis would also have also to be met:

(a) The physical existence of a well-mixed medium such that the probability amplitudes for all particles are spatially constant. The violence of the Earth-planetoid collision and its generated heat seem likely to have caused such mixing.
(b) Hartree two-particle potential functions that have no dependence upon spatial position. Given the effect (a), this is a likely effect. Instead, the potentials are to depend upon the relative particle abundances, which is likewise consistent with (a). The potentials are also to vary with time in the same way as the corresponding biological fitness coefficients. To verify the model, such potentials would have to be found.

### *2.16.4. Apparent Sparsity of Occurrence of Extraterrestrial Life*

However, of course, life has not yet been observed on any astronomical object exterior to the Earth. This is even despite its ability to survive in extremophile conditions (discussed above), and despite our best attempts to sense life, including lunar and martian landers, manned exploration of the moon, and the cosmological "listening" devices of project SETI (Search for Extraterrestrial Intelligence). This suggests that the necessary conditions for life are quite stringent, leading to its occurrence as a very rare event. For example, one analysis concludes there to be about one life form per galaxy (Gordon and Hoover, 2007).

## 3. Vital Role of Order in Living Cells

### 3.1. STABILITY AND RELATIVE STATE OF ORDER/DISORDER

Living systems appear to be relatively stable over long time intervals. The concepts of *order* and change in order, developed next, are essential to maintaining this stability. However, oddly enough, whereas order seems to be an ultimate tool of living systems, they appear to maintain it by monitoring their existing level of *disorder*.

For example, an interesting effect is the remarkable *variation* in *intra*cellular (within-cell) protein ion concentrations when compared to the *extra*cellular space and an equally remarkable *expenditure of energy* by the cell to *maintain this difference*. Thus, if order is the ultimate goal of living systems, this arrangement seems designed to measure *its absence* – the level of disorder. To wit, any perturbation to the cell or its environment will result in *alterations* of its transmembrane gradients.

These changes in transmembrane potentials appear to provide the cell with information that guides its key, global activities (i.e., to move, divide, die, etc.). However, what seems interesting is that even in the presence of an ordered state, the cell allows a range of states of its order to exist. For example, a change in the order of the state can occur from activities *within* it (e.g., messenger proteins carrying DNA information to the nucleus). This requires both energy and information (Gatenby and Frieden, 2010). Or it can occur due to *outside* activities such as sudden change in the environment. In this case, the change in order provides information to the system that allows it to generate a response. This in fact could be a necessary thermodynamic feedback loop that gives living systems their remarkable stability.

### 3.2. INTUITIVE PROPERTIES OF A UNIVERSAL MEASURE OF ORDER

According to the 2nd law of thermodynamics, the level of disorder in the universe is monotonically increasing with time. Yet despite this, we see all around, and within, us systems that are certainly ordered. In fact, a closer (local) look reveals that they are ordered, but on a local level only. In effect, to satisfy the 2nd law, their increased order is at the expense of shipping out extra disorder elsewhere into the universe. Living systems are a paramount example of this tactic. Then, intuitively, it is locally increased order, as distinguished from locally increased information, that best characterizes life. Therefore, we seek a principle of maximum local order. What form should it take? But first, what exactly is "order"?

If one attempts to define "order" as the absence of disorder, it is found to be unsatisfactory. Suppose, as is usual, the level of disorder is measured by the Boltzmann entropy $H$ of the system. Then, how should the "absence" of disorder be expressed – as $-H$ or $1/H$ or even $\exp(-H)$? There is no unique answer by this route, which is simply a cop out, avoiding the issue of what *the direct* measure is. Hence, the question:

Given a system, such as an organism or an organ thereof, whose probability law is $p(x), i = 1, \ldots, N$ and extends over a total region $a \le x \le b$, what is a suitable measure of its level of "order"?

Intuitively, the order describes the degree to which the system exhibits high-frequency details and low random fluctuation. Also, the order should increase with the extension $(b-a)$ of the system. For example, consider a typical apartment house, where $x$ represents the floor number. Each floor has about the same level of order, so two floors should intuitively contain a lot more structure and order than one floor. In fact, according to Eq. 5a below, they contain 4 times (not merely twice) the value of any one. An apt biological example follows below.

## 3.3. THE ORDER MEASURE

The system is regarded as representing *any phenomenon* that degrades in time due to any level of coarse graining (full or partial). Coarse graining may be mathematically expressed as any linear projection of the system that misses or blurs local structure of the $p(x)$. An example is in forming computed tomography (CT) data. In particular, we postulate that whatever form is to measure the order $R$, it should monotonically decrease.

Postulate:

$$\Delta R \le 0 \; for \, \Delta t > 0, \tag{10a}$$

when the system is "coarse grained." Qualitatively, this is whenever it suffers a degradation such as due to viewing on a coarser subdivision. A corollary of postulate (10a) is that an "arrow of time" exists,

$$\Delta R \le 0 \text{ for } t \to t + \Delta t, \Delta t \ge 0 \tag{10b}$$

at each coarse graining. These show decreased order or, conversely, *increased disorder*, with time. This defines what is called the "arrow of time." An "entropy," by definition, indicates the direction of the arrow of time by its monotonic change with increased time. Thus, monotonic property (10b) indicates that the order $R$ (and the Fisher $I$, since $I \propto R$ as seen at (11a and 11b) below) may be regarded as entropies. These are properties of the system, as distinguished from properties of the data. (Fisher informations are usually defined as properties of data (Fisher, 1922, 1959).)

## 3.4. CASES OF PARTIAL COARSE GRAINING

Certain *small system* processes, such as colloidal-particle velocities, orientational states on a quasicrystal, etc., entail an increasingly fine spatiotemporal description of a state of the system (Lemus and García-Colín, 2006). A result is *to actually increase, with time, the number of degrees of freedom* (*d.o.f*) *of the system* (in effect, looking at things with more detail). The introduction into the system of these extra d.o.f tends to increase the level of order. These degrees of freedom probably relate to the extended variables in extended irreversible thermodynamics (Mori, 1965). Thus, although overall the systems suffer decreased order, the reduction in order is diminished by the extra d.o.f. The graining is now only *partial.*

Such *partial* coarse graining also occurs in (1) analyzing biological whole-genome gene expression, or (2) in description of the mesoscopic nonequilibrium thermodynamics (MNET) of small systems. MNET is as an extension of the equilibrium thermodynamics of small systems. Interesting discussions of partial coarse graining may be found in Jou et al. (1985).

In general, partial coarse graining occurs when full-fledged coarse graining operation is constrained by imposed conditions on its permissible perturbations in the system probabilities. An example of such a constraint is a fixed mean-squared value of the perturbations.

### 3.4.1. *Effect of Constraints on Perturbations*

Obviously, the presence of such a constraint must restrain the size of the system perturbations, so that the loss of order in the resulting system is diminished from the loss in the absence of the constraint. In fact, the result below obeys this property.

The order measure $R$ that obeys properties (10a) and (10b), plus constraints, has been found (Frieden and Hawkins, 2010b) to be

$$R = 8^{-1}(b-a)^2\left[I - \sum_{k=1}^{K} I(F_k)\right], \tag{11a}$$

with information measures

$$I(F_k) = \int_a^b dx F_k(x)\frac{p'^2(x)}{p(x)} \text{ and } I \equiv I(1). \tag{11b}$$

Measure (11a) is, as well, unique in the no constraint case $K = 0$. Quantity $I(1)$ is the Fisher information (4a) and (4b). The new measures $I(F_k)$ are "weighted" Fisher informations with specified weight functions $F_k(x)$. The latter were invented in Frieden and Soffer (2010) and define-optimized (minimum *weighted* mean-squared error) parameter estimates, where the weights are $F_k(x)^{-1}$. The minimized weighted error is $I(F_k)^{-1}$, showing the commonsense property of small error for large information. Any state of coarse graining, from constrained to unconstrained ($K = 0$), is allowed in (11a). The unconstrained case, i.e., pure coarse graining, gives a level of order

$$R_0 = 8^{-1}(b-a)^2 I. \tag{11c}$$

From the form of (11a), this defines the maximum possible level of order.

Regarding the *loss* of order after the coarse graining, it follows the form (11a),

$$\Delta R = 8^{-1}\Delta x^2 \left[ I - \sum_{k=1}^{K} I(F_k) \right], \quad \Delta x \ll (b-a). \tag{11d}$$

Since, as informations, all $I(F_k) \geq 0$, and their sum cannot attain value $I$, the loss $\Delta R$ in order is smaller than were there no constraints, $K = 0$. This verifies the intuitive property mentioned in the paragraph preceding Eq. (11a).

*3.4.2. Entropic Properties of Informations and Order*

By (10b) and (11a) and (11b), the measures $R, I(F_k), I_0$ monotonically contract in time. This property defines them as *entropies*, complementing their usual status (Fisher, 1922; Frieden, 2001) as informations. Also, for unitless choices $F_k$, they are unitless, so that the $R$ values for different properties (e.g., protein position, energy, or momentum) may be objectively compared.

Thus, Fisher information $I$, which provides the basis for the EPI principle (3), also provides the quantitative basis for defining order in nature. If the system is purely degraded by coarse graining, the order is just $I$, given by (4a) or (4b). But if the degrading process is *partial* coarse graining, both $I$ and the weighted informations $I(F_k)$ enter in.

All this said, there is one significant difference between information and order. Order is a property of a system whereas information a property of a message from the system. In short, information is order that is transmitted. It is the manifestation of order in observable form.

*Order $R_V$ in a Radially Symmetrical Volume*

The preceding order expressions (11a) to (11c) are for a one-dimensional system $p(x), i = 1, \ldots, N$. The level of order $R_V$ in a three-dimensional volume, described by probabilities $p(x, y, z), i, j, k = 1, \ldots, N$, has yet to be determined. However, for purposes of application to cells, which tend to be roughly spherical, we can approximately find the order in a spherically symmetrical system $p(r)$, as convenient notation, let $0 \leq r \leq r_0$.

The approximation lies in modeling this system as a one-dimensional string that is wound upon itself until it fills all space out to the radius $r_0$. (This in fact approximates the multiple long chain molecules that contribute significant structure to a cell.) Then, the order is approximately the total linear order $R_0$ along the string as given by Eq. (11c). In that expression, $(b-a)$ becomes the total length of the volume-filling string. Let the string have a cross-sectional area $\alpha$. Then, of course, the system volume $\tau$ obeys

$$\tau = (b-a)\alpha = (4/3)\pi r_0^{\,3}, \text{ so that}$$
$$R_V \equiv (b-a)^2 I/8 = \left[(4/3\alpha)\pi r_0^{\,3}\right]^2 I/8 = K_\alpha I r_0^{\,6}, \tag{11e}$$

with the constant $K_\alpha$ defined by the form of (11e). Quantity $R_V$ is the required level of order within the volume. The third equality is by (11c). The order is still linear in the information $I$ but now goes as the sixth power of the volume radius $r_0$, a very steep dependence on size.

## 3.5. HOW DO CRYSTALS AND CELLS DIFFER AS AGENTS FOR LIFE?

A crystal-based origin of life was proposed (Cairns-Smith, 1985) and popularized; see, e.g., Thaxton and Meyer (1998). However, a basic problem with the proposal is that of the storage of genetic information. A definite mechanism has not been found in crystals that would enable them to store the high density of DNA information that a cell does in the form of its A, C, T, and G codons.

Also, a critical difference between crystals and cells lies in the degree of their *spatial heterogeneity* of order – a cell typically exhibits certain finite regions of high order and others of low order. For example, the DNA both in the center of the cell and at the cell membrane at its periphery are both highly ordered structures, but the cytoplasm in between is, by comparison, quite disordered.

Crystals, on the other hand, just add components to the outside edge of their growing structure. The effect is that the entropy per unit volume is more or less constant. By comparison, cells typically bring their building blocks inside their structure, forming molecules with lower entropy that remain constant within boundaries. These boundaries define the alternate regions of high and low order mentioned above.

Perhaps, in this regard, the biggest problem living systems face is the possibility of *too much* order. If some process inside the cell takes small carbon molecules (methane and ethane) from the environment and links them to form long carbon chains that, consequently, cannot leave the membrane, *the cell will rapidly become a static collection of large carbon chains much like a crystal.*

This would be a highly ordered, stable state, which is fine. However, it is also a thermodynamic dead end because it has no ability to reproduce and, thereby, evolve and adapt. Thus, maintaining a relatively disordered cytoplasm or dividing into two cells when order is sufficiently increased may both be necessary to keep the system from becoming a highly ordered, but *adynamic*, state like a crystal. In Sects. 3.8 and 3.9, it is found that such cell division also is preferable on the grounds that it represents a lower state of order, which is beneficial in requiring a reduced level of nutritive inputs.

As discussed in Sect. 3.4, subjecting the cell to partial, or constrained, granularity also rules out this objectionable trend toward a very high, but dead-end, degree of order (also as discussed next).

## 3.6. ACHIEVING REGIONS WITH SPECIFIED LEVELS OF LOCAL ORDER

We now address the preceding question of how the cell can attain high order at its membrane and nucleus while maintaining lower order in between. This can be achieved by use of the effects of partial coarse graining discussed in Sect. 3.4. For example, consider a simple system consisting of a messenger protein moving from cell membrane to cytoplasm to nucleus. As we saw at Eqs. (11a) and (11b), the use of selective constraints on (say) the position variable $x$ of the protein within any given zone of the cell can result in a level of order $R$ there that is as small (or, within bounds, as large) as is required. Therefore, this can be accomplished within each such cell region by means of selective choice of constraints within each. This is the proposed general route to a solution. Unfortunately, to proceed further requires problem specifics and development that are beyond the space limitations of this chapter.

## 3.7. UNIVERSALITY OF MEASURE OF ORDER $R$

We noted that the form for $R$ is unitless, so that the $R$ for different phenomena may be objectively compared. This is an advantageous property, shared as well with the usual Boltzmann entropy measure of information. A single "bit" of entropy represents the same amount of Shannon information, whether in the outcome of flipping a fair coin or the sex of a newborn child. Likewise, *one unit* of order represents the same amount of order, whether in a single cell or an entire elephant (or an inanimate airplane, for that matter).

## 3.8. ORDER IN A MANY-CELLED ORGAN

Many fish, octopus, etc., continually grow in size throughout their lives. Why, then, do their cells *not* likewise grow? The proceeding may provide the basic reason behind this effect.

*Let us see whether nature prefers to produce an organ consisting of a great many tiny cells, or of a smaller number of large cells where the latter are simply scaled up versions of the tiny cells.*

We first evaluate the amount of structural order in a three-dimensional, multiply-celled organ of given size. Then, so as to make obvious the essential difference between them, we purposely compare two extreme cases of cell size, as follows:

(a) *A great many tiny cells.* Here, the organ contains $N$ identical, spherical cells, each of radius $r_0$. The value of $N$ should be quite large, so by the additivity property of the postulate, the resulting $R$ should much exceed the value for one cell. In fact, we found at Eq. (11e) that each spherical cell has an order of level $R_1 = K_\alpha r_0^6 I$. Then, by the additivity property of the information $I$, the total

$$R_N = NR_1 = NK_\alpha r_0^6 I. \tag{12a}$$

As a check, since $N$ is large, the total order in the organ exceeds that in a single cell, as expected.

(b) *A single, huge cell.* The organ has the same volume $Nr_0^3$ as in (i), but now consists of *one* giant cell. Therefore, it is of radius $\rho_0 = N^{1/3} r_0$. Then, by (11e), here the total order is

$$R = K_\alpha I \rho_0^6 = N^2 (K_\alpha I r_0^6) = NR_N. \tag{12b}$$

Comparing this with (12a) shows that, to maintain a fixed level of information $I$, an organ containing one large cell must maintain $N$ times the level of order as the same size organ consisting of $N$ small cells. Does this mean that the $N$ small cells are somehow preferred by nature?

## 3.9. NATURE PREFERS A MULTITUDE OF SMALL CELLS

Some background on the relation between the order $R$ and energy $E$ is first required. The ground state of a system represents that for which the $p(x)$ describing it is maximally spread out or blurred. Then, the Fisher $I$ is a minimum. Also, in being so blurred, it has a minimum of structure, or order $R$, by Eq. (12a). Conversely, a system with high order and structure is in an excited state. Also, excited states are high-energy states. For example, as we saw above, an oscillator in a state $n$ has an energy level $E \propto (n + 1/2)$. Thus, the higher the state, the higher is the order and the higher is the energy.

This implies that it requires less energy, or resource, $E$ for a system to maintain a low-order state than to maintain a high-order state. Therefore, a system (a) of many tiny cells – the less ordered system – requires less energy to maintain its level of order than system (b) of one huge cell. So here, we have a reason why nature prefers many tiny cells over but a few very large ones. The reason is required energy expenditure.

In summary, when an organism grows, it grows in cell *number* rather than in the size of each cell. A growing number of cells requires less energy for maintaining its relatively low total level of order than the corresponding higher-order system with a small number of large cells.

## 3.10. ESSENTIAL NEED FOR CELL REPRODUCTION

Implicit in the foregoing argument is the presence of many cells. This, of course, requires cells to have the ability to reproduce. Thus, life requires not only the growth of each component of a cell but, also, the growth of the number of cells.

Thus, it requires the cells to split off into identical copies. Thus, cell mitosis satisfies information and order requirements for sustaining life.

## 3.11. HOW DOES A CELL ACHIEVE LOCAL INCREASE OF ORDER?

Perhaps the most fundamental property of life is its ability to achieve a local increase in order. According to the 2nd law, order must decrease when viewed globally. Therefore, in order to achieve a local increase, it must be at the expense of a decrease in the order elsewhere, i.e., outside the cell. How does it achieve this binary state of order levels?

By the preceding, system (a) of many cells requires much less energy to achieve its local increase of order than does system (b), since system (b) has such a higher level of order to maintain. Hence, system (a) is much more likely to achieve the required polarization of order levels. Again, it is the process of mitosis, producing many small cell copies, that allows the process to occur.

## 3.12. A PARADOX RESOLVED

This also resolves an apparent paradox. Does a fetus or a grown person provide more cellular information? On the one hand, it must be the grown person, with his much larger number of cells. On the other, the grown person only contains extra copies of the same cells that are in the fetus, so how could these provide extra information? The answer is resolved by distinguishing between the concepts of order and information, as in the preceding subsections. Each cell tends to provide the same level of information, so the same amount of information per unit volume (or cell) exists even as the person grows. However, the total number of cells tends to increase and then saturate, with age, so that the total order follows these trends as well. This is, at least, prior to old age when there is an onset of net cell loss and, therefore, net information loss.

## 3.13. DO LIVING SYSTEMS HAVE AN ULTIMATE GOAL?

Is there a goal to life? An actual, and positive, answer is provided by considering what special significance the EPI principle has to living systems. The answer is not new (as discussed), but the EPI principle now provides powerful evidence for it.

Although no observation is perfect, the EPI principle (3) states that it *provides as much information about the unknown value of the parameter as is available*. Thus, a creature who tries to ascertain the value of some parameter by collecting data from it suffers only minimum loss $I-J$ of information in the data. The result is a relatively accurate estimate of the unknown parameter, in comparison with the accuracy that would accrue were the loss of information not so minimized. See the following example.

This accuracy effect gives aid, and evolutionary advantage, to creatures that depend for their survival on knowing the parameter value. This is perhaps best understood were the opposite true. Suppose the observation contained low information. This means high random noise and therefore inaccurate and unreliable estimates of the parameter whose value is sought. For example, if the prey of a creature were at angular position $\theta$, he would be "seen" instead at (say) angle $\theta + 2.1\theta$. The next time, the estimate might be $\theta - 0.7\theta$, etc. Not only would the prey not be eaten, but worse, no systematic strategy would be suggested by the failures. The creature, as an individual and as a species, would go extinct.

Thus, in providing maximum information, *nature gives maximum advantage to creatures that can formulate strategies of survival* (Vincent, 2004). In fact, the greater the information present, the evermore complex and effective strategies can be devised for increasing the chance of survival.

Dolphins or elephants may, or may not, have their own "aims," but mankind has growing realization of a definite and specific one. We know that once earthly problems of survival (wars, disease, overpopulation, incoming asteroids, etc.) are resolved, there remains one great impediment to survival: the 2nd law of thermodynamics. And perhaps in alliance with it, there is the *dark energy* force that seems to portend the utter stretching apart of all particles of matter toward an End of absolute maximum separation and a state of maximum disorder. This effect would mark the triumph of statistical ignorance and chaos over order.

Worse, such a state is irreversible, since the particles would be so far apart that gravitational attraction could not bring them together again. Therefore, for the sake of his own survival, it is clear that, with all other problems of existence solved, the *ultimate* purpose of man's existence is to solve this remaining problem.

This may be the case for man, but is this life's ultimate purpose? Also, man may not ultimately prove up to the task. Then, future living creatures, perhaps mutations of humans, will try as well. Why should they?

The EPI principle (3) appears to hold universally. The gist of it is that the universe operates out of the principle that maximum information is to be mutually transferred during all interactions of its component systems. (Human observation is but one such interaction.) Such local optimality of information transfer is despite the opposite actions of the 2nd law. This, however, is limited to acting globally, thereby allowing crystals, life, and other locally ordered systems to exist. Evidently, the concept of information is, then, central to all nature. On the other hand, IF the 2nd law of thermodynamics and *dark energy* force ultimately prevail, even so limited an effect as but *local* optimal information transfer would not survive. But, does the universe "want" it to?

Consider the "cosmological principle," which in its most general form states that the presence cosmological scenario in space-time is statistically "typical" of all others. On this basis, the "Big Bang" has happened before and will continue to occur periodically, with intervening "Big Crunches." That is, cosmology is *cyclical,* over a very long time period. (The principle is probably believed by a consensus of hard scientists.)

However, unless these problems of the 2nd law and dark energy are solved, the universe will evolve toward one entirely consisting of nearly infinitely spaced particles. Such a universe is not "typical" of its history and would thus violate the cosmological principle. It also would rule out local optimal information transfer, as discussed above. Therefore, assuming that the cosmological principle is valid, this violation of it is impossible. Hence, *man is simply nature's current "attempt" to ultimately preserve the cosmological principle.* Assuming the principle to be true, and assuming that some form of life will prove up to the task, it will be preserved.

Philosophically, this idea seems included within F. J. Dyson's metaphysics (Dyson, 2004). Also, this use of the cosmological principle has a parallel in the idea of C. J. Peirce (Peirce, 1903; Buchler, 2001) that *habit taking* (i.e., having a homogeneous or habitual appearance) is the supreme universal principle. We thank the philosopher Prof. Vinicius Romanini for these references.

## 4. Dedication

The authors wish to acknowledge our indebtedness to the late Prof. Thomas L. Vincent, a pioneer in the game theory of evolution. To know Tom was to be inspired by his trademark insights and quick wit. Our Friday lunches with him were pleasantly anticipated all week. He was a perpetual fount of ideas – always in the spirit of fun and often leading to good, competitive adventure. This spirit of adventure extended to numerous hikes to the bottom of the Grand Canyon, trips in his sailboat in the Gulf of Lower California, and many spirited games of handball (with BRF) and racquetball (with RAG) over the years. We were privileged to know him, as an esteemed colleague and a sympathetic, reliable friend.

## 5. Bibliography

Allman D, Srivastava B, Lindsley RC (2004) Alternative routes to maturity: branch points and pathways for generating follicular and marginal zone B cells. Immunol Rev 197:147–160

Bernstein H, Byerly H, Hopf F, Michod R, Vemulapali K (1983) The Darwinian dynamic. Q Rev Biol 58:185–207

Born M, Oppenheimer JR (1927) On the quantum theory of molecules. Ann Phys 84: 457–484; in German, see its English translations on the web

Buchler J (2001) The philosophy of Peirce. Routledge, London

Burki J (1999) Transport and cohesive properties of metallic nanocontacts: a free-electron model. Ph.D. dissertation, University of Fribourg, Fribourg, Switzerland

Cairns-Smith AG (1985) Seven clues to the origin of life – a scientific detective story. Cambridge University Press, UK

Cavicchioli R, Thomas T (2000) Extremophiles. In: Lederberg J (ed) Encyclopedia of microbiology, vol 2, 2nd edn. Academic, San Diego, pp 317–337

Chapman CR, Morrison D (1994) Impacts on the Earth by asteroids and comets: assessing the hazard. Nature 367:33–40

Darwin C (1859) On the origin of species by means of natural selection, or the preservation of favoured races in the struggle for life, 1st edn. John Murray, London

Davydov AS (1982) Biology and quantum mechanics. Pergamon, Oxford

Dyson FJ (2004) Infinite in all directions (paperback). Harper Collins, New York

Fisher RA (1922) On the mathematical foundations of theoretical. Philos Trans R Soc Lond A 222:309–368

Fisher RA (1959) Statistical methods and scientific inference, 2nd edn. Oliver and Boyd, London

Frieden BR (1998) Physics from Fisher information: a unification. Cambridge University Press, Cambridge

Frieden BR (2001) Probability, statistical optics, and data testing. Springer, New York

Frieden BR (2004) Science from Fisher information: a unification. Cambridge University Press, Cambridge

Frieden BR, Gatenby RA (2005) Power laws of complex systems from Extreme physical information. Phys Rev E 72:036101

Frieden BR, Gatenby RA (eds) (2007) Exploratory data analysis using Fisher information. Springer, London

Frieden BR, Gatenby RA (2008) Conditions for correspondence between Hartree scattering and biological growth. Phys Rev E 78:041902

Frieden BR, Hawkins RJ (2010a) Asymmetric information and economics. Physica A 389:287–295

Frieden BR, Hawkins RJ (2010b) Quantifying system order for full and partial coarse graining. Phys Rev E 82:066117

Frieden BR, Soffer BH (1995) Lagrangians of physics and the game of Fisher-information transfer. Phys Rev E 52:2274–2286

Frieden BR, Soffer BH (2009) De Broglie's wave hypothesis from Fisher information. Physica A 388:1315–1330

Frieden BR, Soffer BH (2010) Weighted Fisher informations, their derivation and use. Phys Lett A 374:3895–3898

Frieden BR, Plastino A, Plastino AR, Soffer BH (1999) Fisher-based thermodynamics: its Legendre transform and concavity properties. Phys Rev E 60:48–53

Frieden BR, Plastino A, Soffer BH (2001) Population genetics from an information perspective. J Theor Biol 208:49–64

Gatenby RA, Frieden BR (2002) Application of information theory and extreme physical information to carcinogenesis. Cancer Res 62:3675–3684

Gatenby RA, Frieden BR (2004) Information dynamics in carcinogenesis and tumor growth. Mutat Res 568:259–273

Gatenby RA, Frieden BR (2008) Inducing catastrophe in malignant growth. J Math Med Biol 25:267–283

Gatenby RA, Frieden BR (2010) Coulomb interactions between cytoplasmic electric fields and phosphorylated messenger proteins optimize information flow in cells. PLoS One 5(8):e12084. doi:10.1371/journal.pone

Glotzer SC, Paul W (2002) Molecular and mesoscale simulation of polymers. Annu Rev Mater Res 32:401–436

Gordon R, Hoover RB (2007) Could there have been a single origin of life in a big bang universe? Proc SPIE 6694. doi:10.1117/1112.737041

Hartmann WK, Phillips RJ, Taylor GJ (eds) (1986) Origin of the moon. Lunar and planet. Institute, Houston, Texas

Hawkins RJ, Frieden BR (2004) Fisher information and equilibrium distributions in econophysics. Phys Lett A 322:126–130

Jou D, Pérez-Garcia C, Garcia-Colin LS, Lopez de Haro M, Rodriguez RF (1985) Generalized hydrodynamics and extended irreversible thermodynamics. Phys Rev A 31:2502–2508

Lemus HE, García-Colín LS (2006) Non-equilibrium critical behavior: An extended irreversible thermodynamics approach. Journal of Non-equilibrium Thermodynamics 31(4):397–417

Lodish H, Berk A, Zipursky LS et al (2004) Molecular cell biology, 4th edn. Scientific American Books, New York

Lotka AJ (1932) Elements of physical biology. Williams and Wilkins, Baltimore

Mayer AL, Pawlowski CW, Fath BD, Cabezas H (2007) Applications of Fisher information to the management of sustainable environmental systems, Chap 7. In: Frieden BR, Gatenby RA (eds) Exploratory data analysis using Fisher information. Springer, London, pp 217–244

Maynard-Smith J (1989) Evolutionary genetics. Oxford University Press, Oxford

Miller MB, Bassler BL (2001) Quorum sensing in bacteria. Annu Rev Microbiol 55:165–199

Mori H (1965) Transport, collective motion, and Brownian motion. Prog Theor Phys 33:423–450

NASA (2009) At website www.nasa.gov/mission_pages/LCROSS/main/LCROSS_impact.html. Reported at the Lunar Exploration Analysis Group (LEAG) meeting, Houston, TX, 15–19 Nov 2009

O'Hanlon L (2010) New moon finding holds clues to earth's water. Discover magazine, May issue

Orgel L (2000) Origin of life: a simpler nucleic acid. Science 290:1306–1307

Peirce CS (1903) Harvard lectures on pragmatism. Collected papers, Harvard, MA

Plastino A, Plastino AR (2004) Information and thermal physics. In: Frieden BR, Gatenby RA (eds) Exploratory data analysis. Springer, London

Salle M, Smit J, Vink JC (2001) Finite temperature simulations from quantum field dynamics? Nucl Phys B94:431–434

Schiff LI (1955) Quantum mechanics. McGraw-Hill, New York

Schroedinger E (1944) What is life? Cambridge University Press, Cambridge

Seckbach J (2006) Life as we know it. Springer, Dordrecht

Shen K, Teruel MN, Connor JH, Shenolikar S, Meyer T (2000) Molecular memory by reversible translocation of calcium/calmodulin dependent protein kinase II. Nat Neurosci 3:881–886

Sturtevant BD, Corti DS (2008) Computational study of colloidal dynamics in entropic force fields. In Abstract book, the 82nd ACS colloid and surface science symposium, Raleigh, NC

Thaxton C, Meyer SC (1998) Theoretical clay feet. Access Research Network website

Vincent TL (2004) Evolutionary game theory, natural selection and Darwinian dynamics. Cambridge University Press, Cambridge

Volterra V (1926) Fluctuations in the abundance of a species considered mathematically. Nature 118: 558–560

Walker I (1983) The physical dimensions and biological meaning of the coefficients in the Volterra competition equations and their consequences for their possible coexistence. Acta Biotheor 33:93–122

Walker I (1988) Comments on the Manna model. Int Revueges Hydrobiol 73:229–234

Winnewisser G, Walmsley CM (1979) Carbon chain molecules in interstellar clouds. Astrophys Space Sci 65:83–93

Wu YJ, Cheng BM (2010) Synthesis of long carbon chains upon VUV Photolysis of methane in solid neon. www.nsrrc.org.tw/NsrrcWebSystem. YEARLY 2009–2010: 12–14

Biodata of **Thomas Jakschitz**, **Daniel Fitz**, and **Bernd Michael Rode**, authors of "***The Origin of First Peptides on Earth: From Amino Acids to Homochiral Biomolecules***."

**Dipl.-Ing. Dr. Thomas Jakschitz** is a Ph.D. graduate of the Institute of General, Inorganic and Theoretical Chemistry, University of Innsbruck, Austria. He obtained his M.Eng. degree in 2005 from the University of Applied Sciences, Innsbruck, Austria, where he studied process and environment engineering. His research centers on the chemical evolution of peptides toward the origin of life on Earth.

E-mail: **Thomas.Jakschitz@uibk.ac.at**

**Dr. Daniel Fitz** is currently assistant professor at the Institute of General, Inorganic and Theoretical Chemistry at the University of Innsbruck, Austria. He obtained his Ph.D. in 2009 at this institute in the working group of Prof. Rode in the field prebiotic chemistry and the origin of life and continued his research in the same field until January 2011. From February 2011, he is being employed as an analytical chemist at Bionorica Research Company Innsbruck, which produces pharmaceutical products from herbal sources.

E-mail: **Daniel.Fitz@uibk.ac.at**

**Thomas Jakschitz**

**Daniel Fitz**



J. Seckbach (ed.), *Genesis - In The Beginning: Precursors of Life, Chemical Models and Early Biological Evolution*, Cellular Origin, Life in Extreme Habitats and Astrobiology 22, 469–489
DOI 10.1007/978-94-007-2941-4_25, © Springer Science+Business Media Dordrecht 2012

**Professor Dr. Bernd Michael Rode** is the head of the Institute for General, Inorganic and Theoretical Chemistry of the Leopold-Franzens University of Innsbruck, Austria. He obtained his Ph.D. in 1973 from this university and spent postdoctoral stays in Germany, Sweden, and Japan and was guest professor at several European and Asian universities and received four honorary doctorates. His research centers on the theoretical treatment of liquid systems by quantum mechanical simulations on the one hand, and on chemical evolution of biomolecules and related model experiments on the other.

E-mail: **bernd.m.rode@uibk.ac** web: **www.theochem.at**

# THE ORIGIN OF FIRST PEPTIDES ON EARTH: FROM AMINO ACIDS TO HOMOCHIRAL BIOMOLECULES

**THOMAS JAKSCHITZ, DANIEL FITZ, AND BERND MICHAEL RODE**
*Division of Theoretical Chemistry, Institute of General, Inorganic and Theoretical Chemistry, University of Innsbruck, Innrain 52a, A-6020 Innsbruck, Austria*

## 1. Primordial Earth Scenario

Around 4.6 billion years ago, the Earth was formed by accretion of material surrounding the sun. The primarily very hot and molten Earth cooled down to temperatures around the boiling point of water, and water condensed for the first time in the form of droplets in the atmosphere. The presence of liquid water seems to be the most likely condition for a chemical evolution to have started and, therefore, the most basic precondition that life could develop on Earth. Enormous thunderstorms formed hot primordial oceans, lakes and lagoons, which were, however, probably not enduring until the end of the heavy meteoritic bombardment phase around 3.9 billion years ago. In addition to liquid water and a number of different salts, especially sodium chloride, dissolved in it, also a variety of organic molecules was needed for chemical evolution. The base for their formation were atmospheric gases and various types of energy inputs like UV irradiation or thermal energies like volcanic heat and lightnings. The atmospheric gases were mainly supplied by volcanic outgassing. Furthermore, small amounts of oxygen were formed by decomposition of carbon dioxide and water by UV radiation and lightnings (Levine et al., 1982).

## 2. RNA/DNA World vs. Peptide World

Within the past decades, numerous experiments concerning the origin of life have been performed, and lots of theories were discussed. Up to now, there is no secure answer to the question how life emerged on Earth and what the very first initial step was.

Modern life as we know it is a highly complex chemical network of several classes of biomolecules interacting in many ways and depending on each other. The first living systems were for sure much more simple, and the origin of life probably happened with only one single class of biomolecules.

Therefore, we have to consider first of all the most crucial constituents of today's living organisms, i.e. RNA, DNA and proteins. RNA/DNA as the carriers of the genetic information seem to be the most essential class of molecules in modern life forms, what would suggest an RNA/DNA world as a very first step of chemical evolution towards life as a very promising hypothesis (Joyce, 2002; Orgel, 2004). The conditions when life arose on the primitive Earth around 3.8 billion years ago, however, appear rather unfavorable for the formation and the stability of polynucleotides. Geochemists suggest a hot environment with temperatures slightly below the boiling point of water after the first oceans had condensed, and also, some other physical factors like high UV irradiation in the absence of a protecting ozone layer and a high salinity of the hydrosphere appear hostile to most modern life forms and especially to unprotected RNA/DNA. Experiments have shown that RNA is rather decomposed under such conditions within a short time span (Larralde et al., 1995; Lazcano and Miller, 1996; Levy and Miller, 1998; Shapiro, 1995; Cleaves and Miller, 1998).

Proteins, on the other hand, mainly act as catalysts for all kinds of biochemical reactions (besides the important role of structure proteins) and are also deeply involved in the replication process of DNA, which incorporates the genetic plan for the formation of the proteins. It is a classical hen-and-egg problem, no proteins without DNA and no DNA without proteins. Nevertheless, it seems unlikely that life started with both of these molecules and their complex interactions with each other. The question is just if there was a ‘genes first’ scenario with RNA/DNA or a ‘metabolism first’ scenario with peptides/proteins.

From a chemical point of view, amino acids and their polymers would fit much better into the rough primordial environment. They are less sensitive to heat, high salinity and UV radiation than RNA/DNA and their formation, and polymerization seems chemically easier than to connect a nucleobase, ribose and phosphate ion in the correct regiospecific way to form even just one nucleotide. Furthermore, phosphate would rather precipitate in the presence of divalent metal ions like $Ca^{2+}$, and thus be unavailable for RNA synthesis.

In addition to chemical stability, there are still other arguments lending a strong preference to the protein world as initial step. One of these arguments is based on the development of so-called hyper-cycles as proposed by Eigen and Schuster (Eigen and Schuster, 1977). To reproduce an RNA sequence, it is necessary to reduce error probability to such an extent that this can be achieved only by the means of enzyme-controlled mechanisms. For any abiotic RNA formation, the error tolerance possibly achieved is only $10^{-2}$, which would limit the number of constituents of the RNA to 100 elements, what is definitely not enough for any higher-level biochemical process. Further, computer simulations of hyper-cycles and quasi-species lead to a stable system only with a very limited population size and within a very narrow bandwidth, limiting the possibility of a higher organization proceeding through these ‘hyper-cycles’ (Niesert et al., 1981). Another very striking argument was proposed by Dyson (1999) for which it is assumed that in a given system, higher-organized units are formed from discrete monomer chemical species, thus showing a primitive form of metabolism and evolution to more complex species. As can be seen in the formula describing this model, there is an optimal number of chemical species in order to proceed to higher-organized systems with a given preferential binding, which can be compared to specificity of catalysts, of the different monomer species to each other. The inorganic catalysts which should have been the only present on the primitive Earth usually reach a specificity factor between 60 and 100, whereas modern enzymes reach factors up to 5,000. Such a model works well with a number of species involved between 2,000 and 20,000, which guarantees a sufficient complexity. To achieve continuous development to higher forms, the number of basic species should ideally be between 8 and 10, with a smaller number of species like the four basic units making up modern polynucleotides such as RNA and DNA, and this development cannot be achieved with the specific factor of inorganic catalysts. On the other hand, by model conditions of 8–10 basic species, the error tolerance of the system increases to substantial 30% (Dyson, 1999). If this argument is combined with the chemical stability of biopolymers, this leads to the conclusion that four nucleotides cannot have been the first step in the evolution of life, in particular because chemical diversity would have been too low to allow a development to higher organization. Thus, the only alternative to consider the first steps of chemical evolution towards the origin of life is to come back to the models proposed by Oparin (1957), Cairns-Smith (1985) and Lancet and co-workers (Segre and Lancet, 1999), which are sometimes referred to as the 'genes-last model': Starting with ideally ten amino acids, with 2,000–20,000 units for the formation of higher-organized systems, small droplets could be formed as cell precursors. This process would also imply that proteins developed before the genes and that the initial catalysts used might be clay minerals or other inorganic substances like metal-ion complexes with amino acids and/or peptides. The remaining question is whether amino acids and peptides could have provided the necessary conditions which are required in the definition of life processes. Metabolism does not provide any difficulty in this aspect, whereas the questions of carrying the information of self-replication and of autocatalytic processes have to be answered. These processes are well known to occur with RNA and DNA by means of complicated protein-based enzymes, but it had to be shown that it is also possible with peptides alone. In the 1990s, it was demonstrated that there are a number of them able to reproduce themselves from smaller constituents (Isaac and Chmielewski, 2002; Lee et al., 1996; Yao et al., 1998), thus replicating their structure and properties. On the other hand, it was found that amino acids themselves can catalyze peptide-formation processes, which can be seen as a good example of autocatalysis in the production of oligomers (Suwannachot and Rode, 1999).

The question of RNA world or protein world does not seem any more a hen-and-egg problem but a clear answer in favor of a 'protein-first-world', starting from smaller peptides and forming in the end some cell-like organisms with first simple characteristics of life.

## 3. Amino Acid Synthesis

In the early 1950s, Stanley Miller set the starting point for experimental investigations about the origin of life with his famous discharge experiment (Miller, 1953, 1955). More than 40 years earlier, however, very similar experiments had been performed and published as 'The Effect of Silent Discharge on the Reactions of Formamide' by Löb (1906, 1913), but in another context. Miller assumed a reducing primordial atmosphere composed of hydrogen, ammonia, methane and water vapor that he exposed to electric discharges for the simulation of lightning strikes. Within a few days, he detected a tar-like substance at the inner wall of the glass bottle. Its analysis showed a large variety of organic compounds, among others detectable amounts of several amino acids.

In the following years, geochemists and geologists were able, based on geochemical data of up to 3.8 billion years old rocks and even older zirconium crystals, to demonstrate that the primary, reducing atmosphere containing mainly hydrogen escaped rather fast into open space because of the relatively weak gravitation field of the Earth and that ammonia and methane would quickly be decomposed under the influence of high UV irradiation (Sagan and Chyba, 1997; Kasting, 1997). A secondary atmosphere was formed by volcanic outgassing of the Earth's mantle. These gases were very similar to today's volcanic gases consisting of nitrogen, carbon dioxide, water vapor and small amounts of sulfur dioxide and hydrogen sulfide. This neutral atmosphere is extremely different from the one Miller used in his experiments and in subsequent experiments using such neutral gas mixtures in that no formation of amino acids could be detected unless at least minor amounts of a reducing compound (i.e. carbon monoxide) were added. In 2004, however, Plankensteiner et al. demonstrated with advanced analytical methods the possibility of formation of several amino acids in a completely neutral atmosphere of only water vapor, nitrogen and carbon dioxide by electric discharges within just a few weeks (Plankensteiner et al., 2004a, 2006). A short overview of the most important Miller-type experiments and resulting amino acids is presented in Table 1.

Considering the whole primordial atmosphere as a large reaction chamber and the frequent lightnings (and also other energy inputs, like UV irradiation, shock waves, etc.) that would have occurred at the elevated temperatures on the primordial Earth, one can expect, therefore, the accumulation of substantial amounts of amino acids in the hydrosphere over time, even if the conditions were not optimal for the formation of amino acids, due to the absence of reducing atmosphere components.

One more possible provider for amino acids are hydrothermal vents which can be found at the ocean floor in a few thousand meters depth. Under high temperature and high pressure conditions, amino acid precursor molecules as well as amino acids may be formed (Wächtershäuser, 1990, 2000; Cody et al., 2000; Huber and Wächtershäuser, 1997).

**Table 1.** Some examples of Miller-type experiments simulating assumed primordial Earth scenarios.

| Authors | Reactants | Energy source | Results reported |
|---|---|---|---|
| Miller[a] | $CH_4$, $NH_3$, $H_2O$, $H_2$ | Electric discharges | Simple amino acids, organic compounds |
| Abelson[b] | CO, $CO_2$, $N_2$, $NH_3$, $H_2$, $H_2O$ | Electric discharges | Simple amino acids, HCN |
| Bar-Nun et al.[c] | $CH_4$, $NH_3$, $H_2O$ | Shock wave | Simple amino acids |
| Harada and Fox[d] | $CH_4$, $NH_3$, $H_2O$ | Thermal energy | 14 proteinogenic amino acids |
| Lawless and Boynton[e] | $CH_4$, $NH_3$, $H_2O$ | Thermal energy | Glycine, alanine, aspartic acid, β-alanine, N-methyl-β-alanine, β-aminobutanoic acid |
| Groth and Weyssenhoff[f] | $CH_4$, $NH_3$, $H_2O$ | Ultraviolet light | Simple amino acids (low yields) |
| Sagan and Khare[g] | $CH_4$, $C_2H_6$, $NH_3$, $H_2O$, $H_2S$ | Ultraviolet light | Simple amino acids (low yields) |
| Yoshino et al.[h] | $H_2$, CO, $NH_3$, montmorillonite | Thermal energy | Glycine, alanine, glutamic acid, aspartic acid, histidine, lysine, arginine |
| Palm and Calvin[i] | $H_2$, $CH_4$, $NH_3$, $H_2O$ | Election irradiation | Glycine, alanine, aspartic acid |
| Miyakawa et al.[j] | CO, $N_2$, $H_2$, $H_2O$ | Plasma | Glycine, alanine, aspartic acid |
| Plankensteiner et al.[k] | $CO_2$, $N_2$, $H_2O$ | Electric discharges | Glycine, alanine, valine, serine, proline, lysine, histidine |

[a]Miller (1953).
[b]Abelson (1965).
[c]Bar-Nun et al. (1970).
[d]Harada and Fox (1964).
[e]Lawless and Boynton (1973).
[f]Groth and Weyssenhoff (1957).
[g]Sagan and Khare (1971).
[h]Yoshino et al. (1971).
[I]Palm and Calvin (1962).
[j]Miyakawa et al. (1999).
[k]Plankensteiner et al. (2004a).

A third proposed amino acid source are meteorites, especially of carbonaceous chondrite type. Several of these meteorites, which were mainly formed at the same time as our solar system, have been investigated so far, in particular the Murchison meteorite, which fell down in Australia in 1969. This meteorite contains a large number of proteinogenic and non-proteinogenic amino acids, as well as several classes of other organic substances (Kvenvolden et al., 1970; Krishnamurthy et al., 1992; Pizzarello et al., 2004). Suggesting that the amino acid concentration of carbonaceous chondrites is 60 ppm and that 3% of meteorites are from this type, $6 \times 10^{11}$ tons of amino acids could have reached the Earth since it has sufficiently cooled down to enable chemical evolution (Thomas et al., 1998). Summarizing, the availability of amino acids on the primordial Earth can be assumed with high probability and in sufficient amount.

## 4. Peptide Formation

After the formation of amino acids, their polymerization by condensation reactions is an essential step in chemical evolution towards peptides and proteins. This endothermic reaction's equilibrium is far on the side of free amino acids, and in an aqueous solution like the 'primordial soup', the eventuality of a spontaneous polymerization is extremely low. To overcome this problem, a number of hypotheses and possible scenarios have been proposed within the last decades. Among others, peptide-formation reactions can be easily achieved with the help of condensation reagents like cyanates, imidazole, cyanamides, trimetaphosphates, different kinds of polyphosphates, UTP and ATP (Rabinowitz and Hampai, 1985; Weber et al., 1977; Sawai and Orgel, 1975; Sawai et al., 1975; Rabinowitz et al., 1969; Yamanaka et al., 1988), but it is highly implausible that these reactants were available on the primordial Earth in the required concentrations. Another possibility for amino acid polymerization are melting processes, where amino acid mixtures with a high fraction of acidic or basic amino acids are heated up to 180°C to remove the water molecules released during the condensation reaction (Fox and Harada, 1958, 1960; Harada and Fox, 1958). However, the formed polymers have no or only very scarce peptide bonds. Most of the monomers are connected via ester bonds between the amino acids' side chains, and hence, the products called 'proteinoids' are very different from peptides and proteins (Andini et al., 1975).

A further possibility of forming peptides was shown under conditions possibly occurring in hydrothermal vent environments in the presence of aqueous slurries of nickel and iron sulfide, carbon monoxide and dihydrogen sulfide or methyl mercaptane (Huber and Wächtershäuser, 1998; Huber et al., 2003), but the required high pressure of carbon monoxide poses another problem, as well as the restriction of this mechanism to specific locations.

Heterogeneous catalysis by clay minerals like bentonite, kaolinite, montmorillonite or alumina and silica can also form peptides from amino acids, but reactions are restricted to a few amino acids, and the peptide yields are rather low (Bujdak and Rode, 2001; Flores and Bonner, 1974; Bujdak et al., 1995). Nevertheless, clay minerals may have played a significant role in primordial evolution. Their surface provides favorable conditions for protecting adsorbed peptides against hydrolysis and for connecting and thus elongating small peptides. Therefore, clay minerals are very interesting candidates for further evolution steps towards peptides and proteins as they were abundantly present in the primordial Earth environment.

### 4.1. THE SALT-INDUCED PEPTIDE FORMATION REACTION

The salt-induced peptide-formation (SIPF) reaction was discovered in 1988 (Schwendinger and Rode, 1989; Rode and Schwendinger, 1990) on the basis of theoretical investigations of different salt solutions and of metal complexes.

Based on Monte Carlo computer simulations, it could be demonstrated that at sodium chloride concentrations higher than 3M, the sodium ion's first hydration shell is no longer saturated with six water molecules and thus such concentrated salt solutions can act as a dehydration reagent to overcome the thermodynamic barrier of peptide formation also in aqueous solution. For lowering the kinetic barrier, divalent metal ions were assumed to be potential catalysts by forming complexes with amino acids, thereby bringing two reaction partners into close vicinity and preventing the reactive amino group from unfavorable protonation. Tests with different metal cations finally showed that only Cu(II) is suitable to catalyze peptide formation in sufficient yields (Schwendinger and Rode, 1989; Rode and Schwendinger, 1990). Both the ubiquitously present sodium chloride as well as bivalent copper ions were available on the primordial Earth, as indicated by large deposits of copper-containing minerals in so-called greenstone belts (Huston et al., 2000).

Effective sodium chloride concentrations for the dehydrating effect (>3M) can be easily obtained by evaporation of primordial sea water (today's oceans have a sodium chloride concentration of about 0.5M) in lagoons and coastal regions during tide cycles, as well as desiccation of saline lakes due to the hot primordial environment. By rain or high tides, the accumulated concentrated solutions or dry residues would get redissolved, and the whole cycle could start again. Such a scenario can be represented by evaporation cycle experiments, in which dilute solutions of amino acids, sodium chloride and copper chloride are evaporated at 85°C in a heating chamber within 24 h, and the solid residue filled up again with pure water for the next cycle (Saetia et al., 1993). The ideal temperature range of the SIPF reaction of 60–90°C fits perfectly into the hot primordial Earth scenario, concluded from geochemical silicon isotope data (Robert and Chaussidon, 2006).

A further important point is the low dependence of the SIPF reaction on atmospheric gas composition. Only a marginal amount of free oxygen is required to keep the copper ions in the divalent state (Ochiai, 1978). The presence of further inorganic ions like potassium, ammonium or others does not inhibit the reaction but, on the contrary, supports the dehydrating effect of sodium chloride (Eder and Rode, 1994).

These properties and the fact that this reaction works with all amino acids investigated so far let the SIPF reaction appears the most universal and most plausible scenario for the very first steps of chemical peptide formation on the primordial Earth.

### 4.2. REACTION MECHANISM AND PROPERTIES OF THE SIPF REACTION

The crucial species enabling peptide formation via the SIPF reaction is a complex consisting of a central divalent copper ion, two amino acids, a chloride ligand and two axial water molecules (Liedl and Rode, 1992). One of the amino acids is bound

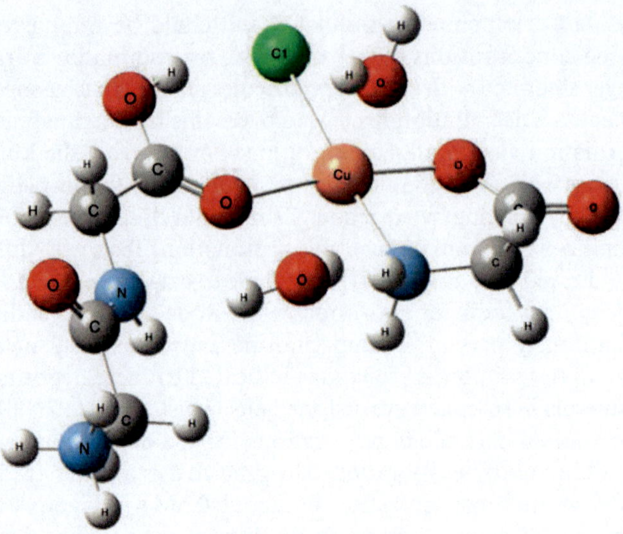


**Figure 1.** Geometry optimized SIPF complex with one chelating glycine ligand, one end-on coordinated diglycine, one chloride ligand and two axial water molecules.

to the copper ion in a chelating way via its carboxy and amino group. The other amino acid or peptide is bound end-on via the carboxy group. The chloride ligand plays an important role by providing suitable electronic and steric preconditions for the peptide-formation process and by preventing the second amino acid from also chelating the copper center, which would result in a too stable and, therefore, unreactive complex. The two axial water molecules complete the first coordination sphere of the Cu(II) ion but stay at elongated distances because of the Jahn-Teller distortion effect. Figure 1 shows a geometry optimized SIPF complex with one glycine and one diglycine ligand. The optimization was carried out at Hartree Fock level with a double zeta basis set for Cu(II), triple zeta basis sets for other atoms and polarizable continuum model for solvation effects.

To form the peptide bond, the coordinated unprotonated amino group binds to the electrophilic carbon atom of the carboxylic group of the end-on bound amino acid. The formed peptide is weaker coordinated to the copper ion than amino acids and preferably released into solution and subsequently replaced by new amino acids to form a new SIPF complex. The SIPF reaction also works with end-on coordinated peptides to provide higher peptides, and only the chelating species has to be an amino acid and will become the carboxylic end of the elongated peptide. Oligopeptides up to at least hexaglycine have been detected after only a few reaction days (Rode and Schwendinger, 1990).

Some specific properties of the SIPF reaction point at a connection to modern life forms. First, it prefers the naturally occurring $\alpha$-amino acids over their $\beta$- and $\gamma$-analogs (Schwendinger et al., 1995), which can be explained by the more favorable five-membered ring of an α-amino acid chelated to the copper ion compared to the less stable six- and seven-membered rings formed by β- and γ-amino acids, respectively.

As mentioned before, the SIPF reaction forms peptides from all amino acids investigated so far, but with varying yields. When binary mixtures of the amino acids A and B are used, one obtains different amounts of A-B and B-A peptides, depending on the activity of the reaction centers influenced by the amino acids' side chains on the one hand, but also on the varying coordination abilities to the Cu(II) ion and the different properties in the highly saline solution. Therefore, specific amino acid sequences are formed with strongly different yields. Consequently, the composition and sequence of smaller and larger polymers formed by this reaction are by no means a statistical mixture of all possible combinations of amino acids but rather follow a certain chemical framework set by the reaction conditions and reaction partners (Rode, 1999; Rode et al., 1997). If the preferably formed A-B and B-A linkages of the SIPF reaction are compared to the most frequently occurring amino acid pairings in ribosomal proteins of archaea and further prokaryota, which are among the oldest and most primitive organisms still being found on Earth, an astonishing coincidence, beyond any statistical probability could be detected (Rode et al., 1997; Rode, 1999). The basic idea of this comparison was that replication mechanisms evolving at a later stage of evolution would have mainly reproduced already existing peptide sequences, thus maintaining them as a kind of 'fingerprint' of the SIPF reaction in existing life forms.

The simplest amino acid glycine, which was certainly also the most abundantly available one on the primordial Earth, not only forms at least hexamers within a few days in the SIPF reaction (Rode and Schwendinger, 1990) but also acts as a catalyst boosting the formation of peptides from most of the other amino acids (Suwannachot and Rode, 1998; Plankensteiner et al., 2002, 2005a, c; Li et al., 2008, 2010), some of which would provide only poor yields when present alone, like valine and leucine. This catalytic effect takes place via the intermediate formation of longer-mixed peptides including glycine and a subsequent hydrolysis, which is one of the side reactions under the relatively harsh conditions of the SIPF reaction and reduces peptide yields in some cases after longer reaction times. Among the other amino acids investigated, only histidine shows a similar catalytic effect (Reiner et al., 2006; Fitz et al., 2008; Li et al., 2008, 2010) and even outperforms glycine in some cases. The highest catalytic factor has been found to result if the catalytic amino acid is present at 1/8 of the other amino acids' concentration.

The simultaneous presence of clays-like montmorillonite and hectorite, and also silica or alumina in the SIPF reaction results in a trend to longer peptides (Son et al., 1998; Rode et al., 1999), as these minerals have been shown to protect adsorbed peptides against hydrolysis and even activate them to induce further chain elongation on their surface. As clay minerals were abundantly available on the primitive Earth, such a scenario indicates a possible next step in chemical evolution towards longer and more complex peptides and proteins.

## 5. Origin of Biohomochirality

Only the L-forms of amino acids, apart of very rare exceptions, act as building blocks of proteins, and in the backbone of RNA/DNA, only D-sugars are incorporated. The proper functioning and reactivity of biomolecules strongly depends on the enantiomeric purity of the building blocks, which can be explained by the structure of proteins. If only one amino acid is replaced by its optical counterpart, destabilization effects are induced (Krause et al., 2000), the structure of α-helices or β-sheets gets distorted, and the protein will not fulfill its tasks properly. Thus, homochiral systems are in general more stable (which is a good argument especially under the rough conditions on the primordial Earth) and mandatory for a wide variety of biochemical reactions. However, the question why L-amino acids and D-sugars and not their mirror images are used in life as we know is still awaiting an answer, although various theories have been and still are discussed concerning the L-preference. Some of them will be outlined shortly as this phenomenon is crucial for a 'protein world'.

### 5.1. THE CHANCE HYPOTHESIS

The simplest but also most unsatisfying explanation is that this form of homochirality arose by pure chance. A certain biomolecule would have achieved the crucial step leading to autocatalytic effects or to other outstandingly advantageous properties so that only this biomolecule and its succession has promoted further biological evolution. If this specific biomolecule by chance consisted of L-amino acids, this could have led to an early decision in favor of this one of the two chemically equivalent forms.

Other possible scenarios to realize the chance hypothesis are provided by various spontaneous resolution mechanisms, like chiral autocatalysis in crystallization processes. Around 5–10% of all chiral molecules are known to crystallize as a mixture of enantiopure crystals, so-called conglomerates. It has been suggested (Welch, 2001) that highly enantioenriched microenvironments could be reached whenever racemates crystallize as conglomerates.

### 5.2. LOCAL CHIRAL INFLUENCES

Another proposed hypothesis for the preference of L-amino acids is an extraterrestrial origin of this phenomenon. In certain locations in the universe, for example, the Orion cloud, circularly polarized infrared and ultraviolet light occurs (Bailey et al., 1998), for example, when light is scattered by non-spherical grains aligned in magnetic fields of high-mass star forming regions. By means of circularly polarized light (CPL), enantioselectivity can be induced in synthetic or degradative chemical reactions. Bonner and co-workers showed that by CPL photolysis of initially racemic leucine an enantiomeric excess of up to 2.5% (Flores et al., 1977) could be obtained. Which stereoisomer is degraded faster depends on the sign of circular polarization. As numerous smaller organic molecules and possibly also amino acids can be found in interstellar clouds, such polarized light could certainly lead to a preference of one of the optical isomers and thereby to a 'chiral success' over the other (Inoue, 1992; Flores et al., 1977). Huge amounts of extraterrestrial material were delivered to the primordial Earth via comets and meteorites (Thomas et al., 1998), and it can be imagined that a part of these compounds was exposed to circularly polarized light during its voyage. Astonishingly, enantiomeric excesses of up to 10% in favor of the L-form for some α-methylated amino acids (Cronin and Pizzarello, 1997) and a D/L ratio of 0.5 for alanine and 0.3 for glutamic acid (Engel and Macko, 1997) have been observed in the Murchison meteorite. If this enantiomeric excess was generated by CPL and the biohomochirality on Earth based on this extraterrestrial enantioenrichment, there is a high possibility that in other regions of the universe where circularly polarized light of opposite sign is predominant, biomolecules or even life forms with opposite chirality are prevailing.

Also on Earth, local chiral influences can be found. The abundant mineral quartz ($SiO_2$) and several further minerals exist in a left- and a right-handed form and enable enantioselective adsorption and reactions on their chiral surfaces. Due to the equal occurrence of both enantiomeric quartz forms on Earth (Frondel, 1978), a possible influence of chiral minerals on the origin of biohomochirality is restricted to local environments where one chiral form might have dominated.

## 5.3. PREDETERMINATION BY FORCES OF NATURE

In 1956, the parity violation in the weak nuclear force was theoretically predicted by Lee and Yang (1956), and in 1957 experimentally proven by showing that electrons ejected together with antineutrinos from an unstable $^{60}Co$ nucleus in the process of β-decay are predominantly left-handed (Wu et al., 1957). The weak force is, in contrast to the other three forces of nature, not invariant against inversion, causing some peculiar effects on chiral molecules: The stabilities, the ground state energies and the reactivities of the two enantiomeric forms are not exactly the same. These findings overthrew the former idea that the forces of nature could not 'distinguish between left and right'. Later, the weak nuclear and the electromagnetic forces could be unified in the electroweak theory (Rubbia, 1985; Weinberg, 1980; Salam, 1980), showing that all atoms are inherently chiral due to parity violation in the weak nuclear force and the inherent chirality of the electrons. Numerous theoretical calculations have been performed on the parity-violating energy differences (PVEDs) between the enantiomers of amino acids and other chiral molecules (Tranter, 1985; Mason and Tranter, 1983b; Laerdahl et al., 2000), predicting that these difference would be extremely small for organic molecules, but the calculations also showed that the inherent chirality is approximately proportional to $Z^5$ or $Z^6$, Z being the atomic number of the most heavy element of the molecule (Mason and Tranter, 1984; Laerdahl and Schwerdtfeger, 1999). For heavier atoms like cesium, however, a slight optical activity could even be experimentally measured (Wood et al., 1997; Guena et al., 2003). The overall optical activity is assembled of weak nuclear and electromagnetic contributions, so the values for the two enantiomers are opposite in sign but have slightly different magnitudes. Theoretical approaches proposed that L-amino acids are stabilized compared to their D-enantiomers by PVED values of $10^{-20}$–$10^{-17}$ hartrees (Tranter, 1985; Mason and Tranter, 1983a). More recent calculations have shown strongly conformation depending values (Berger and Quack, 2000; Laerdahl et al., 2000), but nevertheless, the L-form is mostly the better stabilized one. The shown energy differences would lead to an enantiomeric excess of only $10^{-16}$–$10^{-13}$%, and hence, an appropriate mechanism providing a major amplification of this effect would be needed to lead to a substantial stereoselectivity.

The Vester-Ulbricht hypothesis postulates that longitudinally polarized electrons emitted in the process of β-decay subsequently generate circularly polarized Bremsstrahlung photons which would induce asymmetric reactions in racemic mixtures due to different synthetic or degradative interactions with the two enantiomers (Vester et al., 1959; Ulbricht and Vester, 1962; Ulbricht, 1959). Up to now, all related experiments showed no or just a minimal optical activity within the experimental error range because the high-energetic β-radiation itself acts strongly razemizing.

In 1991, Salam postulated a possible phase transition which is similar to Bose-Einstein condensation below a critical, rather low temperature, to the stabilized enantiomeric L-form (Salam, 1991; Figureau et al., 1995; Wang et al., 2000). No experimental evidence for a Salam phase transition from one enantiomer to another is available to date, although the data of Wang (Wang et al., 2000) can at least prove that phase transitions within enantiomers display parity-violating properties. The question remains if these phase transitions could have had any direct influence on chemical evolution on Earth as the transition temperature of 250 K seems quite low for a favorable evolution scenario.

The salt-induced peptide-formation (SIPF) reaction provides another possible explanation for stereoselectivity leading to L-proteins, which will be presented subsequently.

## 5.4. THE SIPF REACTION AS POSSIBLE ORIGIN OF BIOHOMOCHIRALITY

An initially surprising result were markedly higher di- and trialanine yields when starting from pure L-alanine compared to its D-form (Rode and Suwannachot, 1999). More extensive investigations in this direction showed that dialanine yields are reproducibly around 10% higher from L- than from D-alanine under various starting conditions and reaction times (Plankensteiner et al., 2004b; Fitz et al., 2008).

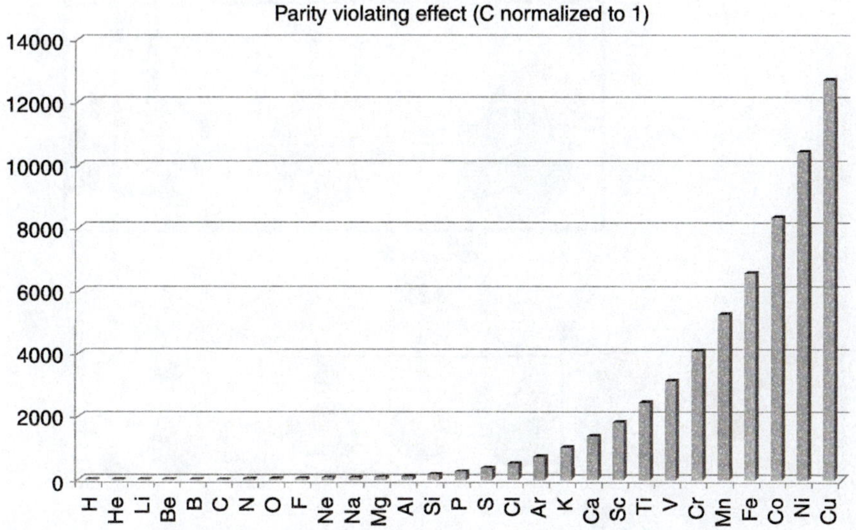


**Figure 2.** Schematic illustration of parity violation effects for the elements H to Cu for a $Z^6$ dependence (C normalized to 1).

Even more pronounced is the stereoselective discrimination in the cases of valine (Plankensteiner et al., 2005b) and isoleucine, where the L-forms produce several-fold higher peptide yields than their D-analogs, while the majority of the other amino acids form the same amounts of dipeptides from both enantiomeric forms within statistical fluctuations. Amino acids with aliphatic side chains seem to take a special position here since such a general enantiospecific preference could not be detected for any other amino acid so far.

In the active SIPF complex, the central copper ion is the heaviest atom ($Z=29$). It can also form a chiral center because of the different ligands coordinating to it. Thus, one could expect much higher PVEDs – by many orders of magnitude as indicated in Fig. 2 – comparing a copper complex containing L-amino acids and their D-counterparts on the one hand and the free amino acids on the other. The combination of the chemical chirality and the inherent chirality based on parity violation in weak nuclear interactions at the copper centre could actually lead to a diastereomer-like behaviour of L- and D-amino acid SIPF complexes, hence providing different chemical properties and reactivities (Fitz et al., 2007).

Quantum mechanical calculations of the geometries of active SIPF complexes with different amino acids have shown an interesting correlation (Fitz et al., 2007). For those amino acids where a stereospecific preference in favor of the L-form in the SIPF reaction has been detected experimentally, the equatorial 'plane' of the amino acids' coordination sites with respect to the Cu(II) centre of the SIPF complex is considerably more distorted towards a tetrahedral conformation than

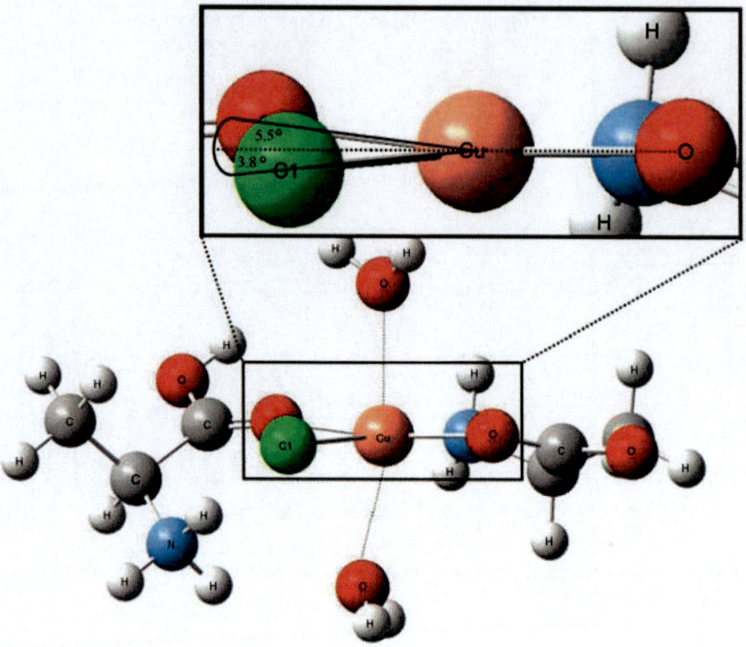


**Figure 3.** SIPF complex with two L-alanine ligands showing the distorted conformation of the central region.

for those amino acids without an enantiomeric preference. Figure 3 shows a complex with two L-alanine ligands and a closer look into the central 'plane' which is considerably distorted. This could mean that the measurable stereoselectivity of SIPF for only a few amino acids might occur due to cumulative interaction of the more or less distorted and thus chiral SIPF complex centre with the inherent chirality of the copper ion, as provided by parity violation in weak interactions. If the SIPF reaction was able to trigger peptide formation to a slight excess of at least some L-amino acids, this would finally lead to a generally L-amino acid–dominated peptide world because of stabilization effects like helix or β-sheet formation, which are only effectively possible for homochiral polymers and which are indispensable for a proper and reproducible chemical reactivity and functioning of a peptide or protein.

## 6. Conclusion

The data presented here indicate a simple and plausible way for the chemical evolution of amino acids, peptides and ultimately proteins from basic atmospheric components and mineral compounds on the primitive Earth. Besides that, the presented form of peptide evolution and the related model experiments also provide a good evidence for the preference of α-amino acids, specific amino acid sequences in proteins and for the natural 'choice' of L-amino acids in biosystems. It is clear that such experiments can only reveal some very fast and rudimentary steps of chemical evolution, but at least, they can form a scientific basis for the understanding of some driving forces leading to life on our planet.

## 7. Acknowledgments

This work was financially supported by the Austrian Science Foundation (Fonds zur Förderung der wissenschaftlichen Forschung, Projekt P19334-N17), which is gratefully appreciated.

## 8. References

Abelson PH (1965) Paleobiochemistry: inorganic synthesis of amino acids. Carnegie Inst Wash Yearbook 55:171–174

Andini S, Benedetti E, Ferrara L, Paolillo L, Temussi PA (1975) NMR studies of prebiotic polypeptides. Orig Life Evol Biosph 6:147–153

Bailey J, Chrysostomou A, Hough JH, Gledhill TM, McCall A, Clark S, Menard F, Tamura M (1998) Circular polarization in star forming regions: implications for biomolecular homochirality. Science 281:672–674

Bar-Nun A, Bar-Nun N, Bauer SH, Sagan C (1970) Shock synthesis of amino acids in simulated primitive environments. Science 168:470–472

Berger R, Quack M (2000) Electroweak quantum chemistry of alanine: parity violation in gas and condensed phase. ChemPhysChem 1:57–60

Bujdak J, Rode BM (2001) Activated alumina as an energy-source for peptide-bond formation – consequences for mineral-mediated prebiotic processes. Amino Acids 21:281–291

Bujdak J, Faybikova K, Eder AH, Yongyai Y, Rode BM (1995) Peptide chain elongation: a possible role of montmorillonite in prebiotic synthesis of protein precursors? Orig Life Evol Biosph 25:431–441

Cairns-Smith AG (1985) Seven clues to the origin of life. Cambridge University Press, Cambridge

Cleaves HJ, Miller SL (1998) Oceanic protection of prebiotic organic compounds from UV radiation. Proc Natl Acad Sci USA 95:7260–7263

Cody GD, Boctor NZ, Filley TR, Hazen RM, Scott JH, Sharma A, Yoder HS (2000) Primordial carbonylated iron-sulfur compounds and the synthesis of pyruvate. Science 289:1337–1340

Cronin JR, Pizzarello S (1997) Enantiomeric excesses in meteoritic amino acids. Science 275: 951–955

Dyson FG (1999) Origins of life. Cambridge University Press, Cambridge

Eder AH, Rode BM (1994) Influence of alkali- and alkaline-earth-metal cations on the 'salt-induced peptide formation' reaction. J Chem Soc Dalton Trans 7:1125–1130

Eigen M, Schuster P (1977) The hypercycle, 'A principle of natural self-organization'. Naturwissenschaften 64:541–565

Engel MH, Macko SA (1997) Isotopic evidence for extraterrestrial nonracemic amino acids in the Murchison meteorite. Nature 389:265–268

Figureau A, Duval E, Boukenter A (1995) Can biological homochirality result from a phase transition? Orig Life Evol Biosph 25:211–217

Fitz D, Reiner H, Plankensteiner K, Rode BM (2007) Possible origins of biohomochirality. Curr Chem Biol 1:41–52

Fitz D, Jakschitz T, Rode BM (2008) The catalytic effect of L- and D-histidine on alanine and lysine peptide formation. J Inorg Biochem 102:2097–2102

Flores J, Bonner WA (1974) On the asymmetric polymerization of aspartic acid polymers by kaolin. J Mol Evol 3:49–56

Flores JJ, Bonner WA, Massey GA (1977) Asymmetric photolysis of (RS)-leucine with circularly polarized ultraviolet light. J Am Chem Soc 99:3622–3625

Fox SW, Harada K (1958) Thermal copolymerization of amino acids to a product resembling protein. Science 128:1214

Fox SW, Harada K (1960) The thermal copolymerization of amino acids common to protein. J Am Chem Soc 82:3745–3751

Frondel C (1978) Characters of quartz fibers. Am Mineral 63:17–27

Groth W, Weyssenhoff HV (1957) Photochemical formation of amino acids from mixtures of simple gases. Naturwissenschaften 44:510–511

Guena J, Chauvat D, Jacquier P, Jahier E, Lintz M, Sanguinetti S, Wasan A, Bouchiat MA, Papoyan AV, Sarkisyan D (2003) New manifestation of atomic parity violation in cesium: a chiral optical gain induced by linearly polarized 6S-7S excitation. Phys Rev Lett 90:143001

Harada K, Fox SW (1958) The thermal condensation of glutamic acid and glycine to linear peptides. J Am Chem Soc 80:2694–2697

Harada K, Fox SW (1964) Thermal synthesis of natural amino acids from a postulated primitive terrestrial atmosphere. Nature 201:335–336

Huber C, Wächtershäuser G (1997) Activated acetic acids by carbon fixation on (Fe, Ni)S under primordial conditions. Science 276:245–247

Huber C, Wächtershäuser G (1998) Peptides by activation of amino acids with CO on (Ni, Fe)S surfaces: implications for the origin of life. Science 281:670–672

Huber C, Eisenreich W, Hecht S, Wächtershäuser G (2003) A possible primordial peptide cycle. Science 301:938–940

Huston DL, Smithies RH, Sun SS (2000) Correlation of the Archaean Mallina – Whim Creek Basin: implications for base-metal potential of the central part of the Pilbara granite-greenstone terrane. Aust J Earth Sci 47:217–230

Inoue Y (1992) Asymmetric photochemical reactions in solution. Chem Rev 92:741–770

Isaac R, Chmielewski J (2002) Approaching exponential growth with a self replicating peptide. J Am Chem Soc 124:6808–6809

Joyce GF (2002) The antiquity of RNA-based evolution. Nature 418:214–221

Kasting JF (1997) Warming early Earth and Mars. Science 276:1213–1215

Krause E, Bienert M, Schmieder P, Wenschuh H (2000) The helix-destabilizing propensity scale of D-amino acids: the influence of side chain steric effects. J Am Chem Soc 122:4865–4870

Krishnamurthy RV, Epstein S, Cronin JR, Pizzarello S, Yuen GU (1992) Isotopic and molecular analyses of hydrocarbons and monocarboxylic acids of the Murchison meteorite. Geochim Cosmochim Acta 56:4045–4058

Kvenvolden K, Lawless J, Pering K, Peterson E, Flores J, Ponnamperuma C, Kaplan IR, Moore C (1970) Evidence for extraterrestrial amino acids and hydrocarbons in the Murchison meteorite. Nature 228:923–926

Laerdahl JK, Schwerdtfeger P (1999) Fully relativistic ab initio calculations of the energies of chiral molecules including parity-violating weak interactions. Phys Rev A 60:4439–4453

Laerdahl JK, Wesendrup R, Schwerdtfeger P (2000) D- or L-alanine: that is the question. ChemPhysChem 1:60–62

Larralde R, Robertson MP, Miller SL (1995) Rates of decomposition of ribose and other sugars: implications for chemical evolution. Proc Natl Acad Sci USA 92:8158–8160

Lawless JG, Boynton CG (1973) Thermal synthesis of amino acids from a simulated primitive atmosphere. Nature 243:405–407

Lazcano A, Miller SL (1996) The origin and early evolution of life: prebiotic chemistry, the pre-RNA world, and time. Cell 85:793–798

Lee TD, Yang CN (1956) Question of parity conservation in weak interactions. Phys Rev 104:254–258

Lee DH, Granja JR, Martinez JA, Severin K, Ghadiri MR (1996) A self-replicating peptide. Nature 382:525–528

Levine J, Augustsson T, Natarajan M (1982) The prebiological paleoatmosphere: stability and composition. Orig Life Evol Biosph 12:245–259

Levy M, Miller SL (1998) The stability of the RNA bases: implications for the origin of life. Proc Natl Acad Sci USA 95:7933–7938

Li F, Fitz D, Fraser DG, Rode BM (2008) Methionine peptide formation under primordial earth conditions. J Inorg Biochem 102:1212–1217

Li F, Fitz D, Fraser DG, Rode BM (2010) Catalytic effects of histidine enationmers and glycine on the formation of dileucine and dimethionine in the salt-induced peptide formation reaction. Amino Acids 38:287–294

Liedl KR, Rode BM (1992) Ab initio calculations concerning the reaction mechanism of the copper(II) catalyzed glycine condensation in aqueous sodium chloride solution. Chem Phys Lett 197:181–186

Löb W (1906) Studien über die chemische Wirkung der stillen elektrischen Entladung. Z Elektrochem 11:282–316

Löb W (1913) Über das Verhalten des Formamids unter der Wirkung der stillen Entladung: Ein Beitrag zur Frage der Stickstoff-Assimilation. Ber Dtsch Chem Ges 46:684–697

Mason SF, Tranter GE (1983a) Energy inequivalence of peptide enantiomers from parity non-conservation. J Chem Soc Chem Commun 1983:117–119

Mason SF, Tranter GE (1983b) The parity-violating energy difference between enantiomeric molecules. Chem Phys Lett 94:34–37

Mason SF, Tranter GE (1984) The parity-violating energy difference between enantiomeric molecules. Mol Phys 53:1091–1111

Miller SL (1953) Production of amino acids under possible primitive earth conditions. Science 117:528–529

Miller SL (1955) Production of some organic compounds under possible primitive earth conditions. J Am Chem Soc 77:2351–2361

Miyakawa S, Kobayashi K, Sawaoka AB (1999) Amino acid synthesis from CO-$N_2$ and CO-$N_2$-$H_2$ gas mixtures via complex organic compounds. Adv Space Res 24:465–468

Niesert U, Harnasch D, Bresch C (1981) Origin of life: between Scylla and Charybdis. J Mol Evol 17:348–353

Ochiai E (1978) The evolution of the environment and its influence on the evolution of life. Orig Life 9:81–91

Oparin AI (1957) The origin of life on Earth. Oliver & Boyd, Edinburgh

Orgel LE (2004) Prebiotic chemistry and the origin of the RNA world. Crit Rev Biochem Mol Biol 39:99–123

Palm C, Calvin M (1962) Primordial organic chemistry. I. Compounds resulting from electron irradiation of $C^{14}H_4$. J Am Chem Soc 84:2115–2121

Pizzarello S, Huang Y, Fuller M (2004) The carbon isotopic distribution of Murchison amino acids. Geochim Cosmochim Acta 68:4963–4969

Plankensteiner K, Righi A, Rode BM (2002) Glycine and diglycine as possible catalytic factors in the prebiotic evolution of peptides. Orig Life Evol Biosph 32:225–236

Plankensteiner K, Reiner H, Schranz B, Rode BM (2004a) Prebiotic formation of amino acids in a neutral atmosphere by electric discharge. Angew Chem Int Ed 43:1886–1888

Plankensteiner K, Righi A, Rode BM, Gargallo R, Jaumot J, Tauler R (2004b) Indications towards a stereoselectivity of the salt-induced peptide formation reaction. Inorg Chim Acta 357:649–656

Plankensteiner K, Reiner H, Rode BM (2005a) Catalytically increased prebiotic peptide formation: ditryptophan, dilysine, and diserine. Orig Life Evol Biosph 35:411–419

Plankensteiner K, Reiner H, Rode BM (2005b) Stereoselective differentiation in the salt-induced peptide formation reaction and its relevance for the origin of life. Peptides 26:535–541

Plankensteiner K, Reiner HA, Rode BM (2005c) Catalytic effects of glycine on prebiotic divaline and diproline formation. Peptides 26:1109–1112

Plankensteiner K, Reiner H, Rode BM (2006) Amino acids on the rampant primordial earth: electric discharges and the hot salty ocean. Mol Divers 10:3–7

Rabinowitz J, Hampai A (1985) Quantitative polyphosphate-induced 'prebiotic' peptide formation in $H_2O$ by addition of certain azoles and ions. J Mol Evol 21:199–201

Rabinowitz J, Flores J, Krebsbach R, Rogers G (1969) Peptide formation in the presence of linear or cyclic polyphosphates. Nature 224:795–796

Reiner H, Plankensteiner K, Fitz D, Rode BM (2006) The possible influence of L-histidine on the origin of the first peptides on the primordial earth. Chem Biodivers 3:611–621

Robert F, Chaussidon M (2006) A paleotemperature curve for the Precambrian oceans based on silicon isotopes in cherts. Nature 443:969–972

Rode BM (1999) Peptides and the origin of life. Peptides 20:773–786

Rode BM, Schwendinger MG (1990) Copper-catalyzed amino acid condensation in water – a simple possible way of prebiotic peptide formation. Orig Life Evol Biosph 20:401–410

Rode BM, Suwannachot Y (1999) The possible role of Cu(II) for the origin of life. Coord Chem Rev 190–192:1085–1099

Rode BM, Eder AH, Yongyai Y (1997) Amino acid sequence preferences of the salt-induced peptide formation reaction in comparison to archaic cell protein composition. Inorg Chim Acta 254:309–314

Rode BM, Son HL, Suwannachot Y, Bujdak J (1999) The combination of salt induced peptide formation reaction and clay catalysis: a way to higher peptides under primitive earth conditions. Orig Life Evol Biosph 29:273–286

Rubbia C (1985) Experimental observation of the intermediate vector bosons $W^+$, $W^-$, and $Z^0$. Rev Mod Phys 57:699–722

Saetia S, Liedl KR, Eder AH, Rode BM (1993) Evaporation cycle experiments – a simulation of salt-induced peptide synthesis under possible prebiotic conditions. Orig Life Evol Biosph 23:167–176

Sagan C, Chyba C (1997) The early faint sun paradox: organic shielding of ultraviolet-labile greenhouse gases. Science 276:1217–1221

Sagan C, Khare BN (1971) Long-wavelength ultraviolet photoproduction of amino acids on the primitive Earth. Science 173:417–420

Salam A (1980) Gauge unification of fundamental forces. Rev Mod Phys 52:525–538

Salam A (1991) The role of chirality in the origin of life. J Mol Evol 33:105–113

Sawai H, Orgel LE (1975) Prebiotic peptide-formation in the solid state. III. Condensation reactions of glycine in solid state mixtures containing inorganic polyphosphates. J Mol Evol 6:185–197

Sawai H, Lohrmann R, Orgel LE (1975) Prebiotic peptide-formation in the solid state. II. Reaction of glycine with adenosine 5′′-triphosphate and P1, P2-diadenosine-pyrophosphate. J Mol Evol 6:165–184

Schwendinger MG, Rode BM (1989) Possible role of copper and sodium chloride in prebiotic evolution of peptides. Anal Sci 5:411–414

Schwendinger MG, Tauler R, Saetia S, Liedl KR, Kroemer RT, Rode BM (1995) Salt induced peptide formation: on the selectivity of the copper induced peptide formation under possible prebiotic conditions. Inorg Chim Acta 228:207–214

Segre D, Lancet D (1999) A statistical chemistry approach to the origin of life. Chemtracts Biochem Mol Biol 12:382–397

Shapiro R (1995) The prebiotic role of adenine: a critical analysis. Orig Life Evol Biosph 25:83–98

Son HL, Suwannachot Y, Bujdak J, Rode BM (1998) Salt-induced peptide formation from amino acids in the presence of clays and related catalysts. Inorg Chim Acta 272:89–94

Suwannachot Y, Rode BM (1998) Catalysis of dialanine formation by glycine in the salt-induced peptide formation reaction. Orig Life Evol Biosph 28:79–90

Suwannachot Y, Rode BM (1999) Mutual amino acid catalysis in salt-induced peptide formation supports this mechanism's role in prebiotic peptide evolution. Orig Life Evol Biosph 29:463–471

Thomas PJ, Chyba CF, McKay CP, Hicks RD (1998) Comets and the origin and evolution of life. Springer, New York

Tranter GE (1985) The parity violating energy differences between the enantiomers of α-amino acids. Mol Phys 56:825–838

Ulbricht TLV (1959) Asymmetry: the non-conservation of parity and optical activity. Q Rev Chem Soc 13:48–60

Ulbricht TLV, Vester F (1962) Attempts to induce optical activity with polarized β-radiation. Tetrahedron 18:629–637

Vester F, Ulbricht TLV, Krauch H (1959) Optical activity and parity violation in β-decay. Naturwissenschaften 46:68–69

Wächtershäuser G (1990) Evolution of the first metabolic cycles. Proc Natl Acad Sci USA 87:200–204

Wächtershäuser G (2000) Origin of life. Life as we don't know it. Science 289:1307–1308

Wang W, Yi F, Ni Y, Zhao Z, Jin X, Tang Y (2000) Parity violation of electroweak force in phase transitions of single crystals of D- and L-alanine and valine. J Biol Phys 26:51–65

Weber LA, Caroon JM, Warden JT, Lemmon RM, Calvin M (1977) Simultaneous peptide and oligonucleotide formation in mixtures of amino acid, nucleoside triphosphate, imidazole and magnesium ion. Biosystems 8:277–286

Weinberg S (1980) Conceptual foundations of the unified theory of weak and electromagnetic interactions. Rev Mod Phys 52:515–523

Welch CJ (2001) Formation of highly enantioenriched microenvironments by stochastic sorting of conglomerate crystals: a plausible mechanism for generation of enantioenrichment on the prebiotic earth. Chirality 13:425–427

Wood CS, Bennett SC, Cho D, Masterson BP, Roberts JL, Tanner CE, Wieman CE (1997) Measurement of parity nonconservation and an anapole moment in cesium. Science 275:1759–1763

Wu CS, Ambler E, Hayward RW, Hoppes DD, Hudson RP (1957) Experimental test of parity conservation in β-decay. Phys Rev 105:1413–1415

Yamanaka J, Inomata K, Yamagata Y (1988) Condensation of oligoglycines with trimeta- and tetrametaphosphate in aqueous solutions. Orig Life Evol Biosph 18:165–178

Yao S, Ghosh I, Zutshi R, Chmielewski J (1998) Selective amplification via auto- and crosscatalysis in a replicating peptide system. Nature 396:447–450

Yoshino D, Hayatsu R, Anders E (1971) Origin of organic matter in the early solar system. III. Amino acids. Catalytic synthesis. Geochim Cosmochim Acta 35:927–938

Biodata of **Martin Klussmann**, author of "***Chemical Models for the Origin of Biological Homochirality***."

**Dr. Martin Klussmann** is currently a research group leader at the Max-Planck-Institut für Kohlenforschung, Muelheim an der Ruhr, Germany. He obtained his Ph.D. at the Technische Universität Darmstadt in 2004, working with Prof. Michael Reggelin on helically chiral polymers. His interest in chirality was further nourished during his stay at Imperial College London, UK, where he worked as a postdoctoral researcher with Prof. Donna G. Blackmond on nonlinear effects and mechanisms of amino acid catalyzed asymmetric reactions. Martin Klussmann's current research interests are in the area of oxidative coupling reactions and the investigation of reaction mechanisms.

E-mail: **klusi@mpi-muelheim.mpg.de**

J. Seckbach (ed.), *Genesis - In The Beginning: Precursors of Life, Chemical Models and Early Biological Evolution*, Cellular Origin, Life in Extreme Habitats and Astrobiology 22, 491–508
DOI 10.1007/978-94-007-2941-4_26, 

# CHEMICAL MODELS FOR THE ORIGIN OF BIOLOGICAL HOMOCHIRALITY

MARTIN KLUSSMANN
*Max-Planck-Institut fuer Kohlenforschung, Kaiser-Wilhelm-Platz 1, 45470 Muelheim an der Ruhr, Germany*

## 1. Introduction

The origin of life on Earth is one of the greatest and most fascinating puzzles encountered by mankind. Compared with our planet as we know it today, the early Earth was a barren place before life existed. Apart from the very different atmosphere, it lacked the complexity found in biological organisms. And yet, over unknown periods of time, chemical systems of ever higher complexity must have evolved that finally made the transition to life (Budin and Szostak, 2010; Luisi, 2006). The question of when exactly such primordial systems would have deserved the designation "life" is in itself an interesting question. A common approach would be to look at present living organisms and try to determine their minimal requirements for existence. Biological homochirality, the fact that biological organisms are made up basically exclusively of L-amino acids and D-sugars and not their mirror images, is one of the less obvious of such requirements. How this exclusiveness could have evolved on the early Earth has intrigued chemists for a long time.

The aim of this chapter is to give a nonconclusive overview of chemical experiments that have been put forward as models for the evolution of biological homochirality. Particular emphasis is given on results concerning amino acids. More detailed surveys of these and other models can be found in review articles (Ávalos et al., 2010; Barron, 2008; Bonner, 1991; Breslow, 2011; Cintas, 2002; Podlech, 2001; Weissbuch et al., 2005).

## 2. Biological Homochirality

Chirality is a property of molecules or objects to be nonsuperimposable with their mirror images. This is true for our right and left hands, from which the property derives its name. For molecules, these two mirror-image forms are called enantiomers. Many natural compounds are chiral, for example, amino acids and sugars which are building blocks of peptides, enzymes, RNA, and DNA, the essential macromolecules of life. These key compounds occur in organisms almost exclusively in one enantiomeric form, amino acids in the L-form and sugars in the D-form.

This phenomenon is called biological homochirality. As a result, a vast number of other natural compounds, made with the help of chiral enzymes, also occur enantiomerically pure in organisms. Examples comprise hormones, toxins and fragrances, or flavors in food.

Biological homochirality strongly affects all biochemical interactions with chiral substances. One enantiomer of a given compound will most likely have a different effect than the other. This is due to the homochirality of receptors, enzymes, and other key parts of an organism that leads to a high degree of chiral discrimination, the ability to differentiate between the two enantiomeric forms of a chiral compound. This can effect smell, taste, and all other kinds of physiological reactions. For example, the terpenoid carvone in its (−)-form smells of mint, whereas its enantiomer, the (+)-form, smells of caraway (Leitereg et al., 1971). The two enantiomers of a pharmaceutically active compound can display different effects; one might show the desired effect, while the other might be inactive or display different, possibly harmful activity (Crossley, 1992; Triggle, 1997).

But more importantly, the biochemical processes in any organism would fail to cooperate in the required way if the functional molecules (e.g., enzymes or RNA) were to consist of building blocks of random chirality. The enantiomer of a whole enzyme would in principle still work with the same efficiency, just in a mirror-image manner, producing enantiomeric products, for example. However, if its chiral amino acid building blocks were to be randomly chosen from either the L- or the D-enantiomeric form, the enzyme would be of completely different shapes. Such a change in the enzyme's secondary or tertiary structure would result in a shutdown or major change in its function. Therefore, it seems obvious that nature "had to make a choice" at one point in its evolution, which enantiomers of the biochemical building blocks to use. Biological homochirality is accordingly seen as a requirement for life, which opens the question of how it was achieved. This seems a remarkable feat, given that chiral molecules on the early Earth were most likely racemic, i.e., consisted of an equal mixture of right- and left-handed enantiomers. The questions of interest to chemists in this context are: is biological homochirality a necessary precondition or an inevitable result of the emergence of life? In other words, did homochirality or something close to it evolve in one place, thus laying the chemical foundation for the origin of life? Or has homochirality been an inevitable by-product of the emergence of life and thus came after or with the origin of life? Furthermore, is it the result of pure chance that all modern organisms are built of L-amino acids and D-sugars, or is there an intrinsic preference for this choice, i.e., was it chance or determinism? Since we are unable to go back in time and find out, most questions concerning homochirality are essentially unanswered and will remain so.

For the "homochirality first" model, usually a stepwise evolution is suggested. At first, a process of "mirror symmetry breaking" is thought to have created an unequal distribution of enantiomers, at least locally. This imbalance was further enhanced by processes called asymmetric amplification which led to a high enantiomeric excess of some compounds. These compounds could have directly constituted the building blocks for the evolution of life like amino acids or sugars, or they transferred their chirality to new products by further chemical reactions. For all these steps, chemical model systems have been put forward that will be discussed in this chapter.

## 3. How Racemic Is the Universe?

If a chiral substance is synthesized by a chemical reaction from achiral or racemic starting material and reagents, the result is normally inevitably a racemate. If chemists want to synthesize only one enantiomer of a substance, they have to resolve the racemate into its components or rely on nature's pool of chiral, enantiopure molecules. These can serve either directly as starting material or as catalysts that induce asymmetry in a reaction leading to a chiral product. After the creation of the universe, first higher elements were formed via nucleosynthesis and only then molecules. Accordingly, one must assume that the universe was racemic at the beginning and still is today – with the exception of the Earth and potentially other planets harboring life. In this chapter, some studies will be presented that cast doubt on this assumption.

An explanation why the universe is not racemic is given by parity violation. This is a phenomenon that violates mirror symmetry in nuclear processes involving weak interactions. It was experimentally verified for the first time by showing that the radioactive β-decay of $^{60}Co$ produces particles with a left-handed spin predominantly (Wu et al., 1957). As a further consequence of parity violation, the two enantiomers of a chiral compound are no longer thermodynamically equal but possess intrinsic differences in energy. This was suggested to explain the dominance of L-amino acids and D-sugars over their enantiomers, the so called Yamagata hypothesis (Yamagata, 1966). But the calculated parity-violating energy differences (PVEDs) are of a very small magnitude, and while some calculations indeed suggest that the natural enantiomers are energetically favored, others call this into question (Wesendrup et al., 2003). More importantly, because PVEDs are so much smaller than any energy barrier usually relevant for chemistry, experimental consequences are generally ruled out and have never been shown to exist beyond doubt (Lahav et al., 2006). Stochastic calculations have also shown that a compound's enantiomeric excess possible due to PVEDs is smaller than the enantiomeric excess due to statistical fluctuations in the number of enantiomers (Lente, 2006; Fuß, 2009). Parity violation is an accepted phenomenon in physics, but the consequence that enantiomeric pairs of molecules have different energies is still a theoretical prediction only (see, e.g., Stoeffler et al., 2011). Whether or not PVEDs will be shown to exist, any connection with biomolecular homochirality on Earth remains unlikely.

While the universe as a whole might be racemic, as discussed, there are indeed observations of *local* imbalances apart from the Earth. In the Orion Nebula, circular polarized light has been found in star-forming regions; in some parts of the nebula right-handed, in others left-handed light dominates (Bailey et al., 1998). Photochemical reactions with circularly polarized radiation have been shown to be able to produce small enantioselectivities, often by degrading one enantiomer faster than the other (Griesbeck and Meierhenrich, 2002; Meinert et al., 2010). Photolysis of racemic leucine with circularly polarized UV light, for example, produced an enantiomeric excess (ee) of 2.5% after degradation of 75% of the material (Flores et al., 1977). It remains to be shown whether circularly polarized UV light also exists in the universe, as the case of the Orion Nebula was found in the infrared. Although its origins are unclear, it is a clear sign that asymmetric forces capable of influencing chemical reactions exist in the universe.

More evidence for a nonracemic universe comes from studies of meteorites. Several organic compounds have been found in the carbonaceous chondrites Murchison and Murray, including amino acids which are significantly enriched in one enantiomer up to 15% ee (Pizzarello and Cronin, 2000; Pizzarello et al., 2003; Strasdeit, 2005). Most of these were nonnatural or very rare amino acids, and their isotope pattern also made it very unlikely that these enantiomeric excesses were of terrestrial origin (Pizzarello et al., 2004). Carbonaceous chondrites have only been a minor contribution to the flux of organic material from space, but in general, dust and meteorites have delivered large amounts of organic material to Earth over time and still do so today (Pasek and Lauretta, 2008).

While the origins of the observations of asymmetry mentioned above remain unclear, they might give us an echo from the time our solar system was formed. The universe as a whole might be racemic, but local imbalances have been found. If the Earth was formed in such a local environment or received organic material from there, its primordial chemistry could have started out with an enantiomeric excess of some compounds already.

## 4. Mirror Symmetry Breaking

"Mirror symmetry breaking" describes cases of the spontaneous creation of an enantiomeric imbalance without driving it with an external bias, for example, a chiral catalyst or reagent. A few of such cases are known, where by chance one enantiomer comes to dominate over the other. Usually, chance does not play a role in chemistry, due to the very large number of molecules present in even the smallest sample. But many examples of mirror symmetry breaking involve crystals, which, as macroscopic objects, can obviously operate in much smaller numbers (Cintas, 2002).

Chiral crystals play an important role in many scenarios on the origin of homochirality. Some chiral substances give a mixture of enantiopure crystals if their racemates are crystallized; this case is called a conglomerate. Each crystal is enantiomerically pure, but all crystals together will be racemic. If by chance a few of these crystals containing predominantly one enantiomer are physically separated from the others, mirror symmetry would have been broken locally. For example, a small pool containing a chiral substance could dry out and lead to conglomerate crystals. Some of these crystals could be blown by wind into another pool which, by chance, could now contain an enantioenriched substance (Welch, 2001).

Crystals can be chiral even if they are built from achiral compounds, quartz being one example. The highly ordered array of molecules within a crystal can transfer chirality by adsorption or by reactions at the surface. In an early experiment, Bonner found small enantiomeric excesses of 1–2% of alanine hydrochloride absorbed on chiral quartz crystals of single handedness (Bonner et al., 1974). More recently, calcite ($CaCO_3$) has been used, which does not form chiral crystals but chiral faces. In careful studies, racemic amino acids were adsorbed on mirror-image faces of calcite. With aspartic acid, one enantiomer showed a small but significant preference for one face of calcite, while the mirror-image face had a preference for the other enantiomer. The average enantiomeric excesses of aspartic acid adsorbed reached 0.2–1.0% ee, while the maximum observed was 5.5% ee (Hazen et al., 2001).

Statistically, chiral crystals and crystal faces should occur with equal amounts of both mirror-image forms on Earth. But chance could have favored enantioenrichment of a compound if there was a *local* dominance of one mirror-image crystal form, as discussed for the conglomerate case above. But there are also ways to generate new crystals in a homochiral fashion.

Sodium chlorate ($NaClO_3$) is an achiral compound that can crystallize with a chiral crystal morphology from water. Under normal circumstances, one will statistically obtain one or the other enantiomorphic crystal and in the sum equal amounts of both. But if a supersaturated solution is stirred while crystallization takes place, crystals of nearly uniform handedness are formed (Kondepudi et al., 1990). Even starting out with a racemic mixture of crystals in a saturated solution, the solid phase can be driven toward chiral purity. If such a mixture of racemic sodium chlorate crystals is stirred with glass balls for an abrasion-grinding process, the solid phase will attain homochirality within hours (Viedma, 2005). Without glass balls, the solid phase stays racemic. The mechanism of this surprising phenomenon has been debated; apparently, the grinding makes large crystals grow at the expense of smaller ones and ultimately makes one enantiomorphic crystal form disappear – statistics determine which one (Blackmond, 2007; Viedma, 2005).

As sodium chlorate itself is achiral, any further chemistry to transfer the solid phase chirality is limited to adsorption phenomena, as discussed above. But the same process was also successfully applied to chiral compounds, for example, the chiral amino acid derivative **1** that crystallizes as a conglomerate. Under racemizing conditions in solution, the solid phase could be driven toward chiral purity, just as in the case of $NaClO_3$ (Fig. 1) (Noorduin et al., 2008).

Here too, one enantiomorphic crystal was transformed into the other one via the solution phase, but now the chiral information was additionally stored in

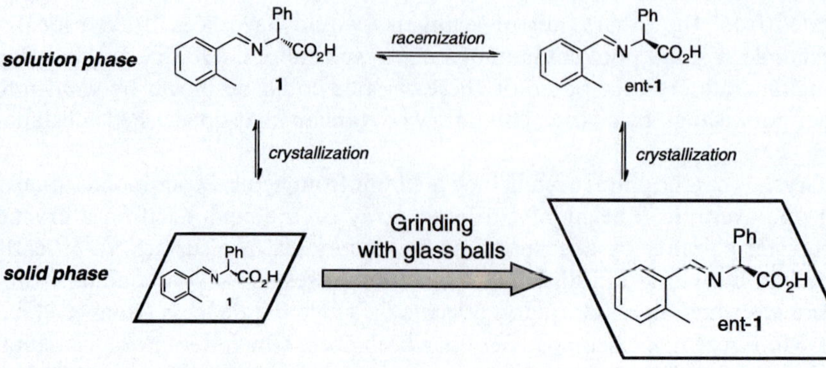


**Figure 1.** Transforming a racemic amino acid derivative to enantiopurity in the solid state by solution racemization and an abrasion-grinding process of the crystals (Noorduin et al., 2008).

the molecular structure and not just in the crystal shape. The same principle was also applied to aspartic acid, one of only two proteinogenic amino acids which crystallize as a conglomerate (Viedma et al., 2008).

## 5. Asymmetric Amplification: From Low to High Enantiomeric Excess

A process or a chemical reaction that produces a high enantiomeric excess starting from a lower one is called asymmetric amplification (Satyanarayana et al., 2009). Relatively many examples for such processes are known, supporting the idea that asymmetric amplification could have provided a route to homochirality after events of mirror symmetry breaking. Most examples known are chemical reactions that have no relevance for prebiotic chemistry, though. The following cases illustrate either the most illustrious examples of asymmetric amplification or examples of significance for prebiotic chemistry because they involve biologically relevant compounds.

### 5.1. ASYMMETRIC AMPLIFICATION IN AUTOCATALYSIS

A theoretical model for spontaneous asymmetric synthesis was developed in 1953 by Frank, combining asymmetric amplification and autocatalysis (Frank, 1953). He proposed that if a catalytic reaction would produce a chiral product which would act as a catalyst for its own formation (autocatalysis), at the same time suppressing the formation of its enantiomer, high enantiomeric excesses could be achieved from nearly racemic starting material. In 1995, Soai and coworkers found the first chemical reaction that corresponded to Frank's model (Fig. 2) (Soai et al., 1995).

The alkylation of the pyrimidyl aldehyde **2** with diisopropylzinc gives a chiral pyrimidyl alcohol **3** that catalyzes its own formation with very high asymmetric amplification. For example, adding a small amount of product **3** with

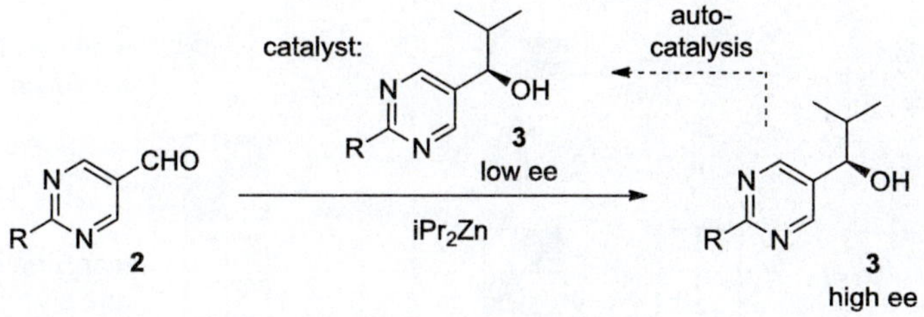


**Figure 2.** Asymmetric autocatalysis, the "Soai reaction." R = H, Me, CCtBu.

an enantiomeric excess of 5% at the beginning of the reaction leads to **3** of 55% ee at the end of the reaction. By using a small amount of this product batch as additive in a repetition of this reaction, nearly enantiopure **3** can be obtained after a few runs. This reaction is so highly sensitive to chiral inductions that basically any source of chirality serves to drive the asymmetric amplification in one direction. Examples include helical hydrocarbons (Sato et al., 2001), chiral quartz crystals (Soai et al., 1999), and substances that are chiral by a selective isotope substitution (Kawasaki et al., 2009; Sato et al., 2000). Moreover, even if the product added at the beginning is completely racemic, enantiomerically enriched product is produced at the end – a case of mirror symmetry breaking. This time, however, the enantiomer produced in excess is determined by chance (Soai et al., 2004). Mechanistic studies revealed that the product alcohol forms dimeric complexes with diisopropylzinc. While the heterochiral dimers of two enantiomeric alcohols are catalytically inactive, the homochiral ones catalyze the reaction, ensuring high enantiomeric excess (Blackmond, 2004; Blackmond et al., 2001).

This so-called Soai reaction has no significance for prebiotic chemistry, due to the structure of its compounds and the fact that it is very sensitive to water. Still, it is the best example illustrating how asymmetric amplification and autocatalysis could have played a role in the evolution of biological homochirality, and there is to date no comparable reaction. Recently, asymmetric autocatalysis has been observed in organocatalytic Mannich reactions (Amedjkouh and Brandberg, 2008; Mauksch et al., 2007; Wang et al., 2010). Some products are structurally similar to natural amino acids, but no asymmetric amplification was observed.

## 5.2. ASYMMETRIC AMPLIFICATION BY CRYSTAL-PHASE BEHAVIOR

We have already seen how chiral crystals and their properties have been evoked for scenarios of mirror symmetry breaking. The cases discussed above all involved conglomerates, substances giving a racemate consisting of enantiopure crystals. But the majority of chiral substances do not crystallize as a conglomerate but belong to the class of so-called racemic compounds. These form racemic crystals containing equal amounts of both enantiomers in a highly ordered fashion

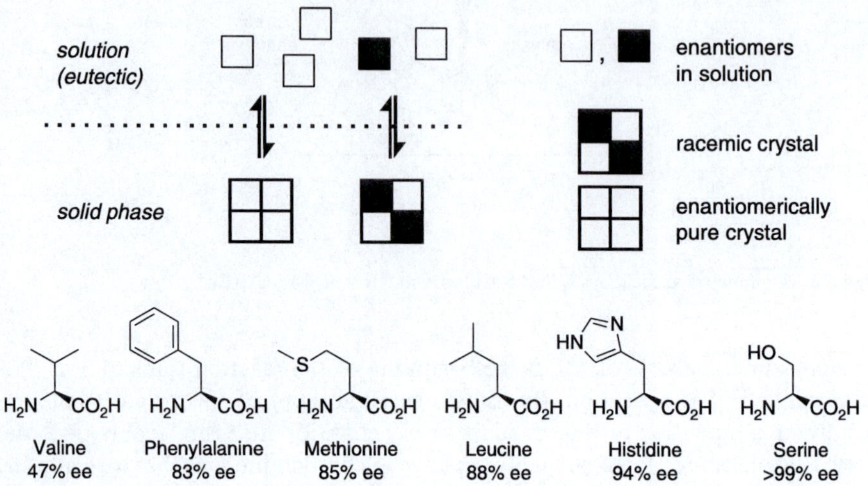


**Figure 3.** Three-phase solid-solution equilibrium (eutectic) of chiral compounds crystallizing as a racemic compounds and enantiomeric excess values at the eutectic of selected amino acids.

(Jacques et al., 1981). If scalemic (nonracemic, nonenantiopure) samples are crystallized, the solid phase will contain the racemic crystals mixed with the excess enantiomer in enantiopure crystals. As these crystals are diastereomorphic, they have different physical properties, for example, melting points and solubilities. Accordingly, if such a mixture of crystals is partially dissolved, the solution will usually have a different enantiomeric composition than the initial one. In a saturated solution, the liquid phase will be in equilibrium with these two different solid phases. Due to the phase rule, such a system has a fixed composition at a given temperature and pressure, called a eutectic point (Fig. 3).

Depending on the compound, the enantiomeric composition at the eutectic can be very high or very low, and cases of very strong asymmetric amplification can occur, if material of low enantiomeric excess is employed. As no molecules are made or destroyed in such a system, the solid phase will be depleted of the excess enantiomer if the solution phase is enriched and the other way around. Striking examples which might be prebiotically relevant involve amino acids. Serine exhibits a eutectic point of >99% ee, so if serine of low enantiomeric excess is only partially dissolved, leaving some of both crystal types (racemic and enantiopure) in the solid phase, the solution will be basically enantiopure (Klussmann et al., 2006a). Other cases of proteinogenic amino acids with relatively high eutectic ee's in water include phenylalanine, methionine, leucine, and histidine (Fig. 3) (Klussmann et al., 2006b).

In the chapter on mirror symmetry breaking, conglomerate crystals played the chief part; now it is the solution phase in systems of racemic compounds that serves as model for prebiotic scenarios (Klussmann et al., 2006a; Morowitz, 1969). The fact that such models are based on systems in thermodynamic equilibrium means that similar scenarios in prebiotic times could have evolved over a long time and would not have required the involvement of chance. For example, an enantiomerically enriched source of a chiral compound like an amino acid could become chirally amplified if part of it were dissolved in a puddle or a small lake. Alternatively, cycles of rain, partial dissolution, and evaporation could lead to a spatially separate deposit of higher ee. Such local enrichment could then be coupled with processes of chirality transfer, for example, by chemical reactions occurring in solution (see the next chapter).

Sublimation processes of chiral compounds also behave similar to solid-solution-phase systems. Such solid-gas-phase systems have sparked similar scenarios, as discussed above (Blackmond and Klussmann, 2007). For example, incomplete sublimation of amino acids in vacuum can lead to asymmetric amplification in the sublimate: sublimation of serine with 3% ee gave condensed batches ranging from 68% to 92% ee (Perry et al., 2007). Similar results were obtained with several other amino acids (Fletcher et al., 2007). It was suggested that such processes could have happened on meteorites in space, where amino acids were enantiomerically enriched by sublimation in vacuum. This material could then have provided a source of chiral, nonracemic compounds for the early Earth, as in the case of the Murchison meteorite.

## 6. Transferring Amplified Chirality

### 6.1. CATALYSIS BY ENANTIOMERICALLY ENRICHED COMPOUNDS

Amino acids, most notably proline, have been shown to be excellent asymmetric organic catalysts for a variety of transformations like aldol and Mannich reactions (Mukherjee et al., 2007). Aldol reactions are particularly interesting for prebiotic scenarios as they can produce carbohydrates, essential building blocks for DNA and RNA, for example. The fact that amino acids are not only the monomers of peptides and enzymes – nature's catalysts – but also catalysts on their own for the formation of carbohydrates has sparked several proposals regarding their possible involvement in the evolution of biological homochirality. Generally, amino-acid-catalyzed reactions are synthetically most useful if performed in organic solvents. A weak point in many prebiotic scenarios is therefore the fact that the performance in water is low due to both a reduced efficiency and stereoselectivity.

As mentioned above, the amino acid isovaline has been found with an ee of up to 15% in meteorites (Pizzarello et al., 2003). Pure L-isovaline has been found to catalyze the synthesis of tetroses – $C_4$ carbohydrates – with up to 11% ee (Pizzarello and Weber, 2004). These experiments illustrate a possible scenario, but of course the modest stereoselectivities of both the product and the catalyst as found in the meteorite would have quickly eroded any enantiomeric excess. If coupled with a process of asymmetric amplification, however, the chirality

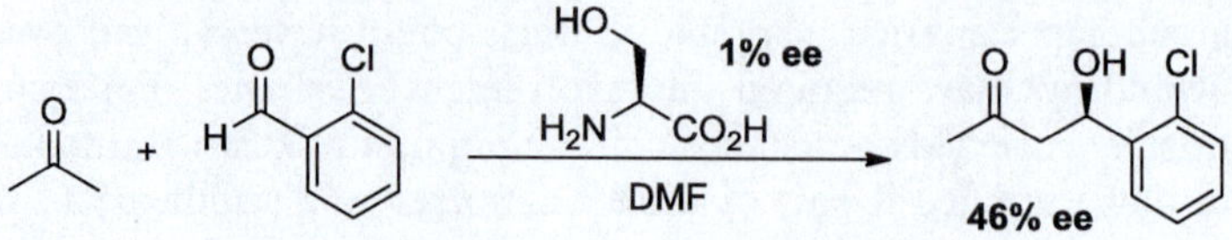


**Figure 4.** Example of asymmetric amplification due to solid-solution-phase behavior in an aldol reaction catalyzed by serine with 1% ee. DMF = N,N-dimethylformamide.

transfer of an amino acid to a new product can become more efficient. For example, serine can be used as an organocatalyst for an aldol reaction in an organic solvent. Enantiomerically pure L-serine catalyzes the reaction shown in Fig. 4 with a product enantioselectivity of 46%. If the reaction is performed with serine of only 1% ee but in a heterogeneous system of solution and undissolved solid serine, the catalyst's ee is amplified in solution (see Fig. 3), and the product is formed again with 46% ee (Klussmann et al., 2006a).

Lower absolute stereoselectivities but similar degrees of asymmetric amplification were achieved in water, too. Still, at present, there is no model reaction generating an enantiomerically pure and prebiotically relevant product using a simple amino acid as a catalyst in water. Alternatively, one could imagine not a direct chirality transfer by amino acid catalysis but first the evolution of improved catalysts from amino acids as building blocks, as discussed in the next part.

## 6.2. CHIRAL MACROMOLECULES

Enzymes are nature's workhorses. Being long-chain polypeptides with defined superstructures and active sites, they are a long way from single amino acids. But already very small oligopeptides can be improved catalysts for the synthesis of carbohydrates under aqueous conditions. For example, alanine catalyzes the aqueous aldol reaction of glycolaldehyde to the tetrose erythrose 5% ee (Pizzarello and Weber, 2004), while the dimer, the dipeptide L-ala-L-ala, gives 33% ee (Fig. 5) (Weber and Pizzarello, 2006). The dipeptide L-val-L-val was found to give even higher stereoselectivities of up to 82% ee (Córdova et al., 2006; Weber and Pizzarello, 2006).

Macromolecules in general possess intriguing features resulting from their superstructure. Polyisocyanates, for example, adopt a helical conformation of the polymer backbone which is chiral. If made from a scalemic mixture of monomers with a chiral side chain, the major enantiomers can overrule the minority of monomers in favor of one helical backbone conformation, a phenomenon called "majority rules" (Green et al., 1999).

The formation of peptides from amino acids under prebiotic conditions is also an interesting question, as the reaction is unfavorable under aqueous conditions (Brack, 2007). A potential prebiotic pathway is the activation of amino acids in the form of *N*-carboxyanhydrides (Kricheldorf, 2006). Similar cooperative effects as in polyisocyanates have been seen in peptides, here too "majority rules."

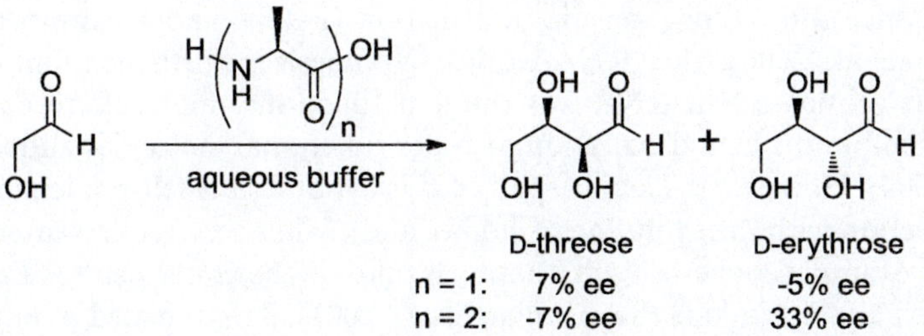


**Figure 5.** Aqueous synthesis of tetrose carbohydrates by amino acid or dipeptide catalysis.

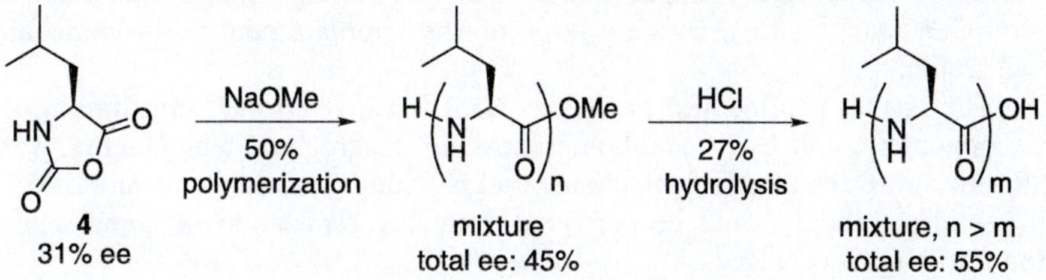


**Figure 6.** Partial polymerization and hydrolysis of leucine *N*-carboxyanhydrides.

When consisting of homochiral monomeric units, peptides have been found to be more stable toward hydrolysis but also more reactive toward the uptake of a monomer of like chirality (Blair and Bonner, 1981). Cycles of chain growth and chain degradation could thus have led to an amplification of chirality in the polymer. For example, partial polymerization of leucine-*N*-carboxyanhydride **4** of 31% ee gave oligopeptides with 45% ee, and subsequent partial hydrolysis gave an oligomer with now 55% ee (Fig. 6) (Blair and Bonner, 1981).

Other results also indicate that homochiral peptide sequences should accumulate from cycles of polymerization and hydrolysis, because they can form stable β-sheets (Brack, 2007).

If one recalls the examples of mirror symmetry breaking and amplification, it seems not unlikely that homochiral L-peptides could have evolved from a pool of amino acids that were just slightly enriched in the L-enantiomers. It is interesting to note in this chapter that chiral quartz crystals have been found to enhance the formation of homochiral oligopeptides from amino acid *N*-carboxyanhydrides (Hitz and Luisi, 2004).

## 7. Self-Replication and Chirality

Self-replication is an essential feature of life and its molecules, specifically the macromolecules RNA and DNA. In many studies, scientists have attempted to create artificial systems capable of self-replication that might serve as models for prebiotic ancestors of our genetic information (Patzke and Kiedrowski, 2007; Robertson et al., 2000). The "RNA-world"-hypothesis is worth mentioning in this chapter. It proposes that RNA was the first functional molecule to come into existence before the actual evolution of other functional molecules and life itself (Orgel, 2004). However, the main aspect of this hypothesis is self-replication, while the connection with the evolution of homochirality has received less attention.

One example concerned with homochirality is the chiral peptide developed by Ghadiri and coworkers (Saghatelian et al., 2001). They created a homochiral 32-mer that could stereoselectively replicate itself from a racemic pool of shorter homochiral fragments. The peptide acts as a template, selectively binding two fragments of the correct configuration by hydrogen bonding, which then undergo a coupling reaction to generate a copy of the template peptide (Saghatelian et al., 2001).

The 32-mer peptide also proved resistant toward "errors": substitution of one monomeric unit by its enantiomer created a heterochiral template that was still active in the catalysis of the homochiral peptide. Thus, the homochirality of all peptides formed would be preserved even if racemization of some single monomeric units occurred.

Similarly, the aqueous polymerization of amino acid *N*-carboxyanhydrides has been shown to yield homochiral β-sheet oligomers which serve as templates for further chain elongations (Rubinstein et al., 2007). The system studied was overall racemic, but, as discussed above, scenarios with events of mirror symmetry breaking prior to the polymerization steps could be plausible.

## 8. Is It an Important Question, After All?

The remarkable features of macromolecules discussed above indicate how complete homochirality might have been achieved and stabilized. The essential role functional macromolecules play in living organisms indicates that their evolution must have been a key part in the evolution of life. But some scientists go even further and suggest that homochirality has been an inevitable outcome during the evolution of functional macromolecules (Siegel, 1998). The aphorism "two equal runners, one tripped" has been coined to describe the idea that two mirror-image systems of homochiral macromolecules evolved that might have led to primordial mirror-image life forms, but one system lost out at some point along the way (Green and Jain, 2010).

Still, even if one is inclined to the proposition that macromolecules have played the chief part in the evolution of biological homochirality, it does not exclude any prior involvement of processes like the ones discussed in the beginning of this chapter. Any event of mirror symmetry breaking and chiral amplification could have helped to provide an enantioenriched pool of building blocks to make the formation of homochiral polymers more feasible. Such events could also have tipped the balance in favor of L-amino acids and D-sugars.

## 9. Conclusions

The puzzle of the evolution of biological homochirality will remain a mystery since we cannot look into the past of our planet – the only known fact is that it did indeed happen. Yet each interesting discovery concerning mirror symmetry breaking or asymmetric amplification adds to the picture. As we collect more of these possible puzzle pieces, some of them will start to fit together and create realistic scenarios.

For example, we have seen indications that the universe is not homogeneously racemic. By collecting nonracemic material from space, the prebiotic Earth could have had a head start, and this could also have decided the "fate" of L-amino acids to become dominant. If these amino acids were concentrated in pools or lakes, crystallization could have amplified their enantiomeric excess in solution. Chemical reactions could have led to small peptides with a preference for homochiral sequences, further enhancing the dominance of L-amino acids. These peptides could have acted as templates for chain elongations or as catalysts producing enantiomerically enriched carbohydrates. Life on Earth might not necessarily have started in these pools, but such scenarios give at least a plausible idea how enantiomerically enriched building blocks like peptides and carbohydrates could have evolved and provided the foundations of life.

## 10. References

Amedjkouh M, Brandberg M (2008) Asymmetric autocatalytic Mannich reaction in the presence of water and its implication in prebiotic chemistry. Chem Commun 44:3043–3045

Ávalos M, Babiano R, Cintas P, Jiménez JL, Palacios JC (2010) Chirality and life. Tetrahedron Asymmetry 21:1030–1040

Bailey J, Chrysostomou A, Hough JH, Gledhill TM, McCall A, Clark S, Ménard F, Tamura M (1998) Circular polarization in star-formation regions: implications for biomolecular homochirality. Science 281:672–674

Barron LD (2008) Chirality and life. Space Sci Rev 135:187–201

Blackmond DG (2004) Asymmetric autocatalysis and its implications for the origin of homochirality. Proc Natl Acad Sci USA 101:5732–5736

Blackmond DG (2007) "Chiral amnesia" as a driving force for solid-phase homochirality. Chem Eur J 13:3290–3295

Blackmond DG, Klussmann M (2007) Spoilt for choice: assessing phase behavior models for the evolution of homochirality. Chem Commun 43:3990–3996

Blackmond DG, McMillan CR, Ramdeehul S, Schorm A, Brown JM (2001) Origins of asymmetric amplification in autocatalytic alkylzinc additions. J Am Chem Soc 123:10103–10104

Blair NE, Bonner WA (1981) A model for the enantiomeric enrichment of polypeptides on the primitive Earth. Orig Life Evol Biosph 11:331–335

Bonner WA (1991) The origin and amplification of biomolecular chirality. Orig Life Evol Biosph 21:59–111

Bonner WA, Kavasmaneck PR, Martin FS, Flores JJ (1974) Asymmetric adsorption of alanine by quartz. Science 186:143–144

Brack A (2007) From interstellar amino acids to prebiotic catalytic peptides: a review. Chem Biodivers 4:665–679

Breslow R (2011) A likely possible origin of homochirality in amino acids and sugars on prebiotic earth. Tetrahedron Lett 52:2028–2032

Budin I, Szostak JW (2010) Expanding roles for diverse physical phenomena during the origin of life. Annu Rev Biophys 39:245–263

Cintas P (2002) Chirality of living systems: a helping hand from crystals and oligopeptides. Angew Chem Int Ed 41:1139–1145

Córdova A, Zou W, Dziedzic P, Ibrahem I, Reyes E, Xu Y (2006) Direct asymmetric intermolecular aldol reactions catalyzed by amino acids and small peptides. Chem Eur J 12:5383–5397

Crossley R (1992) The relevance of chirality to the study of biological activity. Tetrahedron 48: 8155–8178

Fletcher SP, Jagt RBC, Feringa BL (2007) An astrophysically-relevant mechanism for amino acid enantiomer enrichment. Chem Commun 43:2578–2580

Flores JJ, Bonner WA, Massey GA (1977) Asymmetric photolysis of (RS)-leucine with circularly polarized ultraviolet light. J Am Chem Soc 99:3622–3625

Frank FC (1953) On spontaneous asymmetric synthesis. Biochim Biophys Acta 11:459–463

Fuß W (2009) Does life originate from a single molecule? Chirality 21:299–304

Green M, Jain V (2010) Homochirality in life: two equal runners, one tripped. Orig Life Evol Biosph 40:111–118

Green MM, Park J-W, Sato T, Teramoto A, Lifson S, Selinger RLB, Selinger JV (1999) The macromolecular route to chiral amplification. Angew Chem Int Ed 38:3138–3154

Griesbeck AG, Meierhenrich UJ (2002) Asymmetric photochemistry and photochirogenesis. Angew Chem Int Ed 41:3147–3154

Hazen RM, Filley TR, Goodfriend GA (2001) Selective adsorption of L- and D-amino acids on calcite: implications for biochemical homochirality. Proc Natl Acad Sci USA 98:5487–5490

Hitz TH, Luisi PL (2004) Spontaneous onset of homochirality in oligopeptide chains generated in the polymerization of N-carboxyanhydride amino acids in water. Orig Life Evol Biosph 34: 93–110

Jacques J, Collet A, Wilen SH (1981) Enantiomers, racemates and resolution. Wiley, New York

Kawasaki T, Matsumura Y, Tsutsumi T, Suzuki K, Ito M, Soai K (2009) Asymmetric autocatalysis triggered by carbon isotope (13C/12C) chirality. Science 324:492–495

Klussmann M, Iwamura H, Mathew SP, Wells DH Jr, Pandya U, Armstrong A, Blackmond DG (2006a) Thermodynamic control of asymmetric amplification in amino acid catalysis. Nature 441:621–623

Klussmann M, White AJP, Armstrong A, Blackmond DG (2006b) Rationalisation and prediction of solution enantiomeric excess in ternary phase systems. Angew Chem Int Ed 45:7985–7989

Kondepudi DK, Kaufman RJ, Singh N (1990) Chiral symmetry breaking in sodium chlorate crystallization. Science 250:975–976

Kricheldorf HR (2006) Polypeptides and 100 years of chemistry of a-amino acid N-carboxyanhydrides. Angew Chem Int Ed 45:5752–5784

Lahav M, Weissbuch I, Shavit E, Reiner C, Nicholson GJ, Schurig V (2006) Parity violating energetic difference and enantiomorphous crystals-caveats; reinvestigation of tyrosine crystallization. Orig Life Evol Biosph 36:151–170

Leitereg TJ, Guadagni DG, Harris J, Mon TR, Teranishi R (1971) Evidence for the difference between the odors of the optical isomers (+)- and (-)-carvone. Nature 230:455–456

Lente G (2006) Stochastic analysis of the parity-violating energy differences between enantiomers and its implications for the origin of biological chirality. J Phys Chem A 110:12711–12713

Luisi PL (2006) The emergence of life: from chemical origins to synthetic biology. Cambridge University Press, Cambridge

Mauksch M, Tsogoeva SB, Martynova IM, Wei S (2007) Evidence of asymmetric autocatalysis in organocatalytic reactions. Angew Chem Int Ed 46:393–396

Meinert C, Filippi JJ, Nahon L, Hoffmann SV, d'Hendecourt L, de Marcellus P, Bredehöft JH, Thiemann WHP, Meierhenrich UJ (2010) Photochirogenesis: photochemical models on the origin of biomolecular homochirality. Symmetry 2:1055–1080

Morowitz HJ (1969) A mechanism for the amplification of fluctuations in racemic mixtures. J Theor Biol 25:491–494
Mukherjee S, Yang JW, Hoffmann S, List B (2007) Asymmetric enamine catalysis. Chem Rev 107: 5471–5569
Noorduin WL, Izumi T, Millemaggi A, Leeman M, Meekes H, Enckevort WJPV, Kellogg RM, Kaptein B, Vlieg E, Blackmond DG (2008) Emergence of a single solid chiral state from a nearly racemic amino acid derivative. J Am Chem Soc 130:1158–1159
Orgel LE (2004) Prebiotic chemistry and the origin of the RNA world. Crit Rev Biochem Mol Biol 39:99–123
Pasek M, Lauretta D (2008) Extraterrestrial flux of potentially prebiotic C, N, and P to the early Earth. Orig Life Evol Biosph 38:5–21
Patzke V, Kiedrowski GV (2007) Self replicating systems. ARKIVOC 293–310
Perry RH, Wu C, Nefliu M, Cooks RG (2007) Serine sublimes with spontaneous chiral amplification. Chem Commun 43:1071–1073
Pizzarello S, Cronin JR (2000) Non-racemic amino acids in the Murray and Murchison meteorites. Geochim Cosmochim Acta 64:329–338
Pizzarello S, Weber AL (2004) Prebiotic amino acids as asymmetric catalysts. Science 303:1151
Pizzarello S, Zolensky M, Turk KA (2003) Nonracemic isovaline in the Murchison meteorite: chiral distribution and mineral association. Geochim Cosmochim Acta 67:1589–1595
Pizzarello S, Huang Y, Fuller M (2004) The carbon isotopic distribution of Murchison amino acids. Geochim Cosmochim Acta 68:4963–4969
Podlech J (2001) Origin of organic molecules and biomolecular homochirality. Cell Mol Life Sci 58:44–60
Robertson A, Sinclair AJ, Philp D (2000) Minimal self-replicating systems. Chem Soc Rev 29:141–152
Rubinstein I, Eliash R, Bolbach G, Weissbuch I, Lahav M (2007) Racemic beta sheets in biochirogenesis. Angew Chem Int Ed 46:3710–3713
Saghatelian A, Yokobayashi Y, Soltani K, Ghadiri MR (2001) A chiroselective peptide replicator. Nature 409:797–801
Sato I, Omiya D, Saito T, Soai K (2000) Highly enantioselective synthesis induced by chiral primary alcohols due to deuterium substitution. J Am Chem Soc 122:11739–11740
Sato I, Yamashima R, Kadowaki K, Yamamoto J, Shibata T, Soai K (2001) Asymmetric induction by helical hydrocarbons: [6]- and [5]helicenes. Angew Chem Int Ed 40:1096–1098
Satyanarayana T, Abraham S, Kagan HB (2009) Nonlinear effects in asymmetric catalysis. Angew Chem Int Ed 48:456–494
Siegel JS (1998) Homochiral imperative of molecular evolution. Chirality 10:24–27
Soai K, Shibata T, Morioka H, Choji K (1995) Asymmetric autocatalysis and amplification of enantiomeric excess of a chiral molecule. Nature 378:767–768
Soai K, Osanai S, Kadowaki K, Yonekubo S, Shibata T, Sato I (1999) d- and l-quartz-promoted highly enantioselective synthesis of a chiral organic compound. J Am Chem Soc 121:11235–11236
Soai K, Shibata T, Sato I (2004) Discovery and development of asymmetric autocatalysis. Bull Chem Soc Jpn 77:1063–1073
Stoeffler C, Darquie B, Shelkovnikov A, Daussy C, Amy-Klein A, Chardonnet C, Guy L, Crassous J, Huet TR, Soulard P, Asselin P (2011) High resolution spectroscopy of methyltrioxorhenium: towards the observation of parity violation in chiral molecules. Phys Chem Chem Phys 13:854–863
Strasdeit H (2005) New studies on the Murchison meteorite shed light on the pre-RNA world. Chembiochem 6:801–803
Triggle DJ (1997) Stereoselectivity of drug action. Drug Discov Today 2:138–147
Viedma C (2005) Chiral symmetry breaking during crystallization: complete chiral purity induced by nonlinear autocatalysis and recycling. Phys Rev Lett 94:065504
Viedma C, Ortiz JE, de Torres T, Izumi T, Blackmond DG (2008) Evolution of solid phase homochirality for a proteinogenic amino acid. J Am Chem Soc 130:15274–15275
Wang X, Zhang Y, Tan H, Wang Y, Han P, Wang DZ (2010) Enantioselective organocatalytic Mannich reactions with autocatalysts and their mimics. J Org Chem 75:2403–2406

Weber AL, Pizzarello S (2006) The peptide-catalyzed stereospecific synthesis of tetroses: a possible model for prebiotic molecular evolution. Proc Natl Acad Sci USA 103:12713–12717

Weissbuch I, Leiserowitz L, Lahav M (2005) Stochastic "mirror symmetry breaking" via self-assembly, reactivity and amplification of chirality: relevance to abiotic conditions. Top Curr Chem 259:123–165

Welch CJ (2001) Formation of highly enantioenriched microenvironments by stochastic sorting of conglomerate crystals: a plausible mechanism for generation of enantioenrichment on the prebiotic earth. Chirality 13:425–427

Wesendrup R, Laerdahl JK, Compton RN, Schwerdtfeger P (2003) Biomolecular homochirality and electroweak interactions. I. The Yamagata hypothesis. J Phys Chem A 107:6668–6673

Wu CS, Ambler E, Hayward RW, Hoppes DD, Hudson RP (1957) Experimental test of parity conservation in beta decay. Phys Rev 105:1413–1415

Yamagata Y (1966) A hypothesis for the asymmetric appearance of biomolecules on earth. J Theor Biol 11:495–498

Biodata of **Gábor Lente**, author of "***Absolute Asymmetric Synthesis and the Origin of Biological Chirality***."

**Professor Gábor Lente** is currently an associate professor of the Department of Inorganic and Analytical Chemistry at the Faculty of Science of Technology of the University of Debrecen in Hungary. He obtained his Ph.D. in chemistry from the University of Debrecen in 2001 and spent 2 years as a visiting scientist at the Ames Laboratory operated by the Iowa State University in Ames, IA, USA. Professor Lente's scientific interests are in the areas of inorganic reaction mechanisms and stochastic kinetic modeling of chemical reactions.

E-mail: **lenteg@delfin.unideb.hu**

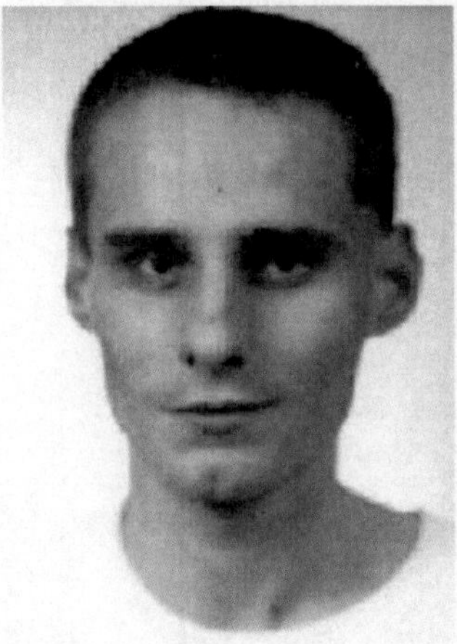



J. Seckbach (ed.), *Genesis - In The Beginning: Precursors of Life, Chemical Models and Early Biological Evolution*, Cellular Origin, Life in Extreme Habitats and Astrobiology 22, 509–523
DOI 10.1007/978-94-007-2941-4_27, 

# ABSOLUTE ASYMMETRIC SYNTHESIS AND THE ORIGIN OF BIOLOGICAL CHIRALITY

**GÁBOR LENTE**
*Department of Inorganic and Analytical Chemistry, University of Debrecen, P.O.B. 10, H-4010 Debrecen, Hungary*

## 1. Introduction

Chiral molecules are quite abundant in nature and play most of the important structural and functional roles in biological systems. It has long been recognized that there is a major imbalance in the natural occurrence of different enantiomers of various molecules. This phenomenon is usually called "homochirality" or "biological chirality" (Pályi et al., 2004) and might be surprising at first sight because the energies of mirror-image enantiomeric pairs are the same. Thermodynamic equilibrium would require the formation of a racemic mixture of enantiomers in all cases. The most common small biomolecules which serve as building blocks of biopolymers are amino acids and carbohydrates, all of which except glycine contain asymmetric carbon atoms and are consequently chiral. Experience shows that L-amino acids and D-carbohydrates are much more common in nature than their mirror-image counterparts.

Present scientific theories of the emergence of Earth and life on it all agree in that there was some initial point in time where molecular chirality could not have existed simply because molecules did not exist. Therefore, molecular asymmetry could not have always existed and must have emerged at some point. Various scientific theories to interpret the origin of biological chirality have been advanced ever since it was recognized. One such theory is that the ultimate cause is absolute asymmetric synthesis (Mislow, 2003; Pályi et al., 2005), which formed a non-racemic mixture of enantiomers at some time during prebiotic evolution. This review summarizes some recent related developments in theoretical and experimental aspects related to absolute asymmetric synthesis and their significance in the origin of biological chirality and life.

## 2. Biological Chirality: Evolutionary Necessity and Random Choice

The basic processes of life known on Earth can only work based on homochiral building blocks (Keszthelyi, 1995; Fujii and Saito, 2004). In more precise wording, enantiomers of the same substance cannot be freely exchangeable in biological systems. This is easily understood based on a simple line of thought. Most biopolymers contain hundreds or thousands of chiral repeating units. In the case of only 200 such repeating units, an imagined biopolymer based on both enantiomers of the building blocks would actually be a varying mixture of potentially $2^{200}$ ($\approx 1.6\times10^{60}$) different molecules, all having the same connections of atoms, but each of the $2^{199}$ ($\approx 8\times10^{59}$) pairs of enantiomers featuring a different geometry. Any reasonable amount of a mixture like this would be extremely unlikely to contain two identically shaped biopolymers. The function of these molecules depends critically on their geometry; therefore, biopolymers based on both enantiomers of the building blocks would be useless. It should be emphasized that the term "homochirality" does not necessarily imply that only one of the two mirror-image enantiomers has biological functions; it only means that the enantiomers are not exchangeable. For example, D-amino acids, which do not have a significant role in the proteins of higher organisms, have well-defined functions in bacterial life (Höltje, 1998; Konno et al., 2007; Lam et al., 2009; Kolodkin-Gal et al., 2010). It is also a widespread opinion that the selective generation of chiral substances must have preceded the appearance of life itself and was an important part of prebiotic chemistry.

When interpreting the origins of biological chirality, two basic questions have to be answered. The first concerns the mechanism how one of two enantiomers could accumulate in large excess over the other despite the apparently symmetric forces of nature. A second and notably different question is why just L-amino acids and D-carbohydrates were selected for their important biological roles by evolution. There are two opposing views in this respect, which will be referred to as the *de lege* and the *de facto* hypothesis following the terminology established earlier (Quack, 2002). The *de lege* (lawlike) view assumes that the present dominance of L-amino acids and D-carbohydrates somehow follows from the laws of nature. The *de facto* (factlike) view posits that there is no such preference law in nature and the present dominance is simply a result of random choice.

Present scientific information seems to prefer the *de facto* view. Currently, there is no known mechanism supported by experimental data that would lead to *de lege* selection. Although there is some fundamental asymmetry in physics called parity violation, its effect on chiral molecules, which is still an experimentally unconfirmed theoretical prediction, is most likely not consistent enough to have any influence on evolution (Crassous et al., 2005). In addition, the predicted effect is smaller than natural fluctuations under normal conditions (Lente, 2006, 2007; Fuß, 2009a, b). The statistical background and the consequences of this fact will be analyzed in some detail in a later section of this review.

An observation-based approach to decide this question would be provided by finding extraterrestrial life. If such life forms depended on D-amino acids or L-carbohydrates, this fact would clearly disprove the *de lege* view. Disproving the *de facto* hypothesis in this way would be much more difficult as it would require studying numerous (preferably in excess of ten) independently evolved life systems where amino acids or carbohydrates have significant roles. Of course, the independent evolution should be proven first in a case like this. Another, more reasonable experimental approach would be to synthesize two batches of polypeptides with enzyme activity on a non-chiral substrate. The first batch would be based on L-amino acids, while the other on D-amino acids exclusively. An experimental confirmation of the expected identical artificial enzyme activities on the non-chiral substrate would be an argument against *de lege* selection as both mirror images of the macromolecule would be shown to function equally well, which would rule out any thermodynamic preference between them.

One might counterargue that the present dominance is a consequence of *de lege* kinetic reasons and not thermodynamic ones connected to the function of these biopolymers. However, the weight of such an argument is seriously decreased by the racemization problem (Bada and Miller, 1987). Amino acids are known to undergo spontaneous racemization in nature, the time scale of which is much shorter than usual geological time scales. Given the right conditions (first of all, reliable information about temperatures), this racemization process can be used for dating certain samples of biological origin on time scales of a few thousand years in a method popularly known as the amino acid clock (Miller et al., 1979; Brooks et al., 1990). The observed spontaneous racemization of amino acids actually is a very strong argument against any *de lege* kinetic interpretation.

## 3. Absolute Asymmetric Synthesis

### 3.1. DEFINITION

The definition of absolute asymmetric synthesis used here follows the recommendation of Mislow (2003): it is a process that forms significant enantiomeric excesses of a chiral product from non-chiral reactants in the absence of any asymmetric external influence. It should be noted that it is quite common in the literature to restrict the chiral influence in the definition to chemical compounds showing molecular chirality. This practice follows earlier recommendations (Bredig et al., 1923; Feringa and van Delden, 1999). In the views of the present author, this earlier definition is quite misleading as there is nothing absolute in the preferred formation of one enantiomer under the influence of a surface of a chiral crystal or intense circularly polarized electromagnetic radiation. In these reactions, the mirror-image pairs of the chiral crystals or oppositely polarized radiation would clearly favor the formation of the opposite enantiomer. Nevertheless, literature examples of this usage are found by the hundreds, only two of which are given here as examples (Nishino et al., 2002; Sakamoto et al., 2004). The fact that the chiral preference depends on asymmetric external influence is often clear from the article titles as well.

The term symmetry breaking is also often used in the context of absolute asymmetric synthesis. This is not necessarily fortunate as it may be misunderstood to imply that there is no symmetry between enantiomers in such processes.

However, the known experimental examples of absolute asymmetric synthesis show stochastic nature. The stochastic distribution of enantiomers is definitely expected to show some symmetry properties: any given enantiomeric excess of the R enantiomer will form with the same probability as the same excess of the S enantiomer. This is actually an important point that is used to make a difference between stochastic nature and irreproducibility (Lente, 2005, 2010). In different words, if one is to repeat the same batch reaction of absolute asymmetric synthesis many times and mixes the final products obtained, this final mixture is expected to be racemic. This is an expression of a symmetry that is different from the usual deterministic idea of a racemic mixture – but it is still a sort of symmetry.

## 3.2. EXPERIMENTAL EXAMPLES

Two significantly different examples of absolute asymmetric synthesis have been reported thus far in the chemical literature. In both cases, the stochastic nature of the process is evident, and some statistical analyses have also been carried out. The analyses most often use the enantiomeric excess, which is given by the following formula:

$$ee = \frac{|r - s|}{r + s} \qquad (1)$$

where $r$ means the number of molecules formed for the R enantiomer and $s$ means the number of molecules formed for the S enantiomer. By definition, $ee\%$ ($= 100 \times ee$) is between 0% and 100% and the actual enantiomer in excess has to be specified for clarity (e.g. 56% S).

The Soai reaction provides the example of absolute asymmetric synthesis that has been studied in most details (Soai et al., 1995, 1996, 2003; Kawasaki et al., 2006, 2009; Soai and Kawasaki, 2009). The chemical reaction itself is the nucleophilic addition of a zinc alkyl to a pyrimidine carbaldehyde. Upon aqueous workup, a chiral alcohol is obtained. In terms of chemical structures, the process can be represented by the following equation:

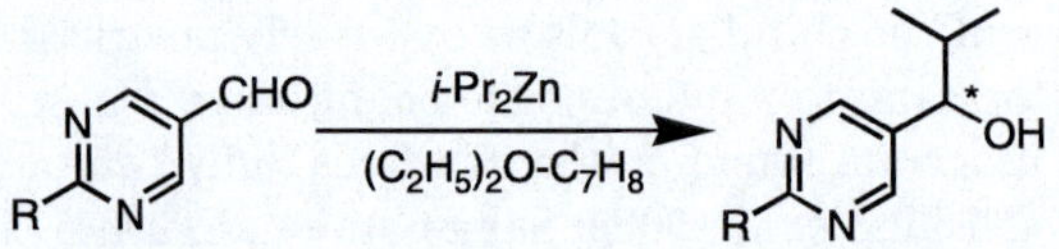


(2)

Group $R$ is variable in Eq. 2. This reaction is actually not as close to the probable conditions of prebiotic processes as one might wish: it has to be carried out with the strict exclusion of water and air because of the sensitive zinc alkyl reagent. Nevertheless, the enantiomeric excess of the product alcohol shows high,

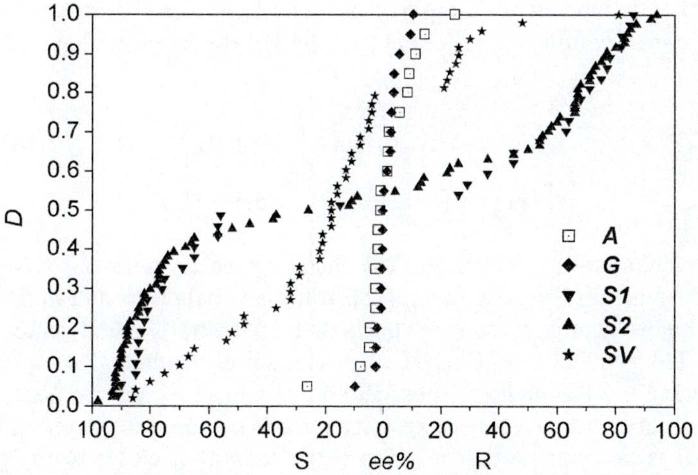


**Figure 1.** Distribution of enantiomeric excesses in reaction (2) (*G*: Gridnev et al., 2003; *S1*: Soai et al., 2003; *S2*: Kawasaki et al., 2006; *SV*: Singleton and Vo, 2002) and reaction (3) (*A*: Asakura et al., 2000).

stochastic fluctuations (Soai et al., 1996, 2003; Kawasaki et al., 2006). Figure 1 shows these stochastic fluctuations by displaying probability distributions. The quantity on the *y* axis is the probability distribution function, which has the enantiomeric excess as the independent variable and gives the probability of the formation of an enantiomeric excess smaller than the independent variable. By definition, the value of the distribution function is 0 at 100% S *ee*, whereas it is 1 at 100% R *ee*. This representation seems much better than histograms, which were often used in earlier articles, as histograms necessarily involve some rather arbitrary categorization of measured points, the dangers of which were pointed out (Lente, 2005). Figure 1 shows two distributions published by the Soai group: series S1 is constructed from 37 identical experiments (Soai et al., 2003), whereas series S2 is a result of 84 repetitions under somewhat different experimental conditions (Kawasaki et al., 2006). Usual *ee* values obtained in these processes are between 40% and 90%.

The basic features of the Soai reaction have now been reproduced in a number of laboratories around the world (Singleton and Vo, 2002, 2003; Gridnev et al., 2003). Distributions observed in two of these studies are also shown in Fig. 1: series G is from 20 identical experiments (Gridnev et al., 2003), whereas series SV was adopted to show an obviously asymmetric distribution based on 48 experiments (Singleton and Vo, 2002). It should be noted that the experiments of series SV were not carried out under strictly identical conditions, and therefore drawing any statistical conclusions would be highly questionable. The interested reader is referred to two recent reviews on experimental progress on the chemistry of the Soai reaction (Soai and Kawasaki, 2009; Gehring et al., 2010).

Another, independent example of absolute asymmetric synthesis involves a trinuclear cobalt complex (Asakura et al., 2000) in the following ligand exchange reaction:

$$\begin{aligned}&\left[(\mathrm{en})_2\mathrm{Co}(\mu-\mathrm{OH})_2\mathrm{Co}(\mu-\mathrm{OH})_2\mathrm{Co}(\mathrm{en})_2\right]^{4+} + 2\mathrm{NH_4Br} + 2\mathrm{H^+} + 2\mathrm{H_2O} = \\ &\qquad 2\left[(\mathrm{en})_2\mathrm{Co}(\mathrm{NH_3})\mathrm{Br}\right]^{2+} + \left[\mathrm{Co}(\mathrm{H_2O})_6\right]^{2+}\end{aligned} \tag{3}$$

In this equation, "en" denotes the chelating bidentate ligand 1,2-diamino-ethane. The reactant trinuclear complex features a cobalt(II) center in the middle, whereas the en ligands coordinate to a pair of cobalt(III) metal centers on the terminus. The product $[(\mathrm{en})_2\mathrm{Co}(\mathrm{NH_3})\mathrm{Br}]^{2+}$ is a chiral octahedral complex whose kinetically inert nature makes it possible to distinguish its enantiomers. The distribution observed in 20 repetitive reactions in this system is displayed in Fig. 1 as series A. It is noteworthy that reaction (3) is somewhat closer to the probable conditions of prebiotic processes as it involves an octahedral complex of a biologically essential metal in aqueous solution exposed to air. Usual *ee* values are quite small in this process; they seldom exceed 10%.

These experimental examples of absolute asymmetric synthesis are extremely important as they transformed the research field from purely theoretical scientific speculations into the realm of experimentally testable hypotheses. Efforts have been made to interpret numerous aspects of the processes including modeling the statistical distributions. Although the distributions shown in Fig. 1 might look different, in fact, all represent the same phenomenon of absolute asymmetric synthesis, but the actual chemical mechanisms or parameters are different. However, further work is needed to demonstrate that multistep chemical mechanisms proposed for these reactions can actually account for the stochastic nature of the processes.

## 3.3. INTERPRETATION: INITIAL FLUCTUATIONS

The mentioned examples of absolute asymmetric synthesis all start from strictly non-chiral conditions and are never under the influence of asymmetric external effects. Therefore, a successful model of these processes also must start from conditions where no chiral material is present. Still, when one thinks about events on a molecular level, it is inevitable to find that it is only the very first chiral molecule that is certain to be formed in a strictly non-chiral environment (Caglioti et al., 2005; 2008). All the rest of the molecules may experience some internal chiral effects because of the pool of chiral molecules already formed. Obviously, the probability of obtaining an S enantiomer as the first chiral molecule is 50%, and obtaining the other enantiomer has the same probability. This line of thought can be extended to higher numbers of molecules, which is usually quite reasonable experimentally as the chiral environment represented by a single molecule (or very few molecules) is unlikely to affect the formation of the next molecules. The statistical description of this process is equivalent to the stochastic view of racemization. As this is necessarily present in the beginning of all absolute asymmetric processes, it is understandable that the enantiomeric distribution of racemization has been analyzed in a number of papers on various levels of sophistication (Mills, 1932; Siegert, 1949; Mislow, 2003; Caglioti et al., 2005, 2006, 2008; Pályi et al., 2005; Caglioti et al., 2006; Lente, 2006, 2007; Barabás et al., 2008a; Caglioti and Pályi, 2008; Fuß, 2009a, b).

A racemization process, irrespective of its chemical mechanism, produces a product mixture that can be characterized with a symmetric binomial distribution. When the total number of molecules formed is $N$, the probability of obtaining a mixture of exactly $r$ molecules of the $R$ enantiomer is given with a simple formula:

$$P(r) = \binom{N}{r} 0.5^N \tag{4}$$

This is a relatively simple distribution whose properties can be found in most introductory statistics textbooks and usually exemplified extensively by coin tossing. The expectations for the number ($0 \le r \le N$) or molar fraction ($0 \le x_R = r/N \le 1$) of $R$ enantiomers are very simply obtained as half of the total number of molecules (for $r$) or 0.5 (for $x_R$).

$$\langle r \rangle = 0.5N \qquad \langle x_R \rangle = 0.5 \tag{5}$$

The standard deviation, which characterizes the average error, can be given for $r$ or $x_R$ as follows:

$$\sigma_r = 0.5N^{0.5} \qquad \sigma_{x_R} = 0.5N^{-0.5} \tag{6}$$

It is more difficult to determine the expectation for the enantiomeric excess as the quantity defined in Eq. 1 is not covered in standard statistics. The first part of the formula (7) below was formulated as a conjecture (Lente, 2010). It can actually be proved by rigorous mathematical methods, but this is beyond the scope of this review.

$$\langle ee \rangle = \frac{1}{N} \prod_{i=1}^{l} \frac{2i+1}{2i} = \frac{(2l+1)!}{l!l!2^{2l} N} = \binom{2l}{l} \frac{2l+1}{2^{2l} N} \approx \sqrt{\frac{2}{\pi N}} \tag{7}$$

In this formula, $l = N/2 - 1$ for even values of $N$, whereas $l = (N-1)/2$ for odd values of $N$. Similarly, an unpublished proof for deriving the standard deviation of the enantiomeric excess shows that it is given by the following equation:

$$\sigma_{ee} = \sqrt{\frac{1}{N} - \frac{1}{N^2} \binom{2l}{l} \binom{2l}{l} \frac{(2l+1)^2}{2^{4l}}} \tag{8}$$

It is well known that a discrete, symmetric binomial distribution can be approximated by a continuous normal distribution for high numbers of $N$. An estimate for $ee_{1/2}$, which is defined as the critical value where the probability function reaches 50%, was introduced based on the properties of the normal distribution (Mills, 1932; Barabás et al., 2008b) and is frequently cited up to today:

$$ee_{1/2} = \frac{0.6743}{\sqrt{N}} \tag{9}$$

Furthermore, it is quite frequently stated that the formation of exactly equal numbers of the two different enantiomers is extremely unlikely in racemic mixtures. Actually, this is impossible when $N$ is odd. When $N$ is even, this "strictly racemic" probability can be obtained from Eq. 4 by simply substituting $r = N/2$. However, this very small probability tends to be a minor point as the other formulas clearly show that it is still the most likely of all individual outcomes, and the expected enantiomeric excesses in racemization for high values of $N$ will be below the detection limit for most or perhaps all practical experimental methods. Figure 2 illustrates this point by displaying the expectation for *ee* as a function of the total number of chiral molecules. Both axes in this figure are shown on a logarithmic scale. The last part of Eq. 7 describes a straight line with a slope of −0.5; this dominates the entire figure as the molecule numbers are changed in an extremely wide range.

Figure 2 clearly shows that expected enantiomeric excesses are extremely low at chemically reasonable amounts of substances (1 μmol and higher: $N > 6 \times 10^{17}$). This figure can also be used to demonstrate that effects resulting in very small asymmetry cannot play a significant role as a basis for a *de lege* mechanism after

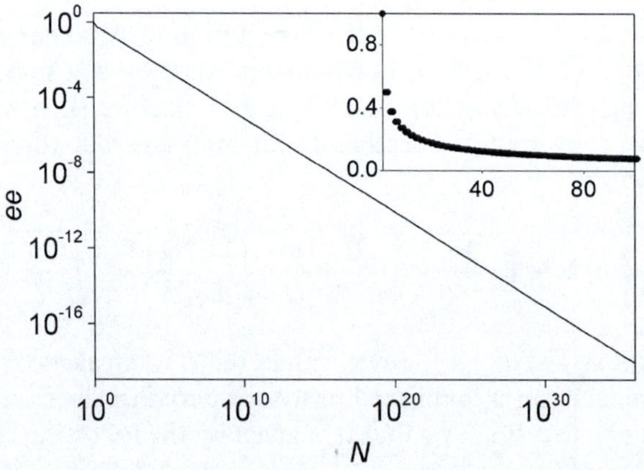


**Figure 2.** Expectation for the enantiomeric excess (*ee*) as a function of the number of molecules (*N*) in racemization as calculated from the binomial distribution. *Inset*: the part of the figure at low molecule numbers magnified (both axis scales are linear).

amplification. For example, the present theoretical estimates of the parity-violating energy difference between enantiomers are about $10^{-13}$ J/mol (Quack, 2002), which translates into an expected enantiomeric excess of about $10^{-17}$ at room temperature (Lente, 2006, 2007). The calculated value should be higher than the *ee* expected from natural fluctuations if it is to play any role at all. Figure 2 shows that this can only occur at $N > 6 \times 10^{33}$ (10 Gmol), which is an extraordinarily high amount of substance, even more so if one also considers that this should be a minute fraction of the chiral material, most of which should be generated in a following amplification process.

The importance of these natural fluctuations in racemization as described by a binomial distribution extends beyond the exclusion of certain *de lege* mechanistic possibilities. For all mechanisms, this process provides an initial imbalance (for the first molecule at the very least) of enantiomers that can be greatly amplified by suitable mechanisms. It also emphasizes the need for elements of stochastic thinking in the interpretation of absolute asymmetric synthesis.

## 3.4. INTERPRETATION: AMPLIFICATION

Small initial enantiomeric excesses can be amplified under certain conditions. This idea probably originates from Frank (1953), who proposed a simple such model, referred to as the Frank model in today's literature, without having any particular chemical system in mind. This model introduced two basic concepts that continue to be productive in this field of theoretical research. The first is the enantioselective autocatalysis, which means that a chiral molecule already formed promotes the formation of further molecules of the same enantiomer without having a similar effect on the opposite enantiomer. This is possible as the existing chiral molecule(s) can provide the asymmetric environment necessary to distinguish between enantiomers. The second important concept was termed mutual antagonism by Frank. This refers to a reaction whereby two enantiomers of opposite chirality react with each other to give a non-chiral product.

Whereas the role of enantioselective autocatalysis (which is not necessarily first order) is obvious in increasing enantiomeric excess, the same may not be so clear for mutual antagonism. A closer look at Eq. 1 reveals important aspects in this regard (Lente and Ditrói, 2009). Enantioselective autocatalysis increases *ee* by overcompensating the increase in the denominator in Eq. 1 by a larger relative increase in the enumerator. Mutual antagonism, on the other hand, leaves the enumerator constant and exerts its effect by decreasing the denominator. This also has the remarkable (but often forgotten) consequence that mutual antagonism decreases the overall amount of chiral material.

A common feature of enantioselective autocatalysis and mutual antagonism is that they can only be effective if at least one of $r$ and $s$ is non-zero and $r \neq s$. Therefore, previous existence of chiral material and an imbalance of enantiomers are both needed. This is why the term amplification is used to describe these processes collectively.

Theoretical research in the last five decades has produced and studied a lot of different mechanism that can amplify minor initial enantiomeric excesses to significant final values. However, most of these models used the deterministic approach to chemical kinetics, which requires the initial small enantiomeric excess as an initial condition (Blackmond, 2000; Kondepudi and Asakura, 2001; hundreds of other related references are not included here to save space). This is rather unsatisfactory from the point of view of interpreting absolute asymmetric synthesis where the essence of the phenomenon is the formation of large enantiomeric excesses from a strictly non-chiral initial state. Attempts could be made by combining the use of the binomial distribution described in the previous section up to the formation of a few molecules with a deterministic amplification model. However, this would have the disadvantage of a sudden switch from one mechanism (racemization) to another one (amplification). In reality, this change must be continuous with conditions under which both types of processes are significant simultaneously. This can be achieved by models that are exclusively stochastic. Some success has been reached in this field recently (Lente, 2004, 2005, 2007, 2010; Saito et al., 2007; Shao and Liu, 2007; Micskei et al., 2008; Sugimori et al., 2008; Lente and Ditrói, 2009; Barabás et al., 2010).

A major drawback of the stochastic kinetic approach is that the simulations cannot be done in a routine way similarly to the simultaneous solution of differential equations in deterministic kinetics. The high number of stochastic variables necessitates the derivation of specific analytical formulas, which seems to be feasible only for very simple chemical kinetic models. For example, a model consisting of the uncatalyzed formation of a racemic product combined with first-order enantioselective autocatalysis was shown to give the following probability distribution for the final product mixture (Lente, 2004, 2005; Shao and Liu, 2007):

$$P(r) = \binom{N}{r} \frac{\prod_{j=0}^{r-1}(0.5+\alpha j) \prod_{j=0}^{N-r-1}(0.5+\alpha j)}{\prod_{j=0}^{N-1}(1+\alpha j)} \tag{10}$$

Parameter $\alpha$ in distribution (10) is characteristic of the ratio of the rates of the uncatalyzed step and the enantioselective autocatalysis. The expectation and the standard deviation for the molar fraction $x_R$ were also determined:

$$\langle x_R \rangle = 0.5 \qquad \sigma_{x_R} = 0.5\sqrt{\frac{N^{-1}+\alpha}{1+\alpha}} \tag{11}$$

Finally, it was also shown that distribution (10) tends to a symmetric beta distribution for large numbers of $N$ (Lente, 2004).

For more complicated and more realistic mechanisms, analytical formulas might be very difficult to find. In such cases, three general strategies have been used thus far. The first is based on carrying out exact stochastic calculations for several different small molecule numbers (Lente and Ditrói, 2009) for which the numerical solution is computationally viable. Some limited conclusions can be drawn by observing the dependence of certain descriptors (such as *ee*) as a function of the molecule number. Another strategy is to use stochastic calculations up to a predetermined low molecule number, after which the deterministic equations are used to extrapolate to a high amount of substance (Lente, 2005). Finally, Monte Carlo methods (sometimes called directed random walk) can also be used to carry out simulations (Saito et al., 2007). It should be emphasized that the computational problems in carrying out full numerical calculations for larger molecule numbers are primarily due to limited memory rather than computer performance problems. This makes it unlikely that future improvements in computer technology will substantially improve the efficiency of such calculations.

It should be noted that statistical analysis of experimental distributions has also been done without any particular chemical mechanism (Lente, 2005, 2007; Barabás et al., 2007, 2009). This is imperative when the symmetry of experimental distributions is tested, which is an important test distinguishing between stochastic nature and irreproducibility.

## 4. Conclusion

With the experimental examples of absolute asymmetric synthesis discovered in the last two decades, it is now possible to model a possible scenario of the origin of biological chirality based on solid observations. These observations clearly confirm the stochastic nature of the formation of significant enantiomeric excesses from non-chiral reactants in the absence of any external asymmetric effects. These results can be readily interpreted using stochastic lines of thought. The statistical description of racemization has special importance in this respect as it provides initial fluctuations that can be amplified by suitable mechanisms in the later part of the processes. The amplifications process can also be modeled using the stochastic approach to chemical kinetics.

## 5. References

Asakura K, Ikumo A, Kurihara K, Osanai S, Kondepudi DK (2000) Random chiral asymmetry generation by chiral autocatalysis in a far-from-equilibrium reaction system. J Phys Chem A 104:2689–2694

Bada JL, Miller SL (1987) Racemization and the origin of optically active organic compounds in living organisms. Biosystems 20:21–26

Barabás B, Caglioti L, Zucchi C, Maioli M, Gál E, Micskei K, Pályi G (2007) Violation of distribution symmetry in statistical evaluation of absolute enantioselective synthesis. J Phys Chem B 111:11506–11510

Barabás B, Caglioti L, Faglioni F, Florini N, Lazzaretti P, Maioli M, Micskei K, Rábai G, Taddei F, Zucchi C, Pályi G (2008a) On the traces of absolute enantioselective synthesis. AIP Conf Proc 963B:1150–1152

Barabás B, Caglioti L, Micskei K, Zucchi C, Pályi G (2008b) Isotope chirality and asymmetric autocatalysis: a possible entry to biological chirality. Orig Life Evol Biosph 38:317–327

Barabás B, Caglioti L, Micskei K, Pályi G (2009) Data-based stochastic approach to absolute asymmetric synthesis by autocatalysis. Bull Chem Soc Jpn 82:1372–1376

Barabás B, Tóth J, Pályi G (2010) Stochastic aspects of asymmetric autocatalysis and absolute asymmetric synthesis. J Math Chem 48:457–489

Blackmond DG (2000) Kinetic aspects of nonlinear effects in asymmetric catalysis. Acc Chem Res 33:402–411

Bredig G, Mangold P, Williams TG (1923) Absolute asymmetric synthesis. Z Angew Chem 36:456–458

Brooks AS, Hare PE, Kokis JE, Miller GH, Enrst RD, Wendorf F (1990) Dating Pleistocene archeological sites by protein diagenesis in ostrich eggshell. Science 248:60–64

Caglioti L, Pályi G (2008) Chiral chemistry of single molecules. Chem Today 26(3):41–42

Caglioti L, Zucchi C, Pályi G (2005) Single-molecule chirality. Chem Today 23(5):38–43

Caglioti L, Hajdu C, Holczknecht O, Zékány L, Zucchi C, Micskei K, Pályi G (2006) The concept of racemates and the Soai-reaction. Viva Origino 34:62–80

Caglioti L, Barabás B, Faglioni F, Florini N, Lazzeretti P, Maioli M, Micskei K, Rábai G, Taddei F, Zucchi C, Pályi G (2008) On the track of absolute enantioselective catalysis. Chem Today 26(5): 30–32

Crassous J, Chardonnet C, Saue T, Schwerdtfeger P (2005) Recent experimental and theoretical developments towards the observation of parity violation (PV) effects in molecules by spectroscopy. Org Biomol Chem 3:2218–2224

Feringa BL, van Delden RA (1999) Absolute asymmetric synthesis: the origin, control, and amplification of chirality. Angew Chem Int Ed Engl 38:3418–3438

Frank FC (1953) On spontaneous asymmetric synthesis. Biochim Biophys Acta 11:459–463

Fujii N, Saito T (2004) Homochirality and life. Chem Rec 4:267–278

Fuß W (2009a) Does life originate from a single molecule? Chirality 21:299–304

Fuß W (2009b) Biological homochirality as result from a single event. Colloids Surf B: Biointerfaces 74:498–503

Gehring T, Busch M, Sclageter M, Weingand D (2010) A concise summary of experimental facts about the Soai reaction. Chirality 22. doi:10.1002/chir.20849

Gridnev ID, Serafimov JM, Quiney H, Brown JM (2003) Reflections on spontaneous asymmetric synthesis by amplifying autocatalysis. Org Biomol Chem 1:3811–3819

Höltje JV (1998) Growth of the stress-bearing and shape-maintaining murein sacculus of *Escherichia coli*. Micribiol Mol Biol Rev 62:181–203

Kawasaki T, Suzuki K, Shimizu M, Ishikawa K, Soai K (2006) Spontaneous absolute asymmetric synthesis in the presence of achiral silica gel in conjunction with asymmetric autocatalysis. Chirality 18:479–482

Kawasaki T, Matsamura Y, Tsutsumi T, Suzuki K, Ito M, Soai K (2009) Asymmetric autocatalysis triggered by carbon isotope (13C/12C) chirality. Science 324:492–495

Keszthelyi L (1995) Origin of the homochirality of biomolecules. Quart Rev Biophys 28:473–507

Kolodkin-Gal I, Romero D, Cao S, Clardy J, Kolter R, Losick R (2010) D-amino acids trigger biofilm disassembly. Science 328:627–629

Kondepudi DK, Asakura K (2001) Kinetic aspects of nonlinear effects in asymmetric catalysis. Acc Chem Res 34:946–954

Konno R, Brückner H, D'Aniello A, Fischer G, Fujii N, Homma H (eds) (2007) D-Amino acids: a new frontier in amino acids and protein research – practical methods and protocols. Nova, New York

Lam H, Oh DC, Cava F, Takacs CN, Clardy J, de Pedro MA, Waldor MK (2009) D-Amino acids govern stationary phase cell wall remodeling in bacteria. Science 325:1552–1555

Lente G (2004) Homogeneous chiral autocatalysis: a simple, purely stochastic kinetic model. J Phys Chem A 108:9475–9478
Lente G (2005) Stochastic kinetic models of chiral autocatalysis: a general tool for the quantitative interpretation of total asymmetric synthesis. J Phys Chem A 109:11058–11063
Lente G (2006) Stochastic analysis of the parity-violating energy differences between enantiomers and its implications for the origin of biological chirality. J Phys Chem A 110:12711–12713
Lente G (2007) The effect of parity violation on kinetic models of enantioselective autocatalysis. Phys Chem Chem Phys 9:6134–6141
Lente G (2010) The role of stochastic models in interpreting the origins of biological chirality. Symmetry 2:767–798
Lente G, Ditrói T (2009) Stochastic kinetic analysis of the Frank model. Stochastic approach to flow-through reactors. J Phys Chem A 113:7237–7242
Micskei K, Rábai G, Gál E, Caglioti L, Pályi G (2008) Oscillatory symmetry breaking in the Soai reaction. J Phys Chem B 112:9196–9200
Miller GH, Hollin JT, Andrews JT (1979) Aminostratigraphy of UK Pleistocene deposits. Nature 281:539–543
Mills W (1932) Some aspects of stereochemistry. Chem Ind (London) 51:750–759
Mislow K (2003) Absolute asymmetric synthesis: a commentary. Collect Czech Chem Commun 68:849–864
Nishino H, Kosaka A, Hembury GA, Aoki F, Miyauchi K, Shitomi H, Onuki H, Inoue Y (2002) Absolute asymmetric photoreactions of aliphatic amino acids by circularly polarized synchrotron radiation: critically pH-dependent photobehavior. J Am Chem Soc 124:11618–11627
Pályi G, Zucchi C, Caglioti L (eds) (2004) Progress in biological chirality. Elsevier, Oxford
Pályi G, Micskei K, Zékány L, Zucchi C, Caglioti L (2005) Racemates and the Soai reaction. Magyar Kém Lapja 60:17–24 (in Hungarian)
Quack M (2002) How important is parity violation for molecular and biomolecular chirality? Angew Chem Int Ed Engl 41:4619–4630
Saito Y, Sugimori T, Hyuga H (2007) Stochastic approach to enantiomeric excess amplification and chiral symmetry breaking. J Phys Soc Jpn 76:044802
Sakamoto M, Kobaru S, Mino T, Fujita T (2004) Absolute asymmetric synthesis by nucleophilic carbonyl addition using chiral crystals of achiral amides. Chem Commun 1002–1003
Shao J, Liu L (2007) Stochastic fluctuations and chiral symmetry breaking: exact solution of Lente model. J Phys Chem A 111:9570–9572
Siegert AJF (1949) On the approach to statistical equilibrium. Phys Rev 76:1708–1714
Singleton DA, Vo LK (2002) Enantioselective synthesis without discrete optically active additives. J Am Chem Soc 124:10010–10011
Singleton DA, Vo LK (2003) A few molecules can control the enantiomeric outcome. Evidence supporting absolute asymmetric synthesis using the Soai asymmetric autocatalysis. Org Lett 5:4337–4339
Soai K, Kawasaki T (2009) Asymmetric autocatalysis. Automultiplication of chiral molecules. Chem Today 27(6, Suppl):3–7
Soai K, Shibata T, Morioka H, Choji K (1995) Asymmetric autocatalysis and amplification of enantiomeric excess of a chiral molecule. Nature 378:767–768
Soai K, Shibata T, Kowata Y (1996) Production of optically active pyrimidylalkyl alcohol by spontaneous asymmetric synthesis. Japan Kokai Tokkyo Koho JP 1997 9–268179. Application date: February 1 and April 18, 1996 (in Japanese)
Soai K, Sato I, Shibata T, Komiya S, Hayashi M, Matsueda Y, Imamura H, Hayase T, Morioka H, Tabira H, Yamamoto J, Kowata Y (2003) Asymmetric synthesis of pyrimidyl alkanol without adding chiral substances by the addition of diisopropylzinc to pyrimidine-5-carbaldehyde in conjunction with asymmetric autocatalysis. Tetrahedron: Asymm 14:185–188
Sugimori T, Hyuga H, Saito Y (2008) Fluctuation induced homochirality. J Phys Soc Jpn 77:064606

Biodata of **Radu Popa** and **Vily Marius Cimpoiasu**, authors of "***Energy-Driven Evolution of Prebiotic Chiral Order (Lessons from Dynamic Systems Modeling)***."

**Dr. Radu Popa** is currently an Associate Professor of Microbiology at the Department of Biology of Portland State University, USA. He obtained his Ph.D. from the University of Cincinnati in 2000 and continued his studies and research at the California Institute of Technology, Jet Propulsion Laboratory and University of Southern California. Professor Popa's scientific interests are in the areas of systems organization during the origin of life and biogeochemical processes on olivine mineral surfaces.

E-mail: **rpopa@pdx.edu**

**Dr. Vily Marius Cimpoiasu** is currently an Associate Professor of Physics in the Biology Department at the University of Craiova, Romania. He obtained his Ph.D. from the University of Bucharest in September 2003 with a thesis entitled 'The role of boron in the physicochemical protobiotical mechanisms'. The actual scientific interests are in the areas of: chirality and origin of life, biophysics of carbohydrates, isotopes and boron biochemistry.

E-mail: **vilycimpoiasu@yahoo.com**

**Radu Popa**

**Vily Marius Cimpoiasu**



J. Seckbach (ed.), *Genesis - In The Beginning: Precursors of Life, Chemical Models and Early Biological Evolution*, Cellular Origin, Life in Extreme Habitats and Astrobiology 22, 525–545
DOI 10.1007/978-94-007-2941-4_28, 

# ENERGY-DRIVEN EVOLUTION OF PREBIOTIC CHIRAL ORDER (LESSONS FROM DYNAMIC SYSTEMS MODELING)

**RADU POPA[1] AND VILY MARIUS CIMPOIASU[2]**
*[1]Department of Biology, Portland State University, Portland, OR, USA*
*[2]Department of Biology, University of Craiova, Craiova, Romania*

## 1. Introduction

The origin of order in early life networks is generally attributed to a combination of factors such as: inherited organization, coupling between entropy disproportionation and energy dissipation, exchange of information with other systems, and adaptive evolution (Kauffman, 1993; Gilat, 2002; Galimov, 2006). The quantitative connection between changes in entropy and energy dissipation was defined in the field of disequilibrium thermodynamics (Nicolis and Prigogine, 1977). The connection between energy and information is still debated (Bennett, 1982; Jarzynski, 1997; Toyabe et al., 2010). Because in complex systems changes in specific information are not predictable from changes in entropy, controlling organization solely through energy dissipation is often very difficult. The predictive relationship between energy and information is not reciprocal. For example, we can calculate how much energy is needed to bring a system to a specific level of order, yet introducing a specific amount of energy in a complex system does not guarantee that a specific type of organization will occur.

Chirality is an interesting form of organization because it has at the same time thermodynamic, isometric, and binary properties. Studying the evolution of chirality is a shortcut toward better understanding of the interplay between various controllers of the origins of prebiotic order. Chiral evolution includes all features of the origins of order (such as changes in entropy, energy, and information) yet without the ambiguity of having more than two choices for each variable. The connection between changes in energy and changes in organization is furthermore straightforward in chiral systems because here, all choices are equivalent, and thus, making a choice about a variable eliminates by default all other liberties of it. This lesser ambiguity of chirality also allows better correlation between changes in energy and changes in information, which simplifies the analysis of the connection between energy dissipation and system organization. In this chapter, we analyze the effect of feedbacks connecting changes in the organization of a chiral system with an external energy flow, the limiting factors of chiral evolution in a precatalytic era, the importance of a system's size in the origins of order, and the need for specific catalysts during the prebiotic evolution of chirality.

The word chirality is often reserved to chemical systems, even though this type of symmetry is common outside chemistry as well. Some theorists attempted generalizing this concept either by via terms such as handedness or mirror symmetry or by using the word chirality in a broader sense that is above and beyond stereochemistry, mirror images, or space dimensions (Mislow and Bickart, 1977; Arnaut, 1997; Gilat, 1999, 2002; Palyi et al., 2005; Micskei et al., 2006a, b; Joshi et al., 2005a, b; Popa et al., 2010). Accepting that chirality is but another example of symmetry, and thus part of a broad mathematical and physical context, helps expand lessons learned in chiral models into the organization of other types of systems. It also helps produce generalizations about drivers, energy, information, and thresholds of the origin of order.

## 2. Controllers of Order in Prebiotic Chiral Systems

Four main factors control the evolution of order in chiral systems:

1. The initial state of the system
2. The dynamics of racemization
3. Connections between chiral disruption and external energy flow
4. Exchanges of information with other systems or phenomena

A variety of dynamic models were used to study chiral evolution (Buhse et al., 1993; Osipov and Stelzer, 2003; Lee et al., 2004; Blackmond and Matar, 2008; Barabas et al., 2009). More has to be done in these computer models to increase their versatility, practicality, and capacity to answer questions about the origin of life. Models are sought that allow studying the evolution of small populations of enantiomers and use integers rather than continuous functions (Caglioti et al., 2006; Micskei et al., 2006b). Also needed is better understanding of the interaction between different controllers of chirality, as well as connections between the evolution of chirality and that of energy flux, information, and the organization of biochemical networks. We produced a basic logical model for analyzing the connection between an external energy flow and the organization of a chiral system. A variety of software platforms can be used to implement this simulation. The example given here is based on the "Stella TM" software, which has the benefits of being easily accessible and user friendly (Stella, 2009). With appropriate changes, this model can be used to ask questions such as: How do different controllers of chiral order influence each other in various stages of prebiotic organization? How much chiral order can a chemical network gain from external sources and how long it takes? What is the energy cost of producing and maintaining chiral order? Was energy a limiting factor during chiral evolution? What was the required sequence of chiral upgrades during the origin of life?

## 3. Working with Intermediates and Integers in Chiral Populations

In most cases, racemization occurs by direct $L \Leftrightarrow D$ interconversion via one or more intermediates, which under defined conditions are produced with predictable probability and have specified half-lives (Wolf, 2008; Eliel et al., 1994). Simulations of very large populations of enantiomers, and simulations where the individual steps are very long (e.g., years), have to take into account that apart from enantiomers, a small subpopulation of LD intermediates always exists. Constantly, a small fraction of the L and D populations become LD intermediates. The contribution of the L and D subpopulations to the pool of LD intermediates is proportional to their size. Upon returning into stable forms, because L and D have similar physical-chemical properties, the probability to form L equals that of forming D and equilibrium (racemic state) is reached when $L = D$.

Although racemization is most often mathematically described via continuous variables, chiral populations only exist as integers (i.e., they only take discrete values). This is little relevance when studying very large populations of enantiomers but becomes important in small populations. Examples where this detail matters include the nucleation of microcrystals, chemistry on chiral surfaces, and small chiral populations trapped in microvesicles. For example, a protocell with a volume of 1 $\mu m^3$ having a chiral chemical at 1 μM concentration will only contain about 600 enantiomers. Chiral populations may contain an even or an odd number of molecules and thus a 50% chance to be chiral disrupted at one given moment due to their size alone. The magnitude of this intrinsic chiral disruption is inversely proportional to the size of the system. A population of 601 molecules, with $L = 301$ and $D = 300$ (or vice versa), has an intrinsic enantiomeric excess (*ee*) of $\sim 1.7 \cdot 10^{-3}$. This is larger than the effect of most chiral asymmetric abiotic factors. Can an autocatalytic mechanism amplify such built in disequilibrium? Is this effect relevant to the evolution of protocells?

Based on the above, simulations of prebiotic chiral evolution have to take into account LD intermediates and integers and also work with small chiral populations.

## 4. Initial System Conditions and the Rate of Racemization

In the simple model from Fig. 1, L or D enantiomers are represented as stocks. The transition between L and D is via an intermediary ("LD interm"). The total number of variable positions "N" is derived from the system's size in moles ("System size moles") and the Avogradro number ("NA"). The rate of racemization is controlled by the difference in the formation of "L" from "D" and "D" from "L" via "LD interm." When initial disequilibrium exists between L and D, for example, when $L > D$, the rate of transformation of L into D ($L \rightarrow D$) is larger than $D \rightarrow L$.

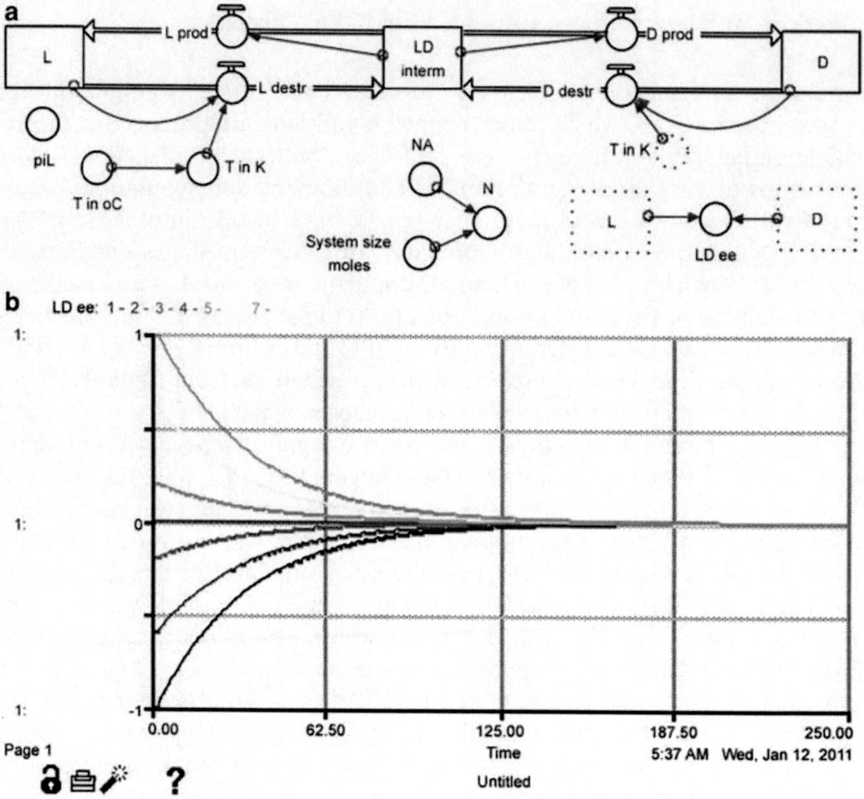


**Figure 1.** *Panel A*: Model of a simple chiral system created in Stella. *Rectangles* are stocks. Flows are *double arrows* with a *circle* in the *middle*. The *simple arrows* are action/connectors, and the *circles* are converters. Values and formulas used in a first-hand simulation are "System size moles" = $10^{-9}$; "piL" = 0, 0.2, 0.4, 0.5, 0.6, 0.8, 1 (= $p_i$ values for L in various initial conditions); "T in C" = 20; "T in K" = T in C + 273.15; "NA" = 6.02214179E23; "N" = NA*System_size_moles; "L" = INT(N*piL); INT is function returning only integers for all values; "D" = INT(N*(1-piL)); "LD interm" = 0; "L destr" = INT(L*T_in_K*0.0001); "D destr" = INT(D*T_in_K*0.0001); "L prod" = INT(LD_interm/2); and "D prod" = INT(LD_interm/2). *Panel B*: Evolution of the *ee* toward racemic state starting from various initial *ee* values, shown as a Stella graph.

The *ee* of L is "LD ee" = (L − D)/(L + D) and has values between −1 and +1. "LD ee" > 0 means excess of L; "LD ee" < 0 means excess of D, while at "LD ee" = 0, the system is racemic. At "LD ee" = 1, the system is fully ordered with 100% L. At "LD ee" = −1, the system is also fully ordered but with 100% D. Temperature is given in °C ("T in C"), then recalculated in K ("T in K") and controls the rate of racemization. A simple model verification of this model can be done by analyzing the evolution toward racemic state starting from various initial conditions (i.e., excess of L or D).

## 5. Coupling Between an External Energy Flow and Enantiomeric Excess

Based on the contribution of liberties (variable elements) from a system, and that of their choices (alternatives of each variable), to the total population of microstates, systems can be described as:

- *Heterogeneous systems* with choices that are unequally shared among variables or with variables not contributing equally to the total population of microstates. An example of heterogeneous system is a solution containing glucose and ribose at different concentrations. These sugars also have different numbers of isomers (i.e., choices). The two variables (glucose and ribose) contribute unequally to the overall information capacity of the system. At the other extreme, *homogeneous systems* have choices that are equally partitioned among variables, and the variables contribute equally to total population of microstates. The type examples of such systems are: a collection of dice tosses, a series of independent lottery numbers, or a solution of two amino acids in equimolar concentration and each amino acid with only two enantiomers (L and D).
- *Isometric systems* (as opposed to *nonisometric systems*) have all choices of each and all variables equally likely. For example, if two choices exist for one variable, each has a $p_i$ value of 0.5; if four choices are present, each has a $p_i$ value of 0.25, *etc.*
- *Binary systems* (as opposed to *nonbinary systems*) contain only two choices for each variable.

Chiral solutions containing a single chemical are typical examples of homogeneous, isometric, and binary systems. To study the transfer of information from an external source (organized system or phenomenon) that may or not be chiral, to a target system that is chiral, we need to be able to compare changes in entropy and information content between systems of dissimilar type (e.g., homogeneous *vs.* heterogeneous, isometric *vs.* nonisometric, and binary *vs.* nonbinary). Also, the study of the coupling between an external energy flow and the internal organization of a system requires using a quantitative relationship between changes in a system's energy and changes in its information. The next section is a review of the basic computational principles that allow such cross-system comparisons possible.

### 5.1. ENTROPY AND INFORMATION

In statistical mechanics (or statistical thermodynamics), entropy does not have a unique value because entropy is a scale dependent parameter, while this discipline can study a system at different scales independently. Moreover, entropy has many aspects which can be analyzed independent of each other. For these reasons, it is often practical to analyze changes in entropy due to one specific scale and particular aspect of a system rather than trying to quantify the total entropy of it. On the other hand, the entropy studied by theory of information has a specified value, because here we can describe the system at its lowest theoretical scale, that is, the scale that deals with one unit of information. Because in information theory energy has no meaning at a superficial glance, it is not possible to connect changes in information with changes in energy. This conundrum has more than one solution, however. First, information theory often studies the probability of virtual numbers, which may or may not represent a physical reality. Therefore, because each microstate is equally likely with any other, no tendency may exist for one microstate to concede ground to another. This lack in tendency of transformation means that the difference in energy between two virtual microstates can be zero. This "on paper" zero-energy content of an information unit from theory of information does not contradict the laws of nature analyzed by statistical mechanics. Second, the scale and the physical context at which entropy is studied also define the energetic cost of a unit of information. These two aspects create discord when information theory and statistical mechanics merge to analyze complex systems (Jaynes, 1957; Jarzynski, 1997; Gough, 2008).

Table 1 summarizes the formulas for entropy applied in different types of systems.

**Table 1.** Entropy in different types of systems based on Gibbs entropy (Gibbs, 1875–1876; 1877–1878). Equation 1 is the general form.

| **Heterogeneous systems** | **Homogeneous systems** | **Homogeneous and isometric systems** |
|---|---|---|
| $S = -k_B \sum_{j=1}^{N} \sum_{i=1}^{M} p_i \ln p_i$ (1) | $S = -k_B N \sum_{i=1}^{M} p_i \ln p_i$ (2) | $S = k_B \ln\left(M^N\right)$ (3) |

$S$ entropy in $J \cdot K^{-1}$, $k_B$ Boltzmann constant $\approx 1.381 \cdot 10^{-23}$ $J \cdot K^{-1}$, $N$ number of variables in the system, $M$ number of choices for each variable, and $p_i$ the probability for a variable to assume the $i$th choice.

To connect energy with information, we analyze the relationship between entropy and the liberties of a system. Relative to the total number of microstates, entropy is:

$$S = k_B \ln \Omega \quad \text{(known as Boltzmann's entropy)}, \tag{4}$$

where $\Omega$ (or $W$) = the total number of microstates for a given macrostate (Boltzmann, 1866).

Formulas for $\Omega$ in different types of systems are given in Table 2.

**Table 2.** Calculating the total number of microstates in systems of different types.

| **Heterogeneous systems** | **Homogeneous systems** | **Homogeneous and isometric systems** |
|---|---|---|
| $\Omega = e^{-\sum_{j=1}^{N} \sum_{i=1}^{M} p_i \ln p_i}$ (5) | $\Omega = e^{-N \sum_{i=1}^{M} p_i \ln p_i}$ (6) | $\Omega = M^N$ (7) |

In chiral systems, Eqs. 3 and 7 are used, yet the other more general forms also apply.

Relative to information, entropy is expressed as:

$$S = k_B \cdot D, \tag{8}$$

where $D$ (= ln Ω) is the amount of decisions needed to order the system (in nat units). A nat is the natural unit for information entropy (Reza, 1994; Comley and Dowe, 2005). 1 nat = 1/ln2 bit = 1.44 bits.

The logarithmic connection between the number of liberties and entropy makes this relationship hard to visualize in heterogeneous and nonisometric systems. Yet, it is easier to see in systems that are homogeneous, isometric, and binary (such as chiral systems), because in such systems it makes sense to express information capacity in bits.

The liberty of enantiomers to be in L or D form is but one aspect of the entropy of a chiral system. Thus, chiral entropy ($S_{LD}$) is but a part of the total entropy ($S$) of a system. $S_{LD}$ is calculated according to Eq. 7, with $M=2$ and $\Omega=2^N$. Each L/D decision about one variable halves the total number of chiral microstates. The total number of binary decisions needed to order a system, or its information capacity, is $H=\log_2(\Omega)$ (Shannon, 1948). This connection allows analyzing the entropy from statistical mechanics relative to the parameter $H$ from theory of information and thus defines entropy as a feature proportional to the energetic cost of making binary decisions about a system:

$$S = k_B \cdot D = k_B \cdot \ln 2 \cdot H = k_B' \cdot H, \tag{9}$$

where $H=\log_2(\Omega)$ = information capacity or Shannon entropy (measured in bits);

$k_B' = k_B \cdot \ln 2 \approx 9.572 \cdot 10^{-24}$ J·Bit$^{-1}$·K$^{-1}$ is a modified Boltzmann constant used to calculate entropy relative to the information capacity measured in bits. $k_B'$ is known in statistical mechanics as "information-theoretic entropy" (Jaynes, 1957; Lindgren, 1988).

If $H$ represents chiral liberties, then Eq. 9 calculates $S_{LD}$.

Equation 9 relates thermodynamic entropy (measurable in energy units) with information capacity (measured in bits). One bit is a binary decision (e.g., yes/no, +/−, 1/0, black/white, L/D, *etc.*) which removes half of the ambiguity (liberties) of a system. The number of binary decisions about a system only depends on its total number of microstates; it does not require individual choices to be binary. Hence, information from all systems, irrespective of their type (heterogeneous or homogeneous, isometric or nonisometric, and binary or nonbinary), can be expressed in bits (see an example in Table 3).

The relationship between $D$ and $H$ ($D$ (in nat) = ln 2 $H$; $H$ (in bit) = 1.434 $D$) helps remove the ambiguity between thermodynamic entropy and information content. In this regard, the entropy from statistical thermodynamics is the energy needed to eliminate a part of the remaining liberties of a system for each increase of temperature by 1°K. How large of a part of a system's liberties can be removed

**Table 3.** Example of comparison of $H$ and $D$ between a system (A) that is binary, isometric, and homogeneous and another system (B) that is nonbinary, isometric, and homogeneous. A similar logic (with different formulas for $\Omega$) also applies to heterogeneous and nonisometric systems.

| System A ($M=2$; $N=9$) | System B ($M=8$; $N=3$) |
|---|---|
| $\Omega=2^9=512$ | $\Omega=M^N=8^3=512$ |
| $H=\log_2(\Omega)=\log_2(512)=9$ (bits) | $H=\log_2(\Omega)=\log_2(512)=9$ (bits) |
| $D_A=\ln\Omega=9\cdot\ln 2=H\cdot\ln 2$ | $D_B=\ln\Omega=3\cdot 3\cdot\ln 2=H\cdot\ln 2$ |

A system with 512 potential microstates requires nine binary decisions to reduce 512 to 1 (hence, $H=9$ bits). This can be seen in the removal of the uncertainty about individual elements as well. The A system has nine variable elements, each with two equivalent choices; thus, a total of nine binary decisions have to be made to make this system unambiguous (one for each variable). In system B, three variables exist, each with eight isometric choices. Here, removing the uncertainty about one variable requires three binary decisions. Again, a total of nine decisions are needed to order the system or to eliminate its ambiguity.

per unit of energy depends on the unit that is used to measure information. To do energy-information analyses in systems that are heterogeneous and nonisometric, it is more practical to express their information in bits. Hence, instead of expressing their information in nat units, it is convenient to compare such systems with other virtual homogeneous, isometric, and binary systems having similar information capacity.

## 5.2. AMPLIFICATION OF INTERNAL ORDER BY COUPLING CHIRAL SYSTEMS WITH EXTERNAL ENERGY FLOWS

The racemic state is the low-energy state, while departure of *ee* from zero requires energy, irrespective of which enantiomer is in excess. Because of continuous racemization, maintaining a system under enantiomeric disequilibrium also requires energy. The purpose of the model shown next is to analyze how chiral equilibrium evolves when feedback regulation mechanisms introduce energy and order simultaneously in a chiral system. In this model, energy comes from an external flow catalyzed by the excess of one enantiomer. This is a simple example of autocatalysis which is not altering the rate of formation of LD intermediates. The feedback may be either positive or negative. Numerous such examples exist in natures that can be modeled. For example, an energy flow may increase or decrease the stability of enantiomers; one enantiomer may be preferentially formed by a catalyst; or under special conditions enantiomers may have dissimilar physical chemical properties. To understand the essence of this phenomenon it suffices to produce a model of a system where the excess of one enantiomer increases the external energy flow, and in turn has positive effects on the formation of more of the same enantiomer.

To construct such model, we need to know the energy equivalence of *ee*. Most scientists agree that energy is needed to produce order, yet the physical meaning of this energy and its magnitude are often controversial. According to Landauer's principle, heat is released when information is lost (Landauer, 1961; Bennett, 2003), with the energy equivalent of one bit of information equal to:

$$E_{bit} = k_B \,.\, T \,.\, \ln 2. \tag{10}$$

This value is also called "Landauer bound" or "Landauer limit." To create *ee*, or to amplify it, an amount of energy has to be introduced in a system proportional to $E_{\text{bit}}$ times the amount of order introduced. Some authors argued that Landauer's principle has not been shown to hold in general (Norton, 2005) or that no precise relationship exists between thermodynamics and logical reversibility (Maroney, 2005). Yet, the Landauer limit is considered by some authors to be a valid and useful physical parameter (Ladyman et al., 2007; Gough, 2008). Irrespective of its exact magnitude, the connection between energy and information in nature is thought to be wide ranging and quantitative (Jarzynski, 1997; Shinitzky et al., 2007). Recent experimental work was used to verify this conjecture (Toyabe et al., 2010). Direct calorimetric measurements showed that heat is released when two chiral disrupted solutions are mixed to produce a solution with lower *ee* (Shinitzky et al., 2007; Shvalb et al., 2010).

Next, we discuss how Landauer's limit can be applied to models of chiral systems. Because enantiomers have similar physical chemical properties, their $p_i$ values at equilibrium are also similar. A chiral change regarding one enantiomer in a direction that is away from racemic equilibrium is equivalent with a nonrandom binary yes/no type of decision. Using $\Omega = 2^N$, Eqs. 4 and 9, we can analyze the $S_{\text{LD}}$ of a chiral system. Chiral choices do not affect the free energy of a system, only its entropy. Hence, the change in energy ($\Delta E$) during chiral disruption is only due to $T\Delta S$. The change in energy associated with $H$ binary decisions that bring a chiral system from racemic state to $ee = 1$ is:

$$\Delta E = T\Delta S = T \,.\, k_{\text{B}} \,.\, \ln 2 \,.\, H, \tag{11}$$

where $H$ = information capacity of the racemic state (in bits).

In binary, homogeneous, and isometric systems, such as chiral solutions, $H = N$ (the number of molecules in the population).

At 20°C, $E_{\text{bit}} \approx 2.8 \cdot 10^{-21}$ J bit$^{-1}$ or ~1.69 kJ (mol·bit)$^{-1}$.

Regarding the physical meaning of $E_{\text{bit}}$, this energy is equivalent with the "discomfort felt by enantiomers" (i.e., their higher energy level associated with their rotational, vibrational, and translational liberties) in a solution with an excess of one enantiomer and relative to the same solution in racemic state.

The model from Fig. 2 contains a couple of upgrades relative to Fig. 1. The production of intermediates is controlled by a counter which allows the system to function even when the chiral population is very small and intermediates are seldom made. The production of enantiomers from the intermediate stock ("Interm LD") may occur via a random path ("Rand prod L&D") or via a nonrandom path ("NonRand prod L&D"). The nonrandom production of enantiomers is controlled by the virtual amounts "NoL" and "NoD." "NoL" and "NoD"

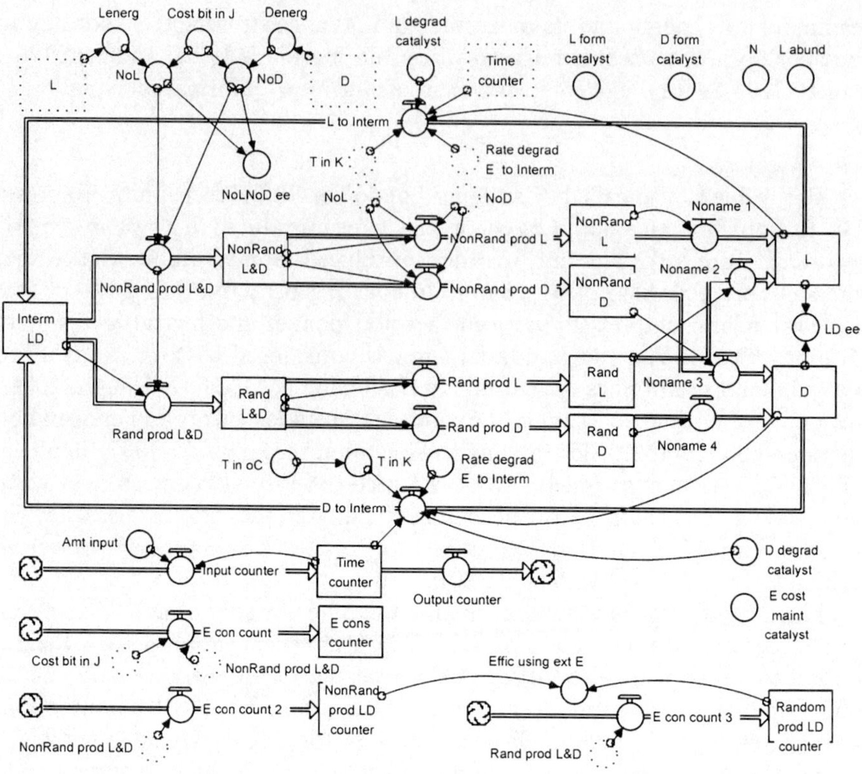


**Figure 2.** Model of a chiral system coupled with an external energy flow via a feedback mechanism. This model can be used to study and compare the combined effects of the initial state of the system, autocatalysis, energy availability, energy return efficiency, rate of racemization, and system size on the evolution of *ee*. A time counter connected with the rate of racemization allows simulating very small populations of enantiomers. Shown below are the model's parts, mathematical relationships between the parts, and initial values for a first-hand simulation.

*Stocks*

Interm LD = stock of LD intermediates = 0.

NonRand L&D = the sum between the L and D enantiomers produced by nonrandom means = 0.

Rand L&D = the sum between the L and D enantiomers produced by random means = 0.

NonRand L = the number of L enantiomers produced by nonrandom means = 0.

NonRand D = the number of D enantiomers produced by nonrandom means = 0.

Rand L = the number of L enantiomers produced by random means = 0.

Rand D = the number of D enantiomers produced by random means = 0.

L = the number of L enantiomers = INT(N*L_abund).

D = the number of D enantiomers = INT(N-L).

Time counter = stock used as in time counter to determine the rate of transformation of enantiomers in intermediary forms = 0.

E cons counter = counter of the energy input invested in the organization of the system during its evolution = 0.

NonRand prod LD counter = counter of the number of nonrandom decisions about the LD intermediates = 0.

Random prod LD counter = counter of the number of random decisions about the LD intermediates = 0.

**Figure 2.** (continued)

*Flows*

NonRand prod L&D = nonrandom production of L and D = NoD + NoL.
Rand prod L&D = random production of L and D = Interm_LD-NonRand_prod_L&D.
NonRand prod L = nonrandom production of L = INT(NonRand_L&D*(NoL/(NoL + NoD))).
NonRand prod D = nonrandom production of D = INT(NonRand_L&D*(NoD/(NoL + NoD))).
Rand prod L = random production of L = INT(Rand_L&D/2).
Rand prod D = random production of D = INT(Rand_L&D/2).
Noname 1 = NonRand_L.
Noname 2 = Rand L.
Noname 3 = NonRand_D.
Noname 4 = Rand D.
L to Interm = the rate of transformation of L in LD intermediates = IF(Time_counter = 0) THEN(INT(L*T_in_K*Rate_degrad_E__to_Interm*L_degradation_catalyst))ELSE(0).
D to Interm = the rate of transformation of L in LD intermediates = IF(Time_counter = 0) THEN(INT(D*T_in_K*Rate_degrad_E__to_Interm*D_degradation_catalyst))ELSE(0).
Input counter = input in the counter = IF(Time_counter = 0)THEN(Amt_input)ELSE(0).
Output counter = output from the counter = 1.
E con count = input in the "E cons counter" = NonRand_prod_L&D*Cost_bit_in_J.
E con count 2 = input in the "Nonrand prod LD counter" = NonRand_prod_L&D.
E con count 3 = input in the "Random prod LD counter" = Rand_prod_L&D.

*Converters*

N = the total number of enantiomers in the system = 1,000,000.
Labund = the initial abundance of L =
LD ee = L- vs. D-dependent enantiomeric excess = (L − D)/(L + D).
NoLNoD ee = the disequilibrium regarding L and D autocatalysis or the LD bias of the external source of energy = (NoL − NoD)/(NoL + NoD).
Lenerg = energy recuperated from the external energy flow and available to produce L from LD intermediates = 2.8E-21*(L).
Denerg = energy recuperated from the external energy flow and available to produce L from LD intermediates = 2.8E-21*(D/2).
Cost bit in J = the energetic equivalent of one bit, given in Joule = 2.8E-21.
NoL = largest amount of L that can be formed using the external source of energy = INT(Lenerg/Cost_bit_in_J).
NoD = largest amount of D that can be formed using the external source of energy = INT(Denerg/Cost_bit_in_J).
T in oC = temperature in degrees C = 20.
T in K = temperature in degrees K = T_in_oC + 273.15.
Rate degrad E to Interm = intrinsic rate of transformation of enantiomers in LD intermediaries = 0.0001.
Amt input = amount reintroduced in the counter when its content becomes zero = 1.
Effic using ext E = the efficiency of using the external source of energy = IF(NonRand_prod_LD_counter = 0)THEN(0)ELSE((NonRand_prod_LD_counter/(NonRand_prod_LD_counter + Random_prod_LD_counter + 0.000001))*100).
L degrad catalyst = the abundance of a catalyst influencing the conversion of L in LD intermediates = 1.
D degrad catalyst = the abundance of a catalyst influencing the conversion of D in LD intermediates = 1.
L form catalyst = the abundance of a catalyst influencing the formation of L from LD intermediates = 1.
D form catalyst = the abundance of a catalyst influencing the formation of D from LD intermediates = 1.
E cost maint catalyst = the energy cost of forming and maintaining a catalyst = 0 (not analyzed here).

represent the largest number of monomers that can be formed using the energy taken by the system from the exterior and taking into account the efficiency of the energy transfer toward the system. If "NoL + NoD" is larger than the total production of intermediates ("D to Interm + L to Interm = Interm LD"), the formation of L and D is 100% controlled. When "NoL + NoD" is smaller than the production of intermediates, part of L and D will form via the random path (i.e., "Rand prod L&D"). The connection with the external energy flow may be fixed (in which cases, we can study the combined effects of external energy flow and internal racemization) or via a feedback (in which case, the amount of each enantiomer controls the return toward the system (which is a form of autocatalysis)). Four potential catalysts/inhibitors are also included that can be used to influence the formation of LD intermediates or of L or D enantiomers.

In the model from Fig. 2, the main controllers of the evolution of the system in terms of $\Delta ee$ are the initial *ee*, the amount of external energy entering the system, the LD bias of the external source of energy, the rate of racemization, and the system's size. A first simulation assumes that the physical chemical properties of the enantiomers are similar. The system is autocatalytic in that the amount of L and D enantiomers that can be formed ("No L" and "No D," respectively) is correlated with the abundance of the L and D stocks, and thus each state (L or D) catalyzes its own amplification.

This model can be used as a platform to answer specific questions about prebiotic chiral evolution. For example, we can use it to analyze:

- Equilibrium between autocatalytic chiral disruption and the rate of racemization
- The effect of the magnitude of an initial *ee* on the evolution of a system
- Effects of stochastic chiral disruption and system size on chiral amplification
- The effect of an asymmetric energy input
- The effect of temperature on the evolution of a chiral system

As shown above, at 20°C, one bit worth of system organization is equivalent with $\sim 2.8 \cdot 10^{-21}$ J. Thus, rate of racemization aside, obtaining $ee = |1|$ in a system with $N = 2.8 \cdot 10^{21}$ ($\approx$ 4.65 mmoles) requires 1 J of energy. Conversely, the amount of energy released from a 1 mol large system as it changes from $ee = |1|$ to $ee = 0$ is ~1.7 kJ. The cost of creating and maintaining chiral order is actually bigger because of constant loss of organization through racemization. If the racemization rate is defined, this model can be used to identify levels of *ee* that can be produced by a specified source of external energy at a given temperature or to determine the amount of energy needed to bring the system to a specified *ee*. As a practical example, we analyzed populations of $\leq 10^6$ enantiomers and found that:

- If "(NoL + NoD) < Interm LD," the initial value of the *ee* excess in the system will not influence the final *ee*, the random production of enantiomers is nonzero, and the following prediction applies about the final state of the system:

  – "LD ee final ≈ [(NoLNoD ee) · (NoL + NoD)]/(Interm LD)"

- If "(NoL + NoD) ≥ Interm LD," the following situations apply:
  - In very large systems (e.g., 1,000,000 enantiomers) and with initial *ee* smaller than the asymmetry of the external energy source (i.e., "NoLNoD ee"), the system will evolve toward $ee = 1$ by autocatalytic amplification of the initial disequilibrium. The rate of change is not constant because during each step of the simulation, the system's "LD ee" tries to reach a value similar with that of a continuously changing "NoLNoD ee."
  - In small systems (e.g., <10,000 enantiomers) and with an initial *ee* smaller than the asymmetry of the external energy source ("NoLNoD ee"), the system will evolve toward an *ee* level that is <1. The smaller the system, the smaller this *ee* threshold is. This effect occurs when the number of L transferred toward "Interm LD" equals the number of L formed by autocatalyis, and reciprocally, the number of D transferred toward "Interm LD" equals the D formed by autocatalyis. This loophole is eliminated by increasing the rate of formation of "Interm LD," which allows the system to conclude its autocatalytic evolution. The "Interm LD" should not increase as much as to produce a situation when "(NoL + NoD) < Interm LD."
  - In systems with initial "LD ee ≥ NoLNoD ee," the "NoLNoD ee" will automatically change with each iteration, and the positive feedback will continue until the final LDee state is reached.

Regarding the role of chance and system size in chiral disruption, the notions that equal chance should not be confused with the populations of L and D enantiomers being of exactly equal size or that true racemates can only exist in mixtures with odd number of enantiomers are long standing in the literature (Pearson, 1898; Caglioti et al., 2006). Because all molecular systems may contain an either odd or even number of enantiomers, 50% of all systems are by default chiral disrupted. In large systems, this intrinsic disequilibrium is not significant yet becomes important in very small systems. For example, a chiral system with 999 molecules cannot be purely racemic at any one given moment, and the state that is closest to the racemic state has an intrinsic $ee = 10^{-3}$. This departure from equilibrium is significant as few physical forces exist in the abiotic nature that can create a larger *ee*. Did early life evolve in micron-size systems for this reason? The simulation of the model from Fig. 2 also shows that if the external controller is LD neutral, the initial asymmetry of a system cannot be amplified.

We can also use this model to define thresholds in a system's size where an intrinsic chiral disequilibrium is large enough to initiate chiral amplification. We analyzed the evolution of a 1-$\mu m^3$-size protocell. At a concentration of a chiral solute of 1 μM (which is likely for prebiotic chemistry), the protocell will contain $6.022 \cdot 10^{17}/10^{15} \approx 6 \cdot 10^2$ chiral molecules. Under these conditions, we found an increase in *ee* and that an upper plateau is reached dictated by the LD bias of the external source of energy ("NoLNoD ee"). Last but not least, the energy controlled amplification of chiral asymmetry is faster if self-catalysis (i.e., the nonrandom formation of enantiomers) is large enough relative to the availability of intermediates (i.e., the rate of racemization).

## 6. The Energy Density of External Sources and the $\Delta E$ of Chiral Disruption

An interesting question about the origin of life is finding a magnitude for an external energy source that is sufficient to introduce chiral order (*ee*) in a protocell. How does the energy needed to make chiral decisions compare to energy sources from the normal environment of protocells? To answer such question, we have to compare the energy equivalent of organization, the rate of formation and lifetime of LD intermediates, and the energy density of external sources. In complex systems, we also have to consider the cost of producing and maintaining chiral catalysts. Shown above, the energetic equivalent of making nonrandom decisions in chemical chiral systems at 20°C is about $2.8 \cdot 10^{-21}$ J bit$^{-1}$. The rate of formation of LD intermediates is derived from the rate of racemization ($k$) which, among all prebiotic monomers, was best studied in amino acids (Schroeder, 1974; Cronin and Pizzarello, 1983, 1997; Bada, 1985). The half-lives of most amino acids in earth's temperature conditions are $10^5$–$10^6$ years (Bada and Miller, 1987). This rate only applies to molecules that racemize faster than they decompose (Bada, 1991). In aspartic acid, the racemization rate in solution at 300 K is $k_{asp} = 5 \cdot 10^{-4}$ year$^{-1}$ (Bada and McDonald, 1995), which is equivalent to a rate of formation (turnover) of the "Interm DL" from Fig. 2 model of:

$$\frac{d(\text{"}Interm\_LD\text{"})}{dt} = 2k_{asp} \cdot (D+L) year^{-1} \tag{12}$$

Simulations on the Fig. 2 model showed that 100% efficient chiral amplification only occurs when the energy density used to make nonrandom decisions is larger than d"Interm DL"/d$t$, and that without specialized catalysts, the largest $\Delta ee$ during prebiotic evolution could not have been faster than the rate of racemization.

A protocell with 1 μm³ volume with 1 mM aspartic acid will contain ~$6 \cdot 10^5$ enantiomers, and at 20°C, it has a rate of "Interm DL" production of ~$6 \cdot 10^2$ bit year$^{-1}$, equivalent to $1.7 \cdot 10^{-18}$ J year$^{-1}$. If the external energy is not limiting, if the efficiency of the external controller to assist the "NonRand prod L&D" path (Fig. 2) is 100%, and if the specificity of the external controller for one of the two enantiomers is 100%, then the energy cost to bring this system from racemic state to $ee = |1|$ is ~$1.7 \cdot 10^{-15}$ J. If the specificity of the external controller is <100%, the time needed to reach equilibrium is longer. This evolution has yet another, larger, energy cost. When the mechanism used to modify the *ee* involves LD intermediates, as it is most often the case, the rate of chiral disruption ($\Delta ee$) is at best similar to the rate of racemization. A population of $6 \cdot 10^5$ molecules of aspartic acid at 20°C will form about $6 \cdot 10^2$ LD intermediates per year (which is about 1.64 per day). These intermediates are generally short lived. If the average life span of the LD intermediates is $\tau_{intLD}$, the energy flux necessary to maintain a population of enantiomers or to increase the *ee* is:

$$E_{ee} = 2.8 \cdot 10^{-21}\ \text{J}\ (\text{s}/\tau_{intLD}).$$

This energy flux can help control a population of 1 s/$\tau_{intLD}$. Under these conditions, the total energy associated with full chiral disruption (which takes about $5.5 \cdot 10^3$ years in aspartic acid) is:

$$2.8 \cdot 10^{-21}/\tau_{intLD} \text{ J } 5{,}500 \text{ year} \sim 4.8 \cdot 10^{-10}/\tau_{intLD} \text{ J s.}$$

Because energy is not always available when LD intermediates are available, the majority of this energy is wasted in the environment as heat. Therefore, the energy necessary to disrupt the system ($E_{ee}$) is larger than the Landauer limit indicates:

$$E_{bit} = 2.8 \cdot 10^{-21} \text{ J bit}^{-1} \text{ (Landauer limit)}$$
$$E_{ee} = 2.8 \cdot 10^{-15} \text{ J bit}^{-1}$$

(this applies to a system forming up to $10^6$ LD intermediates per second and with each LD intermediate existing for about $10^{-6}$ s).

If this energy has to be made available to a 1 μm² cross section (approximately that of a small cell), then the power of the source should be about $2.8 \cdot 10^{-15}$ J $s^{-1}$ $\mu m^{-2} = 2.8 \cdot 10^{-3}$ W $m^{-2}$. We compared this with the energy of three sources common to the environment of early cells. On Earth, the energy of solar light usable for photosynthesis (~400–700 nm) is about $9.5 \cdot 10^{-12}$ J $s^{-1}$ $\mu m^{-2} = 9.5$ W $m^{-2}$ (Gonzalez and Calbo, 2002; Grant and Slusser, 2004). A chemical gradient of 1 mM $cm^{-1}$ $H_2S$, at 20°C will have a rate of diffusion of about $1 \cdot 10^{-19}$ moles $H_2S$ $\mu m^2$ $s^{-1}$. If cells use this $H_2S$ in redox reactions with nitrate as the electron acceptor, the energy available is ~$9.8 \cdot 10^{-15}$ J $s^{-1}$ $\mu m^{-2} = 9.8 \cdot 10^{-3}$ W $m^{-2}$. A protocell with an electroosmotic potential of 60 mV and protons transferred through ATPases (at a density of 100 ATPases per μm²) at ~300 $H^+$ $s^{-1}$ per ATPase have available ~$2.8 \cdot 10^{-12}$ J $s^{-1}$ $\mu m^{-2} = 2.8$ W $m^{-2}$. Obviously, the energy associated with changes in chiral organization is little relative to common environmental energy sources. Most likely, the energy for creating and maintaining *ee* was not in short supply in the precatalytic era of prebiotic evolution. The true limiting factors of chiral evolution seem to be the specificity of chiral synthesis and the small rates of production of "LD intermediates."

## 7. The Effect of Adding Catalysts to the Evolution of Prebiotic Chirality

The objective of this section is to explain the need for catalysis very early during early life chiral evolution. As far as the prebiotic era is concerned, the simple system from Fig. 2 retains a major flaw. Even with unlimited external energy, and biased L vs. D synthesis, the rate of increase of the *ee* (Δ*ee*) cannot be faster than the rate of racemization ("D to interm + L to interm"). In chemicals such as amino acids, this rate is very slow relative to the rates of other cellular processes. Chirality is indispensable to biomolecular systems, but for chiral changes to be relevant in early life evolution, *ee* must be reached rapidly. After racemic molecules are introduced in a cell, they have to change fast and reach the *ee* needed by the host cell.

Biochemical processes cannot wait for enantiomers to meticulously pass through the "LD intermediates" pool in 1,000 years or more. Apparently, the crucial factor in this step of prebiotic chiral evolution was not energy but time. The obvious next evolutionary imperative, after the coupling of L or D production with an external energy source, was to introduce catalysts able to accelerate the rate of production of LD intermediates. Because these intermediates are very short lived, and the energy flux of most common external sources can easily cover the $E_{ee}$ expenses, sizable increases in catalytic efficiency are needed before external energy becomes limiting. To study the consequence of adding catalysis to biochiral evolution, the model from Fig. 2 needs upgrading. Novel parameters that have to be introduced include a stock for the amount (or concentration) of each catalyst, converters indicating catalytic properties, lifetime of catalysts, energy costs for producing and maintaining catalysts, and a means to monitor the turnover of the "Interm LD" stock.

Biochemical systems from modern cells contain a large number of chiral controllers (enzymes). Chiral systems are catalyzed in many ways: autocatalysis, catalysis by chemicals that are external to the chiral system, and increasing the rate of racemization simultaneous with specific L or D synthesis. Autocatalyis of preexisting or stochastic chiral disequilibria was often discussed (Soai et al., 1995, 2004; Soai, 2002, 2004; Micskei et al., 2006a, b). The functioning of chiral enzymes is also well known from the biochemical literature. Irrespective of their type, all mechanisms of chiral catalysis have to increase the turnover through the pool of LD intermediates. The study of the urgency and odds of this evolutionary step requires quantifying shortcomings, advantages, and overall consequences of each form of catalysis as well as the effects of transitions from one form of catalysis to another. Such analysis is extensive and beyond the scope of this study. To the benefit of the users of this model, we only included the different forms of catalysis and the converters representing their energetic costs. Simulation showed that adding simple catalysts to a chiral system will increase the $\Delta ee$ without making energy limitative. Using chiral enzymes is different however. Apart from the information needed to construct enzymes, their energy costs are high and their half-life is short. Arguably, during the evolution of biochirality, energy became limitative at the transition between autocatalysis and catalysis by means of simple chemicals to catalysis by means of enzymes.

## 8. Conclusions

We produced a basic dynamic system model that can be used to study the evolution of prebiotic systems coupled with an external energy flow. Many factors control this evolution including inherited *ee*, rate of racemization, lifetime of intermediates, feedbacks with external energy flow, energy availability, exchange of information with other systems, and catalysis. This model can be used to study patterns of prebiotic chiral evolution and as a basic platform for assembling more complex models. If we interpret chiral disequilibrium as an example of homogeneous, isometric, and binary organization, such simulations can also help obtain insight in the interplay between various controllers of prebiotic order. Unlike mathematical analyses of chiral evolution, which are based on continuous variables and large populations, the model introduced here uses integer values for all stocks of enantiomers and intermediates and allows analyzing chiral evolution in small populations. These populations are comparable to micron-sized protocells containing chiral chemicals in mM range concentration. We exemplified the use of this model by analyzing the effect of an external energy flux on the evolution of *ee* in primitive chiral systems in an early catalytic era when external information did not influence the lifetime of enantiomers. The energy equivalent of chiral order ($E_{bit}$) and the energy flux requirements for producing and maintaining chiral disequilibrium ($E_{ee}$) in chiral systems are small compared to external energy sources common to the environment of protocells (solar light, diffusion combined with redox chemistry, and chemiosmosis). This suggests that prior to the origin of catalysis, the evolution of chiral order was not energy but rate and information limited. During this stage of prebiotic evolution, the rate of *ee* was controlled by the rate of abiotic racemization. Our results indicate that chiral catalysts, particularly those capable of shortening the half-life of enantiomers, were a prerequisite in the organization of information-using molecular networks and in the origin of life. We hypothesize that subsequently during cellular evolution, most of the energy needed to produce and maintain chiral order was defined by the costs of producing and maintaining catalysts.

## 9. Acknowledgment

The authors contributed equally to the work.

## 10. References

Arnaut LR (1997) Chirality in multi-dimensional space with application to electromagnetic characterisation of multi-dimensional chiral and semi-chiral media. J Electromagn Waves Appl 11:1459–1482

Bada JL (1985) Racemization of amino acids. In: Barrett GC (ed) Chemistry and biochemistry of the amino acids. Chapman and Hall, London, pp 399–414

Bada JL (1991) Amino-acid cosmogeochemistry. Philos Trans R Soc Lond B Biol Sci 333:349–358

Bada JL, McDonald GD (1995) Amino acid racemization on Mars: implications for the preservation of biomolecules from an extinct Martian biota. Icarus 114:139–143

Bada JL, Miller SL (1987) Racemization and the origin of optically active organic compounds in living organisms. Biosystems 20:21–26

Barabas B, Caglioti L, Micskei K, Palyi G (2009) Data-based stochastic approach to absolute asymmetric synthesis by autocatalysis. Bull Chem Soc Jpn 82:1372–1376

Bennett CH (1982) The thermodynamics of computation – a review. Int J Theor Phys 21:905–940

Bennett CH (2003) Notes on Landauer's principle, reversible computation and Maxwell's demon. Stud Hist Philos Mod Phys 34:501–510

Blackmond DG, Matar OK (2008) Re-examination of reversibility in reaction models for the spontaneous emergence of homochirality. J Phys Chem B 112:5098–5104

Boltzmann L (1866) Über die Mechanische Bedeutung des Zweiten Hauptsatzes der Wärmetheorie. Wiener Berichte 53:195–220

Buhse T, Lavabre D, Micheau JC, Thiemann W (1993) Chiral-symmetry breaking – experimental results and computer-analysis of a liquid-phase autoxidation. Chirality 5:341–345

Caglioti L, Hajdu C, Holczknecht O, Zekany L, Zucchi C, Micskei K, Palyi G (2006) The concept of racemates and the Soai reaction. Viva Origino 34:62–80

Comley JW, Dowe DL (2005) Minimum message length, MDL and generalised bayesian networks with asymmetric languages. In: Grünwald P, Myung IJ, Pitt MA (eds) Advances in minimum description length: theory and applications. MIT Press, Cambridge

Cronin JR, Pizzarello S (1983) Amino acids in meteorites. Adv Space Res 3:5–18

Cronin JR, Pizzarello S (1997) Enantiomeric excesses in meteoritic amino acids. Science 275:951–955

Eliel EL, Wilen SH, Mander LN (eds) (1994) Stereochemistry of organic compounds. Wiley, New York, pp 153–295

Galimov EM (2006) Phenomenon of life: between equilibrium and non-linearity. Origin and principles of evolution. Geochem Int 44:1–95

Gibbs W (1875–1876; 1877–1878) On the equilibrium of heterogeneous substances. Transactions of the Connecticut Academy III. pp 108–248, Oct. 1875–May 1876, and pp 343–524, May 1877–July 1878

Gilat G (1999) On the biological advantage of chirality. arXiv:physics/9911041v1

Gilat G (2002) Physical chirality, it feeds on negative entropy. In: Pályi G, Zucchi C, Caglioti L (eds) Fundamentals of life. Elsevier, Paris, pp 369–377

Gonzalez JA, Calbo J (2002) Modeled and measured ratio of PAR to global radiation under cloudless skies. Agric Forest Meteorol 110:319–325

Gough P (2008) Information equation of state. Entropy 10:150–159

Grant RH, Slusser JR (2004) Estimation of photosynthetic photon flux density from 368-nm spectral irradiance. J Atmos Ocean Technol 21:481–487

Jarzynski C (1997) Nonequilibrium equality for free energy differences. Phys Rev Lett 78:2690–2693

Jaynes ET (1957) Information theory and statistical mechanics. Phys Rev 106:620–630

Joshi P, Wilkinson AR, Koike T, Fossan DB, Finnigan S, Paul ES, Raddon PM, Rainovski G, Starosta K, Simons AJ, Vaman C, Wadsworth R (2005a) First evidence for chirality in Tc isotopes: spectroscopy of 100Tc. Eur Phys J A 24:23–29

Joshi P, Finnigan S, Fossan DB, Koike T, Paul ES, Rainovski G, Starosta K, Vaman C, Wadsworth R (2005b) Evidence for a new region of chirality around A 104. J Phys G: Nucl Part Phys 31:S1895–S1898

Kauffman SA (ed) (1993) The origins of order. Oxford University Press, New York

Ladyman J, Presnell S, Short AJ, Groisman B (2007) The connection between logical and thermodynamic irreversibility. Stud Hist Philos Mod Phys 38:58–79

Landauer R (1961) Irreversibility and heat generation in the computing process. IBM J Res Dev 5:183–191

Lee K, Shin S, Ka J (2004) Tunneling dynamics of amino-acid: model for chiral evolution? J Mol Struct THEOCHEM 679:59–63

Lindgren K (1988) Microscopic and macroscopic entropy. Phys Rev A 38:4794–4798

Maroney OJE (2005) The (absence of a) relationship between thermodynamic and logical reversibility. Stud Hist Philos Mod Phys 36:355–374

Micskei K, Pota G, Caglioti L, Palyi G (2006a) Empirical description of chiral autocatalysis. J Phys Chem A 110:5982–5984

Micskei K, Maioli M, Zucchi C, Caglioti L, Palyi G (2006b) Generalization possibilities of autocatalytic absolute enantioselective synthesis. Tetrahedron Asym 17:2960–2962

Mislow K, Bickart P (1977) An epistemological note on chirality. Isr J Chem 15:1–6

Nicolis G, Prigogine I (eds) (1977) Self organization in nonequilibrium systems. Wiley, New York

Norton JD (2005) Eaters of the lotus: Landauer's principle and the return of Maxwell's demon. Stud Hist Philos Mod Phys 36:375–411

Osipov MA, Stelzer J (2003) Orientational ordering and chiral symmetry breaking in organic monolayers composed of disklike mesogenic molecules: molecular theory and computer simulations. Phys Rev E Stat Nonlin Soft Matter Phys 67(6 Pt 1):061707

Palyi G, Micskei K, Zekany L, Zucchi C, Caglioti L (2005) Racemates and the Soai reaction. Magy Kem Lapja 60:17–24

Pearson K (1898) Chance or vitalism. Nature 58:495–496

Popa R, Cimpoiasu VM, Scorei RI (2010) Consequences of expanding chirality to include spin isomery (The dilemma of broadening chirality into handedness). Phys AUC 20:64–72

Reza FM (ed) (1994) An introduction to information theory. Dover Publications Inc., New York

Schroeder RA (ed) (1974) Kinetics, mechanism and geochemical applications of amino acid racemization of various fossils. Ph.D. thesis, University of California, San Diego

Shannon CE (1948) A mathematical theory of communication. Bell Syst Tech J 27(379–424):623–659

Shinitzky M, Shvalb A, Elitzur AC, Mastai Y (2007) Entrapped energy in chiral solutions: quantification and information capacity. J Phys Chem 111:11004–11008

Shvalb A, Mastai Y, Shinitzky M (2010) Chiral configuration of the hydration layers of D- and L-alanine in water implied from dilution calorimetry. Chirality 22:587–592

Soai K (2002) Asymmetric autocatalysis and the origin of chiral homogeneity of biologically relevant molecules. In: Pályi G, Zucchi C, Caglioti L (eds) Fundamentals of life. Elsevier, Oxford, pp 426–435

Soai K (2004) Asymmetric autocatalysis, absolute asymmetric synthesis and origin of homochirality of biomolecules. In: Pályi G, Zucchi C, Caglioti L (eds) Progress in biological chirality. Elsevier, Oxford, pp 355–364

Soai K, Shibata T, Marioka H, Choji K (1995) Asymmetric autocatalysis and amplification of enantiomeric excess of a chiral molecule. Nature 378:767–768

Soai K, Sato I, Shibata T (2004) Asymmetric autocatalysis and the origin of homochirality of biomolecules. In: Malhotra SV (ed) Methodologies in asymmetric catalysis, vol 880, ACS symposium series. American Chemical Society, Washington, DC, pp 85–102

Stella TM (2009) http://www.iseesystems.com/softwares/Education/StellaSoftware.aspx (last accessed date Jan 2 2012)

Toyabe S, Sagawa T, Ueda M, Muneyuki E, Sano M (2010) Experimental demonstration of information-to-energy conversion and validation of the generalized Jarzynski equality. Nat Phys 6:988–992

Wolf C (ed) (2008) Dynamic stereochemistry of chiral compounds: principles and applications. RSF Publishing, Cambridge

# PART IV: PANSPERMIA

Gordon
McNichol
Wickramasinghe

Biodata of **Richard Gordon** and **Jesse C. McNichol**, authors of "***Recurrent Dreams of Life in Meteorites.***"

**Dr. Richard Gordon** is an itinerant theoretical biologist with B.Sc. (Math) from the University of Chicago and Ph.D. (Chemical Physics) from the University of Oregon. He is presently Theoretical Biologist, Embryogenesis Center, at the Gulf Specimen Marine Laboratory (http://www.gulfspecimen.org/). He organized Books with Wings (http://bookswithwings.net) which sends books to universities in Afghanistan, and the Embryo Physics Course online in Second Life® (http://embryogenesisexplained.com/course). He writes a blog in Science 2.0 (http://www.science20.com/cosmic_embryo) and is Principal Scientific Advisor for the EvoGrid project (http://www.evogrid.org/), an attempt to simulate the origin of life. He has also worked in diatom nanotechnology, HIV/AIDS prevention, and algorithms for computed tomography.

E-mail: **DickGordonCan@gmail.com**

**Jesse C. McNichol** obtained his honours B.Sc. Degree in Biology from Mount Allison University in 2008 working with Dr. Felix Baerlocher, Dr. Robert Thompson, and Dr. Zoe Finkel. After graduation, Jesse worked for Environment Canada and the National Research Council of Canada studying environmental toxicology and microalgal biofuels. Jesse is currently studying for his Ph.D. with the Woods Hole Oceanographic Institute/MIT joint program under the supervision of Dr. Stefan Sievert. His thesis will be focused on sulfur-oxidizing chemoautotrophs from deep-sea hydrothermal vents and other environments. Jesse's interests outside the lab include foreign languages, botany, and environmental education.

E-mail: **McNichol@mit.edu**

**Richard Gordon**

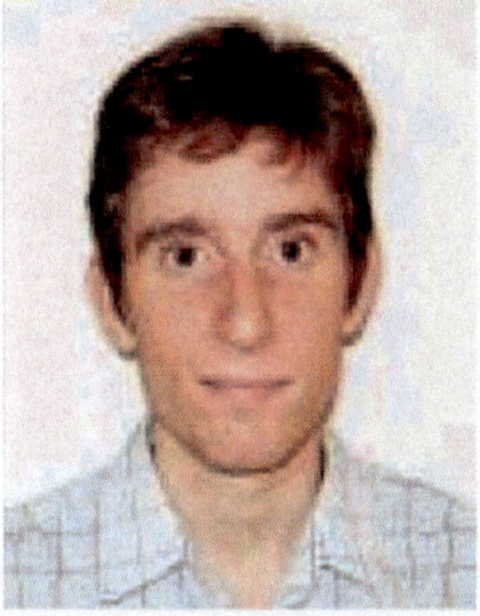

**Jesse C. McNichol**



J. Seckbach (ed.), *Genesis - In The Beginning: Precursors of Life, Chemical Models and Early Biological Evolution*, Cellular Origin, Life in Extreme Habitats and Astrobiology 22, 549–590
DOI 10.1007/978-94-007-2941-4_29, 

# RECURRENT DREAMS OF LIFE IN METEORITES

**RICHARD GORDON[1] AND JESSE C. McNICHOL[2]**
[1]*Embryogenesis Center, Gulf Specimen Marine Laboratories, 222 Clark Drive, Panacea, FL 32346, USA*
[2]*National Research Council of Canada, Institute for Marine Biosciences, 1411 Oxford Street, Halifax, Nova Scotia, Canada B3H 3Z1*

## 1. Introduction: The Iterated Rise, Fall, and Resurrection of Life in Meteorites

> On May 14, 1864, the largest carbonaceous chondrite known until 1950 fell at Orgueil, France, stimulating a controversy that continued for over a century. In the analysis of it published in 1864 by Stanislas Cloez (1817–83) [(Cloez, 1864a, b)], it is stated that it consists of materials "analogous to those of the organic part of several varieties of peats and of lignites."... In 1868, Marcellin Berthelot (1827–1907) reanalyzed the Orgueil meteorite [(Berthelot, 1868, 1869)], reporting materials... "comparable with the oils of petroleum." (p.402 in: (Crowe, 1999))

This was not the first nor last finding suggesting life in meteorites. It is our intention to focus here on the search for visible evidence for life in meteorites, rather than the bulk of the literature, which is on purported chemical signatures, or speculations on or denials of life in meteorites (some reviewed in: Claus and Madri, 1972). Briefly, observation of life in meteorites is now in its eighth round of discovery, denial, and hesitating capitulation. Episodes 2, 4, and 5 have been nicely argued, with a "to be continued" conclusion, by John G. Burke (1986):

1. Life found in a "meteorite" that fell in 1682, proved later to be laden with terrestrial plants and microorganisms (Erhenberg, 1839).
2. With the fall of the Orgueil carbonaceous chondrite of 1864 (Cloez, 1864a, b; Daubrée, 1864a, b; Kesselmeyer, 1864; Laussedat, 1864; Leymerie and Daubrée, 1864; Pisani, 1864), "Louis Pasteur was reported to have investigated unsuccessfully the Orgueil meteorite, seeking evidence of organisms (Morrison and Niehoff, 1979)" (Burke, 1986) (cf. Becquerel, 1924; Galippe and Souffland, 1924; Briggs, 1963a; Claus and Madri, 1972; Campbell and Schulze-Makuch, 2010). We could not find any other early attempts to find life in the Orgueil meteorite. Gümbel sought, but did not find, microfossils in other carbonaceous chondrites (Gümbel, 1878; Claus and Madri, 1972).

3. "The most curious meteoric shower, however, was observed in 1870 in Sweden. Black pieces, consisting almost exclusively of mold, descended on a snow-field, and could thus be easily collected" (Rachel, 1881). Unfortunately, this event apparently was not further documented by Rachel.
4. A flourish of discoveries of fossils in various meteorites by Otto Hahn (1880) was dismissed at the time and in retrospect: "...it is evident from the title and content of Hahn's, 1880 work that he was looking at chondrules" (Burke, 1986). We give details below.
5. "Charles B. Lipman, bacteriologist and graduate dean at the University of California, Berkeley, announced that sixteen sterilized specimens of eight meteorites placed in 24- to 48-hour cultures had produced rods and coccoid cells (Lipman, 1932)" (Burke, 1986), which was soon contradicted (Burke and Wiley, 1937) as "a flight of imagination through space" (Farrell, 1933), despite Lipman's seemingly stringent criteria for avoiding contamination, at least in the context of the state of the art at the time (Lipman, 1932). Lipman's battle with Sherat Kumar Roy (Hudson, 1935; Lipman, 1936; Roy, 1937) is detailed in: Claus and Madri (1972). See also: Nininger (1938).
6. Reexamination of the Orgueil meteorite, which reported "organized elements" (Claus and Nagy, 1962a).
7. The ALH84001 martian meteorite (discussed in McNichol and Gordon, 2012). Further investigation of this meteorite has raised the as yet unanswered question of whether its channels were created by microboring microorganisms (Fisk et al., 2006; McLoughlin et al., 2007; Gibson et al., 2010).
8. Another re-examination of the Orgueil and other carbonaceous chondrites, with the full case a subject of a recent review (Hoover, 2011).

We will focus on morphological evidence of life in meteorites, especially in the Orgueil meteorite. This leaves out analysis of organic molecules and optical activity, for example, that may or may not be bioindicators or terrestrial contamination (Oró and Skewes, 1965; Sephton, 2002), and recent attempts to distinguish terrestrial molecular contaminants from extraterrestrial material (Sephton et al., 1999, 2001; Kissin, 2003). As there are at least 440 publications on the Orgueil meteorite alone, many just published, this is far from a closed story, and, as it is said, "astrobiology is still heavily embroiled in politics" (Rothschild, 2007). Much emotion and bitter feelings have occurred, some expressed in print. But honest disagreement is the grist of science, and conflicts are rarely mere politics, but more fundamental. These range from presently philosophical questions, such as whether the universe is full of life, or we are alone, to assumptions about what can and cannot happen to an unattended museum specimen, and the balance between observations and the weight of authority or consensus.

To begin, it is important to consider the physical state of carbonaceous chondrites:

> Au contact de l'ean froide, la pierre météorique d'Orgueil se délite complétement: elle se divise en particules d'une extrême ténuité qui restent longtemps en suspension dans le liquide, et qui traversent les filtres les plus épais sur lesquels on cherche à les recueillir. Cette action de l'eau est le résultat de la dissolution de matières salines assez abondantes et qui servent en quelque sorte de ciment. (Cloez, 1864b)

Which we translate as:

> On contact of the meteorite with very cold water, the Orgueil meteorite disintegrates completely: it is divided into very fine particles that remained long suspended in the liquid and passed through the filters, the finest that we tried. This action of water is the result of the dissolution of quite abundant saline matter that serves as a kind of cement. (Cloez, 1864b)

And:

> Leur état de porosité est si grand, qu'il n'est pas étonnant que la pyrite qu'elles contiennent se soit altérée en partie dans notre atmosphère pour former des sulfates et des hyposulfites, et pour condenser aussi non-seulement beaucoup d'eau, mais peut-être aussi le peu d'ammoniaque qu'on y trouve. (Pisani, 1864)

Which we translate as:

> Their state of porosity is so great that it is not surprising that the pyrite it contains was altered in part by our atmosphere to form sulfates and hyposulphites, and water and perhaps some ammonia-like compounds may have condensed. (Pisani, 1864)

In our thinking about evidence for lifelike morphologies in the Orgueil meteorite, then, we must take into account the friability, salinity, and porosity of the meteorite samples.

## 2. The Hahn Episode

Given the long history of the panspermia idea, it would seem curious that there is no mention we can find of carbonaceous chondrites by Arrhenius (1908, 1909), whose name is so closely associated with the concept, perhaps because he required light pressure on small, isolated panspermia to push them through space (Claus and Madri, 1972). Nevertheless, life on and in meteorites had certainly been discussed prior to his book (Mason, 1877). We give some detailed quotes here of the Hahn episode, which Arrhenius should have been well aware of, to show the heat, surety, and sometimes equivocation with which conclusions were drawn, providing an historical perspective on present debates:

> It will no doubt be a genuine surprise to all students of nature to learn that a German scientist has found fossil plants and animal forms in most of the meteorites

(chondrites) which he has examined for the purpose. DR. OTTO HAHN,...has, in the usual way, prepared sections of many of these bodies. These he has had photographed and thereby attained a result which is independent of the microscopist's vision. Dr. Hahn claims that they show many forms of plants and animals in a fossil state contained in their mass, of which the highest forms are crinoids, corals and allied species.... We regret that we are unable to endorse this interesting discovery. Professor Whitfield, superintendent of the fossils and minerals in the American Museum of Natural History, has seen Dr. Hahn's drawings [sic] and was unable to verify the presence of the organic forms referred to. He attributed Dr. Hahn's error to a too sanguine temperament, and an "imagination which bodies forth the form of things unknown." (Michels, 1881b)

A review of German publications that support Hahn concludes:

...Hahn has proved it to be: fossil organisms!... The number of the various species of polypi, crinoids, spongiae and algiae which are united by a silicious material, Dr. Weinland estimates after a cursory examination at about fifty.... They tell us of a planet, on which aquatic life was sufficiently developed to produce them and to preserve them after death.... It may be safely claimed that Dr. Hahn's book will prove to be one of the most important contributions to natural science of the present time. (Rachel, 1881)

Michels then vacillated:

An examination which I have since made confirms in every respect the correctness of Dr. Hahn's statement, as to what he saw, and it therefore remains only to decide whether the deductions he made were correct.... I was disappointed to find that high powers failed to develop structure which indicated decisively the nature of these forms, and to show the difficulty of arriving at a correct solution, I may state that the two persons to whom I have so far shown the specimens, differed entirely as to their interpretation; the one pronounced them veritable fossils, and the second was equally sure that they were merely interesting forms of crystallization. I reserve an opinion until the section has been studied with more attention, and comparison made with other specimens now being prepared; in the mean time I shall be pleased to show the section to any person who is interested in this subject, or able by previous study to throw any light on the subject. (Michels, 1881a)

Another reviewer remained more skeptical:

Fossils In Meteors. -Our readers will remember that the late Sir Wyville Thomson, in his presidential address to the British Association at Glasgow, hinted at the possibility that the first germs of life might have been brought to the earth by or on a meteor! More recently a German doctor named Hahn professed,-and professes, for he refuses to believe anything to the contrary,-to have discovered traces of many species of fossils in sections of meteors which he has examined under the microscope. All the meteoric masses hitherto discovered are irony or stony masses, indubitably of igneous origin, and although the microscopic structure of the minerals composing these meteors is often curious and complex, yet no microscopist skilled in the examination of rocks has ever hinted at having seen anything, which by any possibility could be considered organic. Dr. Hahn, however, has been sending specimens and papers describing them over all Europe, and he appears at last to have begun to disseminate his discoveries in the New World. The American Journal called "Science" (Vol. II, p. 410) has an extraordinary account of an interview between Dr. Hahn and Mr. Darwin. Of course no such interview or conversation took place. (Anonymous, 1882)

The reported conversation is:

No sooner had Mr. Darwin peered through the microscope on one of the finest specimens when he started up from his seat and exclaimed: "Almighty God! what a wonderful discovery! Wonderful!" And after a pause of silent reflection he added: "Now reaches life down!" (R, 1881)

Later opinions were more definitive:

Whatever theory be held as to the origin of meteorites, that of volcanic rocks is sufficiently known and universally admitted: they have cooled from a state of fusion. If, therefore, a likeness of internal microscopic structure appears in the two kinds of bodies -besides in many cases a close resemblance in composition- we may safely conclude that meteorites also have passed through a fused condition, and that the structures in question have originated in them as they have in volcanic rocks.... dissenting entirely from the views of Dr. HAHN, as to the presence of organic remains in meteorites. (Weichmann, 1882)

The presence of compounds of carbon, such as result from vegetable decomposition, is curious, but no trace of life or of any organism has ever been seen. (Thornton, 1890)

An excellent complementary account of the Hahn episode, with more references, has been published (Claus and Madri, 1972).

## 3. Another Look in the Latter Half of the Twentieth Century

Surprisingly, despite the long known evidence of organic compounds, apparently no one bothered again to take microscope to meteorite to examine it visually for living or any other structures, until 1953:

Organic substances in meteorites were reported long ago, from 1835 to 1885; abstracts of most of the old references are given in the book of E. Cohen (Cohen, 1894). [Cf. (Cohen, 1903).]... The old literature does not refer to the composition, structure or problems of origin of the carbonaceous matter within the meteorites, and the present research was undertaken to investigate these interesting questions.... the majority are unusually loose in texture, particularly those stones, all without metal, in which the presence or organic substances was established by previous workers using reasonably reliable methods.... Under the binocular microscope, all the stones show uniform, light to dark grey coloration beneath a somewhat darker surface crust, the shade being the same throughout any specimen and even in different stones of the same shower. Thin sections invariably reveal "bituminous" substances, concentrated within the turbid, amorphous ground mass cementing the light coloured chondrites and the occasional phenocrysts. No discrete carbonaceous particles can be resolved in the matrix even under the highest powers of magnification.... It is often implied in popular works that the presence of organic substances in meteorites would indicate the existence of life on the celestial body of their origin.... In the light of modern organic chemical experience, there seems to be no great difficulty in accounting for a non-biological origin of the organic substances. (Mueller, 1953)

The next reported microscopic observation also uncovered nothing morphologically special, though the authors did argue for a biogenic origin of the organic matter, contrary to the previous paper (Mueller, 1953):

> Finely ground particles of the meteorite appeared as irregular shaped, angular, and greenish-brown grains under the petrographic microscope. A few single crystals of elongated to fibrous habit were noted. The irregular particles slowly disintegrated when a drop of water was added to the meteorite powder on the microscope slide. (Nagy et al., 1961)

However, the search for life via microscopy may have not been on the agenda at the time because mineralogy itself was so difficult:

> The mineralogical structure of the carbonaceous meteorites is not known because the very fine grained material and amorphous carbon hinder greatly the microscopic observation of the ground mass. (Boato, 1954)

Nagy next announced, in collaboration with Claus, that simple crushing of the Orgueil meteorite, such as had been done from the beginning (Cloez, 1864b), yielded 1,650–1,700 never before seen "organized elements" per milligram of meteorite, with a contamination rate (most identified as bacteria) of 35–41 per milligram (2%):

> Grains of the meteorite samples were crumbled in water or in glycerol on glass slides. A microscopic examination of the two Orgueil samples and of Ivuna revealed the presence of particles which were dissimilar in morphology to any known mineral form. The "organized elements" resembled, but were not identical to, the morphology of certain species of algae which live in water. (Claus and Nagy, 1961)

One of us (RG) was involved (Anders et al., 1964) as an undergraduate in a 100-year reexamination of the Orgueil meteorite. We uncovered a hoax in which some unknown person placed a piece of coal and the dried bud of a local plant into a sample of the meteorite. I (RG) showed, by analyzing it for the amino acid hydroxyproline, that the mass had been put back together with animal glue. The hoaxster placed the specimen on the local museum shelf, where it sat 100 years without generating the publicity, and the hoaxster died without the attendant glory, such as occurred later with Piltdown Man (Thomson, 1991). The inadvertent timing of the Orgueil hoax discovery (Edward Anders received the hoax sample with all parties assuming it was a legitimate one) may have cast a pale of suspicion from both sides as the 1960s Orgueil episode played out, and at least raised the degree of caution (Botan, 1965). The cause and effect claim was made (see also: Hoover and Rozanov, 2001b) that "...the discovery of a hoax concerning microfossils in the meteorite effectively extinguished all serious scientific investigations seeking evidence for biogenic materials and microfossils in meteorites for over three decades" (Hoover, 2006b) (however, cf. Staplin, 1965b; VanLandingham, 1965a, b; Manten, 1966; Kremp, 1968; Brooks and Shaw, 1969), but, as we have seen, examination of meteorites for traces of life has always been approached in spurts. I can recall Anders' surprise on making this discovery. There was no implication that any more than one specimen of one meteorite had been deliberately tampered with (Anders et al., 1964). Nagy wrote a detailed account, concluding:

> This incident is, in a sense, rather amusing if one can disregard the fact that one of the scarce and valuable Orgueil samples was ruined. Anders et al. (Anders et al., 1964) performed a service to meteorite studies by bringing this episode to light. (Nagy, 1975)

## 4. Organized Elements in the Orgueil Meteorite

Regarding this debate, which echoes to today (see next section), we think it best for the reader to again see the reasoning of the authors pro and con:

> A microscopic examination of the two Orgueil samples and of lvuna revealed the presence of particles which were dissimilar in morphology to any known mineral form. The "organized elements" resembled, but were not identical to, the morphology of certain species of algae which live in water.... One form of the organized elements was morphologically dissimilar to any known terrestrial organism. The organized elements did not show birefringence in polarized light, but they did show fluorescence (yellow) in ultraviolet light. A few specimens of microbiological contaminants (common, airborne bacteria, algae, etc.) were present in the carbonaceous as well as in the non-carbonaceous meteorites... Several of the organized elements appeared to undergo "cell-division".... From these morphological, optical and staining results we have interpreted the organized elements as possible remnants of organisms.... The high concentrations of the organized elements in Orgueil and in Ivuna could have developed only in an environment where moisture was continuously present for a prolonged period of time; yet the meteorite samples were kept in dry museum storage and would have disintegrated if subjected to an aqueous medium. Furthermore, they were outdoors for only a few hours after they had fallen. There are only traces, if any, of capillary water retained in the meteorites.... The organized elements, types 1–4, resemble... species of Dinoflagellates or Chrysomonads.... We are of the opinion that these observations suggest that the organized elements may be micro·fossils indigenous to the meteorite. (Claus and Nagy, 1961)

We can easily note here a number of problems, such as the unproven appeal to dryness of the specimens since they fell, proven bacterial contamination without generalizing this observation to all "organized elements," and identification of eukaryotic algae inconsistent with expected evolutionary progression of life on earth. It is for reasons such as these that in this chapter we do not try to persuade or dissuade the reader about the lifelike nature of the images recorded by meteorite investigators. Some are convincing, others are not, but in any case, finding an "organized element" in a meteorite does not solve how it got there. We will not prove that the organized elements are: alive vs. dead, abiogenic vs. fossil, indigenous vs. invading, prebiotic vs. biotic, minerals vs. organic, real vs. artifacts, prokaryotic vs. eukaryotic, familiar vs. out of this world, etc. Various lines of reasoning lead in all of these contradictory directions.

The above results (Claus and Nagy, 1961) were extended to two more meteorites, with a presumed control of other meteorites, not, however, matched for water content, salinity, and porosity (Claus and Nagy, 1962a). A flurry of papers ensued over the next decades. We list here those dealing with or denying (Orcel and Alpern, 1966) purported fossils. By this time, all live bacteria found in meteorites were presumed contaminants (Hayes, 1967), rather than ancient "living fossils," despite contemporary reports of live bacteria a few hundred million years old (Dombrowski, 1963a, b). The initial salvo occurred in 1962 and was sustained through 1966, followed by another wave of work launched by Richard Hoover in 1996. We record the chief journals involved, suggesting fads in publishing analogous to the stories of polywater (Franks, 1981) and cold fusion (Taubes, 1993) (Table 1) and/or cliques of scientists (Crane, 1972; Hull, 1990). The motivation for hype was there from the start:

> If it can be shown that these hydrocarbons and the "organized elements" are the residue of living organisms indigenous to the carbonaceous chondrites, this would be the most interesting and indeed astounding fact of all scientific study in recent years. (Urey, 1962b)

For the sake of historical completeness, we have attempted to be reasonably thorough in finding and listing the references in Table 1. This is important since this episode is sometimes cast as Edward Anders vs. Bartholomew Nagy, whereas the truth is that there were many independent protagonists on both sides and in the middle despite obvious tensions between these parties (Claus et al., 1963; Urey, 1966). Nagy dedicated his book "...to those investigators who supported or criticized in print the author's early work on carbonaceous chondrites; all considered friends" (Nagy, 1975) while others alluded to "most violent controversies" (Bernal, 1965) and an "acrimonious history" (Chyba and McDonald, 1995). On balance the protagonists appear to have behaved quite well on both sides:

> Although the battle smoke around this fascinating problem has not yet cleared away, and it is still too early to indicate any definite conclusion, the position of the "believers" seems to have strengthened. However this may be, appreciation should be expressed about the careful and thorough way in which Nagy and his group deal with the objections raised against their observations and conclusions. (Manten, 1966)

**Table 1.** Publications on organized elements, fossils, and organically coated particles in meteorites.

| Year(s): #/year | Publication | References |
|---|---|---|
| 1961: 1 | *Nature* | Claus and Nagy (1961) |
| 1962: 26 | *Science* | Anders and Fitch (1962), Morrison (1962), Sagan (1962), Silverman (1962), Urey (1962b) |
| | *Nature* | Bernal (1962), Briggs (1962b, a), Briggs and Kitto (1962), Fitch et al. (1962a), Gregory (1962), Mueller (1962), Nagy (1962), Palik (1962), Pearson (1962), Urey (1962a) |
| | Others | Anonymous (1962), Bairyev and Mamedov (1962), Claus and Nagy (1962a), Deflandre (1962), Fitch et al. (1962b), Gaskell (1962), Nultsch (1962), Staplin (1962a, b), Zigel (1962) |
| 1963: 28 | *Science* | Anders and Fitch (1963), Fitch and Anders (1963b), Mamikunian and Briggs (1963b), Tasch (1963) |
| | *Nature* | Mamikunian and Briggs (1963c), Meinschein (1963), Nagy et al. (1963a, b) |
| | *Annals of the New York Academy of Sciences* | Claus et al. (1963), Fitch and Anders (1963a), Fox and Yuyama (1963), Meinschein et al. (1963), Nagy (1963), Nagy et al. (1963c), Papp (1963), Urey et al. (1963), Vallentyne (1963) |
| | Others | Briggs (1963a, b), Briggs and Mamikunian (1963), Claus and Nagy (1963), Gentner (1963), Krejci-Graf (1963), Mamikunian and Briggs (1963a), Mason (1963), Mueller (1963), Palik (1963), Timofejev (1963) |
| 1964: 15 | *Science* | Anders et al. (1964), Simpson (1964) |
| | *Nature* | Claus and Suba-C (1964), Degens (1964), Fox (1964), Mueller (1964a), Nagy et al. (1964) |
| | *Annals of the New York Academy of Sciences* | Tasch (1964) |
| | Others | Imshenetsky (1964), Mueller (1964b), Nagy and Claus (1964), O'Connell (1964a, b), Vdovykin (1964), Vinogradov and Vdovykin (1964) |
| 1965: 15 | *Science* | Nagy et al. (1965), Oró and Tornabene (1965), Staplin (1965b), Urey (1965) |
| | *Nature* | Claus and Suba-C (1965), Mueller (1965), VanLandingham (1965b) |
| | Others | Abyzov and Imshenetskii (1965), Anonymous (1965), Botan (1965), Briggs and Mamikunian (1965), Fitch and Anders (1965), O'Connell (1965), Staplin (1965a), VanLandingham (1965a) |
| 1966: 9 | *Science* | Urey (1966) |
| | *Annals of the New York Academy of Sciences* | Botan (1966), Hovnanian (1966) |
| | Others | Bitz and Nagy (1966), Manten (1966), Nagy (1966a, b), Orcel and Alpern (1966), Urey and Arnold (1966) |

(continued)

**Table 1.** (continued)

| Year(s): #/year | Publication | References |
|---|---|---|
| 1967: 5 | *Nature* | VanLandingham et al. (1967) |
| | *Geochimica et Cosmochimica Acta* | Hayes (1967) |
| | Others | Nagy (1967), Tan and VanLandhgham (1967), Vdovykin (1967) |
| 1968: 6 | *Geochimica et Cosmochimica Acta* | Hayes and Biemann (1968), Urey et al. (1968) |
| | *Annals of the New York Academy of Sciences* | Claus (1968) |
| | Others | Kremp (1968), Nagy (1968a, b) |
| 1969: 3 | *Nature* | Brooks and Shaw (1969), Nagy and Nagy (1969) |
| | Others | Soffen (1969) |
| 1970–1979: 2.3 | *Science* | Green et al. (1971) |
| | *Annals of the New York Academy of Sciences* | Claus and Madri (1972) |
| | *Nature* | Bauman et al. (1973) |
| | *Geochimica et Cosmochimica Acta* | Bandurski and Nagy (1976), Hayatsu et al. (1977) |
| | Others | Hare (1970), West and Ponnamperuma (1970), Baker (1971), Brooks and Muir (1971), Rossignol-Strick and Barghoorn (1971), Oró (1972), Vdovykin (1972), Alpern and Benkheiri (1973), McCall (1973), Merek (1973), Vdovykin (1973), Folsome et al. (1975), Gregory (1975), Nagy (1975), West et al. (1975), Folsome (1976, 1977), Fox (1977) |
| 1980–1989: 1.1 | *Science* | Zenobi et al. (1989) |
| | Others | Brooks (1981), Horodyski (1981), Dodd (1983), Day (1984), Pflug (1984a, b), Rietmeijer (1985), Hoover et al. (1986), Deamer and Pashley (1989a, b) |
| 1990–1999: 1.9 | *Proceedings of the SPIE* | Hoover (1997), Pflug and Heinz (1997), Folk and Lynch (1998), Hoover et al. (1998), Gorlenko et al. (1999), Hoover and Rozanov (1999), Zhmur and Gerasimenko (1999) |
| | Others | Anders (1991), Buseck and Hua (1993), Chyba and McDonald (1995), Deamer (1997), Zhmur et al. (1997), Alexander et al. (1998), Raulin-Cerceau et al. (1998), Hoover et al. (1999), Kerridge (1999), Murae (1999), Sephton et al. (1999), Zhmur et al. (1999) |
| 2000–2010: 5.1 | *Science* | Nakamura-Messenger et al. (2006) |
| | *Geochimica et Cosmochimica Acta* | Kitajima et al. (2002) |

(continued)

**Table 1.** (continued)

| Year(s): #/year | Publication | References |
|---|---|---|
| | *Proceedings of the SPIE* | Sipiera et al. (2000), Hoover and Rozanov (2001a, b, 2002a, b), Rozanov et al. (2001), Zhmur et al. (2001), Hoover et al. (2003, 2004a, b, c), Rozanov and Hoover (2004), Storrie-Lombardi and Hoover (2004), Hoover (2006a, c, 2007a, b, 2008b), Gordon and Hoover (2007), Mukhopadhyay et al. (2007), Sheldon and Hoover (2008), Wickramasinghe et al. (2010) |
| | Others | Gounelle and Zolensky (2001), Pflug (2001), Sephton (2002), Zhmur (2002), Binet et al. (2004), Lemelle et al. (2004), Llorca (2004), Sephton (2004, 2005), Benzerara et al. (2006), Hoover (2006b, 2008a), Pearson et al. (2006), Line (2007), Mackay (2007), Mimura et al. (2007), Morange (2007), Pearson et al. (2007), Pikuta et al. (2007), Bland et al. (2008), Cody et al. (2008), Bland et al. (2009), Gronstal et al. (2009), Kwok (2009), Campbell and Schulze-Makuch (2010), Rauf et al. (2010), Rozanov (2010) |
| 2011: 17 | *Proceedings of the SPIE* | Hoover and Rozanov (2011), Brandenburg (2011), |
| | Others | Brasier (2011), Cowen (2011), Engel (2011), Godon (2011), Hoover et al. (2011), Hoover (2011), Kaufman (2011), Line (2011), Pikuta (2011), Redfield (2011), Sipiera (2011), Wickramasinghe (2011a, b, c), Youngbull (2011) |

This is in stark contrast to the following account, which we do not endorse:

> The first reports of the detection of microbial fossils in meteorites were published in the columns of *Nature* by Claus, Nagy and others (Claus and Nagy, 1961; Nagy, 1962; Nagy et al., 1963a). No sooner than these publications appeared a vigorous campaign of refutation and denial was mounted by Anders and others, also in the columns of *Nature* (Anders, 1962; Anders and Fitch, 1962; Fitch et al., 1962a). With a ruthless and forceful denigration of these claims on grounds of alleged contamination, the microfossil saga faded from view for a full 20 years. And with a veritable army of nay-sayers braying so stridently the world became convinced that this was a quagmire to which one should never return. The early pioneers of microfossil discovery were thus silenced and had little choice but to recant. I was told by a reliable witness that Claus was ruthlessly bullied into capitulation, and Nagy also retreated somewhat whilst continuing to hint in his writings that it might be so, rather in the manner of Galileo Galilei's whispered '*E pur si mouve*' – and yet it moves. (Wickramasinghe, 2011c)

As we shall see repeatedly, the bottom line is whether or not the possibility of terrestrial contamination has been eliminated. Despite protests that it has (Rozanov et al., 2001 regarding Timofejev, 1963, for example), quantitative work on contamination has just begun.

## 5. Interpreting Organized Elements

The reproducibility of observations of organized elements has been mixed:

> As reported by (Fitch et al., 1962a) a fraction of Orgueil can be obtained that is denser than methylene iodide (p = 3.33) and which on microscopic examination contains numerous approximately spherical particles somewhat resembling the Types I and II 'organized elements' of Claus and Nagy (Claus and Nagy, 1961). These particles are magnetic, and X-ray diffraction studies indicate the presence of troilite and magnetite.... From the results discussed here it must be concluded that the meteorite samples examined did not contain anything identifiable as the 'organized elements' of Nagy et al. (Claus and Nagy, 1961; Nagy, 1962) ... At present there is certainly no convincing evidence for the existence of any material within carbonaceous chondrites of extra-terrestrial, biological origin. (Briggs, 1962b)

> ...the Mokoia meteorite contains complex organic micro-structures of extra·terrestrial origin, a minority of which resemble unicellular organisms, but most are present as unorganized fragments. The evidence is compatible with either a biogenic or an abiogenic origin for the micro-structures.... (Briggs and Kitto, 1962)

> It was found that the majority of the structures claimed as of organic origin could be readily and quite unequivocally identified as mineral grains.... The presence of cores in most of the crystals indicates a process of unmixing of impurities during cooling. (Mueller, 1962)

Claims that some organized elements look like nothing else on earth (Meinschein, 1963) assume a kind of omniscience on the part of the claimant, and represent a challenge to others to find similar terrestrial objects (Tasch, 1963). For example, it was suggested "...that some 'organised elements' (cf. Nagy, 1975; Fig. 81) are similar to artificially produced poorly graphitised carbon particles and therefore not indigenous to the Orgueil (C1) carbonaceous chondrites" (Rietmeijer, 1985). Others have suggested spores (Gregory, 1962; Papp, 1963), algae (Palik, 1962), chrysophytes (Tasch, 1963), bacteria (Tan and VanLandhgham, 1967) and pollen (Deflandre, 1962; Pearson, 1962) (rebutted in: Claus and Nagy, 1963), some adding starch grains (Fitch et al., 1962b) (contradicted in: Claus et al., 1963; B. J. Cholnoky in: Urey et al., 1963) and mitotic figures (Botan, 1965, 1966) (cf. later: "...meteorite microfossils show evidence of life cycle processes (e.g. mitotic cell division, diplococci, chains of cells, trichomes, hormogonia, akinetes and spores)" (Hoover and Rozanov, 2002b)). The discoverers of organized elements claimed to see identifiable microfossils "resembling algae and flagellates" (Claus and Nagy, 1962a): "Types 1–4 appear to be similar to but not identical with known terrestrial species. The fifth type of the organized elements is entirely dissimilar in its morphology to known terrestrial form" (Claus and Nagy, 1961). The latter was then identified as ragweed pollen (Fitch and Anders 1963a, b, Anders in: Urey et al., 1963).

Thus, almost all organized elements that look like microfossils have corresponding terrestrial organisms, and proof hinges on the presumed pristine state of the interior of the meteorites, which we will show may be untenable. There is also the "minor" point that many organized elements are considered eukaryotic fossils in meteorites dated, at $4.56 \times 10^9$ years, older than any rock on earth (Buseck and Hua, 1993), whereas evidence for eukaryotes on earth only goes back to $1.7 \times 10^9$ years (Knoll, 1994; Javaux et al., 2003):

> ...contamination is the likely source of the "organized elements"... especially given the similarities... to plants, lichen, and fungi – organisms which are too advanced and specialized to be expected to exist within an ancient meteorite. (Campbell and Schulze-Makuch, 2010)

Despite the uncertainty about what they are, some authors decided to assign Latin genus and species to many organized elements (Claus and Nagy, 1962b; Staplin, 1962a; Briggs and Mamikunian, 1963; Palik, 1963; Timofejev, 1963; VanLandingham, 1965a). However, "The terrestriality of the organized elements is their most distinctive general characteristic" (Tasch, 1963); "Pierre Bourrelly... believed that the 'organized elements' were definitely organisms.... He was puzzled because they resembled terrestrial forms; he thought they should exhibit greater differences" (Urey et al., 1963). This puzzle is, of course, resolved by assuming that if it looks terrestrial, it is terrestrial, a position no one seems to have clearly stated.

A claim that organized elements may be dividing does not take into account probabilities of random juxtaposition: "...adjoining organized elements have been observed embedded in minerals. This raises the possibility of either copulation or division" (Claus et al., 1963).

If eukaryotes, such as diatoms (Hoover et al., 1986, 1999) or their analogs (Hoover and Rozanov, 2002a) and some acritarchs (Hoover and Rozanov, 2001a) are abundant in space, then why aren't muticellular eukaryotic organisms that can survive desiccation, such as rotifers and tardigrades (Marotta et al., 2010; Persson et al., 2011), not also found in meteorites?

The notion was put forth "...that these 'organized elements' might be earthly forms that had somehow been transferred from the earth to the moon in early geological times and later had returned to earth in carbonaceous meteorites" (Urey in: Urey et al., 1963). Similarly: "...we can not rule out the possibility that they may represent terrestrial microbial life-forms that were ejected by impact phenomena and then contaminated the meteorite parent bodies while in space" (Hoover and Rozanov, 2001b) (cf. Hoover and Rozanov, 2001a). This "outside the box" thinking is all well and good, but, as we shall see, the surface of the moon, for instance, proved sterile.

## 6. The Present Debate on Fossils in Meteorites

There was little debate about the long publication record of NASA astrobiologist Richard B. Hoover on microfossils in meteorites until 2011, when an article appeared in the *Journal of Cosmology* (Hoover, 2011) almost simultaneously with two books (Hoover et al., 2011; Kaufman, 2011). NASA distanced itself from the work based on lack of proof that the meteorites are not contaminated (Anonymous, 2011). The popular press, blog, and television response was immense, raucous, and sometimes irreverent, including much discussion on whether or not the paper was peer-reviewed or NASA-authorized, on Hoover's credentials, and off-the-cuff responses of scientists, mostly wait and see or against (Boyle, 2011; Cowen, 2011), but none of the latter wrote in the peer-reviewed literature, which we can find to date. The following statement was made without any references being given:

> Using state of the art technology Hoover concludes that microbial fossils unambiguously exist in great profusion. The furore that greeted this new publication, with vocal condemnation from Science journals and from NASA chiefs, shows that earlier tactics of rejection by silence have now been replaced by strident ranting and even personal insults. (Wickramasinghe, 2011c)

The editors of the *Journal of Cosmology* took an unusual step, which resulted in wide dissemination, but only 25 comments pro and con, of which we list those that are germane (Brandenburg, 2011; Brasier, 2011; Engel, 2011; Godon, 2011; Line, 2011; Pikuta, 2011; Redfield, 2011; Sipiera, 2011; Wickramasinghe, 2011b; Youngbull, 2011):

> ...Richard Hoover is a highly respected scientist and astrobiologist with a prestigious record of accomplishment at NASA. Given the controversial nature of his discovery, we have invited 100 experts and have issued a general invitation to over 5000 scientists from the scientific community to review the paper and to offer their critical analysis. (Schild, 2011)

Hoover reopened the debate on microfossils in meteorites with a review of the Orgueil organized elements story and observations on a more recent fall, the Murchison Type 2 carbonaceous chondrites of 1969 (Hoover, 1997). He found round and mushroom-shaped bodies, of which the latter could not be attributed to any known terrestrial forms by various authorities, including himself, a highly experienced microbiologist (Hoover, 1997). He argued that these objects were "deeply imbedded":

> The meteorite samples were extracted from the interior of a 1 gram piece of the Murchison meteorite.... The first objects found were smaIl (~2 micron) "mushroom-shaped" structures.... The upper region of the "mushroom" structure appears to be the top of a body which has a supporting tapering stalk. This body was not identifiable as "any known kind myxomycete or protostelid whose assimilative stage of wall-less, amoeboid, protoplasmic strands might have quickly migrated into the meteorite after it fell." High moisture levels are needed for plasmodia to be active. (Hoover, 1997)

Perhaps migration over a few millimeters in nearly three decades should not be considered "quick": "...a huge taxonomic diversity of prokaryotes colonized the Tatahouine meteorite in less than 70 years in the Tatahouine [Sahara desert] sand" (Benzerara et al., 2006). A roughly parallel observation, attributed to extraterrestrial organisms in carbonaceous chondrites instead of terrestrial contaminants, has been made:

> **Evidence for Biogenicity**: Many of the forms have cell-walls and membranes; and are present as colonies, coccoidal clusters and macrocolonies (~ cyanobacteria and purple sulfur bacteria); we have found evidence of life cycle processes including replication (e.g. dividing cells/diplococci, hormogonia, spores & chains of cells); as well as ecologically consistent microbial assemblages and communities (~ Cryoconite Communities from Antarctica – including *Microcoleus spp., Phormidium frigidum, and Nostoc species*). (Hoover and Rozanov, 2001a, 2002a; Hoover et al., 2003)

The relationship between the form of a live and a fossilized microfossil was not discussed, although in one example of experimental silicification shape was preserved (Francis et al., 1978a), and in another not (Kremp, 1968). The effects of desiccation ("collapsed" structures: Staplin, 1965a) before fossilization might also render microscopic forms unrecognizable (Soffen, 1969). An argument is made for a paradigm shift based on the discovery of "extremophiles that live in deep granite and hydrothermal vents and nanofossils in volcanic tuffs" (Hoover, 1997), though it seems that this could also be taken as an argument for the easy penetration of meteorites by terrestrial organisms. In any case, the evidence of our eyes at this level may not be convincing one way or the other: dead form does not necessarily imply prior life. Appeals to authority: "after prolonged examination by many different scientists competent in various fields" (Mamikunian and Briggs, 1963c), "the world's leading authority on ancient microfossils,... leading microbiologists and myxomycologists" (Hoover, 1997), do not settle disputes, nor do committees of them:

> The controversy between Prof. Anders and Prof. Nagy and Dr. Claus... is a matter which clearly requires for its resolution careful comparisons by a panel of impartial experts. (Bernal, 1962)

> On one occasion a single organized element, when it was briefly shown in a microscope to 18 microbiologists, was "identified" as 18 different species of protobionta or organic artifacts. (Claus et al., 1963)

For example, what are we to make of the contrast between the following statements by leading authorities of their day, both making absolute statements that later at least proved equivocal:

> There can be no serious doubt that the Orgeuil material... has been examined in an uncontaminated state. (Bernal, 1961)

> In 1966, the Nobel Laureate, Harold C. Urey (Urey, 1966) provided an extensive review of the scientific evidence for biological materials in meteorites. He observed that the organic substances encountered in meteorites resemble those in ancient terrestrial rocks but not recent contaminants. According to Urey: "*If found in terrestrial objects, some substances in meteorites would be regarded as indisputably biological.*" (Hoover et al., 1998)

More, better, and definitive observations are required. The hint for how to pursue these is given by Hoover's working hypothesis:

> Large numbers of viable microorganisms ejected from diverse planetary bodies by impact ejection/spallation mechanisms may have contaminated the Murchison meteorite and established viable populations. This process may have repeatedly occurred in the past with many different microbial communities from widely dispersed points of origin consuming the remains of organisms long departed. (Hoover, 1997)

> …the biosphere of Earth is open rather than closed. The possibility that biological matter has been transferred into space by the impact of large comets and asteroids onto shallow marine sediments must be considered as well as the possibility that terrestrial microbial extremophiles arrived on Earth during the Hadean bombardment from other regimes of the cosmos. (Hoover, 2008a)

> …asteroids, comets, and carbonaceous meteorites must have been contaminated by the multitude of collisions that have occurred since their formation in the protosolar nebula some 4.6 billion years ago. As a result of this prolonged exposure to solar system debris, it is not logical to consider that any of the meteorites truly remain "pristine" examples of protosolar nebula materials. (Hoover et al., 2003)

This is a testable proposition that implies cross contamination of essentially all bodies in the solar system and perhaps beyond, as the solar system orbits in our galaxy. It complicates the search for origins, but indeed may be the best null hypothesis. For example, once we get samples from Mars, we would anticipate finding organisms quite similar in basic biochemistry to those on earth, and at best, based on molecular clocks, might be able to estimate the date of cross contamination. In terms of independent origins of life, we might find ourselves none the wiser.

## 7. Focus on Motility of Cyanobacteria

Hoover claimed that carbonaceous meteorites (Orgueil, Murchison) contain the fossilized remains of cyanobacteria (Hoover, 2008b), as had been reported earlier (Zhmur et al., 1997) (cf. filaments reported by: Pflug and Heinz, 1997; Zhmur and Gerasimenko, 1999; Zhmur et al., 1999). If we accept his argument that the filamentous objects offer convincing morphological proof of life, we must still consider alternative sources of Earthly contamination. Hoover and Klyce make the assertion that since:

> …"blue-green algae" (cyanobacteria) … are aquatic organisms that grow only during total immersion in liquid water on Earth. If the Orgueil meteorite stones had been immersed in liquid water after landing on Earth, they would have disintegrated. Therefore, the contamination argument against them rings hollow. (Hoover and Klyce, 2010)

Nevertheless, falling in line with the expectation that "…healthily skeptical experts will dream up reasonable mechanisms for these formations" (Youngbull, 2011), we consider it possible that spores of cyanobacteria, landing on a carbonaceous chondrite after its fall, in storage, or in handling (Oró and Skewes, 1965), could have given rise to motile filaments that penetrated the meteorites. In some cases, penetration may be due to the initial impact (Mueller, 1963, 1964a) or air-sucked into the porous meteorite which is in vacuo before entering our atmosphere (Nininger, 1938; Rainer Berger in: Urey et al., 1963). Reports of viable bacteria in meteorites have long been available (Lipman, 1932; Bairyev and Mamedov, 1962; Briggs, 1963a; Nagy, 1966b) and doubted (Roy, 1935, 1937; Mueller, 1963; Staplin, 1965a; Claus and Madri, 1972), although one sample of Orgueil seemed devoid of them under conditions where they were found in the carbonaceous chondrites Murray (F.D. Sisler in: Mason, 1963; Claus and Madri, 1972; Oró and Tornabene, 1965) and Mokoia (Oró and Tornabene, 1965). Cf. (Briggs, 1960, 1962a; Briggs and Kitto, 1962; Claus et al., 1963). However, loss of viability of initial bacterial contaminants has not been either considered nor ruled out:

> The absence of viable organisms in the Orgueil sample may be a reflection of the fact that the sample was well preserved in a sealed container in the Mountaban Museum…. (Oró and Tornabene, 1965)

> To conclude that viable organisms isolated from within meteorites had an extraterrestrial origin would be possible only if the organisms proved unidentifiable and possessed some properties that clearly showed them as unearthly. It is apparent that the evidence… does not meet these criteria and terrestrial contamination is the obvious explanation of most of the positive results. On the other hand, most of the experiments gave negative results and demonstrate the lack of widespread microbiological contamination of stony meteorite interiors during storage under museum conditions. (Briggs, 1963a)

On the contrary, it is quite reasonable that motile microorganisms invaded museum specimens at times of high humidity and subsequently fossilized.

First, we note that filamentous cyanobacteria have evolved to colonize highly erodible materials (Garcia-Pichel and Wojciechowski, 2009), which characterizes carbonaceous chondrites. For example "…the specimen of Orgeuil meteorite in Calcutta Museum… has fallen into fine dust" (Bernal, 1962). Cyanobacteria survive long periods of desiccation in deserts, where they are able to move almost immediately when wetted (Campbell, 1979), and are metabolically active in humid air (Mazor et al., 1996). How dry is "bone dry," the purported condition in carbonaceous chondrites (Brandenburg, 2011)? They have no need for "total immersion" in water. Hoover has acknowledged the presence of motile cyanobacteria in dry environments:

> …motile filamentous cyanobacteria (e.g., *Calothrix*, *Oscillatoria*, *Phormidium*, and *Spirulina*)… grow in geothermal springs and geysers of Earth at temperatures ranging from 320K to 345K and are also found growing in cold polar desert soils. (Hoover, 2006a)

Furthermore, carbonaceous chondrites come with about 20% water in situ (Boato, 1954; Wiik, 1956; Nagy et al., 1964). The force generated by gliding cyanobacteria (Pentecost, 1990) might be sufficient to penetrate a porous carbonaceous chondrite, and perhaps even enlarge existing pores with their microboring capabilities (McLoughlin et al., 2007). New methods of detecting live or dead cells in meteorites could be used to investigate the rate of cell penetration (Lemelle et al., 2004). The very porosity of carbonaceous chondrites, estimated at up to 20% (Anders, 1962; Consolmagno et al., 2008), makes them both a plausible protective habitat for panspermia (Hoover et al., 2004b) and for easy and perhaps quick entrance of earthly motile microorganisms. Many of the cyanobacteria genera tentatively identified in carbonaceous chondrites are motile, multicellular filaments: *Phormidium* (Nultsch, 1968), *Microcoleus* (Whale and Walsby, 1984), and *Lyngba* (Hoiczyk and Baumeister, 1995). A critical analysis of whether the meteorite fossils could be fungi is also lacking (Krumbein, 2009).

When one of us (RG) dissolved a sample of the Orgueil meteorite with HF (1962–1963, in Edward Anders' lab), and looked at the remaining material under a microscope some time later, a microorganism was amusingly seen swimming by. It was either a panspermia or a contaminant (sterile techniques had not been used). In either case, it was clear that Orgueil worked as a culture medium. This was later shown for the Murchison meteorite (Mautner, 2002), though an earlier report found "…these extracts [of]… the Orgueil meteorite… are outright inhibitory" (Claus and Madri, 1972). Thus the inside of a carbonaceous chondrite might be a hospitable environment.

There is a substantial literature on the internal colonization of rocks by microorganisms, and the relationship between rock structure and their behavior (Golubic et al., 1981): "…bacteria can penetrate hard rocks and meteorites" (Oró and Tornabene, 1965). Some rock microborers are cyanobacteria (McLoughlin et al., 2007). Motility speeds of cyanobacteria have been clocked at 5–15 μm/min (Choi et al., 1999) to over 300 μm/min (Gupta and Agrawal, 2006), and filamentous bacteria at 720 μm/min (Ridgway and Lewin, 1988), albeit under water. Given that the five CI1 carbonaceous chondrites that were collected shortly after falling sat in museums since 1806 (Alais), 1864 (Orgueil), 1911 (Tonk), 1938 (Ivuna), and 1965 (Revelstoke), and that the high salt and water content (Hoover, 2011) would suggest hygroscopic behavior as the humidity varied, the time lapsed may be more than enough for penetration by gliding cyanobacteria present on the surface. Note in this regard "…that magnesium sulfate in the [Orgueil] meteorite in New York is $MgSO_4\,4H_2O$; in Chicago it is $MgSO_4\,7H_2O$" (Brian Mason in: Urey et al., 1963). But such long times may not be necessary, as found in this study of organic molecules in one fresh fall:

> It is clear that the organic material found in the surface layer of this meteorite [Pueblito de Allende] is of biological origin and cannot be other than terrestrial contamination acquired even in the short period of time available to the meteorite, namely, between its fall on February 8, 1969, and its collection on February 15, 1969. The speed with which this contamination was acquired, and the diversity of

the material contained in that contamination, make doubtful any interpretation of such organic materials that have been found in meteorites of unknown, or at, least considerably longer, terrestrial history (Nooner and Oro, 1967; Hayes and Biemann, 1968). (Han et al., 1969)

Conditions within carbonaceous chondrites are far from stable:

Carbonaceous chondrites are very porous, with a fractional pore volume of 15 to 20 per cent. They will therefore "breathe" every time the barometric pressure changes, and since many of their mineral constituents, being of very small particle size (DuFresne and Anders, 1961), have large surface areas, considerable adsorption of atmospheric constituents will occur. Carbonaceous chondrites are known to adsorb even atmospheric argon (Stauffer, 1961) and should be even better adsorbents for hydrocarbons. (Anders, 1962)

The free organic components of the Orgueil carbonaceous chondrite have been the subject of much controversy.... Compounds suggestive of terrestrial contamination by essential plant oil derivatives are evident. Their distribution suggests a source such as cleaning products, disinfectant and air-freshener.... (Watson et al., 2003)

Sulfate veins were first described in the Orgueil meteorite in 1961, almost one century after its fall. However, we have observed such veins to form easily during typical sample storage. We suggest that all CI1 sulfate veins formed during the terrestrial residence of these heavily brecciated, porous stones. Reacting with atmospheric water, sulfates originally present in the meteorites dissolved and remobilized, and/or sulfides oxidised, filling the many open spaces offered to them by the very porous rock.... Our reexamination of the original papers reporting on the Alais and Orgueil CI1 meteorites soon after their falls (1806 and 1864, respectively) has revealed that these meteorites landed on the Earth devoid of any white sulfate veins.... an active remobilization of originally extraterrestrial sulfates and/or corrosion of sulfides... led to the filling of the numerous open spaces and to the opening of new fracture networks. (Gounelle and Zolensky, 2001)

This latter notion, that the water-soluble veins could have formed after the Orgueil meteorite fell (Gounelle and Zolensky, 2001), might be testable by methods used to date aqueous vein formation in a martian meteorite (Swindle and Olson, 2002). The suggestion of "new fracture networks" (Gounelle and Zolensky, 2001) may counter arguments that the organized elements "... were firmly embedded in the silicate matrix and thus could hardly have been introduced into the meteorite mechanically, and pollen grains do not grow in such places" (Urey, 1962b) and that "It is improbable at least ["impossible": Hoover, 2008a] that any such massive contamination with terrestrial organisms under the conditions of collection could have occurred [or]... grown under the fairly dry conditions of museums" (Urey, 1962a).

The observation of water-soluble mineral veins formed after the falls of these meteorites not only reopens the question of the aqueous nature of the parent body (DuFresne and Anders, 1961; Briggs, 1962a; Anders, 1963; Nagy et al., 1963c; Chyba and McDonald, 1995; Grossman et al., 2000) but also contradicts a basic assumption: "Nagy has advised us that any biological-like structures firmly embedded in the mineral matrix of the meteorite obviously could not be caused by contamination" (VanLandingham et al., 1967) (cf. Nagy, 1962). This opinion was asserted in more detail as follows:

> Some of the organized elements observed in thin sections have such shapes and conformity with the meteorite texture that they cannot have been introduced in the meteorite at any time on the Earth.... [Some]... particles have no association with any observable vein or pore systems.... A remote possibility exists that... contaminations were introduced into the meteorite some time after its fall, and that the contaminants were since mineralized in the same way as the indigenous particles. Such a possibility is contradicted by the similar shape of these loose particles and those in thin sections and by the fact that they usually cannot be clearly identified as terrestrial on morphological grounds. (Nagy et al., 1963a)

This opinion was pretty well reversed later:

> Nagy reported in (Nagy, 1966b) the results of Volcani and Bruff to the effect that they had successfully cultured quantities of different anaerobic bacteria, supposedly from the interior of the Orgueil meteorite.... The speculation was raised that some of these viable forms might have survived in the films of moisture saturated with the magnesium sulfate of the veins of the stone. In view of the fact that the origin of these veins is uncertain and one cannot exclude the possibility of terrestrial waters participating in their formation, the question of whether the bacteria found were indigenous to the stone was left open. (Claus and Madri, 1972)

Thus, there would seem to have been plenty of opportunity for motile microorganisms to colonize the interior "pristine" centers of these meteorites: "Mason urged... that there might well have been contamination by terrestrial organisms during the long period in which these meteorites have been preserved in collections" (Urey, 1962b). "...it would be truly remarkable if one did not find a variety of terrestrial microorganisms in them" (Mason, 1963). Furthermore, the apparently fossilized nature of the biological structures in carbonaceous chondrites (Hoover and Klyce, 2010) cannot be taken as evidence of their ancient nature, because fossilization can occur rapidly (within a day: Drum, 1968; within 10 weeks: Kremp, 1968; Francis et al., 1978a). In fact, Gerhard Kremp, in the context of discussing alternative explanations of organized elements, mineralized the cyanobacterium (then called a blue-green alga) *Anabaenopsis*: "That the objects may be fossils of blue-green algae type cannot be excluded as a possibility" (Kremp, 1968). The application of detailed studies of fossilization of bacteria (Preston et al., 2011) to meteorite conditions is an area worth pursuing.

Nagy and colleagues seem to here reverse their stand, marking the beginning of the end of their episode:

> Mineralized microstructures, containing seemingly organic residues (that is, the organized elements), have been reported in carbonaceous meteorites. These structures resemble microfossils, although some of them, particularly the ones which are not mineralized, appear to be terrestrial biological contaminations. (Nagy et al., 1965)

> The author is of the opinion that Orgueil contains some type of interesting looking microstructures, but he feels that it may be premature to seriously argue about their identity.... An unhurried and not presupposed evaluation of all evidence and hypotheses seems to be highly desirable to clarify this problem, which in the author's opinion is worthy of further critical consideration. (Nagy, 1966a)

Although we have not reviewed the issue here, a similar conclusion was drawn about the optical activity in the organic matter in meteorites (Meinschein et al., 1966).

We conclude that until criticisms such as those raised above are addressed, the present episode (Hoover, 2011) does not represent "proof" of panspermia in meteorites. On the contrary, we concur that the null hypothesis explaining the results obtained so far is, even if we accept the evidence of cyanobacteria, etc., inside meteorites, that they got there by being surface contaminants that penetrated the friable, saline, hydrated, and porous structure of the intact pieces of meteorites as they sat on museum shelves, under varying humidity, over many years:

> ...these samples have been sitting around in laboratories for between 205 and 73 years. It is well known that microbial contaminants can penetrate deep into such rocks, even during storage. The null hypothesis, therefore, is that many of these objects (e.g., (Hoover, 2011), figs 1, 3) may be prokaryotic contaminants. (Brasier, 2011)

This reasoning is countered by:

> The dominant forms recognized are the mineralized and carbonized remains of cyanobacteria, sulfur bacteria and trichomic mats. Trichomic mats are typically benthic. They would have required the presence of bodies of liquid water for their formation (cyanobacterial mats and the associated anaerobes) and do not form in thin films of water between mineral grains. Since Orgueil landed on Earth, none of the Museum specimens have ever been submerged in bodies or pools of liquid water. Doing so would destroy these fragile meteorites since the cementing evaporite minerals, such as epsomite, readily dissolve in liquid water. (Hoover et al., 2004a)

The possibility of such contamination, one way or the other, is easy to test, by preparing artificial materials resembling these meteorites that are free of dead or alive microorganisms, contaminating their surfaces with gliding microorganisms, and testing how fast they penetrate within as a function of constant or fluctuating humidity. Even better, we need to eventually get fresh samples of CI1 meteorites or their presumed parent bodies, asteroids or comets, under sterile conditions, and test them for organisms within, before any possible residual contaminants could insinuate themselves inside. Contamination controls need to be designed and implemented in advance (Flory et al., 1974) and simulations done to test them (Mueller, 1964a). It would be best to capture fresh material in space, as was done for lunar samples (Barghoorn et al., 1970; Oyama et al., 1970): "None of the meteorites we have studied are devoid of terrestrial contamination, 'falls' and 'finds', alike suffer from the problem" (Swart et al., 1983):

> … all meteorite preparations used in this work contain a variety of rare microstructures in the c. 20-μ size range which exhibit a complex morphology, and yet are not readily identifiable. It is our opinion that most of these objects are unusual mineral grains or terrestrial materials which have contaminated the meteorite during museum storage. However, conclusive identifications are not possible. (Mamikunian and Briggs, 1963c)

Just as "In an effort to resolve contamination issues and explore the possibility of deep microbial life, the Department of Energy launched the subsurface science program" (Hoover, 1997), we would hope that NASA would show similar leadership.

The debate about fossil-organized elements has narrowed from the vast majority of organic coated particles in carbonaceous chondrites (next section) to the "…rich and diverse assemblage of well preserved, mineralized, and carbonized remains [of] large filamentous prokaryotic microorganisms" (Hoover, 2006c), i.e., whether they are invaders from outer space, or from earth. There is now an attempt to find biomarkers that are "unambiguously biological" and "undeniably indigenous" (Hoover, 2007b), though perfection is achieved by an argument based on ignorance: "many [biomarkers]… are not produced by any known abiotic mechanisms and thus must be considered to represent valid chemical biomarkers in carbonaceous meteorites" (Hoover, 2007b).

## 8. Coated Particles

The beginning of the end of the current episode is hinted in the recognition that:

> The Orgueil meteorite contains a number of ovoids and spherical bodies that are of indeterminate biogenicity…. The origin of even the more complex spherical forms (e.g., pyrite or magnetite framboids and ovoids and spheroids with nanofibrils) that are often enclosed within carbon rich or sheath-like films suggestive of a possible biological origin are also grouped with the indeterminate forms since they could have formed by abiotic mechanisms. (Hoover, 2006c)

> The formation of organic microstructures that could be preserved and misinterpreted as biogenic structures in ancient strata has been studied in laboratory experiments (Folsome et al., 1975; Folsome, 1976; Francis et al., 1978b) and several types of biogenic-appearing microstructures occurring in Precambrian rocks and meteorites have been described and interpreted as pseudofossils (Rossignol-Strick and Barghoorn, 1971; Cloud, 1973)…. (Horodyski, 1981)

Unfortunately, we have found no literature that addresses the possible biological alteration of the small organic coated particles.

The small organized elements are often described as "organic matter round a hexagonal mineral nucleus" (Brooks, 1981) (cf. Staplin, 1962a, 1965a; Mueller, 1963, 1964b; Kremp, 1968; Nagy and Nagy, 1969; Brooks and Muir, 1971), some "highly magnetic" (Fitch and Anders, 1963a), "iron with some chlorine and nickel in organized elements" (Nagy et al., 1963a), "silicate micro- and submicrochondrites encased, in most cases, in carbonaceous matter" (Vdovykin, 1964) (cf. Vdovykin, 1972, 1973) and "organic coatings on olivine microchondrules, magnetite and glass globules" (Rossignol-Strick and Barghoorn, 1971). "There is evidence from transmission electron microscopy that graphitic layers coat some mineral grains in Allende (Green et al., 1971)" (Buseck and Hua, 1993). However, Michael Briggs suggests that there are fragments of amorphous high-molecular-weight discharge polymers in meteorites that might coat particles (Briggs, 1963a). These coatings may be fluorescent (Murae, 1999). Possibly, organic coatings have also been found on cosmic dust (Parkin et al., 1962). "Curious stacks of magnetite platelets and 'fuzzy balls' of almost pure elemental iron covered in minute fibrils and encased in thin carbon (probably kerogen) sheaths have been found" (Hoover et al., 2004a). Membrane-forming lipids are found in carbonaceous chondrites (Deamer and Pashley, 1989a, b; Deamer, 1997; Kwok, 2009). In some cases, only the hollow globules presumed to have been formed around "…preexisting ice grains that were protected from radiation processing by the organic mantle" are found (Nakamura-Messenger et al., 2006).

The first three of many steps towards formation of organic matter in meteorites are thought to be: "Molecule formation by a variety of reaction pathways in dense interstellar clouds; Condensation of those molecules onto refractory interstellar grains; Irradiation of organic-rich interstellar-grain mantles…" (Kerridge, 1999). Thus the polymeric organic material may be presolar (Alexander et al., 1998), possibly with "…multiple origins… such as dense interstellar clouds (Kolodny et al., 1980), the solar nebula (Hayatsu and Anders, 1981), and the surfaces and atmospheres of parent asteroids (Peltzer et al., 1984)" (Kitajima et al., 2002). This contrasts significantly with the Miller-Urey scenario using spark discharges (lightning) (Folsome et al., 1975; Folsome, 1977), but it should be possible to distinguish them. Neither mechanism accounts for the apparent active, catalytic role of the mineral grains (Hayatsu et al., 1977).

For all the controversy, from the beginning there has been reasonable agreement that the small organized elements are: "Particles… definitely indigenous…. Their morphology is simple…. The density of most of these particles is high, indicating a predominantly inorganic composition" (Anders and Fitch, 1962), i.e., a combination of organic and inorganic matter worthy of further analysis. The organic coats account for the bulk of the definitively extraterrestrial "insoluble condensed aromatic" organic matter in carbonaceous chondrites (Simmonds et al., 1969):

> The polymeric material occurs mainly as ~1 μm particles containing small mineral nuclei (Alpern and Benkheiri, 1973), and occasionally as surface coatings on larger, spherical or hexagonal grains (Brooks and Muir, 1971; Rossignol-Strick and Barghoorn, 1971). Such selective association with specific grains is suggestive of formation by surface catalysis. Alternative processes, such as condensation of a supersaturated vapor, or reactions of free radicals on grain surfaces, would have given a much more uniform distribution over all mineral grains. (Hayatsu et al., 1977)

In the face of the substantial literature analyzing the kerogen-like, polymer-like aromatic material in carbonaceous chondrites, with abiotic scenarios for its synthesis slowly being worked out (Vinogradov and Vdovykin, 1964; Bitz and

Nagy, 1966; Hayes, 1967; Simmonds et al., 1969; Bauman et al., 1973; Bandurski and Nagy, 1976; Zenobi et al., 1989; Anders, 1991; Sephton et al., 1999; Kitajima et al., 2002; Sephton, 2002; Binet et al., 2004; Pearson et al., 2006; Mimura et al., 2007; Mukhopadhyay et al., 2007; Pearson et al., 2007; Cody et al., 2008; Rauf et al., 2010), the unreferenced statement: "There are no known mechanisms for the production of Kerogens by ABIOTIC PROCESSES" (Hoover et al., 2003) would seem hyperbole.

While size distributions of organized elements are multimodal (Claus and Suba-C, 1964), this does not prove that they "…do not represent just simple mineral grains as suggested (Mueller, 1962; Fitch and Anders, 1963a), but may be the remains of some kind of primitive life indigenous to the meteorites" (Claus and Suba-C, 1964). The size distribution for the inorganic particles was not measured as a control, and anyway:

> Besides, its similarity to the size distribution of magnetite globules and glass globules [(Mueller, 1964b), fig. 8], has been stressed (Mueller, 1965), but the validity of the comparison has been denied (Claus and Suba-C, 1965). These mineral organized elements were previously shown (Fitch and Anders, 1963a), to be soluble in HF-HCl…. Size distribution plots are nondiscriminatory in resolving the question of origin of unknown objects. (Rossignol-Strick and Barghoorn, 1971)

There is also another size class: "…the TEM shows that a considerable portion of the meteorite carbonaceous matter consists of structured particles ranging from 10 to 500 nm in size" (Pflug, 1984b).

Just as speculated for "nanobacteria" (Young and Martel, 2010), "organized elements" may prove to be stepping stones to the origin of life, rather than life itself, perhaps as "a probable precursor of living systems" (Morrison, 1962). It was suggested that organized elements are "prebionta" (Fox and Yuyama, 1963) (cf. Hare, 1970; Merek, 1973; Folsome et al., 1975; Folsome, 1976, 1977; Fox, 1977; Pflug, 1984a, 2001; Nakamura-Messenger et al., 2006), or "protocells" (Rasmussen et al., 2008). The inorganic component may be catalytic for prebiotic chemical syntheses (Sephton, 2004). As such, they may have provided the compartments within which the multitude of "experiments" occurred, with an occasional one resulting in the first viable cells (Gordon, 2008; Damer et al., 2012).

## 9. Next Steps

The common themes in all 8 episodes were: (1) the "organisms" found sooner or later were proved to be similar to organisms already known on earth, raising concerns that that was precisely what they were, and (2) convincing proof that contamination could not have happened was never forthcoming. To break this cycle, we suggest:

1. Deliberate contamination of meteorite specimens with various terrestrial organisms be used as a control (as has been attempted in: Tasch, 1964; Claus, 1968; Claus and Madri, 1972), such as has been done in testing sterility of medical devices (Jacob et al., 1994; Ameen et al., 1996; Luijt et al., 2001; Rasic et al., 2003) or their resistance to infection (Badylak et al., 2003; Brennan et al., 2006)
2. That exploration of the solar system proceed apace to obtain pristine samples and check them for life, under stringent isolating conditions that provably avoid contamination of control specimens
3. That the hypothesis be explored that many organized elements and purported microfossils (including purported extraterrestrial nanobacteria: Folk and Lynch, 1998; Hoover and Rozanov, 1999; Folk and Taylor, 2002; Hoover and Rozanov, 2002b) are analogous to what "nanobacteria" turned out to be: complex structures formed between meteorite "poorly defined high-molecular weight [organic] macromolecules" (Llorca, 2004) and crystals, combinations which "…come in all sizes and shapes and… can mimic biological forms that appear alive" (Young and Martel, 2010)
4. That modern tools of nanotechnology be brought to bear on the study of organized elements, many of which appear to be heteronanocrystals (de Mello Donegá 2011)

The primary question about whether or not the organized elements landed with the meteorites has not been resolved, let alone what they are or were. The cases for and against were clearly stated by him in the Lipman episode when he found live bacteria inside meteorites (Lipman, 1932). While much good science has been done around the question of life or no life in meteorites, fossil or alive, skeptics on both sides have their ground to stand on. It is not a question of integrity, for one of us has not only had "…the chance to hear Hoover personally speak on this subject, [sensing]… his genuine motivation for the truth inside these mysterious type CI1 carbonaceous meteorite" (Youngbull, 2011) (cf. Kaufman, 2011), but has also had the privilege of collaborating with him (Gordon and Hoover, 2007; Gordon et al., 2007). We nevertheless end up by standing firmly in the middle, as have others (Mueller, 1964b; Briggs and Mamikunian, 1965), including Nagy: "…one does not know if the organized elements are contaminations and/or mineral artifacts; or if some of them are indeed indigenous microfossils" (Nagy, 1967), and on the "other side" Anders: "At present, there exists no compelling evidence for extraterrestrial biological activity in these meteorites" (Studier et al., 1968). The half-empty/half-full cup dilemma remains, unresolved by the current episode with the work of Hoover and colleagues, and, we would predict, the dream of life in meteorites is unlikely to be resolved on earth:

> It is a little surprising, and one might even say amusing, that the hypothesis that organized elements may be extraterrestrial microfossils has become perhaps one

of the most thoroughly confirmed and at the same time most exhaustively refuted scientific ideas of the past few years. (Nagy, 1968a)

...some people thought we really believed that biological material has been proven to exist in meteorites, though our colleagues have not interpreted our papers and oral discussion in that way. Perhaps people mistake our hopes for definite conclusions. We think it would be fascinating if we found biological material in the carbonaceous chondrites. We are amazed that so many students of the subject are willing to adopt the negative view immediately without any consideration at all.... in our opinion, all evidence in regard to the question is equivocal; measurements are marginal, the results are due to contamination by biological material after the meteorite arrived on earth, compounds that are thought to be biological can be produced abiologically, etc. Because of these complications, it has never been possible for us to come to a definite conclusion; further, many similarities of the carbonaceous materials of meteorites to terrestrial ancient biological deposits do exist.... we strongly suspect that we will meet the same difficulties when we study the lunar samples. (Urey et al., 1968)

But no similar controversy arose over the lunar samples, once we had them in (sterile) hand (Barghoorn et al., 1970; Meinschein et al., 1970; Oyama et al., 1970): just "...one exception has been a description (Nagy et al., 1970)...of a polymer-like sheet" (Oyama, 1972), one terrestrially contaminated sample (Kozar et al., 2001; Fox, 2002), and one reinterpretation that "...glassy formations are remnant of spiral-homoside trichome" (Zhmur and Gerasimenko, 1999). Urey's hypothesis that organized elements were of terrestrial origin, via the moon (Urey in: Urey et al., 1963), was quietly ignored.

The tantalizing idea that there is actual life in meteorites, not just organic precursors or shells for protocells, is consistent with the concept of panspermia (McNichol and Gordon, 2012) and the idea that the 9 billion years between the Big Bang and the formation of the earth might not have been a lifeless interlude (Gordon and Hoover, 2007; Line, 2007). But that is no proof of the matter.

We summarize with some early wisdom:

At present there is no faultless investigation which is able to prove to existence of cosmobionts in meteorites.... it is quite obvious that even microscopic cracks in a meteorite fallen on the ground will let soil bacteria penetrate easily into the central parts of the meteorite. Unfortunately, this fact was not taken into account during microbiological analyses of meteorites which have been for a few months or even years on the Earth's surface. (Imshenetsky, 1964)

We shall just have to "catch a falling star" (Vance et al., 1957) before it hits the ground to be sure:

One might ask why it is necessary to bother with all this when in ten years time someone will land on an asteroid and settle the question of extraterrestrial life.... (Urey et al., 1963)

1973 has come and gone a long while ago, and it is about time we got on with exploration in space for carbonaceous chondrites. Why bother? Because "...the carbonaceous chondrites may hold the key to the origin of life on our planet" (Staplin, 1965a) or elsewhere.

We shall give Nagy, a careful scientist who stuck his neck out, the last word:

> The origin of the microstructures in carbonaceous meteorites is not known. It seems certain that at least some are terrestrial microbiological contaminations and mineral structures. Yet, there seems to be a growing tendency to consider them indigenous, non-biological organic particles which may be associated with mineral matter. That organized elements are extraterrestrial microfossils is unlikely, even though some investigators believe in this origin. Only further studies can clarify their origin.... Perhaps the wisest attitude is to resist speculating about their origin and intensively study them, instead. (Nagy, 1975)

## 10. Acknowledgment

We would like to thank Cheryl Buors, Neil John Maclean Health Sciences Library, University of Manitoba, for a tremendous job in collecting archaic references.

## 11. References

Abyzov SS, Imshenetskii AA (1965) A technique and some results of meteorite microbiological investigations. Life Sci Space Res 3:155–164

Alexander CMO, Russell SS, Arden JW, Ash RD, Grady MM, Pillinger CT (1998) The origin of chondritic macromolecular organic matter: a carbon and nitrogen isotope study. Meteoritics Planet Sci 33(4):603–622

Alpern B, Benkheiri Y (1973) Distribution de la matière organique dans la météorite d'Orgueil par microscopie en fluorescence [Distribution of organic matter in the Orgueil meteorite by fluorescent microscopy]. Earth Planet Sci Lett 19(4):422–428

Ameen AP, Short RD, Douglas CWI, Johns R, Ballet B (1996) A critical investigation of some of the procedures employed in the surgical use of titanium. J Mater Sci Mater Med 7(4):195–199

Anders E (1962) Meteoritic hydrocarbons and extraterrestrial life. Ann N Y Acad Sci 93(14):651–657

Anders E (1963) On the origin of carbonaceous chondrites. Ann N Y Acad Sci 108(2):514–533

Anders E (1991) Organic matter in meteorites and comets: possible origins. Space Sci Rev 56(1–2):157–166

Anders E, Fitch FW (1962) Search for organized elements in carbonaceous chondrites. Science 138(3548):1392–1399

Anders E, Fitch FW (1963) Erratum: Search for organized elements in carbonaceous chondrites. Science 139(3550):99

Anders E, DuFresne ER, Hayatsu R, DuFresne A, Cavaillé A, Fitch FW (1964) Contaminated meteorite. Science 146(3648):1157–1161

Anonymous (1882) Fossils in meteors. The Midland Naturalist 5:92

Anonymous (1962) Life-forms in meteorites. Curr Sci 31(6):226

Anonymous (1965) Life-forms in meteorites and the problem of terrestrial contamination: a study in methodology – Tasch, P. Psychiatr Q 39(2):382

Anonymous (2011) NASA shoots down alien fossil claims. http://www.abc.net.au/news/stories/2011/03/08/3157645.htm. Accessed 26 June 2011

Arrhenius SA (1908) Worlds in the making; the evolution of the universe. Harper, New York

Arrhenius S (1909) The life of the universe as conceived by man from the earliest ages to the present time, vol 1, 2. Harper & Brothers, New York

Badylak SF, Wu CC, Bible M, McPherson E (2003) Host protection against deliberate bacterial contamination of an extracellular matrix bioscaffold versus Dacron™ mesh in a dog model of orthopedic soft tissue repair. J Biomed Mater Res Part B 67B(1):648–654

Bairyev C, Mamedov S (1962) СЕДБІ ЖИЗНИ ВКАМНЯХ ИЗВ СЕЛЕННО□ [Traces of life in rocks from outer space]. Pravda (June 25):6

Baker BL (1971) Review of organic matter in the Orgueil meteorite. Space Life Sciences 2(4):472–497

Bandurski EL, Nagy B (1976) The polymer-like organic material in the Orgueil meteorite. Geochim Cosmochim Acta 40(11):1397–1406

Barghoorn ES, Phillpott D, Turnbill C (1970) Micropaleontological study of lunar. Science 167(3918):775

Bauman AJ, Devaney JR, Bollin EM (1973) Allende meteorite carbonaceous phase: intractable nature and scanning electron morphology. Nature 241(5387):264–267

Becquerel P (1924) La Vie Terrestre Provient-Elle d'un Autre Monde? [Does life on earth come from another world?]. L'Astronomie 38:393–417

Benzerara K, Chapon V, Moreira D, López-García P, Guyot F, Heulin T (2006) Microbial diversity on the Tatahouine meteorite. Meteoritics Planet Sci 41(8):1249–1265

Bernal JD (1961) Significance of carbonaceous meteorites in theories on the origin of life. Nature 190(4771):129–131

Bernal JD (1962) Comments. Nature 193(4821):1127–1129

Bernal JD (1965) Essay review. In: Middlehurst BM, Kuiper GP (eds) The solar system, vol IV, The Moon, Meteorites and Comets. Sci Prog 53(209):143–146

Berthelot P (1868) Cosmologie. – Sur la matiére charbonneuse des météorites. Comptes rendus hebdomadaires des séances de l'Académie des Sciences, Paris 67:849

Berthelot P (1869) Ueber die kohlige Substanz der Meteoriten. J Prakt Chem 106:254

Binet L, Gourier D, Derenne S, Robert F, Ciofini I (2004) Occurence of abundant diradicaloid moieties in the insoluble organic matter from the Orgueil and Murchison meteorites: A fingerprint of its extraterrestrial origin? Geochim Cosmochim Acta 68(4):881–891

Bitz SM, Nagy B (1966) Ozonolysis of "polymer-type" material in coal, kerogen and in the Orgueil meteorite: a preliminary report. Proc Natl Acad Sci USA 56(5):1383–1390

Bland PA, Howard KT, Cressey G, Benedix CK (2008) The terrestrial component of primitive chondrite alteration. Meteoritics Planet Sci 43(S7):A26

Bland PA, Jackson MD, Coker RF, Cohen BA, Webber JBW, Lee MR, Duffy CM, Chater RJ, Ardakani MG, McPhail DS, McComb DW, Benedix GK (2009) Why aqueous alteration in asteroids was isochemical: High porosity not equal high permeability. Earth Planet Sci Lett 287(3–4):559–568

Boato G (1954) The isotopic composition of hydrogen and carbon in the carbonaceous chondrites. Geochim Cosmochim Acta 6(5–6):209–220

Botan EA (1965) Examination of "organized elements" from the Orgueil meteorite by quantitative fluorescence microscopy. Aerospace Med 36(11):1069–1076

Botan EA (1966) Biotic signatures. Ann N Y Acad Sci 140(A1):307–313

Boyle A (2011) Life in meteorites? Study stirs debate. http://cosmiclog.msnbc.msn.com/_news/2011/03/05/6198177-life-in-meteorites-study-stirs-debate. Accessed 26 June 2011

Brandenburg JE (2011) Second Copernican revolution. The CI and ALH84001 are linked by isotopes and chemistry. J Cosmol 13(March): http://journalofcosmology.com/Life101.html#123

Brasier MD (2011) Life in CI1 carbonaceous chondrites? J Cosmol 13(March): http://journalofcosmology.com/Life101.html#109

Brennan EP, Reing J, Chew D, Myers-Irvin JM, Young EJ, Badylak SF (2006) Antibacterial activity within degradation products of biological scaffolds composed of extracellular matrix. Tissue Eng 12(10):2949–2955

Briggs MH (1960) The origins of life on the earth: a review of the experimental evidence. Sci Cult 26:160–170

Briggs MH (1962a) The nature and origin of meteorite organic matter. Sci Cult 28(8):357–360

Briggs MH (1962b) Properties of organic microstructures of some carbonaceous chondrites. Nature 195(4846):1076–1077

Briggs MH (1963a) Biological problems of meteorites. Tuatara J Biol Soc 11(1):1–16

Briggs MH (1963b) Organic extracts of some carbonaceous meteorites. Life Sci 1:63–68

Briggs MH, Kitto GB (1962) Complex organic micro-structures in the Mokoia meteorite. Nature 193(4821):1126–1127

Briggs MH, Mamikunian G (1963) Organic constituents of the carbonaceous chondrites. Space Sci Rev 1(4):647–682

Briggs MH, Mamikunian G (1965) Trends and problems in exobiology. In: Mamikunian G, Briggs MH (eds) Current aspects of exobiology. Pergamon Press, New York, pp 347–358

Brooks J (1981) Organic matter in meteorites and precambrian rocks: clues about the origin and development of living systems [and discussion]. Philos Trans R Soc Lond Ser A Math Phys Eng Sci 303(1480):595–609

Brooks J, Muir MD (1971) Morphology and chemistry of the organic insoluble matter from the Onverwacht Series Precambrian chert and the Orgueil and Murray carbonaceous meteorites. Grana 11(1):9–14

Brooks J, Shaw G (1969) Evidence for extraterrestrial life: identity of sporopollenin with the insoluble organic matter present in the Orgueil and Murray meteorites and also in some terrestrial microfossils. Nature 223(5207):754–756

Burke JG (1986) Cosmic debris: meteorites in history. University of California Press, Berkeley

Burke V, Wiley AJ (1937) Bacteria in coal. J Bacteriol 34(5):483–488

Buseck PR, Hua X (1993) Matrices of carbonaceous chondrite meteorites. Annual Rev Earth Planet Sci 21:255–305

Campbell SE (1979) Soil stabilization by a prokaryotic desert crust: implications for precambrian land biota. Orig Life Evol Biosph 9(4):335–348

Campbell MR, Schulze-Makuch D (2010) Classics in space medicine: BOTAN EA. Examination of "organized elements" from the Orgueil meteorite by quantitative fluorescence microscopy. Aerosp Med 1965; 36:1069–76. Aviation Space Environ Med 81(7):700–701

Choi JS, Chung YH, Moon YJ, Kim C, Watanabe M, Song PS, Joe CO, Bogorad L, Park YM (1999) Photomovement of the gliding cyanobacterium Synechocystis sp. PCC 6803. Photochem Photobiol 70(1):95–102

Chyba CF, McDonald GD (1995) The origin of life in the solar system: current issues. Annual Rev Earth Planet Sci 23:215–249

Claus P (1968) Studies on terrestrial contaminants of meteorites. Ann N Y Acad Sci 147(9):365–409

Claus G, Madri PP (1972) Studies of terrestrial contaminants in meteorites: Part II Bacteriology of meteorites. Ann N Y Acad Sci 196(9):387–407

Claus G, Nagy B (1961) A microbiological examination of some carbonaceous chondrites. Nature 192(480):594–596

Claus G, Nagy B (1962a) Microfossils, new to science, resembling algae and flagellates, found in meteorites. Pollen et Spores 4(2):339

Claus G, Nagy B (1962b) Taxonomical consideration of certain Incerta Sedes. Phycol Soc Am News Bull 15(1):15–19

Claus G, Nagy B (1963) Discussion de la note de Georges Deflandre sur la présence supposée de micro-organismes d'origine extra-terrestre dans des météorites [Discussion of the note of Georges Deflandre. "Criticisms on the supposed presence of microorganisms extraterrestrial in origin in meteorites]. Rev Agol 6(4):319–323

Claus G, Suba-C EA (1964) Organized element distribution in relation to size in the Orgueil meteorite. Nature 204(495):118–120

Claus G, Suba-C EA (1965) Interpretation of micro-structures in carbonaceous meteorites. Nature 205(4977):1201

Claus G, Nagy B, Europa DL (1963) Further observations on the properties of the "organized elements" in carbonaceous chondrites. Ann N Y Acad Sci 108(2):580–605

Cloez S (1864a) Analyse chimique de la pierre météorique d'Orgueil. Comptes rendus hebdomadaires des séances de l'Académie des Sciences, Paris 59:37–40

Cloez S (1864b) Note sur la composition chimique de la pierre météorique d'Orgueil. Comptes rendus hebdomadaires des séances de l'Académie des Sciences, Paris 58:986–988

Cloud P (1973) Pseudofossils: a plea for caution. Geology (Boulder) 1(3):123–127

Cody GD, Alexander CMO, Kilcoyne ALD, Yabuta H (2008) Unraveling the chemical history of the Solar System as recorded in extraterrestrial organic matter. In: Kwok S, Sandford S (eds) Organic matter in space, proceedings IAU symposium No. 251, vol 4. International Astronomical Union, pp 277–284

Cohen EW (1894) Meteoritenkunde. Heft I. Untersuchungen und Charakteristik der Gemengtheile. E. Schweizerbart'sche Verlagsbuchhandlung, Stuttgart

Cohen EW (1903) Meteoritenkunde. Heft II. Structurformen; Versuche künstlicher Nachbildung von Meteoriten; Rinde und schwarze Adern; Relief der Oberfäche; Gestalt, Zahl und Grösse der Meteorite; Nachträge zu Heft I, vol 2. E. Schweizerbart'sche Verlagsbuchhandlung, Stuttgart

Consolmagno GJ, Britt DT, Macke RJ (2008) The significance of meteorite density and porosity. Chem Erde-Geochem 68(1):1–29

Cowen R (2011) Meteorites may hold fossils from space – or not. ScienceNews:http://www.science-news.org/view/generic/id/70826/title/Meteorites_may_hold_fossils_from_space_—_or_not

Crane D (1972) Invisible colleges, vol 1. The University of Chicago Press, Chicago

Crowe MJ (1999) The extraterrestrial life debate, 1750–1900. Dover, Mineola

Damer B, Newman P, Gordon R, Barbalet T, Norkus R, Karpis M (2012) Cyberbiogenesis: a thought experiment and 21st century grand challenge. In: Seckbach J (ed) In the beginning: Precursors of life, chemical models and early biological evolution. Springer, Dordrecht, pp

Daubrée GA (1864a) Cosmologie. – Note sur les météorites tombées le 14 mai aux environs d'Orgueil (Tarn-et-Garonne). Comptes rendus hebdomadaires des séances de l'Académie des Sciences, Paris 58:984–986

Daubrée GA (1864b) Noveaux renseignements sur le bolide du mai 1864. Comptes rendus hebdomadaires des séances de l'Académie des Sciences, Paris 58:1065–1072

Day W (1984) Genesis on Planet Earth: the Search for Life's Beginning. Yale University Press, New Haven

de Mello DC (2011) Synthesis and properties of colloidal heteronanocrystals. Chem Soc Rev 40(3):1512–1546

Deamer DW (1997) The first living systems: a bioenergetic perspective. Microbiol Mol Biol Rev 61(2):239–261

Deamer DW, Pashley RM (1989a) Amphiphilic components of carbonaceous meteorites. Origins Life Evol Biosph 19:21–38

Deamer DW, Pashley RM (1989b) Amphiphilic components of the Murchison carbonaceous chondrite: surface properties and membrane formation. Orig Life Evol Biosph 19(1):21–38

Deflandre G (1962) Micropaléontologie des météorites. – Remarques critiques sur la présence supposée de microorganismes d'origine extra-terrestre dans des météorites [Micropaleontology of meteorites. – Critical remarks on the supposed presence of microorganisms of extraterrestrial origin of meteorites]. Comptes rendus hebdomadaires des séances de l'Académie des sciences 254(19 Part 3):3405–3407

Degens ET (1964) Genetic relationships between organic matter in meteorites and sediments. Nature 202(4937):1092–1095

Dodd RT (1983) Meteorites. A petrologic-chemical synthesis. Cambridge University Press, London

Dombrowski H (1963a) Bacteria from Paleozoic salt deposits. Ann N Y Acad Sci 108(2):453–460

Dombrowski HJ (1963b) Living bacteria from the Paleozoic. Biol Zentrabl 82(4):477–484

Drum RW (1968) Petrifaction of plant tissue in the laboratory. Nature 218:784–785

DuFresne ER, Anders E (1961) The record in the meteorites-V. A thermometer mineral in the Mighei carbonaceous chondrite. Geochim Cosmochim Acta 23(3–4):200–208

Engel MH (2011) The search for extraterrestrial life. J Cosmol 13(March):http://journalofcosmology.com/Life101.html#102

Erhenberg GC (1839) On a meteoric paper which fell from the sky in the year 1682 at Courland, composed of Confervae and Infusoria. Ann Mag Nat Hist 3:185–186

Farrell MA (1933) Living bacteria in ancient rocks and meteorites. Amer Mus Novit 645:1–3

Fisk MR, Popa R, Mason OU, Storrie-Lombardi MC, Vicenzi EP (2006) Iron-magnesium silicate bioweathering on Earth (and Mars?). Astrobiology 6(1):48–68

Fitch FW, Anders E (1963a) Observations on the nature of the "organized elements" in carbonaceous chondrites. Ann N Y Acad Sci 108(2):495–513

Fitch FW, Anders E (1963b) Organized element: possible indentification in Orgueil meteorite. Science 140(357):1097–1100

Fitch FW, Anders E (1965) Current status of the analysis of organized elements in carbonaceous chondrites. Life Sci Space Res 3:154

Fitch F, Anders E, Schwarcz HP (1962a) 'Organized elements' in carbonaceous chondrites. Nature 193(4821):1123–1125

Fitch F, Schwartz HP, Anders E (1962b) Identification of some organized elements in carbonaceous chondrites. J Geophys Res 67(9):3557–3558

Flory DA, Oró J, Fennessey PV (1974) Organic contamination problems in the Viking molecular analysis experiment. Orig Life Evol Biosph 5(3–4):443–455

Folk RL, Lynch FL (1998) Carbonaceous objects resembling nannobacteria in Allende meteorite. Proc SPIE 3441:115–136

Folk RL, Taylor LA (2002) Nannobacterial alteration of pyroxenes in Martian meteorite Allan Hills 84001. Meteoritics Planet Sci 37(8):1057–1069

Folsome CE (1976) Synthetic organic microstructures and the origins of cellular life. Naturwissenschaften 63(7):303–306

Folsome CE (1977) Reply: Organic microstructures and terrestrial protocells. Naturwissenschaften 64(7):381

Folsome CE, Allen RD, Ichinose NK (1975) Organic microstructures as products of Miller-Urey electrical discharges. Precambrian Res 2(3):263–275

Fox SW (1964) Thermal polymerization of amino-acids and production of formed microparticles on lava. Nature 201(4917):336–337

Fox SW (1977) Organic microstructures and terrestrial protocells. Naturwissenschaften 64(7):380

Fox A (2002) Chemical markers for bacteria in extraterrestrial samples. Anat Rec 268(3):180–185

Fox SW, Yuyama S (1963) Abiotic production of primitive protein and formed microparticles. Ann N Y Acad Sci 108(2):487–494

Francis S, Barghoorn ES, Margulis L (1978a) On the experimental silicification of microorganisms. III. Implications of the preservation of the green prokaryotic alga *Prochloron* and other coccoids for interpretation of the microbial fossil record. Precambrian Res 7(4):377–383

Francis S, Margulis L, Barghoorn ES (1978b) On the experimental silicification of microorganisms. II. On the time of appearance of eucaryotic organisms in the fossil record. Precambrian Res 6(1):65–100

Franks F (1981) Polywater. MIT Press, Cambridge

Galippe V, Souffland G (1924) Recherches sur la présence dans les Météorites, les Pierres dures, les Minerais, le Quartz, le Granite, le Basalte, les Cendres et les laves volcaniques, d'organites susceptibles de reciviscence et sur leur resistance aux hautes températures [Research on presence of recovering micoorganisms in meteorites, hard stones, minerals, quartz, granite, basalt, ash and volcanic lava, and their resistance to high temperatures]. CR Acad Sci, Paris 172:1252–1254

Garcia-Pichel F, Wojciechowski MF (2009) The evolution of a capacity to build supra-cellular ropes enabled filamentous cyanobacteria to colonize highly erodible substrates. PLoS One 4(11)

Gaskell T (1962) Do meteorites reveal life in other worlds? New Scientist 14(2893):458–459

Gentner W (1963) Irdische und Meteoritische Materie [Terrestrial and meteoritic matter]. Naturwissenschaften 50(6):191–199

Gibson EK, McKay DS, Clemett SJ, Thomas-Keprta KL, Pillinger CT, Verchovsky AB, Spencer L (2010) Nature of carbon in martian meteorites. Meteoritics Planet Sci 45(Supplement s1):A63

Godon P (2011) Can a meteorite falling on earth originate from earth?. J Cosmol 13(March): http://journalofcosmology.com/Life101.html#108

Golubic S, Friedmann I, Schneider J (1981) The lithobiontic ecological niche, with special reference to microorganisms. J Sediment Petrol 51(2):475–478

Gordon R (2008) Hoyle's tornado origin of artificial life, a computer programming challenge. In: Seckbach J, Gordon R (eds) Divine action and natural selection: science, faith and evolution. World Scientific, Singapore, pp 354–367

Gordon R, Hoover RB (2007) Could there have been a single origin of life in a big bang universe? Proc SPIE 6694. doi:10.1117/1112.737041

Gordon R, Hoover RB, Tuszynski JA, de Luis J, Camp PJ, Tiffany MA, Nagy SS, Fayek M, Lopez PJ, Lerner BE (2007) Diatoms in space: testing prospects for reliable diatom nanotechnology in microgravity. Proc SPIE 6694:V1–V15. doi:10.1117/1112.737051

Gorlenko VM, Zhmur SI, Duda VI, Osipov GA, Suzina NE, Dmitriev VV (1999) Microbial nature of fibrous kerite of Yolyn. Proc SPIE 3755:83–95

Gounelle M, Zolensky ME (2001) A terrestrial origin for sulfate veins in CI1 chondrites. Meteoritics Planet Sci 36(10):1321–1329

Green HW, Radcliffe SV, Heuer AH (1971) Allende meteorite: a high-voltage electron petrographic study. Science 172(3986):936–939

Gregory PH (1962) Identity of organized elements from meteorites. Nature 194(4833):1065

Gregory PH (1975) Recognition of microscopic objects. Proc R Soc Lond Ser B 189(1095):161–165

Gronstal A, Pearson V, Kappler A, Dooris C, Anand M, Poitrasson F, Kee TP, Cockell CS (2009) Laboratory experiments on the weathering of iron meteorites and carbonaceous chondrites by iron-oxidizing bacteria. Meteoritics Planet Sci 44(2):233–247

Grossman JN, Alexander CMO, Wang JH, Brearley AJ (2000) Bleached chondrules: evidence for widespread aqueous processes on the parent asteroids of ordinary chondrites. Meteoritics Planet Sci 35(3):467–486

Gümbel H (1878) Uber die in Bayern gefundenen Steinmeteoriten [On the stone meteorites found in Bavaria]. Sitzungsber math-phys KI Kgl bayer Akad Wiss München 1:14–72

Gupta S, Agrawal SC (2006) Motility in Oscillatoria salina as affected by different factors. Folia Microbiol 51(6):565–571

Hahn O (1880) Die Meteorite (Chondrite) unde ihre Organismen. Laupp'schen, Tubingen

Han J, Simoneit BR, Burlingame AL, Calvin M (1969) Organic analysis on the Pueblito de Allende meteorite. Nature 222(5191):364–365

Hare F (1970) Évolution de systèmes intervenant au cours de l'abiogenèse [Systems development during abiogenesis]. J Chim Phys Chim Biol 67(9):1681–1704

Hayatsu R, Anders E (1981) Organic compounds in meteorites and their origins. Top Curr Chem 99:1–37

Hayatsu R, Matsuoka S, Scott RG, Studier MH, Anders E (1977) Origin of organic matter in the early solar-system-VII. The organic polymer in carbonaceous chondrites. Geochim Cosmochim Acta 41(9):1325–1339

Hayes JM (1967) Organic constituents of meteorites - a review. Geochim Cosmochim Acta 31(9):1395–1440

Hayes JM, Biemann K (1968) High resolution mass spectrometric investigations of the organic constituents of the Murray and Holbrook chondrites. Geochim Cosmochim Acta 32(2): 239–267

Hoiczyk E, Baumeister W (1995) Envelope structure of four gliding filamentous cyanobacteria. J Bacteriol 177(9):2387–2395

Hoover RB (1997) Meteorites, microfossils, and exobiology. Proc SPIE 3111:115–136

Hoover RB (2006a) Comets, asteroids, meteorites, and the origin of the biosphere. Proc SPIE 6309:J3090–J3090

Hoover RB (2006b) Comets, carbonaceous meteorites, and the origin of the biosphere. Biogeosci Discuss 3:23–70

Hoover RB (2006c) Fossils of prokaryotic microorganisms in the Orgueil meteorite. Proc SPIE 6309:30902

Hoover RB (2007a) Microfossils of cyanobacteria in carbonaceous meteorites. Proc SPIE 6694:69408–69408

Hoover RB (2007b) Ratios of biogenic elements for distinguishing recent from fossil microorganisms. Proc SPIE 6694:D6940

Hoover RB (2008a) Comets, carbonaceous meteorites and the origin of the biosphere. In: Dobretsov N, Kolchanov N, Rozanov A, Zavarzin G (eds) Biosphere origin and evolution. Springer, New York, pp 55–68

Hoover RB (2008b) Microfossils of filamentous prokaryotes in CI1 and CM2 meteorites. Proc SPIE 7097:9703

Hoover RB (2011) Fossils of cyanobacteria in CI1 carbonaceous meteorites: implications to life on comets, Europa, and Enceladus. J Cosmol 13(March):http://journalofcosmology.com/Life100.html

Hoover R, Klyce B (2010) Cosmic ancestry: more evidence for indigenous microfossils in carbonaceous meteorites. http://www.panspermia.org/hoover4.htm. Accessed 8 Nov 2010

Hoover RB, Rozanov AY (1999) Biomorphic structures in Mighei carbonaceous chondrite. Proc SPIE 3755:120–127

Hoover RB, Rozanov AY (2001a) Chemical biomarkers and microfossils in carbonaceous meteorites. Proc SPIE 4495:1–18

Hoover RB, Rozanov AY (2001b) Evidence for biomarkers and microfossils in ancient rocks and meteorites. Proc SPIE 4273:15–32

Hoover RB, Rozanov AY (2002a) Astrobiology: traces of life in the cosmos. Proc SPIE 4765:1–19

Hoover RB, Rozanov AY (2002b) Microfossils, biominerals, and chemical biomarkers in meteorites. Proc SPIE 4939:10–27

Hoover RB, Rozanov AY (2011) Filamentous trichomic prokaryotes in carbonaceous meteorites: indigenous microfossils, minerals, or modern bio-contaminants? Proc SPIE 8152. doi: 10.1117/12.898659

Hoover RB, Hoyle F, Wickramasinghe NC, Hoover MJ, Almufti S (1986) Diatoms on Earth, comets, Europa and in interstellar space. Earth Moon Planets 35(1):19–45

Hoover RB, Rozanov AY, Zhmur SI, Gorlenko VM (1998) Further evidence of microfossils in carbonaceous chondrites. Proc SPIE 3441:203–216

Hoover RB, Hoyle F, Wickramasinghe NC, Hoover MJ, Al-Mufti S (1999) Diatoms on Earth, comets, Europa and in interstellar space. Astrophys Space Sci 268(1–3):197–224

Hoover RB, Jerman GA, Rozanov AY, Davies PCW (2003) Biomarkers and microfossils in the Murchison, Rainbow, and Tagish Lake meteorites. Proc SPIE 4859:15–31

Hoover RB, Jerman G, Rozanov AY, Sipiera PP (2004a) Indigenous microfossils in carbonaceous meteorites. Proc SPIE 5555:1–17

Hoover RB, Pikuta EV, Wickramasinghe NC, Wallis MK, Sheldon RB (2004b) Astrobiology of comets. Proc SPIE 5555:93–106

Hoover RB, Rozanov AY, Jerman G, Costen J (2004c) Microfossils in Cl and CO carbonaceous meteorites. Proc SPIE 5163:7–22

Hoover R, Wickramasinghe CN, Joseph R, Schild R (eds) (2011) The discovery of Alien extra-terrestrial life: the cosmic origins of life. Cosmology Science Publishers, Cambridge

Horodyski RJ (1981) Pseudomicrofossils and altered microfossils from a middle proterozoic shale, Belt Supergroup, Montana. Precambrian Res 16(1–2):143–154

Hovnanian HP (1966) Detection and epistemology of biotic signatures. Ann N Y Acad Sci 140(A1):294–306

Hudson NP (1935) Preface: The question of living bacteria in stony meterorites. Geol Ser Field Museum Nat Hist 6(14):179–180

Hull DL (1990) Science as a process: an evolutionary account of the social and conceptual development of science. University of Chicago Press, Chicago

Imshenetsky AA (1964) Life and space. Life Sci Space Res 2:2–12

Jacob R, Bentolila P, Leroux T, Deschamps P, Bergeron D, Bergeron L, Pelletier J, Douville P, Gaudet D, Levesque P, Pelletier R, Tremblay C, Allard L, Bouchard M, Gilbert G, Laniel P, Bertrand JP, Choquet Y, Girouard Y, Roberge F, Stgeorges G, Roy D, Nootens S, McGregor M (1994) The reuse of single-use cardiac catheters - safety, economical ethical and legal issues. Can J Cardiol 10(4):413–421

Javaux EJ, Knoll AH, Walter M (2003) Recognizing and interpreting the fossils of early eukaryotes. Orig Life Evol Biosph 33(1):75–94

Kaufman M (2011) First contact: scientific breakthroughs in the hunt for life beyond Earth. Simon & Schuster, New York
Kerridge JF (1999) Formation and processing of organics in the early solar system. Space Sci Rev 90(1–2):275–288
Kesselmeyer PA (1864) Der Meteorsteinfall zu Orgueil und Nohic bei Montauban in Südfrankreich, am 14. Mai 1864. Pogg Ann Phys Chem 122:654–658
Kissin YV (2003) Hydrocarbon components in carbonaceous meteorites. Geochim Cosmochim Acta 67(9):1723–1735
Kitajima F, Nakamura T, Takaoka N, Murae T (2002) Evaluating the thermal metamorphism of CM chondrites by using the pyrolytic behavior of carbonaceous macromolecular matter. Geochim Cosmochim Acta 66(1):163–172
Knoll AH (1994) Proterozoic and early Cambrian protists: evidence for accelerating evolutionary tempo. Proc Natl Acad Sci USA 91(15):6743–6750
Kolodny Y, Kerridge JF, Kaplan IR (1980) Deuterium in carbonaceous chondrites. Earth Planet Sci Lett 46(2):149–158
Kozar MP, Krahmer MT, Fox A, Larsson L, Allton J (2001) Lunar dust: A negative control for biomarker analyses of extraterrestrial samples? Geochim Cosmochim Acta 65(19):3307–3317
Krejci-Graf K (1963) Angeblich biogene Stoffe in Meteoriten [Supposedly biogenic substances in meteorites]. Naturwissenschaften 50(16):539–541
Kremp GO (1968) Observations on fossil-like objects in the Orgueil meteorite. J Br Interplanet Soc 21:99
Krumbein WE (2009) Gunflint Chert microbiota revisited - neither stromatolites, nor cyanobacteria. In: Seckbach J, Oren A (eds) Microbial mats: modern and ancient microorganisms in stratified systems. Springer, Dordrecht, pp 53–70
Kwok S (2009) Delivery of complex organic compounds from planetary nebulae to the solar system. Int J Astrobiol 8(3):161–167
Laussedat A (1864) Sur la méthode employée pour déterminer la trajectoire du bolide du 14 mai. Comptes rendus hebdomadaires des séances de l'Académie des Sciences, Paris 58:1100–1105
Lemelle L, Salomé M, Fialin M, Simionovici A, Gillet P (2004) In situ identification and X-ray imaging of microorganisms distribution on the Tatahouine meteorite. Spectrochim Acta Part B Atomic Spectrosc 59(10–11):1703–1710
Leymerie, Daubrée A (1864) Sur l'aérolithe d'Orgueil (Tarn-et-Garonne), tombé le 14 mai 1864, à 8 heures du soit. Comptes rendus hebdomadaires des séances de l'Académie des Sciences, Paris 58:988–990
Line MA (2007) Panspermia in the context of the timing of the origin of life and microbial phylogeny. Int J Astrobiol 6(3):249–254
Line MA (2011) A critical analysis: fossils of cyanobacteria in CII carbonaceous meteorites. J Cosmol 13(March): http://journalofcosmology.com/Life101.html#105
Lipman CB (1932) Are there living bacteria in stony meteorites? Amer Mus Novit 588:1–19
Lipman CB (1936) Bacteria in meteorites. Pop Astron 44:442–446
Llorca J (2004) Organic matter in meteorites. Int Microbiol 7(4):239–248
Luijt DS, Schirm J, Savelkoul PHM, Hoekstra A (2001) Risk of infection by reprocessed and resterilized virus-contaminated catheters: an in-vitro study. Eur Heart J 22(5):378–384
Mackay AL (2007) J. D. Bernal: his legacy to science and to society. J Phys Conf Ser 57:1–16
Mamikunian G, Briggs MH (1963a) A catalog of microstructures observed in carbonaceous chondrites [Technical Report No. 32–398]. Jet Propulsion Laboratory Tech. Rep. Jet Propulsion Laboratory, California Institute of Technology, Pasadena
Mamikunian G, Briggs MH (1963b) "Organized elements" in carbonaceous meteorites. Science 139(355):873
Mamikunian G, Briggs MH (1963c) Some microstructures of complex morphology observed in preparations of carbonaceous chondrites made under sterile conditions. Nature 197(487):1245–1248
Manten AA (1966) Microfossil-like objects in meteorites. Earth Sci Rev 1(4):337–341

Marotta R, Leasi F, Uggetti A, Ricci C, Melone G (2010) Dry and survive: morphological changes during anhydrobiosis in a bdelloid rotifer. J Struct Biol 171(1):11–17

Mason J (1877) Life on meteoric stones. In: Mason J (ed) The year-book of facts in science and art for 1877. Ward, Lock & Co, London

Mason B (1963) Organic matter from space. Sci Am 208(3):43–49

Mautner MN (2002) Planetary resources and astroecology. Planetary microcosm models of asteroid meteorite interiors: electrolyte solutions and microbial growth- space populations and panspermia. Astrobiology 2(1):59–76

Mazor G, Kidron GJ, Vonshak A, Abeliovich A (1996) The role of cyanobacterial exopolysaccharides in structuring desert microbial crusts. FEMS Microbiol Ecol 21(2):121–130

McCall GJH (1973) Meteorites and their origins. Newton Abbot, David and Charles

McLoughlin N, Brasier MD, Wacey D, Green OR, Perry RS (2007) On biogenicity criteria for endolithic microborings on early earth and beyond. Astrobiology 7(1):10–26

McNichol J, Gordon R (2012) Are we from outer space? A critical review of the panspermia hypothesis. In: Seckbach J (ed) Genesis – in the beginning: on prebiotic life, chemical models and early biological evolution. Springer, Dordrecht, pp

Meinschein WG (1963) Benzene extracts of the Orgueil meteorite. Nature 197(487):833–836

Meinschein WG, Hennessy DJ, Nagy B (1963) Evidence in meteorites of former life: the organic compounds in carbonaceous chondrites are similar to those found in marine sediments. Ann N Y Acad Sci 108(2):553–579

Meinschein WG, Frondel C, Laur P, Mislow K (1966) Meteorites: optical activity in organic matter. Science 154(3747):377–380

Meinschein WG, Cordes E, Shiner VJ (1970) Search for alkanes of 15 to 30 carbon atom lenath. Science 167(3918):753–754

Merek EL (1973) Imaging and life detection. Bioscience 23(3):153–159

Michels J (1881a) Editorial, December 24. Science 2(78):605

Michels J (1881b) Editorial, May 14. Science 2(46):217

Mimura K, Okamoto M, Sugitani K, Hashimoto S (2007) Selective release of D and $^{13}$C from insoluble organic matter of the Murchison meteorite by impact shock. Meteoritics Planet Sci 42(3):347–355

Morange M (2007) What history tells us X. Fifty years ago: the beginnings of exobiology. J Biosci 32(6 Supplement 2):1083–1087

Morrison P (1962) Carbonaceous "snowflakes" and the origin of life. Science 135(3504):663–664

Morrison D, Niehoff J (1979) Future exploration of the asteroids. In: Gehrels T, Matthews MS (eds) Asteroids. University of Arizona Press, Tucson, pp 227–250

Mueller G (1953) The properties and theory of genesis of the carbonaceous complex within the cold Bokevelt meteorite. Geochim Cosmochim Acta 4(1–2):1–10

Mueller G (1962) Interpretation of micro-structures in carbonaceous meteorites. Nature 196(4858):929–932

Mueller G (1963) Interpretation of micro-structures in carbonaceous meteorites. In: Colombon U, Hobson GD (eds) Advances in organic geochemistry, proceedings of the first international meeting of the geochemical society, Organic Geochemical Group, Milan, Italy; 10–12 Sept 1962. Pergamon Press, New York, pp 119–140

Mueller G (1964a) 'Impact contamination'of the Mokoia carbonaceous chondrite. Nature 204(4958):567

Mueller G (1964b) Interpretation of micro-structures in carbonaceous meteorites. In: Colombo U, Hobson GD (eds) Advances in organic geochemistry: proceedings of the international meeting in Milan, 1962. Macmillan, pp 119–140

Mueller G (1965) Interpretation of micro-structures in carbonaceous meteorites. Nature 205(4977):1200

Mukhopadhyay PK, Mossman DJ, Ehrman JM (2007) The case for vestiges of Early Solar System biota in carbonaceous chondrites: petroleum geochemical snapshots and possible future petroleum prospects on Mars Expedition. Instruments, Methods, and Missions for Astrobiology X, vol 6694, pp 66940C

Murae T (1999) Fluorescent organic matter in carbonaceous chondrites. Adv Space Res 24(4):469–476
Nagy B (1962) Organic particles embedded in minerals in the Orgueil and Ivuna carbonaceous chondrites. Nature 193:1129–1133
Nagy B (1963) Life-like forms in meteorites and the problems of environmental control on the morphology of fossil and recent protobionta, a conference held by the New York Academy of Sciences, New York, April/May, 1962. Ann N Y Acad Sci 108(2):339–616
Nagy B (1966a) Investigations of the Orgueil carbonaceous meteorite. Geol Fören Stockholm Förh 88(2):235–272
Nagy B (1966b) A study of the optical rotation of lipids extracted from soils, sediments, and the Orgueil carbonaceous meteorite. Proc Natl Acad Sci USA 56(2):389–398
Nagy B (1967) The possibility of extraterrestrial life: ultra-microchemical analyses and electron-microscopic studies of microstructures in carbonaceous meteorites. Rev Palaeobot Palynol 3(1–4):237–242
Nagy B (1968a) Carbonaceous meteorites. Endeavour 27(101):81–86
Nagy B (1968b) Indications of possible biological substances in carbonaceous meteorites. J Astronaut Sci 15(4):161–168
Nagy B (1975) Carbonaceous meteorites. Elsevier Scientific Pub. Co., Amsterdam
Nagy B, Claus G (1964) Mineralized microstructures in carbonaceous meteorites. In: Colombo U, Hobson GD (eds) Advances in organic geochemistry. Pergamon Press, New York, pp 109–114
Nagy B, Nagy LA (1969) Early Pre-cambrian Onverwacht microstructures: possibly oldest fossils on earth? Nature 223(5212):1226–1229
Nagy B, Meinschein WG, Hennessy DJ (1961) Mass spectroscopic analysis of the Orgueil meteorite: evidence for biogenic hydrocarbons. Ann N Y Acad Sci 93(2):27–35
Nagy B, Claus G, Fredriksson K, Percy J, Andersen CA, Urey HC (1963a) Electron probe microanalysis of organized elements in the Orgueil meteorite. Nature 198(487):121–125
Nagy B, Fredriksson K, Kudynowski J, Carlson L (1963b) Ultra-violet spectra of organized elements. Nature 200(4906):565–566
Nagy B, Hennessy DJ, Meinschein WG (1963c) Aqueous, low temperature environment of Orgueil meteorite parent body. Ann N Y Acad Sci 108(2):534–552
Nagy B, Murphy MTJ, Modzeleski V, Rouser G, Claus G, Hennessy DJ, Colombo U, Gazzarrini F (1964) Optical activity in saponified organic matter isolated from the interior of the Orgueil meteorite. Nature 202(4929):228–233
Nagy B, Meinsche WG, Hennessy DJ (1965) Review of earlier work on carbonaceous material and organized elements in meteorites. Science 150(3694):380
Nagy B, Drew CM, Hamilton PB, Modzeles VE, Murphy ME, Scott WM, Urey HC, Young M (1970) Organic compounds in lunar samples: pyrolysis products, hydrocarbons, amino acids. Science 167(3918):770–773
Nakamura-Messenger K, Messenger S, Keller LP, Clemett SJ, Zolensky ME (2006) Organic globules in the Tagish Lake meteorite: Remnants of the protosolar disk. Science 314(5804):1439–1442
Nininger HN (1938) Concerning bacteria in meteorites. Pop Astron 46:214–215
Nooner DW, Oro J (1967) Organic compounds in meteorites-I. Aliphatic hydrocarbons. Geochim Cosmochim Acta 31(9):1359–1394
Nultsch W (1962) Über Funde von Mikroorganismen in Meteoriten [On findings of microorganisms in meteorites]. Dtsch Med Wochenschr 87(39):1972–1973
Nultsch W (1968) Einfluss von Redox-Systemen auf die Bewegungsaktivitat und das phototaktische Reaktionsverhalten von *Phormidium uncinatum* [Effect of redox systems on the motility and the phototactic reactions of *Phormidium uncinatum*]. Arch Mikrobiol 63(4):295–320
O'Connell E (1964a) Solar system science: 1963 literature survey, Part I. Icarus 3(2):172–180
O'Connell E (1964b) Solar system science: 1963 literature survey, Part II. Icarus 3(3):277–284
O'Connell E (1965) Solar system science: 1964 literature survey, Part I. Icarus 4(3):319–333
Orcel J, Alpern B (1966) Étude de la microstructure de la météorite carbonée d'Orgueil [Study of the microstructure of the carbonaceous meteorite of Orgueil]. Comptes Rendus Hebdomadaires des Seances de l'Academie des Sciences Serie D 262(13):1393–1397
Oró J (1972) Extraterrestrial organic analysis. Space Life Sci 3(4):507–550

Oró J, Skewes HB (1965) Free amino-acids on human fingers: the question of contamination in microanalysis. Nature 207(5001):1042–1045

Oró J, Tornabene T (1965) Bacterial contamination of some carbonaceous meteorites. Science 150(3699):1046–1048

Oyama VI (1972) Search for biogenic structures and viable organisms in lunar samples: a review. Space Life Sci 3(4):377–382

Oyama VI, Merek EL, Silverman MP (1970) A search for viable organisms in a lunar sample. Science 167(3918):773–775

Palik P (1962) Further life-forms in the Orgueil meteorite. Nature 194(4833):1065

Palik P (1963) Extraterrestrial taxa and their nomenclature. Taxon 12(8):283

Papp A (1963) Fossil protobionta and their occurrence. Ann N Y Acad Sci 108(2):461–463

Parkin DW, Brownlow AE, Hunter W (1962) Metallic cosmic dust with amorphous attachments. Nature 193(4816):639–642

Pearson R (1962) Life-forms in carbonaceous chondrites. Nature 194(4833):1064–1065

Pearson VK, Sephton MA, Gilmour I (2006) Molecular and isotopic indicators of alteration in CR chondrites. Meteoritics Planet Sci 41(9):1291–1303

Pearson VK, Kearsley AT, Sephton MA, Gilmour I (2007) The labelling of meteoritic organic material using osmium tetroxide vapour impregnation. Planet Space Sci 55(10):1310–1318

Peltzer ET, Bada JL, Schlesinger G, Miller SL (1984) The chemical conditions on the parent body of the Murchison meteorite: some conclusions based on amino, hydroxy and dicarboxylic acids. Adv Space Res 4(12):69–74

Pentecost A (1990) The tractive force generated by a motile filamentous cyanobacterium. Microbios Lett 44(175–176):111–118

Persson D, Halberg KA, Jørgensen A, Ricci C, Møbjerg N, Kristensen RM (2011) Extreme stress tolerance in tardigrades: surviving space conditions in low earth orbit. J Zool System Evolut Res 49:90–97

Pflug HD (1984a) Microvesicles in meteorites, a model of pre-biotic evolution. Naturwissenschaften 71(10):531–533

Pflug HD (1984b) Utrafine structure of the organic matter in meteorites. In: Wickramasinghe NC (ed) Fundamental studies and the future of science. University College Cardiff Press, Cardiff, pp 24–37

Pflug HD (2001) Earliest organic evolution. Essay to the memory of Bartholomew Nagy. Precambrian Res 106(1–2):79–91

Pflug HD, Heinz B (1997) Analysis of fossil-organic nanostructures terrestrial and extraterrestrial. Proc SPIE 3111:86–97

Pikuta E (2011) The discovery of fossil evidence of extraterrestrial life in meteors. J Cosmol 13(March): http://journalofcosmology.com/Life101.html#112

Pikuta EV, Hoover RB, Tang J (2007) Microbial extremophiles at the limits of life. Crit Rev Microbiol 33(3):183–209

Pisani M (1864) Étude chimique et analyse de l'aérolithe d'Orgueil. Comptes rendus hebdomadaires des séances de l'Académie des sciences, Paris 59:132–135

Preston LJ, Shuster J, Fernandez-Remolar D, Banerjee NR, Osinski GR, Southam G (2011) The preservation and degradation of filamentous bacteria and biomolecules within iron oxide deposits at Rio Tinto, Spain. Geobiology 9(3):233–249

R (1881) Mr. Darwin on Dr. Hahn's discovery of fossil organisms in meteorites. Science 2(61):410

Rachel GW (1881) Fossil organisms in meteorites. Science 2(50):275–277

Rasic NF, Friesen RM, Anderson B, Hoban SA, Olson N, Kress J (2003) Prepared endotracheal tubes: are they a potential source for pathogenic microorganisms? Anesth Analgesia 97(4):1133–1136

Rasmussen S, Bedau MA, Chen L, Deamer D, Krakauer DC, Packard NH, Stadler PF (eds) (2008) Protocells: bridging nonliving and living matter. MIT Press, Cambridge

Rauf K, Hann A, Wickramasinghe C (2010) Microstructure and elemental composition of the Tagish Lake meteorite and its astrobiological implications. Int J Astrobiol 9(1):35–43

Raulin-Cerceau F, Maurel MC, Schneider J (1998) From Panspermia to bioastronomy, the evolution of the hypothesis of universal life. Orig Life Evol Biosph 28(4–6):597–612

Redfield R (2011) The evidence is not convincing: review of fossils of cyanobacteria in C11 carbonaceous meteorites. J Cosmol 13(March): http://journalofcosmology.com/Life101.html#124
Ridgway HF, Lewin RA (1988) Characterization of gliding motility in Flexibacter polymorphus. Cell Motil Cytoskeleton 11(1):46–63
Rietmeijer FJM (1985) A poorly graphitized carbon contaminant in studies of extraterrestrial materials. Meteoritics 20(1):43–48
Rossignol-Strick M, Barghoorn ES (1971) Extraterrestrial abiogenic organization of organic matter: the hollow spheres of the Orgueil meteorite. Space Life Sci 3(2):89–107
Rothschild L (2007) Stephen J. Dick; James E. Strick. The Living Universe: NASA and the Development of Astrobiology [book review]. Isis 98(2):423–424
Roy SK (1935) The question of living bacteria in stony meterorites. Geol Ser Field Museum Nat Hist 6(14):180–198
Roy SK (1937) Additional notes on the question of living bacteria in stony meteorites. Pop Astron 45:499–504
Rozanov AY (2010) Pseudomorphic microbial structures. Paleontol J 44(7):819–826
Rozanov AY, Hoover RB (2004) Atlas of bacteriomorphs in carbonaceous chondrites. Proc SPIE 5163:23–35
Rozanov AY, Zhegallo EA, Ushatinskaya GT, Shuvalova YV, Hoover RB (2001) Bacterial paleontology for astrobiology. Proc SPIE 4495:283–294
Sagan C (1962) Summary of a discussion with Erdtman on organized elements in carbonaceous chondrites. Science 137(3530):626
Schild R (2011) Official statement. J Cosmol 13(March): http://journalofcosmology.com/Life100.html
Sephton MA (2002) Organic compounds in carbonaceous meteorites. Nat Prod Rep 19(3):292–311
Sephton MA (2004) Organic matter in ancient meteorites. Astron Geophys 45(2):8–14
Sephton MA (2005) Organic matter in carbonaceous meteorites: past, present and future research. Philos Trans R Soc A Math Phys Eng Sci 363(1837):2729–2742
Sephton MA, Pillinger CT, Gilmour I (1999) Small-scale hydrous pyrolysis of macromolecular material in meteorites. Planet Space Sci 47(1–2):181–187
Sephton MA, Pillinger CT, Gilmour I (2001) Normal alkanes in meteorites: molecular $\delta^{13}C$ values indicate an origin by terrestrial contamination. Precambrian Res 106(1–2):47–58
Sheldon RB, Hoover RB (2008) Cosmological evolution: spatial relativity and the speed of life. Proc SPIE 7097:709716
Silverman SR (1962) Excerpts from letter of 23 May 1962 to Urey from Silverman. Science 137(3530):626–627
Simmonds PG, Rauman AJ, Bollin EM, Gelpi E, Oroó J (1969) Unextractable organic fraction of pueblito de allende metoeorite: evidence for its indigenous nature. Proc Natl Acad Sci USA 64(3):1027–1034
Simpson GG (1964) Nonprevalence of humanoids. We can learn more about life from terrestrial forms than we can from hypothetical extraterrestrial forms. Science 143(3608):769–775
Sipiera PP (2011) A compelling argument: the "fossils" are extra-terrestrial in origin. J Cosmol 13(March): http://journalofcosmology.com/Life101.html#122
Sipiera PP, Hoover RB, Jerman GA (2000) Meteorites and microbes: meteorite collection and ice sampling at Patriot Hills, Thiel Mountains, and South Pole, Antarctica. Proc SPIE 4137:13–21
Soffen GA (1969) Extraterrestrial optical microscopy. Appl Optics 8(7):1341–1347
Staplin FL (1962a) Microfossils from the Orgueil meteorite. Micropaleontology 8(3):343–347
Staplin FL (1962b) Organic remains in meteorites - a review of the problem. J Alberta Soc Petrol Geologist 10(10):575–580
Staplin FL (1965a) Organic remains in meteorites. In: Mamikunian G, Briggs MH (eds) Current aspects of exobiology. Pergamon Press, New York, pp 77–92
Staplin FL (1965b) Possible fossils from Orgueil and other meteorites. Science 150(3694):385

Stauffer H (1961) Primordial argon and neon in carbonaceous chondrites and ureilites. Geochim Cosmochim Acta 24(1–2):70–82

Storrie-Lombardi MC, Hoover RB (2004) Fossil signatures using elemental abundance distributions and Bayesian probabilistic classification. Proc SPIE 5555:18–30

Studier MH, Hayatsu R, Anders E (1968) Origin of organic matter in early solar system-I. Hydrocarbons. Geochim Cosmochim Acta 32(2):151–173

Swart PK, Grady MM, Pillinger CT (1983) A method for the identification and elimination of contamination during carbon isotopic analyses of extraterrestrial samples. Meteoritics 18(2):137–154

Swindle TD, Olson EK (2002) The timing of aqueous weathering on Mars: clues from argon-40-argon-39 analyses of whole-rock samples of the nakhlites Nakhla and Lafayette. Meteoritics Planet Sci 37(S5):A138

Tan WC, VanLandhgham SL (1967) Electron microscopy of biological-like structures in the Orgueil carbonaceous meteorite. Geophys J R Astron Soc 12(3):237

Tasch P (1963) Identity of organized elements in carbonaceous chondrites. Science 142(3589):156–158

Tasch P (1964) Life-forms in meteorites and the problem of terrestrial contamination: a study in methodology. Ann N Y Acad Sci 105:929–950

Taubes G (1993) Bad science: the short life and weird times of cold fusion. Random House, New York

Thomson KS (1991) Piltdown Man: The Great English mystery story. Am Scientist 79(3):194–201

Thornton J (1890) Advanced physiography, 2nd edn. Longmans, Green, and Co, London

Timofejev BW (1963) Lebensspuren in Meteoriten: Resultate einer microphytologischen analyse [Traces of life in meteorites: results of a microphytological analysis]. Grana Palynol 4(1):92–99

Urey HC (1962a) Life-forms in meteorites: origin of life-like forms in carbonaceous chondrites. Nature 193(4821):1119–1123

Urey HC (1962b) Lifelike forms in meteorites. Science 137(3530):623–626

Urey HC (1965) Organic material in meteorites: a review. Science 150(3694):387–388

Urey HC (1966) Biological material in meteorites: a review. Science 151(3707):157–166

Urey HC, Arnold JR (1966) Biological materials in carbonaceous chondrites. In: Pittendrigh CS (ed) Biology and the exploration of Mars: report of a study. National Academies, Washington, DC, pp 114–124

Urey HC, Cholnoky BJ, Berger R, Morrison P, Anders E, Papp A, Bourrelly P, Palmer CM, Nagy B, Hennessy DJ, Claus G, Tasch P, Ross R, Fitch FW, Fox SW, Dombrowski H, Mason B, Bernal JD, Meinschein W (1963) Panel discussion: identity of organized elements. Ann N Y Acad Sci 108(2):606–615

Urey HC, Meinschein WG, Nagy B (1968) Comments on meteoritic hydrocarbons. Geochim Cosmochim Acta 32(6):665

Vallentyne JR (1963) Environmental biophysics and microbial ubiquity. Ann N Y Acad Sci 108(2):342–352

Vance P, Pockriss L, Como P (1957) Catch a falling star [Song]. http://lostpedia.wikia.com/wiki/Catch_a_Falling_Star. Accessed 18 June 2011

VanLandingham SL (1965a) Acid resistant microfossils from the Alais and Orgueil meteorites. Nova Hedwigia Z Kryptogamenk 10(1–2):161–176 + Plates 144–149

VanLandingham SL (1965b) Evidence for microfossils in the Alais and Orgueil carbonaceous meteorites. Nature 208(5014):947–948

VanLandingham SL, Sun CN, Tan WC (1967) Origin of round-body structures in the Orgueil meteorite. Nature 216(5112):252–253

Vdovykin GP (1964) "Organized elements" in carbonaceous chondrites. Geochem Int USSR 4:693–696

Vdovykin GP (1967) Carbon matter of meteorites (organic compounds, diamonds, graphite) [NASA-TT-F-582]. Nauka, Moscow

Vdovykin GP (1972) Meteorites and life. NASA (National Aeronautics and Space Administration) Technical Translation TTF (710):168–193

Vdovykin GP (1973) The Mighei meteorite. Space Sci Rev 14(6):832–879

Vinogradov AP, Vdovykin GP (1964) Multimolecular organic matter of carbonaceous chondrites. Geochem Int USSR 5:831–836

Watson JS, Pearson VK, Gilmour I, Sephton MA (2003) Contamination by sesquiterpenoid derivatives in the Orgueil carbonaceous chondrite. Org Geochem 34(1):37–47

Weichmann FG (1882) Fusion-structures in meteorites. Trans N Y Acad Sc 1:153–155

West MW, Ponnamperuma C (1970) Chemical evolution and origin of life: a comprehensive bibliography. Space Life Sci 2(2):225–288

West MW, Gill ED, Kvenvolden KA (1975) Chemical evolution and origin of life: bibliography supplement 1973. Orig Life Evol Biosph 6(1–2):285–300

Whale GF, Walsby AE (1984) Motility of the cyanobacterium *Microcoleus chthonoplastes* in mud. Br Phycol J 19(2):117–123

Wickramasinghe C (2011a) Bacterial morphologies supporting cometary panspermia: a reappraisal. Int J Astrobiol 10(1):25–30

Wickramasinghe C (2011b) Microfossils in meteors and comet dust: a vindication of panspermia. J Cosmol 13(March): http://journalofcosmology.com/Life101.html#112

Wickramasinghe NC (2011c) Extraterrestrial life and censorship. arXiv:http://arxiv.org/abs/1104.1314

Wickramasinghe C, Wallis MK, Gibson CH, Wallis J, Al-Mufti S, Miyake N (2010) Bacterial morphologies in carbonaceous meteorites and comet dust. Proc SPIE 7819:781913

Wiik HB (1956) The chemical composition of some stony meteorites. Geochim Cosmochim Acta 9(5–6):279–289

Young JD, Martel J (2010) The rise and fall of nanobacteria. Sci Am 302(1):52–59

Youngbull C (2011) Fear of the unknown: do you believe in extraterrestrial life? Definitely maybe! J Cosmol 13(March): http://journalofcosmology.com/Life101.html#103

Zenobi R, Philippoz JM, Buseck PR, Zare RN (1989) Spatially resolved organic analysis of the Allende meteorite. Science 246(4933):1026–1029

Zhmur SI (2002) Possible environments of carbon hondrites accumulation. ESA Special Publications 518:493–494

Zhmur SI, Gerasimenko LM (1999) Biomorphic forms in carbonaceous meteorite Alliende and possible ecological system - producer of organic matter of hondrites. Proc SPIE 3755:48–58

Zhmur SI, Rozanov AY, Gorlenko VM (1997) Lithified remnants of microorganisms in carbonaceous chondrites. Geochem Int 35(1):58–60

Zhmur SI, Gorlenko VM, Gerasimenko LM (1999) Comparative morphology of modern and ancient terrestrial bacteria and microfossils from carbonaceous meteorites. Microbiology 68(6):739–745

Zhmur SI, Duda VI, Roizeman FM (2001) Microfossils in Early Archaen graphites of Aldan shield and some aspects of panspermia. Proc SPIE 4495:19–26

Zigel F (1962) ЖИÐНБ В МЕТЕОРИТЕ [Life in a meteorite, English translation: CFSTI, JPRS-17326]. Ogonek 47:28–29

Biodata of **Jesse C. McNichol** and **Richard Gordon**, authors of "***Are We from Outer Space? A Critical Review of the Panspermia Hypothesis.***"

**Jesse C. McNichol** obtained his honours BSc. Degree in Biology from Mount Allison University in 2008 working with Dr. Felix Baerlocher, Dr. Robert Thompson, and Dr. Zoe Finkel. After graduation, Jesse worked for Environment Canada and the National Research Council of Canada studying environmental toxicology and microalgal biofuels. Jesse is currently studying for his Ph.D. with the Woods Hole Oceanographic Institute/MIT joint program under the supervision of Dr. Stefan Sievert. His thesis will be focused on sulfur-oxidizing chemoautotrophs from deep-sea hydrothermal vents and other environments. Jesse's interests outside the lab include foreign languages, botany and environmental education.

E-mail: **McNichol@mit.edu**

**Dr. Richard Gordon** is an itinerant theoretical biologist with B.Sc. (Math) from the University of Chicago and Ph.D. (Chemical Physics) from the University of Oregon. He is presently Theoretical Biologist, Embryogenesis Center, at the Gulf Specimen Marine Laboratory (http://www.gulfspecimen.org/). He organized Books with Wings (http://bookswithwings.net) which sends books to universities in Afghanistan, and the Embryo Physics Course online in Second Life® (http://embryogenesisexplained.com/course). He writes a blog in Science 2.0 (http://www.science20.com/cosmic_embryo) and is Principal Scientific Advisor for the EvoGrid project (http://www.evogrid.org/), an attempt to simulate the origin of life. He has also worked in diatom nanotechnology, HIV/AIDS prevention and algorithms for computed tomography.

E-mail: **DickGordonCan@gmail.com**

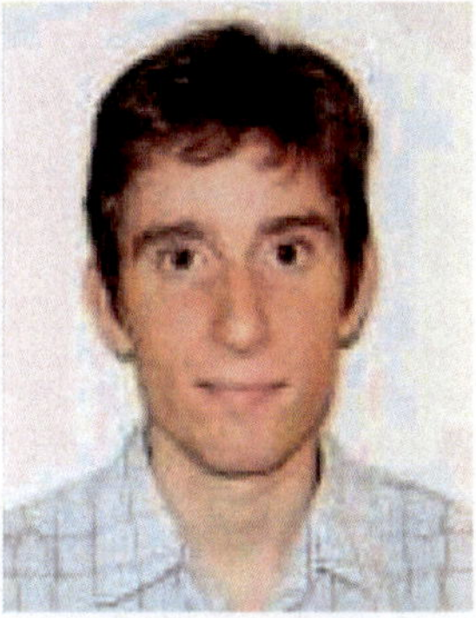

**Jesse C. McNichol**

**Richard Gordon**



J. Seckbach (ed.), *Genesis - In The Beginning: Precursors of Life, Chemical Models and Early Biological Evolution*, Cellular Origin, Life in Extreme Habitats and Astrobiology 22, 591–619
DOI 10.1007/978-94-007-2941-4_30, 

## ARE WE FROM OUTER SPACE?
### *A Critical Review of the Panspermia Hypothesis*


**JESSE C. McNICHOL[1] AND RICHARD GORDON[2]**
*[1]National Research Council of Canada, Institute for Marine Biosciences, 1411 Oxford Street, Halifax, NS, Canada B3H 3Z1*
*[2] Embryogenesis Center, Gulf Specimen Marine Laboratories, 222 Clark Drive, 32346 Panacea, FL, USA*


## 1. Introduction[1]

Aspects of "astrobiological" research, such as searching for exoplanets, characterizing extremophilic microbes, and searching for biosignatures of ancient life undoubtedly offer important insights into the origin and evolution of life on Earth. Because of the important implications of such work, it is hardly surprising that claims of important breakthroughs often inspire heated debate – one famous example was the claim that the Martian meteorite ALH84001 contained fossilized life (McKay et al., 1996). This claim required researchers to critically assess fossil and geochemical evidence with a new level of scrutiny, and it has been suggested that despite the consensus that this claim has been refuted, the academic debate has better prepared scientists to assess future evidence for ET (extraterrestrial) life (Dick and Strick, 2004). While this may be true in a situation where conclusions can be subject to rigorous independent analysis, the field of astrobiology also includes speculative research that is at present difficult, if not impossible, to independently test in the manner of the Martian meteorite (Rinaldi, 2007).

Taken in this context, the hypothesis known as *panspermia* that posits life on Earth was introduced from elsewhere may indeed spur innovative research through healthy debate, but it is equally clear that inconclusive evidence is currently cited by some authors as strong evidence of the panspermia hypothesis (Wickramasinghe et al., 2003).

The concept of panspermia can be traced back to a number of different historical origins, from ancient societies that believed in the universality of life to the musings of the early astronomers (Temple, 2007; Crowe, 2008). Fred Hoyle and Chandra Wickramasinghe are commonly associated with the modern

[1]Abbreviations: ET, extraterrestrial; GC/MS, Gas chromatography/mass spectrometry; LR, labeled release; LUCA, last universal common ancestor; PAHs, polycyclic aromatic hydrocarbons.

incarnation of panspermia, apparently being convinced of the truth of panspermia after the widespread nature of organic compounds in interstellar space was discovered (Hoyle and Wickramasinghe, 1996). Aside from Crick and Orgel's (Crick and Orgel, 1973) discussions of "directed panspermia" (intentional seeding by intelligent beings, proposed earlier: "J.B.S. Haldane… expressed a belief, in 1954 (Haldane, 1954), that life may have… been launched into space by intelligent beings" (Bernal, 1962)), few other authors have developed a hypothesis as extensive in detail as Hoyle and Wickramasinghe (Wickramasinghe, 2010). These authors suggest that microbial life is ubiquitous and persistent in the cosmos and serves as a constant source of inoculum and genetic material for new planets, contributing to evolutionary trends on each world. We use the structure of their hypothesis to inform discussion on the following questions about panspermia:

1. What, if any, evidence is there for life outside Earth?
2. Is it conceivable that life as we know it could be viably transferred through space to other habitats?
3. How likely is it that transferred life would become established and go on to seed neighboring systems, thus amplifying the process?
4. Can panspermia influence evolutionary progress or affect the origin of life debate?

## 2. Question 1: Is There Any Compelling Evidence for Life Outside Planet Earth?

### 2.1. MARS AND THE VIKING EXPERIMENTS

Mars is a planet that has long inspired speculation about extraterrestrial life (Pickering, 1921), and the Viking Mars landers used some of the most sophisticated life-detecting instrumentation yet deployed anywhere outside the Earth to probe Martian soils for signs of microbial life. Three of the four experiments designed to test for the presence of life gave negative or ambiguous results (GC/MS, gas exchange, and pyrolytic release experiments), but one final test, the labeled release (LR) experiment, offered a striking positive response. This test, which aimed to detect the respiration of radioactively labeled organic compounds by microorganisms, yielded a strong positive result, with control experiments offering negative responses. The LR experiment was extensively field-tested on Earth and was found to be sensitive to extremely small quantities of viable cells in extreme Earth environments, the closest we have to a Martian analogue. However, after the fact, this evidence was discounted in favor of an inorganic explanation that theorized that UV-produced peroxides in the Martian soil were responsible for the oxidation (Quinn and Zent, 1999), despite the validity of these theories remaining questioned by the inventor of the apparatus, Gilbert Levin (Levin, 1981, 2007).

Levin's case has been bolstered by a recent investigation that suggested that the GC/MS experiment aboard the Viking would have been insensitive to the organic constituents of Antarctic soils, which are by no means devoid of life (Navarro-González et al., 2006). In addition to this, recent evidence of water ice under the Martian surface (Boynton et al., 2002), in its atmosphere (Pearl et al., 2001), as well as the discovery of seasonal methane emissions (Krashnopolsky et al., 2004) has led Levin to opine: "What would be truly extraordinary in the light of present knowledge would be finding Mars to be sterile" (Levin, 2007). Despite this evidence, recent information obtained from Martian landers suggests an arid, acidic, and oxidizing early Mars (Squyres and Knoll, 2005), and the recent unexpected discovery of perchlorate salts in the Martian soil (Hecht et al., 2009), gives evidence for the presence of an oxidant that may yet convincingly explain the results of the LR experiments as abiotic. Future robotic missions that are slated for the current decade aim to clarify some of these questions by sampling up to 2 m below the Martian surface and may yet offer generally accepted signs of life on Mars (Rinaldi, 2007). Other authors have described instruments that may overcome the limitations of the Viking GC/MS experiments by detecting PAHs (polycyclic aromatic hydrocarbons) with higher sensitivity and distinguishing between enantiomers of amino acids (Bada, 2001), though final confirmation may have to await proposed sample return missions (Schulze-Makuch and Irwin, 2002).

## 2.2. FOSSIL EVIDENCE?

The ALH84001 Martian meteorite caused an intense media stir when it was claimed that it contained proof of past biogenic activity, which spurred then US President Bill Clinton to address the nation on the significance of the finding (Dick and Strick, 2004). Several lines of evidence were used to reach this conclusion, including the presence of PAHs, magnetite and sulfide minerals, and carbonate globules that resembled bacteria, though an order of magnitude smaller than any known Earth cell (McKay et al., 1996). The initial enthusiasm for this claim has largely subsided, based on the fact that the biological production of any one of the lines of evidence used by the authors to argue for this is no more compelling than the argument for a "null" abiotic origin. J. William Schopf addressed some of the limitations of their finding (Dick and Strick, 2004) relating to the "microfossils" discovered within the meteorite:

> Even though they showed multiple examples of the microbe-like objects in question, their suggestion failed because the putative fossils lack cell lumina, carbonaceous cell walls, taphonomically credible variable preservation, and plausible biological size ranges. (Schopf et al., 2005)

Other authors suggested that the geology of the meteorite was inconsistent with a life-supporting environment (Golden et al., 2004; Min and Reiners, 2007; Barrat and Bollinger, 2010) or proposed plausible mechanisms for the abiotic synthesis of organics and magnetite crystals (McCollom, 2003). Despite these criticisms, arguments are still made that some of the magnetite crystals contained in the meteorite represent what "In a terrestrial context…would be interpreted as a biosignature indicative of the presence of past metabolic activity by magnetotactic bacteria" (Thomas-Keprta et al., 2002).

This fierce debate was echoed several years later when the ALH84001 skeptic Schopf found himself on the defensive as Oxford's Martin Brasier raised doubts about the biogenicity and phylogeny of the 3.5-billion-year-old terrestrial Apex chert microfossils, long thought to be conclusive evidence of Archean life (Dalton, 2002). These microscopic fossils, which were originally characterized tentatively as cyanobacteria existing in a shallow sea bottom, were criticized based on their visual appearance as well as their geological context (Brasier et al., 2002). Brasier *et al.* claimed that the fossils had morphology inconsistent with cyanobacteria and that they were in fact carbonaceous artifacts produced in a hydrothermal environment. This debate resulted in many follow-up publications by both Brasier and Schopf describing techniques for testing evidence, such as confirmation of three-dimensional morphology (Schopf and Kudryavtsev, 2005), as well as a number of arguments about the placement of these fossils in a more robust geological context (Brasier et al., 2005).

Although heated at times (Dalton, 2002), the debate served the purpose of clarifying the provenance of some of the oldest microfossils on Earth. Publications by Schopf and colleagues offer evidence to support the biogenicity of these fossils through more in-depth examinations of both morphology as well as the organic constituents of the microfossils (Schopf et al., 2005, 2007; Schopf, 2006). This debate is still far from over – new data suggests the structures are not in fact microfossils, though still indicative of microbial life (Marshall et al., 2011). However, Brasier's most significant achievement has been the reframing of the world where these ancient organisms existed, as well as their phylogenetic affinities – no longer do paleontologists believe that these organisms were undoubtedly cyanobacteria existing in a temperate "early Eden" (Brasier et al., 2004); rather, they appear to have existed in a geologically active hydrothermal environment. This example is a cautionary tale – if Schopf's picture of the early Earth as enshrined in textbooks can be so easily challenged, inferences made about life outside of Earth (that we have even less ability to directly study) may be overturned just as quickly. Purported ET (extraterrestrial) microfossils must be similarly examined with extreme caution with every attempt to confirm the null hypothesis, before claims can be made that they represent unambiguous signs of life.

Another similar debate occurred with the discovery of "nanobacteria," with extraordinary initial claims of an undiscovered type of life ultimately yielding to an abiotic explanation (Cisar et al., 2000; Raoult et al., 2008; Young and Martel, 2010). Despite this, proponents of panspermia have seized on these claims, suggesting a "cosmic prevalence of nanobacteria" (Wickramasinghe and Wickramasinghe, 2008) and detecting their fossils (McKay et al., 1996; Rozanov et al., 2001; Folk and Taylor, 2002; Hoover et al., 2004).

The debate over life in meteorites is a long, repeated, and substantial one, which we delve into in a separate chapter (Gordon and McNichol, 2012).

## 2.3. BACTERIA FROM SPACE?

As these other investigations are proceeding, Wickramasinghe and colleagues have taken a different tack, collaborating with the India Space Research Organization (ISRO) to conduct a series of experiments that seek direct evidence for the panspermia hypothesis. The ISRO has used balloons to sample the upper stratosphere, with the goal of intercepting extraterrestrial microbes filtering down from space (Narlikar et al., 2003; Wainwright et al., 2004). This research has had an unsurprisingly controversial air, with a first experiment isolating several organisms from the upper atmosphere (Wainwright et al., 2003), though a collaborating group disavowed themselves of the results (Willerslev et al., 2003). The organisms that were isolated are common to the terrestrial environment and almost genetically identical, based on 18S rRNA sequences.

Newer work with the same methodology by independent laboratories has isolated new strains from the stratosphere, some of which have been described as new species based on genetic and chemotaxonomic characteristics (Shivaji et al., 2006, 2009; Yang et al., 2009, 2010). Whether or not these species are extraterrestrial in nature is still unproven (see Vaidya, 2009 for some discussion points) since not all possible methods of transfer to the stratosphere from the lower Earth can be discounted – it has been suggested that small particles captured represent Earthly contamination, whereas the larger particles represent incoming biological material (Wainwright et al., 2006), although it should be noted that even small spiders balloon to the upper stratosphere (Filmer, 1991). Supporting an Earthly origin, bacteria and other biological materials are common components of cloud condensation nuclei (Christner et al., 2008), and there have been several mechanisms suggested for the transfer of particles to the upper atmosphere (Rohatschek, 1996; Fromm, 2004; Damoah et al., 2006; Dehel, 2006).

It should also be noted that the discovery of new microbial species is not surprising in itself, and a tolerance to ionizing radiation does not necessarily indicate that these isolates are space-faring – they are by no means extremophiles, growing best under mild conditions (10–30°C). It may be that infrequent events cause bacteria to be transferred to the stratosphere, with only the UV-resistant cells remaining viable for recovery in these microbiological experiments. Alternatively, it may be that the process of transport to the upper atmosphere and consequent return to the surface is enough to select for organisms resistant to ionizing radiation, similar to the argument of Pavlov et al. (2006).

Criticisms aside, these experiments are a realistic method of testing the panspermia hypothesis – as Wainwright (2003) notes. Unless we disavow the methodology completely, one needs to explain the presence of viable bacteria at this height. Even if these isolates can be proven to represent surface life in an extreme environment, it extends the definite reach of microbial life to the upper atmosphere. Such evidence may be strengthened by using culture-independent methods of identifying microbes at this height (Wainwright et al., 2004) or investigating the isotopic composition of any purportedly extraterrestrial biomass and comparing it to known signatures found in cosmic dust (Busemann et al., 2006).

## 2.4. EVIDENCE FOR LIFE IN INTERSTELLAR SPACE

It is now known that molecules essential to life such as water, ammonia (Decin et al., 2010), and carbon compounds (Smolders et al., 2010) are commonly generated in stars and transported to the interstellar medium. The precise identity of this carbonaceous mixture has been variously suggested to be a material similar to kerogen (Papoular, 2001), buckyballs (Chen et al., 2008), or similar compounds formed under conditions common in the circumstellar envelope of certain stars (Kwok, 2004). Certain proponents of panspermia take this ever-increasing inventory of organic molecules as evidence of the ubiquity of life, disputing the previously mentioned processes:

> In view of the impossibility of producing cosmically, by non-biological means, organic material from inorganics like water, carbon monoxide and nitrogen gas, it is clear that the immense amount of interstellar organics must be biological in origin. (Hoyle and Wickramasinghe, 1996)

Recent claims that biological breakdown products are responsible for the signal of organic carbon in the interstellar medium echo this opinion (Rauf and Wickramasinghe, 2010).

While it may indeed turn out that we will discover interstellar carbon-based compounds that are in fact "postbiotic," it is fundamentally difficult to distinguish abiotic from biotic carbon based on spectra alone – more complex direct tests are clearly required as applied to meteoric PAHs (Naraoka et al., 2000) or as have been applied to verify the biogenicity of petroleum reserves on Earth (Glasby, 2006). Equally important to consider is the consistent argument by the proponents of this hypothesis that biological materials more exactly duplicate the emission/absorption spectra collected and therefore should be considered correct (Wickramasinghe, 2011). Few would likely dispute that biological extracts as suggested by these authors create a more complex absorption/emission profile that in some cases may better approximate astronomical data. However, it will be also clear to observers that nonbiological models produce a less complex spectrum because they are a simplification in attempt to explain the data (Horneck, 1995).

Recent investigations of carbonaceous meteorites have discovered many complex organic compounds not of Earthly origin, including several nucleotide bases (Martins et al., 2008). This data has also been claimed as the breakdown products of living organisms, but we do well to caution that simple processes such as UV irradiation of PAHs can produce compounds such as alcohols and quinones (Bernstein et al., 1999), and innumerable papers on the origin of life have demonstrated the abiotic synthesis of compounds such as nucleotides and sugars from precursors present in space (Orgel, 2004; Kim and Benner, 2010; Lambert et al., 2010a, b). While it is not possible to completely discount the possibility of life producing some of these signatures, it strains one's credulity less to explain this evidence in terms of abiotic processes that currently have the balance of evidence in their favor.

## 2.5. EVIDENCE OF OTHER GAIAS? PLANETARY SPECTRAL SIGNATURES

Indirect evidence for extraterrestrial life may soon come from the analysis of exoplanet atmospheres – the identification of certain nonequilibrium gases such as oxygen would strongly argue for life as the most plausible explanation. The detection of seasonal plumes of methane in Mars' atmosphere has ignited debate on the possibility of chemoautotrophic methanogens surviving there (Krashnopolsky et al., 2004), perhaps in habitats similar to those discovered in the high Arctic (Grasby and Londry, 2007). This interpretation is not without controversy, as there are differing theories for the methane's origin (Fiebig et al., 2007). In a similar vein, infrared spectroscopy has also been suggested to identify characteristic chemical bonds in the Europan ice sheets that may indicate biological activity (Dalton et al., 2003).

Outside of our solar system, the ambitious *Darwin Project* aims to track the presence of gases such as $CO_2$, $H_2O$, $CH_4$, and $O_2$ in exoplanets (Cockell et al., 2009). The analysis of the spectrum of a newly discovered exoplanet, a so-called hot Jupiter, has revealed the presence of all the previously mentioned gases in its atmosphere, with the exception of oxygen (Swain et al., 2009). While the presence of methane in large gas giants is not regarded as evidence of life in our own solar system or elsewhere, the recent discovery of several rocky planets dubbed "super Earths" (Charbonneau et al., 2009) gives hope that we will soon discover a rocky planet with an atmosphere indicative of active biology.

Currently, no one line of evidence has offered conclusive proof of extraterrestrial life, let alone panspermia, with controversies that have erupted over fossil evidence exposing our relative lack of understanding about these topics. Clearly, there is mounting evidence that life may survive in the far reaches of space, but the strategy of many proponents of panspermia to make sweeping conclusions from ambiguous evidence is untenable. More cautious authors note that evidence for life such as geochemical biosignatures may well be cryptic (Cockell et al., 2009), perhaps not as clear-cut as on our planet, which has taken many billions of years to reach this level of obvious biological activity.

## 3. Question 2: Is It Conceivable that Life as We Know It Could Be Viably Transferred to Other Habitats?

The existence of numerous lunar and Martian meteorites identified by their unique geology gives strong proof that there is an exchange of material between planets within our solar system. The possibility that these objects could carry viable life has led many to revisit the possibility of panspermia. This topic was ably reviewed by Horneck and Brack (1992), who point out several mechanisms for particles to reach space, including volcanic eruptions, flyby meteorites, and direct meteorite impacts. In these cases, both large pieces of rock as well as smaller particles would reach escape velocity. Large particles may travel to another planet by random chance (lithopanspermia), and smaller particles may be accelerated outwards from their parent star by radiation pressure or carried by interstellar particle clouds (radiopanspermia) (Horneck, 1995). Aside from impacts, small particles may also be directly transported to space (Dehel, 2006) or created through subsequent impacts with other objects (Napier, 2004). In this manner, we can imagine a small number of large meteorites and a larger number of smaller particles exiting a planet's surface, all with the potential to carry resistant forms of life.

A clear barrier to transfer is the sudden shock of temperature and pressure that will occur on both exit and atmospheric reentry. Despite these harsh conditions, simulations indicate that many resistant forms of life such as bacterial endospores can survive (Mastrapa et al., 2001; Stoffler et al., 2007), although there appears to be a critical size below which a meteorite would not be suitable for transfer (Parnell et al., 2008; Foucher et al., 2010) due to heat sterilization at depths of 2–5 cm. This implies that any lithopanspermia event will filter out organisms such as phototrophs that do not exist deep within rocks (Cockell, 2008). While bacteria have been found living in the interior of many substrates (Takai et al., 2001; Wainwright et al., 2009), indicating a random impact will eject organisms to space, the small amount of biological material contained within may reduce the probability of a successful transfer (Clark, 2001).

Noting that many organisms, including some animals (Jonsson et al., 2008), can survive extreme cold and desiccation and remain viable, the largest obstacles to panspermia are the time scale of transfer and the deleterious effect of radiation. Known Martian meteorites took from between ~$6.0 \times 10^5$ and up to ~14 million years to arrive on Earth (Eugster et al., 1997), and radiopanspermia has been suggested to take a minimum of $10^6$ years to reach a new planetary system although considerably less within a solar system (Secker et al., 1996). While bacterial life has been claimed to be potentially viable for possibly up to millions of years (Vreeland et al., 2000), this has been criticized on the basis that DNA is not thermodynamically stable enough to remain intact for this long (Graur and Pupko, 2001) and that there is possibility of contamination during sampling procedures that purport to isolate ancient life (Willerslev et al., 2004; Willerslev and Hebsgaard, 2005). Even if contamination is ruled out, the thermodynamic argument remains strong since DNA is inherently unstable – experiments examining dormant bacteria trapped in Antarctic ice corroborate this line of reasoning (Bidle et al., 2007), showing decreased viability of cells with an increased period of dormancy. In addition, the ability for DNA to remain viable may vary with phylogeny, with calculations suggesting that mesophiles are least likely to retain viable DNA over long periods, while thermophiles have a much better chance (Nicholson, 2003).

Radiation damage will inevitably accelerate this process and, combined with the factors of low temperature and vacuum, may act synergistically to destroy DNA (Fekete et al., 2005), although it has been also suggested that extreme cold tends to retard the destruction of DNA (Demidov et al., 1995). Radiation shielding is plausible with lithopanspermia events, but the small particle sizes necessary for radiopanspermia decrease radiation shielding greatly, although it has been suggested that a thin film of carbon may be sufficient to shelter organisms from inactivating UV radiation (Wickramasinghe et al., 2009).

The radiation environment in which these transfers might take place may also vary – for example, the interior of our galaxy likely experiences higher background radiation from supernovae because the density of stars is greater. The radiation from a supernova may offer a dose sufficient to sterilize particles of the size necessary for radiopanspermia or even small rocks. As Andrew Karam (Karam, 2002) notes, "…radiation may place constraints on the ability of living organisms to be transferred between planets or stellar systems because the mean interval between 'sterilizing' doses is less than the expected transit time." In addition to harmful radiation from an initial blast, supernovae remnants may also make transfer more difficult by increasing the background radiation in interstellar space (Grzedzielski et al., 2010).

Given this, it appears that radiopanspermia transfer events to another solar system will only be possible in certain circumstances. One proposed scenario is during the red giant phase of a star's life where sterilizing UV radiation would be less of an issue (Secker et al., 1996). Experiments involving life exposed to space for nearly 3 years have been proposed (Warmflash et al., 2007), which would further our knowledge about life's resistance to the rigors of space.

Estimates of the probability of successful lithopanspermia events indicate that material potentially containing viable microbes likely was exchanged between Earth, Mars, and Venus in the past, perhaps affecting the early development of the biosphere (Mileikowsky et al., 2000a, b; Schulze-Makuch et al., 2005a, b). In addition, transfer between bodies in a common stellar "nursery" may be likely (Adams and Spergel, 2005), supported by data that "… imply that a substantial fraction of the Oort cloud comets, perhaps exceeding 90%, are from the protoplanetary disks of other stars" (Levison et al., 2010). This is corroborated by the recent understanding gained from the NASA Stardust experiment that material in comets was created from the "mixing on a grand scale" of the original constituents of our solar system (McKeegan et al., 2006; Vaidya, 2009).

In contrast, transfer between stellar systems by lithopanspermia is thought to be extremely unlikely due to the astronomically small probabilities of interstellar meteorites falling on another small planetary body (Melosh, 2003). Some have suggested that panspermia was more likely at the time of our planet's formation since there was a higher number of suitable planets from which an inoculum could arrive (Von Bloh et al., 2003; Franck et al., 2007) or because transfer events would have been more likely due to a larger number of impacts (Mileikowsky et al., 2000a, b). Lineweaver's analysis (Lineweaver, 2001) suggests that most extant terrestrial planets are older than Earth, further supporting the assertion that panspermia may have been more common in the past.

Other, more esoteric mechanisms include the transfer of biological particles from the upper atmosphere to space (Wickramasinghe and Wickramasinghe, 2008), or the ejection of entire "embryonic" planets from solar nebulae, which could go on to support and possibly spread life (Stevenson, 1999). Transfers between galaxies in the same cluster may also be possible when there are ejected jets of material that contain life-supporting material (Gordon and Hoover, 2007).

Nicholson (2009) treats the sequence of probabilities necessary for a successful transfer event in a similar manner to the famous Drake equation, which reminds us that while there may be situations in which viable transfer is possible, we are largely ignorant of the numerous variables that challenge the spread of life. However, even if life was not viably transferred, these events may have an impact – DNA or other prebiotic chemicals could conceivably offer a "kick-start" for the origin of life on sterile planets (Secker et al., 1996).

## 4. Question 3: How Likely Is It That This Transferred Life Would Become Established and Go on to Seed Neighboring Systems, Thus Amplifying the Process?

Discussions of panspermia tend to focus on the mechanistic aspects of viable transfer, but equally important to constraining the probability of successful transfer is the ability of organisms to persist in a new environment. At this time, arguments that life is likely to be based on organic carbon polymers and existing in liquid water (the so-called *carbaquist* perspective (Fry, 2000)) constrain the type of environments where life may exist. Liquid water may exist well below 0°C in some cases (Rothschild and Mancinelli, 2001), and it may be possible for some organisms to have metabolic activity even below this level (Rivkina et al., 2000). With this evidence in mind, Pace (2001) suggests a temperature range from −50°C to 200°C for the existence of life and less stringent requirements on pressure – an interesting experiment has recently demonstrated metabolic activity at gigapascal pressures (Sharma et al., 2002). As evidence begins to accumulate for exoplanets that resemble Earth, Pace (2001) notes that an equally important question becomes the metabolic strategies organisms can potentially exploit to survive in extraterrestrial environments – all organisms, no matter the diversity of their metabolic strategies, require an influx of energy and a means of removing waste products.

If we consider a putative space-faring organism colonizing a lifeless world, it would be a fair assumption that conditions may resemble the early Earth – that is, either with a neutral or partially reducing atmosphere and with a potential influx of organic compounds from comets and interstellar space (Horneck, 1995). If we also assume some form of active geology, as well as light at a wavelength appropriate for photosynthesis, a high diversity of anaerobic strategies may succeed in these environments.

Among the most adaptable microbes thus far discovered are chemosynthetic. These organisms, also known as chemoautotrophs, are able to fix carbon wholly from inorganic chemicals in a process completely independent of photosynthesis. Chemosynthetic organisms exist in the most extreme and poorly understood regions of the Earth, including deep-sea vents where the oxidation of hydrogen sulfide is coupled to carbon fixation (Jannasch, 1985), or deep crustal environments where hydrogen-based, hyperthermophilic subsurface lithoautotrophic microbial ecosystems ("HyperSLiMEs") exist completely independent of the surface biosphere (Takai et al., 2004). Chemoautotrophs appear to be highly diverse (Nakagawa and Takai, 2008), and iron-oxidizing bacteria can apparently survive on unsupplemented extracts of iron meteorites (González-Toril et al., 2005). Speculation about extant life on Mars naturally tends to favor these ecosystems (Nealson et al., 2005) since they do not require a reduced carbon source and can thrive on common inorganic compounds. From the perspective of panspermia, this type of metabolism may be preferentially transferred by meteorite impacts since these organisms are most likely to be living within the interior of rocks.

In contrast to chemoautotrophs, photosynthetic organisms can take advantage of a much more reliable source of energy and thus are able to thrive in a huge diversity of environments on Earth – from Antarctic ice (Fritsen and Priscu, 1998; Thomas, 2005) to the hyperarid deserts of the Atacama (Wierzchos et al., 2006), where cyanobacteria associated with heterotrophic bacteria colonize halite deposits in rocks. While there may be a selective pressure preventing the dispersal of cyanobacteria and other phototrophs (Cockell, 2008), they are interesting to astrobiology for a number of reasons, including high radiation tolerance due to UV-absorbing compounds (Sinha and Hader, 2008) that are increasingly synthesized when exposed to stress such as desiccation (Fleming and Castenholz, 2007). Many cyanobacteria remain viable after long periods of time in dark permafrost soils (Vishnivetskaya et al., 2003), although it appears that the resistant cells, known as akinetes, have limited resistance to the rigors of space unless shielded from radiation (Olsson-Francis et al., 2009). These organisms can survive at extremely low irradiances and extremes of pH and temperature and can switch their photosynthetic metabolism from oxygenic to anoxygenic under differing conditions (Garcia-Pichel and Castenholz, 1990).

Other photosynthetic microbes include strictly anoxygenic phototrophic bacteria, which also thrive under extreme conditions, and may represent the earliest phototrophic life on Earth (Madigan, 2003). These organisms have unique metabolic pathways of carbon fixation (Berg et al., 2007) and may not require solar radiation for photosynthesis – a new strain of green sulfur bacteria has been discovered that exists as an obligate phototroph at hydrothermal vents, relying on purely geothermal radiation to photosynthesize (Beatty et al., 2005). With this knowledge in mind, Cockell and Raven (2004) have concluded that the lack of a UV-absorbing atmosphere would not preclude the colonization of anoxic extrasolar planets by photosynthetic life.

Organisms that are reliant on external sources of reduced carbon are the final category, which can be either aerobic or anaerobic – although it is assumed that when considering the evolution of primitive ecosystems, only organisms that are obligate or facultative anaerobes would be relevant. To produce the reduced compounds necessary for their growth, we can imagine a variety of organic compounds being synthesized through UV irradiation of carbonaceous building blocks in space (Bernstein et al., 1999), and we do know organisms that can survive on completely inorganic compounds such as buckminsterfullerene (Wainwright, 2003) and hydrocarbons such as toluene and naphthalene (Fredrickson et al., 1991). In addition to organisms that survive on simple organic compounds, some are metabolically versatile such as the magnetotactic bacteria that can grow as chemoorganoheterotrophs or chemolithoautotrophs (Thomas-Keprta et al., 2002). Nonetheless, heterotrophic organisms are probably less likely to survive independently on a new planet because they require a constant supply of an external carbon source – something that may be unlikely given the ability of geochemical processes to oxidize organic matter on a relatively short timescale.

## 4.1. CONSTRAINTS ON ALL METABOLIC STRATEGIES

As already mentioned, even with organisms that do not rely on sunlight or reduced organic compounds, there needs to be a constant source of reductant and inorganic carbon; otherwise, their metabolism will grind to a halt (Wilkinson, 2003). Aside from the geochemical processes we are familiar with from Earth, it has been suggested that radiation photolysis of water to $H_2$ may provide this energy in Europa-like systems (Chyba, 2000), and geologically active bodies with sufficient iron may support abiotic $H_2$ production through a process known as serpentinization (Oze and Sharma, 2007).

All life, including photosynthetic organisms, produces waste products that must be removed. In an Earthly context, most organisms do not exist in isolation for this simple reason – syntrophic organisms provide mutually beneficial "services" (see, e.g., Kotsyurbenko, 2005; Guerrero et al., 2002), casting doubt on the possibility of a single strain seeding a new planet. Indeed, theoretical analysis of the supposed "last universal common ancestor" (LUCA) on Earth has suggested that "[the] LUCA does not appear to have been a simple, primitive, hyperthermophilic prokaryote but rather a complex community of protoeukaryotes with RNA genomes, adapted to a broad range of moderate temperatures, genetically redundant, morphologically and metabolically diverse" (Glansdorff et al., 2008). Nonetheless, it may, however, be possible for certain organisms to accomplish multiple ecological functions – see, for example, a recent report of a single-species ecosystem discovered deep in the Earth's crust (Chivian et al., 2008). Wilkinson (2003) has considered this topic in depth and suggests that diverse metabolic strategies are necessary to maintain an active biosphere. This implies that either multiple organisms with different metabolic strategies or one highly adaptable organism would be needed to seed a new world.

Very little is yet known about extrasolar habitats – some guesses have been made regarding a "galactic habitable zone," but there is not yet enough evidence to make any far-reaching conclusions (see Prantzos, 2008). We may yet discover that systems similar to our own are common (Vázquez et al., 2010), but the vast majority of systems discovered include "hot Jupiters," large gas giants orbiting extremely close to their sun. This has led to speculation about the possibility of terrestrial planets existing with these large planets as neighbors, with calculations indicating that these large planets will not necessarily suppress rocky planet formation (Raymond et al., 2005), with other estimates suggesting up to one fourth of known extrasolar planetary systems may be habitable (Menou and Tabachnik, 2003). Other authors have suggested that a host star requires a specific metallicity to effectively form rocky planets like our own (Lineweaver, 2001).

On a similar note, the different elemental composition of planets or oxidation state may be important – it is possible that life evolving on one world may find a new environment toxic due to the abundance of an element rare on its home world. On Earth, cyanobacteria are known to have a low threshold for cadmium toxicity, which is explained by the fact that the environment in which they originally evolved contained less free cadmium due to a different oxidation state (Saito et al., 2003). It is also equally possible that life may be metabolically adaptable in ways we have yet to discover – a new report of bacteria that purportedly use arsenate to replace the biological function of phosphate offers support for this notion (Wolfe-Simon et al., 2011a). However, at the time of writing, the methodology of this study has been widely discussed and criticized (Benner, 2011; Borhani, 2011; Cotner and Hall, 2011; Csabai and Szathmary, 2011; Foster, 2011; Oehler, 2011; Redfield, 2011; Rosen et al., 2011; Schoepp-Cothenet et al., 2011), with the report's claims and response to comments (Wolfe-Simon et al., 2011b) at this point appearing inconclusive at best.

## 4.2. LOCATIONS FOR AMPLIFICATION

Despite the lack of information about the existence of extrasolar habitats, we can make some guesses about the possibility of panspermia on different bodies as modeled on our own solar system. In the case of icy bodies such as Europa, panspermia may be less likely because ice will be less amenable to transfer or UV shielding. Planets with strictly atmospheric life have been suggested as on a contemporary Venus (Schulze-Makuch and Irwin, 2002), and these could offer an excellent method of transfer since the organisms already inhabit a highly extreme environment that is in close exchange with space (Wickramasinghe and Wickramasinghe, 2008).

One of the few types of bodies we can be fairly sure exist throughout the cosmos are comets, and these bodies are an essential part of Hoyle-Wickramasinghe's version of panspermia, in which comets offer a location for the origin of life as well as a "vehicle" for the transfer of life (Hoyle and

Wickramasinghe, 1999). There does appear to be some evidence that comets and potentially asteroids contained liquid water early in their history from the decay of radioactive aluminum (Wilson et al., 1999; Merk and Prialnik, 2006), with frozen water ice still clearly visible in contemporary comets (Sunshine et al., 2006). Wickramasinghe et al. (2009) have also claimed that comets could allow for liquid water near perihelion based on surface temperatures above 300 K. However, the assertion that life evolved and persists in comets has been roundly criticized (Horneck, 1995). Bar-Nun et al. (1981) also point out that comets may not have the structural integrity necessary to contain liquid water and that any organic compounds produced on the surface through interactions with stellar UV radiation would be sublimated at perihelion. In addition, it has been noted that organisms thus far claimed to be inhabitants of comets show no specific adaptations to that niche (Jalasvuori et al., 2009; Vaidya, 2010).

Many questions about comets will be answered by the ROSETTA probe, which is set to intercept comet 67P/Churyumov-Gerasimenko in 2014. Equipped with devices that can measure the chemical composition of the comet from afar, the probe also includes a lander that will be able to measure biologically important characteristics such as the enantiomeric composition of amino acids in the comet (Thiemann and Meierhenrich, 2001), which may offer insight into both solar system formation as well any possible biological activity.

As with comets, asteroids may have been more amenable to life early in the solar system's history, based on the possibility of liquid water (Wilson et al., 1999). Nonetheless, it seems unlikely that life could currently persist in asteroids despite the discovery that asteroids still contain water (Campins et al., 2010), simply because the current temperatures would be too low to support life. Nonetheless, extracts of meteorites have been shown to be amenable to algal growth, with dissolved inorganic salts supporting cell division (Mautner, 2002).

Along with the literature on cometary life, suggestions have been raised that diatoms may inhabit these bodies (Hoover et al., 1999). This is unlikely for a number of reasons. Aside from the previously mentioned criticisms of life in comets, eukaryotic cytoskeletons are unlikely to develop normally in microgravity (Papaseit et al., 2000) (a testable proposition: Gordon et al., 2007), and diatoms are also a recent arrival in the geological record (Sims et al., 2006), which makes their appearance in bodies that originated many billions of years previously extremely unlikely. Other published accounts have attempted to draw a connection between "red rain" events and panspermia (Louis and Kumar, 2006), though no evidence to the authors' knowledge was presented to substantiate this claim.

We agree with Bar-Nun et al. (Bar-Nun et al., 1981), who in reference to the theories of the cosmic abundance of life opine that "…we feel that it is not warranted to extrapolate from these observations to the existence of interstellar or cometary life." Citing inexplicable events such as "red rain" or ambiguous data as "proof" of panspermia will likely only be convincing to the ardent supporters of these hypotheses. Similarly, speculation about how extremophile organisms may thrive in extraterrestrial habitats cannot be taken as proof of life's ability to colonize the universe – there may be unknown factors that prevent life from easily adapting to new environments that we are only just beginning to discover.

## 5. Question 4: Could Panspermia Influence Evolutionary Progress or Affect the Origin of Life Debate?

### 5.1. INFLUENCE OF OUTSIDE GENETIC MATERIAL ON EVOLUTION

Aside from an initial "seeding" playing an essential role in the rise of life, a natural corollary is that "cosmic life forms" could have an ongoing influence on the evolution and diversity of life on Earth. Hoyle and Wickramasinghe have championed the highly controversial suggestion that viruses come from space, believing that the often unpredictable pattern of communicable diseases becomes intelligible in the light of inoculation from above. A paper attributing the recent SARS outbreak to this mechanism (Wickramasinghe, 2003) has been roundly criticized for a number of reasons (Willerslev et al., 2003), and more general problems have been raised in regard to the potential role of viruses in ET environments (Jalasvuori et al., 2009).

The exchange of biological particles with Earth is an essential part of Hoyle-Wickramasinghe's manifestation of the panspermia hypothesis, and they find proof in microbes discovered at high altitudes (see above section):

> The fact that these are similar to terrestrial microbes is no problem; it is fully consistent with panspermia theories in which Earth organisms are derived from cometary organisms that transit through the stratosphere. (Wickramasinghe, 2003)

In this case, the null explanation (that these organisms are terrestrial) is indistinguishable from the "predictions" of their hypothesis, so it is no wonder many scientists remain unconvinced.

The notion that genetic information from panspermia could explain innovation in evolution via galactic horizontal gene transfer (Sheldon and Hoover, 2008) ignores more prosaic hypotheses such as duplication and subsequent specialization of gene batteries (Gordon, 1999) or developmental plasticity.

### 5.2. PANSPERMIA AND THE ORIGIN OF LIFE

Evidence discussed in the previous sections suggests that Earth is unlikely to be uniquely capable of supporting life and therefore may be common throughout the cosmos (Lineweaver and Davis, 2002) – both panspermia and many independent origins of life could produce this inferred pattern. Little is known about the probability of panspermia carrying life over galactic distances, though it has been suggested that:

> On the assumption that this ejection mechanism [radiopanspermia] is common in other planetary systems environmentally capable of supporting life, a "chain

reaction" may seed the disc of the Galaxy within a few billion years. In that case it is unlikely that life originated on Earth. (Napier, 2004)

The belief that effective transfer events are inevitable and the a priori probability of life self-organizing is extremely low has led Hoyle and Wickramasinghe to suggest an independent origin is not relevant to the rise of life on Earth (Hoyle and Wickramasinghe, 1996). While some authors have suggested panspermia events were more likely at the time of Earth's formation (see above sections), the logic that life is an "accident" of prebiotic chemistry and therefore unlikely to have occurred by random chance on Earth has been challenged by other authors (Trevors and Abel, 2004), although considerable uncertainty remains about the probability of an independent origins of life on Earth (Davies, 2003).

Clearly, neither explanation can be proven correct at the moment, but it does appear that a minimum number of multiple origins are necessary if we assume a homogenous distribution of life throughout the universe. Gordon and Hoover (2007) have argued based on a calculated "speed of life" (the maximum speed at which panspermia could move with ejecta) that at least 50,000 unique origins would be necessary to explain universal life (which amounts to at least once per 2,000 galaxies). It is also conceivable that the origin of life is much more probable than Hoyle and Wickramasinghe consider – this would be supported if we were able to discover biochemically distinct forms of life on Earth that are the product of a unique origin. Some authors have suggested searching for "…opposite chirality biomarkers, nonracemic mixtures of biological material, or anomalous ratios of stable isotopes that cannot be explained by abiotic or normal biotic processes" (Davies and Lineweaver, 2005). Other evidence may come from investigating phenomena such as "desert varnish" which appear biological but have not been satisfactorily explained (Cleland, 2007; Davies et al., 2009). The discovery of these hypothesized "shadow biospheres" would offer evidence that the origin of life is highly probable given clement conditions. Any such discovery would, in the opinion of some authors, be more significant than evidence for a successful panspermia event since it would inform the probability of life originating not just on Earth, but throughout the universe (Shapiro and Schulze-Makuch, 2009).

While there is still considerable debate on the interpretation of early microfossils (see above sections and Gordon and McNichol, 2012), it is now generally accepted that life appeared on Earth very quickly after conditions of the Hadean era became more hospitable to life. This early appearance of life, combined with evidence suggesting that many complex biochemical processes were present in the LUCA, has led Line (2002) to consider that an extraterrestrial origin of life was necessary to provide sufficient time for these evolutionary innovations to develop. How much time was available for early life to evolve from a "prebiotic" state to life as we know it is still unknown – it has been recently suggested that early "sterilizing impacts" were not sufficient to wipe out thermophiles (Abramov and Mojzsis, 2009) and that Earth could have been reseeded by impact ejecta (Wells et al., 2003), in which case life may have had more time to develop on Earth than previously supposed. Evidence for this period of Earth's history is missing from the geological record, though it may be preserved as ejected meteorites on the moon or other bodies (Crawford et al., 2008).

Line (2002) has suggested that we may be able to confirm evidence for the evolution of life on Earth elsewhere in the universe if we were to discover life on Mars or Europa that was genetically related to Earthly life. There are naturally pitfalls in any potential interpretation of such data, and Ebach et al. (2008a) caution us to heed the past failures of biogeography and not seek to draw conclusions where data is scanty or lacking. This viewpoint earned criticism from ardent supporters of panspermia as geocentric (Wallis et al., 2008) though it has been noted that at this point, we cannot help but be "geocentric" (Ebach et al., 2008b). The discovery of biochemically related life would be unlikely to quell this debate since it would undoubtedly be difficult to resolve its provenance (Pace, 2001). Indeed, it is likely that the discovery of ET life would only deepen the mystery of our origins.

## 6. Conclusions and Future Research Directions

At first glance, the hypothesis of panspermia appears to be supported by the weight of much biological and astronomical research, but despite a few tantalizing hints, it remains currently in the realm of conjecture. Few would likely disagree that imaginative speculation about the fundamental questions of life's origin has inspired much great research, as it currently inspires research into astrobiology. Unfortunately, science's power to explain these phenomena is constrained by ambiguous, controversial, or nonexistent data. The study of astrobiology therefore is often significantly influenced by opinion – Sir Francis Younghusband's musings on this subject remain relevant even today to discussions of panspermia:

> The onus of proving that nowhere else throughout the whole mighty universe than on this tiny speck of a planet does life appear rests upon those who deny it. (p. 47 in (Younghusband, 1927))

This argument appears central in the writings of ardent supporters of panspermia but cannot in itself prove anything satisfactorily. While proponents of programs optimistic about the existence of extraterrestrial life likely also dream of a day when we have evidence for life outside of this "tiny speck," it is equally clear that, at present, no generally accepted evidence exists.

We do not disavow the possibility of panspermia but believe that a credible case has yet to be made for much of the evidence that seeks to prove a cosmic abundance of life. The debates over the ALH84001 meteorite as well as the Apex Chert microfossils were possible because multiple parties could analyze the same samples critically and have done much to give us a clearer picture of the challenges that will undoubtedly face scientists as we gain more information about life in the universe – the same rigorous debate, with falsifiable hypotheses, is essential for any serious discussion of panspermia. Such debate is now going on regarding the possible substitution of arsenic for phosphorus in DNA, another problem put forward as one of astrobiology (Wolfe-Simon et al., 2011a). Carl Sagan's "Extraordinary claims require extraordinary evidence" (Sagan, 1979), stated in the context of UFOs (unidentified flying objects) has seemed to plague astrobiologists. But in science do we really only require "ordinary" evidence for "ordinary" claims? We suggest not, that is, that all claims in science are open to the same degree of scrutiny. The strongest claims are those that have passed the most scrutiny. The record for panspermia claims in this ordinary science regard is spotty. We are presently faced with the unproven belief that life could be widespread in the universe, fortified by the presence of organic matter in meteorites and elsewhere, contrasted with our certainty that life exists on earth. That should be enough to motivate us to continue the search, with humility:

> I find that as beginners in this business most people are brash but after they have stayed in it a while, they tend to become more humble. I am very humble. Harold C. Urey in: (Laurence, 1961)

Current research on the diversity and limits of microbial life, the discovery and investigation of extrasolar planets, and new scientific missions to bodies in our solar system are most likely to offer clear and uncontroversial evidence for astrobiology. It is, however, important to note that even with unambiguous proof of ET life, controversy over its origin is unlikely to be resolved anytime soon. Whether life on Earth began solely here or was transferred from elsewhere, both panspermia and origin of life research will likely have complementary parts to play in probing these mysteries.

## 7. References


Abramov O, Mojzsis SJ (2009) Microbial habitability of the Hadean Earth during the late heavy bombardment. Nature 459(7245):419–422

Adams FC, Spergel DN (2005) Lithopanspermia in star-forming clusters. Astrobiology 5(4):497–514

Bada JL (2001) State-of-the-art instruments for detecting extraterrestrial life. Proc Natl Acad Sci USA 98(3):797–800

Bar-Nun A, Lazcanoaraujo A, Oro J (1981) Could life have evolved in cometary nuclei? Orig Life Evol Biosph 11(4):387–394

Barrat JA, Bollinger C (2010) Geochemistry of the Martian meteorite ALH 84001, revisited. Meteoritics Planet Sci 45(4):495–512

Beatty JT, Overmann J, Lince MT, Manske AK, Lang AS, Blankenship RE, Van Dover CL, Martinson TA, Plumley FG (2005) An obligately photosynthetic bacterial anaerobe from a deep-sea hydrothermal vent. Proc Natl Acad Sci USA 102(26):9306–9310

Benner SA (2011) Comment on "A bacterium that can grow by using arsenic instead of phosphorus". Science 332(6034)

Berg IA, Kockelkorn D, Buckel W, Fuchs G (2007) A 3-hydroxypropionate/4-hydroxybutyrate autotrophic carbon dioxide assimilation pathway in archaea. Science 318(5857):1782–1786

Bernal JD (1962) Comments. Nature 193(4821):1127–1129

Bernstein MP, Sandford SA, Allamandola LJ, Gillette JS, Clemett SJ, Zare RN (1999) UV irradiation of polycyclic aromatic hydrocarbons in ices: Production of alcohols, quinones, and ethers. Science 283(5405):1135–1138

Bidle KD, Lee S, Marchant DR, Falkowski PG (2007) Fossil genes and microbes in the oldest ice on Earth. Proc Natl Acad Sci USA 104(33):13455–13460
Borhani DW (2011) Comment on "A bacterium that can grow by using arsenic instead of phosphorus". Science 332(6034)
Boynton WV, Feldman WC, Squyres SW, Prettyman TH, Bruckner J, Evans LG, Reedy RC, Starr R, Arnold JR, Drake DM, Englert PAJ, Metzger AE, Mitrofanov I, Trombka JI, d'Uston C, Wanke H, Gasnault O, Hamara DK, Janes DM, Marcialis RL, Maurice S, Mikheeva I, Taylor GJ, Tokar R, Shinohara C (2002) Distribution of hydrogen in the near surface of Mars: Evidence for subsurface ice deposits. Science 297(5578):81–85
Brasier MD, Green OR, Jephcoat AP, Kleppe AK, Van Kranendonk MJ, Lindsay JF, Steele A, Grassineau NV (2002) Questioning the evidence for Earth's oldest fossils. Nature 416(6876):76–81
Brasier M, Green O, Lindsay J, Steele A (2004) Earth's oldest (approximately 3.5 Ga) fossils and the 'Early Eden hypothesis': questioning the evidence. Orig Life Evol Biosph 34(1–2):257–269
Brasier MD, Green OR, Lindsay JF, McLoughlin N, Steele A, Stoakes C (2005) Critical testing of earth's oldest putative fossil assemblage from the similar to 3.5 Ga Apex Chert, Chinaman Creek, western Australia. Precambrian Res 140(1–2):55–1cbrs02
Busemann H, Young AF, Alexander CM, Hoppe P, Mukhopadhyay S, Nittler LR (2006) Interstellar chemistry recorded in organic matter from primitive meteorites. Science 312(5774):727–730
Campins H, Hargrove K, Pinilla-Alonso N, Howell ES, Kelley MS, Licandro J, Mothe-Diniz T, Fernandez Y, Ziffer J (2010) Water ice and organics on the surface of the asteroid 24 Themis. Nature 464(7293):1320–1321
Charbonneau D, Berta ZK, Irwin J, Burke CJ, Nutzman P, Buchhave LA, Lovis C, Bonfils X, Latham DW, Udry S, Murray-Clay RA, Holman MJ, Falco EE, Winn JN, Queloz D, Pepe F, Mayor M, Delfosse X, Forveille T (2009) A super-Earth transiting a nearby low-mass star. Nature 462(7275):891–894
Chen JH, Li MP, Li AG, Wang YJ (2008) On buckyonions as a carrier of the 2175 angstrom interstellar extinction feature. In: Kwok S, Sandford S (eds) Iau: organic matter in space, vol 4, no 251, IAU Symposium Proceedings Series. Cambridge University Press, New York, pp 71–72
Chivian D, Brodie EL, Alm EJ, Culley DE, Dehal PS, DeSantis TZ, Gihring TM, Lapidus A, Lin LH, Lowry SR, Moser DP, Richardson PM, Southam G, Wanger G, Pratt LM, Andersen GL, Hazen TC, Brockman FJ, Arkin AP, Onstott TC (2008) Environmental genomics reveals a single-species ecosystem deep within earth. Science 322(5899):275–278
Christner BC, Morris CE, Foreman CM, Cai RM, Sands DC (2008) Ubiquity of biological ice nucleators in snowfall. Science 319(5867):1214–1214
Chyba CF (2000) Energy for microbial life on Europa - A radiation-driven ecosystem on Jupiter's moon is not beyond the bounds of possibility. Nature 403(6768):381–382
Cisar JO, Xu DQ, Thompson J, Swaim W, Hu L, Kopecko DJ (2000) An alternative interpretation of nanobacteria-induced biomineralization. Proc Natl Acad Sci USA 97(21):11511–11515
Clark BC (2001) Planetary interchange of bioactive material: Probability factors and implications. Orig Life Evol Biosph 31(1–2):185–197
Cleland CE (2007) Epistemological issues in the study of microbial life: alternative terran biospheres? Stud History Philos Sci Part C Stud History Philos Biol Biomed Sci 38(4):847–861
Cockell CS (2008) The interplanetary exchange of photosynthesis. Origins Life Evol Biosph 38(1):87–104
Cockell CS, Raven JA (2004) Zones of photosynthetic potential on Mars and the early Earth. Icarus 169(2):300–310
Cockell CS, Leger A, Fridlund M, Herbst TM, Kaltenegger L, Absil O, Beichman C, Benz W, Blanc M, Brack A, Chelli A, Colangeli L, Cottin H, du Foresto FC, Danchi WC, Defrere D, den Herder JW, Eiroa C, Greaves J, Henning T, Johnston KJ, Jones H, Labadie L, Lammer H, Launhardt R, Lawson P, Lay OP, LeDuigou JM, Liseau R, Malbet F, Martin SR, Mawet D, Mourard D, Moutou C, Mugnier LM, Ollivier M, Paresce F, Quirrenbach A, Rabbia YD, Raven JA, Rottgering HJA, Rouan D, Santos NC, Selsis F, Serabyn E, Shibai H, Tamura M, Thiebaut E, Westall F, White GJ (2009) Darwin-a mission to detect and search for life on extrasolar planets. Astrobiology 9(1):1–22

Cotner JB, Hall EK (2011) Comment on “A bacterium that can grow by using arsenic instead of phosphorus”. Science 332(6034)
Crawford IA, Baldwin EC, Taylor EA, Bailey JA, Tsembelis K (2008) On the survivability and detectability of terrestrial meteorites on the Moon. Astrobiology 8(2):242–252
Crick FHC, Orgel LE (1973) Directed panspermia. Icarus 19(3):341–346
Crowe MJ (2008) The extraterrestrial life debate, antiquity to 1915: a source book. University of Notre Dame Press, Notre Dame
Csabai I, Szathmary E (2011) Comment on “A bacterium that can grow by using arsenic instead of phosphorus”. Science 332(6034)
Dalton R (2002) Microfossils: Squaring up over ancient life. Nature 417(6891):782–784
Dalton JB, Mogul R, Kagawa HK, Chan SL, Jamieson CS (2003) Near-infrared detection of potential evidence for microscopic organisms on Europa. Astrobiology 3(3):505–529
Damoah R, Spichtinger N, Servranckx R, Fromm M, Eloranta EW, Razenkov IA, James P, Shulski M, Forster C, Stohl A (2006) A case study of pyro-convection using transport model and remote sensing data. Atmos Chem Phys 6:173–185
Davies PCW (2003) Does life’s rapid appearance imply a Martian origin? Astrobiology 3(4):673–679
Davies PCW, Lineweaver CH (2005) Hypothesis paper - Finding a second sample of life on earth. Astrobiology 5(2):154–163
Davies PCW, Benner SA, Cleland CE, Lineweaver CH, McKay CP, Wolfe-Simon F (2009) Signatures of a shadow biosphere. Astrobiology 9(2):241–249
Decin L, Agundez M, Barlow MJ, Daniel F, Cernicharo J, Lombaert R, De Beck E, Royer P, Vandenbussche B, Wesson R, Polehampton ET, Blommaert J, De Meester W, Exter K, Feuchtgruber H, Gear WK, Gomez HL, Groenewegen MAT, Guelin M, Hargrave PC, Huygen R, Imhof P, Ivison RJ, Jean C, Kahane C, Kerschbaum F, Leeks SJ, Lim T, Matsuura M, Olofsson G, Posch T, Regibo S, Savini G, Sibthorpe B, Swinyard BM, Yates JA, Waelkens C (2010) Warm water vapour in the sooty outflow from a luminous carbon star. Nature 467(7311):64–67
Dehel T (2006) Uplift and outflow of bacterial spores via electric field. In: Proceedings of the 36th COSPAR Scientific Assembly. Held 16–23 July 2006, in Beijing, China [Meeting Abstracts], vol 36, p 1
Demidov VV, Potaman VN, Solyanina IP, Trofimov VI (1995) Radiation and thermal stabilities of adenine nucleotides. Adv Space Res 3:131–134
Dick SJ, Strick JE (2004) The living universe: NASA and the development of astrobiology. Rutgers University Press, New Brunswick
Ebach MC, Gill AC, Williams DM (2008a) The pitfalls of astrobiogeography. Astrophys Space Sci 317(3–4):143–144
Ebach MC, Gill AC, Williams DM (2008b) Reply, a future for astrobiogeography. Astrophys Space Sci 317(3–4):147–147
Eugster O, Weigel A, Polnau E (1997) Ejection times of Martian meteorites. Geochim Cosmochim Acta 61(13):2749–2757
Fekete A, Modos K, Hegedus M, Kovacs G, Ronto G, Peter A, Lammer H, Panitz C (2005) DNA damage under simulated extraterrestrial conditions in bacteriophage T7. Space Life Sci Astrobiol Steps Toward Origin Life Titan Before Cassini 36(2):303–310
Fiebig J, Woodland AB, Spangenberg J, Oschmann W (2007) Natural evidence for rapid abiogenic hydrothermal generation of $CH_4$. Geochim Cosmochim Acta 71(12):3028–3039
Filmer MR (1991) Southern African Spiders: an identification guide. Struik, Cape Town
Fleming ED, Castenholz RW (2007) Effects of periodic desiccation on the synthesis of the UV-screening compound, scytonemin, in cyanobacteria. Environ Microbiol 9(6):1448–1455
Folk RL, Taylor LA (2002) Nannobacterial alteration of pyroxenes in Martian meteorite Allan Hills 84001. Meteoritics Planet Sci 37(8):1057–1069
Foster PL (2011) Comment on “A bacterium that can grow by using arsenic instead of phosphorus”. Science 332(6034)

Foucher F, Westall F, Brandstätter F, Demets R, Parnell J, Cockell CS, Edwards HGM (2010) Testing the survival of microfossils in artificial Martian sedimentary meteorites during entry into Earth's atmosphere: The STONE 6 experiment. Icarus 207(2):616–630

Franck S, von Bloh W, Bounama C (2007) Maximum number of habitable planets at the time of Earth's origin: new hints for panspermia and the mediocrity principle. Int J Astrobiol 6(2):153–157

Fredrickson JK, Brockman FJ, Workman DJ, Li SW, Stevens TO (1991) Isolation and characterization of a subsurface bacterium capable of growth on toluene, naphthalene, and other aromatic compounds. Appl Environ Microbiol 57(3):796–803

Fritsen CH, Priscu JC (1998) Cyanobacterial assemblages in permanent ice covers on Antarctic lakes: Distribution growth rate, and temperature response of photosynthesis. J Phycol 34(4):587–597

Fromm M (2004) New directions: Eruptive transport to the stratosphere: Add fire-convection to volcanoes. Atmos Environ 38(1):163–165

Fry I (2000) The emergence of life on Earth: a historical and scientific overview. Rutgers University Press, New Brunswide

Garcia-Pichel F, Castenholz RW (1990) Comparative anoxygenic photosynthetic capacity in 7 strains of a thermophilic cyanobacterium. Archiv Microbiol 153(4):344–351

Glansdorff N, Xu Y, Labedan B (2008) The last universal common ancestor: emergence, constitution and genetic legacy of an elusive forerunner. Biol Direct 3:29, http://www.biology-direct.com/content/pdf/1745–6150–1743–1727.pdf

Glasby GP (2006) Abiogenic origin of hydrocarbons: An historical overview. Resource Geol 56(1):83–96

Golden DC, Ming DW, Morris RV, Brearley A, Lauer HV, Treiman AH, Zolensky ME, Schwandt CS, Lofgren GE, McKay GA (2004) Evidence for exclusively inorganic formation of magnetite in Martian meteorite ALH84001. Am Miner 89(5–6):681–695

González-Toril E, Martinez-Frias J, Gomez JMG, Rull F, Amils R (2005) Iron meteorites can support the growth of acidophilic chemolithoautotrophic microorganisms. Astrobiology 5(3):406–414

Gordon R (1999) The hierarchical genome and differentiation waves: novel unification of development, genetics and evolution. World Scientific & Imperial College Press, Singapore/London

Gordon R, Hoover RB (2007) Could there have been a single origin of life in a big bang universe? Proc SPIE 6694, doi:10.1117/1112.737041

Gordon R, McNichol J (2012) Recurrent dreams of life in meteorites. In: Seckbach J, Gordon R (eds) In the beginning: Precursors of life, chemical models and early biological evolution. Springer, Dordrecht, pp

Gordon R, Hoover RB, Tuszynski JA, de Luis J, Camp PJ, Tiffany MA, Nagy SS, Fayek M, Lopez PJ, Lerner BE (2007) Diatoms in space: testing prospects for reliable diatom nanotechnology in microgravity. Proc SPIE 6694:V1–V15. doi:10.1117/1112.737051

Grasby SE, Londry KL (2007) Biogeochemistry of hypersaline springs supporting a mid-continent marine ecosystem: An analogue for Martian springs? Astrobiology 7(4):662–683

Graur D, Pupko T (2001) The Permian bacterium that isn't. Mol Biol Evol 18(6):1143–1146

Grzedzielski S, Bzowski M, Czechowski A, Funsten HO, McComas DJ, Schwadron NA (2010) A possible generation mechanism for the IBEX ribbon from outside the heliosphere. Astrophys J Lett 715(2):L84–L87

Guerrero R, Piqueras M, Berlanga M (2002) Microbial mats and the search for minimal ecosystems. Int Microbiol 5(4):177–188

Haldane JBS (1954) The origins of life. New Biol 16:12–27

Hecht MH, Kounaves SP, Quinn RC, West SJ, Young SMM, Ming DW, Catling DC, Clark BC, Boynton WV, Hoffman J, DeFlores LP, Gospodinova K, Kapit J, Smith PH (2009) Detection of perchlorate and the soluble chemistry of Martian soil at the Phoenix Lander site. Science 325(5936):64–67

Hoover RB, Hoyle F, Wickramasinghe NC, Hoover MJ, Al-Mufti S (1999) Diatoms on Earth, comets, Europa and in interstellar space. Astrophys Space Sci 268(1–3):197–224

Hoover RB, Rozanov AY, Jerman G, Costen J (2004) Microfossils in Cl and CO carbonaceous meteorites. Proc SPIE 5163:7–22

Horneck G (1995) Exobiology, the study of the origin, evolution and distribution of life within the context of cosmic evolution: a review. Planet Space Sci 43(1–2):189–217

Horneck G, Brack A (1992) Study of the origin, evolution and distribution of life with emphasis on exobiology experiments in earth orbit. Adv Space Biol Med 2:229–262

Hoyle F, Wickramasinghe NC (1996) Our place in the Cosmos: the unfinished revolution. Phoenix, London

Hoyle F, Wickramasinghe NC (1999) Comets - A vehicle for panspermia. Astrophys Space Sci 268(1–3):333–341

Jalasvuori M, Örmälä AM, Bamford JKII (2009) On the astrobiological relevance of viruses in extraterrestrial ecosystems. Int J Astrobiol 8(2):95–100

Jannasch HW (1985) Review lecture: the chemosynthetic support of life and the microbial diversity at deep-sea hydrothermal vents. Proc R Soc Lond Ser B Biol Sci 225(1240):277–297

Jonsson KI, Rabbow E, Schill RO, Harms-Ringdahl M, Rettberg P (2008) Tardigrades survive exposure to space in low Earth orbit. Curr Biol 18(17):R729–R731

Karam PA (2002) Terrestrial radiation exposure from supernova-produced radioactivities. Radiat Phys Chem 64(2):77–87

Kim HJ, Benner SA (2010) Comment on "The silicate-mediated formose reaction: bottom-up synthesis of sugar silicates". Science 329(5994)

Kotsyurbenko OR (2005) Trophic interactions in the methanogenic microbial community of low-temperature terrestrial ecosystems. Fems Microbiol Ecol 53(1):3–13

Krashnopolsky VA, Maillard JP, Owen TC (2004) Detection of methane in the Martian atmosphere: evidence for life? Icarus 172(2):537–547

Kwok S (2004) The synthesis of organic and inorganic compounds in evolved stars. Nature 430(7003):985–991

Lambert JB, Gurusamy-Thangavelu SA, Ma KBA (2010a) Response to comment on "The silicate-mediated formose reaction: bottom-up synthesis of sugar silicates". Science 329(5994)

Lambert JB, Gurusamy-Thangavelu SA, Ma KBA (2010b) The silicate-mediated formose reaction: bottom-up synthesis of sugar silicates. Science 327(5968):984–986

Laurence WL (1961) Science; life in space? Analysts of French meteorite say it contains organic matter. New York Times (March 19):E9

Levin G (1981) A search for a nonbiological explanation of the Viking labeled release life detection experiment. Icarus 45(2):494–516

Levin GV (2007) Possible evidence for panspermia: the labelled release experiment. Int J Astrobiol 6(2):95–108

Levison HF, Duncan MJ, Brasser R, Kaufmann DE (2010) Capture of the sun's Oort cloud from stars in its birth cluster. Science 329(5988):187–190

Line MA (2002) The enigma of the origin of life and its timing. Microbiol Sgm 148:21–27

Lineweaver CH (2001) An estimate of the age distribution of terrestrial planets in the universe: quantifying metallicity as a selection effect. Icarus 151:307–313

Lineweaver CH, Davis TM (2002) Does the rapid appearance of life on Earth suggest that life is common in the universe? Astrobiology 2(3):293–304

Louis G, Kumar AS (2006) The red rain phenomenon of Kerala and its possible extraterrestrial origin. Astrophys Space Sci 302(1–4):175–187

Madigan MT (2003) Anoxygenic phototrophic bacteria from extreme environments. Photosynth Res 76(1–3):157–171

Marshall CP, Emry JR, Marshall AO (2011) Haematite pseudomicrofossils present in the 3.5-billion-year-old Apex Chert. Nat Geosci 4(4)):240–243

Martins Z, Botta O, Fogel ML, Sephton MA, Glavin DP, Watson JS, Dworkin JP, Schwartz AW, Ehrenfreund P (2008) Extraterrestrial nucleobases in the Murchison meteorite. Earth Planet Sci Lett 270(1–2):130–136

Mastrapa RME, Glanzberg H, Head JN, Melosh HJ, Nicholson WL (2001) Survival of bacteria exposed to extreme acceleration: implications for panspermia. Earth Planet Sci Lett 189(1–2):1–8

Mautner MN (2002) Planetary resources and astroecology. Planetary microcosm models of asteroid meteorite interiors: electrolyte solutions and microbial growth- space populations and panspermia. Astrobiology 2(1):59–76

McCollom TM (2003) Formation of meteorite hydrocarbons from thermal decomposition of siderite ($FeCO_3$). Geochim Cosmochim Acta 67(2):311–317

McKay DS, Gibson EK Jr, Thomas-Keprta KL, Vali H, Romanek CS, Clemett SJ, Chillier XDF, Maechling CR, Zare RN (1996) Search for past life on Mars: Possible relic biogenic activity in Martian meteorite ALH84001. Science 273(5277):924–930

McKeegan KD, Aleon J, Bradley J, Brownlee D, Busemann H, Butterworth A, Chaussidon M, Fallon S, Floss C, Gilmour J, Gounelle M, Graham G, Guan Y, Heck PR, Hoppe P, Hutcheon ID, Huth J, Ishii H, Ito M, Jacobsen SB, Kearsley A, Leshin LA, Liu MC, Lyon I, Marhas K, Marty B, Matrajt G, Meibom A, Messenger S, Mostefaoui S, Mukhopadhyay S, Nakamura-Messenger K, Nittler L, Palma R, Pepin RO, Papanastassiou DA, Robert F, Schlutter D, Snead CJ, Stadermann FJ, Stroud R, Tsou P, Westphal A, Young ED, Ziegler K, Zimmermann L, Zinner E (2006) Isotopic compositions of cometary matter returned by Stardust. Science 314(5806):1724–1728

Melosh HJ (2003) Exchange of meteorites (and life?) between stellar systems. Astrobiology 3(1):207–215

Menou K, Tabachnik S (2003) Dynamical habitability of known extrasolar planetary systems. Astrophys J 583(1):473–488

Merk R, Prialnik D (2006) Combined modeling of thermal evolution and accretion of trans-neptunian objects - Occurrence of high temperatures and liquid water. Icarus 183(2):283–295

Mileikowsky C, Cucinotta FA, Wilson JW, Gladman B, Horneck G, Lindegren L, Melosh J, Rickman H, Valtonen M, Zheng JQ (2000a) Natural transfer of viable microbes in space – 1. From Mars to Earth and Earth to Mars. Icarus 145(2):391–427

Mileikowsky C, Cucinotta FA, Wilson JW, Gladman B, Horneck G, Lindegren L, Melosh J, Rickman H, Valtonen M, Zheng JQ (2000b) Risks threatening viable transfer of microbes between bodies in our solar system. Planet Space Sci 48(11):1107–1115

Min K, Reiners PW (2007) High-temperature Mars-to-Earth transfer of meteorite ALH84001. Earth Planet Sci Lett 260(1–2):72–85

Nakagawa S, Takai K (2008) Deep-sea vent chemoautotrophs: diversity, biochemistry and ecological significance. Fems Microbiol Ecol 65(1):1–14

Napier WM (2004) A mechanism for interstellar panspermia. Mon Not R Astron Soc 348(1):46–51

Naraoka H, Shimoyama A, Harada K (2000) Isotopic evidence from an Antarctic carbonaceous chondrite for two reaction pathways of extraterrestrial PAH formation. Earth Planet Sci Lett 184(1):1–7

Narlikar JV, Lloyd D, Wickramasinghe NC, Harris MJ, Turner MP, Al-Mufti S, Wallis MK, Wainwright M, Rajaratnam P, Shivaji S, Reddy GSN, Ramadurai S, Hoyle F (2003) Balloon experiment to detect microorganisms in the outer space. Astrophys Space Sci 285(2):555–562

Navarro-González R, Navarro KF, de la Rosa J, Iñiguez E, Molina P, Miranda LD, Morales P, Cienfuegos E, Coll P, Raulin F, Amils R, McKay CP (2006) The limitations on organic detection in Mars-like soils by thermal volatilization-gas chromatography-MS and their implications for the Viking results. Proc Natl Acad Sci USA 103(44):16089–16094

Nealson KH, Inagaki F, Takai K (2005) Hydrogen-driven subsurface lithoautotrophic microbial ecosystems (SLiMEs): do they exist and why should we care? Trends Microbiol 13(9):405–410

Nicholson WL (2003) Using thermal inactivation kinetics to calculate the probability of extreme spore longevity: Implications for paleomicrobiology and lithopanspermia. Orig Life Evol Biosph 33(6):621–631

Nicholson WL (2009) Ancient micronauts: interplanetary transport of microbes by cosmic impacts. Trends Microbiol 17(6):243–250

Oehler S (2011) Comment on "A bacterium that can grow by using arsenic instead of phosphorus". Science 332(6034)

Olsson-Francis K, de la Torre R, Towner MC, Cockell CS (2009) Survival of akinetes (resting-state cells of cyanobacteria) in low earth orbit and simulated extraterrestrial conditions. Origins Life Evol Biosph 39(6):565–579

Orgel LE (2004) Prebiotic adenine revisited: Eutectics and photochemistry. Orig Life Evol Biosph 34(4):361–369

Oze C, Sharma M (2007) Serpentinization and the inorganic synthesis of H-2 in planetary surfaces. Icarus 186(2):557–561

Pace NR (2001) The universal nature of biochemistry. Proc Natl Acad Sci USA 98(3):805–808

Papaseit C, Pochon N, Tabony J (2000) Microtubule self-organization is gravity-dependent. Proc Natl Acad Sci USA 97(15):8364–8368

Papoular R (2001) The use of kerogen data in understanding the properties and evolution of interstellar carbonaceous dust. Astron Astrophys 378(2):597–607

Parnell J, Mark D, Brandstatter F (2008) Response of sandstone to atmospheric heating during the STONE 5 experiment: Implications for the palaeofluid record in meteorites. Icarus 197(1):282–290

Pavlov AK, Kalinin VL, Konstantinov AN, Shelegedin VN, Pavlov AA (2006) Was earth ever infected by Martian biota? Clues from radioresistant bacteria. Astrobiology 6(6):911–918

Pearl JC, Smith MD, Conrath BJ, Bandfield JL, Christensen PR (2001) Observations of Martian ice clouds by the Mars Global Surveyor thermal emission spectrometer: the first Martian year. J Geophys Res Planets 106(E6):2325–12338

Pickering WH (1921) Mars. R. G Badger, Boston

Prantzos N (2008) On the "Galactic Habitable Zone". Space Sci Rev 135(1–4):313–322

Quinn RC, Zent AP (1999) Peroxide-modified titanium dioxide: A chemical analog of putative Martian soil oxidants. Orig Life Evol Biosph 29(1):59–72

Raoult D, Drancourt M, Azza S, Nappez C, Guieu R, Rolain JM, Fourquet P, Campagna B, La Scola B, Mege JL, Mansuelle P, Lechevalier E, Berland Y, Gorvel JP, Renesto P (2008) Nanobacteria are mineralo fetuin complexes. PLoS Pathog 4(2):e41

Rauf K, Wickramasinghe C (2010) Evidence for biodegradation products in the interstellar medium. Int J Astrobiol 9(1):29–34

Raymond SN, Quinn T, Lunine JI (2005) The formation and habitability of terrestrial planets in the presence of close-in giant planets. Icarus 177(1):256–263

Redfield RJ (2011) Comment on "A bacterium that can grow by using arsenic instead of phosphorus". Science 332(6034)

Rinaldi A (2007) Space life holds its breath - Pressured by scepticism, budget cuts and the need to prove itself, astrobiology is coming to a crossroads. Embo Rep 8(5):436–440

Rivkina EM, Friedmann EI, McKay CP, Gilichinsky DA (2000) Metabolic activity of permafrost bacteria below the freezing point. Appl Environ Microbiol 66(8):3230–3233

Rohatschek H (1996) Levitation of stratospheric and mesospheric aerosols by gravito-photophoresis. J Aerosol Sci 27(3):467–475

Rosen BP, Ajees AA, McDermott TR (2011) Life and death with arsenic. Arsenic life: an analysis of the recent report "A bacterium that can grow by using arsenic instead of phosphorus". Bioessays 33(5):350–357

Rothschild LJ, Mancinelli RL (2001) Life in extreme environments. Nature 409(6823):1092–1101

Rozanov AY, Zhegallo EA, Ushatinskaya GT, Shuvalova YV, Hoover RB (2001) Bacterial paleontology for astrobiology. Proc SPIE 4495:283–294

Sagan C (1979) Carl Sagan: "Extraordinary Claims Require Extraordinary Evidence" from: Cosmos, episode 12, "Encyclopaedia Galactica". http://www.youtube.com/watch?v=mRaXvPQ-ayk

Saito MA, Sigman DM, Morel FMM (2003) The bioinorganic chemistry of the ancient ocean: the co-evolution of cyanobacterial metal requirements and biogeochemical cycles at the Archean-Proterozoic boundary? Inorg Chim Acta 356:308–318

Schoepp-Cothenet B, Nitschke W, Barge LM, Ponce A, Russell MJ, Tsapin AI (2011) Comment on "A bacterium that can grow by using arsenic instead of phosphorus". Science 332(6034)

Schopf JW (2006) Fossil evidence of Archaean life. Philos Trans R Soc Lond B Biol Sci 361(1470):869–885

Schopf JW, Kudryavtsev AB (2005) Three-dimensional Raman imagery of Precambrian microscopic organisms. Geobiology 3(1):1–12

Schopf JW, Kudryavtsev AB, Agresti DG, Czaja AD, Wdowiak TJ (2005) Raman imagery: a new approach to assess the geochemical maturity and biogenicity of permineralized precambrian fossils. Astrobiology 5(3):333–371

Schopf JW, Kudryavtsev AB, Czaja AD, Tripathi AB (2007) Evidence of archean life: Stromatolites and microfossils. Precambrian Res 158(3–4):141–155

Schulze-Makuch D, Irwin LN (2002) Reassessing the possibility of life on Venus: Proposal for an astrobiology mission. Astrobiology 2(2):197–202

Schulze-Makuch D, Dohm JM, Fairen AG, Baker VR, Fink W, Strom RG (2005a) Venus, Mars, and the ices on Mercury and the moon: astrobiological implications and proposed mission designs. Astrobiology 5(6):778–795

Schulze-Makuch D, Irwin LN, Lipps JH, LeMone D, Dohm JM, Fairen AG (2005b) Scenarios for the evolution of life on Mars. J Geophys Res Planets 110(E12):E12S23

Secker J, Wesson PS, Lepock JR (1996) Astrophysical and biological constraints on radiopanspermia. J R Astron Soc Can 90(4):184–192

Shapiro R, Schulze-Makuch D (2009) The Search for Alien Life in Our Solar System: Strategies and Priorities. Astrobiology 9(4):335–343

Sharma A, Scott JH, Cody GD, Fogel ML, Hazen RM, Hemley RJ, Huntress WT (2002) Microbial activity at gigapascal pressures. Science 295(5559):1514–1516

Sheldon RB, Hoover RB (2008) Cosmological evolution: spatial relativity and the speed of life. Proc SPIE 7097:709716

Shivaji S, Chaturvedi P, Suresh K, Reddy GSN, Dutt CBS, Wainwright M, Narlikar JV, Bhargava PM (2006) *Bacillus aerius* sp. nov., *Bacillus aerophilus* sp. nov., *Bacillus stratosphericus* sp. nov and *Bacillus altitudinis* sp. nov., isolated from cryogenic tubes used for collecting air samples from high altitudes. Int J System Evolut Microbiol 56:1465–1473

Shivaji S, Chaturvedi P, Begum Z, Pindi PK, Manorama R, Padmanaban DA, Shouche YS, Pawar S, Vaishampayan P, Dutt CBS, Datta GN, Manchanda RK, Rao UR, Bhargava PM, Narlikar JV (2009) *Janibacter hoylei* sp. nov., *Bacillus isronensis* sp. nov and *Bacillus aryabhattai* sp. nov., isolated from cryotubes used for collecting air from the upper atmosphere. Int J System Evolut Microbiol 59:2977–2986

Sims PA, Mann DG, Medlin LK (2006) Evolution of the diatoms: insights from fossil, biological and molecular data. Phycologia 45(4):361–402

Sinha RP, Hader DP (2008) UV protectants in cyanobacteria. Plant Sci 174(3):278–289

Smolders K, Acke B, Verhoelst T, Blommaert J, Decin L, Hony S, Sloan GC, Neyskens P, Van Eck S, Zijlstra AA, Van Winckel H (2010) When an old star smolders: On the detection of hydrocarbon emission from S-type AGB stars. Astron Astrophys 514:L1

Squyres SW, Knoll AH (2005) Sedimentary rocks at Meridiani Planum: Origin, diagetiesis, and implications for life on Mars. Earth Planet Sci Lett 240(1):1–10

Stevenson DJ (1999) Life-sustaining planets in interstellar space? Nature 400(6739):32–32

Stoffler D, Horneck G, Ott S, Hornemann U, Cockell CS, Moeller R, Meyer C, de Vera JP, Fritz J, Artemieva NA (2007) Experimental evidence for the potential impact ejection of viable microorganisms from Mars and Mars-like planets. Icarus 186(2):585–588

Sunshine JM, A'Hearn MF, Groussin O, Li JY, Belton MJS, Delamere WA, Kissel J, Klaasen KP, McFadden LA, Meech KJ, Melosh HJ, Schultz PH, Thomas PC, Veverka J, Yeomans DK, Busko IC, Desnoyer M, Farnham TL, Feaga LM, Hampton DL, Lindler DJ, Lisse CM, Wellnitz DD (2006) Exposed water ice deposits on the surface of comet 9P/Tempel 1. Science 311(5766):1453–1455

Swain MR, Tinetti G, Vasisht G, Deroo P, Griffith C, Bouwman J, Chen P, Yung Y, Burrows A, Brown LR, Matthews J, Rowe JF, Kuschnig R, Angerhausen D (2009) Water, methane, and carbon dioxide present in the dayside spectrum of the exoplanet HD 209458b. Astrophys J 704(2):1616–1621

Takai K, Moser DP, Onstott TC, Spoelstra N, Pfiffner SM, Dohnalkova A, Fredrickson JK (2001) *Alkaliphilus transvaalensis* gen. nov., sp nov., an extremely alkaliphilic bacterium isolated from a deep South African gold mine. Int J System Evolut Microbiol 51:1245–1256

Takai K, Gamo T, Tsunogai U, Nakayama N, Hirayama H, Nealson KH, Horikoshi K (2004) Geochemical and microbiological evidence for a hydrogen-based, hyperthermophilic subsurface lithoautotrophic microbial ecosystem (HyperSLiME) beneath an active deep-sea hydrothermal field. Extremophiles 8(4):269–282

Temple R (2007) The prehistory of panspermia: astrophysical or metaphysical? Int J Astrobiol 6(2):169–180

Thiemann WHP, Meierhenrich U (2001) ESA mission ROSETTA will probe for chirality of cometary amino acids. Origins Life Evol Biosph 31(1–2):199–210

Thomas DN (2005) Photosynthetic microbes in freezing deserts. Trends Microbiol 13(3):87–88

Thomas-Keprta KL, Clemett SJ, Bazylinski DA, Kirschvink JL, McKay DS, Wentworth SJ, Vali H, Gibson EK, Romanek CS (2002) Magnetofossils from ancient Mars: a robust biosignature in the Martian meteorite ALH84001. Appl Environ Microbiol 68(8):3663–3672

Trevors JT, Abel DL (2004) Chance and necessity do not explain the origin of life. Cell Biol Int 28(11):729–739

Vaidya PG (2009) Stardust findings: implications for panspermia. Apeiron 16(2):225

Vaidya PG (2010) Critique on vindication of panspermia – falsification of Hoyle-Wickramasinghe hypothesis? In: Astrobiology Science Conference 2010: Evolution and Life: Surviving Catastrophes and Extremes on Earth and Beyond, vol 1538, held April 26–20, 2010 in League City, TX, p 5135

Vázquez M, Pallé E, Montañés Rodríguez P, Rodríguez PM (2010) The Earth as a distant planet: a Rosetta stone for the search of Earth-like worlds. Springer, New York

Vishnivetskaya T, Spirina E, Shatilovich A, Erokhina L, Vorobyova E, Gilichinsky D (2003) The resistance of viable permafrost algae to simulated environmental stresses: implications for astrobiology. Int J Astrobiol 2(03):171–177

Von Bloh W, Franck S, Bounama C, Schellnhuber HJ (2003) Maximum number of habitable planets at the time of Earth's origin: new hints for panspermia? Orig Life Evol Biosph 33(2):219–231

Vreeland RH, Rosenzweig WD, Powers DW (2000) Isolation of a 250 million-year-old halotolerant bacterium from a primary salt crystal. Nature 407:897–900

Wainwright M (2003) A microbiologist looks at panspermia. Astrophys Space Sci 285(2):563–570

Wainwright M, Wickramasinghe NC, Narlikar JV, Rajaratnam P (2003) Microorganisms cultured from stratospheric air samples obtained at 41 km. FEMS Microbiol Lett 218(1):161–165

Wainwright M, Wickramasinghe NC, Narlikar JV, Rajaratnam P, Perkins J (2004) Confirmation of the presence of viable but non-cultureable bacteria in the stratosphere. Int J Astrobiol 3(01):13–15

Wainwright M, Alharbi S, Wickramasinghe NC (2006) How do microorganisms reach the stratosphere? Int J Astrobiol 5(01):13–15

Wainwright M, Laswd A, Alshammari F (2009) Bacteria in amber coal and clay in relation to lithopanspermia. Int J Astrobiol 8(2):141–143

Wallis M, Wainwright M, Wickramasinghe JT, Wickramasinghe NC (2008) Pitfalls of geocentricism. Astrophys Space Sci 317(3–4):145–146

Warmflash D, Ciftcioglu N, Fox GE, McKay DS, Friedman L, Betts B, Kirschvink J (2007) Living Interplanetary Flight Experiment (LIFE): an experiment on the survivability of microorganisms during interplanetary transfer. In: First International Conference on the Exploration of Phobos and Deimos, Proceedings of the conference held November 5–8, 2007 in Moffett Field, California, vol 1. LPI Contribution No. 1377, p 41

Wells LE, Armstrong JC, Gonzalez G (2003) Reseeding of early Earth by impacts of returning ejecta during the late heavy bombardment. Icarus 162(1):38–46

Wickramasinghe C (2003) SARS - a clue to its origins? Lancet 361(9371):1832–1832

Wickramasinghe C (2010) The astrobiological case for our cosmic ancestry. Int J Astrobiol 9(2):119–129

Wickramasinghe C (2011) Origin of life and panspermia. In: Seckbach J (ed) Genesis - In the beginning: Precursors of life, chemical models and early biological evolution. Springer, Dordrecht, pp

Wickramasinghe NC, Wickramasinghe JT (2008) On the possibility of microbiota transfer from Venus to Earth. Astrophys Space Sci 317(1–2):133–137

Wickramasinghe NC, Wainwright M, Narlikar JV, Rajaratnam P, Harris MJ, Lloyd D (2003) Progress towards the vindication of panspermia. Astrophys Space Sci 283(3):403–413

Wickramasinghe JT, Wickramasinghe NC, Wallis MK (2009) Liquid water and organics in comets: implications for exobiology. Int J Astrobiol 8(4):281–290

Wierzchos J, Ascaso C, McKay CP (2006) Endolithic cyanobacteria in halite rocks from the hyperarid core of the Atacama Desert. Astrobiology 6(3):415–422

Wilkinson DM (2003) The fundamental processes in ecology: a thought experiment on extraterrestrial biospheres. Biol Rev 78(2):171–179

Willerslev E, Hebsgaard MB (2005) New evidence for 250 Ma age of halotolerant bacterium from a Permian salt crystal: Comment and reply. Geology 33(1):e93

Willerslev E, Hansen AJ, Ronn R, Nielsen OJ (2003) Panspermia - true or false? Lancet 362(9381):406–408

Willerslev E, Hansen AJ, Poinar HN (2004) Isolation of nucleic acids and cultures from fossil ice and permafrost. Trends Ecol Evol 19(3):141–147

Wilson L, Keil K, Browning LB, Krot AN, Bourcier W (1999) Early aqueous alteration, explosive disruption, and reprocessing of asteroids. Meteoritics Planet Sci 34(4):541–557

Wolfe-Simon F, Blum JS, Kulp TR, Gordon GW, Hoeft SE, Pett-Ridge J, Stolz JF, Webb SM, Weber PK, Davies PCW, Anbar AD, Oremland RS (2011a) A bacterium that can grow by using arsenic instead of phosphorus. Science 332(6034):1163–1166

Wolfe-Simon F, Blum JS, Kulp TR, Gordon GW, Hoeft SE, Pett-Ridge J, Stolz JF, Webb SM, Weber PK, Davies PCW, Anbar AD, Oremland RS (2011b) Response to comments on "A bacterium that can grow by using arsenic instead of phosphorus". Science 332(6034):10.1126/science.1202098

Yang Y, Yokobori S, Yamagishi A (2009) Assessing panspermia hypothesis by microorganisms collected from the high altitude atmosphere. Biol Sci Space 23(3):151–163

Yang YJ, Itoh T, Yokobori S, Shimada H, Itahashi S, Satoh K, Ohba H, Narumi I, Yamagishi A (2010) *Deinococcus aetherius* sp nov., isolated from the stratosphere. Int J System Evolut Microbiol 60:776–779

Young JD, Martel J (2010) The rise and fall of nanobacteria. Sci Am 302(1):52–59

Younghusband F (1927) Life in the stars: an exposition of the view that on some planets of some stars exist beings higher than ourselves, and on one a world-leader, the supreme embodiment of the eternal spirit which animates the whole. John Murray, London

Biodata of **Chandra Wickramasinghe**, author of "***Origin of Life and Panspermia.***"

**Chandra Wickramasinghe**, Sri Lankan-born mathematician, astrophysicist, and astrobiologist, is one of the pioneers of the modern theories of panspermia. His early work on the organic composition of interstellar dust grains in the 1960s and 1970s has been amply vindicated by modern discoveries of PAHs and organics in the Universe. He did his Ph.D. at Cambridge University under the guidance of the late Sir Fred Hoyle (in 1963 and his Sc.D. in 1973) and has been a collaborator with Hoyle over nearly four decades. He was a Fellow of Jesus College Cambridge from 1962 to 1973 and thereafter Professor at Cardiff University where he set up the Cardiff Centre for Astrobiology in 1989. The recipient of several international honors and honorary degrees, he is a leading figure in modern astrobiology. He has authored or coauthored over 20 books and some 400 scientific papers, with over 70 of these in the journal *Nature*.

He is currently Director of the Buckingham Centre for Astrobiology at the University of Buckingham, UK and Astrobiology Editor of the online journal: JournalofCosmology.com.

E-mail: **ncwick@gmail.com**

J. Seckbach (ed.), *Genesis - In The Beginning: Precursors of Life, Chemical Models and Early Biological Evolution*, Cellular Origin, Life in Extreme Habitats and Astrobiology 22, 621–649
DOI 10.1007/978-94-007-2941-4_31, © Springer Science+Business Media Dordrecht 2012

# ORIGIN OF LIFE AND PANSPERMIA

**CHANDRA WICKRAMASINGHE**
*Buckingham Centre for Astrobiology, The University of Buckingham, Buckingham MK18 1EG, UK*

## 1. Introduction

In the late 1920s, Oparin (1953) and Haldane (1929) introduced the 'Primordial Soup' theory of the origin of life, proposing that the atmosphere of the primitive Earth comprised of a reducing mixture of hydrogen, methane and ammonia and other compounds. Lightening and solar ultraviolet provided the energy to dissociate these molecules, and the radicals so formed recombined through a cascade of chemical reactions to yield biochemical monomers such as amino acids, nucleotide bases and sugars from which life was supposed to arise. In the 1950s, the classic experiments of Miller and Urey (1959) demonstrated the feasibility of the Oparin-Haldane chemistry, and this led to the belief that life could be generated de novo as soon as the biochemical monomers were in place. The formation of the first fully functioning, self-replicating life system with a potential for Darwinian evolution remains an elusive concept, however.

It is generally conceded that the path from chemicals to self-replicating biology must progress through a sequence of organisational steps of ever-increasing complexity. The most popular contender for one possible early step is the RNA world. Here nucleotides polymerise into random RNA molecules that lead to autonomously self-replicating macromolecules (ribozymes) without the need for an intermediary enzyme (Orgel and Crick, 1968). Likewise, other contenders of prebiotic development include the 'iron-sulphur world' (Wachtershauser, 1990), the 'PNA (peptide nucleic acid) world' (Bohler et al., 1995) and the 'clay world' (Cairns-Smith, 1966), the latter involving an inorganic clay system serving as a primitive informational template. In view of the high abundance of silicon in the galaxy, the clay world model might well have a special role to play in a cosmic context. The transition from any of these intermediate systems to a protogene system possessing prescriptive cybernetic information for evolution and finally to DNA-protein-based cellular life is still in the realm of speculation or hypothesis (Abel and Trevors, 2006).

The difficulty of finding unequivocal evidence of the relics of prebiology in the geological record has been a handicap for Earth-based theories of the origin of life. The suite of organics present in interstellar clouds (Kwok, 2009) consistently directs us to possible origins away from Earth to more and more distant parts of the Universe. At the very least, the organic molecules needed for life's origins are much more likely to have been generated in a cosmic context rather than being formed in situ on Earth. Moreover, it is now becoming clear that life arose on Earth almost at the very first moment that it could have survived. During the period from about 3.8 to 4.3 By ago (the Hadean Epoch), the Earth suffered an episode of heavy bombardment by comets and asteroids (Wickramasinghe et al., 2010). Rocks dating back to the tail end of this epoch reveal evidence of an excess of the lighter isotope $^{12}C$ compared with $^{13}C$ pointing to the action of microorganisms that preferentially take up the lighter isotope from the environment (Manning et al., 2006).

The success of the Miller-Urey experiment led to the conviction that it was only a matter of time before the next steps from biochemical monomers to life could be demonstrated in the laboratory. Despite over half a century of effort, this goal has proved stubbornly elusive. If one accepts the calculations showing grotesquely small a priori probabilities for the transition of non-life to life (Hoyle and Wickramasinghe, 1982), it would appear that only two options remain open. The origin of life on Earth was an extremely improbable event that occurred but will effectively not be reproduced elsewhere. Or a very much bigger system than was available on Earth and a longer timescale were involved in an initial origination event, after which life was somehow transferred to Earth. How big or old that system needs to be is still a matter for debate. Arguments by Abel and Trevors (2006) and Abel (2009) suggest that within the framework of Big-Bang-type cosmologies, naturalistic protogene formation still faces almost insuperable difficulties. However, by whatever process life has emerged, this event of origination must be reckoned as unique and its subsequent spread of life throughout the universe assured by the processes of 'panspermia' (Hoyle and Wickramasinghe, 1980, 1981, 1982, 2000).

## 2. Panspermia

The only secure empirical fact relating to the origin of life is encapsulated in the dictum enunciated of Louis Pasteur – *Omne vivum ex vivo* – all life from antecedent life (Pasteur, 1857). If life is always derived from antecedent life in a causal chain, which is clearly manifest in present-day life and through the fossil record, the question naturally arises as to when and where this connection may have ceased. The continuation of the life-from-life chain to a time before the first life appears on our planet and before the Earth itself formed implies the operation of 'panspermia'.

The basic concept of panspermia has a history going back centuries – to the time of classical Greece and even before. It refers in general to the widespread dispersal of the 'seeds of life' in the cosmos (Hoyle and Wickramasinghe, 2000; Arrhenius, 1903, 1908). Whilst the Francis Crick and Leslie Orgel's idea of directed panspermia transfers the problem of origin to another site, possibly invoking intelligent intervention (Crick and Orgel, 1973). Fred Hoyle and the writer have attempted to expand the spatial and temporal domains in which cosmic abiogenesis may have occurred, focussing in particular on totality of comets in our galaxy. Like Crick and Orgel (1973), we were influenced by the super-astronomical odds against the transition from organic molecules to even the most primitive living system – judged for instance by the highly specific complexity of the arrangements of molecules in DNA or enzymes.

The popular view that all extraterrestrial organics arise abiotically – that is to say through non-biologic processes – has no secure empirical basis and is likely to be flawed. On Earth, it is clear that life processes account for almost all the organic molecules on the planet. If biology can somehow be shown to be widespread on a cosmic scale, the detritus of living cells would also be expected to be widely distributed in the cosmos. The bulk of the organic molecules in space would then be explained as break-up products of life molecules (Kwok, 2009). Inorganic processes can scarcely be expected to compete with biology in the ability to synthesise systems of biochemicals resembling the detritus of biology. So wherever complex organics are found in an astronomical setting, one might legitimately infer that biology has spread (Wickramasinghe, 2010). We shall return to this topic in Sect. 6.

## 3. Astronomical Options: Comets

The basic molecules that may be required for prebiotic chemistry – including $H_2O$, simple organics and PAHs – are present in vast quantity in the Galaxy. All that one could hope to achieve in the way of further progress towards biochemistry in interstellar clouds is the production of moderately more complex organic molecules through gas-phase chemistry. These more complex organic molecules must then enter a watery medium in suitably high concentrations to begin the presumptive prebiotic chemistry that may have eventually led to life.

In the formation of a planetary system such as the solar system, the first solid objects to form are the comets. These icy objects would have mopped up the molecules and dust of the parent interstellar cloud, and for a few million years after they condensed would have liquid water interiors, due to the heating effect of radioactive decays. If even the minutest amount of microbial life was already present in the parent interstellar cloud, the newly formed comets could serve to vastly amplify it on a very short timescale.

Prior to life being generated anywhere in the galaxy, primordial comets heated by decay of radioactive nuclides such as $^{26}Al$ and $^{60}Fe$ would have provided trillions of 'warm little ponds' replete with water, organics and nutrients. Their huge numbers would have diminished vastly the improbability hurdle for life to originate. Recent studies of comet Tempel 1 (Fig. 1) have shown evidence of organic molecules including PAHs, clay particles as well as liquid water,

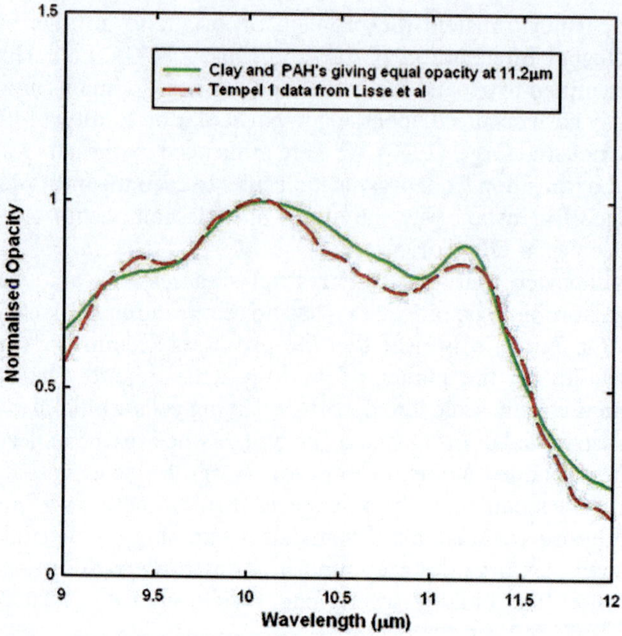


**Figure 1.** Normalised opacities near $\lambda = 10$ μm due to mixtures of clays and biological aromatic molecules (*solid curve*). Comparable contributions at $\lambda = 11.2$ μm from clays and aromatic molecules fit the data (*dashed curve*) best.

providing an ideal setting for the operation of the 'clay theory' of the origin of life (Wickramasinghe et al., 2009; Napier et al., 2007).

Together with Janaki Wickramasinghe and Bill Napier, the writer has argued that a single primordial comet of this kind will be favoured over all the shallow ponds and edges of oceans on Earth by a factor $10^4$, taking into account the total clay surface area for catalytic reactions as well as the timescale of persistence in each scenario. With $10^{11}$ comets, the factor favouring solar system comets over the totality of terrestrial 'warm little ponds' weighs in at a figure of $10^{15}$, and with $10^9$ sun-like stars replete with comets in the entire galaxy, we tot up a factor of $10^{24}$ in favour of a cometary origin life.

The next step in the argument is that once life got started in some comet somewhere, its spread in the cosmos becomes inevitable. The comets themselves are the amplifiers and distributers of life in the Galaxy. Dormant microorganisms are released in the dust tails of comets and propelled by the pressure of starlight to reach interstellar clouds. When a planetary system forms, the newly formed comets in that system provide sites for the amplification of surviving microorganisms that are incorporated in the new system. Transport of microorganisms and spores within the frozen interiors of comets carries only a negligible risk of destruction, but transport in either naked form, within clumps of dust or within meteorites

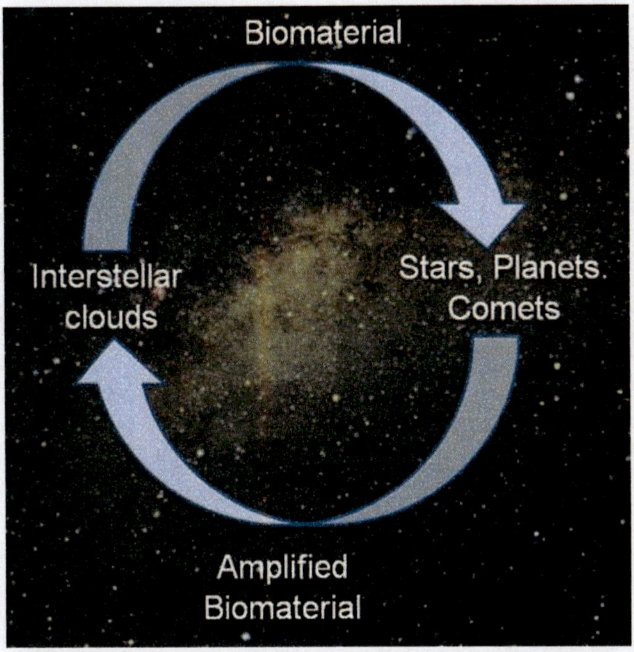


**Figure 2.** Feedback loop of cometary panspermia. Comets amplify and distribute microbial life at every star-/planet-forming event.

entails varying degrees of risk of inactivation by cosmic rays and UV light. However, the successful seeding of life requires only the minutest survival fraction between successive amplification sites. The feedback loop of cometary panspermia is shown schematically in Fig. 2. At the present time, this loop has been traversed some $10^{10}$ times, once for every sun-like star. Of the bacterial particles included in every nascent cometary cloud, only one in $10^{24}$ needs to remain viable to ensure a positive feedback loop for panspermia. All the indications are that this is a very modest requirement that is hard, if not impossible, to violate.

Whilst comets could supply a source of primitive life (archaea and bacteria) to interstellar clouds and thence to new planetary systems, the genetic products of evolved life could also be disseminated on a galaxy-wide scale (Wickramasinghe and Napier, 2008). Our present-day solar system which is surrounded by an extended halo of some 100 billion comets (the Oort cloud) moves around the centre of the galaxy with a period of 240 My. Every 40 million years, on the average, the comet cloud becomes perturbed due to the close passage of a molecular cloud. Gravitational interaction leads to hundreds of comets from the Oort cloud being injected into the inner planetary system, some to collide with the Earth. Such collisions can not only cause extinctions of species (as one impact surely did 65 million years ago, killing the dinosaurs), but they could also result in the expulsion of surface material back into space.

A mechanism thus exists for the genes of evolved Earth life to be transferred to alien planets. A fraction of the Earth debris so expelled survives shock heating and could be laden with viable microbial ecologies as well as genes of evolved life. Such life-bearing Earth material could reach newly forming planetary systems in the passing molecular cloud within a million years of the ejection event. A new planetary system thus comes to be infected with terrestrial microbes and terrestrial genes that can contribute, via horizontal gene transfer, to an ongoing process of local biological evolution. Once life has got started and evolved on an alien planet or planets of the new system, the same process can be repeated (via comet collisions) transferring genetic material carrying local evolutionary 'experience' to other molecular clouds and other nascent planetary systems. If every life-bearing planet transfers genes in this way to more than one other planetary system (say 1.1 on the average) with a characteristic time of 40 My then the number of seeded planets after 9 billion years (lifetime of the galaxy) is $(1.1)^{9,000/40} \sim 2 \times 10^9$. Such a large number of 'infected' planets illustrates that Darwinian evolution, involving horizontal gene transfers, must operate not only on the Earth or within the confines of the solar system but on a truly galactic scale. Life throughout the galaxy on this picture would constitute a single connected biosphere.

## 4. Cosmological Options

How many events of origination of life has there been in the 13.7-By history of the Universe? Can life spread from a single origin in one galaxy to infect the entire universe?

Escape from a galaxy of life-bearing particles, bolides or comets is in principle possible. The value of the escape speed from a galaxy such as ours cannot be determined precisely, but it must surely exceed the orbital velocity of stars in the outer spiral arms, ~250 km/s. Assuming that particles can reach escape speed, say 1,000 km/s (dust could be accelerated by radiation pressure, comets and bolides by gravitational encounters), the distance traversed in the lifetime of our galaxy in $\sim 10^{10}$ years is ~10 Mpc. Diffusion of life over much greater distances will be severely limited by horizon constraints within an expanding universe. Thus, life from a single origin in one galaxy might be thought to be contained within a cluster of galaxies stretching over several Mpc but not beyond.

In standard ΛCDM Big Bang cosmologies, the problem will not be alleviated by going to an earlier epoch of the Universe. In such cosmological models, the first life and its panspermic dispersal can begin only after star formation gets under way and supernovae produce and disperse the heavy elements for life. This happened about 1.3 Gy after the Big Bang according to conformist cosmology, when the universe was about 15% of its present size, and intergalactic distances were also scaled down by the same factor. Transport times, survival constraints and horizon restrictions will still inhibit full-scale cosmological panspermia from a single origin.

The only way to avoid a horizon-limited spread of life would be to adopt a non-conformist cosmological model such as that of Gibson and Schild (2009), which involves the creation of Earth-mass clouds at the plasma recombination stage 300,000 years after the Big Bang. In this model, these cloud fragments condense into frozen planets which are then polluted with heavy elements from the first generation of supernovae. Gibson, Schild and the present author have shown that Earth-mass planets with iron cores, overlain with oceans rich in organic materials and weighed down with extensive H-He atmospheres, provide optimal conditions for a first origin of life. The ocean temperature is close to 647 K, the critical temperature of water, and organic synthesis under these ultra-high-pressure, high-temperature conditions is greatly speeded up (Gibson et al., 2010). The volume of a typical such ocean on a primordial planet is $\sim 10^{10}$ km$^3$, and with $10^{80}$ such planets, a gigantic cosmological 'soup' with a total volume of $10^{90}$ km$^3$ is available for the origin of life. The planetary bodies at this stage are separated by only some tens of AU and with collisional connections established between them, one could imagine a gigantic connected primordial soup in action. These optimal conditions would prevail for 10 million years and cannot be remotely reproduced at any later cosmological epoch.

For judging efficiency in accomplishing an origin of life, the comparison to be made is between the volume of $10^{90}$ km$^3$ for the total set of primordial planets and ~1–10 km$^3$ for all the hydrothermal vents in the Earth's oceans. A factor of nearly of $10^{90}$ in probability is thus gained, compared with the factor of $10^{24}$ we found earlier in going from Earth to comets in the galaxy. The creation of life at one locality in this extremely dense, collision-dominated system in the early universe would lead to the fertilisation of other habitats and the spread of life across the entire primordial universe within a few million years. After some 10 million years, however, ocean temperatures drop with the general cooling down of the Universe, and the primordial planets laden with microbes will freeze. The cosmological legacy of life will thereafter be carried within these frozen primordial planets, which Gibson and Schild (2009) identify as the baryonic dark matter of the Universe. Mergers and disruptions of such life-laden 'giant comets' are associated with every event of star or planet formation occurring in the disc of the galaxy (Figs. 3 and 4).

To conclude this section, it should be stressed that there still remain unsolved problems concerning the origin of prescriptive information (protogene emergence) upon which all life depends. The cybernetic nature of life demands that the processes leading to its emergence be steered and regulated at every metabolic turn. Existing models of abiogenesis are incapable of accounting for such processes, even with the advantage of an enhancement of probability by the factor $10^{90}$ that was discussed above. Abiogenesis modellers still have a great deal of work to do before the ultimate origin of life by naturalistic processes can be fully understood.

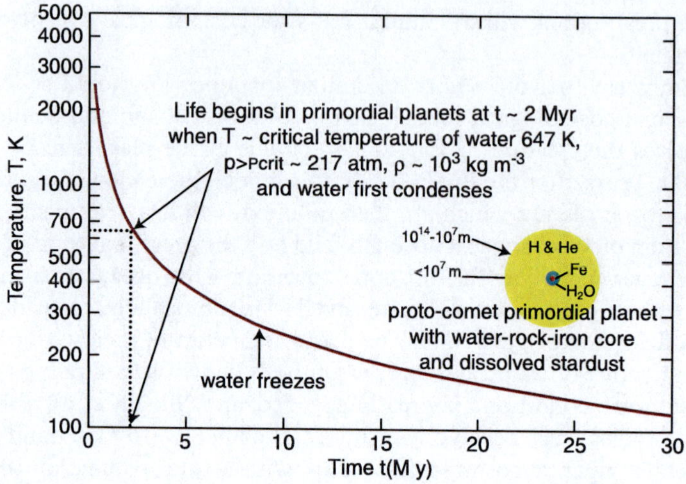


**Figure 3.** Temperature of the Universe as a function of time in million years after the Big Bang, showing stage at which life might optimally arise according to Gibson, Schild, Wickramasinghe model.

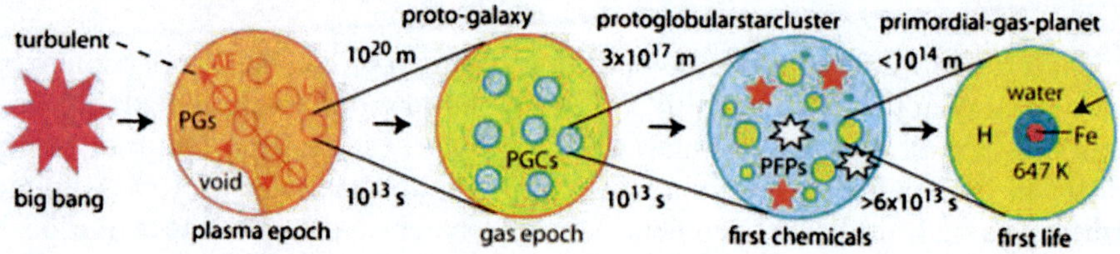


**Figure 4.** Sequence of steps showing the progression to warm primordial planets in which life originates. Courtesy: Carl Gibson.

## 5. Extremophiles and Survival of Cosmic Bacteria

The discovery of bacteria and archaea occupying the harshest environments on Earth provides indirect support for panspermia. Viable transfers of microbial life from one cosmic habitat to another require endurance of high and low temperatures as well as exposure to low fluxes of ionising radiation delivered over astronomical timescales. Survival at some level is needed if panspermia is to work. For microbes deep frozen within kilometre-sized frozen bodies, such as comets or planets, survival for unlimited cosmological timescales would be assured. Individual clumps of microorganisms and microbes embedded in small bolides will, however, have some limits to survivability, but even for naked clumps of microbes survival to some degree during interstellar/interplanetary transit is guaranteed.

A clump of bacteria surrounded with biofilm would inevitably develop a thin outer skin of carbonised material that would provide nearly complete protection

**Table 1.** Survival of spores of *B. subtilis.*

| Mission | Duration of vacuum exposure (days) | Survival fraction at end of exposure (%) |
|---|---|---|
| Spacelab 1 | 10 | 70 |
| EURECA | 327 | 25 |
| LDEF | 2,107 | $1.4 \pm 0.8$ |

Data from Horneck (1993) and Horneck et al. (1995).

from ultraviolet light (Wickramasinghe and Wickramasinghe, 2008). Exposure to ionising radiation such as cosmic rays poses a more serious potential threat, however. The doses of ionising radiation that limit viability appear to be critically dependent on bacterial species. Table 1 shows the percentage of spores of *Bacillus subtilis* surviving after naked exposure to space vacuum in the solar vicinity (Mileikowsky et al., 2000).

Within the solar system, the radiation doses received by an outward-moving bacterium depend critically on the phase of solar activity, being highest at times near the peak of the solar cycle. The average radiation dose received at 1 AU over a timescale comparable to the 11-year period of the solar cycle is less than 10 Mrad, and so viability will not be expected to be drastically reduced. The LDEF experiment shows a reduction to 1% for the extreme case of an unprotected, unfrozen (at ~1 AU) bacterial culture. Freeze-dried clumps of bacteria or bacterial spores are expected to fare significantly better. Far from a star, in a typical interstellar location, a very much lower flux of ionising radiation would be delivered over astronomical timescales, and the total dose delivered is likely to be less relevant than the incident flux.

The estimated doses of ionising radiation intercepted by bacterial clumps in an unshielded interstellar cloud and are in the range 5–15 Mrad in $10^5$ years. Experiments done so far to determine viability as a function of total dose are most likely to be irrelevant for our purpose. Laboratory studies have invariably used very high fluxes of ionising radiation delivered in short pulses, at room temperature in the presence of air. Extrapolation from these studies to the astronomical case involves many uncertainties. In particular, most damage resulting from free-radical formation will be eliminated in an anaerobic environment, and low-flux long-time intervals may not equate to short pulses of high-flux radiation in the laboratory case.

Citing short pulse – high-flux experiments, Horneck et al. (2002) point out that for spores embedded in a meteorite, the timescale for survival to a $10^{-4}$% level (due to primary and secondary CR effects) at a depth of 2–3 m is ~25 My. The same surviving fraction is obtained for shielding at a depth of 1 m after 1 My, and for a naked spore the estimated timescale is ~0.6 My. Over timescales of $3 \times 0.6$ My, the survival fraction is $10^{-18}$. A clump of bacterial spores travelling at 10 km/s would traverse a distance of 18 pc during this time, comfortably less than the mean distance between cometary/planetary systems.

There are strong indications that the survival rates quoted above are an underestimate. They are based on experiments conducted at atmospheric pressure with normal levels of humidity (Lindahl, 1993). There is evidence that bacterial endospores are resistant to inactivation by free-radical damage and chemical processes because their cytoplasm is partially mineralised and their DNA is stabilised (Nicholson et al., 2000). Most significantly, it has been found that bacterial spores (genus Bacillus) in the guts of a bee preserved in amber for 25–40 My could be revitalised (Cano and Borucki, 1995). Similarly, bacterial spores in 250-My-old salt crystals have also been revitalised (Vreeland et al., 2000). In both these studies, the most stringent sterilisation techniques were used to avoid contamination of the samples, and the authors are confident that they have revived bacterial spores of great ages. With a natural background radioactivity (of rocks) of ~1 rad per year, we have here evidence of survival with doses of 25–250 Mrad of ionising radiation.

More recently, Bidle et al. (2007) have examined the most ancient terrestrial ice deposits in the Antarctic and found evidence of viable bacteria/bacterial genes in 8-My-old ice. They estimate an average half-life of bacteria under such conditions of ~1.1 My. The importance of these results in relation to panspermia cannot be denied. To avoid this conclusion, some scientists have used phylogenetic arguments. They point out that the revived organisms bear a close relationship to contemporary species, and so argue that they must be contaminants. However, panspermia theories permit re-introductions of the same organism (stored in comets) separated by millions of years, so no evolutionary divergence is required. Recent phylogenetic studies have indeed shown that 'horizontal' gene transfers across branches of Carl Woese's tree of life befuddle attempts to use the 'tree' as guiding principle or as an evolutionary clock (Woese and Fox, 1977). The phylogenetic modernity of the microorganisms found, as well as their survival over millions of years, consistently point to an extraterrestrial origin.

## 6. Interstellar Dust

### 6.1. INTERSTELLAR EXTINCTION

Identifying the composition of interstellar dust in clouds has been a high priority for astronomical research since the early 1930s (see Wickramasinghe, 1967). The dust absorbs and scatters starlight, causing extinction of the light from stars, and re-emits the absorbed radiation in the infrared. An important clue relating to dust composition follows from studies of extinction of starlight. The total amount of the dust has to be as large as it can be if nearly all the available carbon and oxygen is condensed into grains. The paradigm in the 1960s that the dust was largely comprised of water-ice was quickly overturned with the advent of infrared observations showing absorptions due to CH, OH, C–O–C linkages consistent with organic polymers. The best agreement for a range of astronomical spectra

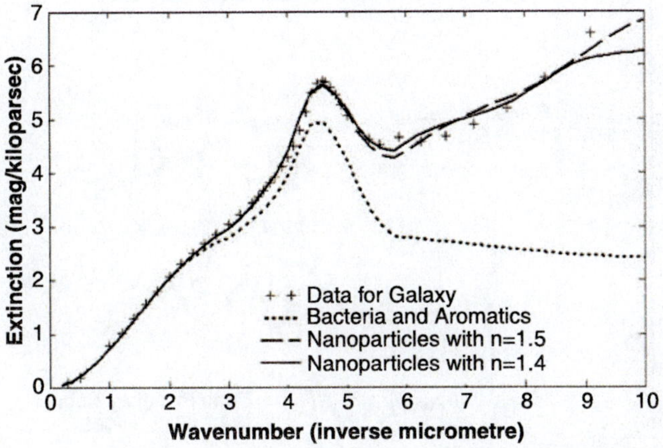


**Figure 5.** Agreement between interstellar extinction (+) and biological models. Mixtures of hollow bacterial grains, biological aromatic molecules and nanobacteria provide excellent fits to the astronomical data. The 2175A hump in the extinction is caused by biological aromatic molecules (See J. Wickramasinghe et al., 2009).

embracing a wide wavelength interval turned out to be material that is indistinguishable from freeze-dried bacteria, and the best overall agreement over the entire profile of interstellar extinction was a mixture of desiccated bacteria, nanobacteria, including biologically derived aromatic molecules as seen in Fig. 5.

## 6.2. THE IDENTIFICATION OF PAH AND BIOLOGICAL AROMATICS

The distribution of unidentified infrared bands (UIBs) between 3.3 and 22 μm is almost identical in their wavelengths in very different emission sources, more or less irrespective of the ambient conditions. Most recent studies of UIBs for a large number of galactic and extragalactic sources have been obtained using the Spitzer Space Telescope (Smith et al., 2007). Figure 6a shows the spectra of a young planetary nebula and a proto-planetary nebula showing characteristic UIBs. Figure 6b shows the average Spitzer spectrum for 13 starburst galaxies displaying the same set of UIBs.

Whilst PAHs (polyaromatic hydrocarbons), presumed to form inorganically, are the favoured model for the UIBs, no really satisfactory agreement with available astronomical data has thus far been shown possible for abiotic PAHs (Hoyle and Wickramasinghe, 1991). This is a particularly serious problem if we require the UIB emitters and the 2175A absorbers to be the same. The latter requirement is of course necessary because it is the starlight energy absorbed in the ultraviolet band that is being re-emitted as UIBs in the infrared. The fact that biological aromatic molecules could act in such a dual role was noticed over two

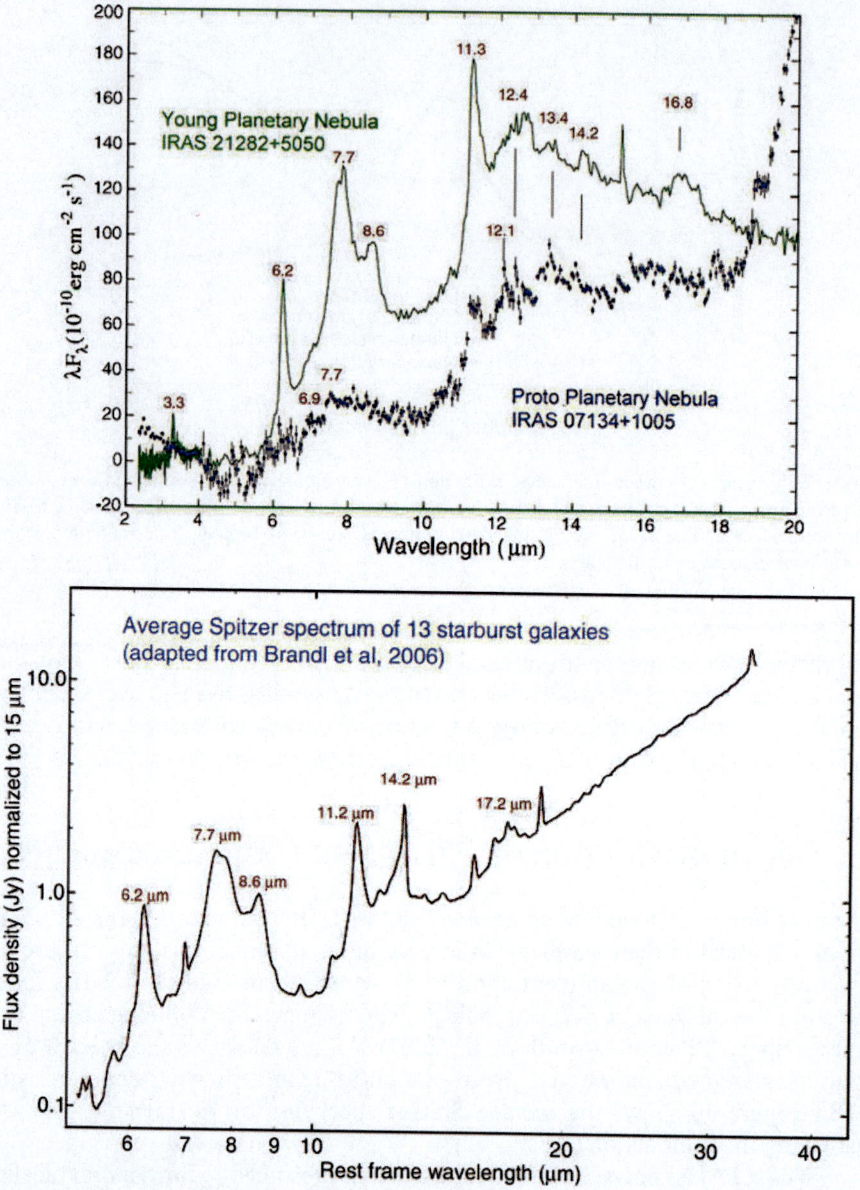


**Figure 6.** **(a)** *ISO SWS01* spectra of the young planetary nebula IRAS 21282+5050 and the proto-planetary nebula IRAS 07134+1005 (Adapted from Kwok, 2009). **(b)** Averaged *Spitzer* IRS spectrum of 13 starburst galaxies (Adapted from Brandl et al., 2006).

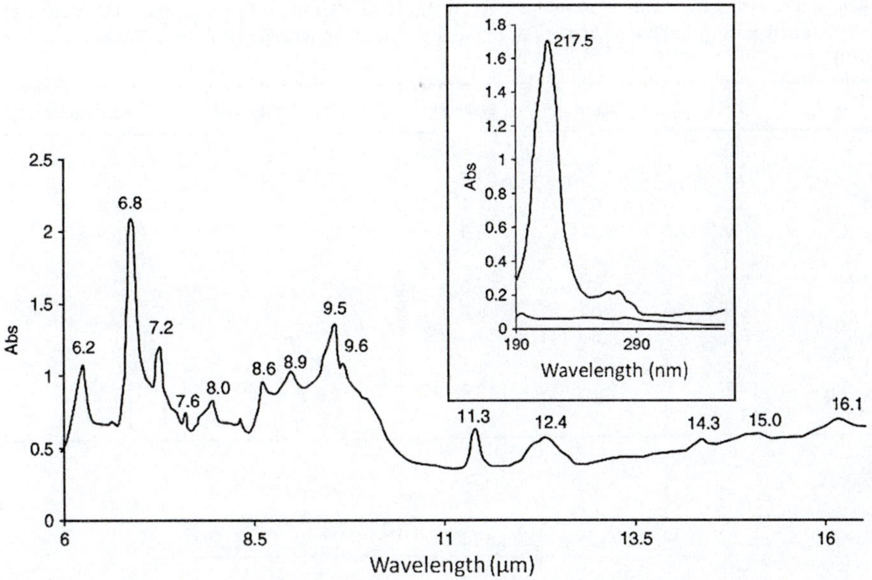


**Figure 7.** IR spectrum of algae in KBr. *Inset* is the spectrum of algae and degradation products in cyclohexane.

decades ago (Wickramasinghe et al., 1989, 1990). Recent laboratory studies by Rauf and the present author have provided further evidence favouring biology as seen for the case of algae in Fig. 7 (Rauf and Wickramasinghe, 2010).

Here we see a range of UIBs and the 2175A extinction feature arising from the same naturally occurring PAH system. A tabulation of the principal infrared bands for biological material for comparison with astronomical data is set out in Table 2 (Rauf and Wickramasinghe, 2010; Gezari et al., 1993).

Figure 8 shows a particularly interesting case where the spectrum of the Antennae galaxies at some 63 million light years distance has been observed to have infrared emissions that match the laboratory spectrum of anthracite (Guillois et al., 1999). Anthracite being a product of biological (bacterial) degradation is again indicative of biological particles, in this case at great distances from the Milky Way. Caltaldo et al. (2002) have shown that aromatic distillates of petroleum, another biological product, exhibit correspondences with the astronomical diffuse infrared bands (UIBs) as well as the λ2175A ultraviolet extinction feature.

## 6.3. OTHER SPECTRAL FEATURES

The most dramatic recent discovery relating to astronomical aromatic molecules is a conspicuous 2175A band in the lens galaxy of the gravitational lens SBS0909+532 which has a red shift of z=0.83 (Motta et al., 2002). The extinction

**Table 2.** Distribution of two astronomical observations (UIBs and proto-planetary nebulae (PPNe)) and major IR absorption bands in laboratory models of terrestrial origin (Rauf and Wickramasinghe, 2010).

| UIBs | PPNe | Algae | Grasses | Bituminous coal | Anthracite coal |
|---|---|---|---|---|---|
| 3.3 | 3.3 | 3.3 | – | 3.3 | 3.3 |
| – | 3.4 | 3.4 | 3.4 | 3.4 | 3.4 |
| 6.2 | 6.2 | 6.0 | 6.1 | 6.2 | 6.2 |
| – | 6.9 | 6.9 | 6.9 | 6.9 | 6.9 |
| – | 7.2 | 7.2 | 7.2 | 7.2 | 7.2 |
| 7.7 | 7.7 | – | 7.6 | – | 7.7 |
| – | 8.0 | 8.0 | 8.0 | – | – |
| 8.6 | 8.6 | 8.6 | – | – | – |
| 11.3 | 11.3 | 11.3 | 11.1 | 11.5 | 11.3 |
| – | 12.2 | 12.1 | 12.05 | 12.3 | 12.5 |
| – | 13.3 | – | – | – | 13.4 |

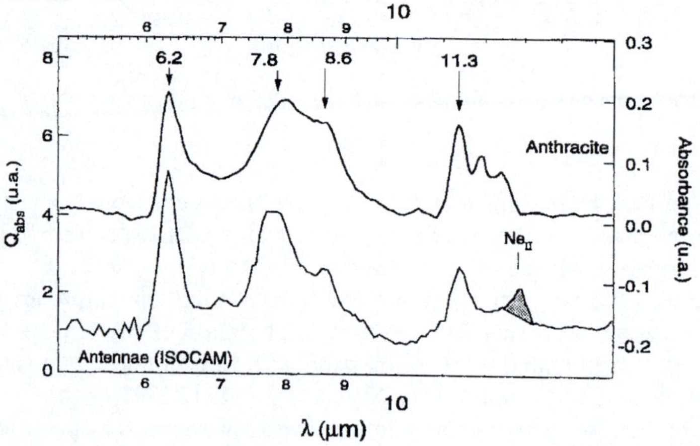


**Figure 8.** ISO spectrum of the Antenna galaxies compared with a spectrum of anthracite (Guillois et al., 1999).

curve for this galaxy is reproduced in Fig. 9, with the dashed curve representing a scattering background attributed to hollow bacterial grains.

The excess absorption over and above a pure scattering background (dashed curve) is normalised to unity at the peak and plotted as the points with error bars in Fig. 10. The curve in this figure shows the absorption of biological aromatic molecules similarly normalised (Wickramasinghe et al., 1989).

The match between the data points and the theoretical curve shows that biological aromatics are a viable explanation of an absorption feature in a galaxy located at a red shift of $z=0.83$. This was an epoch when the universe was nearly half its present radius.

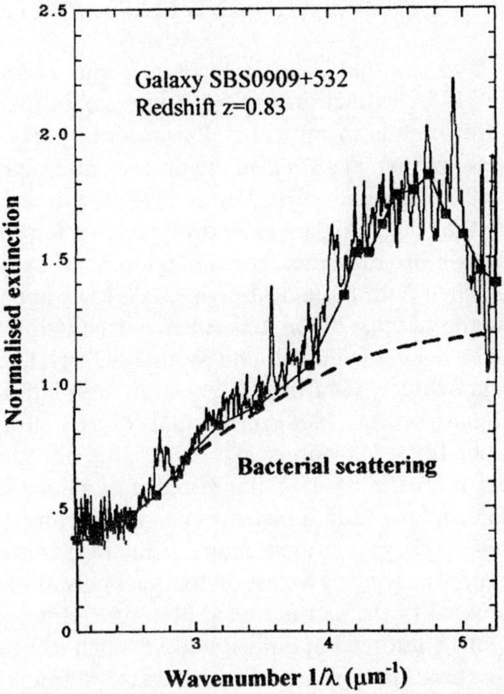


**Figure 9.** The *continuous line* is the extinction curve for the gravitational lens galaxy SBS0909+532 (Motta et al., 2002). The *dashed curve* is the scattering background attributed to bacterial particles.

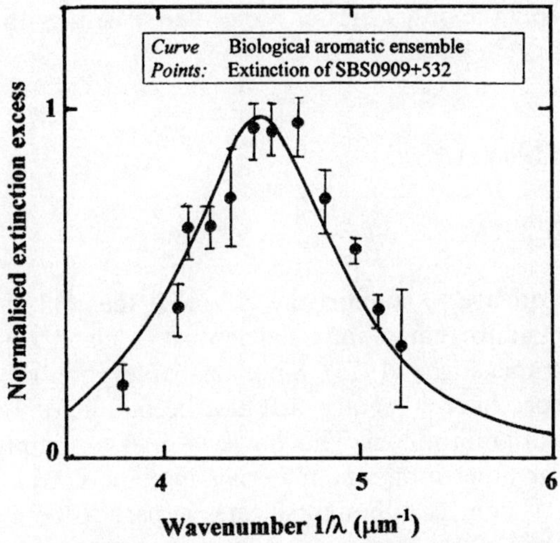


**Figure 10.** The *curve* is the normalised absorption coefficient of an ensemble of 115 biological aromatic molecules. The *points* are observations for the galaxy SBS0909+532 due to Motta et al. (2002), representing the total extinction shown in Fig. 9, from which an underlying bacterial scattering component has been subtracted.

### 6.4. FLUORESCENCE PROPERTIES AND MASS SPECTROSCOPY

In Sects. 6.2 and 6.3, we saw that biologically derived ensembles of aromatics can both UIBs and the 2175A extinction feature in interstellar material. The diffuse interstellar absorption bands in optical stellar spectra, particularly the 4430A feature, also have possible explanations on the basis of molecules such as porphyrins (Hoyle and Wickramasinghe, 1979; Johnson, 1971).

We now show that the so-called extended red emission of interstellar dust (ERE), showing up as a broad fluorescence emission band over the range 5,000–7,500 A, can be matched with biological pigments (Hoyle and Wickramasinghe, 1996; Wickramasinghe et al., 2002). Extended red emission (ERE) has been observed in planetary nebulae (Furton and Witt, 1992), HII regions (Sivan and Perrin, 1993), dark nebulae (Mattila, 1979) and high-latitude cirrus clouds (Szomouru and Guhathakurta, 1998) in the Galaxy as well as in extragalactic systems (Perrin et al., 1995; Darbon et al., 1998). This phenomenon has a self-consistent explanation on the basis of fluorescence of biological chromophores (pigments), e.g. chloroplasts and phytochrome. Competing models based on emission by compact PAH systems are not as satisfactory, as is evident in Fig. 11. Hexa-peri-benzocoronene is one of a class of compact polyaromatic hydrocarbons that have been discussed in the astronomical literature. However, the width and central wavelength of its fluorescent emission leave much to be desired.

Direct mass spectroscopy of interstellar dust in the Stardust mission (Krueger et al., 2004) has also shown the presence of cross-linked hetero-aromatic structures in the degradation products of impacting interstellar grains. Evidence of fragments with atomic mass unit AMU >2,000 consistent with pyrrole, furan substructures and quinones were found (Fig. 12). The fractured components of cell walls are arguably the only types of molecular structures that would survive impacts at speeds of ~30 km/s.

## 7. Comet Dust and Biomaterial

### 7.1. COMET SPECTRA

There is growing evidence to support the view that the composition of dust in comets is very similar to that of interstellar dust, a claim originally made by Vanysek and Wickramasinghe (1975). Since the exploration of comet Halley in 1986, the infrared spectra of cometary dust have been found to be consistent with an organic-biological grain model. This however does not imply the absence of water-ice, silicates or other inorganic materials in comets. Water-ice is indeed a major component of comets although silicates appear to be a relatively minor component.

Spectral features near 19, 24, 28 and 34 μm observed by ISO have been attributed to hydrated silicates in several proto-planetary discs, and in comet

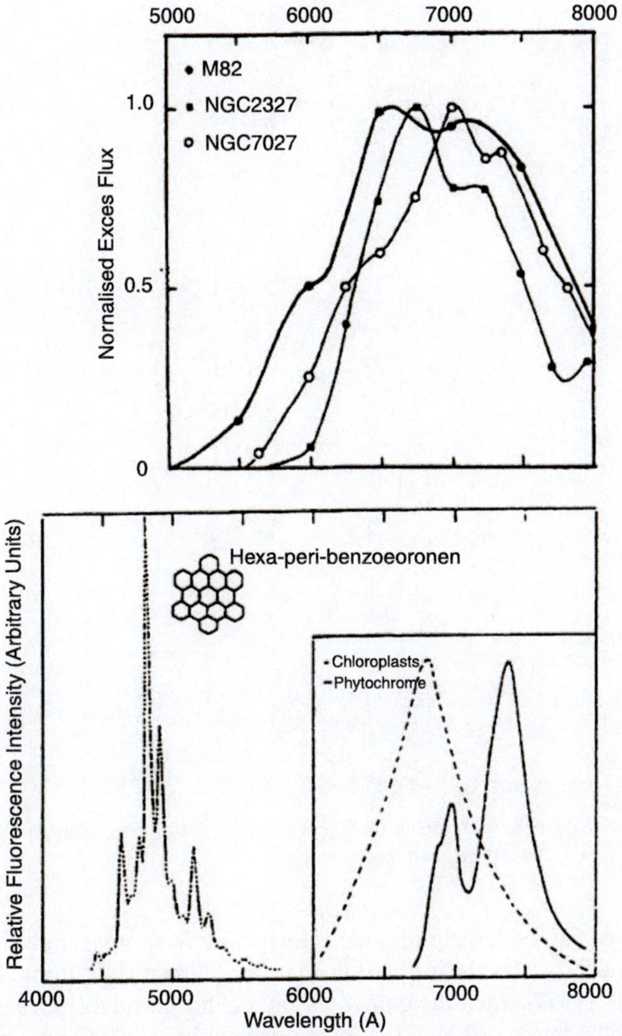


**Figure 11.** The points in the *top panel* show normalised excess flux over scattering continua from data of Furton and Witt (1992) and Perrin et al. (1995). The *bottom right panel* shows relative fluorescence intensity of spinach chloroplasts at a temperature of 77 K. The *dashed curve* is the relative fluorescence spectrum of phytochrome. The *bottom left panel* is the fluorescence spectrum of hexa-peri-benzocoronene.

Hale-Bopp (Crovisier et al., 1997; Waelkens and Waters, 1997). The uniqueness of some of these assignments is still in doubt, but even on the basis of a silicate identification of the principal infrared bands, such material could make up only some few percent of the mass of the dust, the rest being Trapezium-type grains – i.e. grains such as exist in the Trapezium nebula – which also, according to our

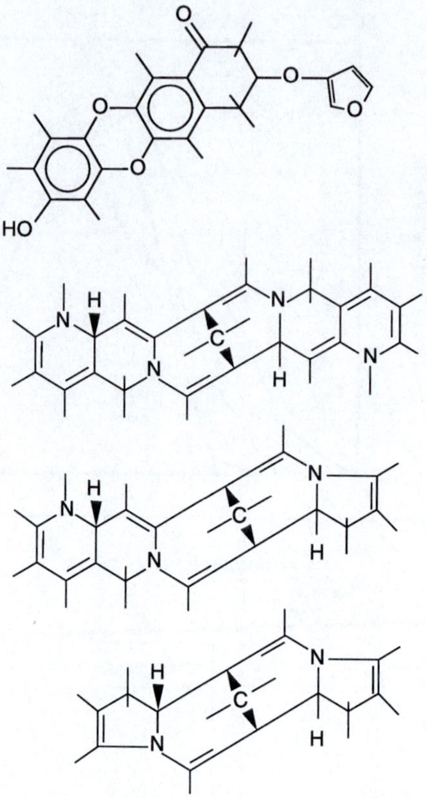

**Figure 12.** Functional groups in the break-up fragments of impacting interstellar dust grains, inferred by Krueger and Kissel (2000) from mass spectroscopy.

interpretations may be largely organic (Hoyle and Wickramasinghe, 1991). This appears to be the case for the infrared flux curve of comet Hale-Bopp, obtained by Crovisier et al. (1997) when the comet was at a heliocentric distance of 2.9 AU.

The dashed curve in Fig. 13 is for a mixed culture of microorganisms containing about 20% by mass in the form of diatoms. Olivine dust, which has a much higher mass absorption coefficient than biomaterial, makes up only 10% of the total mass in this model (Wickramasinghe and Hoyle, 1999). It is interesting to note that the far-infrared emission spectrum of the proto-planetary disc around the star HD100546 has a strikingly similar spectrum to that of comet Hale-Bopp.

## 7.2. DIRECT ANALYSIS OF COMET DUST

The first in situ analysis of comet dust was carried out in 1986 for dust particles from comet Halley as they impacted a mass spectrometer on board the Giotto spacecraft at speeds of ~30 km/s (Kissel and Krueger, 1987). By analysing the

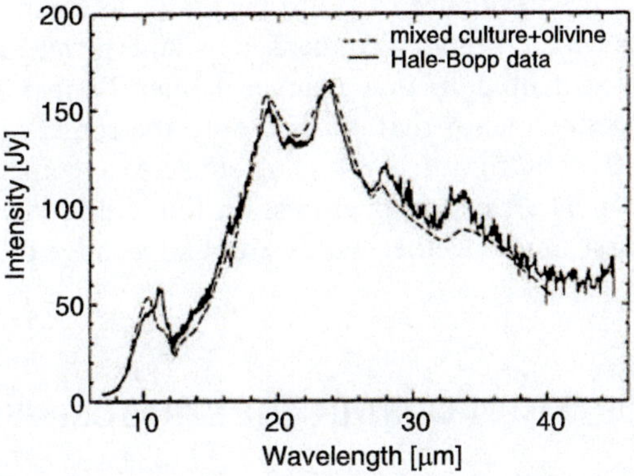


**Figure 13.** The infrared spectrum of comet Hale-Bopp obtained by Crovisier et al. (1997) when the comet was at a heliocentric distance of 2.9 AU (*jagged curve*). The *dashed curve* is for a mixed culture of microorganisms and olivine dust which comprised 10% of the total mass. The diatoms (diatomaceous silica) make up 20% by mass.

masses of break-up products, a large fraction of the grains were found to be organic ('CHON' particles) with compositions that included long-chain hydrocarbons and nitrogenated polyaromatic molecules. These molecular structures are fully consistent with the break-up of bacterial cells, although they do not of course prove such a composition unequivocally.

The claim originally made that bacterial particles are ruled out because the break-up material did not show evidence of the biologically important element phosphorus is flawed. Molecular ion-mass spectra of Kissel and Krueger (1987) could be interpreted as plausible combinations of P with other elements. The break-up of phosphorus groups (as in DNA) could lead to possible mass peaks corresponding to $PO_3^+$ (79), $PO_2^+$ (63) or $PO^+$ (47) rather than $P^+$ (31). That such evidence does indeed exist was shown in an analysis by one of the present authors (Wickramasinghe, 1993). The claim that mass peaks corresponding to two other biologically significant elements Na and K are too low for biology is also open to challenge. Although freeze-dried cultures of vegetative bacteria may be ~100 times richer in Na and K, this is not the case for nutrient-starved bacteria nor for spores (Wickramasinghe, 1993).

The Stardust mission to comet Wild 2 was launched in 1999, and the experiments it carried were planned over the preceding decade at a time when cometary microorganisms were dismissed as an outright impossibility. The protocol used for particle collection would not have permitted the survival of intact cells. In January 2004, the Stardust captured dust from the tail of comet Wild 2 in blocks of aerogel. Each particle impacted the aerogel at an initial speed of ~6 km/s, excavating a track along which break-up debris was deposited. Along the tracks, a wide range of organic molecules including the amino acid glycine was discovered. The detection of hetero-aromatic molecules rich in N and O could be a telltale sign of degraded biomaterial, biology being particularly rich in such structures.

The aerogel also showed evidence of pre-solar grains including a class of highly refractory minerals which probably condensed in supernovae explosions.

The Deep Impact mission that reached comet Tempel 1 on 4 July 2004 involved a high-speed impactor that crashed onto the comet's surface. The crust of the comet was thereby ruptured releasing subsurface cometary material that was analysed in situ. The Deep Impact mission found evidence of clay particles in comets for the first time, together with water and a range of organics (see, for instance, Fig. 1).

## 7.3. CAPTURE OF COMET DUST IN THE STRATOSPHERE

In both the Giotto and Stardust missions, cometary particles were examined and studied after they had endured high-speed impacts. Survival of fragile organic structures (e.g. bacteria) would have been virtually impossible in such conditions.

An obvious place to find fragile particles from comets is the Earth's upper atmosphere. Cometary meteoroids and dust particles are known to enter the atmosphere at a more or less steady rate. Although much of the incoming material burns up as meteors, a significant fraction survives entry. Micron-sized organic grains, arriving as clumps and dispersing in the high stratosphere, would be slowed down gently and would not be destructively heated. The Earth's atmosphere could thus serve as an ideal collector of organic cometary dust.

Techniques for stratospheric collection of cometary dust must of necessity involve procedures for either sifting out terrestrial contamination or for excluding contamination altogether. Such stratospheric dust collections have been carried out from as early as the mid-1960s (Gregory and Monteith, 1967). Balloons and rockets reaching heights well above 50 km were deployed and consistently brought back algae, bacteria and bacterial spores. Although some of the microorganisms thus collected were claimed to exhibit unusual properties, such as pigmentation and radiation resistance, their possible extraterrestrial origin remained in serious doubt at this stage. No DNA sequencing procedure was available at the time to ascertain any significant deviations there might have been from terrestrial species. Moreover, the collection and laboratory techniques in the 1960s left open a high chance of contamination.

In 2001, the Indian Space Research Organisation (ISRO) launched a balloon into the stratosphere carrying devices to collect stratospheric air under aseptic conditions (Harris et al., 2002). The procedure involved the use of cryogenically cooled stainless steel cylinders which were evacuated and fitted with valves that could be opened when they reached a predetermined altitude. Large quantities of stratospheric air at 41 km were thus collected, and the cylinders were brought back for analysis.

The ultra-high-pressure stratospheric air contained within the cylinders was carefully released and passed through a system including millipore membrane

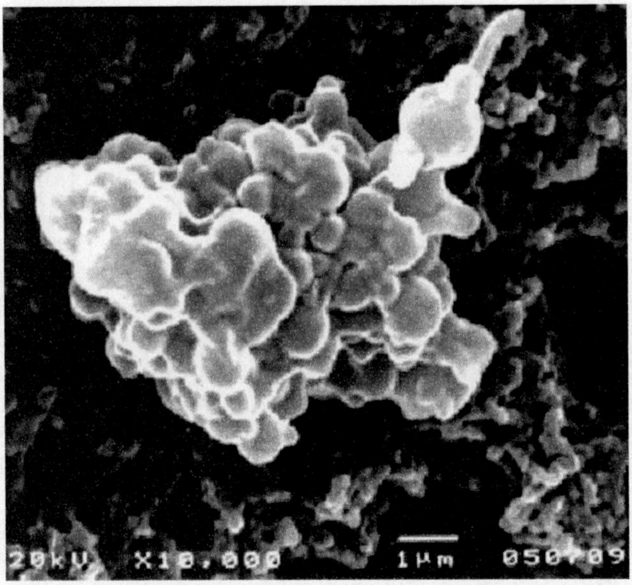


**Figure 14.** A carbonaceous stratospheric particle from 41 km resembling a clump cocci and a rod bacterium.

filters. Upon such filters, stratospheric aerosols were collected, extreme care being taken at every stage to avoid contamination. The particles that were collected fell into two broad classes: (a) mineral grain aggregates, very similar to Brownlee particles, but somewhat smaller; (b) fluffy carbonaceous aggregates resembling clumps of bacteria (see Fig. 14). Typical dimensions were about 10 μm.

The cometary origin of such particles is very strongly indicated, the altitude of 41 km being too high for lofting 10-μm-sized clumps of solid material from the Earth's surface. In addition to structures such as shown in Fig. 14, which can be tentatively identified with degraded bacteria, the stratospheric samples also revealed evidence of similar-sized bacterial clumps that could not be cultured but were nevertheless detected unambiguously by the use of a fluorescent dye. The uptake of the dye revealed the presence of living cells in the clumps.

In a separate series of experiments, a few microbial species were also cultured from stratospheric air samples by Wainwright et al. (2003), and there is tentative evidence that these may also have come from comets. A later balloon flight in 2008 collected more stratospheric material and analysis by S. Shivaji and his colleagues (Shivaji et al., 2009) yield cultures of 3 hitherto unknown microbial species which were highly resistant to ultraviolet light. Although a space origin is tempting to infer, contamination cannot be ruled out entirely and the jury is out on all these matters.

## 8. Microfossils in Meteorites

The topic of microfossils in carbonaceous chondrites has sparked bitter controversy in ever since it was first suggested in the mid-1960s (Claus et al., 1963). Since carbonaceous chondrites are generally believed to be derived from comets, the discovery of fossilised life forms in comets would provide strong prima facie evidence in support of the theory of life in comets and cometary panspermia. However, some contamination was discovered in the original samples, and claims that all the micro-structures (organised elements) discovered in meteorites were artefacts or contaminants led to a general rejection of the microfossil identifications in the 1960s.

The situation remained uncertain until early in 1980 when H.D. Pflug found a similar profusion of 'organised elements' in ultra-thin sections prepared from the Murchison meteorite, a carbonaceous chondrite that fell in Australia on 28 September 1969 (Pflug, 1984). The contaminant free experimental methods adopted by Pflug were beyond reproach. Thin sections of the meteorite were placed on membrane filters, and hydrofluoric acid was used to dissolve out the bulk of the minerals present and the residue examined in an electron microscope.

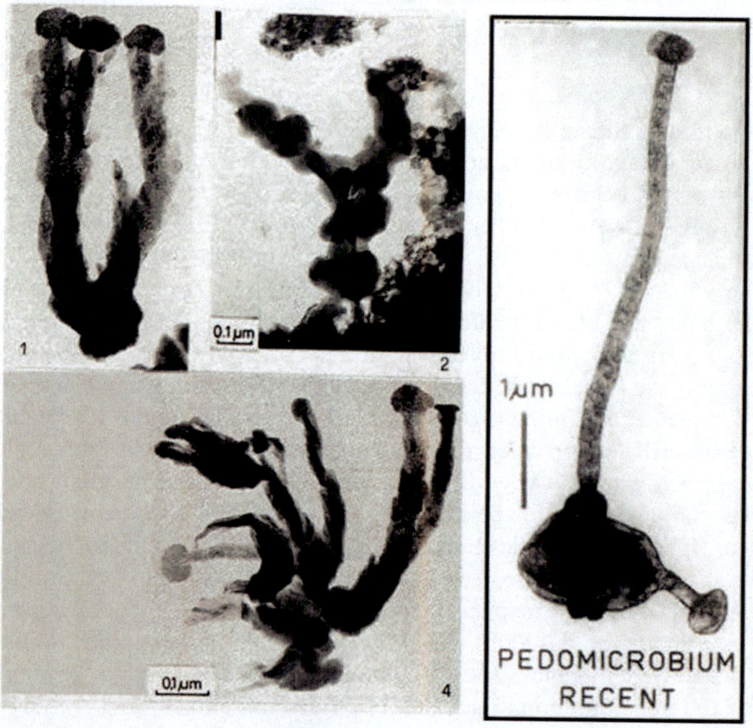


**Figure 15.** Microfossils in the Murchison meteorite discovered by H.D. Pflug (1984).

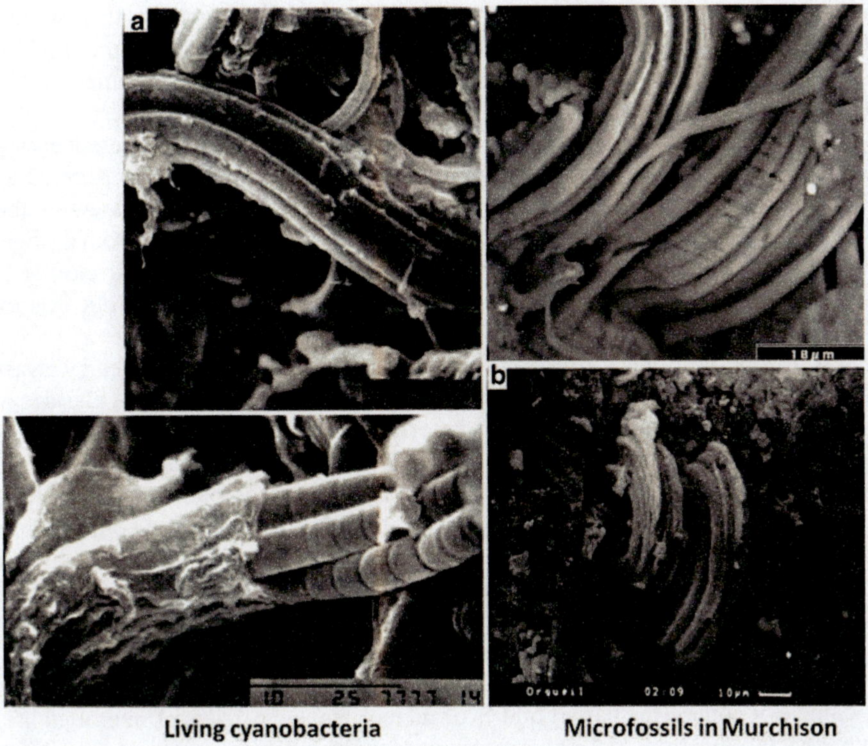


**Figure 16.** Structures in the Murchison meteorite compared with living cyanobacteria (Hoover, 2005).

Figure 15 shows examples of Pflug's findings. Not only were these morphologies strikingly characteristic of particular types of fossil microbes, but laser ion probe analysis yielded compositions of these structures that were also consistent with life. These studies made it very difficult to reject the fossil identification. More recent work by Richard Hoover (2005) and his team now leaves little room for any other interpretation of these structures than that they are microbial fossils (Fig. 16).

## 9. Concluding Remarks

At the end of a century of scientific exploration of the question life's origins from Arrhenius and Oparin to the present day, the problem remains essentially unsolved. If honesty prevails, one has to admit that the precise steps that may have led from organic chemicals to the simplest self-replicating microorganisms are unknown. The transition from non-life to life involves improbabilities of an astronomical or even super-astronomical order. To assert without proof that this must have taken place on our minuscule speck of a planet, Earth, against all the odds, is tantamount to reviving a pre-Copernican philosophy. We argued in this chapter that a first origin of life most probably involved the resources of a large fraction of the observable universe, possibly within the warm interiors of comets or icy planets very early in the history of the universe.

The profusion of biogenically relevant organic molecules that exist over a large part of the observable universe (Kwok, 2009) is incorrectly interpreted as arising from 'prebiotic processes'. The correct interpretation, in the view of the writer, is that panspermic processes continue to distribute living cells from a single cosmic origin of life. What is observed astronomically in interstellar clouds in our galaxy and in external galaxies are molecules representing the detritus of living cells.

Carbonaceous meteorites, such as the Murchison meteorite, also carry a vast array of exceedingly complex organic molecules that could represent the decay products of life. Because amino acids in terrestrial life are left-handed and sugars right-handed, the chirality of meteoritic material would in principle differentiate biotic and abiotic origins. However, chemical changes accompanying the fossilisation tend to switch the handedness and racemise biomaterial. The confirmation of achirality in uncontaminated meteoritic material, with 10–20% left-handed excess (Pizzarello, 2004), presents a serious difficulty for advocates of an entirely abiotic origin of these molecules. The amazingly rich diversity of organics identified in the Murchison meteorite (Schmitt-Kopplin et al., 2010) likewise comes as no surprise in the context of cometary panspermia and is difficult to explain otherwise. If cometary bodies (the parent bodies of meteorites) are carriers of microbial life, one expects a diversity of organic molecules far richer than that which prevails on Earth, generated more via biological than prebiotic processes (Wickramasinghe, 2010). We conceive that extraterrestrial biology carried by comets with diverse initial complements and histories could represent a far greater variety than the minute subset adapted to terrestrial environments and surviving competitive evolution. Even if the question of achirality is ignored, the observed extent of diversity of molecules in the Murchison meteorite – in excess of highly chemically diverse petroleum processes (Schmitt-Kopplin et al., 2010) is improbable to arise from abiotic processes.

To sum up, a wide range of data from astronomy, geology, meteorite studies and microbiology all point to a cosmic origin of life and provide strong support for panspermia. The emerging paradigm is that the origin of life was a unique cosmological event that was accomplished within planetary interiors or within interiors of cometary-type bodies within a few million years of the Big Bang (Gibson et al., 2010).

However, our understanding of the processes involved in this first event of origination still remains far from complete and might well remain so for generations to come (Abel, 2006). What is certain is that a multiplicity of origination events can be excluded from the domain of plausible hypotheses. Comets remain the prime candidates for being the carriers and amplifiers of cosmic life. The Earth was seeded with such life from which we are descended some 4 billion years ago and continues to be seeded even to the present day. On this picture, self-similar life forms must exist abundantly throughout the entire universe. We are by no means alone. Our genetic ancestors are scattered as cosmic dust across the vast reaches of the cosmos.

## 10. References

Abel DL (2009) The universal plausibility metric (UPM) & principle (UPP). Theor Biol Med Model 6(1):27. Open access at http://www.tbiomed.com/content/6/1/27

Abel DL, Trevors JT (2006) Self-organization vs. Self-ordering events in life-origin models. Phys Life Rev 3:211

Arrhenius S (1903) The propagation of life in space. Die Umschau 7:481

Arrhenius S (1908) Worlds in the making. Harper, London

Bidle K, Lee S-H, Marchant DR, Falkowski PG (2007) Fossil genes and microbes in the oldest ice on earth. Proc Natl Acad Sci USA 104(33):13455

Bohler C, Nielson PE, Orgel LE (1995) Template switching between PNA and RNA. Nature 376: 578–581

Brandl BR et al (2006) The mid-infrared properties of starburst galaxies from Spitzer-IRS spectroscopy. Astrophys J 653:1129

Cairns-Smith AG (1966) The origin of life and the nature of the primitive gene. J Theor Biol 10:53

Cano RJ, Borucki M (1995) Revival and identification of bacterial spores in 25- to 40-million-year-old Dominican amber. Science 268:1060

Cataldo F, Keheyan Y, Heymann D (2002) A new model for the interpretation of the unidentified infrared bands (UIBS) of the diffuse interstellar medium and of the protoplanetary nebulae. Int J Astrobiol 1:79

Claus G, Nagy B, Europa DL (1963) Further observations on the properties of the "organized elements" in carbonaceous chondrites. Ann NY Acad Sci 108:580

Crick FHC, Orgel LE (1973) Directed panspermia. Icarus 19:341

Crovisier J, Leech K, Bockelée-Morvan D et al (1997) The spectrum of comet Hale-Bopp (C/1995 O1) observed with the infrared space observatory at 2.9 astronomical units from the Sun. Science 275:1904

Darbon S, Perrin J-M, Sivan J-P (1998) Extended red emission (ERE) detected in the 30 Doradus nebula. Astron Astrophys 333:264

Furton DG, Witt AN (1992) Extended red emission from dust in planetary nebulae. Astrophys J 386:587

Gezari DY, Schmitz M, Pitts PS et al (1993) Catalogue of infrared observations. NASA, Washington

Gibson CH, Schild RE (2009) Hydro-gravitational-dynamics of planets and dark energy. Appl Fluid Mech 2(2):35–41, arXiv:0808.3228

Gibson CH, Schild RE, Wickramasinghe NC (2010) Origin of life from primordial planets. Int J Astrobiol 10(2):83–96

Gregory PH, Monteith JL (eds) (1967) Airborne microbes – symposium of the society for general microbiology, vol 17. Cambridge University Press, Cambridge

Guillois O, Ledoux G, Nenner I et al (1999) Solid interstellar matter: the ISO Revolution. Les Houches, No.11, EDP Sciences, Les Ulis

Haldane JBS (1929) The origin of life. Chatto and Windys, London

Harris MJ, Wickramasinghe NC, Lloyd D et al (2002) Detection of living cells in stratospheric samples. Proc SPIE 4495:192

Hoover RB (2005) In: Hoover RB, Rozanov AY, Paepe RR (eds) Perspectives in astrobiology, vol 366. IOS Press, Amsterdam, p 43

Horneck G (1993) Responses of bacillus subtilis spores to space environment: results from experiments in space. Orig Life Evol Biosph 23:37

Horneck G, Eschweiler U, Reitz G et al (1995) Biological responses to space: results of the experiment "exobiological unit" of ERA on EURECA I. Adv Space Res 16(8):105

Horneck G, Mileikowsky C, Melosh HJ et al (2002) In: Horneck G, Baumstark-Khan C (eds) Astrobiology. The quest for the conditions of life. Springer, Berlin

Hoyle F, Wickramasinghe NC (1979) Diseases from space. J.M. Dent, London

Hoyle F, Wickramasinghe NC (1980) Evolution from space. J.M. Dent, London

Hoyle F, Wickramasinghe NC (1981) Comets - A vehicle for panspermia. In: Ponnamperuma C (ed) Comets and the origin of life, D. Reidel, Dordrecht, p 227

Hoyle F, Wickramasinghe NC (1982) Proofs that life is cosmic. Mem Inst Fund Studies Sri Lanka, No. 1 (www.panspermia.org/proofslifeiscosmic.pdf)

Hoyle F, Wickramasinghe NC (1991) The theory of cosmic grains. Kluwer Academic Press, Dordrecht

Hoyle F, Wickramasinghe NC (1996) Biofluorescence and the extended red emission in astrophysical sources. Astrophys Space Sci 235(343)

Hoyle F, Wickramasinghe NC (2000) Astronomical origins of life: steps towards panspermia. Kluwer Academic Press, Dordrecht/Boston

Johnson FM (1971) Interstellar molecules and cosmochemistry. Ann NY Acad Sci 194:3

Kissel J, Krueger FR (1987) The organic component in dust from comet Halley as measured by the PUMA mass spectrometer on board Vega 1. Nature 326:760

Krueger FR, Kissel J (2000) Stern und Weltraum 5:330

Krueger FR, Werther W, Kissel J et al (2004) Assignment of quinone derivatives as the main compound class composing 'interstellar' grains based on both polarity ions detected by the 'cometary and interstellar dust Analyser' (CIDA) onboard the spacecraft STARDUST. Rapid Commun Mass Spectrom 18:103

Kwok S (2009) Organic matter in space: from star dust to the solar system. Astrophys Space Sci 319:5–21

Lindahl T (1993) Instability and decay of the primary structure of DNA. Nature 362:709

Manning CE, Mojzsis SJ, Harrison TM (2006) Geology, age and origin of supracrustal rocks at Akilia, West Greenland. Am J Sci 306:303

Mattila K (1979) Astron Astrophys 78:253

Mileikowsky C, Cucinotta FA, Wilson JW et al (2000) Icarus 145: 391

Miller SL, Urey HC (1959) Organic compound synthesis on the primitive earth. Science 130:245

Mojzsis SJ, Harrison TM, Pidgeon RT (2001) Oxygen-isotope evidence from ancient zircons for liquid water at the Earth's surface 4,300 Myr ago. Nature 409:178

Motta V, Mediavilla E, Muñoz JA et al (2002) Detection of the 2175 Å extinction feature at z = 0.83. Astrophys J 574:719

Napier WM, Wickramasinghe JT, Wickramasinghe NC (2007) The origin of life in comets. Int J Astrobiol 6:321

Nicholson WL, Munakata N, Horneck G et al (2000) Resistance of Bacillus endospores to extreme terrestrial and extraterrestrial environments. Microbiol Mol Biol Rev 64:548

Oparin AI (1953) The origin of life (trans: Margulis S). Dover Publications, New York

Orgel LE, Crick FHC (1968) Evolution of genetic apparatus, J Mol Biol 38:381–393

Pasteur L (1857) CR Acad Sci 45:913–916

Perrin J-M, Darbon S, Sivan J-P (1995) Observation of extended red emission (ERE) in the halo of M82. Astron Astrophys 304:L21

Pflug HD (1984) In: Wickramasinghe NC (ed) Fundamental studies and the future of science. Univ. College Cardiff Press, Cardiff

Pizzarello S (2004) Orig Life Evol Biosph 34:25–34

Rauf K, Wickramasinghe C (2010) Evidence for biodegradation products in the interstellar medium. Int J Astrobiol 9(1):29–34

Schmitt-Kopplin P et al (2010) High molecular diversity of extraterrestrial organic matter in Murchison meteorite revealed 40 years after its fall. Proc Natl Acad Sci USA 107(7):2763–2768

Shivaji S et al (2009) Janibacter hoylei sp. nov., Bacillus isronensis sp. nov. and Bacillus aryabhattai sp. nov., isolated from cryotubes used for collecting air from the upper atmosphere. Int J Syst Evol Biol 59:2977–2986

Sivan J-P. Perrin J-M (1993) Astrophys J 404:258

Smith JTD, Draine BT, Dale DA et al (2007) The mid-infrared spectrum of star-forming galaxies: global properties of polyaromatic hydrocarbon emission. Astrophys J 656:770–791

Szomouru A, Guhathakurta P (1998) Astrophys J 494:L93

Vanysek V, Wickramasinghe NC (1975) Formaldehyde polymers in comets. Astrophys Space Sci 33:L19

Vreeland RH, Rosenzweig WD, Powers D (2000) Isolation of a 250 million-year-old halotolerant bacterium from a primary salt crystal. Nature 407:897

Wachtershauser G (1990) Evolution of the first metabolic cycles. Proc Natl Acad Sci USA 87(1):200

Waelkens C, Waters LBFM (1997). In: Pendleton YJ, Tielens AGGM (eds) From stardust to planetesimals. ASP conference series, vol 122. Astronomical Society of the Pacific, San Francisco, p 67

Wainwright M, Wickramasinghe NC, Narlikar JV, Rajaratnam P (2003) Are these stratospheric nanoparticles bacteria? FEMS Microbiol Lett 218:161

Wickramasinghe NC (1967) Interstellar grains. Chapman & Hall, London

Wickramasinghe NC (1993) Cosmic Grains 2. Spectroscopic identifications and the organic model, infrared astronomy. In: Mampaso A, Prieto M, Sanchez F (eds) Infrared astronomy. Cambridge University Press, Cambridge, p 303

Wickramasinghe NC (2010) The astrobiological case for our cosmic ancestry. Int J Astrobiol 9(2):119–129

Wickramasinghe JT, Napier WM (2008) Impact cratering and the Oort cloud. MNRAS 387(1): 153–157

Wickramasinghe NC, Wickramasinghe JT (2008) On the possibility of microbiota transfer from venus to earth. Astrophys Space Sci 317:133

Wickramasinghe NC, Hoyle F, Al-Jubory T (1989) Aromatic hydrocarbons in very small interstellar grains. Astrophys Space Sci 158:135

Wickramasinghe NC, Hoyle F, Al-Jubory T (1990) An integrated 2.5–12.5 μm emission spectrum of naturally-occurring aromatic molecules. Astrophys Space Sci 166:333

Wickramasinghe NC, Lloyd D, Wickramasinghe JT (2002) Evidence of photoluminescence of biomaterial in space. Proc SPIE 4495:255

Wickramasinghe JT, Wickramasinghe NC, Wallis MK (2009) Liquid water and organics in comets: implications for exobiology. Int J Astrobiol 8(4):281–290

Wickramasinghe JT, Wickramasinghe NC, Napier WM (2010) Comets and the origin of life. World Scientific Publ, Singapore

Woese C, Fox G (1977) Phylogenetic structure of the prokaryotic domain: the primary kingdoms. Proc Natl Acad Sci USA 74(11):5088

# PART V: THE FIRST STEPS OF CELLULAR EVOLUTION

Agmon
Griffith
Lathe
Mansy
Bianco
Pereira
Rodrigues
Carrapiço
Sapp
Svetina
Trevors
Bej
Elsas

Biodata of **Ilana Agmon**, author of "***The Dimeric Proto-Ribosome Within the Modern Ribosome***."

**Dr. Ilana Agmon** obtained her B.Sc. in Physics and her D.Sc. in crystallography (1990) from the Technion, Israel Institute of Technology, Haifa, Israel. She worked in Ada Yonath's ribosome group at the Weizmann Institute, Rehovot and is currently a visiting scientist in the Institute for Advanced Studies in Theoretical Chemistry, Schulich Faculty of Chemistry, Technion, Haifa and an associate member in the Fritz Haber Research Center for Molecular Dynamics, Hebrew University, Jerusalem, Israel. Her main scientific interests are ribosome evolution and function.

E-mail: **chilana@tx.technion.ac.il; agmon1@gmail.com**

J. Seckbach (ed.), *Genesis - In The Beginning: Precursors of Life, Chemical Models and Early Biological Evolution*, Cellular Origin, Life in Extreme Habitats and Astrobiology 22, 653–668
DOI 10.1007/978-94-007-2941-4_32, 

# THE DIMERIC PROTO-RIBOSOME WITHIN THE MODERN RIBOSOME

**ILANA AGMON**[1,2]
[1]*Schulich Faculty of Chemistry, Institute for Advanced Studies in Theoretical Chemistry, Technion–Israel Institute of Technology, Haifa, Israel*
[2]*Fritz Haber Research Center for Molecular Dynamics, Hebrew University, Jerusalem, Israel*

## 1. Introduction

The evolution of translation is a fundamental issue in the quest for the origin of life. Contemporary translation is an intricate process by which the genetic information is converted into polypeptide chains, which in turn fold into proteins. This process takes place at the ribosome, a complex, universal catalytic machine composed of two subunits. The large subunit is responsible for peptide bond formation and nascent chain progression, while the small subunit hosts the mRNA carrying the genetic code and applies fidelity control. In prokaryotes, the overall mass of the ribosome is 2.3 megadalton, of which RNA constitutes approximately two thirds and proteins a third. The large prokaryotic subunit is named 50S after its sedimentation coefficient and is composed of two RNA chains, 23S and 5S, and over 30 proteins. The small prokaryotic subunit is named 30S and is composed of a single RNA chain, 16S, and over 20 proteins.

Peptide bond formation, the main task of the ribosome, takes place in the peptidyl transferase center (PTC), the active site of the ribosomal large subunit (LSU), which is highly conserved across all of the phylogenetic domains of life. The PTC is composed solely of RNA, classifying the ribosome as a ribozyme whose substrates inhabit its three binding sites: A-site binds an aminoacyl-tRNA (aa-tRNA), P-site binds the peptidyl-tRNA, and E-site binds a deacylated tRNA exiting the ribosome. The amino acids involved in the peptide bond formation reaction are attached to the universally conserved CCA nucleotides at the tRNAs 3′ ends and are poised at the A- and the P-sites within the PTC, related by an approximate twofold symmetry (Nissen et al., 2000). The spatial organization of the reactants results from the primary catalysis which the ribosome exerts on its substrates, i.e., positional catalysis (Gregory and Dahlberg, 2004). The symmetrical positioning places the reactants facing each other in a stereochemistry favorable for peptide bond formation (Bashan et al., 2003; Agmon et al., 2003; Yonath, 2003; Agmon et al., 2005).

In order to enable symmetric accommodation of the substrates, the contemporary ribosome, which – on the whole – is asymmetric, contains a region possessing an approximate twofold symmetry, with the PTC at its center (Agmon et al., 2003, 2005) (Fig. 1a). The symmetry relates two RNA elements, the A- and the P-subregions, each containing about 90 nucleotides, having matching 3D folds and nucleotide conformations but essentially unrelated sequences. The symmetrical region exists in all known structures of the large subunit (Ban et al., 2000; Harms et al., 2001; Schuwirth et al., 2005; Selmer et al., 2006; Korostelev et al., 2006) and is highly conserved throughout kingdoms and organelles, indicating its vital function. Nonsymmetrical extensions radiate from the symmetrical region, together comprising domain V of 23S RNA (Fig. 1a).

The contemporary ribosome is an intricate molecular machine and – as such – is highly unlikely to have emerged in its present state. It is conceivable that the ribosome evolved from a far simpler proto-ribosome, which could catalyze only noncoded peptide bond formation between two small substrates and allowed simple elongation of peptidyl chains. Gradual evolution into the modern ribosome, which has additional functions and an enhanced level of accuracy and efficiency, probably required significant evolutionary changes that blended the original system into the ribosomal bulk. Nevertheless, it was suggested (Agmon et al., 2006; Bokov and Steinberg, 2009; Agmon et al., 2009; Agmon, 2009) that a stand-alone structural entity, a vestige of the original apparatus, can still be inferred from the current structure of the ribosome.

## 2. Identification of the Proto-Ribosome

The proto-ribosome must have been a simple apparatus that could have materialized spontaneously under prebiotic conditions. It would catalyze peptide bond formation by serving as a template to ensure the necessary proximity between the reacting amino acids. An exposition of the plausible assumptions concerning the proto-ribosome was given by Woese (2001): "(1) The primitive translation apparatus had to have been far simpler than its modern counterpart; (2) The primitive apparatus was based upon RNA and RNA interactions; (3) The essence of the primitive apparatus remains at the heart of modern translation."

In the contemporary ribosome, peptidyl transferase activity – the "essence" of the ribosome – is centered in domain V of 23S. Domain V (Fig. 1a) is composed of the symmetrical region, with the PTC at its center, flanked by its nonsymmetrical extensions. These are the helix 72–helix 73 (H72–H73) extension, which connects domain V to the rest of the 23S rRNA; the H75–H79 extension, which forms the L1 arm engaged in the release of E-site tRNA from the ribosome; the H81–H88 extension, which interacts with the P-site tRNA; and the nonsymmetrical parts of H89 and H91, which are associated with accommodating the elongation factors EF-Tu and EF-G.

Aside from their interaction with peripheral ribosomal components, the nonsymmetrical extensions radiating from the symmetrical region interact only with the elongation factors and the modern large substrates but not directly with

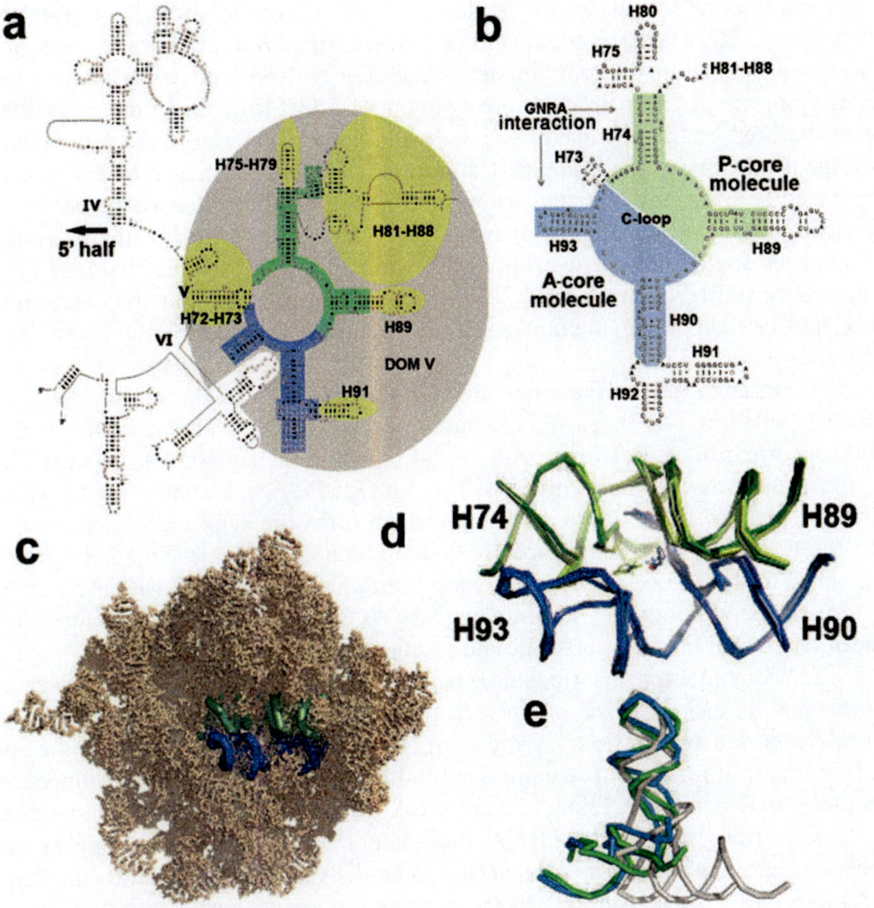


**Figure 1.** **(a)** Secondary structure of 23S 3′ half from *E. coli*. Domain V (*gray background*) is composed of the A (*blue*)- and P (*green*)- subregions of the symmetrical region (the symmetry is not portrayed by this scheme) and the nonsymmetrical extensions (*yellow*). **(b)** Two-dimensional scheme of the symmetrical region from *E. coli* drawn in a manner exhibiting the symmetry. The suggested proto-ribosome is colored. The A- and P-sites and the nonsymmetrical extensions suggested to evolve later are shown with no background. **(c)** The suggested proto-ribosome, embedded within the RNA of *E. coli* LSU (PDB 2AW4). H68–H71 were removed to reveal the proto-ribosome. **(d)** Overlap of the tertiary fold of the suggested proto-ribosome remnant, including models of the amino acid reactants, from the LSU structures of three bacterial ribosomes (*D. radiodurans*, D50S, PDB 1NJP; *E. coli*, PDB 2AW4; *T. thermophilus*, PDB 2WDN) and one archaeon (*H. marismortui*, H50S, PDB 1VQ6). The pocket, obtained through the dimerization of the A- and P-core molecules shown by their RNA backbone, is projected along the symmetry axis. The stem loops of H74, H89, and H90 are suggested to have been lost during evolution. The peptide bond is formed at the bottom of the pocket. **(e)** Overlap of the A- and P-core molecules from the structure of *E. coli* (PDB 2AW4). The projection direction is perpendicular to d. The anticodon helix of a tRNA (PDB 4TRA, *gray*) is overlapped on H89 and H93.

the peptide bond formation site or the reactants. Therefore, these extensions are functional only in the context of the modern ribosome and would have not been required before the programmed translation evolved. The redundancy of the nonsymmetrical extensions in the context of the proto-ribosome is further corroborated by the absence of the majority of these extensions from the RNA of the mitochondrial ribosomes (Cannone et al., 2002), where they are mostly replaced by proteins. Current knowledge suggests that, at least in bacteria, proteins merely play a supporting role in the catalysis performed by the ribosome (Voorhees et al., 2009). The exchange of the nonsymmetrical extensions in the mitochondrial ribosomes from RNA to proteins indicates that the extensions are less critical, and substantiates the possibility of rudimentary function without them.

In order to retrace the initial stage in the evolution of the proto-ribosome, smaller substrates than aa-tRNA should be considered. The current A- and P-loops are not useful for accommodating minimal substrates, because the universally conserved nucleotides G2251 and G2553 (*E. coli* numbering throughout the text) of the modern ribosome, which form the symmetrical base pairs to the P- and A-substrates respectively, are about 21 Å apart. Small substrates attached to these nucleotides would have been unable to interact with each other to form a peptide bond. The contemporary A- and P-sites would therefore have been functional only for substrates no smaller than aa-CA.

In the contemporary ribosome, peptidyl transferase activity takes place in domain V of 23S. Removal of the structural elements concerned solely with the modern coded translation from domain V is therefore expected to point at the remnant of an ancient structural entity that catalyzed the prebiotic noncoded peptide bond formation. Elimination of the A- and P-sites and the nonsymmetrical extensions from the ribosomal RNA of domain V (Fig. 1b) results in a pocket-like entity (Fig. 1c, d), which was suggested to be the vestige of the proto-ribosome (Agmon et al., 2009; Agmon, 2009). A somewhat larger proto-ribosome, composed of the entire symmetrical region attached to the nonsymmetrical portion of H75, was put forward based on the identification of an evolutionary scheme of appending RNA components to the ancient ribosome (Bokov and Steinberg, 2009). Detection of the proto-ribosome in an encircling position around the PTC is consistent with the notion that the region of the large subunit near the PT-origin evolved first and that the parts near the surface are younger (Fox and Naik, 2004; Hsiao et al., 2009).

## 3. Structure of the Dimeric Proto-Ribosome

The suggested proto-ribosome is a pocket-like entity built from a dimer of two L-shaped molecules named the A- and P-ribosomal core molecules. They are assembled in a symmetrical manner around an approximate twofold axis (Fig. 1d) and include the essential ribosomal nucleotides, which interact with the reacting amino acids in the PTC of the modern ribosome. The core molecules, being part of the ribosome's symmetrical region (Agmon et al., 2003, 2005), are similar to each other in fold (Fig. 1e) and nucleotide conformation but have practically unrelated sequences. In *E. coli*, for instance, only 38% of the symmetrical mates have identical nucleotide type compared to 25% expected from random sequences.

Each of the molecules, as found within the ribosome, contains about 60 nucleotides, which are highly conserved in the modern ribosomes. The core molecules are composed of two helices (H90 and H93 in the A-core molecule, H74 and most of H89 in the P-core molecule) joined by an elbow created by a network of hydrogen bonds involving nucleotides from the central loop of domain V (Fig. 1b). Three out of the four helices that compose the vestige of the proto-ribosome within the contemporary ribosome, H74, H89, and H90, lack stem loops (Fig. 1b, d) probably because the ribosome expanded during evolution through the tips of these helices. An L-shaped molecule can either have a single terminal stem loop, where the 3′:5′ ends are found at the edge of one arm (like in the tRNA molecule), or have two loops, where the 3′:5′ ends are located at the elbow region. Completion of the missing stem loops that would have existed in the proto-ribosome results in A- and P-core molecules of 60–70 mer.

In the modern ribosome, association of the two core molecules around the approximate twofold rotational symmetry axis is partially achieved by a highly conserved GNRA interaction motif. The GNRA motif (N = any nucleotide; R = purine) is an abundant non-Watson-Crick RNA-RNA interaction, in which a terminal GNRA stem loop recognizes a helical receptor region (Pley et al., 1994). This motif contains an A-minor contact (Nissen et al., 2001; Battle and Doudna, 2002) involving the GNRA adenosine and a base pair, preferably C:G, from the minor groove of the helical receptor region (Costa and Michel, 1997; Abramovitz and Pyle, 1997). The GNRA interaction is employed in mediating RNA tertiary structure (Jaeger et al., 1994) and in stabilizing symmetrical dimers of RNA molecules (Davis et al., 2005). In the modern ribosome, the GUGA tetra stem loop of H93 interacts with a receptor region comprised of nucleotides from H74. This motif is highly conserved in both the loop and the receptor regions: the GNRA tetraloop nucleotides G2595, R2597, and A2598 are all 100% conserved; the receptor base pair engaged in the A-minor interaction, C2073:G2436, is about 95% conserved; A2435:U2074 is 100% conserved; and U2075 is 99% conserved (Cannone et al., 2002). The presence of an A-minor interaction motif among the dimer-forming contacts evokes the possibility that the initial proto-ribosome dimer may have been associated through two symmetrical A-minor contacts rather than full GNRA motifs. Such a dimerization mode would be compatible with the suggestion that A-minor interactions were used by evolution to add new RNA elements to the expanding proto-ribosome (Bokov and Steinberg, 2009).

This simple pocket-like entity was probably the smallest possible dimeric proto-ribosome, as dictated by the association of the core molecules into dimers through the tips of their stems (Fig. 1b, d). Preservation of the protein synthesis mechanism requires that the structure of the inner part of the pocket, which catalyzes peptide bond formation, should be conserved under any hypothetical structural alteration. The inner void of the pocket is formed by the elbow regions of the two L-shaped constituents of the dimer, and its shape depends on the dimer-forming interactions made by the tips of the stems. The structure of dimers of tectoRNA molecules, obtained via symmetric GNRA interactions, was found to depend critically on the mutual helical twist of the interacting stems (Jaeger et al., 2001). Therefore, conservation of the proto-ribosome's active site requires retention of the interaction motif between the tips of the core molecules. This is only possible if the length of the stems is shortened by a full helical turn, which, for A-form RNA, holds 11 base pairs. Cropping the core molecule stem length by 11 base pairs would eliminate the stems completely (Fig. 1b), implying that the dimeric proto-ribosome was probably the minimal form of this kind of apparatus.

## 4. Formation and Function of the Proto-Ribosome

The ribosomal core molecules are suggested to have emerged in the prebiotic era by spontaneous folding of randomly sequenced RNA chains of about 60–70 nucleotides, which are widely believed to have been available in the primordial world (Pino et al., 2008; Joshi et al., 2009). Alternatively, it could have formed through ligation of two hairpins of about half the length of the full molecule, in analogy with the hypothesis proposed for the prebiotic formation of tRNA molecules (Di Giulio, 1992; Dick and Schamel, 1995; Nagaswamy and Fox, 2003). The core molecule could then have spontaneously dimerized, in a symmetrical manner, with a second core molecule, similar in fold but likely to have a different sequence. The practically unrelated sequences of the A- and the P-core molecule remnants within the ribosome, as well as the infinitesimal probability for encountering a second spontaneously-formed core molecule with identical sequence, support the notion that the proto-ribosome was a heterodimer (Agmon, 2009). The nonenzymatic folding and dimerization of the core molecules resulted in the generation of the pocket-like dimeric proto-ribosomes.

In a successful fraction of the pockets, these apparatuses provided the ability to favorably accommodate single amino acids, other small substrates, or short peptidyl chains obtained during previous cycles of the apparatus' function. Consistent with the expectation for a continuous evolutionary path from the proto-ribosome toward the modern ribosome, the dimeric proto-ribosome is assumed to have harbored the reacting amino acids in the same manner as the modern ribosome does, thus allowing the atoms participating in the nucleophilic attack to interact effectively. When the two prebiotic reactants were accommodated within the dimeric pocket, each attached to one core molecule equivalently to their positioning on the modern ribosome, they were facing each other at a distance and mutual orientation favorable for the formation of a peptide bond.

Subsequently to the peptide bond reaction, the newly obtained dipeptide could have either remained attached to one of the proto-ribosomal sites, or could have fully dissociated and later been reaccommodated on the apparatus. This enabled the positioning of a new amino acid in the vacant site, followed by the formation of a peptide bond, adding another amino acid to the growing peptidyl chain. Reiterations of this dynamic process would have resulted in a primitive mechanism for noncoded elongation, equivalent to the elongation mechanism proposed earlier (Weiner and Maizels, 1987).

The dimeric proto-ribosome may be related to the origin of chiral discrimination in proteins. The PTC of the modern ribosome is suggested to prefer L- to D-amino acids, due to the enhanced steric hindrance exerted on the D conformation (Zarivach et al., 2004; Thirumoorthy and Nandi, 2007). Being proposed to be still embedded in the modern ribosome, the dimeric proto-ribosome may have favored L- over D-amino acids in the prebiotic era as well, indicating a possible relevance to the origin of chirality in proteins.

The dimeric proto-ribosomes would have produced a variety of short, random-sequenced peptides, whose sequence could depend on the affinity of the pocket to specific amino acids (Belousoff et al., 2010) or on the distribution of the concomitant amino acids in the vicinity of a specific proto-ribosome. In turn, these peptidyl chains could have rendered a selective evolutionary advantage to the proto-ribosomes producing them.

## 5. Validating the Hypothesis

To substantiate the hypothesis that the prebiotic proto-ribosome was a simple dimeric RNA enzyme, which is still embedded in the core of the modern ribosomal large subunit, two main aspects have to be examined: (1) the ability of RNA chains of proper length and composition to fold spontaneously into an L-shaped form, similar to the fold of the core molecules within the ribosome, and (2) the ability of two core molecules to dimerize spontaneously, in a symmetrical manner, to form the pocket-like structure suggested to be the vestige of the proto-ribosome. The structural, computational, and experimental aspects of these issues are analyzed below.

Each core molecule is comprised of two helices joined by an elbow region made of single-stranded segments (Fig. 1b). Two examples of comparable L-shaped RNA molecules are known: the tectoRNA molecules with a right-angle motif (Chworos et al., 2004; Jaeger and Chworos, 2006) (Fig. 2b) and the tRNAs (Fig. 2c). TectoRNAs are laboratory designed small RNA building blocks that self-assemble to form modular supramolecular structures (Jaeger et al., 2001). The tectoRNA molecules in general, and the right-angle motif in particular, self-fold into stable units, indicating that RNA chains with the proper sequence can form stable L-shaped RNA molecules spontaneously.

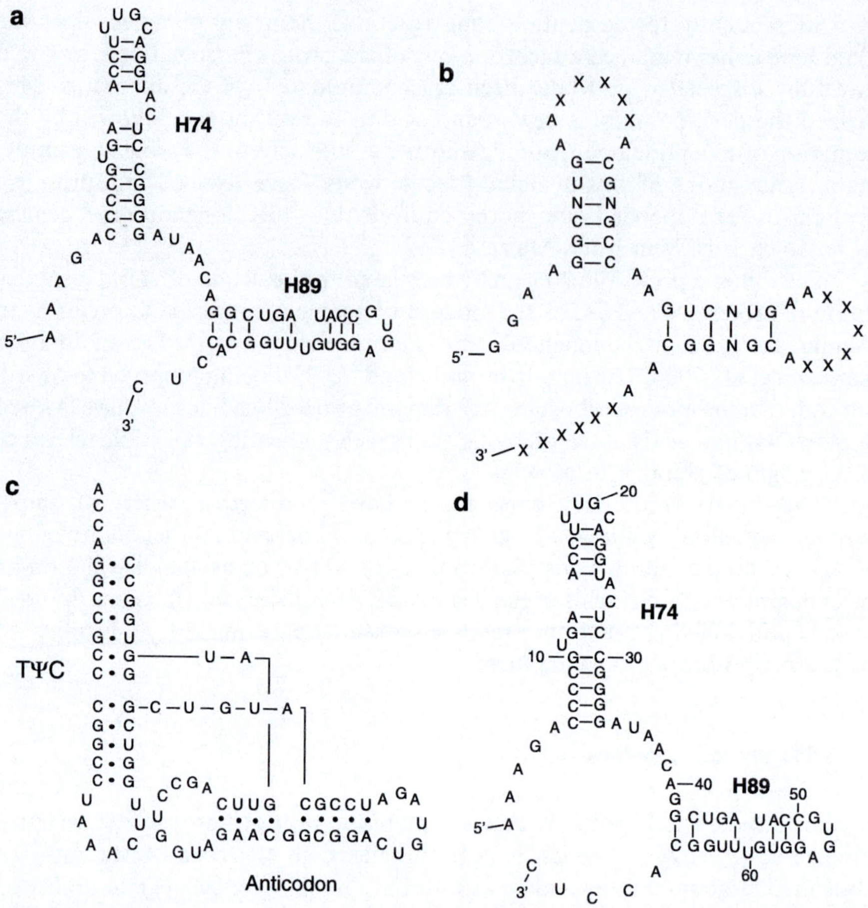


**Figure 2.** Secondary structure of L-shaped RNA molecules: **(a)** P-core molecule from E50S. The stem-loop nucleotides, missing in the ribosome structure, were artificially added to the original sequence to complete truncated helices. **(b)** Right-angle tectoRNA molecule (Jaeger and Chworos, 2006). **(c)** L-shaped scheme of canonical tRNA. **(d)** Mfold prediction for the P-core sequence shown in **(a)**. Folding free energy is $\Delta G = -25.9$ kcal/mol.

Also tRNA, similarly to other small RNAs, forms stable molecules in vitro across a large range of conditions (Russell, 2008). The correspondence between the structure of the tRNA and the core molecule deserves closer scrutiny. Both molecules are L-shaped units of comparable size and outline (Agmon, 2009). If one tRNA helix is overlapped with one of the core molecule helices, their remaining helices point in opposite directions (Fig. 1e). The tRNA is widely accepted to be a relic from the prebiotic world (Eigen et al., 1989; Maizels and Weiner, 1994), while the core molecule was implicated in prebiotic peptide bond formation (Agmon et al., 2006, 2009; Agmon, 2009). These two forms of

L-shaped molecules may have existed, side by side, in the primordial environment, being later recruited for their separate roles in translation. The subsequent recruitment of existing prebiotic molecules for evolving tasks is supported by the variety of roles performed by tRNA-like molecules in replication (Maizels and Weiner, 1994). Both examples indicate that L-shaped RNA molecules can be self-folding and stable, corroborating the possibility that the ancestor of the L-shaped core molecule could have been self-folding and stable enough to emerge in the prebiotic era and to survive evolutionary stresses.

Secondary structures of short RNA sequences, similar in size to the tRNA molecule, can be reliably predicted by free energy minimization (Doshi et al., 2004), providing additional support for conclusions drawn from the 3D structure. Using Mfold (Mathews et al., 1999; Zuker, 2003), thermodynamically favored 2D structures were predicted for the A- and the P-core molecule sequences found in all known structures of the large subunit, i.e., D50S, H50S, and *E. coli* ribosomes (Agmon et al., 2009; Agmon, 2009). The results obtained for *E. coli* are applicable also to T50S because both structures have identical sequence in the investigated region. Although no two sequences were predicted to have identical folds, the results led to a consistent picture for the secondary structure of the core molecule (Fig. 2d). In accord with their fold within the ribosome, the predicted molecules were composed of two helices connected by single-stranded segments. In most cases, the program found a unique solution, suggesting that the likelihood of obtaining alternative secondary structures is low. The free energy predicted for the different folds ranges between $\Delta G = -26 \pm 3$ kcal/mol, supporting the self-folding and stability of the obtained structures. Hence, folding of 60–70 mer RNA segments of proper sequence into an L-shaped conformation is a downhill process from an energetic point of view, confirming that the core molecule could have emerged spontaneously in the prebiotic world.

RNA molecules tend to form stable dimers spontaneously (Kholod, 1999; Roy et al., 2005; Sun et al., 2007). It is therefore likely that two core molecules could have self-assembled under prebiotic conditions to form a dimer. The presence of divalent metal ions might have been a prerequisite because they are crucial for the correct folding, stability, and dimerization of RNA molecules in vitro (Pyle, 2002; Draper, 2004; Davis et al., 2005).

The symmetrical dimerization of the two core molecules took place through the GNRA interaction motifs. This same mode of dimerization is used to construct various forms of tectoRNA dimeric particles (Davis et al., 2005) and was found to contribute to the stability of the molecules involved. Lowering of the free energy by such a mode of dimerization implies that self-assembly of two core molecules into a symmetrical dimer via GNRA interaction is an energetically favorable process that could have occurred spontaneously in a primordial world, generating the dimeric proto-ribosome.

The plausibility of the emergence of a dimeric proto-ribosome was tested experimentally using RNA chains with sequences resembling those observed for the core molecules in the current ribosome. The results indicated that some of the chains were capable of forming dimers that may adopt a pocket-like structure (Davidovich et al., 2009).

Taken together, these structural, computational, and experimental examinations substantiate the feasibility of spontaneous formation of dimeric proto-ribosomes in the prebiotic world.

## 6. Evolution of the Proto-Ribosome

The dimeric proto-ribosome is suggested to have formed through a process of self-organization of small RNA units, i.e., the core molecules. This would have produced proto-ribosome pockets with different levels of protein synthesis efficiency, depending on their diverse sequences, thus having a varying chance of survival. This is a unique application of Darwin's principle of variation and selection, a principle that is usually applied to living organisms. In the scenario suggested here, the sequence variation was due to the randomness of the RNA chains building the proto-ribosomes rather than to mutation and was therefore a prebiotic phenomenon. Nevertheless, selection of the fittest proto-ribosome, which led to its survival, can already be regarded as a life process. The spontaneous appearance of the dimeric proto-ribosome may have been the transition between thermodynamic selection, applicable to chemical reactions, and Darwinian selection, applicable to living organisms, thus forming a crucial step in the emergence of life as we know it.

These proto-ribosomes would have produced a variety of short, random-sequenced peptides that would accumulate in larger amounts in the vicinity of "efficient" RNA dimers. A productive proto-ribosome had an enhanced probability that some of its randomly produced peptidyl chains will fold into a structure possessing weak catalytic abilities of some kind (Agmon, 2009) or will be suitable for stabilizing the proto-ribosomal RNA, thus rendering it a selective evolutionary advantage and improving the prospect of its survival.

The catalytic core of a ribozyme, on its own, provides only a modest level of activity, and it is the peripheral elements which generate a substantial increase in the catalytic rate (Voytek and Joyce, 2007). Improvement of the proto-ribosome efficiency ought to have been achieved through the addition of RNA elements and proteins accompanied by mutational variations. A preliminary step would have probably involved the addition of the A- and P-sites, which could have extended from the far ends of H74 and H90 for P- and A-core molecules, respectively (Fig. 1b). This advanced proto-ribosome would have been capable of accommodating larger substrates via base-pairing, thus better stabilizing the reactants and achieving a higher performance level in noncoded protein synthesis. Further RNA elements would join, possibly through A-minor interactions (Bokov and Steinberg, 2009), perhaps stabilized by the short peptidyl chains produced by the proto-ribosomes. No globular protein exists within the highly conserved ribosomal RNA around the PTC, suggesting that the first proteins to be added to the proto-ribosome were probably short extended peptidyl chains with stabilizing effect, whose remnants are the tails of the present proteins, penetrating into the ribosomal RNA bulk and stabilizing it (Steitz and Moore, 2003; Hsiao et al., 2009).

During the primal stages in the evolution of the proto-ribosome, its two substrate accommodation sites would have been functionally symmetric. Later on, the functional symmetry would have been lost because the procession of amino acid polymerization requires a directional process. In the contemporary ribosome, the A- and P-sites have distinct roles. The A-site accommodates the incoming amino acid while the P-site hosts the peptidyl chain prior to peptide bond formation. Following the reaction, the substrates shift in a specific direction. A-site substrate translocates into the P-site, while the P-site substrate moves into the E-site. The translocation step results in a vacant A-site, allowing a new cycle of amino acid polymerization to commence. Adaptation of the early symmetrical structure to the directional process would have required breaking the symmetry to some extent. Indeed, the current PTC presents an interplay between the overall symmetry required for the symmetrical accommodation of the reactants and structural deviations from it. PTC nucleotides involved in translocation, such as A2602 and U2585, tend to bulge out and break the conformational symmetry (Agmon et al., 2005).

Further structural changes would involve the emergence of the ancestor of the small subunit, which may have promoted the initiation of the mRNA-directed synthesis of proteins. The evolutionary advantage granted to the ribosome by its association with the prebiotic mRNA is not addressed in this chapter. The modern proteins would have been incorporated into the structure, having a supportive role in many aspects of translation, from structural stabilization of the ribosomal RNA and assisting the ribosome assembly (Klein et al., 2004), through increasing the speed and accuracy of the translation, to stabilizing the peptide synthesis reaction, as was recently suggested (Voorhees et al., 2009). External proteins, such as the ancestors of the elongation factors EF-Tu and EF-G, would have coevolved with the ribosome, facilitating steps of translational elongation. This gradual expansion of the system, together with the mutational variations, adapted the proto-ribosome to its advanced functions and has ultimately given rise to the current translation machine, with its modern performance level. Although it is possible to envision how small improvements in translation machinery led to improvement in the proteins produced – and vice versa – the transition from noncoded protein synthesis to coded translation is still incomprehensible.

The evolution of the ribosome described herein, from a small passive template to the current complex molecular machine assembled around it, allows the vision of a continuous path from the primordial world into the present form of life.

## 7. Acknowledgments

Thanks are due to Ada Yonath for initiating the ribosome evolution study and to Amitai Halevi, Noam Adir, Sagi, and Nimrod Agmon for their help. Support was provided by the US National Inst. of Health (GM34360) and the Kimmelman Center for Macromolecular Assemblies.

## 8. References

Abramovitz DL, Pyle AM (1997) Remarkable morphological variability of a common RNA folding motif: the GNRA tetraloop-receptor interaction. J Mol Biol 266:493–506

Agmon I (2009) The dimeric proto-ribosome: structural details and possible implications on the origin of life. Int J Mol Sci 10:2921–2934

Agmon I, Auerbach T, Baram D, Bartels H, Bashan A, Berisio R, Fucini P, Hansen HA, Harms J, Kessler M, Peretz M, Schluenzen F, Yonath A, Zarivach R (2003) On peptide bond formation, translocation, nascent protein progression and the regulatory properties of ribosomes. Eur J Biochem 270:2543–2556

Agmon I, Bashan A, Zarivach R, Yonath A (2005) Symmetry at the active site of the ribosome: structural and functional implications. Biol Chem 386:833–844

Agmon I, Bashan A, Yonath A (2006) On ribosome conservation and evolution. Isr J Ecol Evol 52:359–374

Agmon I, Davidovich C, Bashan A, Yonath A (2009) Identification of the prebiotic translation apparatus within the contemporary ribosome. http://precedings.nature.com/documents/-2921/version/1

Ban N, Nissen P, Hansen J, Moore PB, Steitz TA (2000) The complete atomic structure of the large ribosomal subunit at 2.4 A resolution. Science 289:905–920

Bashan A, Agmon I, Zarivach R, Schluenzen F, Harms J, Berisio R, Bartels H, Franceschi F, Auerbach T, Hansen HA, Kossoy E, Kessler M, Yonath A (2003) Structural basis of the ribosomal machinery for peptide bond formation, translocation, and nascent chain progression. Mol Cell 11:91–102

Battle DJ, Doudna JA (2002) Specificity of RNA-RNA helix recognition. Proc Natl Acad Sci USA 99:11676–11681

Belousoff MJ, Davidovich C, Zimmerman E, Caspi Y, Wekselman I, Rozenszajn L, Shapira T, Sade-Falk O, Taha L, Bashan A, Weiss MS, Yonath A (2010) Ancient machinery embedded in the contemporary ribosome. Biochem Soc Trans 38:422–427

Bokov K, Steinberg SV (2009) A hierarchical model for evolution of 23S ribosomal RNA. Nature 457:977–980

Cannone JJ, Subramanian S, Schnare MN, Collett JR, D'Souza LM, Du Y, Feng B, Lin N, Madabusi LV, Müller KM, Pande N, Shang Z, Yu N, Gutell RR (2002) The comparative RNA Web CRW. Site: an online database of comparative sequence and structure information for ribosomal, intron, and other RNAs. BMC Bioinformatics 3:1–31

Chworos A, Severcan I, Koyfman AY, Weinkam P, Oroudjev E, Hansma HG, Jaeger L (2004) Building programmable jigsaw puzzles with RNA. Science 306:2068–2072

Costa M, Michel F (1997) Rules for RNA recognition of GNRA tetraloops deduced by in vitro selection: comparison with in vivo evolution. EMBO J 16:3289–3302

Davidovich C, Belousoff M, Bashan A, Yonath A (2009) The evolving ribosome: from non-coded peptide bond formation to sophisticated translation machinery. Res Microbiol 160:487–492

Davis JH, Tonelli M, Scott LG, Jaeger L, Williamson JR, Butcher SE (2005) RNA helical packing in solution: NMR structure of a 30 kDa GAAA tetraloop-receptor complex. J Mol Biol 351:371–382. http://www.ncbi.nlm.nih.gov/pubmed/16002091?ordinalpos=3&itool=EntrezSystem2.PEntrez.Pubmed.Pubmed_ResultsPanel.Pubmed_RVDocSum

Di Giulio M (1992) On the origin of the transfer RNA molecule. J Theor Biol 159:199–214

Dick TP, Schamel WA (1995) Molecular evolution of transfer RNA from two precursor hairpins: implications for the origin of protein synthesis. J Mol Evol 41:1–9

Doshi KJ, Cannone JJ, Cobaugh CW, Gutell RR (2004) Evaluation of the suitability of free energy minimization using nearest-neighbor energy parameters for RNA secondary structure prediction. BMC Bioinformatics 5:105

Draper DE (2004) A guide to ions and RNA structure. RNA 10:335–343

Eigen M, Lindemann BF, Tietze M, Winkler-Oswatitsch R, Dress A, von Haeseler A (1989) How old is the genetic code? Statistical geometry of tRNA provides an answer. Science 244:673–679

Fox GE, Naik AK (2004) The evolutionary history of the ribosome. In: de Pouplana LR (ed) The genetic code and the origin of life. Landes Bioscience, Georgetown, pp 92–105

Gregory ST, Dahlberg AE (2004) Peptide bond formation is all about proximity. Nat Struct Mol Biol 11:586–587

Harms J, Schluenzen F, Zarivach R, Bashan A, Gat S, Agmon I, Bartels H, Franceschi F, Yonath A (2001) High resolution structure of the large ribosomal subunit from a mesophilic eubacterium. Cell 107:679–688

Hsiao C, Mohan S, Kalahar BK, Williams LD (2009) Peeling the onion: ribosomes are ancient molecular fossils. Mol Biol Evol 26:2415–2425

Jaeger L, Chworos A (2006) The architectonics of programmable RNA and DNA nanostructures. Curr Opin Struct Biol 16:531–543

Jaeger L, Michel F, Westhof E (1994) Involvement of a GNRA tetraloop in long-range RNA tertiary interactions. J Mol Biol 236:1271–1276

Jaeger L, Westhof E, Leontis NB (2001) TectoRNA: modular assembly units for the construction of RNA nano-objects. Nucleic Acids Res 29:455–463

Joshi PC, Aldersley MF, Delano JW, Ferris JP (2009) Mechanism of montmorillonite catalysis in the formation of RNA oligomers. J Am Chem Soc 131:13369–13374

Kholod NS (1999) Dimer formation by tRNAs. Biochemistry (Mosc) 64:298–306

Klein DJ, Moore PB, Steitz TA (2004) The roles of ribosomal proteins in the structure assembly, and evolution of the large ribosomal subunit. J Mol Biol 340:141–177

Korostelev A, Trakhanov S, Laurberg M, Noller HF (2006) Crystal structure of a 70S ribosome-tRNA complex reveals functional interactions and rearrangements. Cell 126:1065–1077

Maizels N, Weiner AM (1994) Phylogeny from function: evidence from the molecular fossil record that tRNA originated in replication, not translation. Proc Natl Acad Sci USA 91:6729–6734

Mathews DH, Sabina J, Zuker M, Turner DH (1999) Expanded sequence dependence of thermodynamic parameters improves prediction of RNA secondary structure. J Mol Biol 288:911–940

Nagaswamy U, Fox GE (2003) RNA ligation and the origin of tRNA. Orig Life Evol Biosph 33:199–209

Nissen P, Hansen J, Ban N, Moore PB, Steitz TA (2000) The structural basis of ribosome activity in peptide bond synthesis. Science 289:920–930

Nissen P, Ippolito JA, Ban N, Moore PB, Steitz TA (2001) RNA tertiary interactions in the large ribosomal subunit: the A-minor motif. Proc Natl Acad Sci USA 98:4899–4903

Pino S, Ciciriello F, Costanzo G, Di Mauro E (2008) Nonenzymatic RNA ligation in water. J Biol Chem 283:36494–36503

Pley HW, Flaherty KM, McKay DB (1994) Model for an RNA tertiary interaction from the structure of an intermolecular complex between a GAAA tetraloop and an RNA helix. Nature 372:111–113

Pyle AM (2002) Metal ions in the structure and function of RNA. J Biol Inorg Chem 7:679–690

Roy MD, Wittenhagen LM, Kelley SO (2005) Structural probing of a pathogenic tRNA dimer. RNA 11:254–260

Russell R (2008) RNA misfolding and the action of chaperones. Front Biosci 13:1–20

Schuwirth BS, Borovinskaya MA, Hau CW, Zhang W, Vila-Sanjurjo A, Holton JM, Cate JH (2005) Structures of the bacterial ribosome at 3.5 A resolution. Science 310:827–834

Selmer M, Dunham CM, Murphy FV 4th, Weixlbaumer A, Petry S, Kelley AC, Weir JR, Ramakrishnan V (2006) Structure of the 70S ribosome complexed with mRNA and tRNA. Science 313:1935–1942

Steitz TA, Moore PB (2003) RNA, the first macromolecular catalyst: the ribosome is a ribozyme. Trends Biochem Sci 28:411–418

Sun X, Li JM, Wartell RM (2007) Conversion of stable RNA hairpin to a metastable dimer in frozen solution. RNA 13:2277–2286

Thirumoorthy K, Nandi N (2007) Homochiral preference in peptide synthesis in ribosome: role of amino terminal, peptidyl terminal, and U2620. J Phys Chem B 111:9999–10004

Voorhees RM, Weixlbaumer A, Loakes D, Kelley AC, Ramakrishnan V (2009) Insights into substrate stabilization from snapshots of the peptidyl transferase center of the intact 70S ribosome. Nat Struct Mol Biol 16:528–533

Voytek SB, Joyce GF (2007) Emergence of a fast-reacting ribozyme that is capable of undergoing continuous evolution. Proc Natl Acad Sci USA 104:15288–15293

Weiner AM, Maizels N (1987) TRNA-like structures tag the 3′ ends of genomic RNA molecules for replication: implications for the origin of protein synthesis. Proc Natl Acad Sci USA 84:7383–7387

Woese CR (2001) Translation: in retrospect and prospect. RNA 7:1055–1067

Yonath A (2003) Ribosomal tolerance and peptide bond formation. Biol Chem 384:1411–1419

Zarivach R, Bashan A, Berisio R, Harms J, Auerbach T, Schluenzen F, Bartels H, Baram D, Pyetan E, Sittner A, Amit M, Hansen HSA, Kessler M, Liebe C, Wolff A, Agmon I, Yonath A (2004) Functional aspects of ribosomal architecture: symmetry, chirality and regulation. J Phys Org Chem 17:901–912

Zuker M (2003) Mfold web server for nucleic acid folding and hybridization prediction. Nucleic Acids Res 31:3406–3415

Biodata of **Robert W. Griffith**, author of "***Oceanic Arginine Translator: The Origin of Life and Early Evolution of the Genetic Code.***"

**Robert W. Griffith** is a professor emeritus at the Department of Biology, University of Massachusetts Dartmouth in North Dartmouth, Massachusetts. He obtained his Ph.D. from Yale University, New Haven, Connecticut in 1972. He was a post-doctoral scholar at Woods Hole Oceanographic Institution before moving to the University of Massachusetts. His scientific interests include osmoregulatory physiology of fishes, nitrogen metabolism in vertebrates, biology of the coelacanth, and evolution of major groups, notably the origin of gnathostomes, vertebrates, and life. His serious interest in the origin of life began some years ago when he lectured on the topic for the first time in an introductory biology course. When he suggested, rather naively, that he thought the most sensible model was some kind of "macromolecular symbiosis," one of his students, Corey, asked for specifics. This chapter is his belated response.

E-mail: **rgriffith@umassd.edu**

J. Seckbach (ed.), *Genesis - In The Beginning: Precursors of Life, Chemical Models and Early Biological Evolution*, Cellular Origin, Life in Extreme Habitats and Astrobiology 22, 669–690
DOI 10.1007/978-94-007-2941-4_33, © Springer Science+Business Media Dordrecht 2012

# OCEANIC ARGININE TRANSLATOR: THE ORIGIN OF LIFE AND EARLY EVOLUTION OF THE GENETIC CODE

**ROBERT W. GRIFFITH**
*Department of Biology, University of Massachusetts Dartmouth, North Dartmouth, MA 02747, USA*

## 1. Introduction

The question of the origin of life is central to almost all fields of biology. For instance, it provides a starting point for organic evolution and explains the unity of life at the level of molecular genetics and biochemistry. Yet, in spite of its compelling interest and importance, and despite enormous strides that have been made in relevant fields such as molecular genetics, remarkably little real progress has been made toward understanding how life actually arose since Oparin (1938) proposed the model of chemical evolution, later supported experimentally by Miller (1953). To some extent, this is because the classic concept of chemical evolution, with the ocean as an "organic soup," has become viewed as implausible (Miller, 1987; Bada, 2004). This is especially true with the formation of complex monomers and polymers – notably nucleic acids (NA) and peptides – that require high monomer concentrations and conditions that favor condensation. Thus, aside from nonscientific notions of "panspermy" and "ID," most recent origin scenarios involve chemical evolution in isolated locations; mineral surfaces (Deamer et al., 2006; Ferris, 2006), hydrothermal vents (Shock, 1992), organic matrices (Wächtershauser, 1988), aerosols (Tervahattu et al., 2004), or lipid layers or droplets (Segré et al., 2001; Morchio et al., 2001). I believe the latter class of proposals, lipid layers or droplets, represents the most plausible and promising alternative (Griffith, 2009).

The popular view of evolution as a directional process, leading from simplicity to complexity (e.g., Blum, 1951), is implicit in most scenarios for the origin of life. This is the case for models dealing with prebiotic chemical evolution as well as protobiology. I believe this premise is misleading when applied to the origin of life, as indeed it is when applied to contemporary evolutionary processes. I favor the early prebiotic formation of a mix of complex organic molecules (including peptides and NAs, as well as lipids) with chemical and then biological evolution involving natural selection of molecules and combinations of molecules that "worked" together well. Root-Bernstein and Dillon (1997) advocated the importance of interactions among complementary molecules in the origin and early evolution of life, and this is a crucial element of my scenario (Griffith, 2009).

The origin of biological information that now resides in the base sequences of RNA and DNA and the amino acid sequence of proteins represent an issue that is perhaps more enigmatic than the formation of the macromolecules themselves. Currently, there are four kinds of scientific scenarios that attempt to address the issue. Mineral or some other nonbiological but "ordered" matrix could have served as a precursor to NA and peptides (Cairns-Smith, 1982; Wächtershauser, 1988). Metabolism-first scenarios propose that a coherent web of interconnected catalysts (variants of this model utilize random peptides, small molecules, lipids, or molecular aggregates known as composomes) preceded NA and ordered peptides (De Duve, 1991; Morowitz et al., 2000; Shapiro, 2006; Hunding et al., 2006). The nucleic-acid-first (RNA world) scenario is currently the most widely accepted explanation for the origin of biological order (Joyce, 2002; Line, 2002). Independently proposed by Woese (1967), Crick (1968), and Orgel (1968), the RNA world scenario has received experimental support from studies on natural or artificially selected "ribozymes," including recent demonstrations that RNA strands can undergo mutual catalytic replication (Lincoln and Joyce, 2009). The fourth kind of scenario entails the simultaneous prebiotic formation of NA and peptides with mutually supportive function. Such models might logically be called coevolution, but a notable paper by Wong (1975) used that term in a different context, so the phrase molecular mutualism might be preferable. Lahav (1993), Kunin (2000), Nashimoto (2002), and Root-Bernstein (2007) advocate such scenarios, and the premise of molecular mutualism was the basis for my specific proposal for how life originated (Griffith, 2009). In that paper I provided arguments against "molecular scaffolding" and "metabolism first" scenarios and especially against the RNA world (e.g., it is logistically implausible and does not explain how translation or transcription arose). I also provided a number of reasons why molecular mutualism represents an attractive and plausible alternative.

Related to the question of the origin of biological information is how the universal genetic code (based on the assignment of nucleotide triplets to specific amino acids) began and how it subsequently evolved. There are a large number of hypotheses for the code's inception based on the starting premise of the RNA world (e.g., Maizels and Weiner, 1987; Wolf and Koonin, 2007), but most are not very specific, and none has really gained general acceptance. Scenarios based on molecular mutualism of peptides and NA would, on the face of it, seem a more promising approach to the question of the genetic code's origin. In such scenarios, the origin of the code would likely entail direct recognition of a specific amino acid by a specific codon or anticodon nucleotide base sequence. Although this model was only weakly supported by early experimental studies (Ponnamperuma and Hobish, 1984), more recent data provide solid evidence for the phenomenon (Yarus et al., 2005), and it is the basis for several recent mutualism scenarios (Root-Bernstein, 2007; Guimarães et al., 2008; Griffith, 2009). I proposed that the code's inception involved recognition of the amino acid arginine by NA strands bearing its codon and anticodon (Griffith, 2009). Hypotheses for the code's evolution following its initiation constitute a large and disparate literature. Causal explanations for the code's expansion include stereospecific recognition of amino acid classes by NA, ambiguity reduction, frozen accident, and metabolic coevolution (see review by Di Giulio, 2005). The sequence in which codon assignment evolved is contingent upon assumptions of which amino acid was the first coded (e.g., were "simple" amino acids or "functional" amino acids first, and if "functional," what kind of functional role was important) as well as the causal processes by which expansion occurred. In my earlier paper I did not address in detail how I thought the code developed after the original phase of translating the functional amino acid arginine.

In this chapter, I intend to demonstrate that the concept of molecular mutualism can lead to a highly plausible, specific, and testable explanation for the origin of life and the genetic code. In my previous article (Griffith, 2009), I provided arguments for the following general aspects of the scenario: the presence of a surface hydrophobic layer in the ocean, how this surface layer facilitates synthesis and polymerization of biomolecules, the importance of an emulsified zone for molecular accumulation and NA and peptide interactions, why both peptides and NA were present, and why arginine was the first amino acid translated. In that paper I also proposed a specific candidate for the first living system, an autocomplementary hexamer that displays arginine's codons and anticodons, recognizes arginine via stereospecific binding, and incorporates arginine into peptides via phosphate-bond-driven autoaminoacylation. Here, I will give an abbreviated overview of those aspects of my scenario without extensive reiteration of arguments I have already provided, discussing some additional relevant data. Other than the general operative principles of natural selection and molecular mutualism, I did not elaborate the expansion of the genetic code beyond the arginine translator in my earlier paper (Griffith, 2009). Here, I will provide a somewhat longer discussion of that topic.

## 2. The Protobiosphere

### 2.1. PHYSICAL CONDITIONS ON THE PREBIOTIC EARTH

According to current models of Earth's formation, the planet formed by accretion about 4.5 BYA (Lunine, 2006). The "late heavy bombardment" that lasted until around 4.0 BYA may have precluded life's survival (Nisbet and Sleep, 2001; Bada, 2004). My scenario begins soon after the last sterilizing impact with a hot and gradually cooling Earth whose surface was covered with a salty ocean (Knauth, 1998). Since the Earth's crust was recently formed and thin, terrestrial environments were likely limited to isolated volcanic islands (Mason, 1991; Lunine, 2006). One may presume that atmospheric and oceanic dynamic processes (e.g., storms, wind, tides, tectonic, and impact shocks, etc.) were much more severe than they are today, given high temperatures, bolide impact frequency, tectonics, and lunar proximity.

## 2.2. PREBIOTIC ORGANIC SYNTHESIS

Models of organic synthesis and accumulation on the prebiotic Earth are constrained by uncertainty concerning atmospheric gas composition and reducing conditions, the range of physical processes that could induce organic synthesis on the planet, the amount of organic material available from extraterrestrial sources, and the relative rates of synthesis and degradation (Bernstein, 2006). Many chemical evolution models limit themselves to reactions that occur under "moderate" conditions and organic compounds generated through Miller-type simulation experiments or those identified as being present in extraterrestrial bodies, notably the Murchison meteorite (Orgel and Lohrman, 1974). I believe this is a far too conservative approach, given stringent constraints of imagination, scale of simulations, and sensitivity of analytic techniques. The underlying assumptions of my scenario are that life began on Earth (rather than elsewhere) and that prebiotic synthesis and accumulation provided a sufficiently large, complex, and diverse mixture of organic molecules on the planet's surface to allow living systems to arise. Bernstein (2006) supports these assumptions.

## 2.3. ORGANIC PARTITIONING IN THE OCEAN

If one uses the conservative assumption that the bulk of the organic material formed on the prebiotic planet reflect the diversity and proportions of substances produced in simulation experiments and identified from meteorites, this material would include substantial amounts of small polar molecules (including amino acids), insoluble complex "kerogen," and nonpolar hydrocarbons. Certainly, depending upon conditions, hydrocarbons are a major product of Miller-type simulation experiments (Bada, 2004), and they represent a major fraction of organic carbon in meteorites (Mason, 1991). These different organic fractions would be expected to partition into layers within the ocean on the basis of density and solubility. Small polar molecules would dissolve, leading to a dilute "broth" within the water column; kerogen would form dense precipitates on the ocean's floor; low-density hydrocarbons would form a layer on the ocean's surface. Several authors have emphasized that organic matter synthesized on the early Earth could include massive amounts of hydrocarbons that would partition into an "oil slick" that would overlay the Earth's ocean and that the presence of such a slick could have dramatic implications for life's origins (Lasaga et al., 1971; Morchio and Traverso, 1999; Nilson, 2002; Griffith, 2009).

## 2.4. CHEMICAL REACTIONS WITHIN THE SURFACE LIPID LAYER

The superficial nonpolar "oil slick" theoretically represents a highly versatile platform for molecular accumulation and chemical reactions leading toward complex polymers like NA and peptides (Morchio et al., 2001; Griffith, 2009). There is a very substantial literature, most of it in technical reports, on the complex reactions that occur within oceanic oil slicks based upon contemporary oil spills (e.g., Patton et al., 1981). Unfortunately, other than the evolution of oil/water mixtures (mousse), such studies provide little insight into life's origin since conditions are dramatically different nowadays and oil spill researchers tend not to look at components or reactions really relevant to the question. Nonpolar molecules in the primordial slick would likely include heterocyclic molecules (including purines and pyrimidines), metallo-organic compounds (e.g., porphyrins) as well as esters and aliphatic and aromatic hydrocarbons. Aqueous droplets that contain polar solutes would enter the slick from the atmosphere (as rain) and water column (via convective processes), forming "reverse micelles." Among other components, micelles derived atmospherically could contain formaldehyde; those from the water column amino acids. Condensing dehydration reactions could lead to formation of peptides and nucleosides, driven by solar radiation and heating. As noted below, amphiphilic substances would accumulate at interfaces between water and lipid; these might include a variety of organic phosphate compounds (e.g., phospholipids) as well as peptides and nucleosides. Photochemical reactions could be expected to form reactive compounds, ultimately useful in coupled energy-requiring synthetic reactions. Perhaps the "soup theorists" were right after all, but complex reactions like the polymerization of amino acids into peptides did not occur within the aqueous "broth" as might be assumed – they occurred in a layer of fat on the top of the soup!

## 2.5. THE LIPID/AQUEOUS INTERFACE: THE SITE WHERE LIFE BEGAN

In addition to nonpolar substances that would partition at the ocean's surface, prebiotic chemical reactions (including those within the lipid layer) would lead to amphiphiles that possess both polar and nonpolar domains. These would include peptides, nucleosides, and "phospholipids," as well as some of the more hydrophobic carboxylic and amino acids and substances like bile salts. Amphiphiles would associate with the oil/water interface of "reverse micelles" within the lipid layer, and they would also accumulate at the interface between lipid layer and water column and around lipid droplets (micelles). Consequently, an emulsified zone with lipid droplets and vesicles of various sizes would occur, stabilized by amphiphiles. Such droplets and vesicles serve as the basis of the "lipid world" (Segré et al., 2001) and the "composome" scenario of life's origin (Hunding et al., 2006). In my scenario, they serve as the site where life originated; this interface receives a supply of peptides, NA, and reactive compounds from the overlying lipid layer, and amphiphilic peptides aggregate NA at this site, permitting replication and translation.

The ocean's surface layers would be subject to a variety of physical forces that would lead to convective transport of pre- and protobiological material. Internal waves which form at interfaces between liquids of different densities could be of particular significance since they could lead to horizontal and vertical transport of protobiological material and aid replication. I think, it is relevant to the question of life's origins and early evolution that material from oil slicks, including emulsified "mousse," undergoes rapid transport over very long distances, indeed on a global scale (Castanedo et al., 2006). Alternative sites that have been proposed for the origin of life such as mineral surfaces, hydrothermal vents, or isolated pools, aside from being limited in geographic scope, lack comparably extensive convective processes. It is hard to see how they could plausibly account for the origin of life and the development of complexity, notably the elaboration of the genetic code, that surely required experimentation on a grand scale.

## 3. The Arginine Translator

### 3.1. WHY MOLECULAR MUTUALISM WITH BOTH PEPTIDES AND NUCLEIC ACIDS?

The hypothesis that both peptides and NA were present from the beginning is by no means the most widely accepted explanation for how life originated. Most workers advocate either a "metabolism-first" model in which no NA was present or the "RNA world" in which NA (RNA only) but not protein was present. Metabolism-first scenarios do not directly address hereditability, so are generally not favored by biologists. The main argument against an NA plus peptide scenario from RNA world proponents is that such a model involves two complex information-rich molecules, not just one, so it is viewed as less parsimonious than the RNA world (Orgel, 1994). Of course, the RNA world presumes a large number of relatively long, homochiral, folded, and functional RNA strands (Jeffares et al., 1998) which is so implausible, from a logistical standpoint, that an advocate invoked the eternal inflation cosmological model to account for it (Koonin, 2007). Of course, neither metabolism-first nor the RNA world scenarios can easily account for the origin of translation and the genetic code.

The coevolutionary model developed here is based upon the simplest possible mutualistic relationship between peptides and NA; there is only one kind or short, replicating NA, and it interacts with a short peptide containing only one kind of amino acid. The NA/peptide interactions entail peptide-assisted replication, NA-driven translation, energy transformations, and compartmentalization. NA/peptide mutualism can reasonably account for the origin and evolution of the genetic code, and it can also readily explain the chirality of biological molecules, both NA and peptide. Root-Bernstein (2007) advocated the concept of complementarity between NA, amino acids, and peptides and pointed out how this can uniquely explain biological chirality.

## 3.2. WHY ARGININE?

Most schemes for the origin of the genetic code favor the primacy of simple amino acids like glycine, alanine, and glutamic acid that are easily formed in prebiotic simulation experiments and found most abundantly in meteorites (Trifonov, 2004; Copley et al., 2005). Complex amino acids like arginine are presumed to be late additions, following elaboration of synthetic metabolic pathways. The underlying assumption for such scenarios is that early molecular evolution was "progressive" in the sense that it proceeded from simple to complex molecules. Ironically, this view is promulgated by RNA world proponents who argue for a complex array of RNA strands prior to the inception of translation of (simple) amino acids into peptides. It would seem to be logical to me that prebiotic chemical evolution that could plausibly lead to the formation of purines and pyrimidines would also generate arginine, which is of comparable size and composition. This is aside from the fact that the RNA world requires synthesis of pentose sugars, the linkage of bases, sugar, and phosphate into nucleotides and polymerization of these into RNA. The layered ocean model I advocate allows the synthesis and polymerization of arginine (among many other amino acids) as well as the generation of nucleobases, formation of nucleosides and nucleotides and polymerization of these into NA.

My arguments in favor of arginine are based upon functional attributes of this amino acid and of the guanidinium group within peptides. Experimental evidence indicates arginine can be recognized with high specificity by NA aptamers bearing arginine's codon, suggesting that translation via stereospecific recognition is very reasonable for this particular amino acid (Geiger et al., 1996; Yarus et al., 2005). Arginine-enriched peptides bind to the backbone of NA tightly, and this could contribute to maintenance of otherwise water-soluble NA within the ocean's emulsified zone. A considerable literature has developed recently considering the use of oligoarginine peptides as transfection agents, able to encapsulate molecules (DNA, RNA, peptides, and pharmaceuticals) within membranous structures (e.g., Futaki et al., 2007). Research in this field is focused on the practical applications of arginine peptides in therapeutics and biotechnology and is seemingly unaware of possible relevance to origin of life studies. Arginine's guanidinium group forms phosphate bonds that are energetically equivalent to those in ATP. In contemporary organisms, arginine residues participate at active sites of diverse enzymes involved in phosphate transfer, and phosphoarginine is used as a muscle phosphagen in invertebrate animals (Griffith, 2009). In my scenario, phosphate transfer involving arginine residues drives both the elongation and editing of NA strands during replication and the polymerization of amino acids during translation.

## 3.3. GENERAL ASPECTS OF THE ARGININE TRANSLATORS'S COMPOSITION

As discussed above, my model for the first living system contains NA and peptides, specifically peptides enriched with arginine. The NA is a conventional polymer of purine and pyrimidine bases, pentose sugar, and phosphate nucleotides. Thus, my scenario has no need to explain the transition from some nonbiological equivalent (e.g., TNA or PNA) into modern forms of NA. I contend the primordial existence of conventional NA is plausible given my scenario's oil slick chemistry. In my model, both DNA and RNA (and mixed strands) are present – prebiotic chemistry would be expected to generate both ribose and deoxyribose (as well as other sugars) and both thymine and uracil (as well as other bases). Depending upon temperature and other conditions, NA (especially DNA) would form stable double strands; degradation would be much slower for true DNA than for RNA or mixed strands. As noted previously, both NA and peptide would be short because of parsimony; in the case of NA, reliable replication would favor short length.

## 3.4. REPLICATION

In my scenario, NA replication occurs through conventional Watson-Crick base pairing, with strand separation and alignment being assisted by environmental thermal cycles and mediated by arginine-enriched peptides (Griffith, 2009). Temperature cycling could be induced by vertical convection related to internal waves as well as through diurnal cycles. Temperature near the boiling point of water leads to NA melting, with cooling allowing renaturation or replication of new strands on templates. New strand formation commonly entails energetically neutral ligation of preexisting NA segments positioned on complementary sections of a template strand. Experimental studies suggest this is plausible under the conditions of my scenario (James and Ellington, 1997), even in the absence of peptides. Arginine-enriched peptides play a number of roles in replication: arginine-mediated phosphate transfer allows gaps to be filled with nucleotides; arginine peptides hold denatured strands and fragments together, facilitating renaturation and ligation; and arginine peptides protect NA from thermal degradation like the histones of hyperthermophilic archaeans.

## 3.5. TRANSLATION

I proposed that translation began with the direct stereospecific recognition of arginine by a set of complementary NAs; formation of Watson-Crick bonds aligned the amino acid in a suitable position to form a peptide bond with a preexisting arginine-enriched peptide (Griffith, 2009). The preexisting peptide is held to the NA strands through electrostatic bonds to the NA backbone. The energy for peptide bond formation was provided autocatalytically from high-energy guanidinium phosphate bonds present on terminal arginine residues in the peptide. Although energy for peptide bond formation during translation in contemporary organisms does not come directly from phosphate transfer, it does in the ATP-driven nonribosomal synthesis of glutathione (Janowiak and Griffith, 2005). It is worth emphasizing that the ability of arginine to form high-energy phosphate bonds and the length of this amino acid's "R" group make arginine almost uniquely capable of autocatalytic peptide bond formation. In our model, the activation of arginine's guanidinium group was accomplished through transfer from high-energy organic phosphate compounds (or other energy-rich molecules) whose synthesis was facilitated through photochemistry within the oil slick.

The model described above is not inherently able to initiate peptides de novo. When preformed, arginine-enriched peptides were not present. I suggested translation could be initiated through formation of a bond between arginine's amine group and an activated (acylphosphate or thioester) carboxyl group on an amphiphilic molecule. This could help ensure that the peptide/NA complex remained within the emulsified zone.

## 3.6. THE AUTOCOMPLEMENTARY HEXAMER AGCGCT

As a specific and testable model of the core of the first arginine translation system, I proposed the autocomplementary DNA hexamer 5′AGCGCT (Griffith, 2009). Arguments that I used to support this particular kind of NA, size, and sequence include the following: autocomplementary strands are parsimonious and efficient in replication and translation; DNA is favored because the system must function at high temperatures; high G/C content is also favored because of high temperature; a length of six base pairs represents a compromise between speed and reliability of replication and binding affinity for arginine and peptides; the sequence of bases includes the arginine codon CGC and the first two bases of arginine's second codon motif, AGN; because the strand is autocomplementary, it also displays anticodon motifs for arginine. Experimental evidence suggests NA strands that bear arginine codons could display high stereospecific affinity for this amino acid (Yarus et al., 2005). Translation of arginine by NA strands bearing the amino acids codon/anticodon represents a compelling model for the code's origin.

## 3.7. THE PROTOBIONT AS A GLOBAL SUPERORGANISM

Although the discussion thus far has focused on interactions between one particular palindromic NA, arginine, and arginine-enriched peptides, I do not suggest that these components alone represent the first living system or protobiont. Rather, the arginine-translating system serves as the core of a protobiont that encompasses all useful molecules of the environment that support the emergence of life.

These include, but are not limited to, lipid and amphiphilic molecules in the oil slick and emulsified zone, NA that can serve structurally as ribosomes, building blocks during replication, or novel genes, peptides – especially those with nonpolar domains that can stabilize NA/arginine peptide binding to amphiphilic vesicles, light-absorbing molecules including metallo-organics and terpenes in the oil slick and high-energy molecules that can drive arginine phosphorylation. From this perspective, the protobiont may be viewed as a global superorganism. Its evolution is addressed in the next section.

## 4. The Development of Complexity Through Molecular Coevolution

### 4.1. EVOLUTION OF STRUCTURE AND METABOLISM

#### *4.1.1. Variation, Natural Selection, and Cooperativity Drive Molecular Coevolution*

Compared to my hypothetical protobiont's core, a small family of nucleic acid strands that could incorporate only one kind of amino acid into peptides, the simplest prokaryote is an incredibly large and complex biological system. The simplest contemporary cell has a genome with millions of base pairs coding for thousands of distinct functional RNA strands and proteins and the intricate structural framework and metabolic network that these macromolecules direct. The process by which a very simple protobiont (e.g., the arginine translator) evolved the complexity of cellular life certainly included structural and metabolic advancements, but ultimately these required increased information content – NA diversity and elaboration of the genetic code. I propose that the underlying mechanism for the development of complexity in early living systems is molecular coevolution via natural selection. Implicitly, this coevolution requires both a source of variability in genetic material (e.g., novel NA strands, mutation, and recombination) and environmental selective pressures that favor certain NA sequences over others. Selection can work directly upon the structural and functional attributes of the NA or upon those of peptides that are translated. Furthermore, a critical aspect of coevolutionary natural selection in my model is that it involves selection of aggregates of NA strands and other molecules that function cooperatively, not merely for their own benefit.

#### *4.1.2. Composomes with Nucleic Acids and Peptides*

An interesting new concept has been offered by advocates of metabolism-first scenarios who propose that composomes, aggregates of small molecules supported by amphiphiles and lipids, may have constituted a cooperative metabolic web that preceded and eventually led to life based on NA and peptides (Hunding et al., 2006). Proponents of the composome model even suggest these aggregates show attributes of life that include reproduction and the capacity to evolve. The idea that a community of composomes lacking NA replication could evolve into more complex systems has been challenged (Vasas et al., 2010). I do not believe that the pre-NA composome model is necessary or sufficient to account for chemical evolution leading to life based on NA and peptides. Rather, the scenario that I propose here (chemistry in the oil slick and molecular accumulation in the emulsified zone; the latter also being a facet of non-NA composome models) is a more satisfactory model for chemical evolution. However, a community of composomes that possesses complementary NA and peptide (Root-Bernstein, 2007) could provide a powerful platform for early biological evolution – the evolution of cellular structure, metabolism, the translation apparatus, and the genetic code. Thus, I will incorporate the concept and the term composome into my scenario, but I emphasize that these have NA and peptides and they undergo replication and translation homologous with that of extant organisms.

*4.1.3. Evolution of Energy Metabolism*

As already discussed, the sources of bond energy for my arginine translator included preformed polymers, (especially polynucleotides) and high-energy organic phosphate compounds generated in the oil slick by photophosphorylation. Thioesters, generated through photochemistry or geothermal activity, represent another potential source of primordial bond energy (De Duve, 1991). Thus the earliest biological composomes were heterotrophic, relying on extraneous energy sources and arginine-based substrate level phosphorylation. Of course, from the viewpoint of the protobiont as a global superorganism, early life was autotrophic with both photosynthetic and chemosynthetic capabilities. Eventually, evolution of composomes led to prokaryotic cellular organisms that supplemented and then supplanted exogenous sources of bond energy with endogenous phosphorylation, using proton gradients. This achievement would depend upon developing several advanced features: a proton-tight membrane, membrane-bound proteins that can couple proton movement with phosphorylation, organic molecules that can undergo effective photochemical (or chemical) changes, and other molecules that could be used in electron transfer. Certainly, many functional elements that were required for endogenous photophosphorylation were present in the milieu, but their incorporation into a coherent system would have been a fairly late development.

*4.1.4. Evolution of Synthetic Pathways*

The capacity to synthesize building blocks like sugars, amino acids (especially complex ones like arginine), purines, and pyrimidines would become essential as the pool of preformed building blocks was depleted and as abiotic synthesis declined. Given my warm Earth model, I believe the most reasonable scenario for the development of anabolic metabolic pathways can be based upon the retrograde model of Horowitz (1945), where the evolution of synthetic pathways precedes from the terminal steps backward. Instead of a literal interpretation of Horowitz's scheme, I favor a model where enzymatic catalysis began at a pathway's rate limiting step or bottleneck, with the most unfavorable equilibrium constant. The synthesis of complex building blocks from simpler precursors obviously requires a supply of energy, but it may also require the ability to form new carbon-carbon bonds, mechanisms to conduct biological reduction and oxidation reactions, and a set of enzymes to catalyze these and other synthetic steps like amination reactions. Such catalytic versatility requires the ability to translate proteins with a number of functionally distinct amino acids. Thus, the evolution of synthetic pathways evolved in lockstep with the evolution of the translation machinery and the genetic code. Besides enzymes, a variety of sulfur-containing coenzymes, cofactors based upon nucleotides, and other complex accessory molecules would be required for synthetic reactions. Biosynthetic pathways for these "vitamins" would evolve as their availability from the milieu decreased, as we have suggested was the case for the building blocks of NA and proteins.

## 4.2. EVOLUTION OF THE GENETIC CODE

### *4.2.1. Genetic Variability and Genetic Exchange*

I have argued that the protobiont may be viewed as a continuous global superorganism with an open gene pool since there is no effective compartmentalization or barriers to prevent the incorporation of "foreign" NA. This concept is embedded in my model of a community of arginine-translating composomes, but it is by no means a novel idea. Woese (1998) proposed that in early biological evolution, there were minimal barriers to the free exchange of genetic material between living systems. Indeed, any DNA or RNA strands present in the environment can be incorporated into composomes, and if these strands are useful by facilitating replication or translation of useful peptides, they will be kept and amplified. If they are not otherwise useful, they can serve as building blocks during replication and sources of bond energy. By virtue of the synthesis of arginine-enriched peptides, arginine-translating composomes provide a local milieu that favors nucleic acid aggregation, replication, and "editing."

Modification of key NA elements of composomes could occur through several mechanisms. Efficient polymerases were absent, nucleic acid copying in the protobiont would be inherently inaccurate, and base substitutions and various other mutational errors frequent. We have pointed out that replication could entail alignment of partially complementary stands on templates; clearly, this could also lead to elongation. Depending upon the zones of complementarity between the original template strand and other NA fragments, this process could yield a new longer strand with the full template sequence at one end, the full template sequence may be inserted in the middle, or only a portion of the template strand may be incorporated in the new NA. Shorter strands can be produced by aborted replication.

### *4.2.2. The Central Dogma Is Not So Dogmatic*

The central dogma, where molecular information flows in the direction DNA to RNA to protein, would not always be the case early in molecular evolution. In my scenario, the central dogma would be completely ineffectual in regard to RNA and DNA since there are no specific polymerases for transcription and replication. Information flow from RNA to DNA being facile, functionally valuable RNA sequences can be transferred into more stable DNA copies. Indeed, it is possible that the amino acid sequence of functionally useful peptides can affect RNA (and ultimately DNA) base sequence. Reverse translation cannot occur in extant organisms because recognition of amino acids involves adaptor enzymes rather than direct specific binding by tRNA. In my arginine translator, the amino acid is directly recognized by the autocomplementary hexamer, and I argue that direct stereospecific recognition was also the case for the next few amino acids that were translated. In these cases, anticodon sequences would bind to the cognate amino acid. Thus, in the early evolutionary stages of translation, the amino acid sequence of an already formed peptide could direct the alignment of nucleotide bases into anticodon sequences. Replication would then yield complementary codon-bearing mRNA. Zhang and Egli (1994) have proposed a specific mechanism for primordial reverse translation. Peptide-driven organization of nucleotide sequence is an element of code evolution scenarios proposed by Root-Bernstein (2007) and Guimarães et al. (2008).

#### *4.2.3. Selective Pressures Leading Toward a More Complex Genome*

As the arginine translator evolved into a complex living system, selective forces operating on the protobiont's composomes would obviously favor added NAs that were stable and could be replicated easily. The premise of molecular coevolution implies that selection also favors functional attributes of NA that benefit composomes as an entirety, allowing the aggregate to persist in its environment, acquire resources, and reproduce. Hence, NA that (either itself or through translated peptides) increased the overall efficiency and fidelity of nucleic acid replication, promoted acquisition or synthesis of useful molecules, or facilitated the acquisition or transfer of bond energy would be a good neighbor within the composome and would therefore increase in abundance along with other useful composomal constituents.

My scenario's original system translated only one amino acid into peptides and had only one kind of autocomplementary NA strand that served as both tRNA and mRNA. In subsequent sections, I will discuss the expansion of the genetic code, increasing the number of amino acids translated, the equivalent number of NA molecules serving as anticodon-bearing tRNA, and the number and versatility of code-bearing mRNA strands. In addition to the information-related expansion of the genetic code, the evolution of composomal NA also includes improvements in structural elements important in translation – NA that functioned as scaffolds and perhaps catalytically in translation (rRNA) or in the folded sections of tRNA. I suggest the source of these structurally important NA strands was external to the original arginine translator, but incorporation occurred early in composome evolution. Ultimately, the result of the evolution of rRNA would be the various subunits of ribosomal RNA as displayed in living organisms. A number of workers have suggested a simple hairpin loop may have served as the earliest tRNA (e.g., Shimizu, 1995). The ultimate result of tRNA structural evolution would be the complex cloverleaf configuration of extant tRNA molecules that allows the proper positioning of the acylated amino acid for peptide bond formation. Of course, tRNAs possess specific binding sites (the anticodon and adaptor recognition sites) as well as the universal cloverleaf and the 3′ terminus. The utility and versatility of the kinds of amino acids recognized will be the main selection criterion for these exposed functional sections of tRNA strands. Those strands that serve as mRNA should be linear, lacking any significant internal bonding. The principal selective forces on mRNA will, of course, be directed toward the functional properties of peptides that are translated. Discussions of the evolution of mRNA and tRNA strands that participate in the translation of useful peptides are given below.

#### *4.2.4. Scenarios for the Code's Expansion*

The genetic code represents what is perhaps the most intriguing puzzle in science. The pattern by which amino acids and punctuation are assigned codons is neither random nor obviously logical. It invites explanation. The literature related to evolution of the genetic code is large, eclectic in approach, and without consensus. The proposed sequence in which codon assignment evolved is contingent on assumptions of which amino acids were first coded (e.g., were "simple" amino acids or "functional" amino acids first, and if "functional" what kind of functional role was important) as well as the causal processes by which expansion occurred. Among prominent hypotheses for the code's evolution are the frozen accident hypothesis (Crick, 1968), the ambiguity reduction model (Fitch, 1966), the vocabulary expansion scenario (Crick, 1968), the "coevolutionary" hypothesis (Wong, 1975; Di Giulio, 1997), and various stereoselective models, older variants summarized by Ponnamperuma and Hobish, (1984) with some recent models developed by Root-Bernstein (2007), Guimarães et al. (2008), and Yarus et al. (2009). I do not think that one particular model among those already proposed is fully explanatory; rather, I suggest that the elaboration of the genetic code occurred through an eclectic mix of processes that included stereoselective phenomena (particularly early on), vocabulary expansion, ambiguity reduction, and metabolic coevolution. Although the specific causes behind individual codon assignments may have varied and the details of the process may always be obscure because of its contingent nature, the underlying driving force behind the elaboration of the code and the overall development of biological complexity was Darwinian evolution through natural selection.

#### *4.2.5. Early Stereospecific Additions: Nonpolar and Acidic Amino Acids*

The arginine translator provides composomes with numerous useful attributes, but there are limitations on the versatility of such a simple system. There is no obvious way for it to initiate peptide synthesis; rather, it simply sticks arginine onto preformed (and activated) peptides or other carboxylic acids. Furthermore, arginine-enriched peptides are hydrophilic and may not stick tightly to lipids within composomes unless the peptides also possessed hydrophobic domains (to some extent the system accomplishes this by starting with activated hydrophobic acids). I think it likely that the arginine translator quickly incorporated mechanisms to initiate peptides and translate amino acids with hydrophobic residues. Acidic amino acids represent another functional class that would be of compelling structural and functional utility to the protobiont since such amino acids could readily interact with arginine residues. As discussed below, I favor a scenario in which both hydrophobic and acidic amino acids enter translating composomes simultaneously.

Guimarães et al. (2008) suggested that additions of amino acids to the genetic repetoire involved pairs with different complementary properties (e.g., hydrophobic-hydrophilic), and they also suggest that these pairs are coded by complementary NA strands, a given strand displaying the codon of one amino acid of the pair and the anticodon of the other. Even though their arguments were not specifically related to the functional roles of these amino acids, in their scenario, leucine and glutamic acid were added together and were among the earliest additions to the code.

My suggested candidate for a set of complementary NA strands coding for nonpolar and polar (acidic) amino acids maintains the same strand length as the arginine translator and is repetitive. The sequence CUCUCU bears leucine's codon and glutatate's anticodon, and its complement GAGAGA bears glutamate's codon and leucine's anticodon. I suggest that the stereospecific recognition need not be very precise for hydrophobic amino acids and that aspartic acid would serve as well as glutamic acid. Arginine residues in separate peptides might provide energy for peptide bond formation. Alternatively, glutamate and aspartate can undergo internal cyclization, which could lead to formation of peptide bonds through the diketopiperazine reaction (Hartman et al., 1981). Peptides with alternating hydrophobic and acidic residues would be able to associate with lipid components of composomes (leucine residues) and with arginine-enriched peptides (glutamate residues).

Clearly, the three kinds of code-bearing NA hexamers of the current model could be stuck together to form a variety of mRNAs, allowing the translation of longer, complex peptides. The hexamers would serve as tRNA's for the amino acid whose anticodon they bear (e.g., CUCUCU for glutamate, GAGAGA for leucine, and AGCGCT for arginine). The system now is able to develop the capacity to translate a growing array of rather sophisticated peptides incorporating three different types of functional amino acids.

#### *4.2.6. tRNAs and Aminoacyl Transferases*

The next evolutionary step I suggest for the evolution of translation is a transition from linear hexameric "tRNA" strands that hold cognate amino acids stereospecifically, to loop tRNA molecules with internal bonding that hold cognate amino acids through aminoacyl bonds on a terminal stem. These new tRNAs may resemble the hairpin RNAs advocated by Shimizu (1995) in general size and shape. As the protobiont switches from the use of the complementary hexamers as "tRNA" to reliance on looped specific tRNAs, a transitional state could exist where both hexamer and looped tRNAs can work. The new tRNA strands include an anticodon-bearing segment within a loop, a segment with internal bonding, and a stem that ends with an unpaired terminal nucleotide (e.g., adenosine) that holds the cognate amino acid via an aminoacyl bond.

This transition from amino acid stereospecific binding to aminoacylation is contingent upon development of a set of aminoacyl transferases that recognize a specific amino acid and its tRNA and can catalyze aminoacylation. These transferases display three functional zones of significance: an amino acid recognition section, a tRNA (anticodon) recognition section, and a catalytic phosphate transfer section. A logical basis for amino acid recognition is complementarity (e.g., for the amino acids already discussed, nonpolar leucine residues will "recognize" leucine; arginine will be recognized by a core of glutamate residues, glutamate will be recognized by a core of arginine residues). Recognition of tRNA by transferases would involve stereospecific recognition of anticodons within the tRNA's loop, leucine tRNA by leucine residues, glutamate tRNA by glutamate residues, and arginine tRNA by arginine residues (in the case of arginine, the loop includes arginine's codon as well as its anticodon).

It has been proposed that the two classes of transferases, which differ in structure, the manner of interaction with tRNA, and the position of aminoacylation, originated as sense and nonsense strands of the same NA (Rodin and Ohno, 1995). It would be attractive to my model if the amino acids proposed to be the first translated were recognized by transferase enzymes of separate classes. However, arginine, glutamate, and leucine are all recognized by class I transferases. A possibility that will be explored in the next section is that looped tRNAs and aminoacyl transferases for two amino acids entered the protobiont's genome together and that perhaps the two classes of transferases were originally translation products of these tRNA strands.

#### *4.2.7. Filling in the Code*

The sequence in which new amino acids become specified by the code is related to the functional properties of their "R" groups since the capacity to incorporate the new amino acid into protein must provide benefit to composomes in order to be retained. New additions to the code do not use direct recognition by tRNA so that subsequent amino acids or pairs of amino acids need not be complementary to anticodons; transferase enzymes fulfill the recognition role. At this point in my scenario, I suspect that my suggested sequence of additions to the code will become more and more speculative and incapable of providing specific models for testing. I will be brief.

Serine may have been the fourth amino acid to enter the code. Serine would provide the protobiont with an alternative site for NA binding, one with subtler binding affinity. It would enhance the metabolic capacity of the protobiont and would be able to bind phosphate. Serine is recognized by a group II aminoacyl transferase, and I propose it is the incorporation of this amino acid that begins tRNA/transferase-based translation. After this point, triplets rather than hexamers can function as codons and anticodons. When serine enters the translating repertoire, using a looped tRNA and transferase, some of arginine's codons/anticodons (e.g., AGN) also acquire a looped tRNA and transferase. The tRNA's for these two amino acids are complements of each other (e.g., serine tRNA could bear the anticodon AGA, arginine UCU, their shapes would be different as would the 3′ terminal nucleotide). Although it need not have been the case and certainly would not persist, the initial aminoacyl transferases may have been translated from this first set of tRNAs.

I believe it is possible that the "methionine/start" AUG codon was an early specific addition to the composomes' genome. The principle of continuity argues for the conservation of the original start codon and translation efficiency would benefit from imposed directionality. Furthermore, the ability to translate "methionine" provides considerable functional capability, especially the utilization of thioester bond energy. I favor the primacy of homocysteine, methionine's non-methylated precursor, rather than methionine per se. Homocysteine's replacement by methionine would be a late evolutionary step, contingent upon the incorporation of cysteine in the code and other sulfur – containing molecules in metabolism. The sulfhydryl and carboxyl groups of homocysteine react to form a cyclic thiolactone that possesses a high-energy thioester bond. Once incorporated into a polypeptide, homocysteine can catalyze other biosynthetic reactions that require the transfer of groups or bond energy. Furthermore, homocysteine residues can form thioester, disulfide, or thioether bonds with amphiphilic molecules in the milieu or with other homocysteine residues, leading to the folding or aggregation of proteins and linkage with other useful molecules. The incorporation of methionine involves modifications in the tRNA and aminoacyl transferase for arginine or for nonpolar amino acids.

I propose that the next addition(s) to the code involved codons, tRNA, and transferase(s) for nonspecific or "generic" amino acids lacking a truly distinctive "R" group. Such codons would tend to incorporate amino acids that are most abundant in the milieu like glycine and alanine and could provide spacing crucial to protein structure.

With the capacity to incorporate six different kinds of amino acids (arginine, homocysteine, nonpolar, acidic, serine, and "neutral"), it becomes possible to make much more specific adaptor proteins that recognize additional kinds of amino acid. Among the first set of additional amino acids that are specified, I suggest the following: lysine (favored for its catalytic role and allowing better recognition of nucleotide bases by polymerase enzymes), cysteine (takes over several structural and functional roles formerly performed by homocysteine), and aromatic hydrophobic amino acids (e.g., phenylalanine). At this point, new amino acids added to the code would be structurally similar to or metabolic precursors of those already coded, requiring only small modifications of anticodons, codons, and transferases. It seems reasonable at this stage that transferase enzymes could function catalytically in synthesis of their complex cognate amino acids, not just their recognition. Minor changes in codons, anticodons, and transferases could add the substrate amino acid to the genetic code. Thus, the evidence that supports the coevolutionary theory of Wong (1975) would reflect retrograde metabolic evolution. Ultimately, the complete genetic code for 20 amino acids (plus punctuation) evolves as the complexity and sophistication of metabolic pathways grows and the specificity of adaptor proteins and fidelity in translation develops. Structural improvements that facilitate recognition of the tRNA by proteins and aid proper alignment on the mRNA and ribosome lead to the evolution of the characteristic cloverleaf shape of extant tRNA molecules from the simple single loop of our early tRNAs.

## 5. References


Bada JL (2004) How life began on earth: a status report. Earth Planet Sci Lett 226:1–15

Bernstein M (2006) Prebiotic materials from on and off the early earth. Philos Trans R Soc B 361: 1689–1702

Blum HF (1951) Time's arrow and evolution. Princeton University Press, Princeton

Cairns-Smith G (1982) Genetic takeover and the mineral origins of life. Cambridge University Press, Cambridge

Castanedo S, Medina R, Losada I, Vidal C, Mendéz FJ, Osario A, Juanes JA, Puente A (2006) The prestige oil spill in Cantabria (Bay of Biscay). Part I: operational forecasting system for quick response, risk assessment and protection of natural resources. J Coast Res 22:1474–1489

Copley SD, Smith E, Morowitz HJ (2005) A mechanism of association of amino acids with their codons and the origin of the genetic code. Proc Natl Acad Sci USA 102:4442–4447

Crick FHC (1968) The origin of the genetic code. J Mol Biol 38:367–379

De Duve C (1991) Blueprint for a cell: the nature and origin of life. Neil Patterson Publishers, Burlington

Deamer D, Singaram S, Rajamani S, Kompanichenko V, Guggenheim S (2006) Self assembly processes in the prebiotic environment. Philos Trans R Soc B 361:1809–1818

Di Giulio M (1997) On the origin of the genetic code. J Theor Biol 187:573–581

Di Giulio M (2005) The origin of the genetic code: theories and their relationships, a review. Biosystems 80:175–184

Ferris JR (2006) Montmorillonite-catalysed formation of RNA oligomers: the possible role of catalysts in the origin of life. Philos Trans R Soc B 361:1777–1786

Fitch WM (1966) The relation between frequencies of amino acids and ordered trinucleotides. J Mol Biol 16:1–8

Futaki S, Nakase L, Tadokoro A, Takeuchi T, Jones AT (2007) Arginine-rich peptides and their internalization mechanisms. Biochem Soc Trans 35:784–787

Geiger A, Burgstaller P, von der Eltz H, Roeder A, Famulok M (1996) RNA aptamers that bind L-arginine with sub-micromolar dissociation constants and high enitioselectivity. Nucleic Acids Res 24:1029–1036

Griffith RW (2009) A specific scenario for the origin of life and the genetic code based on peptide/oligonucleotide interdependence. Orig Life Evol Biosph 39:517–531

Guimarães RC, Moreira CHC, de Farias ST (2008) A self-referential model for the formation of the genetic code. Theor Biosci 127:249–270

Hartman J, Brand MC, Dose K (1981) Formation of specific amino acid sequences during thermal polymerization of amino acids. Biosystems 13:141–147

Horowitz NH (1945) On the evolution of biochemical synthesis. Proc Natl Acad Sci USA 31:153–157
Hunding A, Kepes F, Lancet D, Minsky A, Norris V, Raine D, Sriram K, Root-Bernstein R (2006) Compositional complementarity and prebiotic ecology in the origin of life. Bioessays 28:399–412
James KD, Ellington AD (1997) Surprising fidelity of template-directed chemical ligation of oligonucleotides. Chem Biol 4:595–605
Janowiak BE, Griffith OW (2005) Glutathione synthesis in *Streptococcus agalachae*: one protein accounts for gamma-glutamylcysteine synthetase and glutathione synthetase activities. J Biol Chem 280:11829–11839
Jeffares DC, Poole AM, Penny D (1998) Relics from the RNA world. J Mol Evol 46:18–36
Joyce GF (2002) The antiquity of RNA-based evolution. Nature 418:214–221
Knauth AP (1998) Salinity history of the earth's early ocean. Nature 395:554–555
Koonin EV (2007) The cosmological model of eternal inflation and the transition from chance to biological evolution in the history of life. Biol Direct 2:15
Kunin V (2000) A system of two polymerases – a model for the origin of life. Orig Life Evol Biosph 30:459–466
Lahav N (1993) The RNA-world and coevolutionary hypotheses and the origin of life: implications, research strategies and perspectives. Orig Life 23:329–344
Lasaga AC, Holland HD, Dwyer MJ (1971) Primordial oil slick. Science 174:53–55
Lincoln TA, Joyce GF (2009) Self-sustained replication of an RNA enzyme. Science 323:1229–1232
Line MA (2002) The enigma of life and its timing. Microbiology 148:21–27
Lunine JI (2006) Physical conditions on the early earth. Philos Trans R Soc B 361:1721–1731
Maizels N, Weiner AM (1987) Peptide-specific ribosomes, genomic tags and the origin of the genetic code. Cold Spring Harb Symp Quant Biol 52:743–749
Mason SF (1991) Chemical evolution: origin of the elements, molecules and living systems. Clarendon, Oxford
Miller SL (1953) A production of amino acids under possible primitive earth conditions. Science 117:528–529
Miller SL (1987) Which organic compounds could have occurred on the early Earth? Cold Spring Harb Symp Quant Biol 52:17–27
Morchio R, Traverso S (1999) The hydrophobic superficial layer: the primordial cradle of life. Biol Forum 92:105–117
Morchio R, Redaelli A, Traverso S (2001) Proteins, nucleic acids and genetic codes. Biol Forum 94: 37–57
Morowitz HJ, Kostelnik JD, Yang J, Cody GD (2000) The origin of intermediary metabolism. Proc Natl Acad Sci USA 97:7704–7708
Nashimoto M (2002) The RNA/protein symmetry hypothesis: experimental support for reverse translation of primitive proteins. J Theor Biol 209:181–187
Nilson FPR (2002) Possible impact of a primordial oil slick on the atmospheric and chemical evolution. Orig Life Evol Biosph 32:247–253
Nisbet EG, Sleep NH (2001) The habitat and nature of early life. Nature 409:1083–1091
Oparin AI (1938) The origin of life (trans: Morgulis S). Dover Reprint, New York
Orgel LE (1968) Origin of the genetic apparatus. J Mol Biol 38:381–393
Orgel LE (1994) The origin of life on the earth. Sci Am 271(4):76–83
Orgel LE, Lohrman R (1974) Prebiotic chemistry and nucleic acid replication. Acc Chem Res 7:368–377
Patton JS, Rigler MW, Boehm PD, Fleet DL (1981) Ixtoc 1 oil spill: flaking of surface mousse on the Gulf of Mexico. Nature 290:235–238
Ponnamperuma C, Hobish MK (1984) The stereochemical approach to studies of the origin of the genetic code. In: Matsuno K, Dose K, Harada K, Rohlfing DL (eds) Molecular evolution and protobiology. Plenum Press, New York, pp 295–312
Rodin SN, Ohno S (1995) Two types of aminoacyl-tRNA synthetases could be originally encoded by complementary strands of the same nucleic acid. Orig Life Evol Biosph 25:565–589
Root-Bernstein RS (2007) Simultaneous origin of homochirality, the genetic code and its directionality. Bioessays 29:689–698

Root-Bernstein RS, Dillon PF (1997) Molecular complementarity I: the complementarity theory of the origin and evolution of life. J Theor Biol 188:447–479
Segré D, Ben-Eli D, Deamer DW, Lancet D (2001) The lipid world. Orig Life Evol Biosph 31:119–145
Shapiro R (2006) Small molecule interactions were central to the origin of life. Q Rev Biol 81:105–125
Shimizu M (1995) Specific aminoacylation of C4N hairpin RNAs with the cognate aminoacyl adenylates in the presence of a dipeptide: origin of the genetic code. J Biochem 117:23–26
Shock E (1992) Chemical environments of submarine hydrothermal systems. Orig Life Evol Biosph 22:67–107
Tervahattu H, Tuck A, Vaida V (2004) Chemistry in prebiotic aerosols: a mechanism for the origin of life. In: Seckbach J (ed) Origins, genesis, evolution and diversity of life, vol 6, Cellular origin and life in extreme habitats and astrobiology. Springer, Dordrecht, pp 153–165
Trifonov EN (2004) The triplet code from first principles. J Biomol Struct Dyn 22:1–11
Vasas V, Szathmáry E, Santos M (2010) Lack of evolvability in self-sustaining autocatalytic networks constraints metabolism-first scenarios for the origin of life. Proc Natl Acad Sci USA 107:1470–1475
Wächtershauser G (1988) Before enzymes and templates: theory of surface metabolism. Microbiol Rev 52:452–484
Woese CR (1967) The genetic code: the molecular basis for gene expression. Harper & Row, New York
Woese CR (1998) The universal ancestor. Proc Natl Acad Sci USA 95:6854–6859
Wolf YI, Koonin EV (2007) On the origin of the translation system and the genetic code in the RNA world by means of natural selection, exaptation and subfunctionalization. Biol Direct 2:14
Wong JT (1975) A co-evolution theory of the genetic code. Proc Natl Acad Sci USA 72:1909–1912
Yarus M, Caporaso JG, Knight R (2005) Origins of the genetic code: the escaped triplet hypothesis. Annu Rev Biochem 74:179–198
Yarus M, Widmann JJ, Knight R (2009) RNA-amino acid binding: a stereochemical era for the genetic code. J Mol Evol 69:406–429
Zhang S, Egli M (1994) A hypothesis: reciprocal information transfer between oligoribonucleotides and oligopeptides in prebiotic molecular evolution. Orig Life Evol Biosph 24:495–505

Biodata of **Richard Lathe**, author of "***Tidal Cycling and the Origin of the Genetic Code: Implications for Cellular Life***."

**Richard Lathe** studied Molecular Biology in Edinburgh, UK and completed his doctoral studies in 1976 at the Department of Molecular Biology at the University of Brussels, Belgium. He has worked in France, Belgium, Germany, and the UK and has held professorships at the Universities of Strasbourg and Edinburgh. He is currently at Pieta Research, a bioscience research and consultancy company in Edinburgh, UK and associate professor at the Institute of Bio-Organic Chemistry, Pushchino, Russian Federation. Current interests in addition to astrobiology include molecular neuroscience, with particular focus on Alzheimer's disease and autism, and the role of physiological dysregulation in brain disorders.

E-mail: **rlathe@pieta-research.org**

J. Seckbach (ed.), *Genesis - In The Beginning: Precursors of Life, Chemical Models and Early Biological Evolution*, Cellular Origin, Life in Extreme Habitats and Astrobiology 22, 691–707
DOI 10.1007/978-94-007-2941-4_34, 

# TIDAL CYCLING AND THE ORIGIN OF THE GENETIC CODE: IMPLICATIONS FOR CELLULAR LIFE

**RICHARD LATHE**[1,2]
[1]*Pieta Research, PO Box 27069, Edinburgh EH10 5YW, UK*
[2]*Institute of Bio-Organic Chemistry, Pushchino, Moscow Region 142290, Russian Federation*

## 1. Introduction

The origin of the genetic code is not understood. Clearly, this was one of the earliest events during the evolution of terrestrial life forms because the same code is essentially present in all organisms. I have argued that the earliest replication of nucleic acids at 3.9 Ga was driven by cycling tidal inundation and drying (Lathe, 2004, 2005). This chapter addresses whether the proposed tidal chain reaction could give insights into the origin of polypeptide synthesis and the genetic code and, speculatively, cellular life. Given space constraints, this paper does not attempt systematic overview, but for recent insightful reviews of the field, see Osawa et al. (1992), Szathmáry (1999), Knight et al. (1999), and Di Giulio (2005).

## 2. Tidal Chain Reaction (TCR)

It has long been held that alignment of nucleotides with a complementary strand can favor the formation of internucleotide bonds, such that one strand can give rise to an accurate complementary copy; the concept is implicit in the structure of DNA (Watson and Crick, 1953). However, as pointed out by Blum (1957), in the absence of any mechanism for removing the complementary strand, the conversion is a dead end. TCR affords a mechanism for cyclic dissociation and reassociation, and therefore for amplification. Briefly, concentration of salts in the early ocean was undoubtedly lower than today. Flooding at high tide produced low salinity, whereas drying of sands, pools, and lakes at low tide rapidly increased both salinity and precursor concentrations.

Cyclic changes in salinity have the following important consequence. The structure of nucleic acid is unusual in that strong hydrogen-bonding interactions (base pairing) alternate with equally strong charge repulsion (opposing phosphate groups). This arrangement has the following properties. At high salt concentration (dry-down at low tide), anions bind to the opposing phosphates, neutralizing their charges, and pairing of complementary strands (or complementary nucleotides)

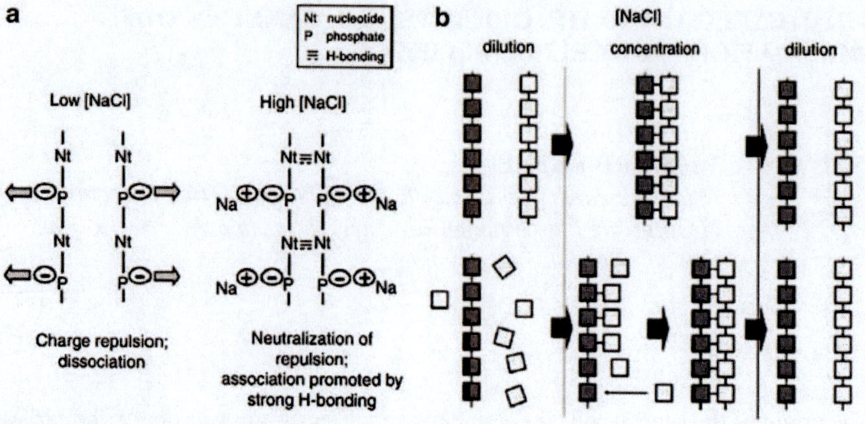


**Figure 1.** (**a**) Salt-dependent association and dissociation of nucleic acids; at low salt (tidal inundation, *left*) the two strands are driven apart by charge repulsion; high salt concentration (low-tide drying phase, *right*) neutralizes charge repulsion and leads to duplex formation. (**b**) Cyclic changes in salt concentration lead to tide-dependent association and dissociation of nucleic acids (*above*), and also cyclic polymerization of precursors on a nucleic acid template, leading to sequence amplification (*below*) (Figure adapted from Lathe, 2006).

can now take place driven by base pairing. Conversely, at low salt concentration (high tide), anion concentration falls, and the two strands are now driven apart by charge repulsion (Fig. 1) (Lathe, 2004).

In the TCR scenario, given sufficient precursor concentrations, cyclic amplification of specific nucleic acid sequences will take place, thereby providing a substrate for Darwinian evolution.

A computer simulation of TCR has lent valuable support to the theory. Using plausible parameters for nucleotide association, dissociation, bond formation, and fragmentation, Fernando et al. (2007) were able to show that tidal cycling accelerated polymer formation by approximately an order of magnitude. The authors drew attention to complex secondary structures in the prototype nucleic acid, emphasizing that, for example, hairpins could have a marked effect on the outcome of the replication pathway.

This chapter now addresses the possibility that TCR could also have driven the evolution of protein-coding sequences and possibly of cell-based life.

## 3. Tidal Cycling, Linkage of Amino Acids to Nucleotides, and the Genetic Code

Cyclic concentration and covalent linkage of nucleic acid chains during the drying phase inevitably took place in the presence of significant concentrations of amino acids. Clearly, during the drying phase, there were important associations between nucleic acids and amino acids and, as first proposed by Gamow (1954) and Woese (1965), these could underlie the emergence of coding. In the following, no rigid distinction is made between DNA and RNA, nor between T and U – both DNA and RNA are likely to have been preceded by a simpler and more variable chemical structure (Lathe, 2005).

Before one can address the present day code, it is expedient to apply some logical simplifications so as to reveal the most primitive code. It is clear that, at the beginning, only a few different amino acids (or groups of amino acids) were encoded. Riddle et al. (1997) point out that only four or five amino acids – for example, A/G, I, E, and K – are necessary to generate a modern complex protein fold. Woese (1965) speculated that early life used "a very few amino acids" and subdivided the spectrum of amino acids into two categories – functional (such as Y, H, K, E, W) and nonfunctional (including F, L, V, A, T). Osawa et al. (1992) envisaged a primitive code determining only three or four categories – hydrophobic (Pho), small uncharged (Sml), and mixed/basic (Mxd/Bas). Fitch and Upper, applying an ambiguity reduction concept, proposed that ultimately the primordial genetic code only differentiated between hydrophobic (Pho) and hydrophilic (Phl) acids (Fitch and Upper, 1987). Görnerup and Jacobi (2010) used "agglomeration" analysis to reduce all amino acids to Bas versus Pho/Sml. At the origin of the code, therefore, it seems likely that very few amino acids (or groups of related amino acids) were encoded.

Importantly, the crucial "coding" nucleotide has been argued to be at the central position of the codon. Woese (1965) pointed out that the central position is the least variable/error prone. Perlwitz et al. (1988) reduced the genetic code to A/G at the central position predominantly encoding Lys/Arg (Bas). Fitch and Upper (1987) suggested a primordial code only based on the central nucleotide that distinguished between –Y– (Y, pyrimidine: C/T/U) encoding Pho, and –R– (R, purine: A/G) encoding Phl. Osawa et al. (1992) envisaged an archetypal genetic code aligned by the codon central position: A, Pho; G, Sml, U and C, Mxd/Bas. Bashford et al. (1998) used a supersymmetry model to regress the code to two or possibly three archetypal codons: –A–, –T–, and possibly –G/C–.

Building on these works, a dendrogram model of genetic code evolution can be established (Fig. 2). The diagram is based on the assumption that purine–purine and pyrimidine–pyrimidine replacements are the most conservative. This assumption is not necessarily all-encompassing because important functional groups are conserved between other nucleotides. With this caveat, the present-day genetic code can be represented as a branching tree that reduces to two parental nucleotide triple codons: (1) –A–, where the dash represents any nucleotide, and specifying a basic amino acid (represented today by lysine, arginine and potential relatives including ornithine), and (2) –Y–, specifying a small or relatively hydrophobic amino acid (Sml/Pho, represented today by leucine, serine and relatives). For further details see Fig. 2 legend. This simplification of the genetic code now permits further observations on the origin of the code.

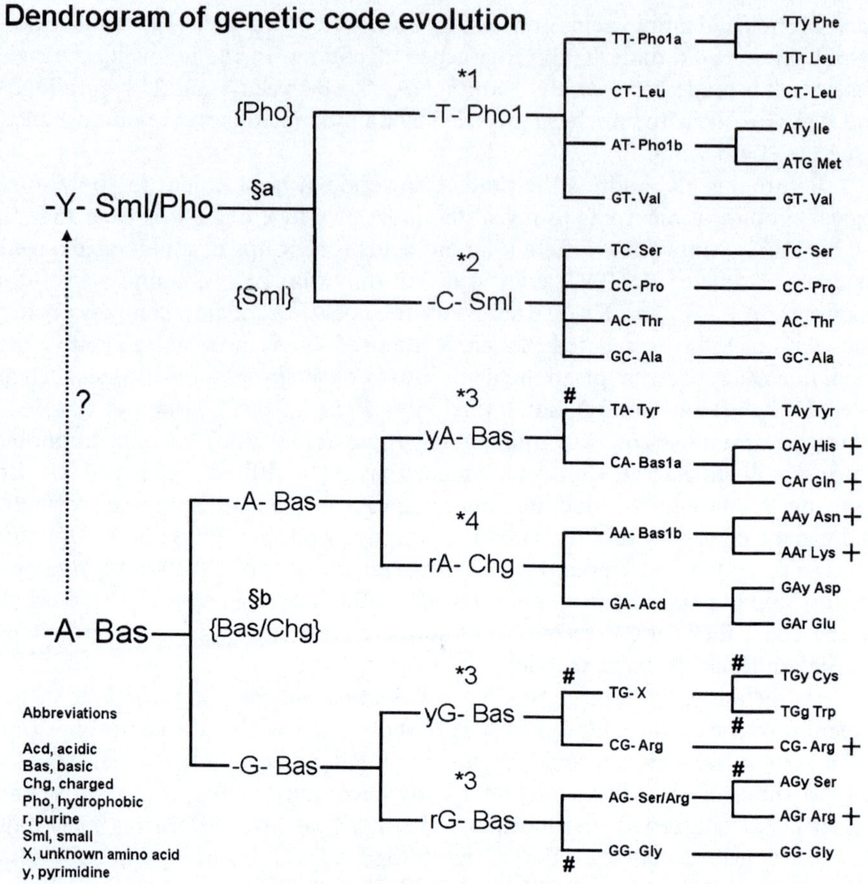


**Figure 2.** Simplification of the genetic code. The modern genetic code is shown far *right*. The diagram suggests possible sequential steps in the divergence from a primordial binary code at the *left*. The *dotted line* linking *–A–* and *–Y–* indicates that it is possible that the earliest code was restricted to the encoding of basic amino acids and that the early binary code (Bas versus Sml/Pho) was a secondary development. The dendrogram is to be compared with Fitch and Upper (1987), Fig. 1, in which the first genetic code was proposed to distinguish only between –R– encoding hydrophilic amino acids (Phl) and –Y– encoding Pho. Perlwitz et al. (1988), Table III, concurs with central position A/G encoding Lys/Arg. Bashford et al. (1998) regress the code by a similar but not identical mechanism to that proposed here, culminating in two or possibly three central nucleotides – A, U/T, and possibly G/C – whereas Fig. 3 of Görnerup and Jacobi (2010) points to two archetypal groups enriched in Bas versus Sml/Pho. *Numbered asterisks* indicate the Riddle groups of amino acids that are sufficient to generate a complex protein fold (Riddle et al., 1997), *§a* and *§b* indicate the two primordial groups proposed by Fitch and Upper (1987); *bracketed* {} groups indicate the archetypal subgroups aligned by central codon position according to Fig. 15 of Osawa et al. (1992); #, Codon recruitment (rather than divergence) that can only have taken place following the emergence of highly structured tRNA (see text); +, codons encoding basic amino acids, and that systematically commence with A or C (see Sect. 3.4).

## 3.1. A CENTRAL ISSUE: PAIR-CODING OR CHARGE-CODING?

In delving deeper into the origins of the code, and how the triplet –A– might encode a basic amino acid (and the reciprocal –Y– a small or hydrophobic amino acid), a central issue must first be resolved – did the code operate via pair-coding or charge-coding? To understand this crucial distinction, one must consider that the sequence of amino acids in a modern protein is governed not by the codon sequence but by specificity of the attachment of amino acids to the 3′ end of the tRNA. If a modern tRNA is charged with a different amino acid, the growing polypeptide chain now incorporates that new amino acid at the position corresponding to the now "incorrect" codon. The central issue of pair-coding and charge-coding is depicted in Fig. 3.

Primordial charge-coding (Fig. 3b) is the only scenario compatible with the inferred evolution of tRNAs and amino acid charging. tRNA charging is now catalyzed by enzymes, but in the first steps towards the evolution of genetic coding, this charging surely took place in the absence of any enzyme, driven purely by chemical affinity and reactivity. An evolutionary scenario for the evolution of tRNA charging is depicted in Fig. 4. This view extends previous work; for example, Szathmáry (1993) proposed a conceptually similar scheme in which amino acids were added to the 3′ ends of precursor tRNAs, but differs from other approaches in which amino acids displayed affinity for codons rather than anticodons.

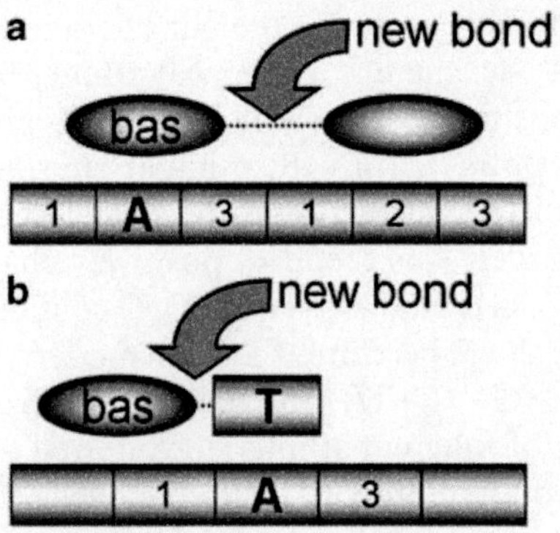


**Figure 3.** The central issue: pair-coding versus charge-coding. (**a**) Pair-coding: affinity between a basic amino acid (*Bas*) and the central triplet nucleotide A favors the incorporation of the Bas residue into a growing polypeptide chain. (**b**) Charge-coding: affinity between a basic amino acid and one or other member of an A–T base pair leads to selective covalent attachment ("charging") of the amino acid to the T nucleotide. The diagram makes no strict assumptions regarding the 5′–3′ orientation of the nucleic acid or N–C orientation of the amino acids. *Square boxes*, nucleic acid bases; *ovals*, amino acids.

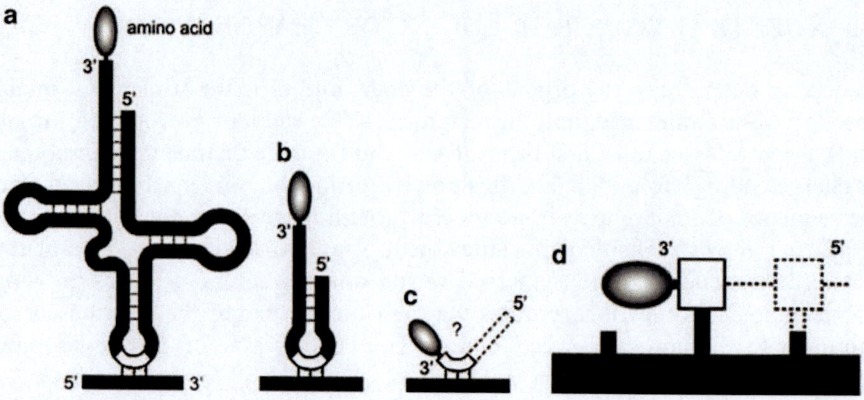


**Figure 4.** Inferred evolution of tRNA and tRNA charging. (**a**) Modern tRNA charged with an amino acid and paired with coding mRNA, and (**b**) inferred primitive tRNA. (**c**) Hypothetical primary precursor to tRNA, where attachment of the amino acid to the 3′ end is catalyzed enzymatically. The ? is to indicate that the existence of a triplet tRNA precursor is unlikely. (**d**) (*Enlarged*): the inferred primordial step where charging of the terminal nucleotide of a short oligonucleotide, perhaps as small as a dinucleotide, takes place in the absence of any enzyme.

## 3.2. EXPLORING THE FIRST CHARGING STEP: DIFFERENTIAL AMINO ACID–NUCLEOTIDE AFFINITIES

It was previously pointed out (Melcher, 1974; Weber and Lacey, 1978) that there is a stereochemical correlation between anticodon nucleotides and their amino acids. This contention is borne out by the considerations discussed below. As presented in Figs. 3b and 4c, there is a chemical affinity between specific amino acids and the central position of the anticodon and, pivotally, between basic amino acids and the nucleotide T (or with a similar but simpler primordial nucleotide such as U). Electrostatic attraction is the strongest contender to explain a specific interaction. In order, A is the most positively charged, with 5 nitrogens (N) and zero oxygens (O); G contains 5N and 1O, C contains 2N and 1O, whereas T (U) contains 2N and 2O. Therefore, the four modern nucleotides decline in charge (basic–acidic) A > G > C > T/U (Melcher, 1974; Weber and Lacey, 1978). In other words, in terms of inherent affinity, the inferred primordial basic amino acid (Bas) has highest bonding potential with the nucleotide T/U and least with A. Figure 5 shows a schematic representation of a linkage between modern lysine and adjacent T (or U).

This model has overt similarities to that proposed by Melcher (1974) with the difference that Melcher modeled affinities for the first two anticodon bases, a contention (if the identity of the first nucleotide is of lesser importance for coding) that is not incompatible with the scenario proposed here. It is therefore proposed that the modern genetic code has, at its origins, the differential affinity for basic versus other amino acids for covalent linkage to the nucleotide T/U, and that the

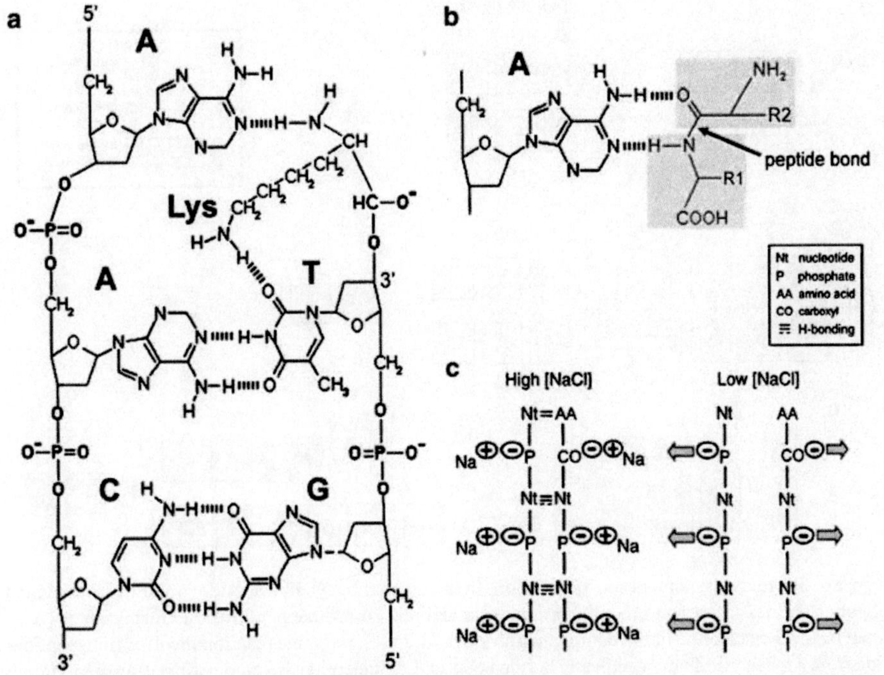


**Figure 5.** (**a**) Pairing of a 5′G-T3′–lysine aminoacylated dinucleotide with a complementary strand and *cis* interaction between the lysine side chain and the unpaired oxygen in T/U. The N-terminus of lysine is shown interacting with the next nucleotide in the complementary strand, noting that the identity of the primordial basic amino acid(s) is unknown, and ornithine (with a shorter side chain) is a contender. (**b**) Interactions between two amino acids and either an A or C base (and only poorly with G or T/U) in the complementary strand stabilize a peptide bond between them and thereby are likely to facilitate peptide bond formation. R1, R2 amino acid side chains. (**c**) The structure generated is compatible with tide-driven association and dissociation; compare with Fig. 1a.

basic elements for first polypeptide chain assembly were amino acids covalently linked to nucleotides or, more likely in view of pairing stability, dinucleotides and longer polymers (that were not involved in coding *per se*). These molecules certainly existed in the primordial soup because, during the drying phase of tidal cycling, increased salt and precursor concentrations no doubt fostered linkage of nucleotides not only to other nucleotides but also to amino acids (Fig. 5).

## 3.3. CATALYSIS OF PEPTIDE BOND FORMATION

In the context of TCR, it is important to note that an aminoacylated dinucleotide (or longer nucleotide) obeys the rules proposed for cyclic amplification driving by oscillating salinity – the negative charge on the C-terminal carboxy group strands in apposition to the negative charge on the phosphate group on the other strand,

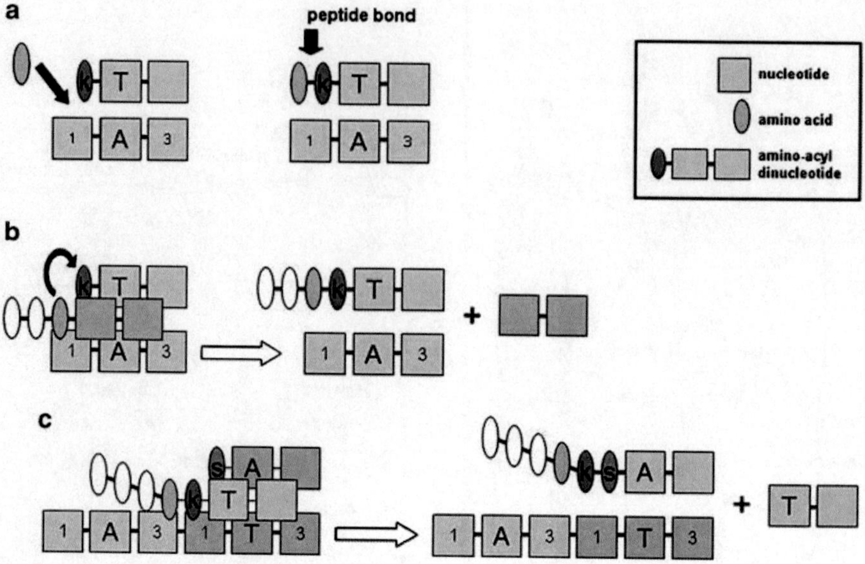


**Figure 6.** Postulated emergence of translation. In (**a**), an aminoacyl dinucleotide (or longer 5′-extended oligonucleotide) accepts an incoming amino acid in a reaction where peptide bond formation is facilitated by the 5′ nucleotide in the complementary strand. In (**b**), the same reaction involves transpeptidation from a loosely and nonspecifically bonded peptidyl-dinucleotide. In (**c**), repetitive transpeptidation allows polypeptide chain synthesis. *K*, lysine; *S*, serine (for illustration). Here, the first position (*1*) is inferred to be either *C* or *A* (*m*).

such that dissociation will take place at low salt concentration (high tide). Conversely, at high salt concentration (low tide), pairing between the N-terminus and the complementary strand (Fig. 5a) permits association and polymerization.

When an aminoacylated nucleotide pairs with another nucleic acid strand, the N-terminus of the nucleotide-attached amino acid can pair with the adjacent nucleotide on the other strand (codon first position). This has an important consequence – depending on the nature and orientation of the adjacent nucleotide, further pairing can take place with a second amino acid; this can promote formation of a peptide bond between the two amino acids, leading to the formation of an amino acid doublet attached to the 3′ nucleotide terminus, simulating modern polypeptide chain synthesis (Fig. 6).

## 3.4. FRAME SELECTION FOR TRANSLATION

These considerations potentially cast light on the selection of the open reading frame. In any triplet coding scheme, translation can read 123–123 (where 1, 2, and 3 are the three triplet positions) but equally 231–231 and 312–312. Modern genes employ a dedicated start codon (ATG) to ensure that only the correct frame is employed. It is possible that the first coding sequences (lacking a dedicated start codon) initiated randomly, and 2/3 of proteins produced were nonfunctional (or, in, e.g., an A-rich sequence, some functionality was preserved in translation products read in the incorrect frame). However, it has been suggested here that only A and C at the first position can catalyze peptide bond formation (Fig. 5) – in view of proximal $-NH_2$ and =N– groups capable of respectively hydrogen bonding to the =O and NH groups of the amino acid (in G, these are $-NH_2$ and –NH– and O= −NH−=O in T). This suggests that the three codon positions of the most primitive codon were mA–, where m ("amino") designates either A or C. This proposal could resolve the issue of frame selection during translation.

Although speculative, the dendrogram of codon evolution (Fig. 2) can be revised to incorporate m at the first position. If the most primitive codon was mA–, subsequent splitting of the mA– tree (lower section of Fig. 5) into yA–, rA–, yG–, and rG– could have been to CA–, AA–, CG–, and AG–. This interpretation derives support from the fact that all four of these codon types, and no others, encode "Bas" in the modern code (indicated by "+" in Fig. 5); this could provide an explanation for why two very different codon types (CG– and AGr) both continue to encode Arg in the present-day code.

### 3.5. A SCENARIO FOR THE FIRST ENCODED POLYPEPTIDE

It is therefore proposed that the primordial tRNA was probably an extended dinucleotide, and charging involved covalent aminoacylation at the 3′ terminus via the C-terminus of the amino acid to generate an aminoacyl dinucleotide or longer oligonucleotide. Bonding between an adjacent A/C residue and two amino acids stabilizes and thereby promotes the formation of a peptide bond between them (Fig. 5b). On this basis, it is possible to envisage a primitive scheme for polypeptide synthesis (Fig. 6). Here, the first position in the "coding" triplet is responsible for aligning the C- and N-termini of two amino acids to promote peptide bond formation (Fig. 6a). Equally, the same interaction is inferred to promote the identical reaction of transpeptidation (Fig. 6b) involving a transient triple-strand structure. This suggests a scheme for primordial protein synthesis (Fig. 6c) driven by cyclic drying and dilution, and explaining the N–C direction of polypeptide synthesis (Bishop et al., 1960). For earlier discussion of double-strand coding and transpeptidation, see Rodin and Rodin (2006) and Weber and Lacey, Jr. (1975), respectively.

## 4. Discussion

Tide-driven cycles of concentration and dilution were previously argued to have generated, and then amplified, primitive nucleic acid polymers. Here, it is suggested that the same cyclic process led to the emergence of template-directed polypeptide assembly. The scheme proposed here has four central aspects.

First, simplification of the genetic code highlights the importance of the central nucleotide in the triplet codon, as previously noted by others. Second, in contrast to other models, the selective interaction between nucleic acids and amino acids takes place, not at pairing with the (*coding*) triplet ("pair-coding"), but instead in covalent linkage to a short complementary (*anticodon*) oligonucleotide precursor to tRNA ("charge-coding") – where, centrally, basic amino acids such as lysine and arginine have a greater affinity for a 3′ T/U (in the anticodon) than for other nucleotides – explaining the preponderance of A at the central position of primordial codon triplets encoding these amino acids. Third, that the adjacent nucleotide (codon first position, predominantly A or C) is stationed in such a way that it can stabilize peptide bond formation and therefore also transpeptidation. Fourth, the binding between an aminoacylated dinucleotide (or longer oligonucleotide) and a complementary strand obeys the same rules as for nucleic acid duplexes – negatively charged groups in apposition (phosphate in nucleic acid and the carboxyl group at the amino acid C-terminus) – such that tidal dilution will lead to dissociation; but also that, in a cyclic process, the charges are neutralized during the drying phase by elevated cation concentrations, promoting duplex and peptide bond formation (Fig. 5c).

The involvement of aminoacylated nucleotides was explored previously (Krampitz and Fox, 1969). Notably, Nakashima and Fox (1972) showed that aminoacyl adenylates can give rise to the formation of peptide polymers and that this reaction is stimulated by a nucleic acid template, supporting the hypothesis that template pairing of aminoacylated nucleotides can contribute to peptide bond formation and transpeptidation.

## 4.1. WHEN DID STRUCTURED tRNAS ARISE?

The foregoing raises the question of how the transition was made from dinucleotide and longer 5′-extended oligonucleotide precursors to modern tRNAs. Clearly, this could only have taken place (i.e., become fixed) once coding was fully established because selective charging of modern tRNA, where the acceptor sites and anticodons are physically separated, requires catalysis by polypeptide enzymes. It is interesting to note that transpeptidation (in contrast to charging) even in modern ribosomes is principally catalyzed by RNA (rather than by polypeptide enzymes), and this adds weight to the notion that transpeptidation preceded the evolution of a fully fledged coding system.

Structured tRNA must have coexisted for a period with dinucleotide (or oligonucleotide) charging – and therefore the first tRNA must have folded back in a complex way to present the new amino acid in the immediate proximity of the codon, placing constraints on the exact structure adopted. However, once the majority of codons were represented by structured tRNAs, the requirement would have relaxed – thereby permitting rapid evolution of modern tRNA molecules. Shimizu (1982) provides elegant observations on tRNA evolution suggesting, in effect, that when tRNA first physically separated the anticodon from the acceptor site, the anticodon must have been an analog of the acceptor site: base preferences at the −4 position, immediately prior to the universal CCA motif, still bear traces of resemblance to the anticodon sequence; moreover, tRNA evolution probably took place when only the second base was firmly established in the code (Rodin and Rodin, 2006).

The emergence of tRNA lifted constraints on triplet coding. New tRNAs and charging enzymes generated by gene duplication could rapidly recruit rare codons to increase the repertoire of amino acids (codon recruitment, Fig. 2). Cases in point are the codons for tryptophan (TGG) and cysteine (TGy): these amino acids are highly unstable and are still among the rarest amino acids in modern proteins. It is presumed that, by tRNA mutation, it was possible to reassign previously dedicated codons to these amino acids (i.e., codon recruitment) and that these are very late additions to the genetic code. The same considerations apply to the dedicated initiator tRNA (tRNA-fMet) and suggest that precise start (and stop) sites for translation are a late addition to the code.

## 4.2. TIDAL CYCLING AND THE EMERGENCE OF CODING

The analysis presented here proposes that physicochemical affinity between amino acids and the central position of the anticodon, leading to selective charging, underlies the emergence of the genetic code. An important feature is that the configuration of the carboxyl linkage and the N-terminus of the amino acid obeys the prerequisite for TCR – strong charge repulsion at low salinity (high tide), but association at high salinity and precursor concentration (low tide) driven by charge neutralization and hydrogen bonding (Lathe, 2004).

Although regression of data regarding the evolution of the Earth–Moon system suggested that tides could have been as fast as every 6 h (Lathe, 2004), this view has been challenged (Varga et al., 2006), and a compromise length of day at 3.9 Ga of ~14 h was suggested (Lathe, 2006), with tides every 7 h. Alternating dilution and concentration on this time scale is proposed to have (a) generated short oligonucleotides selectively charged at their 3′ ends with amino acids according to the primordial code (*anticodon* –T–=Bas) governed by chemical affinity and (b) driven peptide bond formation and transpeptidation.

A scenario thereby arises in which translation emerged in a precellular environment, driven only by physicochemical interactions and tidal cycling. Detailed molecular modeling will be required to test this possibility.

Protein synthesis by this inferred mechanism would have taken place slowly, with only one new polypeptide bond being formed per cycle. This assumption could be incorrect if transpeptidation is driven by stored energy in the amino acid–nucleotide linkage. Moreover, it is inevitable, in the emergence of coding, that primordial polypeptides served in some sense to "benefit" the encoding nucleic acid in an interdependent symbiotic relationship that drove the further evolution of both nucleic acid and protein. Because the inferred primordial code centers on the encoding of basic amino acids, under this scheme, the first proteins were polybasic peptides, perhaps precursors to modern day histones that have strong affinity for nucleic acids and that, even today, protect nucleic acids from environmental stresses.

### 4.3. *TO CROSS A MEMBRANE?* – SPECULATIONS ON THE EMERGENCE OF PROTOCELLS

It is argued that tidal cycling underlies both nucleic acid replication and the emergence of genetic coding. The possibility that such cycling could also underlie the development of cell-based life therefore merits consideration. For future investigations, the following sequence of arguments is offered.

First, what was the most primitive encoded peptide or polypeptide? The arguments presented here suggest that a binary code – Bas versus Sml/Pho (or a derivative tertiary code) – was at the very beginnings of coding. Indirect support derives from consideration of "hot" versus "cold" amino acids. Life undoubtedly first arose at high temperature (>50°C), and although extensive review would be out of place (and this literature is discordant), the consensus is that proteins of heat-adapted (thermophilic) versus cold-adapted (psychrophilic) organisms are selectively enriched in four amino acids: R and E on the one hand and P and L on the other, reiterating the inferred binary/tertiary coding categories.

Second, what could a binary/tertiary coding regime specify? In other words, what was the role of the first polypeptide? Discussion (above) highlighted the potential protective role of histone-like polybasic peptides. In this context, the molecular biologist will of course be aware that polybasic peptides can fulfill a second major biochemical function – the uptake of nucleic acids across cell membranes. Coupling of nucleic acids to, for example, poly-L-lysine (and subsequent counterion provision, notably of $Ca^{2+}$) leads, by mechanisms that are still not understood (but possibly via the formation of local lipid microvesicles), to nucleic acid transfer across membranes.

Third, a polybasic peptide is inefficient. One study demonstrated that a pentapeptide, KKKKK, was unable to catalyze transfer, but a related peptide, KKRPK, was highly effective (Yin et al., 2008); in other words, the fold permitted by proline (CC–) is important [other small candidates (notably glycine, GG–, and serine, AGy and TC–) were not tested in this assay, noting that both serine and arginine stand out because they alone each retain six different codons in the present-day code]. The possibility therefore deserves consideration that a "biased" binary code, only determining majority Bas (R/K) versus minority Sml/Pho (e.g., P, G, or S), could have encoded a primordial and histone-like amino acid polymer that catalyzed nucleic acid transfer into micelles.

Fourth, was nucleic acid transfer across membranes important for the emergence of early life by tidal cycling? To address this question, one must consider the internal versus external environments of lipid micelles. If long-chain fatty acids

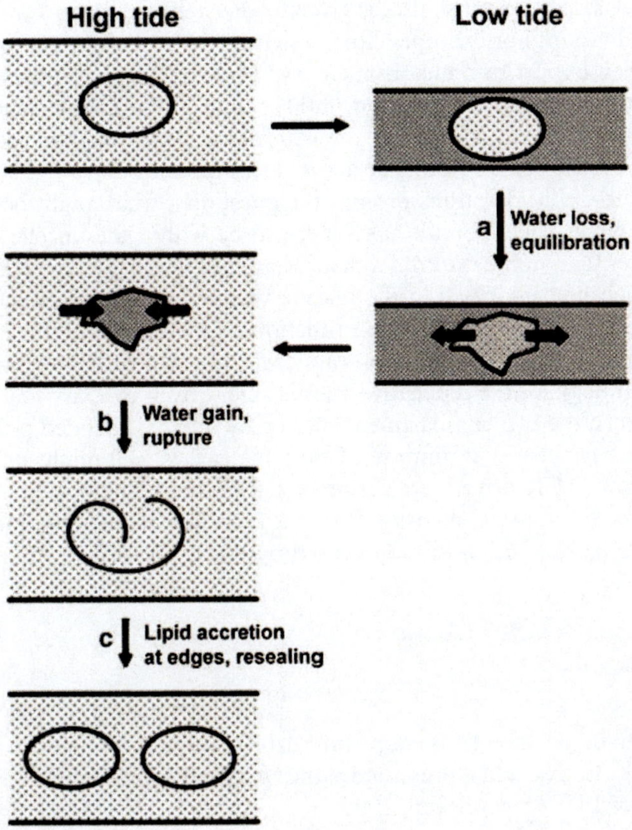


**Figure 7.** Micelle behavior during cyclic concentration and dilution. Disparities in the external versus internal ion concentrations (*shading*) of a micelle (*circle*) drive **(a)** water loss and shrinkage at low tide, followed by **(b)** water gain and rupture on dilution, promoting lipid accretion at exposed edges, and **(c)** micelle "duplication."

were available in the primitive soup, and this seems inevitable, double-membrane structures were undoubtedly generated by shoreline turbulence (discussed in Lathe, 2005). One must therefore consider how such micelles might behave under conditions of tidal dilution and concentration. A feature of (hydrophobic) lipid membranes is that they are only partly permeable to ions but are significantly more permeable to water. Drying down must therefore lead to osmotic water loss, deswelling, and shrinkage of the micelle (Fig. 7a). On subsequent tidal dilution, the shrunken micelle (now with elevated internal ion concentration relative to the external milieu) must gain water, driven by osmosis, and can only rupture (Fig. 7b). Ruptured edges, given lipid availability, inevitably accrete lipid, so permitting resealing and the formation of new micelles (Fig. 7c). In effect, micellar replication is driven by the tide.

Here, consideration must also be given to alternative models of micelle/vesicle growth and division. For example, lipid vesicles have been shown to split spontaneously, driven by intrinsic membrane curvature, under conditions of accretion of new membrane material (Svetina, 2011). Here, the growing micelle has no intermediate phase that is open to the external milieu; this could increase the pressure for nucleic acids to transfer across membranes.

Fifth, these considerations prompt the question – what might be the advantage for the nucleic acid/peptide machinery to be within the micelle during this process, rather than in the external milieu? Rephrased, what is the median concentration of each precursor within the micelle versus the external milieu? – noting that selective uptake/loss of ions as a function of membrane permeability could play a role, notably by influencing intramicellar pH. Or, fundamentally, was the only viable strategy for the replicative nucleic acid, driven by Darwinian selection, to cycle between the two compartments driven by the first encoded polypeptide?

We begin to see a first glimpse of how the earliest cell might have evolved, proceeding from TCR-driven replication of nucleic acids and protein synthesis at ocean borders, to catalyzed incorporation into micelles, all in a hot and turbulent zone between low and high tides governed by the gravitational influence of the early Moon.

## 5. Acknowledgments

I would like to thank John O. Bishop and Yuri Kotelevtsev for helpful discussions. Some of these ideas were first presented at the Nordita Symposium on Astrobiology and Origins of Life, Copenhagen, 2005.

## 6. References

Bashford JD, Tsohantjis I, Jarvis PD (1998) A supersymmetric model for the evolution of the genetic code. Proc Natl Acad Sci USA 95:987–992

Bishop J, Leahy J, Schweet R (1960) Formation of the peptide chain of hemoglobin. Proc Natl Acad Sci USA 46:1030–1038

Blum HF (1957) On the origin of self–replicating systems. In: Rudnick D (ed) Rhythmic and synthetic processes in growth. Princeton University Press, Princeton, pp 155–170

Di Giulio M (2005) The origin of the genetic code: theories and their relationships, a review. Biosystems 80:175–184

Fernando C, von Kiedrowski G, Szathmáry E (2007) A stochastic model of nonenzymatic nucleic acid replication: 'elongators' sequester replicators. J Mol Evol 64:572–585

Fitch WM, Upper K (1987) The phylogeny of tRNA sequences provides evidence for ambiguity reduction in the origin of the genetic code. Cold Spring Harb Symp Quant Biol 52:759–767

Gamow G (1954) Possible relation between DNA and protein structure. Nature 173:318

Görnerup O, Jacobi MN (2010) A model–independent approach to infer hierarchical codon substitution dynamics. BMC Bioinformatics 11:201

Knight RD, Freeland SJ, Landweber LF (1999) Selection, history and chemistry: the three faces of the genetic code. Trends Biochem Sci 24:241–247

Krampitz G, Fox SW (1969) The condensation of adenylates of the amino acids common to protein. Proc Natl Acad Sci USA 62:399–406
Lathe R (2004) Fast tidal cycling and the origin of life. Icarus 168:18–22
Lathe R (2005) Tidal chain reaction and the origin of replicating biopolymers. Int J Astrobiol 4:19–31
Lathe R (2006) Early tides: response to Varga et al. Icarus 180:277–280
Melcher G (1974) Stereospecificity of the genetic code. J Mol Evol 3:121–140
Nakashima T, Fox SW (1972) Selective condensation of aminoacyl adenylates by nucleoproteinoid microparticles (prebiotic–lysine–model system–genetic code). Proc Natl Acad Sci USA 69:106–108
Osawa S, Jukes TH, Watanabe K, Muto A (1992) Recent evidence for evolution of the genetic code. Microbiol Rev 56:229–264
Perlwitz MD, Burks C, Waterman M (1988) Pattern analysis of the genetic code. Adv Appl Math 9:7–21
Riddle DS, Santiago JV, Bray-Hall ST, Doshi N, Grantcharova VP, Yi Q, Baker D (1997) Functional rapidly folding proteins from simplified amino acid sequences. Nat Struct Biol 4:805–809
Rodin SN, Rodin AS (2006) Origin of the genetic code: first aminoacyl–tRNA synthetases could replace isofunctional ribozymes when only the second base of codons was established. DNA Cell Biol 25:365–375
Shimizu M (1982) Molecular basis for the genetic code. J Mol Evol 18:297–303
Svetina S (2011) On the vesicular origin of the cell cycle. In: Seckbach J, Gordon R (eds) Origins: genesis, evolution and diversity of life. Springer, Dordrecht
Szathmáry E (1993) Coding coenzyme handles: a hypothesis for the origin of the genetic code. Proc Natl Acad Sci USA 90:9916–9920
Szathmáry E (1999) The origin of the genetic code: amino acids as cofactors in an RNA world. Trends Genet 15:223–229
Varga P, Rybicki KR, Denis C (2006) Comment on the paper 'Fast tidal cycling and the origin of life' by Richard Lathe. Icarus 180:274–276
Watson JD, Crick FH (1953) Genetical implications of the structure of deoxyribonucleic acid. Nature 171:964–967
Weber AL, Lacey JC Jr (1975) Aminoacyl transfer from an adenylate anhydride to polyribonucleotides. J Mol Evol 6:309–320
Weber AL, Lacey JC Jr (1978) Genetic code correlations: amino acids and their anticodon nucleotides. J Mol Evol 11:199–210
Woese CR (1965) On the evolution of the genetic code. Proc Natl Acad Sci USA 54:1546–1552
Yin S, Fan X, Yu S, Li C, Sy MS (2008) Binding of recombinant but not endogenous prion protein to DNA causes DNA internalization and expression in mammalian cells. J Biol Chem 283: 25446–25454

Biodata of **Cristina Del Bianco** and **Sheref S. Mansy,** authors of "***Heterotrophic Model Protocells.***"

**Cristina Del Bianco** obtained her *Laurea* in chemistry from the University of Rome *La Sapienza* in 2001 and her Ph.D. in structural biology from the University of Florence in 2005. She subsequently worked in the laboratory of Stephen C. Blacklow at Harvard Medical School as a HFSP postdoctoral fellow. She joined the Centre for Integrative Biology at the University of Trento in 2009.

E-mail: **delbianco@science.unitn.it**

**Sheref S. Mansy** obtained his Bachelor's and doctoral degrees from Ohio State University in 1997 and 2003, respectively. After a postdoctoral position in the laboratory of Jack W. Szostak at Harvard Medical School and Massachusetts General Hospital, Sheref was awarded a career development award from the Giovanni Armenise-Harvard foundation. He joined the University of Trento as an assistant professor of biochemistry in 2009.

E-mail: **mansy@science.unitn.it**

**Cristina Del Bianco**

**Sheref S. Mansy**



J. Seckbach (ed.), *Genesis - In The Beginning: Precursors of Life, Chemical Models and Early Biological Evolution*, Cellular Origin, Life in Extreme Habitats and Astrobiology 22, 709–722
DOI 10.1007/978-94-007-2941-4_35, 

# HETEROTROPHIC MODEL PROTOCELLS

**CRISTINA DEL BIANCO AND SHEREF S. MANSY**
*CIBIO, University of Trento, via delle Regole 101, 38060 Mattarello, Italy*

## 1. Introduction

There are many unknowns in the origin of life. Our attempts to remove some of the mysteries of our origins have led to many theories and few experimental data. This is due, in part, to the inherent difficulties in studying a history without record that may never be fully reconstructed. Nevertheless, much can be learned about plausible prebiotic paths to life from experimental research. The resulting data will likely give us much insight into life's past and deepen our understanding of biology.

Of the many unknowns in the origin of life field, one that has enjoyed a lively debate is the question of autotrophic versus heterotrophic beginnings. In short, were Earth's first cells fed complex organic molecules or did they maintain their existence off of simple, small molecules. More specifically, did early cells simply exploit available, premade carbon molecules, such as lipids and nucleotides (heterotrophic lifestyle), or did early life harness sunlight or inorganic molecules to synthesize the organic components of a cell from simple starting material, such as carbon dioxide or methane, that is, an autotrophic lifestyle. A difficulty faced when trying to answer this question is the influences of many additional unknowns. For example, the molecules (including monomers and polymers) and the variety of conditions on prebiotic Earth are unknown. The assumptions made about each of these unknowns influence perspectives and, ultimately, the types of experiments performed in the laboratory. This then leads to experimental research that is often not directly comparable. For example, if it is assumed that prebiotic Earth was only able to provide chemically simple molecules, then research programs based on this hypothesis may tend to focus on demonstrating the emergence of prebiotic chemical cycles that could have been used later as a foundation for primitive, autotrophic, protocellular metabolism. Conversely, if the viewpoint is taken that complex macromolecules existed on prebiotic Earth, then research may be directed toward constructing a model protocell from abiotically constructed, available parts. In other words, the different perspectives rarely converge on similar experiments that can be directly compared in order to evaluate the likelihood of each. The fact that the two vantage points have driven research programs with different goals and types of experiments does not mean that future research efforts have to maintain these rigid perspectives.

Although the concept of a protocellular phase during the emergence of life is old (Oparin, 2003), attempts to build protocells in the laboratory are recent, particularly in comparison with prebiotic synthesis research. Most of the experimental origins of life research relates in some way with a prebiotic soup-type scenario, where the focus is on which molecules can be plausibly synthesized or which chemical cycles are favored under prebiotic conditions. Experimental attempts to construct cell-like structures from the chemical information obtained from these types of prebiotic soup experiments are rarely undertaken. This is understandable for our early attempts in exploring prebiotic chemistry, particularly since such chemistry undoubtedly formed the foundation from which life sprang. Indeed, there is still much to learn from simulating the chemistry of early Earth. However, we now have enough understanding to begin to probe the subsequent steps of prebiotic evolution. It seems that the time is right to begin using our chemical and physical knowledge to explore how complex molecules can assemble into protocells.

Herein we describe paths from abiotic molecules to heterotrophic protocells based on experimental data. These laboratory-constructed models of heterotrophic protocells already have given us much insight into the kinds of functionality achievable from a small number of components and are beginning to define a set of plausible conditions conducive to the emergence of heterotrophic life. It is hoped that similar efforts in building autotrophic model protocell systems will complement the prevalent heterotrophic perspectives.

## 2. The Choice of Heterotrophy

Nearly all the efforts in building laboratory models of protocells have focused on building heterotrophic rather than autotrophic protocells. In large part, the reason is that heterotrophy is simpler. It is easier to build a cell from scratch if it is not necessary to incorporate pathways that chemically transform carbon dioxide, for example, into all of the organic molecules needed to sustain a cell. Arguably, the simplest conception of a protocell is a replicating, self-sustaining chemical system built from lipids and RNA (Szostak et al., 2001). In such a system, lipids and nucleotides would be available for the self-assembled protocell to exploit. Ideally, no protein function would be incorporated, since protein synthesis is a highly complex process, and no alternative, simplified systems are known. Instead, the protocell would consist of catalytic RNA inside of a growing and dividing vesicle. The resulting heterotrophic system likely will not faithfully mimic the protocellular structures that existed in the past, but may do so sufficiently to help reveal how fundamental chemical and physical forces can drive the emergence of biology. At the very least, such experiments are beginning to reveal what is possible. It is worth noting, however, that other conceptions of life that do not focus on reproduction exist (Luisi, 2003; Zepik et al., 2001).

Just as it appears simpler to build a heterotrophic protocell in the laboratory, it may have been simpler for a heterotrophic protocell to emerge on prebiotic

Earth. Since it is unlikely that the complexity observed in modern cells could have originated in a single step, intermediate levels of complexity likely emerged in stages, with the simplest cell-like structures represented by the earliest stages of evolution. Heterotrophic metabolism is less demanding on a cell, whether it is a contemporary cell or a protocell that existed in the past. Less sophisticated heterotrophic protocells could have survived on prebiotic Earth in ways conceptually similar to the many contemporary organisms that live by feeding off of available complex molecules. This simplicity in cellular structure and function does require greater complexity in the surrounding environment. However, this too may not have posed an insurmountable obstacle due to the variety of organic molecules that can be synthesized under prebiotic conditions (Miller, 1953; Oro, 1961; Powner et al., 2009).

Although protocellular research is inherently a bottom-up science that explores prebiotic processes, some insight can be gained from studies on extant cells. In contemporary cells, heterotrophy is associated with a simplification of cellular metabolism in comparison with autotrophic cells. Extreme versions of heterotrophy can be found in parasitic bacteria, since bacteria possess mechanisms that continually remove unneeded regions of DNA. When bacteria enter new environments in which previously needed pathways are no longer necessary, those pathways are typically lost over time (Kuo and Ochman, 2009). Such genetic reductions can lead to significantly smaller genome sizes. For example, the smallest known free-living microorganism is *Mycoplasma genitalium* with 482 genes. Under laboratory conditions, 100 of these genes are nonessential (Glass et al., 2006). Although this organism has a small genome in comparison with typically studied bacteria, for example, *Escherichia coli* has greater than 4,000 genes, natural genetic reductions of the parasitic bacterium *Carsonella ruddii* has resulted in an organism that only requires 182 genes to live (McCutcheon and Moran, 2007). In other words, great simplifications in cellular systems can occur and still sustain a living, heterotrophic cell if the proper environmental conditions are present.

The deconstruction of extant biological systems into the theoretical minimal component parts of life often leads to heterotrophic cells of similar complexity as *C. ruddii*. For example, the Moya laboratory has hypothesized that a minimal cell could be constructed with approximately 200 genes (Gil et al., 2004). Even if such values are impressive in their minimalism relative to commonly studied microorganisms, it still leaves much that is unknown. Genetic sequencing and phylogenetic constructions are not able to reveal the evolutionary processes that occurred for cell-like structures in between zero and 200 genes. This appears to be a limit of top-down approaches in origins of life research. However, at least one further clue presents itself from these considerations. Over one third of the genes needed to sustain both hypothetical minimal cells and microorganisms with greatly reduced genomes are required for protein synthesis. Of all the necessary components of extant life, protein synthesis appears to be one of the most complex and also one of the most highly conserved. If a cell could be built that did not exploit heritable protein function, then much simpler cell-like structures could be envisioned.

Autotrophic origin supporters tend to view the assumptions made by those that attempt to build heterotrophic protocells as fanciful. It is felt that even if heterotrophic protocells could have arisen first, they would have died once their food sources were depleted (Morowitz, 1992). In order for life to have taken hold long enough for evolution to begin to exploit the varied conditions of Earth, the first cells must have been autotrophic. While this is a legitimate viewpoint, autotrophic theories, too, make a series of assumptions in support of autotrophy that may not prove to be true. In terms of protocellular research, they tend to see the appearance of a membrane as a later development and ascribe to all membranes a high degree of impermeability. They also assume that similarities in abiotic and biotic chemistries suggest that modern metabolism evolved from prebiotic geochemical cycles rather than reflecting a convergence on accessible chemistry. This is not to say that autotrophic theories are baseless, but rather that the types of laboratory experiments designed to deepen our understanding of the plausibility of autotrophic beginnings have been narrow in scope. It is hoped that more experiments on autotrophic systems, and in particular on attempts to construct model autotrophic protocells, are forthcoming so that we have a broader spectrum of experimental data to draw on. Attempts to build functioning autotrophic and heterotrophic model protocells will likely uncover plausible scenarios for the emergence of life and allow us to discard incorrect, previously held theories.

## 3. Parts of a Heterotrophic Protocell

There is nothing magical about chemical synthesis nor biological molecules. Just as Friedrich Wöhler synthesized urea and the Venter laboratory synthesized a functioning genome (Gibson et al., 2010), abiologically synthesized molecules behave in the same way as their biologically built counterparts. Further, chemical synthesis is neither confined to the laboratory nor the inside of a cell. Chemistry occurs regularly and continually, providing a range of complex molecules. Thus far, prebiotic routes to the synthesis of amino acids (Miller, 1953), nucleobases (Oro, 1961), nucleotides (Powner et al., 2009), sugars (Ricardo et al., 2004), and lipids (Hargreaves et al., 1977; McCollom et al., 1999) have been demonstrated. More interestingly, many of these molecules have been identified in interstellar ice and meteorites (Pizzarello and Shock, 2010). In other words, fundamental biological molecules can be synthesized in environments devoid of life and from the same molecules that were present when the Earth was formed. Although not every molecule used in contemporary biology has been shown to be made under simulated prebiotic conditions, it appears that the organic complexity provided by natural, abiotic chemistry is capable of providing a wide array of molecules fundamental to life. Therefore, it is plausible that the building blocks of life were present on prebiotic Earth. For a rudimentary heterotrophic cell, all that may have been needed were nucleotides and lipids (Fig. 1). The nucleotides would serve as the starting material for a primitive genomic system, and the lipids would have

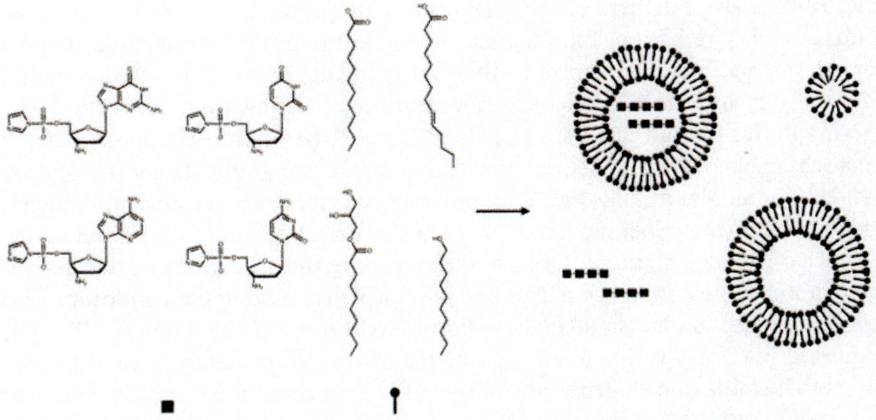

**Figure 1.** The spontaneous assembly of nucleotides and lipids into a protocellular system. The nucleotides shown here are activated with imidazole and amino groups. The single-chain lipids are of the type typically used for laboratory-constructed model systems. Nucleotides and lipids are symbolized by a *square* and a *sphere line*, respectively.

generated a vesicle compartment to house the genome. However, just the existence of the required molecules on prebiotic Earth would not have been sufficient for the emergence of a protocell. The component parts must have been concentrated and assembled in some way.

In order for chemistry to be harnessed for the construction of a protocell, it is important that the chemical components exist at a high enough concentration. A single molecule of each necessary component of a protocell dissolved in the ocean, for example, is unlikely to assemble into a protocell. There are several methods that could be envisioned to bypass this problem. One would be the evaporation of small bodies of water, which would concentrate nonvolatile molecules (Budin and Szostak, 2010). Minerals are capable of binding organic molecules, such as nucleotides (Ferris et al., 1989) and lipids (Hanczyc et al., 2003, 2007), thereby increasing their local concentration. Alternatively, molecules can be concentrated in between ice crystals under eutectic conditions (Monnard et al., 2003). Another particularly interesting mechanism naturally arises from thermal gradients, such as those found within the porous rocks of hydrothermal vents. Convection and thermophoresis within columnar compartments result in significant enrichment of organic molecules, such as monomeric nucleotides, strands of nucleic acids, and lipids (Baaske et al., 2007; Budin et al., 2009).

Higher molecular concentrations lead to increased reactivity of the system, promoting both reactions that form covalent bonds and those that arrange molecules through intermolecular, aggregative forces. Nucleotides have been shown to polymerize under eutectic conditions (Monnard et al., 2003) and on the surfaces of montmorillonite (Ferris et al., 1996). The latter mechanism can generate polymers up to 50 nucleotides long, which is sufficiently large to form aptamers and ribozymes. Further, montmorillonite can protect RNA polymers from damage by UV radiation (Biondi et al., 2007). Although no experimental support for nucleic acid polymerization in thermal gradients exists, the data do suggest that the increased local nucleotide concentrations arising from convection and thermophoresis could facilitate nucleic acid polymerization. Additionally, if the microchannel compartments in which the temperature gradients are present are defined by mineral boundaries, then polymerization could take place directly on the surface of the mineral compartment. Polymerization could also occur in the absence of mineral catalysis. Orgel and others have shown that nucleotide derivatives more reactive than the nucleotide triphosphates used in contemporary cells can be formed under simulated prebiotic conditions (Lohrmann, 1977). The increased reactivity is due to either a better leaving group, such as an imidazole or methyl imidazole (Sleeper and Orgel, 1979), a more nucleophilic center at the 2′ or 3′ position (Zielinski and Orgel, 1985), or both (Schrum et al., 2009). These activated nucleotides can react to form polymers if the nucleotide concentration is sufficiently high (Sleeper and Orgel, 1979), or the nucleotides can condense on a template strand under lower concentration conditions (Lohrmann et al., 1980). The fidelity of the template copying mechanism is sufficiently high to support the maintenance of a protocell, and also low enough to allow for robust evolution (Rajamani et al., 2010).

In addition to nucleic acid synthesis, some of the described processes for concentrating molecules can promote lipids synthesis. For instance, the drying-wetting cycles that facilitate the encapsulation of organic molecules within vesicles (Shew and Deamer, 1985) also can result in lipid synthesis through the condensation of glycerol and fatty acid (Apel and Deamer, 2005). The resulting glycerol esters of fatty acids are potentially useful for a protocell since these molecules facilitate the formation of vesicles (Chen et al., 2005). Mixtures of fatty acid and fatty acid glycerol monoesters also form vesicles with increased stability to salinity (Chen et al., 2005; Monnard et al., 2002) and temperature (Mansy and Szostak, 2008; Maurer et al., 2009) that are additionally better able to acquire complex nutrients from the environment (Mansy et al., 2008).

Much of what has been described occurs under conditions that are conducive to both nucleic acid polymerization and vesicle formation from monoacyl lipids. Essentially, what each mechanism is providing is a way for the local concentration of nucleotides or lipids to be increased, thereby stimulating reactivity. This is perhaps simpler for the case of vesicle formation since lipids aggregate through noncovalent forces. For example, the critical aggregate concentration of oleate is 80 μM (Chen and Szostak, 2004b). Below 80 μM, oleate exists as micelles; however, in a thermal gradient, local concentrations of this same solution can rise above 80 μM and consequently form vesicles (Budin et al., 2009). If the solution contains a mixture of lipids and nucleotides, this mechanism provides for a route that generates vesicles with entrapped contents. Similar mechanisms may operate with clay catalysts. In addition to binding nucleotides, montmorillonite interacts with lipid molecules and catalyzes the formation of vesicles composed of monoacyl lipids (Hanczyc et al., 2007). Since the same catalyst enhances both nucleic acid polymerization and vesicle formation, the presence of this mineral can facilitate the entrapment of RNA strands adsorbed onto the mineral surfaces in vesicles (Hanczyc et al., 2003).

## 4. Toward Compartmentalized Heterotrophic Systems

The assembly of RNA into vesicles does not in itself give life. Nevertheless, even from these few components, the steps of a protocell life cycle begin to emerge (Fig. 2). Many of the life-like processes observed from model protocell systems directly result from the use of vesicles composed of monoacyl vesicles, although interesting examples exist for diacyl lipid systems (Takahashi et al., 2010). In addition to the aggregative processes described above that are dependent upon molecular concentration, morphological transformations required for steps of a protocell life cycle can be induced by chemical factors. For instance, changes in pH greatly influence the types of lipid aggregates formed in solution (Gebicki and Hicks, 1973, 1976; Hargreaves and Deamer, 1978). This plasticity in structure and the spontaneity of vesicle formation and dissolution are the results of the dynamics of the monoacyl lipids (Mansy, 2010). Monoacyl lipids are not only chemically more simple and thus more easily made under prebiotic conditions, the dynamics of the aggregates that they form are highly conducive to processes that typically require complex protein machinery in contemporary biology.

Monoacyl lipids, such as fatty acids, form vesicles that grow and divide. These processes depend on the dynamics of the lipid system. The less dynamic

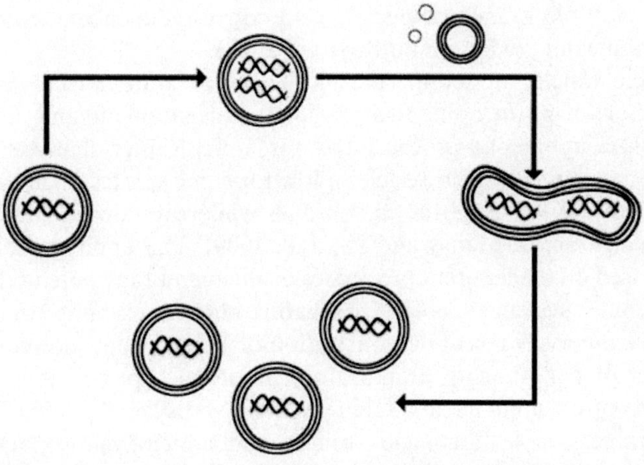

**Figure 2.** A protocell life cycle. One conception of a protocell life cycle includes the replication of a genome inside of a vesicle compartment. Genomic replication induces vesicle growth, which facilitates vesicle division.

diacyl lipids, such as the phospholipids found in extant cells, are not able to exhibit such behavior in the absence of specific protein machinery. The growth of monoacyl vesicles occurs spontaneously upon the mixing of monoacyl micelle and vesicle solutions. Vesicle growth is mediated by the partitioning of the micellar lipids, first into the outer leaflet of the membrane, followed by equilibration with the inner leaflet (Chen and Szostak, 2004a). If multiple layers of membranes are present, then equilibration of the newly adsorbed lipids across multiple membrane layers occurs. Depending upon the permeability characteristics of the solutes in solution, intermediate states during lipid equilibration result in surface area–volume imbalances that transform the spherical vesicles into fragile, elongated structures that easily break apart into multiple daughter vesicles (Zhu and Szostak, 2009). In other words, vesicle growth ultimately results in vesicle division. The division process is highly efficient, retaining nearly all of the entrapped molecules.

The simple scenario of monoacyl vesicles with entrapped RNA leads to preferential growth of nucleic acid replicating systems (Chen et al., 2004). Concentration gradients across membranes generate osmotic forces. These forces are amplified when the solute is a charged polymer, such as RNA, since the concentration gradient of the polymer induces an additional gradient of counter ions to neutralize the negatively charged nucleic acid. Growth of the vesicle compartment reduces the osmotic pressures on the system. In addition to growth mediated by the absorption of micelles, growth of vesicles with entrapped RNA can occur at the expense of empty vesicles. From an evolutionary viewpoint, this leads to a competition between vesicle systems, with nucleic acid containing vesicles surviving at the expense of empty vesicles (Chen et al., 2004). Further, a protocellular system that actively replicates its own genome would consequently induce vesicle growth, thereby coupling two steps of the protocell life cycle, nucleic acid replication and vesicle growth. The thermal stability of monoacyl vesicles is compatible with multiple rounds of nucleic acid copying dependent upon cycles of melting and annealing (Mansy and Szostak, 2008).

Fatty acid vesicles are semipermeable. This is in contrast to the claims often made in support of autotrophic theories for the origin of life and illustrates the usefulness of carrying out protocellular research. Rather than serving as an impenetrable barrier, fatty acid vesicles allow for some solutes, such as small sugars and nucleotides, to cross the membrane, while disallowing others, such as strands of nucleic acids (Mansy and Szostak, 2009). This ability to select for solute uptake based on the inherent properties of the membrane potentially provides for a protocellular system capable of mediating nutrient exchange in the absence of protein machinery. A recent demonstration of this concept showed that with a small number of components, a laboratory model of a protocell could acquire nutrients and copy a small nucleic acid template (Mansy et al., 2008). The system exploited semipermeable fatty acid vesicles and nonenzymatic polymerization reactions to copy the nucleic acid template.

Much has been accomplished in laboratory attempts to build model protocell systems. Model protocells in the laboratory can grow, divide, feed, and copy a nucleic acid template. Each of these steps depends upon heterotrophic assumptions in that the required molecules are added to the system. The protocell does not need to synthesize its own nucleotides, for example. Nevertheless, model heterotrophic protocellular systems require remarkably few components, including fatty acids, nucleotides, and a primer-template pair. In other words, in order for the system to function, it is not necessary to add hundreds of genes or to assemble thousands of different molecules. Further, most of the needed components are either chemically simple or similar to molecules that have been synthesized under simulated prebiotic conditions (Powner et al., 2009; Walde, 2006). Even complications arising from dilute starting materials have many reasonable solutions. The only missing piece thus far seems to be the compilation of each demonstrated reaction into a functional cycle.

## 5. Steps Forward

Putting together each step of a protocell life cycle should be achievable. None of the already demonstrated individual steps require conditions that are known to be incompatible with other steps. Further, some of the reactions naturally couple with other necessary reactions for a protocell life cycle. Encapsulated nucleic acid replication naturally results in an increase in osmotic pressure that induces vesicle growth of fatty acid vesicles (Chen et al., 2004), and vesicle growth naturally results in division of multilamellar vesicles (Zhu and Szostak, 2009). If the protocellular life cycle consists of coupled nucleic acid replication with compartment growth and division, then the cycle appears achievable without the need of building specific regulatory mechanisms to coordinate each step. Nevertheless, not every experiment designed to evaluate each step of the protocell life cycle was performed under identical conditions, and so further experiments are needed to demonstrate the extent of compatibility.

The described growth cycle is fuelled by the replication of a nucleic acid with no particular function other than that of a charged polymer that is efficient at generating an osmotic force. Considering the power of nucleic acids to evolve new function (Lorsch and Szostak, 1996), the described protocellular life cycle seems poised to yield a system with greater complexity and more life-like properties. The fidelity of nonenzymatic replication is conducive to exploring nucleic acid sequence space for new function. Based on the successes of past in vitro nucleic acid selection and evolution experiments, it seems likely that model protocell evolution experiments will yield more robust and efficient protocell-like systems.

Much that has been described herein is an optimistic view of what will be accomplished in the laboratory construction of heterotrophic protocells. However, many challenges still exist. Template-directed nonenzymatic nucleic acid polymerization still requires much advancement in order for it to serve as a robust genomic replication mechanism, and the instability of fatty acid vesicles to high salt conditions limits regions of protocellular emergence to fresh water areas

(Deamer, 1997). Additionally, it may be that our perspectives are too narrow. Not only has protocellular research largely used a heterotrophic perspective, there is an additional bias toward a focus on generating a single, self-replicating protocell. This may not be a realistic possibility for the early Earth. It may not have been possible for a single cell type to emerge alone. Just as it has been argued that it is unlikely that LUCA was a single cell but rather a collection of cooperating cells (Woese, 1998), it may be more realistic to attempt to build a mixture of protocells. A heterogeneous mixture of protocells may more easily form since such a scenario would not require that every needed functionality reside within a single protocell from the beginning. Instead, each protocell would only need to provide a subset of functionality conducive to life.

## 6. Summary

Attempts to build heterotrophic protocells have begun to reveal what is possible from a small number of components. With only a few simple molecules, such as single-chain lipids and activated nucleotides, a protocellular life cycle begins to take shape. The model protocell system can feed, grow, copy nucleic acids, divide, and compete for resources. Mechanisms for the generation and concentration of the needed components to build the protocell have been proposed, although more work is needed to better understand prebiotic nucleic acid chemistry. From the experimental research conducted thus far, the synthesis of heterotrophic protocells appears achievable in the laboratory and perhaps was achievable on prebiotic Earth. However, many challenges remain. More work on autotrophic model systems and on the dynamics of mixtures of protocells will likely contribute greatly to our understandings.

## 7. References

Apel CL, Deamer DW (2005) The formation of glycerol monodecanoate by a dehydration condensation reaction: increasing the chemical complexity of amphiphiles on the early earth. Orig Life Evol Biosph 35:323–332

Baaske P, Weinert FM, Duhr S, Lemke KH, Russel MJ, Braun D (2007) Extreme accumulation of nucleotides in simulated hydrothermal pore systems. Proc Natl Acad Sci USA 104:9346–9351

Biondi E, Branciamore S, Maurel MC, Gallori E (2007) Montmorillonite protection of an UV-irradiated hairpin ribozyme: evolution of the RNA world in a mineral environment. BMC Evol Biol 7(Suppl 2):S2

Budin I, Szostak JW (2010) Expanding roles for diverse physical phenomena during the origin of life. Annu Rev Biophys 39:245–263

Budin I, Bruckner RJ, Szostak JW (2009) Formation of protocell-like vesicles in a thermal diffusion column. J Am Chem Soc 131:9628–9629

Chen IA, Szostak JW (2004a) A kinetic study of the growth of fatty acid vesicles. Biophys J 87:988–998

Chen IA, Szostak JW (2004b) Membrane growth can generate a transmembrane pH gradient in fatty acid vesicles. Proc Natl Acad Sci USA 101:7965–7970

Chen IA, Roberts RW, Szostak JW (2004) The emergence of competition between model protocells. Science 305:1474–1476
Chen IA, Salehi-Ashtiani K, Szostak JW (2005) RNA catalysis in model protocell vesicles. J Am Chem Soc 127:13213–13219
Deamer DW (1997) The first living systems: a bioenergetic perspective. Microbiol Mol Biol Rev 61:239–261
Ferris JP, Ertem G, Agarwal VK (1989) The adsorption of nucleotides and polynucleotides on montmorillonite clay. Orig Life Evol Biosph 19:153–164
Ferris JP, Hill AR, Liu R, Orgel LE (1996) Synthesis of long prebiotic oligomers on mineral surfaces. Nature 381:59–61
Gebicki JM, Hicks M (1973) Ufasomes are stable particles surrounded by unsaturated fatty acid membranes. Nature 243:232–234
Gebicki JM, Hicks M (1976) Preparation and properties of vesicles enclosed by fatty acid membranes. Chem Phys Lipids 16:142–146
Gibson DG, Glass JI, Lartigue C, Noskov VN, Chuang RY, Algire MA, Benders GA, Montague MG, Ma L, Moodie MM et al (2010) Creation of a bacterial cell controlled by a chemically synthesized genome. Science 329:52–56
Gil R, Silva FJ, Pereto J, Moya A (2004) Determination of the core of a minimal bacterial gene set. Microbiol Mol Biol Rev 68:518–537
Glass JI, Assad-Garcia N, Alperovich N, Yooseph S, Lewis MR, Maruf M, Hutchison CA 3rd, Smith HO, Venter JC (2006) Essential genes of a minimal bacterium. Proc Natl Acad Sci USA 103: 425–430
Hanczyc MM, Fujikawa SM, Szostak JW (2003) Experimental models of primitive cellular compartments: encapsulation, growth, and division. Science 302:618–622
Hanczyc MM, Mansy SS, Szostak JW (2007) Mineral surface directed membrane assembly. Orig Life Evol Biosph 37:67–82
Hargreaves WR, Deamer DW (1978) Liposomes from ionic, single-chain amphiphiles. Biochemistry 17:3759–3768
Hargreaves WR, Mulvihill SJ, Deamer DW (1977) Synthesis of phospholipids and membranes in prebiotic conditions. Nature 266:78–80
Kuo CH, Ochman H (2009) Deletional bias across the three domains of life. Genome Biol Evol 1:145–152
Lohrmann R (1977) Formation of nucleoside 5′-phosphoramidates under potentially prebiological conditions. J Mol Evol 10:137–154
Lohrmann R, Bridson PK, Orgel LE (1980) Efficient metal-ion catalyzed template-directed oligonucleotide synthesis. Science 208:1464–1465
Lorsch JR, Szostak JW (1996) Chance and necessity in the selection of nucleic acid catalysts. Acc Chem Res 29:103–110
Luisi PL (2003) Autopoiesis: a review and a reappraisal. Naturwissenschaften 90:49–59
Mansy SS (2010) Membrane transport in primitive cells. Cold Spring Harb Perspect Biol 2:a002188
Mansy SS, Szostak JW (2008) Thermostability of model protocell membranes. Proc Natl Acad Sci USA 105:13351–13355
Mansy SS, Szostak JW (2009) Reconstructing the emergence of cellular life through the synthesis of model protocells. Cold Spring Harb Symp Quant Biol 74:47–54
Mansy SS, Schrum JP, Krishnamurthy M, Tobe S, Treco DA, Szostak JW (2008) Template-directed synthesis of a genetic polymer in a model protocell. Nature 454:122–125
Maurer SE, Deamer DW, Boncella JM, Monnard PA (2009) Chemical evolution of amphiphiles: glycerol monoacyl derivatives stabilize plausible prebiotic membranes. Astrobiology 9:979–987
McCollom TM, Ritter G, Simoneit BR (1999) Lipid synthesis under hydrothermal conditiosn by Fischer-Tropsch-type reactions. Orig Life Evol Biosph 29:153–166
McCutcheon JP, Moran NA (2007) Parallel genomic evolution and metabolic interdependence in an ancient symbiosis. Proc Natl Acad Sci USA 104:19392–19397

Miller SL (1953) A production of amino acids under possible primitive earth conditions. Science 117:528–529

Monnard PA, Apel CL, Kanavarioti A, Deamer DW (2002) Influence of ionic inorganic solutes on self-assembly and polymerization processes related to early forms of life: implications for a prebiotic aqueous medium. Astrobiology 2:139–152

Monnard PA, Kanavarioti A, Deamer DW (2003) Eutectic phase polymerization of activated ribonucleotide mixtures yields quasi-equimolar incorporation of purine and pyrimidine nucleobases. J Am Chem Soc 125:13734–13740

Morowitz HJ (1992) Beginnings of cellular life. Metabolism recapitulates biogenesis. Yale University Press, New Haven

Oparin AI (2003) The origin of life, 2nd edn. Dover Publications, Inc, Mineola

Oro J (1961) Mechanism of synthesis of adenine from hydrogen cyanide under possible primitive earth conditions. Nature 191:1193–1194

Pizzarello S, Shock E (2010) The organic composition of carbonaceous meteorites: the evolutionary story ahead of biochemistry. Cold Spring Harb Perspect Biol 2:a002105

Powner MW, Gerland B, Sutherland JD (2009) Synthesis of activated pyrimidine ribonucleotides in prebiotically plausible conditions. Nature 459:239–242

Rajamani S, Ichida JK, Antal T, Treco DA, Leu K, Nowak MA, Szostak JW, Chen IA (2010) Effect of stalling after mismatches on the error catastrophe in nonenzymatic nucleic acid replication. J Am Chem Soc 132:5880–5885

Ricardo A, Carrigan MA, Olcott AN, Benner SA (2004) Borate minerals stabilize ribose. Science 303:196

Schrum JP, Ricardo A, Krishnamurthy M, Blain JC, Szostak JW (2009) Efficient and rapid template-directed nucleic acid copying using 2′-amino-2′,3′-dideoxyribonucleoside-5′-phosphorimidazolide monomers. J Am Chem Soc 131:14560–14570

Shew RL, Deamer DW (1985) A novel method for encapsulation of macromolecules in liposomes. Biochim Biophys Acta 816:1–8

Sleeper HL, Orgel LE (1979) The catalysis of nucleotide polymerization by compounds of divalent lead. J Mol Evol 12:357–364

Szostak JW, Bartel DP, Luisi PL (2001) Synthesizing life. Nature 409:387–390

Takahashi H, Kageyama Y, Kurihara K, Takakura K, Murata S, Sugawara T (2010) Autocatalytic membrane-amplification on a pre-existing vesicular surface. Chem Commun 46:8791–8793

Walde P (2006) Surfactant assemblies and their varioius possible roles for the origin(s) of life. Orig Life Evol Biosph 36:109–150

Woese C (1998) The universal ancestor. Proc Natl Acad Sci USA 95:6854–6859

Zepik HH, Blochliger E, Luisi PL (2001) A chemical model of homeostasis. Angew Chem Int Ed Engl 40:199–202

Zhu TF, Szostak JW (2009) Coupled growth and division of model protocell membranes. J Am Chem Soc 131:5705–5713

Zielinski WS, Orgel LE (1985) Oligomerization of activated derivatives of 3′-amino-3′-deoxyguanosine on poly(C) and poly(dC) templates. Nucleic Acids Res 13:2469–2484

Biodata of **Luísa Pereira**, **Telma Rodrigues**, and **Francisco Carrapiço**, authors of "***A Symbiogenic Way in the Origin of Life***."

**Luísa Pereira** was born in Lisbon, Portugal (1969) and has a B.Sc. in Biology and Geology (Education) from the University of Lisbon (1996). She teaches at the secondary school Fernando Lopes-Graça, Parede (Portugal), and in 2009 obtained a Master Degree in Earth and Life Sciences. In 2006, she attended the "6th European Workshop of Astrobiology (EANA'06)" in Lyon, France. Recently (2009), she was an invited speaker at the international conference "Evolution today and tomorrow, Darwin evaluated by contemporary evolutionary and philosophical theories," organized by the Center for Philosophy of Sciences of the University of Lisbon, where she presented a communication on the post-neo-Darwinian approach to the origin of life.

E-mail: **luisamadalena@gmail.com**

**Telma Rodrigues** was born in Lisbon, Portugal (1977) and has a B.Sc. in Biology and Geology (Education) from the University of Lisbon (2000) and a Master Degree in Earth and Life Sciences from the same university (2005). She is currently a teacher at a secondary school in Portugal (Dr. José Afonso, Seixal), and simultaneously, she has been developing several projects in astrobiology education and public outreach. She has been responsible for the Astrobiology Project at the same school for the past 5 years. She also has participated in several congresses on astrobiology in Europe and the United States, enhancing the role of education and public outreach in the studies of the origin of life.

E-mail: **telmarodrigues@oniduo.pt**

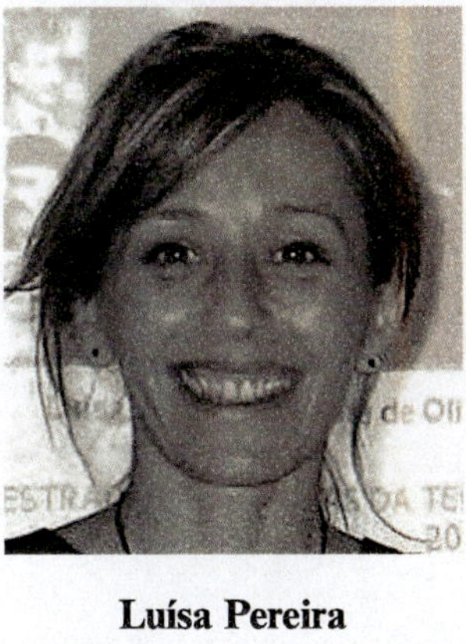

**Luísa Pereira**

**Telma Rodrigues**



J. Seckbach (ed.), *Genesis - In The Beginning: Precursors of Life, Chemical Models and Early Biological Evolution*, Cellular Origin, Life in Extreme Habitats and Astrobiology 22, 723–742
DOI 10.1007/978-94-007-2941-4_36, 

**Dr. Francisco Carrapiço** was born in Lagos (1951), Portugal, and has a B.Sc. in Biology from the University of Lisbon, a Ph.D. in Cell Biology (1985) from the same university, and a postdoctoral from the Arizona State University, USA. Assistant professor at the Faculty of Science of the University of Lisbon, and researcher at the Center for Environmental Biology, his main field of research is Symbiomics, namely, the *Azolla-Anabaena*-bacteria symbiotic system. He has been involved in several projects in Portugal, Europe, Africa, and Latin America. He is also doing research work on the post-neo-Darwinian perspective of evolution, with emphasis on the origin of life and in the symbiogenic approach to this evolutionary process. He also has participated in several congresses on astrobiology in the United States and Europe, focusing on the topic of symbiogenesis and the origin of life.

E-mail: **fcarrapico@fc.ul.pt**

# A SYMBIOGENIC WAY IN THE ORIGIN OF LIFE

**LUÍSA PEREIRA[1], TELMA RODRIGUES[2], AND FRANCISCO CARRAPIÇO[3]**
[1]*Escola Secundária Fernando Lopes-Graça, 2775-200 Parede, Portugal*
[2]*Escola Secundária Dr. José Afonso, 2840-268 Seixal, Portugal*
[3]*Departamento de Biologia Vegetal, Faculdade de Ciências, Centro de Biologia Ambiental e Centro de Filosofia das Ciências, Universidade de Lisboa, 1749-016 Lisboa, Portugal*

## 1. Introduction

The scientific question relating to the origin of life remains, so far, without a consensual answer. The advanced theories essentially highlight a Darwinist or neo-Darwinist approach to the problem, preventing a more global and comprehensive overview, which could be made possible through consideration of other principles, namely symbiogenic ones. The notion of a dynamic and systemic Earth, with natural interdependent and interrelated subsystems, with life in all environments and ecological niches, even in the extreme ones, reveals an evolution of life in close relation with the primitive Earth's geochemical contexts. Under extreme and adverse environmental conditions, life chooses cooperative and synergistic strategies, establishing symbiotic relationships. The extremophile organisms, besides broadening the limits of the known life, instigate new reflections about life beyond Earth. Astrobiology is a privileged eye scoping the very dimensions of the origin, evolution and distribution of life in the Universe. The study presently being carried out, which concerns, among others, extremophile organisms and the search for earthly extreme extraterrestrial environments, analogous to the primitive ones on Earth, is of high importance. Besides characterizing the Earth's primitive environments, it also reinforces the exogenous models of the origin of life.

## 2. Symbiogenesis, the Stepson of Evolution

In the nineteenth century Louis Pasteur raised the fascinating question of the historical origin of life: "Because life can originate only from preexisting life, it has a history and therefore an origin, which must be understood and explained by chemists" (Brack, 2007 in Horneck and Rettberg, 2007, 2). Also, in 1859, Charles Darwin proposed his theory of evolution by natural selection, explaining the origin of all species without the intervention of any supernatural power: current living beings resulted from a long biological evolution, beginning with a very primitive and simple organism. Darwin also developed an approach toward the origins of life and, in 1872, advanced a natural solution based on a prebiotic chemical evolution, considering also the nature of the basic processes involved. Thus, abiogenesis was necessary for an evolutive conception, considering all inanimate matter and life a historical *continuum*, and explaining the emergence of the initial organism(s) (Fry, 2000). Despite Darwin's ideas, it was only in the twentieth century that evolution seriously entered into scientific studies on the origin of life.

Aleksandr Oparin and John Haldane established the guidelines of investigation for the field in the first two decades of the twentieth century. Since then those guidelines remained the main theoretical framework for the majority of models on the origin of life. Their hypothesis and almost all of the life's origin and prebiotic chemistry studies are divided into two contrasting hypotheses: metabolism first or replication first. Until now all research was based on classic Darwinian principles, guided by a strictly competitive and gradual evolution, without any cooperative or synergistic involvement. The Darwinian and neo-Darwinian perspectives have, so far, dominated the scene, but it is clear that they are not able to explain all the mechanisms and processes of evolution (Carrapiço et al., 2007).

Several contradictory solutions and many open questions emerge from the current models on the origin of life. The analysis of the current models and of the recent astrobiological data show underlying difficulties or dilemmas: metabolism *versus* replication, gradual *versus* saltitational process, the smaller time window currently accepted for the emergence of life, endogenous *versus* exogenous building blocks of life, the processes and constraints of the prebiotic chemistry, single *versus* multiple events and terrestrial *versus* extraterrestrial origin.

Oparin and Haldane's first scientific proposal is usually considered as the main source for the development of modern thought considering the origin of life, but in the beginning of the twentieth century, the Russian biologist, Constantin Merezhkowsky made a proposal for a natural origin of life, under the extreme conditions of the primitive Earth. Merezhkowsky was the founder of the symbiogenesis concept in 1909, defined as "the origin of organisms by the combination or by the association of two or several beings which enter into symbiosis" (Sapp et al., 2002, 425). He also highlighted the importance of extremophiles and extreme environments in the early stages and evolution of life. Since then, several authors, have made some innovative approaches including symbiogenic evolutive principles, and highlighting the symbiogenesis significance as a source of evolutive novelty: Barricelli (1957, in Fogel, 2006), Dubos and Kessler (1963), Dyson (1985, 1998), Chapman and Margulis (1998), Margulis (1999), Taylor (1999), Antonelli et al. (2003), Sapp (2003, 2004), Roossinck (2005, 2008), Carrapiço (2006), Ryan (2006), Bouchard (2007), Gontier (2007), Reid (2007) and Watson (2007).

Kutschera (2009a) suggests that Charles Darwin originally employed the term "struggle for life" with two meanings: *competition* but also *cooperation*. It was Heinrich G. Broon, a German paleontologist, who translated Darwin's metaphor exclusively into terms of *competition*. Kutschera's claim is supported by the fact that in a dictionary from 1893 the term *concurrency* used by Darwin had several meanings, ranging from competition to cooperation. In this way, the term *concurrency (struggle for life)* would include competition but also cooperation. Kutschera (2009b) suggests an integrative model of macroevolution, where the Earth's systemic dynamics were responsible for the creation of new habitats and ecological niches. In a global scale, he considers wide environmental changes as responsible for an adaptative evolution, via directional selection. Symbiogenesis, natural selection, and the dynamic of the planet were the key factors causing and determining the course of evolution in an ever-changing Earth. The author considers that this aspect of reality was not considered either by Darwin or by the synthetic theory authors. Kutschera's ideas are also supported by Corning (2005), who considers the theory of evolution by natural selection a functional one, including functional effects (adaptations) of all kinds, and at various levels of the biological organization. This author argues that the functional effects produced, for instance, by an independent set of genomes, as symbionts or coevolving species, with the contingencies of survival and reproduction constitute the directive aspect of natural selection. The notion of "competition via cooperation" is a very important one, because those who cooperate best compete more successfully. Corning (2005) considers the synergies resulting from various kinds of cooperative interactions to be responsible for the evolutionary trend toward greater complexity. In his opinion, several authors recognized that symbiotic and cooperative relationships could provide, under some circumstances, a competitive advantage, to which we add, particularly in the extreme conditions of the primitive Earth. In 1995, the same author considered that complexity should be correlated with the functional synergies it produces.

Considering the role of synergies in the evolution of complexity, we must now turn to symbiosis. Anton de Bary (1878) presented the symbiosis concept as "the common life of different organisms." According to Dubos and Kessler (1963, 8) symbiosis must be considered as a creative manifestation of a general biological phenomenon where "…More remarkable is the fact that many symbiotic systems produce substances and structures that neither one of the two components produces when growing alone." In symbiotic relationships, the central aspect is the creation of evolutive novelty (metabolic, anatomical, and organismal), and the association depends not only on the intrinsic symbionts properties but also on the internal and external system environmental conditions (Carrapiço, 2006). Symbiosis is a robust phenomenon, offering many opportunities for ecological niche occupation in ways that would otherwise be unviable, and also for having highly relevant survival propensities (defense, protection, nutrition, motility, reproduction, communication). We believe symbiosis meets the criteria necessary to overcome the obstacle raised by Theodosius Dobzhansky, who in 1962 stated that "no theory of evolution leaving the mystery of adaptation unraveled should be acceptable" (Corning, 1995, 102). For Corning (1995), the synergies associated to symbiosis perform a leading role in complexity evolutions, being involved in first living organisms' emergence, in primitive prokaryotes mergers, and in the "progressive" temporal coevolution of aquatic and terrestrial ecosystems.

Symbiosis established a new vision of the organism and of evolution, markedly different from the neo-Darwinian view, and was contextualized by symbiomics, a new multi-transdisciplinary evolutive science field with contributions from biology, biochemistry, physiology, genetics, ecology, and evolution of biological systems and their interfaces. It places symbiogenesis in the evolutive context of a post neo-Darwinian perspective (Sapp, 2003), and before it, as a principle of evolution developed by Boris Kozo-Polyansky in 1924 (Kozo-Polyansky, 2010). The concept of symbiogenesis underlines the central role of interactions, in which individuality (new entity) emerges through incorporation (Margulis, 1999 in Carrapiço et al., 2007). Symbiogenesis will be the evolutive mechanism and symbiosis the vehicle through which that mechanism unfolds (Margulis, 1990; Carrapiço et al., 2007). It involves horizontal mergers, which can be rapid, and, usually, discontinuous, creating permanent and irreversible changes, forming the basis for evolutive novelty. Something new arises through merging, resulting in a unique or new metabolism or structure(s), which were not present before symbiosis. Thus, a new entity reveals behaviors previously unobserved in the separate components. The new entity can evolve vertically, but is always preceded by a horizontal merger between two or more entities (Gontier, 2007). Symbiogenesis states that the most probable cause for unexpected complex structures (dead or alive) lies in the association of lesser complicated parts (Dyson, 1997). The evolutionary changes can be explained by an integrated cooperation between organisms, where symbiosis acts as the rule in nature (Carrapiço, 2006).

Corning (2005) also argues that symbiosis has provided many opportunities for organisms to occupy ecological niches that otherwise would not be viable, and that symbiosis typically precipitates the coevolution of various facilitative adaptations, and that frequently involving more than two species. The author mentions three consequences of symbiosis for the science of complexity: fusion via symbiosis as a major mechanism of complexification, evolutionary complexification via fusion includes behavioral dynamic changes that precipitate new options for selection, and the coupling of separately evolved genomes creates a new selective unit that may or may not compete directly with other organisms.

## 3. A Symbiogenic Way to the Origin of Life

A new approach and vision are needed to solve this paradigm. Studies on the origin of life must focus on the physical and chemical principles and naturalistic mechanisms guiding and constraining its emergence. We must use both bottom-up and top-down approaches, and rely on data from several scientific fields, such as geochemical and geophysical data concerning primitive and nowadays' Earth and Solar System, fossil records, current Earth system dynamics, data from living forms interrelations and interdependencies, and the comprehension of the characteristics of complex living systems, among others. This will provide us with a more coherent picture for the understanding of the mechanisms that rule the evolution in nature. We believe metabolic and genetic models explain the different aspects. We must look for the common processes and principles that ruled the prebiotic evolution and question how evolution proceeded in such adverse environmental conditions. We developed our proposal for a symbiogenic understanding of the origin of life based on three perspectives: the consideration of other evolutive mechanisms that would account for a more rapid evolution, the symbiogenesis theory; the leading role of astrobiology, and the notion of a systemic and dynamic planet Earth (Carrapiço et al., 2007).

How can we apply the symbiogenesis theory in order to improve our understanding of the origin of life, and its initial evolution? The symbiogenic concept allows an innovative approach to the origin and evolution of life, with applicability to life's initial stages, and capable of being a fundamental rule in life's establishment and development on Earth and elsewhere. We suggest a naturalistic explanation for the origin of life, through the evolution of complex and adaptative systems, with synergistic, cooperative and symbiogenic mechanisms, guiding chemical evolution through its constraints (Carrapiço et al., 2007). At the turn of the twenty-first century, proposals for universalizing symbiogenesis as an evolutive mechanism were advanced. As far as we know, the first proposal was made by the astrophysical Freeman Dyson, in 1985 (Dyson, 1998). More recently, Gontier (2007) developed the universal symbiogenesis concept. This author considers symbiogenesis as offering a complementary view of evolution, allowing cooperative and synergistic approaches. Gontier (2007) considers a universal and neutral framework for symbiogenesis. "Universal Symbiogenesis is the process whereby new entities are introduced because of interactions between (different) previously independently existing entities. These interactions encompass horizontal mergings and the new entities that emerge because of this are called symbionts. The process is irreversible and discontinuous (…) after symbiogenic processes have occurred, this newly evolved entity can again become the subject of selection" (Gontier, 2007, 174).

We believe this definition allows different phenomena to be regarded as symbiogenic, and that universal symbiogenesis provides us with a general analytical tool for the immense interactions among different entities, namely the prebiotic and biotic entities. The symbiogenesis concept, through universal symbiogenesis, can be applied to the prebiotic evolutive context, beyond the biotic one, as a new paradigm shift in evolution (Carrapiço et al., 2007). The science of complexity defies the selectionist hegemony, by arguing that such a long-term trend toward increasing complexity suggests the presence of additional mechanism(s) and Corning (2005, 94) shares this opinion when he argues: "The evolution of complexity has seemed to require something more than the random point mutations in an amorphous 'gene pool'. The long term toward greater complexity (…) seems to suggest the presence of some additional mechanism or mechanisms."

One main characteristic of the living world is that entities in varied combinations are prodigious novelty generators. Synergies of several kinds have a creative role in evolution, being a source of evolutive novelty. The functional advantages associated with synergistic phenomena constitute a common functional principle as well as a cause for the complex functionally organized systems in "progressive" temporal evolution (Corning, 1995). So, symbiosis is a factor of evolutive change, giving rise rather suddenly to evolutionary novelty, challenging, in our point of view, the Darwinian and neo-Darwinian competition and gradualism (Carrapiço, 2006). Furthermore, synergy refers to the effects that parts (or individuals) cannot attain alone, being interdependent effects, resulting in different "wholes." It resembles an "umbrella" for cooperative interactions, dealing with mutualistics and parasitic combinatorial effects, emerging from the aggregation of multiple incremental contributions of individual parts or organisms (Corning, 1995).

According to Watson (2007), symbiosis allows for the combination of preadapted genetic material enabling an evolutionary process, yet one algorithmically distinct from the Darwinian gradualist framework. In his opinion, there are different underlying algorithmic principles in evolution. Thus, we should expand the framework of evolution to include a greater range of algorithmic possibilities, and break the equivalence between evolution and gradualism. Using evolutionary computation the author shows that compositional evolution is capable of evolving certain kind of complex systems, specifically systems with modular interdependency that would be considered unevolvable under the gradualist framework. Correia (2010) has been developing interesting work in the area of computational evolution. According to the author, evolutionary algorithms (EA) have been used to model and study evolution, using different common evolutionary algorithm models. Some of the models are biologically inspired, including coevolution and, in particular, symbiogenetics. For the author, another area of computational evolution, which has allowed us to study basic principles of evolution and ecology dynamics, is the development of artificial life platforms for open-ended evolution of artificial organisms. According to Dyson (1997), in the mid-1950s, Barricelli proposed a computational symbiogenic model for the origin of life. From his work, Barricelli concluded that symbiogenesis accelerated evolution, by displaying a parallel processing of the genetic code, allowing a more efficient search for the sequences propelling the individual, and species, to a higher level of complexity, the key for evolutive success.

According to Correia (2010), an evolutionary algorithm (EA) is a simplified computational model of evolution. The author considers the implementation of coevolution (competitive or cooperative) in standard models. In cooperative coevolution, there are differentiated species that cooperate. The symbiotic models are based on a strong mutual dependency of two species with different genotypes, and any solution is achieved only by collaboration, that is, symbiosis, which is based on individuals of each species. In the author's opinion, symbiosis and cooperative coevolution in general tend to improve EA performance in difficult problems. According to Correia (2010, 187): "Symbiosis brings the possibility of simultaneous speciation of some symbionts and diversity through the choice of different symbionts available in the populations of the two types. Also, it provides some sort of scaffolding, helping to produce individuals that are able to perform on their own the tasks of former symbiosis."

The majority of autocatalytic and self-organized phenomena are also open to natural selection, with functional synergy being the bridge connecting self-organization and natural selection (Corning, 1995). Although several authors considerer that self-organization and natural selection are opposed to one another, and that the first diminishes the role of the second, others considered that self-organizing systems are not organized by an external force; instead, they believe that the causal dynamics are internal to the parts or participants and their interactions (Corning, 2005). According to the same author, this does not present any inherent conflict with the Darwinian Theory, and natural selection can also be a party to synergistic autocatalytic processes. The author goes further when stating that (Corning, 2005, 103): "It can be party to discontinuous (…) symbiotic functional fusions. And it can be a party, as well, to novelties that create new niches and mitigate competition." Therefore, self-organization is an undisputed fact, and autocatalysis is compelling, especially in the early stages of evolution.

In our point of view, competition and cooperation can coexist in the same scenario of evolution, and probably take place in terms of discontinuous bursts of activity, depending on the internal and external conditions driving evolution. The same population can evolve using competitive and/or cooperative processes in a hypercycle evolutive scenario. A series of synergistic and cooperative effects produced a wide source of creativity and functional advantages that pushed the emergence of complex and functionally integrated biological systems through the evolution of self-organization, self-catalysis and higher complexity. It was only after this biochemical evolution of structures, which gained informational capabilities necessary to self-replication, that the Darwinian competition arose. The complexification in evolution implies the production of new forms (entities) and more powerful synergies. In this context, symbiogenesis creates new important selection units (symbioma) arising through the integration of varied parts and progressive differentiation of the whole, implying a competitive advantage and going beyond the traditional neo-Darwinian units of selection (Carrapiço et al., 2007).

In Corning's (2005) opinion, the order that arises through self-organizing processes probably took place prior to selection. The author believes that the products of evolution were themselves increasingly important co-designers over time, in an interactive process. Thus, and regarding biological complexification, the wholes are more fundamental biological entities than parts. This means that, in the earliest stages of evolution, the parts only had directional consequences and selective value when combined into functional units.

As we already proposed, a symbiogenic scenario could have been responsible for the development of prebiotic processes toward primitive cells. The notion that the synergies associated with symbiosis had a leading role in the complexification was also suggested by other authors such as King (1977), Dyson (1985), Popa (2004), Matthews (1994), Matthews and Minard (2003), Frank (1995), Lee et al. (1997), Fernando (2005), and Carrapiço (2006). We consider that the symbiogenic approach to the origin of life can be associated with Freeman Dyson's hypothesis on the origin of life, which are based on symbiogenic concepts and invoke Lynn Margulis ideas. Dyson (1985) synthesizes the metabolic and genetic visions of life's origin, supported by symbiosis, and proposes a prebiotic evolution accomplished by the independent formation of metabolic systems and self-replicating molecules, which evolved together. In a certain time, the metabolic systems became encapsulated in primitive amphiphilic vesicles, and in a specific time some of the molecules were synthesized first as parasites and later as mutualists and underwent evolution together. Later, the primitive self-replicating molecules joined them, first as parasites and then later as mutualists.

Shenhav et al. (2003) proposed a metabolic system to store and propagate information (compositional information), considering it fundamental to overcome the dogma stating that any transfer information mechanism inferior to base-pair templating is implausible. The authors suggested considering a mesobiotic entity (with primordial molecular assemblies, devoid of long catalytic biopolymers or long informational biopolymers and endowed with molecular and ensemble complexity) embodying primitive features from metabolism and replication. We believe that these mesobiotic entities could be the result of a primitive symbiosis between primitive metabolism and primitive replication molecules, as initially proposed in 1985 by Dyson.

A double origin of life involving metabolism and replication, accomplished in multiple origins on Earth, suggests a metabolism carrying the compositional information, and a replication of sequential information. Symbioses of these two kinds of entities led to a hybridization of their informational content. In a double origin the self-catalyzed and peptide-based metabolic systems emerged through successive symbioses between abiotic components and self-replicating and autocatalyzed molecules. This way, life emerged as two kinds, as metabolism and as genetic information, combined in a symbiotic fashion. The basic organic monomers required different component materials integration. Also, complex polymers were based in monomers fusions. Corning (1995) also considers autocatalysis and self-organization as preceding mutations and the most fundamental novelty source in the formative stages of evolution. Our ideas for a symbiogenic scenario of the origin of life are also supported by Popa (2004) and Woese's (2004) works.

Popa (2004) considers two major symbioses involved in the emergence of life. The first molecular symbiosis in the evolution of life resulted from the association of a frontier (primitive membrane) with a reflexive activity, based in an association or mutual assistance between two distinct characteristics (entities). Metabolism emerges from the coupled entities between a self-assembly process and a reflexive activity that stabilizes it. The reflexive activity was established by different synergisms and mutualisms through autocatalytic cycles and hypercycles that several other authors propose as leading to symbiosis emergence. Eigen et al. (1979, 1981, in Corning, 1995), proposed the hypercycles to explain how organic substances with increasing complexity could overcome the "complexity catastrophe," an example of synergy in evolution. According to King (1977), autocatalytic chemical systems arise spontaneously, in an irradiated aqueous medium. Self-reproducing chemical particles, of any complexity, in a proper environment, have self-regulating properties, allowing for long-term survival. The environmental loss of materials, leading to continuous system decay, can be overcome through physical unions (symbiosis) involving different self-reproducing particles. Those unions create an increasing and irreversible complexity, leading to evolution in the chemical system. The author suggests an evolution through spontaneous successive symbiosis. Thus, successive prebiotic symbioses changed simple autocatalytic particles in this way to primordial cells. According to Fernando (2005), if in a self-reproducing and autocatalytic cycle, with a fundamental organization, two previously independent autocatalytic replicating entities establish a physical union, still capable of autocatalysis, and with a low decay rate, then symbiosis can still be viable in an environment with scarce resources, even if the symbiont growth rate is slower than any of the component parts. The failure in establishing symbiosis can result in deficits, and will be a central problem for metabolism and template replication origins. Bouchard (2007) argues that a careful examination of the evolutionary strategies found in many obligate symbiotic associations validates a view of fitness that is not to be understood in reproductive terms, but in ecological ones. He suggests that complex symbiotic systems can have an emergent fitness value described in terms of differential persistence of the community. In our opinion this vision of persistence is in agreement with the symbiotic relation importance in the withstanding of extreme environmental condition and novel ecological niches occupation (Carrapiço et al., 2007).

We believe in abiotic peptides that assisted a symbiosis with nucleic acids, leading to a primitive translation apparatus emergence. This way, encapsulated entities, analogous to modern cells emerged, based on a nucleic acid organization. According to Woese (2004) an imprecise and primitive small primordial genome replication emerged, with simple and primitive cellular designs, of modular nature, permissive and ill-defined, subjected to an unrestricted Horizontal Gene Transfer (HGT). This HGT would not disturb cellular organization to collapse. The components with certain form/function could be replaced by rough equivalents, the primitive representatives of some of the actual specific enzymes that could have only a specific class reaction. This is a period of ephemera organismal genealogy. For the author, these primitive cells had an essentially communal evolution, with a reticulate and cooperatively collective pattern of evolution. At which point we add hybridization, symbiosis, and synergisms, being the base of a multileveled communication world. This way of looking at the emergence of life supports the notion of multiple origins in strait correlation with the geochemical contexts, natural constraints, and environmental conditions. The progenote proposed consisted in an open community of very diverse primitive cellular entities with intense symbiotic associations, antagonisms, and competition, and with a rapid and reticulate pattern of evolution. Another piece of evidence for the widespread nature of these initial and universal synergistic and cooperative interactions, through symbiogenic mechanisms, can be found in microbial mats and stromatolites. The latter forms one of the first examples of a communal ecological strategy, based on cooperative and synergistic mechanisms, for surviving in the primitive Earth extreme environmental conditions (Carrapiço et al., 2007).

An experiment dealing with symbiotic organisms' survival in space conditions showed that symbiosis, together with synergistic and cooperative principles, is a useful and power strategy for survival in extreme environmental conditions. Experiments performed in the BIOPAN-5 facility, in the European Space Agency (ESA), studied the lichen *Rhizocarpon geographicum* survival fitness when facing space conditions. Ascaso et al. (2006) believe this kind of symbiotic organism could represent model organisms for interplanetary transference. Although they observed several ultrastructural changes, they conclude that fungal and algal cells survive in space conditions. Another example of how symbiotic relations can allow the survival and thriving of life in extreme environmental conditions can be seen in the marine invertebrate *Riftia pachyptila*, a giant tube worm (Annelida) found in the hydrothermal submarine vents.

We are convinced that an origin and initial evolution of life approach should include the symbiogenic perspective, in a more communicational sense. Communal synergistic and cooperative strategies certainly occurred in prebiotic evolution. The symbiogenic approach to the emergence of life contributes to the development of the astrobiological knowledge, as it is able to provide innovative perspectives on the study of life's appearance and development on Earth and elsewhere. We agree with Trainer et al. (2006) that life is an emergent and global phenomenon on Earth and not an isolated one, occurring in a specific place. We think it was also a naturalistic process that occured in several places and times, in a quick step-by-step process with evolutive jumps within a small geological "time window." Its emergence was subject to physical-chemical constraints, intimately correlated with different geochemical contexts and environmental conditions. The symbiogenic universal principals favored an increasing and irreversible diversity and complexity, in balance with the primitive Earth's natural conditions. It was also supported by self-assembly, self-catalysis, liquid water, and basic organic compounds. According to Lehto (2007), we can conceive several locations for the different emergence steps toward life as being locally or temporally separated. This way, several of the different scenarios proposed for life's origin by different theories could be integrated in a convergent approach to the problem (Carrapiço et al., 2007).

Our planet can be considered as an open ecological system, with natural subsystems (lithosphere, atmosphere and hydrosphere), also open and interdependent, with constant exchanges of matter and energy between them and the universe, which evolve and cooperate in order to keep Earth in a dynamical equilibrium. We must consider that this dynamic planet functioning was also present in the primitive Earth. The fourth natural system is the biosphere; it crosses all the others and is supported by their interactions. In contemporary Earth, we observe an immense diversity, biological and environmental, with huge facets, interactions, synergistic phenomena, and antagonisms, in a variety of environmental conditions, some of which are considered extremes. Certainly, in the

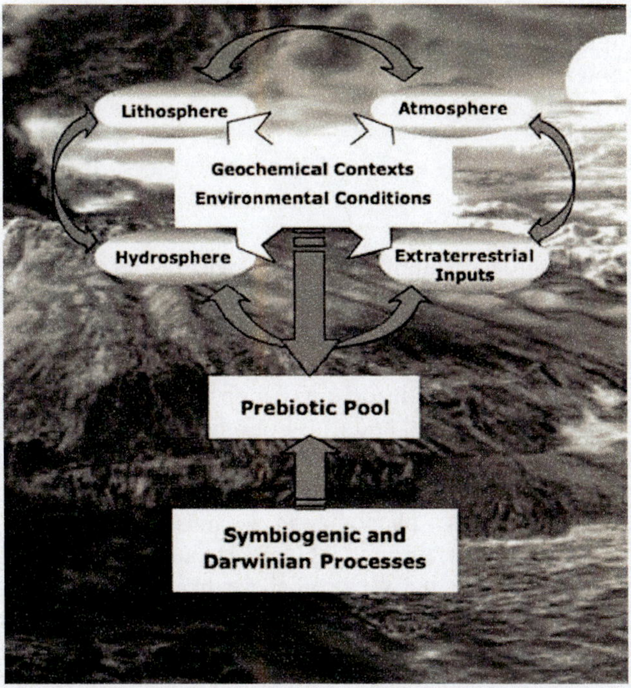


**Figure 1.** Diagram of the possible interactions within the primitive earth prebiotic environment (Carrapiço et al., 2007). (Backup image adapted from http://www2.uol.com.br/sciam/imagens/materia/abre_terra.jpg).

primitive stage, diverse chemical transformations and interrelations between the natural systems also occurred, with important transition phases in their interfaces, translating their current interdependencies (Fig. 1).

The three natural Earth subsystems, synergistically with extraterrestrial sources, provided an abundant input of organic compounds to prebiotic evolution. We assume a diverse reality in the primitive Earth, with different environmental characteristics, and multiple ecological niches, with different organic compounds concentrations. Several authors have already suggested some possible locations for the emergence of the first living systems, namely volcanic fissures, mineral superficies, deep sediments, shallow lagoons, hydrothermal vents and springs, intertidal zones, and ocean surface, aerosols surfaces, and in the subsurface. The existence of several changing habitats in the prebiotic Earth, with semipermeable geographic frontiers, posed environmental challenges that resulted in increasing complex and diverse entities. So, cooperative, synergistic and communicational processes were responsible, using terrestrial and extraterrestrial materials, for the creation of a large prebiotic pool, closely related to geochemical contexts and environmental conditions, and with intense interactions within.

According to Woese (2004) we also believe in life's emergence being accomplished through multiple origins, in different times and environments, displaying a variety of selective contexts, which optimized symbiogenic processes in the promotion of creative novelty (Carrapiço et al., 2007).

## 4. A New Starting Point?

The symbiogenic approach within an astrobiological context is based in several tenets. A currently debated question is the definition of life. According to The Astrobiology Primer (2006), a definition of life based only in extant living forms fails in not considering earlier life forms that preceded current cellular life. Also, NASA's definition of life only considers evolution through competitive processes, excluding symbiogenic processes, and thus not reflecting the natural world's reality. The majority of scientists agree that no single parameter can, by itself, define life, as life emerged from multiple mechanisms acting in concert. According to Kolb (2007), to define life as a phenomenon, we must understand its origin and evolution from abiotic matter, through RNA World and Last Universal Common Ancestor (LUCA). We agree, but we also believe, like Popa (2004, 3, 5, 13, 158), that for the understanding of life as a phenomenon we must consider that life is a complex system, far from equilibrium, delimited from exterior by a frontier, with vast information, with capabilities of self-assembly, self-catalysis, self-organization, self-reproduction, and evolution. It has a metabolism in its own right and homeostasis. It fights against dynamical equilibrium, and uses external sources of energy and nutrients. Its functioning is based in antagonisms, but also in synergistic and cooperative principles, as living beings establish associations and communicate with others.

We believe this proposal for a symbiogenic origin of life has the ability to overcome several of the difficulties highlighted previously. It is a possible way to articulate different theories, valid in specific Earth contexts, and is supported by natural processes, reinforcing the role of the extraterrestrial contributions to the prebiotic chemical pool on Earth, deals with the metabolism-replication dilemma, and puts different primitive environmental conditions in close relation with the existent geochemical contexts. These lines can guide future work, but for that it is necessary to adopt a more cooperative and communicational approach underlying a multidisciplinary scientific work, the reconsidering of positions concerning this field, and the time and place to give, with an open mind, more coherence and meaning to all the work done until now (Carrapiço et al., 2007).

Where to look for future answers? It would appear to be necessary to dedicate more time to the detection of extraterrestrial life, new prebiotic chemistry experimental work, to embrace artificial life studies, and computational simulations with a symbiogenic perspective, extremophile studies, and space exploration. In 2000, Fry said: "It is plausible to speculate that the weight given in the future to metabolic versus genetic models will be largely decided by experimental work." Ten years later, we are facing a possible starting point to the beginning of a new era in the studies on the origin of life. An era that probably will result, according to the Gibson et al. (2010), in a better understanding of the nature of life itself. These authors were responsible for the creation of a bacterial cell controlled by a chemically synthesized genome. The authors designed, synthesized, and assembled a genome starting from digitized genome sequence information and its transplantation into a *Mycoplasma mycoides* cells that were controlled only by the synthetic chromosome.

Gibson et al. (2010) state that the only DNA in the cells was the designed synthetic DNA, including "watermark" sequences and other designed gene deletions and polymorphisms, and mutations acquired during the building process. The new cells have expected phenotypic properties and are capable of continuous self-replication and of logarithmic growth. In the discussion of the results obtained, the authors refer to such a cell controlled by a genome assembled from chemically synthesized pieces of DNA as a "synthetic cell," even though the cytoplasm of the recipient cell is not synthetic. The properties of the cells controlled by the assembled genome are expected to be the same as if the whole cell had been produced synthetically; the authors consider that the DNA software builds its own hardware. According to the authors "if the methods described here can be generalized, design, synthesis, assembly, and transplantation of synthetic chromosomes will no longer be a barrier to the progress of synthetic biology" (Gibson et al., 2010, 56). These authors anticipate that their work will continue to raise philosophical issues, with broad societal and ethical implications. We agree with this point of view, not only with regard for a scientific theory on the nature of life, but also for the symbiogenic origin of life and initial evolution studies.

We think this experimental work illustrates a pratical application of the symbiogenesis concept to the artificial life using live material. This achievement poses several questions regarding the origin of life itself. In the domain of artificial life, we would like to remember the work of Nils Barricelli referred to previously. For this author, the software analog in the living world is a self-replicating DNA molecule. Most of this software is parasitic (symbiotic), depending upon metabolism, which favors the analogies to life. Barricelli also considers that languages are the key to the evolution of self-reproducing structures with greater complexity, through the cooperation of simpler parts. We believe this experimental work represents a very important breakthrough in the origin of life studies. The analogies of software replication (genetic) and hardware metabolism, the fact that they need each other, and their interdependency, are in agreement with the notions of coevolution and symbiogenesis (Dyson, 1997).

According to Correia (2010), when considering artificial life and evolution, perhaps, the most interesting results of artificial life environments are the emergence of evolutionary processes and ecological dynamics equivalent to those observed in nature, such us host-parasite relations and punctuated equilibrium, among others. The author considers it very significant that such macroscopic effects are observed both in natural and artificial environments. We agree with the author opinions when saying (Correia, 2010, 189): "Computational evolution has opened new possibilities of exploring biology by allowing us to liberally build models that may go beyond what we currently know in natural organisms"; "Both EA and artificial life evolution, in software platforms or in robots, can be put to use in testing current or new biological models."

Perhaps we should remember the symbiogenic origin of life theories mentioned previously. Remember that in the primitive Earth there were probably several processes resulting in amphiphilic vesicles and that Matthews (1994), and Matthews and Minard (2003) advanced a symbiogenic idea for the nucleotides endogenous formation on the primitive Earth. According to this model, membrane material – carboxylic acids, carbohydrates, and polypeptides – accumulated in lakes and oceans, while on land polyamidines could have been the original dehydrating agents directing the synthesis of nucleosides and nucleotides from available sugars, phosphates, and nitrogen bases. Most significant would have been the parallel synthesis of polypeptides and polynucleotides arising from the dehydrating action of polyamidines on nucleotides. Metabolic material – hardware – thus arose separately from genetic components – software – as proposed by Freeman Dyson. Subsequent interfacing, perhaps with the help of clays, then produced the first replicating protocells (Carrapiço et al., 2007). On our dynamic planet, this polypeptide-polynucleotide symbiosis mediated by polyamidines may have set the pattern for the evolution of protein-nucleic acid systems controlled by enzymes, the mode characteristic of life today. In 2003, Matthews refers that from hydrolysis of hydrogen cyanide (HCN) polymers, there is some evidence for peptidic structures. Also, hydrolysis/pyrolysis gives rise to nitrogen heterocycles, including purines and pyrimidines found in nucleic acids today. According to the author, some recent experimental results are consistent with this model of HCN polymer chemistry playing an essential role in the origin of life (Carrapiço et al., 2007). Reinforcing this idea, Martins et al. (2008) report the discovery of extraterrestrial nucleobases in the Murchison meteorite.

The role of viruses should also be considered in the origin and initial evolution of life as proposed in 2010 by Villarreal and Witzany. The authors consider that viruses had an evolutionary role and are the predecessors of cells as polyphyletic. Following the ideas of Woese (2004), they propose the coexistence of several LUCAs, as quasispecies based and reticulated. This is a strong argument in favor of our symbiogenic envision since we postulate the existence of several LUCAs, several lineages in the way to life, via symbiotic interactions. The primitive Earth was a natural laboratory where those ephemeral lineages emerged, and where RNA genomes existed, probably, before the DNA ones. This way, varied contributions of the Lipid World and of the RNA World should be considered together in the establishment of the initial roots of the tree of life and with a reticulated pattern of evolution.

## 5. Concluding Remarks

The symbiogenic approach to chemical and biological evolution should be seriously considered. We believe that competition and cooperation can coexist in the same scenario of evolution, and probably take place in terms of discontinuous bursts of activity, depending on the internal and external conditions that drive evolution. It means that the same population can evolve using competitive and/or cooperative processes during the time and space of a hypercycle evolutive scenario. We believe that cooperative and synergistic processes were responsible, using terrestrial and extraterrestrial materials, for the creation of a large prebiotic pool, closely related to geochemical contexts, and intense interactions within. Most likely, a series of synergistic and cooperative effects produced a wide source of creativity, and functional advantages that pushed the emergence of complex and functionally integrated biological systems, through the evolution of self-organization and autocatalysis. It was only after this biochemical evolution of structures, which produced the informational capabilities necessary to self-replication, that the Darwinian mechanisms could arise. This way of perceiving the emergence of life follows the proposals regarding life's initial evolution in which the progenote proposed consisted in an open community of very diverse primitive cellular entities with intense symbiotic associations, antagonisms, and competition, and with a rapid and reticulate pattern of evolution. We believe this symbiogenic approach should be considered in the understanding of chemical and biological evolution. This contributes to the development of astrobiological knowledge, enabling other perspectives on the study of life's appearance and development on Earth and elsewhere. It is also necessary to reconsider symbiosis as a general mechanism in heredity and development, in addition to gene mutations and recombination, as a source of evolutionary innovation, and to hold symbiotic processes as one of the main bases of biodiversity and evolution on Earth. This idea implies also the central role of interactions, in which individuality (new entity) emerges through incorporation. It involves horizontal mergers, which can be rapid, and, usually, discontinuous, creating permanent and irreversible changes, the ground for evolutive novelty. Something new arises through merging, being a unique or new metabolism or structure(s), which was not present before symbiosis (Carrapiço et al., 2007). In this context, we can consider this entity as a new taxonomic novelty or even as a new level of biological organization. Thus, biology must take into account this reality, integrating symbiosis, not only as a factor of evolutionary change, but also as a taxonomic element in the organization of the living world (Carrapiço, 2010).

According to Fry (2000) the "arch and scaffold" model constitutes a metaphor to analyze the origins of life theories. Following this idea we suggest that the synergistic, cooperative, and symbiogenic interactions between the prebiotic entities provided the way for the establishment of the physical and chemical scaffolds the author talks about. In our perspective, those scaffolds or supports were based on symbiogenic principles, which together with physical and chemical prebiotic constraints, the geochemical context compatibility and environmental conditions, channeled evolution, leading to the currently established domains of life. Although there are different models to explain the origin of life, they reveal complementary aspects of the same reality applied to the primitive Earth. Thus, a more cooperative and communicational work is necessary to help reveal the so far hidden answer(s) and for the search of life elsewhere. Astrobiology, being a multi-transdisciplinary science, has the conditions to be the primary actor leading this enterprise and to implement this paradigm shift.

Finally, we believe that an experimental model for the symbiogenic approach to life's origin will come up. Artificial life′s simulations may lead the way: digital primordial soup, the EvoGrid is being developed, through the adaptation of an open source molecular dynamics simulator, to allow artificial nature to take its course, to model the *prebiotic* chemical environment and search for emergent self-organization (Damer et al., 2009).

## 6. Summary

The scientific study of the origin of life remains one of the greatest unanswered scientific questions in the twenty-first century. This study must integrate and articulate the contributions from diverse scientific fields, such as biology, geology, chemistry, geophysics, and astronomy, among others. The transdisciplinary work that is being done in astrobiology becomes fundamental to a scientific comprehension of life's origin, including the conditions and processes of the primitive Earth. Different evolutive mechanisms have been put forward, but the general acceptance of a smaller "time window" for life's emergence has led other researchers, while not ignoring the role of natural selection and the notions of gradualism and competition, to look for other and complementary evolutive mechanisms, which could explain a more rapid evolution. Innovative approaches have eliminated the traditional dichotomy metabolism and replication, such as the double origin of life, developed by the astrophysicist Freeman Dyson, in 1985. More recently, other authors have put forward new evolutive mechanisms, involving cooperation and synergisms. In 2003, Jan Sapp has conceptually integrated symbiogenesis in the symbiomics domain, which became a new evolution theory, one that is post neo-Darwinian. These concepts had their origin in the principles developed by Constantin Merezhkowsky, in 1909, and Kozo-Polyansky, in 1924, which gave symbiosis a determinant role in evolution. These ideas were strengthened by the universal symbiogenesis concept, developed in 2007 by Nathalie Gontier. The inclusion of symbiogenic principles in the comprehension of life emergence, and the contribution from the latest astrobiology developments must be considered in the current life origin studies. A symbiogenic approach toward the origin of life becomes imperative, leading to a broader comprehension of a phenomenon that can be emergent and global.

## 7. Acknowledgments

A special word of gratitude to Benjamin Farrand and Helena Carrapiço for the English revision of the text.

## 8. References

Antonelli PL, Bevilacqua L, Rutz SF (2003) Theories and models in symbiogenesis. Nonlinear Anal Real World Appl 4:743–753

Ascaso C, Los Rios A, Wierzchos J, Risueño C, Horneck G, De La Torre R, Sancho LG (2006) Ultrastructural features of *Rhizocarpon geographicum* symbionts after exposure to space at Biopan-5 facility. In: Abstracts from 6th European workshop on astrobiology, Ecole Normale Supérieure de Lyon, France, October

Bouchard F (2007) What is a symbiotic superorganism and how do you measure its fitness? In: Abstracts of the ISHPSSB meeting, University of Exeter, UK, 25–29 July, pp 45

Brack A (2007) Astrobiology: from the origin of life on earth to life in the universe. In: Horneck G, Rettberg P (eds) Complete course in astrobiology. Wiley-VCH Verlag GmbH & Co.KgaA, Stuttgart, pp 1–22

Carrapiço F (2006) The origins of life and the mechanisms of biological evolution. Proc SPIE 6309:63090O-1–63090O-5

Carrapiço F (2010) How symbiogenic is evolution? Theory Biosci 129(2–3):135–139

Carrapiço F, Pereira L, Rodrigues T (2007) Contribution to a symbiogenic approach in astrobiology. Proc SPIE 6694:669406–1–669406–10

Chapman MJ, Margulis L (1998) Morphogenesis by symbiogenesis. Int Microbiol 1:319–326

Corning PA (1995) Synergy and self-organization in the evolution of complex systems. Syst Res 12(2):89–121

Corning PA (2005) Holistic darwinism. Synergy, cybernetics, and the bioeconomics of evolution. The University of Chicago Press, Chicago

Correia L (2010) Computational evolution: taking liberties. Theory Biosci 129(2–3):183–191

Damer B, Newman P, Gordon R, Barbalet T (2009) The EvoGrid: simulating pre-biotic emergent complexity. http://www.evogrid.org/. Accessed September 2010

De Bary A (1878) Ueber symbiose. – Tageblatt 51. Versamml. Deutscher Naturforscher u. Aerzte. Cassel 1878:121–126

Dubos R, Kessler A (1963) Integrative and disintegrative factors in symbiotic associations. In: Nutman PS, Mosse B (eds) Proceedings of the thirteenth symposium of the Society for General Microbiology, London, pp 1–11

Dyson F (1985) Origins of life. Cambridge University Press, Cambridge

Dyson GB (1997) Darwin among the machines. The evolution of global intelligency. Perseus Books, Cambridge

Dyson F (1998) Evolution, society, science and the universe. Darwin College/Cambridge University Press, Cambridge

Fernando C (2005) The good symbiont. In: Caparrere MS, Freitas AA, Bentley PJ, Johnson CG, Timmis J (eds) ECAL 2005, LNAI, 3630. Springer, Berlin, pp 695–704

Fogel DB (2006) Nils Barricelli – artificial life, coevolution, self-adaptation. IEEE Comput Intell Mag 1(1):41–45

Frank SA (1995) The origin of synergistic symbiosis. J Theor Biol 176:403–410

Fry I (2000) The emergence of life on earth. A historical and scientific overview. Free Association Books, London

Gibson DG, Glass JI, Lartigue C, Noskov VN, Chuang R-Y, Algire MA, Benders GA, Montague MG, Ma L, Moodie MM, Merryman C, Vashee S, Krishnakumar R, Assad-Garcia N,

Andrews-Pfannkoch C, Denisova EA, Young L, Qi Z-Q, Segall-Shapiro TH, Calvey CH, Parmar PP, Hutchison CA III, Smith HO, Venter JC (2010) Creation of a bacterial cell controlled by a chemically synthesized genome. Science 329:52–56

Gontier N (2007) Universal symbiogenesis: an alternative to universal selectionist accounts of evolution. Symbiosis 44:167–181

King GAM (1977) Symbiosis and the origin of life. Orig Life Evol Biosph 8(1):39–53

Kolb VM (2007) On the applicability of the aristotelian principles to the definition of life. Int J Astrobiol 6(1):51–57

Kozo-Polyansky B (2010) Symbiogenesis. A new principle of evolution. Translated from the Russian edition (1924) by Victor Fet. Fet V, Margulis L (eds). Harvard University Press, Cambridge

Kutschera U (2009a) Struggle to translate Darwin's view of concurrency. Nature 458:967

Kutschera U (2009b) Symbiogenesis, natural selection, and the dynamic earth. Theory Biosci 128(3):191–203

Lee DH, Severin K, Yokobayashi Y, Ghadiri MR (1997) Emergence of symbiosis in peptide self-replication through a hypercyclic network. Nature 390:591–594

Lehto K (2007) From molecular evolution to cellular life. In: Horneck G, Rettberg P (eds) Complete course in astrobiology. Wiley-VCH Verlag GmbH & Co.KgaA, Stuttgart, pp 85–120

Margulis L (1990) Words as battle cries: symbiogenesis and the new field of endocytobiology. Bioscience 40(9):673–677

Margulis L (1999) The symbiotic planet. A new look at evolution. Orion Books, London

Martins Z, Botta O, Fogel ML, Sephton MA, Glavin DP, Watson JS, Dworkin JP, Schwartz AW, Ehrenfreund P (2008) Extraterrestrial nucleobases in the Murchison meteorite. Earth Planet Sci Lett 270:130–136

Matthews CN (1994) The HCN world: establishing protein-nucleic acid life. The Alexander Ivanovich Oparin 100th anniversary conference, Miramare, Trieste

Matthews CN, Minard RD (2003) Prebiotic and extraterrestrial chemistry of hydrogen cyanide polymers. In: Abstracts of the workshop on cometary dust in astrophysics, Crystal Mountain, Washington, 2003, pp 46

Popa R (2004) Between necessity and probability: searching for the definition and origin of life. Springer, Berlin

Reid RGB (2007) Biological emergences. Evolution by natural experiment. The MIT Press, Cambridge

Roossinck MJ (2005) Symbiosis versus competition in plant virus evolution. Nat Rev Microbiol 3:917–924

Roossinck MJ (2008) Symbiosis, mutualism and symbiogenesis. In: Evolution of plant virus. Springer, Berlin, pp 157–164

Ryan FP (2006) Genomic creativity and natural selection: a modern synthesis. Biol J Linn Soc 88:655–672

Sapp J (2003) Genesis: the evolution of biology. Oxford University Press, New York

Sapp J (2004) The dynamics of symbiosis: an historical overview. Can J Bot 82:1046–1056

Sapp J, Carrapiço F, Zolotonodov M (2002) Symbiogenesis: the hidden face of Constantin Merezhkowsky. Hist Philos Life Sci 24:421–449

Shenhav B, Segré D, Lancet D (2003) Mesobiotic emergence: molecular and ensemble complexity in early evolution. Adv Complex Syst 6(1):15–25

Taylor TJ (1999) From artificial evolution to artificial life. PhD thesis, Division of Informatics, University of Edinburgh

The astrobiology primer: an outline of general knowledge (2006) – version 1, education paper. Mary Ann Liebert, Inc, New Rochelle

Trainer MG, Pavlov AA, DeWitt HL, Jimenez JL, Mckay CP, Toon OB, Tolbert MA (2006) Organic haze on titan and the early earth. Proc Natl Acad Sci USA 103(48):18035–18042

Villarreal LP, Witzany G (2010) Viruses are essential agents within the roots and stem of the tree of life. J Theor Biol 262:698–710

Watson RA (2007) Compositional evolution and symbiosis. In: Abstracts of the ISHPSSB meeting, University of Exeter, UK, 25–29 July, pp 183–184

Woese CR (2004) A new biology for a new century. Microbiol Mol Biol Rev 68(2):173–186

Biodata of **Jan Sapp**, author of "***Horizontal Gene Transfer and the Tree of Life***."

**Professor Jan Sapp** obtained his Ph.D. from the University of Montreal in 1984. He was professor at the University of Melbourne from 1984 to 1990, and Andrew Mellon Fellow at the Rockefeller University (1991–1992) before his appointment at York University. His scientific interest is in the areas of symbiosis and lateral gene transfer as modes of evolutionary change, microbial phylogeny, and the evolution of complex systems. He has authored many scholarly papers and numerous books, including *Beyond the Gene* 1987; *Evolution by Association: A History of Symbiosis* 1994; *What Is Natural? Coral Reef Crisis* 1999; *Genesis: The Evolution of Biology* 2003; and *The New Foundations of Evolution: On the Tree of Life* (2009).

E-mail: **jsapp@yorku.ca**

J. Seckbach (ed.), *Genesis - In The Beginning: Precursors of Life, Chemical Models and Early Biological Evolution*, Cellular Origin, Life in Extreme Habitats and Astrobiology 22, 743–755
DOI 10.1007/978-94-007-2941-4_37, 

# HORIZONTAL GENE TRANSFER AND THE TREE OF LIFE

**JAN SAPP**
*Department of Biology, York University, 4700 Keele St, Toronto, ON M3J 1P3, Canada*

## 1. Microbes Outside Evolution

The concept of highly conserved characteristics, far removed from the vicissitudes of life, through which one could reconstruct the main course of evolution has been central to evolutionary biology since its earliest beginnings. In *Philosophie Zoologique*, Jean Baptiste Lamarck argued that in ordering organisms in a genealogical series of increasing "complexity of organization," one had to distinguish characteristics that were modified through the influence of environmental conditions, from "the essential system of organs" (Lamarck, 1809, p. 58). Comparisons of the latter could be made only in the higher groupings of animals, not in species or genera; they were less conspicuous in plants, and not at all in the infusoria.

The existence of highly preserved characteristics, which had become integrated within the organism in the course of evolution, and which would unite "large groups of species" was a central to Charles Darwin's concept of descent with modification, too. In *The Origin of Species*, he (1859, pp. 414–415) emphasized the importance of those "essential characteristics" – "organs of high vital or physiological importance." "It might have been thought… that those parts of the structure which determine the habits of life, and the general place of each being in the economy of nature, would be of very high importance in classification. Nothing can be more false…It may even be given as a general rule that the less any part of the organisation is concerned with special habits, the more important it becomes for classification" (Darwin, 1859, p. 414). "Embryological characters," he said, "are the most valuable of all" (Darwin, 1859, p. 479).

Over the next one hundred years, comparative morphology anatomy and the fossil record were the ways to distinguish homology from analogy and to order organisms according to their evolutionary relatedness. But bacteria lacked such morphological complexity as well as a fossil record. One could not tell which traits were homologous and which were analogous, which were ancient and of common ancestry which were more recent adaptations.

Linnaeus had placed all infusoria in one species that he drolly christened *Chaos infusoria* (Dobell, 1932, p. 377). And some bacterial taxonomists of the early twentieth century continued to refer to bacterial classification in terms of chaos because bacteria were not classified into groups based on principles of homology, but rather on as many characteristics as possible. Bacteria were arranged into a nested hierarchy of orders, families, genera, and species based on a plethora of characteristics: cell shape; plane of cell division; ability to form spores and/or colonies; possession of flagella, whether cells were connected, and whether they were branching; staining reactions; relation to temperature and oxygen; pigment production; pathogenicity; and a great diversity of biochemical powers (Sapp, 2009a).

Still, some bacteriologists attempted a classification based on principles of homology and on ever-increasing morphological complexity (Kluyver and van Niel, 1936; Stanier and van Niel, 1941). But by the middle of the twentieth century, they were forced to admit that their efforts were futile (van Niel, 1955). Roger et al. (1957, p. 297) asserted in the first edition of *The Microbial World*: "... it is a waste of time to attempt a natural system of classification for bacteria, ... bacteriologists should concentrate instead on the more humble practical task of devising determinative keys to provide the easiest possible identification of species and genera."

Even demarcating bacteria as a natural grouping apart from other microbes was rife with difficulty. Ernst Haeckel (1866, 1892) had placed bacteria (and blue-green algae) in a division he called "Monera" and, within a third kingdom, Protista. The Monera were supposed to lack the fundamental division of labor – of nucleus and cytoplasm exhibited in "true" cells – and to bridge the gap between the living and the nonliving. Whether the organization of the bacterium fit Haeckel's definition was hotly contested throughout the early twentieth century (Sapp, 2005a, 2009a).

While a few microbiologists proposed that bacteria along with the blue-green algae be assigned to their own kingdom, Monera, (Copeland, 1938; Stanier and van Niel, 1941) critics doubted that it was a genealogically coherent monophyletic grouping. It was far from clear if the blue-green algae really lacked a nucleus, nor was it certain how smaller bacteria such as *Rickettsia* and *Chlamydia* could be distinguished from viruses. "In fact there is no sharp line of distinction between the largest animal viruses and the *rickettsiae*" Stanier et al. (1957, p. 360).

A uniquivocal distinction between the virus and the bacterium would be drawn on the basis of electron microscopy and chemistry: the virus contained only one kind of nucleic acid, either RNA or DNA, enclosed in a coat of protein; it possessed few if any enzymes, and it did not reproduce by division like a cell Lwoff (1957). A few years later, a similar uniquivocal distinction was made between the organization of a bacterium and all the cells of other protists, plants, and animals (Stanier and van Niel, 1962). Borrowing terms from Edouard Chatton (1925), prokaryotic cells (Greek: before karyon or nucleus) were distinguished from eukaryotic cells (Greek: true nucleus). The latter divided by mitosis possessed a membrane-bound nucleus, an intricate cytoskeleton, mitochondria, and, in the case of plants cells, chloroplasts. Prokaryotic cells were smaller and lacked those structures.

> The principle distinguishing features of the procaryotic cell are: 1 absence of internal membranes which separate the resting nucleus from the cytoplasm, and isolate the enzymatic machinery of photosynthesis and of respiration in specific organelles;

2 nuclear division by fission, not by mitosis, a character possibly related to the presence of a single structure which carries all the genetic information of the cell; and 3 the presence of a cell wall which contains a specific mucopeptide as its strengthening element. (Stanier and van Niel, 1962, p. 33)

Although a natural classification could not be achieved for bacteria, it was often assumed that their prokaryotic structure implied a common origin for the grouping. "All these organisms share the distinctive structural properties associated with the procaryotic cell...and we can therefore safely infer a common origin for the whole group in the remote evolutionary past" (Stanier et al., 1963, p. 409). "In fact," they wrote, "this basic divergence in cellular structure, *which separates the bacteria and blue-green algae from all other cellular organisms*, probably represents the greatest single evolutionary discontinuity to be found in the present-day world" (Stanier et al., 1963, p. 85).

## 2. Molecular Phylogenetics and Three Domains

New molecular methods for classification, based on GC content, DNA-RNA hybridization, amino acid sequencing of proteins, and sequencing of ribosomal RNAs, revitalized the aim of a true phylogenetic classification of bacteria in the 1960s and 1970s (Sapp, 2009a).

Sequence analysis of the small subunit RNA of ribosomes, (SSU rRNA) or 16S ribosomal RNA, by Carl Woese and collaborators at the University of Illinois led to an upheaval in bacterial taxonomy. They heralded the revolution in 1980 when providing an outline of a universal phylogenetic tree with a branching pattern unlike any seen before. "For the first time," they said, "a single experimental approach, SSU ribosomal RNA sequence characterization, has been used to develop an overview of phylogenetic relationships in the bacterial world. The technique permits the tracing of relationships back to the common ancestor of all extant life" (Fox et al., 1980, p. 458).

The 16S rRNA sequencing method had remarkable predictive success. It resolved the venerable question of whether mitochondria and chloroplasts arose as symbionts (Sapp, 1994, 2009a). That chloroplasts might have evolved as symbionts was frequently considered since the nineteenth century (Sapp et al., 2002). Some had suggested that the nucleus and cytoplasm reflected a symbiosis – and perhaps centrioles too arose as symbionts (Sapp, 1994). The notion that mitochondria were symbionts was developed by Paul Portier (1918) at the *Institute Oceanographique de Monaco* and by Ivan E Wallin (1927) at the University of Colorado who argued that repeated mitochondrial acquisitions would be the source of new nuclear genes.

Such speculations were generally recognized to be far outside the realm of empirical science. As E. B. Wilson (1925, p. 739) commented in his famed book, *The Cell in Development and Heredity*, "To many, no doubt, such speculations may appear too fantastic for present mention in polite biological society; nevertheless it is within the range of possibility that they may some day call for more serious consideration."

The possible role of symbiosis in the origin of the eukaryotic cell was brought to the fore in the early 1960s with evidence that mitochondria and chloroplasts possessed their own DNA and ribosomes (Sapp, 1994, 2009a). That centrioles possessed DNA remained controversial until disproven in the mid 1990s (Sapp, 1998).

Still, the proof of origin was lacking. Some speculated that one day it might be possible to extract mitochondria, and chloroplasts and even centrioles, and culture them in vitro (Sagan, 1967). But, as opponents of their symbiotic origin noted, those organelle functions were genetically well integrated into nuclear genomes. Discussions of the strength of arguments for exogenous or endogenous origin centered over the issue of parsimony (Sapp, 1994). Others reached the conclusion that the debate over organelle origins remained effectively sterile. W. H. Woolhouse (1967, p. 833) remarked "the time has come to bury this kind of speculation with, by way of an epitaph, a parody of Wittgenstein's well-known remark, 'Whereof one cannot know, thereof one should not speak.'" Roger Stanier (1970, p. 31) decreed that the problem would always remain in the realm of metascience:

> It might have happened thus; but we shall surely never know with certainty. Evolutionary speculation constitutes a kind of metascience, which has the same intellectual fascination for some biologists that metaphysical speculation possessed for some medieval scholastics. It can be considered a relatively harmless habit, like eating peanuts, unless it assumes the form of an obsession; then it becomes a vice.

Such comments were hasty indeed. SSU rRNA comparisons traced mitochondria to alpha proteobacteria and chloroplasts to cyanobacteria (Gray and Doolittle, 1982).

Nothing was more startling than the announcement of an ancient "third form of life": the archaebacteria (Woese and Fox, 1977a). The group comprised methanogens, found in the guts of rumens, extreme halophiles known for rotting salted fish, and thermoacidophiles, that live in conditions that would cook other organisms. These organisms were thought to be completely unrelated before their 16S rRNA analysis brought them together as an ancient primary lineage no more related to typical bacteria than they were to eukaryotes. The group had many characteristics in common: their walls lacked peptidoglycan, the only positive characteristic of the prokaryotes, their tRNAs were unique, the lipids in their membranes were ether-linked not ester-linked, and their transcription enzymes were unlike that of other bacteria, so too their viruses.

The concept of three primary lineages or "urkingdoms" Woese and Fox challenged two assumptions underlying the prokaryote concept: (1) that prokaryotes preceded and gave rise to eukaryotes, and (2) that prokaryotes stem from a common ancestor, itself prokaryotic. (Fox et al., 1980, p. 458) In each urkingdom, there were important differences which suggested that that the three lines of descent may have diverged before the level of complexity of the modern cell was attained: "Genetic control mechanisms seem to differ; RNA polymerase subunit structure differs; rRNA's and tRNA's differ in patterns of post transcriptional modification; cell walls differ in composition, as do lipids, and so on" (Fox et al., 1980, p. 463).

Molecular phylogeneticists who dug into the roots of the universal tree of life typically searched for a universal single prokaryotic-like cell as the most recent common ancestor. Some called it "the cenancestor" (Fitch and Upper, 1987) others "LUCA," the last universal common ancestor (Forterre and Philippe, 1999). Woese sought deeper roots for all three "urkingdoms" in what he and George Fox called "the progenotes" – precellular entities in the throes of developing the relationship between nucleic acids and proteins (Woese and Fox, 1977b; Woese, 1998).

Before the development of the modern translation apparatus, evolution would be driven by a different mode and tempo. Evolution would have been rapid, driven by the high mutation rate of an error – prone genetic system – and by extensive gene transfer of its still relatively loosely connected components. During the progenotes era, there would be no discernable lineages as such. There would be a "ready exchange, a flow, among sub-cellular entities – be they called genes, plasmids, viruses, selfish DNA" at the progenote stage giving rise to "a molecular mosaicism in its descendant lines" (Woese, 1982, p. 13). Primitive systems would be modular and exchange parts freely, and generate enormous diversity very quickly. But, as molecular information processing systems became more accurate, the translation system optimized, organisms with three definable lineages would emerge as those "essential functions" became refractory to horizontal gene transfer.

By 1989, the universal tree based on three primary lineages was rooted by using ancient gene duplications (Iwabe et al., 1989; Gogarten et al., 1989). The ancient stem split archaebacteria and eukaryotes on one side, and typical bacteria on the other, a split that was congruent with other molecular differences among the three taxa. A formal taxonomic proposal of three domains – the Eucarya, the Bacteria, and the Archaea – was made to reflect that phylogenetic ordering (Woese et al., 1990).

Woese (1979, 1980; see also Hartman, 1975) postulated that the first organisms were autotrophs, a view that flew in the face of another widely held doctrine that the first organisms were heterotrophs as articulated in the Oparin-Haldane theory (Oparin, 1938; Haldane, 1929). Accordingly, to the latter theory, the heterotrophs separated from, and then fed on, a rich broth of organic compounds that had evolved over billions of years. Autotrophs came later. Yet, fossil evidence from the 1970s pushed the time of life on Earth back to 3.4 billion years. Simply put, there is no time for the "prebiotic soup." Further, those ancient fossils are thought to be of cyanobacteria, and those organisms possess a highly sophisticated photosynthetic apparatus requiring 100s of million years to evolve. Therefore, one could assume that life on earth evolved very soon after the earth began to cool.

## 3. "Core" Issues

By the end of the last century, *Bergey's Manual* had adopted the phylogenetic framework based on 16S rRNA data. But by that time, doubts arose anew about whether bacterial groups could be organized into a phylogenetic hierarchy (Sapp, 2005b, 2009a).

The problem is horizontal gene transfer (the nongenealogical transfer of genes). Bacterial geneticists of the 1950s had shown that genes could be transferred between bacteria by several means: (1) transformations whereby DNA is picked up from the environment, (2) conjugation where there is a unidirectional transfer of genes between donor and recipient which required physical contact, and (3) transduction where by genes from viruses and their former hosts are transmitted through "infection." Horizontal gene transfer (HGT) could potentially scramble the phylogenetic signal. Concepts of "infectious heredity" accompanied the emergence of bacterial genetics (Lederberg, 1952), but leading bacterial geneticists dismissed HGT as of little evolutionary significance (Stocker, 1955; Luria and Burrous, 1957).

But that perspective was dramatically contradicted with new evidence from genomics during the late 1990s. Horizontal gene transfer was far more pervasive than hitherto expected. Essentially, the data were of two kinds: one based on incongruities of trees generated from different genes and the other on anomalous G+C content and codon usages. The genetic code is highly redundant: all but two amino acids are encoded by more than one codon (e.g., UAU and UAC for tyrosine). There are often species-specific preferences for one of the several codons that encodes the same amino acid. On this latter basis, about 18% of the genes of *E. coli* are relatively recent acquisitions, and all the genes that distinguished *E. coli* from *Salmonella enterica* had been horizontally transferred. (Lawrence and Ochman, 1998). Bacterial "speciation," it seemed, was not driven by point mutations, after all, but rather by leaps, and a high rate of horizontal transfer. Horizontal gene transfer could not only confer antibiotic resistance, but could transform bacteria being benign to pathogenic in a single saltational step.

Debates over the scope and significance of horizontal gene transfer and its ramifications for a tree of life intensified. Are all genes passed around through all bacteria in a haphazard manner in time so as to leave no trace of the genes that defined the organization of the cell? Are there sets of characters to define bacteria taxa in terms of organismal genealogies or was the bacterium merely a collection of genes (or gene clusters), each with its own history (Woese, 1987)? Is there a set of specific kinds of genes that are passed on vertically from cell generation to the next? And if so, can those genes represent the course of organismal evolution in the microbial world?

Many molecular phylogeneticists do indeed adhere to the concept of "the genetic core" as sequences refractory to horizontal gene transfer, and through which true organismal phylogenies can be revealed. This conception sometimes is referred to as the "complexity hypothesis" (Jain et al., 1999). Accordingly, there are two classes of genes: *informational genes*, which are not to be transferred horizontally and *operational* genes, which are so transferred. Informational genes are involved in transcription and translation of DNA and related essential and complex processes that involve the interaction of hundreds of gene products. The complexity of that system would restrict horizontal transfers of genes involved in its individual components. *Operational* genes, whose corresponding enzymes tend to be involved in individual metabolic pathways, such as the synthesis of amino acids, cofactors, cell envelope, and lipids, would have fewer functional constraints.

To be sure, those who adhered to the complexity hypothesis recognize that whatever the universal tree of life is, it is not the conventional tree that Darwin and his followers had sought out (Sapp, 2009b). As Woese (1998, p. 6837) commented, "Its primary branchings reflect the common history of central components of the ribosome, components of the transcription apparatus, and a few other genes. But that is all. In its deep branches, this tree is merely a gene tree." "Vertically generated and horizontally acquired variation" are the "yin and yang of the evolutionary process" Woese (2000, p. 8393). Accordingly, horizontal gene transfer is responsible for novel enzymatic pathways, membrane transport capacities, and so on, whereas, "vertically generated variation is the principal way in which biological complexity, specificity, and cellular integration evolve" (Woese, 2002, p. 8743).

Opponents of the complexity hypothesis insist that the concept of an organismal core refractory to horizontal gene transfer is an unproven and unwarranted assumption. They suggest instead that there are no bacterial histories to be discerned because horizontal gene transfer is the essence of the "prokaryotic" evolutionary process. Bacterial taxa at all ranks are no more real than their supposed genealogies. There is no such thing as an archaeon. Only gene histories were real (Doolittle, 1996, 1999, 2000; Gogarten et al., 2002). HGT, symbiosis, and cell fusions jeopardize the very concept of cellular lineages.

The eukaryotic cell too is chimeric, not just because of the symbiogenetic origin of cytoplasmic organelles mitochondria and chloroplast. The eukaryotic host cell that acquired mitochondria and chloroplasts was formed from a symbiosis of two or three types of organisms. The nuclear genome of ancient eukaryotes comprises informational genes from the Archaea, metabolic genes from the Bacteria, and perhaps other genes from a third extinct line of descent (Hartman and Fedorov, 2002; Sapp, 2009a).

Whereas Woese referred to an organismal core, the "functional essence" of the organism (e.g., translation, transcription, genome replication, and control) and maintained the existence of natural *kinds* to be discovered, detractors argue that all of it embodies an erroneous concept of essentialism. Bacterial taxa, whether defined in terms of a core of genes or in terms of the most shared genes with a common history (majority rule), were merely human inventions:

> It is as if we believed that the words 'eukaryote' and 'prokaryote' named natural kinds whose properties we need only to discover. But in fact they are categories we ourselves invented 30–40 years ago (when our understanding of cell and molecular biology was

> pretty rudimentary) to define organizational grades or identify evolutionary clades. We have not only the right but the obligation to change them now and in future, as our knowledge grows (Doolittle, 1996, p. 8798).
>
> The 'majority-rule' and 'core function' approaches both seem arbitrary, and tinged by the same sort of essentialism that colors our thinking about 'eukaryotes' and 'prokaryotes.' We want to believe that organismal and species lineages do have discrete and definable histories that we can discover, and not that we are choosing, arbitrarily, genes whose phylogeny we will equate with that history. (Doolittle, 1996, p. 8799)

There are no data to indicate that all genes are horizontally transferred. There could indeed be gene sets that define a hierarchy of higher taxa in the bacterial world. At the level of species, there is no stable core of genes. The species concept does not apply. But at higher taxonomic ranks, critics recognize that there could be a stable core of genes that "would truly give the same tree" (Doolittle et al., 2003, p. 47). There could be a stable core for phyla, divisions, domain, or even a universal core. There are about 100 genes at the core of all bacteria – a universal "core" comprised of informational genes (Charlebois and Doolittle, 2004; Doolittle and Bapteste, 2007).

## 4. Concluding Remarks

Darwin said virtually nothing in regard to microbes in the *Origin*, except in the concluding sentence that the power of life "having been originally breathed into a few forms or into one" (Darwin, 1859 p. 490). In subsequent editions, he wrote of "the Creator having breathed." Privately, he said he wished he had not done so. But biologists a century later often quoted from a letter he wrote to his friend Joseph Hooker in (1 February 1871) about a hypothetical "warm little pond" which had the necessary chemicals from which life might evolve. Still, he did not consider a natural origin of life in any of his publications.

Darwin certainly made his views plain in regard to the simile of a bifurcating tree based on descent with modification: "The affinities of all of the beings of the same class have sometimes been represented by a great tree. I believe this simile largely speaks the truth," he said (Darwin, 1859, p. 129). And in his chapter on classification, he famously wrote:

> All the foregoing rules and aids and difficulties in classification are explained, if I do not greatly deceive myself, on the view that the natural system is based on descent with modification; that the characters which naturalists consider as showing true affinity between any two or more species, are those which have been inherited from a common parent, and, in so far, all true classification is genealogical; that community of descent is the hidden bond which naturalists have been unconsciously seeking, and not some unknown plan of creation, or the enunciation of general propositions, and the mere putting together and separating objects more or less alike. (Darwin, 1859, p. 420)

Whether all of microbial taxonomy is actually a matter of "putting together objects more or less alike" or whether there is a core of essential functions through which the main lines of organismal descent may be traced, there is no question that the postgenomics structure of evolutionary theory today differs from all that preceded it. Certainly, there had been discussions of hybridization among some of the architects of the modern synthesis in the 1940s, but make no mistake, hybridization was considered the exception, not the rule (Barton, 2001; Sapp, 2009a).

For those who seek "a universal tree of life," and who maintain the reality of primary lineages, the components of the molecular machinery involved in DNA replication, transcription, and translation are in effect "the essential system of organs" that Lamarck had proposed to understand the evolution of complexity. Those who oppose this view argue accordingly that there are no taxa that can be defined as phylogenetic lineages and that any semblance of a tree of life is an illusion.

## 5. Acknowledgments

This work was partly supported by the Social Sciences and Humanities Research Council of Canada.

## 6. References


Barton NH (2001) The role of hybridization in evolution. Mol Ecol 10:551–568

Charlebois RL, Doolittle WF (2004) Computing prokaryotic gene ubiquity: rescuing the core from extinction. Genome Res 1:2469–2477

Chatton E (1925) *Pansporella Perplexa* reflexions sur la biologie et la phylogenie des protozoaires. Ann Sci Nat Zool VII(10):1–84

Copeland HF (1938) The kingdoms of organisms. The Quarterly Review of Biology 13:383–420

Darwin C (1859) On the origin of species, Facsimile edition of 1859. Harvard University Press, Cambridge 1969

Darwin CR, letter to Hooker JD (1 February 1871) In: Darwin F (ed) The life and letters of Charles Darwin, (1898), vol II. Basic Books, New York, 1959, reprint, pp 202–203

Dobell C (1932) Antony van Leeuwenhoek and his "Little animals". Harcourt, Brace & Co, New York

Doolittle WF (1996) At the core of the archaea. Proc Natl Acad Sci USA 93:8797–8799

Doolittle WF (1999) Phylogenetic classification and the universal tree. Science 284:2124–2128

Doolittle WF (2000) Uprooting the tree of life. Sci Am 282:90–95

Doolittle WF, Bapteste E (2007) Pattern pluralism and the tree of life hypothesis. Proc Natl Acad Sci USA 104:2043–2049

Doolittle WF, Boucher Y, Nesbø CL, Douady CJ, Andersson JO, Roger AJ (2003) How big is the iceberg of which organellar genes in nuclear genomes are but the tip? Philos Trans R Soc Lond B 358:39–58

Fitch WM, Upper K (1987) The phylogeny of tRNA sequences provides evidence for ambiguity reduction in the origin of the genetic code. Cold Spring Harb Symp Quant Biol 52:759–767

Forterre P, Philippe H (1999) The last universal common ancestor (LUCA), simple or complex. Biol Bull 196:373–377

Fox G, Stackebrandt E, Hespell RB, Gibson J, Maniloff J, Dyer TA, Wolfe RS, Balch WE, Tanner RS, Magrum LJ, Zablen LB, Blakemore R, Gupta R, Bonen L, Lewis BJ, Stahl DA, Luerhsen KR, Chen KN, Woese CR (1980) The phylogeny of prokaryotes. Science 209:457–463

Gogarten JP, Kibak H, Dittrich P, Taiz L, Bowman EJ, Bowman BJ, Manolson MF, Poole RJ, Date T, Oshima TJ, Konishi K, Yoshida M (1989) Evolution of the vacuolar H+−ATPase: implications for the origin of eukaryotes. Proc Natl Acad Sci USA 86:6661–6665

Gogarten P, Doolittle WF, Lawrence JG (2002) Prokaryotic evolution in light of gene transfer. Mol Biol Evol 19:2226–2238
Gray MW, Doolittle WF (1982) Has the endosymbiont hypothesis been proven? Microbiol Rev 46:1–42
Haeckel E (1866) Morphologie der Organismen, vol 2. Reimer, Berlin
Haeckel E (1892) The history of creation, vol II, 6th edn. Appleton & co, New York
Haldane JBS (1929) The origin of life. Ration Annu 148:3–10
Hartman H (1975) Speculations on the origin and evolution of metabolism. J Mol Evol 4:359–370
Hartman H, Fedorov A (2002) The origin of the eukaryotic cell: a genomic investigation. Proc Natl Acad Sci USA 99:1420–1425
Iwabe N, Kma K, Hasegawa M, Osawa S, Miyata T (1989) Evolutionary relationship of archaebacteria, eubacteria, and eukaryotes inferred from phylogenetic trees of duplicated genes. Proc Natl Acad Sci USA 86:9355–9359
Jain R, Rivera MC, Lake JA (1999) Horizontal gene transfer among genomes: the complexity hypothesis. Proc Natl Acad Sci USA 96:3801–3806
Kluyver J, van Niel CB (1936) Prospects for a natural system of classification of bacteria. Zentralbl f Bakt 94(2):369–402
Lamarck JB (1809) Zoological philosophy (trans Elliot H). The University of Chicago Press, Chicago 1984
Lawrence JG, Ochman H (1998) Molecular archaeology of the *Escherichia coli* genome. Proc Natl Acad Sci USA 95:9413–9417
Lederberg J (1952) Cell genetics and hereditary symbiosis. Phys Rev 32:403–430
Luria SE, Burrous J (1957) Hybridization between *Escherichia coli* and Shigella. J Bacteriol 74:461–476
Lwoff A (1957) The concept of virus. J Gen Microbiol 17:239–253
Oparin AI (1938) The origin of life. Dover, New York
Portier P (1918) Les Symbiotes. Masson, Paris
Sagan L (1967) On the origin of mitosing cells. J Theor Biol 14:225–274
Sapp J (1994) Evolution by association: a history of symbiosis. Oxford University Press, New York
Sapp J (1998) Freewheeling centrioles. Hist Philos Life Sci 20:255–290
Sapp J (2005a) The prokaryote-eukaryote dichotomy: meanings and mythology. Microbiol Mol Biol Rev 69:292–305
Sapp J (ed) (2005b) Microbial phylogeny and evolution: concepts and controversies. Oxford University Press, New York
Sapp J (2009a) The new foundations of evolution. On the tree of life. Oxford University Press, New York
Sapp J (2009b) Transcending Darwinism: thinking horizontally on the tree of life. Hist Philos Life Sci 31:161–182
Sapp J, Carrapico F, Zolotonosov M (2002) The hidden face of constantin Merezhkowsky, history and philosophy of the life sciences 24:421–449
Stanier RY (1970) Some aspects of the biology of cells and their possible evolutionary significance. In: Charles HP, Knight BC (eds) Organization and control in prokaryotic cells. Twentieth symposium of the society for general microbiology. Cambridge University Press, Cambridge, pp 1–38
Stanier RY, van Niel CB (1941) The main outlines of bacterial classification. J Bacteriol 42:437–466
Stanier RY, van Niel CB (1962) The concept of a bacterium. Archiv Für Mikrobiologie 42:17–35
Stanier R, Doudoroff M, Adelberg E (1957) The microbial world. Prentice-Hall, Englewood Cliffs
Stanier R, Doudoroff M, Adelberg E (1963) The microbial world, 2nd edn. Prentice-Hall, Englewood Cliffs
Stocker BAD (1955) Bacteriophages and bacterial classification. J Gen Microbiol 12:375–381
van Niel CB (1955) Classification and taxonomy of the bacteria and blue green algae. In: Kessel EL (ed) A century of progress in the natural sciences 1853–1953. California Academy of Sciences, San Francisco, pp 89–114
Wallin IE (1927) Symbionticism and the origin of species. William and Wilkins, Baltimore
Wilson EB (1925) The cell in development and heredity. Macmillan, New York
Woese CR (1979) A proposal concerning the origin of life on the planet earth. J Mol Evol 13:95–101

Woese CR (1980) An alternative to the Oparin view of the primeval sequence. In: Halvorson HO, Van Holde KE (eds) The origins of life and evolution. Alan R Liss Inc, New York, pp 65–76
Woese CR (1982) Archaebacteria and cellular origins: an overview. In: Kandler O (ed) Archaebacteria. Proceedings of the 1st international workshop on archaebacteria, Munich, June 27th–July 1st 1981. Gustav Fischer, Stuttgart, pp 1–17
Woese CR (1987) Bacterial evolution. Microbiol Rev 51:221–271
Woese CR (1998) The universal ancestor. Proc Natl Acad Sci USA 95:6854–6859
Woese CR (2000) Interpreting the universal phylogenetic tree. Proc Natl Acad Sci USA 97:8392–8396
Woese CR (2002) On the evolution of cells. Proc Natl Acad Sci USA 99:8742–8747
Woese CR, Fox GE (1977a) Phylogenetic structure of the prokaryotic domain: the primary kingdoms. Proc Natl Acad Sci USA 74:5088–5090
Woese CR, Fox GE (1977b) The concept of cellular evolution. J Mol Evol 10:1–6
Woese CR, Kandler O, Wheelis M (1990) Towards a natural system of organisms: proposal for the domains Archaea, Bacteria, and Eucarya. Proc Natl Acad Sci USA 87(1990):4576–4579
Woolhouse WH (1967) A review of the plastids by JTO Kirk and RAE Tilney-Bassett. New Phytol 66:832–833

Biodata of **Saša Svetina**, author of "***On the Vesicular Origin of the Cell Cycle***."

**Professor Saša Svetina** is currently the Head of the Institute of Biophysics at the Faculty of Medicine, University of Ljubljana and a researcher at the Jožef Stefan Institute, Ljubljana, Slovenia. He obtained his Ph.D. in physics from the University of Ljubljana in 1965. His lifelong work in biophysics revolves around red blood cells and lipid vesicles. Presently, he is focused on the morphology of vesicles and cell organelles as well as on the process of vesicle self-reproduction and its role in the origin of life.

E-mail: **sasa.svetina@mf.uni-lj.si**

J. Seckbach (ed.), *Genesis - In The Beginning: Precursors of Life, Chemical Models and Early Biological Evolution*, Cellular Origin, Life in Extreme Habitats and Astrobiology 22, 757–773
DOI 10.1007/978-94-007-2941-4_38, 

# ON THE VESICULAR ORIGIN OF THE CELL CYCLE

SAŠA SVETINA[1, 2]
[1]*Institute of Biophysics, Faculty of Medicine, University of Ljubljana, Lipičeva 2, SI-1000 Ljubljana, Slovenia*
[2]*Jožef Stefan Institute, SI-1000 Ljubljana, Slovenia*

## 1. Introduction

Life on Earth is based on cells that grow and divide, performing the phenomenon of the cell cycle. Cells are complex organized entities of finite dimensions that are, in general, able to sustain their state under variable external conditions. Cells may exist independently or as parts of multicellular organisms. For several billion years, the number of different cell types was low (Hedges et al., 2004), and it can therefore be assumed that cells evolved reasonably complex structures while still separate entities. It is anticipated that cellular evolution ran in a fashion similar to that of the evolution of species, i.e., by the principle of selection between different populations. Because the prerequisite for producing cell populations is cell division, the cell cycle can be considered as one of the pivotal features of living systems. After entering the cell cycle, a newly born cell proceeds through an ordered set of events (Nurse, 1994). A typical cell cycle involves the S phase, during which a cell doubles its genetic material, and the M phase, during which it spatially separates its contents, and these two phases are preceded by the gap phases G1 and G2, respectively (Murray and Hunt, 1993). The M phase ends by division into two daughters that are replicas of their mother at its birth. How cells coordinate growth and division was (Jorgensen and Tyers, 2004) and is still (Sawin, 2009; Edgar and Kim, 2009) a matter of intensive research.

Vesicles are molecular assemblies characterized by thin, usually flexible, membranes that separate their inner aqueous solution from the aqueous environment. After they were discovered in the mid-1960s (Bangham and Horne, 1964), they rapidly became widely studied (Lipowsky and Sackmann, 1995). Vesicle membranes comprise a variety of amphiphilic molecules, like lipids (Lasic, 1993), surfactants (Marques, 2000), and block copolymers (Discher and Eisenberg, 2002). Depending on the method of preparation, vesicles can be small, large, or giant (Lasic, 1993; Walde et al., 2010), with diameters of ~20 nm, ~0.2 μm, and ~20 μm. They can form spontaneously, as is evidenced by the classical method of their preparation by swelling under hydration (Reeves and Dowben, 1969). For vesicles to be formed, the concentration of their free constituent molecules has to reach a concentration termed as the critical vesicle concentration (CVC). CVC values range from ~$10^{-10}$ M for phospholipid membranes to ~$10^{-3}$ M for fatty acids (Morigaki and Walde, 2007). The concept of lipid involvement in the emergence of life in the form of cells has been suggested on grounds of the probable abundance of lipids in the prebiotic environment and of their capacity to aggregate spontaneously into micelles and vesicles which are able, by division, to transmit their compositional information to their daughters (Segré et al., 2001).

The aim of this chapter is to demonstrate that the process of vesicle self-reproduction provides a possible mechanism, based on physicochemical principles, for the transition from the nonliving to the living world. To this end, we begin by describing the vesicle physical properties relevant to this process. The essential features of the process of vesicle self-reproduction will be revealed by examining the behavior of a simple prototype model of this process and its possible generalization. It will become evident that there are several parallels between the behavior of the cell cycle and the process of vesicle self-reproduction. The idea that the cell cycle is an upgraded version of the process of vesicle self-reproduction will be presented and substantiated by ascribing to the latter the ability to evolve by selection between vesicle populations. The vesicular origin of cellular life will be advocated and the idea contrasted with some other possibilities.

## 2. Some Relevant Vesicle Properties

Here, we shall briefly introduce the properties of vesicles important for their self-reproduction. The behavior of vesicles with fluid membranes depends, to a large extent, on their bending energy ($W$) which is, for a vesicle with a membrane area $A$, expressed as the sum of the local bending energy ($W_b$) and the Gaussian term ($W_G$) (Helfrich, 1973):

$$W = W_b + W_G = \frac{1}{2}k_c\int (C_1 + C_2 - C_0)^2 dA + k_G\int C_1C_2 dA, \quad (1)$$

where $k_c$ is the local bending modulus, $k_G$ is the Gaussian bending modulus, $C_1$ and $C_2$ are the membrane principal curvatures, and $C_0$ is its spontaneous curvature, i.e., the curvature that would be attained by an unstrained piece of membrane. The Gaussian term has the same constant value for vesicles of a given topology and can be disregarded as long as the vesicles remain within a given, e.g., spherical, topology. The Gaussian term has to be taken into account, e.g., in processes of vesicle fission and fusion. The spontaneous curvature $C_0$ reflects transmembrane asymmetry. It is zero for bilayer membranes made of a single component because they are symmetrical. The spontaneous curvature of membranes composed of two or more components, e.g., the components of catanionic vesicles (Jung et al., 2001), in general differs from zero because the membrane leaflets have different compositions. Membrane asymmetry, and consequent nonzero spontaneous curvature, can also result as the consequence of asymmetric membrane inclusions or of different outer and inner solutions, e.g., one of them containing membrane-binding solute(s).

Vesicles attain shapes that correspond to the minimum of the bending energy of their membranes. It was, for instance, demonstrated by studying catanionic vesicles in which, when the bending constant of their membrane is much larger than the average thermal energy, unilamellar spherical vesicles are stabilized by the spontaneous curvature that favors energetically a particular vesicle radius (Jung et al., 2001). In general, when vesicles are formed, they are not necessarily spherical, i.e., they might contain less than the maximum possible amount of internal liquid. In such cases, the minimization of membrane-bending energy defined by Eq. 1 predicts the shape which depends on two parameters, the vesicle reduced volume ($v$), defined as vesicle volume $V$ divided by the volume of the sphere with the membrane area $A$, $V_s$, where $V_s = A^{3/2}/6\pi^{1/2}$, and the reduced spontaneous curvature, defined as $c_0 = R_s C_0$, where $R_s = (A/4\pi)^{1/2}$ is the radius of the sphere with the area $A$ (Deuling and Helfrich, 1976; Seifert et al., 1991).

Due to the dependence of the bending energy on vesicle volume, there is, at $v < 1$, a nonzero pressure difference between the outside and inside of the vesicle given by:

$$\Delta p = -\frac{\partial W_b}{\partial V}. \tag{2}$$

Consequently, water flows through the membrane with the rate:

$$\frac{dV}{dt} = L_p A \Delta p, \tag{3}$$

where $L_p$ is hydraulic permeability coefficient of the membrane.

It is instructive to estimate the time that a vesicle needs to reach the spherical shape if it is initially flaccid, i.e., in that its volume is a fraction ($f$) of its maximum possible volume ($V = fV_s$). For vesicles with a volume close to the volume of the sphere, the pressure difference (Eq. 2) can be expressed analytically as (Ou Yang and Helfrich, 1989):

$$\Delta p_\ell = \frac{2k_c}{R_s^3}(6 - C_0 R_s). \tag{4}$$

The corresponding time interval, obtained by inserting $\Delta p_\ell$ from Eq. 4 for $\Delta p_\ell$ in Eq. 3, is (e.g., for the case $C_0 = 0$) equal to $\Delta t = (1-f)R_s^4/36k_c L_p$. For $f = 0.99$, and by taking as an example the data for SOPC (1-stearoyl-2-oleoyl-sn-glycero-3-phosphocholine), $k_c = 10^{-19}$ J (Waugh et al., 1992), and $L_p = 2.6 \times 10^{-12}$ m$^3$/Ns (Vitkova et al., 2004), $\Delta t$ ranges from much less than a second for small vesicles to much more for giant vesicles. In a typical laboratory experiment lasting for about an hour, small vesicles can be considered to be spherical, while flaccid giant vesicles remain flaccid. It is notable that, for any set of vesicle properties, there is a range of vesicle sizes at which the system's behavior depends on the transmembrane flow of water induced by the membrane bending.

After a vesicle with a given membrane area eventually achieves spherical shape, inflow of water continues until the pressure inside the vesicle becomes equal to that of the outside, with a corresponding increase of membrane tension ($\sigma$). For a spherical vesicle, the general relationship between $\Delta p$ and $\sigma$ is determined by the equilibrium condition that is a generalization of the Laplace equation and reads (Ou Yang and Helfrich, 1989):

$$\Delta p R_s^3 + 2\sigma R_s^2 - k_c C_0 R_s (2 - C_0 R_s) = 0. \tag{5}$$

The membrane tension of an unconstrained vesicle and sufficiently porous membrane at equilibrium and at $\Delta p = 0$ then increases to $\sigma = k_c C_0 (2 - C_0 R_s)/2R_s$.

A spherical vesicle with the smallest local bending energy ($W_b = 0$) has a radius $R_s(0) = 2/C_0$. Spherical vesicles with larger or smaller radii have larger bending energies. From Eq. 4, it follows that a spherical shape is stable only as long as $C_0 R_s < 6$ (Jenkins, 1977; Ou Yang and Helfrich, 1989; for a graphical representation, see Božič and Svetina, 2009). Vesicles have also zero bending energy if their shapes are composites of spheres with radii $R_s(0) = 2/C_0$ connected by infinitesimally narrow necks (Miao et al., 1991).

For the sake of completeness, it has to be stated that the elastic energy of fluid membranes composed of two or more layers that are in contact but can slide laterally, one by the other, involves, in addition to local bending energy terms (Eq. 1), the nonlocal bending energy term (Svetina et al., 1985; Miao et al., 1994; Svetina and Žekš, 1996). This contributes to the membrane elastic energy if the difference between the areas of the two membrane leaflets differs from the preferred area difference, defined as the difference between the equilibrium areas of the two leaflets (Svetina et al., 1985; Božič et al., 1992; Miao et al., 1994). At sufficiently long times, the preferred area difference relaxes by the flip-flop processes to the leaflet area difference, and the nonlocal bending term vanishes (Raphael et al., 2001). The nonlocal bending energy term will therefore be disregarded in our discussions. It was only referred to in order to introduce the preferred area difference, which is the shape determining parameter that acts in a manner similar to but different from that of the spontaneous curvature (Svetina and Žekš, 2002).

Shapes of flaccid vesicles that correspond to the minimum membrane bending energy belong to different symmetry classes (reviewed in Svetina and Žekš, 1996, 2002; Seifert, 1997; Svetina, 2009). Under conditions of continuously varying vesicle volume, membrane area, and membrane spontaneous curvature or preferred area difference, vesicle shape transformations within a given class are continuous. Parameter spaces (in our case, ranges of $v$ and $c_0$ values) of classes are in general bounded, and the boundaries are often represented by the so-called limiting shapes that are combinations of spheres connected by infinitesimally thin necks (Svetina and Žekš, 1989). A typical example of a limiting shape is a sphere connected to a certain number of buds of equal sizes. Some examples of shape transformations of giant phospholipid vesicles that end by the attainment of a

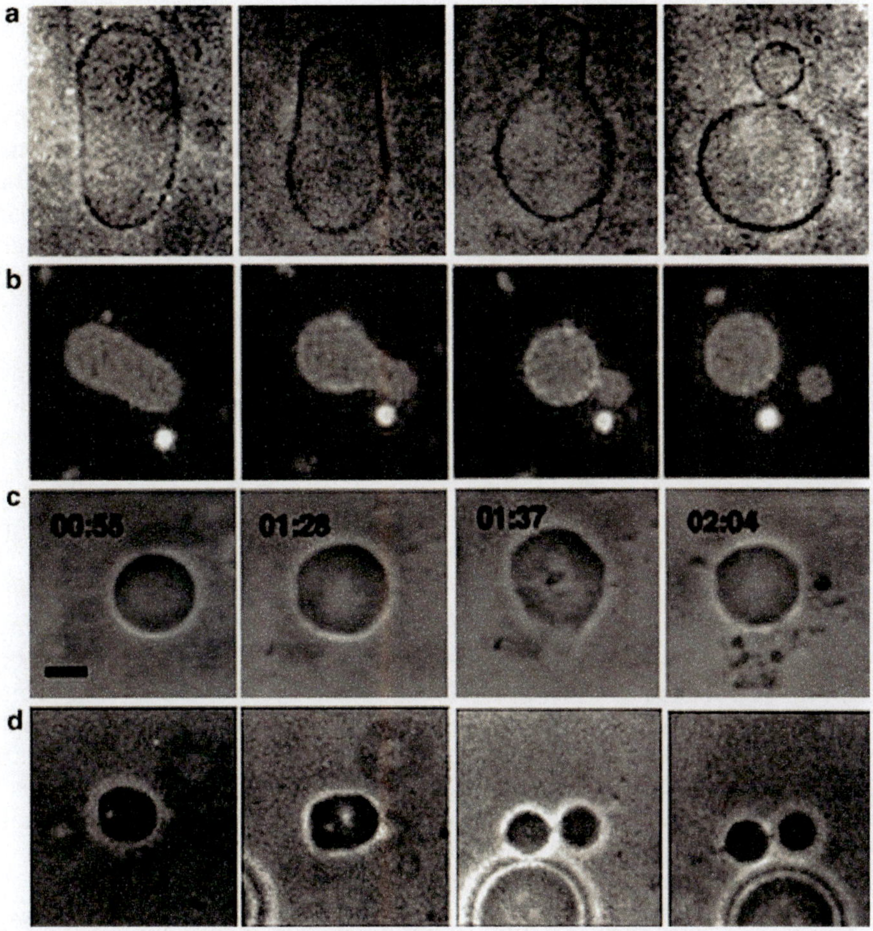


**Figure 1.** Examples of vesicle shape transformations ending with the limiting shapes. (**a**) Budding of a vesicle, made of a single component membrane, caused by increase of temperature (Käs et al., 1993). (**b**) Budding and vesiculation of a vesicle made of a two-component membrane caused by the addition of lysophosphatidylcholine (Inaoka and Yamazaki, 2007). (**c**) A multiple budding that occurred when the membrane area was increasing due to incorporation of oleic acid into the POPC (1-palmitoyl-2-oleoyl-sn-glycero-3-phosphocholine) bilayer (Peterlin et al., 2009). (**d**) Formation of a twin shape, i.e., composed of connected nearly equal spheres, under conditions of increasing membrane area, as in (**c**), and increasing vesicle volume by osmotic swelling (Mally, unpublished).

limiting shape are shown in Fig. 1. It is plausible to assume that a vesicle with such a shape is particularly amenable to splitting, in particular when the necks are strained by external forces. When the limiting shape is a combination of two spheres, breaking of the neck leads to the production of two spherical daughter vesicles.

## 3. Vesicle Self-reproduction

In the process of self-reproduction, a vesicle must double its size and divide. The occurrence of this kind of process has been confirmed experimentally by measuring the increase of the number of vesicles with time under conditions that stimulate the growth of their membranes (Berclaz et al., 2001; Takakura et al., 2003; Chen and Szostak, 2004; Chungcharoenwattana and Ueno, 2005; reviewed in Stano and Luisi, 2010). The possibility of vesicle splitting was indicated by Stano et al. (2006) who, by applying scanning electron microscopy, observed the presence of budded twin shapes within a population of growing vesicles.

The problem in considering the origin of cellular life (Luisi et al., 1999), as well as of designing artificial protocells (Surovtsev et al., 2009; Sole, 2009; Walde, 2010; Stano and Luisi, 2010), is to define the conditions under which the process of vesicle self-reproduction would proceed in a stationary manner, giving rise to exponential growth of the vesicle population. The planning of appropriate experiments should be preceded by a theoretical analysis of the behavior of vesicles under conditions of growth.

The prerequisite for exponential vesicle growth is a constant supply of vesicle components. It has to be assumed that there is more than enough solvent, i.e., water. There must also be a reservoir of the molecules that form membranes in order to keep their solution concentration constantly at the level of their CVC. With respect to the content of the vesicles, the solvent and solutes should be able to cross the membrane at rates which would allow them to double their amount concomitantly with doubling of the membrane area. On the theoretical side, vesicle self-reproduction should be based on the idea that splitting can occur if the vesicle attains the twin shape (Svetina and Žekš, 2002). The attainment of this shape must be synchronized with the doubling of vesicle contents.

To elucidate the basic principles underlying vesicle self-reproduction, we proposed the simplest possible prototype model of this process (Božič and Svetina, 2004, 2009). A vesicle whose flexible membrane exhibits spontaneous curvature $C_0$ was assumed to be suspended in a water solution of membrane-forming molecules. The increase of membrane area due to the continuous incorporation of new molecules was considered as an index of vesicle growth. It was assumed that the rate of incorporation of membrane constituents is proportional to the membrane area so that its growth is exponential, i.e.,

$$A(t) = A_0 2^{\frac{t}{T_d}}, \tag{6}$$

where $t$ is time and $T_d$ is the area doubling time proportional to the reciprocal value of the membrane growth rate. $A_0$ is the initial area which was assumed to correspond to that of the sphere with zero bending energy and radius $R_s(0) = 2/C_0$ where $R_s(0) = (A_0/4\pi)^{1/2}$.

Vesicle volume can only change if its membrane is permeable to the solvent whose net flow is then proportional to the pressure difference given by Eq. 2. Vesicle volume thus changes according to Eq. 3. Because, as shown in the previous section, the pressure inside a nearly spherical vesicle is, according to Eq. 4, lower than the outside pressure, water flows into the vesicle. Depending on the value of the hydraulic permeability coefficient $L_p$, there are two possible consequences. At sufficiently high $L_p$, the water flow predicted by Eq. 3 could, at $\Delta p = \Delta p_\ell$, exceed the volume increase that corresponds geometrically to the increasing membrane area, even if the vesicle were growing as a sphere. Consequently, membrane tension increases, and the pressure difference drops from $\Delta p_\ell$ to the exact value necessary for the water flow defined geometrically according to the increase of the area. The membrane tension attained must satisfy the equilibrium condition Eq. 5. At smaller values of $L_p$, when the water flow defined by Eq. 3 cannot match the rate of volume increase required for spherical growth, a vesicle assumes a nonspherical growth. Even if the vesicle first grows as a sphere, sooner or later it will switch into nonspherical growth. Thus, as membrane area increases, a larger water flow is needed to keep the shape spherical and therefore the corresponding pressure difference must also be larger. At the same time, due to the growth of membrane area, the reduced spontaneous curvature increases and therefore $\Delta p_\ell$ decreases (Eq. 4). For both these reasons, $\Delta p$ sooner or later reaches the value of $\Delta p_\ell$, and from this point on, a vesicle cannot grow anymore as a sphere. The smaller the membrane hydraulic permeability, the larger $\Delta p$ has to be and the sooner in the course of vesicle growth its spherical growth regime will end.

The above reasoning makes it possible to determine at which hydraulic permeability a vesicle would exhibit self-reproduction. At $L_p = 0$, it would deform from its spherical shape at the very beginning of membrane growth. In this case, a possible outcome of continuous membrane growth would be like the last shape of Fig. 1c. As this shape involves many small satellite vesicles, it could actually give rise to a kind of multioffspring self-replication. It was shown (Božič and Svetina, 2004), however, that to reach twin shapes, and thus a type of vesicle self-reproduction similar to cell division, $L_p$ must exceed a certain critical value. Parameter analysis of the behavior of this model showed that it depends on the value of a single constant that can be interpreted as a reduced hydraulic permeability coefficient and is a combination of all parameters of the system, $l_p = T_d L_p k_c C_0^4$.

Figure 2 shows the predicted dependence of the vesicle volume on time for the values of the model parameters for which $l_p > l_{p,cr}$ (where the critical value of $l_p$ was obtained numerically to be $l_{p,cr} = 1.85$). A vesicle first grows as a sphere and, at the point indicated by the first arrow, switches to the phase of nonspherical growth. The last shape reached in this case is that of two connected spheres of different radii. For $l_p = l_{p,cr}$, the vesicle would reach the shape of two equal spheres connected by a narrow neck (shape I), while, for $l_p < l_{p,cr}$, the volume curve would miss to hit the curve of limiting shapes involving a single neck, and the vesicle would grow into a shape with indistinct characteristics.

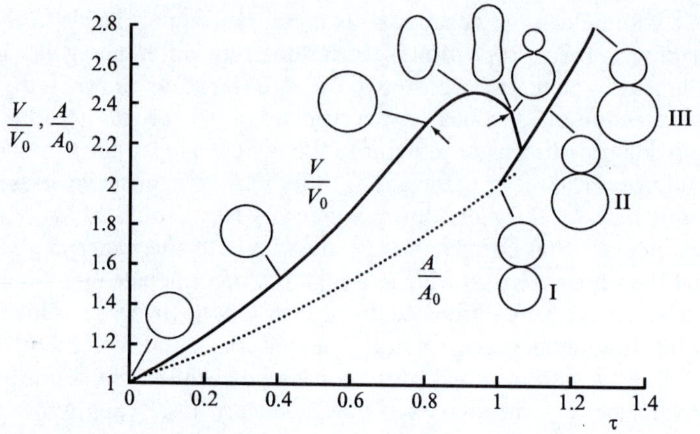


**Figure 2.** The time course of vesicle growth under conditions such that $l_p$ is larger than $l_{p,cr}$. The time $\tau$ is in units of $T_d$. The *dotted curve* shows the exponential growth of membrane area, and the *solid curve* the time dependence of vesicle volume. Membrane area and vesicle volume are in units of a sphere with zero bending energy. Some characteristic shapes are shown. The *first arrow* indicates the transition from spherical to nonspherical growth. The *second arrow* denotes the point of the discontinuous transition from the class of prolate shapes exhibiting equatorial mirror symmetry into the class of prolate shapes without such symmetry. The volume curve ends on the curve of limiting shapes (*bold curve*) where the radii of daughter buds are different. Three examples of limiting shapes are also shown (I–III) (Adapted from Božič and Svetina 2007).

It can be concluded that the process of vesicle self-reproduction resembles cell cycle–like behavior only under the condition $l_p \geq l_{p,cr}$, where for $l_p = l_{p,cr}$, the division is symmetrical, and for $l_p > l_{p,cr}$ it is asymmetrical. In the latter case, both daughters self-reproduce by switching from their spherical growth phase at the same critical size. Thus, the smaller daughter grows as a sphere for a longer time while the larger daughter reaches the critical size sooner (Fig. 3). The average generation time was shown to be larger than the membrane doubling time $T_d$ (Božič and Svetina, 2004).

This prototype model was subsequently generalized by including a solute component which can passively permeate the membrane. The model showed qualitatively analogous behavior; however, there is now an additional component whose vesicle amount must also be doubled in the course of the cycle. The condition for vesicle self-reproduction thus generalizes from a single constant into a function of two variables (Božič and Svetina, 2007). This exemplifies the degree to which the complexity increases on the addition of any new component which has to double. Another interesting observation obtained from the analysis of the generalized system is the change of the size distribution. While in the prototype model, the distribution of vesicle sizes at their birth involved only two different sizes, the size distribution in the generalized case, although still discrete, involves a larger number of possible initial vesicle sizes.

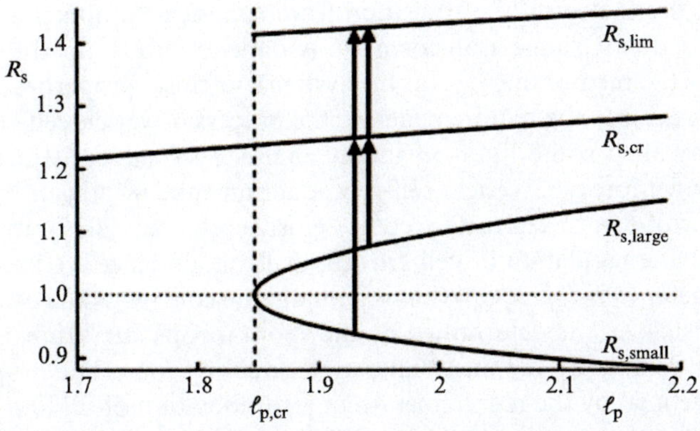


**Figure 3.** The dependence of vesicle radii at birth ($R_{s,small}$ and $R_{s,large}$) and at the critical point ($R_{s,cr}$) on the reduced hydraulic permeability $l_p$. $R_{s,lim}$ is the radius of the sphere with the same area as the area of the limiting shape. The radii are given in units of the radius of the sphere with zero bending energy, $R_s(0)$. The length of arrows at an arbitrarily chosen $l_p$ (~1.935) symbolizes the length of time that a vesicle needs to increase its radius from the lower to the upper value. The small daughter needs more time to reach the critical vesicle size than the larger daughter, but the duration of their division phase is the same (Adapted from Božič and Svetina 2004).

## 4. Cell Cycle and Vesicle Self-reproduction Share Common Features

The cell cycle and vesicle self-reproduction follow the same general goal of producing two daughter entities out of one, which suggests that some general features of these processes could be similar. The most characteristic of these features is certainly the maintenance of the balance between growth and division. The cell cycle can thus, in the most general sense, be considered to consist of a growth phase and a division phase, with definite transition points at which they switch from one to the other. It is in general understood that these transitions serve as the control for different cellular processes to finish before the beginning of the next phase of the cell cycle. In general, the rates of growth of cells during the growth phase depend on the environmental conditions more than the act of cell division, which is more autonomous and dependent primarily on intrinsic cell properties.

The balance between growth and division also has to be maintained in the process of vesicle self-reproduction. Like the cell cycle, this process can also be divided into growth and division phases, with definite transition points. One of the latter is vesicle splitting. The spherical growth that follows it can be considered as the growth phase, being governed essentially only by the rate of membrane growth and membrane hydraulic permeability coefficient. The other transition point is the switch from spherical to nonspherical growth. The critical vesicle size at which this transition occurs is adjusted to the requirement that vesicle shape transformations that follow bring the vesicle into the limiting shape involving narrow neck (Božič and Svetina, 2004, 2007). The course of these shape transformations can be considered as the division phase of the process of vesicle self-reproduction.

It is noteworthy that both the transition from spherical to nonspherical growth and the subsequent shape transformations depend subtly on the mechanical properties of the membrane, i.e., on the system's intrinsic properties.

The insight into the requirements for the process of vesicle self-reproduction to occur may shed some light on the mechanism of the control of cell size. The described prototype vesicle self-reproduction model indicates very clearly why this control can be characterized as "elusive" (Umen, 2005). In this model, all aspects of the regulation of cell growth, division, and size are joined within a single parameter ($l_p = T_d L_p k_c C_0^4$) that is actually the sole regulator of the system's behavior. Vesicle size is determined by the spontaneous curvature, and the rate of growth is defined by the membrane area doubling time. The proper division behavior is defined by the requirement that a combination of all four parameters assumes the required value ($l_p \geq l_{p,cr} = 1.85$). Reasoning analogous to that for the prototype model can also be applied when considering the parameters of its possible generalization (Božič and Svetina, 2007). On extrapolating this notion to cells, one can conclude that their size control very probably involves the coordination of most of their components that has been building up throughout evolution.

The correspondence between the cell cycle and vesicle self-reproduction can be extended to the variability of generation times which, in many cellular systems, is an inherent feature of the interplay between cell growth and division (Di Talia et al., 2007; Tzur et al., 2009). This variability could be ascribed to molecular noise in gene expression, but it was shown, at least for the budding yeast, that it is also the consequence of different sizes of daughter cells and the existence of size control (Di Talia et al., 2007). The coordination between cell growth and the cell cycle occurs in these cells at Start, a short interval in the late G1 phase during which these cells commit to division. The Start can be considered as a critical size threshold enforcing a minimal cell size. Further, the G1 phase of the cell cycle in most other cells is much more variable than the cell cycle phases that follow the cell's commitment to division.

Similarly to that of cells, the growth phase of vesicle self-reproduction is of variable length, whereas the timing of the division phase is strict. The variability of vesicle doubling time occurs in the case of $l_p > l_{p,cr}$, where daughter vesicles attain different initial sizes (Fig. 3). Variability of vesicle generation time is the consequence of variability of the spherical part of its growth, which corresponds to the greater variability of the G1 phase of the cell cycle. Comparison of the prototype and generalized models shows that, with the increase of the number of cell constituents, the need for variable generation times persists because the matching of their doubling becomes more and more difficult.

## 5. Vesicle Self-reproduction Involves the Selectivity Principle

The condition for self-reproduction of the prototype model (Božič and Svetina, 2004), as well as the corresponding conditions of more complex models (Božič and Svetina, 2007), can serve as the selectivity criterion for membrane composition.

This is because they relate the parameters, like the membrane area doubling time, that depend on environmental conditions, to the vesicle material parameters, such as the mechanical properties of their membranes. Such relationships imply the possibility of evolution of vesicle populations.

A possible scenario for vesicle evolution can be envisioned, based on the assumption of a temporally stable primordial soup containing a wide variety of molecules, some of which could form membranes of unilamellar vesicles. The resulting membranes would be expected to have different compositions, giving rise, again by considering the prototype model, to vesicles distributed according to their values of the parameter $l_p$. The most reasonable candidate for the subject of competition between vesicle populations is their rate of growth. The doubling time for growth is shorter for vesicles whose parameter $l_p$ is closer to $l_{p,cr}$, so that the proportion of vesicles with values of $l_p$ closer to $l_{p,cr}$ would increase. Of the vesicle populations that thus reached the optimal $l_p$, those with a shorter $T_d$ and, for instance, larger $L_p$ would be at an advantage with respect to the rate of doubling. The vesicle population with the membrane composition that satisfies the criteria for self-reproduction, but with a shorter doubling time, would outnumber vesicles with other membrane compositions. Vesicle self-reproduction can thus be considered as an evolutionary process that involves the elements of Darwinian selection and can, in a transparent manner, also be related to the laws of physics and chemistry.

## 6. General Implications

The origin of living systems is of continuing interest, and the state of the art in this field is captured comprehensively in several recent review articles (Peretó, 2005; Walde, 2006; Thomas and Rana, 2007; Lancet and Shenhav, 2009; Lazcano, 2008; Wu and Higgs, 2008; Deamer, 2009; Pross, 2009; Tessera, 2009; Fry, 2010). We will therefore focus our discussion on the involvement of vesicles. The appearance of life is considered as a transition from purely chemical states and reactions to autonomous, self-replicating molecular entities capable of evolving by natural selection. At what point the difference between a chemical system and the truly primordial, first organism was established is still an open question (Lazcano, 2008). The evolutionary process is based on the competition between populations of different organisms. Because the first unicellular organisms formed their populations through a process involving cell division, it can be inferred that their initiation was preceded by a system that could self-generate a membrane of composition that would enable reproducible cell division. The main message of this contribution is that the latter may have developed out of the process of vesicle self-reproduction.

Vesicles exhibit several characteristics of living cells. They can form spontaneously and grow by the incorporation of new membrane material. They compartmentalize the space. Assuming that their membranes and internal solution are homogeneous, the two entities produced by vesicle fission inherit the composition of the mother vesicle, in this manner, exhibiting compositional inheritance (Segré et al., 2001; Wu and Higgs, 2008). The proposal that the cell cycle evolved from vesicle self-reproduction is based on the realization that this process involves the principle of selectivity (Božič and Svetina, 2004).

The vesicle version of the origin of the cell cycle implies that after the process of vesicle self-reproduction was properly initiated, it was continued without interruption until now. The question arises about mechanisms that could have transformed a simple vesicle self-reproducing system into extremely complex entities such as cells and other contemporary biological objects. Vesicle systems could become more cell-like only by becoming more autonomous. It is reasonable to consider that diversification of vesicle properties was involved in the increase in metabolic and organizational complexity of early cells. A possible mechanistic scenario for this can be based on the assumption that sufficiently long-lasting stationary environmental conditions could lead to the production of a single large and homogeneous vesicle population. The large number of vesicles obtained in such a manner could, with high probability, stimulate further evolution, for instance, by utilizing their large total membrane area to serve as the substrate for different catalytic reactions that could eventually produce, to their advantage, their own constituents. Vesicle subpopulations could specialize to support separate elements of the common metabolic network. It is easy to envisage (Svetina, 2007) further steps of vesicle evolution based on fusion of vesicles with different acquired, e.g., catalytic, properties thus creating a new, even more successful species.

This vesicle-based approach toward an understanding of the initiation of living structures based on known properties of matter can be contrasted with the often-followed direction to identify molecule(s) with some special properties essential for the establishment of the life process such as the ability of RNA to perform autocatalytic reactions or identifying surfactants that could have existed in the prebiotic environment and could form membranes able to perform selective transmembrane transport and enable energy acquisition (reviewed in Thomas and Rana, 2007). Vesicles are assemblies of molecules, and the presented analysis demonstrates that the vesicular origin of the cell cycle is not so much related to the properties of individual molecules from which vesicle membranes are made but rather to the vesicle shape behavior which is an emergent property of the system as a whole. It has to be stressed that the described shape behavior of vesicles is universal and valid for vesicles with large differences in compositions of their membranes. This is why, in the quest for the origin of life, the term "vesicle world" would be more appropriate than "lipid world" (Svetina, 2007).

As has been suggested, the "vesicle world" of early evolution could have proceeded on the basis of the shortening of the generation times. It is thus possible to draw some conclusions about the properties required of possible candidate molecules for membranes of primordial vesicles. Namely, the evolutionary process must have started with vesicles with relatively long generation times, and for vesicles to exhibit the growth process, the system had to be far from equilibrium. For these reasons, it would appear to be appropriate for vesicle membranes to be formed from molecules with as low CVCs as possible. With molecules with low CVC, they do not have to be at high concentration.

The vesicular origin of the cell cycle can be relevant to the dilemma about which came first, the autopoietic or the hereditary properties of living systems (Peretó, 2005). Vesicle self-reproduction is a process encompassing both these aspects. Namely, vesicle division is, by itself, an autopoietic process giving rise to compositional inheritance. The above-stated dilemma can also be expressed by asking which phase of the cell cycle is primary, the S phase involving molecular replication or the M phase involving also the act of division. Contemporary evolution at the molecular level proceeds by mutations caused by imprecise gene replication. It therefore appears that the most basic part of the cell cycle is the S phase. In principle, chemical evolution could proceed also without the enclosure of its ingredients. However, eventually the system had to be compartmentalized also. In view of the suggested prebiological evolutionary process based on the competition between vesicle populations, it is more plausible to assume that it is the division aspect of the cell cycle that developed first and that the switch into genetics appeared later as an evolutionary improvement.

## 7. Acknowledgments

This work was supported by the Slovenian Research Agency through grants P1-0055 and J3-2268. The author thanks Roger Pain and Peter Walde for the critical reading of the manuscript. Mojca Mally allowed the use of her results before publication, and Bojan Božič helped with the adaptation of Fig. 2.

## 8. References

Bangham AD, Horne RW (1964) Negative staining of phospholipids and their structured modifications by surface active agents as observed in the electron microscope. J Mol Biol 8:660–668

Berclaz N, Müller M, Walde P, Luisi PL (2001) Growth and transformation of vesicles studied by ferritin labeling and cryotransmission electron microscopy. J Phys Chem B 105:1056–1064

Božič B, Svetina S (2004) A relationship between membrane properties forms the basis of a selectivity mechanism for vesicle self-reproduction. Eur Biophys J 33:565–571

Božič B, Svetina S (2007) Vesicle self-reproduction: the involvement of membrane hydraulic and solute permeabilities. Eur Phys J E 24:79–90

Božič B, Svetina S (2009) Comment on "Thermodynamics of vesicle growth and instability". Phys Rev E 80:013401(2)

Božič B, Svetina S, Žekš B, Waugh RE (1992) Role of lamellar membrane structure in tether formation from bilayer vesicles. Biophys J 61:963–973

Chen IA, Szostak JW (2004) A kinetic study of the growth of fatty acid vesicles. Biophys J 87:988–998

Chungcharoenwattana S, Ueno M (2005) New vesicle formation upon oleate addition to preformed vesicles. Chem Pharm Bull 53:260–262

Deamer D (2009) On the origin of systems. EMBO Rep 10:51–54

Deuling HJ, Helfrich W (1976) The curvature elasticity of fluid membranes: a catalogue of vesicle shapes. J Phys Fr 37:1334–1345

Di Talia S, Skotheim JM, Bean JM, Siggia ED, Cross FR (2007) The effect of molecular noise and size control on variability in the budding yeast cell cycle. Nature 448:947–951

Discher DE, Eisenberg A (2002) Polymer vesicles. Science 297:967–973

Edgar BA, Kim KJ (2009) Sizing up cells. Science 325:158–159

Fry I (2010) The role of natural selection in the origin of life. Orig Life Evol Biosph 41:3–16

Hedges SB, Blair IE, Venturi MI, Shoe JI (2004) A molecular timescale of eukaryote evolution and the rise of complex multicellular life. BMC Evol Biol 4:2

Helfrich W (1973) Elastic properties of lipid bilayers: theory and possible experiments. Z Naturforsch 28c:693–703

Inaoka Y, Yamazaki M (2007) Vesicle fission of giant unilamellar vesicles of liquid-ordered-phase membranes induced by amphiphiles with a single long hydrocarbon chain. Langmuir 23:720–728

Jenkins JT (1977) Static equilibrium configuration of a model red blood cell. J Math Biol 4:149–169

Jorgensen P, Tyers M (2004) How cells coordinate growth and division. Curr Biol 14:R1014–R1027

Jung HT, Coldren B, Zasadzinski JA, Iampietro DJ, Kaler EW (2001) The origins of stability of spontaneous vesicles. Proc Natl Acad Sci USA 98:1353–1357

Käs J, Sackmann E, Podgornik R, Svetina S, Žekš B (1993) Thermally induced budding of phospholipid vesicles – a discontinuous process. J Phys II Fr 3:631–645

Lancet D, Shenhav B (2009) Compositional lipid protocells: reproduction without polynucleotides. In: Rasmussen S, Bedau MA, Chen L, Deamer D, Krakauer DC, Packard NH, Stadler PF (eds) Protocells: bridging nonliving and living matter. MIT, Cambridge, pp 233–252

Lasic DD (1993) Liposomes: from physics to application. Elsevier, Amsterdam

Lazcano A (2008) Towards a definition of life: the impossible quest? Space Sci Rev 135:5–10

Lipowsky R, Sackmann E (eds) (1995) Structure and dynamics of membranes, vols 1, 2. Elsevier Science B.V., Amsterdam

Luisi PL, Walde P, Oberholzer T (1999) Lipid vesicles as possible intermediates in the origin of life. Curr Opin Colloid Interface Sci 4:33–39

Marques EF (2000) Size and stability of catanionic vesicles: effects of formation path, sonication, and aging. Langmuir 16:4798–4807

Miao L, Fourcade B, Rao M, Wortis M (1991) Equilibrium budding and vesiculation in the curvature model of fluid lipid vesicles. Phys Rev A 43:6843–6856

Miao L, Seifert U, Döbereiner HG, Wortis M (1994) Budding transitions of fluid-bilayer vesicles: the effect of the area difference elasticity. Phys Rev E 49:5389–5407

Morigaki K, Walde P (2007) Fatty acid vesicles. Curr Opin Colloid Interface Sci 12:75–80

Murray A, Hunt T (1993) The cell cycle: an introduction. Oxford University Press, New York/Oxford

Nurse P (1994) Ordering S phase and M phase in the cell cycle. Cell 79:547–550

Ou Yang Z-C, Helfrich W (1989) Bending energy of vesicle membranes: general expressions for the first, second, and third variation of the shape energy and applications to spheres and cylinders. Phys Rev A 39:5280–5288

Peretó J (2005) Controversies on the origin of life. Int Microbiol 8:23–31

Peterlin P, Arrigler V, Kogej K, Svetina S, Walde P (2009) Growth and shape transformations of giant phospholipid vesicles upon interaction with an aqueous oleic acid suspension. Chem Phys Lipids 159:67–76

Pross A (2009) Seeking the chemical roots of Darwinism: bridging between chemistry and biology. Chem Eur J 15:8374–8381

Raphael RM, Waugh RE, Svetina S, Žekš B (2001) Fractional occurrence of defects in membranes and mechanically driven interleaflet phospholipid transport. Phys Rev E 64:051913(10)

Reeves JP, Dowben RM (1969) Formation and properties of thin-walled vesicles. J Gen Physiol 73:49–60

Sawin KE (2009) Cell division brought down to size. Nature 459:782–783

Segré D, Ben-Eli D, Deamer DW, Lancet D (2001) The lipid world. Orig Life Evol Biosph 31:119–145

Seifert U (1997) Configurations of fluid membranes and vesicles. Adv Phys 46:13–137
Seifert U, Berndl K, Lipowsky R (1991) Shape transformations of vesicles: phase diagram for spontaneous-curvature and bilayer-coupling models. Phys Rev A 44:1182–1202
Sole RV (2009) Evolution and self-assembly of protocells. Int J Biochem Cell Biol 41:274–284
Stano P, Luisi PL (2010) Achievements and open questions in the self-reproduction of vesicles and synthetic minimal cells. Chem Commun 46:3639–3653
Stano P, Wehrli E, Luisi PL (2006) Insights into the self-reproduction of oleate vesicles. J Phys Condens Matter 18:S2231–S2238
Surovtsev IV, Zhang Z, Lindahl PA, Morgan JJ (2009) Mathematical modeling of a minimal protocell with coordinated growth and division. J Theor Biol 260:422–429
Svetina S (2007) The vesicle world: the emergence of cellular life can be related to properties specific to vesicles. Orig Life Evol Biosph 37:445–448
Svetina S (2009) Vesicle budding and the origin of cellular life. ChemPhysChem 10:2769–2776
Svetina S, Žekš B (1989) Membrane bending energy and shape determination of phospholipid vesicles and red blood cells. Eur Biophys J 17:101–111
Svetina S, Žekš B (1996) Elastic properties of closed bilayer membranes and the shapes of giant phospholipid vesicles. In: Lasic DD, Berenholz Y (eds) Handbook of nonmedical applications of liposomes. Theory and basic sciences, vol 1. CRC Press, Boca Raton/New York, pp 13–42
Svetina S, Žekš B (2002) Shape behavior of lipid vesicles as the basis of some cellular processes. Anat Rec 268:215–225
Svetina S, Brumen M, Žekš B (1985) Lipid bilayer elasticity and the bilayer couple interpretation of red cell shape transformations and lysis. Stud Biophys 110:177–184
Takakura K, Toyota T, Sugawara T (2003) A novel system of self-reproducing giant vesicles. J Am Chem Soc 125:8134–8140
Tessera M (2009) Life began when evolution began: a lipidic vesicle-based scenario. Orig Life Evol Biosph 39:559–564
Thomas JA, Rana FR (2007) The influence of environmental conditions, lipid composition, and phase behavior on the origin of cell membranes. Orig Life Evol Biosph 37:267–285
Tzur A, Kafri R, LeBleu VS, Lahav G, Kirschner MW (2009) Cell growth and size homeostasis in proliferating animal cells. Science 325:167–171
Umen JG (2005) The elusive sizer. Curr Opin Cell Biol 17:435–441
Vitkova V, Genova J, Bivas I (2004) Permeability and the hidden area of lipid bilayers. Eur Biophys J 33:706–714
Walde P (2006) Surfactant assemblies and their various possible roles for the origin(s) of life. Orig Life Evol Biosph 36:109–150
Walde P (2010) Building artificial cells and protocell models: experimental approaches with lipid vesicles. Bioessays 32:296–303
Walde P, Cosentino K, Engel H, Stano P (2010) Giant vesicles: preparations and applications. ChemBioChem 11:848–865
Waugh RE, Song J, Svetina S, Žekš B (1992) Local and nonlocal curvature elasticity in bilayer membranes by tether formation from lecithin vesicles. Biophys J 61:974–982
Wu M, Higgs PG (2008) Compositional inheritance: comparison of self-assembly and catalysis. Orig Life Evol Biosph 38:399–418

Biodata of **Jack T. Trevors** and coauthors of "***Hypothesized Microenvironments for the Origin of Microbial Life on Earth.***"

**Professor Jack T. Trevors** is currently professor of microbiology at the University of Guelph, Guelph, Ontario, Canada. A fellow often science academies, an editor for The Journal of Microbiological Methods and Antonie Van Leeuwenhoek Journal of Microbiology, and editor-in-chief of Water, Air, and Soil Pollution. He obtained his Ph.D. from the University of Waterloo, Canada in 1982 and joined the University of Guelph the same year. Professor Trevors' scientific interests are in the areas of bacteriology with an emphasis on metabolism, gene expression in bacteria and soil, genetics, molecular-based methods and biotechnology, membrane fluidity, waste bioremediation, pathogen survival and detection in the environment, bioenergy from wastes, and the origin of microbial life.

E-mail: **jtrevors@uoguelph.ca**

J. Seckbach (ed.), *Genesis - In The Beginning: Precursors of Life, Chemical Models and Early Biological Evolution*, Cellular Origin, Life in Extreme Habitats and Astrobiology 22, 775–795
DOI 10.1007/978-94-007-2941-4_39, 

**Professor Asim K. Bej** is currently professor of biology at the University of Alabama at Birmingham, Birmingham, Alabama, USA. He obtained his Ph.D. from the University of Louisville, Louisville, Kentucky, USA in 1988; he joined the University of Alabama at Birmingham in 1991 and was promoted to full professor rank in 2003. He is on the editorial board of a number of microbiology and molecular biology journals. Professor Bej's research interests are in the areas of biodiversity of Antarctic extremophile microorganisms, gene regulation, cold adaptation, biomedical applications of Antarctic microbial pigments, and diagnostics of microbial pathogens. He has coedited a book on *Polar Microbiology* and is coinventor of a number of microbiological applications to the solution of environmental problems.

E-mail: **abej@uab.edu**

**Asim K. Bej**

**Professor Jan Dirk van Elsas** is currently professor of microbial ecology at the University of Groningen, Groningen, The Netherlands. He is also a fellow of numerous science academies, serves on the editorial board of numerous scientific journals and is an editor of Water, Air, and Soil Pollution and Environmental Biosafety Research. He obtained his Ph.D. from the Federal University of Rio de Janeiro, Brazil in 1983 and worked at that University as well as several research institutes in Wageningen, The Netherlands until he joined the University of Groningen in 2003. Professor van Elsas' scientific interests are in the areas of microbial ecology with an emphasis on ecological interactions and gene expression in the rhizosphere, mycosphere, and soil, gene transfer, molecularly based methods, microbial diversity and biotechnology, pathogen survival and detection in the environment, enzymes involved in production of bioenergy from wastes, and the origin of microbial life.

E-mail: **j.d.van.elsas@rug.nl**

# HYPOTHESIZED MICROENVIRONMENTS FOR THE ORIGIN OF MICROBIAL LIFE ON EARTH

JACK T. TREVORS[1], ASIM K. BEJ[2], AND JAN DIRK VAN ELSAS[3]
[1]*School of Environmental Sciences, Laboratory of Microbiology, University of Guelph, Guelph, ON, Canada N1G 2W1*
[2]*Department of Biology, University of Alabama at Birmingham, 464 Campbell Hall, Birmingham, AL 35294-1170, USA*
[3]*Linneausborgh, University of Groningen, Nijenborgh 7, 9747AG, Groningen, The Netherlands*

## 1. Introduction

The origin of life is still an enigma (Trevors, 2011; Trevors and Masson, 2011). Moreover, there is no generally accepted definition of life, which makes a delineation of the abiotic and biotic world difficult (Cleland and Chyba, 2002). A suite of characteristics is often used to define life. These include the ability of life to synthesize its own biomacromolecules, to generate and convert energy to various forms of cellular work, to perform informational biosynthetic processes to ensure storage of genetic instructions and reproduction; the ability to insulate the internal milieu and control exchanges with the external environment; and the ability to regulate activities and reproduce or multiply (De Duve, 1991). The transition from a lifeless chemosphere on Earth to the beginning of the living biosphere occurred about 4 billion years ago.

A diverse range of viewpoints and hypotheses with some experimentation have shaped origin of life research. Although the emergence of biomolecules from inorganic compounds by purely physicochemical mechanisms is now accepted by some researchers, the transition from a prebiotic state, in which such biomolecules initially occurred unorganized, to a living microbial cell capable of growth and division is a contentious issue. The current immense species diversity that has evolved from simple forms on the basis of only a few nucleotides and 20 L-amino acids has not made it any easier to correctly determine how life originated on the Earth or elsewhere. In particular, one unanswered question is what type of microenvironment did life originate in on the Earth and/or elsewhere? Moreover, there is no established "theory on the origin of life" with supporting observations and experimentation.

Scientists and other researchers who investigate the origin of life generally support either the *metabolism first hypothesis* or the *genetics first hypothesis*. The first hypothesis states that chemical reactions may have led to prebiotic organization of chemical compounds from which an organic genetic code (RNA followed by DNA) emerged. The genetics first hypothesis postulates that replicating genetic polymers (genetic instructions) developed first. Research has also addressed (1) the origin of life on Earth or elsewhere and its transport (Nicholson, 2009) to the Earth or (2) the transition from prebiotic chemistry to microbial cells capable of growth and division. However, there is still a paucity of information on the minimal microenvironment dimensions suitable for the molecular organization of life, in particular, the organic genetic code for cellular instructional storage (RNA, DNA) and the most plausible composition of the microenvironment for the origin of life on the Earth or elsewhere. Therefore, the correct explanation for life's origin is unknown and undecidable at this point in time, as the supporting evidence to form a consensus is yet to be achieved. Although it has been hypothesized that life, most likely bacteria or Archaea, could have been transported from elsewhere to the Earth (panspermia hypothesis), this simply moves the question of the origin of life to a non-Earth location.

In this chapter, we discuss possible physical-chemical microlocations for the origin of life on Earth. Because the first living cell(s) were probably in the magnitude of about a micron in size (and possibly smaller), it can be hypothesized that the first physical-chemical environment conducive to life may also have been a microenvironment, in the order of microns. To what extent such an environment was stable enough and contained the necessary elements and compounds to serve as a site favorable to the origin of life is unknown. However, one can hypothesize that such a location could have been a microlocation of micron-size dimensions.

Numerous possible locations have been hypothesized for the origin of microbial life on the Earth, including hydrothermal vents in oceans (Russell and Hall, 1997), clay and mineral surfaces in aquatic or terrestrial settings as genetic templates and/or scaffolds for life (Cairns-Smith, 1985), organic soups, lipid vesicles (Luisi et al., 1999), the interface between hydrophilic and hydrophobic liquids (Morchio and Traverso, 1999), and more recently, prebiotic films or hydrogels that eventually became living microbial biofilms attached to mineral surfaces (Trevors, 2011) (see Table 1). Whatever the physical location was, the prebiotic organizing molecules in the immediate surrounding microenvironment would need to be sufficiently stable and transitioning to become part of an organizing cell environment, within a higher entropy in the surrounding environment. The microenvironment would also have a nutrient status that provided all necessary minimal nutrients and liquid water, all within pH, pressure, and temperature ranges that permitted prebiotic organization (not decomposition). There would also occur the *transition* or *tipping point* to living microbial cell(s) capable of growth and division.

It has been hypothesized that a hydrogel (Trevors and Pollack, 2005; Trevors, 2011) was a possible component in the origin of life. Here, we expand and refine

**Table 1.** Hypothesized microenvironments for the origin of microbial life on the Earth and/or elsewhere.

| Microlocations | References |
|---|---|
| 1. Hydrothermal vents | McCledon (1999) |
| 2. Mineral surfaces | Cairns-Smith (1985) |
| 3. Lipid vesicles | Tessera (2009) |
| 4. Hydrogels | Trevors (2011)<br>Trevors and Pollack (2005) |
| 5. Hydrogel on surfaces within a hydrophobic-hydrophilic medium (prebiotic film) in a cool/cold environment | This chapter |
| 6. Liquid channels and veins in ice | Price (2009) |
| 7. Alkaline vents | Lane et al. (2010) |

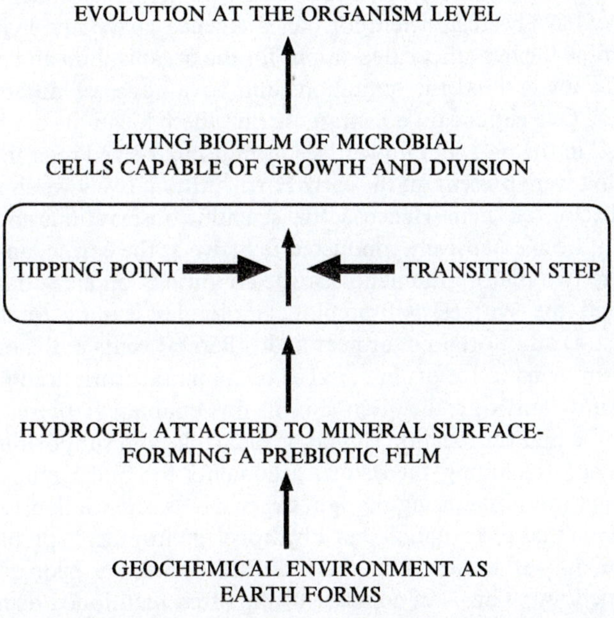


**Figure 1.** Hypothetical pathway from a geochemical environment to the first microbial cells capable of growth and division.

this idea and hypothesize that a prebiotic film of a hydrogel attached to a mineral surface at the interface between a hydrophobic and hydrophilic medium (HM) may have been a plausible environment that allowed life to start. The physical dimensions of the site would need to be in the micron range, as the first prebiotic structures were being organized prior to the transition or tipping point to becoming living cells (see Fig. 1). Since currently new life only arises from preexisting life (with the exception that it occurred at least once from inanimate geochemistry), the nature of this transition event or tipping point is still unknown. For a bacterial cell to divide correctly at its midcell location, it requires specific cell division proteins (Trevors, 2004, 2006). It can only be hypothesized what tipping point event(s) changed a prebiotic structure to become a living cell capable of cell division.

## 2. Microenvironments for the Origin of Microbial Life on Earth

Although numerous locations have been hypothesized for the origin of life on Earth, the correct environment(s) and composition are unknown. Moreover, the spatial dimensions of the location for prebiotic organization and the subsequent emergence of microbial cells capable of growth and division cannot be ignored. Theoretically (because of the micron-size dimensions of microbial cells), the emergence and possible persistence of a life-conducive microenvironment in the same size range may have been sufficient for the emergence of the first living microbial cells. We examine the hypothetical locations for the organization and emergence of life, seeking to understand the limitations and advantages of different hypothesized locations. One significant constraint is that these locations are hypothesized to have existed in the past, but no vestiges or molecular-sized microlocations may still remain that were present on the early Earth. Alternatively we simply have not looked and discovered them. Hence, actual scientific observations and experimentation have not been extensively conducted to arrive at the correct answer.

There are two major arguments for the conditions on early Earth that may have supported the synthesis of prebiotic molecules leading to the origin of microbial life: (a) a hot origin in or near hydrothermal vents in the ocean or (b) a colder origin in veins in ice or on surfaces of minerals. Both arguments present interesting points and warrant discussion in this chapter. A prerequisite in our considerations is that the location should be plausible, and supporting or nonsupporting evidence (including theoretical arguments for and against) should be provided. When examined using thought experiments and available information, one possibility is that life originated in a hydrogel environment (prebiotic film) on a surface like that of minerals, at the interface between a hydrophobic and a hydrophilic medium. This is in contrast to the often-mentioned deep-sea hydrothermal vents, lipid vesicles (Luisi et al., 1999), and mineral surfaces. In the sections below, we discuss why these environments may have been the microlocations for the initial spatial organization of prebiotic life and then the transition (tipping point) from prebiotic to living microbial cells.

### 2.1. HYDROTHERMAL VENTS

Volcanic aquifers have been proposed as locations for the prebiological formation of organic compounds and RNA (Washington, 2000). Moreover, deep-sea, hydrothermal and volcanic vents at temperatures of up to 400°C and about 2-km deep, well beyond the input of sunlight, have been hypothesized as possible sites for the origin of life on Earth (Russell and Hall, 1997). However, these sites are supposedly very turbulent, hot, and not suitable as sources of the minimal necessary prebiotic molecules needed in polymerization reactions for prebiotic structures, followed by the emergence of microbial cells capable of growth and cell division. Moreover, ammonia is required for the generation of nitrogen-containing compounds such as amino acids. There is no evidence that ammonia was present in sufficient amounts at such vents to sustain prebiotic molecule and biopolymer formation. This seems to dismiss the possibility of this location for the origin of life. There is also no evidence that genetic instructions in organic codes such as RNA or DNA would have been able to form and be sustained in such a hydrothermal and turbulent environment. On the other hand, it can be hypothesized that at some distance from the vent cores, in a gradient of decreasing temperatures and at solid (mineral) surfaces, conditions may have existed that permitted the stable occurrence of life-conducive chemical conditions.

Shallow ocean vents, with depths of about 200 m or less, might have been a more suitable origin of life location because sunlight can reach a large part of such depth and act as a source of energy for chemical reactions. What light-capturing pigments might have been present in the early prebiotic mass? Can something be learned from consideration about the evolution of such pigments? Even if the vents were at suitable temperatures for life (10–96°C), the emergence of life at such locations is questionable. In spite of the discovery of some interesting chemistry in the habitat, this still does not explain the emergence of an organic genetic code within a cellular structure, together with the capacity for eventual controlled cell division.

Since virtually all geochemists agree that there was little, if any, free oxygen on the primordial Earth, anaerobic respiration or fermentation seems to have been the most likely primordial energy-generating mechanism. Haldane (1929) supported the latter view, proposing that fermentation was the primordial source of energy on the primitive Earth without oxygen. The “life without oxygen” modus was further explained by Nobel Laureate De Duve (2002, 2005), who argued that life in anaerobic fermentative environments most likely existed on the primitive Earth. The formation of alkaline vents in the primitive Earth’s cooling ocean floor has been suggested by Russell (2007). The scenario of anaerobic fermentative life has been suggested to exist in microenvironments of alkaline vents, in which the main components of essential biomolecules such as amino acids and nucleotides had been formed (Martin and Russell, 2007; Ricardo and Szostak, 2009). Supporting this view, Lane et al. (2010) argued that, based on the knowledge that bacteria and Archaea differ in their gene sequences and crystal structures of enzymes that catalyze the individual steps of fermentation, this process may have evolved in alkaline vents independently and perhaps twice.

The formation of an alkaline vent microenvironment as a location for the formation of key biomacromolecules on the early Earth is a geological process that requires some explanation. Kelley et al. (2001) described that a geothermal process known as serpentinization results in the formation of alkaline vents. Serpentinization is a geochemical reaction that occurs when olivine is hydroxylated in contact with seawater. The hydroxylated rocks expand and rupture, allowing the entry of even more seawater into the crust, driving the reaction again and again (Bach et al., 2006; Sleep et al., 2004). This hydroxylated rock event has occurred on a global scale, as most of the Earth's crust has been in contact with ocean water (Fyfe, 1994). Therefore, one can hypothesize that such mineral-rich alkaline vent microenvironments have existed on an immense scale across the seafloor of the early Earth and were likely sustained for much longer than other hydrothermal vents such as black smokers known to date (Martin and Russell, 2003; Nitschke and Russell, 2009; Russell and Arndt, 2005). Moreover, the serpentinization process is known to generate moderately high temperatures (150–200°C), producing hydrothermal fluids that are strongly alkaline (pH 9–11) and rich in hydrogen. This condition changes to a more favorable temperature of 70°C with alkaline pH and full of dissolved hydrogen gas as the expansion drives the crust up to the seafloor, presenting a scenario for the stability of biomacromolecules that may have been formed in the alkaline vents (Lane et al., 2010). This discovery was important because these alkaline vents in the early Earth's ocean floor provide a plausible scenario for the emergence of chemical conditions that can support the formation of essential life-conducive biomacromolecules.

## 2.2. COLD ORIGIN OF LIFE

The prebiotic Earth, that is, more than 4 Bya, is hypothesized to have been too hot to sustain life. Therefore, it is logical to assume that life started during the Earth's cooling period when a more stable environment facilitated biomolecules necessary for a living cell to form and subsequently to undergo the necessary organization to support life. To that extent, one can hypothesize that the cooling process occurred gradually in pockets of microenvironments rather than as a sudden global event. The counterargument to the "cooling microenvironment hypothesis" that facilitated the first living cells on Earth is the "hot origin of life." The strongest support for this origin of life hypothesis has been presented by Woese (1987) with the phylogenetic tree rooting from several thermophilic microbial ancestors. However, molecular analysis of bacterial phylogeny has provided some contrasting evidence, namely, that early microbial life may have been rooted from the phylum Planctomycetes, which includes a number of unique psychrophilic bacteria. This has prompted the alternative hypothesis of the "cold origin of life" on ancient Earth (Doolittle, 1999; Brochier and Philippe, 2002). In support of this hypothesis, Nisbet and Fowler (1999) have elaborated that the Precambrian biosphere probably experienced a significantly lower temperature microenvironment during the Paleoproterozoic period when cyanobacteria appeared, and subsequently in the Neoproterozoic era when severe glaciation led to the appearance of "snowball" Earth. A counterargument is that, from 2.5 to 4 Bya, during the

Archaean aeon, the Earth was at least three times warmer than today (Nisbet and Sleep, 2001). Based on stromatolite fossil records, the Archaean aeon period was the estimated time when life originated on Earth (Nisbet and Sleep, 2001).

The alternative possibility is that at the time the Earth's surface was generally hot, the deep ocean water was cooler. Therefore, life may have flourished, instead of at the "hot Earth's surface," in the cooler deep ocean water. Genetic data and the stability of biomolecules in organisms at the root of the tree of life, especially nucleic acids, may hold the clue as to the origin and existence of ancient life, possibly in "colder microenvironments" rather than in widespread hot surfaces of the Archaean aeon.

If life originated on a cooling Earth, then one has to consider that essential prebiotic biomolecules and their precursors must have formed and been stable in colder conditions. Hydrogen cyanide (HCN) is considered to be the earliest naturally occurring biologically relevant precursor compound, polymerization of which has been described as the hallmark for the formation of amino acids and nucleic acids on ancient Earth (Matthews, 2004). As early as 1875, Pflüger described that cyanogen and its components have the tendency to polymerize and can form labile protein when reacted with water, oxygen, and natural salts. Later, Miller (1953) and Miller and Urey (1959) demonstrated that when methane, ammonia, and water in combination were subjected to electric discharge, the formation of α-amino acids occurred. This led to a long-accepted concept of the formation of amino acids on primitive Earth as occurring by the condensation of these monomers in yet to be known suitable conditions. Later, Matthews and Moser (1966) reported that instead of being the direct product of the monomers as shown by Miller and Urey (1959), α-amino acids on early Earth may have been formed as a by-product during polymerization of HCN. Another series of pioneering experiments by Oro et al. (1980) showed the synthesis of α-amino acid and adenine from HCN oligomers in ammonium hydroxide solution. Subsequently, Minard et al. (1998) reported that HCN polymers are the primary candidates for the abiogenesis of the building blocks of nucleic acids (the purines and the pyrimidines) and polypeptides (amino acids). The source of HCN on Earth has been proposed to be extraterrestrial, which may have occurred during frequent bolide bombardment causing the prebiotic ancient Earth atmosphere to be covered with this compound (Matthews, 2004). This raises the question of whether HCN and its derivative biomolecules stably existed on a "cold" or "hot" Earth. Miyakawa et al. (2002a, b) reported that, based on its physical and chemical characteristics, it is unlikely that HCN polymerized in a warm, primitive ocean. Therefore, the prediction is that a eutectic microenvironment may have occurred, at least on some part of Earth, to concentrate HCN sufficiently for it to polymerize, leading to the formation of the prebiotic compounds that are necessary for living cells.

Which molecules came first? This part of the debate is concerned with the question whether proteins or nucleic acids came first on the prebiotic Earth. Based on the experiments by Miller (1953), Miller and Urey (1959) and Oro et al. (1980), it was hypothesized that proteins formed first. In addition, the catalytic activity of proteins, which was originally thought to be unique, supported the idea of proteins first. However, the discovery of ribozymes (a group of RNA species with catalytic activity) by Zhang and Cech (1997, 1998) prompted an alternative argument, that is, RNA might have been the first type of biomolecule that emerged during chemical organization on early Earth, yielding biologically active macromolecules in the biotic world of the Archaean aeon about 3.5 billion years ago. This led to the concept of the "RNA world," which was described by Gilbert (1986) as life with polyribonucleotides instead of polydeoxyribonucleotides present on the early Earth. This "RNA world" concept has merit because (a) RNA has the properties to function as genetic material that can self-replicate, be altered or mutate, and undergo its own selection process, enabling it to cope with a changing environment, and (b) it has catalytic activity that may have been utilized to promote phosphodiester transfer and phosphodiester hydrolysis in RNA. In addition, rRNA in ribosomes involved in catalytic activity in protein synthesis found in modern living cells is considered to be the remnant of the ancient biomolecular interaction, thereby supporting the "RNA world" hypothesis.

An argument in support of a hot origin of life is that rRNA sequences seem to show that the first (bacterial) phylum that emerged in the tree of life was hyperthermophilic (DiGiulio, 2003). However, in an earlier study, Miller and Lazcano (1995) suggested that hyperthermophilic bacteria may have outcompeted older mesophilic ancestors. Using a more reliable analysis of bacterial phylogeny, Brochier and Philippe (2002) reported that hyperthermophilic bacteria did not emerge first. Instead, Planctomycetes, a phylum that includes many psychrophiles found in polar regions, could be hypothesized as the last universal common ancestor (LUCA) to all life forms.

A correlation between the optimal growth temperature of prokaryotes (Bacteria and Archaea) and the G+C nucleotide content of ribosomal RNA (rRNA) led Galtier et al. (1999) to infer that the existence of life forms was incompatible with survival at high temperatures on the early Earth. The model concluded that ancestral RNA had only a moderate G+C content, well below that of all known hyperthermophiles and consistent with organisms that live at moderate temperatures (Vogel, 1999). This inference was based on the model that predicts nucleotide sequence evolution leading to an estimated ancestral base composition, with the assumption of the existence of varying G+C content among lineages and unequal substitution rates. In the same year, Arrhenius et al. (1999) supported the study and extended the argument, stating that LUCA may have had biochemical properties similar to those of a contemporary bacterium. If correct, then the predecessors of LUCA may have adapted to varying environmental temperature fluctuations at least once or possibly multiple times.

Although photochemical reactions, which occur at high temperature, may have supported the origin of life on the ancient "hot" Earth, it is now known that, unless there is strong adaptation such as found in extreme thermophiles, biological macromolecules are highly unstable when subjected to prolonged exposure to high temperatures (Bada and Lazcano, 2002). Therefore, it is unlikely that RNA-based life persisted in hot ocean waters or hot surface environments (Moulton et al., 2000).

In contrast, it is known that nucleic acids are stable and can be preserved for a long period of time at low temperatures. For example, Priscu and Christner (2004) reported the recovery of DNA from ancient glacial ice. Perreault et al. (2008) studied heterotrophic and autotrophic sulfur-oxidizing bacteria from cold springs and permafrost of the Canadian High Arctic. Vincent et al. (2000) and Vincent (2007) described oscillatorians, which resemble ancient fossilized cyanobacteria and are found in abundance in both Antarctic and Arctic icy environments. This indicates that life on early Earth could have existed and evolved in cold microenvironments rather than at high temperatures, as hypothesized by some researchers.

Woese (1987) reported that the early Bacteria and Archaea dominated the hot Archaean Earth; most of these organisms were thermophiles. Although this contention may be true, several other reports have shown that many psychrophilic Bacteria and Archaea (primarily the Crenarchaeota) are predominantly inhabitants of Antarctic and Arctic icy environments as well as the deep frigid ocean (Karner et al., 2001; Massana et al., 1998; Vincent, 2007; Vincent et al., 2004a, b). Therefore, the ancestors of these Prokarya living in high-temperature environments are also commonly found in cold and even subzero-temperature environments. This suggests that the essential macromolecules for life could have been formed or remained stable in a relatively cold geological location. The existence of geological conditions favoring the principle biomacromolecules on the early Earth, supporting the "cold origin of life" hypothesis, has been presented by some investigators. The first step to confirm this hypothesis is to establish how an icy microenvironment in the early history of Earth may have supported the synthesis of nucleic acids, which subsequently became the organic genetic code of life in a biologically active living cell. Price (2009) presented three different cold microenvironments that could have potentially supported the origin of life. They are as follows: (a) liquid veins in ice, (b) individual ice crystals in polycrystalline ice, and (c) permafrost with mineral grain surfaces. The first argument supporting the cold origin of life is based on the prediction that the reduced luminosity of the young sun and low greenhouse gases in the atmosphere contributed to the oceans remaining frozen during the early history of Earth (Bada et al., 1994). It was proposed that only 300 m of the ocean surface were frozen, leaving cold water trapped hundreds of meters below. The icy surface may have shielded the cold liquid water from solar radiation, giving biological macromolecules the opportunity to form and organize. Later, heavy volcanic activity resulted in a greenhouse effect, leading to the melting of the ocean's ice cap. Therefore, microbial life may have originated in cold ocean water before the surface ice melted (Kasting and Catling, 2003). The 16S rRNA gene–based phylogenetic tree suggests that the first phyla that emerged were of hyperthermophilic origin (DiGiulio, 2003). However, since RNA is thermally unstable, it is difficult to put the parts of the puzzle together, that is, to understand that a minimal RNA genome was the first organic genetic instruction set in life on this planet.

If the cold origin of life hypothesis is a plausible one, the cellular mechanisms that support microbial survival, growth, and adaptation on an icy Earth must be explained. The survival of early bacteria in an icy environment could be explained from modern-day discoveries of the functions of ice-binding, antifreeze, and cold-adaptive proteins in microorganisms that occur in low-temperature environments. Antifreeze and ice-binding proteins have the ability to inhibit the recrystallization of ice, thereby hindering freeze damage to the cells (Venketesh and Dayananda, 2008). It has been proposed that the antifreeze or ice-binding proteins create a microenvironment of liquid water by preventing growth of ice crystals surrounding microorganisms in ice cores, thereby keeping a constant supply of nutrients and minerals for the growth and division of such cells (Bej and Mojib, 2009). A bacterial strain isolated from a 3,519-m deep ice core just above liquid water (the accreted ice) of the subglacial Lake Vostok exhibited the presence of a 54-kDa ice-binding protein (Raymond et al., 2008). A psychrophilic yeast isolated from an Arctic pond exhibited a secreted 26-KDa ice-binding protein with thermal hysteresis and recrystallization inhibition (Lee et al., 2010). These ice-binding proteins showed high sequence identity with ice-binding proteins found in fungi, diatoms, and bacteria, clustering within the class of ice-active proteins.

Antifreeze proteins are structurally diverse polypeptides that have been discovered in many cold-adapted organisms. These proteins generate bipyramidal ice crystals and exhibit thermal hysteresis in aqueous suspension at subzero temperatures (Muryoi et al., 2004; Xiao et al., 2010). In contrast, cold-adaptive proteins are relatively low-molecular-weight proteins that function as RNA chaperones, assisting cells in their maintenance of physiological functions at low temperatures. These cold-adaptive proteins are common in microorganisms isolated from polar environments (Bej et al., 2000; Bej and Mojib, 2009; Panicker et al., 2002, 2010). A recent study by Rice et al. (2009) showed that bacterial teichoic acids help to establish long thin liquid water channels surrounding ice crystals at −40°C. Bacterial cells remain sequestered in these liquid water channels and pockets, suggesting that they could remain viable and physiologically active in frozen ice at very low temperatures for a long period of time. The microchannel in the ice core is in fact the bacterial microenvironment in which active growth and cell division take place.

In addition, many bacteria in cold polar environments can secrete "slimy" exopolysaccharides (high-molecular-weight carbohydrate polymers) to establish an "ice-free" microenvironment, in which water retention and nutrient availability are conducive to slow but steady cellular metabolism, growth, and division (Nichols et al., 2005; Panicker et al., 2006). These cellular survival mechanisms found in microorganisms in recent Earth polar environments support the possible origin and existence of life in a cold and even icy early Earth environment.

## 2.3. MINERAL SURFACES

An essential element like phosphorus may have been concentrated on metal sulfide minerals, which concentrated organic compounds, used for the organization of prebiotic and then primitive living cells (Bebie and Schoonen, 1999). The warm/hot subsurface of the Earth with the internal Earth heat and possibly hydrogen driving early metabolism (Morita, 2000) is a possible hypothesis for the location of the first reactions of a primitive cell. Morita (2000) suggested hydrogen as the potentially universal energy source for bacteria. Hydrogen in an anoxic subsurface may have been the gaseous substrate that supported early metabolism until energy generation from a pathway like glycolysis emerged. Coliform bacteria can use hydrogen as an energy source, but not for active growth and division, which are now served by better substrates. It is interesting that a hydrogen-based subsurface microbial community dominated by methanogens was suggested as the basis of production in a primitive ecosystem on both planet Mars and the Jupiter moon, Europa (Chapelle et al., 2002). Mineral surfaces offer the advantage of being stable, abundant, and ordered. Moreover, they provide some capacity to concentrate diluted solutes. However, by themselves, they do not offer the hydrogel-type environment that is similar to cellular cytoplasm (or structured water) in present-day living organisms.

## 2.4. LIPID VESICLES

Lipid vesicles are organized by simple chemistry, without requiring any instructions necessary for assembly. They mimic a diffusion membrane, come in numerous sizes, but are somewhat fragile. However, lipid bilayers are permeable to amino acids and phosphate (Chakrabarti and Deamer, 1992), which has significant implications in the organization of the first stable and functional membranes and all subsequent fluid bilayer membranes.

Goltsov and Barsukov (2002) hypothesized a scenario whereby a membrane reorganized from a micelle to a vesicle under conditions of increasing temperature, which caused melting. Cycles of heating and cooling have probably been common on the early Earth. These temperature cycles may have been of central importance to the organization and transition of simple micelles to stable vesicles composed of bilayer membranes. At the higher temperature, the simple membrane would have acted like a fluid bilayer. As the temperature decreased, the membrane would undergo a transition to a more gel-like or rigid bilayer that may even have fragmented. As the temperature increased again, the fragments may have fused to form longer membranes. At some point, the temperature values between heating and cooling cycles would become closer and membrane fragmentation would cease and the stable vesicle or precells would remain intact. It is also noteworthy that the rate of cooling could also have had an influence on the vesicles (Souzu, 1989). For example, low-temperature stress can remove water from membranes.

The organization of micelles and their transition to a continuous membrane would have been a pivotal event in bringing organization at the interface between the primitive aqueous and hydrocarbon environment, if this actually happened. As a mixture of compounds containing fatty acids cools down, the fatty acids can aggregate to form a milky suspension of micellar and vesicle structures (Deamer, 1997).

## 2.5. HYDROPHILIC/HYDROPHOBIC INTERFACES

Gold (2001) hypothesized that methane and oil were present in the subsurface of the Earth when it formed. It is plausible that prebiotic life organized at anaerobic subsurface interfaces between an oily hydrophobic environment and an aquatic environment. Morchio and Traverso (1999) hypothesized that intense ultraviolet (UV) radiation (also unfiltered by minimal ozone in the atmosphere) caused the polymerization of atmospheric methane. The resulting condensed hydrocarbons over time could have covered parts, or all, of Earth, creating a thick hydrophobic layer. Such a layer would only need to contain a suitable microenvironment for the origin of life. Morchio and Traverso (1999) also speculated that a hydrophobic medium may have been the primordial cradle of life. This hypothesis fits with the hypothesis that the first cells originated by soap-like chemistry, with soaps bridging the gap between oil and water (Schopf, 1999). The primordial soup may have been a warm, water-oil emulsion or an interface between the two with still unknown surfactant-like compounds present. Hydrophobic organic compounds would aggregate while hydrophilic compounds would be in the aqueous layer. This oil/water chemistry would be conducive to the first compartments or organizing closed membranes that provided a prebiotic biostructure with an inside and outside. The interface between an aqueous and hydrophobic environment would also permit elements and compounds of varying water solubilities to be in the same physical, molecular location, where they could be used in prebiotic organization.

## 2.6. HYDROGELS AND PREBIOTIC FILMS

Possibly, the warm primordial soup referred to in some origin of life literature was actually a hydrogel environment attached to mineral surfaces, that is, a prebiotic film. Such an environment may have provided an immense number of molecular gel sites on mineral surfaces contained with a larger hydrophobic/hydrophilic medium (HM). This brings forward the question – what was the hydrogel composed of and how did it attach and remain on the mineral surface? For example, a 2% by weight clay mineral added to water results in a semisolid gel-like material at room temperature (Seydel et al., 2008). Is it plausible that the soup was composed of some large unknown number of microscopic mineral surfaces embedded in a gel in an HM? The composition of the gel can only be hypothesized on at this time and is addressed latter. The surfaces could have been provided by charged clays, products of weathering on Earth. If a gel environment was possible, it provides many attributes, including the all-important stability (Trevors, 2011).

A definition of a gel is that the elastic modulus is greater than the viscous modulus. Unlike liquids, gels do not flow as a result of steady shear. A phenomenological definition is that gels are solid or solid-like materials that consist of two or more components, one of which is a liquid of some abundance (Almdal et al., 1993). One feature of hydrogels, that is, gels in water, is that the solvent is not ordinary bulk water. As dipoles, water molecules will adsorb onto hydrophilic surfaces, and onto one another, forming networks. When enmeshed in a polymer network, such layered or "structured" water holds together the network, forming a stable gel (Pollack, 2001). The hydrogen bonding in water can hold together even polymers that are not cross-linked to one another, as happens in common colloidal gels such as gelatin (Bromberg and Ron, 1998).

Trevors and Pollack (2005) hypothesized that the first microbial cell(s) on Earth was/were assembled in a hydrogel environment capable of retaining water, oily hydrocarbons, solutes, and gas bubbles. These cells, under these conditions, were capable of carrying out numerous functions in a confined environment, even in the absence of a membrane. Thus, a gel-like environment may have conferred distinct advantages to the assembly of the first prebiotic cells. It is known that the bacterial cell is a gel (Pollack, 2001), and a gel as a primitive active cytoplasm appears to be a suitable environment for the assembly of precells, followed by cells capable of growth and division. It is hypothesized that a hydrogel microenvironment attached to a mineral surface may have been a type of film, in which prebiotic cells organized prior to these gaining the capacity to grow and divide. A hydrogel attached to a mineral surface retains its integrity even in the absence of a cytoplasmic membrane. Primitive organelles, nutrients, ions, proteins, and nucleic acids could have remained ordered and in continuous and close molecular proximity within the gel, without the danger of dispersing as would have happened in an aqueous environment with free diffusion. The question of how the precell, without an intact membrane and with only a small mass, could retain its integrity needs not be an issue if the cytoplasm is viewed as a cohesive hydrogel. In addition, numerous basic functions performed by the cell or precell can be carried out by gels themselves. A hydrogel attached to a mineral surface would have the advantages conferred by both the hydrogel and the mineral surface, such as stability, abundant presence of surface area, order, and the possibility of concentration of dilute solutes. The transition of a prebiotic gel or prebiotic film to a living gel cell may be more plausible than the origin of microbial life events in a thermal hydrovent or fragile vesicle.

## 3. Cell Division and the Origin of Life

The transition or tipping point event, at which cell(s) became capable of cellular division, is an enigma. Some minimal amount of core structural and molecular biology must have been present that allowed controlled cell division to proceed (Trevors, 2004). This would have been a cell with a minimal stable size, two genome copies, sufficient cytoplasm for two offspring cells, and the capability to partition the cell into two offspring cells. However, the early cells may not have divided equally if their division mechanism was not as complex as in present-day bacteria. This type of cell would possess complex regulated molecular functions and may have been capable of a complete cell cycle (period from one cell division to the next cell division). The initial cellularization of life allowed the cell to emerge as the basic unit of life. The genome needed to replicate at least once, so two genomes would be available for each cell during division. Moreover, cells cannot commence DNA synthesis unless they have sufficient cytoplasm to generate two cells. This means that evolving bacterial cells had to extract energy from their environment and had to develop their biological functions to a minimal threshold that eventually permitted cell division while in a higher entropy surrounding environment (Trevors, 2004).

## 4. Summary

A number of intriguing scientific hypotheses have been presented during the past decades, indicating that the early Earth has experienced a cold environment (cold environment is defined as relatively cold, which could possibly be the current mesophilic conditions on parts of the recent Earth). One possible location for the origin of life is the ocean bottom, where the chemical organization for biologically active macromolecules may have formed the first cell(s). Among the supporting scientific information is the probability that RNA, and in particular rRNA, is more stable under cool conditions. The most likely scenario that supports the cold origin of life hypothesis seems to be the appearance of localized stable microenvironments during the cooling period of the Earth, in which key biological macromolecules assembled and underwent meaningful modifications to support life.

From a thermodynamic perspective, an aqueous microenvironment for the origin of life may not have been suitable for the formation of peptide and nucleotide polymers, which are unstable due to their tendency to hydrolyze in water. Moreover, without any mechanism to confine molecules to a certain space, the essential molecules for life to emerge may have diffused away in an open water scenario.

The biofilm mode of attached growth is common in present-day microorganisms. We therefore hypothesize that a primitive hydrogel (not simply unbound water)-type environment attached to a mineral surface may have been the microenvironment that has supported the organization of prebiotic life. A hydrogel environment thus provides advantages in terms of stability of the environment, not only for the accretion of polymeric mass, but also for subsequent cell division and evolution. Hydrogels have the capacity to carry out various energy-requiring tasks, and there is evidence that phase transitions are used today in eukaryotic cells (Verdugo et al., 1992; Tasaki, 1999; Pollack, 2001). Furthermore, the hydrogel hypothesis offers a solution to the conundrum of how prebiotic structures remained in close physical proximity before the appearance of the continuous cell membrane. We suggest that the hydrogel-like nature of current-day cells needs to be further investigated; in the event, it provides a better understanding of how the first microbial cells organized and divided on the primitive Earth.

Possibly, a set of scientifically plausible explanations for the origin of life will be pieced together in future research. Central to the understanding of life's origin is the mechanism whereby an organic genetic code carried by nucleic acids came to specify the correct protein catalysts that make life possible, assembling cells that are capable of metabolism, growth, and cell division. From the foregoing arguments, it appears that the hypothetical micron-size environment, that is, a prebiotic film composed of a hydrogel on mineral surfaces (contained within a larger hydrophobic/hydrophilic medium) in a cooling environment where organizing macromolecules remained stable, is one possibility. This resembles a primitive biofilm mode of microbial life. Given the fact that oil-in-water emulsions can be removed by bentonite organoclay (Moazed and Viraraghavan, 2002), this may indicate that clay surfaces had a role in concentrating emulsions on the early Earth, providing the necessary concentrations of substances for prebiotic films and cells. The active concentration of the molecules needed for the organization of prebiotic films and then living cells solves the problem of local depletion of elements and compounds, as the latter might have been replenished in the film. Finally, atmospheric gases such as $N_2$, $H_2$, and $CO_2$ were likely ubiquitous on primitive Earth. These gases can diffuse through membranes, hydrogels, and lipid vesicles. Alkaline vents in the early Earth's ocean floor may have provided a suitable anoxic microenvironment rich in hydrogen, with an alkaline pH, minerals, and suitable temperatures for biomacromolecules to form and later be involved in the transition to living cells.

From a prebiotic organizational perspective, the more numerous the hydrogel microenvironments for prebiotic macromolecular organization, the higher the probability of success for living cells to emerge from a prebiotic gel. A dual-phase aqueous-hydrophobic interface may have provided an immense number of microenvironments with a diverse supply of substances, that is, water-soluble, water-insoluble, and with varying solubilities in both phases for initial membrane organization. A cool environment also favors decreased chemical decomposition of biopolymers.

Humans may never know the origin of life. However, from an origin-of-life and subsequent evolutionary perspective, it is rational to discuss all primordial Earth and non-Earth environments and piece together the most plausible environmental conditions that were needed for the emergence of living microbial cells. A hydrogel-type gel and then a transition to a living biofilm present on mineral surfaces in a cool anoxic environment is a possibility for the origin of life.

## 5. Acknowledgments

Research by J.T.T. is supported by the NSERC (Canada) Discovery Program; A. Bej's work was supported by the UAB Faculty Development Award; Tawani Foundation (Col. James Pritzker), Chicago; NASA/NSSTC/VCSI Antarctic expedition (Marty Kress); NCAOR, India (Rasik Ravindra) and AARI/RAE, Russia.

## 6. References


Almdal K, Dyre J, Hvidt S, Kramer O (1993) Towards a phenomenological definition of the term "gel". Polym Gels Netw 1:5–17

Arrhenius G, Bada JL, Gerald FJ, Lazcano A, Miller S, Orgel LE (1999) Origin and ancestor: separate environments. Science, New Series 283:792

Bach W, Paulick H, Garrido CJ (2006) Unraveling the sequence of serpentinization reactions: petrography, mineral chemistry, and petrophysics of serpentinites from MAR 15 degree N (ODP Leg 209, Site 1274). Geophys Res Lett 33:L13306

Bada JL, Lazcano A (2002) Origin of life. Some like it hot, but not the first biomolecules. Science 296:1982–1983

Bada JL, Bigham C, Miller SL (1994) Impact melting of frozen oceans on the early earth: implications for the origin of life. Proc Natl Acad Sci USA 91:1248–1250

Bebie J, Schoonen MAA (1999) Pyrite and phosphate in anoxia and an origin-of-life hypothesis. Earth Planet Sci Lett 171:1–5

Bej AK, Mojib N (2009) Cold adaptation in Antarctic biodegradative microorganisms. In: Bej AK, Aislabie J, Atlas RM (eds) Polar Microbiology:The ecology, biodiversity and bioremediation potential of microorganisms in extremely cold environments. CRC Press (Taylor & Francis Group), Boca Raton, pp 157–177

Bej AK, Saul D, Aislabie J (2000) Cold tolerance of alkane-degrading bacteria isolated from soil near Scott Base. Antarct Polar Biol 23:100–105

Brochier C, Philippe H (2002) Phylogeny: a non-hyperthermophilic ancestor for bacteria. Nature 417:244

Bromberg LE, Ron ES (1998) Temperature-responsive gels and thermogelling polymer matrices for protein and peptide delivery. Adv Drug Deliv Rev 31:197–221

Cairns-Smith AG (1985) Seven clues to the origin of life: a scientific detective story. Cambridge University Press, Cambridge

Chakrabarti AC, Deamer DW (1992) Permeability of lipid bilayers to amino acids and phosphate. Biochim Biophys Acta 1111:171–177

Chapelle FH, O'NeillK BPM, Methe BA, Ciufo SA, Knobel LL, Lovley DR (2002) A hydrogen based subsurface microbial community dominated by methanogens. Nature 415:312–315

Cleland CE, Chyba CF (2002) Defining life. Orig Life Evol Biosph 32:387–393

De Duve C (1991) Blueprint for a cell. Neil Patterson Publishers, Burlington

De Duve C (2002) Life evolving: molecules mind and meaning. Oxford University Press, New York

De Duve C (2005) Singularities: landmarks on the pathways of life. Cambridge University Press, Cambridge

Deamer DW (1997) The first living systems: a bioenergetic perspective. Microbiol Mol Biol Rev 61:239–261

DiGiulio M (2003) The universal ancestor and the ancestor of bacteria were hyperthermophiles. J Mol Evol 57:721–730

Doolittle WF (1999) Phylogenetic classification and the universal tree. Science 284:2124–2129

Fyfe WS (1994) The water inventory of the earth: fluids and tectonics. Geol Soc Lond (Special Publications) 78:1–7

Galtier N, Tourasse N, Gouy M (1999) A nonhyperthermophilic common ancestor to extant life forms. Science 283:220–221

Gilbert W (1986) The RNA world. Nature 319:618

Gold T (2001) The deep hot biosphere. Copernicus Books, New York

Goltsov AN, Barsukov LI (2002) Synergetics of the membrane self-assembly: a micelle-to-vesicle transition. J Biol Phys 26:27–41

Haldane JBS (1929) The origin of life. Ration Annu 3:3–10

Karner MB, DeLong EF, Karl DM (2001) Archaeal dominance in the mesopelagic zone of the Pacific Ocean. Nature 409:507–510

Kasting JF, Catling D (2003) Evolution of a habitable planet. Annu Rev Astron Astrophys 41:429–463
Kelley DS, Karson JA, Blackman DK (2001) An off-axis hydrothermal vent field near the Mid-Atlantic Ridge at 30 degrees N. Nature 412:145–149
Lane N, Allen JF, Martin W (2010) How did LUCA make a living? Chemiosmosis in the origin of life. Bioessays 32:271–280
Lee JK, Park KS, Park S, Park H, Song YH, Kang SH, Kim HJ (2010) An extracellular ice-binding glycoprotein from an Arctic psychrophilic yeast. Cryobiology 60:222–228
Luisi PL, Walde P, Oberholzer T (1999) Lipid vesicles as possible intermediates in the origin of life. Curr Opin Colloids Interface Sci 4:33–39
Martin W, Russell M (2003) On the origins of cells: a hypothesis for the evolutionary transitions from abiotic geochemistry to chemoautotrophic prokaryotes, and from prokaryotes to nucleated cells. Philos Trans R Soc Lond B 358:59–85
Martin W, Russell MJ (2007) On the origin of biochemistry at an alkaline hydrothermal vent. Philos Trans R Soc Lond B 367:1887–1925
Massana R, Taylor LT, Murray AE, Wu KY, Jeffrey WH, DeLong FF (1998) Vertical distribution and temporal variation of marine planktonic archaea in the Gerlache Strait, Antarctica, during early spring. Limnol Oceanogr 43:607–617
Matthews CN (2004) The HCN world. In: Seckbach J (ed) Origins: genesis, evolution and diversity of life. Kluwer Academic Publisher, Dordrecht
Matthews CN, Moser RE (1966) Prebiological proteins synthesis. Proc Natl Acad Sci USA 56:1087–1094
McCledon JH (1999) The origin of life. Earth Sci Rev 47:71–93
Miller SL (1953) Production of amino acids under possible primitive earth conditions. Science 117:528
Miller SL, Lazcano A (1995) The origin of life – did it occur at high temperatures? J Mol Evol 41:689–692
Miller SL, Urey HC (1959) Organic compound synthesis on the primitive earth. Science 130:245
Minard RD, Hatcher PG, Gourley RC, Matthews CN (1998) Structural investigations of hydrogen cyanide polymers: new insights using TMAH thermochemolysis/GC-MS. Orig Life Evol Biosph 28:461–473
Miyakawa H, Celeaves HJ, Miller S (2002a) The cold origin of life: A. Implications based on the hydrologic stabilities of hydrogen cyanide and formamide. Orig Life Evol Biosph 32:195–208
Miyakawa S, Cleaves H, Miller S (2002b) The cold origin of life: B. Implications based on pyrimidines and purines produced from frozen ammonium cyanide solutions. Orig Life Evol Biosph 32:209–218
Moazed H, Viraraghavan T (2002) Coalescence/filtration of an oil-in-water emulsion in a granular organo-clay/anthracite mixture bed. Water Air Soil Pollut 138:253–270
Morchio R, Traverso S (1999) The hydrophobic superficial layer: the primordial cradle of life. Biol Forum 92:105–117
Morita RY (2000) Is $H_2$ the universal energy source for long-term survival? Microb Ecol 38:307–320
Moulton V, Gardner PP, Pointon RF, Creamer LK, Jameson GB, Penny D (2000) RNA folding argues against a hot-start origin of life. J Mol Evol 51:416–421
Muryoi N, Sato M, Kaneko S, Kawahara H, Obata H, Yaish MW, Griffith M, Glick BR (2004) Cloning and expression of afpA, a gene encoding an antifreeze protein from the arctic plant growth-promoting rhizobacterium *Pseudomonas putida* GR12–2. J Bacteriol 186:5661–5671
Nichols CA, Guezennec J, Bowman JP (2005) Bacterial exopolysaccharides from extreme marine environments with special consideration of the southern ocean, sea ice, and deep-sea hydrothermal vents: a review. Mar Biotechnol (NY) 7:253–271
Nicholson WL (2009) Ancient micronauts: interplanetary transport a microbe's cosmic impacts. Trends Microbiol 17:243–250
Nisbet EG, Fowler CMR (1999) Archaean metabolic evolution of microbial mats. Proc R Soc Lond 266:2375–2382
Nisbet EG, Sleep NH (2001) The habitat and nature of early life. Nature 409:1083–1091
Nitschke W, Russell MJ (2009) Hydrothermal focusing of chemical and chemiosmotic energy, supported by delivery of catalytic Fe, Ni, Mo, Co, S and Se forced life to emerge. J Mol Evol 69:481–496

Oro J, Holzer G, Lazcano-Araujo A (1980) The role of HCN and its derivatives in prebiotic evolution. In: Vennesland B, Conn EE, Knowles CJ, Westley J, Wissing F (eds) Cyanide in biology. Academic, New York, pp 517–541
Panicker G, Aislabie J, Saul D, Bej AK (2002) Cold tolerance of *Pseudomonas* sp. 30–3 isolated from oil-contaminated soil. Antarct Polar Biol 25:5–11
Panicker G, Aislabie J, Bej AK (2006) Analysis of aggregative behavior of *Pseudomonas* sp. 30–3 isolated from Antarctic soil. Soil Biol Biochem 38:3152–3157
Panicker G, Mojib N, Nakatsuji T, Aislabie J, Bej AK (2010) Occurrence and distribution of *cap*B in Antarctic microorganisms and study of its structure and regulation in the Antarctic biodegradative Pseudomonas sp. 30/3. Extremophiles 14:171–183
Perreault NN, Greer CW, Andersen DT, Tille S, Lacrampe-Couloume G, Lollar BS, Whyte LG (2008) Heterotrophic and autotrophic microbial populations in cold perennial springs of the high arctic. Appl Environ Microbiol 74:6898–6907
Pflüger E (1875) Arch Ges Physiol 10:641 (see Oro et al., 1980)
Pollack GH (2001) Cells, gels and the engines of life: a new, unifying approach to cell function. Ebner & Sons, Seattle
Price PB (2009) Microbial genesis, life and death in glacial ice. Can J Microbiol 55:1–11
Priscu JC, Christner BC (2004) Earth's icy biosphere. In: Bull AT (ed) Microbial diversity and bio-prospecting. American Society for Microbiology, Washington, DC, pp 130–145
Raymond JA, Christner BC, Schuster SC (2008) A bacterial ice-binding protein from the Vostok ice core. Extremophiles 12:713–717
Ricardo A, Szostak JW (2009) Life on earth. Sci Am 9:54–61
Rice CV, Harrison W, Kirkpatrick K, Brown ED (2009) Cryoprotection from bacterial teichoic acid. Proc SPIE 7441:74410M
Russell MJ (2007) The alkaline solution to the emergence of life: energy, entropy and early evolution. Acta Biotheor 55:133–179
Russell MJ, Arndt NT (2005) Geodynamic and metabolic cycles in the Hadean. Biogeosciences 2:97–111
Russell MJ, Hall AJ (1997) The emergence of life from iron monosulphide bubbles at a submarine hydrothermal redox and pH front. J Geol Soc (Lond) 154:377–402
Schopf JW (1999) Cradle of life. Princeton University Press, Princeton
Seydel T, Wiegart L, Juranyi F, Struth B, Schober H (2008) Unaffected microscopic dynamics of macroscopically arrested water in dilute clay gels. Phys Rev E 78:061403
Sleep NH, Meibom A, Fridriksson T (2004) $H_2$-rich fluids from serpentinization: geochemical and biotic implications. Proc Natl Acad Sci USA 101:12818–12823
Souzu H (1989) Changes in chemical structure and function in *Escherichia coli* cell membranes caused by freeze-thawing. I. Change of lipid state in bilayer vesicles and in the and in the original membrane fragments depending on rate of freezing. Biochim Biohys Acta 978:105–111
Tasaki I (1999) Evidence for phase-transition in nerve fibers, cells and synapses. Ferroelectrics 220:305–316
Tessera M (2009) Life began when evolution began. a lipidic vesicle-based scenario. Orig Life Evol Biosph 39:559–564
Trevors JT (2004) Evolution of cell division in bacteria. Theory Biosci 123:3–15
Trevors JT (2006) The Big Bang, Superstring Theory and the origin of life on the earth. Theory Biosci 124:403–412
Trevors JT (2011) Hypothesized origin of microbial life in a prebiotic gel and the transition to a living biofilm and microbial mats. C R Biol 334:269–272
Trevors JT, Masson L (2011) Quantum microbiology. Curr Issues Mol Biol 13:43–50
Trevors JT, Pollack GH (2005) Hypothesis: the origin of life in a hydrogel environment. Prog Biophys Mol Biol 89:1–8
Venketesh S, Dayananda C (2008) Properties, potentials, and prospects of antifreeze proteins. Crit Rev Biotechnol 28:57–82
Verdugo P, Deyrup-Olsen I, Martin AW, Luchtel DL (1992) Polymer gel-phase transition: the molecular mechanism of product release in mucin secretion? In: Karalis TK (ed) Mechanics of swelling, vol H64, NATO ASI series. Springer, Berlin

Vincent WF (2007) Cold tolerance in cyanobacteria and life in the cryosphere. In: Seckbach J (ed) Algae and cyanobacteria in extreme environments. Springer, Heidelberg, pp 287–301

Vincent WF, Gibson JA, Pienitz R, Villeneuve V, Broady PA, Hamilton PB, Howard-Williams C (2000) Ice shelf microbial ecosystems in the High Arctic and implications for life on snowball earth. Naturwissenschaften 87:137–141

Vincent WF, Mueller DR, Bonilla S (2004a) Ecosystems on ice: the microbial ecology of Markham Ice Shelf in the High Arctic. Cryobiology 48:103–112

Vincent WF, Mueller D, Van Hove P, Howard-Williams C (2004b) Glacial periods on early earth and implications for the evolution of life. In: Seckbach J (ed) Origins: genesis, evolution and diversity of life. Kluwer Academic Publishers, Dordrecht, pp 481–501

Vogel G (1999) RNA study suggests cool cradle of life. Science, New Series 283:155–157

Washington J (2000) The possible role of volcanic aquifers in prebiological genesis of organic compound and RNA. Orig Life Evol Biosph 30:53–79

Woese C (1987) Bacterial evolution. Microbiol Rev 51:221–271

Xiao N, Suzuki K, Nishimiya Y, Kondo H, Miura A, Tsuda S, Hoshino T (2010) Comparison of functional properties of two fungal antifreeze proteins from *Antarctomyces psychrotrophicus* and *Typhula ishikariensis*. FEBS J 277:394–403

Zhang B, Cech TR (1997) Peptide bond formation by in vitro selected ribozymes. Nature 390:96–100

Zhang B, Cech TR (1998) Peptidyl-transferase ribozymes: trans reactions, structural characterization and ribosomal RNA-like features. Chem Biol 5:539–553

# PART VI: THE DISTRIBUTION OF LIFE IN THE SOLAR SYSTEM AND ELSEWHERE IN THE COSMOS

**Lineweaver**
**Chopra**
**Raulin**
**Raven**
**Cockell**
**Kaltenegger**
**Seckbach**
**Chela-Flores**

Biodata of **Charles H. Lineweaver** author (with co-author **Aditya Chopra**) of "***What Can Life on Earth Tell Us About Life in the Universe?***"

**Dr. Charles H. Lineweaver** is the convener of the Australian National University's Planetary Science Institute and holds a joint appointment as an associate professor in the Research School of Astronomy and Astrophysics and the Research School of Earth Sciences. He obtained an undergraduate degree in physics from Ludwig Maximilians Universitat, Munich, Germany and a Ph.D. in astrophysics from the University of California at Berkeley (in 1994) He was a member of the COBE satellite team that discovered the temperature fluctuations in the cosmic microwave background. Before his appointment at ANU, he held post-doctoral positions at Strasbourg Observatory and the University of New South Wales where he taught one of the most popular general studies courses "Are We Alone?" His research areas include cosmology, exoplanetology, and astrobiology.

Personal homepage: **http://www.mso.anu.edu.au/~charley**
E-mail: **charley@mso.anu.edu.au**

**Aditya Chopra** is undertaking Ph.D. research at the Australian National University's Planetary Science Institute. He obtained an undergraduate degree in chemistry from the University of Western Australia in 2007 and in 2008 graduated with 1st class honours in astronomy at the Australian National University. His research areas include astrobiology and planetary science with the focus of his Ph.D. research being the examination of the elemental abundances in different life forms and their environments to gain insight into the origin and evolution of life.

E-mail: **aditya@mso.anu.edu.au**

**Charles H. Lineweaver**

**Aditya Chopra**



J. Seckbach (ed.), *Genesis - In The Beginning: Precursors of Life, Chemical Models and Early Biological Evolution*, Cellular Origin, Life in Extreme Habitats and Astrobiology 22, 799–815
DOI 10.1007/978-94-007-2941-4_40, © Springer Science+Business Media Dordrecht 2012

# WHAT CAN LIFE ON EARTH TELL US ABOUT LIFE IN THE UNIVERSE?

**CHARLES H. LINEWEAVER AND ADITYA CHOPRA**
*Planetary Science Institute, Research School of Astronomy and Astrophysics and the Research School of Earth Sciences, Australian National University, Canberra, Australia*

## 1. Quirkometry and Terrestrial Life as a Model Organism

The universe is filled with stars similar to our Sun (Robles et al., 2008), rocky planets similar to our Earth (Lineweaver and Grether, 2003; Ida and Lin, 2004; Mordasini et al., 2009), water like our oceans (Kuchner, 2003; Léger et al., 2004), amino acids like our proteins, and all the other ingredients for life (Pizzarello, 2007). But is the universe filled with life? If it is, what kind of life is it? We argue that if there is life out there at all, its basic features are likely to be a subset of the features common to all terrestrial life.

Here on Earth, we can distinguish generic features common to all terrestrial life from quirky features unique to only one group of organisms. Sometimes, the distinction between generic and quirky is easy. For example, all terrestrial life is based on carbon, but only a tiny subset barks at passing cars. From this, we can infer that on other terrestrial planets, we are more likely to find carbon-based extraterrestrials than we are to find dogs. All terrestrial life has DNA, but only a tiny subset has naked mole rat DNA. From this, we can infer that on other planets, we are more likely to find extraterrestrials with DNA than we are to find naked mole rats.

Sometimes, the distinction between generic and quirky is more difficult. Multicellularity, sexual reproduction, and encephalization are not common to all terrestrial life. But are they common enough (or adaptive enough) that we should expect extraterrestrials to be multicellular, sexual reproducers with heads? Quirkometry – distinguishing the quirky, uniquely terrestrial, from the terrestrially generic and possibly universal – is a young science.

Biologists study model organisms such as mice (*Mus musculus*), fruit flies (*Drosophila melanogaster*), nematode roundworms (*Caenorhabditis elegans*), yeast (*Saccharomyces cerevisiae*), slime mold (*Dictyostelium discoideum*), colon bacteria (*Escherichia coli*), and mustard plants (*Arabidopsis thaliana*) not because the features of these species are more generic or representative of other life forms, or because these species are more quirky and interesting than other species, but because they are easier to study. Among the life forms in the universe, terrestrial life can be viewed as a model organism, because it is the easiest for us to study. But how representative is terrestrial life of life elsewhere?

When a *Drosophila* expert learns all that is known about *Drosophila*, the next step – the most useful step as far as other biologists are concerned – is to figure out how much of what is known about *Drosophila* applies to other organisms. How generic and how quirky are the various *Drosophila* features? To make these distinctions, we simply compare *Drosophila* to many other species. But even in the absence of other species, one could make some good guesses. For example, basic *Drosophila* biochemistry is likely to be shared by other life forms, but the exact shape of the wings is not. Details are different from fundamentals.

*Drosophila* embryology can help us understand other species which had a common ancestor with *Drosophila* hundreds of millions of years ago. For example, segmentation occurs very early on in the embryogenesis of *Drosophila,* and because ontogeny tends to recapitulate phylogeny (Gould, 1977), we could make an educated guess that other species, with whom *Drosophila* shared a common ancestor a few hundred million years ago, would also be segmented. Fundamental features of an individual member can provide information about the group because the most fundamental features appear earliest in the embryonic development of an individual, and these features are often phylogenetically the most deeply rooted. Thus, they are likely to be shared by ancient ancestors and their other descendants. Similarly, the earliest "embryonic" ontogeny – the earliest steps of development of the single example of terrestrial life we know – may recapitulate the paths that all life in the universe has to follow during its earliest evolution.

Here is an example of our reasoning. Suppose we did not know that chimpanzees existed, but we wondered – or even suspected – that there were beings out there in the jungle with whom we shared a common ancestor about 6 million years ago. Without having detected chimpanzees, the best way to study these possibly existent beings would be to study the fossils of early hominids that lived about 6 million years ago. This is because these early ancestors are (and were) more closely related to chimps than we are today. In the absence of SETI detections or visits from ET, the earliest branchings of the terrestrial phylogenetic tree of life (Figs. 3 and 4) are the closest we can get to the earliest branchings of life elsewhere. Thus, the study of the earliest and most fundamental features of terrestrial life is probably one of the best ways to study possibly existent extraterrestrial life.

## 2. Evolution: From Deterministic to Quirky

Terrestrial life emerged from nonlife. Quirky biology emerged from deterministic physics and chemistry. If this assumption is correct, then the first steps of molecular evolution are deterministic or quasi-deterministic. De Duve (1995) has argued that this initial determinism makes life a "cosmic imperative" built into the chemistry of the universe.

The earliest life in warm little ponds or hydrothermal vents adapted to abiotic environmental challenges such as variations in temperature, pH, salinity, solar radiation, and humidity. These adaptations to environmental conditions do not have strong feedback. The temperature of the environment does not get higher or lower when tolerance to hotter and colder temperature evolves. The pH of the environment does not get higher or lower when an autocatalytic cycle or proto-organism learns how to handle free protons.

However, as proto-life evolved into life, an increasingly important part of an organism's environment became other organisms. Life forms became dependent on each other for nutrients and metabolism. Increasingly, life had to adapt to life. Thus, evolution became more self-referential and quirky, similar to the nonlinear screech of feedback when a microphone comes too close to a loudspeaker. The screech is an arbitrary random noise being amplified into a dominant signal. In a similarly positive feedback loop, DNA is pushed by selection pressure from other DNA to fitness peaks on an adaptive landscape created by the same DNA. Predator/prey arms races, sexual selection, and the selective pressure from conspecifics and other life forms provide the feedback that makes current evolution self-referential, divergent, and contingent. Biology is an historical science. The further life diverges, the more its direction is determined by the accumulated quirks of past history (Gould, 1989, 2002; McShea and Brandon, 2010).

Picture a tree with a trunk that splits into branches and branches that split into twigs. If one accepts the idea that life has evolved from a deterministic trunk into quirky twigs and if one accepts the idea that quirkiness increases with time, then the first divergences of life, the thick branches closest to the trunk, are more likely to be representative of life elsewhere than will be the quirky twigs. Our best guesses about the nature of extraterrestrials will come from a study of the features common to all life on Earth, for it is there that the deterministic initial stages can be read. When we look at the earliest divergences in the tree of life, we are looking at divergences that took place when life was closer to its deterministic roots. In other words, the divergent paths that the early evolution of life took are likely to be more relevant to early life elsewhere, than the quirky contingent paths that life took further downstream (Fig. 3).

The most fundamental features of mice are the features they share with all mammals. The most fundamental features of mammals are the features they share with all eukaryotes. The most fundamental features of eukaryotes are the features they share with all terrestrial life. And among the most fundamental features of terrestrial life forms are the features they share with all life in the universe. So what are the most fundamental features of terrestrial life?

## 3. The Most Fundamental Features of Terrestrial Life

The most fundamental features of terrestrial life have been discussed in some detail (e.g., Feinberg and Shapiro, 1980; Pace, 2001; Benner et al., 2004; De Duve, 2007). These features are often used to attempt to define life (e.g., Sagan, 1970; Joyce, 1994; Cleland and Chyba, 2002). Here, we briefly review some major fundamental features including liquid water as the solvent, carbon as the scaffold for biochemistry, the stoichiometry of the major bioelements, the LEGO principle, homochirality, free energy from thermodynamic disequilibria, and the Darwinian evolution of inheritable molecules.

### 3.1. LIQUID WATER AS SOLVENT

Life is made of molecules dissolved in a solvent. It seems reasonable that any life would need to be based on a liquid because of the way biomolecules, atoms, and ions need to move freely and interact, both in the liquid and on the liquid–solid boundary (Benner et al., 2004; Bains, 2004). In solids, molecules lack mobility. In gases, molecules lack structure.

Whether life elsewhere can be based on liquids besides water is an open question. Bains (2004) and Benner et al. (2004) have pointed out that if life can be based on other liquids, then that liquid determines what the biochemistry will be. The elements out of which water-based life is made, hydrogen, oxygen, carbon, nitrogen, phosphorus, and sulfur (HOCNPS), are the most common elements in the universe (Pace, 2001; Lodders et al., 2009). These abundances do not preclude life based on other liquids and elements, but they do suggest that if there is some life that is not based on liquid water, it would not be based on HOCNPS molecules and therefore would not be as abundant as water-based life. It would be limited by the availability of nutrients just as terrestrial life is limited by the low abundance and correspondingly low availability of phosphorus. Thus, most extraterrestrial life, like terrestrial life, should be based on water as a solvent. Hence, in its search for extraterrestrial life, NASA's "follows the water" (Hubbard et al., 2002) makes sense.

### 3.2. CARBON AS SCAFFOLD

Light elements are more abundant in the universe than heavier elements. This is at least a partial explanation for why life is made of lighter elements. For example, there are ~20 carbon atoms for every silicon atom in the universe. Thus, based on abundance alone, if silicon were just as likely as carbon to be the basis of life (Benner et al., 2004), there would be ~20 life forms based on carbon for each life form based on silicon.

### 3.3. THE STOICHIOMETRY OF THE MAJOR BIOELEMENTS: HOCNPS

Terrestrial life is made of HOCNPS at the 98% (wt%) level. The bulk elements hydrogen, oxygen, carbon, and nitrogen make up 97%. Phosphorus and sulfur make up another 1%. The remaining 2% is dominated by potassium, sodium, calcium, magnesium, and chlorine.

Trace elements like iron, copper, and cobalt make up ~0.03% (Chopra et al., 2010). Another way to look at the chemical composition of life is as follows. Life is made of about 70% water. If we remove that 70%, the rest is 65% carbon, 10% oxygen, 7% hydrogen, and 7% nitrogen. These stoichiometric percentages are fairly stable and common to all life on Earth, much as Redfield ratios C:N:P ≈ 106:16:1 (Redfield, 1934) are fairly stable among phytoplankton and more generally marine biomass.

Terrestrial life's Redfield ratios (based on bacteria and humans) are approximately 106:13:2 (slightly less N and twice as much P as Redfield's marine biomass) (Chopra et al., 2010). Our educated guess is that if life elsewhere is based on water, then it too will be made of the most abundant elements in the universe, HOCNPS and possibly with stoichiometric ratios that correlate strongly with the cosmic (i.e., solar) ratios H:O:C:N:P:S ≈ 430,000:230:106:31:0.1:6 (Lodders et al., 2009).

## 3.4. THE LEGO PRINCIPLE

Since the elemental ingredients for life are the most common elements in the universe, it is not surprising that the molecular ingredients of life are common. Terrestrial life is built of subunits called monomers such as amino acids, fatty acids, sugars, and nitrogenous bases. Amino acids link together to form proteins. Fatty acids link together to form lipids. Sugars link to form carbohydrates. And nitrogenous bases combine with sugar and phosphate to make nucleotide monomers, which link together to form RNA/DNA. Life links available monomers together to make polymers, much as we link these words together to make these sentences. McKay (2004) has dubbed this monomer-linkage-strategy the "LEGO Principle." Importantly, life does not make all possible polymers. From a much larger pool of possibilities, life links a very limited set of specific monomers together to produce a very specific set of polymers. For example, out of the ~100 amino acids found in the Murchison meteorites, only 8 are part of the 20 protein amino acids that life on Earth is made of (Schmitt-Kopplin et al., 2010).

McKay (2004) and Davies et al. (2009) pointed out that abiotic processes usually produce a broad spectrum of complex molecules, while life produces a specific set of molecular polymers that shows up like a fingerprint in a molecular mass spectrometer. Since the LEGO principle is common to all terrestrial life, it may be common to all life. This is one of the biosignatures that upcoming missions to Mars will be looking for.

## 3.5. HOMOCHIRALITY

Take a dozen children, each speaking a different language and put them in a room. After a while, they will start speaking the same language. Or take a country in which horse-drawn carts travel down the middle of the roads. Add more traffic and maybe a few cars. After a while, people will be driving on the right side or the left side of roads. The symmetry is broken by the necessity to get along – by the necessity to choose one side only. It does not matter which side, as long as there is general agreement. When such correlated behavior is adaptive, statistical fluctuations can get amplified into a consensus.

In terrestrial biology, the polypeptide backbones of proteins are made exclusively from homochiral (L) amino acids. Carbohydrates and nucleic acids are made with homochiral (D) sugars. Amino acids and sugars produced abiotically are usually racemic (see however Pizzarello, 2007). This distinction between racemic abiotically produced molecules and nonracemic biotically produced molecules makes homochirality a fundamental feature of terrestrial life and one of our expectations about life elsewhere. If aliens have roads and cars, they need agreement on which side to drive on, but there would be no way to guess whether it is on the right or left. Similarly, if extraterrestrial life is made of chiral molecules, an educated guess would be that these molecules are homochiral, but there is no way to guess whether their monomers will be L or D.

## 3.6. FREE ENERGY FROM THERMODYNAMIC DISEQUILIBRIA

The observation of planetary atmospheres out of chemical equilibrium has been proposed as a way to distinguish live planets from dead planets (Lovelock, 1975; Catling and Bergsman, 2010). Life needs to do something for a living, and this living depends on extracting free energy from an environment out of thermodynamic equilibrium. This extraction is based on absorbing photons and catalyzing redox reactions (Kleidon, 2010).

If we generalize the traditional definitions of life to include all far-from-equilibrium dissipative structures (Schneider and Sagan, 2005; Lineweaver, 2006; Lineweaver and Egan, 2008), then stars, volcanoes, hurricanes, convection cells, and fires would be considered forms of life. Stars are based on a nuclear disequilibrium, volcanoes and hurricanes are based on a pressure and thermal disequilibrium, convection cells are based on thermal disequilibrium, and fires are based on a chemical disequilibrium. Thus, some kind of disequilibrium is a common feature of all life, even with the most general definition of life.

## 3.7. DARWINIAN EVOLUTION OF INHERITABLE MOLECULES

Life has been defined as "a chemical system capable of Darwinian evolution" (Joyce, 1994). All terrestrial life forms (including viruses) have this in common. If we accept this as a definition, then, tautologically, extraterrestrial life will be capable of Darwinian evolution. However, to identify Darwinian evolution in extraterrestrial life, we need to identify the channel of inheritance.

None of the far-from-equilibrium dissipative structures listed above seem to have channels of inheritance, in which information is passed from one generation to the next. One could argue that hurricanes may be influenced by temperature, moisture, and pressure gradients set up in the tracks of previous hurricanes, but there is no identifiable unit of information that is built upon through generations of hurricanes. We expect extraterrestrial life to have some form of inheritance since, otherwise, we would not consider it alive.

In the proto-biological RNA world (Gilbert, 1986; Joyce, 2002), bits of RNA stuck to each other. Some combinations were stable and survived while others fell apart. Strands of RNA, ribozymes, catalyzed reactions and acted as a template for self-replication (Cech, 1985; Chen et al., 2007; Orgel, 2004). Inexact copying, point mutations and random conjugation were the source of variations, and these variations were inheritable. The correlated survival of some genes and not others, and their isolation in cells, was the beginning of Darwinian evolution and life on Earth (e.g., Martin and Russell, 2003). Thus, we can expect extraterrestrial life to have inheritable molecules.

## 4. Evolutionary Transitions: Early and Generic vs. Late and Quirky

The phylogenetic tree of life has many branches. Every extant species sits at the end of a branch – an evolutionary pathway – that one can follow back in time (Dawkins, 2004). If we follow the branches back in time, they converge two at a time into thicker branches which meet at the common trunk of the tree – the common origin of all extant life, known as the Last Universal Common Ancestor (LUCA). The fundamental features common to all life on Earth have their origins in the roots and trunk of the tree. Following the trunk and branches forward in time, many familiar species were identical (in the trunk) for the first few billion years, and then recently, they diverged into different branches. For example, animals and fungi diverged about a billion years ago after sharing the same evolutionary pathway for ~3 billion years. Vertebrates and invertebrates diverged about 500 million years ago after being identical for ~3.5 billion years. Because of this pattern of divergence, the most fundamental "major transitions" are the earliest transitions – the ones that occurred in the trunk and are common to all extant terrestrial life.

Some features that are not common to all terrestrial life are often interpreted as "major transitions" in evolution (Smith and Szathmáry, 1995) and are sometimes thought to be so universally adaptive that we should expect these transitions to have occurred in the evolution of extraterrestrial life. Eukaryogenesis, multicellularity, sexual reproduction, and encephalization are among these. These candidates for being features of extraterrestrial life are less secure because they are based on subjective notions of universal fitness, not on features common to all terrestrial life (Lineweaver, 2005, 2009).

"Major transitions" (particularly recent ones) are arbitrary because a series of major transitions can be identified along the evolutionary pathway to any extant species. Evolution is path-dependent. There is no linear progression.

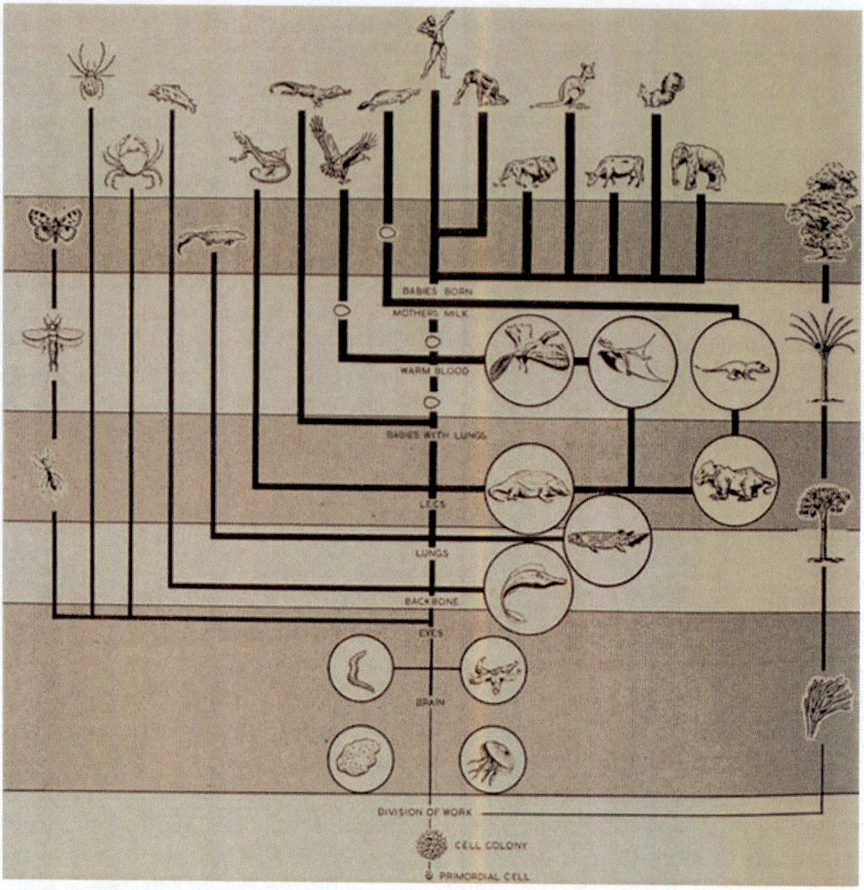


**Figure 1.** The Schwarzeneggerization of Life. Illustration from Gatland and Dempster (1957) in which some of the transitions that led to a male Caucasian *Homo sapiens* bodybuilder are portrayed as major transitions central to the trends of evolution. Plants are marginalized to the far *right*. Fungi and prokaryotes are *left out*. Jellyfish and starfish appear in the *central lower right* of the diagram but, for some reason, do not evolve to the present day. For a detailed discussion of what is wrong with such "Great Chain of Being" illustrations, see Gould (1989).

Figure 1 is a misleading diagram showing how seductive the linear-progression-Great-Chain-of-Being interpretation of evolution can be. However, Fig. 2, the front cover of a more academically respectable book, "The Major Transitions in Evolution" by Smith and Szathmáry (1995), shows much the same thing. On the right of Fig. 2, the arrows summarize the major transitions that Smith and Szathmáry (1995) have identified. Gould (1989) has written a concise criticism of what is wrong with such subjective interpretations of biological evolution.

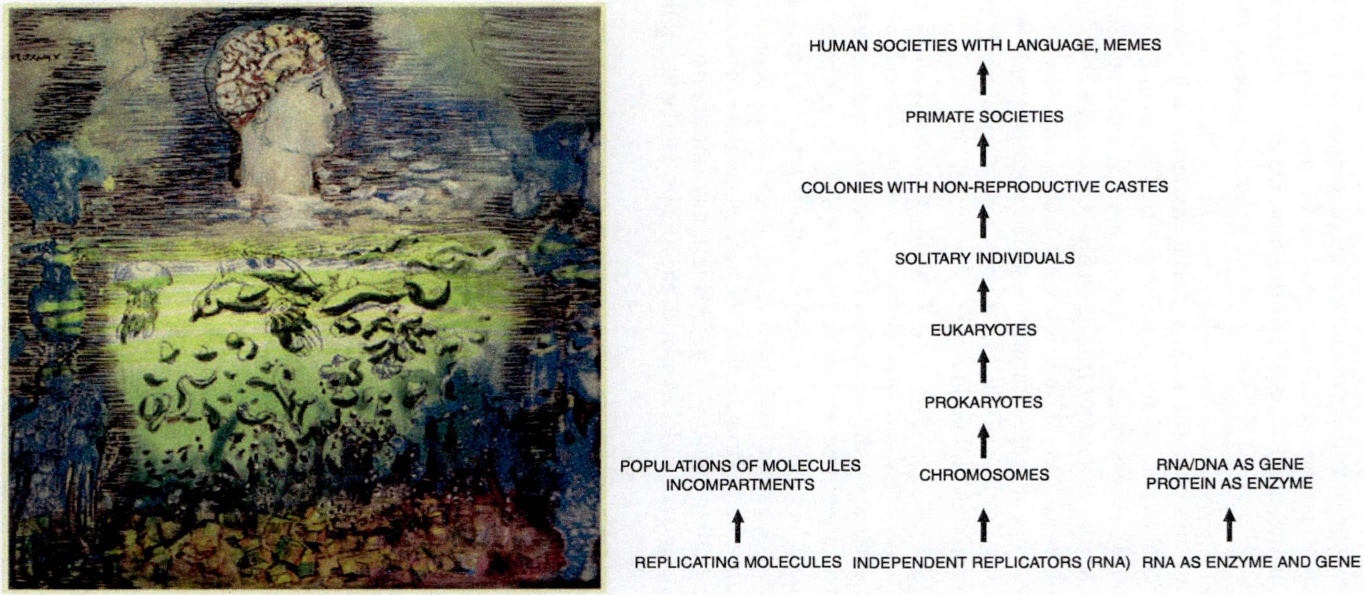


**Figure 2.** The Macedonification of Life. Illustration from the cover of Smith and Szathmáry (1995) "The Major Transitions in Evolution" in which some of the transitions that led to a skinless Macedonian brain are portrayed as major transitions central to the evolution of life. The major transitions of life described in the book are shown on the *right*. Unlike Fig. 1, where Schwarzenegger shares the spotlight with a dozen other creatures, the skinless Macedonian brain is alone at the top.

Social insects do not need big brains. Moles do not need eyes. Legs are not a major transition on your way to becoming a tuna or a sequoia tree. Vertebrae are not a major transition on the evolutionary path to worms. And multicellularity and sexual reproduction are not major transitions in the evolutionary path to the marine bacterium *Pelagibacter*, the most abundant organism on Earth (Morris et al., 2002).

The major transitions that led to any particular terrestrial species (e.g., *Homo sapiens*) have no objective claim to being major transitions in the evolution of extraterrestrial life. We may consider the transition from a primate society to our specific human societies as a major transition for us, but this was not a major transition in the evolutionary paths that let to macaques, gibbons, or chimps. Smith and Szathmary's list of major transitions would be appropriate for a bird except that the transition from "walking" to "flying" would replace the "primate societies" to "human societies" transition.

If we can make guesses about how terrestrial life transitioned, then we have some idea of what it transitioned from earlier on in its history. The earliest transitions of terrestrial life inform our best guesses at the nature of extraterrestrial life. On this reasoning, we suggest that extraterrestrial life had an emergence similar to our emergence in an RNA world – a viral world of replicating inheritable molecules, molecules like RNA ribozymes, both enzymes and genes, functioning in metabolism and passing on information to the next generation. These are the features listed at the bottom right of Fig. 2, before any transitions take place. With regard to an extraterrestrial RNA world, the "RNA" can be thought of as any inheritable molecule.

## 5. Hyperthermophilia and the Deepest Branches of Terrestrial Life

The deepest roots of the 16S rRNA phylogenetic tree are hyperthermophilic (Fig. 3, see also Lineweaver and Schwartzman, 2005; Wong et al., 2007). Extant organisms with the shortest branches are hyperthermophiles able to tolerate temperatures above 90°C. These organisms – Aquifex, Thermotoga, Nanoarchaeota, and Korarchaeota – seem to be the best representatives of the Last Universal Common Ancestor of all terrestrial life. Hyperthermophilic organisms closest to the root suggest that LUCA was hyperthermophilic and by extension that the origin of life on Earth was hyperthermophilic.

Although hyperthermophilia is not now a feature common to all terrestrial life, it may have been ~4 billion years ago. Thus, we might expect extraterrestrial life to have originated as a hyperthermophilic RNA world. What it evolved into from there is much less certain and quirky. However, life everywhere may still be embedded in its origin, much as we are still embedded in our viral (RNA) world.

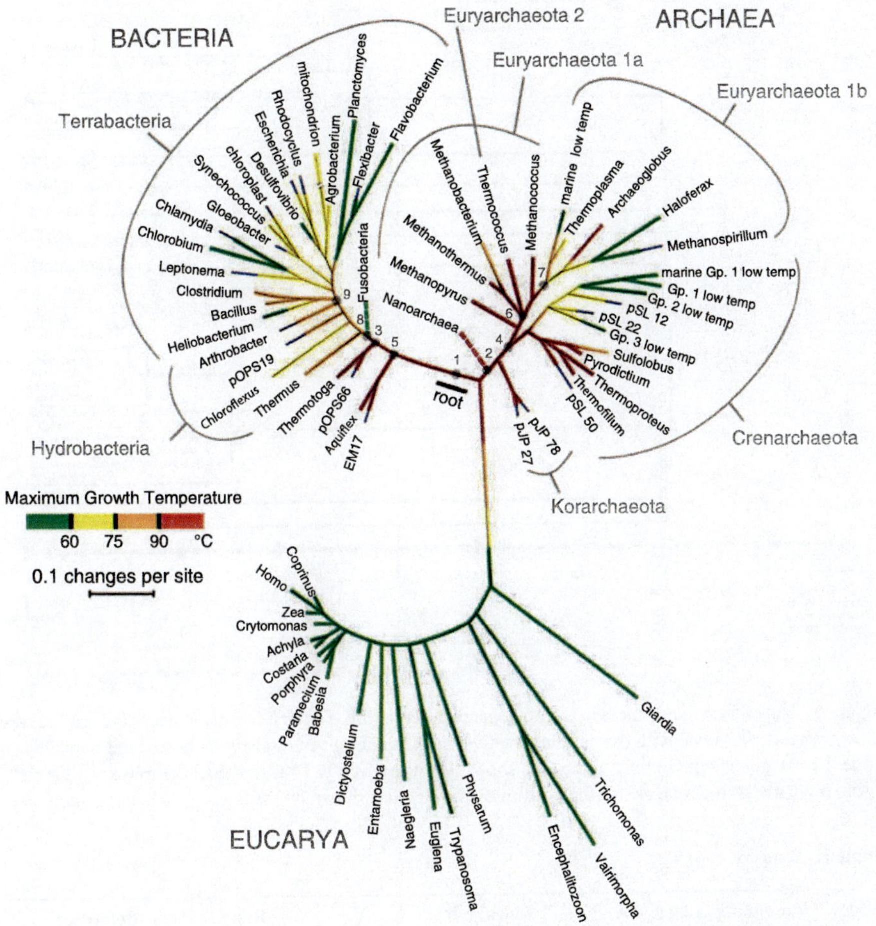


**Figure 3.** The 16S rRNA tree of life suggests that life started out hyperthermophilic (*red*) and then learned how to live at lower temperatures too. The color coding refers to the maximum temperature able to be tolerated by a given clade. For example, eukaryotes all require temperatures less than 60°C. Plants, animals, and fungi are represented in the *lower left* by *Zea* (Corn), *Homo* (humans), and *Coprinus* (mushrooms), respectively. The position of the root is from ancient gene duplication studies (Iwabe et al., 1989; Brown and Doolittle, 1995; Wong et al., 2007; Gaucher et al., 2010). Numbered nodes are described in Fig. 4 and Table 1. Figure modified from Lineweaver and Schwartzman (2005) modified from Pace (1997).

## 6. Terrestrial Ubiquity as a Frozen Accident Uninformative of ET

We have argued that universal features of life are our best guess for the nature of extraterrestrial life. However, just because a feature is universal in terrestrial life does not necessarily mean that it is a feature of extraterrestrial life. Ubiquity

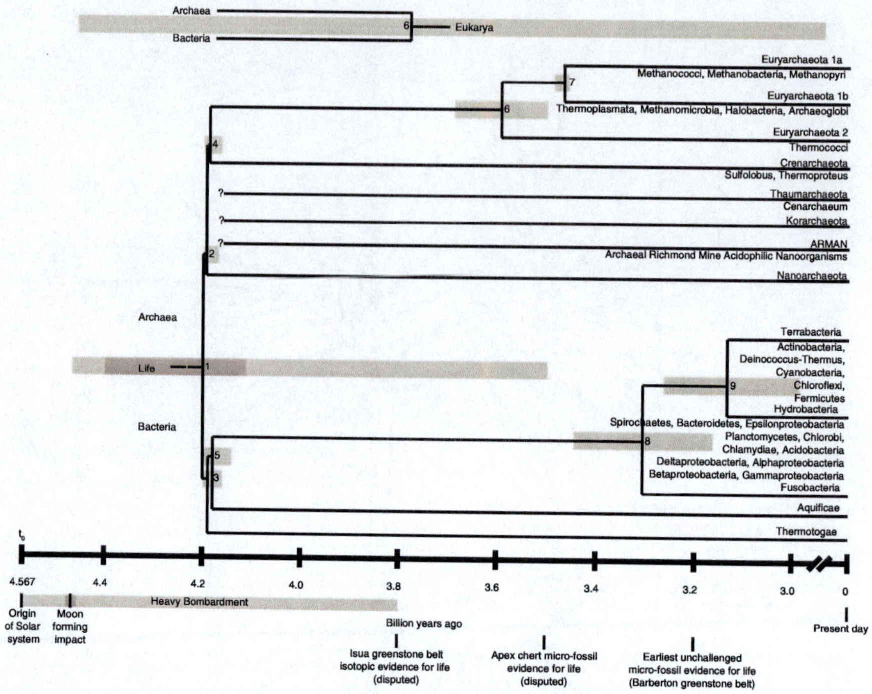


**Figure 4.** The earliest divergences in the phylogenetic tree of life. The first five divergences (nodes 1–5) are shown here at ~4.2 Gya, with overlapping relative uncertainties given by the gray horizontal bars at each node. The larger absolute uncertainties are shown for node 1 only. That is, the whole tree can shift right and left relative to the time axis at the bottom by the amount shown by the horizontal bar at node 1.

**Table 1.** Data for Fig. 4.

| Node | Time (*Mya*) | Lineage A | Lineage B | Range (*Mya*) | Reference |
|---|---|---|---|---|---|
| 1 | 4,200 | Bacteria | Archaea | 4,400–4,112[a] | Hedges (2009) |
| Archaea | | | | | |
| 2 | 4,193 | Nanoarchaeota | Crenarchaeota and Euryarchaeota | 4,200–4,176 | Battistuzzi and Hedges (2009a) |
| 4 | 4,187 | Crenarchaeota | Euryarchaeota | 4,199–4,163 | |
| 6 | 3,594 | Euryarchaeota2 | Euryarchaeota1 | 3,691–3,503 | |
| 7 | 3,468 | Euryarchaeota1a | Euryarchaeota1b | 3,490–3,460 | |
| Bacteria | | | | | |
| 3 | 4,189 | Thermotogae | Aquificae and Fuso. and Hydro. and Terra. | 4,200–4,159 | Battistuzzi and Hedges (2009b) |
| 5 | 4,179 | Aquificae | Fuso. and Hydro. and Terra. | 4,197–4,141 | |
| 8 | 3,306 | Fusobacteria | Hydro. and Terrabacteria | 3,447–3,165 | |
| 9 | 3,134 | Hydrobacteria | Terrabacteria | 3,265–2,987 | |

[a]This range is the dark gray, relative uncertainty (Hedges, 2009). The larger, light gray absolute uncertainty is based on the Moon-forming impact at ~4,470 Mya (Halliday, 2008; Sleep et al., 1989) and the earliest fossil evidence for life ~3,500 Mya (Hedges, 2009). Origin of solar system: 4,567 ± 2 Mya (Amelin and Krot, 2007).

among extant terrestrial life forms does not necessarily imply ubiquity among all terrestrial life that has ever lived or among all extraterrestrial life. The tree of life has been pruned by both selection pressure and random events.

Most species that have ever existed are extinct. It is possible that a large percentage of phyla, kingdoms, and even domains of life that have ever existed are extinct (e.g., Davies and Lineweaver, 2005). Thus, whatever is common to all life today could be the result of a severe and possibly arbitrary pruning of the tree of life – a pruning based on fitness but also on luck.

The case of DNA is illustrative. DNA is common to all terrestrial life forms. But the code for translating base pairs into amino acids is largely arbitrary (e.g., Tlusty, 2010; Freeland et al., 2000). That arbitrariness is what makes it an abstract symbolic *code* rather than a model or a pictograph system.

Crick (1968) suggested that the genetic code was universal in all organisms, and the result of a "frozen accident," unable to evolve further even if the current state were suboptimal. Frozen accidents, even when common to all extant and extinct life, would not be expected to be common to all extraterrestrial life.

## 7. Summary

Terrestrial life can be seen as a model organism representing extraterrestrial life. We have briefly reviewed some of the fundamental features common to all terrestrial life including: liquid water as the solvent, carbon as the scaffold for biochemistry, the stoichiometry of the major bioelements, the LEGO principle, homochirality, free energy from thermodynamic disequilibria, and the Darwinian evolution of inheritable molecules. We have argued that our best guesses for the features of extraterrestrial life are a subset of the features on this list. Other frequently espoused candidate features (e.g., multicellularity, sexual reproduction, heads) are less secure because they are based on subjective notions of universal fitness.

By identifying the more deterministic processes that led to the origin and evolution of life on Earth and by tracking the earliest divergences, we are identifying the possibly universal processes which may have led to life elsewhere. One prediction of this reasoning is that the life forms on other worlds will have hyperthermophilic roots and will still be embedded in their earliest ancestors, much as we are still embedded in our viral (RNA) world.

## 8. References

Amelin Y, Krot A (2007) Pb isotopic age of the Allende chondrules. Meteorit Planet Sci 42:1321–1335

Bains W (2004) Many chemistries could be used to build living systems. Astrobiology 4:137–167

Battistuzzi FU, Hedges SB (2009a) Archaebacteria. In: Hedges SB, Kumar S (eds) The timetree of life. Oxford University Press, Oxford/New York, pp 101–105

Battistuzzi FU, Hedges SB (2009b) Eubacteria. In: Hedges SB, Kumar S (eds) The timetree of life. Oxford University Press, Oxford/New York, pp 106–115

Benner SA, Ricardo A, Carrigan MA (2004) Is there a common chemical model for life in the universe? Curr Opin Chem Biol 8:672–689

Brown JR, Doolittle WF (1995) Root of the universal tree of life base on ancient aminoacyl-tRNA synthetase gene duplications. Proc Natl Acad Sci USA 92:2441–2445
Catling DC, Bergsman DS (2010) On detecting exoplanet biospheres from atmospheric chemical disequilibrium. Astrobiology science conference 2010 abstract #5533
Cech TR (1985) Self-splicing RNA: implications for evolution. In: Bourne GH, Danielli JF, Jeon KW (eds) Int Rev Cytol Acad Press 93:3–22
Chen X, Li N, Ellington A (2007) Ribozyme catalysis of metabolism in the RNA world. Chem Biodivers 4:633–655
Chopra A, Lineweaver CH, Brocks JJ, Ireland TR (2010) Palaeoecophylostoichiometrics: searching for the elemental composition of the last universal common ancestor. In: Short W,. Cairns I (eds) Australian space science conference series: 9th conference proceedings. NSSA full refereed proceedings CD, National Space Society of Australia Ltd, Sydney
Cleland C, Chyba C (2002) Defining life. Orig Life Evol Biosph 32:387–393
Crick FHC (1968) The origin of the genetic code. J Mol Biol 38:367–379
Davies PCW, Lineweaver CH (2005) Finding a second sample of life on earth. Astrobiology 5:154–163
Davies PCW, Benner SA, Cleland CE, Lineweaver CH, McKay CP, Wolfe-Simon F (2009) Signatures of a shadow biosphere. Astrobiology 9:241–249
Dawkins R (2004) The ancestor's tale: a pilgrimage to the dawn of life. Weidenfeld and Nicholson, London
De Duve C (1995) Vital dust: the origin and evolution of life on earth. Basic Books, New York
De Duve C (2007) Chemistry and selection. Chem Divers 4:574–583
Feinberg G, Shapiro R (1980) Life beyond earth: the intelligent earthling's guide to life in the universe. William Morrow, New York
Freeland SJ, Knight RD, Landweber LF, Hurst LD (2000) Early fixation of an optimal genetic code. Mol Biol Evol 17:511–518
Gatland KW, Dempster DD (1957) The inhabited universe: an enquiry staged on the frontiers of knowledge. McKay, New York
Gaucher EA, Kratzer JT, Randall RN (2010) Deep phylogeny – how a tree can help characterize early life on earth. Cold Spring Harb Perspect Biol 2:a002238
Gilbert W (1986) Origin of life: the RNA world. Nature 319:618
Gould SJ (1977) Ontogeny and phylogeny. Harvard University Press, Cambridge
Gould SJ (1989) Implications of an iconography. In: Wonderful life: the burgess shale and the nature of history. Norton & Company, New York
Gould SJ (2002) The structure of evolutionary theory. Harvard University Press, Cambridge
Halliday AN (2008) A young moon-forming giant impact at 70–110 million years accompanied by late-stage mixing, core formation and degassing of the earth. Philos Trans R Soc A 366:4163–4181
Hedges SB (2009) Life. In: Hedges SB, Kumar S (eds) The timetree of life. Oxford University Press, Oxford/New York, pp 89–98
Hubbard GS, Naderi FM, Garvin JB (2002) Following the water, the new program for mars exploration. Acta Astron 51:337–350
Ida S, Lin DNC (2004) Toward a deterministic model of planetary formation. I. A desert in the mass and semimajor axis distributions of extrasolar planets. Astrophys J 604:388–413
Iwabe N, Kuma K-I, Hasegawa M, Osawa S, Miyata T (1989) Evolutionary relationship of archaebacteria, eubacteria, and eukaryotes inferred from phylogenetic trees of duplicated genes. Proc Natl Acad Sci USA 86:9355–9359
Joyce GF (2002) The antiquity of RNA based evolution. Nat Insight 418:214–221
Joyce GF (1994) In: Deamer DW, Fleischacker GR (eds) Origins of life: the central concepts. Jones and Bartlett Publishers, Boston, pp xi–xii
Kleidon A (2010) Life, hierarchy, and the thermodynamic machinery of planet earth. Phys Life Rev 7(4):424–460
Kuchner MJ (2003) Volatile-rich earth-mass planets in the habitable zone. Astrophys J 596: L105–L108
Léger A, Selsis F, Sotin C et al (2004) A new family of planets? "Ocean-planets". Icarus 169:499–504

Lineweaver CH (2005) Intelligent life in the universe book review of "Intelligent life in the universe: from common origins to the future of humanity" by Peter Ulmschneider, review published in Astrobiology 5:658–661

Lineweaver CH (2006) We have not detected extraterrestrial life, or have we? In: Seckbach J, Walsh M (eds) Life as we know it: cellular origins and life in extreme habitats and astrobiology. Springer Life Sciences, Dordrecht, p 445

Lineweaver CH (2009) Paleontological tests: human-like intelligence is not a convergent feature of evolution. In: Seckbach J, Walsh M (eds) From fossils to astrobiology, cellular origins and life in extreme habitats and astrobiology, vol 12. Springer, Dordrecht, pp 353–368

Lineweaver CH, Egan C (2008) Life, gravity and the second Law of thermodynamics. Phys Life Rev 5:225–242

Lineweaver CH, Grether D (2003) What fraction of sun-like stars have planets? Astrophys J 598:1350–1360

Lineweaver CH, Schwartzman (2005) Cosmic thermobiology: thermal constraints on the origin and evolution of life in the universe. In: Seckbach J (ed) Origins: cellular origins and life in extreme habitats and astrobiology, vol 6. Springer, Dordrecht, pp 233–248

Lodders K, Palme H, Gail H-P (2009) Abundances of the elements in the solar system. In: JE Trumper (ed) Landolt-Bornstein, new series, astronomy and astrophysics. vol VI/4B, Chapter 4.4, Springer, Berlin, pp 560–630

Lovelock JE (1975) Thermodynamics and the recognition of alien biospheres. Proc R Soc Lond B 189:167–181

Martin W, Russell MJ (2003) On the origins of cells: a hypothesis for the evolutionary transitions from abiotic geochemistry to chemoautotrophic prokaryotes, and from prokaryotes to nucleated cells. Philos Trans R Soc Lond B Biol Sci 358:59–83

McKay CP (2004) What is life – and how do we search for it on other worlds? PLoS Biol 2:1260–1263

McShea DW, Brandon RN (2010) Biology's first law: the tendency for diversity and complexity to increase in evolutionary systems. University of Chicago Press, Chicago

Mordasini C, Alibert Y, Benz W, Naef D (2009) Extrasolar planet population synthesis. II. Statistical comparison with observations. Astron Astrophys 501:1161–1184

Morris RM, Rappe MS, Connon SA, Vergin KL, Siebold WA, Carlson CA, Giovannoni SJ (2002) SAR11 clade dominates ocean surface bacterioplankton communities. Nature 420:806–810

Orgel LE (2004) Prebiotic chemistry and the origin of the RNA world. Crit Rev Biochem Mol Biol 39:99–123

Pace N (1997) A molecular view of microbial diversity and the biosphere. Science 276:734–740

Pace N (2001) The universal nature of biochemistry. Proc Natl Acad Sci USA 98:805–808

Pizzarello S (2007) The chemistry that preceded life's origin: a study guide from meteorites. Chem Biodivers 4:680–693

Redfield AC (1934) On the proportions of organic derivations in sea water and their relation to the composition of plankton. In: Daniel James RJ (ed) Johnstone memorial volume. University Press of Liverpool, Liverpool, pp 177–192

Robles JA, Lineweaver CH, Grether D et al (2008) A comprehensive comparison of the sun to other stars: searching for self-selection effects. Astrophys J 684:691–706

Sagan C (1970) "Life" in the encyclopedia britannica. 14th edition

Schmitt-Kopplin P, Gabelica Z, Gougeon RD et al (2010) High molecular diversity of extraterrestrial organic matter in Murchison meteorite revealed 40 years after its fall. Proc Natl Acad Sci USA 107:2763–2768

Schneider ED, Sagan D (2005) Into the cool: energy flow, thermodynamics, and life. The University of Chicago Press, Chicago/London

Sleep NH, Zanhnle KJ, Kasting JF, Morowitz HJ (1989) Annihilation of ecosystems by large asteroid impacts on the early earth. Nature 342:139–142

Smith JM, Szathmáry E (1995) The major transitions in evolution. Oxford University Press, Oxford

Tlusty T (2010) A colorful origin for the genetic code: information theory, statistical mechanics and the emergence of molecular codes. Phys Life Rev 7:362–376

Wong JT-F, Chen J, Mat W-K, Ng S-K, Xue H (2007) Polyphasic evidence delineating the root of life and roots of biological domains. Gene 403:39–52

Biodata of **François Raulin**, author of "***Potential for Life in the Saturn System.***"

**François Raulin** is Full Professor at UPEC (University Paris Est Créteil) and develops his researches at LISA, Laboratoire Interuniversitaire des Systèmes Atmosphériques, in the GPCOS Group. F. Raulin was Director of LISA (1995–2005) and Director of the federation of CNRS laboratories "Groupement de Recherche" in Exobiology (1999–2006).

His scientific fields of interest are related to planetology and exo/astrobiology: studies of organic chemistry in extraterrestrial environments (Titan, giant planets, comets, and Mars) using complementary approaches: laboratory experiments (experimental simulations, IR et UV spectral data, analytical techniques by GC, GC-MS, and Pyr-GC-MS); theoretical modeling (using microphysics, kinetics, et thermodynamics); and observational data: remote sensing and in situ space exploration. F. Raulin is IDS (InterDisciplinary Scientist) of the Cassini-Huygens mission and Co-Investigator of the CIRS (Cassini), ACP, and GC-MS (Huygens) experiments. He is Co.I of the COSAC and COSIMA experiments of the Rosetta European cometary mission and deputy Team leader of the MOMA experiment on ExoMars. He was Chair of COSPAR Commission F (Life Sciences), Vice Chair of COSPAR Planetary Protection Panel, 1st Vice President of ISSOL and Chair of the exo/astrobiology working group of CNES. He is currently Chair of the Planetary Protection Working Group of ESA and President of SFE (Société Française d'Exobiologie).

He is the author of 300 scientific papers and nine books related to the field of the origins of Life and Exobiology.

E-mail: **Francois.Raulin@lisa.u-pec.fr**

J. Seckbach (ed.), *Genesis - In The Beginning: Precursors of Life, Chemical Models and Early Biological Evolution*, Cellular Origin, Life in Extreme Habitats and Astrobiology 22, 817–833
DOI 10.1007/978-94-007-2941-4_41, © Springer Science+Business Media Dordrecht 2012

# POTENTIAL FOR LIFE IN THE SATURN SYSTEM

**FRANÇOIS RAULIN**
*LISA-IPSL, CNRS/UPEC & Univ. Paris Diderot,*
*61 Avenue Général de Gaulle, F-94000 Créteil, France*

## 1. Introduction

The potential for life on a planetary body first requires the presence of an efficient prebiotic chemistry able to transform carbonaceous materials into replicating macromolecules encapsulated in a cell-like structure. Another possibility is the interplanetary transfer of living systems to this planetary body. Then, the planetary environment must be or must become suitable for life, that is to say habitable (Lammer et al., 2009). Habitable zone used to be defined as a zone in a planetary system where the physical conditions are compatible with the presence of liquid water on a planetary object. From such a restricted definition, the habitable zone in the Solar system would be limited to a region comprised roughly between the orbits of Venus and Mars.

Now, there are several planetary objects in the solar system which are not in this zone, and have liquid water, not on their surface but in their internal structure. This is the case with three of the four Galilean satellites of Jupiter: Ganymede, Callisto, and specially Europa, the internal ocean of which may be close to the surface and its bottom in contact with the bedrock. This is also the case in the Saturn system, at least with Titan, its largest satellite, and Enceladus. Organic chemical processes are present in the environment of these two Kronian satellites, as recently clearly shown by the Cassini-Huygens mission. The dense atmosphere of Titan is a gigantic organic reactor, which produces volatile and complex organic compounds, accumulating on Titan's surface. Moreover, Titan presents many analogies with the Earth, in particular, the early Earth, another important astrobiological aspect of this planetary body. Several organic compounds have been detected in the plumes of Enceladus which indicates that an active organic chemistry is also going on in the internal structure of this satellite. And the Cassini observations of several other icy bodies of the Saturn system suggest the presence of organics on their surface.

This chapter describes the several astrobiological aspects of the Kronian system, especially Titan and Enceladus, on the basis of the recent data from the Cassini-Huygens mission, associated to laboratory experiments and models.

## 2. The Saturn System and the Cassini-Huygens Mission

### 2.1. THE SATURN SYSTEM

Saturn (Fig. 1), king of the rings, with its magnificent planetary ring system, is the second giant planet of the solar system by its size and distance to the sun, as well as by its number of cataloged satellites – more than 50.

Located at about 10 AU from the sun, it receives a solar flux of about 1% that received by the Earth, high enough however to initiate a complex photochemistry in its atmosphere and that of its largest satellite, Titan. With an obliquity of 27°, Saturn has seasons which last for more than 7 terrestrial years, and so have its satellites. Two of those are of prime interest for exobiology and even for habitability: Titan and Enceladus. But several others, such as Dione, Hyperion, Iapetus, and Phoebe, have also interesting properties, due to their surface composition. Many of the data of the Saturn system were obtained from the Voyager mission. However most of the new data came and are currently coming from the Cassini-Huygens mission.

**Figure 1.** Photomontage of Saturn and several of its satellites. Dione is in the forefront, Enceladus on the *left*, close to the rings, and Titan on the *right upper side* of the picture (Credit: NASA/JPL (ref PIA01482)).

## 2.2. THE CASSINI-HUYGENS MISSION

After the flyby of Saturn and several of its satellites, especially Titan, by the Voyager mission in the early 1980s and the discovery of many important properties of Titan, it became of prime importance to go back to the Saturn system. This became a reality with the Cassini-Huygens mission (JPL/NASA, 2010; ESA, 2010; Russell, 2003, 2005; Lebreton and Matson, 2002; Brown et al., 2009; Lebreton et al., 2009).

The Cassini-Huygens mission, developed in collaboration between NASA and ESA, includes a Saturn orbiter (Cassini) and a Titan atmospheric probe (Huygens). It was launched in October 1997 and reached the Saturn system in 2004. The Cassini spacecraft (Fig. 2) became artificial satellite of Saturn on July 1, 2004. Cassini carries 12 instruments (Table 1), which have provided a tremendous amount of data on the Kronian system. It was also carrying the Huygens probe (Lebreton and Matson, 2002), which was released on December 25, 2004, and entered Titan's atmosphere on January 14, 2005. Huygens carried six scientific instruments (Table 1), which explored and analyzed Titan's environment during 2 h and 20 min of descent of the probe in its atmosphere, and provided also more than 1 h of data on Titan's surface after landing.

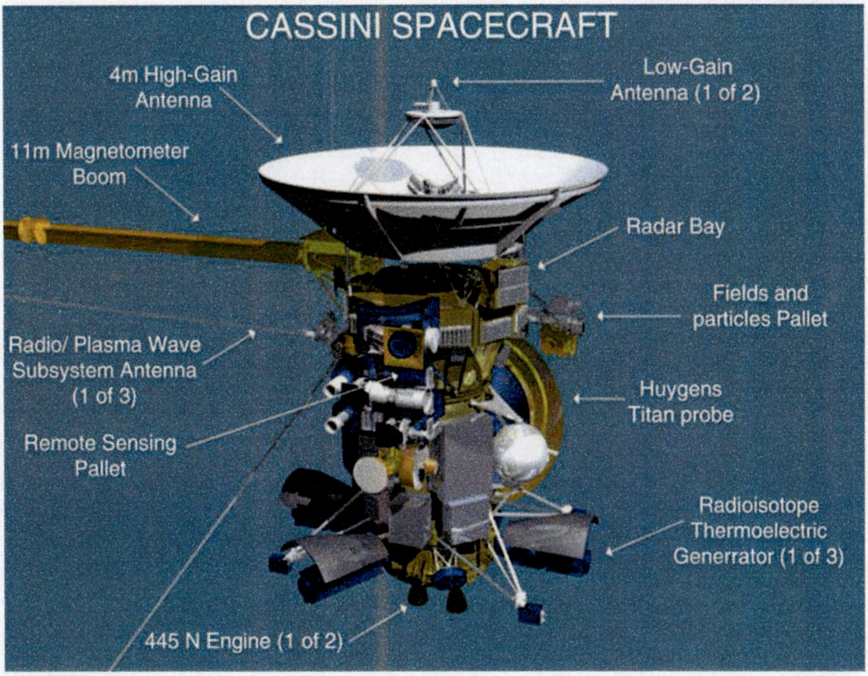


**Figure 2.** The Cassini spacecraft and some of its instruments (Credit NASA/JPL/Caltech (IMG001943)).

**Table 1.** Cassini-Huygens: scientific instruments, their associated principal investigators (PI) or team leaders, and interdisciplinary scientists (IDS).

| Cassini-Huygens instruments and interdisciplinary programs | P.I., team leader or IDS | |
|---|---|---|
| **Cassini probe** | | |
| *Optical remote sensing instruments* | | |
| Composite Infrared Spectrometer (CIRS) | V. Kunde/M. Flasar, | USA |
| Imaging Science Subsystem (ISS) | C. Porco, | USA |
| Ultraviolet Imaging Spectrograph (UVIS) | L. Esposito | USA |
| Visual & I.R. Mapping Spectrometer (VIMS) | R. Brown | USA |
| *Fields particles and waves instruments* | | |
| Cassini Plasma Spectrometer (CAPS) | D. Young | USA |
| Cosmic Dust Analysis (CDA) | E. Grün/R. Srama, | Germany |
| Ion & Neutral Mass Spectrometer (INMS) | H. Waite | USA |
| Magnetometer (MAG) | D. Southwood / M. Dougherty | U.K. |
| Magnetospheric Imaging Instrument (MIMI) | S. Krimigis | USA |
| Radio & Plasma Wave Spectrometer (RPWS) | D. Gurnett | USA |
| *Microwave remote sensing* | | |
| Cassini Radar (Radar) | C. Elachi | USA |
| Radio Science Subsystem (RSS) | A. Kliore | USA |
| *Interdisciplinary scientists* | | |
| Magnetosphere and plasma | M. Blanc | France |
| Rings and dust | J.N. Cuzzi | USA |
| Magnetosphere and plasma | T.I. Gombosi | USA |
| Atmospheres | T. Owen | USA |
| Satellites and asteroids | L.A. Soderblom | USA |
| Aeronomy & solar wind interaction | D.F. Strobel | USA |
| **Huygens probe** | | |
| *Scientific instruments* | | |
| Gas Chromatograph-Mass Spectrometer (GC-MS) | H. Niemann | USA |
| Aerosol Collector & Pyrolyser (ACP) | G. Israël | France |
| Huygens Atmospheric Structure Instrument (HASI) | M. Fulchignoni | Italy |
| Descent Imager & Spectral Radiometer (DISR) | M. Tomasko | USA |
| Doppler Wind Experiment (DWE) | M. Bird | Germany |
| Surface Science Package (SSP) | J. Zarnecki | U.K. |
| *Interdisciplinary scientists* | | |
| Aeronomy | D. Gautier | France |
| Atmosphere/surface interactions | J.I. Lunine | USA |
| Chemistry and exobiology | F. Raulin | France |

After 4 years of nominal mission, the Cassini-Huygens mission was extended for 2 years, until July 2010, and became the Cassini Equinox mission. The later was again extended for 7 more years, with the Solstice mission. The end of the mission is planned in 2017, with the entry of the Cassini orbiter into Saturn atmosphere.

## 3. Titan

Discovered by Huygens in 1655, Titan is the largest satellite of Saturn and the second of the solar system, after Ganymede, by the radius of its solid body (2,575 km). One of its major particularities is the presence of a dense atmosphere: it is the only satellite of the solar system in that case.

The Voyager mission has shown that this atmosphere is mainly made of dinitrogen, with noticeable fraction of methane, inducing a complex atmospheric chemistry dominated by organic compounds. The atmosphere also includes hazes which mask the surface of the satellite in the visible range (Fig. 3a). However, the atmosphere becomes almost transparent in spectral windows in the near IR and near UV, allowing the observation of Titan's surface (as seen on Fig. 3b, resulting from images of Titan taken by Cassini ISS in the near UV).

The many data obtained from Cassini observations (Brown et al., 2009) clearly show many important aspects of astrobiological interest for Titan. Those include several analogies with the Earth, and the primitive Earth in particular, the presence of a complex organic chemistry in all the different parts of what could

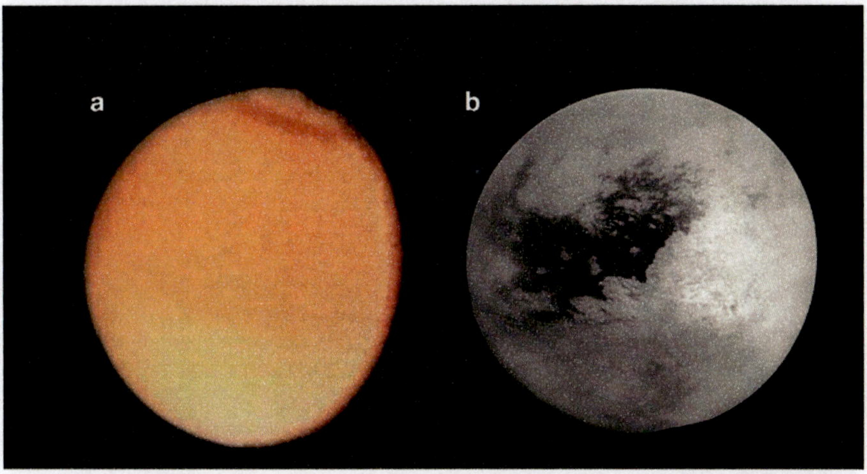


**Figure 3.** (**a**) Titan's picture taken by the Voyager 2 camera, from a distance of approximately 2.3 million kilometers, on August 1981. Several well-defined bands can be seen in the haze and clouds which mask the surface (Credit NASA/JPL-Caltech (PIA01532)). (**b**) Mosaic image of Titan from 16 individual images taken with Cassini narrow-angle camera at distances ranging from approximately 226,000–242,000 km in October 2004.The large bright white area is named Xanadu (Credit: NASA/JPL/Space Science Institute (PIA06185)).

be called "Titan's geofluid" (atmosphere, solid aerosols, and solid and liquid surfaces), and the potential for life, with the presence of habitable conditions, even for life as we know it.

## 3.1. ANALOGIES WITH THE EARTH

When looking at Titan's properties, one can observe many analogies with the Earth. First of all, both have a dense atmosphere, dominated by dinitrogen. From all planetary objects with an atmosphere, Titan is even the closest to the Earth conditions, not only because of its atmospheric $N_2$, but also because of its surface pressure: 1.5 bar for Titan (Fig. 4), close to the 1 terrestrial bar. Both atmospheres have many other similarities. Their vertical profiles look qualitatively the same, especially in the low atmosphere, with a troposphere and a stratosphere (Fig. 4).

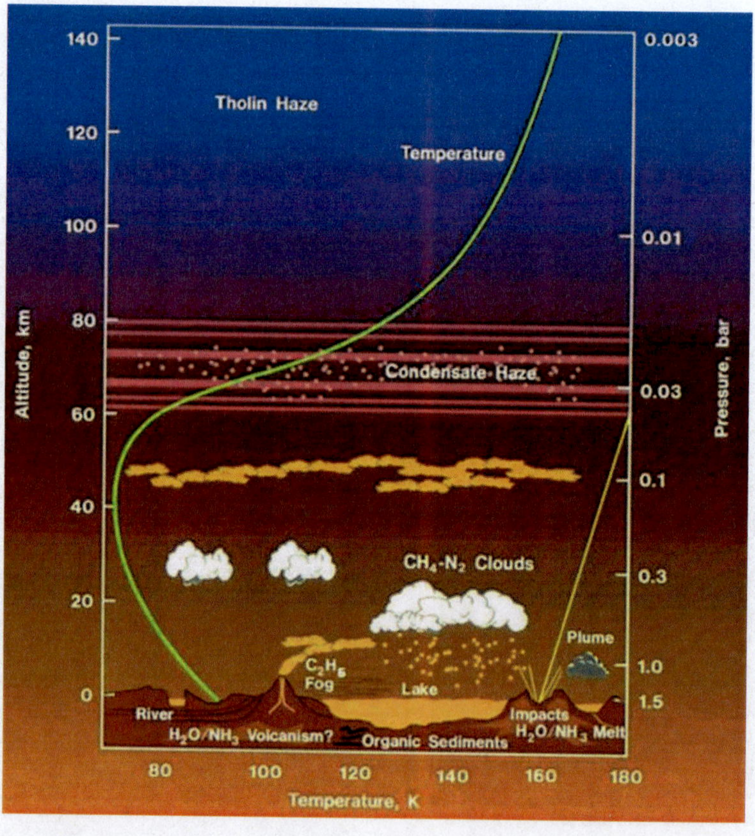


**Figure 4.** Atmospheric profile of the low atmosphere of Titan and potential surface features (Credit: ESA (ID number: SEM49YXO4HD)).

Both include greenhouse gases: condensable $CH_4$ and noncondensable $H_2$ for Titan, analog of condensable $H_2O$ and noncondensable $CO_2$ for the Earth. Both include antigreenhouse elements: aerosols and clouds. Both have a polar vortex playing a role in the atmospheric chemistry (south pole atmosphere chemistry for Titan, ozone chemistry for the Earth). Both atmospheres include $^{40}Ar$ as major noble gas (although in smaller abundance in Titan's atmosphere: about 30 ppm as detected by Huygens GCMS and Cassini INMS). This indicates that these atmospheres are of secondary origin, $^{40}Ar$ being produced by the radioactive decay of $^{40}K$ and degassed from the interior of the planetary body to the atmosphere. The very low concentration of primordial gases (the abundance of $^{36}Ar$ is about 1% of that of $^{40}Ar$) also indicates that the current atmosphere is a secondary one, and that nitrogen was initially on the form of ammonia. Ammonia was then photolyzed and transformed into $N_2$.

On Titan, methane seems to play the role on water on Earth, forming clouds, precipitation, liquid surfaces. There is a methane cycle on Titan which has many analogies with the water cycle on Earth (Atreya et al., 2006). There are indeed clouds in Titan's atmosphere. Cassini-Huygens observation strongly suggests the occurrence of rain and shows many traces of fluvial erosion on Titan's surface such as the dendritic structures seen by Huygens before it lands and the pebbles observed near its landing site (Fig. 5). But these pebbles are likely to be made of dirty water ice instead of rocks: on Titan, water replaces silicates on Earth. There

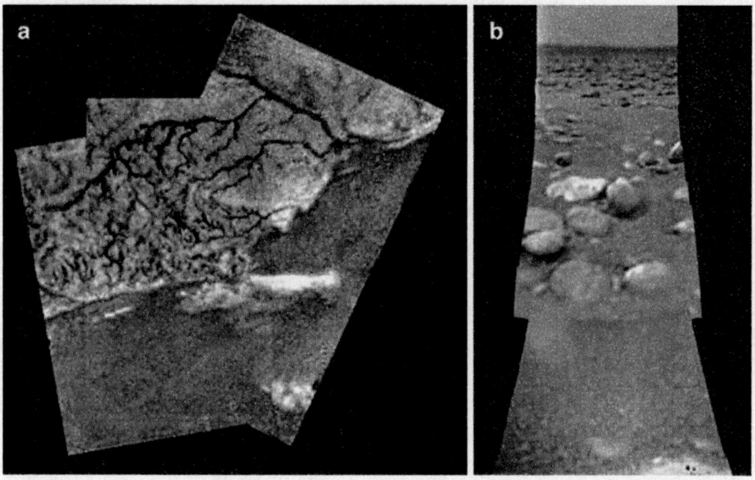


**Figure 5.** Titan's surface seen by DISR camera on Huygens (Credit: ESA/NASA/JPL/University of Arizona). (**a**) Dendritic structures indicating a flow down into a major river channel (ID number: SEMNL681Y3E). (**b**) On the landing site: pebbles on a surface looking like a dry lakebed (ID number: SEM9TWULWFE).

**Figure 6.** A large lake seen by Cassini radar in the north polar region of Titan. The big island in the *middle* is about 90 by 150 km across (Credit: NASA/JPL (PIA09180)).

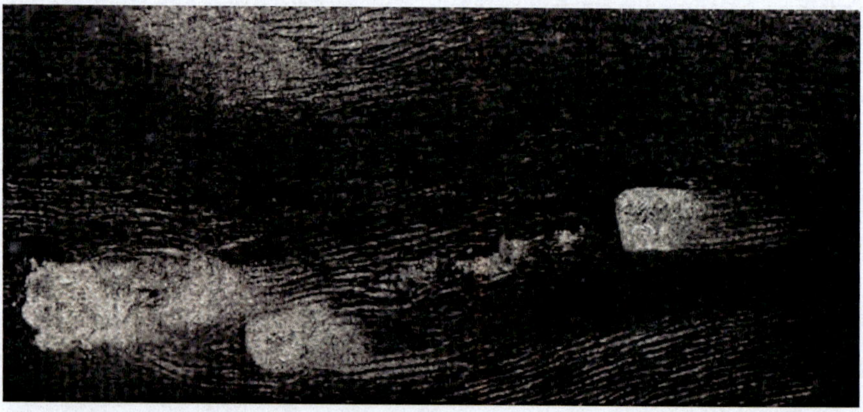

**Figure 7.** Titan's dunes in low latitude regions seen by the Cassini radar. The surface covered by this image is about 160 by 325 km. The dunes are up to 3 km spaced (Credit NASA/JPL (PIA08738)).

are many lakes seen by Cassini in the polar region of Titan (Fig. 6). Those lakes are indeed filled with a liquid as demonstrated by the detection of the IR signature of liquid ethane (Brown et al., 2008) and by the observation of a specular reflexion (Stephan et al., 2010). This liquid should be mainly composed of ethane and methane (Cordier et al., 2008; Raulin et al., 2009).

Several other geological features, similar to the terrestrial ones have been observed on Titan. There are many huge dune fields (Fig. 7) especially in the low latitude regions, mountains, cryovolcanoes, signs of tectonic activity where again, water ice plays the role of silicates on Earth.

## 3.2. COMPLEX ORGANIC CHEMISTRY

Another similarity, especially with the primitive Earth, is the presence of an active organic chemistry which is occurring in the atmosphere of Titan but also involving its surface.

The coupled chemistry of $N_2$ and $CH_4$ in the high atmosphere of Titan, induced by energetic electrons of Saturn's magnetosphere and solar VUV photons, allows the formation of many species, which recombine, evolve, and produce a large diversity of hydrocarbons and nitriles. Before the Cassini arrival in the Saturn system, several of these organics were already detected in Titan's atmosphere: alkanes up to C3, alkenes and alkynes up to C4, benzene, and several nitriles, HCN, $CH_3CN$, $C_2N_2$, $C_4N_2$, and $HC_3N$. The presence of HCN – the most abundant nitrile detected in gas phase – and $HC_3N$, because of their important role in terrestrial prebiotic chemistry, is by itself of astrobiological importance. But organic chemistry is also involved in the condensed phases, and particularly in the haze particles. It was assumed that Titan's aerosol is made of complex organic matter, similar to the laboratory so-called tholins; these solid organics produced when simulating experimentally Titan's atmosphere chemistry. These laboratory Titan's tholins have interesting properties; in particular, they can release amino acids after hydrolysis.

The ACP experiment on Huygens provided the first direct in situ chemical analysis of Titan's aerosols. It collected haze particles in Titan's stratosphere and troposphere, heated them at different temperatures, and sent the produced gases for analysis to the GCMS instrument. The obtained results indicate that the aerosols are made of a refractory nucleus, composed of H, C, and N atoms (Fig. 8), and produce $NH_3$ and HCN after pyrolysis at 600°C (Israël et al., 2005). This strongly supports the tholins hypothesis. It also strongly suggests that Titan's aerosols may evolve once in contact with water ice on Titan's surface, and may produce a variety of organics of biological interest, such as amino acids (Neish et al., 2010; Ramirez et al., 2010).

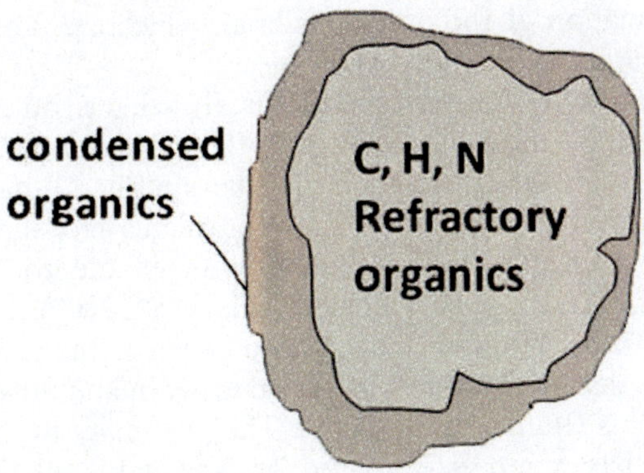


**Figure 8.** Composition of Titan's aerosols.

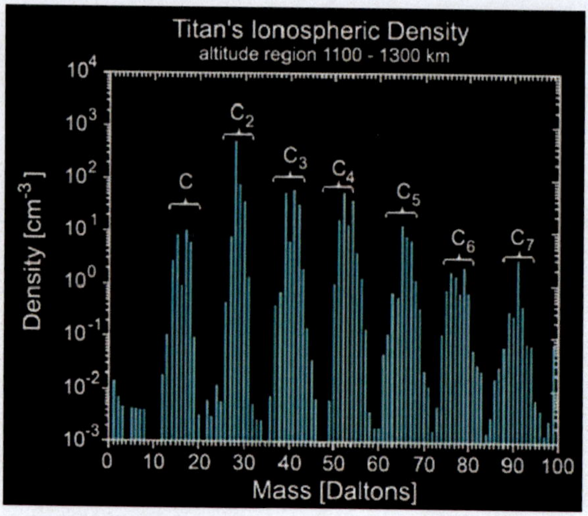


**Figure 9.** Mass spectrum of Titan's ionosphere in the 1,100–1,300-km altitude region, taken by INMS instrument on Cassini. $C_n$ is a chemical group with n (C + N) atoms (Credit NASA/JPL/University of Michigan (PIA07865)).

One of the most surprising results related to the chemistry of Titan's atmosphere was provided by the INMS and CAPS instruments on the Cassini orbiter. Direct mass spectrometric measurements of the ionosphere by INMS show the presence of a large variety of organics (Waite et al., 2007). Their molecular mass extends up to more than 100 Da (the higher mass of the instrument) and includes various hydrocarbons and nitriles (Fig. 9). CAPS instrument detected positive and negative ions, the latter with mass reaching around 10,000 Da! This indicates that the complex organic chemistry starts already in the ionospheric regions at very high altitude (up to ~1,300 km) and may directly participate in the formation of tholin-like materials which are present in the lower atmospheric regions, as detected by ACP.

The end products of this chemistry, mainly ethane, other low molecular organics, including benzene and nitriles, and the macromolecular organic compounds, tholin-like materials, accumulate on Titan's surface. Ethane with methane can condense and fill the polar lakes. Thermodynamic modeling of these lakes indicates that they should be mainly composed of ethane, with several 10% of methane and propane (Cordier et al., 2008; Raulin et al., 2009), and a large variety of minor species. Those should include many organics, with much higher concentration in the lakes than in the atmosphere and easily quantitatively analyzable by in situ measurements (Raulin et al., 2009). The possibility for life in these cold hydrocarbon lakes has even been envisaged (McKay and Smith, 2005).

## 3.3. HABITABILITY

The mean volumic mass of Titan is 1.88 kg·l$^{-1}$ suggesting that it is made approximately of 50% ices, mainly water ice (~1 kg·l$^{-1}$) and 50% of silicates (~3 kg·l$^{-1}$). Models of the formation of Titan from the Saturn subnebula, and its following thermal evolution up to now, suggest that its internal structure includes a water-ammonia ocean, sandwiched between two water ice layers, with a rocky core (Fig. 10). Recent data from the RSS experiment on Cassini indicate that the solid layers are poorly differentiated and should be a mixture of water ice and silicates (Iess et al., 2010).

This internal ocean could be located at about 30–40 km below the surface (Béghin et al., 2009) and could include up to about 15% ammonia. This corresponds to a pH around 11.5. The pressure reaches ~5 kbar at 200-km depth, and it could include hot spots reaching −20°C. Such conditions are not incompatible with life as we know it on Earth (Fortes, 2000; Raulin, 2008; Raulin et al., 2009). Thus, the possibility of life in this ocean cannot be excluded. Moreover, models also predict that during the first ten millions of years after Titan's formation, the ocean was in direct contact with the atmosphere on one side and with the bedrock

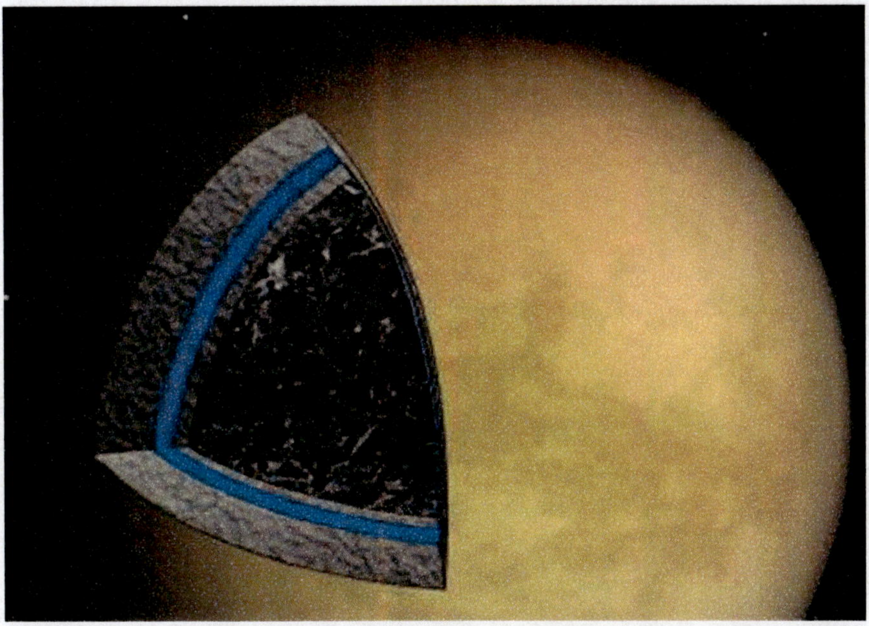

**Figure 10.** Artist view of the internal structure of Titan: the models predict the presence of an internal water-ammonia ocean between two large layers of water ice mixed with rocks (Credit NASA/JPL (PIA12843)).

on the other side. This could have provided conditions very favorable for an efficient prebiotic chemistry toward the emergence of life, with the possible involvement of hydrothermal vents. Thus, the internal ocean of Titan not only is habitable but could be habited.

## 4. Enceladus

One of the biggest not expected discovery of the Cassini-Huygens mission is Enceladus activity (Fig. 11).

Discovered by William Herschel in 1789, Enceladus is about ten times smaller than Titan, with a radius of 250 km. During the first close flyby of Enceladus by Cassini, the ISS narrow-angle camera discovered the presence of large plumes ejected from the south pole region of the satellite. The images of Enceladus surface in the same region show large faults, now called "tiger stripes" since they look like. Infrared observations of these stripes by CIRS show that they are several 10°K warmer than the rest of the surface, strongly suggesting that the plumes are ejected from these faults, coming from a warm water reservoir.

In situ mass spectrometric analysis of the plumes by INMS shows that they are mainly composed of water but include many organic compounds, such as methane, acetylene, and propane. Ammonia has also been recently detected by INMS, which opens the possibility of N-organics present also in the plume (Waite et al., 2009).

Among the models which have been proposed to explain these plumes, one suggests the presence of a reservoir of pressurized liquid water in the internal structure of Enceladus (Matson et al., 2007; Waite et al., 2009). This liquid would

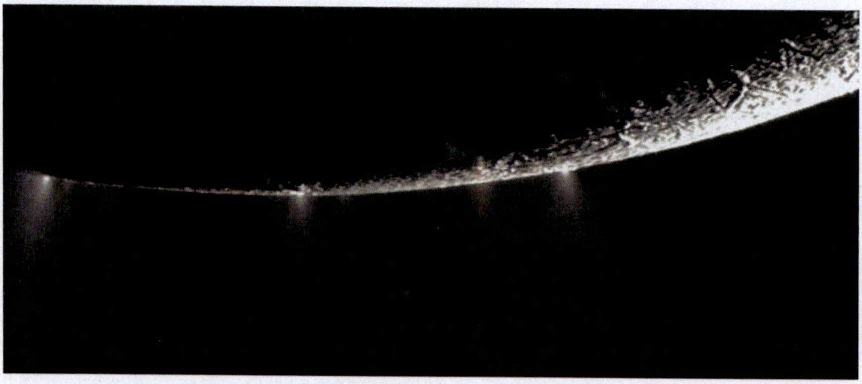

**Figure 11.** This picture of Enceladus taken by Cassini narrow-angle camera on November 2009 shows many large plumes ejected from the tiger stripes, fissures located in the south pole of Enceladus (Credit: NASA/JPL/Space Science Institute (PIA11688)).

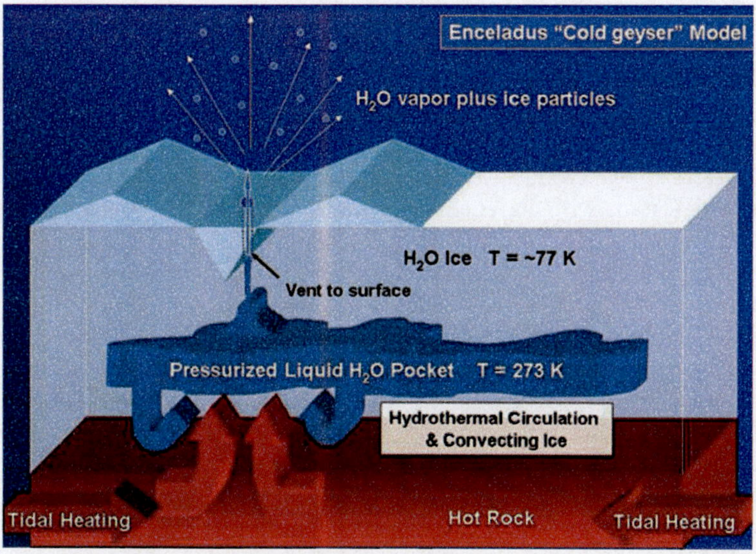


**Figure 12.** A model of Enceladus internal structure proposed to explain the plumes (Credit: NASA/JPL/Space Science Institute (PIA07799)).

be in contact with a hot bedrock (Fig. 12), providing, again, interesting conditions for prebiotic chemistry and habitability.

The possible presence of life in this liquid reservoir has been considered recently (McKay et al., 2008) and cannot be excluded, in spite of the violence of the plumes and the restlessness of this planetary environment.

## 5. Other Satellites

VIMS observation of several other satellites of Saturn, in the near infrared region, shows that their surface is covered by dark materials. This is particularly the case with Dione, but also with Phoebe, Iapetus, Hyperion, Epimetheus and even with the F-ring (Clarke et al., 2008). This dark material could be made of cyanide compounds and could be of cometary origin (Clarke et al., 2008).

Spectral signatures of hydrocarbons have also been found from VIMS data on Iapetus and Phoebe suggesting the presence of organic compounds such as PAH's, kerogen- or coal-like structure compounds (Cruishank et al., 2008).

Some of these satellites may also have internal liquid water pockets and thus may present the requisite essential for the emergence and development of life: liquid water and organic compounds. The solstice mission may be able to discover such properties and thus extend drastically the list of planetary bodies of important astrobiological interest in the Saturn system.

## 6. Conclusions

Although already considered as a target of prime interest for exo/astrobiology, because of the dense and chemically reducing atmosphere of its largest satellite Titan, with the many new data obtained from the Cassini-Huygens mission, the Saturn system appears today even much more interesting for this field.

Indeed, Titan, with its chemically active atmosphere, appears more and more as a giant organic factory. Recent modeling of the primitive Earth atmosphere and especially its exosphere strongly suggests that it was much more reducing than expected (Tian et al., 2005) and thus closer to Titan's atmosphere. Looking at Titan, atmospheric chemistry now is thus a way to explore the chemistry and other processes in the primitive atmosphere of the Earth. The presence of an internal water ocean on Titan, although it has not yet been fully demonstrated by Cassini observations, is more and more likely. If life is present on Titan, for some exobiologists, its search and discovery could be even more important for our understanding of chemical evolution, the origin of life and extraterrestrial life, than finding life on Mars or Europa (Shapiro and Schulze-Makuch, 2009). Although more speculative, the presence of Life in Enceladus is also an attractive possibility.

The Cassini-Huygens mission is not finished, but it is likely that it will not be able to detect clear traces of Life in those two planetary objects. Such a detection would require a new mission to the Saturn system, with more powerful in situ techniques, especially to perform a molecular analysis of the surface of the satellites and of the particles of the plumes. Such missions are already envisaged, and could be developed under fully international collaboration, involving in particular ESA and NASA, such as the TSSM mission (Lebreton et al., 2009).

## 7. Acknowledgments

The author wishes to thank CNES and ESA for their support in his Cassini-Huygens participation as an IDS of the mission, making possible the writing of this chapter.

## 8. References

Atreya SK, Adams EY, Niemann HB et al (2006) Titan's methane cycle. Planet Space Sci 54:1177–1187

Béghin C, et al (2009) New insights on Titan's plasma-driven Schumann resonance inferred from Huygens and Cassini data. Planet Space Sci 57:1872–1888

Brown RH, et al (2008) The identification of liquid ethane in Titan's Ontario Lacus. Nature 454:607–610

Brown R, Lebreton JP, Waite H (eds) (2009) Titan from Cassini-Huygens. Springer, Dordrecht

Clark RN, et al (2008) Compositional mapping of Saturn's satellite Dione with Cassini VIMS and implications of dark material in the Saturn system. Icarus 193:372–386

Cordier D, Mousis O, Lunine JI, Lavvas P, Vuitton V (2008) An estimate of the chemical composition of Titan's lakes. Astrophys J Lett 689:61–65

Cruikshank DP, et al (2008) Hydrocarbons on Saturn's satellites Iapetus and Phoebe. Icarus 193:334–343

ESA (2010). Web site on Huygens probe: http://sci.esa.int/huygens/

Fortes AD (2000) Exobiological implications of a possible ammonia-water ocean inside Titan. Icarus 146:444–452

Iess L, Rappaport N, Jacobson R, Racioppa P, Stevenson D, Tortora P, Armstrong J, Asmar S (2010) Gravity field, shape, and moment of inertia of Titan. Science 327:1367–1369

Israël G, et al (2005) Evidence for the presence of complex organic matter in Titan's aerosols by in situ analysis. Nature 438:796–799

JPL/NASA (2010). Web site on Cassini mission: http://saturn.jpl.nasa.gov/index.cfm

Lammer H, Bredehöft JH, Coustenis A, Khodachenko ML, Kaltenegger L, Grasset O, Prieur D, Raulin F, Ehrenfreund P, Yamauchi M, Wahlund J-E, Grießmeier J-M, Stangl G, Cockell CS, Kulikov YuN, Grenfell JL, Rauer H (2009) What makes a planet habitable? Astron Astrophys Rev 17:181–249

Lebreton JP, Matson D (2002) The Huygens probe: science, payload and mission overview. Space Sci Rev 104:59–100

Lebreton JP, Coustenis A, Lunine J, Raulin F, Owen T, Strobel D (2009) Results from the Huygens probe on Titan. Astron Astrophys Rev 17:149–179

Matson DL, Castillo JC, Lunine JI, Johnson TV (2007) Enceladus' Plume: compositional evidence for a hot interior. Icarus 187:569–573

McKay CP, Smith HD (2005) Possibilities for methanogenic life in liquid methane on the surface of Titan. Icarus 178:274–276

McKay CP, Porco CC, Altheide T, Davis WL, Kral TA (2008) The possible origin and persistence of life on Enceladus and detection of biomarkers in the plume. Astrobiology 8:909–919

Neish CD, Somogyi A, Smith MA (2010) Titan's Primordial soup: formation of amino acids via low-temperature hydrolysis of tholins. Astrobiology 10:337–347

Ramirez SI, Coll P, Buch A, Brassé C, Poch O, Raulin F (2010) The fate of aerosols on the surface of Titan. Faraday Discuss 147:419–427

Raulin F (2008) Astrobiology and habitability of Titan. Space Sci Rev 135:37–48

Raulin F, McKay CP, Lunine JI, Owen T (2009) Titan's astrobiology. In: Brown R, Lebreton J-P, Waite H (eds) Titan from Cassini–Huygens. Springer, New York, pp 215–233

Russell CT (2003) The Cassini-Huygens mission: overview objectives and Huygens instrumentarium. Kluwer Academic Publishers, Dordrecht

Russell CT (2005) The Cassini-Huygens mission: orbiter remote sensing investigations. Kluwer Academic Publishers, Dordrecht

Shapiro R, Schulze-Makuch D (2009) The search for alien life in our solar system: strategies and priorities. Astrobiology 9:335–343

Stephan K, et al (2010) Specular reflection on Titan: liquids in Kraken Mare. Geophys Res Lett 37:L07104–L07109

Tian F, Toon OB, Pavlov AA, De Sterck H (2005) A hydrogen-rich early earth atmosphere. Science 308:1014–1017

Waite JH, Young DT, Cravens TE, Coates AJ, Crary FJ, Magee B, Westlake J (2007) The process of tholin formation in Titan's upper atmosphere. Science 316:870–875

Waite JH, et al (2009) Liquid water on Enceladus from observations of ammonia and $^{40}Ar$ in the plume. Nature 460:487–490

Biodata of **John A. Raven, Charles S. Cockell, and Lisa Kaltenegger**, authors of "***Energy Sources for, and Detectability of, Life on Extrasolar Planets.***"

**John A. Raven FRS FRSE** is an Emeritus Professor of Biology at the University of Dundee. He obtained his PhD in plant biophysics from the University of Cambridge (UK) in 1967. He has widespread interests including the bioenergetics, biogeochemistry and ecophysiology of aquatic and terrestrial photosynthetic organisms, paleoecology, and, most recently, astrobiology. Professor Raven has published over 330 scientific papers and more than 50 book chapters; he has written a book on energetic and transport in aquatic plants and coauthored a book on aquatic photosynthesis as well as edited multiauthor volumes and special issues of journals. He is currently an associate editor of five journals.

E-mail: **j.a.raven@dundee.ac.uk**

J. Seckbach (ed.), *Genesis - In The Beginning: Precursors of Life, Chemical Models and Early Biological Evolution*, Cellular Origin, Life in Extreme Habitats and Astrobiology 22, 835–857
DOI 10.1007/978-94-007-2941-4_42, © Springer Science+Business Media Dordrecht 2012

**Charles S. Cockell** is Professor of Astrobiology at the University of Edinburgh, and is Director of the UK Centre for Astrobiology. He obtained his DPhil in molecular biophysics from the University of Oxford (UK). He has wide-ranging interests in Earth Sciences (how microbes contribute to rock weathering and nutrient release, subsurface biogeochemistry) and Astrobiology (how microbes acquire energy and nutrients, and the potential for life beyond Earth). Professor Cockell has published over 120 peer-reviews publications, has written two popular science books and is regularly engages withy the news media in promoting astrobiology and Earth science.

E-mail: **c.s.cockell@ed.ac.uk**

**Lisa Kaltenegger** is a Group Leader at the Max Planck Institute for Astronomy and a lecturer at Harvard University. She obtained her PhD on the search for extrasolar planets at the Karl Franzen University of Graz (Austria). Her research on extrasolar planets includes the study of rocky planets and super-Earths, the habitable zone and spectral fingerprints of exoplanets. Dr Kaltenegger has published over 50 peer-reviewed publications and is involved in outreach and the presenting science to lay audiences.

E-mail: **lkalte@cfa.harvard.edu, kaltenegger@mpia.de**

# ENERGY SOURCES FOR, AND DETECTABILITY OF, LIFE ON EXTRASOLAR PLANETS

**JOHN A. RAVEN[1], CHARLES S. COCKELL[2,3], AND LISA KALTENEGGER[4,5]**

[1]*Division of Plant Sciences, University of Dundee at TJHI, The James Hutton Institute, Invergowrie, Dundee DD2 5DA, UK*
[2]*Centre for Earth, Planetary and Space Research, Open University, Walton Hall, Milton Keynes MK7 6AA, UK*
[3]*School of Physics and Astronomy, James Clerk Maxwell Building, University of Edinburgh, Mayfield Road, Edinburgh EH9 3JZ, UK*
[4]*Astronomy Department, Harvard University, 60 Garden, St. MS20, Cambridge, MA 02138, USA*
[5]*MPIA, Koenigtuhl 17, 69115 Heidelberg, Germany*

## 1. Introduction

Life in the accessible biosphere on Earth today is energised in primary productivity mainly by solar radiation through photolithotrophy with a minor role for chemolithotrophy based on reductants from hydrothermal vents and volcanoes, as well as reductant from photolithotrophy (Johnston et al., 2009; Raven, 2009a). Food webs downstream of primary producers involve chemo-organotrophs which regenerate chemical resources for photolithotrophy and chemolithotrophy. In addition to these energy sources for growth and maintenance, there are also a range of other potential energy sources for growth and maintenance of organisms (Muller and Schulze-Makuch, 2006), with varying extents of experimental support for their occurrence. The analysis of these alternative energy sources involves not only the likelihood of their occurrence when the energy source is available but also the availability of these energy sources on Earth. These analyses permit estimates of how globally significant the alternative energy sources could be. In the context of astrobiology, we review and extend previous attempts to quantify the possible extent of photolithotrophy and chemolithotrophy, and of the alternative energy sources, as well as the possibilities of remote sensing of these processes on Earth.

We use our own planet as a test case for the analysis of the detectability of primary productivity on extrasolar planets. This has been possible for about half of the lifetime of our planet (Kaltenegger et al., 2007), mostly from oxygenic photolithotrophy in the context of other chemicals in our atmosphere. Detectability of technological life on extrasolar planets will require big telescopes to detect man-made chemicals in the atmosphere that are trace gases in Earth's present atmosphere, or technology could potentially be detected in a very different way due to a positive outcome of SETI.

## 1.1. DETECTING HABITABILITY REMOTELY

The spectrum of the planet can contain signatures of atmospheric species, what creates its spectral fingerprint (Owen 1980). The trade-off between contrast ratio and design is not discussed here, but leads to several different configurations for space-based mission concept. Figure 1 shows observations and model fits to spectra of the Earth in three wavelength ranges (see Kaltenegger et al., 2007 for details). The data shown in Fig. 1 (left) are the visible Earthshine spectrum (data by Woolf et al., 2002), (centre) is the near-infrared Earthshine spectrum (data by Turnbull et al., 2006), and (right) is the thermal infrared spectrum of Earth as measured by a spectrometer en route to Mars (data by Christensen and Pearl, 1997). The data are shown in black, and the SAO atmosphere model is shown in red. In each case, the constituent gas spectra in a clear atmosphere are shown in the bottom panel, for reference.

The interferometric systems suggested operate in the mid-IR (6–20 μm) and observe the thermal emission emanating from the planet. The coronagraph and occulter concepts detect the reflected light of a planet and operate in the visible and near-infrared (0.5–1 μm). The viewing geometry results in different flux contributions of the overall detected signal from the bright and dark side, for the reflected light, and the planet's hot and cold regions for the emitted flux. The reflected spectrum will exhibit phases depending on the part of the lit surface that can be seen from the observer's point of view, while in the mid-infrared, the day and nightside of the planet roughly emit at the same temperature, and the whole planetary surface emits throughout the planet's orbit.

Figure 2 shows the Earth's spectrum in transmission, the spectrum a remote observer would see if the Earth were to transit the Sun; see Montanes-Rodriguez et al. (2007); Kaltenegger and Traub (2009) and Pallé et al. (2009) for details on Earth's transmission spectrum. Figure 2 shows the effective height in the atmosphere of the transiting Earth as a function of wavelength. Absorption features in an atmosphere increase the apparent size of a transiting planet at that wavelength because they absorb the light at characteristic wavelengths, making the planet appear bigger. Adding the effective height of the atmosphere to the planet's radius results in the detected size of the planet versus wavelength. Kaltenegger (2010) discusses transit spectroscopy of potentially habitable moons. Both transmission and emergent spectra contain the signature of atmospheric gases that may indicate habitable conditions and, possibly, the presence of a biosphere: $CO_2$, $H_2O$, $O_3$, $CH_4$ and $N_2O$ in the thermal infrared and $H_2O$, $O_3$, $O_2$, $CH_4$ and $CO_2$ in the visible to near-infrared. The presence or absence of these spectral features (detected individually or collectively) will indicate similarities or differences with the atmospheres of terrestrial planets and moons, and its astrobiological potential.

Our search for signs of life is based on the assumption that extraterrestrial life shares fundamental characteristics with life on Earth, in that it requires liquid water as a solvent and has a carbon-based chemistry (see, e.g. Brack, 1993; Des Marais et al., 2002). Life based on a different chemistry is not considered here

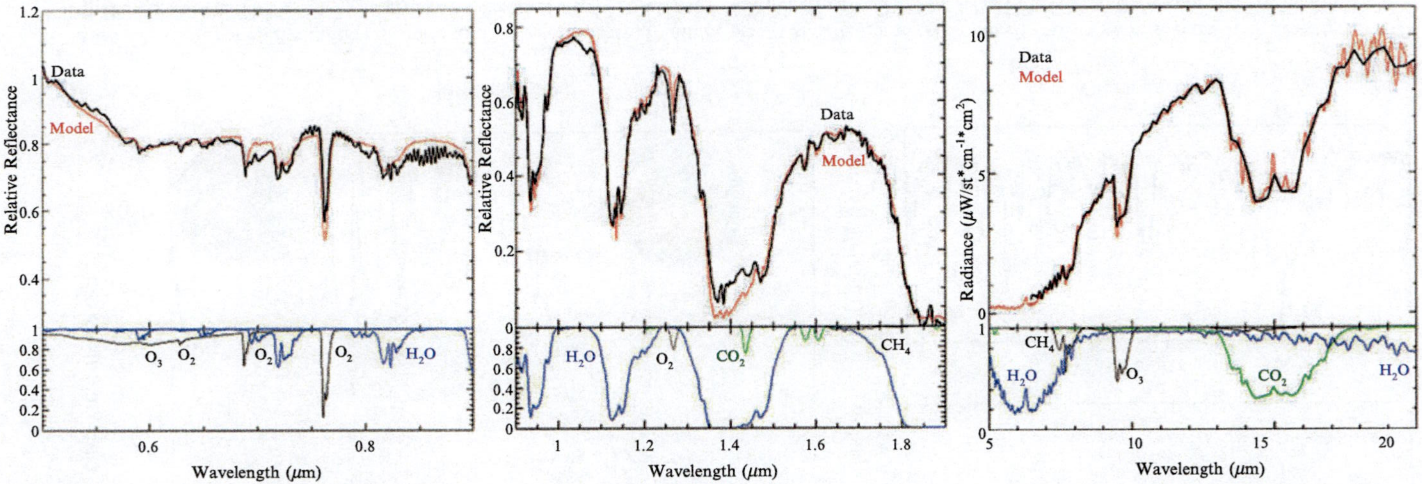


**Figure 1.** Observed reflectivity spectrum in the visible (Woolf et al., 2002) (*left*), near-infrared (Turnbull et al., 2006) (*middle*) and emission spectrum in the infrared (Christensen and Pearl, 1997) (*right panel*) of the integrated Earth, as determined from Earthshine and space, respectively. The data are shown in black, and the SAO model is in red (Adapted from Kaltenegger et al., 2007). The reflectivity scale is arbitrary.

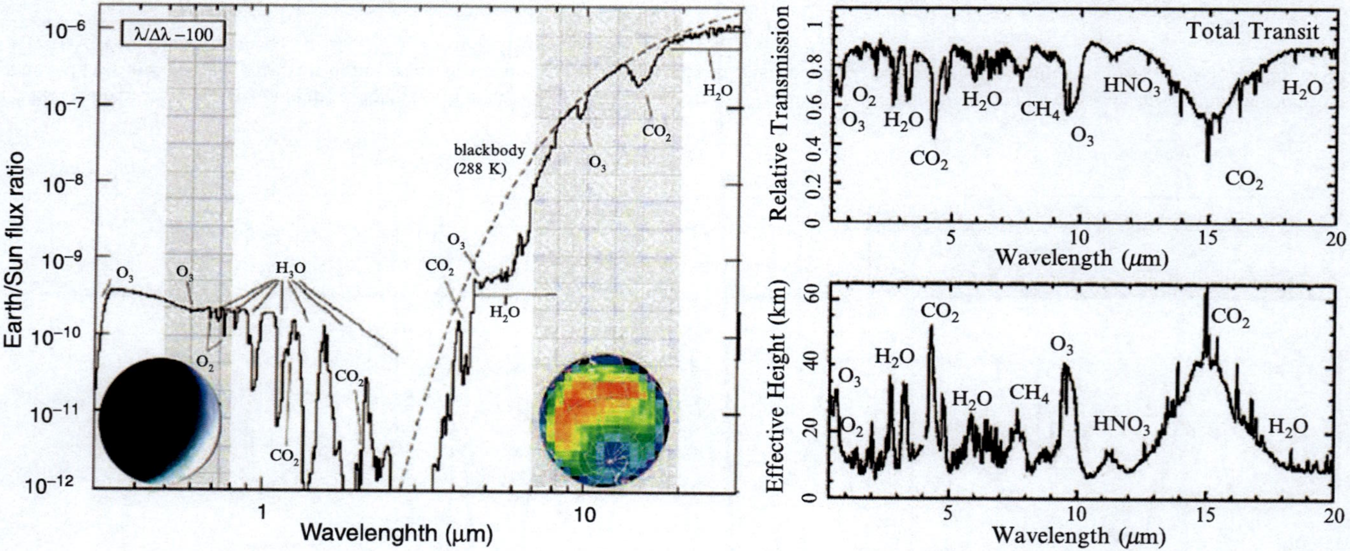


**Figure 2.** (*Left panel*) Synthetic reflection and emission spectra (Kaltenegger et al., 2010) and (*right panel*) transmission spectra of the Earth from UV to IR are shown. The intensity is given as a fraction of solar intensity (*left panel*) as well as the relative height in the atmosphere. The atmospheric features are indicated (Adapted from Kaltenegger and Traub, 2009; Kaltenegger, 2010).

because the vast possible life-forms produce signatures in their atmosphere that are so far unknown. Therefore, we assume that extraterrestrial life is similar to life on Earth in its use of the same input and output gases and that it necessarily exists out of thermodynamic equilibrium with its environment (Lovelock, 1975). 'Biomarkers' is used here to mean detectable species, or set of species, whose presence at significant abundance strongly suggests a biological origin (e.g. couple $CH_4+O_2$, or $CH_4+O_3$, Lovelock, 1975; Pavlov et al. 2000, 2003). Bio-indicators are indicative of biological processes but can also be produced abiotically, e.g. $O_2$ that can be produced by UVB photolysis of water but is estimated to have only yielded an atmospheric content before oxygen photosynthesis evolved on Earth that is $10^{-14}$ of the present value (Buick, 2008). It is their quantities, and detection along with other atmospheric species, and in a certain context (for instance the properties of the star and the planet) that points towards a biological origin. Note that if the presence of biogenic gases such as $O_2/O_3+CH_4$ may imply the presence of a massive and active biosphere, their absence does not imply the absence of life. Life existed on Earth before the interplay between oxygenic photosynthesis and carbon cycling produced an oxygen-rich atmosphere. $O_2$, $O_3$ and $CH_4$ are good biomarker candidates that can be detected by a low-resolution (spectral resolution $\lambda/d\lambda<50$, where $\lambda$ denotes the wavelength) spectrograph throughout half of Earth's geological history (see, e.g. Kaltenegger and Traub 2009; Kaltenegger et al. 2010). Remote detectability of different geochemical cycles and cryptic biota are discussed in detail in, e.g. Kaltenegger and Sasselov (2010) and Cockell et al. (2009a) respectively.

## 2. The Range of Energy Sources Which Are, or Could Be, Used to Power the Growth and Maintenance of Primary Producers on Earth

Table 1 lists the known or possible energy inputs to primary producers on Earth and, where appropriate, estimates of their contributions to past and present global primary production in the relatively readily observable biosphere and in the deep biosphere.

### 2.1. PHOTOCHEMICAL ENERGY CONVERSION

#### *2.1.1. Oxygenic Photolithotrophy*

Solar radiation is the main energy source allowing primary producers on Earth to convert inorganic compounds into organic matter by photolithotrophy. As is indicated in Table 1, oxygenic photolithotrophy is by far the quantitatively predominant photolithotrophic, and indeed autotrophic, process occurring in the known biosphere (see also Raven, 2009a). Since essentially all of the oxygen in the atmosphere has been produced by oxygenic photosynthesis, all of the other (chemolithotrophic) examples of autotrophy which consume oxygen depend on oxygenic photosynthesis; these are considered below. The organic matter produced in oxygenic

**Table 1.** Known and possible energy inputs to primary producers on Earth.

| Energy input: output | Process | Quantitative significance on Earth/ Pg C $m^{-2}$ $a^{-1}$ | References |
|---|---|---|---|
| Light: redox, electrochemical gradient →ATP | Oxygenic photolithotrophy | 45–>50 in the ocean ~60 on continents | Raven (2009a) |
| Light: redox, electrochemical gradient → ATP | Anoxygenic photolithotrophy | <0.033–0.077 in the ocean today | Johnston et al. (2009), Raven (2009a) |
| | | <1.2–<5 in the ocean 3.8 Ga ago | Kharecha et al. (2005), Canfield et al. (2006) |
| Inorganic redox couple: redox, electrochemical gradient →ATP | Chemolithotrophy | <0.40 in the ocean today | Raven (2009a) |
| | | <1.2–<5 in the ocean 3.8 Ga ago | Kharecha et al. (2005), Canfield et al. (2006) |
| | | 0.00074 in the ocean 3.8–4 Ga ago | Sleep and Bird (2008) |
| Light: electrochemical gradient →ATP | Chemo-organotrophy supplemented by light energy transformed using bacteriochlorophyll-based photochemistry in aerobic anoxygenic bacteria or by halo- or proteorhodopsin in archaea and bacteria, respectively | Not readily quantified; economise on organic C by replacing some respiration | Raven (2009a, b) |
| Variations in external osmolarity: electrochemical gradient →ATP | Possibility of supplementation of energy supply for organisms in estuarine environments, and in the interaction of the thermohaline circulation with surrounding seawater | No real evidence for use of varying osmolarity as a supplementary energy source | Muller and Schulze-Makuch (2006), Kaufmann (2009) |

| | | | |
|---|---|---|---|
| Variations in speed of water flow over benthic organism: ATP | Possibility of stretch relaxation in breaking waves could drive actomyosin-based mechanochemical motors in the direction of ATP synthesis. Water flow in rivers, marine currents, thermohaline circulation could use rotary motors based on bacterial flagella and $F_O$–$F_1$ ATP synthases, with appropriate gearing | No real evidence for the use of this mechanism as a supplementary energy source | Raven (1988), Kaufmann (2009) |
| Temperature gradients: thermo-osmotic energy transformation | Possibility of thermo-osmotic energy transformation: problems with small temperature gradients tolerable for organisms and the maximum efficiency of a heat cycle, and high thermal conductivity of water and biological membranes | No real evidence for the use of this mechanism as a supplementary energy source | Muller and Schulze-Makuch (2006), Kaufmann (2009) |

Where estimates have been made, the quantitative contributions to past and present productivity in the (relatively) readily observable biosphere and the deep biosphere.

photolithotrophy is the reduced carbon source for almost all the non-autotrophic life on Earth using chemo-organotrophy, and the great majority of chemo-organotrophs use oxygen, from oxygenic photolithotrophy, as their oxidant.

The accumulation of oxygen (and hence ozone), and the potential for spectroscopic remote sensing of oxygenic photosynthesis, requires burial of organic carbon which could otherwise recombine biologically or otherwise with the oxygen produced in its oxygenic photosynthesis and also to offset tectonic input of reductant into the surface biosphere. Atmospheric features are remotely detectable at low resolution if they exist in sufficient concentration to influence the planet's atmosphere (see, e.g. Kaltenegger et al., 2010 for a detailed discussion). Species like water, carbon dioxide, oxygen and ozone are readily observable on a current Earth with low resolution, while man-made trace gases like freons require high resolution as thus big telescopes to be remotely detectable.

The other possibility for detecting oxygenic photolithotrophy by remote sensing is to focus on surface features. Surface features can be distinguished remotely due to characteristic reflection versus wavelength of different substances like rocks, vegetation and water (see, e.g. Kaltenegger et al., 2007).

On current Earth, the marked increase in reflectance at about 700 nm (the 'vegetation red edge' or VRE), with a low reflectance in the 'photosynthetically active radiation' range 400–700 nm (rarely 730 nm) and a high reflectance at longer wavelengths amounts to a few percent of the overall detectable flux of the planet, a much lower signal than atmospheric species, but addressing a very different aspect of habitability (e.g. if plants or plant-like reflecting surface could be present). The difference in the level of detectability of surface features versus atmospheric features results from areal coverage of surface features like vegetation – e.g. current Earth is covered by about 70% of ocean and the 30% of land area is not uniformly covered by vegetation. In addition, clouds obscure part of the surface, reducing the overall detectable signal. With high spectral as well as temporal resolution, clouds could be distinguished from a surface that rotates with the planet's rotation period.

The occurrence of some reflection (and transmission) in the 400–700 (−730) nm range rather than complete absorption can be related to trade-offs with other evolutionary requirements (Key et al., 2010; Raven, 1984, 2011). On Earth, around 440 million years ago (see, e.g. Gensel, 2008; Rubinstein et al., 2010; Tomescu et al., 2009), an extensive land plant cover developed, generating the red chlorophyll edge in the reflection spectrum between 700 and 750 nm. The high infrared and red absorptance of water means that most aquatic oxygenic photolithotrophs cannot be a remotely sensed by this method because they are covered by a layer of water, leaving only surface photolithotrophs that can be sampled (Cockell et al., 2009a).

The VRE feature is readily observed in terrestrial photolithotrophs other than those that are not being screened by overlying mineral material as a means of restricting water loss and UVB damage: this unscreened vegetation accounts for the great majority of the photosynthetic pigments on the land surface of

Earth today (Cockell et al., 2009a, b). Individual leaves show a strong red-edge reflection feature, but on a disc-averaged spectrum, what we will detect for an extrasolar planet, where all information will be collected in one pixel, the red edge is a small feature compared to chemical absorption features in the atmosphere (see, e.g. Kaltenegger et al., 2007; Kaltenegger and Traub, 2009). Averaged over a spatially unresolved hemisphere of Earth, the additional reflectivity of this spectral feature is typically only a few percent as discussed above (see also Kaltenegger and Traub, 2009; Montañés-Rodriguez et al., 2005). Several groups (Arnold et al., 2002; Christensen et al., 1997; Montañés-Rodriguez et al., 2005; Turnbull et al., 2006; Woolf et al., 2002) have measured the integrated Earth spectrum via the technique of Earthshine, using sunlight reflected from the non-illuminated, or 'dark', side of the moon. Earthshine measurements have shown that detection of Earth's VRE is feasible if the resolution is high and the cloud coverage is known, but is made difficult owing to its broad, essentially featureless spectrum and cloud coverage.

The low-resolution method to detect surface features is making measurements of reflectance at two wavelengths in the blue-green part of the spectrum. Such a method only detects photosynthetic pigments in the top metre or two of a water body (Cockell et al., 2009a, b) or on the surface. This means that any deep chlorophyll maximum at the bottom of the upper mixed water layer (Tett, 1990; Cockell et al., 2009a) cannot be remotely detected.

The rationale for the large decrease in absorptance by vegetation at about 700 nm (rarely 730 nm) has recently been analysed in energetic and evolutionary terms (Bjorn et al., 2009: Milo, 2009; Raven, 2009b, 2011; see also Raven and Wolstencroft, 2004; Kiang et al., 2007a, b; Stomp et al., 2007; Wolstencroft and Raven, 2002). In summary, the 'red edge' results from two constraints. One is the energetics of carbon dioxide reduction and water oxidation in relation to the energy per photon at 680–700 nm using a mechanism involving two photons per electron transferred from water to carbon dioxide. The other is maximising the capacity to use photons at lower wavelengths down to just below 400 nm where absorption by protein and the potential for photodamage both increase with decreasing wavelength. The use of just below 400–700 nm for photosynthesis, with the primary photochemistry involving the energetic equivalent of photons at 680 and 700 nm in the two photochemical reactions, involves the energetically downhill transfer of excitation energy from shorter (high energy per photon) to longer wavelengths, and includes the wavelength of maximum photon, as opposed to energy, output of the sun at about 600 nm incident on the Earth's atmosphere and some 685 nm at the Earth's surface as a result of the $O_3$ Chappuis band, 500–700 nm, and just before the decrease due to $O_2$ at 687.5 nm (Kiang et al., 2007a, p. 241; see also Raven, 2011). Some plants from higher altitudes have higher absorptance in the near-infrared than plants from lower altitudes (Purohit et al., 2008), with the possibility of a less pronounced 'red edge' in the high-altitude plants. Such differences in absorptance have been related to the warming of leaves in cooler high-altitude environments, although it must be remembered that the allocation of absorbed radiation between sensible heat (warming the leaf) and latent heat (evaporating water) represents a trade-off between regulation of the temperature of the plant and of its water content.

On a cloud-free Earth, the diurnal flux variation in the visible caused by different surface features rotating in and out of view could be high, assuming hemispheric inhomogeneity. When the planet is only partially illuminated, a more concentrated signal from surface features could be detected as they rotate in and out of view on a cloudless planet (see, e.g. Ford et al., 2001; William and Gaidos, 2008). Earth has an average of 60% cloud coverage, which prevents easy identification of the features without knowing the cloud distribution. Clouds are an important component of exoplanet spectra because their reflection is high and relatively flat with wavelength. Clouds reduce the relative depths, full widths and equivalent widths of spectral features, weakening the spectral lines in both the thermal infrared and visible (Kaltenegger et al., 2007). In the thermal infrared, clouds emit at temperatures that are generally lower than the surface, while in the visible the clouds themselves have different spectrally dependent albedo that further influence the overall shape of the spectrum.

If one could record the planet's signal with a very high time resolution (a fraction of the rotation period of the planet) and signal-to-noise ratio (SNR), one could determine the overall contribution of clouds to the signal (Cowan et al., 2010; Pallé et al., 2008). During each of these individual measurements, one has to collect enough photons for a high individual SNR per measurement to be able to correlate the measurements to the surface features, what precludes this method for first-generation missions that will observe a minimum of several hours to achieve a SNR of 5–10. For Earth (Cowan et al., 2010; Pallé et al., 2008), these measurements show a correlation to Earth's surface feature because the individual measurements are time-resolved as well as have an individual high SNR, making it a very interesting concept for future generations of missions. Earth's hemispherical integrated vegetation red-edge signature is very weak, but planets with different rotation rates, obliquities, land-ocean fraction and continental arrangement may have lower cloud cover and higher vegetated fraction. Knowing that other pigments exist on Earth and that some minerals can exhibit a similar spectral shape around 750 nm (Seager et al., 2005), the detection of the red edge of the chlorophyll on exoplanets, despite its interest, will not be unambiguous.

### *2.1.2. Anoxygenic Photolithotrophy*

Oxygenic photolithotrophy uses the electron source of last resort, the thermodynamically and kinetically recalcitrant weak reductant water. There is also anoxygenic photolithotrophy, in which stronger reductants supply electrons to photochemistry (Cockell and Lee, 2002), permitting autotrophic $CO_2$ fixation. Today, these are restricted to habitats which are illuminated to a level at least as great as moonlight at the surface of the ocean at full moon (0.3–5.0 nmol photon $m^{-2}$ $s^{-1}$: Overmann et al., 1992; Raven and Cockell, 2006) and are also oxygen-free. In practice, these habitats result from sedimentation of particulate organic matter at rates in excess of that at which ventilation provides $O_2$, so that microbial oxidation of the organic carbon uses electron acceptors other than $O_2$. In seawater, the major possible oxidant is sulphate, which yields much less energy per unit organic carbon oxidised than would $O_2$. This habitat has negligible $O_2$ and enrichment in inorganic carbon species and sulphide and other reduced sulphur species resulting from the reduction of sulphate, i.e. the growth conditions appropriate to purple sulphur and purple non-sulphur photosynthetic bacteria. Even though there are chemical subsidies (sulphide, carbon dioxide) generated indirectly from the products of oxygenic photosynthesis, the global magnitude of anoxygenic photolithotrophy today is very small compared to that of oxygenic photosynthesis (Table 1; Johnston et al., 2009; Raven, 2009a).

The wavelength range for photosynthesis by the anoxygenic photolithotrophs is less energetically constrained than is that of oxygenic photolithotrophy. Only one kind of photochemical reaction, using bacteriochlorophyll for photochemistry and (with carotenoids) photon harvesting, is found in any one anoxygenic photolithotroph, with the infrared absorption maximum related to minima in the absorption spectrum of water (Kiang et al., 2007a; Stomp et al., 2007). The range of wavelengths used by these organisms for photosynthesis is from 400–760 nm to 400–1,100 nm, i.e. 'red edges' range from 760 to 1,100 nm.

As for remote detection, the absence of oxygen production by the anoxygenic photolithotrophs means that detection from atmospheric changes is not possible even if the global magnitude of this form of autotrophy were greater. Reflectance spectroscopy of photosynthetic pigments on Earth is rendered difficult by the great range of 'infrared edge' wavelengths in these organisms (Kiang et al., 2007a; Stomp et al., 2007), the small area occupied by these organisms and the difficulty in distinguishing the variable 'infrared edge' from mineral spectra. Furthermore, their requirement for anaerobic conditions for photolithotrophic growth generated by mineralisation of organic matter faster than oxygen can diffuse to the mineralising microorganisms, which generally means a considerable depth of water and/or sediment over the anoxygenic photolithotrophs which greatly restricts remote sensing of their pigments.

### *2.1.3. Non-solar Energy for Photolithotrophy?*

In addition to the direct use of solar radiation, there is also the possibility that photochemical energy conversion on Earth could be powered by other sources of electromagnetic radiation in the range 400–700 nm (Cockell and Raven, 2004; Cockell et al., 2009a, b; Raven and Cockell, 2006; Raven et al., 2000; Wolstencroft and Raven, 2002). One source is the short wavelength tail of infrared radiation from hydrothermal vents, i.e. a geological (tectonic) energy source. Chemiluminescence from abiological sulphide oxidation is another radiation source, with sulphide from hydrothermal vents or from reduction of sulphate by mineralisation of sedimented organic carbon in anoxic sediments and photosynthetically produced oxygen as oxidant. A third energy source is sonoluminescence, i.e. the collapse of vey small gas bubbles derived from hydrothermal activity. A further radiation source is the indirect use of solar radiation after reflection from the moon. The maximum land or sea surface irradiance from sunlight reflected from a full moon is similar to the minimum irradiance needed for photolithotrophic growth of the most low light-adapted organisms, so very little global primary productivity could result from moonlight (Raven and Cockell, 2006). The maximum radiation incident on the land of sea surface of Earth from stars other than the Sun is about six orders of magnitude less than that from the full moon (Munz and McFarland, 1973), so starlight is not a plausible source of energy for photolithotrophy (Raven and Cockell, 2006). On Earth, these sources of energy could only make a very small contribution to global energy flow and carbon cycling. There is very little possibility of remote detection of the very small quantity of photosynthetic pigments involved and no chance for those associated with sediments and hydrothermal vents. Only a portion of any such photolithotrophy is oxygenic, meaning that the remote detection of this photolithotrophy through atmospheric composition is not feasible. A final radiation source could be the biota itself. Bioluminescent bacteria generate light, the function of which is still unelucidated. This could be a source of light for phototrophs; however, such a theoretical possibility is unlikely to drive large-scale phototrophy, not least because the energy for bioluminescence almost all comes ultimately from autotrophy based on solar energy. More plausible is the direct detection of light generated by bioluminescent organisms. The bioluminescent signature could be detected if the areal extent was great enough. On the Earth, bioluminescence can cover areas of tens of thousands of square kilometres, and this phenomenon is amenable to satellite observation from space (Miller et al., 2005).

*2.1.4. Aerobic Anoxygenic Photochemistry in Chemo-organotrophs*

The other biologically catalysed photochemistry that influences global productivity on Earth is mainly carried out by predominantly chemo-organotrophic organisms. This photochemistry involves either bacteriochlorophyll or an energy-converting rhodopsin (Raven, 2009a, b; Zubkov, 2009). Unlike the autotrophic bacteria with photochemistry catalysed by bacteriochlorophyll, the anoxygenic aerobic bacteria are tolerant of oxygen, and occur near the surface of water bodies. The anoxygenic aerobic bacteria lack autotrophic inorganic carbon-assimilating pathways, and the contribution of these organisms to the energetics and carbon economy of aquatic ecosystems is in supplementing the respiratory energy conversion by the bacteria in their growth and maintenance in the light by generating transmembrane ion gradients and coupled ADP phosphorylation to ATP (Table 1; Raven, 2009a; Zubkov, 2009). The archaea and non-cyanobacterial bacteria with energy-converting rhodopsins have a similar function in global energetics and carbon cycling to that of the aerobic anoxygenic bacteria (Table 1; Raven, 2009a, b). As Raven (2009a) points out, it is very difficult to quantify the global role of these two pathways. However, a role has recently been shown in a *Vibrio* species for energy-transducing proteorhodopsin in prolonging survival in the absence of usable organic substrates (Gómez-Consanau et al., 2010). The role of the energy-converting rhodopsins in the few oxygenic photolithotrophs in which they have been found is no better understood (Raven, 2009a, b).

The wavelength range for energy-converting photochemistry is within the range 400–700 nm for the energy-converting rhodopsins (Sabeh et al., 2007) and 400–900 nm for the aerobic anoxygenic photosynthetic bacteria (Kiang et al., 2007a; Stomp et al., 2007). Despite the widespread occurrence of these organisms in surface waters of aquatic habitats (Raven, 2009a), there are problems with their remote sensing of them and molecular genetic methods (polymerase chain reaction) based on environmental samples are used to determine their abundance (Jiao et al., 2007).

## 2.2. CHEMOLITHOTROPHY

This mode of autotrophy involves an inorganic chemical reaction as the energy source (Table 1). The reductant for chemolithotrophy on Earth can be generated in biogeochemical cycles. Examples are nitrification oxidising ammonium to nitrite and nitrite to nitrate with oxygen as oxidant, and sulphide oxidation to sulphate using oxygen or nitrate as oxidant, so that the reductant as well as the oxidant is a direct or indirect outcome of oxygenic photolithotrophy on Earth today (Raven, 2009a). The alternative source of reductant is geological (tectonic), e.g. the sulphide or ferrous iron from hydrothermal vents, of sulphide from basalt-hosted aquifers (Raven, 2009a). The values in Table 1 indicate that chemolithotrophy is a very minor component of primary productivity today. Before oxygenic photosynthesis became significant there were anoxygenic photosynthetic bacteria dominating primary production in the surface plankton 2.5–3.8 Ga ago. The sum of anoxygenic photolithotrophy and chemolithotrophy was an order of magnitude greater than today, although before the evolution of photosynthesis estimates of chemolithotrophy are low (Table 1; Canfield et al., 2006).

In some cases, the function of chemolithotrophs in their natural environment involves a vertical migration through a chemocline, with sulphide-oxidising bacteria migrating up from high-sulphide zones to a more oxidised zone where they take up nitrate, which is carried down and acts as oxidant in sulphide oxidation. Analogous migrations could, as will be seen below, facilitate the use of other, much less well-characterised, possible energy sources for life on Earth.

The remote detection of chemolithotrophy on Earth is limited to the atmospheric detection of the fraction of atmospheric methane derived from chemolithotrophy, and possibly the occurrence of satellite-detectable sulphur particles in surface ocean water off Namibia. Methanogenesis can be regarded as chemolithotrophy when it uses hydrogen as electron donor and carbon dioxide as electron acceptor for generation of an electrochemical potential, as well as the source of organic carbon for growth by reducing carbon dioxide to the redox level of carbohydrate (Raven, 2009a). The satellite-detectable sulphur particles at the ocean surface (e.g. off Namibia) are the result of the oxidation to elemental sulphur of upwelled hydrogen sulphide produced by sulphate-reducing chemoorganotrophic bacteria in sediments mentioned above (Weeks et al., 2002) and from deeper in the sediment (Brüchert et al., 2009); the upwelled sulphide has escaped the attentions of the dense populations of sulphide-oxidising bacteria (Schulz et al., 1999). While much of the sulphide oxidation in the water column is abiotic, chemolithotrophic sulphide-oxidising bacteria can also be involved (Lavik et al., 2009). In the case examined by Lavik et al. (2009) off Namibia the chemolithotrophic oxidation of sulphide to sulphur and sulphate with nitrate as electron acceptor occurred in subsurface waters, and so the blooms of chemolithotrophic bacteria would not have been detectable by remote sensing.

## 2.3. OTHER POSSIBLE ENERGY SOURCES FOR AUTOTROPHY

### *2.3.1. Variations in External Salinity*

In principle, variations in external salinity could act as a source of energy for organisms (Muller and Schulze-Makuch, 2006). In a high-salinity environment, there is a driving force for ion entry, and ions could enter cells through rectifying channels. When these ion-loaded cells are exposed to low-salinity environments there is a driving force for ion loss, and ions exiting through transporters couple to nutrient uptake and or to ADP phosphorylation, thus contributing energy input to cells. This mechanism could operate for a benthic organism in a tidal estuary with large semi-diurnal changes in salinity. It is not clear what other environments could support this mechanism since other spatial differences are relatively small, e.g. across typical haloclines or between the thermohaline circulation and surrounding water. In both cases, the organism must move between the two salinity regimes. In addition, energy conversion by this mechanism gives obvious outcomes similar to those from anoxygenic aerobic bacteria and energy-conserving rhodopsins in archaea and bacteria, i.e. no redox products unless the ion-gradient-driven ion fluxes are coupled to endergonic electron transfer. The per unit biomass and global magnitude of transformation of energy useful to the organism by this mechanism is likely to be low. There would accordingly be little chance of remote detection of such metabolism even if there were redox products in view of their small global magnitude.

### *2.3.2. Variations of Water Flow over Benthic Organisms*

Raven (1988) indicates that the varying rates and directions of water movement over benthic intertidal and immediate subtidal organisms could drive actomyosin-based mechanochemical motors in the direction of ATP synthesis. This could only apply to relatively tall (mm – tens of mm or more) organisms that can have parts exposed to rapid water flow rather than just inhabiting the boundary layer. The use of essentially constant water flow over a benthic organism to generate intermediates from which metabolism can be energised could occur through variants on rotary molecular motors such as bacterial flagella or ATP synthetase (Nicholls and Ferguson, 2002), with suitable additions which interact with the water, and gearing. Again, there are questions of the extent to which redox reactions could be energised, the limited per organism and global magnitude of the energy conversion, and hence the possibility of detection of the products of any coupled redox reaction.

*2.3.3. Temperature Differences and Thermo-osmosis*

Muller and Schulze-Makuch (2006) give a detailed account of the possibilities of thermo-osmosis as an early mechanism of energising early life on Earth. The mechanism suffers from the relatively small temperature differences and the thermal conductivity of water and biological membranes (see Tsong and Kingsley, 1975). It is unlikely that thermo-osmosis is a significant contributor to the energetics of extant organisms. As for the two mechanisms considered in the last two paragraphs, there are questions of the extent to which redox reactions could be energised, the limited per organism and global magnitude of the energy conversion, and hence the possibility of detection of the products of any coupled redox reaction. In view of the small global magnitude of any contribution that temperature differences makes it would be difficult to detect, even if the organisms produced a significant quantity of spectroscopically detectable gas with a significant atmospheric half-life per unit of metabolism.

### 2.4. POTENTIAL ENERGY SOURCES FOR THE DEEP BIOSPHERE

The concept of a deep biosphere can be traced to the suggestion by Gold (1992) that there life occurs down to 6 m in the Earth's crust. This was followed by attempts to quantify the biomass and explore the phylogenetic diversity of the chemolithotrophic, including methanogen (Chapelle et al., 2002) deep subsurface biosphere, as well as to estimate what appears to be low metabolic rates on cell and a volume of habitat basis (Jørgensen and D'Hondt, 2006). There has also been culture work (Finster et al., 2009) and assembly of a genome from environmental genomic data indicating a single-species ecosystem (Chivian et al., 2008) to characterise two deep subsurface bacteria. A major energy source for chemolithotrophy by archaea and bacteria in the deep biosphere is thought to be hydrogen produced by radiolysis of water driven by the decay of potassium, thorium and uranium (Lin et al., 2004). Davies et al. (2009) discuss aspects of the remote sensing of the deep biosphere. Whether methane from the deep biosphere escapes to the atmosphere in sufficient quantities for remote detection needs further investigation. Methane can also be produced by outgassing from the planet interior, making detectable methane concentration in an atmosphere a very interesting but not unique signature to life.

## 3. Possibilities of Remote Detection of Autotrophy on Earth

Remote (satellite) sensing of oxygenic photolithotrophy on Earth involves spectroscopic detection of oxygen and ozone in the atmosphere, and of vegetation on land by the 'red edge', is possible since about 2.3 billion years and about

500 million years, respectively (see, e.g. Kaltenegger et al., 2007). A bigger collecting area is needed to detect for surface features than atmospheric absorption features. Note that oxygen and ozone could also be produced in detectable amounts by 'cryptic photosynthesis', indicating organisms that would not show any distinct surface features (Cockell et al., 2009a).

While oxygenic photolithotrophs in the ocean can be remotely sensed as spectral reflectance at two blue-green wavelengths with extremely high resolution using satellites in Earth's orbit for Earth's oceans, this can only detect pigments at a few metres depth at best. Such pigments can be those of non-oxygenic photosynthetic bacteria, so ground truthing is necessary here. This is also the case for major aggregations of oxygenic photolithotrophs such as coccolithophores and coral reefs whose detection by satellites was for Earth, where detectability is based on light scattering by calcite (coccolithophores) and aragonite (coral reefs) and more detailed closer observation is needed to demonstrate that the aggregations contain, and are produced by, oxygenic photolithotrophs.

The supplementation of energy from chemo-organotrophy by energy-transforming photochemistry is, as we have seen, difficult to quantify based on sampling of water bodies, and spectroscopic detection of the pigments by remote sensing of surface waters would be difficult to distinguish from the quantitatively predominant pigments of oxygenic photolithotrophs. This trophic mechanism has no specific gaseous products.

For chemolithotrophs on Earth, the global primary productivity is relatively low, decreasing the chance of spectroscopic detection even if the gas has a long atmospheric lifetime and is potentially detectable when present at a higher enough partial pressure detectable. The most likely candidate organism here is the subset of methanogens that use hydrogen as their reductant and carbon dioxide as their oxidant for chemiosmotic energy conversion (Lane et al., 2010; Nicholls and Ferguson, 2002) and as their carbon source for producing carbon skeletons for cell material, and so are autotrophs. Raven (2009a) suggested that half of global biological methanogenesis results from this kind of metabolism rather than from organic carbon sources, though this is probably an upper limit. As was pointed out above, the remote detection of the organisms responsible is unlikely, and even if they were detected their identification as chemolithotrophs would be next to impossible.

The other possible trophic modes on Earth, involving at least partial energisation by gradients of salinity, water flow rates and temperature, have not been documented so far. Even if they occurred, their global quantitative importance would, as discussed above, be small, and not detectable by remote sensing either as the organisms or a specific gaseous product. The negligible role for these mechanisms of Earth today does not, of course, rule out a role for them in the very early evolution of life on Earth (Kaufmann, 2009; Muller and Schulze-Makuch, 2006), although a chemolithotrophic origin involving some chemiosmotic energy conversion seems more likely as an early mean of energising life (Lane et al., 2010).

### 3.1. POSSIBILITIES OF REMOTE DETECTION OF AUTOTROPHY ON OTHER PLANETS AND MOONS

Within our solar system, there are analyses of possible energy supplies for any life on Mars from electromagnetic radiation for photolithotrophy (Cockell and Raven, 2004) and for chemolithotrophy in a deep biosphere (Fisk and Giovannoni, 1999) on Mars, and for Europa's ocean (Schulze-Makuch and Irwin, 2002). Of these, the only possibility for remote detection with current or plausibly projected technology is that of photolithotrophy, and this could only have applied to early Mars rather than today (Cockell and Raven, 2004). For planets outside the solar system, the only possibility for detection appears to be oxygenic photolithotrophy (Cockell et al., 2009a, b, c) in context with other atmospheric gases that put the detected oxygen/ozone in chemical context. A combination of $O_2/O_3$ and a reducing gas like $CH_4$ are good indicators for life that can be detected by a low-resolution (Resolution <50) spectrograph. Note that if the presence of biogenic gases such as $O_2/O_3 + CH_4$ may imply the presence of a massive and active biosphere, their absence does not imply the absence of life. Life existed on Earth before the interplay between oxygenic photosynthesis and carbon cycling produced an oxygen-rich atmosphere. The context of the detection of biomarkers is very important, e.g. on a hot planet, water can easily be photolyzed. Because the lighter gas $H_2$ is being lost to space preferentially, a build-up of oxygen in the planetary atmosphere could occur over a few million years and be remotely detectable. Such a situation would not indicate biology and can be excluded by knowing the planet's temperature. An important point is that the 'red edge' for pigments of oxygenic photolithotrophs would not be at 700–730 nm if the photolithotrophy involved three photochemical reactions rather than two, for any Earth-like planets in the continuously habitable zone of M stars with their maximum photon emission at longer wavelengths than of our sun, a G2V star (Wolstencroft and Raven, 2002; Cockell and Raven, 2004; Raven and Wolstencroft, 2004; Kiang et al., 2007b; Cockell et al., 2009a, b, c).

## 4. Conclusion

The emerging field of extrasolar planet search has shown an extraordinary ability to combine research by astrophysics, chemistry, biology and geophysics into a new and exciting interdisciplinary approach to explore planetary environments. It is relatively straightforward to remotely ascertain that Earth is a habitable planet, replete with oceans, a greenhouse atmosphere, global geochemical cycles and life – if one has data with arbitrarily high SNR and spatial and spectral resolution. The interpretation of observations of other planets with limited signal-to-noise ratio and spectral resolution as well as absolutely no spatial resolution, as envisioned for the first-generation instruments, will be far more challenging and implies that we need to gather information on the planet environment to understand what we will see. The spectrum will be used to identify the compounds of the planetary atmosphere, and constrain the potential habitable environment. In that context, we can then test if we have an abiotic explanation of all compounds seen in the atmosphere of such a planet. Oxygenic photolithotrophy is the means of energising a biosphere on other planets that is more likely to be remotely detected than are other means of energising biospheres.

If we do not have an abiotic explanation of the oxygen detected in an exoplanet's atmosphere, we can work with the exciting biotic hypothesis.

## 5. Acknowledgments

The University of Dundee is a registered Scottish Charity, No. SC 010596.

L. Kaltenegger acknowledges the support of the Harvard Origins of Life Initiative and the NASA Astrobiology Institute.

## 6. References

Arnold L, Gillet S, Lardière O, Riaud P, Schroeder J (2002) A test for the search for life on extrasolar planets. Astron Astrophys 352:231–237

Bjorn LO, Papageorgiou GC, Blankenship RE, Govindjee (2009) Why chlorophyll *a*? Photosynth Res 99:85–98

Brack A (1993) Liquid water and the origin of life. Orig Life Evol Biosph 23:3–10

Brüchert V, Curie B, Beard KR (2009) Hydrogen sulphide and methane emission of the central Namibian shelf. Prog Oceanogr 83:169–178

Buick R (2008) When did oxygenic photosynthesis evolve? Philos Trans R Soc Lond B Biol Sci 363:2731–2743

Canfield DE, Rosing MT, Bjerrum C (2006) Early anaerobic metabolisms. Philos Trans R Soc Lond B Biol Sci 361:1819–1836

Chapelle FH, O'Neill K, Bradley M, Methé BA, Ciufo SA, Knobel LL, Lowley DR (2002) A hydrogen-based subsurface microbial community dominated by methanogens. Nature 415:312–315

Chivian D, Brodie EL, Alm EJ, Culley DE, Dehal PS, SeSantis TZ, Gihring TM, Lapidus A, Lin LH, Moser DP, Richardson PM, Sotham G, Wanger G, Pratt LM, Anderson GL, Hazen TC, Brockman FJ, Arkin AP, Onstott TC (2008) Environmental genomics reveals a single-species ecosystem deep with earth. Science 322:275–278

Christensen PR, Pearl JC (1997) Initial data from the mars global surveyor thermal emission spectrometer experiment: observations of the earth. J Geophys Res 102:10875–10880

Cockell CS, Lee M (2002) Interstellar predation. J Br Interplanet Soc 55:8–29

Cockell CS, Raven JA (2004) Zones of photosynthetic potential on mars and the early earth. Icarus 169:300–310

Cockell CS, Kaltenegger L, Raven JA (2009a) Cryptic photosynthesis – extrasolar planetary oxygen without a surface biotic signature. Astrobiology 9:623–626

Cockell CS, Raven JA, Kaltenegger L, Logan RC (2009b) Planetary targets in the search for extrasolar oxygenic photosynthesis. Plant Ecol Divers 2:207–219

Cockell CS, Léger A, Fridlund M, Herbst TM, Kaltengger L, Absil C, Beichman C, Benz W, Blanc M, Brack A, Chelli A, Colangeli L, Cottin H, Coudé de Foresto F, Danchi WC, Desfrère D, den Herder J-W, Eiroa C, Greaves J, Henning T, Johnston KC, Jones H, Labardie L, Lammer H, Launhardt R, Lawson P, Lay OP, LeDuigou J-M, Liseau R, Malbet F, Martin SR, Mawet D,

Mourard D, Moutou C, Mugnier LM, Ollivier M, Paresce F, Quirrenbach A, Rabbia YD, Raven JA, Rottgering HJA, Rouan D, Santos NC, Selsis F, Serabyn A, Shibai H, Tamura M, Thiébaut E, Westal F, White GJ (2009c) Darwin – a mission to detect and search for life on extrasolar planets. Astrobiology 9:1–22

Cowan NB, Agol E, Meadows VS, Robinson T, Livengood TA, Deming D, Lisse CM, A'Hearn MF, Wellnitz DD, Seager S, Charbonneau D, the EPICA Team (2010) Alien maps of an ocean-bearing world. Astrophys J 700:915–923

Davies PCW, Benner SA, Cleland CE, Lineweaver CH, McKay CP, Wolfe-Simon F (2009) Signatures of a shadow biosphere. Astrobiology 9:241–249

Des Marais DJ, Harwit MO, Jucks KW, Kasting JF, Lin DNC, Lunine JI, Schneider J, Seager S, Traub WA, Woolf NJ (2002) Remote sensing of planetary properties and biosignatures on extrasolar terrestrial planets. Astrobiology 2:153–181

Finster KW, Cockell CS, Voytek MA, Gronstal AL, Kjedsen KJ (2009) Description of *Tetratococcus profundi* sp. nov., a deep-subsurface actinobacterium isolated from a Chesapeake impact crater drill (940 m depth). Antonie von Leuwenhoek 96:515–526

Fisk MR, Giovannoni SJ (1999) Sources of nutrients and energy for a deep biosphere on mars. J Geophys Res Planets 104:11805–11815

Ford E, Seager S, Turner EL (2001) Characterization of extrasolar terrestrial planets from diurnal photometric variability. Nature 412:885–887

Gensel PG (2008) The earliest land plants. Annu Rev Ecol Evol Syst 39:439–477

Gold T (1992) The deep, hot biosphere. Proc Natl Acad Sci USA 89:6045–6049

Gómez-Consanau L, Akram N, Lindell K, Petersen A, Neutze R, Milton DL, González JM, Pinhoussi J (2010) Proteorhodopsin phototrophy promotes survival of marine bacteria during starvation. PLoS Biol 8:e1000358

Jiao N, Zhan Y, Zeng Y, Hong N, Liu R, Chen F, Wang P (2007) Distinct distribution pattern of abundance and diversity of aerobic anoxygenic bacteria in the global ocean. Env Microbiol 9:3091–3099

Johnston DT, Wolfe-Simon F, Pearson A, Knoll AH (2009) Anoxygenic photosynthesis modulated proterozoic oxygen and sustained earth's middle age. Proc Natl Acad Sci USA 106:16925–16929

Jørgensen BB, D'Hondt S (2006) A starving majority deep beneath the seafloor. Science 314:932–934

Kaltenegger L (2010) Characterizing habitable exo-moons. Astrophys J 711:L1–L6

Kaltenegger L, Sasselov D (2010) Detecting planetary geochemical cycles in exoplanets: atmospheric signatures and the case of $SO_2$. Astrophys J 708:1162–1167

Kaltenegger L, Traub W (2009) Transits of earth-like planets. Astrophys J 698:519–527

Kaltenegger L, Traub WA, Jucks KW (2007) Spectral evolution of an earth-like planet. Astrophys J 658:598–616

Kaltenegger L, Selsis F, Fridlund M, Lammer H, Beichman C, Danchi W, Lvoa C, Henning T, Herbst T, Léger A, Liseau R, Lurine J, Paresce F, Penny A, Quirrenbach A, Röttgering H, Schneider J, Stam D, Tinetti G, White GD (2010) Characterization of terrestrial exoplanets and detection of biomarkers. Astrobiology 10:89–102

Kaufmann M (2009) On the free energy that drove primordial anabolism. Int J Mol Sci 10:1853–1871

Key T, McCarthy A, Campbell DA, Six C, Roy S, Finkel ZV (2010) Cell size trade-offs govern light-strategies in marine phytoplankton. Environ Microbiol 12:95–104

Kharecha P, Kasting J, Siefert J (2005) A coupled atmosphere-ecosystem model of the early Archean Earth. Geobiology 3:53–76

Kiang NY, Siefert J, Govindjee, Blankenship RE (2007a) Spectral signatures of photosynthesis. I. Review of earth organisms. Astrobiology 7:222–251

Kiang NY, Segura A, Tinetti G, Govindjee, Blankenship RE, Cohern M, Siefert J, Crisp D, Meadows VS (2007b) Spectral signatures of photosynthesis. Astrobiology 7:252–274

Lane N, Allen JF, Martin W (2010) How did LUCA make a living? Chemiosmosis in the origin of life. Bioessays 30:271–280

Lavik G, Stührman T, Brüchert V, Van der Plas A, Mohrholz V, Lam P, Muβmann M, Fuchs BM, Amann R, Lass U, Kuypers MMM (2009) Detoxification of sulphidic African shelf waters by blooming chemolithotrophs. Nature 457:581–584

Lin L-H, Slater GF, Lolar BS, Lacrampe-Coulume G, Onstott TC (2004) The yield and isotopic composition of radiolytic energy source for the deep subsurface biosphere. Geochim Cosmichim Acta 69:893–903

Lovelock JE (1975) Thermodynamics and the recognition of alien biospheres. Proc R Soc Lond B Biol Sci 189:167–180

Miller SD, Haddock SHD, Elvidge SV, Lee TF (2005) Detection of a bioluminescent milky sea from space. Proc Natl Acad Sci USA 102:14181–14184

Milo R (2009) What governs that reaction center excitation wavelength of photosystems I and II? Photosynth Res 101:59–67

Montañés-Rodriguez P, Pallé E, Goode PR, Hickey J, Koonin SE (2005) Globally integrated measurements of the earth's visible spectral albedo. Astrophys J 629:1175–1182

Montañés-Rodriguez P, Palle E, Goode PR (2007) Measurements of the surface brightness of the earthshine with applications to calibrate lunar flashes. Astronom J 134:1145–1149

Muller AWJ, Schulze-Makuch D (2006) Thermal energy and the origin of life. Orig Life Evol Biosph 36:177–189

Munz FW, McFarland WN (1973) The significance of spectral position in the rhodopsins of tropical marine fish. Vision Res 13:1829–1874

Nicholls DG, Ferguson SJ (2002) Bioenergetics 3. Academic Press, London, pp. xviii + 297

Overmann J, Cypionka H, Pfennig N (1992) An extremely low-light-adapted phototrophic sulphur bacterium from the Black Sea. Limnol Oceanogr 37:150–155

Owen T (1980) The search for early forms of life in other planetary systems - future possibilities afforded by spectroscopic techniques. In: Papagiannis MD (ed) Strategies for the search of life in the universe. Reidel, Dordrecht, pp 177–185

Pallé E, Ford EB, Seager S, Montañés-Rodríguez P, Vazquez M (2008) Identifying the rotation rate and the presence of dynamic weather on extrasolar earth-like planets from photometric observations. Astrophys J 676:1319–1329

Pallé E, Zapatero Osorio MR, Barrena R, Montañés-Rodríguez P, Martín EL (2009) Earth's transmission spectrum from lunar eclipse observations. Nature 459:814–816

Pavlov AA, Kasting JF, Brown LL, Rages KA, Freedman R, Greenhouse R (2000) Greenhouse warming by $CH_4$ in the atmosphere of early earth. J Geophys Res 105:981–992

Pavlov AA, Hurtgen MT, Kasting JF, Arthur MA (2003) Methane-rich proterozoic atmosphere? Geology 31:87–92

Purohit AN, Nautiyal AR, Taylor P (2008) Leaf optical properties of an alpine perennial herb *Sedum vaginatum* Clarke grown at two altitudes. Biol Plant 30:373–378

Raven JA (1984) A cost-benefit analysis of photon absorption by photosynthetic unicells. New Phytol 98:593–625

Raven JA (1988) Algae on the move. Bot J Scotl 36:505–515

Raven JA (2009a) Contributions of anoxygenic and oxygenic phototrophy and chemolithotrophy to carbon and oxygen fluxes in aquatic environments. Aqu Microb Ecol 56:177–192

Raven JA (2009b) Functional evolution of photochemical energy transformation in oxygen-producing organisms. Funct Plant Biol 36:505–515

Raven JA (2011) The cost of photoinhibition. Physiol Plant 142:87–104

Raven JA, Cockell CS (2006) Influence on photosynthesis of starlight, moonlight, planetlight, and light pollution (reflections on photosynthetically active radiation in the universe). Astrobiology 6:668–675

Raven JA, Wolstencroft RD (2004) Constraints on photosynthesis in earth and earth-like planets. In: Norris RP, Stootman FH (eds) Bioastronomy 2002, life among the stars. International astronomical union symposium 213. Astronomical Society of the Pacific, San Francisco, pp 305–308

Raven JA, Kübler JE, Beardall J (2000) Put out the light, and then put out the light. J Mar Biol Assoc UK 80:1–25

Rubinstein CV, Gevienne P, de la Puente GS, Actini RA, Steeman P (2010) Early middle Ordovician evidence for land plants in Argentina (Eastern Antarctica). New Phytol 188:365–369

Sabeh G, Kirkup BC, Rosenberg M, Stambler N, Polz NF, Béjà O (2007) Adaptation and spectral tuning in divergent marine proteorhodopsins from the eastern marine proteorhodopsin for the eastern Mediterranean and the Sargasso Sea. ISME J 1:48–55

Schulz HN, Brinkhoff T, Ferdelman TG, Hernández Mariné M, Teske A, Jørgensen BB (1999) Dense populations of giant sulphur bacteria in Namibian shelf sediments. Science 284:493–495

Schulze-Makuch D, Irwin LN (2002) Energy cycling and hypothetic organisms in Europa's ocean. Astrobiology 2:105–121

Seager S, Turner EL, Schafer J, Ford EB (2005) Vegetation's red edge: a possible spectroscopic biosignature of extraterrestrial plants. Astrobiology 5:372–390

Sleep NH, Bird DK (2008) Evolutionary ecology during the rise of dioxygen in the earth's earliest biosphere. Philos Trans R Soc Lond B Biol Sci 363:2651–2664

Stomp M, Huisman J, Stal LJ, Matthijs HCP (2007) Colorful niches of phototrophic microorganisms shaped by vibrations of water molecules. ISME J 1:271–282

Tett P (1990) The photic zone. In: Herring PJ, Campbell K, Whitfield M, Maddock L (eds) Light and life in the sea. Cambridge University Press, Cambridge/New York

Tomescu AMF, Pratt LM, Rothwell GW, Strother PK, Nadon GC (2009) Carbon isotopes support the presence of extensive land floras pre-dating the origin of vascular plants. Palaeogeogr Palaeoclimatol Palaeoecol 283:45–59

Tsong TY, Kingsley E (1975) Hemolysis of human erythrocyte induced by a rapid temperature jump. J Biol Chem 256:786–789

Turnbull MC, Traub WA, Jucks KW, Woolf NJ, Meyer MR, Gorlova N, Skrutskie MF, Wilson JC (2006) Spectrum of a habitable world: earthshine in the near-infrared. Astrophys J 644:551–559

Weeks SJ, Currie B, Bakun A (2002) Massive emissions of toxic gases in the Atlantic. Nature 415:493–494

William DM, Gaidos E (2008) Detecting the glint of starlight on the oceans of distant planets. Icarus 195(2):927–937

Wolstencroft RD, Raven JA (2002) Photosynthesis: likelihood of occurrence and possibility of detection of earth-like planets. Icarus 17:535–548

Woolf NJ, Smith PS, Traub WA, Jucks KW (2002) The spectrum of earthshine: a pale blue dot observed from the ground. Astrophys J 574:430–442

Zubkov MV (2009) Photoheterotrophy in marine prokaryotes. J Plankt Res 31:935–958

Biodata of **Joseph Seckbach**, editor of this volume and the author (with **Julian Chela-Flores**) of the chapter "***Habitable Environments by Extremophiles on Earth, the Solar System, and Elsewhere.***"

**Professor Joseph Seckbach** is the founder and chief editor of the book series ***Cellular Origins, Life in Extreme Habitats and Astrobiology*** (**'COLE'**). See **www.springer.com/sereis/5775**. He is the author of several chapters in this series. Dr. Seckbach earned his Ph.D. from the University of Chicago, Chicago, IL (1965) and spent his postdoctoral years in the Division of Biology at Caltech (Pasadena, CA). Then, he headed at the University of California at Los Angeles (UCLA) a team for searching for extraterrestrial life. He has been appointed to the faculty of the Hebrew University (Jerusalem, Israel) and spent his sabbatical periods in UCLA, Harvard University, and served at Louisiana State University (LSU, Baton Rouge, LA) as the first chair for the Louisiana Sea Grant. Dr. Seckbach obtained two DAAD fellowships (German fellowships for exchange academicians) in Tübingen and in Ludwig Maximilians University in Munich.

Among his publications are books, scientific articles, edited and translated of popular books. Dr. Seckbach is the co-author (with R. Ikan) of the *Chemistry Lexicon* (1991, 1999) and a co-editor of ***Proceeding of Endocytobiology VII Conference*** (Freiburg, Germany, 1998) and the ***Proceedings of Algae and Extreme Environments meeting*** (Trebon, Czech Republic, 2000). His edited volume (with Richard Gordon) entitled ***Divine Action and Natural Selection: Science, Faith, And Evolution*** has been published by World Scientific Publishing Company (2008). His recent interest is in the field of enigmatic microorganisms, life in extreme environments and astrobiology.

E-mail: **seckbach@huji.ac.il**

J. Seckbach (ed.), *Genesis - In The Beginning: Precursors of Life, Chemical Models and Early Biological Evolution*, Cellular Origin, Life in Extreme Habitats and Astrobiology 22, 859–870
DOI 10.1007/978-94-007-2941-4_43, 

**Professor Julian Chela-Flores** was born in Caracas, Venezuela and studied in the University of London, England, where he obtained his Ph.D. in quantum mechanics (1969). He was a researcher at the Venezuelan Institute for Scientific Research (IVIC) and Professor at Simón Bolívar University (USB), Caracas until his retirement in 1990. During his USB tenure, he was the Dean of Research for 6 years. He is a Fellow of The Latin American Academy of Sciences, The Academy of Sciences of the Developing World, the Academy of Creative Endeavors (Moscow) and a corresponding member of the Venezuelan 'Academia de Física, Matemáticas y Ciencias Naturales'. His current positions are staff associate of the Abdus Salam International Centre for Theoretical Physics (ICTP), Trieste, research associate, Dublin Institute for Advanced Studies (DIAS) and professor titular, Institute of Advanced Studies (IDEA), Caracas. His particular area of expertise is astrobiology, in which he is the author of numerous papers. He organised a series of conferences on Chemical Evolution and the Origin of Life from 1992 to 2003. In 2001, he published the book: *The New Science of Astrobiology From Genesis of the Living Cell to Evolution of Intelligent Behaviour in the Universe* (Kluwer), and in 2009 he published the book: *A Second Genesis: Stepping-Stones Towards the Intelligibility of Nature* (World Scientific Publishers).

E-mail: **chelaf@ictp.it**

# HABITABLE ENVIRONMENTS BY EXTREMOPHILES ON EARTH, THE SOLAR SYSTEM, AND ELSEWHERE

**JOSEPH SECKBACH[1] AND JULIAN CHELA-FLORES[2,3]**
[1]*Hebrew University of Jerusalem, P.O.B. 113, 220 Mevo Hadas, Efrat 90435, Israel*
[2]*The Abdus Salam ICTP, Strada Costiera 11, 34151 Trieste, Italy*
[3]*Instituto de Estudios Avanzados, IDEA, Caracas 1015A, República Bolivariana de Venezuela*

## 1. Introduction

We know that life exists on Earth in almost every ecological niche. One of the prerequisites for life is the availability of liquid water, sources of energy, and a reasonable supply of organic molecules. From our experience with the Earth biota, wherever there is water, there is a good opportunity of finding living organisms.

The search for extraterrestrial life is encouraged by a comparison between organisms living in severe environmental conditions on Earth and the physical and chemical conditions that exist on some Solar System bodies. The extremophiles that could tolerate more than one factor of harsh conditions are called polyextremophiles. There are unicellular and even multicellular organisms that are classified as hyperthermophiles (heat lovers), psychrophiles (cold lovers), halophiles (salt lovers), barophiles (living under high pressures), and acidophiles (living in media of the lower scale of pH). At the other end of the pH scale, they are called alkaliphiles (namely, microbes that live at the higher range of the pH scale). Thermoacidophilic microbes thrive in elevated thermoenvironments with acidic levels that exist ubiquitously in hot acidic springs.

*Cyanidium caldarium* is a classical example of an acidothermophilic red alga that thrives in places such as hot springs (<57° and in the range 0.2–4 pH). This algal group shows a higher growth rate (expressed as number of cells and higher oxygen production when cultured with a stream of pure $CO_2$ rather than when bubbled with a stream of air (Seckbach, 2010). It has been reported that *Cyanidium* cells resisted being submerged in sulfuric acid (1N $H_2SO_4$). This is a practical method for purifying cultures in the laboratory and eliminating other microbial contamination (Allen, 1959). The psychrophiles thrive in cold environments, such as within the territories found in the Siberian permafrost, around the North Pole in Arctic soils, and they may also grow in Antarctica.

Barophilic microorganisms can tolerate a pressure of 1,000 atm on the seafloor, while other barophilic microorganisms have been detected in the subsurface of dry land. In hypersaline areas (such as the Dead Sea, Israel), we find halophilic bacteria (Arahal et al., 1999) and algae that can balance the osmotic pressure of hypotonic external solutions (Oren, 1988).

*Chroococcidiopsis* is one of the most primitive cyanobacterium known so far. This microbe survives in a wide range of extreme habitats that are hostile to most other forms of life. *Chroococcidiopsis* grows in hot springs, in hypersaline habitats, in a number of hot, arid deserts throughout the world, as well as in the frigid Ross Desert in Antarctica (Fewer et al., 2002).

Recently, the segmented microscopic animals tardigrades (0.1–1.5 mm) have been under investigations (Goldstein and Blaxter, 2002; Horikawa, 2008). These "water bears" are polyextremophilic and are able to tolerate a temperature range from about 0°C up to +151°C (much more that other known microbial prokaryotic extremophiles, Bertolani et al., 2004). But even in low Earth orbit, extreme temperatures are possible: tardigrades can survive being heated for a few minutes to 151°C, or being chilled for days at −200°C, or for a few minutes at −272°C, 1° warmer than absolute zero (Jönsson et al., 2008). These extraordinary temperatures were discovered by an ESA project of research into the fundamental physiology of the tardigrade, named TARDIS. Tardigrades are also known to resist high radiation, vacuum, and anhydrous condition for a decade in a dehydrated stage and can tolerate a pressure of up to 6,000 atm. These aquatic creatures are ideal candidates for extraterrestrial life and for withstanding long periods in space. They have already been used in space and have survived such stress.

For further information see in Google the images of this animal (http://en.wikipedia.org/wiki/Tardigrade), and in Tardigrade—New World Encyclopedia (http://www.newworldencyclopedia.org/entry/Tardigrada).

Several chapters dealing with relevant topics as this chapter have been published in the *Cellular Origin, Life in Extreme Habitats and Astrobiology* series www.springer.com/series/5775.

## 2. Some Microbes May Be Dormant for Long Periods in Harsh Conditions

Some microorganisms live in desiccation conditions (in dormant stage and at minimum metabolic rates). In such circumstances, these microbes, or their spores, could last in a dormant stage for millions of years before being revived. Recently, researchers have breathed new life into bacteria trapped deep under glacial ice in Greenland for over 120,000 years. In addition, scientists found an ancient ecosystem below the "Blood Falls" in an Antarctic glacier (Mikucki and Priscu, 2007; Mikucki et al., 2009). It was determined that this community had survived millions of years in a salty pool without light or oxygen. J. Grom (2009) describes the phenomenon as follows:

> Scientists have found life in an ecosystem trapped underneath a glacier in Antarctica for nearly 2 million years. The microbes, they suggest, are surviving the dark, oxygen-free waters by drawing energy from sulfur and iron. The findings provide insight into how life may have survived "Snowball Earth"—periods when some scientists speculate that the planet was entombed in ice—and hint at the possibility of life in other inhospitable environments, such as Mars and Jupiter's icy moon Europa.

Among the biological samples, there was a diversity of bacteria that thrive in cold, salty water loaded with iron and sulfur. The water averages minus 10°C (the high salt concentration prevents the water from freezing). These bacterial cells convert iron and sulfur into their nourishment (chemosynthesis). The fact that life can thrive in one of the most extreme environments on Earth supports the hypothesis that it has emerged in extraterrestrial ecosystems as well. R. Cano and M. Borucki (1995) reported the revival and identification of bacterial spores within the intact body of a bee that was trapped in amber for 25–40 million years. R. H. Vreeland et al. (2000) have claimed to reawaken bacteria from spores inside a 250-million-year-old salt crystal. There has been some controversy regarding these ancient extant bacteria (Nickle et al., 2002; Oard, 2001), or whether they may represent the most ancient life forms on Earth (Hoyle, 2001). Other authors await confirmation of the Vreeland et al. results in different salt deposits (Adam, 2000).

## 3. Astrobiology

Does life exist beyond our planet and, if so, is it comparable to what we know here on Earth? These are some of the most fascinating questions facing science today, particularly astrobiology, the study of the origin, evolution, distribution, and destiny of life in the universe. Three strategies have been devised for the search for extraterrestrial life: firstly, the study of the cellular makeup of exotic organisms on Earth; secondly, the search for organic matter and living microorganisms beyond Earth; and last but not least, the use of radio telescopes to detect signals of intelligent behavior in the universe. The first strategy has focused on understanding how life began on Earth. Research has shown exotic organisms living in inhospitable environments, such as the seafloor, the Antarctic glacial sheets, and volcanic lava streams—all of which display temperatures and pressures that may have been present during the process of the Earth's evolution. Perhaps one of the most unexpected recent discoveries has been that there are underground ecosystems, which, to a large extent, are independent of sunlight, extending our old concept of what was a habitable zone in a given Solar System. Research into our own origins not only broadens our appreciation of the ability of extremophiles to conquer every accessible niche (cf., Part I in this book), but such investigation also helps us understand the environmental extremes tolerated by simple organisms.

The second strategy for deciding if we are not alone in the universe is to search for the simplest forms of organic matter—amino acids or proteins—that may be embedded in ancient rocks of planets, comets, or meteorites, or even suspended in interstellar clouds. The search has focused elsewhere in the Solar System: Mars, Europa (a moon of Jupiter), and Titan and Enceladus (satellites of Saturn). The discovery of meteorites from Mars suggests that all the terrestrial planets (Mercury, Venus, Earth, Moon, and Mars) at one stage in the past may have been in biological intercourse. There is compelling evidence that liquid water has flowed in the geologically recent past on Mars (or may even be flowing now).

The third strategy used in the search of life beyond Earth is the most relevant one to the subject matter of the present book. It relies on radio telescopes such as the very large one at the National Astronomy and Ionosphere Center in Arecibo, USA. These "dishes" actually have two roles to play: first and foremost, they help to examine wavelengths that cannot be seen by the human eye, for example, radio waves and microwaves. Such information has proven to be essential for understanding the movement and behavior of planets and stars. Secondly, radio telescopes also seek anomalies in microwaves and radio waves wafting across the universe. Such anomalies may represent the imprint of intelligent life. Thus far, astronomers have been scanning the radio and microwave spectrum for over half a century with no reliable signal yet from an extraterrestrial civilization. But this initial difficulty does not imply that the initiative is likely to be abandoned (Ekers et al., 2002).

## 4. Is Extraterrestrial Life a Possibility that Can Be Tested?

Data and photographs transmitted by the Voyagers revealed previously unknown details about each of the giant planets and their moons. Close-up images from the spacecraft uncovered a variety of phenomena in the Jupiter System: most surprising among them is the volcanic activity on Io, one of its so-called Galilean satellites. Among the Voyager discoveries, one of the most significant was the icy surface of the Jovian satellite Europa. The excitement surrounding Europa is due to the data retrieved by the subsequent Galileo mission (1995–2003). This mission has changed the way we look at the Solar System and especially Europa, and even to the possibility of exomoons that may be inhabitable (Kipping, 2009). This mission was the first to conduct long-term observations of the Jovian System from orbit. It found evidence of subsurface saltwater on Europa, Ganymede, and Callisto and further revealed the intensity of volcanic activity on Io. From the similarity of the processes that gave rise to planets and satellites, we may expect that hot springs may lie at the bottom of the ocean triggered by tidal stresses and radiogenic energy.

It has been assumed in the past that Jupiter's protonebula must have contained many organic compounds. Organisms similar to thermophiles could possibly exist at the bottom of Europa's ocean. However, given the incomplete understanding of the evolution of early life on Earth, at present, we should allow for the possibility that microorganisms are a possible Europan biota (cf., Sect. 4.2). We may add that up to the present time, we do not fully understand the evolution of the earliest ancestor of all life on Earth. Indeed, plate tectonics has obliterated fossils of the early organisms from the crust of the Earth, which would constitute the only record of the evolution of early life.

Testing whether life is extant on the icy satellites of the Outer Solar System depends on the choice of the right instrumentation. The British Penetrator

Consortium is developing an attractive possibility. Penetrators are being suggested for the exploration of the Solar System. There was a related proposal for ESA that had the name of LunarEX (Smith et al., 2008). It consisted of small projectiles that could be delivered at high velocity to reach just beneath the surface of satellites for probing samples of surficial chemical elements. They would be appropriate for in situ chemical laboratories. Eventually, these preliminary studies that may not be implemented have nevertheless encouraged the development of suitable instrumentation that could be tested on the icy surface of Europa and Ganymede in our next visit to the Jovian System (cf., Sect. 4.2 and Gowen et al., 2010).

## 4.1. MARS

The present view on the surface of Mars shows that Mars in the past was wet and warmer than today. Images show several contours of water bodies carved in the surface, such as tunnels, deep rivers, lakes, canyons, and other structures where water used to run. Therefore, our nearest neighbor planet is a candidate for having supported life in the past. We cannot exclude its presence in some isolated environments. The possibility of extending the biosphere deep into the silicate crust in another terrestrial planet deserves special attention. The emergence of life on Mars is pertinent to astrobiology, since we cannot exclude the analogy with organisms that have been found to inhabit deep in the silicate crust of the Earth. Such microbes may have been deposited with the original sediment. Life, in these conditions may have evolved during an early "clement period" that may have occurred contemporary with our own Early Archean: the Noachian Epoch, or Early Hesperian in Mars stratigraphy, according to the standard terminology (Sleep, 1994). Possible candidates for sites in which life may have evolved are located in the Tharsis region located on Mars' equator, at the western end of Valles Marineris, where volcanic activity has taken place since, by analogy with the Earth, the heat from underground magma may have produced hot springs, which are known to be possible sources of hyperthermophilic microorganisms (cf., Parts I and II). Knowledge of these possible locations raises the question whether life may have survived till the present confined to regions where pockets of liquid water may occur.

In the short term, the Mars missions that are being planned now will add valuable insights as to the possibility of extant or extinct life on Mars. For instance, the Mars Science Laboratory (MSL), a NASA rover is expected to be launched in 2011. The MSL rover will include instruments for the analysis of soil samples. It will also investigate the past or present ability of Mars to support microbial life. On the other hand, ExoMars (exobiology in Mars) is planned for 2018. This will be collaboration between ESA and NASA. A robotic rover will be sent to the surface of Mars. It will deploy a rover carrying analytical instruments dedicated astrobiology and geology. The ExoMars spacecraft will consist of an orbiter, the carrier module. The rover itself will carry a scientific payload.

One of its aims is to search for possible biosignatures of Martian life, past or present. Especially significant will be the European orbiter, whose aim includes tracking down sources of methane that have been detected in the past (Lefèvre and Forget, 2009). Confirming the presence of methane is a high priority; nevertheless, an abiogenic origin is thought to be equally plausible (Atreya et al., 2007).

## 4.2. EUROPA

The Galileo Space Mission (1995–2003) has provided ample evidence for an ocean on Europa underneath its frozen icy surface. One of the primary goals of astrobiology is to determine whether life ever existed in places other than the Earth and, if so, what were the environmental conditions that made it possible. The discovery of an independent life form, a separate tree of life from our own on Europa, would not only be fascinating in its own right, but it would shed revealing light on the microbes that inhabited the Earth more than 4 billion years ago, when higher temperatures were common and the continual bombardment of meteorites and comets made the surface of the Earth a hostile environment.

From the point of view of the possibility of the existence of life on Europa, we should consider a lake called Vostok (Karl et al., 1999), which is the largest of about 80 subglacial lakes in Antarctica. Its surface is of approximately 14,000 $km^2$, and its volume is 1,800 $km^3$. Indeed this Ontario-sized lake in Eastern Antarctica is also deep, with a maximum depth of 670 m. On the other hand, from the point of view of microbiology, the habitat analogue provided by Lake Vostok for the Europa environment seems appropriate. Lake Vostok appears to be harboring hydrothermal vents beneath the water surface. This is suggestive of what may be occurring on Europa. The circulation of pure water in Lake Vostok will be driven by the differences between the density of meltwater and lake water. Geothermal heating will warm the bottom water to a temperature higher than that of the upper layers.

The water density will decrease with increasing temperature resulting in an unstable water column. This leads to vertical convective circulation in the lake, in which cold meltwater sinks down the water column and water warmed by geothermal heat ascends up the water column (Siegert et al., 2001). Similarly, Europa may also have geothermally heated warm water under its ice crust. Processes of the type that occur in Lake Vostok may be taking place on Europa, where biogenic sulfur may be reaching the surface (Singer, 2003).

Early discussions considered the possibility exploring Europa's habitability in terms of direct use of a submersible called a hydrobot (Horvath et al., 1997). This question is still relevant a decade later, in terms of new NASA tests of an autonomous underwater vehicle (AUV) called "endurance" for the Astrobiology Science and Technology for Exploring Planets (ASTEP) program (Doran et al., 2007), a worthy successor of our cryobot-hydrobot early planning.

In relation with the distribution and destiny of life in the universe, we have argued that if the experiments on evolution were to be successful, the science of the distribution of life in the universe would lie on solid scientific bases, for instance, direct verification of whether the transition prokaryote-eukaryote has taken place within the Solar System bodies (Chela-Flores, 1998, 2000).

Beyond Galileo, a return to the Jupiter System is being considered in the next decade with the Europa Jupiter System Mission (EJSM), a worldwide collaboration that will point mainly on Europa and Ganymede, the largest satellite in the Solar System. The mission consists of two flight elements operating in the Jovian System: the NASA-led Jupiter Europa Orbiter (JEO) and the ESA-led Jupiter Ganymede Orbiter (JGO). JEO and JGO will explore Europa and Ganymede, respectively (Grasset et al., 2009). Possible biomarkers accessible to EJSM have been discussed recently (Chela-Flores and Kumar, 2008; Chela-Flores, 2010).

## 4.3. IS HABITABILITY POSSIBLE ELSEWHERE?

Titan is a potential cradle for life. After the early successes of the Cassini-Huygens Mission, many interesting questions have been raised, including the source of methane, and possible ammonia-water ocean, inside this large satellite that resembles the Archaean Earth in some respects. For instance, Titan has a nitrogen atmosphere, and so does the Earth, including its atmosphere before life. Titan has organics that are almost certainly supplied in the absence of life. Not all the Earth's prebiotic ingredients are present on Titan though, because the Earth probably had $CO_2$ unlike Titan (Coustenis and Taylor, 2008).

Other reasons for focusing on Enceladus is that Cassini flew within 175 km in 2005 confirming in detail the presence of an atmosphere: their instruments found that the atmosphere contains water vapor comprising up to about 65%, with molecular hydrogen at about 20%. The rest is mostly carbon dioxide and some combination of molecular nitrogen and carbon monoxide. Another Cassini instrument showed that the south pole is warmer than near the equator. The poles should be colder because the Sun shines so obliquely there. However, in small areas of the pole, concentrated near the fractures known as the "tiger stripes," the temperatures can reach temperatures of well under −110 K (−261 F). This should be compared with the equatorial temperature of approximately 80 K.

Cassini has also confirmed that icy jets shooting up to 500 km are ejected from Enceladus, a tiny satellite of Saturn. The presence of liquid water in its interior raises this moon to a prime candidate for the search for life. Saturn's moon Enceladus emits plumes of water vapor and ice particles from fractures near its south pole, suggesting the possibility of a subsurface ocean. The water plumes could be caused by a liquid ocean many kilometers underground rather than by geysers erupting from a salty ocean just beneath the moon's surface (Postberg et al., 2009).

## 5. Conclusions

All the above facts encourage the intensification of search for life on Earth's closest neighbors (terrestrial planets and satellites in the Outer Solar System). Further investigations, and especially the forthcoming missions to Europa and the various missions to Mars, will shed additional light on the potential extraterrestrial life on or inside Mars and Europa. A central role would be played by appropriately chosen instrumentation, and we have described the high hopes that penetrators promise to yield in our search for life in Solar System exploration. We conclude that if life can thrive in some of the most extreme environments on earth, perhaps it can meet the challenges of existing elsewhere in the Solar System, including a Europan biota under its icy surface sustained by chemosynthesis on hydrothermal vents at the seafloor.

## 6. Acknowledgment

The senior author (JS) thanks the Israeli and Hungarian Academies of Sciences and Humanities for their kindly support towards and during the conference of Astronomy and Civilization in Budapest (Aug. 2009). Special appreciation is due to Anna-Teresa Tymieniecka—President of The World Phenomenology Institute—for permission to reprint this chapter in the forthcoming volume *Origins: Genesis, Evolution and Diversity of Life* (2nd Edition) Springer 2011.

## 7. References

Adam D (2000) Hardcore hibernation. Published online (19 October), Nature. doi:10.1038/news 001019-9

Allen MB (1959) Studies with *Cyanidium caldarium*, an anomalously pigmented chlorophyte. Arch Mikrobiol (Berlin, Heidelberg) 32:270–277

Arahal DR, Marquex MC, Volcani BE, Schleifer KH, Ventosa A (1999) *Bacillus marismortui* sp. nov., a new moderately halophilic species from the Dead Sea, Int J Syst Evol Microbiol 49:521–530

Atreya SK, Mahaffy PR, Wong AS (2007) Methane and related trace species on Mars: origin, loss, implications for life, and habitability. Planet Space Sci 55:358–369

Bertolani R, Guidetti R, Jönsson KI, Altiero T, Boschini D, Rebecchi L (2004) Experiences with dormancy in tardigrades. J Limnol 63(Suppl 1):16–25

Cano R, Borucki M (1995) Revival and identification of bacterial spores in 25 to 40 million year old Dominican amber. Science 268:1060–1064

Chela-Flores J (1998) A search for extraterrestrial eukaryotes: physical and biochemical aspects of exobiology. Orig Life Evol Biosph 28:583–596. http://www.ictp.trieste.it/~chelaf/searching_for_extraterr.html

Chela-Flores J (2000) Testing the Drake Equation in the solar system. In: Lemarchand GA, Meech K (eds) A new era in astronomy. ASP conference series, vol 213. Astronomical society of the Pacific, San Francisco, pp 402–410. http://www.ictp.trieste.it/~chelaf/TestingDrakeEq.html

Chela-Flores J (2010) Instrumentation for the search of habitable ecosystems in the future exploration of Europa and Ganymede. Int J Astrobiol 9:101–108. http://www.ictp.it/~chelaf/jcf_IJA_2010.pdf

Chela-Flores J, Kumar N (2008) Returning to Europa: can traces of surficial life be detected? Int J Astrobiol 7:263–269 (copyright holder: Cambridge University Press). http://www.ictp.it/~chelaf/JCFKumar.pdf

Coustenis A, Taylor FW (2008) Titan exploring an earthlike world, 2nd edn. WSP, Singapore, 412 pp

Doran PT, Stone W, Priscu J, McKay Ch, Johnson A, Chen B (2007) Environmentally Non-Disturbing Under-Ice Robotic ANtarctiC Explorer (ENDURANCE), American Geophysical Union, Fall meeting, abstract #P52A-05

Ekers RD, Kent Cullers D, Billingham J, Scheffer LK (2002) SETI 2020: a roadmap for the search for extraterrestrial intelligence. SETI Press, Mountain View, 549 pp

Fewer D, Friedl T, Büdel B (2002) Chroococcidiopsis and heterocyst differentiating cyanobacteria are each other's closest living relatives. Mol Phylogenet Evol 23:82–90

Goldstein B, Blaxter M (2002) Quick guide: Tardigrades. Curr Biol 12:R475

Gowen RA, Smith A, Fortes AD, Barber S, Brown P, Church P, Collinson G, Coates AJ, Collins G, Crawford IA, Dehant V, Chela-Flores J, Griffiths AD, Grindrod PM, Gurvits LI, Hagermann A, Hussmann H, Jaumann R, Jones AP, Joy A, Sephton KH, Karatekin O, Miljkovic K, Palomba E, Pike WT, Prieto-Ballesteros O, Raulin F, Sephton MA, Sheridan MS, Sims M, Storrie-Lombardi MC, Ambrosi R, Fielding J, Fraser G, Gao Y, Jones GH, Kargl G, Karl WJ, Macagnano A, Mukherjee A, Muller JP, Phipps A, Pullan D, Richter L, Sohl F, Snape J, Sykes J, Wells N (2010) Penetrators for in situ sub-surface investigations of Europa. Adv Space Res 48(4):725–742

Grasset O, Lebreton J-P, Blanc M, Dougherty M, Erd C, Greeley R, Pappalardo B, the Joint Science Definition Team (2009) The Jupiter Ganymede Orbiter as part of the E J. SA/NASA Europa Jupiter System Mission (EJSM), EPSC abstracts, 4: EPSC2009–784, European Planetary Science Congress

Grom J (2009) Ancient ecosystem discovered beneath Antarctic glacier. Science Now Daily News 16 April

Horikawa DD (2008) The Tradigrade Ramazzottium varieornatus as a model animal for Astrobiological studies. Biol Sci Space 22(3):93–98

Horvath J, Carsey F, Cutts J, Jones J, Johnson E, Landry B, Lane L, Lynch G, Jezek K, Chela-Flores J, Jeng T-W, Bradley A (1997) Searching for ice and ocean biogenic activity on Europa and Earth. In: Hoover RB (ed) Instruments, methods and missions for investigation of extraterrestrial microorganisms. Proceedings of SPIE, vol 3111. SPIE, Bellingham, pp 490–500. http://www.ictp.it/~chelaf/searching_for_ice.html

Hoyle B (2001) Ancient bacteria may be oldest life form. Am Soc Microbiol News 67. http://newsarchive.asm.org/jan01/topic4.asp

Jönsson KI, Rabbow E, Schill RO, Harms-Ringdahl M, Rettberg P (2008) Tardigrades survive exposure to space in low earth orbit. Curr Biol 18(17):R729–R731

Karl DM, Bird DF, Björkman K, Houlihan T, Shackelford R, Tupas L (1999) Microorganisms in the accreted ice of Lake Vostok, Antarctica. Science 286:2144–2147

Kipping D (2009) Transit timing effects due to an exomoon. Mon Not R Astron Soc 392:181–189

Lefèvre F, Forget F (2009) Observed variations of methane on Mars unexplained by known atmospheric chemistry and physics. Nature 460:720–723

Mikucki JA, Priscu JC (2007) Bacterial diversity associated with blood falls, a subglacial outflow from the Taylor Glacier, Antarctica. Appl Environ Microbiol 73:4029–4039

Mikucki JA, Pearson A, Johnston DT, Turchyn AV, Farquhar J, Schrag D, Panbar AD, Priscu JC, Lee PA (2009) A contemporary microbially maintained subglacial ferrous 'ocean'. Science 324:397–400

Nickle DC, Learn GH, Rain MW, Mullins JI, Mittler JE (2002) Curiously modern DNA for a '250 million-year-old' bacterium. J Mol Evol 54:134–137

Oard MJ (2001) Aren't 250 million year old live bacteria a bit much? Creation Ministries International, article 2415

Oren A (1988) The microbial ecology of the Dead Sea. In: Marshall KC (ed) Advances in microbial ecology, vol 10. Plenum Publishing Company, New York, pp 193–229

Postberg F, Kempe S, Schmidt J, Brilliantov N, Beinsen A, Abel B, Buck U, Srama R (2009) Sodium salts in E-ring ice grains from an ocean below the surface of Enceladus. Nature 459:1098–1101

Seckbach J (2010) Overview of Cyanidian biology. In: Seckbach J, Chapman D (eds) Red algae in genome age. Cellular Origin, Life in Extreme Habitats and Astrobioloy. Springer, Dordrecht, 13:343–356

Siegert MJ, Ellis-Evans JC, Tranter M, Mayer C, Petit J, Salamantin A, Priscu JC (2001) Physical, chemical and biological processes in Lake Vostok and other Antarctic subglacial lakes. Nature (London) 414:603–609

Singer E (2003) Vital clues from Europa. New Sci Mag 2414:22–23. http://www.ictp.it/~chelaf/VitalClues.pdf

Sleep NH (1994) Martian plate tectonics. J Geophys Res 99:5639–5655

Smith A, Crawford IA, Gowen RA, Ball AJ, Barber SJ, Church P, Coates AJ, Gao Y, Griffiths AD, Hagermann A, Joy KH, Phipps A, Pike WT, Scott R, Sheridan S, Sweeting M, Talboys D, Tong V, Wells N, Biele J, Chela-Flores J, Dabrowski B, Flannagan J, Grande M, Grygorczuk J, Kargl G, Khavroshkin OB, Klingelhoefer G, Knapmeyer M, Marczewski W, McKenna-Lawlor S, Richter L, Rothery DA, Seweryn K, Ulamec S, Wawrzaszek R, Wieczorek M, Wright IP, Sims M (2008) LunarEX – a proposal to cosmic vision. Exp Astron.doi: 10.1007/s10686–008–9109–6. http://www.ictp.it/~chelaf/Penetrator.pdf

Vreeland RH, Rosenzweig WD, Powers DW (2000) Isolation of a 250 million-year-old halotolerant bacterium from a primary salt crystal. Nature 470:1075–1077

# PART VII: HISTORY AND FRONTIERS OF ASTROBIOLOGY AND THE HUMANITIES

Aretxaga-Burgos
Chela-Flores
Raulin-Cerceau
Strick

Biodata of **Roberto Aretxaga-Burgos** and **Julian Chela-Flores,** co-authors of the chapter "***Cultural Implications of the Search and Eventual Discovery of a Second Genesis.***"

**Dr. Roberto Aretxaga-Burgos** was born in Bilbao, Spain. He studied in the University of Deusto, Bilbao, Spain, where he obtained his Ph.D. in Philosophy in 1998. He is Universitary Specialist in Science, Technology and Society (STS). Currently he is a researcher in the Department of Philosophy, University of Deusto. His special area of expertise is J.D. García-Bacca's thought, in which he is the author of numerous studies, outstanding among which is *La filosofía de la técnica de J. D. García Bacca* (Deusto University Press, 1999). The Venezuelan Science and Technology Ministry used this work for some of its programs in 2004. He was an Organizing Committee Member of the International Philosophy Congress: *Centenario del Nacimiento de J. D. García Bacca* (Bilbao, 2001), and co-editor of its Congress Proceedings (2003). He was a contributor (by invitation of the Juan David García-Bacca Foundation) to the Commemorative book *Juan David García Bacca: "Vivir dos veces despierto" 1901/1992* (Caracas, 2005), Venezuela's National Book Award 2005. His research interests are Philosophical Anthropology, Philosophy of Mind, and Philosophy of Astrobiology, in which he is the author of numerous papers. He is the Promoter, Director, and Co-ordinator of the Multidisciplinary Series *Astrobiologia y Filosofía* in the Deusto University's journal *Letras de Deusto*, as well as the homonymous Section in Cibernous (Philosophy in Network Educative Project: www.cibernous.com). He also collaborates with the Mexican Society of Astrobiology (SOMA). He is the author of *Filosofía de la Astrobiología* website (http://astrobiologiayfilosofia.blogia.com).

E-mail: **roarebur@hotmail.com**

J. Seckbach (ed.), *Genesis - In The Beginning: Precursors of Life, Chemical Models and Early Biological Evolution*, Cellular Origin, Life in Extreme Habitats and Astrobiology 22, 873–890
DOI 10.1007/978-94-007-2941-4_44, 

**Professor Julian Chela-Flores** was born in Caracas, Venezuela, and studied in the University of London, England, where he obtained his Ph.D. in quantum mechanics (1969). He was a researcher at the Venezuelan Institute for Scientific Research (IVIC) and Professor at Simón Bolívar University (USB), Caracas, until his retirement in 1990. During his USB tenure he was Dean of Research for 6 years. He is a Fellow of The Latin American Academy of Sciences, The Academy of Sciences of the Developing World, the Academy of Creative Endeavors (Moscow), and a Corresponding Member of the Venezuelan "Academia de Física, Matemáticas y Ciencias Naturales." His current positions are Staff Associate of the Abdus Salam International Center for Theoretical Physics (ICTP), Trieste, Research Associate, Dublin Institute for Advanced Studies (DIAS), and Profesor-Titular, Institute of Advanced Studies (IDEA), Caracas. His particular area of expertise is astrobiology, in which he is the author of numerous papers. He organized a series of Conferences on Chemical Evolution and the Origin of Life from 1992 till 2003. In 2001 he published the book *The New Science of Astrobiology From Genesis of the Living Cell to Evolution of Intelligent Behavior in the Universe* (Kluwer) and in 2009 he published the book *A Second Genesis: Stepping-Stones Towards the Intelligibility of Nature* (World Scientific Publishers).

E-mail: **chelaf@ictp.it**

# CULTURAL IMPLICATIONS OF THE SEARCH AND EVENTUAL DISCOVERY OF A SECOND GENESIS

**ROBERTO ARETXAGA-BURGOS[1]**
**AND JULIAN CHELA-FLORES[2,3]**
[1]*Department of Philosophy, University of Deusto, Bilbao, Spain*
[2]*The Abdus Salam ICTP, Strada Costiera 11, 34014 Trieste, Italy*
[3]*Instituto de Estudios Avanzados, IDEA, Caracas 1015A, República Bolivariana de Venezuela*

## 1. Introduction: Implications of a Second Genesis Elsewhere in the Universe

The question we wish to address in this chapter concerns one of the most rewarding scientific and cultural activities that are possible at present. The frontier of science and humanities is the exciting territory we will discuss in this chapter. The fast development of technology has allowed us to address the issue of a possible "second Genesis" interpreted as the emergence of biological complexity elsewhere in the universe (McKay, 2001).

The origin of life is not fully understood. However, the general outline of chemical evolution of the precursors of the biomolecules has greatly advanced. The outline includes the likely pathways that nature may have followed during the molecular evolution that preceded the Darwinian evolution of the living cell. The seminal work of Charles Darwin in the nineteenth century established the basis for the second stage in the discussion of astrobiology.
Two questions emerge out of the pioneering work in astrobiology:

Firstly, we consider the implications of *the search* for a second Genesis: Since its search is already a fact, its implications are already observable in different aspects of our culture, such as in science, philosophy, and technology.

Secondly, in this chapter, in a second stage, we shall later discuss the implications of an eventual *discovery* of a second Genesis. If the search for a second Genesis is a fact, its eventual discovery is only a possibility, but a scientifically sound possibility, which places the study of its implications in the domain of heuristic interest.

Regarding the second stage of our discussion, we recall that the theory of Darwinian evolution is not a predictive theory. In order to get further insights as to what can be the eventual destiny of life in the universe that might not be evident from our current knowledge of biology, we should search for alternative manifestations of the living process. Having assumed that Darwinian evolution is a universal process, we may argue in favor of the inevitability of the origin and evolution of life, including intelligence.

In the wider context of general philosophical implications of the Science of Astrobiology, a shorter presentation of some of the topics below has been published in a recent book (Chela-Flores, 2011).

## 2. Cultural Implications of the Scientific Search of a Second Genesis

### 2.1. THE PLURALITY OF INHABITED WORLDS: FROM SPECULATION TO SCIENCE

The debate in Western philosophy on the existence of other inhabited worlds goes back, at least, to the early Greek philosophers, being carried on since that time up to the middle of the twentieth century in the realm of speculation. Only in the last few decades we have begun to dedicate effort, knowledge, and resources to settle this question through observation and experiments. Astrobiology emerges from such undertakings.

In other occasions, we have discussed extensively the scientific bases of astrobiology, reaching the conclusion that, either by the nature of its assumptions and its theoretical framework, astrobiology, understood as the search for a second Genesis, is indeed a valid scientific research (Aretxaga, 2006, 2008; Chela-Flores, 2003, 2006, 2008). What was originally a question in the realm of philosophy now has been shifted radically to the center of science. One example that will be discussed below illustrates our point: Astrobiology can explore in strictly scientific terms the possibility of whether the evolution of intelligent behavior is inevitable in an evolving cosmos.

This change of point of view experienced by the question of the plurality of inhabited worlds in our culture inserts the human beings, for the first time in history, facing the possibility of giving a definite answer to one of the most fascinating questions ever formulated.

### 2.2. ASTROBIOLOGY AS A TRANSDISCIPLINARY SCIENCE

It is not unusual to consider astrobiology as an intrinsically multidisciplinary or more precisely interdisciplinary activity. It is well known that its study requires a collaboration of knowledge and methods of a diversity of disciplines, in order to tackle its diverse relevant questions, so that astrobiology forces upon us bring together different areas of knowledge that often are far apart from each other. Nevertheless, some scientists maintain that astrobiology implies a deeper level of interaction between different disciplines: *transdisciplinary investigations*. This means a new approach to conceive the research activity, as well as its results.

Transdisciplinary science (TDS) has been defined as the integration of theoretical and methodological perspectives drawn from different disciplines, in order to produce new pathways for the investigation of a specific problem from a novel point of view (Stokols et al., 2003). In this sense, astrobiology is an outstanding example, since the problem of the search for a second Genesis may be approached in order of increasing complexity by different disciplines.

According to the Spanish Engineer J. A. Rodríguez-Manfredi, astrobiology exploits the synergy generated by different underlying disciplines, in order to create its own methods, procedures, and objectives. In this manner, it goes beyond the mere exchange of information between different fields of specialization, since research requires the learning of a common language and learning new mental approaches that were initially foreign to the separate disciplines. Consequently, astrobiology is contributing to the progress of scientific research with new challenges (Rodríguez-Manfredi, 2008). To sum up, astrobiology, besides transforming the question of the plurality of inhabited words from philosophy to science, is now contributing to a new way of doing science, transdisciplinarity.

## 2.3. THE SEARCH OF A SECOND GENESIS AS A PROMOTER OF COLLABORATION BETWEEN THE TWO CULTURES

Ever since the 1959 Rede Lecture by Charles Percy Snow (Snow, 1993), the suggestive term "two cultures" has called attention to the wide gap between science and humanities. These two intrinsically different cultural activities largely address the same questions. For this, we should all persevere in establishing a constructive dialogue. Biologist Edward Osborne Wilson has attempted to bridge the gap between "the two cultures" with his seminal 1998 book "Consilience: The Unity of Knowledge" (Wilson, 1998). In this book, Wilson discusses methods that have been used to unite the sciences and might in the future unite them with the humanities. The term *consilience* describes the synthesis of knowledge from different specialized fields of human endeavor.

The difficulty in finding a common dialogue is deep in our culture: Giordano Bruno was a pioneer in this intercultural dialogue (Bruno, 2000). In 1584, he wrote his first three Italian dialogues during a visit to England. These writings were stimulated by debates at the University of Oxford. In his third dialogue "*On the infinite universe and worlds,*" Bruno introduced concepts that are still at the center of astrobiology.

The lesson that we may draw from both history and our present experience is that the progress of human knowledge cannot be achieved, neglecting either of the "two cultures" that were pointed out by Snow. Astrobiologists themselves often have requested the presence of humanists in this field, so as to achieve together better and deeper insights in the scientific search for a second Genesis.

## 2.4. ASTROBIOLOGY AS A PROMOTER OF A NEW VISION OF REALITY AND HUMAN BEINGS

The Spanish engineer Rodríguez-Manfredi considers the consequences of the following remark: The instruments for observation model reality through its interaction, so that the aspect of a phenomenon will depend on the kind and stage of development of the artifacts employed for its observation—as in the case of the double nature of light: corpuscular and wavelike (Rodríguez-Manfredi, 2008). He concludes that astrobiology in so far as it contributes to new technologies and methods is already contributing to a change of perception that could lead to a revised version of both "life" and "universe."

E. Mayz-Vallenilla has gone somewhat deeper in this question. The Venezuelan philosopher considers that a human being has access in a natural manner to reality through a cognitive pattern elaborated with categories modeled predominantly by the spectrum of visible light (*optical-luminous logos*). Nevertheless, there are artifacts of a "meta-technical" nature, such as radio telescopes, scanning microscopes, or particle accelerators, that are in contrast to traditional telescopes and microscopes. The latter have been invented by outstanding minds (*meta-technical logos*). The fact that astrobiology, among other branches of modern science, requires for its implementation such advanced models, methods, and techniques (*meta-technical* technologies) implies that the basic concepts such as "physical universe" (*physis*) or "life" (*bios*) are profoundly transformed (Mayz-Vallenilla, 2003). Taking the importance of these concepts in the elaboration of other equally fundamental for the construction of a culture, such as that of mind, body, or intelligence, we could assert that its transformation shall reflect itself on the very definition of culture. To such a transformation, we could assign the label "Copernican." Just as our perception from time immemorial of the Solar System was radically transformed due to Copernicus and Galileo, also astrobiology is making it possible to conceive the reality of living organisms from a new point of view, which earlier we have called *sub specie Universorum*: Under this category, life appears as a phenomenon intimately connected with the universe, as a natural tendency of its matter-energy substratum. The immediate consequence of this new conception is that the word "life" no longer refers to an *exceptional* phenomenon, to signify a local episode, a special case of a universal phenomenon (Aretxaga, 2008). This is the central problem of astrobiology, namely to investigate whether life on Earth is but one example of a ubiquitous phenomenon.

Today, we are aware that both Copernicus and Galileo were correct (at least partially) and that their discoveries transformed profoundly human thoughts and feeling to reality (knowing and experiencing it), as well as thinking and feeling in it (to know oneself and to live in it). As a result of that change, our lives continue to evolve, without being aware of it, in a world constructed on its consequences in a wide spectrum: science, art, literature, ideas, projects, in our daily lives. In the same way, we can foresee that confirming our view of life *sub specie Universorum* by an eventual success in the scientific search for a second

Genesis may have consequences when we consider related phenomena, such as mind, intelligence, and even other aspects of our general knowledge. We can even imagine that all of these changes will, in turn, have an effect in altering how we conceive the world (*Weltanschauung*), including the perception that we have of ourselves, of our position in the tree of life, and of our place and responsibility in the cosmos. As pointed out by Lederberg, out of all the objectives of space exploration, "the study of life beyond the earth, which we term 'exobiology,' is the most subtle and demanding, for it insists: 'Know thyself'" (Lederberg, 1963).

## 2.5. THE SCIENCE OF ASTROBIOLOGY AND SCIENCE COMMUNICATION

Even though the debate in the Western world related to the plurality of inhabited worlds goes back at least 2,500 years, it only started to become popular topic of discussion in the nineteenth century. There were several causes that contributed, for instance, the observations and speculations of some astronomers—Schiaparelli, Lowell, or Flammarion—or the rising popularity of science fiction that encouraged by the possibility in principle of interplanetary travel, discovered in the topic of alien life a source of inspiration.

A key fact for understanding the rise in popularity of the search for extraterrestrial life during the twentieth century was the creation of NASA in 1958, during the cold war. Soon after its foundation, the American space agency created the *Office of Life Sciences* in 1960. This action demonstrated to the public opinion the priority that politicians and scientists granted to the relation between the studies of the origin of life and the technological challenge represented by interplanetary exploration. This intriguing relation between both of these fields, now formally established, contributed to a general perception that extraterrestrial life and its scientific quest began to be perceived socially as a serious effort within culture, but at the same time, the search began to excite popular imagination, occasionally excessively on this particular research line.

The sociocultural context in which astrobiology was born encouraged a constantly fluid and mutually beneficial interaction between this discipline and science fiction, but at the same time, it contributed to many misunderstandings as to the true nature and possibilities of such research. At present, even at the level of the school curricula, much confusion arises due to a misconception of the real limits of science and the humanities. Indeed, science is an activity that is essentially experimental and supported by a reasonable theoretical background that is often presented in detailed mathematical language that is not easily understood by everyone. The discussion of astrobiology with the general public requires a detailed knowledge of science, philosophy, and theology. It is difficult for single individuals, either scientists or humanists, to be able to approach reliably all aspects of contemporary culture. But it is nevertheless desirable to take additional modest steps in this challenging direction. Some experience in all aspects of culture is essential for a constructive, comprehensive, and interdisciplinary discussion that dares to consider the implications of science in philosophy and religion; for a mutual understanding, yet another chapter on the popularization of science would be justified.

On the other hand, the close temporal proximity between the emergence and acceptance of astrobiology as a new science turns this subject into an excellent playground to analyze and better comprehend the inevitable clash of interests and interactions between science and technology as well as between the latter two and other sectors of human knowledge. These additional areas of influence included also the humanities and theology, not excluding the relation of astrobiology with other areas of influence, such as the social, cultural, political, or economic ones. Due to the technologic challenges and its sense of adventure, astrobiology presents an excellent opportunity to advance in several fronts: pedagogical, science communication, and the new technologies (Alcibar, 2008).

## 3. Cultural Implications of the Discovery of a Second Genesis

### 3.1. WHAT CAN BE REASONABLY EXPECTED TO BE DISCOVERED

Often, astrobiologists have to face the criticism of doing science without a subject matter. Currently, we do not have a definition of life that is unanimously accepted. This is in fact an open question in both philosophy and science. However, our present knowledge of life on Earth allows us to elaborate a reasonably well-accepted explanatory model. Putting these remarks together with our recent insights into life in extreme environments induces us to think that a similar phenomenon—life—could have taken place elsewhere in the universe in the favorable conditions that the increasing repertoire of exoplanets begin to suggest.

Extrapolating elsewhere in the universe our present understanding for the emergence of life does provide us with an object for astrobiology, since it allows us:

(a) To set some theoretical bases for the possible existence of the main object of our research (a second Genesis)
(b) To suggest the elaboration of a coherent and realistic strategy, namely with possibilities of success, of the search and obtaining reliable data of extant, of extinct life or of simply the identification of reliable biomarkers

Consequently, astrobiology can be understood as the *scientific exploration* of the universe searching for a second Genesis (Aretxaga, 2008).

We are gradually discovering other places in the universe where such conditions exist, suggesting that it may be possible to detect the presence of bacterial life. As a consequence of the exciting discoveries of Cassini that come more than a decade after the epoch making discoveries of the Jovian moons by the Galileo mission, our general view of the conditions favorable for a second Genesis in our own Solar System, our cosmic backyard, have been greatly enhanced. The conditions for habitability are more widespread than were thought to be possible at the end of the twentieth century.

In our Solar System, there are several cases of some interest, where life can reasonably expected to be discovered:

- Europa may be a typical biofriendly body among extrasolar planets. The main problem is how to select appropriate experiments in situ after surface landers are able to filter melt water from Europa's frozen surface.
- Evolutionary tests are also suitable in potential water sources under the Martian surface, such as in its north pole. These environments may contain microorganisms, an exciting possibility that was raised by the Allan Hill meteorite. Although the possibility of having detected life in this meteorite was not subsequently confirmed, it was nevertheless an important contribution that stimulated discussion about the possibility of life elsewhere in our Solar System. Cases of possible habitability to test in the future have been discussed in the literature (Chela-Flores, 2010).

From the above examples, we have learned that searching for a second Genesis is feasible in the foreseeable future. The ultimate aim of a biology experiment in Solar System exploration is to develop robotic tests that are compatible with the necessarily reduced dimension of landers. In the case of a Martian mission to subterranean pockets of liquid water, we find a possibility that has attracted wide attention both by scientists and by the popular press. For this purpose, miniaturization of instrumentation is essential. Many difficulties though are inherent in the eventual design of a test that would intend to identify microbes, robotically, in any extraterrestrial environment. This question begins to be important, in view of the decisions that have to be made in the selection of biological experiments that should be performed in situ on Europa and on Mars.

The discovery on Earth of unicellular microorganisms (eukaryotes) in an analogous habitat to Mars—the Tinto River, Spain—suggests the possibility of the presence of eukaryotic microorganisms elsewhere in our Solar System. Since there is no evidence for inhibiting the origin and evolution of such microorganisms beyond the Earth in analogous habitats, it seems reasonable to discuss the hypothesis of the universality of eukaryogenesis (Chela-Flores, 2003). Finally, eukaryoticity seems to be a necessary condition for the emergence of a neuron (with the capability of setting up a series of action potentials that characterize it), and eventually, multicellular complex nervous systems seem inevitable. The lowest multicellular organisms where action potentials have been documented are the eukaryotic sponges (Villegas et al., 2000). In multicellular prokaryotes, such as stromatolitic colonies of cyanobacteria, action potentials seem unnecessary and, in fact, have not been detected to the best of our knowledge. These arguments reinforce the hypothesis that has been assumed anyway by radio astronomers that there will be other intelligences that would give away their presence by radio waves emitted from some exoplanet or exomoon (the SETI project).

## 3.2. THE BIOLOGY OF THE POSSIBLE: ARTIFICIAL LIFE?

We should not rule out other possibilities, and indeed some further possibilities have been discussed by Dick, such as postbiological entities (Dick, 2008). In this context, artificial life is a terminology used referring to attempts at developing systems with the properties of all biological organisms, for instance, just to name a few, self-reproduction, homeostasis, adaptability and mutational variation. The term is often associated with computer simulation of artificial life. This term has gained support since unlike robotics it emphasizes reprogramming, and other digital characteristics (Langton, 1990). In common with astrobiology this line of research can attempt to answer the question: "what is the simplest possible living system?"

## 3.3. RELEVANCE AND IMPLICATIONS OF THE DISCOVERY OF A SECOND GENESIS

Both the implications and the degree of importance of the eventual discovery of a second Genesis will depend on the type of life that is discovered. Considering our human condition, the discovery with the largest impact and relevance at all levels of our society would be the discovery of intelligent life, especially if it were technologically advanced. But the majority of scientists would be of the opinion that the discovery of the simplest extraterrestrial microbial life would be an event of historic proportions. In the present work, we shall distinguish implications at different levels.

### *3.3.1. Straightforward Consequences If Life Is Discovered*

In this context, we have two different aspects:

(a) The setting up of different protocols for our actions to be taken for communicating the news to society in general as well as preventing undesirable ethical, political, or scientific consequences, both for life on Earth as well as for the life that would be discovered (Race, 2008).
(b) Confirming the universality of biology, sharpening the definition of life, and the fall of biogeocentism: The sharp distinction between chance (contingency) and necessity (natural selection as the main driving force in evolution) is relevant for astrobiology. For this reason, it is important to document the phenomenon of evolutionary convergence at all levels, in the ascent from stardust to brain evolution.

The universality of biochemistry suggests that in Solar System missions, biomarkers should be selected from standard biochemistry. Given the importance of deciding whether the evolution of intelligent behavior has followed a convergent evolutionary pathway, and given the intrinsic difficulty of testing these ideas directly, we can alternatively begin testing the lowest stages of the evolutionary pathway within the Solar System. Within a few years, we will be in a position to search directly for evolutionary biomarkers on Europa, the Jovian satellite.

We have considered that if extant microorganisms were to be encountered, a possible set of evolutionary biomarkers may be considered. In addition, if only landers are possible in the foreseeable future, it is possible with the help of available techniques in biogeochemistry to decide on the possible biomarkers by sampling regions of the icy surface of Europa where there is abundant presence of non-ice elements such as, for example, sulfur (Chela-Flores, 2010).

Testing evolutionary biomarkers clearly lies in the distant future. However, we expect that technology will allow us to be in a position to undertake a variety of experiments in situ on the surface of other solar system bodies, such as the Jovian moon Europa. Given the length of time before we can test reliable biomarkers directly, a full discussion at the present time of the feasibility of carrying out a proper test is timely. In this sense, the discussion of biomarkers is reasonable at the present stage, since the Galileo mission has already provided us with a wealth of information about the chemical non-water-ice elements on the icy surface. The careful interpretation of such information might conceivably lead us to reliable biomarkers without actually penetrating Europa's icy surface. Testing directly the Europan icy surface with a lander is a possibility that can be taken into account.

The discovery of other solar systems suggests that their formation seems to be analogous to ours. This is compatible with the current extensive knowledge of interstellar matter. From the assumed universality of biology, it seems inevitable that intelligent behavior will emerge in the cosmos, provided certain conditions favorable to the presence of continuous life on a given planet (or satellite) are maintained. One of these conditions is that early stages in the formation of a solar system are characterized by a heavy bombardment period.

This period would end after a few hundred million years. Consequently, planetary conditions over geologic time are likely to allow the continuous presence of life, as it has already occurred on our own planet, once the heavy bombardment ceased. Observational techniques continually improve (the Terrestrial Planet Finder and the large telescopes of the European Space Observatory are now under construction). These new instruments will allow us to estimate the duration of the initial heavy bombardment in other solar systems as well as the subsequent life-favorable quiescent period. These future observations should give us a more precise idea of the temporal constraints that allow the continuous presence of life on a given planet. In our own Solar System, the most attractive site for the search for life has already been explored.

The Galileo mission to Jupiter and the Galilean satellites completed an 8-year period of continuous exploration in the year 2003. This mission focused our attention on Europa. In the seventeenth century, Galileo discovered Europa together with three other satellites Io, Ganymede, and Callisto. However, Europa remains the leading contender for being the host of an independent evolutionary line. A second Genesis could, in principle, be brought to our attention in the foreseeable future if the funding of the space agencies allows this technical possibility.

Ganymede is the largest satellite in the Solar System, bigger than the planets Mercury and Pluto. In fact, the Galileo mission already has given us data to suggest that Ganymede, Callisto, and Europa may harbor large oceans underneath their icy surfaces. Not only is there strong evidence for the internal oceans in the Jovian system, but also Jupiter's large icy moons appear to have three ingredients essential for the origin and sustained evolution of life, namely, water, energy, and the necessary chemical molecules.

It is still premature to attempt the exploration of extraterrestrial oceans of the Jovian system by means of submersibles. But for learning whether a second Genesis has occurred, probably, a lander may be sufficient. We should recall some aspects of the icy Europan surface: It is possible that matter from the interior of the satellite may be raised to the surface itself. The Galileo mission has led to the discovery of a phenomenon called "lenticulae" that are interpreted as surface areas on Europa, whose origin is matter from its deep interior. A lander, without penetrating the icy shell, may be sufficient to retrieve information that might shed some light on its subsurface ocean that may be pregnant with life.

Although there are still many questions to be answered, at present, it seems possible (although not an easy matter) to probe the oceans of the iced Galilean satellites, even if indirectly, by reaching their icy surfaces with landers or micropenetrators.

The preliminary steps taken by science, philosophy, and natural theology assumed that humans had a privileged position in the cosmos. With Copernicus, geocentricism was abandoned. Natural theology has incorporated this view, after an initial opposition to Galileo's inexorable conclusions. After Darwin's theory of evolution was formulated, anthropocentrism was abandoned, although more than a century of research was needed, culminating with the solid genetic bases of neo-Darwinism. Natural theology is also beginning to incorporate this second step into its discussions. We have called, for lack of a better word in the English language, "biogeocentrism" the concept that life is exclusive to the Earth (Aretxaga, 2004; Chela-Flores, 2001, 2009). The underlying difficulty for many scientists arises from a confrontation between the physical and biological sciences. Firstly, everyone agrees that the Newton's theory of gravitation can be extrapolated without any difficulty throughout the universe, except for the small corrections required by Einstein's theory of relativity. One example regards the orbits of Jupiter-like planets. Secondly, the case of extrapolating the theory of biological evolution throughout the cosmos requires more care and is still an open problem. Arguments against the hypothesis of biogeocentrism can now be formulated, thanks to progress in our understanding of Darwinian evolution (cf., Sect. 4). The role of randomness has to be qualified since Darwin's time. Although chance is implicit in *The Origin of Species* and Monod captured the essence of Darwinism with the suggestive contrast between chance and the necessary filtering of natural selection, we have also seen that molecular biology greatly constrains chance. Convergence will be a further factor to take into account. To sum up, Darwinian contingency is constrained, and evolution tends to converge on similar solutions when natural selection acts on similar environments. Cosmochemistry and planetary science suggest that the environments where life can originate are limited. We already are gathering information on a significant number of Jupiter-like planets around stars arising from protonebulae that are likely to grant them an array of satellites. In our Outer Solar System, this can be confirmed. Each of our giant planets has a large number of satellites.

Such is the case in the Outer Solar System. Factors giving rise to atmospheres in Jovian satellites are known. Titan, for instance, has an atmosphere produced by outgassing, combined by seeding of volatiles by comets carrying a fraction of water-ice. Evidence is leaning in favor of the existence of Jovian planets with masses larger than Jupiter; hence, tidal heating responsible for Io's volcanic eruptions could be even more efficient in other solar systems. On Europa, it is not clear that tidal heating may produce hydrothermal vents capable of giving rise to life, unlike the case of satellites orbiting around large Jupiter-like planets. To sum up, natural selection will be working on a finite number of similar environments. According to cosmochemistry, similar chemical elements will enter into pathways of chemical evolution. We have also learned that there are no laws in chemical evolution that are specific to the Earth; it is reasonable to hypothesize that biological evolution will follow chemical evolution. Today, Darwinism cannot be seen as a simple dichotomy between chance and necessity, but rather, we must build into its fabric constrained chance and convergent evolution. Thus, in environments similar to Earth, we expect analogous pathways that lead from chemical evolution to the evolution of intelligent behavior.

These speculations are made all the more exciting and timely by the new discovery of over 1,200 new planets, some of which are hypothesized to be habitable-zone candidates. The Kepler mission of NASA was launched in March 2009 (Fridlund et al., 2010). Kepler can remain pointed toward the same point in the sky for years and it is expected to make observations for a period of 3–4 years having the potential to detect small long-period planets. The observed star field is in the constellations Cygnus and Lyra. With a significant field of view, it will follow about 100,000 stars. Six planets were discovered early made of a mix of rock and gases orbiting a yellow dwarf star (2,000 light years away), known as Kepler-11 (Lissauer et al., 2011). This is an extraordinary solar system in which five inner planets have orbital periods less than 50 days around the dwarf star, while an additional one has a period of 118 days. We are witnessing these extraordinary discoveries, but must wait for a revision of the data to establish, or refute, one of the major discoveries in astrobiology. With these results the catalogue of known exoplanets is expected to explode way beyond the preliminary 500 that were known in the first decade of the 21st century.

### *3.3.2. Consequences of the Specific Nature of Exolife*

The consequences and implications of the specific nature of life to be discovered elsewhere in the Solar System, or eventually on exoplanets or exomoons, are the main topics of this section. They shall depend on the type of life and the stage of the evolution of the life to be identified (Race, 2008). Taking as a point of reference the terrestrial model, we can conceive the following possibilities: unicellular life (both prokaryotic and eukaryotic), multicellular life capable of coordination via the emission of action potentials, multicellular life that has reached the stage of intelligent behavior, and possibly, a stage further where the intelligent behavior manifests itself at the social and cultural levels which we commonly associate with civilization. We should not rule out other possibilities, and indeed, some further possibilities have been discussed by Dick, such as postbiological entities (Dick, 2008).

While the discovery of eukaryotic microorganisms would confirm the hypothesis of the universality of eukaryogenesis, reinforcing the possibility of the existence of intelligent life elsewhere in the universe, the detection of technologically advanced civilizations would have more profound consequences, for instance, the incorporation of the category of *person* assigned to deserving beings different from humans (Aretxaga, 2006).

We should also consider the possibility that the messages received by Earth-bound radio telescopes could provide information on other forms of thinking and comprehending reality in its multiple aspects, such as science, philosophy, art, or religion, that would induce us to revise our theories and knowledge related to the origin, structure, and function of mind and intelligence, not forgetting the origin, evolution, and contents of its most significant creations: society and culture (Billingham et al., 1994; NASA, 2000; Tough, 2000). Faced by these possibilities, the Spanish theologian E. Miret-Magdalena (Miret-Magdalena, 2008) recommends an open mind in searching all the positive common elements that might be learned after the discovery, especially including the idea of the new beings of God and other ideas of religious significance. These reflections led him to suggest a cosmic ecumenism. In this context, it is also worth mentioning the concept of *cosmotheology* (Dick, 2000, 2005).

## 4. Discussion: Some of the Larger Philosophical Issues Arising from a Second Genesis

Although astrobiology is far from reaching its maturity, it lies squarely within the frontiers of traditional research. Some of the deepest questions raised within astrobiology lie close to those raised within the humanities. From the point of view of philosophy and theology, it is conceivable to view conventional science as one aspect of a wider empiricism that would take into account such facts as the intelligibility of the universe. Some humanists feel that it is possible that a search for a rich empiricism could bring within human rationalization what lies beyond the scientific approach.

The implications of a second Genesis are likely to be relevant, not only to astrobiology, but also ethics and especially for philosophy and theology. Such a discovery would provide significant additions to our insights on the intelligibility of nature. Intelligible means the capability of being understood, or comprehended.

Alternatively, intelligible can signify to be apprehensible by the intellect alone. A third aspect of intelligible, closer to the significance of the term in the context of the present work, is related to something that is beyond perception. An "intelligible universe" can be the starting point of a prolonged and systematic discussion among astrobiologists as well as philosophers and theologians (Chela-Flores, 2009).

The Belgian Nobel Laureate Christian de Duve, at the end of a review on the origin and evolution of life, asks himself the question: "What does it all mean?" not only in science (De Duve, 2002), but also in philosophy (Dear, 2006) and in theology (Russell, 2001). The intelligibility of the universe raises questions that lie on the frontier between science and the humanities; and we need an open all-embracing approach.

Considerations of intelligibility return to the often quoted, but less frequently debated statement of Steven Weinberg, the American Physics Nobel Laureate: The more the universe seems comprehensible, the more it also seems pointless (Weinberg, 1977). Such pessimism seems to leave no room for a first Genesis on Earth as a special case of a wide and meaningful phenomenon of second Genesis elsewhere. We should clarify that the quoted statement reflected a specific philosophical trend, or attitude, characteristic of the first half of last century. Hence, if that were the case, the statement should not discourage the general dialogue at the frontier of science and religion. The common approach from both ends of the academic discourse to discovering a second Genesis anywhere else in the universe should contribute, as we shall see below, to the progress of the philosophy and theology. Reciprocally, the questions raised in any field of the humanities arising from the discovery of a second Genesis could, in turn, enrich the search for the place of humans in the universe. The above quotation can be best understood in the context of a philosophical trend called Existentialism.

We should also ask: Which role would astrobiology play in the systematic search for meaning beyond Hegel's Absolute Idealism? In other words, what would be the implications of a new philosophical approach for the meaning of human life?

The eventual success of astrobiology in its search for life elsewhere in the universe would force upon us a reconsideration of the question of intelligibility and comprehension of nature, a subject that is the very nucleus of Western thought.

From this new perspective, we could profit by returning to concepts that can be retraced to Lawrence J. Henderson, especially those of fitness and biocentrism: The fitness of the cosmos for the origin and evolution of life is discussed in Henderson's "The Fitness of the Environment" (Henderson, 1913), where the shift forms two separate discussions, one for the cosmic evolution and the other for biological evolution (Darwinism), begin to merge. This influential book was published almost a century ago, not as a response to the crisis in the systematic search for meaning that we have attempted to review in this chapter. Henderson was a graduate of Harvard and a professor of the University (Chela-Flores, 2007). Henderson's main interests ranged widely: He was a physiologist, chemist, biologist, philosopher, and sociologist. He discussed the question of teleology in biochemistry to give some rationale to the question of fitness of the environment for the evolution of life. For many chemical compounds he discussed, the difficulties that the evolution of life would have encountered had these compounds not been freely available in the environment. Water was one example. Its search, even today, is a main objective of the exploration of the Solar System. Henderson concludes at the end of his book that:

> *The properties of matter and the course of cosmic evolution are now seen to be intimately related to the structure of the living being and to its activities; they become, therefore, far more important in biology than has been previously suspected. For the whole evolutionary process, both cosmic and organic, is one, and the biologist may now rightly regard the universe in its very essence as biocentric.*

Today we should search the roots of Henderson's "biocentrism" at the molecular level. In fact, fine-tuning in biochemistry is represented by the strength of the chemical bonds that gives rise to the "dictionary" that contains the correspondence between amino acids and the monomers of RNA. This dictionary is called the genetic code. Both transcription and translation of the messages coded in RNA and DNA would not be possible if the strength of the bonds had different values. Hence life, as we understand it today, would not have arisen.

We have argued in favor of fitness of the cosmos for the origin and evolution of life without touching on the question of teleology. In this sense, we approach the subject without restricting ourselves exclusively to biological evolution in the universe, but rather, we also include the evolution of the structure of the cosmos itself.

The contrast between different views on the intelligibility of the universe, as discussed in this chapter, illustrates the need for philosophical understanding of meaning with fresh approaches in philosophy at the frontier of astrobiology. We need especially to understand the influence of philosophical doctrines that will tend to encourage any future constructive dialogue between the new physical sciences and the humanities—especially philosophy—pointing toward unified two-culture knowledge, aiming at consilience (with the Edward Wilson fortunate choice of word).

## 5. Chart of the Implications of a Second Genesis Elsewhere

| | |
|---|---|
| I. Cultural implications of searching for a second genesis (Secs. 1–2) | *From speculations to science* |
| | *Astrobiology as a transdisciplinary science* |
| | *Collaboration between the two cultures* |
| | *Towards insights into reality from the point of view of astrobiology* |
| | *To promote societal interest in science and technology* |
| II. Cultural Implications of finding a second genesis (Sec. 3) | *Straightforward consequences* |
| | *Implications that depend on the nature of what is discovered* |
| III. Deeper philosophical issues (Sec. 4) | *Ontological issues* |
| | *Towards the end of biogeocentirsm* |

## 6. References

Alcibar M (2008) Astrobiología, cultura popular y comunicación social de la ciencia y la tecnología. Letras de Deusto 38(118):76–85

Aretxaga R (2004) Astrobiology and biocentrism. In: Seckbach J, Chela-Flores J, Owen T, Raulin F (eds) Life in the universe, vol 7, Cellular origin and life in extreme habitats and astrobiology. Kluwer Academic Publishers, Dordrecht, pp 345–348

Aretxaga R (2006) La ciencia astrobiológica: Un nuevo reto para el humanismo del siglo XXI. Letras de Deusto 36(110):12–20

Aretxaga R (2008) Astrobiología: entre la ciencia y la exploración. Letras de Deusto 38(118):13–27

Billingham J, Heins R, Milne D, Doyle S, Klein M, Heilbron J, Ashkenazi M, Michaud M, Lutz J, Shostak S (eds) (1994) Social implications of the detection of an extraterrestrial civilization. SETI Press, SETI Institute, Mountain View

Bruno G (2000) *De l'infinito, universo e mondi*, Venice, 1584. [English translation: On the infinite universe and innumerable worlds, Cambridge, 1650]. A more comprehensive bibliographic reference: Giordano Bruno 1548–1600, Biblioteca di Bibliografia Italiana 164, Leo S. Olschki Editore, Roma, pp 105–106

Chela-Flores J (2001) The new science of astrobiology from genesis of the living cell to evolution of intelligent behavior in the universe. Kluwer Academic Publishers, Dordrecht

Chela-Flores J (2003) Marco cultural de la Astrobiología. Letras de Deusto 33(98):199–215

Chela-Flores J (2006) La evolución del comportamiento inteligente: ¿Existe evidencia de proposito en las moleculas complejas de la química? Letras de Deusto 36(110):21–36

Chela-Flores J (2007) Fitness of the cosmos for the origin and evolution of life: from biochemical fine-tuning to the Anthropic principle. In: Barrow JD, Conway Morris S, Freeland SJ, Harper CL (eds) Fitness of the cosmos for life: biochemistry and fine-tuning. Cambridge University Press, Cambridge, pp 151–166

Chela-Flores J (2008) La posibilidad de la existencia de vida extraterrestre inteligente, su búsqueda científica e interés filosófico. Astrobiologia y Filosofia (III), Letras de Deusto 38(118):38–47

Chela-Flores J (2009) A second genesis: stepping stones towards the intelligibility of nature. World Scientific Publishers, Singapore, 248 pp

Chela-Flores J (2010) Instrumentation for the search of habitable ecosystems in the future exploration of Europa and Ganymede. Int J Astrobiol 9(2):101–108

Chela-Flores J (2011) The science of astrobiology a personal point of view on learning to read the book of life, 2nd edn. Cellular origin, life in extreme habitats and astrobiology, Springer, Dordrecht, Chapters 13, 14 and 16. http://www.ictp.it/~chelaf/ss220.html

De Duve C (2002) Life evolving molecules mind and meaning. OUP, New York

Dear P (2006) The intelligibility of nature how science makes sense of the world. The University of Chicago Press, Chicago, 254 p

Dick SJ (2000) Cosmotheology. In: Dick SJ (ed) Many worlds: the new universe, extraterrestrial life, and its theological implications. Templeton Press, Philadelphia

Dick SJ (2005) Cosmotheology revisited: theological implications of extraterrestrial life. In: Cons-Ciências: actas do Fórum Internacional Ciência, Religião e Consciência. Porto, Edições Universidade Fernando Pessoa, CTEC, 2005, pp 287–301

Dick SJ (2008) The Postbiological Universe. Acta Astronautica 62(8–9):499–504

Fridlund M, Eiroa C, Henning T, Herbst T, Kaltenegger L, Léger A. Liseau R, Lammer H, Selsis F, Beichman C, Danchi W, Lunine J, Paresce F, Penny A, Quirrenbach A, Röttgering H, Schneider J, Stam D, Tinetti G, White GJ (2010) A roadmap for the detection and characterization of other Earths. Astrobiology 10:113–119

Henderson LJ (1913) The fitness of the environment an enquiry into the biological significance of the properties of matter. Peter Smith, Gloucester, 1970, p 312

Langton CG (1990) Computation at the edge of Chaos: Phase-Transitions and Emergent Computation. Ph.D. Thesis, University of Michigan

Lederberg J (1963) Life beyond the earth. Stanford Today, Stanford University, California, Series 1, No 7, December 1963, 1–8

Lissauer JJ et al. (2011) A closely packed system of low-mass, low-density planets transiting Kepler-11. Nature 470:53–58

Mayz-Vallenilla E (2003) Astrofísica y Meta-técnica. Letras de Deusto 33(98):217–224 [English translation: (2000) Astrophysics and Meta-technics. In: Chela-Flores et al (eds) Astrobiology: origin from the big bang to civilization. Kluwer, Dordrecht, pp 235–240]

McKay CP (2001) The search for a second genesis in our solar system. In: Chela-Flores J, Owen T, Raulin F (eds) The first steps of life in the universe. Kluwer Academic Publishers, Dordrecht, pp 269–277

Miret-Magdalena E (2008) Ciencia astrobiológica y pensamiento cristiano. Letras de Deusto 38(118): 58–67

Race MS (2008) Communicating about the discovery of extraterrestrial life: different searches, different issues. Acta Astronaut 62:71–78

Rodríguez-Manfredi JA (2008) La innovación tecnológica en la astrobiología como motor de cambio en la mentalidad actual. Letras de Deusto 38(118):68–75

Russell R (2001) Life in the universe: philosophical and theological issues. CTNS Bull Center Theol Nat Sci 21:3–9 (Spring 2001)

Snow CP (1993) The two cultures. Cambridge University Press, London/New York

Stokols D, Fuqua J, Gress J, Harvey R, Phillips K, Baezconde-Garbanati L, Unger J, Palmer P, Clark M, Colby S, Morgan G, Trochim W (2003) Evaluating transdisciplinary science. Nicotine Tob Res 5(S-1):S21–S39

NASA Technical Memorandum (2000) Workshop on the societal implications of astrobiology. Final report, 15 December 2000

Tough A (ed) (2000) When SETI succeeds: the impact of high-information contact. Foundation for the Future, Washington, DC

Villegas R, Castillo C, Villegas GM (2000) The origin of the neuron: the first neuron in the phylogenetic tree of life. In: Chela-Flores J, Lemarchand GA, Oro J (eds) Astrobiology: origins from the big bang to civilization. Kluwer Academic Publishers, Dordrecht, pp 195–211

Weinberg S (1977) The first three minutes. Fontana/Collins, London

Wilson EO (1998) Consilience the unity of knowledge. Knopf, New York, p 70

Biodata of **Florence Raulin-Cerceau,** author of "***The Concept of Chemical Evolution Before Oparin.***"

**Florence Raulin-Cerceau** is an associate Professor at the Natural History Museum of Paris (France). She obtained her Ph.D. in Astronomy and Space Techniques at the University Pierre et Marie Curie (Paris 6) in 1983. Her thesis, entitled "Studies of infrared spectra of nitrogenous compounds likely to be present in the atmosphere of Titan" dealt with Exobiology. These previous works were implicated in the preparation of the Cassini-Huygens mission launched in 1997.

Since 1988, she has a permanent position at the Museum of Paris where she contributed to the creation of the *Grande Galerie de l'Evolution* in the *Jardin des Plantes*. Since 1997, her research field concerns the history and epistemology of Exo/Astrobiology.

Florence Raulin Cerceau is author of several books (in French) dealing with the history of ideas about the origin of life and the search for life elsewhere in the universe.

E-mail: **raulin@mnhn.fr**

J. Seckbach (ed.), *Genesis - In The Beginning: Precursors of Life, Chemical Models and Early Biological Evolution*, Cellular Origin, Life in Extreme Habitats and Astrobiology 22, 891–906
DOI 10.1007/978-94-007-2941-4_45, © Springer Science+Business Media Dordrecht 2012

# THE CONCEPT OF CHEMICAL EVOLUTION BEFORE OPARIN

**FLORENCE RAULIN-CERCEAU**
*Centre Alexandre Koyré (UMR 8560/CNRS/EHESS/MNHN)*
*Muséum national d'Histoire naturelle, 57 rue Cuvier,*
*75005 Paris, France*

## 1. Introduction

> *The process of chemical evolution could be considered to have taken place in three stages: the inorganic, the organic, and the biological. In the molecular approach to the problem of the origin of life, attempts have been made to elucidate how compounds of biological significance could have arisen on Earth before the first life appeared.* (Ponnamperuma and Gabel, 1968)

During the later part of the nineteenth century, the question of the origin of life faced two problems. The theory of spontaneous generation was refuted by Pasteur's works (Pasteur, 1861), while the hypothesis of panspermia (Richter, 1865), maintaining that life was eternal, put back the problem toward the origin of the universe itself. Darwin's famous publication *The Origin of Species* (Darwin, 1859) had shown that, according to the principle of natural selection, large timescales with extremely slow processes were needed for biological evolution. However, Darwin's theory of evolution had left the question of the origin of life unanswered, at least in his 1959 major publication. In a letter dated 1 February 1871 sent to his friend Hooker, Darwin timidly proposed in very few words that life could have arisen from inorganic compounds in a very distant past (Darwin, 1871):

> But if (and Oh! what a big if!) we could conceive in some warm little pond, with all sorts of ammonia and phosphoric salts, light, heat, electricity, etc., present, that a protein compound was chemically formed ready to undergo still more complex changes, at the present day such matter would be instantly devoured or absorbed, which would not have been the case before living creatures were formed.

Anyway, the consequences of this view of natural processes acting along with time did not have an immediate effect on the problem of the origin of life, among the scientific community. A first difficulty had to be surmounted: the assumed barrier between living and nonliving matter. The question of the origin of life was then reformulated: was there an impassable "hiatus" between animate and inanimate nature (Oparin, 1938)?

The centuries preceding Darwin's theory offered a strong inheritance in favor of the specificity of living matter. The idea that life could be animated by a "vital force" which could not be explained by physicochemical forces was accepted until the beginning of the nineteenth century. The first syntheses of organic compounds (Wöhler, 1828), supposed until then to be produced only by living organisms, weakened this hypothesis. In the midst of such a context, a new experimental field was born: the chemistry of organic compounds. If molecules belonging to living matter could be formed in laboratory from man's hand, why nature could not do likewise with the help of natural processes? The particular circumstances of the evolutionary thought existing at that time led this question to clearly emerge in the 1870s in the mind of the supporters of the Darwinian theory. A materialistic concept of the origin of life was then proposed. Evolutionists such as Huxley, Tyndall, Spencer, or Haeckel began to realize that the problem of the origin of life was a specific one: it had to be placed in the peculiar physical conditions prevailing on Earth in a distant past. An "abiogenic" view of evolution[1] taking place on the primitive Earth became necessary to complete the one of biological evolution. However, the problem which still remained to be solved was to define the different stages of evolutionary processes without knowing some important parameters: first, the conditions of the primitive environment, secondly the timescale needed to turn inanimate matter into living matter.

This last point was particularly unclear. It was not obvious in many minds that *abiogenic* processes asked for long time span. A certain form of spontaneous generation involving inorganic compounds was still accepted at that time. In fact Pasteur refuted one form of spontaneous generation: the generation coming from self-organization of organic matter (itself coming from living matter such as the processes of fermentation). And during the 1860s, the debate (at least in France with the Pasteur/Pouchet controversy) was above all focused on that last topic. And then, one form of spontaneous generation was generally left aside: the generation coming from inorganic matter, i.e., the *abiogenic* view of life. This view was confronted to many experimental as well as ideological difficulties. While it became conceivable that inorganic compounds could combined themselves in order to form living matter, it was generally supposed that the processes involved in that particular form of generation could needed relatively short timescale. Accordingly, the idea of a chemical evolution before a biological evolution could only have emerged after the acceptance of the following points:

- A materialistic view of the origin of life involving *abiogenic* processes
- Long timescales necessary to allow the natural "action" of these *abiogenic* processes
- Long series of progressive events leading to life
- A specific environment on the primitive Earth, favorable to organic chemical syntheses

---

[1] *Abiogenesis* was a term coined by Huxley to designate the formation of living matter from simple compounds available on the primitive Earth and then which did not come from living matter itself (for instance, in state of decay) (see Sect. 2).

## 2. The Materialistic View of the Origin of Life

A materialistic view of the processes leading to the origin of life appeared in the 1870s following Charles Darwin's theory of biological evolution. As a result, the idea of progressive transformation of inorganic matter into living matter by means of evolutionary steps logically appeared as a consequence of the Darwinian theory. Through materialistic concepts, the whole history of life on Earth could be explained by natural causes. This position was particularly followed by the advocates of the Darwinian theory during the second part of the nineteenth century.

Living matter coming from inorganic matter without the intervention of any specific vital forces was then strongly supported by the advocates of the Darwinian theory. While the consideration that living and nonliving matter could obey the same laws of evolution was emerging, evolutionists were wondering what could have been the circumstances favorable to natural processes leading to life. Two parameters seemed to have played a particular role: environment and time. The problem of the origin of life was placed in an evolutionary context in which the question of the difference between organic and inorganic matter was a central one. Among Darwin's followers, the British naturalist Thomas Huxley (1825–1895) stated in 1870 that *protoplasm* (a substance he considered as the "physical substratum of all vital phenomena") might have evolved from nonliving matter at some stage during Earth's history. Protoplasm could be the product of natural physicochemical processes involving water, carbonic acid, and ammonia (Huxley, 1870). Huxley coined the term *abiogenesis* to designate this generation of living matter from nonliving matter (Kamminga, 1980). Such a viewpoint did not maintain a gap between living and nonliving matter. Another Darwin's British follower, the physicist John Tyndall (1820–1893), argued that all natural phenomena could be explained in terms of the properties of matter (Tyndall, 1871). The origin of life was included in this assumption, and then living matter should have emerged from inorganic compounds by purely physical causes. Inorganic compounds combined into living matter during the first steps of the primitive Earth, as the planet cooled. Tyndall's viewpoint was very closed to Huxley's one. However, no detailed process of *abiogenesis* emerged from Huxley or Tyndall's proposals. In fact, inorganic compounds were supposed to organize themselves with no explicit intermediary steps (Kamminga, 1980).

The idea of a long series of progressive events leading to life and occurring on the primitive Earth is present in the works of the British philosopher Herbert Spencer (1820–1903). Spencer's philosophy supposed that everything from the stars to the embryo, from civilizations to the individual, was in process of development, interaction, change, growth, and progress. Spencer was then a strong advocate of evolutionary processes and thought that gradual transitions were necessary to go from nonliving to living matter (Spencer, 1898). But Spencer's view about this last point was not widely followed at that time. One reason was probably the controversy dealing with the age of the Earth, which seemed to imply insufficient time for the evolution of living organisms to have occurred (Farley, 1974). Another reason was perhaps the fact that, in the 1870s, the concept of spontaneous generation was not completely abandoned in spite of Darwin and Pasteur's conclusions. In reality, it reappeared a little bit transformed. Even among the supporters of the Darwinian theory, the possibility of a "spontaneous generation" born by *abiogenesis* in relatively short time, nowadays and in specific environments (such as the depths of the oceans), remained a solution to the origin of life.

In Germany, one of the most fervent supporters of the Darwinian concepts was the naturalist Ernst Haeckel (1834–1919). In this country, scientific materialism was already present since the first years of the 1850s, as a consequence of several theoretical works such as those of Ludwig Büchner (1824–1899) and Karl Vogt (1817–1895). Following this trend in materialism, Haeckel proposed an explanation of the universe, including geological and biological processes, completely founded on materialistic considerations. His philosophical view on the unity of nature led him to accept that there was no fundamental distinction between living and nonliving world. As a result, the origin of life was, according to Haeckel, a step in evolution no more complicated than other transformations in the universe. His theory on the origin of life (Haeckel, 1876) was based on a very restrictive definition of life (this opinion has been often submitted to criticism):

- Life is ruled by physicochemical processes.
- Life is the result of the combination of element carbon with H, O, N, P, and S.
- Life is a particular state of matter, a semifluid state of aggregation of these atoms.

In spite of this quite reducing position, Haeckel highlighted many points which are nowadays considered as very important (as we will see in Sect. 4). Among them was the significance of the initial conditions needed to start the processes leading to living matter.

## 3. The First Primitive Syntheses in Laboratory

The first experimental organic syntheses dated back from 1828 when the German chemists Friedrich Wöhler (1800–1882) and Hermann Kolbe (1818–1884) respectively obtained urea and acetic acid in laboratory. These early results on organic chemistry demonstrated that the chemical components produced by living beings (such as urea or acetic acid) were not specific to life. From now on, it was possible to distinguish the organic matter coming from living beings[2] and the organic

[2] Until the end of the nineteenth century, and despite these organic syntheses, the term "organic" was generally reserved to compounds or matter coming from living beings.

matter coming from inanimate matter. In Germany, the strong materialistic movement (as we have seen in Sect. 2) led to various attempts to put into practice the fabrication in laboratory of basic organic compounds. Techniques in organic chemistry were improved during this part of the nineteenth century, for instance, with the German physiologist Emil du Bois-Reymond (1818–1896) or the French chemist Marcellin Berthelot (1827–1907) (Raulin-Cerceau, 2009). It should be noted that, as early as 1861, the Russian chemist Alexander Butlerov (1828–1884) showed how many different sugars (including sugars found in DNA and RNA obviously unknown at that time) could be synthesized from concentrated formaldehyde solutions containing mineral catalysts such as carbonate and alumina (Ponnamperuma and Gabel, 1968; Lazcano and Bada, 2003). Experimental works were also carried out at the beginning of the twentieth century by the chemist Walther Löb (1872–1916) on the synthesis of glycine, as mentioned by Bada and Lazcano (2002) and Brack (2003).

Considering the evolutionary view of the history of life, the idea that primitive syntheses could be possible in laboratory emerged as early as the 1870s. It became conceivable to make an attempt at producing in laboratory the "key components" of living matter. But what were exactly, at that time, these "key components"? One of the main questions that had to be answered in order to solve the problem of the origin of life concerned indeed the precise nature of living matter. Following Huxley's hypothesis on *protoplasm* (1870), living matter was generally considered as made of a unique material (*protoplasm*) whose own composition was supposed to be only proteinic or albuminoid-based compounds. However, the question persisted: what were the detailed components of this basic living material? Pioneers experimental works, based on what the basic components of life could have been, are detailed below.

We can consider that the first proposals to synthesize, in an evolutionary view, the characteristic compounds of living matter appeared in 1875 with the works of the German physiologist Eduard Pflüger (1829–1910) (Tirard, 1996). Pflüger supported a "cyanic theory" in which the cyanogen-radical (CN-radical) played a crucial role among living processes (Pflüger, 1875). According to Pflüger, the CN-radical would be the key compound to solve the problem of the origin of life. Starting from what was known about the constitution of proteins, he thought that the protoplasm – supposed to be made of proteins and accepted at that time as the main component of living matter – contained "living" proteins. Pflüger thought that two categories of proteins existed in organisms: "living" (or "protoplasmic") proteins and "dead" (or "storage") proteins (Kamminga, 1980). The fundamental difference between these two types of proteins was located at the chemical level. The living property of proteins would have come from the CN-radical whose instability would have constituted the basis for chemical transformations in living matter (Raulin-Cerceau, 2009). At that time, proteins (or *protoplasm*) were considered to be the essence of living processes, an opinion also defended by Pflüger. He concluded from his hypothesis that living proteins were characterized by the presence of the CN-radical. If this radical could be regarded as the link between nonliving and living world, the problem was then to understand how it could be generated. To this point, experiments in laboratory had demonstrated that cyanogen compounds were formed under high temperature conditions. Pflüger's hypothesis defended the fact that the synthesis of cyanide compounds would have taken place during an early period of incandescence of the primitive Earth. His main idea was that "life originated from fire" and fire could be considered as "the force that has produced the constituents of proteins by synthesis" (Pflüger, 1875). Pflüger thought that the energy required for the transformation of elementary compounds into "living" compounds was provided by fire. The polymerization could have started when the Earth's surface was cooler and reactions with oxygen first, and secondly with water and salts, could have undergone under more clement conditions.

Pflüger wondered about the initial conditions of the primitive Earth and about the "key" compounds that could be essential to the emergence of living matter. The idea of long timescales needed to gradually transform the compounds was also present in Pflüger's hypothesis. Protein synthesis could not have occurred in a single step (Kamminga, 1980). However, one point of his theory was opened to criticism: the specific role attributed to the CN-radical for the reason that it implied a reducing view of the nature of life. The naturalist Ernst Haeckel shared this last view during the 1870s since he credited carbon atom with having a central role in the origin of life. And Haeckel too was criticized by his contemporaries for this assumption. As a final point, we could consider that the concept of chemical evolution was born with Pflüger's theory, even if Pflûger himself did not use, with his own words, this specific term.

At the very end of the nineteenth century, many experiments were attempted in laboratory to obtain some of the characteristic compounds of all living beings. The German chemist Hermann Emil Fischer (1852–1919) carried out numerous experimental works on sugars (between 1884 and 1894). One of his greatest successes was the synthesis of glucose, fructose, and mannose in 1890, starting from glycerol. Fischer coined the term purine and synthesized several purines such as adenine, xanthine, and caffeine. Between 1899 and 1908, Fischer contributed also to the knowledge of proteins. He searched for effective analytic methods of separating and identifying the individual amino acids and discovered a new type of them, the cyclic amino acids (proline and oxyproline). He also studied the synthesis of proteins. He identified the peptide bond as a bond capable of connecting the amino acids between them to obtain polypeptides. These important results contributed to develop the experimental field of the primitive syntheses aimed to define the main life components. Of course, great difficulties were still to be surmounted in order to simulate prebiotic chemistry with an evolutionary view, as it will be later attempted by Stanley Miller.

## 4. Birth of the Concept of Chemical Evolution

At the turn of the twentieth century, more and more scientists wondered about a first primitive generation coming from inorganic materials. This view of the origin of life asked new questions. What conditions are necessary, not to say essential, to the emergence of life? What are the simple chemical molecules involved in the reactions producing the main life compounds? Finally, how to explain a primordial genesis with the help of simple materials available on the primitive Earth? Extending these processes to other planets, is chemical evolution a banal process in the universe?

The Belgium botanist and physiologist Leo Errera (1858–1905) stated in his *Essais de Philosophie Botanique* (*Essays on Botanical Philosophy*) published in 1899 that a "spontaneous generation" from inert compounds (based on C, O, H, and N atoms) available on the primitive Earth should have been necessary for the formation of the first organisms (Errera, 1899). In the same way, his book *The Evolution Theory* published in 1904, the German biologist August Weismann (1834–1914) asked himself about the conditions required for a first spontaneous generation and rose the question dealing with the experimental conditions employed to recreate life in laboratory:

> *After the fiery Earth had so far cooled that its outermost layer had hardened to a firm crust, and after water had condensed to a liquid form, there could at first only have been inorganic substances in existence. In order to prove spontaneous generation, therefore, it would be necessary to try to find out from what mingling of inorganic combinations organisms could arise; to prove that spontaneous generation could never have been possible is out of the question... It would be impossible to prove by experiment that spontaneous generation could never have taken place; because each negative experiment would only prove life does not arise under the conditions of the experiments. But this by no means excludes the possibility that it might arise under other conditions* (Weismann, 1904).

In Germany, Haeckel's theory of origin of life, initially proposed in his book *The History of Creation* (1876 – firstly published in German in 1868), integrated progressive chemical steps of growing complexity. As seen in Sect. 2, Haeckel asserted that living matter could have been formed by combinations between carbon and specific atoms. These combinations would have produced aggregates, which would become ever more and more complex and eventually would develop into the simplest organisms known to Haeckel, the *Monera* (Haeckel, 1866). The formation process of the *Monera* was then supposed to be the result of physicochemical forces inherent in matter present in the primeval oceans (Kamminga, 1980). In his famous book *The History of Creation*, Haeckel named this process of *abiogenesis*, "autogony" (Haeckel, 1876). However, until the very beginning of the twentieth century, he accepted the idea that *Monera* could repeatedly arise by autogony in the bottom of the oceans, as what he called "repeated acts of spontaneous generation" (Haeckel, 1876). Haeckel's suggestion that life was the result of a spontaneous "crystallization" of simple carbon compounds into protoplasm was seen by some of his contemporaries as a flagrant oversimplification (Kamminga, 1980).

For instance, Carl Nägeli and August Weismann did not accept that autogony could be just a single physicochemical process. Both Nägeli and Weismann asserted that *abiogenesis* was a logical necessity, but they thought that living matter must have developed gradually over immense period of time (Kamminga, 1980).

In his following books, Haeckel integrated Nägeli and Weismann's criticisms, and the simplicity of autogony disappeared in his book *The Riddle of the Universe* (1900 – firstly published in German in 1899) and in *The Wonders of Life* (1904). Haeckel developed in this last book an entire chapter on the question of the origin of life. More and more doubts are present about the "repeated acts of spontaneous generation." In order to avoid any ambiguity with the so-called spontaneous generation, Haeckel proposed the term "primordial generation" to designate the first (and unique) processes of formation of living matter on Earth. Above all, Haeckel developed in this chapter rudimentary stages of chemical evolution including Pflüger's works that came to consolidate his *Monera* theory. Ever since then he considered that the generation of life must have been an extremely slow process, taking place over very long time span. His hypotheses of archigonia (autogony occurring on the primitive Earth) are detailed in this last book with a description of every evolutionary stage (Haeckel, 1904):

- Formation of hydrocarbons and cyanogen compounds on the molten Earth.
- Transformation of simple cyanogen compounds in a series of nitrocarbures on the cooled Earth, when water had condensed and the atmosphere was rich in carbonic acid. Nitrocarbures finally produced albuminoid compounds.
- Formation of larger and larger molecular structures (micelles) thanks to chemical affinities with albuminoid compounds.
- Association of micelles leading to the formation of homogenous plasma (plassonelles).
- Development and division of plassonelles leading to the formation of spheres of homogeneous plasma (*Monera*).
- Formation of a solid envelop (membrane) and a soft medium (central grain) after chemical differentiation. Formation of cytodes.
- Transformation of these simple cytodes in simple cells with nucleus after concentration of the hereditary material.

The first three stages could be typically considered as stages of chemical evolution with the formation of albuminoid compounds (regarded at that time as a basic material characteristic of living matter, in the same way as proteins or *protoplasm*) from very simple inorganic molecules such as cyanogen compounds. These (theoretical) stages of matter evolution leading to life prepared to a new way to study the origin of life: defining progressive steps of *abiogenesis* occurring in a specific environment of the primitive Earth (Raulin Cerceau, 2009). Beside this, the very few attempts carried out at that time in laboratory to produce the different stages of *abiogenesis* did not provide any positive result. This fact strengthened the assumption that the conditions needed for life to appear had definitively disappeared from the surface of the Earth (and above all, these negative results underlined that these conditions were completely unknown). This point was particularly important since it seemed to prove that the origin of life was a nonrepetitive phenomenon throughout the history of life.

Haeckel's *Monera* theory and Haeckel's theory on the origin of life had a large influence on the supporters of the materialistic view of life. In France, the biologist Félix Le Dantec (1869–1917), a Pasteur's pupil, supported Haeckel's *Monera* theory in his book *Théorie Nouvelle de la Vie* (*New Theory of Life*) published in 1896. In this writing, he mentioned an "elementary life" starting from *Monera* and involving stages of chemical evolution having appeared on the primitive Earth:

> *Within special conditions realized only once, a "monerian" molecule*[3] *would have been synthesized and elementary life would have appeared* (Le Dantec, 1896)

In very few words, Le Dantec underlined that life would have appeared on Earth under very specific and nonrepetitive conditions. Le Dantec's assumptions showed that at that time some scientists had already realized that a process of chemical evolution – considered as a one-way process depending on singular conditions – would have occurred on the primitive Earth.

We can say eventually that at the turn of the twentieth century, the scientific field of the origin of life began to be built around the idea of chemical evolution, even if at that time no detailed mechanism was formulated to explain in chemical terms the formation of proteins from inorganic matter. Nevertheless, two important parameters were considered from then on:

- Extremely long time spans allowing evolutionary processes of inorganic matter (time was acting as well as forces)
- Specific environmental conditions on the primitive Earth, very different from those existing today

Taking these parameters into consideration gave a complete new insight into the problem of the origin of life: life have had a unique origin on Earth. Namely, have occurred on Earth only one *abiogenesis*, one generation of living matter. Life had a starting point located in a very distant past without any possibility to go back to the initial conditions. Life was a historical process.

## 5. Chemical Evolution as an Established Fact

The concept of chemical evolution was already firmly established during the 1910s. The British biochemist Benjamin Moore (1867–1922) published in 1913 a book entitled *The Origin and Nature of Life*, in which he proposed a theory about matter evolution from atoms to living organisms. Starting from the genesis of chemical elements, he developed a reflection on chemical evolution in the stars. With the

---

[3]The term "molecule" has not here the same meaning than the one we know today. It designed a very small unit of living matter.

help of spectroscopy, stellar chemistry have shown the existence of a chain of transformation of elements which was included in an evolutionary process involving energy in a whole cosmic scheme. Moore distinguished several types of evolution stages. The origin of life was also included in this progression and was not a unique event which would have only occurred on our planet:

> *The stellar evidences of the synthesis of matter and the creation of worlds, leaves the position entirely untenable, that life once came by a single creative unrelated act upon this earth, or any other earth from which it was carried to ours. Life is at present originating in countless other worlds as daily it is originating upon our own, and given a certain stage in evolution when matter has reached a certain complexity in structure, and become tenanted by certain types of energy, life must come, and having come must evolve into higher and higher forms.* (Moore, 1913)

Besides this, Moore would have perceived that the photochemical synthesis of formaldehyde using sunlight and catalysts, could be the starting point for the organic syntheses at the origin of the formation of life compounds.

While chemical evolution was incorporated into the ways of envisaging the processes of the origin of life, primitive syntheses were developed in laboratory. In the 1910s, the French chemists Daniel Berthelot and Henry Gaudechon (Berthelot and Gaudechon, 1912) carried out the synthesis of carbon hydrates from carbon dioxide and water, under the influence of ultraviolet radiation (Dauvillier, 1958). Berthelot also demonstrated that formaldehyde (HCHO) was a starting point for the formation of many carbon hydrates, and he established that this compound had a tendency to polymerization and then to the formation of glucides (glucose, saccharose, starch, cellulose) according to Würtz reaction:

$$\text{formaldehyde} \rightarrow \text{alcohols-aldehydes} \rightarrow \text{glucides.}$$

As early as 1913, Walther Löb had obtained glycine from a gaseous mixture of carbon dioxide, ammoniac, and water under electrical discharges. However, according to Brack, the objective of Löb's experiment was not approaching the problem of life's origin but rather getting a better understanding of the assimilation of atmospheric nitrogen by plants (Brack, 2003). The published related paper remained relatively unknown until the famous Miller's experiment in 1953.

The French biologist Pierre Gavaudan (1905–1985) brought the works of the Italian physiologist Ermanno Giglio Tos (1865–1926) out of oblivion. According to Gavaudan, Giglio Tos formulated in 1910 the fundamental principles of a "theory of biogenetic evolution" (Gavaudan, 1967). In this theory, the biological cell (that he perceived as too complex to have appeared following very simple processes) would have been the result of a long series of biochemical stages in which "biomolecules" born in a specific organic environment would have played an important part in the evolutionary processes. First, Giglio Tos supposed that acetylene (coming from mineral reactions) could have been at the origin of many organic compounds. Various types of biomolecules would have been endowed of a primitive metabolism depending on heterotrophy. Thanks to secondary biochemical activities depending on autotrophy, these biomolecules would have been able to survive in an environment poorer and poorer in organic molecules. New biomolecules insoluble in water would have appeared, and then they would have joined together thanks to cohesion forces. This "symbiotic" association would have created again new substances, named by Giglio Tos "biomores," and which would have represented the basic structures constituting the first living cells. According to Gavaudan, Giglio Tos's theory prefigured Oparin's coacervates hypothesis (Gavaudan, 1967).

Approaching the 1920s, more and more scientists adopted the concept of chemical evolution. The American geologist and paleontologist Henry Fairfield Osborn (1857–1935) was one of those who supported the hypothesis suggesting that the first mechanisms leading to life would have followed gradual stages in the specific environment of the primitive Earth. His theory on the origin of life was developed in his book *The Origin and Evolution of Life* published in 1917. Osborn provided here relatively detailed evolutionary stages starting from the simplest compounds up to the more complex ones (Osborn, 1917):

- Assembling one by one the elements essential to life: hydrogen, oxygen, nitrogen, carbon, phosphorus, sulphur, potassium, calcium, magnesium, iron. The four first elements (the most important ones) would have been available thanks to nitrogenous compounds coming either from eruptive emanations or from an atmosphere very rich in nitrogen and carbon dioxide.
- By mutual attraction, these elements would have joined together and would have formed a new form of unit in the universe: an organic unit, a biological individual endowed with a coordinated functioning, in a word an *organism*, distinct from the aggregates of inorganic matter whose they came from by an assembling process.
- This assembling occurred in a colloidal state – a dynamic state of matter proposed by the British chemist Thomas Graham (1805–1869) in 1861 (in the same way *protoplasm* was considered as a colloid).
- Specific actions and reactions appeared in the new living organisms along with new biochemical compounds.
- Evolution and specialization of various chemical agents such as catalysts, have accompanied step by step the evolution of plants and animals functions.
- Formation of a colloidal state, synthesis of new compounds and chemical coordination were completed by another process: competition and rudimentary natural selection.

In the 1920s, other viewpoints on chemical evolution were developed in spite of the need of more observations and experimental methods to put into practice this new scientific area. In France, the naturalist Edmond Perrier (1844–1921) proposed similarly to Osborn a theory of the origin of life including stages of chemical evolution. Informed about Berthelot's and Gaudechon's works, Perrier admitted that primitive syntheses of numerous organic compounds would have been possible, as a starting point in time as well as in space of the origin of life (Perrier, 1920). According to Perrier, life was not the result of the evolution of only one substance but rather of a mixture and collection of many chemical compounds (as it was the case with Osborn). The solution to solve the problem of the origin of life would be, as for him, to carry out appropriate experiments in laboratory. He suggested searching for processes able to synthesize banal substances ordinarily inert, which could give birth to life if they are mixed and associated in particular proportions (Perrier, 1920). He underlined this last point (the correct proportions of selected substances) which was, according to him, one of the main difficulties when experiments on chemical evolution were undertaken in laboratory. Beside this, Perrier attached a particular importance to the function of chlorophyll in increasing "the first living masses" coming from the fortuitous assembly of the mixed substances. The first living organisms would have been then classified in the plant kingdom. Perrier also confirmed one of the main difficulties remaining when one was confronted to the problem of the origin of life: it was a historical science without any possibility to return to the initial conditions, conditions that were unknown and that could only be reconstructed by thought (Perrier, 1920).

## 6. Conclusion

The first proposals dealing with a process of chemical evolution occurring before biological evolution came from evolutionists as early as the 1870s. This idea was consecutive to the one of *abiogenesis* formulated in 1870 by Huxley, asserting the formation of living matter (*protoplasm*) from inorganic compounds. However, *abiogenesis* was rapidly confronted to some problems. The supposed simplicity of the characteristic materials of the living matter (*protoplasm*, albuminoid compounds) was in contradiction with the nonobservation of *abiogenesis* in laboratory. At the end of the nineteenth century, two problems emerged following this statement:

1. The basic components of life were probably more complex than previously thought.
2. *Abiogenesis* was certainly a process asking for large time spans.

*Abiogenesis* was not observable either in nature. Protagonists of an evolutionary view of life accepted that there were primitive stages of complexity of matter form inorganic to living organisms. They finally realized that the processes involved in these transformations were only possible under specific circumstances on the primeval Earth. The question was then focused on the description of these primitive stages and the primordial syntheses implicating simple and "inert" compounds available in a primeval environment, completely unknown. In the meantime, developments in organic chemistry emphasized that it was expectable to produce in laboratory – not the living matter itself but – some of the building bricks of life. One of the main points highlighted at the turn of the twentieth century concerned these primitive syntheses which should have been carried out in an evolutionary view, and above all in the conditions of the primitive environment. Different "scenarios" explaining the first stages of the formation of living matter were proposed at that time, even if these scenarios were more or less detailed.

The contribution of these early works was very important to establish the concept of chemical evolution as a historical process. Of course, a lot of data were missing at that time dealing with the nature of the fundamental components of life (nucleic acids particularly) and the environment on the primeval Earth. We have seen that Pflüger's, Haeckel's, Giglio Tos's, and Osborn's theories attempted to offer an idea of the possible stages from very simple atmospheric compounds to biomolecules. It could be considered that these theoretical proposals could represent the earliest "scenarios" of chemical evolution, even if this last term was not explicitly used to describe this process. Between the 1870s and the 1920s, the concept of chemical evolution became a scientific question integrated in evolutionary models proposed within the context of materialistic views. Therefore, the idea of chemical evolution was present in many works before the first Oparin's publication on the origin of life (Oparin, 1924). It should be noted that Oparin paid tribute to the main forerunners of chemical evolution in his famous book published in English in 1938 (Oparin, 1938), especially in the chapter entitled "Theories of the Origin of Life at Some Distant Period of the Earth's Existence." Anyway, Oparin provided the most complete contribution on this topic in giving accurate stages of long evolution from atmospheric compounds and organic syntheses to coacervates formation. Oparin's publication represents without reservation a decisive turning point in the studies of the origin of life.

## 7. References


Bada JL, Lazcano A (2002) Miller revealed new ways to study the origins of life. Nature 416:475

Berthelot D, Gaudechon H (1912) Sur la longueur d'onde des radiations actives dans la synthèse photochimique des composés ternaires. Comptes rendus de l'Académie des Sciences, séance du 24 juin 1912, pp 1803–1805

Brack A (2003) La chimie de l'origine de la vie. In: Gargaud M, Despois D, Reisse J (eds) Les Traces du Vivant. Presses Universitaires de Bordeaux, Bordeaux

Darwin Ch (1859) On the origin of species by natural selection, or the preservation of favoured races in the struggle for life. J. Murray, London

Darwin Ch (1871) Letter to J. Hooker. In: Darwin F (ed) The life and letters of Charles Darwin. J. Murray, London

Dauvillier A (1958) L'origine photochimique de la vie. Masson, Paris

Errera L (1899) Essais de philosophie botanique. Revue de l'Université de Bruxelles, Tome V, pp 545–565

Farley J (1974) Spontaneous generation controversy from Descartes to Oparin. The Johns Hopkins University Press, Baltimore/London

Gavaudan P (1967) Aperçu sur l'évolution du problème de l'origine de la vie dans la première moitié du XXe siècle. In: La Biogenèse – La recherche chimique en France, Numéro special annuel de Physique et Chimie, Paris

Haeckel E (1866) Generelle Morphologie der Organismen. G. Reimer, Berlin

Haeckel E (1876) The history of creation. D. Appleton, New York

Haeckel E (1900) The riddle of the universe. Harper, New York

Haeckel E (1904) The wonders of life. Harper, New York

Huxley TH (1870) Biogenesis and abiogenesis, reprinted In: Collected essays, London, vol 8, 1894, pp 229–271

Kamminga H (1980) Studies in the history on the origin of life from 1860. Thesis for the degree of Doctor of Philosophy in the University of London, Department of History and Philosophy of Science, Chelsea College, London

Lazcano A, Bada JL (2003) The 1953 Stanley L. Miller experiment: fifty years of prebiotic organic chemistry. Orig Life Evol Biosph 33:235–242

Le Dantec F (1896) Théorie nouvelle de la vie. Alcan, Paris

Moore B (1913) The origin and nature of life. Williams and Norgate, London

Oparin AI (1924) Proiskhozhdenie zhizny, Moscow

Oparin AI (1938) The origin of life. The Macmillan Company, New York

Osborn HF (1917) The origin and evolution of life: on the theory of action reaction and interaction of energy. Charles Scribner's Sons, New York

Pasteur L (1861) Mémoire sur les corpuscules organisés qui existent dans l'atmosphère. Examen de la doctrine des generations spontanées. Annales des sciences naturelles (partie zoologique), 4ème série, XVI, pp 5–98

Perrier E (1920) La Terre avant l'histoire. Les origines de la vie et de l'homme. La Renaissance du Livre, Paris

Pflüger E (1875) Ueber die phyisiologische Verbrennung in den lebendigen Organismen. Archiv Ges Physiol 10:251–367

Ponnamperuma C, Gabel NW (1968) Current status of chemical studies on the origin of life. Orig Life Evol Biosph 1(1):64–96

Raulin-Cerceau F (2009) Les origines de la vie, Histoire des idées. Ellipses, Paris

Richter HE (1865) Zür Darwin'schen Lehre. Schmidt'sJahrb Ges Med 126:243–249

Spencer H (1898) Principles of biology. Williams and Norgate, London

Tirard S (1996) Les travaux sur l'origine de la vie de la fin du XIXe siècle jusqu'aux années 1970. Thèse de doctorat d'épistémologie et histoire des sciences, Presses du Septentrion, Paris

Tyndall J (1871) Fragments of science for unscientific people: a series of detached essays, lectures, and reviews. D. Appleton and Company, New York

Weismann A (1904) The evolution theory. Edward Arnold, London

Wöhler F (1828) On the Artificial Production of Urea. Annalen der Physik und Chemie 12:253–256

Biodata of **James E. Strick**, author of "***A History of Origin of Life Ideas from Darwin to NASA.***"

**Professor James E. Strick** earned his Ph.D. from Princeton University in 1997. He is Associate Professor and chair of the Department of Earth and Environment and the Program in Science, Technology, and Society at Franklin and Marshall College, Lancaster, Pennsylvania, USA. He is a historian of biology and medicine, interested in the history of ideas and experiments on the origin of life, 1750–present. His earlier works have included *Sparks of Life: Darwinism and the Victorian Debates over Spontaneous Generation* (Harvard, 2000) and, coauthored with Steven Dick, *The Living Universe: NASA and the Development of Astrobiology* (Rutgers, 2004).

E-mail: **jstrick@fandm.edu**

J. Seckbach (ed.), *Genesis - In The Beginning: Precursors of Life, Chemical Models and Early Biological Evolution*, Cellular Origin, Life in Extreme Habitats and Astrobiology 22, 907–921
DOI 10.1007/978-94-007-2941-4_46, © Springer Science+Business Media Dordrecht 2012

# A HISTORY OF ORIGIN OF LIFE IDEAS FROM DARWIN TO NASA

**JAMES E. STRICK**
*Department of Earth and Environment, Franklin and Marshall College, 637 College Avenue, Lancaster, PA 17603, USA*

## 1. Nineteenth Century Ideas: Spontaneous Generation

Charles Darwin had remarkably ambivalent feelings about spontaneous generation—the belief, dating back at least to Aristotle, that life could emerge from nonliving matter. In the first edition of *On the Origin of Species*, Darwin appears to have carefully avoided the topic through almost the entire book. Then in a single throwaway line near the very end, he upset many of his supporters by adopting this language: "Therefore I should infer from analogy that probably all the organic beings which have ever lived on this earth have descended from some one primordial form, *into which life was first breathed*" (Darwin, 1859, p. 484, italics mine).

In the second edition, released just seven weeks later on 31 December 1859, Darwin made it "...breathed by the Creator." Darwin received much criticism, from both supporters and opponents of evolution, about dissembling in this passage. In the 1862 third edition, he simply dropped the entire phrase after "one primordial form" and left it that way through all remaining editions of the *Origin* in his lifetime, essentially taking a public position of "no comment" on the implications of his theory for the origin of life on Earth. In private, to close confidantes, he said "I have long regretted that I truckled to public opinion, and used the Pentateuchal term of creation, by which I really meant 'appeared' by some wholly unknown process. It is mere rubbish, thinking at present of the origin of life; one might as well think of the origin of matter" (Darwin, 1863). He thought Pasteur's experiments much more convincing than Pouchet's, in their famous debates over spontaneous generation before the French Academy of Sciences from 1861 to 1864 (Darwin, 1862; Geison, 1995; Farley and Geison, 1974). But to others, for example, Ernst Haeckel, Darwin said it would be a great thing if spontaneous generation could be proven, as it would undergird his theory in an important way (Darwin, 1872b, 1873). His reluctance to discuss the matter in 1859 seems to have become genuine ambivalence on the topic.

If Darwin was trying to walk a middle road in how he described the origin of life in public, he may perhaps have been trying to keep liberal-minded Christians from rejecting his evolution theory, as long as he did not challenge a Creator bringing the first life into existence. Many critics of his theory accused it of being a stalking horse for a materialistic worldview. When his supporters seemed equally upset by his "into which life was first breathed" language, he may have simply decided the best course was to avoid the subject altogether.

On the British scientific scene in the 1860s and the 1870s, the most entrenched advocates of spontaneous generation, and those most resistant to the germ theory of disease, were among the medical community (Strick, 2000). At that time, despite Pasteur's suggestive research on silkworm diseases, British doctors strongly favored a theory of contagious disease as analogous to fermentation via chemical catalysis, a theory first promulgated by Justus Liebig in the early 1840s. This was in part due to a brief but highly publicized theory of 1849 that cholera was caused by a fungus. When the supposed cholera-fungus was soon discovered to be a common mold contaminant, many British medical men who might have had leanings toward a germ theory felt that it had been discredited (Pelling, 1978). Thus, when Pasteur's claims of the mid-1860s reached Britain, most British doctors were skeptical that living microbes were the sources of contagion, and they continued to favor a chemical poison, viewing the microbes as some kind of product or concomitant of the disease process rather than its cause. Thus, when the physicist John Tyndall (1870) gave a famous lecture on "Dust and Disease" in January 1870, arguing that doctors must accept the germ theory of Pasteur in order to have any really scientific approach to disease, many British doctors took offense. Tyndall was known for his brash public persona and as a general spokesman for science in London society circles. In addition to his quite haughty tone, he was trained as a physicist and certainly had no clinical experience of disease or of the role of patients' constitutions in causing their susceptibility to be so highly varied. So, it was not surprising that doctors accused Tyndall of being an interloper in biology and medicine, an area where his opinions had no weight. They thought he was totally unaware of the cholera-fungus hoax of 1849 and of other important technical reasons for medical skepticism about an oversimplified germ theory. Tyndall's version of the germ theory denied any role to the "constitution": it compared patients to so many identical tubes of infusion. If germ-laden dust particles fell into the patient, he would get sick, Tyndall said. Those in hospitals or towns with epidemics who did not get sick were those who, such particles, or "germ clouds," passed by.

Chief among the medical professionals who opposed Tyndall was Henry Charlton Bastian, professor of pathological anatomy at London's University College Medical School. Bastian was an avowed supporter of Darwin's and Spencer's writings on evolution and did much experimental work to try to show that microorganisms could arise by spontaneous generation, or biogenesis as he at first called it. Bastian, like Alfred Russel Wallace and many others interested in natural selection, thought at that time that Darwin's theory required spontaneous generation to explain in a nonmiraculous way where the original common ancestor of all species came from (Strick, 2000). Wallace wrote excitedly to Darwin (1872b) and in a review in *Nature* in August 1872, that Bastian's findings could help Darwinian evolution out of the time crunch presented by William Thomson's (later Lord Kelvin) challenge, that the age of the Earth—a few tens of millions of years at most, by his physics calculations—was utterly inadequate to allow the hundreds of millions of years of evolution Darwin's gradualist theory required to get from single-celled creatures to complex vertebrates by a single, unbroken chain of descent (Wallace, 1872a, b; Burchfield, 1990). Darwin (1872a) was tempted but still ambivalent. By this time, he had written to T. H. Huxley and J. D. Hooker that he certainly believed that life at first emerged from nonlife by natural processes in "some warm little pond" (Darwin, 1871; Hardin, 1950). However, Bastian's idea of spontaneous generation going on all the time, everywhere, up to the present meant many different possible lines of descent with new ones beginning constantly, a model similar to Lamarck's. Darwin was still deeply intellectually committed to a single branching tree of descent, which would require only one (or a small handful) of progenitor(s) created by an original emergence of life entirely confined to the very earliest period of life on Earth. Furthermore, Darwin (1871) now argued for the first time that once a first organism emerged and multiplied, its kind would gobble up the precursors of further organisms before they had the chance to fully develop into complete living forms. Darwin followed closely the debates over Bastian's experiments, as did all Darwinians who had come to know Bastian as one of their own, indeed one of their brightest rising young stars in experimental science.

Bastian also thought that bacteria in diseased patients resulted from spontaneous generation, as by-products of the disease process. Between 1870 and 1875, he had published the results of hundreds of experiments in which he showed that bacteria could be found in tubes of various infusions boiled for periods varying from a few minutes up to an hour (Bastian, 1872, 1874, 1875). Attempting to refute Bastian's work from 1875 to 1877, Tyndall devised an ingenious dust-free cabinet in which to carry out the experiments.

Tyndall's close friend T. H. Huxley was at first interested in Bastian's experiments, but soon concluded that the pathologist must be mistaken in his conclusions. He declared that Bastian "had gotten out of his tubes exactly what he put into them," i.e., that organisms must have gotten into the tubes as contaminants (Huxley, 1870). Both Huxley and Tyndall were also evolutionists but found it much easier to believe in living things able to survive boiling somehow (or, even more believable, that Bastian was a sloppy experimenter) than to believe that organisms as complex as protozoa could come from anything other than the "seeds" or "germs" of other like organisms. They lobbied widely among scientific colleagues to convince others that Bastian was at best a poor experimenter and at worst a fraud and a cheat. As mentioned above, Bastian, for his part, found it much more difficult to believe that life could survive the boiling temperature for more than a minute or two, than to believe (especially since he saw it as a logical necessity of continuity in nature and of evolutionary science) that the transition in stages from nonliving to living matter should be possible. A noted physiologist, John Burdon Sanderson, observed Bastian's technique carefully and reported that it was up to high scientific standards (Strick, 2000).

Huxley, in a famous address to the British Association for Advancement of Science meeting at Liverpool in September 1870, moved to gain the rhetorical upper hand in the debate by defining the terms. He defined "biogenesis" to mean life only from other life. The opposite belief he termed "abiogenesis" and began to argue that it very well might be possible, even probable, but only in the conditions of the primitive Earth. This convenient dualistic terminology was rapidly picked up by contemporaries and propagated in textbooks. Given this, it is more than a little ironic that Huxley had hijacked the term "biogenesis" from Bastian, who was using it up until that time to mean exactly the opposite, i.e., spontaneous generation (Strick, 2000). Also important, it is from this time that all discussions of the origin of life question began to be pushed more and more into the distant past. Only after Huxley's talk (1870), for example, did Darwin advance the argument that if spontaneous generation had occurred in the Earth's distant past, it would no longer be possible after the evolution of the first heterotrophs, since they would consume any organic molecules that formed before those could assemble into a new organism.

Those opposed to both evolution and spontaneous generation had a simpler solution. The physicist William Thomson, Lord Kelvin, for example, responded to both Huxley and Bastian in his own Presidential Address to the BAAS in 1871 (Thomson, 1871). He suggested that the germs of life might have originally been brought to Earth from another world via a meteorite. To those demanding a completely naturalistic origin of life like the Darwinians, of course, this merely pushed the question back a step to some other planet. It did not escape their notice that Thomson was a deeply religious man.

Much of the public interested in evolution followed the debates closely. At stake was whether evolution would be inseparably linked to belief in spontaneous generation (Bastian's position), or whether the official Darwinian position would be the "once-only, in the distant past" version of the origin of life, which Huxley tried to encapsulate in the term "abiogenesis." As if to vindicate Darwin's fear that his theory had enough enemies even without broaching the origin of life question, Frederick Barnard, president of Columbia College in New York and an influential American intellectual, published a striking account of where the debate stood in 1873. Bastian's evidence seemed strong, said Barnard, but if it were true, he would be forced to conclude that the new evolutionary science had completely disproven any supernatural Creator. And with that, gone would also be any hope of an afterlife. Faced with that choice, Barnard showed unusual frankness in confessing:

> If these doctrines are true, all talk of creation or methods of creation becomes absurdity; for just as certainly as they are true, God himself is impossible... But we are told it is unphilosophical, in the pursuit of truth, to concern ourselves about consequences. We should accept the truth with gladness,... and let consequences take care of themselves. To this canon I am willing to subscribe up to a certain point. But if, in my study of nature, I find the belief forced upon me that my own conscious spirit... is but a mere vapor, which appeareth for a little time and then vanisheth away forever, that is a truth... for which I shall never thank the science which has taught it me. Much as I love truth in the abstract, I love my hope of immortality still more; and

> if the final outcome of all the boasted discoveries of modern science is to disclose to men that they are more evanescent than the shadow of the swallow's wing upon the lake,... give me then, I pray, no more science. Let me live on, in my simple ignorance, as my fathers lived before me, and when I shall at length be summoned to my final repose, let me still be able to fold the drapery of my couch about me, and lie down to pleasant, even though they be deceitful dreams. (Barnard, 1873, p. 79)

Darwin was right, apparently, in imagining the explosive potential of discussing the origin of life too directly.

Studies on the life cycles of protozoa, made with painstaking continuous microscopic observation of live cultures over days and weeks on end, showed a number of protozoa produced heat- and desiccation-resistant cysts or spores (Dallinger and Drysdale, 1873, 1874, 1875). Pleomorphist theories of bacteria having no fixed morphology and transforming fluidly into fungi and back again also came under attack, for example, by Robert Koch (Bulloch, 1938). Both of these changes tended to undermine the framework of ideas about microbes in which spontaneous generation made sense. Emerging support for the germ theory of disease—after Pasteur's successful demonstration of an anthrax vaccine in 1881 and Koch's identification of the causative microbe of tuberculosis in 1882—also drastically undercut belief in the possibility of spontaneous generation by the early 1880s.

But let us back up to the experimental debate on microbes in boiled infusions. Beginning in the autumn of 1876, Tyndall began to have difficulty with the experiments carried out in his dust-free cabinet. Infusions which had been sterilized by only five minutes' boiling a year earlier now could not be sterilized even after *hours* of boiling. In 1876, the discovery by the German botanist Ferdinand Cohn, that certain species of *Bacillus*, especially common in hay and in cheese (and in infusions made from these that showed microbial growth after extended boiling), were capable of producing heat-resistant endospores. This has been taken to explain why Bastian's infusions were full of microbial growth after boiling, and upon being handed a copy of Cohn's (1876) article, Tyndall immediately adopted this stance to explain his own recent difficulties (Tyndall, 1876, 1877b). Darwin appears to have finally been convinced by Tyndall's experiments that present-day spontaneous generation was impossible and that Huxley's view of abiogenesis, only once under the different conditions of primitive Earth, was the correct answer to the origin of life (Darwin, 1877).

Tyndall struggled for several months and finally discovered that sterility could be achieved by repeated short boilings, followed by a period of allowing the remaining spores to germinate. This process came to be known as "fractional sterilization" or "Tyndallization" (Tyndall, 1877a). This was largely superceded with the development of the autoclave in the 1880s. Equally interesting from a historical point of view is that though this rapid about-face also showed that Tyndall and Huxley had been wrong that Bastian must be a sloppy, incompetent experimenter, that is precisely the version of Bastian (and most spontaneous generation advocates) that has usually gone into the textbooks until quite recently. The sole exception was Pasteur's student Emile Duclaux, the only contemporary writer among the winners who credited Bastian's perseverance in sticking to his experimental results with the value that it truly had, i.e., that without the goad of Bastian and his popularity, Pasteur, Cohn, and Tyndall might never have discovered that they were wrong about the existence of bacteria in boiled hay infusions, and thus been led to realize that heat-resistant spores exist and require autoclaving or fractional sterilization to guarantee their being killed (Duclaux, 1896). Pasteur was a devout Catholic and never believed, to his dying day, the viewpoint favored by many modern origin of life hypotheses, that random or chance organization of materials could have formed life, even the very first living organisms. (Despite this, he did hold private beliefs that spontaneous generation might be possible—he had even secretly tried to prove it experimentally—but only by bringing to bear some kind of "cosmic asymmetric force," related to the ability of life to select so exquisitely among stereoisomers of organic molecules (Geison, 1995)). It is important to realize that believers in "spontaneous generation" were an important minority faction *within* the Darwinian evolutionary camp until the debates of the 1870s. In those struggles between opposing factions of Darwinians, the Huxley and Tyndall–led group won out with their "abiogenesis" view, that the origin of life was indeed a naturalistic process, but could only have occurred once under the very different conditions of the primitive Earth.

## 2. Twentieth Century Ideas

The origin of life has continued to be a subject of active interest in the twentieth century, and, despite avoidance of the term "spontaneous generation" since 1880, some modern concepts have resembled older ideas more than superficially. Swedish chemist Svante Arrhenius published (1908) a revived and much more detailed version of Thomson's argument for transport of microbial life through the cosmos via meteorites and comets, what he termed "panspermia." Beginning with Bastian (1905, 1907), who revived experiments on spontaneous generation from 1900 to 1915, a number of investigators came to believe that the basic chemistry specific to the living was the chemistry of colloids. A burst of research in colloid chemistry from 1910 until the 1930s included many workers who believed that this avenue would lead to understanding the simplest combination of materials to cross the boundary from nonlife to living matter (Morgan, 1990). Among these were biochemists Benjamin Moore (1912) and Albert Mary, bacteriologist Arthur I. Kendall (1925), the Mexican biologist Alfonso Herrera (1917, 1928, 1942), and medical doctor and researcher George Crile (Crile et al., 1932). A revival of pleomorphist theories of microbes from 1910 to 1940 also helped revive older ideas that the boundary between the simplest living cells and the largest nonliving colloid molecules was not so wide, nor so difficult to cross (Amsterdamska, 1991; Wainwright, 1997). These ideas were supported by some bacteriologists such as Kendall and Theobald Smith, but attacked ferociously by others, especially Hans Zinsser (Coulter, 1994). In the end, the development of the ultracentrifuge and the emergence of the macromolecule concept put an end by 1940 or so to belief that colloids were responsible for the crucial properties of life (Olby, 1994). During the Depression Era, when the Rockefeller Foundation was practically alone as a source of large-scale funding for life sciences research, the director of the Natural Sciences division, Warren Weaver, believed deeply in the new macromolecular approach (Kohler, 1989). The resulting asymmetry of funding hastened the demise of colloidalist theories of the origin of life, while accelerating the development of physicochemical reductionism in the life sciences.

Geneticist H. J. Muller (1929) suggested that the first chemical assembly of a gene should be considered the origin of life, though this still suggested a sudden origin of life from nonlife, perhaps the most central idea implicit in the doctrine of spontaneous generation. The Russian biochemist A. I. Oparin (1924) proposed instead a gradual chemical evolution of life, probably proceeding through the intermediate stage of coacervates. This idea proposed a chemically reducing atmosphere on the primitive Earth, with lightning and heat providing the energy to drive synthesis of organic molecules from methane, ammonia, water, and hydrogen. Even if random chemistry between such monomers might require billions of years to assemble a first protocell, this was nonproblematic since Earth had probably been cool enough for 3.8 or 3.9 billion years and the oldest fossil life forms known were much less than 1 billion years old. British geneticist J. B. S. Haldane published very similar ideas (1929); the organic precursors would accumulate in the ocean forming a "hot, dilute soup," Haldane surmised. This "soup" would serve as food for heterotrophs, the simplest original organisms. Research into life's origin had lost none of its larger associations: the issue still attracted researchers with anticlerical leanings (Herrera was one) as well as affiliations with radical, even Communist political beliefs. Haldane and J. D. Bernal did research in the field and were avowed Communists, as was Muller until the late 1930s. Furthermore, Oparin and Wilhelm Reich (1938) explicitly credited the philosophy of dialectical materialism as having been crucial to their origin of life research programs (Graham, 1987).

Once Oparin's book was translated into German in 1936 and English (1938), Oparin's ideas of chemical evolution became widely influential, inspiring, for example, Reinhard Beutner (1938) as well as Norman Horowitz's work (1945) on successive evolution of the steps of metabolic pathways. Thereafter, Oparin's and Haldane's ideas were usually referred to together as the "Oparin-Haldane" hypothesis or the "heterotroph hypothesis."

Research on viruses, particularly Wendell Stanley's crystallization of tobacco mosaic virus (1935), lent credence for a time to the idea that a virus might be the simplest "living molecule" (Podolsky, 1996). In addition, after the rise of molecular biology and the discovery of the Watson-Crick DNA structure, many considered Muller's idea of "the gene as the origin of life" more likely. The Urey-Miller experiment, published just 3 weeks after Watson and Crick's paper on DNA, lent further support to the notion that the building blocks of life could have assembled very rapidly. In that experiment, an electrical discharge passed through a mixture of steam, hydrogen, methane, and ammonia in a sealed container produced amino acids and other organic compounds after only a few days (Miller, 1953). Throughout the 1950s and the 1960s, such views were in tension with those of Oparin and his students in the Soviet Union, that the process of chemical evolution was a more gradual one, with no single decisive "living molecule." Biophysicist J. D. Bernal also took this view of a more gradual, evolutionary process, for which he later coined the term "biopoesis" (1959). Cold War and anti-Lysenkoist hostilities were involved, as any scientific theory openly lauding dialectical materialism (as Oparin and Bernal both did) was seen as suspect in the West in the aftermath of Lysenko's takeover of Soviet genetics (Farley, 1977). Bernal (1949) also first directed attention to the possibility that the surfaces of clay minerals might have served as important catalytic sites for the early stages of chemical evolution, the "scaffold" upon which the "arch" of the first protolife was constructed, a line of research much further developed by Graham Cairns-Smith (1965).

Origin of life experiments have included a tradition of synthesizing "model protocells" in the tradition of Oparin and Herrera. Prominent representatives of this approach have included Sidney Fox (1965), Krishna Bahadur (Smith et al., 1981; Grote, 2011), and Claire Folsome (Folsome and Morowitz, 1969). Others, in the mode of Urey, Miller (1953), Oró (1961), and Horowitz (1945), have pursued precise chemical analysis of hypothesized early reactions leading to chemical precursors of life. A chicken-and-egg problem exists, because of the complete interdependence in all known cells of metabolism and enzyme machinery with the nucleic-acid-based machinery of replication and information storage. The advocates of the synthetic "protocell model" approach have tended to believe that metabolism in a membrane-enclosed structure must have come first. Those emphasizing precise chemical analysis have tended to believe in a "gene-first" scenario, though that has become complicated by the understanding of how dependent genes are upon enzymes for their successful replication and expression. The discovery of ribozymes, RNA molecules capable of replication and of enzymatic activity (Cech, 1986), suggested to many researchers that an intermediate "RNA World" must have existed, to bridge the gap before modern cells with the two functions inseparably mutually dependent. Since RNA monomers are very difficult to synthesize, and quite unstable, in most primitive Earth simulation experiments, what came before RNA remains unsolved (Orgel, 1998).

The involvement of NASA as a major patron for origin of life research, beginning in 1960, led to the funding of a wide variety of approaches, including gradual evolution models as well as sudden, gene-based models: "protein-first" or "metabolism-first" as well as "nucleic acid-first" or "information-first" models. This work took place under the rubric of "exobiology" (broadened and renamed "astrobiology" in the late 1990s). The sudden influx of funding from a patron on a large scale led to rapid growth of origin of life research and linking that research to the search for life on other planets (Strick, 2004; Dick and Strick, 2004). Historian Iris Fry has referred to this time as the "Era of Optimism" in modern origin of life research, because, operating under the Oparin-Haldane "heterotrophs first, reducing atmosphere" paradigm until 1980 or so, researchers made many breakthroughs and felt the solution of the origin of life problem might be near at hand (Fry, 2000). However, if the Oparin-Haldane hypothesis counts as a "paradigm" in Thomas Kuhn's sense (1970), then a number of the new discoveries from the late 1960s until 1980 count as anomalies that began to undermine that paradigm.

For example, the discovery of continually older microfossils greatly narrowed the time window for prebiotic chemistry to take place, to 1 billion years or less—even half that, if one accepts Schopf's (1993) claim of 3.45-billion-year-old Apex chert microfossils, recently questioned by Brasier et al. (2002). Random chemistry appears increasingly unlikely to be able to accomplish the entire task in the time allotted at the Earth's surface. Furthermore, by 1980, the consensus among planetary and atmospheric scientists was that Earth's early atmosphere was probably not reducing, but rather neutral or perhaps even mildly oxidizing. This leads to far slower organic synthesis—orders of magnitude slower—in Urey-Miller-type simulations. Other lines of evidence suggest the chemistry may be nonrandom; rather, it is biased strongly in favor of organic synthesis. Detection by radio astronomy of giant clouds of organic molecules in interstellar space, and discovery of extraterrestrial compounds such as amino acids on meteorites that reached Earth, both suggest the possibility that some steps of organic synthesis could have taken place beyond the Earth, so that precursor molecules could have been delivered via exogenous delivery, short-cutting how much of the process had to occur on Earth's surface (Kvenvolden et al., 1970; Cronin and Pizzarello, 1997). (While not the same as Arrhenius or Kelvin's belief in transport of entire organisms, some have considered this at least a partial vindication of "panspermia" notions.) Discovery of thriving hydrothermal vent ecosystems in 1977 suggested the possibility of a high-temperature, high-pressure origin of life, not at Earth's surface (Corliss et al., 1981). Brock's (1967) discovery of extreme thermophiles in Yellowstone hot springs and Woese's subsequent recognition of the Archaea as a distinct kingdom, perhaps containing some of the most primitive types of cells (Woese, 1987), seemed to support the high-temperature ideas, since many of the Archaea were thermophiles capable of metabolizing methane, sulfur, etc. Life could exist, it was realized by 1980, in a far wider, more hostile range of conditions than previously imagined—even the conditions beneath the surface of Mars and/or on a warmer, wetter early Mars, some imagine (Brock, 1985). This despite the fact that two 1976 Viking spacecraft seemingly proved life nonexistent on the Martian surface.

Thus, the Oparin-Haldane hypothesis may have been wrong about most of its fundamental assumptions. But it has still been a valuable guide and impetus to research, greatly expanding knowledge of planetary science and of the early stages of origin and evolution of life. Furthermore, NASA Exobiology Program funding supported and made possible other related discoveries, such as Lynn Margulis's (1970) serial endosymbiosis theory of the origin of eukaryotic cells, James Lovelock's "Gaia hypothesis" and the use of far-from-equilibrium atmospheric chemistry as the basis for the search for Earthlike planets in planetary systems of other stars, etc. Much of Woese's work was supported by this NASA funding (Strick, 2004); thus, it has made possible his newer theoretical work on what horizontal gene transfer means about the nature of the "last common ancestor" of all life on Earth (Sapp, this volume). Most of all, the era of large-scale NASA funding greatly expanded origin of life research and fused it purposefully, along with all the other disciplines required to search for life on other planets, into a new interdisciplinary scientific discipline called exobiology, now renamed astrobiology (Strick, 2004; Dick and Strick, 2004).

"Spontaneous generation" has been considered a dead-end, disproven belief since 1880 or so, most of all its implication that life could come very suddenly into existence from nonlife in a few hours or days. But, despite the dropping of the older term, the conceptual continuities between modern and older origin of life ideas, particularly for abiogenesis in the Earth's distant past, are in some ways as interesting as the discontinuities. A few modern researchers, such as Sidney Fox, continued to use the term "spontaneous generation" into the late 1950s to emphasize those continuities. The continued association of these ideas with larger philosophical and political concerns is also a feature of discussions well into the twentieth century, particularly during the Cold War. While no single dominant paradigm has replaced the Oparin-Haldane model since 1980, the field of origin of life research remains vibrant, exploring many productive new avenues.

## 3. References


Amsterdamska O (1991) Stabilizing instability: the controversy over cyclogenic theories of bacterial variation during the interwar period. J Hist Biol 24:191–222

Arrhenius S (1908) Worlds in the making: the evolution of the universe (Eng trans Borns H). Harper & Bros., New York

Barnard F (1873) The germ theory of disease and its relations to hygiene. Public Health Rep Pap 1:70–87

Bastian HC (1872) The beginnings of life, 2 vols. Macmillan, London; republished (as are all the Bastian sources cited here) in Strick JE (ed) (2001) Evolution and the spontaneous generation debate, 6 vols. Thoemmes Press, Bristol

Bastian HC (1874) Evolution and the origin of life. Macmillan, London

Bastian HC (1875) Introductory paper to "Discussion on the germ theory of disease". Trans Path Soc (Lond) 26:255–345

Bastian HC (1905) The nature and origin of living matter. Fisher Unwin, London

Bastian HC (1907) The evolution of life. Methuen & Co., London

Bernal JD (1949) The physical basis of life. Proc Phys Soc Lond 62:537–558; later revised and expanded as a monograph (Routledge and Kegan Paul, London, 1951)

Bernal JD (1959) The problem of stages in biopoesis. In: Clark F, Synge RLM (eds) Proceedings of the first international conference on the origin of life, Moscow, 19–24 August 1957. Pergamon Press, New York, pp 38–53

Beutner R (1938) Life's beginnings on the earth. Williams and Wilkins, Baltimore

Brasier M, Green O, Jephcoat A, Kleppe A, Van Kranendonk M, Lindsay J, Steele A, Grasineau N (2002) Questioning the evidence for earth's oldest fossils. Nature 416:76–81 (7 March)

Brock TD (1967) Life at high temperatures. Science 158:1012–1019 (24 Nov)

Brock TD (1985) Life at high temperatures. Science 230:132–138 (11 Oct)
Bulloch W (1938) A history of bacteriology. Oxford University Press, Oxford
Burchfield JD (1990) Lord Kelvin and the age of the earth, 2nd edn. University of Chicago Press, Chicago
Cairns-Smith AG (1965) The origin of life and the nature of the primitive gene. J Theor Biol 10:53–88
Cech T (1986) RNA as an enzyme. Sci Am 255:64–75 (Nov)
Cohn F (1876) Untersuchungen über Bacterien IV. Beit z Biol Pflanzen 2(2):249–276
Corliss JB, Baross J, Hoffman SE (1981) An hypothesis concerning the relationship between submarine hot springs and the origin of life. Oceanol Acta Sp:59–69
Coulter H (1994) Divided legacy. Part IV, twentieth century medicine: the bacteriological era. North Atlantic Books, Berkeley
Crile G, Telkes M, Rowland A (1932) Autosynthetic cells. Protoplasma 15:337–360
Cronin J, Pizzarello S (1997) Enantiomeric excesses in meteoritic amino acids. Science 275:951–955 (14 Feb)
Dallinger W, Drysdale W (1873) Researches on the life history of a cercomonad: a lesson in biogenesis. Mon Microsc J 10:53–58
Dallinger W, Drysdale W (1874) Further researches into the life history of the Monads. Mon Microsc J 11:7–10, 69–72, 97–103
Dallinger W, Drysdale W (1875) Further researches into the life history of the monads. Mon Microsc J 13:185–197
Darwin C (1859) On the origin of species. John Murray, London
Darwin, C (1862) Letter to Henry Holland, April, in Correspondence of Charles Darwin, v. 10, Cambridge U. Press, Cambridge, UK, pp 141–142
Darwin, C (1863) Letter to Joseph Hooker, 29 March, in Correspondence of Charles Darwin.,v. 11, Cambridge U. Press, Cambridge, UK, pp 277–278
Darwin C (1871) Letter to Joseph Hooker, 1 Feb. In: Darwin papers (unpublished ms.). American Philosophical Society Library, Philadelphia (forthcoming in Darwin Correspondence, vol 19)
Darwin C (1872a) Letters to A.R. Wallace, 28 Aug., 2 Sept. In: Marchant J (ed) Alfred Russel Wallace: letters and reminiscences, vol 1. Harper & Bros., New York, 1916, pp 225–226, 228
Darwin C (1872b) Letter to Ernst Haeckel, 2 Sept. In: Darwin papers, unpublished ms. American Philosophical Society Library, Philadelphia
Darwin C (1873) Letter to Ernst Haeckel, 25 Sept. In: Darwin papers, unpublished ms. B: D25p. American Philosophical Society Library, Philadelphia
Darwin C (1877) Letter to G.J. Romanes, 23 May. In: Darwin papers unpublished ms., Carroll #513. American Philosophical Society Library, Philadelphia
Dick SJ, Strick JE (2004) The living universe: NASA and the development of astrobiology. Rutgers University Press, New Brunswick
Duclaux E (1896) Pasteur: Histoire d'un Esprit. English trans Smith E, Hedges F Pasteur: the history of a mind. W.B. Saunders, Philadelphia
Farley J (1977) The spontaneous generation controversy from Descartes to Oparin. Johns Hopkins University Press, Baltimore
Farley J, Geison GL (1974) Science, politics and spontaneous generation in nineteenth century France: the Pasteur-Pouchet debate. Bull Hist Med 48:161–198
Folsome C, Morowitz H (1969) Prebiological membranes: synthesis and properties. Space Life Sci 1:538–544
Fox S (ed) (1965) The origins of prebiological systems and their molecular matrices. Academic, New York
Fry I (2000) The emergence of life on earth. Rutgers University Press, New Brunswick
Geison GL (1995) The private science of Louis Pasteur. Princeton University Press, Princeton
Graham L (1987) Science, philosophy and human behavior in the Soviet Union. Columbia University Press, New York, chapter 2
Grote M (2011) Jeewanu, or the 'particles of life': the work of Krishna Bahadur in 20th century origin of life research. J Biosci 36(4):563–570

Haldane JBS (1929) The origin of life. Ration Annu 148:3–10; reprinted in Bernal JD (1967) The origin of life, World Pub, New York, pp 242–249
Hardin G (1950) Darwin and the heterotroph hypothesis. Sci Mon 70:178–179
Herrera AL (1917) Los Protobios 1917. In: Castellanos I (ed) La Plasmogenia, Imprenta de Rambla, Havana, 1921, pp 127–150
Herrera AL (1928) Plasmogeny. In: Alexander J (ed) Colloid chemistry, theoretical and applied, vol 2: biology and medicine. Chemical Catalog Co, New York, pp 81–91; reprinted Cleaves HJ (ed) The works of Alfonso Herrera in translation. Springer, New York, forthcoming
Herrera AL (1942) A new theory of the origin and nature of life. Science 96:14
Horowitz N (1945) On the evolution of biochemical synthesis. Proc Natl Acad Sci 31:153–157
Huxley TH (1870) "Biogenesis and abiogenesis," presidential address to BAAS meeting in Liverpool. Nature 2:400–406 (15 Sept); also pub. in (1901) The scientific memoirs of Thomas Henry Huxley, vol 3. Macmillan, London, pp 572–594; and in (1925) Discourses biological and geological, essays. Appleton, New York, pp 232–274
Kendall AI (1925) Bacteria as colloids. In: Holmes HN (ed) Colloid symposium monograph, vol 2. Chemical Catalog Co., New York, pp 195–203
Kohler R (1989) Partners in science. University of Chicago Press, Chicago
Kuhn T (1970) The structure of scientific revolutions, 2nd edn. University of Chicago Press, Chicago
Kvenvolden K, Lawless J, Pering K, Peterson E, Flores J, Ponnamperuma C, Kaplan IR, Moore C (1970) Evidence for extraterrestrial amino acids and hydrocarbons in the Murchison meteorite. Nature 228:923–926
Margulis L (1970) Origin of eukaryotic cells. Yale University Press, New Haven
Miller S (1953) A production of amino acids under possible primitive earth conditions. Science 117:528–529
Moore B (1912) The origin and nature of life. Williams and Norgate, London
Morgan N (1990) The strategy of biological research programmes: reassessing the 'Dark age' of biochemistry, 1910–1930. Ann Sci 47:139–150
Muller HJ (1929) The gene as the basis of life. In: Proceedings of the 4th International Congress of Plant Biology, Menasha, WI, 1929, vol 1, pp 897–921 (paper presented 19 Aug. 1926)
Olby R (1994) The path to the double helix, 2nd edn. Dover, New York
Oparin AI (1924) Proiskhozhdenie Zhizny. Moscow Worker pub., Moscow. English trans: Bernal JD (1967) The origin of life. World Pub, New York, pp 199–241
Oparin AI (1938) The origin of life. Macmillan, New York
Orgel L (1998) The origin of life: a review of facts and speculations. Trends Biochem Sci 23:491–495
Oró J (1961) Mechanism of synthesis of adenine from hydrogen cyanide under possible primitive earth conditions. Nature 191:1193–1194
Pelling M (1978) Cholera, fever and English medicine. Oxford University Press, Oxford
Podolsky S (1996) The role of the virus in origin of life theorizing. J Hist Biol 29:79–126
Reich W (1938) Die Bione. Sexpol Verlag, Oslo. English trans (1979) The Bion experiments on the origin of life. Farrar, Straus, New York
Schopf JW (1993) Microfossils of the Early Archean Apex Chert: new evidence of the antiquity of life. Science 260:640–646 (30 April)
Smith AE, Folsome C, Bahadur K (1981) Nitrogenase activity of organo-molybdenum microstructures. Experientia 37:357–359
Stanley W (1935) Isolation of a crystalline protein possessing the properties of tobacco-mosaic virus. Science 81:644–645 (28 June)
Strick JE (2000) Sparks of life: Darwinism and the Victorian debates over spontaneous generation. Harvard University Press, Cambridge
Strick JE (2004) Creating a cosmic discipline: the crystallization and consolidation of exobiology, 1957–1973. J Hist Biol 37:131–180
Thomson W (1871) Presidential address to the BAAS. Nature 4:262–270 (3 Aug)
Tyndall J (1870) Dust and disease. In: Essays on the floating matter of the air. Johhnson, New York, reprint, 1966; originally published as (1870) On haze and dust. Nature 1:339–342 (27 Jan)

Tyndall J (1876) Optical deportment of the atmosphere in relation to putrefaction and infection. Philos Trans R Soc Lond 166:27–74

Tyndall J (1877a) On heat as a germicide when discontinuously applied. Proc R Soc Lond 25:569

Tyndall J (1877b) Further researches on the deportment and vitality of putrefactive organisms from a physical point of view. Philos Trans R Soc Lond 167:149–206

Wainwright M (1997) Extreme Pleomorphism and the bacterial life cycle: a forgotten controversy. Perspect Biol Med 40:407–414

Wallace AR (1872a) Review of Bastian's the beginnings of life. Nature 6:284–287, 299–303 (8 and 15 August)

Wallace, A.R. (1872b) Letters to Charles Darwin, 4 Aug. and 31 Aug. In: Marchant J (ed) Alfred Russel Wallace: letters and reminiscences, vol 1. Harper & Bros, New York, 1916, pp 223, 226

Woese CR (1987) Bacterial evolution. Microbiol Rev 51:221–271

# SUBJECT AND ORGANISM INDEX





J. Seckbach (ed.), *Genesis - In The Beginning: Precursors of Life, Chemical Models and Early Biological Evolution*, Cellular Origin, Life in Extreme Habitats and Astrobiology 22, 923–932
DOI 10.1007/978-94-007-2941-4, 

**T**

# AUTHOR INDEX





J. Seckbach (ed.), *Genesis - In The Beginning: Precursors of Life, Chemical Models and Early Biological Evolution*, Cellular Origin, Life in Extreme Habitats and Astrobiology 22, 933–934
DOI 10.1007/978-94-007-2941-4, 